REINFORCED CONCRETE STRUCTURES

(VOLUME I)

* 1992 *

[CONTAINING 28 CHAPTERS]

REINFORCED CONCRETE STRUCTURES

(VOLUME I)

By

Dr. B.C. PUNMIA
Formerly,
Professor and Head, Deptt. of Civil Engineering, &
Dean, Faculty of Engineering
M.B.M. Engineering College,
Jodhpur

Er. ASHOK KUMAR JAIN
Director,
Arihant Consultants,
Jodhpur

Dr. ARUN KUMAR JAIN
Assistant Professor
M.B.M. Engineering College,
Jodhpur

(AS PER IS : 456-1978)

SEVENTH EDITION

THOROUGHLY REVISED AND ENLARGED INCORPORATING WORKING STRESS AND LIMIT STATE DESIGN METHODS

LAXMI PUBLICATIONS (P) LTD

(An ISO 9001:2008 Company)

BENGALURU • CHENNAI • COCHIN • GUWAHATI • HYDERABAD
JALANDHAR • KOLKATA • LUCKNOW • MUMBAI • RANCHI • NEW DELHI
BOSTON (USA) • ACCRA (GHANA) • NAIROBI (KENYA)

REINFORCED CONCRETE STRUCTURES

Printed and bound in India
First Edition : April 1975, *Second Edition* : Jan. 1981, *Third Edition* : Feb. 1983
Fourth Edition : April 1985, *Fifth Edition* : May 1988, *Sixth Edition* : May 1990
Seventh Edition : May 1992 *Reprint* : Aug. 1993, July 1994, Sept. 1997, May 1999, Sept. 2000, July 2002, March 2003, July 2004, Aug. 2006, June 2008, Aug. 2008, Aug. 2009, Feb. 2010, May 2011, July 2012, Oct. 2013, May 2016
ISBN : 978-81-318-0644-9

PUBLISHED IN INDIA BY

LAXMI PUBLICATIONS (P) LTD
(An ISO 9001:2008 Company)
113, GOLDEN HOUSE, DARYAGANJ, NEW DELHI-110002, INDIA
Telephone : 91-11-4353 2500, 4353 2501
Fax : 91-11-2325 2572, 4353 2528
www.laxmipublications.com info@laxmipublications.com

Branches

✆ **Bengaluru**	080-26 75 69 30
✆ **Chennai**	044-24 34 47 26
✆ **Cochin**	0484-237 70 04, 405 13 03
✆ **Guwahati**	0361-254 36 69, 251 38 81
✆ **Hyderabad**	040-27 55 53 83, 27 55 53 93
✆ **Jalandhar**	0181-222 12 72
✆ **Kolkata**	033-22 27 43 84
✆ **Lucknow**	0522-220 99 16
✆ **Mumbai**	022-24 91 54 15, 24 92 78 69
✆ **Ranchi**	0651-220 44 64

C—11421/016/02

Printed at : Ajit Printing Press, Delhi.

Preface

Reinforced Concrete occupies a leading position in modern construction. Proper reinforced concrete construction depends upon thorough understanding of the action of the structure and upon the knowledge of characteristics and limitations of those materials that are used in its construction. The personal element — the care with which work is excecuted in the field — is also of major importance in concrete construction. The text book, in its two volumes, presents modern methods of design for both ordinary and prestressed concrete structures. Volume I has 19 chapters and deals with the design of more common and elementary structures. Volume 2 incorporates the design principles of advanced structures, including prestressed concrete structures.

In the present volume, the subject matter has been introduced gradually, beginning with properties and characteritics of the materials used in the reinforced concrete. Chapter 2, 3, 4 and 5 deal with the basic design principles of various types of beams and slabs while chapter 6 gives their detailed designs. In chapter 7, the designs of various types of stairs, including cinema balcony, are given. Chapter 8, 9, 10 and 11 present the detailed design procedures for various types of roof slabs. Axially loaded columns are given in chapter 12 while problems of combined direct and bending stresses are discussed in chapter 13. Chapters 14 and 15 deal with reinforced concrete foundations and reinforced concrete retaining walls respectively. Practically all the common types of foundations and retaining walls have been included. Reinforced concrete domes are given in chapter 16. The design of water tanks have been divided into two parts : (*i*) tank resting on ground surface, and (*ii*) under–ground and overhead tanks. The first has been given in chapter 17. The underground and overhead tanks have been given in volume 2 of the book. Chapter 18 describes the formwork of various concrete structures. The tests for cement and concrete have been described in detail in chapter 19.

Each chapter begins with clear statements of pertinent definitions, design principles and theories, and the design procedures. The basic principles are supplemented with numerous design examples and illustrations, along with the detailed drawing. The book is a text for students of civil engineering discipline. The designing and construction engineers will also find it a useful guide.

Various designs in this volume are based on latest Indian Standards. The tables or curves giving permissible stresses and/or design coeffcients, as well as the basic rules for design, taken from the various Indian Standards are gratefully acknowledged. In spite of every care taken to check the numerical work, some errors may remain, and I shall be obliged for any intimation of these readers may discover.

Jodhpur
1-4-75

B.C. PUNMIA

Preface to the Second Edition

In the Second Edition, the subject matter has been thoroughly updated and the revised Indian Standard Code (IS : 456-1978) has been introduced. The revised Indian Standard Code differs from the earlier Code (IS : 456-1964) in two aspects : (1) some design procedures have been basically changed, and (2) the Code uses SI units. The Second Edition of the book incorporates both the aspects. However, since the change over from the MKS to SI units has to be gradual, and since the revised Code has yet not been included in the teaching curriculum of all the Institutions, the design procedures based on earlier Code have been retained. The design procedures laid down in the revised Code have been given at the end of each chapter and the illustrative examples using these new procedures have been solved in SI units. Further suggestions will be greatly appreciated.

Jodhpur
1-11-80

B.C. PUNMIA

Preface to the Third Edition

In the Third Edition of the book the subject matter has been thoroughly revised and updated. The revised Indian Standard Code IS : 456-1978 has now been incorporated in all the chapters. The notable addition to the book is the introduction of the development length requirements at the supports of beams and slab to limit the maximum diameter of steel reinforcement envisaged in the equation

$\frac{M_1}{V} + L_0 > L_d$. Similarly, in Chapter 11 on Flat Slabs, simple tables have been provided with the help of which flexural stiffness coefficients for columns and slabs (with or without drops) can be computed with ease, thus taking into account the effect of pattern of loading. This would make the design of flab slab easier and simpler. The IS Code on water tanks has not been revised as yet. However, the changes introduced in the permissible stresses in steel reinforcement in the Code in its *fourth reprint* (May 1982) have been incorporated in the book. Further suggestions will be greatly appreciated.

Jodhpur
11.2.83

B.C. PUNMIA

Preface to the Fourth Edition

In the Fourth Edition, the book has been completely rewritten, using SI units and incorporating the recommendations of the Latest Indian Standards. The basic text has now been given new orientation. The design requirements laid down in Indian Standard Code IS : 456-1978 have been explained in greater details through illustrative examples. Special attention has been paid to the shear and development length requirements of the Code, for beam, slabs, flat slabs, footings and retaining walls. The useful suggestions given by many readers have been incorporated. Further suggestions will be greatly appreaciated.

Jodhpur
21.3.85

B.C. PUNMIA

Preface to the Fifth Edition

In the Fifth Edition, the subject matter has been thoroughly revised and enlarged. In all the chapters, a large number of illustrative design examples have been added, using Fe 415 Steel. This will enable the designers to use high yield strength deformed bars (HYSD bars). Further suggestions will be greatly appreciated.

Jodhpur
1st Jan. 1988

B.C. PUNMIA

Preface to the Sixth Edition

In the Sixth Edition, the subject matter has been thoroughly revised and enlarged. The book has now been divided into two parts. Part I deals with the Working Stress Method of Design. Part II, containing Eight chapters is on Limit State Method of Design and is completeley a new addition. This has been done to acqaint the students with the modern trends of Designs in which preference is given to the Limit State Method of Design. In many Institutions and Universities, both the methods are taught simultaneously, so that the students can compare both the methods in each design. It is hoped, this new addition of Part II will be useful to the students teachers and the Field/Practising Engineers.

Jodhpur
6th May 1990

B.C. PUNMIA
ASHOK KUMAR JAIN

Preface to the Seventh Edition

In the seventh Edition, the book has been thoroughly **revised, enlarged** and **updated. The number of chapters have been increased from 27 to 31.** The book has been divided into **five parts.** *Part I*, containing 12 chapters, is on *Beams and Slabs. Part II* on *Compression Members* has 2 chapters. *Part III*, containing four chapters is on *Foundations and Retaining Structures. Part IV*, consisting of Four chapters is on *Water Tanks and Miscellaneous structures*. Lastly, *Part V* having 9 chapters is devoted to the *Limit State Design.* **Thus, in the first volume of the book, the student will be introduced to both the Working Stress Method as well as the Limit State Method.** It is hoped, with these changes, the **New Edition** will be more useful to both the students, as well as Field Engineers.

B.C. PUNMIA
ASHOK KUMAR JAIN
ARUN KUMAR JAIN

Jodhpur
15.5.92

Contents

Part I
BEAMS AND SLABS

PART II
COMPRESSION MEMBERS

PART III
FOUNDATIONS AND RETAINING STRUCTURES

CHAPTER 15. CONTINUOUS AND ISOLATED FOOTINGS

CHAPTER 16. COMBINED FOOTINGS

CHAPTER 17. PILE FOUNDATIONS

CHAPTER 18. RETAINING WALLS

PART IV
WATER TANKS AND MISCELLANEOUS STRUCTURES

PART V

LIMIT STATE DESIGN METHOD

SYMBOLS

A = Total area of section.

A_b = Equivalent area of helical reinforcement.

A_c = Area of compressive steel.

A_e = Equivalent area of section.

A_k = Area of concrete core.

A_m = Area of steel or iron core.

A_{sc} = Area of longitudinal reinforcement (comp.)

A_{st} = Area of steel (tensile).

A_l = Area of longitudinal torsional reinforcement.

A_{sv} = Total cross-sectional area of stirrup legs or bent up bars within distance s_v.

A_w = Area of web reinforcement.

A_{Φ} = Area of cross-section of one bar.

a = Lever arm.

a_c = Area of concrete.

B = Flange width of T-beam.

b = Width.

b_r = Width of rib.

C = Compressive force.

c = Compressive stress in concrete.

c' = Stress in concrete surrounding compressive steel.

c_s = Permissible tensile stress in concrete.

c_1 = Compressive stress at the junction of flange and web.

D = Depth.

d = Effective depth.

d_c = Cover to compressive steel.

d_s = Depth of slab.

d_t = Cover to tensile steel.

e = Eccentricity.

= Compressive steel depth factor ($= d_c/d$).

F = Shear force,

F_r = Radial shear force.

f = Stress (in general).

f_{ck} = Characteristic compressive stress.

f_y = Characteristic strength of steel.

H = Height.

I = Moment of inertia.

I_e = Equivalent moment of inertia of section.

j = Lever arm factor.

K_a = Coefficient of active earth pressure.

K_p = Coefficient of passive earth pressure.

k = Neutral axis depth factor (n/d).

L = Length.

L_d = Development length.

l = Effective length of column ; Length ; Bond length.

M = Bending moment ; moment.

M_r = Moment of resistance ; Radial bending moment.

M_t = Torsional moment.

M_θ = Circumferential bending moment.

m = Modular ratio.

n = Depth of neutral axis.

n_c = Depth of critical neutral axis.

$\Sigma 0$ = Sum of perimeter of bars.

P_a = Active earth pressure.

P_p = Passive earth pressure.

P_u = Ultimate Load.

p = Percentage steel.

p' = Reinforcement ratio (A_{st}/bd).

p_a = Active earth pressure intensity.

p_e = Net upward soil pressure.

p_a = Passive earth pressure intensity.

Q = Shear resistance.

q = Shear stress (due to bending).

q' = Shear stress due to torsion.

R = Radius ; Resistance factor $\left(=\frac{1}{2}cjk\right)$.

r = Radius ; cost ratio of steel and concrete ; L/B ratio.

s = Spacing of bars ; standard deviation.

s_a = Average bond stress.

s_b = Local bond stress.

T = Tensile force ; Thickness of wall ; Torsional moment.

t = Tensile stress in steel.

t_c' = Compressive stress in compressive steel.

W = Point load ; Total load.

X = Co-ordinate.

Z = Distance : Co-ordinate.

Y, y = Co-ordinate

Z_B, Z_L = Bending moment coefficients.

α = Inclination ; coefficient.

β = Surcharge angle.

γ = Unit weight of soil.

γ' = Submerged unit weight of soil.

σ_{cc} = Permissible stress in concrete (direct comp).

σ_{cc}' = Direct compressive stress in concrete.

σ_{cbc} = Permissible compressive stress in concrete due to bending.

σ_{cu} = Ultimate compressive stress in concrete cubes.

σ_{sc} = Permissible compressive stress in bars.

σ_{sh} = Permissible stress in helical reinforcement.

σ_{sp} = Permissible punching shear stress.

σ_{st} = Permissible stress in steel in tension.

σ_{st} = Permissible tensile stress in shear reinforcement.

σ_{st} = Permissible tensile stress in main reinforcement.

σ_{sy} = Yield point compressive stress in steel.

μ = Coefficient of friction.

Φ = Diameter of bar ; angle of internal friction

τ_{bd} = Design bond stress,

τ_c = Shear stress in concrete

τ_{cmax} = Max. shear stress in concrete with shear reinforcement.

τ_v = Nominal shear stress.

1

Introduction

1.1 CEMENT CONCRETE

Concrete is a product obtained artificially by hardening of the mixture of cement, sand, gravel and water in predetermined proportions. When these ingredients are mixed, they form a plastic mass which can be poured in suitable moulds, called *forms*, and set on standing into hard solid mass. The chemical reaction of cement and water, in the mix, is relatively slow and requires time and favourable temperature for its completion. This time, known as *setting time* may be divided into three distinct phases. The first phase, designated as time of *initial set*, requires from 30 minutes to about 60 minutes for completion. During this phase, the mixed concrete decrease its plasticity and develops pronounced resistance to flow. The second phase, known as *final set*, may vary between 5 to 6 hours after the mixing operation. During this phase, concrete appears to be relatively soft solid without surface hardness. The third phase consists of progressive hardening and increase in strength. The process is rapid in the initial stage, until about one month after mixing, at which time the concrete almost attains the major portion of its potential hardness and strength.

Depending on the quality and proportions of the ingredients used in the mix, the properties of concrete vary almost as widely as different kinds of stones. Concrete has enough strength in compression, but has little strength in tension. Due to this, concrete as such is weak in bending, shear and torsion. Hence the use of *plain concrete*, described above, is limited to applications where great compressive strength and weight are the principal requirements and where tensile stresses are either totally absent or are extremely low. However, to use cement concrete for common structures such as beams, slabs, retaining structures etc, *steel bars* may be placed at tensile zones of the structure which may then be concreted. The steel bars, known as *steel reinforcement*, embedded in the concrete, takes the tensile stresses. The concrete so obtained is termed as *reinforced cement concrete,* commonly abbreviated as R.C.C.

1.2. CLASSIFICATION AND COMPOSITION OF CEMENT

1. *Classification*

Cement may be classified into five groups : (1) Portland Cement, (2) High Alumina Cement, (3) Super Sulphated Cement, (4) Natural Cements and (5) Special Cements, with the following subdivisions :

1. *Portland Cements*
   - (a) Ordinary Portland Cement
   - (b) Rapid Hardening Cement
   - (c) Extra Rapid Hardening Cement
   - (d) Low Heat Portland Cement
   - (e) Portland Blast furnace Slag Cement
   - (f) Portland-Puzzolana Cement
   - (g) Sulphate Resisting Portland Cement
   - (h) White Portland Cement
   - (i) Coloured Portland Cement
2. *High Alumina Cement*
3. *Super Sulphated Cement*
4. *Natural Cements*
5. *Special Cements*
   - (a) Masonry Cement
   - (b) Trief Cement
   - (c) Expansive Cement
   - (d) Oil Well Cement

2. *Composition of Portland Cement*

The principal raw materials used in the manufacture of cement are :

(a) Argillaceous or silicates of alumina in the form of clays and shales.

(b) Calcareous or calcium carbonate, in the form of lime stone, chalk and marl which is a mixture of clay and calcium carbonate.

The ingradients are mixed in the proportion of about two parts of calcareous material to one part of argillaceous material and then crushed and ground in ball mills in a dry state or mixed in a wet state. The dry powder or the wet slurry is then burnt in a rotary kiln at a temperature between 1400° to 1500°C. The clinker obtained from the kiln is first cooled and then passed on to ball mills where gypsum is added and it is ground to the requisite fineness according to the class of product.

The chief chemical constituents of Portland cement are as follows.

Lime (CaO)	60 to 67%
Silica (SiO_2)	17 to 25%
Alumina (Al_2O_3)	3 to 8%
Iron Oxide (Fe_2O_3)	0.5 to 6%
Magnesia (MgO)	0.1 to 4%
Sulphur Trioxide (SO_3)	1 to 3%
Soda and/or Potash (Na_2O+K_2O)	0.5 to 1.3%

The above constituents forming the raw materials undergo chemical reactions during burning and fusion, and combine to form the following compounds (called Bogue compounds) in the finished products :

Compound	*Abbreviated designation*
1. Tricalcium silicate ($3CaO.SiO_2$)	C_3S
2. Dicalcium silicate ($2CaO.SiO_2$)	C_2S
3. Tricalcium aluminate ($3CaO.Al_2O_3$)	C_3A
4. Tetracalcium alumino-ferrite ($4CaO.Al_2O_3.Fe_2O_3$)	C_4AF

The proportions of the above four compounds vary in the various Portland cements. Tricalcium silicate and dicalcium silicates contribute most to the eventual strength. Initial setting of Portland cement is due to the tricalcium aluminate. Tricalcium silicate hydrates quickly and contributes more to the early strength. The contribution of dicalcium silicate takes place after 7 days and may continue for upto 1 year. Tricalcium aluminate hydrates quickly, generates much heat and makes only a small contribution to the strength within the first 24 hours. Tetracalcium alumino-ferrite is comparatively inactive. All the four compounds generate heat when mixed with water, the aluminate generating the maximum heat and the dicalcium silicate generating the minimum. Due to this, tricalcium aluminate is responsible for most of the undesirable properties of concrete. Cement having less C_3A will have higher ultimate strength, less generation of heat and less cracking. Table 1.1 gives the composition and percentage of the four compounds for normal, rapid hardening and low heat portland cement.

TABLE 1.1

COMPOSITION AND COMPOUND CONTENT OF PORTLAND CEMENT
(After Lea)

	Normal	*Rapid hardening*	*Low heat*
(a) **Composition : Percent**			
Lime	63.1	64.5	60
Silica	20.6	20.7	22.5
Alumina	6.3	5.2	5.2
Iron Oxide	3.6	2.9	4.6
(b) **Compound : Percent**			
C_3S	40	50	25
C_2S	30	21	45
C_3A	11	9	6
C_4AF	12	9	14

3. *Ordinary Portland Cement (IS : 269)*

The properties of various types of portland cement differ because of relative proportions of the four compounds and the fineness to which the cement clinker is ground. The *Ordinary Portland Cement* or the *Setting Cement* is the basic Portland cement and is manufactured in larger quantities than all the others. It is admirably suited for use in general concrete construction where there is no exposure to sulphates in the soil or in ground water.

4. *Rapid Hardening Portlund Cement (IS : 269)*

This cement is also known as *high-early strength cement.* It is similar to ordinary Portland cement except that it is ground finer, possesses more C_3S and less C_2S than the ordinary Portland Cement. The magnitude of the increase in strength is gauged from the fact that the strength developed at the age of 3 days is about the same as 7-day strength of ordinary Portland cement with the same water-cement-ratio. The main advantage of a rapid hardening cement is that shuttering may be removed much earlier, thus saving considerable time and expenses. Similarly, in the concrete products in industry, moulds can be released quicker. Rapid hardening cement is also used for road work where it is imperative to open the road traffic with the minimum delay.

5. *Extra Rapid Hardening Cement*

Extra rapid hardening cement is obtained by inter-grinding calcium chloride with rapid Hardening Portland cement. The normal addition of $CaCl_2$ is 2% (of the commercial 70% $CaCl_2$) by weight of the rapid hardening cement. The addition of $CaCl_2$ also imparts quick setting properties. Hence this cement should be placed and fully compacted within 30 minutes of mixing.

6. *Low Heat Portland Cement (IS : 269)*

When concrete is poured in any structure, an increase in temperature occurs and a certain amount of heat is generated. This is due to the chemical reaction that takes place while the cement is setting and hardening. Low heat Portland cement is used in massive constructions like abutments, retaining walls, dams, etc., where the rate at which the heat can be lost at the surface is lower than at which the heat is initially generated. The heat generated in ordinary cement at the end of 3 days may be of the order of 80 calories per gram of cement, while in low heat cement it is 50 calories per gram.

It has low percentagae of C_3A and relatively more C_2S and less C_3S than ordinary Portland cement. This is achieved by restricting the amount of calcium and increasing the silicates present in the raw materials of manufacture. Therefore, it has low rate of gain of strength, but the ultimate strength is practically the same.

7. *Portland Blast Furnace Cement (IS : 455)*

This cement is made by intergrinding Portland cement clinker and blast furnace slag, the proportion of the slag being not less than 25% or more than 65% by weight of cement, as prescribed by IS : 455. The slag should be granulated blast furnace slag of high lime content, which is produced by rapid quenching of molten slag obtained during the manufacture of pig iron in a blast furnace. It is usual for the Portland cement clinker to be ground with a slag, a small percentage of gypsum being added to regulate the setting time. The blending of the Portland cement clinker with the slag, by no means detracts from any desired property of cement. Indeed, it confers upon it some additional advantage. This is because the granulated slag itself possesses latent hydraulic properties which are tremendously activated when the slag is crystalised and integrated with Portland cement clinker. In general, blast furnace cement will be found to gain strength more slowly than the ordinary Portland cement. It has less heat of hydration than ordinary Portland cement. From the point of view of a builder and the structural engineer, Portland blastfurnace slag cement may be used for all purpose for which ordinary Portland cement is used. In addition, in view of its low heat evolution, it can be used in mass concrete structures such as dams, retaining walls, foundations and bridge abutments.

8. *Portland Pozzolana Cement (IS : 1489)*

Portland Pozzolana cement is manufactured either by intergrinding Portland cement clinker and pozzolana or by intimately and uniformly blending Portland cement and fine pozzolana. While intergrinding presents no difficulty, *blending* tends to result in a non-uniform product and Indian Standard is specific in specifying that the latter method should be confined to factories and other such works where intimate blending can be ensured through mechanical means. As per Indian Standard, the proportion of pozzolana may vary from 10 to 25% by weight of cement. The pozzolana used in the manufacture of Portland Pozzolana cement in India is, at present, burnt clay or shale, or fly ash. Although pozzolanas have no cementing value themselves, they have the property of combining with free lime to produce a stable lime pozzolana compound which has definite cementitious properties. This cement has higher resistance to chemical agencies and to attack by sea water, because of absence of free lime. Portland Pozzolana cement also has a lower heat of evolution. Portland Pozzolana cement is frequently stated to have a lower rate of development of strength than ordinary Portland cement. However, when the pozzolana is selected with care and is refined and ground with Portland cement clinker under controlled conditions, the compressive strengths reached by Portland pozzolana cement are comparabale with those reached by ordinary Portland cement. This can be seen from the following table which compares the strength at different ages of Portland Pozzolana cement and Ordinary Portland cement manufactured at the cement works of the Associated Cement Companies (ACC) Ltd., India :

	Compressive Strength, N/mm²	
Age in days	*Portland Pozzolana Cement*	*Ordinary Portland Cement*
3	20 - 22	19 - 23
7	26 - 33	27 - 32
28	37 - 48	36 - 52

9. *Sulphate Resisting Cement*

In sulphate resisting Portland cement, the quantity of tricalcium aluminate is strictly limited. They are normally ground finer than Portland cement. The action of sulphates is to form sulpho-aluminates which have expansive properties and so cause disintegration of the concrete. Sulphate resistisng cement should be allowed to harden in the air for as long as possible to allow a resistant skin to be formed through carbonation by the action of atmospheric carbon dioxide.

10. *White and Coloured Portland Cements*

The greyish colour of Portland cements is due to the presence of iron oxide. White Portland cement is manufactured in such a way that the percentage of iron oxide is limited to less 1%. To achieve this, superior raw materials, such as chalk and lime stone having low percentage of iron, and white clay (China clay) are used. Sodium aluminium fluoride (cryolite) is added to act as flux in the absence of iron oxide. Oil fuel is used in place of pulverised coal in the kilning process in order to avoid contamination by coal ash. Coloured Portland cements are usually obtained by adding strong pigments, upto 10% to the ordinary or white cement, during grinding of clinker. The essential requirements of a good pigment are that it should be permanent and should be chemically inert when mixed with cement.

11. *High Alumina Cement*

High Alumina cement, also known as *aluminous cement foundu* is manufactured in entirely different way from that of Portland cements. The raw materials used for its manufacture are chalk and bauxite which is a special clay of extremely high alumina content. The manufacture of this type of cement is more expansive than the Portland cements, though it has many advantages over other types of cements. High alumina cement is characterised by its dark colour, high early strength, high heat of hydration and resistance to chemical attack. It thus produces concrete of far greater strength and in considerably less time even than Rapid - Hardening Portland cement, allowing earlier removal of the formwork. Its rapid hardening properties arise from the presence of calcium aluminate chiefly monocalcium aluminate ($Al_2O_3 . CaO$) as the predominant compound in place of calcium silicates of Portland cement and after setting and hardening there is no free hydrated lime as in the case of Portland cement. However, great care should be taken in the use of high alumina cement, and it must not be mixed with any other type of cement, since the heat given off on setting is greater than with other cement.

12. *Super Sulphated Cement (IS : 6909)*

Super sulphated cement is made from well granulated blastfurnace slag (80 to 85%), calcium sulphate (10 to 15%) and Portland cement (1 to 2%) and is ground finer than the Portland cement. One of its most important properties is its low total heat of hydration. It is, therefore, very suitable for construction of dams and mass concreting work. Concrete made from super sulphated cement may expand if cured in water and may shrink if the concrete is cured in air. Another big advantage of super sulphated cement is its comparatively high resistance to chemical attack.

13. *Natural cement*

Natural cements are those cements which are manufactured from naturally occurring *cement rocks* which have compositions similar to the artificial mix of argillaceous and calcareous materials from which Portland cement is manufactured. However, the natural cement rocks are burned at somewhat lower temperatures than those used for the production of Portland cement clinker. The properties of such cement depend upon the composition of the natural cement rock.

14. *Masonry Cement*

For a long time, lime gauged with sand was used for mortar for laying brick work. However, in order to increase the strength and rapidity of gaining strength it became common to mix Portland cement with the lime. The usual proportions of cement : lime : sand may range from 1 : 1 : 6 for heavy loads to 1 : 3 : 12 for light loads. Cement sand mortars are too harsh, while lime makes the mortar easier to work. In order to avoid the necessity for mixing cement and lime, masonry cements have recently been introduced. According to Wuerpel, most successful masonry cement are composed of Portland cement clinker, lime stone, gypsum and air-entraining agent. These constituents are ground to an even greater fineness than that of high early strength Portland cement. The plasticity and workability of masonry cements are imparted by the lime stone and air-entraining agents. The ease of working masonry cements and their water retentive properties help to increase their adhesion to bricks or other building units and this is further assisted by the fact their shrinkage is fairly low.

15. *Trief Cement*

Trief cement is practically the same as blastfurnace cement except that the blastfurnace slag is ground wet and separately from the cement. Wet grinding results in a fine product, with a specific surface of at least 3000 cm^2/g. Due to this, the slow rate of gain of strength normally associated with blastfurnace cement is avoided and strength from early ages equal to those of ordinary Portland cement are obtained. This cement has smaller shrinkage and a smaller heat of evolution while setting than ordinary Portland cement.

16. *Expansive Cement*

Expansive cement expands while hardening. Ordinary concrete shrinks while hardening, resulting in shrinkage cracks. This can be avoided by mixing expansive cement with the normal cements in the concrete, which will neither shrink nor expand. Another useful application of expansive cement is in repair work where the opened up joints can be filled with this cement so that after expansion a tight joint is obtained. Expansive cements have been used in France for underpinning and for the repair of bomb damaged arch bridges.

17. ***Oil Well Cements***

In the drilling of oil wells, cement is used to fill the space between the steel lining tube and the wall of the well, and to grout up porous strata and to prevent water or gas from gaining access to oil-bearing strata. The cement used may be subject to very high pressure, and the temperature may rise to 400°F. Cement used must be capable of being pumped for up to about 3 hours. It must also harden quickly after setting. These properties can be achieved by (*a*) adjusting the composition of the cement and (*b*) by adding retarders to ordinary Portland cement. In case (*a*), the proportion of Fe_2O_3 is adjusted so that it is above that required to combine with all the Al_2O_3 to form tetra calcium alumino-ferrite $4CaO.Al_2O_3.Fe_2O_3$. The proportion of tricalcium aluminate $3CaO.Al_2O_3$ formed is therefore very small and the setting time is accordingly increased. Setting times of upto 4 hours at a temperature of 200°F and 6 hours at a temp. of 70°F can be obtained with a Portland cement containing no tricalcium aluminate. By the use of retarders setting time of upto $6\frac{1}{2}$ hours at temperatures of upto 220°F can be obtained.

1.3 SPECIFICATIONS FOR PORTLAND CEMENT

For the quality control of Portland cement used for plain and reinforced concrete, the Indian Standard Institution has recommended the following specifications and tests : (1) Chemical composition (2) fineness (3) soundness (4) setting time (5) compressive strength, and (6) heat of hydration.

The detailed procedures for the above tests are given in a separate chapter at the end of the book.. The specification for the above requirements, recommended by the Indian Standard are given below.

1. ***Chemical Requirements***

When tested in accordance with the methods given in IS : 4032–1968 (Methods of chemical analysis of hydraulic cement), ordinary cement and rapid hardening Portland cement shall comply with the following chemical requirements :

(*a*) Ratio of percentage of lime to percentages of silica, alumina and iron oxide; when calculated by the formula	Not greater than 1.02. and not less than 0.66.

$$\frac{CaO - 0.7\,SO_3}{2.8\,SiO_2 + 1.2Al_2O_3 + 0.65\,Fe_2O_3}$$

(*b*) Ratio of percentage of alumina to that of iron oxide	Not less than 0.66.
(*c*) Weight of insoluble residue	Not more than 2 per-cent
(*d*) Weight of magnesia	Not more than 6 per-cent
(*e*) Total sulphur content calculated as sulphuric anhydride(SO_3)	Not more than 2.75 per-cent
(*f*) Total loss on ignition	Not more than 4 per-cent

When tested in accordance with methods given in IS : 4032-1968, Low Heat Portland cement shall comply with the following requirements as to its chemical composition :

The percentage of lime, after deduction of that necessary to combine with the sulphuric anhydride present, shall be :

not more than $2.4(SiO_2)+1.2(Al_2O_3)+0.65(Fe_2O_3)$

and not less than $1.9(SiO_2)+1.2(Al_2O_3)+0.65(Fe_2O_3)$.

Each symbol in brackets refers to the percentage (by weight of total cement) of the oxide, excluding any contained in the insoluble residue. In all other respects, low heat Portland cement shall comply with requirements specified in *(b)*, *(c)*, *(d)*, *(e)* and *(f)* above.

2. *Fineness*

When tested for fineness in terms of specific surface, by Blaine's air permeability method as described in IS : 4031-1968 (Methods of chemical analysis of hydraulic cements), the cement shall comply with the following requirements :

Type of Cement	*Specific Surface*
Ordinary	Not less than 2250
Rapid-hardening	Not less than 3250
Low heat	Not less than 3200

3. *Soundness*

When tested by the 'Le Chatelier' method described in IS : 4031–1968 (see chapter on 'Tests for cement and concrete'), unaerated ordinary rapid hardening and low heat Portland cement shall not have an expansion of more than 10 mm.

In the event of the cement failing to comply with the above requirements, a further test shall be made by the 'Le Chatelier' method, from another portion of the same sample after aeration, by being spread out to a depth of 75 mm at a relative humidity of 50 to 80 percent for a total period of 7 days, when the expansion of each of the three types of cements mentioned above shall not be more than 5 mm.

When specified by purchaser at the time of placing the order, unaerated ordinary, rapid hardening and low heat Portland cements shall not have an expansion of more than 0.3 percent when tested by the autoclave test described in IS : 4031–1968.

All cements having a magnesia content more than 2 percent shall be tested for soundness by autoclave test and shall comply with the requirements specified in the above para.

4. *Setting time*

The setting time of the cements, when tested by Vicat apparatus method shall conform to the following requirements :

	Ordinary	*Rapid hardening*	*Low heat*
(*a*) Initial Setting time in minutes, *not less than*	30	30	60
(*b*) Final setting time in minutes, *not more than*	600	600	600

5. *Compressive strength*

The average compressive strength of atleast three mortar cubes (area of face 50 cm^2) composed of one part of cement, three parts of standard sand (conforming to IS : 650–1966) by weight and $\frac{P}{4} + 3.0$ percent (combined weight of cement plus sand) water, and prepared, stored and tested in the manner described in IS : 4031–1968, shall be as follows :

	Ordinary (N/mm^2)	*Rapid hardening* (N/mm^2)	*Low heat* (N/mm^2)
(*a*) 24 h±30 min., *not less than*	—	16	—
(*b*) 72 h±1 h, *not less than*	16	27.5	10
(*c*) 168 h ±2 h, *not less than*	22	—	16
(*d*) 672 h ±4 h, *not less than*	—	—	35

Alternatively, the cement may be accepted based on the compressive strength limits indicated in the para below :

The average compressive strength of atleast three mortar cubes (area of face 50 cm^2) composed of one part of cement, three parts of sand by weight, and $\frac{P}{4} + 3.5$ percent (of combined weight of cement plus sand) water, and prepared, stored and tested in the manner described in IS : 4031–1968 shall be as follows :

	Ordinary (N/mm^2)	*Rapid hardening* (N/mm^2)	*Low heat* (N/mm^2)
(*a*) 24 h±30 min., *not less than*	—	11.5	—
(*b*) 72 h±1 h, *not less than*	11.5	21	7
(*c*) 168 h ±2 h, *not less than*	17.5	—	11.5
(*d*) 672 h ±4 h, *not less than*	—	—	26.5

Where P is the percentage of water required to produce a paste of standard consistency, to be used for the determination of the water content of mortar for the compressive strength tests and for the determination of soundness and setting time shall be obtained by the method described in IS : 4031–1968.

6. *Heat of Hydration*

The requirement shall apply only to low heat cement. When tested according to the method described in IS : 4031–1968, the heat of hydration of low heat Portland cement shall be as follows :

TABLE 1.2. INDIAN STANDARD SPECIFICATIONS FOR CEMENTS : PHYSICAL REQUIREMENTS

	Type of Cement	Ordinary Portland	Rapid Hardening Portland	Low heat Portland	Portland blast Furnace Slag	Portland Pozzalana	High alumina	Super Sulphated
	Indian Standard	269-1967	269-1967	269-1967	455-1967	1489-1967	6452-1972	6909-1973
1. Fineness	Residue by Wt. on 90-micron IS sieve	$\ngtr$ 10 %	$\ngtr$ 5 %	—	$\ngtr$ 10 %	$\ngtr$ 5 %	—	—
	Min. specific surface (Blain's air permeability) cm^2/gm.	2250	3253	3200	2250	3000	2250	4000
2. Soundness	Explansion[1] (Lechatelier)	$\ngtr$ 10mm	$\ngtr$ 10 mm	$\ngtr$ 10 mm	$\ngtr$ 10 mm	$\ngtr$ 10 mm	$\ngtr$ 5mm	$\ngtr$ 5mm
	Autoclave test[2]	$\ngtr$ 0.8%	$\ngtr$ 0.8%	$\ngtr$ 0.8%	$\ngtr$ 0.8%	$\ngtr$ 0.8%	—	—
3. Setting time	Initial	$\nless$ 30 min	$\nless$ 30 min	$\nless$ 60 min	$\nless$ 30 min	$\nless$ 30 min	$\nless$ 30 min	$\nless$ 30 min
	final	$\ngtr$ 600	$\ngtr$ 600	$\ngtr$ 600	$\ngtr$ 600	$\ngtr$ 600	$\ngtr$ 600	$\ngtr$ 600
4. Min. compressive strength[4] (N/mm^2)	24 hours[3]	—	16	—	—	—	30	—
	72 hours[3]	16	27.5	10	16	—	35	15
	168 hours[3]	22	—	16	22	32	—	22
	762 hours	—	—	35	—	31	—	30
				Heat of hydration to be not more than 65 cal/gm at 7 days and 75 cal/gm at 28 days		Average drying shrinkage of mortars to be not more than 0.15%		

Notes : 1. In the event of the cements failing to comply with this requirement, the test is repeated with another portion of the same sample after aeration for 7 days, when the expansion for the first five types shall not be more than 5 mm.

2. When specified by purchaser and for all cements having a magnesia contents more thant 3 per cent.

3. There shoud be a progressive increase in strength from the strength at 24., 72 or 168 hours.

4. Alternatively, see amendment 1 of IS : 269-1967, amendment 2 of IS : 455-1967 and amendment 2 of IS : 1489- 1967 tor revised compressive strength

TABLE 1.3. INDIAN STANDARD SPECIFICATIONS FOR CEMENTS : CHEMICAL REQUIREMENTS

Type of Cement	Ordinary and rapid hardening Portland	Low heat Portland	Portland blast furnaoe Slag	Portland pozzalana	High alumina	Super sulphated
Indian Standard	**269-1967**	**269-1967**	**455-1967**	**1489-1972**	**6452-1972**	**6909-1973**
	Max-permissible values					
1. Insoluble residue	2.0%	2.0%	$(x'+0.025)+2.0$	$x+2.0\frac{(100-x)}{100}$	—	4%
2. Magnesia (MgO)	6%	6%	8%	6%	—	10%
3. Sulphur						
as sulphuric anhydride (SO_3)	2.75%	2.75%	3%	2.75	—	6%
as sulphide (S)	—	—	1.5%	—	—	1.5%
4. Manganic oxide (Mn_2O_3)	—	—	2%	—	—	
5. Loss on ignition	4%	4%	4%	5%	—	—

where : x is the declared percentage of pozzalana in the given cement.
x' is the declared percentage of granulated blast furnace slag in the cement.

(a) *7 days : not more than* 65 Calories per gram, and

(b) *28 days : not more than* 75 calories per gram

Comparison of properties of various type of cements

Table 1.2 gives the *physical requirements* of various common types of cements, as specified by the Indian Standards. Similarly, Table 1.3 gives the chemical requirements for various types of cements.

1.4 AGGREGATES

Aggregates is a general term applied to those inert or chemically inactive materials which, when bonded together by cement, form concrete. Most of the aggregates used are naturally occurring aggregates such as crushed rock, gravel and sand. Artificial and processed aggregates may be broken brick of crushed air-cooled blastfurnace slag. Light weight aggregates, such as pumice, furnace clinker, coke, breeze, sawdust, foamed slag, expanded slates, expanded vermiculite etc. are also used for the production of concrete of low density.

Classification

Aggregates may be divided into two groups : (*a*) coarse aggregates and (*b*) fine aggregate. Aggregates less than 4.75 mm are known as *fine aggregates* while those more than 4.75 mm in size are known as *coarse aggregate.* For large and important works it has become usual to separate the coarse aggregate also into two or more sizes, and these fractions are kept separate until the proper quantity of each has been weighed out for a batch of concrete. *All in aggregate,* that is to say, aggregate as it comes from the pit or river bed, is sometimes used for unimportant works.

Quality of aggregate

Natural aggregate used for concrete construction is required to comply with the norms laid down in IS : 383–1970 '*specification for coarse and fine aggregates from natural sources for concrete*'

Some of the important characteristics of aggregates are : (1) strength (2) size (3) particle shape (4) surface texture (5) grading (6) impermeability (7) cleanliness (8) chemical inertness (8) physical and chemical stability at high temperatures (10) coefficient of thermal expansion, and (11) cost.

Aggregate should be chemically inert, strong, hard, durable, of limited porosity, free from adherent coatings clay lumps, coal, and coal residues and should contain no organic or other admixture that may cause corrosion of the reinforcement, or impair the strength or durability of the concrete. The limits of the content of deleterious materials in aggregate are given in Table 1.4.

The strength of concrete depends upon the strength of aggregate. The strength of an aggregate depends upon the nature of the base rock. Granite aggregate provides greater strength than pumice or burnt clay aggregates. Strength quality of an aggregate is usually specified in terms of (*i*) crushing strength (*ii*) aggregate crushing value, or (*iii*) impact value.

Crushing strength is determined on a rock sample of some specific size. A minimum of 80 N/mm^2 as crushing strength is normally acceptable for good concrete aggregate.

The *aggregate crushing value* is given by :

$$\text{Aggregate crushing value} = 100\frac{W_2}{W_1}$$

where W_1 = Weight of surface dry sample of aggregate passing 12.5 mm IS sieve and retained on 10 mm IS sieve.

W_2 = Weight of the material passing through 2.36 mm IS sieve after a specified load is applied in a specified manner.

A *higher* aggregate crushing value indicates a weak aggregate. Its value should not exceed 45 for plain cement concrete, 40 for R.C.C. for buildings and 30 for prestressed concrete.

The *aggregate impact value* is given by :

$$\text{Aggregate impact value} = 100\frac{W_2}{W_1}$$

Where W_1 = Weight of oven dry sample of aggregate passing through 12.5 mm IS sieve but retained on 10 mm IS sieve.

W_2 = Weight of fractured aggregate finer than 2.36 mm IS sieve, formed after 15 impacts of a special hammer falling through 300 mm.

A higher aggregate impact value indicates weaker concrete. An impact value of 30 indicates good quality aggregate, though a value upto 45 may sometimes be accepted.

The size of coarse aggregate used depends upon the nature of work. The coarse aggregate must be small enough to enable it to be worked between and around all reinforcements and into all corners of the work. For R.C.C. work, the maximum size of aggregate is limited to 20 mm to 25 mm. A coarse aggregate may have three shapes : rounded, irregular and angular. For a concrete of given workability, rounded aggregates require least water cement ratio while angular aggregates require highest water cement ratio. The particle shape is thus very important, since the water cement ratio governs greatly the strength of concrete. Similarly a concrete made with aggregates of rough surface is stronger than with smooth one. Grading of aggregates greatly affects strength and imperviousness of concrete. If the coarse and fine aggregates are well graded, the percentage of voids is considerably reduced. The voids of the fine aggregates are then occupied by the cement paste while the voids of coarse aggregate are filled with the mortar consisting sand, cement and water. The imperviousness of aggregates is an essential requirement, especially when the concrete is used for water retaining structures.

TABLE 1.4 : LIMITS OF CONTENT OF DELETERIOUS MATERIALS *
(IS : 383 — 1970)

Deleterious substances	*Fine Aggregate*		*Coarse Aggregates*	
	uncrushed	*crushed*	*uncrushed*	*crushed*
Coal and lignite	1.00	2.90	1.00	1.00
Clay lumps	1.00	1.00	1.00	1.00
Soft fragments	—	—	3.00	—
Materials finer than 75μ IS sieve	3.00	15.00	3.00	3.00
Shale	1.50	—	—	—
Total of percentage of all deleterious materials**	1.00	2.00	5.00	5.00

* Percentage by weight of aggregate.

** Mica is excluded

This is also essential in other R.C. works of permanence, otherwise air and moisture would penetrate with the result that outer concrete would spall out. Aggregates must be clean and free from clay, silt, fine dust etc. so that proper mixing is possible. Dirt or other adherent coating would weaken the adhesion between the individual particles in a hardened concrete. Impurities, such as traces of sulphur or unburnt coal etc. may cause swelling due to chemical action, or may attack the reinforcement. The aggregate should have a thermal expansion similar to that of cement matrix. To summarise, the aggregate should be composed of inert mineral matter, should have high resistance to attrition, should be clean, free from any adhering coating, dense, durable and sufficiently strong to enable the full strength of the cement matrix to be developed.

Coarse aggregate

The material retained on 4.75 mm sieve is termed as coarse aggregate. Crushed stone and natural gravel are the common materials used as coarse aggregate for concrete. Natural gravels can be quarried from pits where they have been deposited by alluvial or glacial action, and are normally composed of flint, quartz, schist and igneous rocks. Coarse aggregates are obtained by crushing various types of granites (such as syenites, dolerites, diorites, quartzites etc.), schist, gneiss, crystalline hard lime stone and good quality sand stones. When very high strength concrete is required, a very fine-grained granite is perhaps the best aggregate. Coarse grained rocks make harsh concrete, and need high proportion of sand and high water-cement ratio to get reasonable degree of workability. Harder types of sand stones, having fine grained texture, are suitable as coarse aggregate, but softer varieties should be used with caution. Concrete made with sand stone aggregate give trouble due to cracking, because of high degree of shrinkage. Similarly hard and close-grained crystalline lime stones are very suitable for aggregate, is cheap, but should be used only in plain concrete. The bricks should be clean, hard, well burnt and free from mortar and should not contain more than half percent of soluble sulphates. It should not be used for reinforced concrete work, since it is porous and may corrode the reinforcement. Blast furnace slag, coal ashes, coke-breeze etc. may also be used as aggregate to obtain light weight and insulating concrete of low strength.

Fine aggregate

The material smaller than 4.75 mm size is called fine aggregate. Natural sands are generally used as fine aggregate. Sand may be obtained from pits, river, lake or sea shore. When obtained from pits, it should be washed to free it from clay and silt. Sea shore sand may contain chlorides which may cause efflorescence, and may cause corrosion of reinforcement. Hence it should be thoroughly washed before use. Similarly, if river sand contains impurities such as mud etc. it should be washed before use. Angular grained sand produces good and strong concrete, because it has good interlocking property, while round grained particlels of sand do not afford such interlocking

Grading of aggregates

Gradation of the aggregates is almost as important as its quality is. The grading of the aggregates has marked effect on the workability, uniformity, and finishing qualities of concrete. The grading of coarse aggregate may be varied through wider limits than that of sand without appreciably affecting the workability of concrete. The following Indian Standard sieves are generally used for grading aggregates :

Fine mesh wire cloth : 3.35 mm, 2.36 mm, 1.18 mm, 600 micron, 300-micron, 150-micron, 75-micron.

Square hole perforated plate : 80-mm, 63-mm, 50-mm, 40-mm, 31.5 mm, 25-mm, 20-mm, 16-mm, 12.5 mm, 10-mm, 6.3 mm, 4.75-mm

Table 1.5 gives the grading limits for fine aggregate as specified in IS : 383–1970. Table 1.6 and 1.7 give, respectively, the grading limits of coarse aggregate as well as for all-in-aggregate as specified in IS : 383–1970.

TABLE 1.5 : GRADING LIMITS FOR FINE AGGREGATES (IS : 383—1970)

IS Sieve designation	*Percentage Passing*			
	Grading Zone I	*Grading Zone II*	*Grading Zone III*	*Grading Zone IV*
10 mm	100	100	100	100
4.75 mm	90–100	90–100	90–100	95–100
2.36 mm	60–95	75–100	85–100	95–100
1.18 mm	30–70	55–90	75–100	90–100
600 micron	15–34	35–39	60–79	80–100
300 micron	5–20	8–30	12–40	15–50
150 micron	0–10	0–10	0–10	0–15

According to IS : 383–1970, the fine aggregate (*i.e.* sand) is graded into four zones, as shown in Fig. 1.1. A fine aggregate whose grading falls outside the limits of any particular grading zones or sieves other than 600-micron I.S. Sieves is regarded as falling within that grading zone provided the total of the excess on the other four sieves is not greater than 5%. Sands of zones 1 to 3 may

TABLE 1.6. GRADING LIMITS FOR COARSE AGGREGATES
(IS : 383—1970)

IS sieve designation	Percentage passing, for single size aggregat of a nominal size						Percentage passing, for graded aggregate of nominal size			
	63 mm	40 mm	20 mm	16 mm	12.5mm	10 mm	40 mm	20 mm	16mm	12.5 mm
80 mm	100	—	—	—	—	—	100	—	—	—
63 mm	85–100	100	—	—	—	—	—	—	—	—
40 mm	0-30	85–100	100	—	—	—	95–100	100	—	—
20 mm	0-5	0-20	85–100	100	—	—	30–70	90–100	100	100
16 mm	—	—	—	85–100	100	—	—	—	90–100	—
12.5 mm	—	—	—	—	85-100	100	—	—	—	90–100
10 mm	0-5	0-5	0-20	0–30	0–45	85–100	10–35	25–25	30–70	40–85
4.75 mm	—	-	0-5	0–5	0-10	0-20	0–10	0–10	0–10	0–10
2.36 mm	—	—	—	—	—	0-5	—	—	—	—

be used for concrete work. Zone-4 sand should not be used except in special mixes.

TABLE 1.7. GRADING LIMIT FOR ALL IN AGGREGATE (IS : 383—1970)

IS Sieve designation	*Percentage passing, for all in aggregate of*	
	40 mm nominal size	*20 mm nominal size*
80 mm	100	–
40 mm	95–100	100
20 mm	45–57	95–100
4.75 mm	25–45	30–50
600 micron	8–30	10–25
150 micron	0–6	0–6

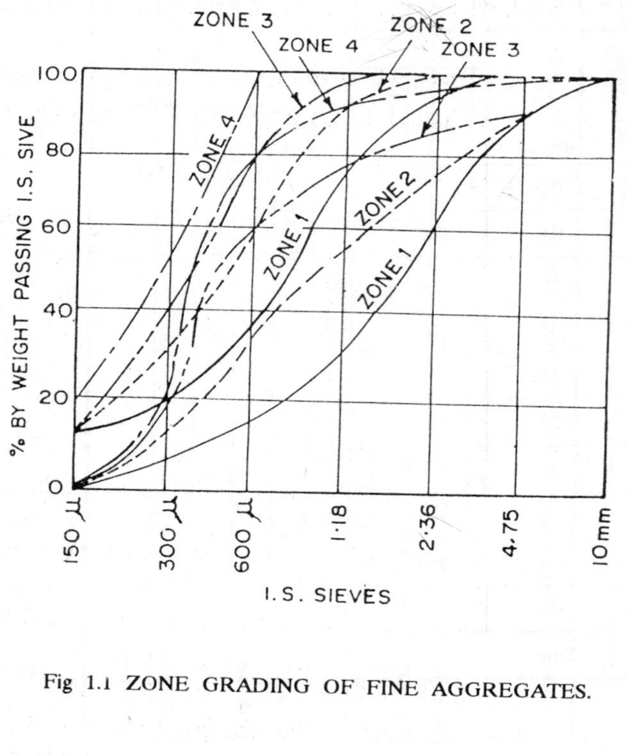

Fig 1.1 ZONE GRADING OF FINE AGGREGATES.

Fineness Modulus

The *fineness modulus* of an aggregate is an index number which is roughly proportional to the average size of the particles in the aggregate. The coarser the aggregate, the higher the fineness modulus. The fineness modulus is obtained by adding the percentage of the weight of material retained on the following I.S. sieves and dividing it by 100 :

80 mm, 40 mm, 20 mm, 10 mm, 4.75 mm, 2.36 mm, 1.18 mm, 600-micron, 300-micron and 150-micron (total 10 sieves).

Table 1.8(*a*) illustrates the method of determining fineness modulus of both coarse and fine aggregates. It has been found that certain values of fineness moduli for the fine and coarse aggregates give good workability, with a minimum quantity of cement. The limits of fineness moduli are given in table 1.8(*b*)

TABLE 1.8. (*a*) DETERMINATION OF FINENESS MODULUS

IS Sieve	*Coarse aggregate (10 kg)*			*Fine aggregate (1 kg)*		
	weight retainded (kg.)	*Total wt. retainded (kg.)*	*% weight retainded*	*weight retained (kg.)*	*Total wt. retained (kg.)*	*% weight retained*
80 mm	0.0	0.0	0.0	0.0	0.0	0.0
40 mm	0.0	0.0	0.0	0.0	0.0	0.0
20 mm	3.5	3.5	35.0	0.0	0.0	0.0
10 mm	3.0	6.5	65.0	0.0	0.0	0.0
4.75 mm	2.8	9.3	93.0	0.0	0.0	0.0
2.36 mm	0.7	10.0	100.0	0.1	0.1	10.0
1.18 mm	0.0	10.0	100.0	0.25	0.35	35.0
600 micron	0.0	10.0	100.0	0.35	0.70	70.0
300 micron	0.0	10.0	100.0	0.20	0.90	90.0
150 micron	0.0	10.0	100.0	0.10	1.00	100.0
	Sum : 693.0			Sum : 305.0		
Fineness modulus	$\frac{693.0}{100} = 6.93$			$\frac{305.0}{100} = 3.05$		

TABLE 1.8 (*b*) LIMITS OF FINENESS MODULI

Maximum size of aggregate		*Fineness modulus*	
		max.	*min.*
(a) Fine aggregate		2.0	3.5
(b) Coarse aggregate	(*i*) 20 mm	6.0	6.9
	(*ii*) 40 mm	6.9	7.5
	(*iii*) 75 mm	7.5	8.0
	(*iv*) 150 mm	8.0	8.5
(c) Mixed aggregate	(*i*) 20 mm	4.7	5.1
	(*ii*) 25 mm	5.0	5.5
	(*iii*) 32 mm	5.2	5.7
	(*iv*) 40 mm	5.4	5.9
	(*v*) 75 mm	5.8	6.3
	(*vi*) 150 mm	6.5	7.0

1.5 WATER

Water acts as lubricant for the fine and coarse aggregates and acts chemically with cement to form the binding paste for the aggregate and reinforcement. Water is also used for curing the concrete after it has been cast into the forms.

Water used for both mixing and curing should be free from injurious amount of deleterious materials. Potable waters are generally considered satisfactory for mixing and curing of concrete. If water contains any sugar or an access of acid, alkali or salt, it should not be used. As a guide, the following concentrations represent the maximum permissibles :

(*a*) To neutralize 200 ml sample, it should not require more than 2 ml 0.1 Normal NaOH.

(*b*) To neutralize 200 ml sample, it should not require more than 10 ml of 0.1 Normal NaCl.

(*c*) Percentagae of solids should not exceed the following :

	Percent
Organic	0.02
Inorganic	0.30
Sulphates	0.05
Alkali Chlorides	0.10

Table 1.9 gives tolerable concentration of some impurities in water.

Carbonates and bicarbonates of sodium and potassium. Sodium Carbonate may cause very rapid setting while carbonates may either accelerate or retard the setting. They may also reduce the strength of concrete, if present in large concentrations.

Chlorides and Sulphates. They are normally present in brackish water. Water is harmless if sulphates do not exceed 3000 ppm or chlorides do not exceed 10000 ppm.

Calcium Chloride. They accelerate both setting and hardening. The tolerable concentration is 2 percent by weight of cement in non-prestressed concrete.

Other Inorganic Salts. Salts of manganese, tin, zinc, copper and lead (nitrate) cause a marked influence on the reduction in the strength of concrete specially the last three salts are the most active. The action of lead nitrate is completely destructive. Sodium sulphide has detrimental effect and a concentration of even 100 ppm is undesirable. Salts of sodium, *i.e.* sodium iodate, sodium phosphate, sodium arsenate and sodium borate reduce the initial strength to a very large extent.

Turbidity. The turbidity in water due to presence of silt, is limited to 2000 ppm.

Sea water. Sea water containing upto 3.5% salts may be used for un-reinforced concrete. However, it is undesirable to use sea water for reinforced concrete structures exposed to air, for risk of corrosion of reinforcement. Sea water should never be used for prestressed concrete.

Acidic and Alkaline Waters The tolerable limits of hydrochloric, sulphuric and other common inorganic acids is 10,000 ppm. Water containing concentration of sodium hydroxide of higher than 0.5% by weight of cement may reduce the strength of concrete.

TABLE 1.9. TOLERABLE CONCENTRATION OF SOME IMPURITIES IN MIXING WATER

Impurity	Tolerable Concentration
1. Sodium and potassium carbonates and bicarbonates	1000 ppm (total). If this is exceeded it is advisable tc make tests both for setting times and 28-days Strength.
2. Chlorides	10,000 ppm.
3. Sulphuric anhydride	3,000 ppm.
4. Calcium chloride	2 percent by weight of cement in non-prestressed concrete.
5. Sodium iodate, sodium phosphate, sodium arsenate, sodium borate.	Very low
6. Sodium sulphide	Even 100 ppm warrants testing.
7. Sodium hydroxide	0.5 per cent by weight of cement, provided quick set is not induced.
8. Silt and suspended particles	2000 ppm. Mixing water with a high content of suspend solids should be allowed to stand in setting basin before use.
9. Total dissolved salts	15,000 ppm.
10. Organic material	3,000 ppm. Water containing humic acid or such organic acids may adversely affect the hardening of concrete : 780 ppm of humic acid are reported to have seriously impaired the strength of concrete. In the case of such waters, therefore, further testing is necessary.
11. pH	4.5 to 8.5

Algae. Algae in water may cause a marked reduction in strength of concrete. Algae present on the surface of aggregates also weakens the bond between them and cement paste.

1.6 MEASUREMENT OF MATERIALS

The materials used for preparation of concrete are (*i*) cement (*ii*) fine aggregate (*iii*) coarse aggragate and (*iv*) Water. Their accurate measurement before mixing is very important so that the required quantities in the proportion of the concrete mix are obtained.

1. ***Cement.***

It is preferable to measure cement in terms of its weight, and not in terms of volume. The volume of cement changes with the conditions of measurement. In our country, cement is supplied in bags, each bag weighing 50 kg. Under normal conditions, the volume of cement in the bag is considered equivalent to 34.5 Litres. However, if the same cement is shovelled, the bag may measure upto 42 litres. Before mixing, therefore, cement is measured in terms of weight.

2. ***Fine aggregates.***

Fine aggregate (*i.e.* sand) may be measured by weight, for accurate works and by volume for ordinary works. However, when dry sand absorbs water from atmosphere, or when water is mixed to it artificially, its volume increases. This increase in volume due to moisture in sand is known as 'bulking of sand'. Water particles lubricate the

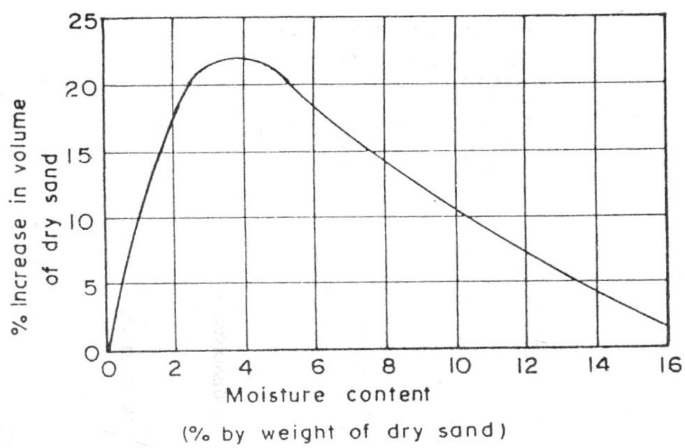

FIG 1.2. BULKING CURVE FOR SAND

sand particles, causing surface tension, and due to this, particles are pulled apart. Thus increase in volume results. This increase in volume depends on the gradation of sands, but may be taken to be maximum at a moisture content of about 4% by weight of dry sand. Further increase in moisture results in decrease in the percent increase of volume, as shown in Fig. 1.2. The bulking increases with fineness, and may be about 25% by volume.

Due to this, if sand is measured by volume, bulking should be properly accounted for. In order to determine unit weight of sand and percentage bulking corresponding to a given moisture content a measuring cylinder, 15 cm in diameter and 87 cm high is used. The cylinder may first be weighed empty. It is then filled with water and weighed. The difference of the two weights would give the volume of the cylinder.

To determine the unit weight of loose dry sand, fill it gently with the sand, without tamping, and level the top surface by striking off the excess sand. Weigh the cylinder. The difference between this weight and the empty weight would give the weight of loose dry sand. Dividing this with the volume of cylinder, the unit weight of loose dry sand is known. To know unit weight of dense dry sand, fill the sand in the cylinder in three equal layers, each layer being compacted 25 times with 16 mm dia. rod 60 cm long, provided at its lower end. The tamping rod should penetrate well below the layer being compacted. After filling, the level of sand is struck to the level of upper end of the cylinder, and the cylinder is weighed. Dividing the weight of compacted sand in the cylinder to the volume, unit weight of dry compacted sand is known.

To determine the percent bulking at various moisture contents, take about 6 litres of dry compacted sand, weigh it and dump it in a mixing pan. Mix it with x percent (say), by weight of dry sand in water thoroughly. Fill the cylinder with this wet sand, without compacting, and strike off the excess sand. Weigh the cylinder, to get the weight of wet sand in the cylinder. Repeat the experiment with various values x, varying from 0 to 20 percent.

Let W_1 = Weight of one cubic metre of compacted dry sand (*i.e.* unit weight of dry compacted sand).

W_3 = Weight of one cubic metre of loose wet sand.

x = percentage of water added.

W_2 = Weight of dry sand contained in one cubic metre of wet loose sand.

Now W_3 = Weight of dry sand + weight of water

$$= W_2 + \frac{x}{100} \times W_2 = W_2 \left[1 + \frac{x}{100}\right]$$

Hence $$W_3 = \frac{W_2}{1 + \frac{x}{100}} \qquad ...(1.1)$$

Therefore, percentage bulking = $\dfrac{\text{increase in volume}}{\text{original volume}}$

$$= \frac{W_1 - W_2}{W_2} \times 100 \qquad ...(1.2)$$

Bulking factor $$= \frac{W_1}{W_2} \qquad ...(1.3)$$

With the help of Eqs. 1.1 and 1.2 , curve of Fig 1.1 can be plotted at various moisture contents.

In order to determine the bulking of a given sand sample in the field, fill a graduated cylinder with the sand, upto a certain mark. Now fill the cylinder to the top, with water. The sand will settle in the cylinder to another mark. Determine the decrease in the volume. To determine the bulking in the laboratory, first determine the volume V_1 of sand corresponding to a mark in the graduated cylinder. Dry the sand and determine its decreased volume V_2.

The percentage bulking $$= \frac{V_1 - V_2}{V_2} \times 100 \qquad (1.4)$$

Thus, knowing the percentage bulking at the site, actual volume of corresponding dry sand can be estimated by subtracting from the measured volume of sand the increase in volume due to bulking. For accurate and large scale works, sand is always measured by weight and necessary allowance is made for the hygroscopic moisture in the sand.

3. ***Coarse aggregate.***

There is no problem of bulking in coarse aggregate, and hence it may be measured either by volume or by weight. However, the weight of a given volume of aggregate is influenced by the size of the measuring box. Hence for accurate and large scale works, measurement should be done by weight. The unit weight of coarse aggregate in loose and dry state is found exactly in the same manner as for fine aggregate, except that a bigger container is used. Since the size of container has effect on the determination, Indian Standards specify the following container for carrying out the tests :

1. Maximum size of aggregate 5 mm to 40 mm : 15 litre capacity cylinder of 25 cm diameter.
2. Maximum size of aggregate over 40 mm : 40 litre capacity cylinder of 35 cm diameter.

4. ***Water.***

Water is normally measured by volume, and specified as so many litres per bag of cement. For a given quantity of water to be mixed in concrete, adjustment should be made for the amount of water present in sand and aggregate. The amount of the water present in the aggregate, due to hygroscopic action etc. should be subtracted from the total required quantity of water. However, if the aggregate is dry, and found to absorb water, extra water should be *added* to account for this. The percentage absorption should be determined first.

1.7 WATER CEMENT RATIO

Water-cement ratio is the ratio of volume of water mixed in concrete to volume of cement used. The strength and workability of concrete depend to a great extent on the amount of water used. For a given proportion of the materials, there is an amount of water which gives the greatest strength. Amount of water less than this optimum water decreases the strength and about 10 percent less may be insufficient to ensure complete setting of cement. Similarly, more than the optimum water increases the workability but decreases the strength. An increase in 10% above the optimum may decrease the strength approximately by 15%, while an increase in 50% may decrease the strength to one half. The use of an excessive amount of water not only produces low strength but increases shrinking and decreases density and durability.

According to Abram's water-cement ratio law, for any given conditions of test the strength of workable concrete mix is dependent only on the water- cement ratio. Lesser the water-cement ratio in a workable mix greater will be its strength. From Abram's law, it follows that provided the concrete is fully compacted, the strength is not affected by aggregate shape, type or surface texture, or the aggregate grading, the workability and the richness of the mix.

According to Powers, cement does not combine chemically with more than half the quantity of water in the mix. Cement requires about 1/5 to 1/4 of its weight of water to become completely hydrated. This suggests that if water cement ratio is less than 0.4 to 0.5, complete hydration will not be secured. Some practical values of water cement ratio for structural reinforced concrete are about 0.45 for 1 : 1 : 2 concrete, 0.50 for 1 : $1\frac{1}{2}$: 3 concrete and 0.55 to 0.60 for 1 : 2 : 4 concrete. However, concrete vibrated by efficient mechanical vibrators require less water cement ratio, and hence have more strength. Sometimes, plasticising agents may be mixed to increase the workability of the mix. For such concrete, therefore, water-cement ratio is reduced, resulting in an increase in the strength.

1.8 PROPERTIES AND TESTS ON CONCRETE

The important properties of concrete, which govern the design of a concrete mix are (*i*) strength (*ii*) durability (*iii*) workability and (*iv*) economy. The aim of proportioning a concrete mix will be to find the economic proportions of cement, coarse aggregate, fine aggregate and water so as to get a mix of given strength, proper workability and durability. The various tests conducted on concrete are discussed in detail in a separate chapter 'Tests on cement and concrete'.

Durability of Concrete

Durability is the property of concrete by virtue of which it is capable of resisting its disintegration and decay. The concrete should be durable with proper regard to the various weathering conditions such as action of atmospheric gases, moisture changes, temperature variations. Disintegration and decay of concrete may be due to the following reasons :

(1) Use of unsound cement, which due to some delayed chemical reactions, undergo volume changes after the concrete has hardened.

(2) Use of less durable aggregate, which may either react with cement, or may be reacted upon by atmospheric gases.

(3) Entry of harmful gases and salts through excessive pores and voids present in unsound concrete, causing its disintegration.

(4) Freezing and thawing of water sucked through the cracks or crevices, by capillary action causing its disintegration.

(5) Expansion and contraction resulting from temperature changes or alternate wetting and drying.

As stated earlier, water required for chemical reaction is about 25% of the weight of cement. Hence excess water present in concrete later evaporates, leaving voids and pores. These pores or voids are later responsible for decay of concrete. Hence for durable concrete, water cement ratio should be as small as possible to get a workable mix. A well compacted concrete has less voids and pores and has more durability. The entrainment of air in concrete has been found to increase very considerably the resistance of concrete to freezing and thawing. The improvement in this respect is due to relief, occassioned by the minute dispersed air bubbles which act as expansion chambers, of stresses and pressures, caused by temperature and moisture changes and by expansion of the moisture contained in concrete on freezing. Vinsol resin is sometimes mixed with concrete to have the property of entrapping innumerable minute air bubbles in concrete. In order to prevent Vinsol resin reacting chemically with the cement, and to make it soluble in water, it is first neutralised by the addition of sodium hydroxide which converts it into a soap. The quantity of resin required for such purpose is extremely small ranging from 0.005 to 0.05 of 1 percent of the weight of cement.

Workability of concrete

It is difficult to properly define and measure the 'workability' of concrete, despite its being the most important property. In its simplest form, the term 'workability' may be defined as the ease with which concrete may be mixed, handled, transported, placed in position and compacted. According to Indian Standard (IS : 1199–1959), *workability of concrete is that property of concrete which determines the amount of internal work necessary to produce full compaction.* The greatest single factor affecting the workability is the amount of water in the mix. A workable concrete does not show any bleeding or segreggation. Bleeding of concrete takes place when excess of water in the mix comes up at the surface, causing small pores through the mass of concrete. Segregation is caused when coarse aggregate separate out from the finer materials, resulting in large voids, less durability and less strength.

Several tests which have been developed to measure the workability of concrete are : (1) Slump test (2) Compacting factor test (3) Vee-Bee test and (4) Vibro-workability test. These have been described in detail separately.

Slump test is probably the simplest and commonly used test, though it is not the true guide to workability. In this test, concrete is compacted in a vessel of the shape of the frustum of a cone and open at both the ends. Concrete is compacted with the help of standard tamping rod, in four equal layers. Immediately after the vessel is filled, it is raised vertically, without giving any jerks etc. The concrete in the vessel becomes free and therefore slump. The vertical settlement, measured in mm, is termed as slump. The following table gives a rough guide of workability of concrete, in terms of slump for various types of work :

	Type of work	*Slump (mm)*
1.	Concrete for road work	20 to 30
2.	Ordinary R.C.C. work for beams and slabs etc.	50 to 100
3.	Columns, retaining wall and thin vertical sections	75 to 150
4.	Vibrated concrete	12 to 25
5.	Mass concrete	25 to 50

The compaction factor test measures the workability of concrete in terms of internal energy required to compact the concrete fully. In this test concrete is compacted in a lower cylindrical mould by making it to fall through two vertically placed hoppers. The weight of concrete in mould is determined. The theoretical weight of materials, required to fill the mould without air voids is also calculated from the knowledge of the proportions of the mix. The compacting factor is then calculated by dividing the observed weight of concrete in the mould by the theoretical weight. A concrete of low workability is represented by a compaction factor of about 0.85, of medium workability for a compaction factor of 0.92 and of good workability for a compaction factor of 0.95.

The various factors which influence the workability of concrete are (1) water in the mix (2) maximum size of particles (3) ratio of coarse and fine aggregates (4) particles interference (5) particle interlocking and (6) admixtures. Out of these, water in the mix is greatest single factor affecting the workability. Addition of water increases workability. The larger the maximum aggregate size and coarser the grading the smaller is the amount of water required for a given workability. In general, the grading requiring the least amount of water for a given workability will be that which gives the smallest surface area for a given amount of aggregate. A smooth rounded aggregate requires less water for a given workability than the irregular shaped aggregate. For a given aggregate-cement ratio, if the quantity of coarse aggregate is increased, the total surface area is reduced and hence more water would be available for lubrication, for a constant water cement ratio, resulting in increase in workability.

1.9 METHODS OF PROPORTIONING CONCRETE MIXES

1. ARBITRARY METHOD

This method is adopted only for work of small magnitude or of moderate importance. The combined aggregate should be dense and should have least voids. For this, the quantity of fine aggregates should be sufficient to fill the voids of coarse aggregate. The ratio of coarse aggregate to fine is found to lie between $1\frac{1}{2}$ to $2\frac{1}{2}$, for a dense mix of aggregates. However, a common practice is to take the quantities of fine and coarse aggregate in the proportion of 1 : 2, and hence to express the quantities of cement, sand and coarse aggregate in the proportions of $1 : n : 2n$ by volume. The ratios of 1 : 1 : 2 and 1 : 1.2 : 2.4 are considered suitable for very high strength concrete, the ratios $1 : 1\frac{1}{2} : 3$ and 1 : 2 : 4 are used for normal reinforced concrete work and ratios 1 : 3 : 6 and 1 : 4 : 8 are used for foundations and mass concrete work. The amount of water to be used in the above mixes is decided on the basis of workability of the mix. The workability depends upon the type of work and the method of compaction. In this method, there is no rigid control over the strength of the mix. However, because of simplicity in the design, the method is widely used for all works of small magnitude.

2. MINIMUM VOIDS METHOD

In this method, the voids of coarse aggregate and fine aggregate are determined separately. The quantity of sand used should be such that it completely fills the voids of the coarse aggregate. Similarly, the quantity of cement used should be such that it fills the voids of sand, so that a dense mix having minimum voids is obtained. However, in actual practice, the quantity of sand used in the mix is kept 10% more than the voids in the coarse aggregate and the quantity of cement is taken 15% more than the voids in the sand. To the mix of cement, sand and coarse aggregate so obtained, sufficient water is added to make the mix workable. However, this method does not give satisfactory result because the presence of cement, sand and water separates the constituents of the coarse aggregate, thereby increasing its voids determined previously in absence of sand and cement. Similalrly, the voids of sand are increased due to the addition of cement and water. Hence we do not always get a dense concrete. At the same time, the grading of aggregates has not been done so as to require least amount of water (and hence least w/c ratio) resulting in higher strength.

3. MAXIMUM DENSITY METHOD

This method of minimum voids was later improved by Fuller, to get a grading of materials to get maximum density. Based on wide scale experiments, he gave the following expression for the grading of materials :

$$P = 100\left(\frac{d}{D}\right)^{\frac{1}{2}} \qquad (1.4)$$

where D = maximum size of aggregate

P = percentage by weight, of material finer than diameter d.

The coarse and fine aggregates should be fully graded according to the above rule. For example, let the maximum size of coarse aggregate be 20 mm and the maximum size of fine aggregate be 4 mm, the percentage of material finer than 4 mm is given by

$$P = 100\left(\frac{4}{20}\right)^{\frac{1}{2}} = 44.7\%$$

i.e. 44.7 Kg of the aggregate, including the weight of cement, are to be mixed with 55.3 kg of coarse aggregate. The quantity of various intermediate sizes should also correspond to this formula.

Let us prepare the mix having a ratio of cement to aggregates (fine + Coarse) as 1 : 6 by weight.

$\therefore$ Quantity of cement in 100 kg. of mix $= \dfrac{100}{7} = 14.3$ kg

$\therefore$ Quantity of sand $= 44.7 - 14.3 = 30.4$ kg.

Hence the ratio of cement, sand and coarse aggregate by weight will be 14.3 : 30.4 : 55.3. Let us assume the unit weights of cement, fine aggregate and coarse aggregate as 1440, 1750 and 1600 kg per cubic metre respectively. Then the ratio of the three constituents, by volume will be

$$\frac{14.3}{1440} : \frac{30.4}{1750} : \frac{55.3}{1600}$$

or

$$1 : \left(\frac{30.4}{1750} \times \frac{1440}{14.30}\right) : \left(\frac{55.3}{1600} \times \frac{1440}{14.30}\right)$$

or

$$1 : 1.75 : 3.48$$

or

$$1 : 1\tfrac{3}{4} : 1\tfrac{1}{2} \text{ (nominally)}$$

After having decided the proportions of various materials, sufficient quantity of water is added to make the mix workable. Table 1.10 gives the grading of mixed aggregate for 40 mm and 20 mm maximum size of aggregate. The method is not so popular since grading cannot be accurately achieved in field, and there is no control over the strength.

TABLE 1.10 GRADING OF MIXED AGGREGATE

Max. Size of coarse aggregate	*Percentage passing the I.S size*								
	40 mm	*20 mm*	*10 mm*	*4 mm*	*2 mm*	*1 mm*	*500 micron*	*250 micron*	*125 micron*
40 mm	100	71	50	32	22	16	11	8	6
20 mm	—	100	71	44	32	22	16	11	8

4. FINENESS MODULUS AND WATER CEMENT RATIO METHOD

(*a*) *Fineness modulus.*

It has been observed that strength of mix is dependent wholly on the water cement ratio while the grading of the particles is important from workability and economic point of view. The grading of particles by Fullers formula, to get maximum density, is difficult and sometimes uneconomical to achieve in practice. Fineness modulus method essentially is a substitute for Fuller maximum density method, aimed at standardisation of the grading of aggregates. The term *fineness modulus*, suggested by Abram, is a numerical index of fineness of both fine as well as coarse aggregates.

Certain values of fineness modulus for mixed aggregates are found to give the best result. Let p be the desired fineness modulus for a mix of fine and coarse aggregates. If p_1 and p_2 are the fineness moduli of fine and coarse aggregates respectively, than the proportion R of finer aggregate to the combined aggregate, by weight is given by :

$$R = \frac{p_2 - p}{p - p_1} \times 100 \qquad \text{...(1.5)}$$

For example of Table 1.8 (*a*) if the desired fineness modulus of the combined aggregate is 5.3, we have

$$R = \frac{6.93 - 5.3}{5.3 - 3.05} \times 100 = 72.5\%$$

(*b*) *Abram's water-cement ratio law.*

Abram's water-cement ratio (w/c ratio) law states that for any given conditions of test, the strength of workable concrete mix is dependent only on the water-cement ratio. It follows from this law that provided the concrete is fully compacted, the strength is not affected by aggregate shape, type or surface texture, or the aggregate grading, the workability and the richness of mix. We know that workability of mix (defined as 'that property of the concrete which determines the amount of useful internal work necessary to produce full compaction') is dependent on the amount of water in the mix. But amount of water in the mix, corresponding to a given strength, is governed by the water-cement ratio law. Hence the only way to increase the quantity of water to increase the workability of the mix is to increase the amount of cement also. However, as the grading of aggregate does not effect the strength of the concrete directly, the object must be to choose the grading to give the best workability with lowest water content. The grading of the particles should, therefore, be such that in the fully compacted state the total surface area of particles of aggregates as well as voids in them are the least. This means that larger the maximum aggregate size and the coarser the grading, the smaller is the amount of water required for a given workability. However, beyond a certain limit, the further increase in the maximum size and coarseness of grading results in harsh and under-sanded mixtures causing honeycombing, thus requiring more cement for smoothness. On the other hand, increase in the proportion of fine aggregate (*i.e.* sand) gives smooth mix but requires more cement and hence results, in uneconomical mix. Between these two limits lies the grading, which can be either determined by Fuller's maximum optimum density method or the fineness modulus method.

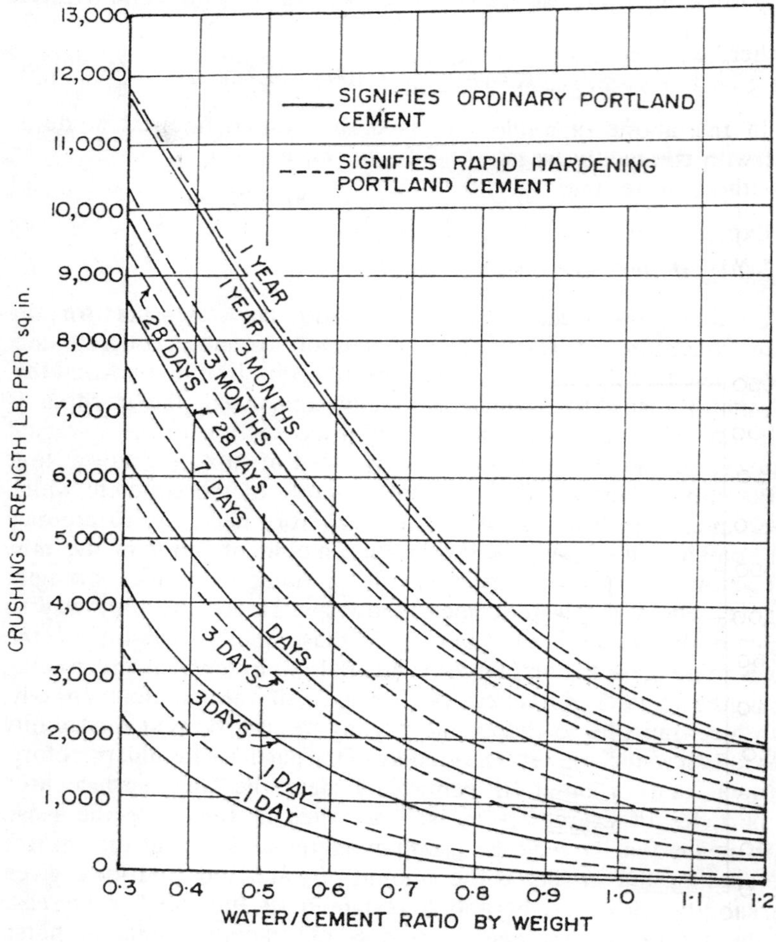

FIG 1.3. RELATION BETWEEN CRUSHING STRENGTH AND WATER-CEMENT RATIO FOR FULLY COMPACTED CONCRETE.

According to Abram's law, the strength of mix increases with the decrease in the water cement ratio. In terms of crushing strength after 7 days curing, the law can be expressed as follows :

$$p_7 = \frac{984}{7^x} \qquad ...(1.6\,a)$$

where p_7 = cylinder crushing strength, in kg/cm^2, after 28 days curing ; x = water cement ratio by volume.

In the above expression, the constant 984 and 7 may vary slightly with the quality of aggregates and cement, method of curing and method of testing.

Expressed in terms of strength after 28 days curing, the law can be written as :

$$p_{28} = \frac{984}{4^x} \qquad ...(1.6\,b)$$

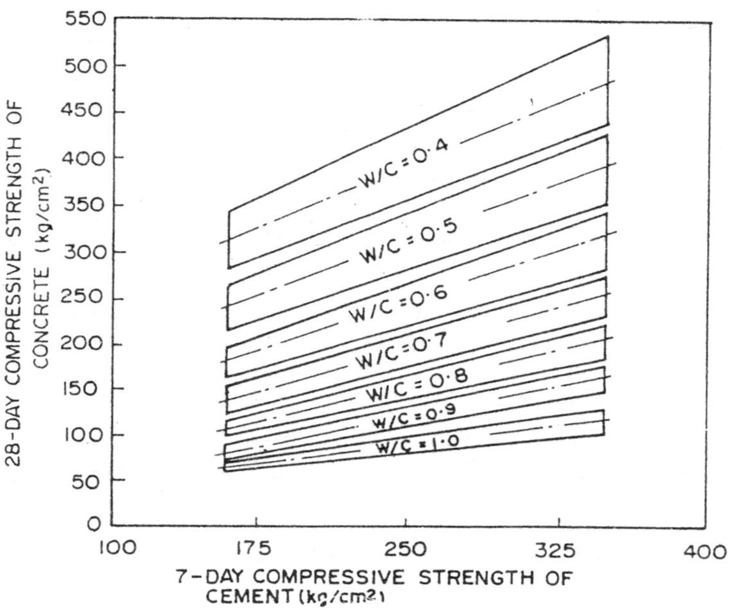

FIG 1.4 RELATION SHIP BETWEEN COMPRESSIVE STRENGTH OF CONCRETE MIXES WITH DIFFERENT WATER-CEMENT RATIO AND THE 7 DAY COMPRESSIVE STRENGTH OF CEMENT.

Fig 1.3 shows the relations between the crushing strength and water cement ratio (by weight) for various periods of curing. Fig 1.4 gives the relationship between 28 days compressive strength(cube) of concrete mixes with different water-cement ratios by weight and the 7 days compressive strength of cement (IS : 456–1964). Both Figs. 1.3 and 1.4 may be used as a guide for the selection of proper water cement ratio for a mix of given strength.

(*c*) ***Procedure for design of mix.***

The procedure for the design of mix can be summarised as follows:

1. For the requirements of strength of the mix, choose suitable water cement ratio from Fig 1.3 and 1.4.

2. Determine the maximum size of the aggregate available. Also determine the fineness modulus of both coarse and fine aggregates.

3. Determine the grading of the aggregates by Fuller's maximum density method. If this does not correspond to the grading of available materials, try to improve that grading to make it similar to that obtained by Fuller's method. If this is difficult to achieve, design the grading by fineness modulus method. For that, choose suitable fineness modulus of combined aggregate and determine the ratio of fine aggregates to coarse aggregatess by Eq 1.5.

4. Determine the workability of the mix in terms of slump or compaction factor, required for the work.

5. Fix the ratio of cement to that of combined asggregate, mix quantity of water corresponding to the water cement ratio determined in step 1, and determine the workability of mix. Change the cement-aggregate ratio by trial to get the desired workability.

6. Determine the actual proportions of cement, fine aggregates, coarse aggregate and water, from the above steps, so that economical concrete of desired strength and workability is obtained.

1.10 GRADES OF CONCRETE AND CHARACTERISTIC STRENGTH : IS : 456–1978

Indian Standard IS : 456–1978 specifies seven grades of concrete designated as M 10, M 15, M 20 , M 25, M 30, M 35 and M 40. In the designation of concrete mix, letter M refers as the mix and the number to the specified *characteristic compressive strength*

(f_{ck}) of 15 cm cube at 28 days, expressed in N/mm^2. The *characteristic strength* is defined as the strength of the material below which not more than 5% of the test results are expected to fall.

It should be noted that 1 N/mm^2 $\approx$ 10 kg/cm^2. Thus 15 N/mm^2 $\approx$ 150 kg/cm^2. Hence M 15 grade of concrete of IS : 456–1978 corresponds to M 150 grade concrete of IS : 456–1964.

Increase in strength with age

Where it can be shown that a member will not receive its full design load/stress within a period of 28 days after the casting of the member (for example, in foundation and lower columns in multi–storeyed buildings), the characteristic compressive strength can be increased. In general, the strength of concrete at any age, can be expressed by the relation

$$f_T = \frac{T}{a + bT} \cdot f_{28}$$

Where f_T = Strength of conrete at T days

f_{28} = Strength of concrete at 28 days

T = No. of days

a,b = Empirical constant.

TABLE 1.11. GRADES OF CONCRETE

IS : 456-1964		*IS : 456-1978*	
Grade Designation	*Specified 28 days work cube compressive strength (kg /cm^2)*	*Grade Designation*	*Specified characteristic compressive strength f_{ck} at 28 days (N/mm^2)*
M 100	100	M 10	10
M 150	150	M 15	15
M 200	200	M 20	20
M 250	250	M 25	25
M 300	300	M 30	30
M 350	350	M 35	35
M 400	400	M 40	40

Note 1 : M 5 (M 50) and M 7.5 (M 75) grades of concrete may be used for lean concrete bases and simple foundations for masonry walls. These mixes need not be designed.

Note 2 : Grade of concrete lower than M 15 (M 150) shall not be used in reinforced concrete.

As per IS : 456–1978, the characteristic strength given in Table 1.11 may be increased by multiplying by the factors given below.

Minimum age of member when full design load/stress is expected (Months)	*Age factor*
1	1.00
3	1.10
6	1.15
12	1.20

Note 1 : No increase in respect of age at loading should be allowed where high alumina cement concrete is used.

Note 2 : Where members are subjected to lower direct load during construction they should be checked for stresses resulting from combination of direct load and bending during construction.

Note 3 : The permissible stress or design strength shall be based on the increased value of compressive strength.

Tensile Strength of concrete

The flexural and split tension strengths shall be obtained as described in IS : 516 – 1959 and IS : 5816 – 1970 respectively. When the designer wishes to use an estimate of the flexural tensile strength from the compressive strength, the following formula may be used (IS : 456–1978) :

Flexural tensile strength $f_{ck} = 0.7\sqrt{f_{ck}}$ N/mm^2

where f_{ck} is the *characteristic compressive strength of concrete* in N/mm^2.

Modulus of Elasticity

The modulus of elasticity is primarily influenced by the elastic properties of the aggregate and to a lesser extent by the conditions of curing and age of the concrete, the mix proportions and the type of cement. The modulus of elasticity is normally related to the compressive strength of concrete. In the absence of test data, the modulus of elasticity for structural concrete may be assumed as follows (IS : 456 – 1978) :

$$E_c = 5700\sqrt{f_{ck}}$$

where E_c is the short term static modulus of elasticity in N/mm^2 and f_{ck} is the characteristic cube strength of concrete in N/mm^2.

DESIGN MIX CONCRETE

The mix shall be designed to produce the grades of concrete having the required workability and a characteristic strength not less than appropriate values given in Table 1.11. As long as the quality of the materials does not change, a mix design done earlier may be considered adequate for later work.

Nominal mix concrete

The nominal mix concrete may be used for concrete of grades M 5, M 7.5, M 10, M 15 and M 20. The proportions of material for nominal mix concrete shall be in accordance with Table 1.12.

The cement content of the mix specified in Table 1.12 for any nominal mix shall be proportionately increased if the quantity of water in a mix has to be increased to overcome the difficulties of placement and compaction, so that the water- cement ratio as specified is not exceeded.

TABLE 1.12. PROPORTIONS FOR NOMINAL MIX DESIGN

<table>
<tr><th>Grade of concrete IS : 456-1978</th><th>Nominal mix</th><th>Limiting proportion of fine aggregate to coarse aggregate (by mass)</th><th>Total quantity of dry aggregate by mass per 50 kg of cement, to be taken as sum of the individual masses of fine and coarse aggregates (max.)</th><th>Quantity of water per 50 kg of cement (liters)</th></tr>
<tr><td>M 5</td><td>1 : 5 : 10</td><td rowspan="5">Generally 1 : 2 but subject to an uper limit of $1 : 1\frac{1}{2}$ and a lower limit of $1 : 2\frac{1}{2}$</td><td>800</td><td>60</td></tr>
<tr><td>M 7.5</td><td>1 : 4 : 8</td><td>625</td><td>45</td></tr>
<tr><td>M 10</td><td>1 : 3 : 6</td><td>480</td><td>34</td></tr>
<tr><td>M 15</td><td>1 : 2 : 4</td><td>350</td><td>32</td></tr>
<tr><td>M 20</td><td>$1 : 1\frac{1}{2} : 3$</td><td>250</td><td>30</td></tr>
</table>

Note : The proportions of the fine to coarse aggregate should be adjusted from upper limit progressively as the aggregates become fine and the maximum size of coarse aggregates becomes larger. Graded coarse aggregates should be used.

Optional test requirements of the concrete

In order to get a relatively quicker idea of quality of concrete, optional tests on beams for modulus of rupture at 72 ± 2 hours or at 7 days or compressive strength tests at 7 days may be carried out in addition to 28 day compressive strength tests. For this purpose, the values given in Table 1.13 may be taken for general guidance in the case of concrete made with ordinary Portland cement. In all cases, the 28 day compressive strength specified in Table 1.11 shall alone be the criterion for acceptance or rejection of the concrete. If, however, from tests carried out in a particular job over a reasonably long period, it has been established to satisfaction of the engineer in charge that a suitable ratio between 28 days compressive strength and the modulus of rupture at 72 ± 2 hours of 7 days or compressive strength at 7 days may be accepted, the engineer in charge may suitably relax the frequency of 28 days compressive strength specified below, provided the expected strength values at the specified early age are consistently met.

TABLE 1.13 OPTIONAL TESTS REQUIREMENTS OF CONCRETE

Grades of concrete (IS: 456-1978)	*Compressive strength on 15 cm cubes min. at 7 days N/mm²*	*Modulus of rupture by beam test, min.*	
		at 72 ± 2h N/mm²	*at 7 days N/mm²*
M 10	7	1.2	1.7
M 15	10	1.5	2.1
M 20	13.5	1.7	2.4
M 25	17.0	1.9	2.7
M 30	20.0	2.1	3.0
M 35	23.5	2.3	3.2
M 40	27.0	2.5	3.4

Frequency of sampling

A random sampling procedure shall be adopted to ensure that each concrete batch shall have a reasonable chance of being tested, that is, the sampling should be spread over the entire period of concreting and cover all mixing units. The minimum frequency of sampling concrete of each grade shall be in accordance with the following :

Quantity of concrete in the work (m^3)	*No of samples*
1–5	1
6–15	2
16–30	3
31–50	4
51 and above	4 plus one additional sample for each 50 m^3 or part thereof.

Test strength of sample and standard deviation

The test strength of the sample shall be average of the strength of three specimens. The individual variation should not be more than ± 15 percent of the average.

The total number of test results required to constitute an accepted record for calculation of standard deviation shall be not less than 30. Attempts should be made to obtain the 30 test results as early as possible when a mix is used for the first time. The standard deviation of concrete of a given grade shall be calculated using the following formula from the results of individual test of concrete of that grade obtained as specified above :

$$\text{Estimated standard deviation, } s = \sqrt{\frac{\Delta^2}{n-1}}$$

where Δ = deviation of the individual test strength from the avaerage strength of n samples, and

n = number of sample test results.

Where sufficient tests results for a particular grade of concrete are not availabale, the value of standard deviation given below may be assumed (IS : 456–1978).

Grade of concrete	*Assumed standard deviation (N/mm^2)*
M 10	2.3
M 15	3.5
M 20	4.6
M 25	5.3
M 30	6.0
M 35	6.3
M 40	6.6

Acceptance Criteria (IS : 456–1978)

(*A*) The concrete shall be deemed to comply with the strength requirements if :

(*a*) Every sample has a test strength not less than the chasracteristic value, or

(*b*) The strength of one or more samples though less than the characteristic value, is in each case not less than the greater of :

(i) the characteristic strength minus 1.35 times the standard deviation, and

(ii) 0.8 times the characteristic strength ; and the average strength of all samples is not less than the characteristic strength plus

$$\left[1.65 - \frac{1.65}{\sqrt{No.\ of\ samples}}\right] \text{ times the standard deviation.}$$

(*B*) The concrete shall be deemed not to comply with the strength requirements if :

(*a*) The strength of any sample is less than the greater of :

(i) The characteristic strength minus 1.35 times the standard deviation, and

(ii) 0.8 times the characteristic strength, or

(*b*) the average strength of all the samples is less than the characteristic strength plus

$$\left[1.65 - \frac{3}{\sqrt{No.\ of\ samples}}\right] \text{ times the standard deviation.}$$

Concrete which does not meet the strength requirements as specified in (*A*), but has a strength greater than that required by (*B*) may, at the discretion of the designer, be accepted as being structurally adequate without further testing.

1.11. PERMISSIBLE STRESSES IN CONCRETE (IS : 456–1978)

1. DIRECT TENSILE STRESS

The direct tensile stress f_t is calculated from the expression

$$f_t = \frac{F_t}{A_c + m A_{st}}$$

where F_t = total tension on member minus pretension in steel, if any, before concreting :

A_c = cross sectional area of concrete, excluding any finishing material and reinforcing steel

m = Modular ratio, and

A_{st} = Cross-sectional area of reinforcing steel in tension.

For members in direct tension, when full tension is taken by the reinforcement alone, the tensile stress (f_t) shall be not greater than the values given below :

TABALE 1.14 PERMISSIBLE DIRECT TENSILE STRESS

Grade of concrete	M 10	M 15	M 20	M 25	M 30	M 35	M 40
Tensile stress N/mm²	1.2	2.0	2.8	3.2	3.6	4.0	4.4

2. COMPRESSIVE STRESS AND BOND STRESS

The permisible compressive stresses and average bond stresses shall be as given in table 1.15.

TABLE 1.15. PERMISSIBLE STRESSES IN CONCRETE (IS : 456–1978)

Grade of concrete	*Permissible stress in compression* (N/mm^2)		*Permissible stress in Bond (Average) for plain bars in tention* (N/mm^2) τ_{bd}
	Bending σ_{cbc}	*Direct* σ_{cc}	
M 10	3.0	2.5	—
M 15	5.0	4.0	0.6
M 20	7.0	5.0	0.8
M 25	8.5	6.0	0.9
M 30	10.0	8.0	1.0
M 35	11.5	9.0	1.1
M 40	13.0	10.0	1.2

The bond stress given in Table 1.15 shall be increased by 25 percent for bars in compression.

In the case of deformed bars conforming to IS : 1139–1966 and IS : 1786–1979, the bond stress given in Table 1.15 may be increased by 40 percent.

3. SHEAR STRESS

The permissible shear stress in concrete in beams without shear reinforcement shall be as given in Table 1.16.

TABLE 1.16 PERMISSIBLE SHEAR STRESS IN CONCRETE (IS : 456–1978)

$\frac{100A_s}{bd}$	*Permissible shear stress in concrete τ_c, N/mm² for grades of concrete*					
	M 15	*M 20*	*M 25*	*M 30*	*M 35*	*M 40*
0.25	0.22	0.22	0.23	0.23	0.23	0.23
0.50	0.29	0.30	0.31	0.31	0.31	0.32
0.75	0.34	0.35	0.36	0.37	0.37	0.38
1.00	0.37	0.39	0.40	0.41	0.42	0.42
1.25	0.40	0.42	0.44	0.45	0.45	0.46
1.50	0.42	0.45	0.46	0.48	0.49	0.49
1.75	0.44	0.47	0.49	0.50	0.52	0.52
2.00	0.44	0.49	0.51	0.53	0.54	0.55
2.25	0.44	0.51	0.53	0.55	0.56	0.57
2.50	0.44	0.51	0.55	0.57	0.58	0.60
2.75	0.44	0.51	0.56	0.58	0.60	0.62
3.00 and above	0.44	0.51	0.57	0.60	0.62	0.63

Note : A_s is that area of longitudinal tension reinforcement which continues at least one effective depth beyond the section being considered except at supports where the full area of tension reinforcement may be used.

4. MODULAR RATIO

The modular ratio m has the value $\frac{280}{3\,\sigma_{cbc}}$ where σ_{cbc} is the permissible compressive stress due to bending in concrete in N/mm^2 as specified in Table 1.15. The above expression for m partially takes into account long term effects such as creep. Based on the above expression, the values of m (rounded off to the nearest integral value) for various grades of concrete will be as given in Table 1.17.

TABLE 1.17 MODULAR RATIO

Grade of concrete	M 10	M 15	M 20	M 25	M 30	M 35	M 40
Modular ratio m	31 (31.11)	19 (18.67)	13 (13.33)	11 (10.98)	9 (9.33)	8 (8.11)	7 (7.18)

Note : The figures in the bracket are the exact values. However, it is recommended to use only the rounded-off values for the design purposes.

5. *INCREASE IN PERMISSIBLE STRESSES*

Where stresses due to wind (or earthquake) and temperature effects are combined with those due to dead, live and impact load, the stresses specified in Table 1.14, 1.15, 1.16 and 1.17 may be exceeded upto a limit of $33\frac{1}{3}$ percent. Wind and seismic forces need not be considered as acting simultaneously.

1.12 MIXING, COMPACTING AND CURING CONCRETE

1. *Mixing of Concrete.*

The operation of manufacture of concrete is called 'mixing'. The mixing can be done by two methods : (*a*) hand mixing (*b*) mixing by concrete mixers.

(*a*) **Hand mixing.** In this method, mixing is done manually on a steel plate, 2m × 2m in size or on a clean hard surface. The method is resorted to when the quantity of concrete needed for the work is small. Sand and cement in appropriate proportions are mixed first in a dry state. The coarse aggregate is then added and the whole mixture is mixed thoroughly with the help of shovels. The predetermined amount of water is then sprinkled over the mix. The mass is then turned till the colour of concrete obtained is homogeneous and workable mix is obtained.

(*b*) **Machine mixing.** Concrete should normally be mixed in a mechanical mixer. The main part of mixer is a drum in which the ingredients are mixed thoroughly by mechanically rotating the drum. The drum is made of steel plates, with a number of blades put in inclined position in the drum. As the drum rotates the materials encounter resistance to rotation from the blades and this disturbing effect helps in a good mixing of the ingradients. The mixers are either operated electrically or else are driven by oil engines attached to them. Coarse aggregate should be fed first, then sand and lastly cement. In the revolving state, when the three get thoroughly mixed, water should be added either with the help of a can or automatically through the pipe attached to the mixer. Mixing should be continued until there is a uniform distribution of the materials and the mass is uniform in colour and consistency, but in no case shall the mixing be done for less than two minutes.

Concrete mixers may be of two types : (i) Batch mixers and (ii) Continuous mixers. Batch type mixers are employed for work of relatively small magnitude. Batch type mixers can either be of tilting drum type or closed drum type. In the tilting drum type, drum rotates about a trunnion axis and is so arranged that it is quite easy to rotate and tilt it when it is empty as well as when full. In the closed drum type, the drum remains rotating in one direction and is emptied by means of the hopper which tilts to receive the discharge. Continuous mixers are used in mass concreting work where a large and continuous flow of concrete is required. In these mixers, processes of feeding, mixing and emptying go on continuously without break.

2. ***Transporting concrete.***

Concrete should be handled from the place of mixing to the place of final deposit as rapidly as practical by methods which will prevent the segregation or loss of any of the ingradients. If the segregation does occur during transport, the concrete should be remixed before being placed. During hot or cold weather, concrete should be transported in deep containers, on account of their lower ratio of surface area to mass, reduce the rate of loss of water by evaporation during hot weather and loss of heat during cold weather.

3. ***Placing concrete.***

The concrete should be placed and compacted before setting commences and should not subsequently be disturbed. Method of placing should be such as to preclude segreggation. Before concrete is placed, it should be ensured that the forms are rigid, in their correct position, well cleaned and oiled. Oiling of the forms will prevent the concrete from sticking to it, and it will then be easier to remove the forms when they are no longer required. Concrete should not be poured into the forms only at one point, but should be uniformly spread on all the sides for better compaction. When the work has to be resumed on a surface which has hardened, such surface should be roughened. It should then be swept clean, thoroughly wetted and covered with a 13 mm layer of mortar composed of cement and sand in the same ratio as the cement and sand in the concrete mix. This 13 mm layer of mortar should be freshly mixed and placed immediately before the placing of the concrete. Where the concrete has not fully hardened, all laitance should be removed by scrubbing the wet surface with wire or bristle brushes, care being taken to avoid dislodgement of particles of aggregate. The surface should be thoroughly wetted and all free water removed. The surface should then be coated with neat cement grout. The first layer of concrete to be placed on this surface should be well rammed against old work, particular attention being paid to corners and close spots.

4. ***Compacting Concrete.***

The removal of entrapped air during production of concrete and the uniform, dense arrangement of the constituents of concrete are effected during the compaction of concrete. The density and, consequently, the strength and durability of concrete depend upon this operation. Concrete should be thoroughly compacted during the operation of placing and thoroughly worked around the reinforcement, around embedded fixtures and into corners of the form work. Concrete is compacted by vibration during which the vibrator communicates rapid vibrations of low amplitude to the particles, as a result of which the concrete becomes fluid, that is to say, its mobility is increased, and the particles, in movement, under the force of gravity occupy a more stable position, with which volume of concrete is least. Vibrators are of three general types (i) internal vibrators (ii) external vibrators, and (iii) surface vibrators. Internal or immersion vibrator consists of a vibrating element enclosed in a casing which is immersed in fresh concrete and transmit vibrations through the vibrator body. External or form vibrators are fastened to the form work by a clamping device and transmit vibrations, to the concrete through the form. In precast members of concrete, the vibrating tables are very helpful. These tables vibrate the entire mass of concrete uniformly. Surface vibrators, set up on the concrete surface after placing, transmit vibrations to it through a working platform. They are generally employed in concrete road construction.

Mechanical method of compacting the concrete is used only when the mix is stiff. Over-vibration or vibration of very wet mixes is harmful and should be avoided. Alternatively, concrete may be compacted manually by rodding, tamping or hammering. Rodding is generally done to compact thin vertical members while temping is done for compacting concrete for slabs etc.

5. *Curing concrete.*

Curing is one of the most essential operation in which concrete is kept continuously damp for some days to enable the concrete to gain more strength. Curing replenishes the loss of moisture from the concrete due to evaporation, absorption and heat of reactions. The period of curing depends upon atmospheric conditions such as temperasture, humidity and wind velocity. The normal period is between 7 to 10 days. There are several methods of curing the concrete, the more common being the following : (i) Covering the exposed surface with a layer of sacking, canvas, hessian or similar absorbant materials, and keeping them continuously wet, (ii) thoroughly wetting the surface of concrete and then keeping it covered with a layer of suitable water proof material (iii) impounding water in earthen to sandy bunds in squares over the flooring (iv) curing with the help of steam or hot water, resulting in rapid development of strength.

1.13. STEEL REINFORCEMENT

Steel reinforcement used in reinforced concrete may be of the following types :

(*a*) 1. Mild steel bars conforming to IS : 432 (Part I)–1966.

2. Hot rolled mild steel deformed bars conforming to IS : 1139–1966.

(*b*) 1. Medium tensile steel conforming to IS : 432 (Part I)–1966.

2. Hot rolled medium tensile steel deformed bars conforming to IS : 1139–1966.

(*c*) 1. Hot rolled high yield strength deformed bars (HYSD bars) conforming to IS : 1139–1966.

2. Cold-worked steel high strength deformed bars conforming to IS : 1786–1979 (Grade Fe 415 and Grade Fe 500).

(*d*) 1. Hard drawn steel wire fabric comforming to IS : 1566-1967.

2. Rolled steel made from structural steel conforming to IS : 226–1975.

The most important characteristic of a reinforcing bar is its *stress strain curve* and the important property is the *yield stress* or 0.2% *proof stress*, as the case may be. The permissible stresses in steel reinforcement as per IS : 456–1978 are given in Table 1.18, in which columns 2, 3 and 4 correspond to types (*a*), (*b*) and (*c*) respectively, mentioned above. The modulus of elasticity E for these steels may be taken as 2×10^5 N/mm^2.

Type (*a*) bars have yield strength of 250 N/mm^2. Hence it is sometimes known as Fe 250 steel, having characteristic strength of 250 N/mm^2.

Type (*c*) bars are high yield strength bars, also known as, HYSD bars. These may be *hot rolled* high yield strength bars or *cold worked* steel high strength deformed bars. The latter are also known as CTD bars (cold twisted deformed bars) or *Tor Steel*, and are available in two grades : (i) Fe 415 (or Tor 40) and (ii) Fe 500 (Tor 50) having 0.2% proof stress as 415 N/mm^2 and 500 N/mm^2 respectively.

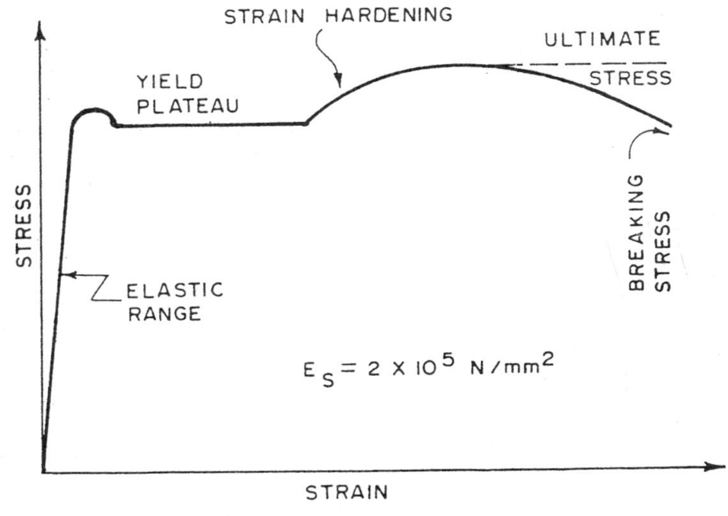

FIG 1.5 STRESS STRAIN CURVE FOR MILD STEEL.

The hot rolled mild steel bars, in common use for nearly six decades, has low yield stress but has significant *yield plateau* having plastic strain nearly 10 times the limiting strain at yield point. Due to such an excessive yielding, there is considerable readjustment of stress in concrete so as to maintain equilibrium of internal stresses, resulting in excessive cracking and deformations in the structural member. The idealised stress strain curve for mild steel bars assumed by IS : 456–1978, is shown in curve 1 of Fig. 1.6.

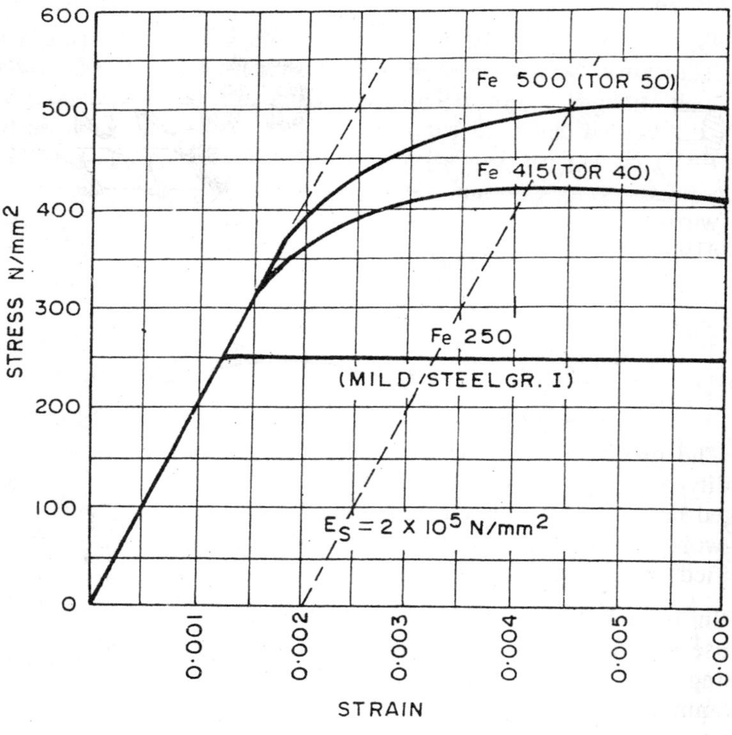

FIG. 1.6 STRESS STRAIN CURVE FOR CTD BARS

The *yield plateau* can be eliminated by straining the bar beyond the yield plateau either by twisting or *stretching* and then unloading as shown in Fig. 1.6. This process is known as cold working, and the *Tor– Steel reinforcement bars,* available in India are of this type. By this process, higher yield stress can be attained with much smaller strains. A *twisted bar* has considerable increased yield stress, about 50% to 100% more than the ordinary mild steel bars. Their use can permit higher working stress and hence considerable saving in quantity of steel can be achieved. Bond between concrete and steel can be improved by use of *deformed bars.* A deformed bar is a bar of steel provided with lugs or ribs deformation on surface of the bar to minimise the slippage of the bar in concrete, as shown in Fig. 1.7.

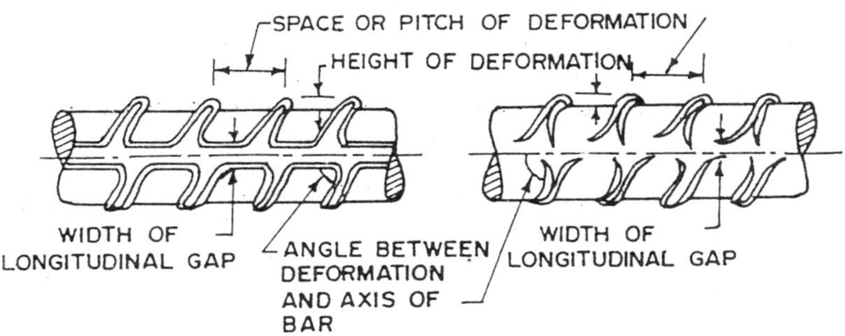

FIG. 1.7 DEFORMED BARS

The process of cold workings has the disadvantage that useful ductility of mild steel is reduced. It is to be noted that if the cold worked bar is reheated, the original yield plateau is recovered. Since cold-worked bars lack a well-defined yield plateau, yield stress is specified at 0.2% *proof stress.*

In reinforced concrete, a long time trend is evident towards the use of high strength materials, both steel and concrete. In big housing projects, *High yield strength deformed bars* (HYSD bars) are in common use, having yield stress (0.2 percent proof stress) equal to 230 to 415 N/mm^2 (4150 kg/cm^2) and permissible stress eequal to 230 N/mm^2 (2300 kg/cm^2) for grade Fe 415. For grade Fe 500 HYSD bars, the yield stress is as high as 500 N/mm^2 while the permissible tensile stress is equal to 275 N/mm^2. *Modern high strength steels* of Fe 415 and Fe 500 grades require the simultaneous use of high strength concrete of M 20 grade and higher.

TABLE 1.18. PERMISSIBLE STRESSES IN STEEL REINFORCEMENT

Type of stress in steel reinforcement	*Permissible stresses*		
	***(a)* Mild steel bars conforming to Grade I of IS : 432 (Part I) 1966 or deformed mild steel bars conforming to IS : 1139-1966**	***(b)* Medium tensile steel conforming to IS : 432 (Part I) 1966 or Deformed medium tensile stel bars conformly to IS : 1139-1966**	***(c)* High yield strength deformed bars (HYSD bars) conforming to IS : 1139-1966 or IS : 1786-1979 (Grade Fe 415)(CTD bars)**
(1)	*(2)*	*(3)*	*(4)*
1. Tension (σ_{st} or σ_{sv})		Half the guaranteed yield stress subject to a maximum of 1900 kg/cm^2 (190 N/mm^2)	
(a) up to and including 20 mm	1400 kg/cm^2 (140 N/mm^2)		2300 kg/cm^2 (230 N/mm^2)
(b) over 20 mm	1300 kg/cm^2 (130 N/mm^2)		2300 kg/cm^2 (230 N/mm^2)
2. Compression in column bars (σ_{sc})	1300 kg/cm^2 (130 N/mm^2)	1300 kg/cm^2 (130 N/mm^2)	1900 kg/cm^2 (190 N/mm^2)
3. Compression in bars in a beam or a slab where the compressive resistance of the concrete is taken in to account.	The calculated compressive stress in the surrounding concrete multiplied by 1.5 times the modular ratio or σ_{sc} whichever is lower.		
4. Compression in bars in a beam or a slab where compressive resistance of the concrete is taken into account.		Half the guaranteed yield stress subject to a maximum of 1900 kg/cm^2 (190 N/mm^2)	
(a) Upto and including 20 mm.	1400 kg/cm^2 (140 N/ mm^2)		1900 kg/cm^2 (190 N/mm^2)
(b) Over 20 mm.	1300 kg/cm^2 (130 N/mm^2)		1900 kg/cm^2 (190 N/mm^2)

Note 1. For high yield strength deformed bars of Grade Fe 500, the permissible stress in direct tension and flexural tension shall be 0.55 f_y. The permissible stresses for shear and compression reinforcement shall be as for Grade Fe 415.

Note 2. For welded wire fabric conforming to IS : 1566–1967, the permissible value in tension is 2300 kg/cm^2 (230 N/mm^2)

Note 3. For the purpose of this code, the yield stress of steels for which there is no clearly defined yield point should be taken to be 0.2 percent proof stress.

Note 4. When mild steel conforming to Grade II of IS : 432 (Part I)–1966 is used the permissible stresses shall be 90 percent of permissible stress in column 2, or if the design details have already been worked out on the basis of mild steel conforming to Grade I of IS : 432 (Part I)–1966, the area of reinforcement shall be increased by 10 percent of that required for Grade I steel.

Percent Elongation. Minimum requirement of *% elongation* should be as follows :

	M.S. bars Grade I	*Medium tensile steel bars*	*Medium tensile steel bars*	*Medium steel deformed bars*	*HYSD and CTD bars*
Dia<10mm	20	17	23	20	14.5
Dia≥10 mm	23	20	23	20	14.5

Characteristic Strength of Steel Reinforcement

The *characateristic strength* (f_y) means that value of strength below which not more than 5% of the test results are expected to fall. Until the relevant Indian Standard specifications for reinforcing steel are modified to include the concept of characteristic strength, the characteristic value shall be assumed as the minimum yield/0.2 percent proof stress specified in the relevant Indian Standard Codes.

The values of characteristic strength for three common types of reinforcements are given in Table 1.19.

TABLE 1.19. CHARACTERISTIC STRENGTH OF STEEL REINFORCEMENT

Type of reinforcement	*Yield stress or 0.2% proof stress* (N/mm²)	*Characteristic strength f_y* (N/mm²)	*Permisible tensile strength σ_{st}* (N/mm²)
1. Mild steel bars conforming to Grade I of IS : 432 (Part I) or deformed m.s. bars of I.S. 1139	250 (average)	250	140 (upto 20 mm dia.) 130 (over 20 mm dia.)
2. High yield strenth deformed bars conforming to IS : 1109 or Grade Fe 415 of IS : 1786—1979	415	415	230
3. High yield strength deformed bars of Grade Fe 500 of IS 1786—1979	500	500	275

PROBLEMS

1. Discuss in brief various types of cements used for construction.
2. Discuss salient specifications for Portland cement used for reinforced concrete.
3. Write a note on 'aggregates' used for cement concrete.
4. Explain the following:
   (i) Determination of fineness modulus
   (ii) Grading aggregates
   (iii) Quality of water for cement concrete
   (iv) Durability of concrete
   (v) Workability of Concrete
   (vi) Water cement Ratio
5. Explain in brief various types of reinforcement used for R.C.C.
6. Discuss the salient properties of various grades of concrete as per IS : 456–1978. What are the corresponding nominal mixes ?

2

Theory of Reinforced Beams and Slabs

2.1. INTRODUCTION

Though plain cement concrete has high compressive strength, its tensile strength is relatively low. Normally, the tensile strength of concrete is about 10 to 15% of its compressive strength. Hence, if a beam is made of plain cement concrete, it has a very low load carrying capacity since its low tensile strength limits its overall strength. It is, there, *reinforced* by placing steel bars in the tensile zone of

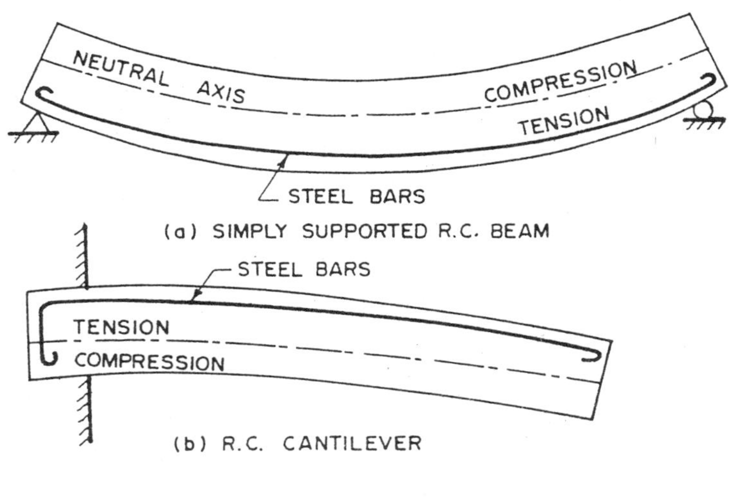

FIG. 2.1. REINFORCED CONCRETE BEAM.

the concrete beam so that the compressive bending stress is carried by concrete and tensile bending stress is carried entirely by steel reinforcing bars.

Fig 2.1 (*a*) shows a simply supported reinforced concrete beam subjected to transverse loads, bending it downwards. The reinforcement, consisting of steel bars, is placed at a suitable depth below the *neutral axis*. Similarly, Fig. 2.1 (*b*) shows a cantilever bending downwards. Since the tensile zone is above the neutral axis in this case, the steel bars are provided at some suitable height above the neutral axis. In both the cases, steel reinforcement is provided in the tensile zone only ; such beams are known as singly reinforced beams. However, if reinforcement is provided in compressive zone also, to carry the compressive stresses, it is known as a *doubly reinforced section* (see Chapter 4).

The concrete, reinforced by steel bars, is called *reinforced cement concrete* and is abbreviated as R.C.C. Reinforced concrete is thus a rational union of concrete and steel combined to act jointly. This joint action of steel and concrete in a reinforced concrete section is dependent on the following important factors :

(*i*) Bond between concrete and steel bars.

(*ii*) Absence of corrosion of steel bars embedded in the concrete.

(*iii*) Practically equal thermal expansion of both concrete and steel.

2.2 MODULAR RATIO

Fig. 2.2 shows a composite section consisting of two bars, one of concrete and other of steel, well bonded together with the help of end plugs, and subjected to a compressive load *P*.

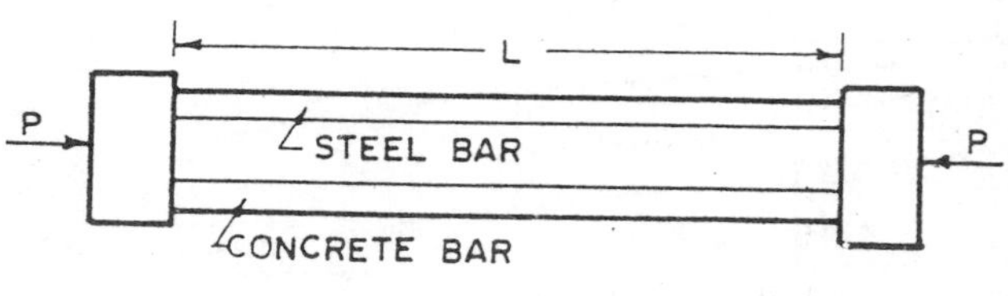

FIG. 2.2 COMPOSITE SECTION

Let E_s = Modulus of elasticity of steel

E_c = Modulus of elasticity of concrete.

Let the area of cross-section of each bar be equal to A. Let P_s be the load carried by steel and P_c be the load carried by concrete. Hence from statics:

$$P_S + P_C = P \qquad ...(1)$$

Also, from compatibility,

$$\Delta_S = \Delta_C = \Delta$$

where Δ is the shortening of each bar under the external load.

Hence $$\frac{P_S L}{A E_S} = \frac{P_C L}{A E_C}$$

or $$\frac{P_S}{P_C} = \frac{E_S}{E_C} \qquad ..(2)$$

From (1) and (2) we get

and $$P_S = \frac{P \,.\, E_S}{E_C + E_S} = \frac{mP}{1+m} \qquad ...(2.1)$$

$$P_C = \frac{P \,.\, E_C}{E_C + E_S} = \frac{P}{1+m} \qquad ...(2.1(a))$$

where $$m = \frac{E_S}{E_C} = \text{modular ratio.}$$

From Eqs. 2.1 and 2.1(a), it is seen that the load carried by steel is m times the load carried by concrete.

Equivalent areas of composite section

Fig 2.3 shows a reinforced concrete column having area of cross-section A_S of steel bar and a_C of that of concrete such that

$$A = A_S + a_C$$

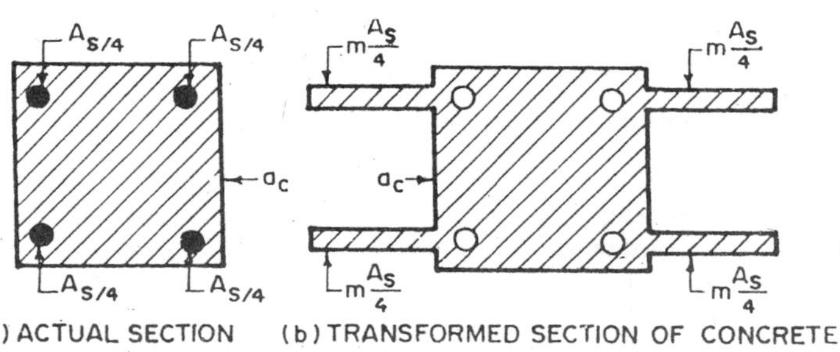

FIG. 2.3

If p_c is the stress set up in concrete and p_s that in steel, under an external compressive load P, we have

$$p_s \,.\, A_s + p_c \,.\, a_c = P \qquad ...(1)$$

Again, if we assume a perfect bond between steel and concrete then the compressive strains in both will be equal. Hence :

$$\frac{p_s}{E_s} = \frac{p_c}{E_c}$$

or $$p_s = \frac{E_s}{E_c} \,.\, p_c = m . p_c \qquad ...(2)$$

which again shows that stress in steel is equal to m times the stress in concrete.

Substituting the value of p_s in (1), we get

$$A_s \,.\, m \,.\, p_c + a_c \,.\, p_c = P$$

$$p_c = \frac{P}{a_c + m \,.\, A_s} = \frac{P}{A_e} \qquad ...(2.2)$$

In Eq. 2.2, the expression$(a_c + m\,A_s)$ is called the *equivalent area of the section in terms of concrete.* In other words, the area of steel A_s can be replaced by an equivalent area of concrete equal to $m\,A_s$ as shows in Fig. 2.2(a).

Again $$A_e = a_c + mA_s$$

$$= (A - A_s) + m\,A_s$$

$$= A + (m - 1)A_s \qquad ...(2.3)$$

where A is the total cross sectional area of the original column section.

2.3. SINGLY REINFORCED BEAM

(*a*) *Fundamental Assumptions*

Reinforced concrete beams are non-homogeneous in nature and, therefore, an exact theory of bending cannot be developed. The stress-strain curve for reinforced concrete is not a straight line, since plastic deformations accompany the elastic strains. It thus follows that Hooke's law applies only in a narrow range of stress. There are two fundamental theories on which analysis of reinforced concrete can be done : (*i*) *straight-line theory* or the *elastic theory,* and (*ii*) *ultimate load theory* and *limit state theory.* In the straight line theory the ultimate compressive strength of concrete and the yield point stress of steel are divided by the same or different appropriate safety factors to determine the allowable stresses that may be permitted in the two materials under working conditions. In the ultimate load theory, the resistance of the beam to pure bending is determined either by the ultimate strength of concrete or by the yield point stress of steel, and then this resistance can be divided by proper safety factor to determine the bending resistance that can be relied upon under working conditions. We shall first take up the straight line theory or the elastic theory. The ultimate load theory and the limit state design have been discussed seperately in detail.

The application of elastic theory to the beams is based on the following assumptions :

1. At any cross-section, plane section before bending remain plane after bending. This means that unit strain above and below the neutral axis are proportional to the distance from the neutral axis.

2. The concrete and steel reinforcement are well bonded. This means that the tensile strain in concrete surrounding the steel is equal to the tensile strain in the steel reinforcement.

3. All tensile stresses are taken by the reinforcement alone and none by the concrete. This means that while calculating the moment of resistance of the beam, the contribution of the concrete in the tensile zone is to be completely neglected.

4. Modulus of elasticity of concrete is constant at all stresses and is not a function of duration of stress. Young's modulus of concrete is generally found by secant method from the stress-strain curve.

5. Steel reinforcement is free from initial stresses when it is embedded in concrete.

Fig. 2.4(*a*) shows a singly reinforced concrete beam section. At very small loads both concrete and steel resist tension. However, as the load increases, the concrete at the bottom of the section will reach a tensile stress at which cracks occur, and thus the steel bars alone resists the tension. Fig. 2.4 (*b*), (*c*) and (*d*) shows the general of distribution of compressive stress in concrete as the load increases. Since the stress-strain diagram for concrete [Fig.2.4(*f*)] is not linear, the stresses in the concrete do not increase in direct proportion to increase in strain. Therefore, the stress diagram is slightly curved. The straight line theory assumes that the stress distribution in the concrete at working stresses of approximately half the ultimate compressive strength of concrete (σ_{cu}) is like that of Fig. 2.4 (*c*) but the slight curvature in the compression diagram is neglected.

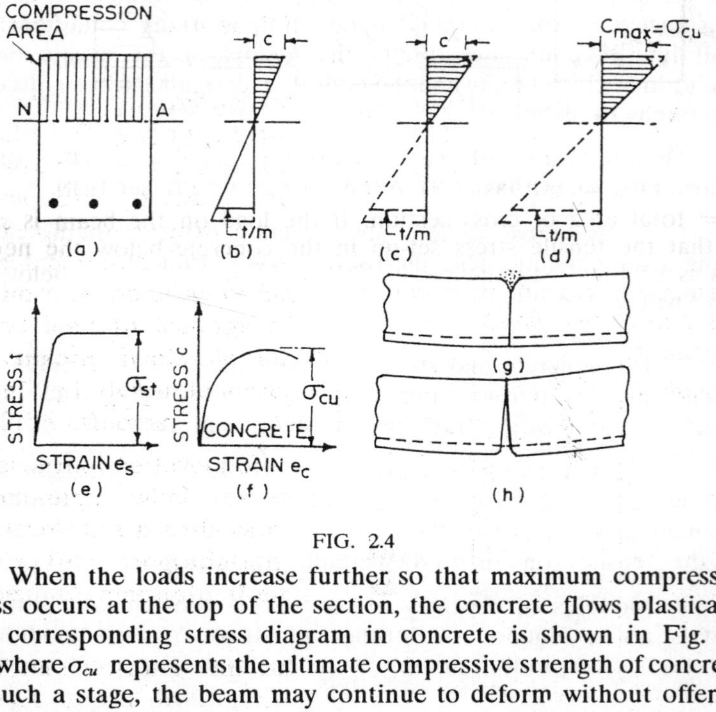

FIG. 2.4

When the loads increase further so that maximum compressive stress occurs at the top of the section, the concrete flows plastically. The corresponding stress diagram in concrete is shown in Fig. 2.4 (*d*), where σ_{cu} represents the ultimate compressive strength of concrete. At such a stage, the beam may continue to deform without offering proportionately increased resistance, and a small increase in load may cause complete failure of beam, without sufficient warning. Such compression failure is indicated in Fig. 2.4 (*g*). However, if the tensile stress in steel reaches its ultimate value first, the beam would "pop open" [Fig 2.4(*h*)] and the beam will fail in tension or by local crushing of concrete at top of the concrete. However, upto the elastic limit of steel, the reinforcing bars resist tension in proportion to their strain.

(*b*) ***Equivalent Area***

(*i*) **Uncracked section.** Fig. 2.5 (*a*) shows a R.C. beam section having A_{st} = area of steel reinforcement, a_c = area of concrete and

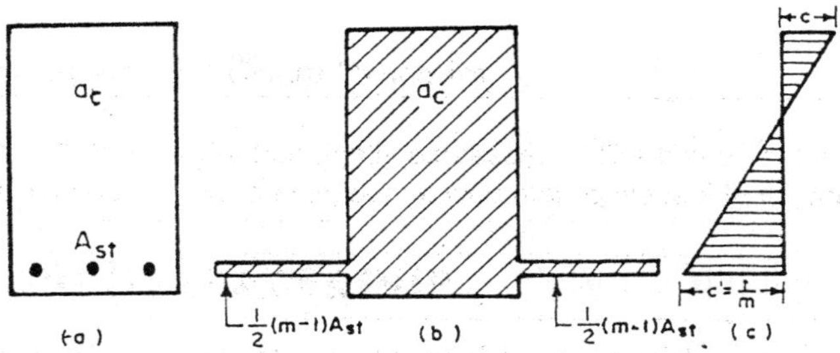

FIG. 2.5 EQUIVALENT AREA : UNCRACKED SECTION.

A = total area of cross section. If the load on the beam is *small* so that the tensile stress set-up in the concrete below the neutral axis is smaller than the permissible ; the concrete area below the N.A. will not crack. In that case,

$$e_s = e_c \text{ at the centre of the reinforcement}$$

$$\therefore \quad \frac{t}{E_s} = \frac{c'}{E_c}$$

or $$t = \frac{E_s}{E_c} c' . = m . c' \quad ...(2.4)$$

or $$c' = \frac{t}{m} \quad ...[2.4(a)]$$

where e_s = strain in steel

e_c = strain in concrete surrounding steel

t = stress in steel

c' = stress in concrete surrounding steel, assuming that concrete is not cracked.

c = compressive stress in concrete at the extreme fibre.

Eq. 2.4 suggests that stress in steel is m times the stress in concrete, or conversely, the stress in concrete is $\frac{1}{m}$ times the stress in steel. Fig 2.5(c) shows the stress distribution in the beam section consisting of an equivalent concrete sectin of Fig. 2.5(b). if a_c' is the additional area of concrete equivalent to steel area, we have Load carried by steel = Load carried by equivalent concrete or $$A_{st} \cdot t = a_c' \, c'$$

$$a_c' = \frac{t}{c'} \cdot A_{st} = m\,A_{st} \left(\text{since } \frac{t}{c'} = m\right) \quad ...(2.5)$$

However, *net* increase in the equivalent concrete area (uncracked) is

$$a_{ce} = a_c' - A_{st} = m\,A_{st} - A_{st} = (m-1)\,A_{st} \quad ...[2.5(a)]$$

(ii) Cracked Section.

If the load on the beam is increased, the tensile stress in the concrete below the N.A. will increase and will crack it. Even under normal loading conditions, the tensile stresses set up in concrete will be more than the permissible one. Hence it is assumed that the concrete area in tension zone will be *ineffective*. In that case, the compressive stress will be taken up by the concrete above N.A. and the tensile stresses will be taken up by the steel reinforcement. Fig. 2.6 (*c*) shows the stress distribution in the *equivalent concrete section* shown in Fig. 2.6(*b*). Since concrete is assumed to have cracked, the equivalent area of concrete will be such that the load

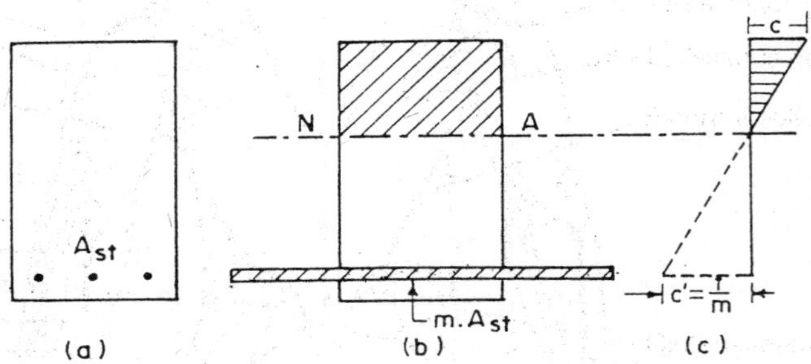

FIG. 2.6 EQUIVALENT AREA : CRACKED SECTION

carried by steel is the same as the load carried by an equivalent concrete area a_{ce} (strong enouch to carry that tensile load),

i.e. $$A_{st} \cdot t = a_{ce} \cdot c'$$

$$a_{ce} = \frac{t}{c'} A_{st} = m\,A_{st} \quad ...(2.6)$$

where c' is the tensile stress in the *equivalent concrete area.*

The difference between the stress diagrams of Fig. 2.5(c) and 2.6(c) must be clearly borne in mind. In Fig. 2.5(c), the tensile stress distribution is all over the portion of the concrete below the N.A., while in Fig. 2.6 (c), the tensile stress $c' \left(= \frac{t}{m} \right)$ is in the equivalent concrete area concentrated at the level of steel reinforcement.

2.4 NEUTRAL AXIS OF BEAM SECTION

In order to determine the position of the neutral axis in a reinforced concrete beam section, we shall consider two cases : (i) when the stresses developed in the concrete section are known and (ii) when the dimensions of the section are known but stresses are not known. Fig. 2.7 (a) shows the reinforced concrete beam section having steel reinforcement placed in the tension zone. Such a section is known as *singly reinforced section.*

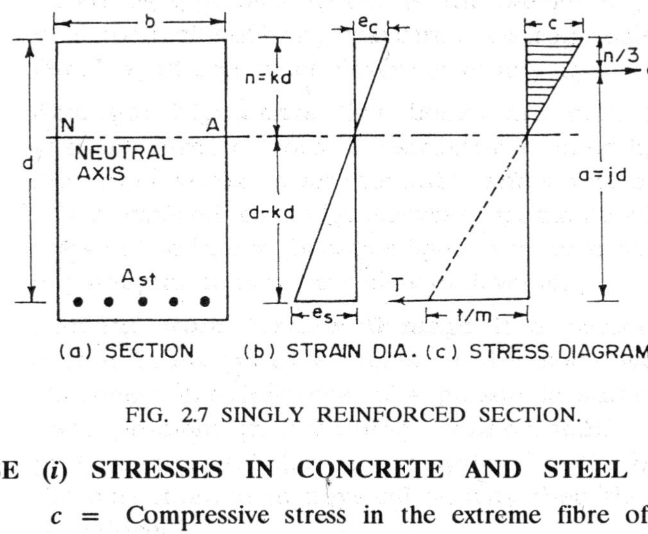

FIG. 2.7 SINGLY REINFORCED SECTION.

CASE (i) STRESSES IN CONCRETE AND STEEL KNOWN

Let c = Compressive stress in the extreme fibre of concrete
t = tensile stress in steel reinforcement
b = breadth of beam
d = depth to the centre of reinforcement (known as the *effective depth)*
n = kd = depth of N.A. below the top of the beam
k = Neutral axis depth factor = n/d
A_{st} = Area of tensile reinforcement,

Since there is no resultant force across the section,

TOTAL COMPRESSION = TOTAL TENSION.

Neglecting the tensile stress in concrete, we get

$$\frac{1}{2}c\,.\,b\,.\,kd = t\,.\,A_{st}$$

$$\frac{A_{st}}{b\,.\,d} = k\,.\,\frac{c}{2t} \qquad ...(2.7)$$

If t is the stress in steel, the stress c' in the equivalent concrete area will be equal to $\frac{t}{m}$ as proved in Eq. 2.4. Hence from the diagram [Fig. 2.7 (*c*)], we get

$$\frac{c}{k\,.\,d} = \frac{t}{m}\,.\,\frac{1}{d - kd} = \frac{t}{md\,(1 - k)}$$

$$\frac{c}{k} = \frac{t}{m\,(1 - k)}$$

or $$mc - mck = kt$$

From which $$k = \frac{mc}{mc + t} \qquad ...(2.8)$$

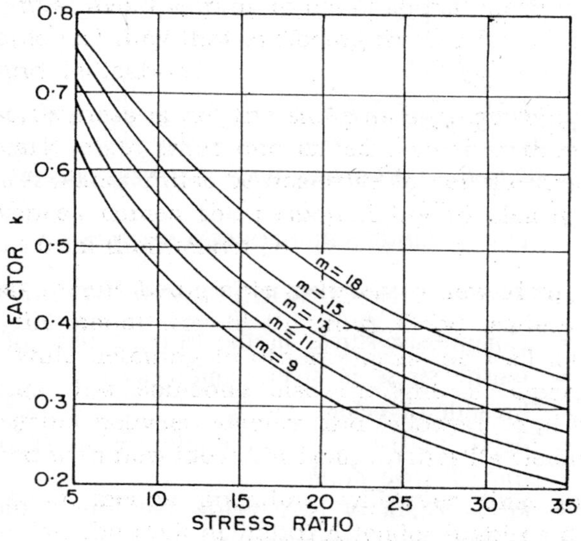

FIG. 2.8. DETERMINATION OF *k*

$$k = \frac{1}{1 + \frac{1}{m}\frac{t}{c}} = \frac{1}{1 + \frac{r}{m}} \qquad \text{...[2.8(a)]}$$

where r = stress ratio = t/c

Eq. 2.8 gives the location of the neutral axis when the stresses in concrete and steel are known. If the stress ratio $r\left(=\frac{t}{c}\right)$ and the modular ratio m are known, N.A. can be located from Eq. 2.8 (a). .Fig. 2.8 shows the graphical representation of Eq. 2.8(a) for various values of stress ratio and modular ratio.

Let σ_{st} = Permissible tensile stress in reinforcement (*i.e.* permissible value of t)

σ_{cbc} = Permissible compressive stress in concrete due to bending (*i.e.* permissilble value of c)

n_c = Depth of critical neutral axis, corresponding to permissible values of stresses in steel and concrete.

Then

$$n_c = k \,.\, d = \frac{mc}{mc + t} \,.\, d = \frac{m\,\sigma_{cbc}}{m\sigma_{cbc} + \sigma_{st}} \,.\, d = \frac{1}{1 + \frac{r}{m}} d \qquad \text{...[2.8(b)]}$$

where r = stress ratio $= \dfrac{\sigma_{st}}{\sigma_{cbc}}$

The section in which the stresses in concrete and steel reach their corresponding permissible values simultaneously is called a *balanced section* (*See* § 2.6).

Note. As per IS : 456–1978, the modular ratio is given by

$$m = \frac{280}{3\,\sigma_{cbc}}$$

$$\therefore \qquad m \,.\, \sigma_{cbc} = \frac{280}{3} = 93.33$$

Hence Eq. 2.8 reduces to the following form

$$k = \frac{m\,\sigma_{cbc}}{m\sigma_{cbc} + \sigma_{st}} = \frac{93.33}{93.33 + \sigma_{st}} \qquad \text{...[2.3(c)]}$$

or

$$k = \frac{1}{1 + 0.0107\,\sigma_{st}} \qquad \text{...[2.8(d)]}$$

This suggests that value of k does not depend upon the grade of the concrete. It depends only on the permissible stress in reinforcement See Table 2.1.

The area of steel for the section is given by Eq. 2.7. If p is the *percentage steel*, we have

$$p = \frac{A_{st}}{b\,.\,d} \times 100$$

But $$\frac{A_{st}}{b\,.\,d} = \frac{k\,.\,c}{2t} \quad \text{(from Eq. 2.7)}$$

$$\therefore \quad p = \frac{k\,.\,c}{2t} \times 100 \qquad ...(2.9)$$

$$= 50\,\frac{c}{t} \cdot \frac{mc}{mc + t}$$

or $$p = \frac{50\,mc^2}{t\,(mc + t)} \qquad ...[2.9(a)]$$

Expressed in terms of the permissible values of the stresses, the above equation may be written as :

$$p = \frac{50\,m\sigma_{cbc}^{\;2}}{\sigma_{st}\,(m\,\sigma_{cbc} + \sigma_{st})} \qquad ...[2.9(b)]$$

CASE (*ii*) DIMENSIONS OF THE BEAM COMPLETELY KNOWN

Let us now take the case when the dimensions of the section, including the reinforcement are known, but the stresses in concrete and steel are *not known.* Refer Fig. 2.7. The neutral axis of a homogeneous beam is the horizontal line that passes through the centre of gravity of the cross-section. Hence equating the moment of area in compression to the moment of the *equivalent area* in tension about the N.A., we get.

$$b\,.\,kd\,.\,\frac{kd}{2} = mA_{st}\,(d - kd)$$

or $$k^2\,.\,\frac{b\,d^2}{2} + m\,.\,kd\,A_{st} - m\,.\,dA_{st} = 0$$

or $$k^2 + 2k\,.\,m\,.\,\frac{A_{st}}{b\,d} - 2m\,\frac{A_{st}}{b\,d} = 0$$

puting $\frac{A_{st}}{b\,d} = p'$ = reinforcement ratio, we get

$$k^2 + 2k\,mp' - 2mp' = 0,$$

From which.

$$k = \frac{-2mp' \pm \sqrt{4m^2p'^2 + 8mp'}}{2}$$

or $$k = \sqrt{2mp' + (mp')^2} - mp' \qquad ...(2.10)$$

Thus if p' is known (*i.e.* if the dimensions b and d, and the area of steel A_{st} are known) the neutral axis depth factor k can be calculated from Eq. 2.10 and the N.A. can be located.

2.5 MOMENT OF RESISTANCE

Since there is no resultant force across the section, the total compressive force acting at the centre of gravity of the compressive forces is equal to the total tensile force acting at the centre of gravity of the steel reinforcement. Hence the moment of resistance of the reinforced concrete beam section is equal to the moment of the couple consisting of the compressive force and the tensile force. The total compressive force $C = \frac{1}{2} c \, . \, k \, . \, b \, . \, d \; (= \frac{1}{2} \sigma_{cbc} \, . \, kd \, . \, b)$ and acts at a distance of $kd/3$ from the top of the section. Similarly, the total tensile force $T = t \, . \, A_{st} \; (= \sigma_{st} A_{st})$ and acts at a disdtance d from the top of the section. If jd is the lever arm (*i.e.* the distance between the compressive force and the tensile force), we have

Lever arm $$a = jd = d - \frac{kd}{3} = d\left(1 - \frac{k}{3}\right)$$

or $$j = 1 - \frac{k}{3} \qquad ...(2.11)$$

Eq. 2.11 gives the *lever arm factor* j in terms of the neutral axis depth factor k.

Now, moment of resistance M_r is given by

M_r = force (either compressive or tensile) × lever arm

$$\therefore \quad M_r = \tfrac{1}{2} c \, . \, kd \, . \, b \, (jd) = (\tfrac{1}{2} cjk)\, bd^2 \qquad ...[2.12(a)]$$

or $$M_r = Rbd^2 \qquad ...(2.12)$$

where $$R = \tfrac{1}{2} cjk = \tfrac{1}{2} \sigma_{cbc} \, . \, jk \qquad ...(2.13)$$

Table 2.1 gives the values of design constants k, j, R and p for balanced design.

TABLE 2.1. VALUES OF DESIGN CONSTANTS

Grade of concrete	→	M 15	M 20	M 25	M 30
Modular ratio m	→	18.67	13.33	10.98	9.33
σ_{cbc} N/mm^2	→	5.0	7.0	8.5	10.0
$m\sigma_{cbc}$		93.33	93.33	93.33	93.33
(a) σ_{st}=140 N/mm^2 (Fe 250)	***k***	0.400	0.400	0.400	0.400
	j	0.867	0.867	0.867	0.867
	R	0.867	1.214	1.474	1.734
	p(%)	0.714	1.000	1.214	1.429
(b) σ_{st}=190 N/mm^2	***k***	0.329	0.329	0.329	0.329
	j	0.890	0.890	0.890	0.890
	R	0.732	1.025	1.244	1.464
	p(%)	0.433	0.606	0.736	0.866
(c) σ_{st}=230 N/mm^2 (Fe 415)	***k***	0.289	0.289	0.289	0.289
	j	0.904	0.904	0.904	0.904
	R	0.653	0.914	1.110	1.306
	p(%)	0.314	0.440	0.534	0.628
(d) σ_{st}=275 N/mm^2 (Fe 500)	***k***	0.253	0.253	0.253	0.253
	j	0.916	0.916	0.916	0.916
	R	0.579	0.811	0.985	1.159
	p(%)	0.230	0.322	0.391	0.460

From Eq. 2.12, the dimensions of the beam section can be determined to develop a given moment of resistance. *It should be noted that for a given type of concrete and steel reinforcement, factor R is a constant and does not depend upon the beam dimensions.*

Also $M_r = t \,.\, A_{st} \,.\, jd = \sigma_{st} \,.\, A_{st} \,.\, jd$ ...(2.14)

Hence

$$A_{st} = \frac{M_r}{t\,.\,jd} = \frac{M}{t\,.\,jd} = \frac{M}{\sigma_{st}\,.\,jd} \qquad ...(2.15)$$

where M is the external bending moment and σ_{st} is permissible stress in steel reinforcement.

Eq 2.15 is used for determining the cross sectional area of steel reinforcement for given moment of resistance and effective depth.

If the section of a beam, including the steel reinforcement, is completely known, its moment of resistance is calculated both by Eqs. 2.12(a) and 2.14, using maximum permissible values of stresses c ($= \sigma_{cbc}$) and t ($= \sigma_{st}$). Thus the values of M_r will be obtained and if the two values do not agree, it will be concluded that both concrete and steel do not develop their allowable stresses simultaneously. The moment of resistance of the beam will then be equal to the lesser of the two values.

2.6 BALANCED, UNDER-REINFORCED AND OVER-REINFORCED SECTION

(*a*) BALANCED SECTION

In a beam section, if the area of steel reinforcement A_{st} is of such magnitude that the permissible stresses c ($= \sigma_{cbc}$) and t ($= \sigma_{st}$) in concrete and steel respectively, are developed simultaneously, the section is known as the balanced section, critical section or *economical section.*

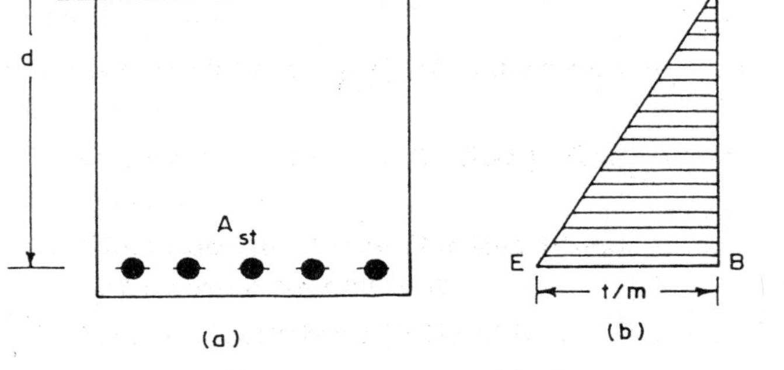

FIG. 2.9 BALANCED SECTION.

For such a section (Fig 2.9), the neutral axis factor k is determined by Eq. 2.8 (*b*), using c ($= \sigma_{cbc}$) and t ($= \sigma_{st}$) as the maximum permissible stress in the two materials and depth n_c of the critical N.A. is taken equal to k times d. For such a balanced section, the moment of resistance obtained from the compressive force will be equal to the moment of resistance obtained from the tensile force.

The percent area of reinforcement for a balanced section is obtained from Eq. 2.9 (*a*) :

$$p = \tfrac{1}{2} k \,.\, \frac{c}{t} \times 100 = \tfrac{1}{2} k \,.\, \frac{\sigma_{cbc}}{\sigma_{st}} \times 100 \quad ...(2.9)$$

or $$p = \frac{50\,k}{r} \quad ...(2.9a)$$,where r is the stress ratio.

For example, if $c = \sigma_{cbc} = 5 \text{ N/mm}^2$, $t = \sigma_{st} = 104 \text{ N/mm}^2$ and $m = 19$, we have

$$k = \frac{1}{1 + \frac{1}{m}\frac{t}{c}} = \frac{1}{1 + \frac{140}{19 \times 5}} = 0.404$$

$$\therefore \quad p = \tfrac{1}{2} \times 0.404 \times \frac{5}{140} \times 10 = 0.72\%$$

Thus for a reinforced concrete having the above properties if the reinforcement provided is 0.72% of the cross-section ($b \times d$), a balanced section will be obtained. The N.A. of a balanced section is called the *critical neutral axis..*

(*b*) UNDER REINFORCED SECTION

An under reinforced section is the one in which the percentage steel provided is *less* than that given in Eq 2.9 and therefore full strength of concrete in compression is not developed. The actual neutral axis of such a section will fall *above* the critical neutral axis of a balanced section.

Fig 2.10 (*a*) shows the under-reinforced section in which the actual N.A. is above the critical N.A. Let $c = \sigma_{cbc}$ be the permissible stress in concrete and $t = \sigma_{st}$ be the permissible stress in steel. If full permissible compressive stress $c = \sigma_{cbc} = AD_1$ is permitted to be developed in concrete, the corresponding stress t_1 in steel will be such that $BE_1 = t_1/m$. Evidently, since $t_1 > t$, full compressive stress c cannot develop in concrete. Instead, a compressive stress $c_1 = AD$ is developed of such a magnitude that $BE = t/m = \frac{\sigma_{st}}{m}$ is obtained. Thus, in a under-reinforced concrete, the concrete is not

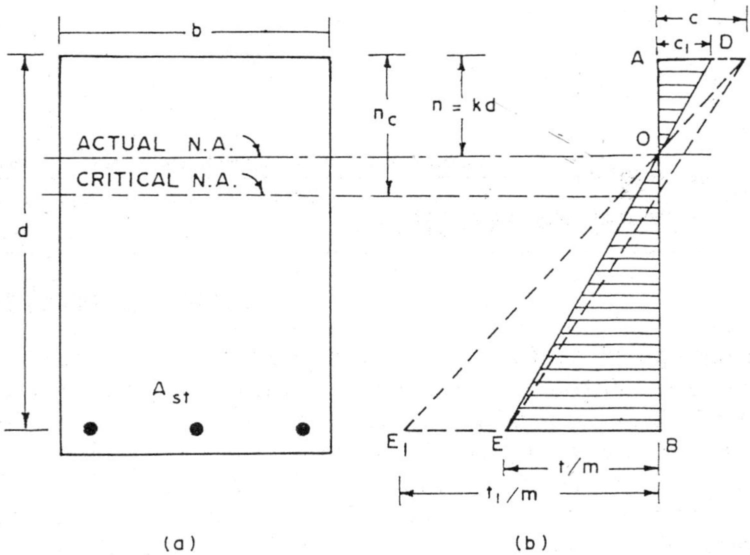

FIG. 2.10 UNDER-REINFORCED SECTION.

fully stressed to its permisible value when stress reaches its maximum value of $t = \sigma_{st}$.

The moment of resistance of an under-reinforced section is, therefore, computed on the basis of the tensile force in steel :

$$M_r = t A_{st} \cdot jd = \sigma_{st} \cdot A_{st} \cdot jd \qquad \text{...(Eq. 2.14)}$$

(c) OVER-REINFORCED SECTION

In an over-reinforced section, the reinforcement provided is more than critical one and therefore the actual N.A. of such a section falls below the critical N.A. of a balanced section.

Fig 2.11(a) shows a over-reinforced section in which the actual N.A. falls below the critical one. If stress in steel is permitted to be equal to maximum permissible value $t = (\sigma_{st})$ the corresponding stress in concrete will be equal to $AD_1 = c_1$ which is greater than permissible value $c = (\sigma_{cbc})$in concrete. Hence the stress distribution will be governed by line DOE (and not $D_1 O E_1$) , giving rise to a stress t_1 in steel which is lesser than t. Thus, in a over-reinforced section, steel reinforcement is not fully stressed to its permissible

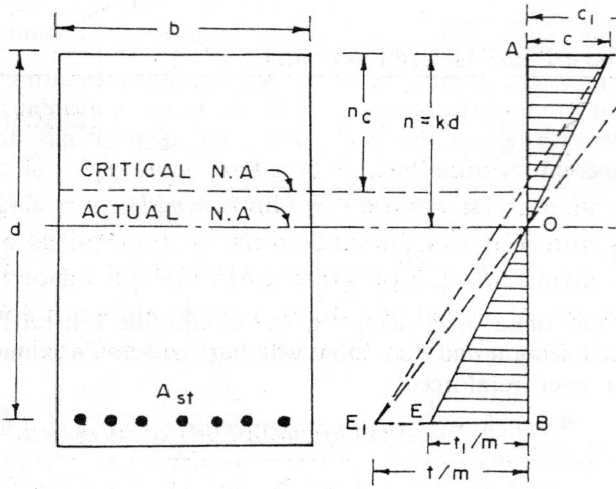

FIG. 2.11. OVER-REINFORCED SECTION.

value and the moment of resistance is determined on the basis of compressive force developed in concrete :

$$M_r = \frac{1}{2} c \,.\, kd \,.\, b \,.\, jd = \frac{1}{2} c \,.\, j \,.\, k \,.\, bd^2 = \frac{1}{2} \sigma_{cbc} \,.\, j \,.\, k \,.\, bd^2 \quad ...[2.12(a)]$$

or $$M_r = \frac{1}{2} c \,.\, n \,.\, \left(d - \frac{n}{3}\right) b = \frac{1}{2} \sigma_{cbc} \,.\, n \left(d - \frac{n}{3}\right) b \quad ...[2.12(b)]$$

where n = depth of the neutral axis

2.7 TYPES OF PROBLEMS IN SINGLY-REINFORCED BEAMS

There may be three types of problems in singly reinforced concrete beams :

Type I : Determination of the moment of resistance of the given section

Type II : Determination of stresses in the given section subjected to a given bending moment.

Type III : Design of section to resist a given bending moment.

TYPE I : DETERMINATION OF MOMENT OF RESISTANCE

In this type of problem, the section including the steel reinforcement is completely defined and it is required to determine the moment of resistance corresponding to a given set of permissible stresses in concrete and steel. The procedure for solution is as under :

1. Determine the position of the neutral axis by equating the moment of the compressive area of the concrete to equivalent area of steel in tension, about the N.A.. Thus, the neutral axis depth factor k is given by Eq. 2.10

$$k = \sqrt{2mp' + (mp')^2} - mp' \qquad \text{...(2.10)}$$

where $p' = \dfrac{A_{st}}{bd}$ = reinforcement ratio.

2. Find the value of k, using Eq. 2.8(a), for the critical N.A. of the balanced section and determine whether the section is under-reinforced or over reinforced.

3. Find the moment of resistance M_r by either using Eq. 2.12(a) or Eq. 2.14, whichever may be the case.

4. If it is required to find the uniformly distributed load or point load or any other type of the load on a simply-supported beam, calculate the maximum B.M. in terms of the unknown load and equate it to the moment of resistance determined in step 3.

TYPE II : DETERMINATION OF STRESSES IN THE SECTION

In this type of problem, the dimensions of the section, including the reinforcement, are completely given and it is required to find the stresses developed in steel and concrete, when subjected to a given bending moment. The solution is done in the following steps:

1. Determine the positon of neutral axis by equating the moment of the compressive area of the concrete to equivalent area of steel in tension, about the N.A. Thus, the N.A. depth factor k is given by

$$k = \sqrt{2mp' + (mp')^2} - mp' \qquad \text{...(2.10)}$$

where $p' = \dfrac{A_{st}}{bd}$.

2. Knowing k, determine the lever arm factor j from Fig 2.11

$$j = 1 - \frac{k}{3}.$$

3. Determine the moment of resistance of the section in terms of tensile force in the reinforcement (Eq. 2.14) and equate it to the external moment M, Thus

$$A_{st} \cdot t \cdot jd = M \qquad ...(2.16)$$

Thus the stress t in steel is known.

4. Calculate the stress in concrete from the relation (Fig. 2.7)

$$\frac{c}{kd} = \frac{t}{m} \cdot \frac{1}{d - kd}$$

or

$$c = \frac{kt}{m(1-k)} = \frac{nt}{m(d-n)} \qquad ...(2.17)$$

(where $n = kd$ = depth of the neutral axis).

TYPE III : DESIGN OF THE SECTION

This is more common type of problem in which it is required to determine the dimensions of the section, including the reinforcment, to develop a given moment of resistance corresponding to a given set of permissible stresses in concrete and steel. The design is done in the following steps :

1. Determine the position of the critical neutral axis of the balanced section from Eq. 2.8 (*a*) or Eq 2.8 (*b*)

$$k = \frac{1}{1 + \frac{1}{m}\frac{t}{c}} = \frac{1}{1 + \frac{1}{m}\frac{\sigma_{st}}{\sigma_{cbc}}} = \frac{1}{1 + \frac{r}{m}}$$

where r = stress ratio

Also determine $j = 1 - \frac{k}{3}$.

2. Determine the factor $R = \frac{1}{2} c\, j\, k = \frac{1}{2} \sigma_{cbc}\, j \cdot k$

3. Assume a suitable $\frac{b}{d}$ ratio for the beam section and calculate its moment of resistance from Eq. 2.12 :

$$M_r = R \cdot b\, d^2.$$

4. Equate this to be maximum bending moment to which the beam section is subjected. To compute the maximum bending moment, the approximate self-weight of the beam of any suitable assumed section should also be taken into account. Thus the dimensions b and d are known.

5. Determine the area of steel reinforcement from Eq. 2.15 :

$$A_{st} = \frac{M}{t \cdot jd} = \frac{M}{\sigma_{st} \cdot jd}$$

The detailed design features of a singly reinforced beam are given in chapter 7.

Example 2.1 *For a balanced rectangular section ($b \times d$) of a singly reinforced beam, determine (i) depth of neutral axis (ii) moment of resistance and (iii) percentage of steel using M 15 concrete and using (a) mild steel (b) Fe 415 steel.*

If b = 200 mm and d = 300 mm, determine the numerical values of n, M_r and A_{st}.

Solution (*a*) Mild steel bars.

Given $c = \sigma_{cbc} =$ N/mm², $m \approx 19$ for M 15 concrete, and

$t = \sigma_{st} = 140$ N/mm² for mild steel bars.

(*i*) For a balanced section, we have, from Eq. 2.8

$$k = \frac{mc}{mc + t} = \frac{19 \times 5}{19 \times 5 + 140} = 0.404$$

∴ Depth of N.A. $= n = kd$

$= \mathbf{0.404\ d.}$

(*ii*) From Eq. 2.12 (*a*), moment of resistance is given by

$$M_r = \frac{1}{2} cjk \cdot b\,d^2 = R \cdot b\,d^2$$

where

$$j = 1 - \frac{k}{3} = 1 - \frac{0.404}{3} = 0.865$$

$$\therefore \quad R = \frac{1}{2} cjk = \frac{1}{2} \times 5 \times 0.865 \times 0.404 = 0.874$$

$$\therefore \quad M_r = R\,bd^2 = \mathbf{0.874\,bd^2}$$

(where M is in N-mm units)

(*iii*) From Eq. 2.9 percentage steel is given by

$$p = \frac{50\,mc^2}{t\,(mc + t)} = \frac{50 \times 19\,(5)^2}{140\,(19 \times 5 + 140)} = 0.722\%$$

Alternatively $p = \dfrac{k \cdot c}{2t} \times 100 = \dfrac{0.404 \times 5}{2 \times 140} \times 100 = 0.722\%$

For the given numerical values,

$n = 0.404\ d = 0.404 \times 300 = \mathbf{121.2\ mm}$

$M_r = 0.874\,b\,d^2 = 0.874\,(200)\,(300)^2$

$$= 15.73 \times 10^6 \text{ N} - \text{mm} = \mathbf{15.73 \text{ kN}-m}$$

$$A_{st} = p \cdot \frac{b\,d}{100} = \frac{0.722 \times 200 \times 300}{100} = \mathbf{433.2 \text{ mm}^2}$$

(b) Fe 415 Steel

We have $c = \sigma_{cbc} = 5 \quad \text{N/mm}^2,\ m \approx 19$

and $\quad t = \sigma_{st} = 230 \text{ N/mm}^2$ for Fe 415 steel

(i) For balanced section,

$$k = \frac{mc}{mc + t} = \frac{19 \times 5}{19 \times 5 + 230} = 0.292$$

$\therefore$ Depth of N.A.$=n=kd=\mathbf{0.292\ d}$

(ii) From Eq. 2.12 (a)

$$M_r = \tfrac{1}{2} cjk \,.\, b\,d^2 = R \,.\, b\,d^2$$

where

$$j = 1 - \frac{k}{3} = 1 - \frac{0.292}{3} = 0.903$$

$$\therefore \quad R = \tfrac{1}{2} cjk = \tfrac{1}{2} \times 5 \times 0.903 \times 0.292 = 0.659$$

$$\therefore \quad M_r = R\,bd^2 = \mathbf{0.659\,bd^2}$$

(where m is in N-mm units)

(iii) From Eq 2.9 percentage of steel is given by

$$p = \frac{50\,mc^2}{t\,(mc + t)} = \frac{50 \times 19\,(5)^2}{230\,(19 \times 5 + 230)} = 0.318\%$$

Alternatively, $p = \dfrac{k \,.\, c}{2t} \times 100 = \dfrac{0.292 \times 5}{2 \times 230} \times 100 = 0.318\%$

For the given numerical values,

$$n = 0.292\ d = 0.292 \times 300 = \mathbf{87.6 \text{ mm}}$$

$$M_r = 0.659\,bd^2 = 0.659\,(200)\,(300)^2$$

$$= 11.86 \times 10^6\,N{-}mm = \mathbf{11.86 \text{ kN}-m}$$

$$A_{st} = p \cdot \frac{b\,d}{100} = \frac{0.318 \times 200 \times 300}{100} = \mathbf{190.8 \text{ mm}^2}$$

Thus it is seen that by using HYSD bars, the moment of resistance of the beam has decreased.

For a given external B.M., greater depth of the beam is required if HYSD bars are used in place of mild steel bars. However, great saving in steel area is achieved by using HYSD bars.

Example 2.2 *Determine the moment of resistance of a singly reinforced beam 160 mm wide and 300 mm deep to the centre of reinforcement, if the stresses in steel and concrete are not to exceed 140 N/mm² and 5N/mm². The reinforcement consists of 4 bars of 16 mm diameter. Take m=18.*

If the above beam is used over an effective span of 5m, find the maximum load the beam can carry, inclusive of its own weight.

Solution $A_{st} = 4 \times \frac{\pi}{4}(16)^2 = 804 \text{ mm}^2$

$c = \sigma_{cbc} = 5 \text{ N/mm}^2 ; t = \sigma_{st} = 140 \text{ N/mm}^2 ;$

$m = 18.$

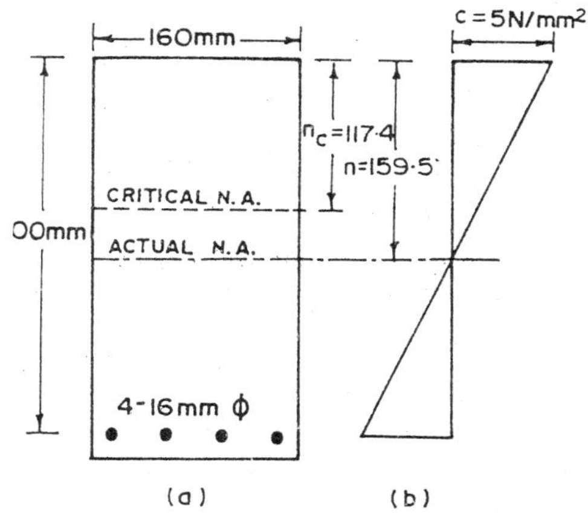

FIG. 2.12.

Here the dimensions of the section, including the tensile reinforcement are directly given.

Equating the moment of area in compression to the moment of the equivalent area in tension, about the N.A., we get

$$b \times n \times \frac{n}{2} = m A_{st} (d-n)$$

$$\therefore \qquad \frac{160 \times n^2}{2} = 18 \times 805\,(300 - n)$$

or $\qquad n^2 + 181.125\,n - 54337.5 = 0$

From which $\qquad n = 160$ mm

If n_c is the depth of critical neutral axis, we have, from Eq. 2.8

$$n_c = k \cdot d = \frac{mc}{mc + t}\,d.$$

$$= \frac{18 \times 5}{(18 \times 5) + 140} \times 300 = 117.4 \text{ mm}$$

Since the depth of actual neutral axis is more than the critical one, the section is over reinforced.. Thus, concrete reaches its maximum stress earlier to steel. Hence the moment of resistance is found on the basis of compressive force developed in concrete.

Lever arm $\qquad = d - \frac{n}{3} = 300 - \frac{159.5}{3} = 246.8$ mm

$$\therefore M_r = \tfrac{1}{2} c \cdot n \cdot b \cdot \left(d - \frac{n}{3}\right) = \tfrac{1}{2} \times 5 \times 159.5 \times 160\,(246.8)$$

$$= 15.75 \times 10^6 \text{ N-mm} = 15.75 \text{ kN-m}$$

Let $\qquad w$ = uniformly distributed load in kN/m inclusive of the self-weight of the beam.

Maximum B.M. $\qquad = \frac{w L^2}{8} = \frac{w \times 25}{8}$ kN − m

Equating this to the moment of resistance of the beam we get

$$w = \frac{15.75 \times 8}{25} = \mathbf{5.04 \text{ kN/m.}}$$

Example 2.3 *The cross-section of a simply supported reinforced beam is 200 mm wide and 300 mm deep to the centre of the reinforcement which consists of 3 bars of 16 mm dia. Determine from the first principles the depth of N.A. and the maximum stress in concrete when steel is stressed to 120 N/mm². Take m=19.*

Solution. Here, the dimensions are fully given.

$$A_{st} = 3 \times \frac{\pi}{4}\,(16)^2 = 603.2 \text{ mm}^2.$$

Let the depth of N.A. be n. Equating the moment of the compressive area to the moment of equivalent area of steel, about N.A., we get

$$b \,.\, n \,.\, \frac{n}{2} = m\, A_{st}\,(d - n)$$

or $$100\, n^2 = 19 \times 603.2\,(300 - n)$$

or $$n^2 + 114.6\, n - 34382 = 0$$

From which n = **136.8 mm**

Fig. 2.13 (*b*) shows the stress diasgram.

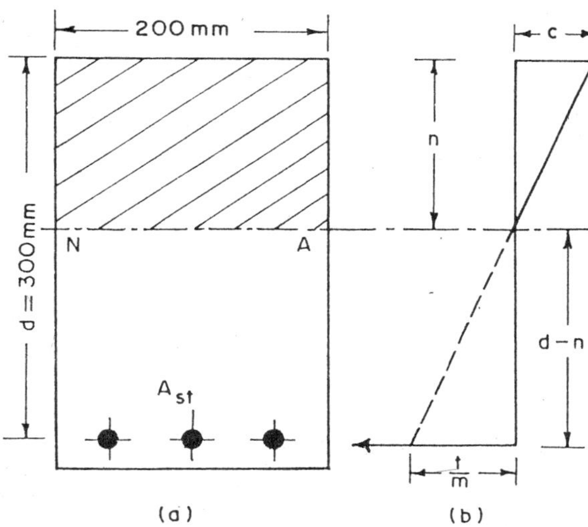

FIG. 2.13

Stress in steel = t = 120 N/mm² (Given)

$$\therefore \qquad \frac{t}{m} = \frac{120}{19} = 6.316$$

From the stress diagram

$$\frac{c}{n} = \frac{t/m}{d-n}$$

or $$\frac{c}{136.8} = \frac{6.316}{300 - 136.8} = 0.0387$$

From which c = **5.29 N/mm²**

Example 2.4 *A rectangular, singly reinforced beam, 300 mm wide and 500 mm effective depth is used as a simply supported beam over an effective span of 6 m. The reinforcement consists of 4 bars of 20 mm diameter. If the beam carries a load of 12 kN/m, inclusive of the self-weight, determine the stresses developed in concrete and steel. Take m=19.*

Solution Let n = depth of N.A.

$$A_{st} = 4 \times \frac{\pi}{4}(20)^2 = 1256.6 \text{ mm}^2.$$

Equating the moments of two areas about N.A. we get

$$b \times n \times \frac{n}{2} = m A_{st}(d-n)$$

$$\therefore \quad 300 \times \frac{n^2}{2} = 19 \times 1256.6\,(500-n)$$

or $n^2 + 159.2n - 79587 = 0$

From which $n = 213.5$ mm

$$\text{Lever arm} = a = d - \frac{n}{3} = 500 - \frac{213.5}{3} = 428.8 \text{ mm}$$

$$\text{Maximum B.M.} = \frac{wl^2}{8} = \frac{12(6)^2}{8} = 54 \text{ kN}-\text{m}$$

$$= 54 \times 10^6 \text{ N}-\text{mm}$$

Let c be the compressive stress in concrete.

$$\therefore \quad M_r = \tfrac{1}{2}c.n.b \times a = \tfrac{1}{2}c \times 300 \times 213.5 \times 428.8$$

$$= 13.732 \times 10^6 c \text{ N}-\text{mm}$$

Equating M_r to the external B.M. we get

$$13.732 \times 10^6 c = 54 \times 10^6$$

$$\therefore \quad c = \frac{54}{13.732} = \mathbf{3.93\ N/mm^2}$$

If t is the corresponding stress in steel, we get

$$\frac{c}{n} = \frac{t/m}{d-n}$$

$$\therefore \quad t = \frac{mc}{n}(d-n) = \frac{19 \times 3.93}{213.5}(500 - 213.5)$$

$$= \mathbf{100.2\ N/mm^2}$$

Alternatively, $M_r = A_{st}.\,t\,a = 1256.6 \times t \times 428.8$

$$\therefore \quad 1256.6 \times 428.8\,t = 54 \times 10^6$$

$$t = \frac{54 \times 10^6}{1256.6 \times 428.8} = 100.2 \text{ N/mm}^2$$

Example 2.5(*a*) *Design a reinforced concrete beam subjected to a bending moment of 20 kN-m. Use M 15 concrete, and mild steel reinforcement. Keep the width of the beam equal to half the effective depth.*

Solution. For M 15 concrete, $c = \sigma_{cbc} = 5$ N/mm^2 and $m \approx 19$
For mild steel bars, $t = \sigma_{st} = 140$ N/mm^2.

The beam is to be designed in such a way that maximum allowabale stresses in both the materials reach simultaneously.

Thus, for a balanced section

$$k = \frac{m\sigma_{cbc}}{m\sigma_{cbc} + \sigma_{st}} = \frac{19 \times 5}{19 \times 5 + 140} = 0{\cdot}4043$$

$$j = 1 - \frac{k}{3} = 1 - \frac{0.4043}{3} = 0.865$$

$$R = \frac{1}{2}\sigma_{cbc} \cdot j \cdot k = \frac{1}{2} \times 5 \times 0.865 \times 0.4043 = 0.874$$

Now $$M_r = R \cdot b\,d^2 = R\left(\frac{d}{2}\right)(d^2) = \frac{R\,d^3}{2}$$

$$= \frac{0.874}{2} d^3 = 0.437\, d^3 \text{ N}-\text{mm}$$

Given B.M. $$M = 20 \text{ kN}-\text{m} = 20 \times 10^6 \text{ N}-\text{mm}$$

Since $$M_r = M$$

$\therefore$ $$0.437\, d^3 = 20 \times 10^6$$

or $$d = \left(\frac{20 \times 10^6}{0.437}\right)^{\frac{1}{3}} \approx \mathbf{358\ mm}..$$

$\therefore$ $$b = \frac{1}{2} d = \mathbf{179\ mm.}$$

Area of steel is given by

$$A_{st} = \frac{M}{\sigma_{st}\, jd} = \frac{20 \times 10^6}{140 \times 0.865 \times 358} = \mathbf{461.3\ mm^2}$$

Thus provide a beam of 179 mm width and 358 mm depth (effective) with area of tensile reinforcement equal to 461.3 mm^2.

Example 2.5 (*b*) *Redesign the beam of example 2.5 (a) using M 20 concrete and Fe 415 steel.*

Solution. For M 20 concrete, $c = \sigma_{cbc} = 7$ N/mm², and $m \approx 13$. For Fe 415 steel, $\sigma_{st} = 230$ N/mm². For a balanced section,

$$k = \frac{m\sigma_{cbc}}{m\sigma_{cbc} + \sigma_{st}} = \frac{13 \times 7}{13 \times 7 + 230} = 0 \cdot 283$$

$$j = 1 - \frac{k}{3} = 1 - \frac{0.283}{3} = 0.906$$

$$R = \frac{1}{2}\ c.j.k = \frac{1}{2} \times 7 \times 0.906 \times 0.283 = 0.898$$

Now $$M_r = R\,.\,b\,d^2 = R\left(\frac{d}{2}\right)(d^2) = \frac{R\,d^3}{2}$$

$$= \frac{0.898}{2} d^3 = 0.449\, d^3 \text{ N-mm}$$

Given B.M. $= M = 20$ kN $-$ m $= 20 \times 10^6$ N $-$ mm.

Now $M_r = M$

$\therefore$ $0.449\, d^3 = 20 \times 10^6$

or $$d = \left(\frac{20 \times 10^6}{0.446}\right)^{\frac{1}{3}} \approx \mathbf{354\ mm.}$$

$\therefore$ $b = \frac{1}{2} d = \mathbf{177\ mm.}$

Area of steel is given by

$$A_{st} = \frac{M}{\sigma_{st}\, jd} = \frac{20 \times 10^6}{230 \times 0.906 \times 354} = \mathbf{271.1\ mm^2}$$

Thus provide a beam of 177 mm width and 354 mm effective depth, with area of tensile reinforcement equal to 271.1 mm².

Example 2.5(*c*) *Redesign the beam of example 2.5(a) using M15 concrete and Fe 415 steel.*

Solution. For M15 concrete, $c = \sigma_{cbc} = 5$ N/mm², and $m \approx 19$. For Fe 415 steel, $\sigma_{st} = 230$ N/mm². For a balanced section,

$$k = \frac{m\sigma_{cbc}}{m\sigma_{cbc} + \sigma_{st}} = \frac{19 \times 5}{19 \times 5 + 230} = 0 \cdot 2923$$

$$j = 1 - \frac{k}{3} = 1 - \frac{0.2923}{3} = 0.9026$$

$$R = \tfrac{1}{2}\sigma_{cbc}\,.j.k = \tfrac{1}{2} \times 5 \times 0.9026 \times 0.2923 = 0\cdot6596$$

Now $$M_r = R\,.\,b\,d^2 = R\left(\frac{d}{2}\right)(d^2) = \frac{R\,d^3}{2}$$

$$= \frac{0.6596}{2}d^3 = 0.3298\,d^3.$$

Given $$\text{B.M.} = M = 20\text{ kN} - \text{m} = 20 \times 10^6\text{ N} - \text{mm}$$

Now $$M_r = M$$

$\therefore$ $$0.3298\,d^3 = 20 \times 10^6$$

Which gives $d \approx$ **392 mm**

$$b = \frac{d}{2} = \textbf{196 mm.}$$

Area of steel is given by

$$A_{st} = \frac{M}{\sigma_{st}\,jd} = \frac{20 \times 10^6}{230 \times 0.9026 \times 392} = \textbf{245.8 mm}^2$$

Thus provide a beam of 196 mm width and 392 mm effective depth, with area of tensile reinforcement equal to 245.8 mm^2.

Example 2.6 *A reinforced concrete beam 200 mm×400 mm effective depth is used over an effective span of 5 m. It is subjected to a uniformly distributed load of 5 kN/m inclusive of its own weight. Find the necessary steel reinforcement at the centre of the span. Take allowable stresses in steel and concrete as 130 N/mm² and 4 N/mm² respectively, and m=16.*

Solution. Given : $c = \sigma_{cbc} = 4$; $t = \sigma_{st} = 130$ and $m = 16$.

$$\text{Maximum B.M.} = \frac{w\,l^2}{8} = \frac{5 \times 25}{8}\text{ kN} - \text{m}$$

$$= 15.625\text{ kN} - \text{m} = 15.625 \times 10^6\text{ N} - \text{m}$$

$$M_r = R\,.\,b\,d^2 = 15\cdot625 \times 10^6$$

or $$R = \frac{15\cdot625 \times 10^6}{200\,(400)^2} = 0\cdot488$$

Hence $$M_r = 0\cdot488\,b\,d^2 \qquad ...(i)$$

For the balanced section,

$$k = \frac{m\,c}{m\,c + t} = \frac{16 \times 4}{16 \times 4 + 130} = 0\cdot33.$$

$$j = 1 - \frac{k}{3} = 1 - 0 \cdot 11 = 0 \cdot 89.$$

$$\therefore \quad R = \tfrac{1}{2} c\, j\, k = \tfrac{1}{2} \times 4 \times 0.89 \times 0.33 = 0 \cdot 587$$

$$\therefore \quad M_r = 0 \cdot 587\, b\, d^2 \qquad \text{... } (ii)$$

The moment of resistance of the given beam has to be less than the moment of resistance of the critical section. Hence the steel reinforcement will be corresponding to an under-reinforced beam. For such a section, the stress in steel will reach the maximum value of 130 N/mm^2.

Let n = depth of N.A. of the actual section. Then the corresponding stress in concrete is given by

$$\frac{c}{t/m} = \frac{n}{d-n}$$

$$\therefore \quad c = \frac{t}{m} \cdot \frac{n}{d-n} = \frac{130}{16} \times \frac{n}{400-n} \qquad \text{...}(iii)$$

Now $$M_r = \tfrac{1}{2} c \,.\, n \,.\, b \left(d - \frac{n}{3} \right)$$

$$\therefore \quad 15 \cdot 625 \times 10^2 = \tfrac{1}{2} \left\{ \frac{130}{16} \times \frac{n}{400-n} \right\} n \times 200 \left(400 - \frac{n}{3} \right)$$

or $$n^2 \frac{(1200 - n)}{400 - n} = 57700$$

This is a cubic equation in n. Solving it by trial and error, we get

$$n = 121.8 \text{ mm}$$

Hence from (iii) $$c = \frac{130}{16} \times \frac{121.8}{400 - 121.8} = 3 \cdot 56 \text{ N/mm}^2$$

Now, Total compression = Total tension,

$$\therefore \quad \tfrac{1}{2} cnb = A_{st} \,.\, t$$

or $$A_{st} = \frac{1}{2} \frac{cnb}{t} = \frac{1}{2} \frac{3 \cdot 56 \times 121 \cdot 8 \times 200}{130}$$

or $$= \mathbf{333 \text{ mm}^2} = 3.33 \text{ cm}^2$$

Check. For the actual section,

$$k = \frac{m\,c}{m\,c + t} = \frac{16 \times 3 \cdot 56}{16 \times 3 \cdot 56 + 130} = 0 \cdot 3047.$$

$$j = 1 - \frac{0 \cdot 3047}{3} = 0 \cdot 8984$$

$$R = \tfrac{1}{2} c\, j\, k = \tfrac{1}{2} \times 3 \cdot 56 \times 0 \cdot 8984 \times 0 \cdot 3047$$

$$= 0 \cdot 4873$$

$$M_r = 0 \cdot 4873 \times 200 \times (400)^2 = 15 \cdot 625 \times 10^6 \text{ N} - \text{mm}$$

Also $M_r = A_{st}\, t\, j\, d = 3 \cdot 33 \times 130 \times 0.8984 \times 400$

$$= 15 \cdot 6 \times 10^6 \text{ N} - \text{mm}.$$

Hence O.K.

Example 2.7 *A reinforced concrete beam 200 mm × 400 mm effective depth is used over an effective span of 5 m. It is subjected to a uniformly distributed load of 7 kN/m inclusive of its own weight. Find the necessary steel reinforcement at the centre of the span. Take allowable stresses in steel and concrete as 130 N/mm² and 4 N/mm² respectively and m=16.*

Solution. This example is exactly the same as the previous example except that w has been increased to 7 kN/m.

$$c = \sigma_{cbc} = 4 \text{N/mm}^2 \;; t = \sigma_{st} = 130 \text{ N/mm}^2 \text{ and } m = 16.$$

$$\text{Maximum B.M.} = \frac{w L^2}{8} = \frac{7 \times 25}{8} = 21 \cdot 88 \text{ kN} - \text{m}$$

$$= 21 \cdot 88 \times 10^6 \text{ N} - \text{mm}$$

$$\therefore \quad R = \frac{M_r}{b d^2} = \frac{M}{b d^2} = \frac{21 \cdot 88 \times 10^6}{200 \times (400)^2} = 0 \cdot 684$$

Hence $M_r = 0 \cdot 684\, b\, d^2$ ...(*i*)

For the balanced section

$$k = \frac{m c}{m c + t} = \frac{16 \times 4}{16 \times 4 + 130} = 0 \cdot 33.$$

$$j = 1 - \frac{0 \cdot 33}{3} = 1 - 0 \cdot 11 = 0 \cdot 89.$$

$$R = \tfrac{1}{2} c\, j\, k = \tfrac{1}{2} \times 4 \times 0.89 \times 0.33 = 0 \cdot 587$$

$M_r = 0 \cdot 587\, b\, d^2$ ..(*ii*)

Since the moment of resistance of the given beam is greater than the moment of resistance of the critical section, the steel reinforcement will be corresponding to an over-reinforced beam. For such a section, the stress in concrete will reach the maximum value of 4 N/mm².

Now, $M_r = \frac{1}{2} c\, n\, b \left(d - \frac{n}{3}\right)$

$\therefore \quad 21 \cdot 88 \times 10^6 = \frac{1}{2} \times 4\, n \times 200 \left(400 - \frac{n}{3}\right)$

$\therefore \quad n^2 - 1200\, n + 164100 = 0$

$\therefore \quad n = 1042 \text{ or } 157 \cdot 5$

Hence $n = 157 \cdot 5$ mm

(**Note.** For critical section, $n_c = k\, d = 0 \cdot 33 \times 400 = 132$ mm).
The stress in steel is given by

$$\frac{t/m}{c} = \frac{d - n}{n}$$

$$\therefore \quad t = \left(\frac{d - n}{n}\right) m\, c = \frac{400 - 157 \cdot 5}{157 \cdot 5} \times 16 \times 4$$

$$= 98.5 \text{ N/mm}^2$$

Now, total tension = total compression

$$\therefore \quad t \,.\, A_{st} = \frac{1}{2} cn \,.\, b$$

or

$$A_{st} = \frac{1}{2} \frac{c\, n\, b}{t} = \frac{1}{2} \frac{5 \times 157 \cdot 5 \times 200}{98 \cdot 5}$$

$$= \mathbf{639.6\ mm^2}$$

Check. $M = A_{st} \,.\, t\, j\, d = A_{st} \,.\, t \left(d - \frac{n}{3}\right)$

$$= 639 \cdot 6 \times 98 \cdot 5 \left[400 - \frac{157.5}{3}\right]$$

$$= 21.89 \times 10^6 \text{ N–mm.}$$

Note. In the above example of an over-reinforced section, the stress in steel is only 98.5 N/mm^2 which is very much less than the permissible value of 130 N/mm^2. Thus the strength of steel is not fully utilized. Such a design is undesirabale and uneconomical. In circumstances where the dimensions of the beam are limited, and where the section has to develop greater moment of resistance than that of the balanced section, it is always desirable to design the section as *doubly reinforced* (see Chapter 4). In doubly reinforced section, steel reinforcement is placed in the compression zone also, which increases its moment of resistance.

Hence a doubly reinforced section is always prefered over an over-reinforced section.

Example 2.8. (*a*) *The moment of resistance of a rectangular reinforced concrete beam, of breadth b cm and effective depth d cm is 0.9 bd*^{2}. *If the stresses in the outside fibre of the concrete and in the steel do not exceed 5 N/mm*^{2} *and 140 N/mm*^{2} *respectively, and the modular ratio equals 18, determine the ratio of depth of the neutral axis from the outside compression fibre to the effective depth of the beam and the ratio of area of tension steel to the effective area of the beam. The beam is reinforced for tension only.*

Solution. Given : $c = \sigma_{cbc} = 5$; $t = \sigma_{st} = 140$ and $m = 18$. Let the ratio of depth of the neutral axis to the effective depth be k.

Now $$R = \frac{1}{2} c j k = 0 \cdot 9 \text{ (given)}$$

$$\therefore \quad \frac{1}{2} c \left(1 - \frac{k}{3}\right) k = 0 \cdot 9$$

or $$\left(1 - \frac{k}{3}\right) k = \frac{1 \cdot 8}{c} = \frac{1 \cdot 8}{5} = 0 \cdot 36$$

or $$k^2 - 3k + 1 \cdot 08 = 0$$

From which, $k = 0 \cdot 43$.

For the balanced section

$$k = \frac{mc}{mc + t} = \frac{18 \times 5}{18 \times 5 + 140} = 0.392$$

Since the actual N.A. is below the critical N.A., the section is *over-reinforced*. Hence the stress in concrete reaches a maximum value of 5 N/mm^2. The corresponding value of stress in steel is given by

$$\frac{c}{t/m} = \frac{kd}{d - kd}$$

$$\therefore \quad t = \frac{mcd\,(1 - k)}{kd} = \frac{mc\,(1 - k)}{k}$$

$$= \frac{18 \times 5\,(1 - 0 \cdot 43)}{0 \cdot 43} = 119 \text{ N/mm}^2$$

Now $$p' = \frac{A_{st}}{bd} = \frac{M}{tjd} \cdot \frac{1}{bd} = \frac{0 \cdot 9\,bd^2}{tjbd^2} = \frac{0 \cdot 9}{t \cdot j}$$

Here $$j = 1 - \frac{k}{3} = 1 - \frac{0 \cdot 43}{3} = 0 \cdot 857$$

$$t = 119 \text{ N/mm}^2$$

$$\therefore \quad p' = \frac{0.9}{tj} = \frac{0.9}{119 \times 0.857} = 0.00882$$

or $$p = 100\,p' = \mathbf{0.882\%}$$

Note. For a balanced section $p = \dfrac{50\,k\,\sigma_{cbc}}{\sigma_{st}} = 50 \times \dfrac{0.395 \times 5}{140} = 0.7\%$

Example 2.8 (*b*) *Solve example 2.8 (a) if M 20 grade concrete and Fe 415 grade steel reinforcement is used for the beam.*

Solution

Given $$c = \sigma_{cbc} = 7 \text{ N/mm}^2$$

$$m = 13\,;\ \sigma_{st} = 230 \text{ N/mm}^2$$

$$R = \tfrac{1}{2}cjk = 0{\cdot}9 \text{ (given)}$$

$$\therefore \quad \tfrac{1}{2}c\left(1 - \frac{k}{3}\right)k = 0{\cdot}9$$

or $$\left(1 - \frac{k}{3}\right)k = \frac{1{\cdot}8}{c} = \frac{1{\cdot}8}{7} = 0{\cdot}2571$$

$$\therefore k^2 - 3k + 0{\cdot}7714 = 0$$

From which $k = \mathbf{0{\cdot}284}$

For the balanced section

$$k = \frac{mc}{mc + t} = \frac{13 \times 7}{13 \times 7 + 230} = 0.2825 \approx 0.284$$

Thus the beam is nearly balanced. Hence the stresses in concrete and steel reach their permissible values simultaneously.

For a balanced section.

$$p = 50\frac{kc}{t} = 50 \times \frac{0{\cdot}284 \times 7}{230} = \mathbf{0{\cdot}432\%}$$

Check. $$p = \frac{A_{st}}{b\,d} \times 100 = \frac{M}{tjd} \times \frac{1}{bd} \times 100$$

$$= \frac{0{\cdot}9\,b\,d^2}{t\,j\,bd^2} \times 100 = \frac{90}{tj}$$

Here $$j = 1 - \frac{k}{3} = 1 - \frac{0{\cdot}284}{3} = 0{\cdot}9053$$

$$\therefore \qquad p = \frac{90}{230 \times 0 \cdot 9053} = 0 \cdot 432\%$$

Example 2.8(c) *Solve example 2.8(a) if M 15 grade concrete and Fe 415 grade steel reinforcement is used for beam.*

Solution Given $c = \sigma_{cbc} = 5 \text{ N/mm}^2$

$$m = 19 \text{ and } t = \sigma_{st} = 230 \text{ N/mm}^2$$

$$R = \tfrac{1}{2} cjk = 0 \cdot 9 \text{ (given)}$$

$$\therefore \qquad \tfrac{1}{2} c \left(1 - \frac{k}{3}\right) k = 0 \cdot 9$$

or

$$\left(1 - \frac{k}{3}\right) k = \frac{1 \cdot 8}{c} = \frac{1 \cdot 8}{5} = 0 \cdot 36$$

$$\therefore \qquad k^2 - 3k + 1 \cdot 08 = 0$$

From which $k = \mathbf{0 \cdot 43}$

For the balanced section

$$k = \frac{m\,\sigma_{cbc}}{m\,\sigma_{cbc} + \sigma_{st}} = \frac{19 \times 5}{19 \times 5 + 230} = 0.2923$$

Since the actual N.A. is below the critical N.A., the section is *over-reinforced*. Hence the stress in concrete reaches a maximum value of 5 N/mm^2. The corresponding value of stress in steel is given by

$$\frac{c}{t/m} = \frac{kd}{d - kd}$$

$$\therefore \qquad t = \frac{mcd\,(1 - k)}{kd} = \frac{mc\,(1 - k)}{k}$$

$$= \frac{19 \times 5\,(1 - 0 \cdot 43)}{0 \cdot 43} = 125 \cdot 93 \text{ N/mm}^2$$

Now

$$p' = \frac{A_{st}}{b\,d} = \frac{M}{t\,j\,d} \cdot \frac{1}{bd} = \frac{0 \cdot 9\,bd^2}{tjbd^2} = \frac{0 \cdot 9}{t\,.\,j}$$

where

$$j = 1 - \frac{k}{3} = 1 - \frac{0 \cdot 43}{3} = 0 \cdot 857$$

and

$$t = 125 \cdot 93 \text{ N/mm}^2$$

$$\therefore \qquad p' = \frac{0 \cdot 9}{125 \cdot 93 \times 0 \cdot 857} = 0 \cdot 00834$$

or $$p = 100\,p' = \mathbf{0\cdot834\%}$$

Note. For a balanced section

$$p = \frac{50\,k\,\sigma_{cbc}}{\sigma_{st}} = 50 \times \frac{0\cdot2923 \times 5}{230} = 0\cdot318\%$$

2.8. VARIATION OF M_r WITH p

We have seen that a *balanced section* is the one in which the stresses in steel and concrete reach the permissible values simultaneously. The percentage steel for a balanced section is given by Eq. 2.9 :

$$p = p_c = \frac{k_c \,.\, \sigma_{cbc}}{2\,\sigma_{st}} \times 100 \qquad ...(2.9b)$$

The moment of resistance of a balanced section is computed from Eq. 2.12. If, however, the section is under-reinforced, M_r is computed on the basis of tensile force in steel (Eq 2.14). If the section is over-reinforced M_r is determined on the basis of compressive force developed in concrete (Eq 2.12*a*).

For a beam with given dimensions (*i.e.* b, d, p') the neutral axis depth factor k is computed from Eq. 2.10. In general,

$$M_r = R \,.\, b\, d^2$$

Hence M_r depends upon R, and the variation of M_r is directly proportional to variation of R.

$$\therefore \qquad R = \frac{M_r}{b\,d^2}\ \text{N/mm}^2$$

(where M_r is in N-mm, b in mm and d in mm)

For illustration purposes, let us study the variation of R (or M_r) with p for M 15 concrete using (*a*) Mild steel bars (Fe 250) having $\sigma_{st} = 140$ N/mm^2 and (*b*) HYSD bars (Fe 415) having $\sigma_{st} = 230$ N/mm^2.

(*a*) **Fe 250** : $\sigma_{st} = 140$ N/mm^2

$$k_c = \frac{m\,\sigma_{cbc}}{m\,\sigma_{cbc} + \sigma_{st}} = \frac{1}{1 + 0\cdot0107\,\sigma_{st}} \qquad [2.8(d)]$$

$$= \frac{1}{1 + 0\cdot0107 \times 140} = 0\cdot400$$

$$j_c = 1 - \frac{0\cdot400}{3} = 0\cdot867$$

$$p_c = \frac{k_c \,.\, \sigma_{cbc}}{2\,\sigma_{st}} \times 100 = \frac{0\cdot4 \times 5}{2 \times 140} \times 100 = 0\cdot714$$

$$R_c = \frac{1}{2}\sigma_{cbc} \cdot j_c k_c = \frac{1}{2} \times 5 \times 0\cdot867 \times 0\cdot4$$

$$= 0\cdot867 \text{ N/mm}^2$$

For any other value of k is given by Eq. 2.10

$$k = \sqrt{2mp' + (mp')^2} - mp'$$

where $p' = \frac{p}{100} = \frac{A_{st}}{b\,d}$ = reinforcment ratio

and m = 18.67 for M 15 concrete

For $p < p_c$ $\quad M_r = \sigma_{st} \cdot A_{st} \cdot jd = \sigma_{st}(p' \cdot bd)\, jd.$

$$= (\sigma_{st} \cdot p'j)\, b\, d^2 = Rb\, d^2$$

where $R = \sigma_{st} \cdot p' \cdot j.$

For example, when $p = 0\cdot4$, $p' = 0\cdot004$

$$k = \sqrt{(2 \times 18\cdot67 \times 0\cdot004) + (18\cdot67 \times 0\cdot004)^2} - 18\cdot67 \times 0\cdot004$$

$$= 0\cdot319$$

$$j = 1 - \frac{k}{3} = 0\cdot894$$

$$R = \sigma_{st} \cdot p' \cdot j = 140 \times 0\cdot004 \times 0\cdot894 = 0\cdot5$$

Similarly, $p > p_c$, $M_r = \frac{1}{2}\sigma_{cbc} \cdot j \cdot k \cdot bd^2$

$$= R\, bd^2$$

where $R = \frac{1}{2}\sigma_{cbc} \cdot j \cdot k.$

For example, when $p = 0\cdot8$, $p' = 0\cdot008$

$$k = \sqrt{(2 \times 18\cdot67 \times 0\cdot008) + (18\cdot67 \times 0\cdot008)^2} - 18\cdot67 \times 0\cdot008$$

$$= 0\cdot417$$

$$j = 1 - \frac{k}{3} = 0\cdot861$$

$$R = \frac{1}{2}\sigma_{cbc} \cdot j \cdot k = \frac{1}{2} \times 5 \times 0\cdot861 \times 0\cdot417 = 0\cdot898$$

Values of $k, j \left(= 1 - \frac{k}{3}\right)$ and R for various values of p are tabulated below :

p	0.4	0.6	0.714	0.8	1.0	1.2
p'	0.004	0.006	0.00714	0.008	0.01	0.012
k	0.319	0.374	0.4	0.417	0.452	0.482
j	0.894	0.875	0.867	0.861	0.849	0.839
R	0.5	0.735	0.867	0.898	0.960	1.011

(*b*) **Fe 415** : $\sigma_{st} = 230$ N/mm²

$$k_c = \frac{1}{1 + 0 \cdot 0107 \times 230} = 0 \cdot 289$$

$$j_c = 1 - \frac{k_c}{3} = 0 \cdot 904$$

$$p_c = \frac{0 \cdot 289 \times 5}{2 \times 230} \times 100 = 0.314$$

$$R = \frac{1}{2} \times 5 \times 0 \cdot 904 \times 0 \cdot 289 = 0 \cdot 653$$

The values of k, j and R for various values of p are tabulated below :

p	0.1	0.2	0.314	0.4	0.5	0.6
k	0.175	0.238	0.289	0.319	0.349	0.374
j	0.942	0.920	0.904	0.894	0.884	0.875
R	0.217	0.423	0.653	0.713	0.772	0.818

The variation of $R\left(= \dfrac{M_r}{bd^2}\right)$ with p, for both the types of bars are shown in Fig. 2.14. It is seen that upto $p = p_c$ the moment of resistance increases almost *linearly*. Provision of reinforcement in excess of p_c does not lead to proportionate increase in M_r and is not economical.

As per IS : 456–1978, the minimum reinforcement is given by

$$p' = \frac{A_{st}}{b\,d} = \frac{0 \cdot 85}{f_y} \quad \text{(see chapter 7)}$$

or

$$p = \frac{A_{st}}{b\,d} \times 100 = \frac{85}{f_y} \text{ per cent}$$

For Mild steel bars having $f_y = 250$ N/mm²

$$p = \frac{85}{250} = 0 \cdot 34\%$$

For HYSD bars having $f_y = 415$ N/mm²

$$p = \frac{85}{415} = 0.205\%$$

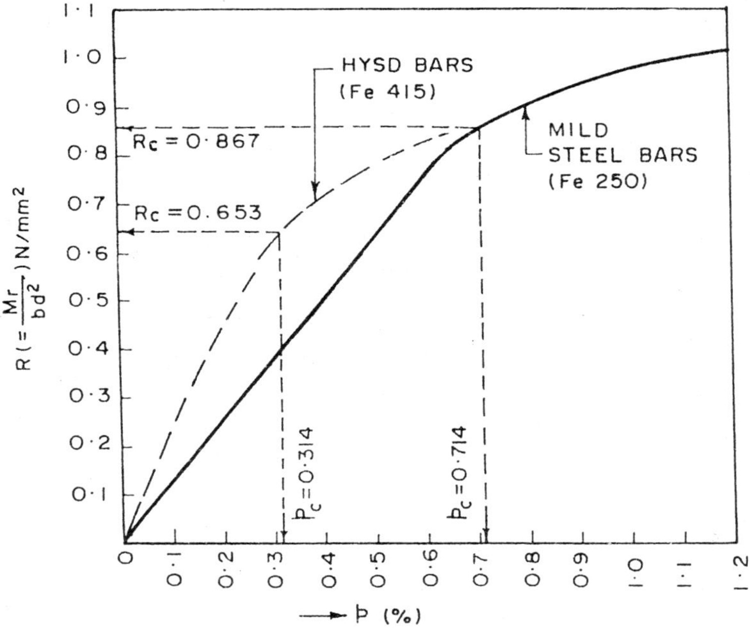

FIG. 2.14 VARIATION OF R AND M_r WITH p.

2.9 BEAM OF TRIANGULAR SECTION

(a) *Balanced Section*

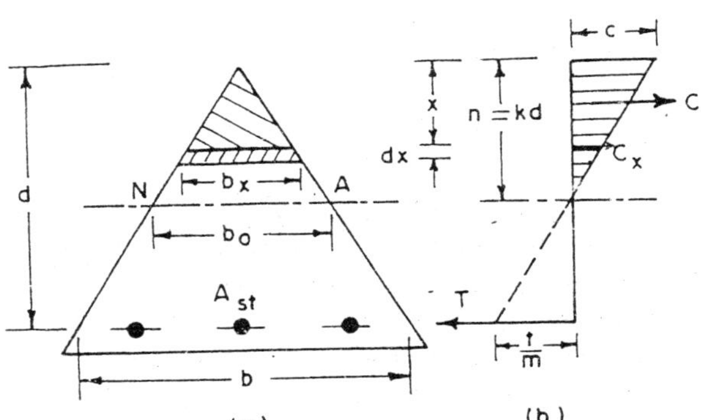

FIG. 21.5

Fig. 2.15 (a) shows a triangular section having effective depth d and width b at the level of the reinforcement. Let the N.A. be at a depth n below the apex. At depth x below the apex, the width b_x and stress c_x are given by

$$b_x = \frac{b}{d}x$$

$$c_x = \frac{kd - x}{kd}c$$

Hence the compressive force dC_x of elementary strip of thickness dx is given by

$$dC_x = b_x \, . \, c_x \, . \, dx$$

The total compressive force is given by

$$C = \int dC_x = \int_0^{kd} b_x \, . \, c_x \, . \, dx$$

$$= \int_0^{kd} \frac{b}{d}x \, . \, \frac{kd - x}{kd} c \, dx$$

$$= \frac{cb}{k\,d^2} \int_0^{kd} x \, . \, (kd - x) \, dx$$

$$= \frac{cb}{k\,d^2}\left[kd \, . \frac{x^2}{2} - \frac{x^3}{3} \right]_0^{kd} = \frac{k^2\,bd}{6} c \quad ...(2.18)$$

Taking $c = \sigma_{cbc}$ and $k = k_c$ for a balanced design

$$C = \frac{k_c^2\,bd}{6} \, . \, \sigma_{cbc} \quad ...[2.18(n)]$$

Total compressive force $T = A_{st} \, . \, \sigma_{st}$

Equating the two, for equilibrium,

$$A_{st} \, . \, \sigma_{st} = \frac{k_c^2 \, . \, bd}{6} \, . \, \sigma_{cbc}$$

or

$$A_{st} = \frac{k_c^2 \, . \, bd}{6} \, . \, \frac{\sigma_{cbc}}{\sigma_{st}} \quad ..(2.19)$$

Eq. 2.18 will be useful in determining reinforcement for a *balanced section*, where k_c is given by

$$k_c = \frac{m\,\sigma_{cbc}}{m\,\sigma_{cbc} + \sigma_{st}} = \frac{1}{1 + 0{\cdot}0107\,\sigma_{st}} \quad ...[2.18(b)]$$

In order to find the moment of resistance of the section, take the moment of compressive force about the centre of tensile steel.

$$\therefore \qquad M_r = \int_0^{kd} (d - x)\, dC_x$$

$$= \int_0^{kd} (d - x)\, b_x \,.\, c_x\, dx$$

$$= \int_0^{kd} (d - x) \frac{b}{d} x \,.\, \frac{kd - x}{kd} \,.\, c\, dx$$

$$= \frac{bc}{k d^2} \int_0^{kd} x \,.\, (kd - x)(d - x)\, dx$$

$$= \frac{bc}{kd^2} \left[kd^2 \,.\, \frac{x^2}{2} - d(k+1) \,.\, \frac{x^3}{3} + \frac{x^4}{4} \right]_0^{kd}$$

$$M_r = \frac{bc}{kd^2} \left[\frac{kd^2}{2} \,.\, (kd)^2 - \frac{d(k+1)}{3} (kd)^3 + \frac{1}{4} (kd)^4 \right]$$

or
$$M_r = \frac{c \,.\, k^2 (2 - k)}{12} bd^2 \qquad ...(2.20)$$

Taking $c = \sigma_{cbc}$ and $k = k_c$ for a balanced section,

$$M_r = \frac{k_c^2 (2 - k_c)}{12} \sigma_{cbc} \,.\, bd^2 = Q\, bd^2 \qquad ...(2.21)$$

where
$$Q = \frac{k_c^2 (2 - k_c)}{12} \,.\, \sigma_{cbc} = K \,.\, \sigma_{cbc} \qquad ...[2.21(a)]$$

and
$$K = \frac{1}{12} k_c^2 (2 - k_c) \qquad ...[2.21(b)]$$

and k_c is given by Eq. 2.8 (*d*)

The values of K and Q are given in the table (next page).

(b) *Determination of M_r, and Stresses*

If the dimensions of the section (*i.e.* b, d and A_{st}) are given the M_r, and the actual stress can be determined by first locating the N.A. This can be done by equating the moment of the compressive area about N.A. to the moment of the equivalent tensile area about N.A.

σ_{st} *(N/mm²)*	140 (Fe 250)	190	230 (Fe 415)	275 (Fe 500)
$k = k_c$	0.400	0.329	0.289	0.253
K	0.0213	0.0151	0.0119	0.0094
Q for				
σ_{cbc} = 5 (M 15)	0.107	0.076	0.060	0.047
σ_{cbc} = 7 (M 20)	0.149	0.106	0.083	0.066
σ_{cbc} = 8.5 (M 25)	0.181	0.128	0.101	0.080
σ_{cbc} = 10.0 (M 30)	0.213	0.151	0.119	0.094

Thus, trom Fig. 2.15, width b_0 at the N.A. is given by

$$b_0 = \frac{b}{d} . kd \quad = bk = \frac{bn}{d}$$

∴ Moment of compressive area about N.A.

$$= \left(\frac{1}{2} b_0 . n \right) \frac{n}{3} = \frac{1}{2} \times \frac{bn}{d} . n . \frac{n}{3}$$

$$= \frac{1}{6} \frac{b}{d} n^3 .$$

Moment of equivalent tensile area about N.A.

$$= m A_{st} (d - n)$$

Equating the two, we get

$$\frac{1}{6} \frac{b}{d} n^3 = m A_{st} (d - n)$$

or
$$n^3 + \frac{6\, md\, A_{st}}{b} n - \frac{6\, md^2 A_{st}}{b} = 0$$

This is a cubic equation, the solution of which (by trial) would yield the value of n.

The depth of critical N.A. is given by

$$n_c = \frac{m\, \sigma_{cbc}}{m\, \sigma_{cbc} + \sigma_{st}} . d \quad = k_c . d$$

(*i*) if $n = n_c$, the section is *balanced*, and the M_r is obtained by Eq. 2.21.

(*ii*) If $n < n_c$, the section is *under-reinforced*, and the stress in steel will reach its maximum value first.

Thus $t = \sigma_{st}$.

Hence the corresponding stress c in concrete is given, from the stress diagram (Fig. 2.15*b*), by

$$\frac{c}{n} = \frac{\sigma_{st}/m}{d-n}$$

or
$$c = \frac{\sigma_{st}}{m} \cdot \frac{n}{d-n} \qquad ...(2.17)$$

Knowing c, the moment of resistance of the section is found from Eq. 2.20.

Thus
$$M_r = c \cdot \frac{k^2(2-k)}{12} bd^2 \qquad ...(2.20)$$

where
$$k = \frac{n}{d}$$

(*iii*) if $n > n_c$, the section is *over reinforced* and the stress in concrete will reach its maximum value first.

Thus
$$c = \sigma_{cbc}$$

The corresponding stress in steel is given by

$$\frac{\sigma_{cbc}}{n} = \frac{t/m}{d-n}$$

or
$$t = \sigma_{cbc} \cdot \frac{m(d-n)}{n} \qquad ...[2.17(a)]$$

Also, M_r is given by Eq. 2.20 by substituting $c = \sigma_{cbc}$

Thus,
$$M_r = \sigma_{cbc} \cdot \frac{k^2(2-k)}{12} bd^2$$

where
$$k = \frac{n}{d}$$

2.10. BEAM OF TRAPEZOIDAL SECTION

(*a*) *Balanced Section*

The moment of resistance for a trapezoidal section can be found by taking the sum of M_r for middle rectangular and M_r for triangular sections.

Thus,
$$M_r = M_{r1} + M_{r2}$$

Where M_{r1} = moment of resistance of triangular portion

$$= \frac{ck^2(2-k)}{12}(b-b_1)d^2 \qquad ...(\text{Eq. } 2.18)$$

and M_{r2} = moment of resistance of triangular portion

$$= \frac{1}{2} c \left(1 - \frac{k}{3}\right) k\, b_1 d^2 \qquad \text{...[Eq.2.12(b)]}$$

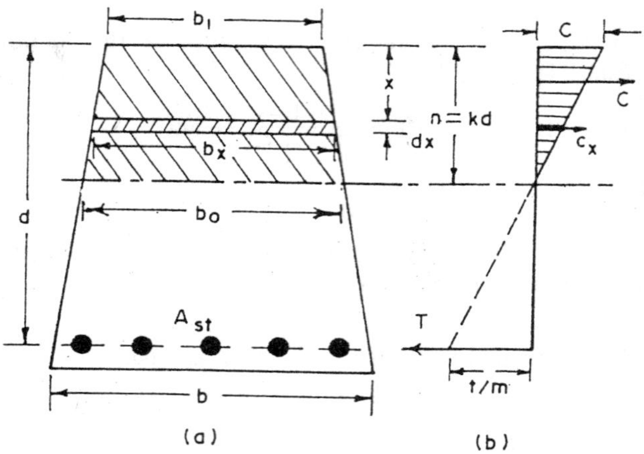

FIG. 2.16

$$\therefore \quad M_r = \left[\frac{ck^2(2-k)}{12}(b-b_1) + \frac{1}{2} c \left(\frac{1-k}{3}\right) k\, b_1\right] d^2 \qquad \text{...(2.22)}$$

For a balanced section, $c = \sigma_{cbc}$ and $k = k_c$

$$M_r = \left[\sigma_{cbc} \frac{k_c^2(2-k_c)}{12}(b-b_1) + \frac{1}{2}\sigma_{cbc}\left(1 - \frac{k_c}{3}\right) k_c\, b_1\right] d^2$$

or $$M_r = [Q(b-b_1) + Rb_1] d^2 \qquad \text{...[2.22(a)}$$

where $$Q = \frac{k_c^2(2-k_c)}{12} . \sigma_{cbc} \qquad \text{...[2.21(a)]}$$

$$R = \frac{1}{2}\sigma_{cbc} . \left(1 - \frac{k_c}{3}\right) . k_c$$

and $$k_c = \frac{m\,\sigma_{cbc}}{m\sigma_{cbc} + \sigma_{st}} = \frac{1}{1 + 0.0107\,\sigma_{st}}$$

The total compressive force is given by

$$C = C_1 + C_2$$

where C_1 = compressive force for triangular portion

$$= \frac{k^2}{6} c (b - b_1) d \quad \text{...(Eq. 2.18)}$$

and C_2 = compressive force for rectangular portion

$$= \frac{1}{2} c b_1 k d$$

$$\therefore \quad C = \frac{k^2}{6} c (b - b_1) d + \frac{1}{2} c b_1 k d$$

$$C = \frac{kd}{2} \left[\frac{k (b - b_1)}{3} + b_1 \right] c \quad \text{...(2.23)}$$

For a balanced section, $c = \sigma_{cbc}$ and $k = k_c$

$$\therefore \quad C = \frac{k_c d}{2} \left[b_1 + \frac{k_c (b - b_1)}{3} \right] \sigma_{cbc} \quad \text{...[2.23(a)]}$$

Total tensile force, for a balanced section is

$$T = A_{st} \cdot \sigma_{st}$$

Equating the two, we get

$$A_{st} = \frac{k_c \cdot d}{2} \left[b_1 + \frac{k_c}{3} (b - b_1) \right] \frac{\sigma_{cbc}}{\sigma_{st}} \quad \text{...(2.24)}$$

(b) Determination of M_r and stresses

If the dimensions of the section are given, the M_r and actual stresses can be determined by first locating N.A. This can be done by equating the moment of compressive area about N.A. to the moment of equivalent tensile area about N.A.

Thus, from Fig 2.16, moment of compressive area about N.A.

$$= \frac{1}{6} \frac{b - b_1}{d} \cdot n^3 + \frac{b_1 n^2}{2}$$

Moment of equivalent tensile area about N.A.

$$= m \cdot A_{st} (d - n)$$

Equating the two, we get

$$\frac{1}{6} \frac{b - b_1}{d} \cdot n^3 + \frac{b_1 n^2}{2} - m \cdot A_{st} (d - n) = 0$$

This is cubic equation, the solution of which (by trial) would yield the value of n.

Hence $\quad k = \frac{n}{d}$

The depth of critical N.A. is given by

$$n_c = \frac{m \sigma_{cbc}}{m \sigma_{cbc} + \sigma_{st}} \cdot d = k_c d .$$

(*i*) If $n = n_c$, the section is balanced, and M_r is obtained by Eq. 2.22 (*a*).

(*ii*) If $n < n_c$, the section is *under-reinforced* and stress in steel will reach its maximum value first. Thus $t = \sigma_{st}$. The corresponding stress c in concrete is given by

$$c = \frac{\sigma_{st}}{m} \cdot \frac{n}{d-n} \quad \text{(from stress diagram)}$$

Knowing c, M_r is found from Eq. 2.22.

(*iii*) If $n > n_c$, the section is *over-reinforced*, and stress in concrete will reach its maximum value first. Thus $c = \sigma_{cbc}$. The corresponding stress t in steel is given by

$$t = \sigma_{cbc} \cdot \frac{m(d-n)}{n} \quad \text{(from stress diagram)}$$

And M_r is given by Eq. 2.22, by taking $c = \sigma_{cbc}$.
Thus,

$$M_r = \left[\frac{k^2(2-k)}{12}(b-b_1) + \frac{1}{2}\left(1-\frac{k}{3}\right)kb_1\right]\sigma_{cbc} \cdot d^2$$

$\left(\text{where } k = \frac{n}{d}.\right)$

(*c*) ***Trapezoidal section with $b_1 > b$***

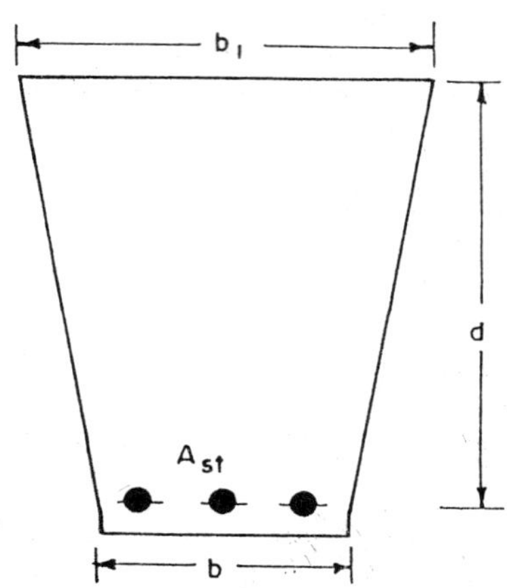

FIG. 2.17.

Since the area of concrete in tension zone is not useful, it is more economical to keep b_1 greater than b. All the equations (*i.e.* 2.22, 2.23, 2.24) will be valid, except that the term containing $(b-b_1)$ factor will be negative. Alternatively these equations can be *rearranged* in the following form :

$$M_r = \left[\frac{1}{2}c\left(1-\frac{k}{3}\right)kb_1 - \frac{ck^2(2-k)}{12}(b_1-b)\right]d^2 \quad \text{...(2.25)}$$

$$M_r = [\,Rb_1 - Q(b_1-b)\,]\,d^2 \quad \text{...[2.25(a)]}$$

$$C = \frac{kd}{2}\left[b_1 - \frac{k}{3}(b_1-b)\right]c \quad \text{...(2.26)}$$

$$C = \frac{k_c d}{2}\left[b_1 - \frac{k_c}{3}(b_1-b)\right]\sigma_{cbc} \quad \text{...[2.26(a)]}$$

$$A_{st} = \frac{k_c d}{2}\left[b_1 - \frac{k_c}{3}(b_1-b)\right]\frac{\sigma_{cbc}}{\sigma_{st}} \quad \text{...(2.27)}$$

(*d*) *Inverted triangular section* (Fig. 2.18)

Such a section can be analysed or designed by taking $b \approx 0$ at the level of reinforcement, and using the equations of case (*c*) of the trapezoidal section. The corresponding equations can be rearranged in the following form.

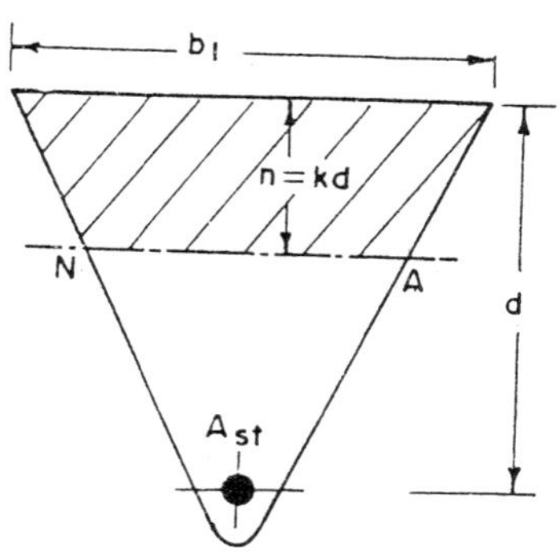

FIG. 2.18.

$$M_r = \left[\frac{1}{2}c\left(1-\frac{k}{3}\right)k - \frac{ck^2(2-k)}{12}\right]b_1 d^2 \quad \text{...(2.28)}$$

$$M_r = [\,R - Q\,]\,b_1 d^2 \quad \text{..(2.29)}$$

$$C = \frac{kd(3-k)}{6}b_1 c \quad \text{...(2.30)}$$

$$C = \frac{k_c \,.\, d(3-k_c)}{6}b_1 \,.\, \sigma_{cbc} \quad \text{...[2.30(a)]}$$

$$= \frac{(0.404)^2 (2 - 0.404)}{12} \times 5 = 0.1085$$

$$\therefore \quad M_r = 0.1085 \times 300\,(600)^2$$

$$= 11.72 \times 10^6 \text{ N-mm} = \mathbf{11.72 \text{ kN-m.}}$$

Example 2.11 *Determine the moment of resistance of the beam section of Fig. 2.19 if the beam is inverted and reinforcement is provided at the apex. Also, determine the reinforcement and moment of resistance corresponding to the balanced section.*

Solution. Assume width at the c.g. of reinforcement as zero. Hence $b = 0$

$$A_{st} = 4 \times \frac{\pi}{4}(12)^2 = 452.4 \text{ mm}^2.$$

The width b_0 at the N.A. is given by the relation

$$\frac{b_0}{b_1} = \frac{d - n}{d} \text{ (approx.).}$$

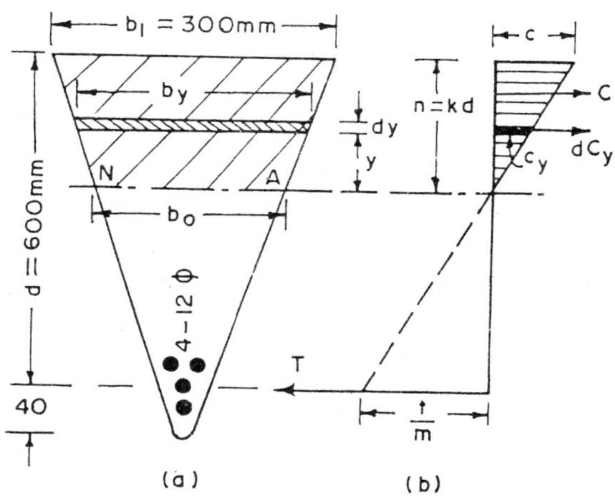

FIG. 2.20.

$$\therefore \quad b_0 = \frac{d - n}{d} \cdot b_1 = \frac{600 - n}{600} \times 300$$

$$= \frac{600 - n}{2}$$

Area of compression zone $= \dfrac{b_0 + b_1}{2} \times n$

$$= \frac{1}{2}\left[\frac{600-n}{2}+300\right]n$$

$\bar{x}$ from N.A. $= \dfrac{2b_1 + b_0}{b_1 + b_0} \times \dfrac{n}{3}$

$$= \frac{600 + \dfrac{600-n}{2}}{300 + \dfrac{600-n}{2}} \times \frac{n}{3}$$

$\therefore$ Moment of compression area about N.A.

$$= \frac{1}{2}\left[\frac{600-n}{2}+300\right]n \times \frac{\dfrac{600-n}{2}+600}{\dfrac{600-n}{2}+300} \times \frac{n}{3}$$

$$= \frac{n^2}{6}\left[600 + \frac{600-n}{2}\right] \qquad \text{...(1)}$$

Moment of equivalent tensile area about N.A.

$$= m\,A_{st}\,(d-n)$$

$$= 19 \times 452 \cdot 4\,(600-n) \qquad \text{...(2)}$$

Equating the two, we get

$$\frac{n^2}{6}\left[600 + \frac{600-n}{2}\right] = 19 \times 452 \cdot 4\,(600-n)$$

or $\quad n^3 - 1800\,n^2 - 103147\,n + 61888320 = 0$

Solving this by trial, we get

$$n = 165 \text{ mm} \;;$$

Hence $\quad k = \dfrac{165}{600} = 0 \cdot 275$

For the balanced section, $k_c = 0 \cdot 404$

$\therefore \quad n_c = 0 \cdot 404 \times 600 = 242 \cdot 4$ mm.

Since $n > n_c$ the section is *under-reinforced*, and the stress in steel will reach its max. permissible value first.

Hence $\quad t = \sigma_{st} = 140$ N/mm^2

$$\therefore \quad c = \frac{\sigma_{st}}{m} \cdot \frac{n}{d-n} = \frac{140}{19} \times \frac{165}{600-165}$$

$$\doteq 2 \cdot 795 \text{ N/mm}^2.$$

In order to find the moment of resistance, consider a strip of thickness dy at a distance of y from N.A. Width b_y is given by

$$\frac{b_y}{b_1} = \frac{d-n+y}{d}$$

$$\therefore \qquad b_y = \frac{d-n+y}{d} b_1 = \frac{(435+y)}{2}$$

The stress c_y at the level of strip is given by

$$\frac{c_y}{c} = \frac{y}{n}$$

or
$$c_y = \frac{y}{n} . c = \frac{2 \cdot 795}{165} y$$

$$\therefore \qquad dC_y = b_y . c_y \, dy$$

$$= \frac{435+y}{2} \times \frac{2 \cdot 795}{165} y \, dy$$

$$= 0 \cdot 00847 \, (435+y) \, y \, dy$$

$$\therefore \qquad C = \int_0^n dC_y = \int_0^{n=165} 0.00847 \, (435+y) \, y \, dy$$

$$= 0.00847 \left[\frac{435 y^2}{2} + \frac{y^3}{3} \right]_0^{n=165} \qquad \text{...(I)}$$

$$= 62838 \text{ N}$$

Also
$$M_r = \int_0^n (dC_y) \, (d-n+y)$$

$$= \int_0^{165} 0.00847(435+y)y \, [600-165+y] dy$$

$$= 0.00847 \int_0^{165} (435+y)^2 y \, dy$$

$$= 0.00847 \left[189225 \frac{y^2}{2} + \frac{y^4}{4} + 870 \frac{y^3}{3} \right]_0^{165} \text{...(II)}$$

$$= 34.42 \times 10^6 \text{ N-mm} = \mathbf{34.42 \text{ kN-m}}$$

Check. From Eq. 2.28,

$$M_r = \left[\frac{1}{2} \left(1 - \frac{k}{3} \right) k - \frac{k^2 (2-k)}{12} \right] c b_1 d^2$$

$$= \left[\frac{1}{2} \left(1 - \frac{0.275}{5} \right) \times 0.275 - \frac{(0.275)^2 (2-0.275)}{12} \right] 2.795 \times 300 \, (600)^2$$

$= 34.42 \times 10^6$ N-mm.

This is about *three times more* than the one of the example 2.9, with the same dimensions but placement reversed.

(*b*) For the Balanced Section

$$k_c = 0.404\,;\, n_c = 0.404 \times 600 = 242.4 \text{ mm}$$

$$c_y = \frac{y}{n_c} \cdot \sigma_{cbc} = \frac{5}{242.4} \cdot y = 0.02062\,y$$

$$b_y = \frac{d - n_c + y}{d} b_1 = \frac{600 - 242.4 + y}{600} \times 300$$

$$= \frac{357.6 + y}{2}$$

$$\therefore \qquad dC_y = \frac{357.6 + y}{2} \times 0.02062\,y\,dy$$

$$= 0.01031\,(357.6 + y)\,y\,dy$$

$$C = \int_0^{n_c} dC_y = \int_0^{242.4} 0.01031(357.6 + y)y\,dy$$

$$= 0.01031 \left[\frac{357.6 y^2}{2} + \frac{y^3}{3}\right]_0^{242.4}$$

$$= 157264 \text{ N.}$$

Equating this to total tensile area, we get

$$A_{st} = \frac{C}{\sigma_{st}} = \frac{157264}{140} = \mathbf{1123.3\ mm^2}.$$

Check : From Eq. 2.31

$$A_{st} = \frac{k_c d\,(3 - k_c)}{6} b_1 \frac{\sigma_{cbc}}{\sigma_{st}}$$

$$= \frac{0.404 \times 600\,(3 - 0.404)}{6} \times 300 \times \frac{5}{140}$$

$$= 1123.7 \text{ mm}^2$$

Also

$$M_r = \int_0^{n_c} (d\,C_y)\,(d - n_c + y)$$

$$= \int_0^{242.4} 0.01031\,(357.6 + y)\,y\,(600 - 242.4 + y)\,dy$$

$$= 0.01031 \int_0^{242.4} (357.6 + y)^2 y\,dy$$

$$= 0.01031 \left[\frac{127878y^2}{2} + \frac{y^4}{4} + \frac{715.2y^3}{3} \right]_0^{242.4}$$

$$= 82.64 \times 10^6 \text{ N–mm} = \mathbf{82.64 \text{ kN–m}} .$$

Check : From Eq. 2.29

$$M_r = (R - Q)\, b_1 d^2$$

where $$R = \frac{1}{2} \times 5 \left(1 - \frac{0.404}{3}\right) \times 0.404 = 0.874$$

$$Q = \frac{k_c^2 (2 - k_c)}{12} \sigma_{cbc}$$

$$= \frac{(0.404)^2 (2 - 0.404)}{12} \times 5 = 0.1085$$

$$\therefore M_r = (0.874 - 0.1085) \times 300 (600)^2$$

$$= 82.67 \times 10^6 \text{ N–mm} .$$

Example 2.12. *A simply supported beam of trapezoidal shape is shown in Fig. 2.21(a), Taking σ_{cbc}= 5 N/mm², σ_{st}= 140 N/mm² and m=18, determine the moment of resistance of the beam section and the actual stresses in the section.*

Solution. Let n be the depth of the N.A.

width $$ab = 500 - \frac{200}{600} n = 500 - \frac{n}{3}$$

Width b at reinforcement level

$$= 500 - \frac{200}{600} \times 560 = 313.3 \text{ mm} .$$

The trapezium *abcd* of compression zone can be divided into rectangle *abhg* and two triangle *acg* and *bhd*.

Width $$cg = hd = \frac{1}{6} n$$

Hence moment of the compressive area about N.A. is

$$= n \left(500 - \frac{n}{3}\right) \frac{n}{2} + \frac{1}{2} n \times \frac{1}{3} n \times \frac{2}{3} n$$

$$= 250 n^2 - \frac{n^3}{6} + \frac{n^3}{9} = 250 n^2 - \frac{n^3}{18} \quad ...(i)$$

$$A_{st} = 5 \times \frac{\pi}{4} (20)^2 = 1570.8 \text{ mm}^2$$

$\therefore$ Moment of equivalent tensile area, about N.A.

$$= 18 \times 1570.8 (560 - n) \quad ...(ii)$$

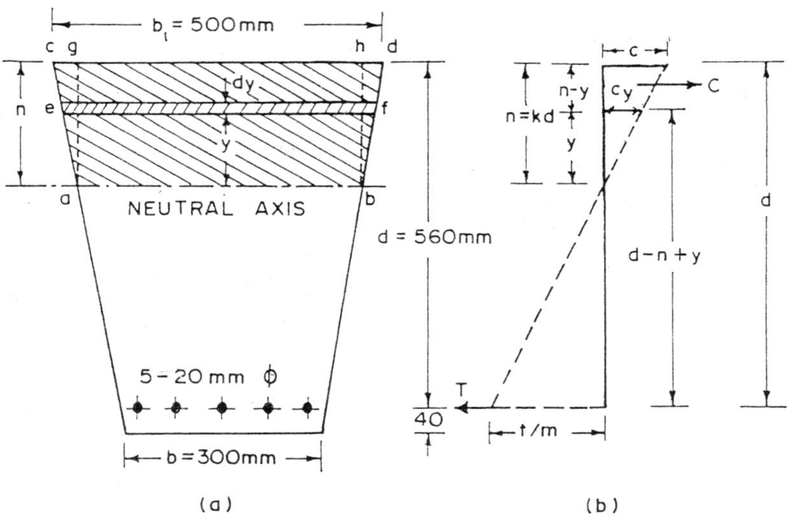

FIG. 2.21

Equating the two area moments, we get

$$250\,n^2 - \frac{n^3}{18} = 18 \times 1570.8\,(\,560 - n\,)$$

or
$$n^3 - 4500\,n^2 - 508940\,n + 285 \times 10^6 = 0$$

Solving it by trial and error, we get

$$n = 205 \text{ mm}\,;\, k = \frac{n}{d} = \frac{205}{560} = 0.366$$

$$k_c = \frac{m\,\sigma_{cbc}}{m\,\sigma_{cbc} + \sigma_{st}} = \frac{18 \times 5}{18 \times 5 + 140} = 0.391$$

$$n_c = 0.391 \times 560 = 219 \text{ mm}\,.$$

Since the actual N.A. is *above* the critical N.A., the section is *under-reinforced*, and the stress in steel will reach the maximum value earlier.

$\therefore$ $$t = \sigma_{st} = \mathbf{140\ N/mm^2}.$$

Corresponding stress in concrete is given by

$$c = \frac{t}{m} \times \frac{n}{d-n} = \frac{140}{18} \times \frac{205}{560 - 205}$$

$$= 4.49 \text{ N/mm}^2$$

In order to find the moment of resistance of the beam, consider an elementary strip of area $b_y.dy$ at a distance y from the N.A.

Width of strip $ef = b_y = 500 - \frac{n - y}{3} = 500 - \frac{205 - y}{3}$

Stress c_y at $ef = c \,.\, \frac{y}{n} = \frac{4.49}{205} y = 0.0219 y$

$$\therefore \quad dC_y = b_y \,.\, dy \,.\, c_y$$

$$= \left(500 - \frac{205 - y}{3}\right)(0.0219\, y)\, dy$$

$$= (1295 + y) \times 0.0073\, y\, dy$$

Moment of the force about the centre of steel area is

$$d M_y = b_y \,.\, dy \,.\, c_y (d - n + y)$$

$$= 0.0073 (1295 + y) y\, dy (560 - 205 + y)$$

$$M_r = \int_0^n 0.0073 (1295 + y)(355 + y) y\, dy$$

$$= 0.0073 \int_0^{205} (y^3 + 1650 y^2 + 459275 y)\, dy$$

$$= 0.0073 \left[\frac{y^4}{4} + \frac{1650 y^3}{3} + \frac{459725 y^2}{2}\right]_0^{205}$$

$$= 108.33 \times 10^6 \text{ N–mm}$$

$$= \mathbf{108.33 \text{ kN–m}}$$

Altermatively, from Eq. 2.85,

$$M_r = \left[\frac{1}{2}\left(1 - \frac{k}{3}\right) k b_1 - \frac{k^2(2 - k)}{12}(b_1 - b)\right] cd^2$$

$$= \left[\frac{1}{2}\left\{1 - \frac{0.366}{3}\right\} \times 0.366 \times 500 - \frac{(0.366)^2 (2 - 0.366)}{12}(500 - 313.3)\right] \times 4.49 (560)^2$$

$$= 108.3 \times 10^6 \text{ N–mm} = 108.3 \text{ kN–m}$$

Example 2.13 *A singly reinforced concrete beam 50 cm wide and 60 cm deep has a 10 cm wide and 10 cm deep groove in the compression side, as shown in Fig. 2.22. The reinforcement consists of 4 bars of 20 mm diameter. Determine (i) position of the neutral axis (ii) maximum compressive and tensile stresses when it is subjected to a bending moment of 80 kN-m. Take m=15.*

Solution

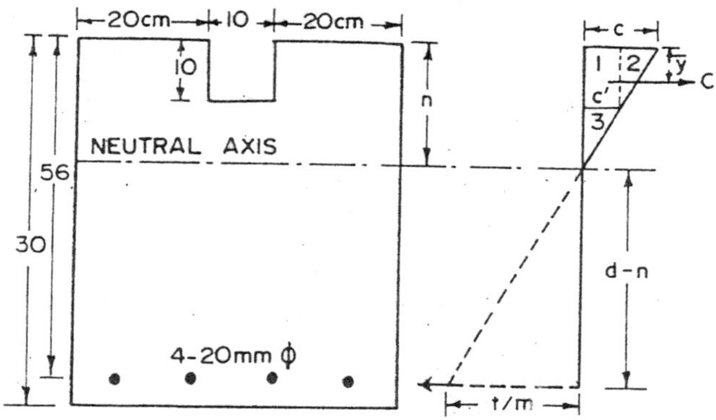

FIG. 2.22.

Area of steel $= 4 \times \frac{\pi}{4}(20)^2 = 1256.6 \text{ mm}^2$

Let the depth of the N.A. be n.

$\therefore$ Moment of tensile area about N.A.

$$= 15 \times 1256.6\,(560 - n) \quad \text{...(i)}$$

Moment of compressive area about N.A.

$$= 400\,n \times \frac{n}{2} + 100\,(n - 100)\,\frac{(n - 100)}{2}$$

$$= 200\,n^2 + 50\,(n^2 - 200\,n + 10000)$$

$$= 250\,n^2 - 10000\,n + 500000 \quad \text{...(ii)}$$

Equating he two we get

$$250\,n^2 - 10000\,n + 50000 = 15 \times 1256.6\,(560 - n)$$

or $n^2 + 35.4\,n - 42022 = 0$

From which n = 188 mm

Stress c' in concrete at the bottom level of groove is

$$c' = c\,\frac{n - 100}{n} = c\,\frac{188 - 100}{188} = 0.468\,c$$

Let $\bar{y}$ be the distance of C.G. of total compressive force, from top fibre. Let us divide the total compressive force into three parts as marked in the stress diagram.

Force $C_1 = 400 \times 100\, c' = 40000 \times 0.468\, c$

$= 18720\, c$, acting at 50 mm from top.

Force $C_2 = 400 \times 100 \times \frac{1}{2}(c - c') = 20000 \times 0.532\, c$

$= 10640\, c$, acting at $\frac{100}{3}$ m from top

Force $C_3 = 500\,(188 - 100)\,\frac{1}{2} c' = 22000 \times 0.468\, c$

$= 10296\, c$ acting at $100 + \frac{1}{3}(188 - 100)$

$=$ 129 mm from the top.

Hence taking the moments of individual forces about the top we get

$$\bar{y} = \frac{(18720\, c \times 50) + (10640\, c \times 33.3) + (10296\, c \times 129)}{18720\, c + 10640\, c + 10296\, c}$$

$= 66$ mm.

Lever arm $= d - \bar{y} = 560 - 66 = 494$ mm.

Now $M = A_{st}\,.\,t \times \text{Lever arm} = A_{st}\,.\,t\,a$

$\therefore \quad t = \frac{M}{a\,.\,A_{st}} = \frac{80 \times 10^6}{494 \times 1256.6}$

$= \mathbf{128.9\ N/mm^2}$

From the stress diagram,

$$\frac{c}{t/m} = \frac{n}{d - n}$$

$\therefore \quad c = \frac{t}{m} \cdot \frac{n}{d - n} = \frac{128.9}{15} \times \frac{188}{560 - 188}$

$= \mathbf{4.34\ N/mm^2}$.

2.11. SLAB SPANNING IN ONE DIRECTION

Fig. 2.23(a) shows a room fo size $l \times L$, i ι which size L is much greater than l. The roof of the room consists of a reinforced concrete slab. When uniformly distributed load is applied to the slab, the bending generally takes place about two axes, parallel to directions of L and l. However, if the ratio is greater than 1.5, 84% of the total load is carried by the shorter span (l) and 16% by the longer span (L). Hence if the ratio $\frac{L}{l}$ is the greater than 1.5, it is usual to consider the bending of the slab to take place in the span l, as shown in Fig. 2.23(b). Such a slab is, therefore, known as a *slab spanning in one direction*. If however, $\frac{L}{l}$ is less than 1.5, bending takes place in both the directions. The analysis of slabs spanning in both the directions have been given in Chapter 10.

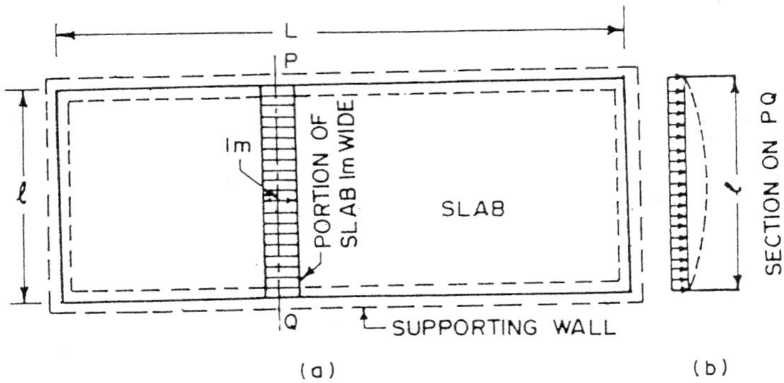

FIG. 2.23

The analysis of the slab spanning in one direction is done by assuming it to be a beam of 1 m width, as shown in Fig. 2.23(*a*) though the slab width is continuous and is not composed of individual beams of 1 m width. The reinforcement etc. are calculated for 1 m width and the bars are distributed accordingly. The formulae for determination of N.A., moment of resistance and steel reinforcement are the same as found for singly reinforced beam, except that b is taken to be equal to 1000 mm.

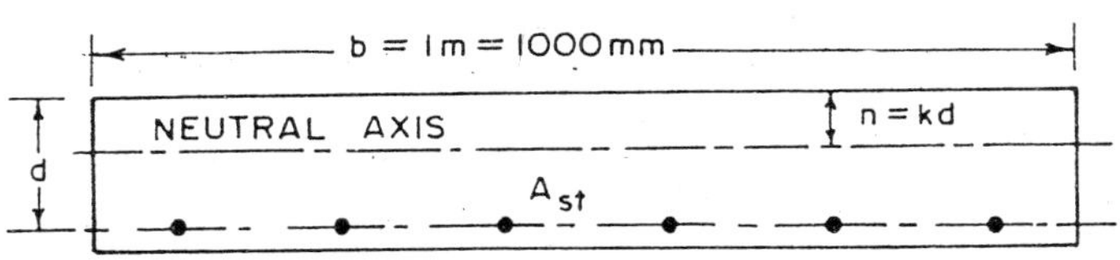

FIG. 2.24.

Fig. 2.24 shows a slab having b = 1 m =1000 mm and an effective depth d. For a balanced design the depth of critical N.A. is given by

$$k = \frac{m\,\sigma_{cbc}}{m\,\sigma_{cbc} + \sigma_{st}}$$

$$j = 1 - \frac{k}{3}$$

$$\therefore \quad R = \frac{1}{2}\sigma_{cbc} \cdot j \cdot k$$

$$\therefore \quad M_r = Rbd^2$$

Equating this to the external B.M. M, we get

$$d = \sqrt{\frac{M}{bR}} = \sqrt{\frac{M}{1000\,R}} \qquad \text{...(2.32)}$$

Thus the effective depth d of the slab can be known.

The area of steel per 1000 mm width is given by

$$A_{st} = \frac{M}{\sigma_{st} \cdot j \cdot d} \qquad \text{...(2.33)}$$

Let Φ = diameter of the bar chosen (mm)

A = area of one bar (mm^2)

s = spacing of bars (mm), c/c

Then No. bars $= \dfrac{1000}{s}$

Total area $= \dfrac{1000}{s} \cdot A_\Phi$

Equating this to the total area (A_{st}) required,

$$A_{st} = \frac{1000}{s} A_\Phi$$

or

$$s = \frac{1000\,A_\Phi}{A_{st}}$$

The complete procedure of design of slabs spanning in one direction is given in chapter 8

For a slab of given dimensions d and A_{st}, the N.A. can be located by equating the moments of equivalent areas about the N.A.

Thus, if kd is the depth of N.A., we get

$$b \cdot kd \frac{kd}{2} = m\,A_{st}\,(d - kd)$$

or

$$k^2 + 2\,km\frac{A_{st}}{bd} - 2m\frac{A_{st}}{bd} = 0$$

or

$$k^2 + 2\,km\frac{A_{st}}{1000\,d} - 2m\frac{A_{st}}{1000\,d} = 0 \qquad \text{...(2.35)}$$

From which k can be found.

Knowing k, stresses in the section can be determined exactly in the same manner, as adopted for beams.

Example 2.14 *A reinforced concrete slab has an effective span of 4 m, and carries a uniformly distributed load of 6 kN/m² inclusive of its own weight. Determine (i) effective depth of the slab (ii) steel reinforcement. Use M 15 concrete and (a) Fe 250 steel, (b) Fe 415 steel.*

Solution. Consider 1 m width of the slab.

$$w = 6 \times 1 \times 1 = 6 \text{ kN/m length of slab}$$

$$M = \frac{wL^2}{8} = \frac{6 \times 4 \times 4}{8} \text{ k N–m}$$

$$= 12 \text{ k N–m} = 12 \times 10^6 \text{ N–mm}$$

(*a*) Mild steel (Fe 250) bars

Given $\sigma_{cbc} = 5$ N/mm² ; $\sigma_{st} = 140$ N/mm² and $m \approx 19$

$$\therefore \quad k = \frac{19 \times 5}{19 \times 5 + 140} = 0.404$$

$$j = 1 - \frac{k}{3} = 0.865$$

$$R = \frac{1}{2}\sigma_{cbc} \cdot j \cdot k = \frac{1}{2} \times 5 \times 0.865 \times 0.404 = 0.874$$

Hence $$d = \sqrt{\frac{M}{1000R}} = \sqrt{\frac{12000000}{1000 \times 0.874}} = \mathbf{117.2 \text{ mm}}$$

Keep $$d = 120 \text{ mm}$$

$$A_{st} = \frac{M}{\sigma_{st}\, jd} = \frac{12000000}{140 \times 0.865 \times 120} = \mathbf{826.7 \text{ mm}^2}$$

Choosing 12 mm diameter bars,

$$A_\Phi = \frac{\pi}{4}(12)^2 = 113.1 \text{ mm}^2$$

$$\therefore \quad s = \frac{1000 A_\Phi}{A_{st}} = \frac{1000 \times 113.1}{826.7} = 136.8 \text{ mm}$$

Hence provide 12 mm dia. bars @ 135 mm c/c.

(*b*) Fe 415 Steel

$\sigma_{cbc} = c = 5$ N/mm², $m = 19$ and $\sigma_{st} = 230$ N/mm²

$$k = \frac{19 \times 5}{19 \times 5 + 230} = 0.2923$$

$$j = 1 - \frac{0.2923}{3} = 0.9026$$

$$R = \frac{1}{2}\sigma_{cbc}\, jk = \frac{1}{2} \times 5 \times 0.9026 \times 0.2923 = 0.6596$$

$$d = \sqrt{\frac{M}{1000R}} = \sqrt{\frac{12 \times 10^6}{1000 \times 0.6596}} = 134.9$$

Keep $d = 135$ mm

$$A_{st} = \frac{M}{\sigma_{st} jd} = \frac{12 \times 10^6}{230 \times 0.9026 \times 135} = \mathbf{428.2\ mm^2}$$

Choosing 10 mm dia. bars.

$$A_\Phi = \frac{\pi}{4}(10)^2 = 78.54 \text{ mm}^2$$

$\therefore$ Spacing
$$s = \frac{1000 A_\Phi}{A_{st}} = \frac{1000 \times 78.54}{428.2}$$
$$= 183.4 \text{ mm}$$

Hence provide 10 mm dia. bars at 180 mm c/c. It is seen that by using HYSD bars, though the depth of slab is increased, there is considerable saving in steel reinforcement.

Example 2.15. *Redesign the slab of example 2.14, using M 20 concrete and Fe 415 steel reinforcement.*

Solution. $\sigma_{cbc} = 7 \text{ N/mm}^2$, $m \approx 13$ and $\sigma_{st} = 230 \text{ N/mm}^2$

$$\therefore \quad k = \frac{13 \times 7}{13 \times 7 + 230} = 0.2835$$

$$j = 1 - \frac{0.2835}{3} = 0.9055$$

$$R = \frac{1}{2} \sigma_{cbc} jk = \frac{1}{2} \times 7 \times 0.9055 \times 0.2835 = 0.898$$

$$M = 12 \times 10^6 \text{ N–mm as before}$$

$$\therefore \quad d = \sqrt{\frac{M}{1000 R}} = \sqrt{\frac{12 \times 10^6}{1000 \times 0.898}} = 115.6 \text{ mm}$$

Keep $d = 116$ mm

$$A_{st} = \frac{M}{\sigma_{st} jd} = \frac{12 \times 10^6}{230 \times 0.9055 \times 116} = \mathbf{496.7\ mm}$$

Choosing 10 mm dia. bars,

$$A_\Phi = \frac{\pi}{4}(10)^2 = 78.54 \text{ mm}^2$$

$\therefore$ Spacing
$$s = \frac{1000 A_\Phi}{A_{st}} \quad \frac{1000 \times 78.54}{496.7}$$
$$= 158 \text{ mm}$$

Hence provide 10 mm dia. bars at 150 mm c/c.

PROBLEMS

1 (a) Discuss in brief basic assumptions of straight line theory.
(b) Drive expressions for the position of neutral axis and moment of resistance of balanced rectangular section.
(c) How do you find the moment of resistance of a beam section.

2. Working from the first principles, determine the moment of resistance and percentage of steel in a singly reinforced rectangular section ($b \times d$) for a balanced design, if the stresses in concrete and steel are not to exceed 5 N/mm^2 and 140 N/mm^2 respectively. Take m = 18.

3. Design a R.C. beam to carry a load of 6 kN/m inclusive of its own weight on an effective span of 6 m. Keep the breath to be 2/3rd of effective depth. The permissible stresses in concrete and steel are not to exceed 5 N/mm^2 and 140 N/mm^2 respectively. Take m = 18.

4. The cross-section of singly reinforced beam is 320 mm wide and 450 mm deep to the centre of the reinforcement. The reinforcement consists of 4 bars of 16 mm. Determine the moment of resistance of the beam section. The permissible stresses in concrete and steel are 5 N/mm^2 and 140 N/mm^2 respectively. Take m = 18.

5. Determine the moment of resistance, per meter width of a R.C. slab having the following details:
Overall thickness: 100 mm
Reinforcement: 12 Φ @ 100 mm c/c
Cover to the centre of reinforcement : 20 mm
Take the permissible stresses in concrete and steel as 6 N/mm^2 and 140 N/mm^2 respectively, and m = 15.

6. A R.C.C. beam 300 mm wide and 600 mm deep has 4 bars of 20 mm diameter as tension reinforcement, the centre of the bars being 50 mm from the bottom of the beam. Determine the uniformly distributed load (exclusive of its own weight) the beam can carry over an effective simply supported span of 6 m. Take the permissible stresses in concrete and steel as 5.2 N/mm^2 and 126 N/mm^2 respectively and m = 18.

7. A R.C. beam has a section 400 mm wide and 600 mm deep. Find the necessary reinforcement to carry a load of 10 kN/m. In addition to its own weight, over an effective span of 6 m. The beam has an effective cover of 50 mm to the centre of reinforcement. The permissible stresses in concrete and steel may be taken as 5 N/mm^2 and 140 N/mm^2 respectively. Take m = 18.

8. Solve problem 7 if the beam has size 400 mm×450 mm.

9. The moment of resistance of a rectangular reinforced concrete beam, of breadth b (mm) and effective depoth d (mm) is 0.7 bd^2 N–mm. If the stresses in the outside fibre of concrete and in the steel do not exceed 4.2 N/mm^2 and 140 N/mm^{2}respectively, and the modular ratio equals 18, determine the ratio of the neutral axis from the outside compression fibres to the effective depth of the beam and the ratio of the area of tension steel to the effective area of the beam.

10. A reinforced concrete beam has section 300 mm wide and 600 mm deep. The reinforcement consists of mild steel bars of 25 mm diameter with a cover of 50 mm to the centre of reinforcement. If it is subjected to a bending moment of 120 kN-m, determine the stresses developed in steel and concrete. Take m = 15.

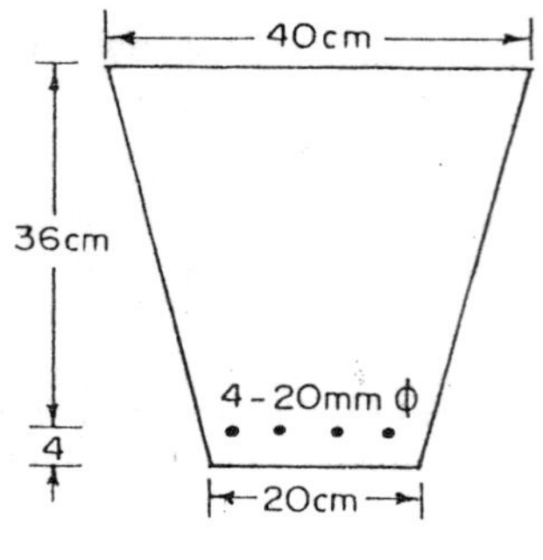

FIG. 2.25.

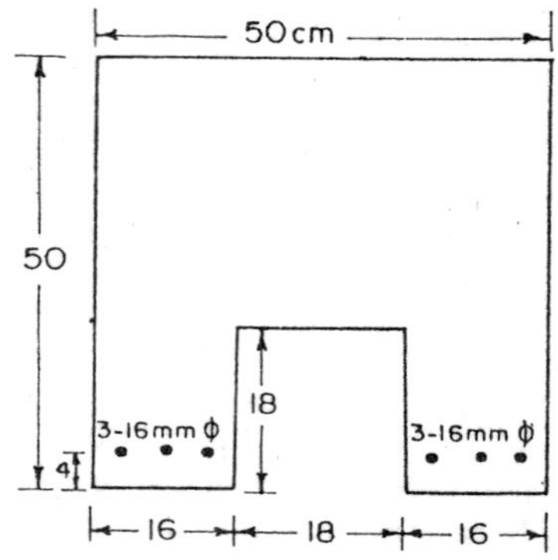

FIG. 2.26

11. Determine the moment of resistance of the beam section shown in Fig. 2.25. Take $\sigma_{cbc} = 5$ N/mm^2 and $m = 18$.
12. Determine the moment of resistance of the section shown in Fig. 2.26. Take $\sigma_{cbc} = 5$ N/mm^2 and $\sigma_{st} = 140$ N/mm^2 and $m = 18$.

ANSWERS

2. 0.85 bd^2; 0.698%
3. $b = 240$ mm ; $d = 360$ mm ; $A_{st} = 616$ mm^2
4. 44.9 kN-mm.
5. 8.5×10^6 N-mm
6. 12.38 kN/m
7. 1040 mm^2
8. 1970 mm^2
9. $k = 0.38$; $p = 0.65\%$
10. 6.57 N/mm^2; 106 N/mm^2.
11. 45.5 kN-m
12. 68.5 kN-m.

3

Shear and Bond

3.1. SHEAR STRESS IN R.C. BEAMS

When a beam is loaded with transverse loads, the bending moment varies from section to section. Shearing stresses in the beam are caused by this variation of bending moment along the span. Fig. 3.1(a) shows two vertical sections mn and $m_1 n_1$ distant dx apart, of a homogeneous beam. The B.M. at mn is M (say) and that at $m_1 n_1$ is $M + \delta M$. Hence the bending stresses at section $m_1 n_1$ are

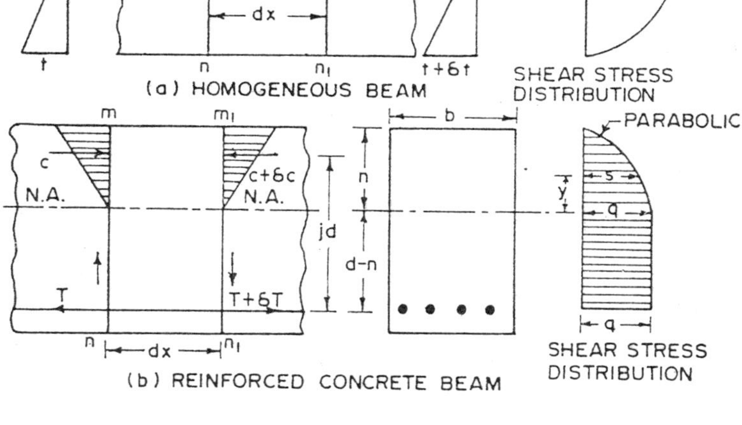

FIG. 3.1. SHEAR STRESSES IN BEAMS

greater than those at section *mn*. Thus, at the *same fibre*, there are unequal bending stresses at two cross-sections distant *dx* apart. This inequality of stresses produces a tendency in each fibre to slide over adjacent fibre in a horizontal plane, causing horizontal shear stress which is accompanied by the complementary shear stress in the vertical direction.

In the case of homogeneous beams, the shear stress q at any plane is given by

$$q = \frac{V}{Iz}(A\bar{y})$$

where V = shear force at the section

I = moment of inertia of the beam section

z = width of the section

$A\bar{y}$ = moment of area above the section, about N.A.

Fig. 3.1(b) shows two sections mn and m_1n_1 distant dx apart, of a R.C. beam. Let the bending moment at mn be M and that at m_1n_1 be $M + \delta M$. Due to this variation of bending moment in a length δx, the compressive forces in concrete at these two sections will be C and $C + \delta C$, and the tensile forces in steel will be T and $T + \delta T$ as shown. Let us assume that the shear force V remains constant in the length dx.

Let us assume that concrete does not take any tension. Consider any plane between N.A. and centre of steel reinforcement. Let q be the intensity of shear at the plane.. Hence total horizontal shear $=q.b.\delta x$ at that layer. Also, the total horizontal force that tends to slide this layer past the adjacent one is equal to $(T+\delta T)-(T)=\delta T$. Hence equating the two, we get

$$q \,.\, b \,.\, \delta x = \delta T$$

or
$$q = \frac{\delta T}{b \,.\, \delta x} \qquad ...(i)$$

In order to get the value of q, it is first essential to determine the value of δT. For equilibrium $\Sigma M = 0$,

$$\therefore \quad (T + \delta T)\,jd = Tjd + V \,.\, \delta x$$

$$\therefore \quad \delta T = \frac{V \,.\, \delta x}{jd} \qquad ...(ii)$$

Substituting in (i), we get

$$q = \frac{V}{bjd} \qquad ...(3.1)$$

This value of q is evidently the same for any layer between the N.A. and the centre of steel. Thus the shear stress distribution below the N.A. is rectangular.

The shear stress distribution above the N.A. is parabolic, as in the case of homogeneous beams. The intensity of shear stress s at any layer distant y above the N.A. is given by

$$s = \frac{V}{Ib}\int_{y}^{n} b \, . y \, . dy$$

or
$$s = \frac{V}{2I}(n^2 - y^2) \qquad ...(3.2)$$

where I is the moment of inertia of the R.C. beam section about the N.A. and can be found the following expression:

$$I = \frac{1}{3} b \, . n^3 + m A_{st} (d - n)^2 \qquad ...(3.3)$$

The complete shear stress distribution diagram is shown in Fig. 3.1(b). The intensity of maximum shear (q) can also be found be equating the external shear force to the shearing resistance of the beam, as under :

$$V = (\text{ Area of shear stress diagram }) \times b$$

$$= \left\{ \left(\frac{2}{3} qn\right) + q(d - n) \right\} b$$

$$= \left(\frac{2}{3} qn + qd - qn\right) b$$

$$= qb\left(d - \frac{n}{3}\right) = q \, . bjd$$

$$q = \frac{V}{b \, . jd} \qquad ...(3.1)$$

Eq. 3.1, obtained from the stress based approach, gives the value of maximum shear stress q at N.A. and for all layers below N.A. in a cracked reinforced beam. Though in pure flexure, we neglect the concrete area below N.A., the same concrete between the cracks is needed for shear transfer between steel reinforcement and the compression zone. Also the flexural cracks (assumed vertical in pure flexure) are not vertical in the shear-flexural zone under consideration. We have also neglected the shear deformations, which are really responsible for the mobilization of aggregate interlock at the cracks and dowel forces in the reinforcement. Due to these reasons, the *stress-based approach* followed above in the derivation of Eq. 3.1 may not represent the real behaviour of the beam in shear and flexure. Hence Eq. 3.1 , representing conventional shear stress formula $q = V/bjd$, has now been abandoned. Indian Standard Code (IS : 456–1978) recommends to use *nominal shear stress* given by the expression.

$$\tau_v = \frac{V}{bd} \qquad ...(3.4)$$

where τ_v = nominal shear stress

V = shear force at the section, due to design loads.

Thus in Fig. 3.2(*b*), the maximum shear stress $q(=V/bjd)$ obtained from elastic theory, is greater than the *nominal shear stress* (or average

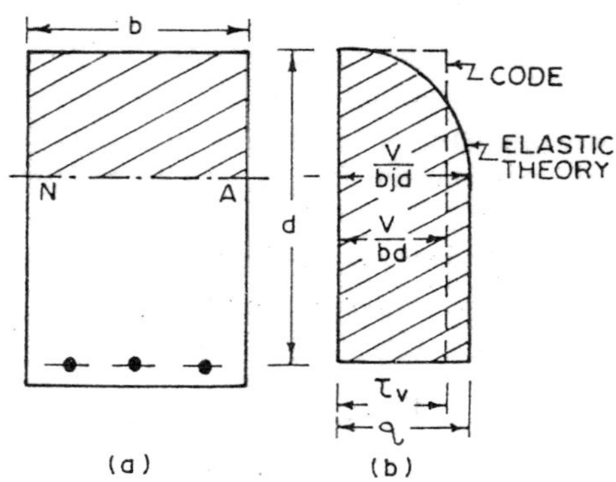

FIG. 3.2.

shear stress) τ_v suggested by the code.

3.2. EFFECTS OF SHEAR : DIAGONAL TENSION

We know that the bending stress and the shearing stress vary across the cross-section of R.C. beam. Below the N.A., the bending stres (f) is tensile while the shear stress q is constant. Fig. 3.3(*b*) shows a small element, taken from the portion below the N.A. This element is subjected to a longitudinal tensile stress f and horizontal shear stress q alongwith the vertical complementary shearing stress q.

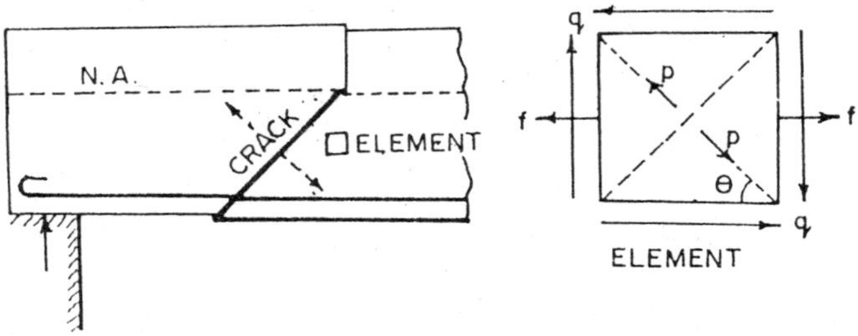

FIG. 3.3. DIAGONAL TENSION.

$$p = \frac{f}{2} \pm \sqrt{\left(\frac{f}{2}\right)^2 + q^2} \qquad ...(3.5)$$

and the inclination of the principal plane with horizontal is given by

$$\tan 2\theta = -\frac{2q}{f} \qquad ...[3.5(a)]$$

The major principal stress p_1 is tensile and is given by

$$p_1 = \frac{f}{2} \pm \sqrt{\left(\frac{f}{2}\right)^2 + q^2} \qquad ...[3.5(b)]$$

At the supports, where bending stress f is practically zero, the value of the principal (tensile) stress is equal to the shear stress q, and it is inclined at 45° to the horizontal, *i.e.* it acts diagonally. Hence it is known as *diagonal tension.* At the centre of the beam, where shearing stresses are practically negligible, the principal tensile stress is equal to the longitudinal tensile stress f, its directioon will be horizontal and the cracks will be vertical.

Concrete is quite strong in shear, but the diagonal tension, which is caused by the combined action of longitudinal tension and the transverse shearing stresses, is to be resisted by the provision of *shear reinforcement or diagonal (or inclined tension) reinforcement or web reinforcement.*

3.3. REINFORCEMENT FOR DIAGONAL TENSION

As stated above, proper reinforcement must be provided to resist the diagonal tension. Fig 3.4 shows a portion of the beam near its support, reinforced with inclined bars 1, 2, 3 etc. spaced at distance s apart.

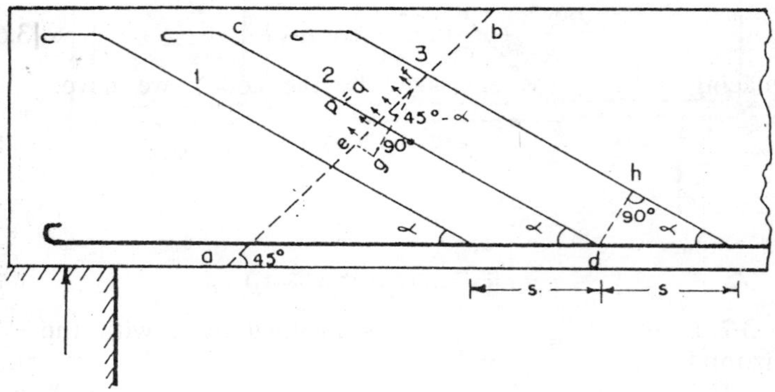

FIG. 3.4 WEB REINFORCEMENT.

Let the web reinforcement, be inclined at an angle α with the horizontal and the direction of diagonal tension be at 45° with the horizontal. Near the support, it is asssumed that the value of the diagonal tension is equal to the maximum shearing stress q. The beam is likely to crack along the line ab inclined at 45° ; hence total tension in one inclined bar (say bar 2) must be such that its vertical component balances the vertical component of the diagonal tension force along the length ef centre to centre between the bars.

Let A_{sv} = cross-sectional area of bar 2 = web reinforcement

σ_{sv} = permissible tensile stress in shear reinforcement

Vertical component to the force in the bar

$$= \sigma_{sv} \,.\, A_{sv} \sin \alpha$$

Vertical component of the diagonal tension in the length ef

$$= q \times b\,(ef) \sin 45°$$

Equating the two, we get

$$\sigma_{sv} \,.\, A_{sv} \sin \alpha = q \,.\, b\,(\,ef\,) \sin 45° \qquad ...(i)$$

In order to get the value of the length ef, draw fg, perpendicular to bar 2 ; also draw eg perpendicular to fg.

Evidently, $fg = hd = s \sin \alpha$

$$\therefore \qquad ef = fg \,.\, \sec efg = fg \,.\, \sec\,(\,45° - \alpha\,)$$

or $$ef = s \sin \alpha \,.\, \sec\,(\,45° - \alpha\,)$$

Substituting the value of ef in (i), we get

$$\sigma_{sv} . A_{sv} \sin\alpha = (q . b . \sin 45°) \times s \sin\alpha \sec(45° - \alpha)$$

$$s = \frac{\sigma_{sv} A_{sv}}{q . b . \sin 45°} \cos(45° - \alpha)$$

$$s = \frac{\sigma_{sv} A_{sv}}{q . b} (\sin\alpha + \cos\alpha) \qquad ...[3.6(a)]$$

Replacing q by τ_v (as envisaged by the code), we have

$$s = \frac{\sigma_{sv} A_{sv}}{\tau_v . b} (\sin\alpha + \cos\alpha) \qquad ...(3.6)$$

But $\tau_v = \frac{V}{bd}$

$$\therefore \quad s = \frac{\sigma_{sv} A_{sv} . d}{V} (\sin\alpha + \cos\alpha) \qquad ...(3.7)$$

Eq. 3.7 gives the spacing of bars inclined at α with the horizontal.

If the web reinforcement is vertical, *i.e.* if$\alpha = 90°$ we get

$$s = \frac{\sigma_{sv} A_{sv} . d}{V} = \frac{\sigma_{sv} A_{sv}}{\tau_v . b} \qquad ...(3.8)$$

If the bars are inclined at 45°,the spacing is given by

$$s = \frac{\sigma_{sv} A_{sv} . d}{V} \sqrt{2} = \frac{\sigma_{sv} A_{sv}}{\tau_v . b} \sqrt{2} \qquad ...(3.9)$$

Types of Shear reinforcement

Shear reinforcement will be necessary if the nominal shear stress exceeds the allowable shear stress τ_c. The code recommendations are discussed is § 2.6.

However, in general, shear reinforcement may be provided in the following three forms :

(*i*) Shear reinforcement in the form of vertical bars, known as *stirrups.*

(*ii*) Shear reinforcement in the form of inclined bars.

(*iii*) Shear reinforcement in the form of combination of stirrups and inclined bars.

The total external shear V is jointly resisted by concrete as well as shear reinforcement, and is represented by the following :

$$V = V_c + V_s$$

where V_c = shear strength of concrete $= \tau_c . bd$

V_s =shear strength of shear reinforcement given by rearranging Eq. 3.7 in the following form :

$$V_s = \frac{\sigma_{sv} . A_{sv} . d}{s} (\sin\alpha + \cos\alpha) \qquad ...(3.10)$$

3.4 VERTICAL STIRRUPS

Shear reinforcement in the form of vertical *stirrups* consists of 5 mm to 15 mm diameter steel bars bent round the tensile reinforcement where it is anchored to 6 to 12 mm dia. anchor bars or holding bars. Depending upon the magnitude of the shear stress to be resisted, a stirrup may be one legged, two legged, four legged or multi-legged, as shown in Fig. 3.5.

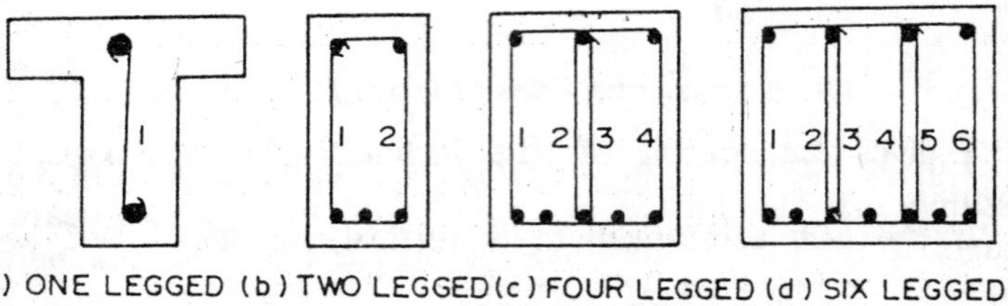

FIG. 3.5. FORMS OF VERTICAL STIRRUPS

The spacing of vertical stirrups will depend upon the net shear force to be resisted at the section, determined from the expression:

$$F = V - V_c = V_s$$

Evidently, the net shear force F should not exceed the shear strength V_s of stirrup. Hence from Eq. 3.8, replacing V by F, we get

$$s = \frac{\sigma_{sv} \cdot A_{sv} \cdot d}{F} \quad \text{...(3.11)}$$

Alternatively, when vertical stirrups are used in conjunction with other forms of shear reinforcement the strength of vertical stirrups, spaced at s is given by

$$V_s = \frac{\sigma_{sv} \cdot A_{sv} \cdot d}{s} \quad \text{...[3.8(a)]}$$

In the above expressions,

A_{sv} = cross-sectional area of stirrups

= (Area of cross-section of stirrups bar) × number of legs

$= A_{\Phi} \cdot n$

s = centre to centre spacing of stirrups.

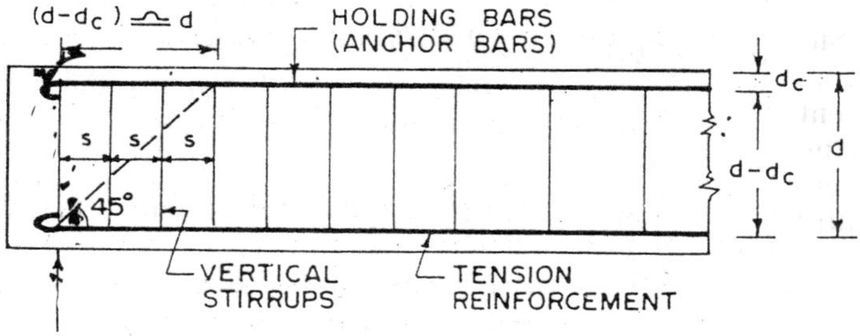

FIG. 3.6. SPACING OF VERTICAL STIRRUPS

Eq. 3.11 can also be *alternatively* obtained from Fig 3.6. Let us assume that in absence of shear reinforcement, the beam fails in diagonal tension, the inclination of the tension crack being at 45° to the axis of the beam and extended up to a horizontal distance equal to $(d - d_c) \approx d$

$\therefore$ No. of stirrups resisting shear force $= \dfrac{d}{s}$

$$\therefore \quad \begin{bmatrix}\text{Shear force } F \\ \text{resisted by stirrups}\end{bmatrix} = \begin{bmatrix}\text{force resisted} \\ \text{by each stirrup}\end{bmatrix} \times \begin{bmatrix}\text{No. of} \\ \text{stirrups}\end{bmatrix}$$

or
$$F = \sigma_{sv} \,.\, A_{sv} \times \frac{d}{s}$$

or
$$s = \frac{\sigma_{sv} \,.\, A_{sv} \,.\, d}{F} = \frac{\sigma_{sv} \,.\, A_{sv} \,.\, d}{V - V_c} \qquad \text{...(3.11)}$$

where F = net S.F. to be resisted by stirrups $= V - V_c$

Spacing diagram for vertical stirrups. The spacing of the stirrups at any section along the length of the beam can be determined from Eq. 3.6, since the net shear force F at that section can be

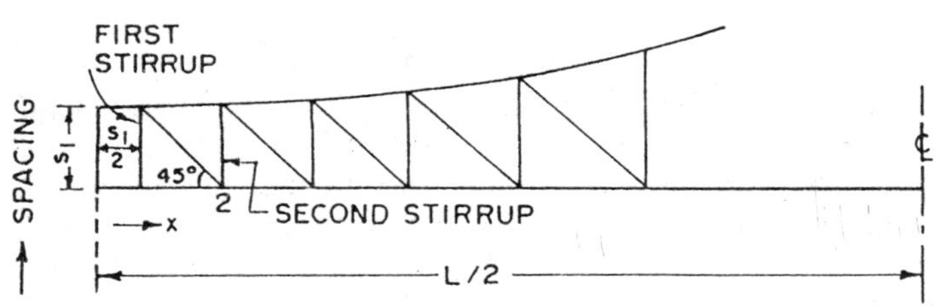

FIG. 3.7 SPACING DIAGRAM.

easily determined. This will give variable spacings all along the length of the beam. These variable spacings can be *averaged* out by the construction of *spacing diagram.*

Procedure. (1) Determine net shear force at various sections distant x_1, x_2, x_3 etc. from the support and calculate the spacings at these sections from Eq. 3.11.

(2) Plot the spacing diagram with the calculated spacing as ordinate (Fig. 3.7).

(3) Calculate the spacing of the first stirrup, on the basis of net S.F. at the support. Set the first stirrup at half the distance of this spacing, from the support.

(4) Where this ordinate at (1) intersects the spacing diagram, set a line at 45°, meeting the span line in (2). This will give the position of the second stirrup. Repeat this procedure to get the position of other stirrups.

3.5. INCLINED BARS

Some of the main reinforcing bars, provided to resist the tensile bending stress, can be bent up near the supports where they are no longer required since bending moment is very much reduced near the support. Such bent up bars resist the diagonal tension. To maintain symmetry about the vertical axis of the section, bars are generally bent up in *pairs.* These bars are bent usually at 45°.

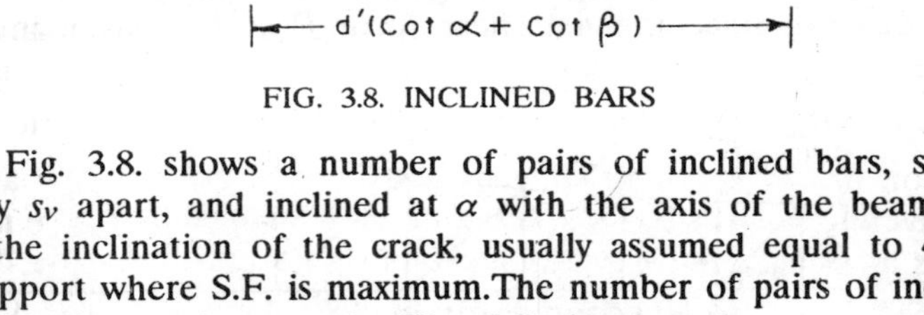

FIG. 3.8. INCLINED BARS

Fig. 3.8. shows a number of pairs of inclined bars, spaced equally s_v apart, and inclined at α with the axis of the beam. Let β be the inclination of the crack, usually assumed equal to 45° at the support where S.F. is maximum.The number of pairs of inclined bars, crossing such a crack (Fig. 3.8) is given by

$$n = \frac{(d - d_c)(\cot\alpha + \cot\beta)}{s_v}$$

where $\qquad d_c =$ cover from top

Taking $d - d_c = d'$

$$n = \frac{d' \; (\cot\alpha + \cot\beta)}{s_v} \qquad ...(3.12)$$

The vertical component (V_s) of the tensile force on inclined bars crossing the diagonal crack is given by

$$V_s = n \,.\, A_{sv} \,.\, \sigma_{sv} \,.\, \sin\alpha \qquad ...(3.13)$$

Substituting the value of n,

$$V_s = A_{sv} \,.\, \sigma_{sv} \,.\, (\cos\alpha + \sin\alpha \cot\beta) \frac{d'}{s_v} \qquad ...(3.14)$$

If $\quad n = 1$ (*i.e.* only one pair of inclined bars)

$$s_v = d' (\cot\alpha + \cot\beta) \qquad ...[3.15(a)]$$

and $$V_s = A_{sv} \,.\, \sigma_{sv} \,.\, \sin\alpha \qquad ...[3.15(b)]$$

Taking $\alpha = 45°$, $V_s = 0.707 \,.\, A_{sv} \,.\, \sigma_{sv}$ ...(3.15)

If $\quad n = 2$ (*i.e.* two pairs of inclined bars)

$$s_v = \tfrac{1}{2}\, d' (\cot\alpha + \cot\beta) \qquad ...(3.16)$$

and $$V_s = 2 A_{sv} \,.\, \sigma_{sv} \,.\, \sin\alpha \qquad ...(3.17)$$

The above expressions are also valid in the case of *inclined stirrups*, in which case A_{sv} is the total area of the all the legs of each inclined stirrups. The above analysis helps in understanding the *truss analogy* or *lattice girder effect.*

Lattice girder effect

The resistance of shear at any section of a beam reinforced with inclined bars may be calculated on the assumption that the inclined bars form the tension members of a system of the lattice girder in which the concrete forms the compression members. There are two systems of lattice girders : (*i*) single system (*ii*) double system

(*i*) *Single System* (Fig. 3.9)

In this system, the bars are bent at 45° and the imaginary compression members (known as *struts*) *AB, CD* etc. are assumed to be at $67\frac{1}{2}°$ with horizontal. Member *AC* is an imaginary horizontal portion of the inclined bar acting as *tension chord* and is assumed to have the same cross-sectional area as of inclined bars. The member *BD* is an imaginary *compression chord.* It is assumed that the distance between the top compression chord and bottom tension chord is equal to lever arm. In the triangle *ABC, BC* and *AC* are tension members while *AB* is compression member. From geometry of triangle *ABC,*

$$AC = BC = a\sqrt{2} = 1.414\, a$$

Thus in order that the inclined bars are to be effective, the bars nearest to the support are bent at 45° at a distance of 1.414*a* from the support.

Since the triangle *ABC* is isosceles, the total force in the inclined bar *BC* is equal to the total force in the horizontal bar. Thus, the force in the inclined bar is $A_{sv}\,\sigma_{sv}$ where A_{sv} is the area of cross-section of the bent up bars. The vertical component of this force to resist the shear force is equal to :

$$V_s = A_{sv} \,.\, \sigma_{sv} \,.\, \sin 45° = 0.707\, A_{sv} \,.\, \sigma_{sv} \qquad ...(3.15)$$

The inclined bar is effective for a length equal to lever arm on either side measured along the N.A.

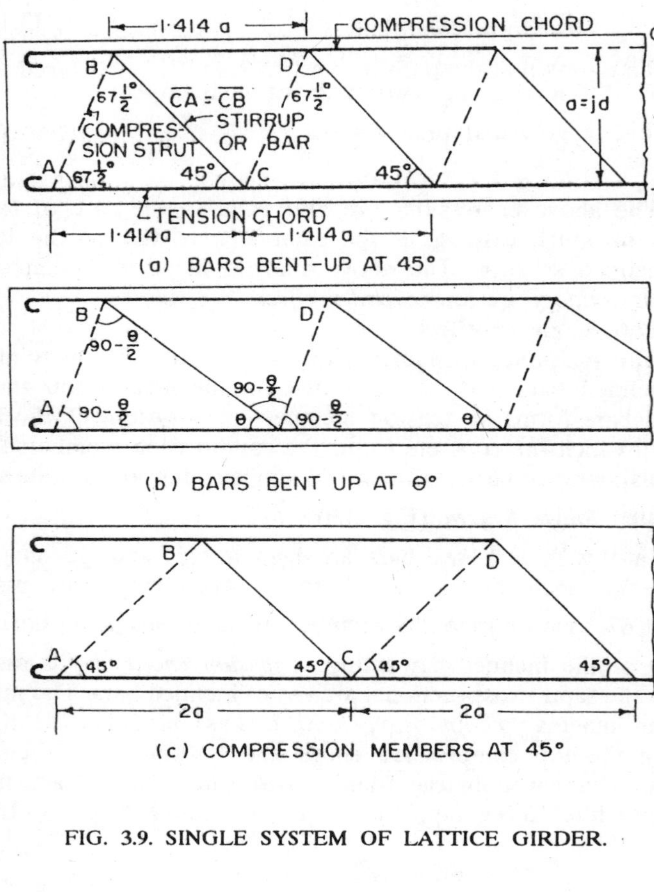

FIG. 3.9. SINGLE SYSTEM OF LATTICE GIRDER.

If, however, the bars are bent up at an angle θ, the compression member is assumed to be inclined at $90° - \frac{\theta}{2}$ with the horizontal as shown in Fig. 3.9(*b*).

Sometimes, the imaginary compression members are assumed to be inclined at 45° as shown in Fig. 3.9(*c*). In this case, the lengths *AB* and *BC* wil! be equal, and the spacing will be equal to 2*a*. The total force in the bar *BC* can be calculated from the force triangle *ABC* as under :

$$\frac{\text{Force in } BC}{\text{Force in } AC} = \frac{\sin 45°}{\sin 90°}$$

$$\therefore \qquad \text{Force in } BC = \frac{A_{sv} \,.\, \sigma_{sv} \,.\, \sin 45°}{\sin 90°} = A_{sv} \frac{\sigma_{sv}}{\sqrt{2}}$$

This shows that the permissible stresses in *BC* is reduced to $\frac{\sigma_{sv}}{\sqrt{2}}$.

This ensures that the stress in the horizontal portion of the inclined bars is equal to the permissible value of σ_{sv}.

Hence the vertical S.F. resisted by *BC*

$$= V_s = A_{sv} \,.\, \frac{\sigma_{sv}}{\sqrt{2}} \,.\, \sin 45°$$

or

$$V_s = 0.707 \,.\, A_{sv} \,.\, \frac{\sigma_{sv}}{\sqrt{2}} \qquad ...(31.8)$$

(*ii*) ***Double system*** **(Fig. 3.10)**

In this system, the bars are bent up at spacing equal to half the spacing of the single system. The compression members are assumed to be inclined at $67\frac{1}{2}°$, but the spacing between them is assumed to be equal to 1.414 *a*, while the spacing between the inclined bars is kept equal to 0.707*a*, as shown in Fig. 3.10.

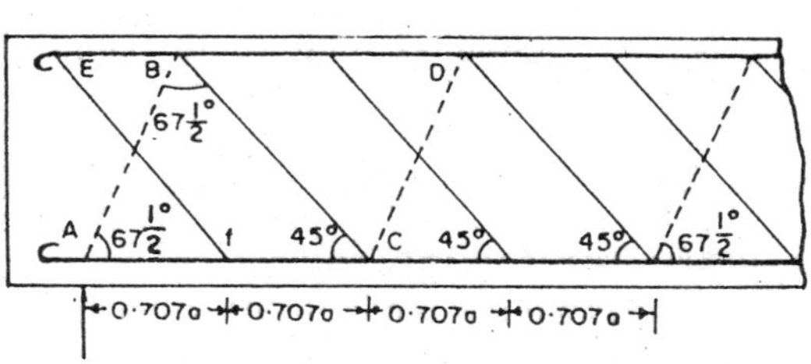

FIG 3.10. DOUBLE SYSTEM.

The shear resistance of such a system will evidently be equal to twice that of the single system *i.e.*, it will be equal to

$$2 \times 0.707 A_{sv} . \sigma_{sv}$$

Combination of vertical stirrups and inclined bars. If the bent up bars are not sufficient to resist the shear, additional vertical stirrups are also provided to act in combination with the inclined bars. In such a case, the total shear resistance is taken equal to the sum of the resistances of the vertical stirrups and the inclined bars.

3.6. INDIAN STANDARD CODE RECOMMENDATIONS (IS : 456–1978)

1. NOMINAL SHEAR STRESS

The revised Indian Standard Code (IS : 456–1978) uses a concept of *nominal shear stress* (τ_v) rather than *shear stress* (q) given by Eq. 3.1. The nominal shear stress in beams or slabs of uniform depth shall be calculated by the following equation :

$$\tau_v = \frac{V}{bd} \qquad ...(3.4)$$

where V = shear force due to design loads

b = breadth of the member, which for flanged section shall be taken as the breadth of web (b_r)

and d = effective depth

In the case of beams of varying depth. the equation shall be modified as :

$$\tau_v = \frac{V \pm \frac{M}{d} \tan \beta}{b \, . \, d} \qquad ...[3.4(a)]$$

where τ_v ,V, b and d have the same meanings as above and is the angle between the top and the bottom edges of the beam. The negative sign in the formula applies when the bending moment M increases numerically in the same direction as the effective depth increases, and the positive sign when the moment decreases numerically in this direction.

(2) DESIGN SHEAR STRENGTH OF CONCRETE

(*a*) *Without Shear Reinforcement*

The permissible shear stress (τ_c) in concrete in beams without shear reinforcement is given in Table 3.1. The permissible shear stress depends upon two factors : grade of concrete and percentage reinforcement $(100 A_s/bd)$.

TABLES 3.1. PERMISSIBLE SHEAR STRESS (τ_c) IN CONCRETE (IS : 456–1978)

$\frac{100A_s}{bd}$	*Permissible shear stress in concrete τ_c N/mm^2 for grade of concrete*					
	M 15	*M 20*	*M 25*	*M 30*	*M 35*	*M 40*
0.25	0.22	0.22	0.23	0.23	0.23	0.23
0.50	0.29	0.30	0.31	0.31	0.31	0.32
0.75	0.34	0.35	0.36	0.37	0.37	0.38
1.00	0.37	0.39	0.40	0.41	0.42	0.42
1.25	0.40	0.42	0.44	0.45	0.45	0.46
1.50	0.42	0.45	0.46	0.48	0.49	0.49
1.75	0.44	0.47	0.49	0.50	0.52	0.52
2.00	0.44	0.49	0.51	0.53	0.54	0.55
2.25	0.44	0.51	0.53	0.55	0.56	0.57
2.50	0.44	0.51	0.55	0.57	0.58	0.60
2.75	0.44	0.51	0.56	0.58	0.60	0.62
3.00 and above	0.44	0.51	0.57	0.60	0.62	0.63

For solid slab the permissible shear strees in concrete shall be $k \cdot \tau_c$ where k has the value given in Table 3.2.

TABLE 3.2 FACTOR *k*

Overall depth of slab (mm)	300 or more	275	250	225	200	175	150 or less
k	1.00	1.05	1.10	1.15	1.20	1.25	1.30

The above requirement does not apply to flat slabs.

Note. A_s is that area of longitudinal tensile reinforcement which continues at least one effective depth beyond the section being considered except at supports where the full area of tension reinforcement may be used.

(b) With Shear Reinforcement

When shear reinforcement is provided the nominal shear stress τ_v in beam shall not exceed $\tau_{c.max}$ given in Table 3.3.

For slabs, τ_v shall not exceed half the value of given in Table 3.3.

TABLE 3.3. MAXIMUM SHEAR STRESS : $\tau_{c.max}$ N/mm^2 (IS : 456–1978)

Concrete grade	M 15	M 20	M 25	M 30	M 35	M 40
τ_{cmax} *(N/mm^2)*	1.6	1.8	1.9	2.2	2.3	2.5

(3) MINIMUM SHEAR REINFORCEMENT

When τ_v is less than τ_c given in Table 3.1, minimum shear reinforcement shall be provided as explained below:

Minimum shear reinforcement in the form of stirrups shall be provided such that

$$\frac{A_{sv}}{b \cdot s_v} > \frac{0.4}{f_y} \qquad \text{...(3.19)}$$

where A_{sv} = total cross-sectional area of stirrup legs effective in shear.

τ_v = stirrup spacing along the length of the member

b = breadth of beam or breadth of the web of flanged beam and,

f_y = characteristic strength of the stirrup reinforcement, in N/mm^2, which shall not be taken greater than 415 N/mm^2. The values of f_y for three types of steel reinforcement are given in Table 1.19.

However, in members of minor structural importance such as lintels or where the maximum shear stress calculated is less than half the permissible value, this provision need not be complied with.

The term *characteristic strength* means that value of the strength below which not more than 5 percent of the test results are expected to fall. *Until the relevant Indian Standard specifications* for reinforcing steel are modified to include the concept of characteristic strength, the characteristic value shall be assumed as the minimum yield/0.2

percent proof stress specified in the relevant Indian Standard specifications.

Eq. 3.19 can be rearranged in the following form to give the maximum spacing of stirrups :

$$s_v \leq \frac{2.5\, A_{sv} \cdot f_y}{b} \qquad ...[3.19(a)]$$

Eq. 3.19 may also be written in the form

$$\frac{A_{sv} \cdot f_y \cdot d}{s_v} = 0.4\, bd \qquad ...[3.19(b)]$$

In the above expression, the left hand term $\frac{A_{sv} \cdot f_y \cdot d}{s_v}$ is equal to the shear resistance V_s of the vertical stirrups while the right hand term 0.4 bd represent $\tau_c \cdot bd$ (*i.e.* shear reinforcement V_c for concrete).

Thus Eq. 3.19 ensures provision of reinforcement for a net shear stress of 0.4 N/mm^2.

The *minimum shear reinforcement* for two types of steel bars for b=300 mm works out as under

(a) *Mild steel bars*

Taking f_y = 250 N/mm^2

$$s_v = \frac{2.5\, A_{sv} \times 250}{300} = 2.083\, A_{sv} \text{ mm}$$

Taking 8 mm dia bars,

$$A_{sv} = 2 \times \frac{\pi}{4}(8)^2 = 100.5 \text{ mm}^2$$

Hence $s_v = 2.083 \times 100.5 = 209$ mm.

(b) *HYSD bars*

Taking f_y = 415 N/mm^2

$$s_v = \frac{2.5\, A_{sv} \times 415}{300} = 3.458\, A_{sv} \text{ mm}$$

For 8 mm dia. bars,

$$s_v = 3.458 \times 100.5 = 348 \text{ mm.}$$

(4) DESIGN OF SHEAR REINFORCEMENT

When τ_v exceeds τ_c given in Table 3.1, shear reinforcement shall be provided in any of the following forms :

(i) Vertical stirrups

(ii) Bent-up bars along with stirrups and

(*iii*) Inclined stirrups.

Where bent up bars are provided, their contribution towards shear resistance shall not be more than half that of the total shear reinforcement.

Shear reinforcement shall be provided to carry a shear equal to $V - \tau_c \,.\, bd$. The strength of shear reinforcement V_s shall be calculated as below :

(*i*) For vertical stirrups

$$V_s = \frac{\sigma_{sv} \,.\, A_{sv} \,.\, d}{s_v} \qquad ...(3.20)$$

(*ii*) For inclined stirrups or a series of bars bent-up at different cross-section :

$$V_s = \frac{\sigma_{sv} \,.\, A_{sv} \,.\, d}{s_v} (\sin \alpha + \cos \alpha) \qquad ...(3.21)$$

(*iii*) For single bar or single group of parallel bars, all bent up at the same cross-section

$$V_s = \sigma_{sv} \,.\, A_{sv} \,.\, \sin \alpha \qquad ...(3.22)$$

where :

A_{sv} = total cross-sectional area of stirrups legs or bent up bars within a distance s_v

s_v = spacing of the stirrups or bent up bars along the length of the member

τ_c = design shear strength of the concrete.

b = breadth of the member which for flanged beams, shall be taken as breadth of the web b_w.

σ_{sv} = permissible tensile stress in shear reinforcement which shall not be taken greater than 230 N/mm^2.

α = angle between the inclined stirrups or bent up bar and the axis of the member, not less than 45° and

d = effective depth.

Note : Where more than one type of shear reinforcement is used to reinforce the same portion of the beam, the total shear resistance shall be computed as the sum of the resistance of the various types separately. The area of stirrups shall not be less than the minimum specified in para 3 above.

(5) MAXIMUM SPACING OF SHEAR REINFORCEMENT

The maximum spacing of shear reinforcement measured along the axis of the member shall not exceed 0.75 *d* for vertical stirrups and *d* for inclined stirrups at 45°, where *d* is the effective depth of the section under consideration. In no case shall the spacing exceed 450 mm.

(6) SUMMARY

(*i*) If $\tau_v \le \frac{1}{2}\tau_c$

or $V \le \frac{1}{2}V_c$ (where $V_c = \tau_c \cdot bd$)

No shear reinforcement is really needed.

(*ii*) If $\frac{1}{2}V_c \le V \le V_c$

Provide nominal reinforcement given by Eq. 3.19.

However, the maximum spacing is restricted to 0.75 *d* or 450 mm, whichever is less.

(*iii*) If $V_c \le V \le V_{cmax}$

Design the transverse reinforcement for the net S.F.

$$F = (V - V_c).$$

If ΣV_s is the sum of the resistances of various types of shear reinforcements (Eq. 3.20, 3.21, 3.22) then ΣV_s should be equal to or greater than $(V - V_c)$.

However, the maximum spacing (s_v) is restricted to 0.75 *d* or 450 mm, whichever is less. Also the contribution of bent-up bars should not be more than half of the total shear reinforcement and the area of stirrups should not be less than the one provided by Eq. 3.19.

(*iv*) If $V \ge V_{c.max}$ redesign the web of the section such that $V_{c.max}$ becomes equal to or more than external shear V.

Example 3.1 *A reinforced concrete beam 250 mm wide and 400 mm effective depth is subjected to a shear force of 95 kN at the supports. The tensile reinforcement at the support is 0.5 percent. Find the spacing of 12 mm diameter 2-legged stirrups to resist the shearing stress at supports, for M 15 concrete. Use (a) mild steel reinforcement (b) HYSD bars. Also, design the minimum shear reinforcement at the mid-span of the beam.*

Solution.

Given $p = \dfrac{100 A_{st}}{bd} = 0.5\%$

Hence from Table 3.1, for M 15 concrete, we get

$$\tau_c = 0.29 \text{ N/mm}^2$$

Also from Table 3.3, maximum shear stress

$$\tau_{c.max} = 1.6 \text{ N/mm}^2$$

The nominal shear stress in the beam is

$$\tau_v = \frac{V}{bd} = \frac{95 \times 1000}{250 \times 400} = 0.95 \text{ N/mm}^2$$

This is greater than the permissible shear stress τ_c but less than the maximum shear stress $\tau_{c.max}$.

The shear reinforcement is provided to carry a shear V_s equal to $V - \tau_c . bd$.

$$\therefore \quad V_s = V - \tau_c bd = 95000 - 0.29 \times 250 \times 400$$

$$= 66000 \text{ N}$$

(*a*) **Mild steel bars** (Fe 250) $\sigma_{sv} = 140$ N/mm^2

The spacing of stirrups at supports is given by Eq. 3.20, rearranged as follows :

$$s_v = \frac{\sigma_{sv} \, d \, A_{sv}}{V_s}$$

$$= \frac{140 \times 400}{66000} A_{sv} = 0.848 \, A_{sv} \text{ mm.}$$

Using 12 mm Φ bars two legged stirrups,

$$A_{sv} = 2 \frac{\pi}{4} (12)^2 = 226.2 \text{ mm}^2$$

$$\therefore \quad s_v = 0.848 \times 226.2 = 191.9 \text{ mm}$$

Hence provide stirrups at spacing of 190 mm *c/c* at the supports.

For nominal (minimum) shear reinforcement at mid span, the spacing at 12 mm Φ 2-lgd stirrups is given by Eq. 3.19(*a*).

$$s_v = \frac{2.5 \, A_{sv} f_y}{b}$$

$$= \frac{2.5 \times 226.2 \times 250}{250} = 565 \text{ mm.}$$

However, max. spacing=0.75 $d = 0.75 \times 400 = 300$ mm.

Thus, use of 12 mm dia stirrups will be uneconomical. Using 10 mm dia bars for stirrups,

$$A_{sv} = 2 \times \frac{\pi}{4} (10)^2 = 157.1 \text{ mm}^2$$

$$\therefore \quad s_v = \frac{2.5 \, A_{sv} . f_y}{b}$$

$$= \frac{2.5 \times 157.1 \times 250}{250} = 393 \text{ mm.}$$

Subject to a maximum of $0.75 \, d = 0.75 \times 400 = 300$ mm.

Hence use 10 mm dia 2-lgd. stirrups at 300 mm spacing at the location where nominal reinforcement is required.

(*b*) HYSD bars

$$s_v = \frac{\sigma_{sv} \, . \, d \, . \, A_{sv}}{V_s}$$

$$= \frac{230 \times 400}{66000} A_{sv} = 1.3939 \, A_{sv} \text{ mm.}$$

Using 12 mm Φ bars two legged stirrups,

$$A_{sv} = 2 \times \tfrac{\pi}{4}(12)^2 = 226.2 \text{ mm}^2$$

$$\therefore \quad s_v = 1.3939 \times 226.2 = 315.3 \text{ mm}$$

subject to a maximum of $0.75\, d = 0.75 \times 400 = 300$ mm.

Hence provide the stirrups @ 300 mm c/c.

For nominal (minimum) shear reinforcement, let us use 8 mm dia bars.

$$A_{sv} = 2 \times \tfrac{\pi}{4}(8)^2 = 100.53 \text{ mm}^2$$

The spacing is given by Eq. 3.19 (*a*)

$$s_v = \frac{2.5\, A_{sv} \, . \, f_y}{b}$$

$$= \frac{2.5 \times 100.53 \times 415}{250} = 417 \text{ mm.}$$

subject to a maximum spacing equal to $0.75d = 300$ mm

Hence provide 8 mm dia. 2 *lgd* stirrups at 300 mm spacing at the location where nominal reinforcement is required.

Example 3.2 *A reinforced concrete beam has 300 mm width and 500 mm effective thickness. The shear reinforcement consists of 8 mm diameter 4-legged stirrups spaced at 100 mm centre to centre at the supports. If the beam is subjected to a shear force of 120 kN at the ends, calculate the maximum shear stress developed in the shear reinforcement. The beam has 0.5% reinforcement at the ends.*

Solutin From Table 3.1, for $\frac{A_s}{bd} \times 100 = 0.5$,

$$\tau_c = 0.29 \text{ N/mm}^2$$

$$\therefore \quad V_c = \tau_c \, . \, bd = 0.29 \times 300 \times 500 = 43500 \text{ N}$$

$$\text{Hence} \quad V_s = V - V_c = 120000 - 43500 = 76500 \text{ N}$$

Now, for four legged stirrups.

$$A_{sv} = 4 \times \frac{\pi}{4}(8)^2 = 201.1 \text{ mm}^2.$$

Let t_{sv} = stress developed in the stirrups.

Then, $$V_s = \frac{A_{sv} \cdot t_{sv} \cdot d}{s_v}$$

or $$t_{sv} = \frac{V_s \cdot s_v}{A_{sv} \cdot d} = \frac{76500 \times 100}{201.1 \times 500}$$

$$= \mathbf{76 \ N/mm^2.}$$

Example 3.3 *A simply supported beam, 300 mm wide and 500 mm effective depth carries a uniformly distributed load of 50 kN/m including its own weight, over an effective span of 4 metres. Design the shear stirrups in the form of vertical stirrups. Use M 15 concrete. Take $\sigma_{st} = \sigma_{sv} = 140$ N/mm² and $f_y = 250$ N/mm². Assume that the beam contains 0.75% reinforcement throughout the length.*

Solution The nominal shear stress in the beam is given by

$$\tau_v = \frac{V}{bd} = \frac{V}{300 \times 500} = \frac{V}{150000} \text{ N/mm}^2$$

At the ends, $$V = \frac{wL}{2} = \frac{50000 \times 4}{2} = 100000 \text{ N}$$

$$\tau_v = \frac{100000}{150000} = 0.67 \text{ N/mm}^2$$

From Table 3.1, for $\frac{100 A_s}{bd} = 0.75$ %, permissible shear stress $\tau_c = 0.34$ N/mm². Since the nominal shear stress is more than this, shear reinforcement is necessasry at the end section.

Using 10 mm Φ 2-lgd vertical stirrups,

$$A_{sv} = 2\frac{\pi}{4}(10)^2 = 157 \text{ mm}^2$$

Shear resistance to be provided by the stirrups at end section is

$$V_s = V - \tau_c bd = 100000 - 0.34 \times 300 \times 500$$
$$= 49000 \text{ N}$$

$$\therefore \quad s_v = \frac{\sigma_{sv} \cdot A_{sv} \cdot d}{V_s} = \frac{140 \times 157 \times 500}{49000} = 224.3 \text{ mm}$$

$$\approx 220 \text{ mm.}$$

Maximum spacing of stirrups is given by

$$s_v = \frac{2.5\, A_{sv} \,.\, f_y}{b} = \frac{2.5 \times 157 \times 250}{300} = 327 \quad ...(1)$$

$$\approx 320 \text{ mm.}$$

In no case should the spacing exceed $0.75\, d = 0.75 \times 500$ $= 375$ mm.

Thus, the spacing is to be varied from 220 mm at end section to 320 mm at a section distant x (say) from the mid span. Let us locate this section, where shear is equal to V_x where

$$V_x = \frac{V}{2}(x) = \frac{100000}{2} x = 50000\, x$$

$$\therefore \quad V_s = V_x - \tau_c\, bd = 50000\, x - 0.34 \times 300 \times 500$$

$$= 50000\, x - 51000$$

$$\therefore \quad s_v = \frac{\sigma_{sv} \,.\, A_{sv} \,.\, d}{V_s} = \frac{140 \times 157 \times 500}{50000\, x - 51000} \quad ...(2)$$

Equating (1) and (2) we get

$$\frac{140 \times 157 \times 500}{50000\, x - 51000} = 320$$

From which $x \approx 1.7$ m. Thus vary the spacing from 220 mm c/c at supports to 320 mm c/c at 1.7 m from the mid span. For the remaining length, provide the stirrups at 320 mm c/c.

Example 3.4 *Redesign the shear stirrups of the beam of example 3.3, using HYSD bars, taking* $\sigma_{st} = \sigma_{sv} = 230$ *N/mm*^{2} *and* $f_y = 415$ *N/mm*^{2}.

Solution.

$$V = \frac{w\,L}{2} = \frac{50000 \times 4}{2} = 100000 \text{ N}$$

$$\tau_v = \frac{V}{bd} = \frac{100000}{300 \times 500} = 0.67 \text{ N/mm}^2$$

For $\frac{100\, A_s}{bd} = 0.75\ \%$, we get $\tau_c = 0.34$ N/mm^2 from Table 3.1. Hence shear reinforcement is necessary at the ends. Using 8 mm dia. 2-*lgd* vertical stirrups,

$$A_{sv} = 2\,\frac{\pi}{4}(8)^2 = 100.53 \text{ mm}^2$$

Now

$$V_s = V - \tau_c\, bd = 100000 - 0.34 \times 300 \times 500$$

$$= 49000 \text{ N}$$

$$\therefore \quad s_v = \frac{\sigma_{sv} \,.\, A_{sv} \,.\, d}{V_s} = \frac{230 \times 100.53 \times 500}{49000}$$

$$= 236 \text{ mm } (= 230 \text{ mm, say})$$

Maximum spacing of stirrups is given by

$$s_v = \frac{2.5 A_{sv} \cdot f_y}{V_s} = \frac{2.5 \times 100.53 \times 415}{300} = 348 \text{ mm}$$

$$\approx 340 \text{ mm} \qquad ...(1)$$

In no case should the spacing exceed $0.75\,d = 0.75 \times 500$

$$= 375 \text{ mm.}$$

Hence the spacing is to be varied from 230 mm at end section to 340 mm at a section distant x m from the mid span where minimum shear reinforcement, given by eq. (i) is to be provided.

At that section,

$$V_x = \frac{V}{2}(x) = \frac{100000}{2} x = 50000\,x$$

$$\therefore \quad V = V_x - \tau_c\, bd = 50000\,x - 0.34 \times 300 \times 500$$

$$= 50000\,x - 51000$$

$$\therefore \quad s_v = \frac{\sigma_{sv} \cdot A_{sv} \cdot d}{V_s} = \frac{230 \times 100.53 \times 500}{50000\,x - 51000} \qquad ...(2)$$

Equating (1) and (2) , we get

$$\frac{230 \times 100.53 \times 500}{50000\,x - 51000} = 340$$

From which we get $x = 1.7$ m

Thus vary the spacing from 230 mm c/c at supports to 340 mm c/c at 1.7 m from mid-span. For the remaining length, provide the stirrups at 340 mm c/c.

Example 3.5 *A singly reinforced concrete beam has an effective depth of 600 mm and width of 300 mm and is reinforced with 8 bars of 20 mm diameter at the centre of the span. From the requirements of bending stresses, it is known that four of these bars can be safely bent up at a distance of 1.6 m from the support. The maximum S.F. at the ends is 160 kN and the distance from the either support shear reinforcement is needed is 1.3 m. Explain how you would use the bent up bars as diagonal tension reinforcement. Also design the stirrup reinforcement required. Take* σ_{sv} *= 140 N/mm*^{2}*. Use M 15 concrete.*

Solution. For M 15 concrete

$$j = 0.867 \text{ with m.s. bars}$$

$\therefore$ Lever arm $a = jd = 0.867 \times 600 = 520$ mm

Let us bend two bars at a distance

$$= 1.414\,a = 1.414 \times 520 = 735 \text{ mm from the support.}$$

Shear resistance of the two bars bent at 45° is

$$V_{s1} = A_{sv} \cdot \sigma_{sv} \sin 45° = 0.707 \cdot A_{sv} \cdot \sigma_{sv}$$

$$= 0.707 \left\{ 2 \times \frac{\pi}{4} (20)^2 \right\} \times 140$$

$$= 62190 \text{ N.}$$

However, maximum contribution towards shear resistance shall not be more than half the total shear reinforcement.

$$\text{Also, } \frac{100 A_s}{bd} = \frac{100 \times 4 \left(\frac{\pi}{4}\right) (20)^2}{300 \times 600} = 0.7\%$$

Hence from Table 3.1, $\tau_c = 0.33 \text{ N/mm}^2$

$$\therefore \quad V_c = \tau_c \cdot bd = 0.33 \times 300 \times 600 = 59400 \text{ N.}$$

$$\therefore \quad V_s = V - V_c = 160000 - 59400 \text{ N} = 100600 \text{ N}$$

Max. Contribution V_{s1} due to bent up bars $= \frac{1}{2} V_s$

$$= \frac{1}{2} \times 100600 = 50300 \text{ N.}$$

V_{s2} from stirrups $=50300$ N.

Using 8 mm Φ two-legged stirrups

$$A_{sv} = 2 \frac{\pi}{4} (8)^2 = 100.5 \text{ mm}^2$$

$$s_v = \frac{\sigma_{sv} \cdot A_{sv} \cdot d}{V_{s2}} = \frac{140 \times 100.5 \times 600}{50300} = 167.8 \text{ mm.}$$

This spacing should not be more than s_v corresponding to minimum reinforcement requirement.

i.e. $$s_v \leq \frac{2.5 \cdot A_{sv} \cdot f_y}{b} = \frac{2.5 \times 100.5 \times 250}{300} = 209 \text{ mm.}$$

Also, spacing should not be more than 0.75 d ($=0.75 \times 600=450$ mm) nor more than 450 mm.

Hence provide 2-*lgd* 8 mm dia. stirrups @ 160 mm c/c.

Let us bend the other two bars at a distance of $2 \times 1.414\, a = 2 \times 735 = 1470$mm from the support. This can safely be done since from B.M. point of view, we can bend them even at a distance of 1600 mm from the support. Effective length of these bars resisting shear $= 2 \times 1.414\, a = 1470$ mm. This is more than the length of 1300 mm in which shear reinforcement is required. Hence no extra shear reinforcement is required. However, provide minimum shear reinforcement, in the form of 8 mm Φ 2-*lgd* stirrups @ 209 mm (≈ 200 mm, say) c/c for the remaining length. The arrangement is shown in Fig 3.11.

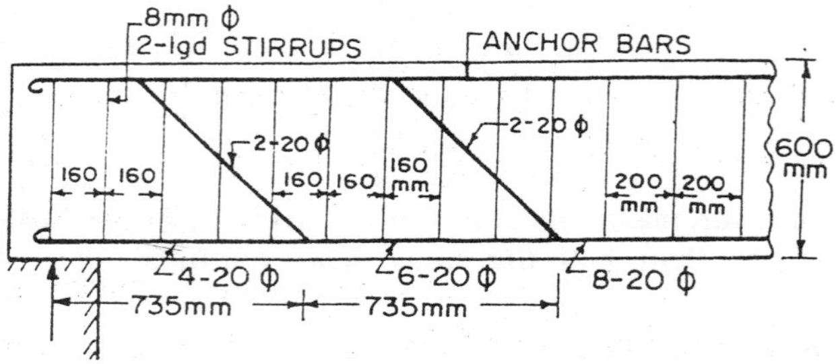

FIG. 3.11

Example 3.5 *A simply supported beam, 300 mm wide and 600 mm effective depth carries a uniformly distributed load of 70 kN/m including its own weight, over an effective span of 6 m. The reinforcement consists of 5 bars of 25 mm diameter. Out of these, two bars can be safely bent at 1 m distance from the support. Design suitable shear reinforcement for the beam. Use M 15 concrete. Take* $\sigma_{st} = \sigma_{sv} = 140$ *N/mm*^{2} *and* $f_y = 415$ *N/mm*^{2}.

Solution.

Max. S.F. $V = \frac{wL}{2} = \frac{70000 \times 6}{2} = 210000$ N

Normal shear stress $= \frac{V}{bd} = \frac{210000}{300 \times 600} = 1.17$ N/mm^2

Area of reinforcement $= (5-2)\frac{\pi}{4}(25)^2 = 1473$ mm^2

$$p = \frac{1473}{300 \times 600} \times 100 \approx 0.82\%$$

Hence from Table 3.1, for $p = 0.82\%$ and M 15 concrete, we get

$$\tau_c = 0.35 \text{ N/mm}^2$$

From Table 3.3, $\tau_{cmax} = 1.6$ N/mm^2

Since the nominal shear stress is less than this, the beam is safe.

For M 15 concrete, let $j = 0.87$

Lever arm $jd = 0.87 \times 600 = 522$ mm.

Let us bend 2 bars at 45° at a distance $= 1.414 \times jd = 1.414 \times 522 \approx 730$ mm from the support.

Area of cross-section of 2 bent up bars

$$= 2\frac{\pi}{4}(25)^2 = 982 \text{ mm}^2$$

Shear resistance of concrete $= V_c = \tau_c bd$

$$= 0.35 \times 300 \times 600 = 63000 \text{ N}$$

Shear resistance to be provided by shear reinforcement

$$= 210000 - 63000 = 147000 \text{ N}$$

Shear resistance of two bent up bars $= \sigma_{sv} . A_{sv} . \sin\alpha$

$$= 140 \times 982 \times 0.707 \approx 97173 \text{ N}$$

However the maximum shear resistance to be assigned to the bent up bars should not exceed half the resistance to be provided by shear reinforcement, *i.e.* it should not exceed $\frac{1}{2} \times 147000 = 73500$ N.

Hence, the remaining resistance to be provided by the vertical stirrups = 73500 N

Thus V_s = 73500 N

Hence, the spacing of vertical stirrups is given by

$$s_v = \frac{\sigma_{sv} . A_{sv} . d}{V_s} = \frac{140 \times A_{sv} \times 600}{73500}$$

or $s_v = 1.143 A_{sv}$.

Using 10 mm Φ *2lgd* stirrups, having

$$A_{sv} = 2\frac{\pi}{4}(10)^2 = 157 \text{ mm}^2$$

$$s_v = 1.143 \times 157 \approx 180 \text{ mm}$$

The maximum spacing of stirrups is given by

$$s_v = \frac{2.5 . A_{sv} . f_y}{b} = \frac{2.5 \times A_{sv} \times 250}{300} = 2.083 A_{sv}$$

or $s_v = 2.083 \times 157 = 325$ mm for 10 mm Φ 2 *lgd* stirrups.

At distance greater than 730 mm (0.73 m) from the supports, bent up bars are not available and the sheaı resistance is available only from the stirrups as well as from concrete.

At 0.73 m from supports (or 2.27 m from mid span), the S.F. is given by

$$V = \frac{210000}{3} \times 2.27 = 158900 \text{ N}$$

$$A_{st} = 5\frac{\pi}{4}(25)^2 = 2454 \text{ mm}^2$$

$$\therefore \qquad p = \frac{2454}{300 \times 600} \times 100 \approx 1.36\%$$

Hence from Table 3.1, $\tau_c = 0.41\ \text{N/mm}^2$.

$\therefore$ Shear resistance of concrete

$$= \tau_c\, bd = 0.41 \times 300 \times 600 = 73800\ \text{N}$$

Hence shear resistance to be provided by stirrups is given by

$$V_s = V - \tau_c\, bd = 158900 - 73800 = 85100\ \text{N}$$

Thus, the spacing of 10 mm Φ 2-*lgd* stirrups is

$$s_v = \frac{\sigma_{sv} \,.\, A_{sv} \,.\, d}{V_s} = \frac{140 \times 157 \times 600}{85100} = 155\ \text{mm.}$$

$$\approx 150\ \text{mm (say).}$$

To locate the section where maximum spacing of 325 mm is provided, proceed as follows :

Let the distance of the section be x m from mid-span.

$$V = \frac{210000}{3} \times x = 70000x\ \text{N}$$

Shear resistance of concrete.

$$= \tau_c\, bd = 0.41 \times 300 \times 600 = 73800\ \text{N}$$

$$\therefore \qquad V_s = 70000x - 73800$$

$$s_v = \frac{\sigma_{sv} \,.\, A_{sv} \,.\, d}{V_s} = \frac{140 \times 157 \times 600}{70000x - 73800}\ \text{mm.}$$

Equating this to 325 mm we get

$$\frac{140 \times 157 \times 600}{70000x - 73800} = 325\ \text{mm.}$$

which gives $x = 1.63$ m from mid-span (or 1.37 m from supports).

Thus, the shear reinforcement may be summarised as follows after slight adjustment in the spacings :

(a) At supports : 10 mm Φ 2-*lgd* stirrups @ 180 mm c/c plus 2 bent up bars from the main reinforcement.

(b) At 0.73 m from supports to 1.37 m from supports : vary the spacing of the 10 mm Φ 2-*lgd* stirrups from 150 mm c/c to 305 mm c/c.

(c) Beyond 1.37 m from supports : 10 mm Φ 2-*lgd* stirrups @ 325 mm c/c.

Provided 2-12 mm Φ holding bars at top.

Note. At locations where τ_v is less than half τ_c, the provisions of minimum shear reinforcment need not be complied with, as per IS : 456–1978.

Fig. 3.12 shows the shear reinforcement.

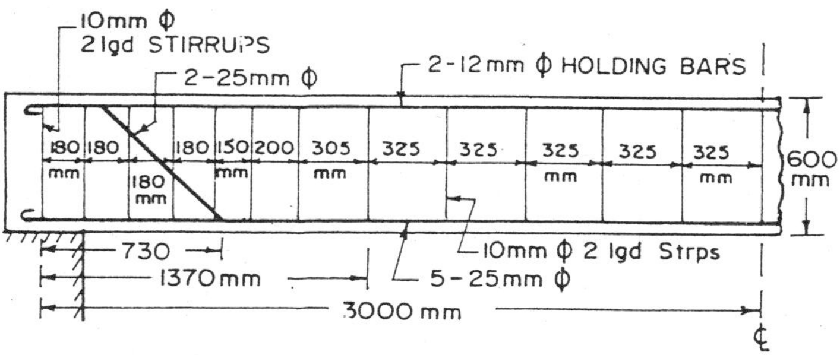

FIG. 3.12

3.7. BOND, ANCHORAGE AND DEVELOPMENT LENGTH

One of the most important assumptions in the behaviour of reinforced concrete structures is that there is proper bond between concrete and reinforcing bars. When steel bars are embedded in concrete, the concrete, after setting, *adheres* to the surface of the bars and thus resist any force that tends to pull or push this rod. The intensity of this adhesive force is called *bond stress*. Thus the term 'bond' is used to describe the means by which slip between steel and concrete is prevented. The *bond stresses* are the *longitudinal shearing stresses* acting on the surface between the steel and concrete.

The bond between steel and concrete comprise of three resistances :

(*i*) Pure adhesive resistance,

(*ii*) Frictional resistance, and

(*iii*) Mechanical resistance,

Pure adhesive resistance is provided by the relatively weak adhesion of the 'chemical gum' produced by concrete during setting. Mechanical friction is provided by the shrinkage of the concrete thereby gripping the steel bars. When loaded the adhesive resistance is broken first. But even after the adhesion is broken, friction between the two materials continues to provide a considerable bond resistance. The hooks and other means of *anchoring* steel bar do not prevent the initial slipping ; but they delay the collapse of structure as the member acts as a tied arch with the uncracked concrete representing the arch and the anchored bars as the tie rods. The frictional resistance is low for a smooth bar surface, such as that of cold-rolled steel, and is higher for rough surfaces. Hence it is essential that the bars must be clean amd free from dirt, grease, scale and loose rust so that pure adhesive resistance and frictional resistance can be developed fully. Anything that destroys the ability of the concrete to grip the steel may prove to be serious because it will prevent the stress in the latter from being fully developed.

The *mechanical resistance* is provided by *deformed bars* only, and not by the plain bars. The deformed bars have lugs, or corrugations, and give higher bond resistance by providing an interlock between steel and concrete. In the case of deformed bars, adhesion and friction become minor elements, and the bond strength is primarily dependent on the bearing (compressive) strength of the concrete against the lugs or corrugations. The plain bars slip easily and cause localised serious cracking while the deformed bars tend to distribute the deformation in the concrete in a series of closely spaced hair cracks.

Bond stress may be visualised as *interface shear*. The bond stress in reinforced concrete *members* arises due to two distinct situations (*a*) from the change in tensile force carried by the bar, along its length, due to *change in the bending moment* along the length of the member, and (*b*) from the *anchorage of bar* in case of tension or compression. The first type is known as *flexural bond stress* (τ_{bf}) and the second type is known as *anchorage bond stress.* (τ_{ba} *or* τ_{bd})

FLEXURAL BOND STRESS

Fig. 3.13 shows two section *ab* and *cd*, spaced *dx* apart of a reinforced concrete beam. Let *M* be the bending moment at section *ab* and $M + dM$ be the bending moment at section *cd*. Similarly let *T* and $T + dT$ be the tensile forces developed in steel reinforcement at *ab* and *cd* respectively.

Now $$M = T \, . \, jd$$

and $$M + dM = (T + dT) \, jd$$

Hence $$dM = dT \, . \, jd$$

or $$dT = \frac{dM}{jd} \qquad ...(i)$$

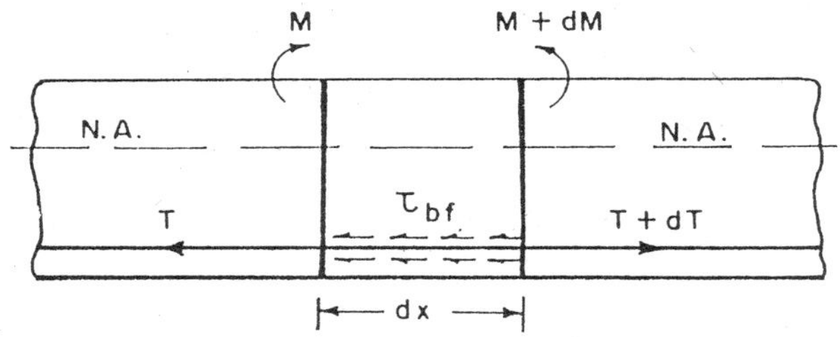

FIG. 3.13

The difference in the tensile force (dT) in steel bars between sections is resisted by bond stress developed in the length dx.

Let τ_{bf} = bond stress developed in concrete around the steel reinforcement.

$\Sigma 0$ = sum of the perimeters of the steel bars resisting tension.

Total surface area of bars $= dx\,\Sigma\,0$.

Hence $\tau_{bf}\,[dx\,\Sigma\,0] = dT$ ...(*ii*)

Substituting the value of dT in (*i*), we get.

$$\tau_{bf}\,(dx\,\Sigma 0) = \frac{dM}{jd}$$

or

$$\tau_{bf} = \frac{dM}{dx}\cdot\frac{1}{jd\,\Sigma\,0}$$

But $\dfrac{dM}{dx}$ = shear force = V

Hence $$\tau_{bf} = \frac{V}{jd\,\Sigma 0} = \frac{V}{a\,\Sigma 0} \quad ...(3.23)$$

The stress τ_{bf} at a particular section is called *local bond stress*

General expression for bond stress

Eq 3.23 is applicable only when all the reinforcing bars placed are equidistant from the neutral axis. However, there are cases, such as the case of a beam of circular section, in which it is not possible to place the bars equidistant from the N.A. Also, Eq. 3.33 is not applicable for finding bond stress in the reinforcement provided in the compression zone of the beam. In such cases, the expression derived below can be used to find bond stress in the individual bars.

Let a reinforcing bar of diameter Φ be placed at a distance y from the N.A. Let dM be the change in bending moment in a length dx.

$\therefore$ Change in the stress in the bar due to change in the bending moment dM is equal to $m. \frac{dM}{I}.y$, where I the moment of inertia of the equivalent section and m is the modular ratio.

$\therefore$ Force causing slip = stress × area

$$= \frac{\pi}{4}\Phi^2 . m\frac{dM}{I}.y \qquad ...(i)$$

Bond force resisting the slip $= \tau_{bf} . \pi\,\Phi . dx.$ ...(*ii*)

Equating the two, we get

$$\tau_{bf} . \pi\,\Phi . dx = \frac{\pi}{4}\Phi^2 . \frac{dM}{I}y$$

or

$$\tau_{bf} = \frac{dM}{dx}\,\frac{my}{I}.\frac{\Phi}{4}$$

But

$$\frac{dM}{dx} = \text{shear force} = V$$

$\therefore$

$$\tau_{bf} = \frac{V}{4I}\,my . \Phi \qquad ...(3.24)$$

ANCHORAGE BOND STRESS : DEVELOPMENT LENGTH

Eq. 3.23 and 3.24 give the localised flexural bond stress at a *particular section*. However, bond stresses are *developed* over a specified length and any attempt to visualise bond on the basis of stresses at a particular cross-section is bound to be erroneous. In the case of bond, it has been recently concluded that the concept of *anchorage bond* is more meaningful than concept of flexural bond stresses of localised nature. *Current design methods disregard such*

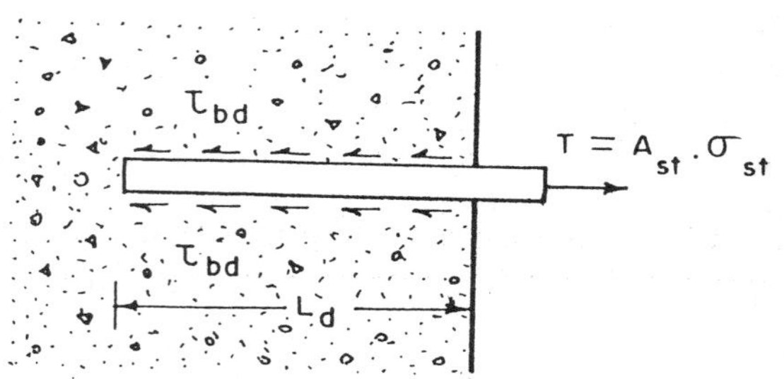

FIG. 3.14

local bond stresses, even though they may result in localized slip between steel and concrete adjacent to the cracks. Thus, the concepts of *development length* (in the case of flexure) and *anchorage length* (in case of tension or compression) have gained prominence.

Fig. 3.14 shows a steel bar embedded in concrete, and subjected to a tensile force *T*. Due to this force, there will be a tendency of the bar to slip out and this tendency is resisted by the bond stress developed over the perimeter of the bar, along its length of embedment. Let us assume that average, uniform bond stress is developed along the length. The required length, necesssary to develop full resisting force, is called *anchorage length* in case of axial tension(or compression) and *development length* in the case of flexural tension, and is designated by symbol L_d.

Hence if Φ is the nominal diameter of the bar, we have

$$\frac{\pi}{4}\Phi^2 . \sigma_{st} = \tau_{bd} . \pi \Phi . L_d$$

or

$$L_d = \frac{\Phi . \sigma_{st}}{4\,\tau_{bd}} \qquad ...(3.25)$$

The permissible bond stress τ_{bd} is specified in the Code (Table 3.4) for various grades of concrete.

Eq. 3.25 shows that a bar must extend a length L_d beyond any section at which it is required to develop its full strength so that sufficient bond resistance is mobilized.

For M 15 concrete, $\tau_{bd} = 0.6$ N/mm^2 while for mild steel bars, $\sigma_{st} = 140$ N/mm^2. Hence from Eq. 3.25

$$L_d = \frac{140\,\Phi}{4 \times 0.6} = 58.3\,\Phi$$

The *development length* for other grades of concrete, and for mild steel and HYSD bars, is given in Table 3.5.

HOOKS AND BENDS

Quite often space available at the end of beam is limited to accommodate the full development length L_d. In that case, hooks or bends are provided. The anchorage value (Le) of hooks or bend is accounted as contribution to the development length L_d.

Fig. 3.15 (ai) shows a semi-circular hook, fully dimensioned, with respect to a factor K. The value of K is taken as 2 in the case of mild steel conforming to IS : 432–1966, (specifications for

Mild-Steel and Medium Tensile Steel bars and Hard-Drawn steel wires for concrete reinforcement) or IS : 1139–1959, (specifications for 'Hot rolled mild steel and medium tensile steel deformed bars for concrete reinforcement'). The hook with $K = 2$ is shown in Fig. 3.15(*aii*) with equivalent horizontal length of the hook. For

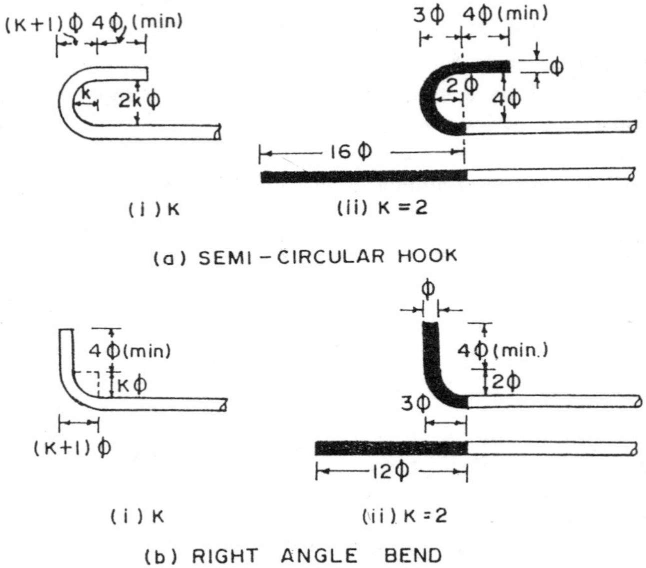

FIG. 3.15

the case of Medium Tensile Steel conforming to IS : 432–1966 or IS : 1139–1959, K is taken as 3. In the case of cold worked steel conforming to IS : 1986–1961, (specifications for cold twisted steel bars for concrete reinforcement), K is taken as 4. In the case of bars above 25 mm, however, it is desirable to increase the value of K to 3, 4 and 6 respectively.

Fig. 3.15 (*b*) shows a right angled bend, with dimensions in terms of K, the value of which may be taken as 2 for ordinary mild steel for diameters below 25 mm and 3 for diameters above 25 mm.

In the case of deformed bars, the value of bond stress for various grades of concrete is greater by 40% than the plain bars. Hence deformed bars may be used without hooks, provided anchorage requirements are adequately met with. Where hooks are formed in deformed bars, the internal radius of the bend should be at least three times the diameter of the bar. The length of straight bar beyond the end of the curve should be atleast four times the diameter of the bars.

3.8 IS CODE ON BOND AND ANCHORAGE REQUIREMENTS (IS : 456–1978)

(a) PERMISSIBLE STRESS IN BOND

The permissible stress in bond (average) for plain bars in tension are given in table 1.15. However, the values are reproduced in Table 3.4 for ready reference.

TABLE 3.4 PERMISSIBLE BOND STRESS τ_{bd} (IS : 456–1978)

Grade of concrete	*M 10*	*M 15*	*M 20*	*M 25*	*M 30*	*M 35*	*M 40*
Permissible stress in bond τ_{bd} *(N/mm²)*	–	0.6	0.8	0.9	1.0	1.1	1.2

Note 1. The bond stress given shall be increased by 25 percent for bars in compression.

Note 2. In the case of deformed bars conforming to IS : 1139–1966 and IS : 1786–1979, the bond stress given above may be increased by 40%.

(b) DEVELOPMENT OF STRESS IN REINFORCEMENT

The calculated tension or compression in any bar at any section shall be developed on each side of the section by the appropriate development length of end anchorage or by a combination thereof.

Development length of bars. The development length L_d is given by

$$L_d = \frac{\Phi \cdot \sigma_s}{4\tau_{bd}} \quad \text{...(3.25)}$$

where Φ = nominal diameter of the bar

σ_s = stress in bar at the section considered at design load,

and

τ_{bd} = design bond stress given in Table 3.4

The values of L_d for plain mild steel bars ($\sigma_s = \sigma_{st} = 140$ N/mm^2) and high yield strength deformed bars (HYSD bars having

$\sigma_s = \sigma_{st} = 230$ N/mm^2), for various grades of concrete are summarised in Table 3.5.

TABLE 3.5. DEVELOPMENT LENGTH IN TENSION

Grade of Concrete	*Plain m.s. bars*		*HYSD bars*	
	τ_{bd} *(N/mm^2)*	L_d/Φ	τ_{bd} *(N/mm^2)*	L_d/Φ
M 15	0.6	58	0.84	68
M 20	0.8	44	1.12	51
M 25	0.9	39	1.26	46
M 30	1.0	35	1.40	41
M 35	1.1	32	1.54	37
M 40	1.2	29	1.64	34

When the bars are in compression, the bond stress given in Table 3.4 shall be increased by 25%. Hence the development length in compression is given by

$$L_{dc} = \frac{\Phi\, \sigma_{sc}}{5\, \tau_{bd}} \qquad \text{...[3.25(a)]}$$

Note 1. The development length includes anchorage values of hooks in tension reinforcement.

Note 2. For bars of sections other than circular, the development length should be sufficient to develop the stress in the bar by bond.

Bars bundled in contact. The development length of each bar or bundled bars shall be that for the individual bar, increased by 10 percent for the two bars in contact, 20 percent for three bars in contact and 33 percent for four bars in contact.

(c) ANCHORING REINFORCING BARS

(i) *Anchoring Bars in Tension*

Deformed bars may be used without end anchorages provided development length requirement is satisfied. Hooks should normally be provided for plain bars in tension.

The anchorage value of bend shall be taken as 4 times the diameter of the bar for each 45° bend subject to a maximum of 16 times the diameter of the bar.

The anchorage value of a standard U-type hook shall be equal to 16 times the diameter of the bar.

(ii) Anchoring Bars in Compression

The anchorage length of straight bar in compression shall be equal to the development length of bars in compression. The projected length of hooks, bends and straight lengths beyond bends if provided for a bar in compression, shall be considered for development length.

(iii) Anchoring Shear Reinforcement

Inclined bars. The development length shall be as for bars in tension ; this length shall be measured as under : (1) in tension zone from the end of the sloping or inclined portion of the bar, and (2) in the compression zone, from mid depth of the beam.

Stirrups. Notwithstanding any of the provisions of this standard, in case of secondary reinforcement, such as stirrups and transverse ties, complete development lengths and anchorage shall be deemed to have been provided when the bar is bent through an angle of atleast 90° round a bar of atleast its own diameter and is continued beyond the end of the curve for a length of atleast eight diameters, or when the bar is bent through an angle of 135° and is continued beyond the end of curve for a length of atleast six bar diameters or when the bar is bent through an angle of 180° and is continued beyond the end of the curve for a length atleast four bar diameters.

(d) DIAMETER OF REINFORCEMENT AND DEVELOPMENT LENGTH REQUIREMENTS AT THE SUPPORT

The code stipulates that at the simple supports (and at the point of inflection), the positive moment tension reinforcement shall be *limited* to a diameter such that

$$L_d \leq \frac{M_1}{V} + L_0 \qquad \text{...(3.26)}$$

where L_d = development length computed for σ_{st} from Eq 3.25.

M_1 = moment of resistance of the section, assuming all reinforcement at the section to be stressed to σ_{st}.

V = shear force at the section due to design loads

L_o = sum of anchorage beyond the centre of support and the equivalent anchorage value of any hook or mechanical anchorage at the simple support (at the point of inflection, L_o is limited to d or 12Φ which ever is greater).

Note. For the computation of L_o, the support width should be known. Fig. 3.16 shows a beam with end support, in which x' is the side cover and x_o is the distance of the *beginning* of the hook from the centre line of the support. Now L_o = sum of anchorage beyond the centre of the support and the equivalent anchorage value.

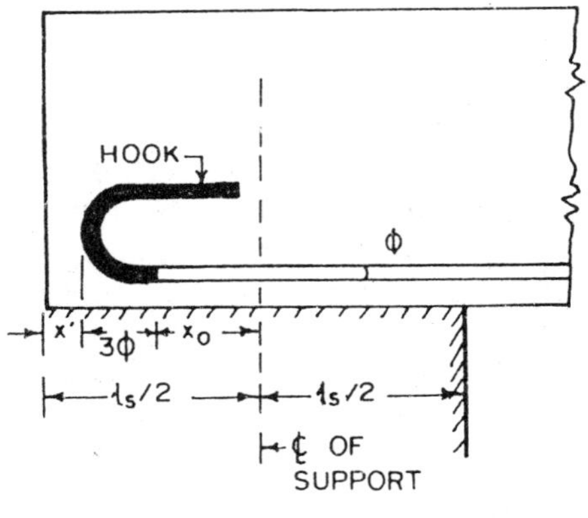

FIG. 3.16.

The dark portion shows the hook which has an anchorage value of 16Φ [Fig 3.15(*a*)]. The distance of the beginning of the hook from its apex of the semi-circle can be taken to be equal to 3 Φ. Let l_s be the length of support.

$$\therefore \quad L_0 = x_0 + 16\,\Phi \quad \text{where} \quad x_0 = \frac{l_s}{2} - x' - 3\Phi$$

$$\therefore \quad L_0 = \left(\frac{l_s}{2} - x' - 3\,\Phi\right) + 16\,\Phi$$

or

$$L_0 = \frac{l_s}{2} - x' + 13\,\Phi \qquad ...(3.27)$$

If no hook is provided, as in case of deformed bars,

$$L_0 = \frac{l_s}{2} - x'$$

The code further recommends that the value of M_1/V in Eq. 3.26 may be increased by 30 percent when the ends of the reinforcement are confined by a compressive reaction. This condition of *'confinement'* of reinforcing bars may not be available at all the types of simple supports.

Four types of simple supports are shown in Fig. 3.17. In Fig. 3.17(*a*), the beam is simply supported on a wall which offers a compressive reaction which confines the ends of reinforcement. Hence a factor of 1.3 will be applicable with the term M_1/V. Similarly in Fig. 3.17(*b*), the slab is simply supported on a wall and hence factor 1.3 will be applicable . However, in Fig. 3.17(*c*) and (*d*), though a simple support is available, the reaction does not confine the ends of the reinforcement and hence the factor 1.3 will not be applicable with M_1/V term.

Similarly, for the case of slab connected to a beam [Fig. 3.17(*e*)] or for the case of a secondary beam connected to a main beam [3.17(*f*)] tensile reaction is induced and hence factor 1.3 will not be available.

Thus, at simple supports, where the compressive reaction confines the ends of reinforcing bars, we have

$$L_d \leq 1.3\frac{M_1}{V} + L_0 \qquad ...[3.26\ (a)]$$

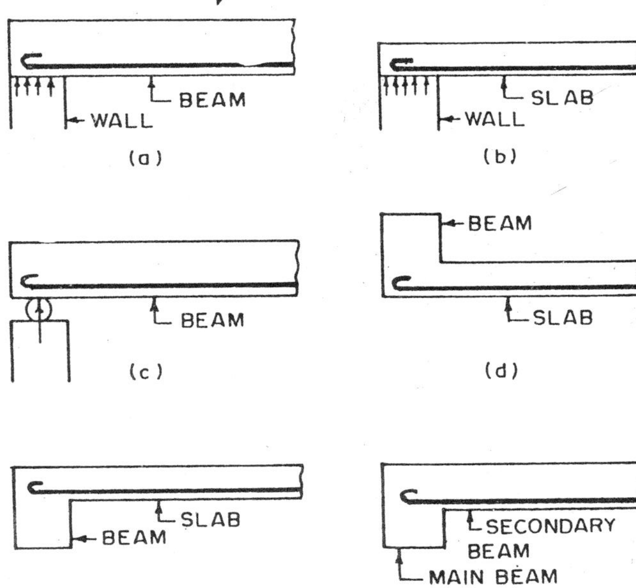

FIG. 3.17.

In Eq. 3.26 or 3.26(*a*), M_1 is the moment of resistance of bars available at the support.

(*e*) REINFORCEMENT SPLICING

The reinforcing bars are normally manufactured in specified lengths. If in any R.C. structure, the length of the bar required is more than the available length, splicing of bars is done. In such a case some minimum lap length is required so that the force can be transferred from one bar to the other bar.

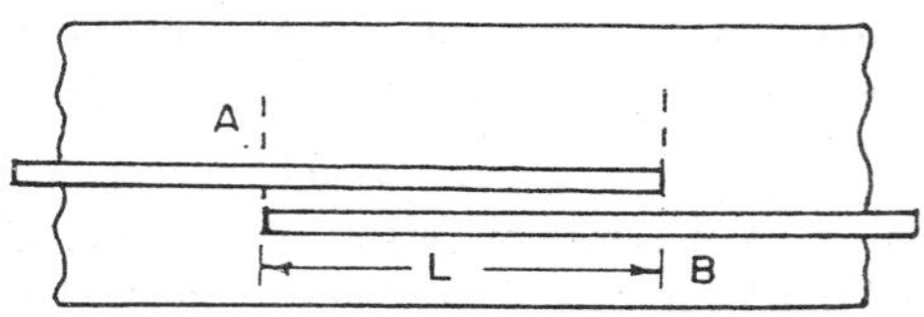

FIG. 3.18.

The following are I.S. Code recommendations.

1. Where splices are provided in the reinforcing bars, they shall, as far as possible be away from the sections of maximum stress and be staggered. It is recommended that splices in flexural members should not be at sections where the bending moment is more than 50 percent of the moment of resistance ; and not more than half the bars shall be spliced at a section.

Where more than half of the bars are spliced at a section or where splices are made at points of maximum stress, special precautions shall be taken, such as increasing the length of lap and/or using spirals or closely spaced stirrups around the length of the splice.

2. Lap splices shall not be used for bars larger than 36 mm ; for larger diameters, bars may be welded ; in cases where welding is not practicable, lapping of bars larger than 36 mm may be permitted, in which case additional spirals should be provided around the lapped bars.

3. Lap splices shall be considered as staggered if the centre to centre distance of the splices is not less than 1.3 times the lap length as described in para 4 below.

4. Lap length including anchorage value of hooks in flerxural tension shall be L_d or 30Φ whichever is greater and for direct tension $2L_d$ or 40Φ whichever is greater. The straight length of the lap shall not be less than 15Φ or 20 cm.

5. The lap length in compression shall be equal to L_d in compression [Eq. 3.25(*a*)], but not less than 24Φ.

6. When bars of two different diameters are to be spliced, the lap length shall be calculated on the basis of diameter of smaller bar.

7. When splicing of welded wire fabric is to be carried out, lap splices of wires shall be not less than the spacing of cross wires plus 10 cm.

8. In case of bundled bars, lapped splices of bundled bars shall be made by splicing one bar at a time ; such individual splices within a bundle shall be staggered.

3.9. CURTAILMENT OF TENSION REINFORCEMENT

Some bars of tensile reinforcement can be curtailed if not required from flexure point of view. The distance x, measured from the centre of the span, at which N_x number of bars (out of that N_c bars) can be curtailed, is given by Eq. 3.28.

$$x = \frac{L}{2}\sqrt{\frac{N_x}{N_c}} \qquad \text{...(3.28)}$$

Alternatively, a simple graphical method can be used for such curtailment. Fig. 3.19 shows the bending moment diagram for half the span of the beam , in which $N_c(=4$, say) number of bars are required at the mid span. Each bar has area equal to A_{st}/N_c , where A_{st} is the total area of tensile reinforcement. The stress in the bars is close to the permissible value (σ_{st}) at the midspan, but decreases

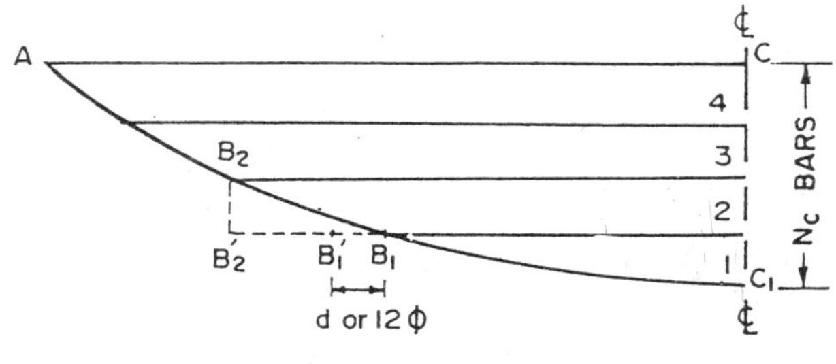

FIG. 3.19. CURTAILMENT OF BARS.

with the B.M. towards the point B_1 at which one of the bars can be curtailed and the remaining bars are stressed to the permissible value. However, one of the bars can be terminated at point B_1 only. if the length $C_1 B_1$ is equal to or greater than the development length L_d for that bar. Similarly, the second bar can be curtailed at point B_2 if $B_1 B_2'$ is equal to or greater than L_d for the second bar. The Code specifies that the reinforcement shall extend the point at which it is no longer required to resist flexure, for a distance equal to effective depth d or $12\,\Phi$ whichever is greater. Thus, for the first bar, if $B_1 C_1 > L_d$, it should extend , beyond B_1, to a point B_1' such that $B_1 B_1'$ is equal to d or 12Φ whichever is more.

Another important recommendation of the code is that in simple beams, one third of the positive reinforcement should be continued to the supports, and the diameter should be so chosen that Eq. 3.26 is satisfied.

PROBLEMS

1. A R.C. beam supports a total uniformly distributed load of 80 kN inclusive of its own weight and has 16 cm width and 32 cm effective depth. Determine (*i*) maximum shear stress (*ii*) spacing of 10 mm diameter two legged stirrups near the support. Take $\sigma_{st} = \sigma_{sv} = 140$ N/mm^2, $\sigma_{cbc} = 5$ N/mm^2 and m=18.
2. A beam has an effective depth of 40 cm. At a particular section, it has 8 mm Φ 2 lgd stirrups provided at 20 cm centre to centre. Calculate the sheare resistance of the stirrups. Take $\sigma_{sv} = 140$ mm^2 and j=0.87.
3. A hook made from 12 mm diameter rod is embedded in a concrete roof, for a distance of 100 mm. Calculate the maximum load that can be suspended from the hook. Take permissible bond stress = 0.6 N/mm^2.
4. A R.C. beam supports a total uniformly distributed load of 160 kN. The beam has 20 cm width and 36 cm effective depth. Determine the spacing of 10 mm diameter two legged vertical stirrups, in addition to two main reinforcing bars of 16 mm bent up at 45° to resist stear. Take $\sigma_{cbc} = 5$ N/mm^2, $\sigma_{st} = \sigma_{sv} = 140$ N/mm^2, and m=18.

4

Torsion

4.1. INTRODUCTION

Axial loads, flexure, shear and torsion are the basic loading situations for which independent theories have been developed, and the more complicated interactive loading situations have been analysed as combinations of these basic effects. A common example of torsional loading is that of *peripheral beams* in each floor of a multistoreyed building, in which the slab is cast monolithic with the beam giving rise to L-beam configuration (see chapter 7). Another example of torsional loading is that of a ring beam provided at the bottom of an elevated circular water tank. Such a ring beam is subjected to bending moment, shear force and torsional moment. The beams supporting cantilevered canopy slabs are also subjected to significant torsional loading. Other prominent examples of torsional loadings are edge beams of concrete shell roofs, and helicoidal staircases.

Fig. 4.1(*a*) shows the loading on the peripheral beam received from the slab. Due to monolithic beam-slab construction, the end

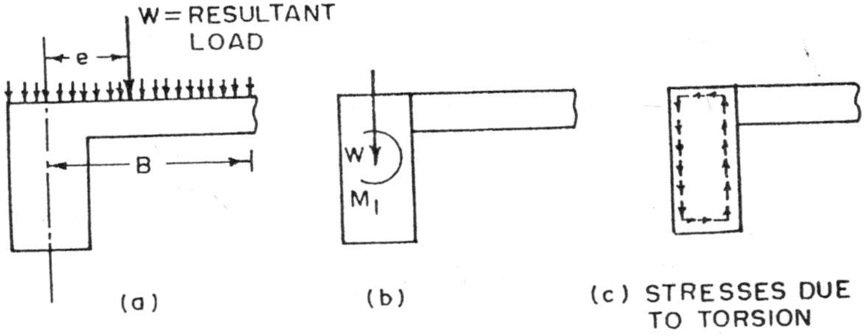

FIG. 4.1. TORSIONAL LOADING IN L-BEAMS.

beam has a flange to one side, giving rise to L-beam configuration. A portion of the slab, of length B, acts as flange of the L-beam. Evidently, such an L-beam is not symmetrically loaded. Due to load on the flange, an eccentric load W is transferred to the beam. This eccentric load may be considered as a central load W plus a moment M_t, as shown in Fig. 4.1(b). The central load W causes bending moment in the beam, which is jointly resisted by the rectangular portion of the beam as well as the flange (of width B). The torsional moment M_t causes torsion in the beam which is resisted by the rectangular portion alone and the flange does not contribute to any torsional moment of resistance.

4.2. TORSIONAL RESISTANCE : ELASTIC BEHAVIOUR

When a shaft of circular section is subjected to torsional moment (or torque) M_t, the maximum shearing stress induced in the material is given by the classical formula :

$$\frac{M_t}{J} = \frac{N\theta}{l} = \frac{f_s}{R} \qquad \text{...(4.1)}$$

where M_t = applied torque.

J = polar moment of inertia.

N = shear modulus.

l = length of the shaft.

f_s = maximum shear stress at the extreme fibres.

R = radius of circular shaft.

However, circular sections are very rarely used in reinforced concrete construction. When a beam of rectangular and other non-circular section is subjected to torsion, considerable warping of the cross-section takes place, and plane sections do not remain plane.

For a beam of rectangular section subjected to torsional moment M_t, the maximum torsional shear stress occurs at the middle of the *long side*, and its magnitude is given by

$$\tau_l = \tau_{max} = \frac{M_t}{\alpha\, b^2 D} = \frac{M_t}{Z_t} \qquad \text{...(4.2)}$$

The torsional shear stress at the middle of the *short side* of the rectangular section is given by

$$\tau_s = \eta\, \tau_{max}\,. \qquad \text{...(4.3)}$$

where D = depth or long side of the section.

b = short side of the section.

Z_t = polar or torsional section modulus

α, η = coefficient depending upon $\frac{D}{b}$ ratio (Table 4.1)

TABLE 4.1

$\frac{D}{b}$	1.0	1.5	2.0	3.0	6.0	10.0	∞
α	0.208	0.231	0.246	0.267	0.299	0.313	0.333
η	1.00	0.859	0.795	0.753	0.743	0.742	0.742

For the design purposes, the *torsional section modulus* for rectangular sections may be taken as :

$$Z_t = \alpha b^2 D \qquad \text{...[4.4(a)]}$$

$$Z_t \approx \frac{b^2 D}{3 + \frac{2b}{D}} \qquad \text{...[4.4(b)]}$$

For the T and L-sections, Cowan gives the following expressions for maximum torsional shear stress :

$$\tau_{max} = \frac{3 M_t . t_i}{\Sigma b_i t_i^3} \qquad \text{...(4.5)}$$

where t_i is t_1 or t_2 .

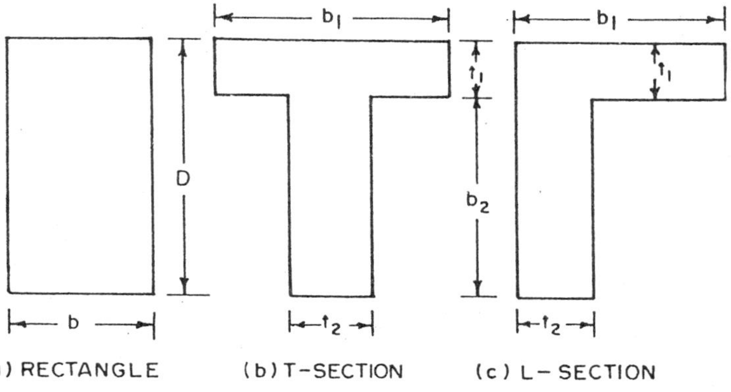

FIG. 4.2.

4.3. INDIAN STANDARD RECOMMENDATIONS ON DESIGN FOR TORSION

When a beam (either rectangular or flanged) is subjected to torsional moment, shear stresses are developed in it. Shear stresses resulting from torsional moment can be accurately assessed by membrane analogy or by any other recognised theory. However, such an analysis is beyond the scope of the present book.

The Indian Standard (IS : 456–1978) recommendations given in para 2 and 3 below are applicable to beams of solid rectangular cross- section. However, these clauses may also be applied to flanged beams, by substituting b_w for b, in which case they are generally conservative ; therefore, specialist literature may be referred to.

1. CRITICAL SECTION.

Section located less than a distance d from the face of the support may be designed for the same torsion as computed at the distance d where d is the effective depth.

2. SHEAR AND TORSION

(i) Equivalent shear

Equivalent shear V_e shall be calculated from the formula :

$$V_e = V + 1.6\,\frac{T}{b} \qquad ...(4.6)$$

where V_e = equivalent shear

V = S.F.

T = torsional moment, and

b = breadth.

The *equivalent nominal shear stress* (τ_{ve}) *shall be calculated from the expression*

$$\tau_{ve} = \frac{V_e}{bd}$$

(where $b = b_w$ for flanged sections).

(*i*) The value of τ_{ve} shall not exceed that values of $\tau_{c\,max}$ given in Table 3.3. (If it exceeds, the section should be redesigned by increasing the concrete area).

(*ii*) If the equivalent nominal shear stress τ_{ve} does not exceed τ_c given in Table 3.1, minimum shear reinforcement shall be provided in the form of stirrups, such that

$$\frac{A_{sv}}{b \cdot s_v} \geq \frac{0.4}{f_y} \qquad ...(4.17)$$

The above equation can be rearranged in the following form to give the maximum spacing of stirrups :

$$s_v \leq \frac{2.5\, A_{sv} f_y}{b} \qquad ...[4.17(a)]$$

where A_{sv} is the total cross-sectional area of stirrups legs effective in shear, s_v is the stirrup spacing and f_y is the characteristic strength of the stirrup reinforcement, in N/mm^2, which shall not be taken greater than 415 N/mm^2.

(*iii*) If τ_{ve} exceeds τ_c given in Table 3.1, both longitudinal and transverse reinforcement shall be provided in accordance with para 3 below.

3. REINFORCEMENT OF MEMBER SUBJECTED TO TORSION

Reinforcement in torsion, when required, shall consist of longitudinal and transverse reinforcement.

(*a*) ***Longitudinal reinforcement***

The longitudinal reinforcement shall be designed to resist an equivalent bending moment M_{e1} given by

$$M_{e1} = M + M_T$$

where M = B.M. at the cross-section, and

$$M_T = T \frac{\left(1 + \frac{D}{b}\right)}{1.7} \qquad ...(4.7)$$

where T is the torsional moment, D is the overall depth of the beam and b is the breadth of the beam.

If the numerical value of M_T as defined above exceeds the numerical value of M, longitudinal reinforcement shall be provided on the flexural compression face, such that beam can also withstand an equivalent moment M_{e2} given by

$$M_{e2} = M_T - M$$

the moment M_{e2} being taken as acting in the opposite sense to the moment M.

(*b*) ***Transverse reinforcement***

Two legged closed hoops enclosing the corner longitudinal bars shall have an area of cross-section A_{sv} given by

$$A_{sv} = \frac{T \cdot s_v}{b_1 d_1 \sigma_{sv}} + \frac{V \cdot s_v}{2.5\, d_1 \sigma_{sv}} \qquad \text{...(4.8)}$$

However, the total transverse reinforcement shall not be less than

than $$\frac{(\tau_{ve} - \tau_c)\, b \cdot s_v}{\tau_{sv}}$$

where T = torsional moment

V = shear force

s_v = spacing of the stirrup reinforcement

b_1 = centre to centre distance between corner bars in the direction of width.

d_1 = centre to centre distance between corner bars in the direction of depth.

b = breadth of member

σ_{sv} = permissible tensile stress in shear reinforcement

τ_{ve} = equivalent shear stress

τ_c = shear strength of the concrete as specified in Table 3.1.

Note 1. ***Distribution of torsional reinforcement*** :

When a member is designed for torsion, torsion reinforcement shall be provided as below.

(*a*) The transverse reinforcement for torsion shall be rectangular closed stirrups placed perpendicular to the axis of the member. The spacing of the stirrups shall not exceed the least of x_1, $\frac{x_1 + y_1}{4}$ and 300 mm, where x_1 and y_1 are respectively the short and long dimensions of the stirrup.

(*b*) Longitudinal reinforcement shall be placed as close as is practicable to the corners of the cross-section and in all cases, there shall be at least one longitudinal bar in each corner of the ties. When the cross-sectional dimension of the member exceeds 450 mm, additional longitudinal bars shall be provided to satisfy the requirements of minimum reinforcement and spacing given in Note 2.

Note 2. ***Side face reinforcement*** :

Where the depth of the web in a beam exceeds 450 mm. the *side face reinforcement* shall be provided along the two faces. The total area of such reinforcement shall be not less than 0.1 percent of the web area and shall be distributed equally on two faces at a spacing not exceeding 300 mm or web thickness whichever is less.

Design Example. *A rectangular beam 400 mm wide is subjected to the following at a section : (i) B.M. of 45 kN-m (ii) S.F. of 30 kN, and (iii) Torsional moment of 20 kN-m. Design the section. Take the following values :*

(i) Permissible stress in steel in tension and shear : 140 N/mm^2

(ii) Permissible stress in concrete : 5 N/mm^2

(iii) m =18

Solution.

(1) *Design constants.*

For the given set of stresses ($c = \sigma_{cbc} = 5$ N/mm^2, $t = \sigma_{st} = 140$ N/mm^2 and $m=18$) we have $k = 0.39$, $j = 0.87$ and $R = 0.85$.

(2) *Design of section for B.M.*

b = 400 mm ; M = 45 kN-m = 45$\times 10^6$ *N-mm*

$$\therefore \quad d = \sqrt{\frac{M}{Rb}} = \sqrt{\frac{45 \times 10^6}{0.85 \times 400}} = 326 \text{ mm.}$$

Keep total $D = 400$ mm. Using 16 mmΦ bars and a clear cover of 30, mm available $d = 400 - 30 - 8 = 362$ mm.

(3) *Equivalent shear*

$$V_e = V + 1.6 \times \frac{T}{b}$$

$$= 30000 + 1.6 \times \frac{20 \times 10^6}{400} = 110000 \text{ N}$$

Hence equivalent shear is given by

$$\tau_{ve} = \frac{V_e}{bd} = \frac{110000}{400 \times 362} = 0.76 \text{ N/mm}^2$$

For M15 concrete, τ_c for 0.5% reinforcement is 0.29 N/mm^2 and τ_c for 1.0% reinforcement is 0.37 N/mm^2 (Table 3.1). Though the percentage reinforcement is not known, it is certain that permissible shear stresss τ_c will be less than τ_{ve}. Hence shear reinforcement is necessary.

Also, from Table 3.3, $\tau_{c\,max}$ for M 15 concrete is 1.6 N/mm^2 . Thus τ_{ve} is less than $\tau_{c.max}$. Hence the dimensions of the beam are acceptable.

(4) *Longitudinal reinforcement*

Since τ_{ve} exceeds τ_c, both longitudinal and transverse reinforcement are necessary. The longitudinal reinforcement is designed to resist an equivalent bending moment M_{e1}, given by

$$M_{e1} = M + M_T$$

where M =B.M.=45×10^6 N-mm

$$M_T = T\frac{(1 + D/b)}{1.7}, \quad \text{where } T\text{=torisional moment}$$

$$= 20 \times 10^6 \left(\frac{1 + 400/400}{1.7}\right) = 23.53 \times 10^6 \text{ N-mm}$$

$$\therefore \ M_{e1} = 45 \times 10^6 + 23.53 \times 10^6 = 68.53 \times 10^6 \text{ N-mm}$$

$$\therefore \ A_{st} = \frac{M_{e1}}{\sigma_{st} \cdot jd} = \frac{68.53 \times 10^6}{140 \times 0.87 \times 362} = 1554 \text{ mm}^2$$

Using 16 mm Φ bars, $A_\Phi = \frac{\pi}{4}(16)^2 = 201 \text{ mm}^2$

$\therefore$ No. of 16 mm Φ bars $= \frac{1554}{201} \approx 8$

(5) *Transverse reinforcement*

Transverse reinforcement is provided in the form of vertical stirrups. Provide two corners bars of 16 mm Φ, at the top face, at a clear cover of 30 mm, to hold the stirrups, as shown in Fig. 4.3..

$\therefore$ b_1 =centre to centre distance between corner bars in the direction of width = 400−60−16=324 mm

d_1 =centre to centre distance between corner bars in the direction of depth = 400 − 60 − 16 = 324 mm

The area of cross-section A_{sv} of the stirrups is given by :

$$A_{sv} = \frac{T\,s_v}{b_1 d_1 \sigma_{sv}} + \frac{V \cdot s_v}{2.5\, d_1 \sigma_{sv}}$$

Using 12 mm Φ 2-lgd stirrups,

$$A_{sv} = 2\frac{\pi}{4}(12)^2 = 226 \text{ mm}^2$$

$$\therefore \ 226 = \left[\frac{20 \times 10^6}{324 \times 324 \times 140} + \frac{3000}{2.5 \times 324 \times 140}\right] s_v.$$

From which s_v =153 mm.

However, the spacing should not exceed the least of x_1, $\frac{x_1+y_1}{4}$ and 300 mm, where,

x_1 =short dimension of stirrup = 324 + 16 + 12 = 352 mm

$y_1 = 324+16+12=352$

$$\therefore \quad \frac{x_1 + y_1}{4} = \frac{352 + 352}{4} = 176 \text{ mm}$$

Thus s_v = 153 mm is permissible.

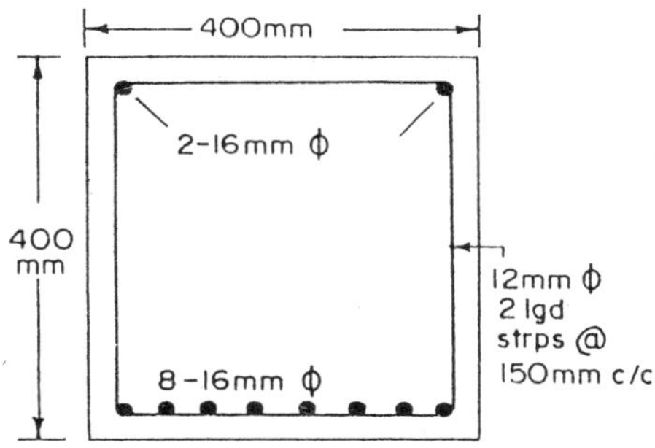

However, provide 12 mm Φ two-legged stirrups @ 150 mm centre to centre. The section of the beam is shown in Fig. 4.3.

FIG. 4.3.

PROBLEMS

1. Discuss I S Code provisions for design for torsion.
2. A rectangular beam 300 mm wide is subjected to the following at a given section
   (i) B.M. : 30 kN-m
   (ii) S.F. : 20 kN
   (iii) Torsional moment 40 kN-m.

   Design the section. Use (*a*) M 15 concrete and mild steel reinforcement (*b*) M 20 concrete and Fe 415 steel, and (*c*) M 15 concrete and Fe 415 steel.

5

Doubly Reinforced Beams

5.1. INTRODUCTION

We have seen that the balanced section of a beam is the most economical section from the requirement of steel point of view. If the area of tensile steel reinforcement is doubled, the moment of resistance of the beam is increased only by about 22%. The moment of resistance of a balanced section is equal to Rbd^2. If a beam of specified dimensions ($b \times d$) is required to resist a moment much greater than Rbd^2, there are two alternatives: (*i*) to use an over-reinforced section, or (*ii*) to use doubly reinforced section. An over reinforced section is always uneconomical since the increase in the moment of resistance is not in proportion to the increase in the area of tensile reinforcement since the concrete, having reached maximum allowable stress, cannot take more additional load without adding compression steel. The other alternative is to provide reinforcement in the compression side of the beam and thus to increase the moment of resistance of the beam beyond the value Rbd^2 for a singly reinforced balanced section. Sections reinforced with steel in compression and tension are known as *doubly reinforced sections.*

Apart from the requirement of developing more resistance than Rbd^2, doubly reinforced sections are used in the following circumstances :

(1) When the members are subjected to alternate external loads and the bending moment in the sections reverses, such as in concrete piles, etc.

(2) When the members are subjected to loading eccentric to either side of the axis, such as in columns subjected to wind loads.

(3) When the members are subjected to accidential lateral loads, shock or impact.

The steel reinforcement provided in the compression zone is subjected to compressive stress. However, concrete undergoes creep strains due to continued compressive stress, with the result that the strain in concrete goes on increasing with time. This increases compressive strain in steel also in addition to creep strain, in compressive steel. *Thus the total compressive strain in compressive steel will be much greater than the strain in surrounding concrete due to flexure alone.* According to IS : 456–1978, the compressive stress in steel should be calculated by multiplying the stress in the surrounding concrete by 1.5 m. In other words, *the modified modular ratio* (m_C) to be used for compression steel is 1.5 times the modular ratio normally used (*i.e.* $m_c = 1.5\ m$). The stress in compression steel so found should not exceed the permissible value (σ_{sc}) specified in Table 1.17.

5.2. LOCATION OF NEUTRAL AXIS

Fig. 5.1. show a doubly reinforced section. The theory of doubly reinforced section is based on the same assumptions as for singly reinforced section. The neutral axis of a doubly reinforced section can be found by finding the centre of gravity of the combined section consisting of concrete in compression only and steel in compression and tension.

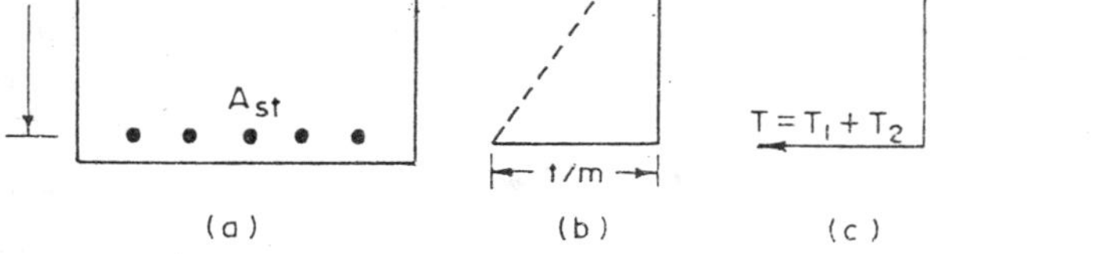

FIG. 5.1. DOUBLY REINFORCED SECTION.

Let b = breadth of the beam.

d = effective depth of the beam

d_c = depth of the centre of the compressive steel$=e.d$

e = compressive steel depth factor $= \dfrac{d_c}{d}$

c = maximum stress in concrete

t = maximum stress in steel

c' = stress in concrete surrounding compressive steel

t_c = stress in compressive steel $= m_c \,.\, c' = 1.5\, m \,.\, c'$

A_{st} = area of tensile steel

A_{sc} = area of compressive steel

m_c = modular ratio for compression zone = 1.5 m

From Fig. 5.1 (b) we get

$$\frac{c}{c'} = \frac{kd}{kd - ed} = \frac{k}{k-e}$$

$$\therefore \qquad c' = \frac{k-e}{k} c$$

Hence stress in steel $t_c = m_c c' = m_c \dfrac{k-e}{k} c = 1.5m \dfrac{k-e}{k} . c$

If maximum stresses in concrete and tensile steel are known the N.A. can be located from stress diagram [Fig. 5.1(a)] exactly in the same manner as in singly reinforced section :

$$\frac{c}{t/m} = \frac{n}{d-n} = \frac{kd}{d-kd} = \frac{k}{1-k}$$

or
$$k = \frac{mc}{mc+t} = \frac{m\,\sigma_{cbc}}{m\,\sigma_{cbc} + \sigma_{st}}$$

where σ_{cbc} and σ_{st} are the permissible stresses in concrete and steel respectively. Thus the presence of steel in compression zone does not affect the depth of N.A.

Again neglecting the concrete in the tensile zone and equating the moment of the compressive area about N.A. to the moment of the tensile area about N.A. we get

$$\frac{bn^2}{2} + m_c \,.\, A_{sc}(n-d_c) - A_{sc}(n-d_c) = m\, A_{st}(d-n)$$

or
$$\frac{bn^2}{2} + (m_c - 1) A_{sc}(n-d_c) = m\, A_{st}(d-n) \qquad \text{...[5.1(a)]}$$

or
$$\frac{bn^2}{2} + (1.5m - 1) A_{sc}(n-d_c) = m\, A_{sc}(d-n) \qquad \text{...(5.1)}$$

where $(m_c - 1) A_{sc}$ or $(1.5\, m - 1) A_{sc}$ is the equivalent concrete area to compressive steel. In the above equation, all the terms except

n is known. Hence the N.A. can be located. The above equation can also be expressed in terms of the neutral axis depth factor k and compresssive steel depth factor e as follows :

$$\frac{b}{2} k^2 d^2 + (m_c - 1) A_{sc} (k - e) d = m A_{st} (1 - k) d$$

or
$$\frac{b k^2 d}{2} + (m_c - 1) A_{sc} (k - e) = m A_{st} (1 - k) \qquad [5.1(b)]$$

where $m_c = 1.5\ m$

5.3. MOMENT OF RESISTANCE

The moment of resistance of a doubly reinforced section can be determined by taking the moment of the compressive force about the centre of the tensile reinforcement.

Let M = total bending moment to be resisted by the section

= moment of resistance of the section $= M_1 + M_2$

M_1 = moment that can be developed by the balanced section, without compressive reinforcement= moment of the compressive force in concrete about the centre of tensile reinforcement.

M_2 = moment, in excess of M_1, that is to be provided by the compression steel = moment of the compressive force in steel, about the centre of the tensile reinforcement.

A_{st} = total tensile reinforcement $= A_{st1} + A_{st2}$.

A_{st1} = tensile steel for the balanced section, corresponding to the moment M_1.

A_{st2} = additional tensile steel necessary to develop the remaining moment M_2

Refer Fig. 5.1.

Taking the moments of the compressive forces about the centre of the tensile steel, we get

$$M = b\, n \frac{c}{2} \left(d - \frac{n}{3} \right) + (m_c - 1) A_{sc} \,.\, c' \,(d - d_c) \qquad ...(5.3)$$

(where c may be equal to or less than. σ_{cbc}. See note 1).

Expressed in terms of k and c, we get

$$M = b\, kd \frac{c}{2} \left(d - \frac{kd}{3} \right) + (m_c - 1) A_{sc} \,.\, c' \,(d - ed)$$

or $$M = \frac{1}{2} jck \,.\, bd^2 + (m_c - 1) A_{sc} \,.\, c' \, d\, (1-e)$$

or $$M = R\, b\, d^2 + c'\, d\, (m_c - 1) A_{sc}\, (1-e) \qquad \text{...[5.3(a)]}$$

$$\downarrow \qquad\qquad \downarrow$$

or $$M = M_1 + M_2 \qquad \text{...[5.3(b)]}$$

Now $$A_{st1} = \frac{M_1}{t \times \text{lever arm}} = \frac{M_1}{t\, j\, d} = \frac{M_1}{\sigma_{st} \,.\, jd} \qquad \text{...(5.4)}$$

and

$$A_{st2} = \frac{M_2}{t \times \text{lever arm}} = \frac{M_2}{t\, (d - d_c)} \qquad \text{...(5.5)}$$

(where t may be equal to or less than σ_{st}. See Note 3 below)

The additional tensile area A_{st2} is actually required to balance the compressive steel. Hence A_{st2} can also alternatively be determined by equating the moment of the equivalent concrete area of the compressive steel to the moment of equivalent concrete area of the tensile steel A_{st2} about N.A.. Thus,

$$(m_c - 1) A_{sc} (n - d_c) = m\, A_{st2}\, (d - n)$$

or $$A_{st2} = A_{sc} \frac{m_c - 1}{m} \left(\frac{n - d_c}{d - n} \right) = A_{sc} \frac{m_c - 1}{m} \left(\frac{k - e}{1 - k} \right) \qquad \text{...(5.6)}$$

(where $m_c = 1.5\, m$)

The total area A_{st} of tensile steel will be equal to $A_{st1} + A_{st2}$.

Note 1. The value of c in Eq 5.3 will be equal to the permissible value of stress σ_{cbc} in concrete if the total area of steel A_{st} is so provided that it balances both the compressive force in concrete as well as compressive force in steel. However, there are *two cases* in which the value of c in Eq. 5.3 may be less than the working or permissible stress. *Case (i)* : In this case the tensile steel may be insufficient even to balance the compressive concrete, with the result that neutral axis will be above the critical one and the stress in steel will reach its permissible value first. *Case (ii)* : In this case the compressive steel A_{sc} may be more than amount required to balance the additional tensile reinforcement A_{st2} (Eq 5.6) with the result that the N.A. will again be above the critical N.A. and the stress in tensile steel will reach its maximum permissible value first. In *both these cases*, the value of c has to be first determined from the known value of t from the following :

$$c = \frac{t}{m} \frac{n}{d - n} = \frac{\sigma_{st}}{m} \cdot \frac{n}{d - n}$$

Except in these two cases, the value of c to be substituted in Eq. 5.3 will be equal to its permissible value σ_{cbc}.

Note 2. The stress in concrete (c') surrounding the compression steel can also be found from the following relation easily derived from the stress diagram [Fig. 5.1(b)] :

$$\frac{c'}{c} = \frac{n - d_c}{n}$$

or

$$c' = c\,\frac{n - d_c}{n} = c\,\frac{k - e}{k} \qquad \text{...(5.7)}$$

In this equation, the value of c should be determined as explained in note 1.

Note 3. If the additional tensile reinforcement A_{st2} required to balance the compressive reinforcement cannot be provided in the same tier or layer as that of A_{st1}, the value of stress t, to be used in Eq. 5.5, will be slightly less than the permissible value σ_{st} and will have to be first determined from the stress diagram.

5.4 STEEL BEAM THEORY

We have seen that the total moment of resistance of a doubly reinforced beam consists of two terms : the moment of resistance M_1 of the compression concrete and the corresponding tensile steel (A_{st1}) to balance it, and the moment of resistance M_2 of the compression steel (A_{sc}) and the remaining tensile steel (A_{st2}) to balance it. In the steel beam theory, the concrete is completely neglected and the moment of resistance is taken equal to the moment of steel couple, taking the permissible value of stresses in compressive steel equal to permissible value in tensile steel, σ_{st}.

$$M = A_{st}\,.\,t\,(d - d_c) = A_{st}\,.\,\sigma_{st}\,(d - d_c) \qquad \text{...(5.8)}$$

Thus, in the steel beam theory, the concrete serves only as a web and the compressive and tensile steels as the flanges of an imaginary steel joist having equal area of both the flanges. However, there seems to be no justification in neglecting the compressive area of concrete and permitting the stress in compressive steel equal to stress in tensile steel, as the stress in concrete would pass the limit of the safe working stress. However, the method can be used to find the approximate value of the moment of resistance of the section specially when the area of compression steel is equal to or more than the area of the tensile steel and when these areas are within 2 to 3 percent of the cross-sectional area.

5.5 TYPES OF PROBLEMS IN DOUBLY REINFORCED BEAMS

There are three types of problems in doubly reinforced beams. The methods of solving these problems are practically similar to those of singly reinforced beams.

TYPE 1. DETERMINATION OF MOMENT OF RESISTANCE

In this type of problem, the dimensions of the section, including the area of tensile and compression reinforcement is completely known and it is required to determine the moment of resistance of the beam, corresponding to a given set of permissible stresses in concrete and steel. The procedure for solution is as under :

1. Determine the position of the neutral axis by equating the moments of area of compressive concrete and the area of compressive steel expressed in equivalent concrete to the moment of the area of the concrete equivalent to the area of steel in tension, about the N.A.. Thus, from Eq. 5.1,

$$\frac{bn^2}{2} + (m_c - 1)A_{sc}(n - d_c) = mA_{st}(d - n) \quad ...(5.1)$$

or

$$\frac{bk^2d}{2} + (m_c - 1)A_{sc}(k - e) = mA_{st}(1 - k) \quad ...[5.1(a)]$$

From which either n or k can be determined.

2. Find the positon of the critical neutral axis from the relation

$$n_c = \frac{mc}{mc + t} \cdot d = \frac{m\sigma_{cbc}}{m\sigma_{cbc} + \sigma_{st}} \cdot d$$

3. If the actual N.A. falls below the critical neutral axis, the stress c in concrete reaches its maximum permissible value σ_{cbc} first.

If, however, the actual N.A. falls above the critical neutral axis, the stress t in tensile steel reaches its maximum permissible value σ_{st} first (*i.e.* t = 140 N/mm^2). From this value of t, calculate the value of c from the stress diagram :

$$\frac{c}{n} = \frac{t/m}{d - n}$$

or

$$c = \frac{nt}{m(d - n)} = \frac{kt}{m(1 - k)}$$

Having known the value of c in either case, find the value of c' from the stress diagram :

$$c' = c\frac{n - d_c}{n} = c\frac{k - e}{k} \quad ...(5.7)$$

4. Having known c and c', find the moment of resistance of the beam by taking the moment of all the forces about the tensile steel. Thus from Eq. 5.3.

$$M = bn\frac{c}{2}\left(d - \frac{n}{3}\right) + (m_c - 1)A_{sc} . c'(d - d_c) \quad ...(5.3)$$

or

$$M = \frac{1}{2}cjkbd^2 + (m_c - 1)A_{sc} . c' d(1 - e) \quad ...[5.3(a)]$$

TYPE II. DETERMINATION OF STRESSES IN THE SECTION

In this type of problem, the dimensions of the section are completely known, including the tensile and compressive reinforcement, and it is required to find the stresses developed in concrete, tensile reinorcement and compressive reinforcement, when subjected to a given bending moment. The solution is done in the following steps :

(1) Determine the position of the neutral axis as explained in previous case [*i.e.*, from Eq. 5.1 or 5.1(a)].

(2) Determine the value of stress c' in the concrete surrounding the compressive reinforcement, in terms of maximum stress c in concrete, from the stress diagram. :

$$c' = \frac{c(n - d_c)}{n} = c\frac{k - e}{k} \quad ...(5.7)$$

(3) Determine the moment of resistance of the beam by taking the moments of forces about the centre of the tensile steel, and equate it to the external moment M. Thus :

$$M = bn . \frac{c}{2}\left(d - \frac{n}{3}\right) + (m_c - 1)A_{sc} . c'(d - d_c)$$

or

$$M = bn . \frac{c}{2}\left(d - \frac{n}{3}\right) + (m_c - 1)A_{sc} . \left\{\frac{c}{n}(n - d_c)\right\}(d - d_c) \quad ...(5.8)$$

(where $m_c = 1.5\,m$)

This equation contains only one unknown *i.e.*, stress c in concrete, which can be determined.

(4) Determine the stress t in tensile steel from following relation :

$$t = mc\left(\frac{d - n}{n}\right)$$

Similarly, determine the stress t in the compressive steel from Eq. 5.7 :

$$t_c = m_c c' = \frac{1.5\,m\,.\,c}{n}(d - d_c) \qquad ...(5.9)$$

TYPE III. DESIGN OF THE SECTION

This is more common type of problem in which it is required to determine the dimensions of the section, to develop a *given* moment of resistance corresponding to a given set of permissible stresses in concrete and steel. Generally, the size of the beam (b and d) is known and it is required to determine the area of tensile and compressive reinforcement. The design is done in the following steps.

(1) Determine the position of the critical neutral axis of the section for the given set of stress :

$$n = \frac{mc}{mc + t} d = \frac{m\sigma_{cbc}}{m\sigma_{cbc} + \sigma_{st}} . d$$

(2) Determine the area of tensile reinforcement A_{st1} and the moment of resistance M_1 corresponding to the singly reinforced balanced section :

$$M_1 = Rbd^2 = \frac{1}{2} cjk \,.\, bd^2 = bn \frac{c}{2}\left(d - \frac{n}{3}\right)$$

and $$A_{st1} = \frac{M_1}{tjd} = \frac{M_1}{\sigma_{st} \,.\, jd} \qquad ...(5.4)$$

where $c = \sigma_{cbc}$ and $t = \sigma_{st}$

Note. If the external moment M is greater than M_1, the beam is designed as doubly reinforced beam. This test should always be applied before step 3.

(3) Determine the additional area of tensile reinforcement A_{st2} for the remaining bending moment M_2 from Eq. 5.5.

$$A_{st2} = \frac{M_2}{t\,(d - d_c)} = \frac{M - M_1}{t\,(d - d_c)} \qquad ..(5.5)$$

where $t = \sigma_{st}$ if A_{st2} and A_{st1} are provided at the same level.

Thus, the total area of tensile reinforcement will be equal to

$$A_{st} = A_{st1} + A_{st2}$$

(3) Determine the area of compressive steel A_{sc} by equating the moments of equivalent areas of compressive steel (A_{sc}) and additional tensile area (A_{st2}). Thus from Eq. 5.6 :

$$A_{sc} = \frac{m(d - n)}{(m_c - 1)\,(n - d_c)} A_{st} = \frac{m(1 - k)}{(m_c - 1)\,(k - e)} A_{st2} \qquad ...(5.6)$$

Note. If A_{sc} comes out to be equal to or more than A_{st} the beam should redesigned using steel beam theory.

Example 5.1 *A beam section, 300 mm wide and 560 mm deep is reinforced with 4 bars of 25 mm diameter in the tensile zone and 4 bars of 12 mm diameter in the compression zone. The cover to the centre of both the reinforcement is 40 mm. Determine the moment of resistance of the section, if (a) M 15 concrete and mild steel bars are used (b) M 20 concrete and HYSD bars are used.*

Solution

(a) M 15 concrete and mild steel bars.

Given $\sigma_{cbc} = 5 \text{ N/mm}^2$,

$\sigma_{st} = 140 \text{ N/mm}^2$,

$$A_{sc} = 4 \times \frac{\pi}{4}(12)^2 = 452.4 \text{ mm}^2$$

$$A_{st} = 4 \times \frac{\pi}{4}(25)^2 = 1963.5 \text{ mm}^2$$

From Eq. 5.1

$$\frac{b\,n^2}{2} + (m_c - 1)A_{sc}(d - d_c) = m\,A_{st}(d - n)$$

Here $b = 300$ mm ;

$d = 560 - 40 = 520$ mm.

$d_c = 40$ mm ;

$m_c = 19 \times 1.5 = 28.5$

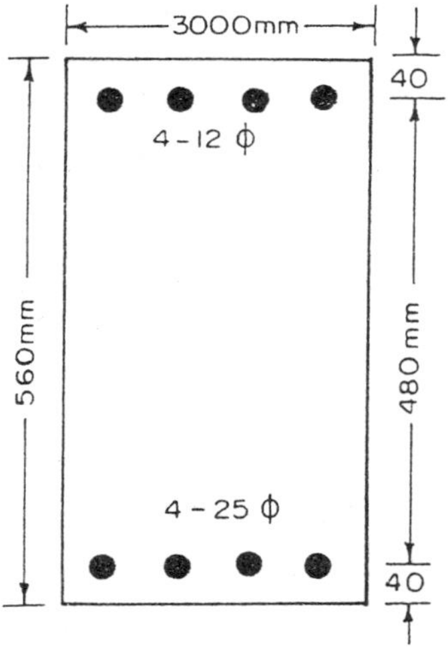

FIG. 5.2

$$\therefore \quad \frac{300}{2}n^2 + (28.5 - 1) \times 452.5\,(n - 40) = 19 \times 1963.5\,(520 - n)$$

or $$n^2 + 331.7\,n - 132647 = 0$$

From which $n = 234$ mm

The critical N.A. is given by

$$n_c = \frac{m\sigma_{cbc}}{m\,\sigma_{cbc} + \sigma_{st}}\,d = \frac{19 \times 5}{19 \times 5 + 140} \times 520 = 210.2 \text{ mm.}$$

Since the actual N.A. is below the critical N.A., the stress in concrete will reach its maximum permissible value first. Thus, $c = \sigma_{cbc} = 5 \text{ N/mm}^2$. Hence the stress in concrete surrounding compressive steel, given by Eq. 5.7 is

$$c' = c \cdot \frac{n - d_c}{n} = \frac{5}{234}(234 - 40) = 4.15 \text{ N/mm}^2$$

The moment of resistance is given by Eq. 4.3

$$M_r = b \cdot n \cdot \frac{c}{2}(d - \frac{n}{3}) + (m_c - 1) A_{sc} \cdot c' (d - d_c)$$

or

$$M = 300 \times 234 \times \frac{5}{2}\left(520 - \frac{234}{3}\right) + (28.5 - 1)\, 452.4 \times 4.15\,(520 - 40)$$

$$= 77.57 \times 10^6 + 24.78 \times 10^6 \text{ N-mm}$$

$$= 102.35 \times 10^6 \text{ N-mm} = \mathbf{102.25 \text{ kN-m.}}$$

Note. $t_c = m_c\, c' = 28.5 \times 4.15 = 118.3$ N/mm^2 which is less than the permissible value σ_{sc} $(=130$ N/mm$^2)$.

(b) M 20 Concrete and HYSD bars

$$\sigma_{cbc} = 7 \text{ N/mm}^2 : \sigma_{st} = 230 \text{ N/mm}^2 \;;\; m = 13$$

$$A_{sc} = 452 \text{ mm}^2 \text{ and } A_{st} = 1963.5 \text{ mm}^2, \text{ as before.}$$

$$m_c = 13 \times 1.5 = 19.5$$

From Eq. 5.1

$$\frac{bn^2}{2} + (m_c - 1) A_{sc} (n - d_c) = m A_{st} (d - n)$$

or

$$\frac{300}{2} n^2 + (19.5 - 1) \times 452.4 (n - 40) = 13 \times 1963.5 (520 - n)$$

or $n^2 + 226\, n - 90720 = 0$

From which $n = 208.7$ mm

The critical N.A. is given by

$$n_c = \frac{m\sigma_{cbc}}{m\sigma_{cbc} + \sigma_{st}}\, d = \frac{13 \times 7}{13 \times 7 + 230} \times 520$$

$$= 147.4 \text{ mm}$$

Since the actual N.A. is below the critical N.A., the streses in concrete will reach its maximum permissible value first.

$$\therefore \quad c = \sigma_{cbc} = 7 \text{ N/mm}^2$$

$$c' = c\, \frac{n - d_c}{n} = 7 \times \frac{208.7 - 40}{208.7} = 5.658 \text{ N/mm}^2$$

The moment of resistance is given by Eq. 5.3

$$M_r = bn\, \frac{c}{2}\left(d - \frac{n}{3}\right) + (m_c - 1) A_{sc} \cdot c' (d - d_c)$$

$$= 300 \times 208.7 \times \frac{7}{2}\left(520 - \frac{208.7}{3}\right) + (19.5 - 1) \times 452.4 \times 5.658\,(520 - 40)$$

$$= 98.71 \times 10^6 + 22.73 \times 10^6 \text{ N-mm}$$

$$= 121.44 \times 10^6 \text{ N-mm} \quad = \mathbf{121.44 \text{ kN-m}}$$

Note $t_c = m_c c' = 19.5 \times 5.658 = 110.3 \text{ N/mm}^2$ which is less than the permissible value $\sigma_{sc}\,(= 190 \text{ N/mm}^2)$

Example 5.2 *A doubly reinforced rectangular beam of breadth b and effective depth d has tensile reinforcement equal to 1.2 percent and compressive reinforcement also 1.2 percent. The compressive reinforcement is at 0.1d below the top. Determine the moment of resistance of the section if the permissible streseses in concrete and steel are 5 N/mm² and 140 N/mm² respectively, and m = 18. What will be the moment of resistance by the steel beam theory ?*

Solution. Given : $\sigma_{cbc} = 5 \text{ N/mm}^2$; $\sigma_{st} = 140 \text{ N/mm}^2$; $m \approx 18$

From Eq. 5.1

$$\frac{bn^2}{2} + (m_c - 1)A_{sc}\,(n - d_c) = m\,A_{st}\,(d - n)$$

$$\frac{bn^2}{2} + (1.5 \times 19 - 1)\frac{1.2}{100}\,bd\,(n - 0.1\,d) = 19 \times \frac{1.2}{100}\,bd\,(d - n)$$

or
$$n^2 + 1.116\,n\,d - 0.522\,d^2 = 0$$

which gives $n = 0.355\ d.$

The critical N.A. is given by

$$n_c = \frac{m\sigma_{cbc}}{m\sigma_{cbc} + \sigma_{st}}\,d = \frac{19 \times 5}{19 \times 5 + 140}\,d = 0.404\,d.$$

The actual N.A. is above the critical one ; hence the stress in tensile steel reaches its maximum value of 140 N/mm² before the stress in concrete can reach the value of 5 N/mm². Corresponding to $t = 140 \text{ N/mm}^2$, c and c' can be found as under :

$$c = \frac{t}{m}\cdot\frac{n}{d - n} = \frac{140}{19} \times \frac{0.355\,d}{0.645\,d} = 4.06 \text{ N/mm}^2$$

$$c' = \frac{t}{m}\cdot\frac{n - d_c}{d - n} = \frac{140}{19}\left(\frac{0.355 - 0.1}{0.645}\right) = 2.91 \text{ N/mm}^2.$$

The moment of resistance is given by Eq. 5.3

$$M_r = b\,.\,n\,\frac{c}{2}\left(d - \frac{n}{3}\right) + (m_c - 1)\,A_{sc}\,.\,c'\,(d - d_c)$$

$$= b \times 0.355\, d \times \frac{4.06}{2}\left(d - \frac{0.355\, d}{3}\right) + (1.5 \times 19 - 1) \times \frac{1.2\, bd}{100} \times 2.91\,(d - 0.1\, d)$$

$$= 0.653\, bd^2 + 0.864\, bd^2$$

$$= \mathbf{1.517\, bd^2\ N - mm}$$

(where b and d are in mm)

Using the *steel beam theory*

$$M_r = A_{st} \,.\, \sigma_{st}\,(d - d_c)$$

$$= \frac{1.2\, bd}{100} \times 140\,(d - 0.1\, d) = \mathbf{1.512\ bd^2\ N\text{-}mm}$$

Thus it is seen that the moment of the resistance found by the approximate theory is nearly the same as that calculated by the exact theory, the difference being 0.3%.

Example 5.3. *A doubly reinforced concrete beam is 400 mm wide and 600 mm deep to the centre of tensile reinforcement. The compression reinforcement consists of 4 bars of 16 mm diameter, and is placed with its centre at a depth of 40 mm from the top. The tensile reinforcement consists of 4 bars of 20 mm diameter. The section is subjected to a bending moment of 100 kN-m. Detemine the stresses in concrete and steel. Take m = 16.*

Solution.

$$A_{sc} = 4 \times \frac{\pi}{4}(16)^2 = 805\ \text{mm}^2\ ;\ A_{st} = 4 \times \frac{\pi}{4}(20)^2 = 1250\ \text{mm}^2$$

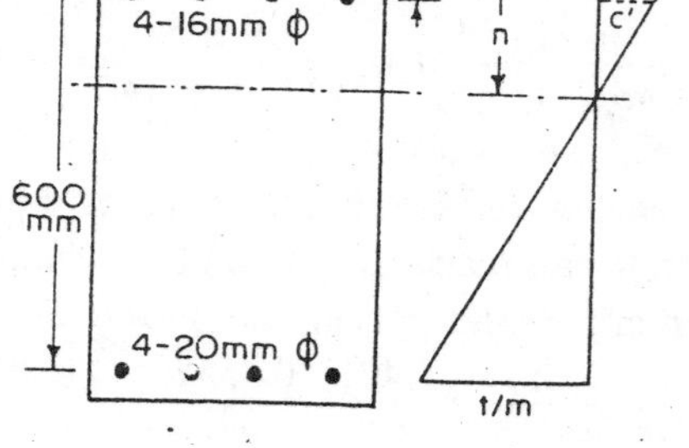

FIG. 5.3

Taking moment of equivalent areas, about N.A.

$$\frac{bn^2}{2} + (m_c - 1)A_{sc}\,(n - d_c) = m\, A_{st}\,(d - n)$$

or $$\frac{400}{2}n^2 + (1.5 \times 16 - 1)\,805(n - 40) = 16 \times 1250\,(600 - n)$$

or $$n^2 + 192.6\,n - 63700 = 0$$

From which $n = 173.8$ mm.

Let the maximum stress in concrete be c. The stress c' in concrete surrounding the compresesion steel is given by :

$$c' = \frac{n - d_c}{n}c = \frac{173.8 - 40}{173.8}c = 0.77\,c.$$

From Eq. 5.3 we get

$$M_r = b\,n\,.\frac{c}{2}\left(d - \frac{n}{3}\right) + (m_c - 1)A_{sc}\,.\,c'\,(d - d_c)$$

$$100 \times 10^6 = 400 \times 173.8\,\frac{c}{2}\left(600 - \frac{173.8}{3}\right) + (1.5 \times 16 - 1)\,805 \times 0.77c\,(600 - 40)$$

From which c = **3.73 N/mm²**

$$c' = 0.77\,c = 0.77 \times 3.73 = 2.87 \text{ N/mm}^2$$

The stress in compressive steel = 1.5 mc'

$$= 1.5 \times 16 \times 2.87 = \mathbf{68.9\ N/mm^2}$$

Stress in tensile steel is

$$t = \frac{cm\,(d - n)}{n} = \frac{3.73 \times 16\,(600 - 173.8)}{173.8}$$

$$= \mathbf{146.3\ N/mm^2}$$

Example 5.4 (a) *A doubly reinforced rectangular beam is 240 mm wide and 500 mm deep. If the limiting stresses in concrete and steel are 5 N/mm² and 140 N/mm² respectively determine the steel areas for bending moment of 80 kN-m. Assume that steel is buried on both faces with its centre 40 mm from either face. Take m = 19.*

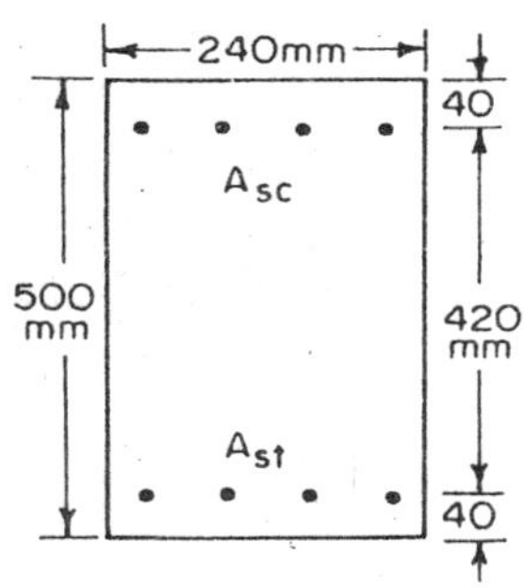

FIG. 5.4

Solution. Given $c = \sigma_{cbc} = 5$ N/mm² ;

$$t = \sigma_{st} = 140 \text{ N/mm}^2$$

Then depth of critical N.A. is given by

$$n = n_c = \frac{mc}{mc + t}d$$

$$= \frac{19 \times 5}{19 \times 5 + 140} \times 460 \approx 186 \text{ mm.}$$

Max. stress in concrete = c = 5 N/mm²

The stress c' in concrete surrounding the compression steel is

$$c' = \frac{n - d_c}{n} c = \frac{186 - 40}{186} \times 5 = 3.925 \text{ N/mm}^2$$

From Eq. 5.3

$$M_r = b \cdot n \cdot \frac{c}{2}\left(d - \frac{n}{3}\right) + (m_c - 1) A_{sc} \cdot c' (d - d_c)$$

or $$80 \times 10^6 = 240 \times 186 \times \frac{5}{2}\left(460 - \frac{186}{3}\right) + (1.5 \times 19 - 1) A_{sc} \times 3.925 (460 - 40)$$

From which $A_{sc} =$ **785 mm²**.

Now, total compression = total tension.

$$\therefore \quad bn \frac{c}{2} + (m_c - 1) A_{sc} \cdot c' = A_{st} \cdot t$$

or $$240 \times 186 \times \frac{5}{2} + (1.5 \times 19 - 1)\, 785 \times 3.925 = A_{st} \times 140$$

which gives A_{st} = **1402 mm²**.

Alternative Solution

$$n = n_c = 186 \text{ mm, as found above.}$$

Lever arm $a = d - \frac{n}{3} = 460 - 62 = 398$ mm.

Now, M_1 for singly-reinforced section.

$$= bn \frac{c}{2}\left(d - \frac{n}{3}\right) = 240 \times 186 \times \frac{5}{2}\left(460 - \frac{186}{3}\right)$$

$$= 44.42 \times 10^6 \text{ N-mm} = 44.42 \text{ kN-m.}$$

For this M_1,

$$A_{st1} = \frac{M_1}{t \times a} = \frac{44.42 \times 10^6}{140 \times 398} = 797 \text{ mm}^2$$

Balance of B.M. $= M_2 = M - M_1 = (80 - 44.42)\, 10^6$

$= 35.58 \times 10^6$ N-mm.

$\therefore$ Tensile steel required for this B.M. is given by

$$A_{st2} = \frac{M_2}{t\,(d - d_c)} = \frac{35.58 \times 10^6}{140\,(460 - 40)} = 605 \text{ mm}^2$$

Compression steel to balance this is given by Eq. 5.6 :

$$A_{sc} = \frac{m\,(d-n)}{(m_c-1)\,(n-d_c)} A_{st2} = \frac{19\,(460-186)}{(1.5 \times 19-1)\,(186 - 40)} \times 605$$

$=$ **785 mm²**.

Total tensile steel $A_{st} = A_{st1} + A_{st2} = 797 + 605 =$ **1402 mm²**

Example 5.4 (*b*). *Solve example 5.4 (a) if HYSD bars are used in place of mild steel bars.*

Solution.

Given $c = \sigma_{cbc} = 5\ \text{N/mm}^2\ ;\ t = \sigma_{st} = 230\text{N/mm}^2$

$$n = n_c = \frac{mc}{mc + t} d$$

$$= \frac{19 \times 5}{19 \times 5 + 230} \times 460 = 134.5 \text{ mm.}$$

Max. stress in concrete $= c = 5\ \text{N/mm}^2$,

$$c' = \frac{n - d_c}{n} . c = \frac{134.5 - 40}{134.5} \times 5 = 3.513 \text{ N/mm}^2.$$

From Eq. 5.3,

$$M_r = b . n . \frac{c}{2} \left(d - \frac{n}{3}\right) + (m_c - 1) A_{sc} . c' (d - d_c)$$

or $$80 \times 10^6 = 240 \times 134.5 \times \frac{5}{2} \left(460 - \frac{134.5}{3}\right) + (1.5 \times 19 - 1) A_{sc} \times 3.513 (460 - 40)$$

From which A_{st} = **1146 mm²**.

Now, total compression = total tension

$$\therefore\ bn \frac{c}{2} + (m_c - 1) A_{sc} . c' = A_{st} . t$$

or $$240 \times 134.5 \times \frac{5}{2} + (1.5 \times 19 - 1) \times 1146 \times 3.513 = A_{st} \times 230$$

From which A_{st} = **832 mm²**.

Alternative solution.

$n = n_c = 134.5$ mm, as found above.

Lever arm $a = d - \frac{n}{3} = 460 - \frac{134.5}{3} = 415.2$ mm.

Now M_1 for singly reinforced section

$$= b . n . \frac{c}{2} \left(d - \frac{n}{3}\right) = 240 \times 134.5 \times \frac{5}{2} \left(460 - \frac{134.5}{3}\right)$$

$$= 33.5 \times 10^6 \text{ N-mm} = 33.5 \text{ kN-m}$$

For this M_1 ,

$$A_{st1} = \frac{M_1}{t \times a} = \frac{33.5 \times 10^6}{230 \times 415.2} = 350.8 \text{ mm}^2.$$

Balance of B.M. $= M_2 = M - M_1 = (80 - 33.5)\, 10^6$

$$= 46.5 \times 10^6 \text{ N-mm.}$$

$\therefore$ Tensile steel required for this B.M. is given by

$$A_{st2} = \frac{M_2}{t\,(d - d_c)} = \frac{46.5 \times 10^6}{230\,(460 - 40)} = 481.4 \text{ mm}^2.$$

Compression steel to balance this is given by Eq. 5.6

$$A_{sc} = \frac{m\,(d-n)}{(m_c-1)\,(n-d_c)} \times A_{st2} = \frac{19\,(460-134.5)}{(1.5 \times 19-1)\,(134.5-40)} \times 481.4$$

$$= \mathbf{1145.6 \text{ mm.}}$$

Total $A_{st} = A_{st1} + A_{st2} = 350.8 + 481.4 = \mathbf{832.2 \text{ mm}^2}$.

Note. It is found here that $A_{sc} > A_{st}$. Such a design is uneconomical because the HYSD bars (A_{sc}) provided in compression zone are highly under-stressed. For such a circumstance, it is better to re-design the beam, using steel beam theory, by providing equal areas of steel in tension and compression zones. Thus,

$$A_{sc} = A_{st} = \frac{M}{\sigma_{st}\,(d - d_c)} = \frac{80 \times 10^6}{230\,(460 - 40)} = 828 \text{ mm}^2$$

Thus, total steel area = 828+828 = 1656 mm^2, against a total area of 1145.6 + 832.2 = 1977.8 mm^2 found by balanced design.

The reader may now find the actual stresses developed in concrete, compression steel and tensile steel, with the above provision of equal steel areas.

Example 5.5. *A reinforced concrete section is subjected to a reversal of bending moment of equal magnitude of 120 kN-m in either direction. Design the section if the permissible stresses in concrete and steel are 6 N/mm^2 and 140 N/mm^2 respectively, and m = 15. Assume b = 0.6 d and effective cover to steel equal to 0.1 d.*

Solution. Given $c = \sigma_{cbc} = 5$ N/mm^2 and $t = \sigma_{st} = 140$ N/mm^2.

Since the moment can act in either direction (*i.e.*, hogging as well as sagging), we will have to provide equal reinforcement at the top as well as bottom. Hence $A_{sc} = A_{st} = A_s$ (say)

Now $$n = n_c = \frac{mc}{mc + t} = d = \frac{15 \times 6}{15 \times 6 + 140}\,d = 0.391\,d$$

Equating the moment of equivalent areas about the N.A., we get

$$\frac{bn^2}{2} + (m_c - 1)\,A_{sc}\,(n - d_c) = m\,A_{st}\,(d - n)$$

$$\therefore \frac{0.6\,d\,(0.391d)^2}{2} + (1.5 \times 15 - 1)A_s\,(0.391 - 0.1)d = 15\,A_s\,(d - 0.391d)$$

or $$0.046\,d^2 + 6.26\,A_s = 9.14\,A_s$$

$$\therefore \quad A_s = A_{sc} = A_{st} = 0.016\,d^2 \text{ (mm}^2\text{)} \qquad \text{...(1)}$$

Now $$M_r = bn\frac{c}{2}\left(d - \frac{n}{3}\right) + (m_c - 1)\,A_{sc}\,.\,c'\,(d - d_c)$$

where $c' = \frac{n - d_c}{n} c = \frac{0.391 - 0.1}{0.391} \times 6 = 4.46 \text{ N/mm}^2$

$$A_{sc} = 0.016\, d^2 \text{ mm}^2 ; M_r = M = 120 \times 10^6 \text{ N-mm}$$

$$\therefore \quad 120 \times 10^6 = 0.6\, d \times 0.391 d \times \frac{6}{2}\left(d - \frac{0.391}{3} d\right) + (1.5 \times 15 - 1) \times 0.016\, d^2 \times 4.46\, (d - 0.1\, d)$$

or $120 \times 10^6 = 0.612\, d^3 + 1.381\, d^3 = 1.993\, d^3$

From which $d \approx$ **392 mm**

$$b = 0.6 \times 392 = \mathbf{235\ mm}$$

$$A_{sc} \approx A_{st} = 0.016\,(392)^2 = \mathbf{2460\ mm^2}$$

Cover to steel $= 0.1\, d \approx$ **40 mm.**

Example 5.6. *Determine the moment of resistance of a doubly reinforced beam of trapezoidal section shown in Fig. 5.5. Take* $\sigma_{cbc} = 5$ *N/mm*^{2} , $\sigma_{st} = 140$ *N/mm*^{2} *and* $m = 18$.

Solution. Let n be the depth of N.A. To get the position of N.A., take the moments of the equivalent areas about the N.A. The concrete area consisting of trapezium *abdc* can be divided into the rectangles *abhg* and two triangles *acg* and *bhd* [Fig.5.5(*b*)].

$$\text{Width } ab = 500 - \frac{200}{600} n = 500 - \frac{n}{3}$$

$$\text{Width } cg = hd = \frac{1}{6} n$$

$$A_{st} = 5 \times \frac{\pi}{4} (12)^2 = 565.5 \text{ mm}^2$$

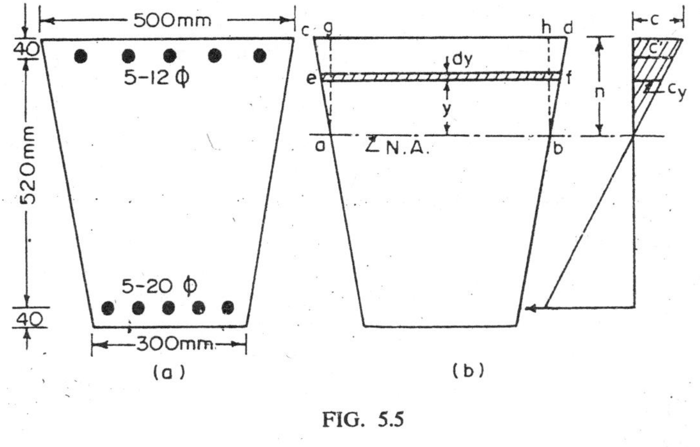

FIG. 5.5

$$A_{st} = 5 \times \frac{\pi}{4}(20)^2 = 1570.8 \text{ mm}^2$$

Equating the moment of equivalent areas about the N.A., we get

$$n\left[500 - \frac{n}{3}\right]\frac{n}{2} + \frac{1}{2}n \times \frac{n}{3} \times \frac{2n}{3} + (m_c - 1)A_{sc}[n - d_c]$$
$$= m \times A_{st}(d - n)$$

$$250n^2 - \frac{n^3}{18} + (1.5 \times 18 - 1) \times 565.5\,[n - 40] = 18 \times 1570.8 \times (560 - n)$$

$$n^3 - 4500n^2 - 773593n + 2.956 \times 10^8 = 0$$

Solving this by trial and error, we get

$$n = 187 \text{ mm}$$

For the given stresses, the position of critical N.A. is given by

$$n_c = \frac{m\,\sigma_{cbc}}{m\,\sigma_{cbc} + \sigma_{st}}\,d = \frac{18 \times 5}{18 \times 5 + 140} \times 560 = 219 \text{ mm.}$$

Since the actual N.A. is above the critical N.A. the stress in tensile steel will reach its maximum permissible value first. Thus $t = \sigma_{st} = 140\ \text{N/mm}^2$.

The corresponding stress in concrete is given by

$$c = \frac{t}{m}\,\frac{n}{d-n} = \frac{140}{18} \times \frac{187}{560 - 187}$$
$$= 3.9\ \text{N/mm}^2$$

The stress in concrete surrounding compressive steel is given by

$$c' = \frac{n - d_c}{n} \times c = \frac{187 - 40}{187} \times 3.9 = 3.07\ \text{N/mm}^2$$

The moment of resistance, M_r of the section is found by taking the moment of the compressive forces about the centre of tensile steel.

Let M_{r1} = moment of compressive force in concrete, about the centre of tensile steel,

M_{r2} = moment of compressive force in steel, about the centre of tensile steel,

so that $M_r = M_{r1} + M_{r2}$

To find M_{r1}, consider an elementary strip of thickness dy at a distance y from the N.A. [Fig. 5.5(b)] and let the stress at the level be c_y.

Breadth of strip $ef = b_y = 500 - \frac{500 - 300}{600}[n - y]$

$$= 500 - \frac{1}{3}[187 - y]$$

Stress $c_y = c \times \frac{y}{n} = \frac{3.9}{187} y = 0.0209\, y$

Force in elementary strip $= b_y \,.\, dy \,.\, c_y$

Moment of this force about the centre of steel area is given by

$$d\,M_{r1} = b_y \,.\, dy \,.\, c_y\,(d - n + y)$$

$$\therefore \quad M_r = \int_0^N b_y \,.\, c_y\,(d - n + y)\, dy$$

$$= \int_0^{187} \left\{500 - \tfrac{1}{3}(187 - y)\right\} 0.0209\, y\,(560 - 187 + y)\, dy$$

$$= \frac{0.0209}{3} \int_0^{187} (y^3 + 1686\, y^2 + 489740\, y)\, dy$$

$$= \frac{0.0209}{3}\left[\frac{y^4}{4} + \frac{1686}{3} y^3 + \frac{489749}{2} y^2\right]_0^{187}$$

$$= 87.39 \times 10^6 \text{ N-mm.}$$

Moment of compressive forces in the compression reinforcement is given by

$$M_{r2} = (m_c - 1) A_{sc} \,.\, c'\,(d - d_c)$$

$$= (1.5 \times 18 - 1)\, 565.5 \times 3.07\,(560 - 40)$$

$$= 23.47 \times 10^6 \text{ N-mm.}$$

Hence total moment of resistance is given by

$$M_r = M_{r1} + M_{r2} = 87.39 \times 10^6 + 23.47 \times 10^6 \text{ N-mm}$$

$$= 111.86 \times 10^6 \text{ N-mm}$$

$$= \mathbf{111.86 \text{ kN-m.}}$$

5.6. SHEAR STRESS, BOND STRESS AND DEVELOPMENT LENGTH

(*a*) MAXIMUM SHEAR STRESS

Fig. 5.6 (*b*) shows the shear stress distribution in a doubly reinforced beam. The distribution of the shear stress at any plane, distant y *above* the N.A. is given by $q = \frac{VA\bar{y}}{bI}$. There will be sudden increase in the shear stress below the level of the compression steel. This is so because the quantity $A\bar{y}$ suddenly increases for all planes below the level compression steel.

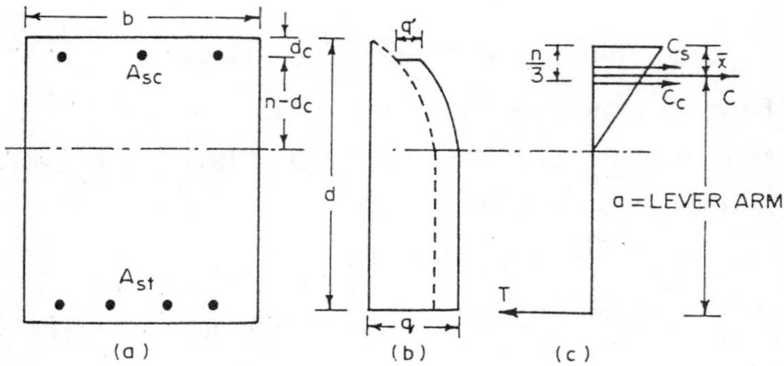

FIG. 5.6. SHEAR STRESS DISTRIBUTION.

Let q' be the increase in the shear stress just below the level of the compression steel. Then

$$q' = \frac{V}{bI} \times (\text{ increase in } A\bar{y})$$

or

$$q' = \frac{V}{bI} \times (m_c - 1) A_{sc} (n - d_c) \quad ..(5.11)$$

where I is the moment of inertia of the equivalent section and $m_c = 1.5\ m$.

For all planes below the N.A., the shear stress will be maximum and of a constant magnitude given by

$$q = \frac{V}{ba} \quad ...(5.12)$$

where a = Lever arm

= distance between tensile force and the centre of the compressive force.

Fig. 5.6 (*c*) shows the tensile and compressive forces.

Let C_s = compressive force in steel, acting at d_c below the top of the beam.

C_c = compressive force in concrete, acting at $\frac{n}{3}$ below the top of the beam.

C = resultant compressive force acting at $\bar{x}$ below the top of the beam.

Now $C_s = (m_c - 1) A_{sc} . \frac{n - d_c}{n} c$, where $m_c = 1.5\, m$

$C_c = b \times n \times c/2$

$$\therefore \quad \bar{x} = \frac{C_s \times d_c + C_c . \frac{n}{3}}{C_s + C_c}$$

$$\bar{x} = \frac{(m_c - 1) A_{sc} . \frac{n - d_c}{n} . c . d_c + bn \frac{c}{2} \times \frac{n}{3}}{(m_c - 1) A_{sc} . \frac{n - d_c}{n} . c + \frac{bnc}{2}}$$

$$\therefore \quad a = d - \bar{x} = d - \frac{\frac{bn^2}{6} + (m_c - 1) A_{sc} . \frac{n - d_c}{n} d_c}{\frac{bn}{2} + (m_c - 1) A_{sc} . \frac{n - d_c}{n}} \qquad ...(5.13)$$

Alternatively , lever arm can also be found from the following relation if M and A_{st} are known.

$$M = t . A_{st} . a$$

or

$$a = \frac{M}{t A_{st}} = \frac{M}{\sigma_{st} . A_{st}} \qquad ...(5.14)$$

(b) NOMINAL SHEAR STRESS

IS code (IS : 456–1978) recommends the use of *nominal stress* in the beam, given by

$$\tau_v = \frac{V}{bd} \qquad ...(5.15)$$

where τ_v = nominal shear stress

V = S. F. at the section

d = effective depth of the beam

b = width of the section.

(c) FLEXURAL BOND STRESS

In the case of doubly reinforced beam, flexural bond stress is determined both in the tensile steel as well as the compressive steel, though the bond stress around the compressive steel will be of smaller magnitude. The bond stress around tensile steel may be found by Eq. 3.23.

$$\tau_{bf} = \frac{V}{j\, d\, \Sigma 0} = \frac{V}{a \Sigma 0} \qquad ...(5.16)$$

where a is the lever arm given by Eq. 5.13 or 5.14.

The bond stress around compression steel can be found from expression derived below.

Consider a small length δx of the beam and let δM be the change in the bending moment in the length. Knowing that $p = \frac{M}{I} y$ from elementary theory, the variation in the stress δt_c in compression steel is given by

$$\delta t_c = \frac{\delta M}{I} (n - d_c) m_c \quad \text{where } m_c = 1.5\, m$$

$\therefore$ Variation in force (δC_s) in compression steel, in a length δx is given by

$$\delta C_s = \frac{\delta M}{I} (n - d_c) m_c \,.\, A_{sc}$$

Let τ'_{bf} = bond stress around compression steel.

$\Sigma 0'$ = sum of the perimeters of the steel bars of the compression reinforcement.

Then $$\tau_{bf}' \,\delta x \times \Sigma 0' = \frac{\delta M}{I} (n - d_c) m_c \,.\, A_{sc}$$

or $$\tau_{bf}' = \frac{\delta M}{\delta x} \cdot \frac{n - d_c}{I} \cdot \frac{m_c \,.\, A_{sc}}{\Sigma 0'}$$

or $$\tau_{bf}' = \frac{F}{I \Sigma 0'} (n - d_c) m_c A_{sc} \qquad \text{...(5.7)}$$

where $$I = \frac{bn^3}{3} + (m_c - 1) A_{sc} (n - d_c)^2 + m \,.\, A_{st} (d - n)^2 .$$

(*d*) DEVELOPMENT LENGTH

Fig. 5.16 and 5.17 give the localized bond stress at a *particular section*. However, bond stresses are *developed* over a specified length and any attempt to visualise bond on the basis of stress at a particular cross-section is bound to be erroneous. In the case of bond, it has been recently concluded that the concept of *anchorage bond* is more meaningful than the concept of flexural bond stress of localised nature. *Current design methods disregard such local bond stresses even though they may result in localised slip between steel and concrete adjacent to the cracks.*. Thus the concept of *development length* has gained prominence.

As per IS : 456–1978, the development length for tension reinforcement is given by

$$\underset{\text{tension}}{(L_d)} = \frac{\Phi\, \sigma_s}{4\, \tau_{bd}} \qquad ...(5.18)$$

where σ_s = stress in bar at the section considered at the design load.

For bars in compression, the permissible bond stress is increased by 25%. Hence

$$\underset{\text{comp.}}{(L_d)} = \frac{\Phi\, .\, \sigma_{sc}}{5\, \tau_{bd}} \qquad ...(5.19)$$

where σ_{sc} = *compressive stress* in the bars, at the section considered, at the design load.

PROBLEMS

1. A doubly reinforced rectangular beam is 300 mm wide and 450 mm deep and is subjected to a bending moment of 50 kN-m. If the limiting stresses in concrete and steel are 5 N/mm^2 and 140 N/mm^2, determine the steel areas. Assume that steel is buried on both faces, with its centre 40 mm from either face. Take m = 18.

2. A rectangular beam of breadth b and effective depth d is reinforced with 1.2% tensile and 1.2% compressive steel. Determine the moment of resistance of the section(a) by accurate method and (b) by steel beam theory, neglecting concrete. Take compression reinforcement to be placed at 0.1 effective depth below top. Take the permissible stress in steel and concrete as 140 N/mm^2 and 5.2 N/mm^2 respectively, and m = 18.

3. A beam 250 mm × 500 mm in section is reinforced with 3 bars of 14 mm diameter at top and 5 bars of 20 mm diameter at the bottom, each at an effective cover of 40 mm. Determine the moment of resistance of the beam section. Take the permissible stress in steel and concrete as 126 N/mm^2 and 5.2 N /mm^2 respectively, and m = 18.

4. A doubly reinforced concrete beam is 250 mm wide and 500 mm deep from the compression edge to the centre of tensile steel. The area of compression and tensile steel are both 1300 mm^2 each. The centre of compression steel is 50 mm from the compression edge. If the beam is subjected to a total bending moment of 70 kN-m, determine the stress in concrete, and tension and compression steel. Take m = 18.

5. Determine the moment of resistance of a doubly reinforced beam of trapezoidal section shown in Fig. 5.7. What will be the moment of resistance using steel beam

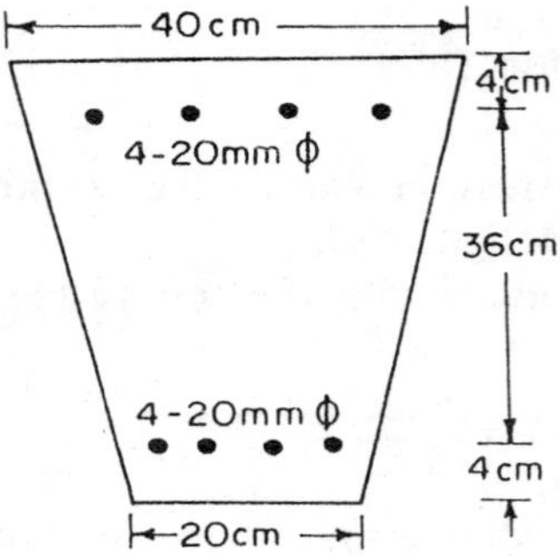

FIG. 5.7

theory ? Take permissible stress in concrete and steel as 5 N/mm^2 and 140 N/mm^2 and m = 18.

6. A pile 300 mm × 300 mm is reinforced with 4 bars of 20 mm diameter with concrete cover of 40 mm over the centre of steel. When being hauled up to the frame, it is subjected to a bending moment of 25 kN-m. Determine the maximum stresses set up in concrete and steel. Take m = 18.

ANSWERS

1. A_{st} = 1770 mm^2; A_{sc} = 1300 mm^2

2. (*a*) 1.594 bd^2 N-mm
   (*b*) 1.598 bd^2 N-mm

3. 73.5 kN-m

4. c = 3.56 N/mm^2
   t = 123.8 N/mm^2
   t_c = 68.3 N/mm^2

5. (*a*) 56.7 kN-m (*b*) 56.2 kN-m

6. c = 3.19 N/mm^2; t = 88.9 N/mm^2

6

T and L-Beams

6.1 INTRODUCTION

We have seen that the concrete below the N.A. does not resist any bending moment but simply serves to embed the tensile steel. Also, the portion of the concrete just above the N.A. carries only very little compressive force since the intensity of compressive stress there, is of a very small magnitude. This suggests that the section of the beam should be such that it has greater width at the top (compression side) in comparison to the width below the neutral axis. Such a section is known as a T- section as shown in Fig. 6.1 (*aii*).

In actual practice, a beam of T-section, known as T-beam is more common than the rectangular section. Fig. 6.1 (*b*) shows the plan of a big size room. The roof system of the room may be of two types : (*i*) a number of beams of rolled steel section with a concrete roof slab, both acting independently, as shown in Fig 6.1 (*c*) or (*ii*) a number of R.C. beams with a roof slab acting monolithically [Fig. 6.1 (*d*)]. The second system is more common because in this system a portion of the slab of width *B* acts monolithically with the rectangular section of the beam, thus giving rise to T-beam. In fact, the slab forms the compression flange of the T-beam, while the comparatively narrow rectangular section forms the web of the T-beam

In such a system, the roof slab is designed to bend over the span *BD* and the reinforcement for positive and negative bending moment is provided in the direction at right angles to the direction *AB* of the beam. The beam, on the other hand, takes uniformly distributed load from the slab, and bends over the span *AB*. The

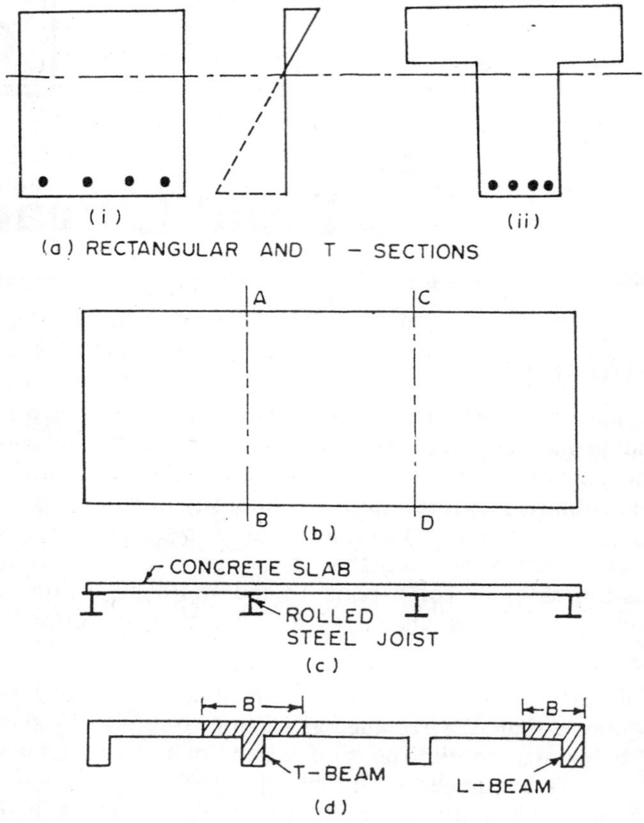

FIG 6.1

slab, which forms the upper part of the T-beam, is stressed laterally in compression, but this does not reduce its capacity to carry longitudinal compression as a part of the beam.

6.2 DIMENSIONS OF A T-BEAM

Fig. 6.2 shows the following four important dimensions of a T-beam section :

(*i*) Breadth of the flange (b_f)

(*ii*) Thickness of the flange (D_f)

(*iii*) Breadth of the rib (b_w)

(*iv*) Depth of the rib (d_w)

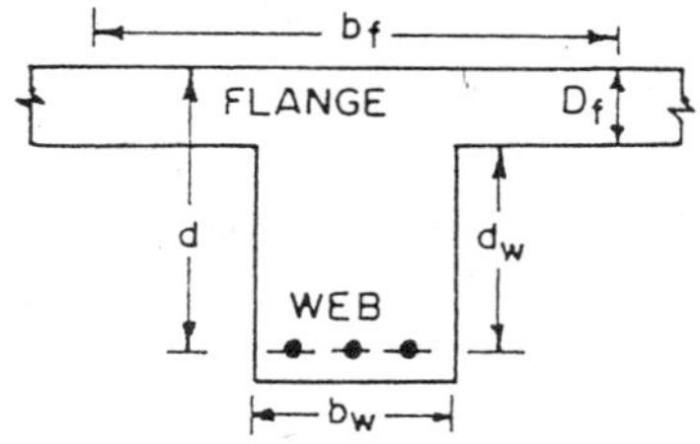

FIG. 6.2

(i) ***Breadth of the flange*** **(b_f)**

The breadth of the flange of the T-beam is that portion of the slab which acts monolithic with the beam and which resists the compressive stresses.

According to Indian standard code (IS : 456–1978), a slab which is assumed to act as a flange of a T-beam (or of a L-beam) shall satisfy the following :

(*a*) The slab be cast integrally with the web, or the web and the slab shall be effectively bonded together in any other manner, and

(*b*) If the main reinforcement of the slab is parallel to the beam, transverse reinforcement shall be provided. Such reinforcement shall not be less than 60% of the main reinforcement at the mid-span of the slab.

Note. Reinforcement in flanges of T-beams (and also of L-beams) shall satisfy the above requirement 1 (*b*). Where flanges are in tension, a part of the main tension reinforcement shall be distributed over the effective flange width or a width equal to one length of the span ,whichever is smaller. If the effective flange width exceeds one length of span, nominal longitudinal reinforcement shall be provided in the outer portion of flange.

In absence of more accurate determinations, the effective width b_f of the flange may be taken as the least of the following :

(a) $b_f = \dfrac{l_0}{6} + b_w + 6\,D_f$

(b) $b_f = b_w + \dfrac{1}{2}$ (sum of clear distances to the adjacent beams on either side)

(c) for isolated beams the flange shall be obtained as below but in no case greater than actual width :

$$b_f = \frac{l_0}{\left(\frac{l_0}{b}\right) + 4} + b_w$$

where $b_f =$ effective width of flange

$l_0 =$ distance between points of zero moments in the beam

$b_w =$ breadth of web

$D_f =$ thickness of flange and

$b =$ actual width of flange.

Note : For continuous beams and frames, l_0 may be assumed as 0.7 times the effective span.

(ii) *Thickness of the flange (D_f)*

The thickness of the flange is taken equal to the total thickness of the slab, including cover. The slab spans in a direction transverse to the span of the beam, and, therefore, the thickness of the slab is determined on the basis of its bending in transverse direction. As far as design of the T-beam is concerned, D_f is the fixed dimension known from the considerations of bending etc. of the slab.

(iii) *Breadth of the rib or web (b_w)*

The breadth of the rib is equal to width of the portion of the beam in the tensile zone. This width should be sufficient to accommodate the tensile reinforcement with proper spacing between the bars. The width of the beam should also be sufficient to provide lateral stability ; from this point of view, it should atleast be equal to one-third of the height (depth) of the rib. From architectural point of view, the breadth of the rib should be equal to the width of the column supporting the beam.

(iv) *Depth of rib or web (d_w)*

The depth of rib is the vertical distance between the bottom of the flange and the centre of the tensile reinforcement, and is dependent on the effective depth (d) of the beam. The effective depth (d) of the T-beam is the distance between the top of the flange and the centre of the tensile reinforcement. Thus $d = d_w + D_f$. The assumed overall depth of the T-beam may be taken an 1/12 to 1/15 of the span when it is simply supported at the ends. When it is continuous, the assumed overall depth may be taken as 1/15 to 1/20 of span for light loads, 1/12 to 1/15 of span for medium loads and 1/10 to 1/12 of span for heavy loads.

6.3 POSITION OF NEUTRAL AXIS

The neutral axis in a T-beam section may fall in two alternative locations : (*i*) within the flange [Fig. 6.3(*a*)], or (*ii*) outside the flange [Fig. 6.3 (*b*)]. Case (*i*) may happen when the slab is comparatively thicker. In that case, the N.A. may be located exactly in the same manner that is adopted in singly reinforced beam :

$$\frac{b_f n^2}{2} = m A_{st} (d - n) \qquad ...(6.1)$$

If, however, the N.A. falls exactly at the junction of flange and rib, the above equation reduces to :

$$\frac{b_f D_f^2}{2} = m A_{st} (d - D_f) \qquad ...(6.2)$$

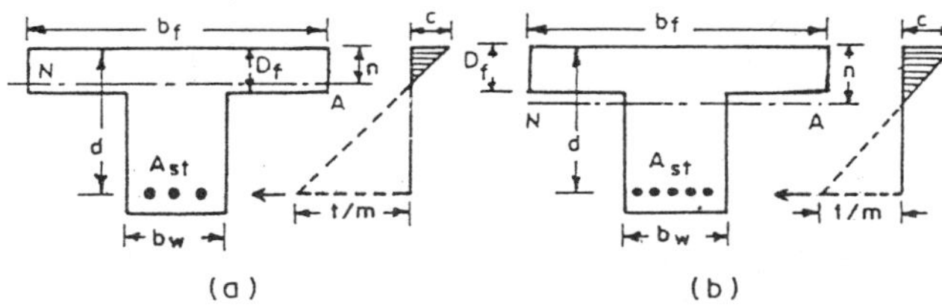

FIG. 6.3 LOCATION NEUTRAL AXIS.

In the second position, which is the most common one, the N.A. falls outside the flange. In that case, N.A. can be located by equating the moments of compressive and tensile areas about the N.A. :

$$b_f . D_f \left(n - \frac{D_f}{2}\right) + \frac{b_w(n - D_f)^2}{2} = m A_{st} (d - n) \qquad ...(6.3)$$

In the above equation, the term $\frac{b_w (n - D_f)^2}{2}$ is usually very small and it is customary to reject this term. In that case, the above expression reduces to :

$$b_f D_f \left(n - \frac{D_f}{2}\right) = mA_{st} (d - n) \qquad ...(6.4)$$

Thus, if the dimensions of the T-beam section are completely known, the N.A. can be located either by Eq. 6.1 or by Eq. 6.4. If by application of Eq. 6.1, n comes out to be more than D_f, calculations should be repeated by using Eq. 6.4. Similarly, if Eq. 6.4 is used first and n comes out to be less than D_f calculations should be repeated by using Eq. 6.1.

If, however, the maximum stresses in concrete ($c = \sigma_{cbc}$) and steel ($t = \sigma_{st}$) are *known* , the N.A. is located by using the stress diagram :

$$\frac{n}{d-n} = \frac{c}{t/m} \qquad ...(6.5)$$

6.4 LEVER ARM AND MOMENT OF RESISTANCE

(*a*) *Lever Arm*

Fig. 6.4 (*b*) shows the stress diagram for a T-beam section in which the N.A. falls outside the flange, which is more general case.

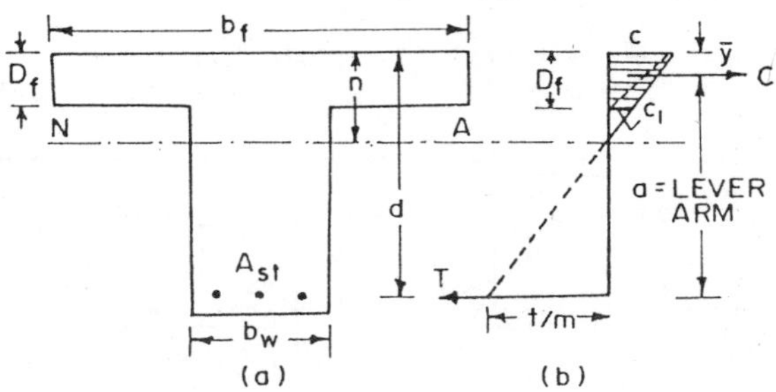

FIG. 6.4 LEVER ARM.

Let c be the stress in the concrete at the top (extreme fibre) and c_1 be the stress in the concrete at the junction of flange with the rib. Since the compression area lying between the junction and the N.A. is very small, the compressive force of this area is *neglected.* Therefore, the shaded area in Fig. 6.4(*b*) shows the net compressive stress diagram, the centroid of which is at a depth $\bar{y}$ (say) below the top. The value of $\bar{y}$ can be found by dividing the trapezoidal stress diagram into two triangles by a diagonal (shown dotted) and taking moments at the top :

$$\bar{y} = \frac{\left[c \cdot \frac{D_f}{2} \cdot \frac{D_f}{3}\right] + \left[c_1 \cdot \frac{D_f}{2} \cdot \frac{2}{3} D_f\right]}{\left[c \cdot \frac{D_f}{2}\right] + \left[c_1 \frac{D_f}{2}\right]}$$

or
$$\bar{y} = \left[\frac{c + 2c_1}{c + c_1}\right] \frac{D_f}{3} \qquad ...(6.6)$$

But from stress diagram,

$$\frac{c_1}{n - D_f} = \frac{c}{n}$$

or $$c_1 = \frac{n - D_f}{n} c \qquad ...(6.7)$$

Substituting this value of c_1 in Eq. 6.6, we get

$$\bar{y} = \frac{1 + 2\left\{\frac{n - D_f}{n}\right\}}{1 + \left\{\frac{n - D_f}{n}\right\}} \times \frac{D_f}{3} = \frac{3n - 2D_f}{2n - D_f} \times \frac{D_f}{3} \qquad ...[6.7(a)]$$

Lever arm $\bar{a} = d - \bar{y}$

$$= d - \frac{D_f}{3}\left\{\frac{1 + 2\left\{\frac{n - D_f}{n}\right\}}{1 + \frac{n - D_f}{n}}\right\}$$

$$= d - \frac{D_f}{3}\left\{\frac{3n - 2D_f}{2n - D_f}\right\}$$

$$= d - \frac{D_f}{3}\left\{\frac{3}{2} - \frac{D_f}{2(2n - D_f)}\right\}$$

or $$a = d - \frac{D_f}{2} + \frac{D_f^{\,2}}{6(2n - D_f)} \qquad ...(6.8)$$

If the actual N.A. is above the critical N.A., the section is under-reinforced, the stress (t) in steel reaches its maximum permissible value σ_{st} first and the stress c in concrete should be found from Eq. 6.5 first, before substituting Eq. 6.6. If, however, the actual N.A. falls at or below the critical N.A., the section is over-reinforced and the stress (c) in concrete reaches its maximum permissible value (σ_{cbc}) first.

(b) *Moment of Resistance*

The moment of resistance of the T-beam is found by multiplying the total compression and the lever arm.

If the N.A. lies in the flange ,

$$M_r = b_f \cdot n \cdot \frac{c}{2}\left(d - \frac{n}{3}\right) \qquad ...[6.9(a)]$$

If the N.A. lies in the web, the moment of resistance is given by

$$M_r = b_f \cdot D_f \cdot \frac{c + c_1}{2} \times a \qquad ...[6.9(b)]$$

(Neglecting the compression of the rib)

Hence $$M_r = \frac{b_f \cdot D_f \cdot c}{2}\left[1 + \frac{n - D_f}{n}\right]\left[d - \frac{D_f}{2} + \frac{D_f^2}{6(2n - D_f)}\right] \qquad ...(6.9)$$

Also, $$M_r = t \cdot A_{st} \cdot a = \sigma_{st} \cdot A_{st} \cdot a. \qquad ...(6.10)$$

6.5 MOMENT OF RESISTANCE TAKING COMPRESSION OF RIB INTO ACCOUNT

When the compression of the rib is taken into account, the neutral axis can be located by :

$$b_f \, . \, D_f \, . \left(n - \frac{D_f}{2}\right) + \frac{b_w}{2}(n - D_f)^2 = m\,A_{st}\,(d - n) \qquad ...(6.11)$$

The moment of resistance of the beam can be found by two methods :

Method 1.

In this method, the distance $\bar{y}$ of the C.G. of compressive force is found first and then lever arm is determined. The moment of resistance is then found by multiplying the total compressive force by the lever arm.

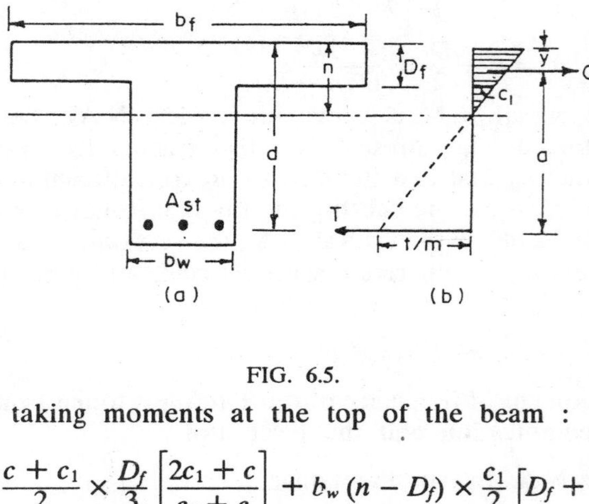

FIG. 6.5.

Thus, taking moments at the top of the beam :

$$\bar{y} = \frac{b_f \, . \, D_f \, . \, \frac{c + c_1}{2} \times \frac{D_f}{3}\left[\frac{2c_1 + c}{c_1 + c}\right] + b_w\,(n - D_f) \times \frac{c_1}{2}\left[D_f + \frac{n - D_f}{3}\right]}{b_f \, . \, D_f \, . \, \frac{c + c_1}{2} + b_w\,(n - D_f)\,\frac{c_1}{2}} \qquad ...(6.12)$$

where $$c_1 = \frac{n - D_f}{n} \, . \, c$$

$\therefore$ Lever arm $a = d - \bar{y}$

Hence $$M_r = \left\{b_f \, . \, D_f \, . \, \frac{c + c_1}{2} + b_w\,(n - D_f)\,\frac{c_1}{2}\right\} a \qquad ...(6.13)$$

Method 2.

The moment of resistance of the beam is the sum of the moment of resistance due to force in flange and the moment of resistance due to force in rib. Thus

$M_r = M_{r1} + M_{r2}$

where M_{r1} = moment of resistance of the force in flange

or $$M_{r1} = \frac{b_f \cdot D_f \cdot c}{2}\left[1 + \frac{n - D_f}{n}\right]\left[d - \frac{D_f}{2} + \frac{D_f^2}{6(2n - D_f)}\right] \quad ...(6.14)$$

and M_{r2} = moment of resistance of the force in rib.

$$= b_w (n - D_f) \times \frac{c_1}{2}\left[d - n + \frac{2}{3}(n - D_f)\right]$$

$$= \frac{b_w}{2}(n - D_f)\left\{\frac{n - D_f}{n} c\right\}\left[d - \frac{n}{3} - \frac{2}{3} D_f\right]$$

$$= \frac{c \cdot b_w}{2n}(n - D_f)^2\left[d - \frac{n + 2D_f}{3}\right] \quad ...(6.15)$$

5.6 DEPTH OF BALANCED SECTION OF T-BEAM

Let us find the depth of balanced section of T-beam, corresponding to external moment M.

The critical neutral axis depth factor k is given by :

$$k = \frac{mc}{mc + t} = \frac{m\sigma_{cbc}}{m\sigma_{cbc} + \sigma_{st}} \quad ...(i)$$

where $c = \sigma_{cbc}$ = permissible stress in concrete

or $t = \sigma_{st}$ = permissible stress in steel

The depth of balanced section of T-beam is determined on the assumption that the compression of the rib is negligible and can be ignored. Hence the compressive force (C) of beam is equal to the compressive force in flange, given by

$$C = b_f \cdot D_f \cdot \frac{1}{2}(c + c_1)$$

or $$C = b_f \cdot D_f \cdot \frac{1}{2}\left[c + \frac{n - D_f}{n} c\right] = \frac{b_f \cdot c \cdot D_f}{2}\left[\frac{2n - D_f}{n}\right]$$

$$= \frac{b_f \cdot c \cdot D_f}{2}\left[\frac{2kd - D_f}{kd}\right] \quad ...(ii)$$

$$\text{Lever arm} = a = d - \frac{D_f}{2} + \frac{D_f^2}{6(2n - D_f)} \quad ...(\text{Eq. } 6.8) \quad ...(iii)$$

$$M = M_r = \text{compressive force } C \times \text{lever arm } a$$

or $$M_r = \frac{b_f \cdot c \cdot D_f}{2}\left[\frac{2kd - D_f}{kd}\right]\left[d - \frac{D_f}{2} + \frac{D_f^2}{6(2\,kd - D_f)}\right]$$

or $$M = \frac{b_f \cdot c \cdot D_f}{12\,kd} [\,12\,k\,d^2 - 6\,d \cdot D_f - 6\,k\,d\,D_f + 3\,D_f^2 + D_f^2\,]$$

or $$M = \frac{b_f \cdot c \cdot D_f}{12\,kd} [12\,k\,d^2 - 6\,d\,D_f\,(1+k) + 4\,D_f^2]$$

or $$M = b_f \cdot c \cdot D_f \left[d - \frac{D_f(1+k)}{2k} + \frac{D_f^2}{3\,k\,d}\right] \qquad ...(6.16)$$

In the above expression, all quantities except d are known. Hence d can be calculated from the solution of the above quadratic equation.

Alternatively, d can be computed by *assuming* the lever arm a equal to $0.9\,d$, though its actual value may vary from $0.88\,d$ to $0.95\,d$. Then :

$$M = M_r = C \times a$$

$$= \frac{b_f \cdot c \cdot D_f}{2} \left[\frac{2k - D_f}{k \cdot d}\right] \times 0.9\,d \qquad ....[6.17(a)]$$

$$M = 0.45\,b_f \cdot c \cdot D_f \left[\frac{2kd - D_f}{k}\right] \qquad ...(6.17)$$

This is a linear equation in terms of the unknown d.

6.7 ECONOMICAL DEPTH OF T - BEAM

The balanced depth found above is not economical. This is because of the fact that in T-beam large compression area is provided by the flange and consequently balanced depth is comparatively reduced and area of steel is increased. The cost of steel is several times more than the cost of concrete and thus overall cost of T-beam, having balanced depth is increased. To find an expression for the economical depth of beam, we will have to take into account the *cost-ratio* of steel and concrete.

Let R_s = cost of steel, per cubic mm = $r \cdot R_c$

R_c = cost of concrete, per cubic mm

r = cost ratio = $\dfrac{\text{cost of steel}}{\text{cost of concrete}}$ per unit volume

$= \dfrac{R_s}{R_c}$

d_c = depth of cover, measured below the centre of the tensile steel of T-beam.

The volume of concrete provided by the flange is *fixed,* and is, therefore, not included in the cost calculations.

Volume of concrete, per mm length, in the web $= b_w(d - D_f + d_c)$

$\therefore$ cost of concrete in the rib, per mm length

$$= b_w (d - D_f - d_c) R_c$$

Similarly, cost of steel, per mm length of beam

$$= A_{st} R_c = r . R_c . A_{st}$$

But $A_{st} = \frac{M}{tjd}$, where $t = \sigma_{st}$ = permissible stress in steel

$\therefore$ Total cost (Q) of the beam, per mm length is given by

$$Q = R_c . b_w (d - D_f + d_c) + r . R_c . \frac{M}{tjd}$$

or

$$Q = R_c . \left[b_w (d - D_f + d_c) + \frac{r . M}{tjd} \right]$$

In the above expression, b_w, D_f and d_c are fixed dimensions while depth d is variable. For the cost to be *minimum* ,

$$\frac{dQ}{dd} = 0 = R_c \left[b_w - \frac{r M}{tjd^2} \right]$$

$$\therefore \quad d^2 = \frac{r . M}{tjb_w}, \text{ where } t = \sigma_{st}$$

or

$$d = \sqrt{\frac{r . M}{tj\, b_w}} = \sqrt{\frac{r . M}{\sigma_{st} j\, b_w}} \quad \text{...(6.18)}$$

Taking $t = \sigma_{st}$ = 140 N/mm^2 and j = 0.9, the above expression reduces to

or

$$d = \sqrt{\frac{r . M}{126\, b_w}} \quad \text{...[6.18(a)]}$$

Similarly, taking σ_{st} = 230 N/mm^2 and j = 0.9, we have

$$d = \sqrt{\frac{r . M}{207\, b_w}} \quad \text{...[6.18(b)]}$$

In the above expression, the cost of shuttering has not been taken into account. For deeper beams, the cost of shuttering may be appreciable. In that case the depth of beam found from Eq. 6.18 may be further reduced by 10 to 15% to account for the additional cost of shuttering.

The width of web (b_w) is kept a certain minimum, so as to accommodate tensile steel. Sometimes, steel may be provided in two or three layers, but this reduces the lever arm. Suitable value of b_w is also selected on the consideration of shear.

6.8 SHEAR, BOND AND DEVELOPMENT LENGTH

1. *Shear stress distribution*

The distribution of shear stress in T-beam can be found exactly in the same way as for a rectangular beam. Fig. 6.6 (*b*) shows the shear stress distribution across the depth.

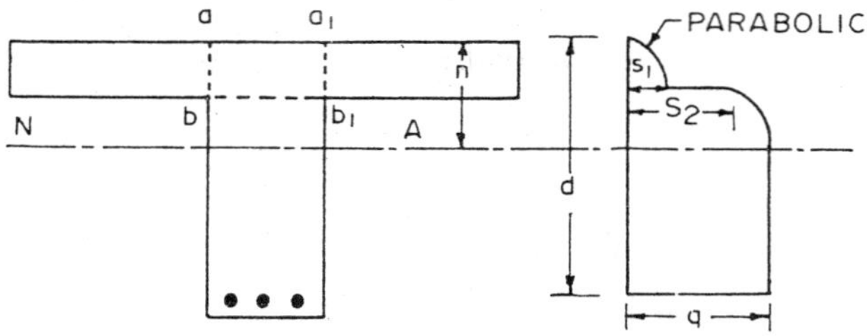

FIG. 6.6 SHEAR STRESS DISTRIBUTION.

The maximum shear (q) is induced in all fibres between the N.A. and the centre of tensile reinforcement and its magnitude is given by Eq. 3.1,

$$q = \frac{V}{b_w \cdot a} \quad ...(6.19)$$

where a = lever arm $= d - \frac{D_f}{2} + \frac{D_f^2}{6\,(2n - D_f)}$

The shear stress (s) at any layer y above the N.A. is given by Eq. 3.2.

$$s = \frac{V}{2I}(n^2 - y^2). \quad ...(6.20)$$

where I is the moment of inertia of the T-beam section, about the N.A. The ratio of stress s_1 in the flange to stress s_2 in the rib, at the junction is evidently given by

$$\frac{s_1}{s_2} = \frac{b_w}{b_f}$$

The complete shear stress distribution across the section is shown in Fig. 6.6 (*b*).

(*b*) *Nominal Shear Stress*

Indian Standard Code (IS : 456–1978) recommends the use of nominal shear stress (τ_v) in design given by the expression,

$$\tau_v = \frac{V}{b_w \cdot d} \quad ...(6.21)$$

(c) ***Flexural bond and development length***

The flexural bond stress around main reinforcement is given by Eq. 3.23 :

$$\tau_{bf} = \frac{V}{a\,\Sigma 0}$$

However, in the case of bond, it has been recently concluded that the concept of *anchorage bond* is more meaningful than the concept of flexural bond stresses of localised nature. Current design methods disregard such local bond stresses, and recommend the use of development length given by Eq. 3.25.

$$L_d = \frac{\Phi\,.\,\sigma_{st}}{4\,\tau_{bd}}$$

6.9 TYPES OF PROBLEMS IN T–BEAM

There may be three types of problems in T-beam :

Type I. Determination of moment of resistance of the given section.

Type II. Determination of stresses in the given section subjected to a given bending moment.

Type III. Design of section to resist a given bending moment.

TYPE I. DETERMINATION OF MOMENT OF RESISTANCE

In this type of problem, the section including the steel reinforcement is completely defined and it is required to determine the moment of resistance corresponding to a given set of permissible stresses in concrete and steel. The procedure for solution is as under:

1. Determine the position of the neutral axis by equating the moment of the compressive area of concrete to the area of concrete equivalent to the area of steel in tension. Assuming that the N.A. lies in the flange, we get from Eq. 6.1 :

$$\frac{b_f\,.\,n^2}{2} = m\,.\,A_{st}\,(d - n)$$

If n comes out to be greater that D_f , N.A. falls in the rib and Eq. 6.4 is used :

$$b_f\,.\,D_f\left[n - \frac{D_f}{2}\right] = m\,.\,A_{st}\,(d - n)$$

Thus n is known.

2. Determine the position of critical neutral axis, on the basis of permissible stresses and m, using Eq. 6.5 :

$$\frac{n_c}{d - n_c} = \frac{c}{t/m} = \frac{\sigma_{cbc}}{\sigma_{st}/m}$$

or $$n_c = kd = \frac{d}{1 + t/mc} = \frac{d}{1 + \sigma_{st}/m \,.\, \sigma_{cbc}}$$

where $c = \sigma_{cbc}$ = permissible stress in concrete

and $t = \sigma_{st}$ = permissible stress in steel.

3. If the actual neutral axis is *above* the critical N.A., the section is *under-reinforced* and steel reaches maximum stress (σ_{st}) first. The actual stress c in concrete is calculated from the relation

$$c = \frac{t}{m} \cdot \frac{n}{d - n}$$

(where $t = \sigma_{st}$ = permissible stress in steel)

If the N.A. lies in the flange, the moment of resistance is given by

$$M_r = b_f \,.\, n \frac{c}{2} \left(d - \frac{n}{3}\right)$$

If the N.A. lies in the web, the moment of resistance is given by

$$M_r = \frac{b_f \,.\, D_f \,.\, c}{2} \left[1 + \frac{n - D_f}{n}\right] \left[d - \frac{D_f}{2} + \frac{D_f^2}{6\,(2n - D_f)}\right]$$

Alternatively, the moment of resistance may be calculated from Eq. 6.10.

$$M_r = t \,.\, \sigma_{st} \,.\, a = t A_{st} \left[d - \frac{D_f}{2} + \frac{D_f^2}{6\,(2n - D_f)}\right]$$

4. If the actual neutral axis is below the critical N.A., the section is *over-reinforced,* and concrete reaches maximum stress (σ_{cbc}) first.

If the N.A. falls in the flange, the moment of resistance is given by Eq. 6.9 (a).

$$M_f = b_f \,.\, n \,.\, \frac{c}{2} \left[d - \frac{n}{3}\right]$$

(where $c = \sigma_{cbc}$ = permissible stress in concrete)

If the N.A. falls in the web, the moment of resistance is given by Eq. 6.9.

$$M_r = \frac{b_f \,.\, D_f \,.\, c}{2} \left[1 + \frac{n - D_f}{n}\right] \left[d - \frac{D_f}{2} + \frac{D_f^2}{6\,(2n - D_f)}\right]$$

5. If it is further required to find the uniformly distributed load or point load or other type of load on a simply supported

beam, calculate the maximum B.M. in terms of the unknown load and equate it to the moment of resistance determined in step 3 or 4.

TYPE II. DETERMINATION OF STRESSES IN THE SECTION

In this type of problem, the dimensions of the section, including the reinforcement, are completely known and it is required to find the stresses developed in steel and concrete, when subjected to a given bending moment. The solution is done in the following steps :

1. Determine the position of the neutral axis by equating the moments of equivalent areas about the N.A. Assuming that the N.A. lies in the flange, we get from Eq. 6.1.

$$\frac{b_f \, . \, n^2}{2} = m \, A_{st} \, (d - n).$$

If n comes out to be greater than D_f, N.A. falls in the rib and Eq. 6.4 is used,

$$b_f \, . \, D_f \left(n - \frac{D_f}{2}\right) = m \, . \, A_{st} \, (d - n)$$

Thus n is known.

2. Equate the moment of resistance of the beam to the external moment M, and determine the compressive stress in concrete.

Thus, if the N.A. falls in the flange, we have

$$M = M_r = b_f \, . \, n \, . \, \frac{c}{2} \left(d - \frac{n}{3}\right)$$

from which c can be calculated.

If the N.A. falls in the rib, we have from Eq. 6.9,

$$M = M_r = \frac{b_f \, . \, D_f \, . \, c}{2} \left[1 + \frac{n - D_f}{n}\right] \left[d - \frac{D_f}{2} + \frac{D_f^2}{6\,(2n - D_f)}\right]$$

Thus maximum compressive stress c in concrete is known.

3. Calculate the tensile stress in steel by the stress relationlship

$$t = mc \, \frac{d - n}{n}$$

TYPE III. DESIGN OF THE SECTION

This is more common type of problem in which it is required to determine the dimensions of the rib including tensile reinforcement to develop a given moment of resistance corresponding to a given set of permissible stresses in concrete ($c = \sigma_{cbc}$) and steel

($t = \sigma_{st}$). The dimensions of the flange (*i.e.* b_f and D_f) are treated as known.

1. Assume some suitable value of width of rib based on criteria discussed in § 6.2.

2. Determine the effective width b_f of the flange, as discussed in § 6.2.

3. Determine the efffective depth of the beam on three considerations : (*i*) balanced section (*ii*) economy (*iii*) shear stress.

For balanced section, the effective depth is given by Eq. 6.16

$$M = b_f \, . \, c \, . \, D_f \left[d - \frac{D_f(1+k)}{2k} + \frac{D_f^2}{3kd} \right]$$

Alternatively, taking lever arm to be equal to 0.9*d*, and use Eq. 6.17.

$$M = 0.45 \, b_f \, . \, c \, . \, D_f \left[\frac{2kd - D_f}{k} \right]$$

For economical depth, use Eq. 6.18

$$d = \sqrt{\frac{rM}{t j b_w}}$$

For shear considerations, use Eq. 6.21

$$\tau_v = \frac{V}{b_w \, . \, d}$$

or

$$d = \frac{V}{b_w \, . \, \tau_v}$$

where τ_v may be taken equal to $\tau_{c.max}$ given in Table 3.3.

The *maximum* value of d obtained from the above three criteria is taken as the effective depth of the beam.

Take the approximate lever arm $= d - \frac{D_f}{2}$, and determine the area of tensile reinforcement

$$A_{st} = \frac{M}{t \times a} = \frac{M}{\sigma_{st} \times a}$$

Arrange suitably the steel reinforcement.

5. Determine the actual N.A. and check the stresses in concrete and steel.

6. Check the section for shear and provide shear reinforcement if necessary.

7. Check for bond.

The complete design procedure for T-beam is given in Chapter 7.

Example 6.1 *An isolated T-beam, simply supported over a span of 6 m has following dimensions : Width of flange 750 mm, thickness of flange 125 mm ; over all depth of 400 mm ; width of web 260 mm ; effective cover to tensile reinforcement, 40 mm. The beam is reinforced with 4 bars of 20 mm dia. Determine the moment of resistance of the beam if (a) mild steel bars are used (b) Fe 415 steel bars are used. Take* $\sigma_{cbc} = 5\ N/mm^2$ *and* $m = 19$ *in each case.*

Solution. The section of the T-beam, is shown in Fig. 6.7

Assume that N.A. lies in flange.

$$\text{Area of steel} = A_{st} = 4 \times \frac{\pi}{4}(20)^2 = 1250.6\ \text{mm}^2$$

For an isolated T-beam, effective flange width

$$b_f = \frac{l_0}{\frac{l_0}{b} + 4} + b_w \quad (\text{where } l_0 = 6000\ \text{mm})$$

$$\therefore \quad b_f = \frac{6000}{\frac{6000}{750} + 4} + 260 = 760\ \text{mm}$$

Available $b = 750$ mm.

$\therefore \quad b_f = b = 750$ mm

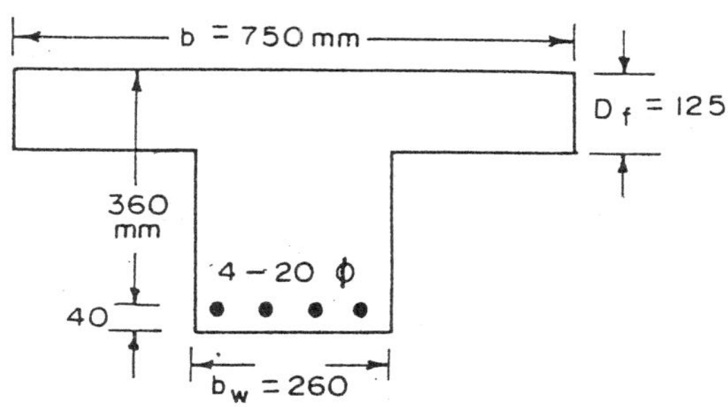

FIG. 6.7

Equating the moments of equivalent areas about the N.A., we get

$$750 \times \frac{n^2}{2} = 19 \times 1250.6\,(360 - n)$$

or $$n^2 = 22920 - 63.67\,n$$

or $$n^2 + 63.67\,n - 22920 = 0$$

$$n = 122.9\ \text{mm}$$

This is less than D_f (=125 mm).

Hence the N.A. just lies inside the flange.

(a) Mild steel reinforcement.

The depth of critical N.A. is given by

$$n_c = \frac{d}{1+\dfrac{\sigma_{st}}{m\sigma_{cbc}}} = \frac{360}{1+\dfrac{140}{19\times 5}} = 145.5 \text{ mm}$$

Since the actual N.A. falls *above* the critical N.A., the section is *under-reinforced* and the stress in steel reaches its maximum value first. Thus, $t = \sigma_{st} = 140$ N/mm². The corresponding stress in concrete is given by

$$c = \frac{t}{m}\frac{n}{d-n} = \frac{140}{19}\times\frac{122.9}{360-122.9} = 3.82 \text{ N}/mm^2$$

The moment of resistance of the beam is given by

$$M_r = b_f \cdot n \cdot \frac{c}{2}\left(d-\frac{n}{3}\right)$$

$$= 750\times 122.9\times\frac{3.82}{2}\left(360-\frac{122.9}{3}\right)$$

$$= 56.17\times 10^6 \text{ N-mm} = \mathbf{56.17 \text{ kN-m.}}$$

(*b*) Fe 415 Steel.

The depth of critical N.A. is given by

$$n_c = \frac{d}{1+\dfrac{\sigma_{st}}{m\sigma_{cbc}}} = \frac{360}{1+\dfrac{230}{19\times 5}} = 105.2 \text{ mm}$$

Hence $n > n_c$.

Since the actual N.A. falls *below* the critical N.A., the section is *over-reinforced* and the stress in concrete reaches its maximum value first. Thus $c = \sigma_{cbc} = 5$ N/mm². The corresponding stress in steel is given by

$$t = \frac{m\cdot c\,(d-n)}{n} = \frac{19\times 5\,(360-122.9)}{122.9} = 183.3 \quad \text{N/mm}^2$$

The moment of resistance of the beams is given by

$$M_r = b_f \cdot n \cdot \frac{c}{2}\left(d-\frac{n}{3}\right)$$

$$= 750\times 122.9\times\frac{5}{2}\left(360-\frac{122.9}{3}\right)$$

$$= 73.52\times 10^6 \text{ N-mm} = \mathbf{73.52 \text{ kN-m}}\ .$$

Example 6.2. *An isolated T-beam has a flange width of 1000 mm, flange thickness of 80 mm and effective depth of 400 mm. The rib is 240 mm wide and reinforced with 5 bars of 20 mm diameter. Determine the moment of resistance of the section, if the permissible stresses in concrete and steel are 5.2 N/mm² and 140 N/mm² and m = 18. Neglect the compressive force in the web. The beam is simply supported over a span of 4 m.*

Solution.

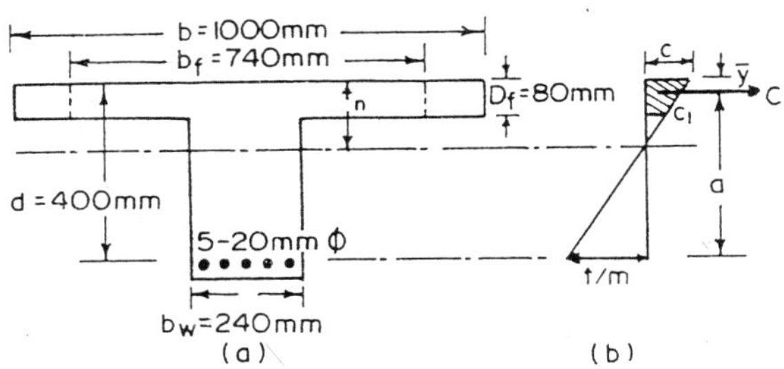

FIG. 6.8.

The effective width $b_f = \dfrac{l_0}{\left(\dfrac{l_0}{b}\right) + 4} + b_w$

where l_o=effective span =4000 mm.

$$b_f = \frac{4000}{\left(\frac{4000}{1000}\right) + 4} + 240 = 740 \text{ mm}$$

Thus the effective width of flange is very much less than the actual flange width.

Area of steel $A_{st} = 5\frac{\pi}{4}(20)^2 = 1570 \text{ mm}^2$

Equating the moments of equivalent areas about N.A. and assuming that N.A. falls in the flange, we get

$$b_f \cdot \frac{n^2}{2} = m \cdot A_{st}(d - n)$$

or $$740 \times \frac{n^2}{2} = 18 \times 1570\,(400 - n)$$

or $$n^2 + 76.4n - 30550 = 0$$

From which n=141 mm > D_f

Hence N.A. falls in the web.

Again, equating the moments of equivalent areas, and neglecting the compression of the web, we get

$$b_f \, . \, D_f \left(n - \frac{D_f}{2}\right) = m \, A_{st} \, (d - n)$$

or $$740 \times 80 \, (n - \frac{80}{2}) = 18 \times 1570 \, (400 - n)$$

or $$1.477 \, n = 230.9$$

From which $n = 156.4$ mm

Taking $\sigma_{cbc} = 5.2$ N/mm^2, $\sigma_{st} = 140$ N/mm^2 and $m = 18$, depth of critical N.A. is given by

$$n_c = \frac{d}{1 + \frac{\sigma_{st}}{m\sigma_{cbc}}} = \frac{400}{1 + \frac{140}{18 \times 5.2}} = 160.3 \text{ mm}$$

Since the actual N.A. falls above the critical N.A., the stress in steel reaches the maximum value first. Hence $t = \sigma_{st} = 140$ N/mm^2. The corresponding stress in concrete at the outer fibre is given by

$$c = \frac{t}{m} \frac{n}{d - n} = \frac{140}{18} \times \frac{156.4}{400 - 156.4} = 4.99 \text{ N/mm}^2$$

$$c_1 = \frac{n - D_f}{n} c = \frac{156.4 - 80}{156.4} \times 4.99 = 2.44 \text{ N/mm}^2$$

Hence $$\bar{y} = \left[\frac{c + 2c_1}{c + c_1}\right] \frac{D_f}{3} = \left[\frac{4.99 + 4.88}{4.99 + 2.44}\right] \frac{80}{3} = 35.4 \text{ mm.}$$

Lever arm $a = d - \bar{y} = 400 - 35.4 = 364.6$ mm

$$M_r = b_f \, d_f \frac{c + c_1}{2} \times a = 740 \times 80 \frac{4.99 + 2.44}{2} \times 364.6$$

$$= 80.1 \times 10^6 \text{ N-mm}$$

$$= \mathbf{80.1 \text{ kN-m.}}$$

Alternatively,

$$M_r = \sigma_{st} \, A_{st} \, . \, a = 140 \times 1570 \times 364.6$$

$$= 80.1 \times 10^6 \text{ N-mm} = 80.1 \text{ kN-m.}$$

Example 6.3 *Solve Example 6.2 taking web compression into account.*

Solution. The N.A. will lie in the web at a distance n below the top of the flange.

Equating the moments of equivalent areas about the N.A.,

$$b_f \, . \, D_f \left(n - \frac{D_f}{2}\right) + \frac{b_w}{2} (n - D_f)^2 = m \, A_{st} \, (d - n)$$

$$740 \times 80(n - 40) + 120 \, (n - 80)^2 = 18 \times 1570 \, (400 - n)$$

or $$493.3 \, n - 19733 + n^2 + 6400 - 160 \, n = 94200 - 235.5n$$

or $$n^2 + 568.8n - 107533 = 0$$

From which $n = 149.7$ mm.

As before $n_c = 160.3$ mm.

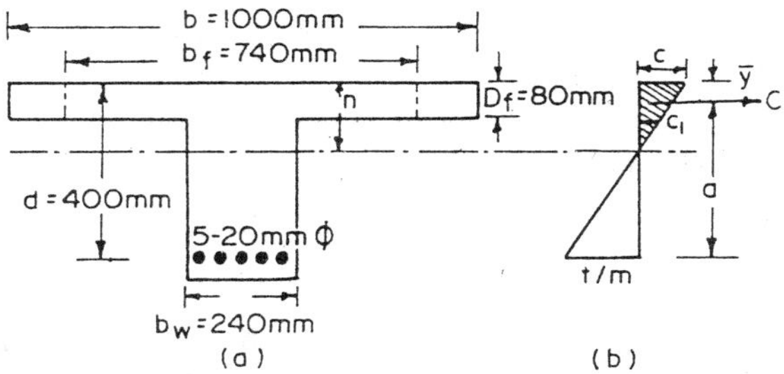

FIG. 6.9.

The actual N.A. falls above the critical one and hence the section is under-reinforced. The stress in steel reaches its maximum value first. Thus, $t = \sigma_{st} = 140$ N/mm^2. The corresponding value of stress in concrete at the outer fibre is given by

$$c = \frac{t}{m} \cdot \frac{n}{d-n} = \frac{140}{18} \times \frac{149.7}{400 - 149.7} = 4.65 \quad \text{N/mm}^2$$

$$c_1 = \frac{n - D_f}{n} \cdot c = \frac{149.7 - 80}{149.7} \times 4.65 = 2.16 \quad \text{N/mm}^2$$

$$\bar{y} = \frac{b_f \cdot D_f \cdot \frac{c + c_1}{2} \times \frac{D_f}{3}\left\{\frac{2c_1 + c}{c_1 + c}\right\} + b_w\,(n - D_f) \times \frac{c_1}{2}\left\{D_f + \frac{n - D_f}{3}\right\}}{b_f \cdot D_f \cdot \frac{c + c_1}{2} + b_w\,(n - D_f)\,\frac{c_1}{2}}$$

$$= \left[740 \times 80 \left\{\frac{4.65 + 2.16}{2}\right\} \times \frac{80}{3}\left\{\frac{4.32 + 4.65}{2.16 + 4.65}\right\}\right.$$

$$\left. + 240 \left\{149.7 - 80\right\} \times \frac{2.16}{2}\left\{80 + \frac{149.7 - 80}{3}\right\}\right]$$

$$\div \left[740 \times 80 \left\{\frac{4.65 + 2.16}{2}\right\} + 240 \left\{149.7 - 80\right\} \frac{2.16}{2}\right]$$

$$= 40.8 \text{ mm}$$

$\therefore \quad$ L.A. $= a = d - \bar{y} = 400 - 40.8 = 359.2$ mm

$$M_r = \left\{ b_f \,.\, D_f \frac{c + c_1}{2} + b_w (n - D_f) \frac{c_1}{2} \right\} a$$

$$= \left\{ 740 \times 80 \left(\frac{4.65 + 2.16}{2} \right) + 240 (149.7 - 80) \frac{2.16}{2} \right\} 359.2$$

$$= 79 \times 10^6 \text{ N–mm} = \mathbf{79 \text{ kN–m}}$$

Alternatively,

$$M_r = \sigma_{st} \,.\, A_{st} \,.\, a = 140 \times 1570 \times 359.2$$

$$= 79 \times 10^6 \text{ N-mm} = 79 \text{ kN-m}.$$

Example 6.4. *Given the T-beam of the dimensions of Ex. 6.2 except that the reinforcement consists of 4 bars of 25 mm diameter. Determine the maximum uniformly distributed load, inclusive of its self weight, that the beam can carry over an effective span of 4 m. Neglect the compression of the rib.*

Solution.

$$A_{st} = 4 \frac{\pi}{4} (25)^2 = 1963 \text{ mm}^2.$$

Let the N.A. be in the web at a distance n below the top of the flange. Equating the moments of the equivalent areas about the N.A., we get

$$b_f D_f \left(n - \frac{D_f}{2} \right) = m A_{st} (d - n)$$

or $\quad 740 \times 80 \left(n - \dfrac{80}{2} \right) = 18 \times 1963 (400 - n)$

from which $\quad n = 174.6$ mm.

But $n_c = 160.3$ mm as before.

Since the actual N.A. falls below the critical N.A., the section is over-reinforced and hence the stress in concrete reaches its maximum value of $c = \sigma_{cbc} = 5.2$ N/mm^2 first.

$\therefore \quad c_1 = \dfrac{n - D_f}{n} c = \dfrac{174.6 - 80}{174.6} \times 5.2 = 2.82$ N/mm^2

and $\quad t = mc \dfrac{d - n}{n} = 18 \times 5.2 \dfrac{400 - 174.6}{174.6} = 120.83$ N/mm^2

$$\bar{y} = \frac{c + 2c_1}{c + c_1} \times \frac{D_f}{3} = \frac{5.2 + 5.64}{5.2 + 2.82} \times \frac{80}{3} = 36 \text{ mm}$$

L.A. $\quad a = d - \bar{y} = 400 - 36 = 364$ mm

$$M_r = b_r \,.\, D_f \,.\, \frac{c + c_1}{2} \times a = 740 \times 80 \,\frac{5.2 + 2.82}{2} \times 364$$

$$= 86.40 \times 10^6 \text{ N-mm} = 86.40 \text{ kN-m}$$

(*Alternatively* $M_r = t \,.\, A_{st} \,.\, a = 120.83 \times 1963 \times 364 = 86.4 \times 10^6$ N-mm)

Let w kN/m be the safe uniformly distributed load, inclusive of the safe weight, that the beam can carry,

$$\therefore \quad M_{max} = \frac{wL^2}{8} = \frac{w \times 16}{2} = 2w \text{ kN-m.}$$

Equating this to the moment of resistance of the beam.

$$2w = 86.4$$

$$\therefore \quad w = \mathbf{43.2 \ kN/m.}$$

Example 6.5. *The flange of a T-beam is 120 mm thick and 1200 mm wide. The web of the beam is 300 mm wide and depth upto the centre of tensile reinforcement is 700 mm. The reinforcement consists of 5 bars of 25 mm diameter arranged in one row. If the beam section is subjected to a B.M. of 200 kN-m, determine the stresses developed in concrete and steel. Take m = 18. The beam is simply supported over a span of 6 m.*

Solution.

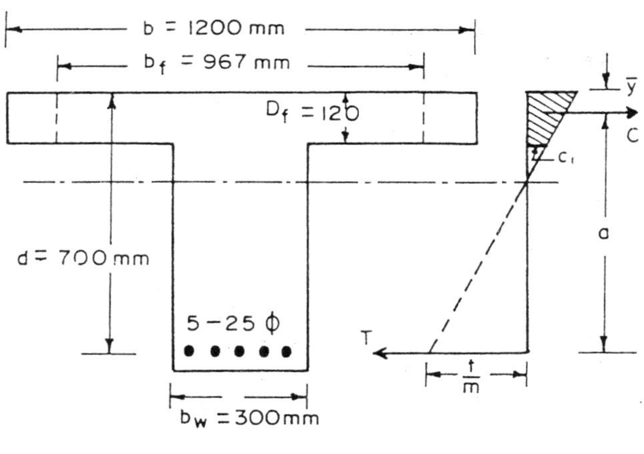

FIG. 6.10.

The effective width $b_f = \dfrac{l_0}{\left(\dfrac{l_0}{b}\right) + 4} + b_w$

$$b_f = \frac{6000}{\frac{6000}{1200}+4} + 300 = 967 \text{ mm}$$

This is less than the actual width (b) of the flange.

$$A_{st} = 5 \times \frac{\pi}{4}(25)^2 \approx 2460 \text{ mm}^2$$

Let the depth of N.A. be n. Equating the moments of equivalent areas above N.A. and neglecting the compression of web, we get

$$b_f . D_f \left[n - \frac{D_f}{2}\right] = m\,A_{st}\,(d-n)$$

or $\quad 967 \times 120\,(n - 60) = 18 \times 2460(700 - n)$

From which $\quad n = 236.8$ mm

Let c be the maximum compressive stress in concrete at the top of the flange and c_1 be the compressive stress at the bottom of the flange.

$$c_1 = \frac{n - D_f}{n} . c = \frac{236.8 - 120}{236.8} \times c = 0.493\,c$$

$$\therefore \quad \bar{y} = \left[\frac{c + 2c_1}{c + c_1}\right]\frac{D_f}{3} = \left[\frac{1 + 0.986}{1 + 0.493}\right] \times \frac{120}{3} = 53.2 \text{ mm.}$$

L.A. $\quad a = D - \bar{y} = 700 - 53.2 = 646.8$ mm.

$$\therefore \quad M_r = b_f . D_f . \frac{c+c_1}{2} \times a$$

$$= 967 \times 120 \frac{1 + 0.493}{2} c \times 646.8$$

External B.M.=200 kN-m

Equating the two, we get

$56.03\ c = 200$

or $\quad$ **$c = 3.57$ N/mm²**

The stress in steel is given by

$$t = mc\frac{d-n}{n} = 18 \times 3.57 \frac{700 - 236.8}{236.8}$$

$$= \mathbf{125.68 \ N/mm^2}$$

Alternatively,

$$t = \frac{M}{A_{st} \times a} = \frac{200 \times 10^6}{2460 \times 646.8}$$

$$= 125.68 \text{ N/mm}^2.$$

Example 6.6. *A reinforced concrete T-beam has a flange 1200 mm wide and 100 mm thick. The effective depth of the beam is 400 mm and width of web is 250 mm. Determine the moment of resistance of the beam and the area of steel in tension, if m = 19, and the permissible stresses in concrete and steel are 5 N/mm² and 140 N/mm² respectively. Assume that the beam is simply supported over a span of 6 m.*

Solution.

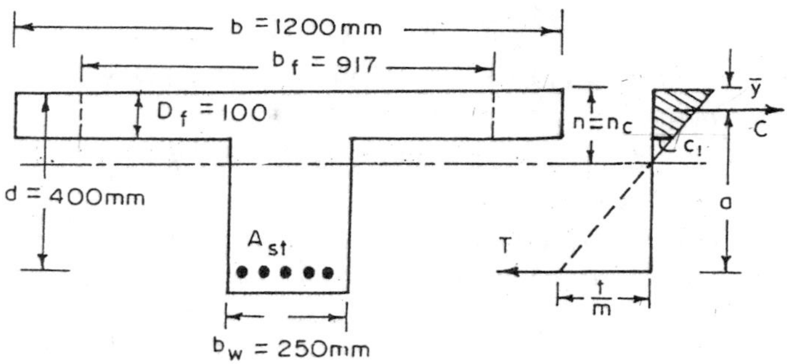

FIG 6.11

For an isolated T-beam,

$$b_f = \frac{l_0}{\frac{l_0}{b} + 4} + b_w = \frac{6000}{\frac{6000}{1200} + 4} + 250$$

$$\approx 917 \text{ mm.}$$

The area of tensile steel is to be found for balanced section, for which the depth of critical N.A. is given by

$$n_c = \frac{d}{1 + \frac{\sigma_{st}}{m\sigma_{cbc}}} = \frac{400}{1 + \frac{140}{19 \times 5}} \approx 162 \text{ mm}$$

Equating the moments of equivalent areas about the N.A. and neglecting the compressive area of the web, we get

$$b_f \,.\, D_f \left(n_c - \frac{D_f}{2}\right) = m\, A_{st}\, (d - n_c)$$

or

$$917 \times 1000 \left(162 - \frac{100}{2}\right) = 19 \times A_{st}(400 - 162)$$

From which A_{st} = **2271 mm²**

The moment of resistance of the beam is given by

$$M_r = b_f \, . D_f \left(\frac{c + c_1}{2}\right) \times \text{lever arm}$$

Now $$c_1 = \frac{n_c - D_f}{n_c} . c = \frac{162 - 100}{162} \times 5 = 1.914 \text{ N/mm}^2$$

$$\bar{y} = \left(\frac{c + 2c_1}{c + c_1}\right) \frac{D_f}{3} = \left(\frac{5 + 2 \times 1.914}{5 + 1.914}\right) \frac{100}{3} = 42.6 \text{ mm}$$

$\therefore$ $$a = d - \bar{y} = 400 - 42.6 = 357.4 \text{ mm}$$

$\therefore$ $$M_r = 917 \times 100 \left(\frac{5 + 1.914}{2}\right) \times 357.4$$

$$\approx 113.4 \times 10^6 \text{ N-mm} = \mathbf{133.4 \text{ kN-m}}$$

Alternatively ,

$$M_r = \sigma_{st} \, . A_{st} \, a = 140 \times 2271 \times 357.4$$

$$\approx 113.6 \times 10^6 \text{ N-mm}$$

Example 6.7. *A T-beam roof consists of 100 mm thick reinforced concrete slab cast monolithic with 300 mm beams spaced 4 metres centre to centre . The super-imposed load over the slab is 3.9 kN/m², and the effective span of the beam is 5 metres. Determine the depth and area of steel in one of the intermediate beams. Take permissible stresses in concrete and steel as 5 N/mm² and 140 N/mm² respectively and m = 18.*

Solution. Given D_f= 100 mm and b_w= 300 mm.

The effective width of the flange is given by

$$b_f = \frac{l_0}{6} + b_w + 6D_f$$

$$= \frac{5000}{6} + 300 + 6 \times 100 = 1733 \text{ mm}$$

This is less than b_w plus half the sum of clear distances to the adjacent beams on either side.

The effective depth d can be determined on the basis of two criteria : (*i*) B.M. or maximum allowable compressive stress in concrete (*ii*) shear considerations and (*iii*) economy.

For the balanced section, taking $c = \sigma_{cbc} = 5$ N/ mm² and $t = \sigma_{st} =$ 140 N/mm², we get

$$k = \frac{mc}{mc + t} = \frac{18 \times 5}{11 \times 5 + 140} = 0.301$$

Assume a=0.9d for preliminary calculations.

Hence from Eq. 6.17 (*a*).

$$M_r = b_f \,.\, c \,.\, \frac{D_f}{2}\left[\frac{2kd - D_f}{kd}\right] \times a, \text{ where } a = 0.9d$$

$$\therefore \quad M_r = 1733 \times 5 \times 50 \left[\frac{2 \times 0.391\,d - 100}{0.391\,d}\right] \times 0.9\,d$$

or $\quad M_r = 779850\,[0.782\,d - 100]$ N–mm

$$\approx 0.78\,[0.782\,d - 100] \text{ kN–m}$$

(*Alternatively*, we can use Eq. 6.16 in which it is not necessary to assume a).

Unit weight of concrete = 25000 N/mm^3

Now, load of slab per sq. m $= \dfrac{(25000) \times 100}{1000} = 2500$ N

Super-imposed load per sq. m. $= 3900$ N

Total $= 6400$ N

$\therefore$ Load per metre run of beam $= 6400 \times 4 = 25600$ N

Weight of web, per metre run $= 2400$ N

Total = 28000 N = 28 kN

$$\text{B.M. } M = \frac{wL^2}{8} = \frac{28 \times 25}{8} = 87.5 \text{ kN–m}$$

Equating this to the moment of resistance, we get

$$0.78\,[0.782\,d - 100] = 87.5$$

From which $d \approx 272$ mm

$$\text{S.F.} = V = \frac{wl}{2} = \frac{28 \times 5}{2} = 70 \text{ kN}$$

From shear considerations,

$$\tau_v = \frac{V}{bd} = \frac{V}{b_w \,.\, d}$$

Taking $\tau_v = \tau_{c.max} = 1.6$ N/mm^2, we get

$$d_{min.} = \frac{V}{b_w \,.\, \tau_v} = \frac{70000}{300 \times 1.6} \approx 146 \text{ mm.}$$

Economical depth is given by Eq. 6.13

$$d = \sqrt{\frac{r \,.\, M}{\sigma_{st} \,.\, jb_w}}, \text{ where } M = 87.5 \times 10^6 \text{ N-mm}$$

Taking $r = 60$

$$d = \sqrt{\frac{60 \times 87.5 \times 10^6}{140 \times 0.9 \times 300}} = 372 \text{ mm}$$

Adopt d=370 mm. Lever arm $a = 0.9 \times 370 = 333$ mm

Area of steel, $A_{st} = \dfrac{M}{\sigma_{st} \times a} = \dfrac{87.5 \times 10^6}{140 \times 333} = 1867 \text{ mm}^2$

Using 5-22 mm Φ bars, actual area provided is

$$A_{st} = 5 \times \frac{\pi}{4}(22)^2 = 1900 \text{ mm}^2$$

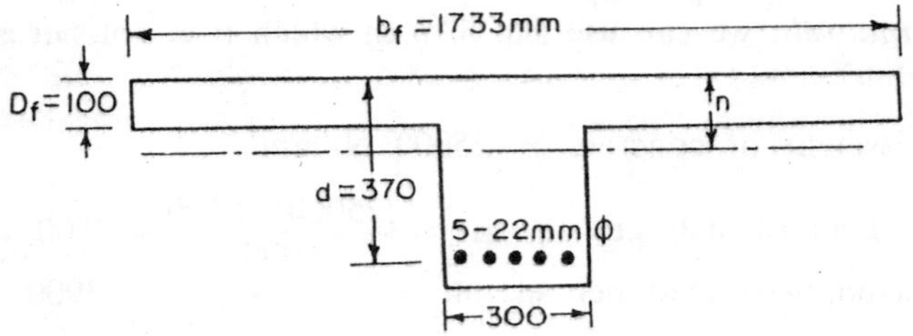

FIG. 6.12.

Thus the section is completely known (Fig 6.12). However, we have to determine the actual stresses induced in the section.

To find the position of the N.A. , we have

$$b_f \, . \, D_f \, . \left(n - \frac{D_f}{2}\right) = m \, A_{st}\,(d - n)$$

or $\quad 1733 \times 100\,(n - 50) = 18 \times 1900\,(370 - n)$

or $\quad 1.197\, n = 123$

$\therefore \quad n = 102.7$ mm

$$\bar{y} = \frac{3n - 2D_f}{2n - D_f} \times \frac{D_f}{3} = \frac{3 \times 102.7 - 200}{2 \times 102.7 - 100} \times \frac{100}{3} = 34.2 \text{ mm}$$

$\therefore \quad a = d - \bar{y} = 370 - 34.2 = 335.8$ mm

(against the assumed value of 333 mm assumed earlier).

$$M_r = b_f \, . \, D_f \frac{c + c_1}{2} a = b_f D_f \frac{c}{2}\left[1 + \frac{n - D_f}{n}\right] \times a$$

$$= b_f \, . \, D_f \, \frac{c}{2} \, \frac{2n - D_f}{2} \times a$$

$$= 1733 \times 100 \times \frac{c}{2} \, \frac{2 \times 102.7 - 100}{102.7} \times 335.8$$

$$= 29.86 \times 10^6 c \text{ N–mm} = 29.86\, c \text{ kN–m}$$

Equating this to the external B.M. of 87.5 kN–m

$$29.86 \text{ c} = 87.5$$

$$c = \mathbf{2.93 \ N/mm^2}$$

The corresponding stress in steel is given by

$$t = mc\frac{d-n}{n} = 18 \times 2.93\frac{370 - 102.7}{102.7}$$

$$= 137.3 \text{ N/mm}^2$$

6.10 DOUBLY REINFORCED T-BEAMS

Sometimes, the dimensions of the web are restricted from architectural or other considerations. If the beam is expected to carry more bending moment than it can take as a balanced section, compression reinforcement is provided to increase its moment of resistance. Such a beam is known as doubly reinforced T-beam.

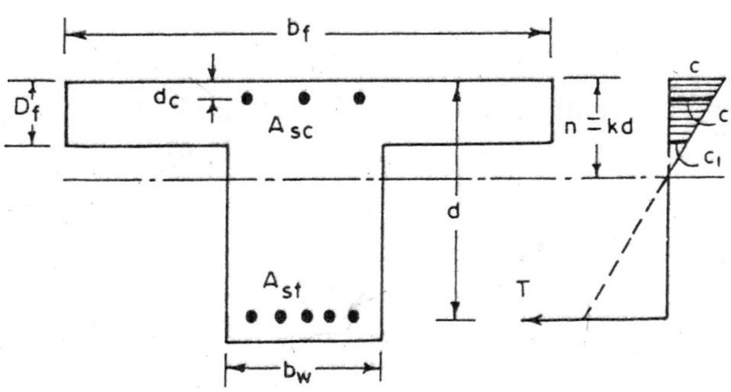

FIG. 6.13

Let A_{sc} = area of compressive reinforcement

d_c = cover for the compressive reinforcement

c' = compressive stress in concrete surrounding compressive steel.

The design of such a beam is done in the following steps :

1. Find the moment of resistance M_1 of the T-beam as the balanced section. Using Eq. 6.9 :

$$M_1 = b_f \cdot D_f \cdot \frac{c}{2}\left[1 + \frac{kd - D_f}{kd}\right]\left[d - \frac{D_f}{2} + \frac{D_f^2}{6\,(2kd - D_f)}\right] \quad ...(6.22)$$

2. Calculate the moment of resistance (M_2) to be provided by the compression steel.

$M_2 = M - M_1$, where M = total B.M.

But $M_2 = (m_c - 1) A_{sc} \times c'\,(d - d_c)$

where $c' = \frac{kd - d_c}{kd} . c$ and $m_c = 1.5\,m$

$$\therefore \quad M_2 = (m_c - 1)\,A_{sc} \times c \times \frac{kd - d_c}{kd}\,(d - d_c) \qquad ...(6.23)$$

Thus, A_{sc} is known from this equation.

3. Find the tensile reinforcement as under :

$$A_{st1} = \frac{M_1}{\sigma_{st} \times a} \quad (\text{where } t = \sigma_{st}) \qquad ...(6.24)$$

where
$$a = d - \frac{D_f}{2} + \frac{D_f^2}{6\,(2kd - D_f)}$$

$$A_{st2} = \frac{M_2}{\sigma_{st}(d - d_c)} \qquad ...(6.25)$$

Total area of steel $A_{st} = A_{st1} + A_{st2}$ ...(6.26)

For a doubly reinforced T-beam of given dimensions, the neutral axis can be located by equating the moments of equivalent areas about the neutral axis. Thus :

$$b_f . D_f \left(n - \frac{D_f}{2}\right) + (m_c - 1)\,A_{sc}(n - d_c) = m\,A_{st}\,(d - n) \qquad ...(6.27)$$

(Assuming the N.A. to lie in the web)

The moment of resistance of such a beam, neglecting the compressive force in the rib, is given by

$$M_r = b_f . D_f \left(\frac{c + c_1}{2}\right) \times a + (m_c - 1)\,A_{sc} \times c \times \frac{n - d_c}{n}\,(d - d_c) \qquad ...(6.28)$$

Taking a equal to $d - \frac{D_f}{2} + \frac{D_f^2}{6\,(2n - D_f)}$, the above expression reduces to :

$$M_r = b_f . D_f . c \left[\frac{2n - D_f}{2n}\right] \left[d - \frac{D_f}{2} + \frac{D_f^2}{6\,(2n - D_f)}\right] + (m_c - 1)\,A_{sc} . \frac{c(n - d_c)}{n}\,(d - d_c) \qquad ...(6.29)$$

where $c = \sigma_{cbc}$ = permissible compressive stress in concrete.

Example 6.8. *An isolated T-beam carries a uniformly distributed load of 40 kN/m run, inclusive of its own weight, over an effective span of 6 m. The beam has the following dimensions : width of flange : 800 mm ; thickness of flange : 100 mm; effective depth of the beam : 480 mm and width of rib 300 mm. Determine the necessary areas of tensile and compressive reinforcement. Take* $\sigma_{cbc} = 5$ *Nmm*^{2}, $\sigma_{st} = 140$ *N/mm*^{2} *and m=18.*

Solution. Given $\sigma_{cbc} = 5$ N/mm^2 ; $t = \sigma_{st} = 140$ N/mm^2.

The effective width of an isolated T-beam is given by

$$b_f = \frac{l_0}{\left(\frac{l_0}{b}\right) + 4} + b_w = \frac{6000}{\left(\frac{6000}{800}\right) + 4} + 300 = 822 \text{ mm.}$$

But this cannot be more than $b(=800$ mm).

Hence $b_f = 800$ mm,

$$\text{B.M.} = \frac{wL^2}{8}$$

$$= \frac{40 \times 36}{8}$$

$$= 180 \text{ kN-m}$$

$$= 180 \times 10^6 \text{ N-mm}$$

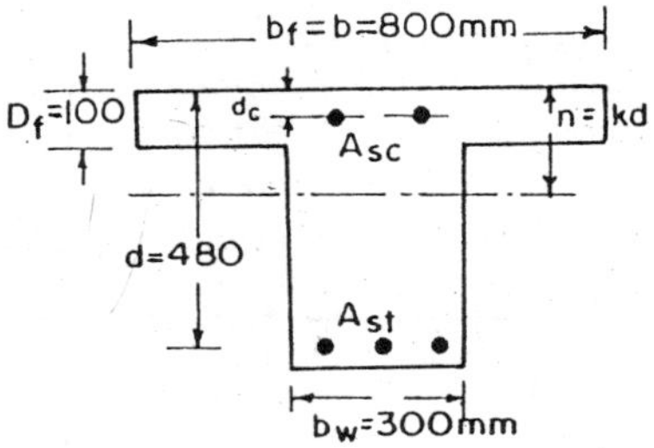

FIG. 6.14

For balanced section,

$$n = kd = \frac{mc}{mc + t} d$$

$$= \frac{18 \times 5}{18 \times 5 + 140} \times 480 = 187.8 \text{ mm.}$$

The moment of resistance of the balanced section is

$$M_r = b_f \,.\, D_f \frac{c}{2} \left[\frac{2kd - D_f}{kd}\right] \left[d - \frac{D_f}{2} + \frac{D_f}{6(2\,kd - D_f)}\right]$$

$$= 800 \times 100 \times \frac{5}{2} \left[\frac{2 \times 187.8 - 100}{187.8}\right]$$

$$\times \left[480 - 50 + \frac{100 \times 100}{6\,(2 \times 187.8 - 100)}\right]$$

$$= 293504 \times [436] \text{ where } 436 = a = \text{L.A.}$$

$$= 128 \times 10^6 \text{ N-mm}$$

$$A_{st1} = \frac{M_1}{t \times a} = \frac{128 \times 10^6}{140 \times 436} = 2097 \text{ mm}^2$$

$$M_2 = M - M_1 = (180 - 128)\, 10^6 = 52 \times 10^6 \text{ N-mm}$$

$$\therefore \quad 52 \times 10^6 = (m_c - 1) A_{sc} \times c \left[\frac{kd - d_c}{kd}\right] (d - d_c)$$

Assuming $d_c = 40$ mm,

$$A_{sc} = \frac{(52 \times 10^6) \times (187.8)}{(1.5 \times 18 - 1) \times 5(187.8 - 40)(480 - 40)}$$

$$= 1155 \text{ mm}^2,$$

and $$A_{st2} = \frac{M_2}{t\,(d - d_c)} = \frac{52 \times 10^6}{140(480 - 40)} = 844 \text{ mm}^2$$

$\therefore$ Total tensile steel A_{st} = 2097 + 844 = **2941** mm^2

and Comp. steel A_{sc} = **1155** mm^2.

Example 6.9 *Fig. 6.15 shows the section of a doubly reinforced T-beam section. The tensile reinforcement consist of 5 bars of 24 mm diameter while the compressive reinforcement consists of 3 bars of 20 mm diameter. Taking $\sigma_{cbc} = 5$ N/mm^2, $\sigma_{st} =$ 140 N/mm^2 and m = 18, determine the moment of resistance of the section (a) neglecting compression of web (b) taking compression of web into account. The beam is used over an effective span of 5 m.*

Solution. The effective width of flange is given by

$$b_f = \frac{l_0}{(l_0/b) + 4} + b_w = \frac{5000}{(5000/1200 + 4)} + 300 = 912 \text{ mm}$$

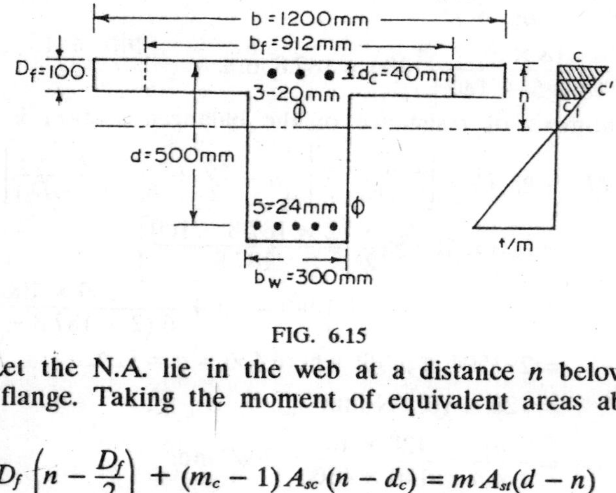

FIG. 6.15

Let the N.A. lie in the web at a distance n below the top of the flange. Taking the moment of equivalent areas about N.A. we get

$$b_f D_f \left(n - \frac{D_f}{2}\right) + (m_c - 1) A_{sc} (n - d_c) = m A_{st}(d - n)$$

where $A_{sc} = 3\frac{\pi}{4}(20)^2 = 945 \text{ mm}^2$ and $A_{st} = 5\frac{\pi}{4}(24)^2 = 2260 \text{ mm}^2$

$\therefore$ $$912 \times 100(n - 50) + (1.5 \times 18 - 1)945(n - 40) = 18 \times 2260(500 - n)$$

or $$n - 50 + 0.269\,n - 10.78 = 223 - 0.446\,n$$

from which $$n = 165.5 \text{ mm}$$

Depth of critical N.A. is given by

$$n_c = \frac{m\sigma_{cbc}}{m\sigma_{cbc} + \sigma_{st}} d = \frac{18 \times 5}{18 \times 5 + 140} \times 500 = 196 \text{ mm}$$

Since the actual N.A. falls above the critical N.A., the section is under-reinforced and the stress in steel will reach its maximum value first. Hence $t = \sigma_{st} = 140 \text{ N/mm}^2$.

$$\therefore \qquad c = \frac{t}{m}\,\frac{n}{d-n} = \frac{140}{18} \times \frac{165.5}{500-165.5} = 3.848 \text{ N/mm}^2$$

The moment of resistance is given by

$$M_r = b_f \, . \, D_f \frac{c}{2}\left[1 + \frac{n - D_f}{n}\right]\left[d - \frac{D_f}{2} + \frac{D_f^2}{6(2n - D_f)}\right]$$
$$+ (m_c - 1) A_{sc} \, . \, c \times \frac{n - d_c}{n}(d - d_c)$$

$$= 912 \times 100 \times \frac{3.848}{2}\left[1 + \frac{165.5 - 100}{165.5}\right] \times$$
$$\left[500 - 50 + \frac{(100)^2}{6(2 \times 165.5 - 100)}\right] + (1.5 \times 18 - 1) \times$$
$$945 \times 3.848 \frac{165.5 - 40}{165.5}(500 - 40)$$

$$= 111.98 \times 10^6 + 32.98 \times 10^6 = 144.96 \times 10^6 \text{ N–mm}$$
$$= 144.96 \text{ kN–m}.$$

(b) Taking compressive force of web into account

The N.A. is given by

$$b_f \, . \, D_f\left(n - \frac{D_f}{2}\right) + \frac{b_w(n - D_f)^2}{2} + (m_c - 1) A_{sc}(n - d_c) = m A_{st}(d - n)$$

or $912 \times 100(n - 50) + 150(n - 500)^2 + (1.5 \times 18 - 1)945(n - 40)$
$= 18 \times 2260\,(500 - n)$

or $608n - 30400 + n^2 + 10000 - 200n + 164n - 6552 = 135600 - 271n$

or $\qquad n^2 + 843\,n - 162\,552 = 0$

From which $n = 161.8$ mm

As found earlier, $n_c = 196$ mm.

The section is, therefore, under-reinforced, and

$$t = \sigma_{st} = 140 \text{ N/mm}^2.$$

The corresponding stresses c, c' and c_1 are as follows :

$$c = \frac{t}{m}\,\frac{n}{d-n} = \frac{140}{18} \times \frac{161.8}{500 - 161.8} = 3.72 \text{ N/mm}^2.$$

$$c' = \frac{t}{m} \cdot \frac{n - d_c}{d - n} = \frac{140}{18} \times \frac{161.8 - 40}{500 - 161.8} = 2.80 \text{ N/mm}^2.$$

$$c_1 = \frac{t}{m} \cdot \frac{n - D_f}{d - n} = \frac{140}{18} \times \frac{161.8 - 100}{500 - 161.8} = 1.42 \text{ N/mm}^2.$$

The moment of resistance of the beam will be equal to the sum of the moment of resistance due to (*i*) compressive force of flange (*ii*) compressive force of rib, and (*iii*) compressive force of steel.

or $$M_r = M_{r1} + M_{r2} + M_{r3}$$

Now $$M_{r1} = \text{M.R. due to compression in flange}$$

$$= b_f \cdot D_f \left(\frac{c + c_1}{2}\right)(d - \bar{y})$$

where $$\bar{y} = \left(\frac{c + 2c_1}{c + c_1}\right)\frac{D_f}{3} = \left(\frac{3.72 + 2.84}{3.71 + 1.42}\right)\frac{100}{3} = 42.54 \text{ mm}$$

$$\therefore \quad M_{r1} = 912 \times 100 \left(\frac{3.72 + 1.42}{2}\right)(500 - 42.54)$$

$$= 107.22 \times 10^6 \text{ N-mm}$$

$$M_{r2} = \text{M.R. due to compressive force in rib}$$

$$= b_w (n - D_f)\frac{c_1}{2}\left[d - D_f - \frac{1}{3}(n - D_f)\right]$$

$$= 300(161.8 - 100)\frac{1.42}{2}\left[500 - 100 - \frac{1}{3}(161.8 - 100)\right]$$

$$= 4.99 \times 10^6 \text{ N-mm}$$

$$M_{r3} = \text{M.R. due to compressive force in steel.}$$

$$= (m_c - 1) A_{sc} \cdot c' (d - d_c)$$

$$= (1.5 \times 18 - 1)\, 945 \times 2.80\, (500 - 40)$$

$$= 31.65 \times 10^6 \text{ N–mm}$$

$$\therefore \text{ Total } M = (107.22 + 4.99 + 31.65)\, 10^6$$

$$= 143.86 \times 10^6 \text{ N–mm} = \mathbf{143.86 \text{ kN–m.}}$$

Example 6.10. *A T-beam has the following dimensions : width of flange 1000 mm, thickness of flange 120 mm, width of web 250 mm ; effective depth 500 mm. The beam is reinforced with 2400 sq. mm of steel in tension and 1200 sq. mm of steel in compression. Effective cover of compressive steel = 40 mm. Determine the following:*

(*i*) *depth of N.A.*

(*ii*) *stress in compressive and tensile steel, if maximum stress in concrete at the top edge is 4 N/mm*^{2}.

(*iii*) *moment of resistance of the section.*

(*iv*) *lever arm.*

Take m = 16 and assume that the beam is simply supported over a span of 12.5 m.

Solution. For an isolated T-beam,

$$b_f = \frac{l_0}{(l_0/b) + 4} + b_w = \frac{12500}{(12500/1000 + 4)} + 250$$

$$= 1007.6 \text{ mm} > b.$$

Hence $b_f = b = 1000$ mm.

Let the depth of N.A. be n below the top of flange. Taking moments of the equivalent areas about the N.A., we get

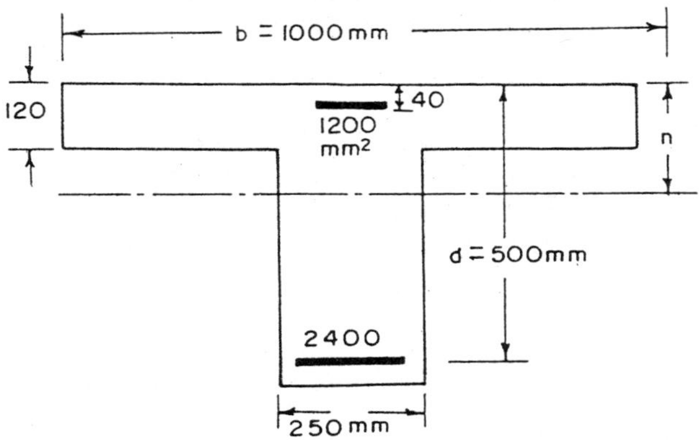

FIG. 6.16

$$b_f \cdot D_f\left(n - \frac{D_f}{2}\right) + \frac{b_w(n - D_f)^2}{2} + (m_c - 1)A_{sc}(n - d_c) = m\,A_{st}(d - n)$$

$$\therefore \quad 1000 \times 120\left(n - \frac{120}{2}\right) + \frac{250(n - 120)^2}{2} + (1.5 \times 16 - 1)$$
$$\times 1200\,(n - 40) = 16 \times 2400\,(500 - n)$$

or $120000n - 7200000 + 125n^2 + 1800000 - 30000n + 27600n - 110400$
$$= 19200000 - 38400n$$

or $n^2 + 1248n - 205632 = 0$

From which n = **147 mm.**

Maximum compressive stress in concrete = 4 N/mm^2 (given)

$\therefore$ Stress in tensile steel $= t = mc\,\dfrac{d - n}{n}$

$$= 16 \times 4 \times \frac{500 - 147}{147} = \mathbf{153.7\ N/mm^2}$$

$$c' = c\,\frac{n - d_c}{n}$$

$$= 4 \times \frac{147 - 40}{147} = \mathbf{2.91\ N/mm^2}$$

$\therefore$ Stress in comp. steel $= t_c = m_c \,.\, c' = 1.5 \times 16 \times 2.91$

$$= \mathbf{69.8\ N/mm^2}$$

Stress $c_1 = c\,\frac{n - D_f}{n} = 4 \times \frac{147 - 120}{847} = 0.735\ \text{N/mm}^2$

In order to find the *lever arm,* let us find distance $\bar{y}$ of the centroid of the compressive force. Fig 6.17 shows the compressive stress diagram in which the total compressive force has been split

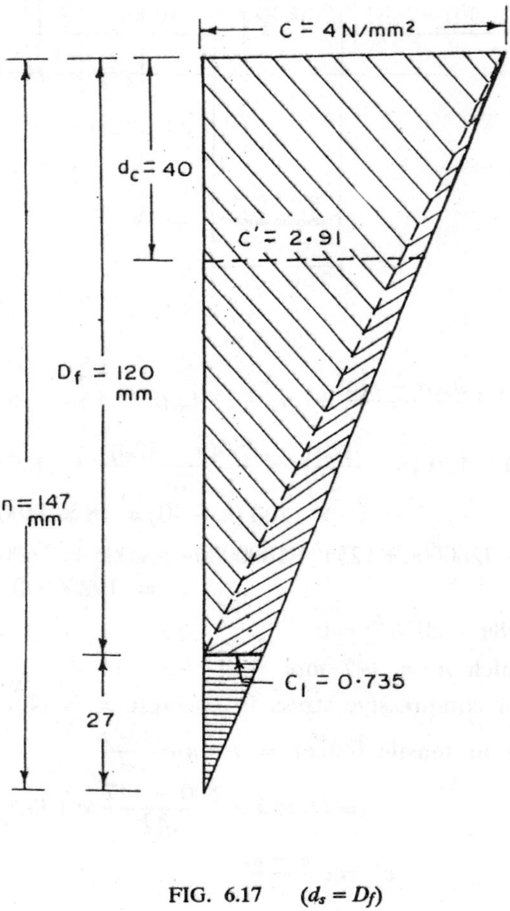

FIG. 6.17 $(d_s = D_f)$

into four components: (*i*) compressive force in steel (*ii*) compressive force in rib (*iii*) compressive force in flange, due to upper triangle and (*iv*) compressive force in flange, due to lower triangle.

Total compressive force

$$C = A_{sc}(m_c - 1)c' + \tfrac{1}{2}c_1(n - D_f)b_w + b_f \cdot D_f \cdot \frac{c}{2} + b_f \cdot D_f \cdot \frac{c_1}{2}$$

$$= [1200(1.5 \times 16 - 1) \times 2.91] + [\tfrac{1}{2} \times 0.735\,(147 - 120)\,250] + \left[1000 \times 120 \times \frac{4}{2}\right] + \left[1000 \times 120 \times \frac{0.735}{2}\right]$$

$$= 80316 + 2481 + 240000 + 44100 = 366897 \text{ N}$$

$$\bar{y} = \frac{(80316 \times 40) + 2481\left[120 + \frac{27}{3}\right] + \left[240000 \times \frac{120}{3}\right] + \left[44100 \times \frac{120 \times 2}{3}\right]}{366897}$$

or $\bar{y} = 45.4$ mm

$\therefore$ Lever arm $a = d - \bar{y} = 500 - 45.4 = 454.6$ mm

$\therefore$ $M_r =$ compressive force × lever arm

$= 366897 \times 454.6 \approx 167 \times 10^6$ N-mm

$=$ **167 kN-m**

Check $M_r = A_{st} \cdot \sigma_{st} \cdot a = 2400 \times 153.7 \times 454.6$

$\approx 167 \times 10^6$ N-mm

6.11 L-BEAM

In the case of T-beam, the rectangular beam has flanges on both the sides and hence load is transferred to it in vertical plane passing through the middle of its width. However, in the case of monolithic beam-slab construction, the end beams have flanges to one side only, as shown in Fig. 6.18(*a*). Such a beam having the shape of letter 'L' is known as L-beam.

Fig. 6.18 (*b*) shows the loading on the beam received from the slab. Evidently, the beam is not symmetrically loaded. Due to this, an eccentric load *W* is transferred to the beam. This eccentric load may be considered as central load *W* plus a moment M_t, as shown in Fig. 6.18(*c*). The central load *W* causes bending moment in the beam, which is jointly resisted by the rectangular portion of the beam as well as the flange (of width *B*), similar to that of a T-beam. The moment M_t causes *torsion* in the beam, and is known as the torsional moment which is resisted by the rectangular portion alone, and the flange does not contribute to any torsional moment of resistance. Separate torsional resistance has to be provided.

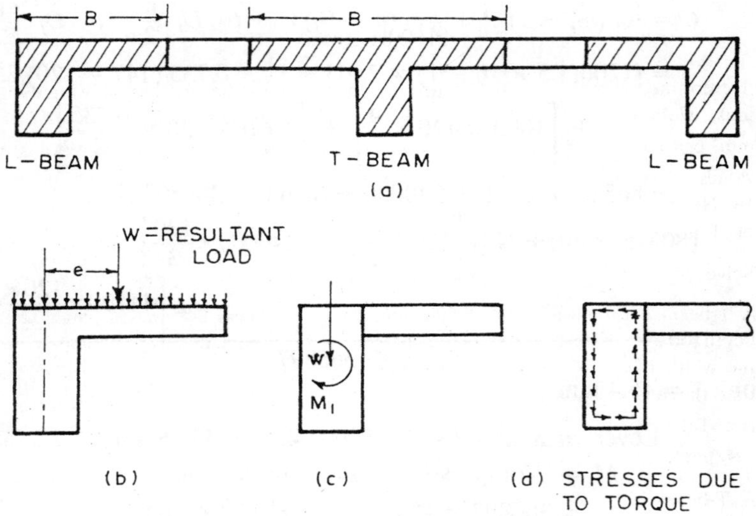

FIG. 6.18. L-BEAM.

According to Indian Standard Code (IS : 456–1978), the width B (or b_f) of the slab, acting monolithic with the beam, forming the flange of the L-beam is taken as the least of the following:

(*a*) $b_f = \frac{l_0}{12} + b_w + 3\,D_f$

(*b*) For isolated beam

$$b_f = \frac{0.5\,l_0}{\frac{l_0}{b} + 4} + b_w$$

(but in no case greater than the actual width of flange)

where l_0 is the distance between points of zero moments in the beam. For continuous beams and frames, l_0 may be assumed as 0.7 times the effective span.

The location of the N.A. and the calculations of the moment of resistance of L-beam are done exactly in the same way as that for T-beam, and all the equation developed for T-beam are also applicable for L-beam. The design for bending moment is also done exactly in the way as that for T-beam. However, additional reinforcement, as discussed in chapter 4, has to be provided to resist shear induced by torsional moment.

The design procedure for L-beams is given chapter 7.

PROBLEMS

1. Determine the moment of resistance of T-beam section having the following dimensions: width of flange 600 mm; thickness of slab 80 mm: tensile steel-area 3200 mm^2 centred at 500 mm from the top ; width of web 250 mm. Take the permissible values of stresses in concrete and steel as 4.2 N/mm^2 and 126 mm^2 and m = 18. Neglect the compression of the rib. The beam is used over an effective span of 4 m.

2. Solve problem 1, taking into account compressive force in the rib.

3. A T-beam has flange width of 800 mm, flange thickness 80 mm and an effective depth of 500 mm. The reinforcement consist of 5 bars of 30 mm diameter and the width of rib is 300 mm. Determine the maximum uniformly distributed load that the beam can carry over an effective span of 5 m. Neglect compression of rib. Take permissible stress in concrete and steel as 5 N/mm^2 and 140 N/mm^2 and m = 18.

4. A T-beam has a flange of 1000 mm width and 80 mm thickness. The effective depth is 300 mm and the width of rib is 200 mm. Taking $\sigma_{cbc} = 5\ N/mm^2$, $\sigma_{st} = 140\ N/mm^2$ and m = 18, determine (*a*) area of steel in tension and (*b*) moment of resistance of the beam if it is used over an effective span of 3 m.

5. A R.C. T-beam has the following dimensions : width of flange 1300 mm ; thickness of flange 100 mm; effective depth 560 mm ; width of rib 300 mm ; reinforcement : 6 bars of 25 mm diameter. Determine the stresses developed in the concrete and steel, when the beam is subjected to a bending moment of 180 kN-m. The beam has an effective span of 5m.

6. A T-beam roof consists 100 mm thick reinforced concrete slab cast monolithic with 250 mm wide beams spaced 3 metres centre to centre, the effective span of each beam being 4 metres. Design one of the intermediate beams for a super imposed load of 3 kN/m^2 over the slab Take $\sigma_{cbc} = 5\ N/mm^2$, $\sigma_{st} = 140\ N/mm^2$ and m = 18. The ratio of cost of steel to that of concrete may be taken as 50.

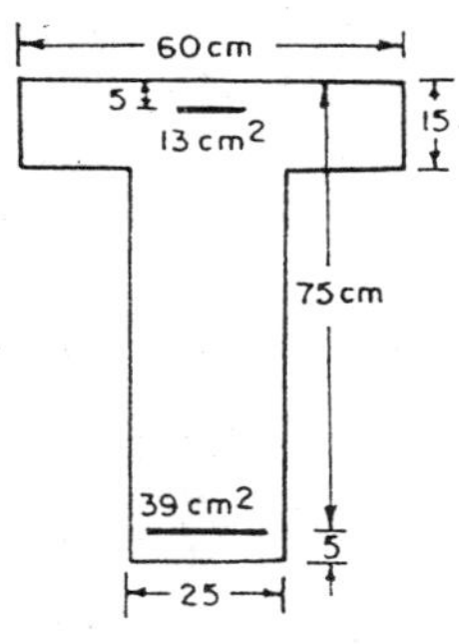

FIG. 6.19

7. The sectional dimensions of a T-beam are shown in Fig. 6.19. The beam is reinforced width 3900 sq. mm of steel in tension and 1300 sq. mm

in compression. Taking $c = 7$ N/mm^2 and $m = 15$, calculate from the first principles:

(*a*) The depth of N.A.
(*b*) The stress in the soffit of the top slab in terms of the top edge stress of $c = 7$ N/mm^2
(*c*) Compressive stress in the top steel
(*d*) Moment of resistance of the section
(*e*) Lever arm.

Assume that the beam is simply supported over a span of 5 m.

ANSWERS

1. 80 kN-m
2. 101 kN-m
3. 32.47 kN-m
4. (*a*) 1185 mm^2
   (*b*) 44.25 kN-m
5. $c = 4.79$ N/mm^2 ; $t = 127.4$ N/mm^2
7. (*a*) 282 mm
   (*b*) 3.28 N/mm^2
   (*c*) 174.3 N/mm^2
   (*d*) 460 kN-m
   (*e*) 677 mm.

7

Design of Beams and Slabs

7.1 DESIGN OF BEAMS

(*a*) BASIC RULES FOR DESIGN (IS : 456–1978)

We have already discussed the theory of singly reinforced beams. While designing the beams, following important rules must be kept in mind :

1. *Effective span*

Unless otherwise specified, the effective span of a member shall be as follows :

(*a*) *Simply supported beam or slab.* The effective span of a member that is not built integrally with its supports shall be taken as clear span plus the effective depth of slab or beam or centre to centre of supports, whichever is less.

(*b*) *Continuous beam or slab.* In case of continuous beam or slab, if the width of the support is less than 1/12 of the clear span, the effective span shall be as in (*a*) above. If the supports are wider than 1/12 of the clear span or 600 mm whichever is less, the effective span shall be taken as under :

(*i*) For the end span with one end fixed and the other continuous or for intermediate spans, the effective span shall be clear span between supports, and

(*ii*) For the end span with one end free and the other continuous, the effective span shall be equal to the clear span plus half the effective depth of the beam or slab or the clear span plus half the width of the discontinuous support, whichever is less.

Note. In the case of spans with roller or rocker bearings, the effective span shall always be the distance between the centres of bearings.

(*c*) *Frames.* In the analysis of a continuous frame, centre to centre distance shall be used.

2. *Control of deflection*

The deflection of a structure or part there of shall not adversely affect the appearance or efficiency of the structure or finishes or partitions.

For beams and slabs, the vertical deflection limits may generally be assumed to be satisfied provided that the span to depth ratios are not greater than the values obtained as below :

(*a*) Basic values of span to effective depth ratios for span upto 10 m :

Cantilever	7
Simply supported	20
Continuous	26

(*b*) For spans above 10 m, the values in (*a*) may be multiplied by 10/span in metres, except for cantilever in which case deflection calculations should be made.

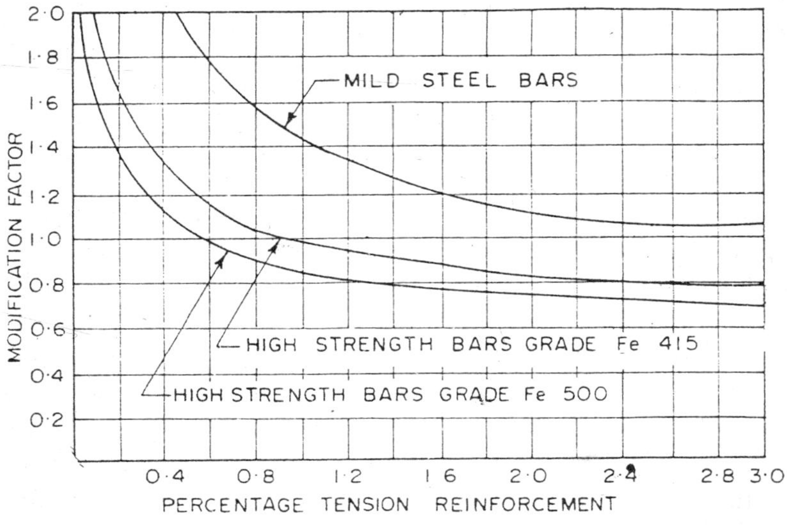

(c) Depending on the area and the type of steel for tension reinforcement, the values in (*a*) or (*b*) shall be modified as per Fig.7.1.

(*d*) Depending on the area of compression reinforcement, the value of span to depth ratio be further modified as per Fig. 7.2.

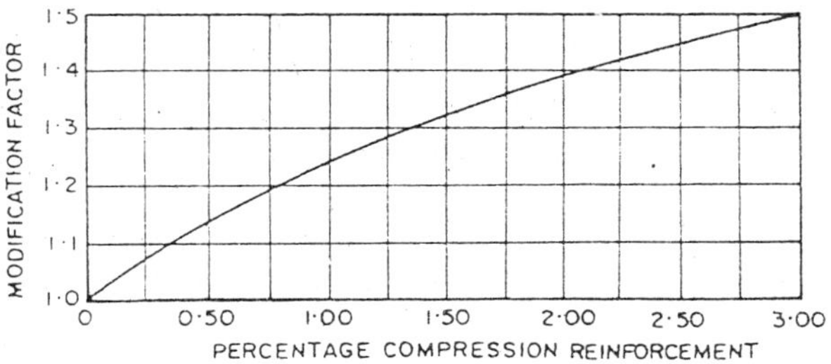

FIG. 7.2. MODIFICATION FACTOR FOR COMPRESSION REINFORCE-

(*e*) For flanged beams, the values of (*a*) or (*b*) be modified as per Fig. 7.3 and the reinforcement percentage for use in Fig 7.1 and 7.2 should be based on area of section of equal to$b_f d$.

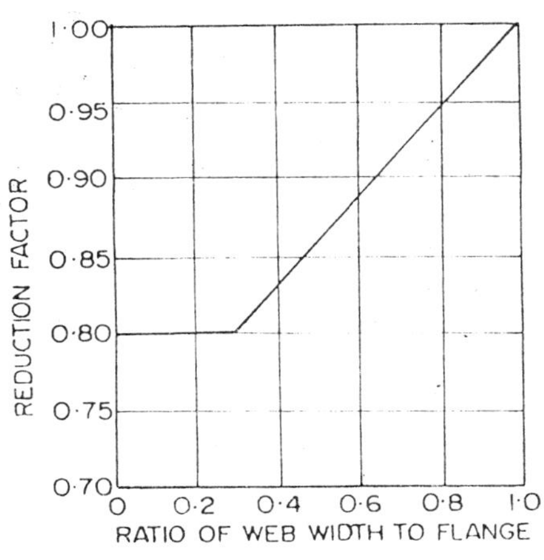

FIG. 7.3 REDUCTION FACTORS FOR FLANGED BEAMS.

Note 1. For slabs spanning in two directions, the shorter of the two spans should be used for calculating the span to effective depth ratios.

Note 2. For two-way slabs of small spans (up to 3.5 m), with mild steel reinforcement, the span to overall depth ratio given below may generally be assumed to satisfy vertical deflection limits for loading class upto 3000 N/mm^2.

Simply supported slabs : 35
Continuous slabs : 40

For high strength deformed bars of grade Fe 415, the values given above should be multiplied by 0.8.

3. *Slenderness limits for beams to ensure lateral stability*

A simply supported on continuous beam shall be so proportioned that the clear distance between the lateral restraints does not exceed 60 b or 250 b^2/d whichever is less, where d is the effective depth of the beam and b the breadth of the compression face midway between the lateral restraints.

For a cantilever, the clear distance from the free end of the cantilever to the lateral restraint shall not exceed 25 b or 100 b^2/d whichever is less.

4. *Critical sections for moments and shear*

For monolithic construction, the moments computed at the face of the supports shall be used in the design of the member at those sections. For non-monolithic construction the design of the member shall be done keeping in view para 1.

Critical section for shear. The shear computed at the face of the support shall be used in the design of the member at that section except as in case mentioned below.

When the reaction in the direction of the applied shear introduces compression into the end region of the member, sections located at the distance less than d from the face of the support may be

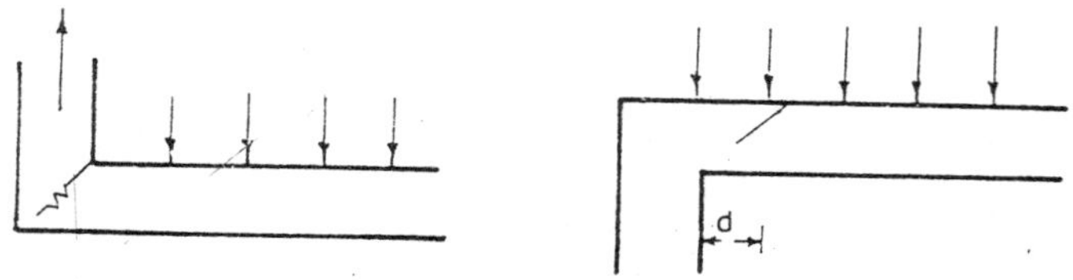

(a) Critical Section at Face of the Support

(b) Critical Section at Distance d from the Face of the Support

FIG. 7.4. CRITICAL SECTIONS FOR SHEAR.

designed for the same shear as that computed at distance d (see Fig. 7.4.

5. *Reinforcement in beams*

(a) Tension reinforcement

(*i*) *Minimum reinforcement.* The minimum area of the tension reinforcement shall not be less than that given by the following :

$$\frac{A_{st}}{bd} = \frac{0.85}{f_y}$$

where f_y is the characteristic strength of reinforcement in N/mm^2.

(ii) *Maximum reinforcement.* The maximum area of tension reinforcement shall not exceed 0.04 bD.

(b) Compression reinforcement

The maximum area of compression reinforcement shall not exceed 0.04 bD. Compression reinforcement in beams shall be enclosed by stirrups for effective lateral restraint.

(c) Side face reinforcement

Where the depth of the web in a beam exceeds 750 mm, side face reinforcement shall be provided along the two faces. The total area of such reinforcement shall be not less than 0.1 percent of the web area and shall be distributed equally on two faces at a spacing not exceeding 300 mm or web thickness whichever is less.

(d) Transverse reinforcement in beams for shear and torsion

The transverse reinforcement in beams shall be taken around the outer-most tension and compression bars. In T-beams and L-beams, such reinforcement shall pass around longitudinal bars located close to the outer face of the flange.

(e) Maximum spacing of shear reinforcement

The maximum spacing of shear reinforcement measured along the axis of the member shall not exceed 0.75 d for vertical stirrups and d for inclined stirrups at 45° , where d is the effective depth of the sections under consideration. In no case shall the spacing exceed 450 mm.

(f) Minimum shear reinforcement

Minimum shear reinforcement in the form of stirrups shall be provided such that

$$\frac{A_{sv}}{b \, . \, s_v} \geq \frac{0.4}{f_y}$$

where A_{sv} = total cross-sectional area of stirrup legs effective in shear

s_v = stirrup spacing along the length of member

b = breadth of the beam or breadth of the web of flanged beam, and

f_y = characteristic strength of the stirrup reinforcement in N/mm^2 which shall not be taken greater than 415 N/mm^2.

Maximum diameter. The diameter of the reinforcing bars shall not exceed one eighth of the total thickness of the slab.

6. *Curtailment of tension reinforcement in flexural member*

(*a*) For curtailment, reinforcement shall extend beyond the point at which it is no longer required to resist flexure for a distance equal to the effective depth of the member or 12 times the bar diameter, whichever is greater except at simple support or end of cantilever. In addition, clauses (*b*) to (*f*) given below shall be satisfied.

Note. A point at which reinforcement is no longer required to resist flexure is where the resistance moment of the section, considering only the continuing bars, is equal the design moment.

(*b*) Flexural reinforcement shall not be terminated in a tension zone unless any one of the following conditions is satisfied :

(*i*) The shear at the cut-off point does not exceed two thirds that permitted, including the shear strength of web reinforcement provided.

(*ii*) Stirrup area in excess of the required for shear and torsion is provided along each terminated bar over a distance from the cut-off point equal to three fourths the effective depth of the member. The excess stirrup area shall be not less than 0.4 bs/f_y , where b is the width of the beam, s is the spacing and f_y is the characteristic strength of reinforcement in N/mm^2. The resulting spacing shall not exceed $d/8\beta_0$ where β_0 is the ratio of the area of bars cut off to the total area of bars at the section, and d is the effective depth.

(*iii*) For 36 mm and smaller bars, the continuing bars provide double the area required for flexure at the cut-off point and the shear does not exceed three-fourth that permitted.

(c) *Positive moment reinforcement*

(*i*) At least one-third the positive moment reinforcement in simple members and one-fourth the positive moment reinforcement in continuous members shall extend along the same face of the member into the support, to a length equal to $L_d/3$.

(*ii*) When a flexural member is part of the primary lateral load resisting system, the positive reinforcement required to be extended into the support as described in (*i*) above shall be anchored to develop its design stress in tension at the face of the support.

(*iii*) At simple supports and at point of inflection, positive moment tension reinforcement shall be limited to a diameter such that L_d computed for σ_{st} $\left(\text{by the relation } L_d = \dfrac{\Phi\,\sigma_{st}}{4\,\tau_{bd}}\right)$ does not exceed

$$\left\{\frac{M_1}{V} + L_0\right\}$$, where

M_1 = moment of resistance of the section assuming all reinforcement at the section to be stressed to σ_{st}.

V = S.F. at the section due to design loads

L_0 = Sum of the anchorage beyond the centre of the support and the equivalent anchorage value of any hook or mechanical anchorage at simple support ; and at a point of inflection, L_0 is limited to the effective depth of the member or 12 Φ whichever is greater, and

Φ = diameter of bars.

The value of M_1/V in the above expression may be increased by 30 percent when the ends of the reinforcement are confined by a compression reaction.

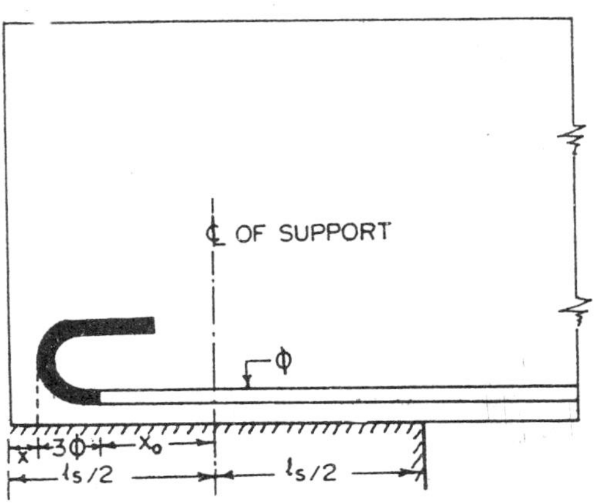

FIG. 7.5

Note. For the computation of L_0, the support width should be known. Fig 7.5 shows a beam with end support, in which x' is the side cover and x_0 is the distance of the *beginning* of the hook from the centre line of the support. Now L_0 = sum of anchorage beyond the centre of the support and the equivalent anchorage value.

The dark portion shows the hook which has an anchorage value of 16 Φ [Fig. 3.15(*a*)]. The distance of the beginning of the hook from its apex of the semi-circle can be taken to be equal to 3 Φ. Let l_s be the length of the support.

$$\therefore \quad L = x_0 + 16\,\Phi$$

where $$x_0 = \frac{l_s}{2} - x' - 3\,\Phi$$

$$\therefore \quad L_0 = \left(\frac{l_s}{2} - x' - 3\,\Phi\right) + 16\,\Phi$$

or $$L_0 = \frac{l_s}{2} - x' + 13\,\Phi.$$

for HYSD bars, no hooks are provided.

Hence, $$L_0 = \frac{l_s}{2} - x'$$

(*d*) ***Negative moment reinforcement***

At least one third of the total reinforcement provided for negative moment at the support shall extend beyond the point of inflection for a distance not less than the effective depth of the member or 12 Φ or one-sixteenth of the clear span whichever is greater.

(*e*) ***Curtailment of bundled bars***

Bars in a bundle shall terminate at different points spaced apart by not less than 40 times the bar diameter except for bundles stopping at support.

7. ***Spacing of reinforcement***

For the purpose of the clause, the diameter of a round bar shall be its nominal diameter and in the case of bars which are not round, or in case of deformed bars or crimped bars ; the diameter shall be taken as the diameter of a circle giving an equivalent effective area. Where spacing limitations and minimum concrete cover (see clause 9) are based on bar diameter, a group of bars bundled in contact shall be treated as a single bar of diameter derived from the total equivalent area.

(*a*) *Minimum distance between individual bars*

The following shall apply for spacing of bars :

(*i*) The horizontal distance between two parallel main reinforcing bars shall usually be not less than the greatest of the following :

(1) The diameter of the bar if the diameters are equal (2) the diameter of the larger bar if the diameters are unequal, and (3) 5 mm more than the nominal maximum size of coarse aggregate.

Note. This does not preclude the use of larger size of aggregates beyond the congested reinforcement in the same member ; the size of aggregates may be reduced around congested reinforcement to comply with this provision.

(*ii*) Greater horizontal distance than the minimum specified in (*i*) should be provided wherever possible. However when needle vibrators are used the horizontal distance between bars of a group may be reduced to two-thirds the nominal maximum size of the coarse aggregate, provided that sufficient space is left between groups of bars to enable the vibrator to be immersed.

(*iii*) Where there are two or more rows of bars, the bars shall be vertically in line and the minimum vertical distance between the bars shall be 15 mm, two-thirds the nominal maximum size of aggregate or the maximum size of bar whichever is the greatest.

(*b*) *Maximum distance between bars in tension*

Unless the calculation of crack widths shows that a greater spacing is acceptable , the following rules shall be applied to flexural members in normal internal or external conditions of exposure.

The horizontal distance between parallel reinforcement bars, or groups, near the tension face of a beam shall not be greater than the value given in Table 7.1 depending on the amount of redistribution carried out in analysis and the characteristic strength of the reinforcement.

TABLE 7.1. CLEAR DISTANCE BETWEEN BARS

f_y	*Percentage redistribution to or from section considered*				
	−30	*−15*	*0*	*+15*	*+30*
	Clear distance between bars				
N/mm²	*mm*	*mm*	*mm*	*mm*	*mm*
250	215	260	360	300	300
415	125	155	180	210	235
500	105	130	150	175	135

Note. The spacings given in the Table are not applicable to members subjected to particularly aggressive environments unless in the calculation of the moment of resistance, f_y has been limited to300 N/mm^2 in limit state design andσ_{st} limited to165 N/mm^2 in working stress design.

8. ***Cover to reinforcement***

(*a*) Reinforcement shall have concrete cover and thickness of such cover (exclusive of plaster or other decorative finish) shall be as follows :

(*i*) At each end of reinforcing bar not less than 25 mm, nor less than twice the diameter of such bar ;

(*ii*) For longitudinal reinforcing bar in a beam, not less than 25 mm, nor less than diameter of such bar ;

(*iii*) For any other reinforcement not less than 15 mm, nor less than the diameter of such bar.

(*b*) Increased cover thickness may be provided when surfaces of concrete members are exposed to the action of harmful chemicals (as in the case of concrete in contact with earth faces contaminated with such chemicals), acid, vapour, saline atmosphere, sulphurous smoke (as in the case of steam operated railways), etc., and such increase of cover may be between 15 mm to 50 mm beyond the figures given in (*a*) above, as may be specified by the engineer-in-charge.

(*c*) For reinforced concrete members, totally immersed in sea-water, the cover shall be 40 mm more than that specified in (*a*) above.

(*d*) For reinforced concrete members, periodically immersed in sea water or subjected to sea spray, the cover to concrete shall be 50 mm more than that specified in (*a*).

(*e*) For concrete of grade M 25 and above, the additional thickness of cover specified in (*b*) to (*d*) may be reduced to half.

(*f*) In all such cases the cover should not exceed 75 mm.

9. Anchoring reinforcing bars

(*a*) *Anchoring bars in tension*

(*i*) Deformed bars may be used without end anchorages provided development length requirement is satisfied. Hooks should normally be provided for plain bars in tension.

(*ii*) *Bends and hooks.* The anchorage value of bend shall be taken as 4 times the diameter of the bar for each 45° bend subject to a maximum of 16 times the diameter of the bar. The anchorage value of a standard U-type hook shall be equal to 16 times the diameter of the bar.

(*b*) ***Anchoring bar-in-compression***

The anchoring length of straight bar in compression shall be equal to the development length of bar in compression. The projected length of hooks, bends straight length beyond bends if provided for a bar in compression shall be considered for development length.

10. ***Moment and shear coefficients for continuous beams and slabs***

(*a*) Unless more exact estimates are made, for beams and slabs of uniform cross-section which support substantially uniformly distributed loads over three or more spans which do not differ by more than 15 percent of the longest, the bending moments and shear forces used in design may be obtained by using coefficients given in Table 7.2 and Table 7.3 respectively. For moment at supports where two unequal spans meet or in case where the spans are not equally loaded, the average of the two values for the negative moment at the support may be taken for design.

(*b*) Where a member is built into a masonry wall which develops only partial restraint, the member shall be designed to resist a negative moment at the face of the support of $\frac{Wl}{24}$ where W is the total design load and l is the effective span, or such other restraining moment as may be shown to be applicable. For such a condition, shear coefficient given in Table 7.3 at the end support may be increased by 0.05.

TABLE 7.2 BENDING MOMENT COEFFICIENTS

Type of load	*Span moment*		*Support moment*	
	Near middle of end span	*At middle of interior span*	*At support next to the support*	*At other interior supports*
1. Dead load and imposed load (fixed)	$+\frac{1}{12}$	$+\frac{1}{24}$	$-\frac{1}{10}$	$-\frac{1}{12}$
2. Imposed load (not fixed)	$+\frac{1}{10}$	$+\frac{1}{12}$	$-\frac{1}{9}$	$-\frac{1}{9}$

Note. For obtaining the B.M. , the coefficient shall be multiplied by the total design load and effective span.

TABLE 7.3. SHEAR FORCE COEFFICIENTS

Type of load	*At end support*	*At support next to the end support*		*At all other Interior supports*
		Outer side	*Inner side*	
1. Dead load and imposed load (fixed)	0.4	0.6	0.55	0.5
2. Imposed load (not fixed)	0.45	0.6	0.6	0.6

Note. For obtaining the S.F., the coefficient shall be multiplied by the total design load.

11. ***Unit weight of concrete***

Indian Standard Code (IS : 456–1978) specifies that *unless more accurate calculations are warranted*, the unit weights of plain concrete and reinforced concrete made with standard gravel or crushed natural stone aggregate may be taken as 24000 N/m^3 (24 kN/mm^3) and 25000 N/m^3 (25 kN/mm^3) respectively.

In the opinion of the Author, the above figures are on the higher side. The average *unit mass* (or specific mass) of *heavy* reinforced concrete vibrated into place is 2500 kg/m^3; this corresponds to a *unit weight* (or specific weight) of $2500 \times 9.81 = 24525$ N/m^3. The average *unit mass* (or specific mass) of heavy reinforced concrete placed without vibration is 2400 kg/m^3 ; this corresponds to a unit weight (or specific weight) of $2400 \times 9.81 = 23544$ N/m^3. Most of the countries of the world use unit weight of concrete as 2400 kg(f)/m^3 only, which corresponds to 23544 N/m^3.

(*b*) DESIGN PROCEDURE

The design of a singly reinforced beam is done in the following steps :

1. **Calculation of constants.** For tne given set of stresses, determine k, j and R.

2. **Calculation of bending moment..** Assume suitable values of overall depth and breadth of the beam, and determine the effective span. Calculate the self weight of the beam, and add it to the given superimposed load to get the total U.D.L. per metre length of the beam. Calculate the maximum bending moment in the beam.

3. **Design of the section.** Calculate the effective depth of the beam by the expression :

$$d = \sqrt{\frac{M}{R \cdot b}}$$

Find out the total depth of the beam by adding suitable cover to the above calculated effective depth. If this is nearly equal to the value of depth assumed in (1) it is alright. If the difference is more, recalculate the self-weight on the basis of the revised depth of the beam. Recalculate the depth on the basis of this new bending moment.

4. **Steel Reinforcement.** Calculate the area of steel by the formula

$$A_{st} = \frac{M}{\sigma_{st} \cdot jd}$$

where σ_{st} = permissible tensile stress in steel.

Choose suitable diameter and number of bars. Arrange them suitably in the width.

Curtailment of Reinforcement

Curtail the reinforcement at other locations where it is no longer required from the bending moment point of view.

Let M = bending moment at the centre of the span.

M_x = bending moment at other section distance x from the centre

N = numbers of bars at the centre of the span

N_x = number of bars curtailed at the section distance x from the centre

$N - N_x$ = number of bars at the given section distant x from the centre

Since A_{st} is proportional to the bending moment, we have

$$\therefore \quad \frac{N - N_x}{N} = \frac{M_x}{M}$$

or

$$1 - \frac{N_x}{N} = \frac{M_x}{M}$$

or

$$\frac{N_x}{N} = 1 - \frac{M_x}{M} = \frac{M - M_x}{M} \qquad ...(i)$$

If the beam carries U.D.L., $M = \frac{wL^2}{8}$ and $M_x = \left(\frac{wL^2}{8} - \frac{wx^2}{2}\right)$

$$\therefore \qquad M - M_x = \frac{wx^2}{2}$$

or
$$\frac{M - M_x}{M} = \frac{Wx^2}{2} / \frac{wL^2}{8} = \frac{4x^2}{L^2} \qquad ...(ii)$$

Equating (*i*) and (*ii*),we get

$$\frac{N_x}{N} = \frac{4x^2}{L^2}$$

or
$$x = \frac{L}{2}\sqrt{\frac{N_x}{N}} \qquad ...(7.1)$$

Thus, the above expression gives the distance x from the centre where N_x bars can be curtailed out of total bars N. However, the bars have to develop full tension at the centre of span, and hence the distance x should be equal to or greater than the development length L_d of the bar which is to be curtailed. The Code also specifies that the reinforcement shall extend the point at which it is no longer required to resist flexure, for distance equal to effective depth d or 12 Φ whichever is greater. Hence the bar should be curtailed at a distance $x+d$ (or12Φ) from the centre ; and this distance should be greater than L_d. See § 3.9 and Fig. 3.18 also for graphical method.

Another important recommendation of the Code is that in simple beams, one third of the positive reinforcement should be continued to the supports, and the diameter should be so chosen that Eq. 3.26 or 3.26(*a*) is satisfied.

However, if the bar is to be bent up, it may be safely bent up at a distance x given by Eq. 7.1.

5. **Shear reinforcement**. Calculate the maximum shear force in the beam. Find the nominal shear stress in the beam by the expression

$$\tau_v = \frac{V}{bd}$$

Also, determine (or estimate) design stress τ_c for the centre using Table 3.1

(*i*) If $\tau_v \leq \frac{1}{2}\tau_c$

or $V < \frac{1}{2}V_c$ (where $V_c = \tau_c . bd$)

No shear reinforcement is really necessary.

(*ii*) If $\frac{1}{2} V_c \leq V \leq V_c$

Provide nominal shear reinforcement given by Eq. 3.19 :

$$\frac{A_{sv}}{b \cdot s_v} \geq \frac{0.4}{f_y}$$

Choosing bar diameter, A_{sv} is known, and hence s_v can be computed from the above expression. However, the maximum spacing is restricted to 0.75 d or 150 mm, whichever is less.

(*iii*) If $V_c \leq V \leq V_{c(max)}$ [where $V_{c(max)} = \tau_{c(max)} \cdot bd$]

Design the transverse reinforcement for the net S.F. $= V - V_c$.

If ΣV_s is the sum of the resistances of various types of shear reinforcement (Eq. 3.20, 3.21 and 3.22) then ΣV_s should be equal to or greater than $(V - V_c)$. However, the contribution of bent up bars should not be more than half, the area of stirrups should not be less than the one given by Eq. 3.19, and maximum s_v should be restricted to 0.75 d (or 450 mm).

(iv) If $V \geq V_{c(max)}$, redesign the web of the section such that $V_{c.max}$ becomes equal to or more than external shear V.

6. ***Check for development length at the end.*** Check for development length of discussed in para $a(6)$.

i.e.
$$L_d \leq \frac{M_1}{V} + L_0$$

See Fig 7.5 for computation of L_o.

Design Example 7.1 (*a*) *A reinforced concrete beam is supported on two walls 750 mm thick, spaced at clear distance of 6 metres. The beam carries a superimposed load of 9.8 kN/m. Using M 15 concrete, design the beam. Take the permissible tensile and shear stress in steel as 140 N/mm² for mild steel bars.*

Solution.

1. ***Calculation of Design Constants.*** For M 15 concrete, $c = \sigma_{cbc} = 5$ N/mm² and $m \approx 19$.

Also, $t = \sigma_{st} = 140$ N/mm².

$$\therefore \quad k = \frac{mc}{mc + t} = \frac{19 \times 5}{19 \times 5 + 140} = 0.404$$

$$j = 1 - k/3 = 1 - 0.404/3 = 0.865$$

$$R = \frac{1}{2} cjk = \frac{1}{2} \times 5 \times 0.865 \times 0.404 = 0.874$$

2. ***Calculation of B.M.***

Let the effective depth of beam $= \frac{1}{10}$ span $= \frac{6000}{10} = 600$ mm

$\therefore$ Total depth of beam = 600 mm (say)

Let the width of the beam$= \frac{1}{2} d = 300$ mm (say)

Self load of beam per metre run $= (0.6 \times 0.3 \times 1)\ 25000$

$= 4500$ N.

External load=9800 N/m

$\therefore$ Total load per metre run= 9800 + 4500 = 14300 N

Effective span $= L = l + d = 6 + 0.6 = 6.6$ m

(This is smaller than the centre to centre distance of 6.75 m between the supports).

$$M = \frac{w\,l^2}{8} = \frac{14300(6.6)^2}{8} = 77870 \text{ N-m} = 77.87 \times 10^6 \text{ N-mm}$$

3. ***Design of section***

$$d = \sqrt{\frac{M}{Rb}} = \sqrt{\frac{75.77 \times 10^6}{0.874 \times 300}} = 545 \text{ mm}$$

Let us take $d = 545$ mm and $D = 580$ mm.

Revised self load of beam $= 0.3 \times 0.58 \times 25000 = 4350$ N/m

$\therefore \quad w = 9800 + 4350 = 14150$ N/m

Effective span $= 6 + 0.545 = 6.545$ m

$$M = \frac{14150\,(6.545)^2}{8} \approx 75770 \text{ N-m} = 75.77 \times 10^6 \text{ N-mm}$$

$$d = \sqrt{\frac{75.77 \times 10^6}{0.874 \times 300}} = 538 \text{ mm}$$

Assuming that 20 mm Φ bars will be used, and a clear cover of 25 mm (min.), $D = 538 + 25 + 10 = 573$ mm.

However, keep $D = 580$, so that available $d = 580 - 25 - 10 = 545$ mm

4. ***Steel reinforcement***

$$A_{st} = \frac{M}{\sigma_{st} \cdot jd} = \frac{75.77 \times 10^6}{140 \times 0.865 \times 545} = 1150 \text{ mm}^2$$

A_Φ for 20 mm Φ bars $= \frac{\pi}{4}(20)^2 = 314\ mm^2$

$$\therefore \quad \text{No. of bars} = \frac{1150}{314} = 3.66$$

Hence provide 4 bars, having $A_{st} = 4 \times 314 = 1256\ mm^2$.

Keeping 25 mm cover outside, clear spacing between bars will be $= \frac{1}{3}(300 - 25 \times 2 - 4 \times 20) = 56.7$ mm, which is much more than the diameter of the bar.

Min. reinforcement is given by

$$\frac{A_{st}}{bd} = \frac{0.85}{f_y}$$

Taking $f_y = 250\ N/mm^2$,

$$A_{st} = \frac{0.85\ (300 \times 545)}{250} = 556\ mm^2$$

Since the actual A_{st} provided ($=1256\ mm^2$) is much more than this, the design is O.K.

5. ***Check for shear and design of shear reinforcement***

The reaction at the wall supports will be uniformly distributed over the full width. Hence the shear force will be maximum of the edge of the support.

$$\text{Max. } V = \frac{wl}{2} = \frac{14150 \times 6}{2} = 42450\ N$$

$$\tau_v = \frac{V}{bd} = \frac{42450}{300 \times 545} = 0.26\ N/mm^2.$$

Assuming that out of 4 bars of main reinforcement, 2 bars will be bent up near the support, only 2 bars will be available.

$$\therefore \quad \frac{100 A_s}{bd} = \frac{100}{300 \times 545} \times 628 = 0.38\%$$

Hence from Table 3.1, permissible shear (τ_c) for M 15 concrete, for 0.38% steel $\approx 0.26\ N/mm^2$, which is equal to the nominal shear stress. Hence no shear reinforcement is required, However, provide minimum shear reinforcement governed by the expression:

$$\frac{A_{sv}}{bs_v} \geq \frac{0.4}{f_y}$$

Using 8 mm Φ 2 lgd stirrups, $A_{sv} = 2\frac{\pi}{4}(8)^2 = 100.5\ mm^2$.

Taking $f_y = 250\ N/mm^2$, the spacing of the stirrups is given by

$$s_v = \frac{2.5\, A_{sv} \cdot f_y}{b} = \frac{2.5 \times 100.5 \times 250}{300}$$

$$= 209 \text{ mm.}$$

However provide the stirrups @ 200 mm c/c. Provide 2-12 mm Φ holding bars.

6. ***Check for development length at supports.***

The code stipulates that at the simple supports, where the reinforcement is confined by a compressive reaction the diameter of the reinforcement be such that

$$1.3 \frac{M_1}{V} + L_0 \geq L_d$$

Assuming that 2 bars are bent up and 2 bars are available at the supports

$$A_{st} = 2 \frac{\pi}{4} (20)^2 = 628 \text{ mm}^2$$

$\therefore$ M_1 = moment of resistance of the section, assuming all reinforcement stressed to σ_{st}

$$= 140 \times 628 \times 0.865 \times 545 = 41.45 \times 10^6 \text{ N-mm}$$

$$V = 42450 \text{ N}$$

L_0 = sum of anchorage value of hooks.

Let us provide a support equal to the width of the wall, *i.e.* 600 mm. Let the clear side cover $x' = 40$ mm. For a U hook having anchorage value of 16 Φ , we have

$$\therefore \quad L_0 = \left(\frac{L_s}{2} - x' - 3\Phi\right) + 16\Phi$$

or

$$L_0 = \frac{L_s}{2} - x' + 13\Phi$$

$$= \frac{600}{2} - 40 + 13 \times 20 = 520 \text{ mm}$$

$$\therefore \quad 1.3 \frac{M_1}{V} + L_0 = 1.3 \times \frac{41.45 \times 10^6}{42450} + 520$$

$$= 1269 + 520 = 1789 \text{ mm}$$

Development length $L_d = \dfrac{\Phi\, \sigma_{st}}{4\, \tau_{bd}}$

$$= \frac{20 \times 140}{4 \times 0.6} = 1167 \text{ mm}$$

Alternatively, $L_d \approx 58\,\Phi$ (Table 2.5)

$$= 58 \times 20 = 1160 \text{ mm.}$$

Thus $\left(1.3\frac{M_1}{V}+L_0\right) > L_d$

Hence code requirements are satisfied.

7. Details of reinforcement.

Due the partial fixidity that may be caused at the supports, some reinforcement is always provided at the top of the beam near the ends. Let us bend two bars, one at a distance x_1 from the support and the other at a distance x_2 from the support.

$$\text{B.M. at } x_1 = \frac{wL}{2}x_1 - \frac{wx_1^2}{2}$$

This should be $\frac{3}{4}$ of the maximum B.M.

$$\therefore \quad \frac{wL}{2}x_1 - \frac{wx_1^{\ 2}}{2} = \frac{3}{4}\frac{wL^2}{8}$$

This gives $x_1 = 0.25\,L = 0.25 \times 6.545 = 1.64$ m.

However, bend one bar at a distance of 1.60 m from the support.

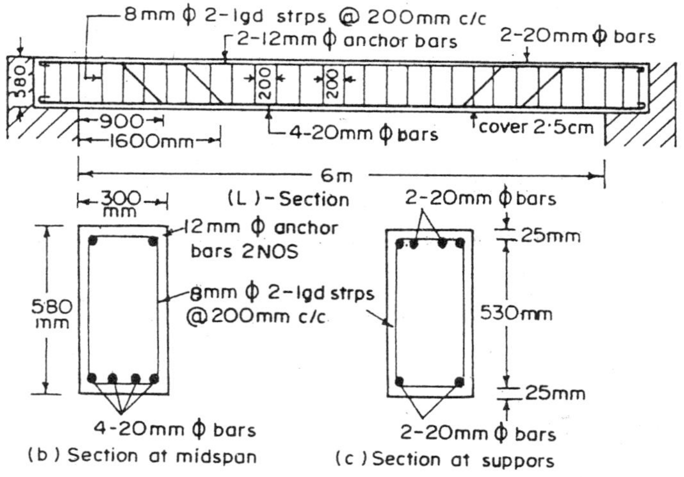

FIG. 7.6

Similarly, for the second bar,

$$\frac{wLx_2}{2} - \frac{wx_2^2}{2} = \frac{1}{2}\frac{wL^2}{8}$$

This gives $x_2 = 0.14L \approx L/7 \approx 0.14 \times 6.545 = 0.92$ m. However, bend the second bar at a distance of 0.9 m from the face of the support. The remaining 2 bars shall be taken straight into the support.

Fig 7.6 shows the longitudinal section and the cross-section of the beam.

Design Example 7.1 (*b*) *Redesign the beam of example 7.1(a) using M 20 concrete mix and HYSD bars.*

Solution.

(i) Calculation of design constants.

For M 20 concrete, $c = \sigma_{cbc} = 7\ \text{N/mm}^2$ and $m=13$. For HYSD bars, $t = \sigma_{st} = 230\ \text{N/mm}^2$

$$k = \frac{mc}{mc+t} = \frac{13 \times 7}{13 \times 7 + 230} = 0.2835$$

$$j = 1 - \frac{k}{3} = 1 - \frac{0.2835}{3} = 0.9055$$

$$R = \frac{1}{2}cjk = \frac{1}{2} \times 7 \times 0.9055 \times 0.2835 = 0.898.$$

(ii) Calculation of B.M.

Let the effective depth of beam

$$= \frac{1}{10}\text{span} = \frac{6000}{10} = 600\ \text{mm}$$

∴ Total depth of beam = 640 mm (say)

Let the width of beam $= \frac{1}{2}d = 300$ mm (say)

Self load of beam per metre run

$$= (0.6 \times 0.3 \times 1) = 25000$$

$$= 4500\ \text{N/m}$$

External Load = 9800 N/m

∴ Total load per metre run

$$= 9800 + 4500 = 14300\ \text{N/m}$$

Effective span $= L = l + d = 6 + 0.6 = 6.6$ m, which is smaller than the centre to centre distance of 6.75 m between the supports:

$$M = \frac{wl^2}{8} = \frac{14300(6.6)^2}{8}$$

$$= 77870\ \text{N-m} = 77.87 \times 10^6\ \text{N-mm}$$

(iii) Design of section

$$d = \sqrt{\frac{M}{Rb}} = \sqrt{\frac{77.87 \times 10^6}{0.898 \times 300}} = 553\ \text{mm.}$$

Let us take $d = 555$ mm and $D = 590$ mm

Revised self load of beam

$$= 0.3 \times 0.59 \times 25000 = 4425 \text{ N/m}$$

$$\therefore \quad w = 9800 + 4425 = 14225 \text{ N/m}$$

$$M = \frac{14225\,(6.555)^2}{8}$$

$$= 76402 \text{ N-m} \approx 76.4 \times 10^6 \text{ N-mm}$$

$$d = \sqrt{\frac{76.4 \times 10^6}{0.898 \times 300}} = 532 \text{ mm}$$

Keep total D=570 mm. Using 20 mm dia. bars and 25 mm clear cover, $d = 570 - 25 - 10 = 535$ mm.

(iv) Steel reinforcement

$$A_{st} = \frac{M}{\sigma_{st}\, jd} = \frac{76.4 \times 10^6}{230 \times 0.9055 \times 535} = 686 \text{ mm}^2$$

$$A_{\Phi} \text{ for 20 mm dia. bars} = \frac{\pi}{4}(20)^2 = 314 \text{ mm}^2$$

$$\therefore \text{ No. of bars} = \frac{686}{314} = 2.18$$

If 16 mm dia. bars are used,

$$A_{\Phi} = \frac{\pi}{4}(16)^2 = 201 \text{ mm}^2$$

$$\therefore \text{ No. of bars} = \frac{686}{201} = 3.41.$$

Hence provide 4 Nos. of 16 mm dia. bars.

Actual $A_{st} = 4 \times 201 = 804 \text{ mm}^2$. Keeping 25 mm cover (outside) clear spacing between bars will be

$$= \frac{1}{3}(300 - 25 \times 2 - 4 \times 16) = 62 \text{ mm},$$

which is much more then the diameter of the bar.

Minimum reinforcement is given by

$$\frac{A_{st}}{bd} = \frac{0.85}{f_y}, \text{ where } f_y = 415 \text{ N/mm}^2$$

$$\therefore \quad A_{st} = \frac{0.85\,(300 \times 535)}{415} = 329 \text{ mm}^2$$

Since the actual A_{st} provided ($= 804 \text{ mm}^2$) is much more than this, the design is O.K.

(*v*) ***Check for shear and design of shear reinforcement***

$$\text{Max.} \quad V = \frac{wl}{2} = \frac{14225 \times 6}{2} = 42675 \text{ N.}$$

$$\tau_v = \frac{V}{bd} = \frac{42675}{300 \times 535} = 0.266 \text{ N/mm}^2$$

Assuming that out of 4 bars of main reinforcement, 2 bars will be bent up near the support, only 2 bars will be available.

$$\frac{100 A_s}{bd} = \frac{100 \times 402}{300 \times 535} = 0.25 \%$$

Hence from Table 3.1, permissible shear (τ_c) for M 20 concrete $=0.22 \text{ N/mm}^2$.

Since $\tau_v > \tau_c$ shear reinforcement is necessary.

$$V_c = \tau_c . bd = 0.22 \times 300 \times 535 = 35310$$

Hence $$V_s = V - V_c = 42675 - 35310 = 7365 \text{ N}$$

This is quite small and the two bent up bars willl take care of this. However, nominal shear reinforcement, given by the following expression is necessary

$$\frac{A_{sv}}{b . s_v} \geq \frac{0.4}{f_y}$$

Using 8 mm Φ 2-lgd stirrups,

$$A_{sv} = 2 \frac{\pi}{4} (8)^2 = 100.5 \text{ mm}^2$$

∴ Spacing of stirrups is given by

$$s_v = \frac{2.5 A_{sv} . f_y}{b} = \frac{2.5 \times 100.5 \times 415}{300}$$

$$= 347 \text{ mm.}$$

This spacing should not be more than $0.75d$ ($= 0.75 \times 535 = 401$ mm) or 450 mm, whichever is less.

Hence provide the stirrups @ 340 mm c/c. Provide 2-10 mm Φ holding bars.

(*vi*) ***Check for development length at supports***

The code stipulates that at simple supports, where the ends of bars are confined by compressive reaction the diameter of the reinforcement should be such that

$$1.3 \frac{M_1}{V} + L_0 \geq L_d .$$

Now $$A_{st} = 2 \frac{\pi}{4} (16)^2 = 402 \text{ mm}^2$$

∴ M_1 = moment of resistance of the section, assuming all reinforcement stressed to σ_{st}

$$= \sigma_{st} . A_{st} . jd = 230 \times 402 \times 0.9055 \times 535$$

$$= 44.79 \times 10^6 \text{ N-mm}$$

$$V = 42675 \text{ N}$$

L_0 = sum of anchorage beyond centre line of support

$$= \frac{l_s}{2} - x'$$

Let us provide a support equal to the width of the wall (*i.e.* 600 mm). Let the side cover $x' = 40$ mm.

$$\therefore \quad L_0 = \frac{600}{2} - 40 = 260 \text{ mm.}$$

$$\therefore \quad 1.3\frac{M_1}{V} + L_0 = 1.3 \times \frac{46.47 \times 10^6}{42675} + 468$$

$$= 1416 + 260 = 1676 \text{ mm}$$

Development length $L_d = \dfrac{\Phi . \sigma_s}{4\,\tau_{bd}}$

For HYSD bars, $\sigma_s = \sigma_{st} = 230$ N/mm^2

For M 20 concrete and HYSD bars,

$$\tau_{bd} = 1.40 \times 0.8 = 1.12 \text{ N/mm}^2$$

$$\therefore \quad L_d = \frac{16 \times 230}{4 \times 1.12} = 821 \text{ mm}$$

Thus $\left(1.3\dfrac{M_1}{V} + L_0\right) > L_d$.

Hence code requirements are satisfied

(viii) Details of reinforcement

The bending of bars, near supports, will be done in the same manner as explained in Example 7.1(*a*).

7.2 DESIGN OF LINTEL BEAMS

Lintels are provided over the openings of doors, windows, almirahs, etc. Generally, they support the load of the wall over it, and sometimes also the live load are transferred by the sub-roof of the room. Following five cases may arise from point of view of distribution of load over the lintels :

(1) When the length of wall on each side is more than half the effective span of the lintel.

(2) When the length of wall on the side is less than half the effective span.

(3) When the length of wall on both sides is less than half the effective span.

(4) When there are openings over the lintel.

(5) When there is load carrying slab over the lintel.

Case I. ***When the length of wall on each side is more than the effective span of the lintel.***

This is most general case. Because of arch-action in the masonry, all the load of the wall above the lintel is not transferred to the lintel. It is assumed that the load transferred is in the form of equilateral triangle, and the load on the lintel is equal to the weight of the masonry in the triangular portion, as shown in Fig 7.7(*a*). If, however, the height of the wall above the lintel is insufficient (*i.e.* if the

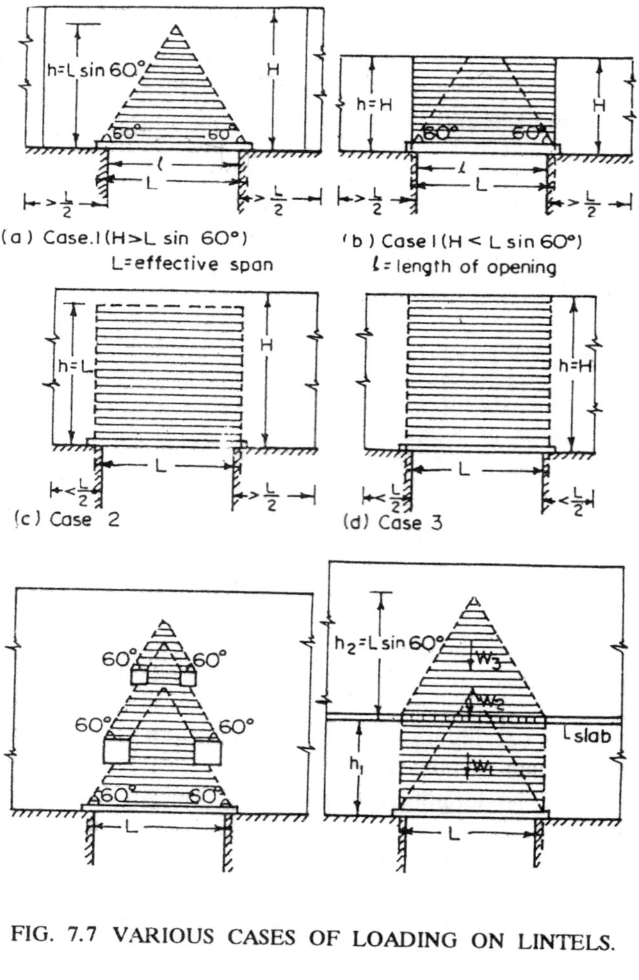

FIG. 7.7 VARIOUS CASES OF LOADING ON LINTELS.

apex of the triangle falls above the top of the wall), whole of the rectangular load above the lintel is taken to act on the lintel, as shown in Fig. 7.7(*b*).

Let H = height of wall above the lintel.

l = actual opening.

L = effective length of the opening.

h = effective height of masonry, above the lintel.

Thus, $h = L \sin 60° = L \cdot \frac{\sqrt{3}}{2} = \frac{L}{2}\sqrt{3}.$

$\therefore$ Area of triangle $\frac{1}{2} \times L \times \frac{L}{2}\sqrt{3} = \frac{L^2}{4}\sqrt{3}.$

Case 2. ***When the length of the wall to one side is less than half the effective span.***

Fig 7.7 (*c*) shows the situation where the length of wall to one side is less than half the effective span, but the length to the other side is more than half the effective span. In that case, the load transferred to the lintel will be equal to the weight of masonry contained in the rectangle of height h equal to the effective span.

Case 3. ***When the length of the walls to each side is less than half the effective span.***

This is shown in Fig. 7.7(*d*). The load acting on the lintel will be equal to the weight of the masonry contained in the rectangle of height h equal to the full height H of the wall.

Case 4. ***When there are openings on the lintel.***

If there are openings, due to the provision of ventilators etc. and if these openings are intersected by the 60° lines, the loading will be calculated by allowing dispersion lines at 60° from the top edges of the openings, as shown in Fig. 7.7(*e*). The total load on the lintel will be equal to the weight of the masonry contained in the shaded area.

Case 5. ***When there is load carrying slab falling within the dispersion triangle [Fig. 7.7(f))].***

If the roof slab is provided at a level well above the apex of the dispersion triangle, the uniformly distributed load carried by the slab is *not* transferred to the lintel. If, however, the slab intersects the dispersion triangle, three types of loads are transferred to the lintel :

(*i*) Load W_1 due the weight of the masonry contained in the rectangle of height h_1 equal to the height of the slab above the lintel.

(*ii*) Load W_2 carried by the slab, in a length L.

(*iii*) Load W_3 due to the weight of masonry contained in the equilateral triangle above the slab.

The design of the lintel is similar to that of R.C. beam discussed in § 7.1. The width of lintel is normally kept equal to width of the wall.

The max. bending moment M_1 due to the triangular load is taken as

$$M_1 = \frac{WL}{6}$$

where W = total weight of masonry contained in the equilateral triangle.

The max. bending moment M_2 due the uniformly distributed self-weight w of the lintel is

$$M_2 = \frac{wL^2}{8}$$

$$\therefore \quad \text{Total} \quad M = \frac{WL}{6} + \frac{wl^2}{8}$$

The maximum shear will act at the edge of the opening and is given by

$$F = \frac{W}{2} + \frac{wl}{2}$$

Example 7.2. *Design a R.C. lintel to be provided over an opening of 2 m wide. The height of masonry above the lintel is 3 m. The opening is centrally located in a long wall, 300 mm thick. The unit weight of masonry may be taken as 1.9 kN/m³. Take the following permissible stresses :*

$c = \sigma_{cbc} = 5$ N/mm² ; $t = \sigma_{st} = 140$ N/mm² ; $m = 19$.

Solution.

1. Calculation of constants

$$k = \frac{mc}{mc + t} = \frac{19 \times 5}{19 \times 5 + 140} = 0.404$$

$$j = 1 - k/3 = 0.865$$

$$R = \frac{1}{2} c\, j\, k = \frac{1}{2} \times 5 \times 0.865 \times 0.404 = 0.874$$

2. *Calculation of bending moment*

Let the total depth of lintel = 24 cm.

Width of lintel = 30 cm.

Self weight of lintel, per m length $= 0.24 \times 0.30 \times 1 \times 25000$
$= 1800$ N/m

Effective span will be the minimum of the following :

(*i*) Centre to centre of bearing : Providing a bearing 20 cm,
$$L = 2 + 0.2 = 2.2 \text{ m}$$

(*ii*) Clear span + effective depth : Assuming a cover of 3 cm,
$$d = 24 - 3 = 21 \text{ cm}$$

$$\therefore \quad L = 2 + 0.21 \text{ m} = 2.21 \text{ m.}$$

Hence adopt $L = 2.2$ m.

Height of equilateral triangle, assuming 60° dispersion
$$= L \sin 60° = \frac{L}{2}\sqrt{3} = \frac{2.2}{2} \times \sqrt{3} = 1.9 \text{ m.}$$

This is less than the height of the wall above the lintel. Hence load on the lintel will be equal to the weight of the masonry contained in the triangular portion.

$$\therefore \quad W = \left(\frac{1}{2} \times 2.2 \times 1.9\right)(0.3 \times 19000) = 11913 \text{ N}$$

$$\text{B.M.} \quad M = \frac{WL}{6} + \frac{wL^2}{8}$$

$$= \left[\frac{11913 \times 2.20}{6} + \frac{1800 \times (2.20)^2}{8}\right]$$

$$= 5457 \text{ N-m} = 5457000 \text{ N-mm}$$

3. *Design of section*

$$\therefore \quad d = \sqrt{\frac{M}{Rb}} = \sqrt{\frac{5457000}{0.874 \times 300}} = 144 \text{ mm}$$

Adopt $d = 150$ mm. Using 10 mm Φ bars and a clear cover of 25 mm, the total depth = 150 + 25 + 5 = 180 mm. This is less than the assumed value. However, recalculation of the load is not necessary, because self-weight of lintel is comparatively smaller than the super-imposed load.

4. *Reinforcement*

$$A_{st} = \frac{M}{\sigma_{st}\, jd} = \frac{5457000}{140 \times 0.865 \times 150} = 300.4 \text{ mm}^2$$

Provide 10 mm diameter bars.

No. of bars required $= \dfrac{300.4}{\frac{\pi}{4}(10)^2} = 3.8$

Hence provide 4 Nos. of 10 mm Φ bars.

Total area provided $= 4 \times \frac{\pi}{4}(10)^2 = 314.2 \text{ mm}^2$

Minimum reinforcement is given by

$$\frac{A_{st}}{bd} = \frac{0.85}{f_y}$$

or
$$A_{st} = \frac{0.85}{250} \times (300 \times 150) = 153 \text{ mm}^2$$

Since the actual steel provided (=314.2) is more than this, the design is O.K.

Bend two of the bars at a distance of $\frac{L}{7} = \frac{2200}{7} \approx 300$ mm from the face of each support. Keep clear cover = 24 mm.

5. ***Check for shear and design of shear reinforcement***

$$V = \frac{W}{2} + \frac{wl}{2} = \frac{11913}{2} + \frac{1800 \times 2}{2}$$

$$= 7757 \text{ N.}$$

$$\therefore \qquad \tau_v = \frac{V}{bd} = \frac{7757}{300 \times 150} = 0.172 \text{ N/mm}^2$$

At supports, $\dfrac{100 A_s}{bd} = \dfrac{100}{300 \times 150}(\frac{1}{2} \times 314.2) = 0.35\ \%$

From Table 3.1, for M 15 concrete, for 0.35%, $\tau_c \approx 0.25$ N/mm². Hence only nominal shear reinforcement is required, given by the relation :

$$\frac{A_{sv}}{b \cdot s_v} \geq \frac{0.4}{f_y}$$

Using 8Φ-two legged stirrups, $A_{sv} = 2 \times \frac{\pi}{4}(8)^2 = 100.5 \text{ mm}^2$

$$\therefore \qquad s_v = \frac{2.5\ A_{sv}}{b} \times f_y = \frac{2.5 \times 100.5 \times 250}{300}$$

$$= 209 \text{ mm}$$

Hence provide these @ 200 mm c/c throughout. Provide 2–8Φ anchor bars (holding bars) throughout the length.

6. ***Check for development length at supports***

At the supports, $1.3\dfrac{M_1}{V} + L_0 \geq L_d$

$$A_{st} = \frac{1}{2} \times 314.2 = 157.1 \text{ mm}^2$$

$\therefore \quad M_1 = A_{st} \,.\, jd \,.\, \sigma_{st} = 157.1 \times 140 \times 0.865 \times 150$

$$= 2.854 \times 10^6 \text{ N-mm}$$

$$V = 7757 \text{ N.}$$

$$L_0 = \frac{L_s}{2} - x' + 13\Phi$$

Let $L_s = 200$ mm

$$x' = \text{side cover} = 25 \text{ mm}$$

$$L_0 = \frac{200}{2} - 25 + 13 \times 10 = 205 \text{ mm}$$

$$\therefore \quad 1.3\frac{M_1}{V} + L_0 = 1.3 \times \frac{2.854 \times 10^6}{7757} + 205$$

$$= 478 + 205 = 683 \text{ mm.}$$

$$L_d = \frac{\Phi\,\sigma_{st}}{4\,\tau_{bd}} = \frac{10 \times 140}{4 \times 0.6} = 583 \text{ mm.}$$

Hence code requirements are *satisfied.*

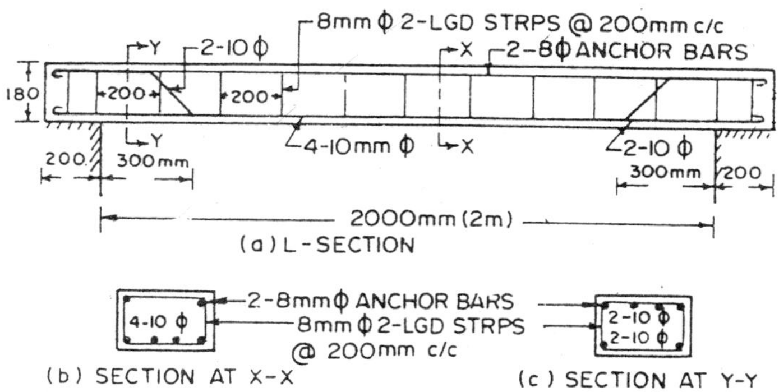

FIG. 7.8

7. *Details of reinforcement*

Fig. 7.8 shows the longitudinal section and cross-section of the lintel beam.

Example 7.3 *Fig 7.9(a) shows the cross-section of a wall of room, 5 m wide and 18 m long from inside. Design the lintel for the window having a opening of 2 m. The reinforced concrete slab, 16 cm thick, may be assumed to transfer half the load to the wall shown. Use the following data for design :*

(*i*) *Weight of lime concrete terracing* $= 19 \text{ kN/m}^3$

(ii) *Weight of masonry* = *19.2 kN/m³*

(iii) *Weight of reinforced concrete* = *24 kN/m³*

(iv) *Super-imposed load on the slab* = *1.6 kN/m²*

Use M-15 concrete and take σ_{st}= 140 N/m². The window may be assumed to be centrally located along the length of the room.

Solution

1. Loading and bending moment.

Let us assume the overall depth of the lintel to be 300 mm, and a bearing of 300 mm on either side. Hence effective span of lintel will be equal to 2 + 0.3 = 2.3 m. Assuming the dispersion to be along 60° lines, the height of load triangle, measured above the top of the lintel $= 2.3 \sin 60° = \frac{2.3 \times \sqrt{3}}{2} \approx 2$ m, which is more than height (1.5m) of the slab above the lintel. Hence the load of the slab will also be transferred to lintel.

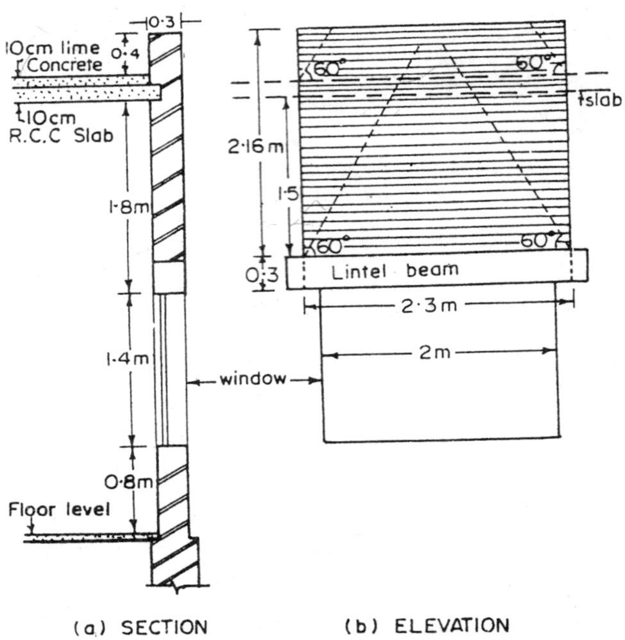

FIG. 7.9

Fig. 7.9. (*b*) shows the elevation, showing all the heights. If we construct equilateral triangle above the top of the slab, its apex will fall very much above the top of the parapet wall. Hence the weight of the whole wall above the lintel will be transferred to the lintel. Thus , the loads per metre length of the lintel will consist of the following :

(*i*) *Weight of the wall*

Weight per metre run $= 0.3 \times 2.16 \times 19000 = 12300$ N

(*ii*) *Load transferred by the slab*

Consider 1 m strip of the slab of 5 m span.

Live load on this strip $= \frac{1}{2}(5 \times 1) \times 1600 \qquad = 4000$ N

Dead weight of slab $= \frac{1}{2}\left(5 \times 1 \times \frac{16}{100}\right)24000 \qquad = 6900$N

(16 cm thick)

Dead weight of lime concrete $= \frac{1}{2}\left(5 \times 1 \times \frac{10}{100}\right) 19000$

$= 4750$ N

∴ Total load per metre run of the lintel $= 18350$ N

(*iii*) *Self-weight of lintel*

Weight per metre run $= (0.3 \times 0.3 \times 1)\, 24000 = 2160$ N

Hence total U.D.L. per meter run of lintel

$= 12300 + 18350 + 2160 = 32810$ N

Max. bending moment $M = \frac{wL^2}{8}$

$$= \frac{32810 \times (2.3)^2}{8} = 2.17 \times 10^4 \text{ N–m}$$

$$= 21.7 \times 10^6 \text{ N–mm}$$

2. *Calculation of constants*

$$k = \frac{mc}{mc + t} = \frac{19 \times 5}{19 \times 5 + 140} = 0.404$$

$$j = 1 - k/3 = 0.865$$

$$R = \frac{1}{2}cjk = 5 \times 0.865 \times 0.404 = 0.874$$

3. *Design of section*

Effective depth $d = \sqrt{\frac{M}{R \, . \, b}} = \sqrt{\frac{21.7 \times 10^6}{0.874 \times 300}} = 288$ mm

≈ 290 mm

Using 10 mm Φ bars and keeping a clear cover of 25 mm, total depth =290 + 5 + 25 = 320, against the assumed value of 300 mm. This will slightly increase the dead load, but since the dead load of lintel is very small in comparison to the other loads, revision of calculations with new dimensions is not necessary.

4. *Steel reinforcement*

$$A_{st} = \frac{M}{\sigma_{st} \cdot jd} = \frac{21.7 \times 10^6}{140 \times 0.874 \times 290}$$

$$= 612 \text{ mm}^2$$

$$\text{No. of 10 mm } \Phi \text{ bars} = \frac{612}{\frac{\pi}{4}(10)^2} \approx 7.8$$

Hence use 8 nos. 10 mm Φ bars.

Total $A_{st} = 8 \times \frac{\pi}{4}(10)^2 = 628 \text{ mm}^2$

Out of these, bend 4 bars up at a distance of $\frac{L}{7} = \frac{2300}{7} \approx 300$ mm from the edge of each support.

5. *Check of shear and design of shear reinforcement*

Max. S.F. $V = \frac{w \cdot l}{2} = \frac{32810 \times 2}{2} = 32810$ N

(At the edge of the support)

$$\therefore \quad \tau_v = \frac{V}{bd} = \frac{32810}{300 \times 290} = 0.38 \text{ N/mm}^2$$

At support, $A_{st} = \frac{1}{2} \times 628 = 314 \text{ mm}^2$

$$\therefore \quad \frac{100 A_s}{bd} = \frac{100 \times 314}{300 \times 290} = 0.36\%$$

Hence from Table 3.1. $\tau_c \approx 0.26$ N/mm^2

Since τ_v is more than τ_c, shear reinforcement in necessary.

$V_c = \tau_c \cdot bd = 0.26 \times 300 \times 290 = 22620$ N

$\therefore \quad V_s = V - V_c = 32810 - 22620 = 10190$ N

Shear resistance V_{s1} of bent up bars in given Eq. 3.22

$V_{s1} = \sigma_{sv} \cdot A_{sv} \sin \alpha = 0.707\, \sigma_{sv} \cdot A_{sv}$

$= 0.707 \times 140 \times 314 = 31080$ N

But V_{s1} assigned to inclined oars cannot be more than $\frac{1}{2} V_s$ $= \frac{1}{2} \times 10190 = 5045$ N.

Hence vertical stirrups are to be provided by a shear V_{s2} = 5045 N only.

Using 8 mm Φ 2-lgd stirrups,

$$A_{st} = 2\frac{\pi}{4}(8)^2 = 100.5 \text{ mm}^2$$

Hence from Eq. 3.20.

$$s_v = \frac{\sigma_{sv} . A_{sv} . d}{V_{s2}} = \frac{140 \times 100.5 \times 290}{5045}$$

$$= 800 \text{ mm. This is too large.}$$

However, max. spacing corresponding to nominal shear stirrups is given by Eq. 3.19(*a*).

$$s_v \le \frac{2.5 A_{sv} . f_y}{b} = \frac{2.5 \times 100.5 \times 250}{300}$$

$$\le 209 \text{ mm.}$$

Hence provide 8 mm Φ 2-lgd stirrups @ 200 mm c/c. Provide 2-8 mm Φ holding bars at top.

6. *Check for development length at the end*

At the support, the code requirements are that

$$1.3\frac{M_1}{V} + L_0 \ge L_d$$

$$A_{st} = \frac{1}{2} \times 628 = 314 \text{ mm}^2$$

$$\therefore \quad M_1 = A_{st} . \sigma_{st} . jd = 314 \times 140 \times 0.865 \times 290$$

$$= 11.03 \times 10^6 \text{ N-mm}$$

$$V = 32810 \text{ N}$$

$$L_0 = \frac{L_s}{2} - x' + 13\,\Phi$$

Let L_s = 300 mm ;

x' = side cover = 25 mm

$$\therefore \quad L_0 = \frac{300}{2} - 25 + 13 \times 10 = 255 \text{ mm.}$$

$$\therefore \quad 1.3\frac{M_1}{V} + L_0 = 1.3 \times \frac{11.03 \times 10^6}{32810} + 255 = 436 + 225 = 661 \text{ mm}$$

$L_d \approx 58.3\,\Phi = 583$ mm.

Hence the code requirement are satisfied.

7. ***Details of reinforcements.*** Shown in Fig. 7.10.

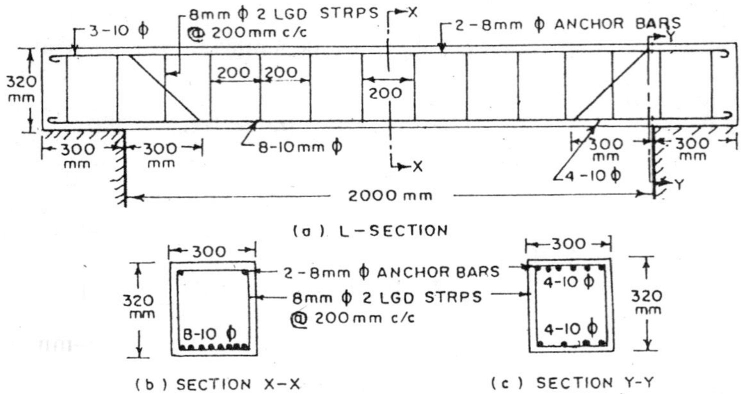

FIG. 7.10.

7.3. DESIGN OF CANTILEVER

When a cantilever beam or slab is subjected to downward loads, it bends downwards with its convexity upwards. Due to this, the upper fibres of the section are subjected to tensile stress and the lower fibres are subjected compressive stresses. Hence the main reinforcement is provided at the *upper face,* and not the lower face. The bending moment varies from a zero value at the free section to maximum value at the fixed end (support). Hence the depth of section and the steel reinforcement is varied as per requirements of bending moment. In order that the whole cantilever does not overturn, it should be properly fixed in the wall. The reinforcement should be embedded in the support for a distance of 58 diameters from the edge of the support.

When curtailment of bar is done, it should be kept in mind that the embedment of the bar is not less than L_d from the point it is needed the most, nor less that 12 diameters (or *d*) from the point it can be spared from the bending moment point of view.

Example 7.4. *A cantilever (beam) projects 2.5 m beyond the fixed end, and carries a superimposed load of 12 kN per metre run. Design the cantilever for the following values of permissible stress :* $c = \sigma_{cbc} = 5\ N/mm^2$; $t = \sigma_{st} = 140\ N/mm^2$; $m = 18$; *Take unit weight of concrete as* $24\ kN/m^3$.

Solution.

1. ***Calculation of bending moment***

Let the section of the cantilever be 250 mm wide and 500 mm deep at the fixed end, and 250 mm and 200 mm deep at the free end.

$\therefore$ Weight of beam $= \frac{1}{2}(0.5+0.2) \times 0.25 \times 0.25 \times 24000 = 5250$ N

Acting at $\dfrac{0.5 + 2 \times 0.2}{0.5 + 0.2} \times \dfrac{2.5}{3} = 1.07$ m from the fixed end.

$\therefore$ Maximum B.M.

$$M = \left[(5250 \times 1.07) + \frac{12000 \times 2.5^2}{2}\right]$$

$$= 4.31 \times 10^4 \text{ N–m} = 43.1 \times 10^6 \text{ N–mm}$$

S.F. at the edge of the support $= 5250 + 12000 \times 2.5$

$= 35250$ N

2. ***Design constants***

$$k = \frac{mc}{mc + t} = \frac{18 \times 5}{18 \times 5 + 140} = 0.391$$

$$j = 1 - k/3 = 0.87$$

$$R = cjk = \frac{1}{2} \times 5 \times 0.87 \times 0.391 = 0.85$$

3. ***Design of section***

Effective depth $d = \sqrt{\dfrac{M}{R \, . \, b}} = \sqrt{\dfrac{43.1 \times 10^6}{0.85 \times 250}}$

Using 16 mm Φ bars, and 25 mm clear cover, total depth $= 450 + 8 + 25 = 483$ mm. However, keep total depth $= 490$ mm. Then available effective depth $= 490 - 25 - 8 = 457$ mm. Keep total free end depth at 200 mm.

4. ***Steel reinforcement***

$$A_{st} = \frac{M}{\sigma_{st} \, . \, jd} = \frac{43.1 \times 10^6}{140 \times 0.87 \times 457}$$

$$= 774 \text{ mm}^2$$

No. of 16 mm Φ bars $= \dfrac{774}{\frac{\pi}{4}(16)^2} = 3.85$

Hence provide 4 Nos. 16 mm Φ bars giving total $A_{st} = 804$ mm^2.

Since the B.M. decreases to zero at the end, let us curtail few bars. Let 2 bars be *curtailed* at a distance x from the free end. Assuming the B.M.D. to be parabolic, the B.M. at this section may be approximately taken to be equal to $\left(\frac{x}{2.5}\right)^2 \times 43.1 \times 10^6$.

$= 6.9 \times 10^6 x^2$ N-mm. Area of 2-16 Φ bars = 402 mm².

$$\therefore \quad 402 = \frac{6.9 \times 10^6 x^2}{140 \times 0.87\, d_x} \quad ...(i)$$

where d_x = effective depth at that section.

Total depth at the section $= \left[200 + \frac{490 - 200}{2.5} x\right]$

$$\therefore \quad d_x = \left[200 + \frac{290}{2.5} x\right] - [8 + 25] = 167 + 116x \quad ..(ii)$$

Substituting in (i),

$$402 = \frac{6.9 \times 10^6 x^2}{140 \times 0.87 \times (167 + 116x)}$$

From which x = 1.57 m

Minimum embedment requirement beyond this = 12Φ

= 12×16 = 192 mm, or equal to d_x = 167+116×1.57 = 350 mm whichever is more.

∴ Two bars may be curtailed at 2.5−1.57+0.35 = 1.28 m from the edge of the support. This distance should be greater than $L_d \approx 58.3\,\Phi = 58.3 \times 16 = 933$ mm = 0.933 m. Hence the bars can be curtailed at a distance of 1.28 m from the support.

5. *Check for shear and design of shear reinforcement*

$$V = 35250 \text{ N}$$

$$\therefore \quad \tau_v = \frac{35250}{250 \times 457} = 0.31 \text{ N/mm}^2$$

For the given values of σ_{cbc} and m, the concrete may be assumed to be of M 15 grade

$$\frac{100 A_s}{bd} = \frac{100 \times 804}{250 \times 457} = 0.7\%$$

Hence from Table 3.1, $\tau_c \approx 0.33$ N/mm².

Hence shear reinforcement is not required. However, provide nominal shear reinforcement given by the expression

$$s_v \le \frac{2.5 A_{sv} . f_y}{b} \le \frac{2.5 \times 250 A_{sv}}{250}$$

or $$s_v \le 2.5 A_{sv}$$

Using 8 mm Φ-2 lgd stirrups,

$$A_{sv} = 2\frac{\pi}{4}(8)^2 = 100.5 \text{ mm}^2$$

$$\therefore \quad s_v \leq 2.5 \times 100.5 = 251 \text{ mm}$$

Hence provide the stirrups @ 250 mm c/c. Provide 2-10 Φ holding bars.

6. *Check for development length at the support*

Since the ends of the bars are confined by compressive reaction,

$$1.3\frac{M_1}{V} + L_0 \geq L_d$$

Here, $M_1 = 804 \times 140 \times 0.87 \times 457 = 44.75 \times 10^6$ N–mm

$V = 35250$

Let L_s = width of support = 30 mm

x' = side cover = 30 mm

$$\therefore \quad L_0 = \frac{L_s}{2} - x' + 13\,\Phi = \frac{500}{2} - 30 + 13 \times 16 = 428 \text{ mm.}$$

$$\therefore \quad 1.3\frac{M_1}{V} + L_0 = 1.3\frac{44.75 \times 10^6}{35250} + 428 = 1651 + 428 = 2079 \text{ mm}$$

$$L_d \approx 58.3\Phi = 58.3 \times 16 = 933 \text{ mm.}$$

Hence code requirements are satisfied. However, it is preferable to anchor the bars by embedding as shown in Fig. 7.11.

7. *Details of reinforcement* : As shown in Fig. 7.11.

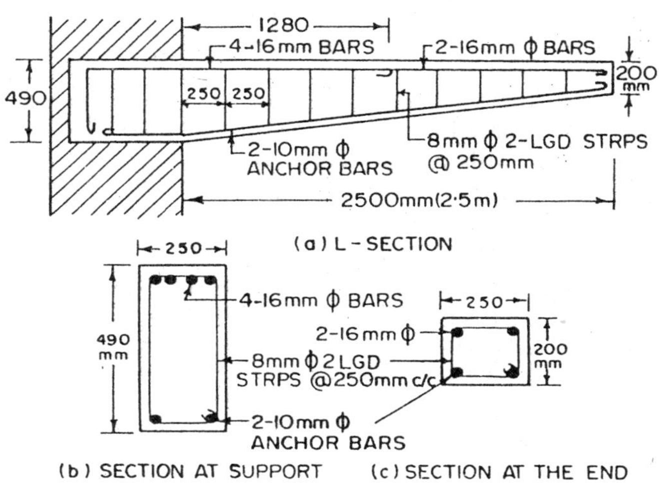

FIG. 7.11.

7.4. DESIGN OF ONE-WAY SLAB

(a) LONGITUDINAL AND TRANSVERSE REINFORCEMENTS

The analysis of slab spanning in one direction is done by assuming it to be a beam of 1 m width. The reinforcement etc. are calculated for 1 m width and bars are distributed accordingly.

Thus, $$d = \sqrt{\frac{M}{1000\,R}}, \text{ where } M \text{ is in N-mm}$$

$$A_{st} = \frac{M}{\sigma_{st}\,.\,jd}, \text{ (where } t = \sigma_{st}\text{)}$$

and spacing $$s = \frac{1000\,A_{\Phi}}{A_{st}} \text{ mm}$$

where A_{Φ} is the cross-sectional area of one bar.

In addition to the main reinforcement (called longitudinal reinforcement), *transverse reinforcement* (also called distribution reinforcement) is also provided in a direction at right angles to the span of the slab. The transverse reinforcement is provided to serve the following purposes

(*i*) It distributes the effects of point load on the slab more evenly and uniformly.

(*ii*) It distributes the shrinkage and temperature cracks more evenly.

(*iii*) It keeps the main reinforcement in position.

The amount of transverse reinforcement may vary from a minimum of 0.15% of gross concrete area for ordinary slab to 0.3% for bridge slabs or for slab where temperature variations are high. The distribution reinforcement is placed in contact with the main reinforcement, and is also provided with hooked ends.

(b) BASIC RULES FOR DESIGN (IS : 456–1978)

1. *Effective span*

The basic rules for the effective span of a slab are the *same* as given by beams (clause *a*-1 of § 7.1). Thus, in the case of freely supported slab, the effective span is taken equal to the distance between centre to centre of supports, or the clear distance between the supports plus the effective depth of the slab whichever is less.

2. *Control of deflection*

The same rules apply, as given for beams, in clause *a*-2 of § 7.1.

Thus, the *basic* values of span of effective depth ratios, for span upto 10 m are as follows :

Cantilever	7
Simply supported	20
Continuous	26

3. ***Minimum reinforcement***

The reinforcement in either direction in slabs shall not be less than 0.15% of the total cross-sectional area. However, this value can be reduced to 0.12% when high strength deformed bars or welded wire fabric are used.

4. ***Maximum diameter***

The diameter of the reinforcing bars shall not exceed one eighth of the total thickness of the slab.

5. ***Spacing of bars***

The horizontal distance between parallel main reinforcement bars shall not be more than three times the effective depth of a solid slab or 450 mm, whichever is smaller. The horizontal distance between parallel reinforcement bars provided against shrinkage and temperature shall not be more than five times the effective depth of the slab or 450 mm whichever is smaller. The horizontal distance between two parallel main reinforcing bars shall usually be not less than the greatest of the following :

(*i*) The diameter of the bar if the diameters are equal.

(*ii*) The diameter of the larger bar if the diameters are unequal, and

(*iii*) 5 mm more than the nominal maximum size of coarse aggregate.

6. ***Cover to reinforcement***

(*a*) At the end of reinforcing bar not less than 25 mm, nor less than twice the diameter of such bar.

(*b*) For tensile, compressive, shear and other reinforcement in a slab, not less than 15 mm, nor less than the diameter of such bar, and

(*c*) For any other reinforcement, not less than 15 mm, nor less than the diameter of such bar.

7. ***Curtailment of tension reinforcement***

The same rules, as given for beams in clause *a* 6 of § 7.1 apply for slabs also. Thus, for curtailment, reinforcement shall extend beyond the point at which it is no longer required to resist flexure, for a distance equal to effective depth or 12 times the bar diameter whichever is greater, except at supports or end of cantilever.

At least one-third the positive moment reinforcement in simple members and one-fourth of the positive moment reinforcement in continuous members shall extend along the same face of the member into the support, to a length equal $L_d/3$.

At simple supports, and at point of inflexion, positive moment tension reinforcement shall be limited to such a diameter that

$$\left(\frac{M_1}{V} + L_0\right) \geq L_d.$$

The value of M_1/V the above expresion may be increased by 30% when the ends of the reinforcement are confined by a compressive reaction.

At least one-third of the total reinforcement provided for negative moment at the support shall extend beyond the point of inflexion, for a distance not less than the effective depth of the member or 12 Φ or one-sixteenth of the clear span whichever is greater.

Design Example 7.5. (*a*) *Design a R.C. slab for a room having inside dimensions 3 m × 6 m. The thickness of the supporting wall is 300 mm. The slab carries 100 mm thick lime concrete at its top, the unit weight of which may be taken as 19000 N/m³. The live load on the slab may be taken as 2500 N/m². Assume the slab to the simply supported at the ends. Use M 15 concrete. Take permissible tensile stress in steel as 140 N/mm². Take unit weight of concrete as 24 kN/m³.*

Solution.

1. ***Calculation of constants***

For M 15 concrete $c = \sigma_{cbc} = 5$ N/mm^2 and $m \approx 19$

Also, $t = \sigma_{st} = 140$ N/mm^2

$\therefore$ $k = 0.404$, $j = 0.865$ and $R = 0.874$ as in the previous example.

2. ***B.M. and S.F.***

The slab will evidently bend over a clear span of 3 m. Let the overall depth of slab be 120 mm, and the approximate effective depth be $120 - 20 = 100$ mm

$\therefore$ Effective span $= 3+0.1 = 3.1$ m, which is evidently less than the centre to centre distance between the supports.

(*i*) Load due to self weight of slab $= (1\times1\times0.12\times24000)$
$= 2880$ N

(*ii*) Load due to lime concrete finish $= 1\times1\times0.1\times19000$
$=900$ N

(*iii*) Superimposed live load $= 1 \times 1\times 2500= 2500$ N

Total U.D.L. (w), per metre run of slab $=7280$ N

$$\therefore \quad M = \frac{wl^2}{8} = \frac{7280(3.10)^2}{8} = 8745 \text{ N-m/m width}$$

$$V = \frac{wL}{2} = \frac{7280 \times 3}{2} = 10920 \text{ N at the face of the support.}$$

3. ***Design of section***

$M = 8745$ N–m $= 8.745 \times 10^6$ N-mm per m or 1000 mm width.

$$d = \sqrt{\frac{M}{bR}} = \sqrt{\frac{8.745 \times 10^6}{1000 \times 0.874}} \approx 100 \text{ mm}$$

Using 10 mm bars Φ and a minimum cover of 15 mm, over all depth $=100+5+15=120$ mm.

From stiffness (or deflection) point of view, span/effective depth ratio= 20 (see clause 2).

For a balanced design, percentage reinforcement $= \dfrac{k\,\sigma_{cbc}}{2\,\sigma_{st}} \times 100$
$= \dfrac{0.404 \times 5}{2 \times 140} \times 100 = 0.72\%$ for M 15 concrete. Hence from Fig. 7.1, modification factor (using mild steel bars) $\approx$ 1.6.

$$\therefore \quad \frac{\text{span}}{d} = 20 \times 1.6 = 32$$

$$\therefore \quad d = \frac{\text{span}}{32} = \frac{3100}{32} = 96.9 \text{ mm.}$$

Hence keep d = 100 mm, as required from bending point of view. Thus total depth = 120 mm.

4. ***Main reinforcement***

$$A_{st} = \frac{M}{\sigma_{st} \cdot jd} = \frac{8.745 \times 10^6}{140 \times 0.865 \times 100} = 722 \text{ mm}^2$$

This is more than minimum reinforcement = 0.15%

$$= \frac{0.15 \times 120 \times 1000}{100} = 180 \text{ mm}^2.$$

Using 10 mm Φ bars ; area of each bar, $A_\Phi = \dfrac{\pi}{4}(10)^2 = 78.54$ mm²

$$\therefore \text{ Pitch } s = \frac{1000\,A_\Phi}{A_{st}} = \frac{1000 \times 78.54}{722} = 109.4 \text{ mm} \approx 110 \text{ mm.}$$

$$\text{Actual } A_{st} = \frac{1000 \times 78.54}{110} = 714 \text{ mm}^2$$

$$\% \text{ reinforcement} = \frac{714}{120 \times 1000} \times 100 \approx 0.6\ \%$$

It is to be noted that the above pitch of 110 mm is less than 3 times the effective depth of the slab.

Bend every third bar up at the support, at a distance of $\frac{l}{5} = \frac{3.10}{5}$ $= 0.6$ m from the edge of the support.

A_{st} at the support $= \frac{2}{3} \times 714 = 476 \text{ mm}^2$ or $= 0.4\%$ which is more than the minimum of 0.15%.

5. ***Check for development length at the support.***

The code stipulates that at simple supports, the diameter of the reinforcement be such that

$$1.3\frac{M_1}{V} + L_0 \geq L_d$$

where $M_1 = 140 \times 472 \times 0.865 \times 100 = 5.72 \times 10^6$ N–mm

$V = 10920$ N

Let us assume that a clear cover of 25 mm is provided at the side (end) and the slab has a support of 300 mm width. Providing U-hook,

$$L_o = \frac{l_s}{2} - x' + 13\,\Phi$$

$$= \frac{300}{2} - 25 + 13 \times 10 = 255 \text{ mm}$$

$$1.3\frac{M_1}{V} + L_0 = \frac{1.3 \times 5.72 \times 10^6}{10920} + 255 = 681 + 255$$

$$= 936 \text{ mm}$$

$$L_d = \frac{\Phi\,\sigma_{st}}{4\tau_{bd}} = \frac{10 \times 140}{4 \times 0.6} = 583 \text{ mm}$$

(Alternatively, $L_d \approx 58.3\,\Phi = 583$ mm)

$$\therefore \quad 1.3\frac{M_1}{V} + L_0 \geq L_d$$

Hence anchorage requirements are satisfied.

Note : The reinforcement should extend by a length equal to $\frac{L_d}{3} = \frac{583}{3} = 200$ mm beyond the face of the support. This suggests that the width of the support should not be less than $200+25=225$ mm. In the present case, the support width is 300 mm.

$$\tau_{bd} = \frac{V}{bd} = \frac{10920}{1000 \times 100} \approx 0.11 \text{ N/mm}^2.$$

This is much less than the permissible value of $\tau_c = 1.3 \times 0.26 = 0.34$ N/mm^2 for M 15 concrete for $p' = 0.4\%$ (Table 3.1 and 3.2) Hence safe.

7. ***Distribution reinforcement***

$$A_{sd} = \frac{0.15\, bD}{100} = \frac{0.15 \times 1000 \times 120}{100} = 180 \text{ mm}^2$$

Using 6 mm Φ bars, each having $A_\Phi = 28.3$ mm^2.

pitch $$s = \frac{1000\, A_\Phi}{A_{sd}} = \frac{1000 \times 28.3}{180} = 157 \text{ mm}.$$

This is less than five times the effective depth of the slab. Hence provide 6 mm Φ bars @ 150 mm c/c. Near the edge of

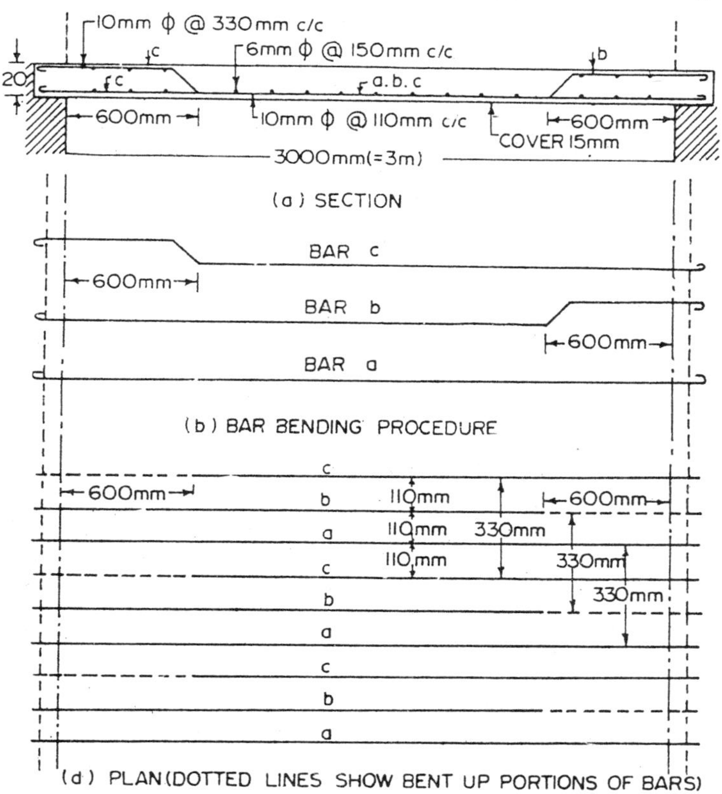

FIG. 7.12.

the support, the distribution reinforcement may be provided both at top as well as the bottom.

The details of reinforcement etc. are shown in Fig. 7.12.

Design Example 7.5 (*b*) : *Redesign the slab of Example 7.5(a), using Fe 415 steel in place of mild steel Use M 15 concrete.*

Solution

1. ***Calculation of constants***

For M 15 concrete, $c = \sigma_{cbc} = 5 \text{ N/mm}^2$ and $m \approx 19$

For Fe 415 steel, $t = \sigma_{st} = 230 \text{ N/mm}^2$

$$\therefore \quad k = \frac{mc}{mc + t} = \frac{19 \times 5}{19 \times 5 + 230} = 0.2923$$

$$j = 1 - \frac{0.2923}{3} = 0.9026$$

$$R = \frac{1}{2} \times 5 \times 0.9026 \times 0.2923 = 0.6596$$

2. ***B.M. and S.F.***

It is to be noted that R has decreased from 0.874 to 0.6596. This will result in increased value of depth of slab.

Let $D = 140$ mm, and $d \approx 120$ mm

Effective span $= 3 + 0.12 = 3.12$ m

(*i*) Self weight of slab $= 1 \times 1 \times 0.14 \times 25000 = 3500$ N

(*ii*) Load due to lime concrete finish $= 1900$ N (as before)

(*iii*) Superimposed load $= 1 \times 1 \times 2500 = 2500$ N

Total load, w, per m run of slab $= 7900$ N

$$M = \frac{wl^2}{8} = \frac{7900\,(3.12)^2}{8} = 9613 \text{ N-m/m width}$$

$$= 9.613 \times 10^6 \text{ N-mm}$$

$$V = \frac{wL}{2} = \frac{7900 \times 3}{2} = 11850 \text{ N at the face of the support}$$

3. ***Design of the section***

$$d = \sqrt{\frac{M}{bR}} = \sqrt{\frac{9.613 \times 10^6}{1000 \times 0.6596}} = 120.7 \text{ mm}$$

Using 8 mm dia. bars and a minimum cover of 15 mm, overall depth $= 120.7 + 4 + 15 = 139.7$ mm

From stiffness (or deflection) point of view, span/effective depth ratio = 20.

For a balanced design, % reinforcement $= \dfrac{k\,\sigma_{cbc}}{2\cdot\sigma_{st}} \times 100$

$= \dfrac{0.2923 \times 5}{2 \times 230} \times 100 \approx 0.32\ \%$. Hence from Fig. 7.1, modification factor (using Fe 415 bars) ≈ 1.6

$$\therefore \quad d = \frac{\text{span}}{20 \times 1.6} = \frac{3120}{32} = 97.5 \text{ mm}$$

Hence keep d = 120.7 mm, as required from B.M. point of view. Keep D = 140 mm, and hence $d = 140-4-15 = 121$ mm.

4. ***Main reinforcement***

$$A_{st} = \frac{M}{\sigma_{st}\,jd} = \frac{9.613 \times 10^6}{230 \times 0.9026 \times 121} = 382.7 \text{ mm}^2$$

This is more than the minimum reinforcement = 0.12 %

$$= \frac{0.12 \times 140 \times 1000}{100} = 168 \text{ mm}^2$$

Using 8 mm dia. bars, $A_{\Phi} = \dfrac{\pi}{4}(8)^2 = 50.26 \text{ mm}^2$

$$\therefore \text{ pitch} \qquad s = \frac{1000\,A_{\Phi}}{A_{st}} = \frac{1000 \times 50.26}{382.7} = 131.3 \text{ mm}$$

Keep pitch = 130 mm

$$\text{Actual} \quad A_{st} = \frac{1000 \times 50.26}{130} = 386.6 \text{ mm}^2$$

$$\%\ \text{reinforcement} = \frac{386.6}{140 \times 1000} \times 100 = 0.276\ \%$$

It is to be noted that the above pitch of 130 mm is less than 3 times the effective depth of the slab.

Bend every third bar up at the support, at a distance of $\dfrac{l}{5}$ $= \dfrac{3.12}{5} \approx 0.6$m from the edge of the support.

A_{st} at the support $= \dfrac{2}{3} \times 386.6 = 257.7 \text{ mm}^2$, or equal to 0.184 % which is more than minimum of 0.12 %.

5. ***Check for development length at the supports***

The code stipulates that at simple supports, the diameter of the reinforcement be such that

$$1.3\,\frac{M_1}{V} + L_0 \geq L_d$$

where $\quad M_1 = 230 \times 257.7 \times 0.9026 \times 121 = 6.473 \times 10^6$ N–mm

$$V = 11850 \text{ N}$$

Let us assume that a clear cover of 25 mm is provided at the side (end) and the slab has a support of 300 mm width.

$$L_0 = \frac{l_s}{2} - x' = \frac{300}{2} - 25 = 125 \text{ mm}$$

$$1.3\frac{M_1}{V} + L_0 = 1.3 \times \frac{6.473 \times 10^6}{11850} + 125 = 710 + 125 = 835 \text{ mm}$$

$$L_d = \frac{\Phi\, \sigma_{st}}{4\, \tau_{bd}} = \frac{8 \times 230}{4\,(0.6 \times 1.4)} \approx 548 \text{ mm}$$

(Alternatively, $L_d \approx 68.4\, \Phi = 548$ mm)

$$1.3\frac{M_1}{V} + L_0 > L_d$$

Hence anchorage requirements are satisfied.

Note : The reinforcement should extend by a length equal to $\frac{L_d}{3} = \frac{548}{3} = 183$ mm beyond the face of the support. This suggests that the width of the support should not be less than 183 + 25 = 208 mm. In the present case, the support width is 300 mm.

6. ***Check for shear***

$$\tau_v = \frac{V}{bd} = \frac{11850}{1000 \times 121} \approx 0.098 \text{ N/mm}^2$$

This is much less than the permissible value of $\tau_c = 1.3 \times 0.22 = 0.29$ N/m m^2 for M15 concrete for $p' = 0.184\,\%$ (Table 3.1 and 3.2)

7. ***Distribution reinforcement***

$$A_{sd} = \frac{0.15\, bD}{100} = \frac{0.15 \times 1000 \times 40}{100} = 210 \text{ mm}^2$$

Using 6 mm dia. mild steel bars, having $A_\Phi = 28.3$ mm^2,

$$\text{pitch } s = \frac{1000\, A_\Phi}{A_{sd}} = \frac{1000 \times 28.3}{210} = 134.8 \text{ mm.}$$

Hence provide 6 mm dia. mild steel bars @ 130 mm c/c. This is less than five times the effective depth of the slab. Near the edge of the support, the distribution reinforcement may be provided both at top as well as bottom.

The details of the reinforcement etc. will be similar to the one shown in Fig. 7.12.

7.5 DESIGN OF CANTILEVER CHAJJA

A cantilever slab bends downwards, with the result that tension is developed at the other face. Hence reinforcement is also provided at the upper face. The span of the slab is taken equal to the actual length of overhang. If the width of the cantilever is long, 1 metre width of the cantilever is taken for the design purposes. However, if the width of the cantilever is short, whole of the width may be taken as the width of the slab for the design purposes.

Design Example 6.6. *Design a cantilever slab having an overhang of 1.25 m. Take live load intensity of 2000 N/m² on the cantilever. Use M 15 concrete and mild steel bars having* $\sigma_{st} = 140\ N/mm^2$.

Solution.

(i) ***Design constants***

For $\sigma_{cbc} = 5$ N/mm², $\sigma_{st} = 140$ N/mm² and $m \approx 19$, we have $k = 0.404$; $j = 0.865$ and $R = 0.874$

(ii) ***B.M. and S.F.***

Assume the cantilever to be of average total thickness of 60 mm.

$$\text{Dead weight, per m}^2 = \frac{60}{1000} \times 1 \times 1 \times 25000 = 1500 \text{ N}$$

Live load per m² = 2000 N

∴ Total $w = 1500 + 2000 = 3500$ N

$$\therefore \quad M = \frac{wL^2}{2} = \frac{3500(1.25)^2}{2} = 2734 \text{ N-m}$$

$$= 2.734 \times 10^6 \text{ N-mm}$$

$$V_{max} = wL = 3500 \times 1.25 = 4375 \text{ N}$$

(iii) ***Design of Section***

$$d = \sqrt{\frac{M}{Rb}} = \sqrt{\frac{2.734 \times 10^6}{0.874 \times 1000}} = 56 \text{ mm}$$

Use a total depth of 80 mm at the fixed end and reduce it to 60 mm at the free end. Using 8 mm Φ bars and 15 mm clear cover, available $d = 80 - 15 - 4 = 61$ mm at the fixed support.

(iv) ***Reinforcement***

$$A_{st} = \frac{M}{\sigma_{st}\, jd} = \frac{2.734 \times 10^6}{140 \times 0.865 \times 61} = 370 \text{ mm}^2.$$

Choosing 8 mm Φ bars, $A_\Phi = \frac{\pi}{4}(8)^2 = 50.3$ mm².

$\therefore \quad \text{pitch } s = \dfrac{1000 A_\Phi}{A_{st}} = \dfrac{1000 \times 50.3}{370} = 136 \text{ mm}$

Hence provide 8 mm Φ bars @ 130 mm c/

Actual $\quad A_{st} = \dfrac{1000 \times 50.3}{136} = 370 \text{ mm}^2.$

(v) Check for development length at the support

The code stipulates that at simple supports, where the ends of the bars are confined by compressive reaction the diameter of the reinforcement be such that

$$1.3 \frac{M_1}{V} + L_0 \geq L_d$$

where $\quad M_1 = 140 \times 370 \times 0.865 \times 61 = 2.734 \times 10^6 \text{ N–mm}$

$V = 4375 \text{ N}$

$L_0 = \dfrac{L_s}{2} - x' + 13\,\Phi$

$L_s = \text{width of support} = 200 \text{ mm (say)}$

$x' = \text{side cover} = 25 \text{ mm}$

$\therefore \quad L_0 = \dfrac{200}{2} - 25 + 13 \times 8 = 179 \text{ mm}$

$$1.3 \frac{M_1}{V} + L_0 = 1.3 \times \frac{2.734 \times 10^6}{4375} + 179 = 812 + 179 = 991 \text{ mm}$$

$$L_d \approx 58.3\,\Phi = 58.3 \times 8 = 466 \text{ mm.}$$

Hence the code requirements are satisfied.

(vi) Check for shear

$$\tau_v = \frac{V}{bd} = \frac{4375}{1000 \times 61} = 0.072 \text{ N/mm}^2$$

This is much less than the permissible value of $\tau_c = 1.3 \times 0.31 = 0.4 \text{ N/mm}^2$ for M 15 concrete for $p' = \dfrac{100 A_s}{bd} = \dfrac{100 \times 370}{1000 \times 61} = 0.6\%$.

Hence safe.

(vii) Distribution reinforcement

$$A_{sd} = \frac{0.15\, bD}{100} = \frac{0.15 \times 1000\, D}{100} = 1.5\, D \text{ mm}^2$$

$$= 1.5 \times 80 = 120 \text{ mm}^2.$$

Using 6 mm Φ bars, each having $A_\Phi = 28.3 \text{ mm}^2$

$\text{pitch } s = \dfrac{1000 A_\Phi}{A_{sd}} = \dfrac{1000 \times 28.3}{120} = 236 \text{ mm}$

However, provide these @ 200 mm c/c.

The section of the cantilever slab is shown in Fig. 7.13.

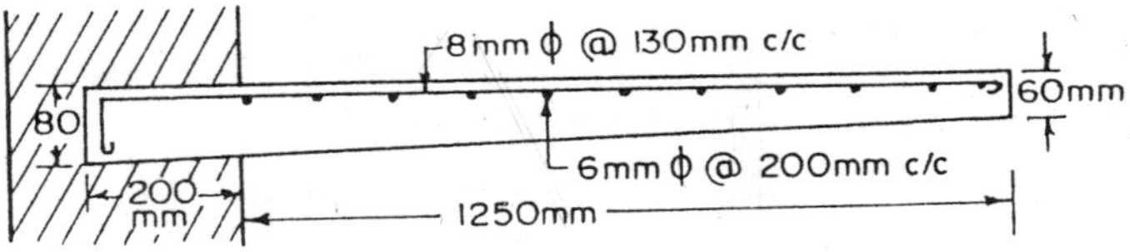

FIG. 7.13.

7.6. DESIGN OF LINTEL WITH SUNSHADE

Design Example 7.7 *A lintel, covering a verandah opening of 3 m, carries a sunshade, 1 m wide, monolithic with it. The bottom of the sunshade and the lintel are at the same level, as shown in Fig. 7.14. The verandah roof carries a live load of 2000 N/m². Design the lintel and the sunshade. Use M 15 concrete and take $\sigma_{st} = 140$ N/mm².*

Solution.

Design constants : Given σ_{cbc}=5 N/mm² ; σ_{st}=140 N/mm²; $m \approx 19$

$$k = \frac{m\,\sigma_{cbc}}{m\,\sigma_{cbc} + \sigma_{st}} = \frac{19 \times 5}{19 \times 5 + 140} = 0.404$$

$$j = 1 - k/3 = 0.165$$

$$R = \frac{1}{2}\sigma_{cbc}\,jk = \frac{1}{2} \times 5 \times 0.865 \times 0.404 = 0.874$$

1. *Design of sunshade*

Let us design the sunshade for the same live load that is carried by the verandah roof.

Let the total thickness of the sunshade be 50 mm (average). Consider one metre length of the sunshade.

Live load, per m run = 2000 N.

Dead load, per m run $= \frac{1 \times 1 \times 50}{1000} \times 2500 =$ 1250 N.

$\therefore \quad w = 2000 + 1250 = 3250$ N/m.

$$M = \frac{wL^2}{2} = \frac{3250 \times 1 \times 1}{2} = 1625 \text{ N–m} = 1625 \times 10^3 \text{ N–mm.}$$

$$\therefore \quad d = \sqrt{\frac{M}{Rb}} = \sqrt{\frac{1625 \times 10^3}{0.874 \times 1000}} = 43 \text{ mm}$$

Adopt 65 mm overall depth at the support (lintel) and reduce it to 45 mm at free end. Average total thickness = 55 mm.

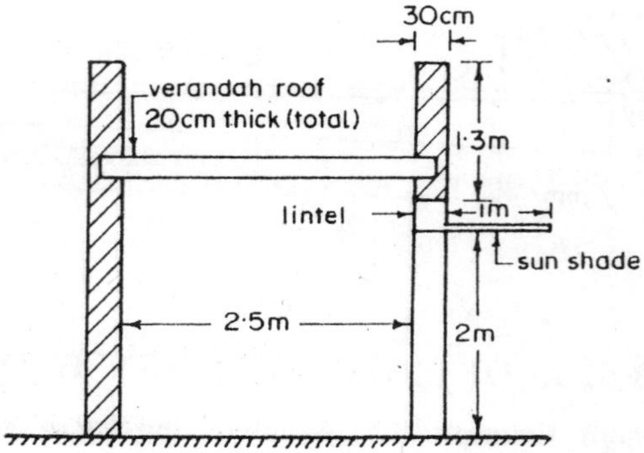

FIG. 7.14

Using 8 mm dia. bars, and 15 mm clear cover, available effective depth = 65 – 15 – 4 = 46 mm.

$$A_{st} = \frac{M}{\sigma_{st}\, jd} = \frac{1625 \times 10^3}{140 \times 0.865 \times 46} = 292 \text{ mm}^2.$$

Choosing 8 mm Φ mm bars, $A_\Phi = \frac{\pi}{4}(8)^2 = 50.3 \text{ mm}^2$.

$$\text{pitch } s = \frac{1000\, A_\Phi}{A_{st}} = \frac{1000 \times 50.3}{292} = 172 \text{ mm}.$$

Max. spacing = $3d$ = 3× 46 = 138 mm.

Hence provide 8 mm Φ bars @ 130 mm c/c. Anchor the bars inside the lintel as shown.

Transverse reinforcement

$$A_{sd} = 0.15\,\% = \frac{0.15}{100} \times bD = \frac{0.15}{100} \times 1000\, D$$

$$= 1.5\, D = 1.5 \times 55 = 82.5 \text{ mm}^2.$$

(D = average total depth = 55 mm)

Using 6 mm Φ bars having $A_\Phi = 28.27 \text{ mm}^2$

$$\text{Spacing} = \frac{1000\, A_\Phi}{A_{sd}} = \frac{1000 \times 28.27}{82.5} = 342 \text{ mm}.$$

Max. spacing = $5d = 5 \times 36 = 180$ mm.

Hence provide 6 mm Φ distribution bars @ 180 mm c/c.

Check for shear and development length at the support

$V = wL = 3250$ N

$$s_v = \frac{V}{bd} = \frac{3250}{1000 \times 46} = 0.07 \text{ N/mm}^2.$$

Hence safe.

At the support, $1.3 \frac{M_1}{V} + L_0 \geq L_d$.

Here $M_1 \approx M = 1625 \times 10^3$

$V = 3250$

$L_s = 300$ mm ; $x' =$ side cover $= 25$ mm

$$\therefore \quad L_0 = \frac{L_s}{2} - x' + 13\,\Phi = \frac{300}{2} - 25 + 13 \times 8 = 229 \text{ mm}$$

$$\therefore \quad 1.3 \frac{M_1}{V} + L_0 = 1.3 \times \frac{1625 \times 10^3}{3250} + 229 = 650 + 229 = 879 \text{ mm}$$

$L_d \approx 58.3\,\Phi = 58.3 \times 8 = 466$ mm.

Hence code requirements are satisfied.

2. *Design of Lintel beam*

Let the width of Lintel = 300 mm.

Depth of Lintel $= \frac{1}{10}$ span = 400 mm (say)

Consider one metre length of the lintel.

(*i*) Dead load of verandah roof $= \frac{1}{2}(2.5 \times 1 \times 0.2 \times 25000)$
$= 6250$ N

(*ii*) Live load from verandah roof $= \frac{1}{2}(2.5 \times 1 \times 2000)$
$= 2500$ N

(*iii*) Dead weight of masonry $= 0.3 \times 1.3 \times 1 \times 20000$
$= 7800$ N

(Assuming unit wt. of masonry = 20 kN/m^3)

(*iv*) Self weight of lintel $= 0.3 \times 0.4 \times 1 \times 25000 = 3000$ N

(*v*) Dead weight of sunshade $= 1 \times 0.055 \times 1 \times 25000 = 1375$ N

Total of (*i*) to (*v*) : Load per m run = 20925 N
≈ 21000 N/m.

Note : The live load on the sunshade is not likely to be there simultaneously with live load on varandah roof and hence has not been included. Similarly, torsional moment due to load on the sunshade will be extremely small and hence been neglected.

Assume a bearing of 400 mm on either end.

Effective span $= L = 3 + 0.4 = 3.4$ m.

$$M = \frac{wL^2}{8} = \frac{21000\,(3.4)^2}{8} = 30340 \text{ N–m} = 30.34 \times 10^6 \text{ N–mm}$$

$$\therefore \quad d = \sqrt{\frac{M}{Rb}} = \sqrt{\frac{30.34 \times 10^6}{0.874 \times 300}} = 340 \text{ mm.}$$

Using 16 mm Φ bars and keeping a clear cover of 25 mm, total depth = 340 + 8 + 25 = 373 mm.

However, keep total depth = 400 mm, so that available d = 400 − 8 − 25 = 367mm.

$$A_{st} = \frac{M}{\sigma_{st}\, jd} = \frac{30.34 \times 10^6}{140 \times 0.865 \times 367} = 683 \text{ mm}^2.$$

No. of 16 mm Φ bars $= \dfrac{683}{\frac{\pi}{4}(16)^2} = 3.4 \approx 4$

No. of 12 mm Φ bars $= \dfrac{683}{\frac{\pi}{4}(12)^2} \approx 6.$

Hence provide 6-12Φ bars. Out of these, bend 2 bars up, at a distance of $\frac{L}{7} = \frac{3.4}{7} \approx 0.5\ m$ from the edge of each support.

Remaining $A_s = 4 \times \frac{\pi}{4}(12)^2 = 452.4 \text{ mm}^2.$

Check for shear and development length

$$V = \frac{wl}{2} = \frac{21000 \times 3}{2} = 31500 \text{ N}$$

$$\tau_v = \frac{V}{bd} = \frac{31500}{300 \times 367} = 0.286 \text{ N/mm}^2,$$

$$\frac{100 A_s}{bd} = \frac{100 \times 452.4}{300 \times 340} = 0.44\ \%.$$ Hence from Table 3.1.

$\tau_c \approx 0.28 \text{ N/mm}^2$. Hence shear reinforcement is necessary.

$$V_c = \tau_c \,.\, bd = 0.28 \times 300 \times 340 = 28560 \text{ N}.$$

$$\therefore V - V_c = 31500 - 28560 = 2940 \text{ N}.$$

Thus, even if the shear resistance of bent up bars is neglected, the remaining shear ($V - V_c$) is very small. Hence provide only nominal shear stirrups, determined from the expression :

$$s_v \le \frac{2.5 A_{sv} f_y}{b} \le \frac{2.5 \times 250}{300} A_{sv} \le 2.08 A_{sv}$$

Using 8 mm Φ 2-lgd stirrups, having $A_{sv} = 2 \times \frac{\pi}{4}(8)^2 = 100.5 \text{ mm}^2$,

$$s_v \le 2.08 \times 100.5 = 209 \text{ mm} \approx 200 \text{ mm}.$$

Hence provide there @ 200 mm c/c. Provide 2–10 Φ anchor bars. Also, at the support, the code requirements are :

$$1.3\frac{M_1}{V} + L_0 \ge L_d$$

Here $M_1 = A_{st} \,.\, \sigma_{st} \,.\, jd = 452.4 \times 140 \times 0.865 \times 367$

$= 21.11 \times 10^6$ N−mm

$$V = 31500 \text{ N} \quad L_s = 400 \text{ mm}$$

$$x' = \text{side cover} = 30 \text{ mm (say)}$$

$$\therefore \quad L_0 = \frac{L_s}{2} - x' + 13\,\Phi = \frac{400}{2} - 30 + 13 \times 12 = 326 \text{ mm.}$$

$$\therefore \quad 1.3\frac{M_1}{V} + L_0 = 1.3 \times \frac{21.11 \times 10^6}{31500} + 326 = 871 + 326 = 1197 \text{ mm}$$

$$L_d = 58.3\,\Phi = 58.3 \times 12 = 700 \text{ mm.}$$

Hence code requirements are satisfied.

The details of reinforcement etc are shown in Fig. 7.15.

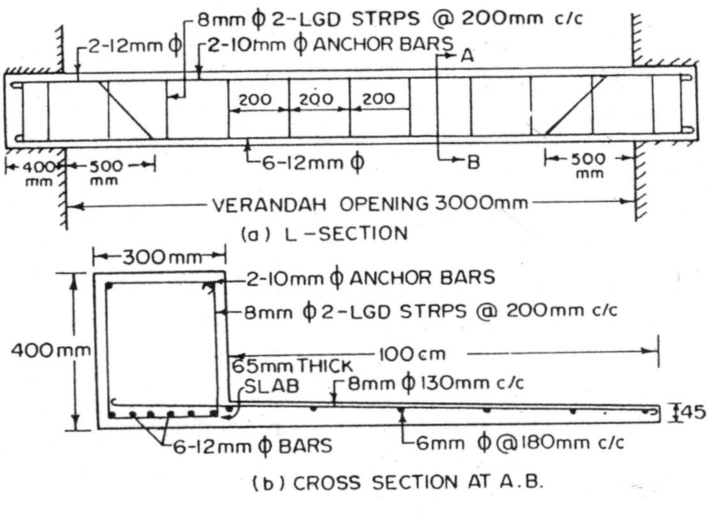

FIG. 7.15

7.7 DESIGN OF CONTINUOUS SLAB

When a slab is continuous over several spans, negative (*i.e*; hogging) bending moment is induced over the intermediate supports. and hence reinforcement has to be provided at the top of the slab portion over the intermediate supports. These intermediate supports may either be monolithic with the slab, or may be of some different material such as rolled steel joist etc. Following are some of the *basic* rules for design, as per IS : 456–1978.

1. *Effective span.* In the case of continuous slab, if the width of the support is less than 1/12 of the clear span, the effective span shall be the distance between centres of supports, or the clear distance between supports plus the effective depth of the slab, whichever is smaller. If the supports are wider than 1/12 the clear span or 600 mm, whichever is less, the effective span shall be taken as under :

(*a*) For end span with one end fixed and the other continuous, or for intermediate spans, the effective span shall be the clear span between supports ; and

(*b*) For end span with one end free and the other continuous the effective span shall be equal to the clear span plus half the effective depth of the slab or the clear span plus half the width of the continuous support, whichever is less.

Note : In the case of spans with roller or rocker bearings, the effective span shall always be the distance between the centres of bearings.

2. *Limiting stiffness.* The ratio of span to effective depth should not exceed 26.

3. *Reinforcement* Same rules as for simply supported slabs.

4. *Moment and shear coefficients.* Same rules as for continuous beams, given in clause *a*-10 of § 7.1. Thus, the B.M. values are indicated on Fig. 7.16.

5. *Arrangement of reinforcement.* The reinforcement is arranged as illustrated in Fig. 7.17.

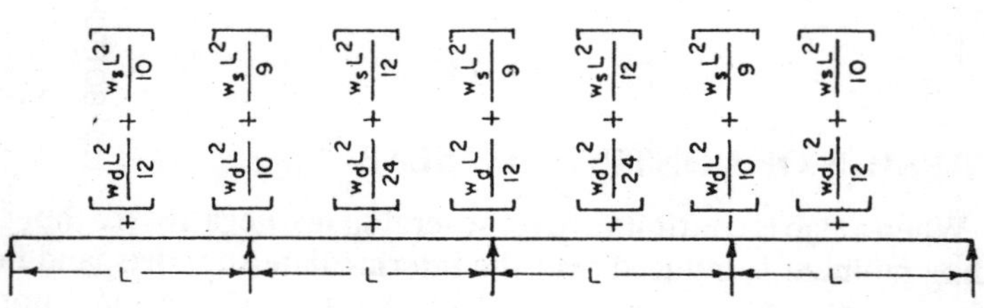

FIG. 7.16

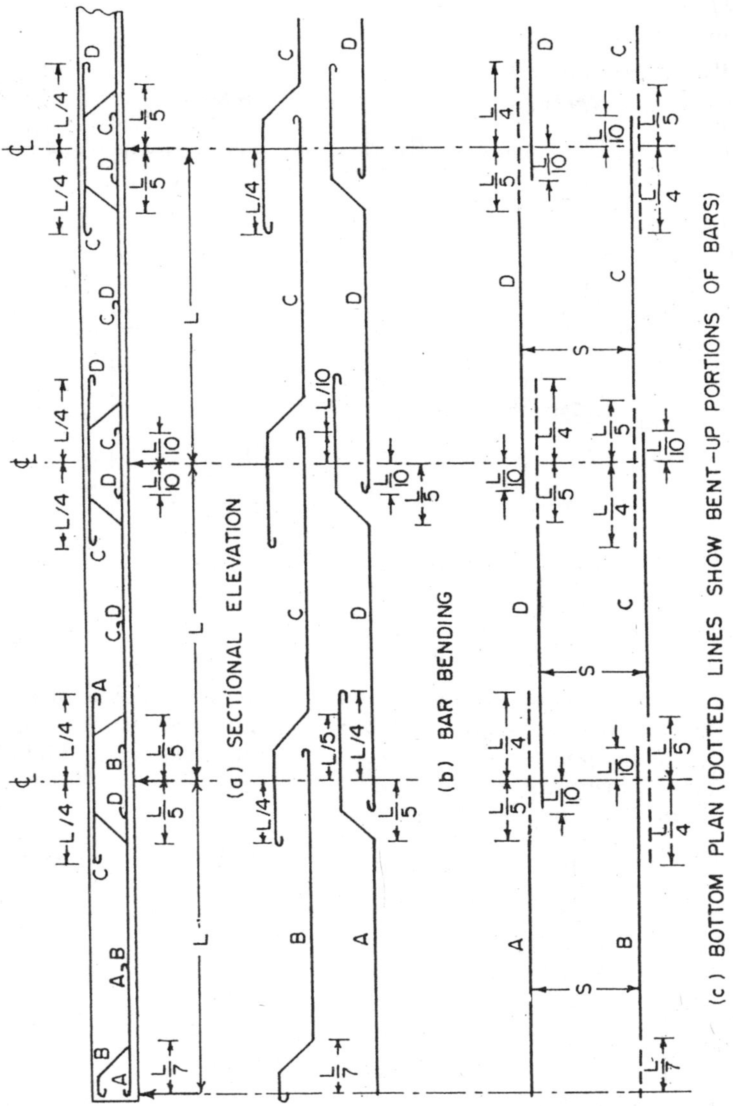

FIG. 7.17

Design Example 7.8. *Design a continuous R.C. slab for a hall 6.5 m wide and 13.5 long. The slab is supported on R.C.C. beams, each 200 mm wide. The end bearing of the slab on the walls is 120 mm. The super- imposed load is 2500 N/m^2. The floor carries a finishing weighing 600 N/m^2. Use M 15 concrete and mild steel bars.*

Solution.

1. ***Design constants***

We have $\sigma_{cbc} = 5\ \text{N/mm}^2$, $m \approx 19$; $\sigma_{st} = 140\ \text{N/mm}^2$,

$$\therefore \quad k = \frac{m\,\sigma_{cbc}}{m\,\sigma_{cbc} + \sigma_{st}} = \frac{19 \times 5}{19 \times 5 + 140} = 0.404$$

$$j = 1 - k/3 = 0.865$$

$$R = \frac{1}{2} \times 5 \times 0.865 \times 0.404 = 0.874$$

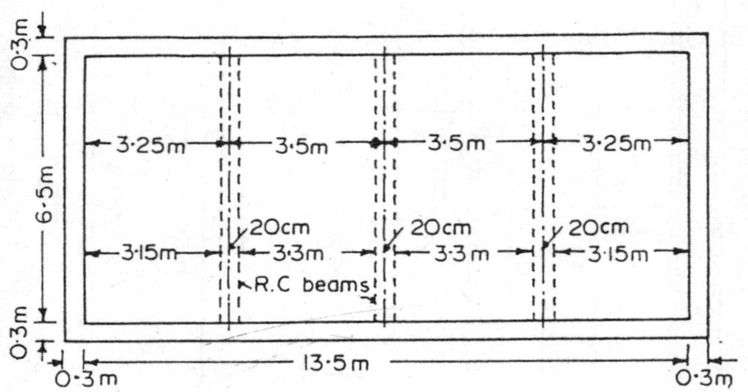

FIG. 7.18

2. ***Effective spans.*** Let us arrange the spans as shown in Fig. 7.18. Let the effective depth of slab = 100 mm. The width of the intermediate supports is less than 1/12 of the clear span. Hence the effective span will be as under :

(*i*) For end spans, effective span :

L = centre to centre of supports $= 3.25 + \dfrac{0.20}{2} = 3.35$ m

or L = clear distance $+ d = 3.15 + 0.10 = 3.25$ m

whichever is less.

Hence $L = 3.25$ m.

(*ii*) For intermediate spans :

L = centre to centre of supports = 3.5 m

or $L = 3.3 + 0.1 = 3.4$ m, whichever is less.

Hence $L = 3.4$ m.

Thus, in this case, the lengths of effective span is not equal. However, the difference is very much less than 15% of the longer spans; hence they may be treated as being equal for the purposes of using the bending moment coefficients given in the previous article.

3. *Calculation of bending moment and shear force*

Assume overall depth of slab = 120 mm. Consider 1 m width of slab.

(*i*) Self weight of slab = 0.12 × 1 × 1 × 25000 = 3000 N

(*ii*) Weight of finishing = 600 N

∴ w_d = 3600 N/m

Super-imposed load = w_s = 2500 N/m run.

For end spans

$$M \text{ (near the centre) } = \frac{w_d L^2}{12} + \frac{w_s L^2}{10}$$

$$= \frac{3600\,(3.25)^2}{12} + \frac{2500\,(3.25)^2}{10} = 5800 \text{ N-m}$$

$$= 5.8 \times 10^6 \text{ N/mm} \qquad \text{...}(i)$$

$$M \text{ (at support next to end support) } = -\left(\frac{w_d L^2}{10} + \frac{w_s L^2}{9}\right)$$

$$= -\left[\frac{3600\,(3.25)^2}{10} + \frac{2500\,(3.25)^2}{9}\right]$$

$$= -6737 \text{ N-m} = -6.737 \times 10^6 \text{ N-mm} \qquad \text{...}(ii)$$

S.F. at end support $= 0.4\, w_d \,.\, L + 0.45\, w_s \,.\, L$

$= (0.4 \times 3600 + 0.45 \times 2500)\ 3.25$

$= 8336$ N

S.F. (at the exterior face of the support next to end support)

$= 0.6\, w_d \,.\, L + 0.6\, w_s \,.\, L$

$= (0.6 \times 3600 + 0.6 \times 2500)\ 3.25 = 11895$ N

For intermediate spans

$$M \text{ (at the middle) } = \frac{w_d L^2}{24} + \frac{w_s L^2}{12}$$

$$= \frac{3600\,(3.4)^2}{24} + \frac{2500\,(3.4)^2}{12} = 4142 \text{ N-m}$$

$$= 4.142 \times 10^6 \text{ N/mm.} \qquad \text{...}(iii)$$

$$M \text{ (at interior suppori } = -\left(\frac{w_d L^2}{12} + \frac{w_s L^2}{9}\right)$$

$$= -\left[\frac{3600\,(3.4)^2}{12} + \frac{2500\,(3.4)^2}{9}\right] = -6679 \text{ N-m}$$

$$= 6.679 \times 10^6 \text{ N-mm}$$

F (interior faces) $= 0.5\, w_d \,.\, L + 0.6\, w_s \,.\, L$

$= (0.5 \times 3600 + 0.6 \times 2500)\, 3.4 = 11220$ N

4. *Design of section*

The effective depth will be calculated on the basis of maximum bending moment out of four values calculated above.

$$M_{max.} = 6.737 \times 10^6 \text{ N-mm}$$

$$\therefore \quad d = \sqrt{\frac{6.737 \times 10^6}{0.874 \times 1000}} = 88 \text{ mm}$$

Keep overall depth = 110 mm, and a clear cover equal to 15 mm. Using 12 mm Φ bars, available effective depth = 110 − 15 − 6 = 89 mm.

5. *Main reinforcement*

The maximum bending moments in the end spans (*ii*) and intermediate span (*iv*) are nearly equal. Hence same reinforcement will be provided in all spans.

$$M_{max} \text{ for design} = 6.737 \times 10^6 \text{ N-mm}$$

$$\therefore \quad A_{st} = \frac{M}{\sigma_{st}\, jd} = \frac{6.737 \times 10^6}{140 \times 0.865 \times 89} = 625 \text{ mm}^2.$$

Use 12 mm Φ bars, having $A_\Phi = \frac{\pi}{4}(12)^2 = 113.1 \text{ mm}^2$

$$\therefore \text{ pitch} \quad s = \frac{1000\, A_\Phi}{A_{sd}} = \frac{1000 \times 113.1}{625} = 181 \text{ mm}$$

Hence keep pitch = 180 mm. Alternate bars from the bottom can be bent up near supports to bear the negative bending moment.

Actual A_{st} provided per metre width $= \frac{100 \times 113.1}{180} = 628 \text{ mm}^2$

6. *Distribution reinforcement*

$A_{st} = 1.5D = 1.5 \times 110 = 165 \text{ mm}^2$

Using 6 mm Φ bars having $A_\Phi = \frac{\pi}{4}(6)^2 = 28.3 \text{ mm}^2$,

$$\text{pitch} \quad s = \frac{1000\, A_\Phi}{A_{sd}} = \frac{1000 \times 28.3}{165} = 171.5 \text{ mm}$$

Hence provide pitch at 170 mm c/c , which is less than five times the effective depth of the slab.

7. *Check for development length at support*

The code stipulates that the diameter of the bars should be such that

$1.3 \frac{M_1}{V} + L_0 \geq L_d$ at the end support.

$M_1 = A_{st}\, \sigma_{st}\, jd = (\frac{1}{2} \times 628)(140 \times 0.865 \times 89) = 3.49 \times 10^6$ N–mm

$$V = 11080 \text{ N}$$

$$L_s = 120 \text{ mm} \; ; \; x' = 25 \text{ mm (say)}$$

$$L_0 = \frac{L_s}{2} - x' + 13\Phi = \frac{120}{2} - 25 + 13 \times 12 = 191 \text{ mm.}$$

$$1.3\frac{M_1}{V} + L_0 = 1.3 \times \frac{3.49 \times 10^6}{8336} + 191 = 544 + 191 = 735 \text{ mm.}$$

$$L_d \approx 58.3\,\Phi = 58.3 \times 12 = 700 \text{ mm.}$$

Thus code requirements are satisfied.

8. ***Check for shear***

Maximum shear = 11895

$$\tau_v = \frac{V}{bd} = \frac{11895}{1000 \times 89} = 0.134 \text{ N/mm}^2$$

$$\frac{100A_s}{bd} = \frac{1000 \times 324}{1000 \times 89} = 0.364\%$$

Hence from table 3.1 and 3.2, $\tau_c = 0.25 \times 1.3 = 0.325$ N/mm^2 for M 15 concrete. Thus the slab is safe in shear.

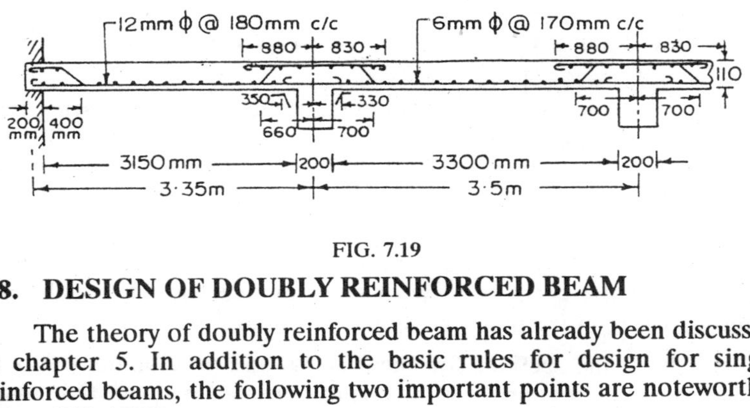

FIG. 7.19

7.8. DESIGN OF DOUBLY REINFORCED BEAM

The theory of doubly reinforced beam has already been discussed in chapter 5. In addition to the basic rules for design for singly reinforced beams, the following two important points are noteworthy. (IS : 456–1978).

1. ***Percentage of compression reinforcement***

The maximum area of compression reinforcement shall not exceed 0.04 *bD*. Compression reinforcement in beams shall be enclosed by stirrups for effective lateral restraint.

2. ***Anchoring bars in compression***

The anchoring length of straight bar in compression shall be equal to the development length of bar in compression. The projected length of hooks, bends and straight length beyond bends if provided for a bar in compression shall be considered for development length.

Design procedure

1. Determine the position of the critical neutral axis of the section of the given set of stresses :

$$n = \frac{m\,\sigma_{cbc}}{m\,\sigma_{cbc} + \sigma_{st}}\,d$$

where d is the known effective depth of the beam, σ_{cbc}=permissible stress in concrete, and σ_{st}=permissible tensile stress in steel.

2. Determine the area of tensile reinforcement A_{st1} and the moment of resistance M_1 corresponding to the singly reinforced balanced section :

$$M_1 = Rbd^2 = \frac{1}{2}\sigma_{cbc}\,jk\,.\,bd^2 = bn\,\frac{\sigma_{cbc}}{2}\left(d - \frac{n}{3}\right)$$

and $$A_{st} = \frac{M_1}{\sigma_{st}\,.\,jd}.$$

3. Determine the additional area of tensile reinforcement A_{st2} for the remaining bending moment M_2 :

$$A_{st2} = \frac{M_2}{\sigma_{st}\,(d - d_c)} = \frac{M - M_1}{\sigma_{st}\,(d - d_c)}$$

Thus, the total area of tensile reinforcement will be equal to

$$A_{st} = A_{st1} + A_{st2}$$

4. Determine the area of compressive steel A_{sc} by equating moments of equivalent areas of compressive steel (A_{sc}) and additional tensile area (A_{st2}) Thus

$$A_{sc} = \frac{m(d - n)}{(m_c - 1)(n - d_c)}\,.\,A_{st2} = \frac{m(1 - k)}{(m_c - 1)(k - e)}\,.\,A_{st2}$$

where $m_c = 1.5\,m$ and $e - d_c/d$

5. From the point of view of economy, curtail the tensile and compressive reinforcement at suitable sections.

6. Check the sections for shear stress and development length. Design shear reinforcement if the beam is unsafe for shear. Otherwise provide nominal shear reinforcement.

Design Example 7.9. *Design a doubly reinforced beam to carry a supper-imposed load of 60 kN per meter run. The overall depth and width of the beam are restricted to 840 mm and 300 mm respectively. The beam has a clear span of 5m and a bearing of 50 cm on each end. Take* $c = \sigma_{cbc} = 5\ N/mm^2$; $t = \sigma_{st} = \sigma_{sv} = 140\ N/mm^2$ *and* $m=18$.

Solution

1. ***Design constants***

$$k = \frac{m.c}{mc + t} = \frac{18 \times 5}{18 \times 5 + 140} = 0.391$$

$$j = 1 - k/3 = 0.87$$

$$R = \frac{1}{2} cjk = \frac{1}{2} \times 5 \times 0.87 \times 0.391 = 0.85$$

2. ***Bending moment and shear force***

Effective span $= 5 + 0.5 = 5.5$ m

Wt. of beam per m run $= \frac{30 \times 84}{100 \times 100} \times 25000 = 6300$ N

Supper-imposed load per m run $= 60000$ N

Total U.D.L. per m run $= 66300$ N

$$\therefore \quad M = \frac{wL^2}{8} = \frac{66300 \times (5.5)^2}{8} = 250700 \text{ N-m.}$$

$$= 250.7 \times 10^6 \text{ N-m}$$

Maximum S.F. at the support is given by

$$F = \frac{wl}{2} = \frac{66300 \times 5}{8} = 165750 \text{ N.}$$

3. ***Moment of resistance M_1 and reinforcement A_{st1}***

Let us assume that the centre of the tensile reinforcement will be at 65 mm above the bottom fibre.

$$\therefore \quad d = 840 - 65 = 775 \text{ mm}$$

$$n = kd = 0.391 \times 775 = 303 \text{ mm.}$$

For a singly reinforced balanced section,

$$M_1 = Rbd^2 = 0.85 \times 300\,(775)^2 = 153 \times 10^6 \text{ N-mm.}$$

Area of tensile reinforcement is given by

$$A_{st1} = \frac{M_1}{\sigma_{st}\, jd} = \frac{153 \times 10^6}{140 \times 0.87 \times 775} = 1622 \text{ mm}^2.$$

4. ***Moment of resistance M_2 and reinforcement A_{st2}***

$$M_2 = M - M_1 = (250.7 - 153)10^6 = 97.7 \times 10^6 \text{ N-mm.}$$

This remaining B.M. has to be resisted by a couple provided by the tensile and compressive reinforcements. Let the centre of compressive reinforcement be placed at 40 mm below the top fibre *i.e.* $d_c = 40$ mm.

$$\therefore \quad A_{st2} = \frac{M_2}{\sigma_{st}\,(d - d_c)} = \frac{97.7 \times 10^6}{140(775 - 40)} = 949 \text{ mm}^2$$

$\therefore$ Total area of tensile reinforcement is

$$A_{st} = A_{st1} + A_{st2} = 1622 + 949 = 2571 \text{ mm}^2.$$

5. *Compressive reinforcement A_{sc}*

$$A_{sc} = \frac{m(d-n)}{(m_c-1)(n-d_c)} A_{st2}, \text{ where } n = 0.391 \times 775 = 303$$

$$= \frac{18(775-303)}{(1.5 \times 18 - 1)(303-40)} \times 949 = 1179 \text{ mm}^2$$

6. *Reinforcing bars*

$$A_{st} = 2571 \text{ mm}^2.$$

Using 20 mm Φ bars, having $A_\Phi = 314 \text{ mm}^2$

$$\therefore \quad \text{No. of bars} = \frac{2571}{314} = 8.2$$

Hence provide 9-20 mm Φ bars. Place 5 bars in the bottom tier and 4 bars in the top tier, keeping a clear distance of 25 mm between the two tiers. Keep a clear cover of 25 mm. Use 25 mm Φ spacers bars @ 1 m c/c. $A_{sc} = 1179 \text{ mm}^2$.

$$\text{No. of 20 mm Φ bars} = \frac{1179}{314} = 3.8$$

Hence provide 4-20 Φ bars at the top, in one tier, keeping a clear cover of 25 mm.

7. *Curtailment of reinforcement*

The bending at any point distant x metres from the centre of the span is given by

$$M_x = \left(\frac{wL^2}{8} - \frac{wx^2}{2}\right) 1000 = \left(M - \frac{wx^2}{2}\right) 1000$$

(where the moments M_x and M are in N-mm units).

At the point where compressive reinforcement is not required, the bending moment should be equal to M_1

$$\therefore \quad M_1 = M - \frac{wx^2}{2} \times 1000$$

$$\therefore \quad x = \sqrt{\frac{2(M_1 - M)}{1000\, w}} = \sqrt{\frac{2M_2}{1000\, w}} = \sqrt{\frac{2 \times 97.7 \times 10^6}{1000 \times 66300}}$$

$$\approx 1.72 \text{ m.}$$

Hence at $x = 1.70$ m from the centre, compressive reinforcement is no longer required and it may, therefore, be curtailed. However, curtail only 2 bars and continue 2 bars upto the support, to hold the shear stirrups.

At this section, the bending moment is M_1 only. Hence tensile reinforcement required $= A_{st1} = 1622 \text{ mm}^2$, which will need only 6 bars. Hence curtail 3 bars of the upper tier at this point and continue rest of the bars to the support.

8. *Shear reinforcement*

Near the support, where the S.F. is maximum, the section is singly reinforced (since the three compressive reinforced bars serve as holding bars of the shear stirrups).

Available effective depth $d = 840 - 25 - 10 = 805$ mm

$$\tau_v = \frac{V}{bd} = \frac{165750}{300 \times 805} = 0.67 \text{ N/mm}^2.$$

Available $A_s = 6 \times 314 = 1884 \text{ mm}^2$

$$\therefore \quad \frac{100 A_s}{bd} = \frac{100 \times 1884}{300 \times 805} = 0.78\%$$

Hence from table 3.1, $\tau_c \approx 0.34$ N/mm^2.

Hence shear reinforcement is necessary.

$V_c = \tau_c \, . \, bd = 0.34 \times 300 \times 805 = 82110$ N

$\therefore \quad V_s = V - V_c = 165750 - 82110 = 83640$ N.

The shear stirrups are governed by Eq. 3.20

$$s_v = \frac{\sigma_{sv} \, . \, d \, . \, A_{sv}}{V_s} = \frac{140 \times 805}{82110} A_{sv} = 1.373 \, A_{sv}$$

Using 10 mm Φ 2-lgd stirrups, $A_{sv} = 2 \frac{\pi}{4} (10)^2 = 157 \text{ mm}^2$

$\therefore \quad s_v = 1.373 \times 157 = 215$ mm.

Maximum spacing is given by Eq. 3.19(*c*)

$$s_v = \frac{2.5 f_y}{b} \, . \, A_{sv} = \frac{2.5 \times 250}{300} A_{sv} = 2.08 \, A_{sv}$$

$$= 2.08 \times 157 = 327 \text{ mm}.$$

Hence provide 10 mm Φ 2-lgd stirrups @ 200 mm c/c near the support. This spacing can be increased to 300 mm toward the mid span of the beam.

9. *Check for development length at supports*

At the supports $1.3 \frac{M_1}{V} + L_0 \geq L_d$.

Here $\quad M_1 = \sigma_{st} \, . \, A_{st} \, . \, jd = (140 \times 6 \times 314 \times 0.87 \times 775)$

$$= 177.8 \times 10^6 \text{ N–mm}$$

$V = 165750$ N

$L_s = 500$ mm ; $x' = 30$ mm (say).

$$L_0 = \frac{L_s}{2} - x' + 13\Phi = \frac{500}{2} - 30 + 13 \times 20 = 480 \text{ mm}$$

$$\therefore\ 1.3\frac{M_1}{V} + L_0 = 1.3 \times \frac{177.8 \times 10^6}{165750} + 480 = 1395 + 480 = 1875 \text{ mm}$$

$$L_d \approx 58.3\,\Phi = 58.3 \times 20 = 1166 \text{ mm.}$$

Hence code requirements are satisfied. The details of reinforcement etc. are shown in Fig. 7.20.

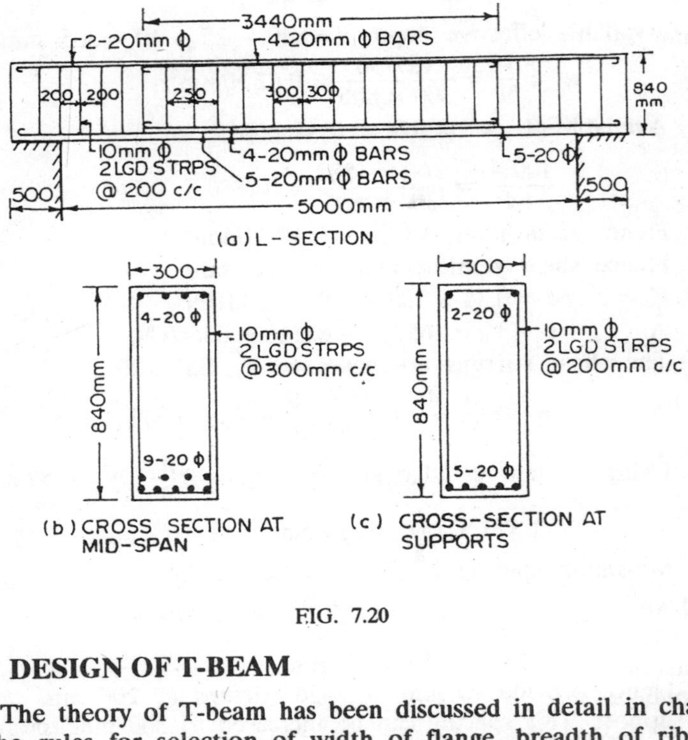

FIG. 7.20

7.9. DESIGN OF T-BEAM

The theory of T-beam has been discussed in detail in chapter 6. The rules for selection of width of flange, breadth of rib and depth of rib have been in given in that chapter. The general notes given in § 7.1 for rectangular beam also apply for this beam. The design of T-beam is done in the following steps :

1. Assume some suitable value of width of web based on criteria discussed § 6.2.

2. Determine the effective width b_f of the flange as discussed in § 6.2.

3. Assume effective span of the beam and calculate the bending moment and S.F.

4. Determine the effective depth of the beam on any one of the three considerations: (*i*) balanced section (*ii*) economy (*iii*) shear stress.

For balanced section, the effective depth is given by Eq. 6.16:

$$M = b_f . \sigma_{cbc} . D_f \left[d - \frac{D_f(1+k)}{2k} + \frac{D_f^2}{3kd}\right]$$

Alternatively, taking lever arm to be equal to $0.9d$, we have from Eq. 7.17.

$$M = 0.45\, b_f . \sigma_{cbc} . D_f \left[\frac{2kd - D_f}{k}\right]$$

For economical depth, use Eq. 6.18,

$$d = \sqrt{\frac{rM}{\sigma_{st}\, j\, b_w}}$$

For shear consideration, use Eq. 6.18,

$$d_{min} = \frac{V}{b_w . \tau_v} = \frac{V}{b_w . \tau_{c.max}}$$

where τ_v may be taken equal to $\tau_{c.cmax}$ given in Table 3.3

The *maximum* value of d obtained from the above three criteria is taken as the effective depth of the beam.

5. Take the approximate lever arm $a = d - \frac{D_f}{2}$ and determine the area of tensile reinforcement :

$$A_{st} = \frac{M}{\sigma_{st} . a}$$

Arrange suitably the steel reinforcement.

6. Determine the actual N.A. and check the stresses in concrete and steel.

7. Check the section for shear and provide shear reinforcement, if necessary. Design of shear reinforcement is done in the same manner as for rectangular beams.

8. Check for development length.

Design Example 7.10. *The floor of a hall measures 16 m× 6 m to the faces of the supporting walls. The floor consists of three beams spaced at 4 m centre to centre, and the slab thickness is 120 mm. The floor carries a uniformly distributed load of 5 kN/m², inclusive of the floor finishes. Design the intermediate beam. The following value of allowable stresses may be taken :* $c = \sigma_{cbc} = 5$ *N/mm²;* $t = \sigma_{st} = \sigma_{sv} = 140$ *N/mm²;* $m = 19$. *The ratio of cost of steel to cost of concrete may be taken equal to 60 . The support width may be assumed equal to 500 mm.*

Solution.

1. ***Calculation of constants***

$$k = \frac{m\,\sigma_{cbc}}{m\,\sigma_{cbc} + \sigma_{st}} = \frac{19 \times 5}{19 \times 5 + 140} = 0.404$$

$$j = 1 - k/3 = 0.865$$

$$R = \frac{1}{2} \times 5 \times 0.865 \times 0.404 = 0.874$$

2. ***Determination of b_w and b_f***

The width of web should be sufficient to accomodate the tensile reinforcement. Assume that the reinforcement consists of 6 bars of 20 mm diameter. Keeping the clear distance between bars equal to 20 mm, and clear cover at the end to be 20 mm, the total width $b_w = (6 \times 20) + (5 \times 20) + (2 \times 20) = 260$ mm.

Keep $b_w = 300$ mm

The width of flange b_f is taken the least of the following

(*a*) $$b_f = \frac{l_0}{6} + b_w + 6\,D_f$$

where $l_0 = 6 + 0.5 = 6.5 = 6500$ mm

$$\therefore \quad b_f = \frac{6500}{6} + 300 + 6 \times 120 \approx 2100 \text{ mm}$$

(*b*) $b_f = b_w + \frac{1}{2}$ (sum of clear distances to adjacent beams on either side)

$$= 300 + \frac{1}{2}[2\{4000 - 300\}] = 4000 \text{ mm.}$$

Hence adopt b_f = 2100 mm.

3. ***Bending moment and S.F.***

Effective span $= 6 + 0.5 = 6.5$ m.

Dead load of slab per m^2 $= 0.12 \times 1 \times 1 \times 25000 = 3000$ N

super-imposed load per m^2 $= 5000$ N

Total load $= 8000$ N/m^2

$\therefore$ Load per metre run of beam $= 8000 \times 4 = 32000$ N/m

Weight of web per metre run $= 3000$ N (say)

Total load w $= 35000$ N/m

$$\therefore \quad M = \frac{wL^2}{8} = \frac{35000(6.5)^2}{8} = 184840 \text{ N-m} = 184.84 \times 10^6 \text{ N-mm}$$

$$\text{S.F. } V = \frac{wl}{2} = \frac{35000 \times 6}{2} = 105000 \text{ N}$$

4. *Effective depth of beam*

For balanced section, we have from Eq. 6.17

$$M = 0.45\, b_f \,.\, \sigma_{cbc} \,.\, D_f \left(\frac{2kd - D_f}{k}\right)$$

$$\therefore\; 184.84 \times 10^6 = 0.45 \times 2100 \times 5 \times 120 \left[\frac{2 \times 0.404\, d - 120}{0.404}\right]$$

From which $d = 312$ mm

$$d_{min} = \frac{V}{b_w \,.\, \tau_{c\,.\,max}}$$

For M 15 concrete, $\tau_{c\,.\,\max} = 1.6$ N/mm^2 (Table 3.3)

$$\therefore\quad d_{min} = \frac{105000}{300 \times 1.6} = 219 \text{ mm}$$

From economic consideration, taking $j \approx 0.9$

$$d = \sqrt{\frac{r\,.\,M}{\sigma_{st}\,.\,j\,.\,b_w}} = \sqrt{\frac{60 \times 184.84 \times 10^6}{140 \times 0.9 \times 300}} \approx 540 \text{ mm}$$

Hence adopt $d = 540$ mm.

5. *Area of steel*

Approx. lever arm $a = d - \frac{D_f}{2} = 540 - \frac{120}{2} = 480$ mm

(Alternatively, $a = 0.9d = 0.9 \times 540 = 486$ mm)

Adopt a=480 mm for preliminary calculations.

$$A_{st} = \frac{M}{\sigma_{st}\,.\,a} = \frac{184.84 \times 10^6}{140 \times 480} = 2751 \text{ mm}^2.$$

Choosing 25 mm Φ bars, No. of bars $= \frac{2751}{490.9} = 5.6$

Hence choose 6 Nos. of 25 mm Φ bars.

Actual A_{st} provided $= 6 \times 490.9 = 2945$ mm^2

Keep total depth of beam = 580 mm, and a clear cover of 25 mm. Hence available $d = 580 - 25 - 25/2 \approx 542$ mm.

6. *Determination of actual N.A. and stresses in steel and concrete*

Assuming that the N.A. falls below the flange, we have

$$b_f \,.\, D_f \left(n - \frac{D_f}{2}\right) = m\, A_{st}\, (d - n)$$

or $$2100 \times 120 \left(n - \frac{120}{2}\right) = 19 \times 2945\, (542 - n)$$

From which $n = 148$ mm

$$\bar{y} = \frac{3n - 2D_f}{2n - D_f} \times \frac{D_f}{3}$$

$$= \frac{3 \times 148 - 2 \times 120}{2 \times 148 - 120} \times \frac{120}{3} = 46.4 \text{ mm}$$

$\therefore$ Lever arm $a = d - \bar{y} = 542 - 46.4 \approx 496$ mm

(against an assumed value of 480 mm)

$$\therefore \text{ Stress in steel } = \frac{M}{A_{st} \cdot a} = \frac{184.84 \times 10^6}{2945 \times 496} = 126.5 \text{ N/mm}^2$$

Corresponding stress in concrete is given by

$$c = \frac{t}{m} \cdot \frac{n}{d - n} = \frac{126.5}{19} \times \frac{148}{542 - 148} = 2.5 \text{ N/mm}^2$$

Alternatively, $M = b_f \cdot D_f \dfrac{c + c_1}{2} \cdot a$

where $c_1 = \dfrac{n - D_f}{n} c = \dfrac{148 - 120}{148} c = 0.189\, c$

$$\therefore \quad M = b_f \cdot D_f \left(\frac{1 + 0.189}{2}\right) c \times a$$

$$\therefore \quad c_1 = \frac{184.84 \times 10^6}{2100 \times 120 \times 0.5945 \times 496} \approx 2.5 \text{ N/mm}^2$$

7. *Shear reinforcement*

$$\tau_v = \frac{V}{b_w \cdot d} = \frac{105000}{300 \times 542} = 0.646 \text{ N/mm}^2$$

At supports, assuming to continue only 3 bars up to supports

$$\frac{100 A_s}{b_w d} = \frac{100 \times 3 \times 490.9}{300 \times 542} \approx 0.9\%$$

Hence from Table 3.1, for M 15 concrete,

$$\tau_c \approx 0.36 \text{ N/mm}^2.$$

Hence shear reinforcement is necessary.

$$V_c = \tau_c \cdot bd = 0.36 \times 300 \times 542 = 58540 \text{ N}$$

$$\therefore \quad V_s = V - V_c = 10500 - 58540 = 46460 \text{ N.}$$

Let us bend 3 bars up at 45°, at a distance $1.414\, a = 1.414 \times 496 \approx 700$ mm from the edge of support. Shear resistance of the 3-bent up bars is

$$V_{s1} = A_{sv} \cdot \sigma_{sv} \sin 45^o = 0.707 A_{sv} \cdot \sigma_{sv}$$

$$= 0.707\,[3 \times 490.9] \times 140 = 145768 \text{ N.}$$

However, max. resistance assigned to bent up bars

$$= \tfrac{1}{2} V_s = \tfrac{1}{2} \times 46460 = 23230 \text{ N.}$$

Hence provide vertical stirrups for the remaining 23230 N shear ($=V_{s2}$).

Using 8 mm Φ-2-lgd. stirrups, $A_{sv} = 2 \dfrac{\pi}{4} (8)^2 = 100.5 \text{ mm}^2$

$$\therefore \quad s_v = \frac{\sigma_{sv} . A_{sv} d}{V_{s2}} = \frac{140 \times 100.5 \times 542}{23230}$$
$$= 328 \text{ mm.}$$

The maximum spacing corresponding to nominal stirrups is given by

$$s_v \leq \frac{2.5 A_{sv} . f_y}{b} = \frac{2.5 \times 100.5 \times 250}{300} = 209 \text{ mm}$$

Also, the spacing should not be more than 0.75 d = 0.75 × 542 = 406 mm or more than 450 mm.

Hence provide 8 mm Φ 2 lgd-stirrups @ 200 mm c/c throughout.

8. *Check for development length at supports*

The Code envisages that the diameter of bars should be so selected that following relation is satisfied at the supports :

$$1.3 \frac{M_1}{V} + L_0 \geq L_d$$

Here,
$$M_1 = 3 \times 490.9 \times 140 \times 496 = 1.023 \times 10^8 \text{ N-mm}$$
$$V = 105000 \text{ N}$$
$$L_s = 500 \text{ mm}$$
$$x' = 30 \text{ mm (say)}$$
$$L_0 = \frac{L_s}{2} - x' + 13\,\Phi = \frac{500}{2} - 30 + 13 \times 25 = 545 \text{ mm}$$

$$\therefore 1.3 \frac{M_1}{V} + L_0 = 1.3 \times \frac{1.023 \times 10^8}{105000} + 545 = 1267 + 545 = 1812 \text{ mm}$$

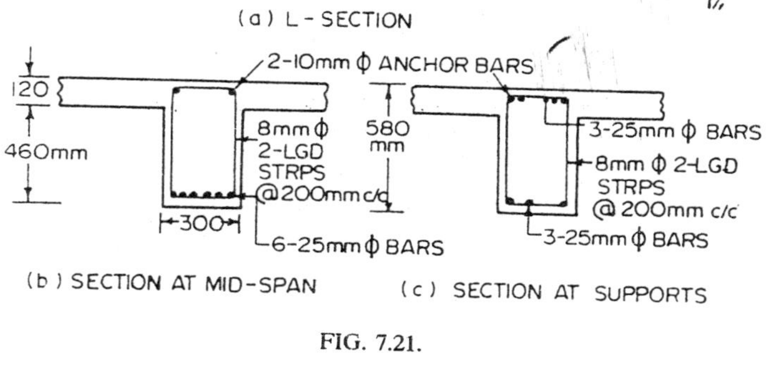

FIG. 7.21.

$L_d \approx 58.3\,\Phi = 58.3 \times 25 = 1458$ mm

Hence the code requirements are satisfied. The details of reinforcement etc. are shown in Fig. 7.21.

7.10. DESIGN OF CYCLE STAND SHADE

Design Example 7.11. *A cycle stand shade consists of a R.C. slab which cantilevers 3 metres on each side of central R.C. beam and is monolithic with it. The R.C. beam is simply supported on columns 400 mm wide, at the ends, over a clear span of 6 metres. Design the shade for a superimposed load of 2000 N/m². Use M 15 concrete and take* $\sigma_{st} = 140$ *N/mm². Take unit weight of concrete as 24 kN/m³.*

Solution.

For M 15 concrete we have $c = \sigma_{cbc} = 5$ N/mm² and $m = 19$.

Also, $t = \sigma_{st} = 140$ N/mm².

$\therefore\ k = 0.404$; $j = 0.865$ and $R = 0.874$

(*A*) Design of cantilevering slab

For stiffness the ratio of span to effective depth of cantilever =7. For a balanced design, the percentage reinforcement for M 15 concrete =0.72%. Hence from Fig. 7.1 modification factor, using mild steel bars≈1.6.

$$\therefore \qquad \frac{\text{span}}{d} = 7 \times 1.6 = 11.2$$

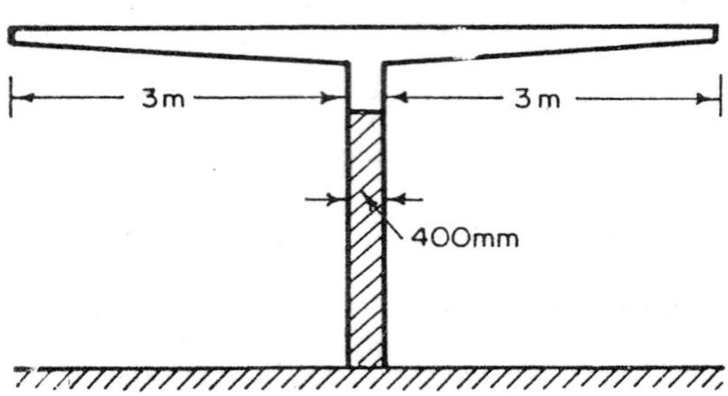

FIG. 7.22.

$$\therefore \quad \text{Effective depth} = \frac{\text{span}}{11.2} = \frac{3000}{11.2} = 268 \text{ mm.}$$

Keep the total depth of 300 mm and reduce is to 100 mm at the ends. Mean thickness of slab $= (300 + 100)\frac{1}{2} = 200$ mm.

Dead load of slab per sq. metre

$= 0.2 \times 1 \times 1 \times 24000 = 4800$ N

Superimposed load = 2000 N/m^2

$\therefore$ Total load/m^2 = 2000 + 4800 = 6800 N/m^2

$$\therefore \quad M = \frac{wL^2}{2} = \frac{6800 \times 3 \times 3}{2} = 30600 \text{ N-m} = 30.6 \times 10^6 \text{ N-mm/m}$$

$$V = wL = 6800 \times 3 = 20400 \text{ N/m width.}$$

Considering b =1 m=100 mm,

$$d = \sqrt{\frac{M}{Rb}} = \sqrt{\frac{30.6 \times 10^6}{0.874 \times 1000}} = 187 \text{ mm}$$

However, from stiffness (deflection) point of view, keep d=268 mm. Using 12 mm Φ bars and keeping a minimum clear cover of 15 mm, D = 268 + 15 + 6 = 289 mm. Hence keep D=290 mm, so that available d = 290 − 15 − 6 = 269 mm. At the end, reduce the total depth to 100 mm so that available d = 100 − 15 − 6 = 79 mm.

$$A_{st} = \frac{M}{tjd} = \frac{30.6 \times 10^6}{140 \times 0.865 \times 269} = 939 \text{ mm}^2$$

For 12 mm Φ, $A_\Phi = \frac{\pi}{4}(12)^3 = 113$ mm^2

$$\therefore \text{ pitch } s = \frac{1000 A_\Phi}{A_{st}} = \frac{1000 \times 113}{939} = 102.3 \text{ mm.}$$

Hence provide 12 mm Φ bars @ 120 mm c/c.

Actual area of steel provided

$$= \frac{1000 A_\Phi}{s} = \frac{1000 \times 113}{120} = 942 \text{ mm}^2.$$

Let us curtail alternate bars at half the cantilever length.

$$\text{B.M. at the point} = \frac{wx^2}{2} = \frac{6800\,(1.5)^2}{2} \times 1000 = 7.65 \times 10^6 \text{ N-mm}$$

$$\text{Effective depth of slab} = \frac{269 + 79}{2} = 174 \text{ mm}$$

$$\therefore A_{st} \text{ required } \frac{7.65 \times 10^6}{140 \times 0.865 \times 174} = 363 \text{ mm}^2$$

Actual A_{st} available $= \frac{1}{2} \times 942 = 471$ mm^2. Hence O.K.

Check for shear

$$\tau_v = \frac{V}{bd} = \frac{20400}{1000 \times 269} \approx 0.08 \text{ N/mm}^2$$

Distribution reinforcement

$$A_{st} = \frac{0.15bD}{100} = \frac{0.15 \times 1000 \times 290}{100} = 435 \text{ mm}^2$$

Using 8 mm Φ bars, A_Φ=50.3 mm^2

$$\therefore \quad \text{pitch } s = \frac{1000\,A_\Phi}{A_{sd}} = \frac{1000 \times 50.3}{100} = 116 \text{ mm}$$

Hence provide 8 mm Φ bars @ 110 mm c/c.

At half the cantilever length, average depth = $(290 + 110)\frac{1}{2}$ =195 mm.

$$\therefore \quad A_{sd} = \frac{0.15bD}{100} = \frac{0.15 \times 1000}{100}D = 1.5\,D = 1.5 \times 195 = 293 \text{ mm}^2$$

$$\text{pitch } s = \frac{1000 \times 50.3}{293} \approx 170 \text{ mm c/c}$$

(*B*) Design of Beam

The beam will be designed as T-beam with tapered flanges. However we can approximately assume the tapered flanges to be equivalent to flanges of uniform average thickness.

Width of web = (b_w) = width of column = 400 mm.

$$\text{Average depth of flange, } D_f = \frac{290 + 100}{2} = 195 \text{ mm} = 0.195 \text{ m}$$

Total width of flange = b = 6.4 m = 6400 mm.

Let the effective span beam= $l_0 \approx 6 + 0.4 = 6.4$ m

For an isolated T-beam, the width of the flange is

$$b_f = \frac{l_0}{\left(\frac{l_0}{b}\right) + 4} + b_w = \frac{6400}{\left(\frac{6400}{6400}\right) + 4} + 400$$

=1680 mm.

Assume the overall depth of beam = 400 mm, for the purpose of calculating the dead load.

(*i*) Dead load of slab	=	(0.195 × 6.0 ×)24000 = 28080 N
(*ii*) Dead load of beam	=	0.4 × 0.4 × 1 × 24000 = 3840 N
(*iii*) Superimposed load	=	2000(3 + 3 + 0.4) × 1 = 12800 N
	Total w =	44720 N

$$M = \frac{wl_0^2}{8} = \frac{44720(6.4)^2}{8} \times 1000 = 229 \times 10^6 \text{ N-mm}$$

Effective depth of the beam

For a balanced section, we have from Eq. 5.17

$$M = 0.45\, b_f c \,.\, D_f \left(\frac{2kd - D_f}{k}\right)$$

or $229 \times 10^6 = 0.45 \times 1680 \times 5 \times 195 \left[\frac{2 \times 0.404\,d - 195}{0.404}\right]$

From which $d = 395$ mm.

Using 30 mm Φ bars arranging them in two tiers with a clear vertical distance of 30 mm between them, and a clear cover of 30 mm, total $D = 395 + 15 + 30 + 30 = 470$ mm.

Revised dead load of beam $= 0.4 \times 0.47 \times 1 \times 24000 = 4512$

$\therefore$ Total $w = 28080 + 4512 + 12800 = 45392$ N/m

$$M = \frac{45392\,(6.4)^2}{8} \times 1000 = 232.4 \times 10^6 \text{ N–mm}$$

Approximate L.A. $= a = d - \frac{D_f}{2} = 395 - \frac{195}{2} \approx 298$ mm

$$A_{st} = \frac{M}{\sigma_{st} \times a} = \frac{232.4 \times 10^6}{140 \times 298} = 5570 \text{ mm}^2$$

For 30 mm Φ bars $A_\Phi = 707 \text{ mm}^2$

$\therefore$ No. of 30 mm Φ bars $A_\Phi = \frac{5560}{707} \approx 8.$

Actual $A_{st} = 707 \times 8 = 5656 \text{ mm}^2$.

Determination of N.A. and actual stresses

Assuming that the N.A. falls in the flange, we have

$$\frac{b_f n^2}{2} = m \,.\, A_{st}\,(d - n)$$

$$\therefore \qquad \frac{1680}{2}(n)^2 = 19 \times 5656\,(395 - n)$$

or $n^2 + 128\,n - 50534 = 0$

From which $n \approx 170$ mm

$$\therefore \qquad \text{L.A.} = d - \frac{n}{3} = 395 - \frac{170}{3} = 338.3 \text{ mm}$$

Average stress in steel $= \frac{M}{A_{st} \times a} = \frac{232.4 \times 10^6}{5656 \times 338.3} = 121.5 \text{ N/mm}^2$

Corresponding stress in concrete is given by

$$c = \frac{t}{m} \times \frac{n}{d - n} = \frac{121.5}{19} \times \frac{170}{390 - 1760} = 4.93 \text{ N/mm}^2 \text{ (Safe)}$$

Stress in extreme layer of steel $= \frac{121.5}{395} \times (395 + 30)$

$= 130.8 \text{ N/mm}^2$

Shear reinforcement

$$V = \frac{wL}{2} = \frac{45392 \times 6.0}{2} = 136176 \text{ N}$$

$$\therefore \qquad \tau_v = \frac{136176}{400 \times 395} = 0.86 \text{ N/mm}^2.$$

Assuming that at the ends, four bars will be bent up, out of 8 bars, available $\frac{100 A_s}{bd} = \frac{100 \times 2828}{400 \times 395} = 1.79\%$. Hence from Table 3.1, for M 15 concrete, we get $\tau_c = 0.44 \text{ N/mm}^2$. Since $\tau_v > \tau_c$ shear reinforcement is necessary. Also from Table 3.3, for M 15 concrete, $\tau_{c.max} = 1.6 \text{ N/mm}^2$. Since τ_v is less than this, the section is O.K.

Shear reinforcement will be provided both in the form of inclined bars as well as in the form of vertical stirrups.

Out of 8 bars, let us bend 4 bars as follows :

(i)Two bars bent up at a distance of $\sqrt{2} \,.\, a = \sqrt{2} \times 314.2 \approx 480$ mm from the edge of the support.

(*ii*) Two bars bent up at a distance of $2a\sqrt{2} \approx 960$ mm from the edge of the support.

It can be shown that the above bending will be permissible from the B.M. point of view.

Area of cross-section of 2 bent-up bars,

$$= 2\frac{\pi}{4}(30)^2 = 1414 \text{ mm}^2$$

$\therefore$ Shear resistance of two bent up bars $= V_1$

$= \sigma_{sv} \,.\, A_{sv} \,.\, \sin\alpha = 140 \times 1414 \times 0.707 \approx 139960$ N.

This is more than the S.F. itself. Though shear resistance of concrete ($V_2 = \tau_c \,.\, bd$) is additionally available, we will have to provide nominal shear stirrups.

Using 10 mm Φ- 2 lgd stirrups

$$A_{sv} = 2\frac{\pi}{4}(10)^2 = 157 \text{ mm}^2.$$

The maximum spacing of stirrups is given by,

$$s_v = \frac{2.5 A_{sv} f_y}{b} = \frac{2.5 \times 157 \times 250}{400} = 245 \text{ mm}.$$

Hence provide the stirrups @ 240 mm c/c upto 960 mm from the edge of the support.

At 960 mm (=0.96 m) from the edge of the support.

Shear force $\frac{136176}{3}(3-0.96) = 92600$ N.

Shear resistance of concrete $= \tau_c . b_w . d = 2.44 \times 400 \times 395$
$= 69520$ N.

Hence shear to be resisted by the stirrups is given by

$V_s = V - \tau_c . b_w . d = 92600 - 69520 = 23080$

The spacing of 10 mm Φ-2 lgd stirrups is

$$s_v = \frac{\sigma_{sv} . A_{sv} . d}{V_s} = \frac{140 \times 157 \times 395}{23080} = 376 \text{ mm.}$$

But maximum spacing cannot exceed 240 mm. Hence provide the stirrups @ 240 mm c/c throughout the length of the beam.

Check for development length at the end.

The code requires that

$$1.3\frac{M_1}{V} + L_0 \geq L_d$$

A_{st} available at end=2828 mm^2

$\therefore \quad M_1 = 2828 \times 140 \times 0.865 \times 395 = 135.3 \times 10^6$ N−mm.

V=136176 N.

Let us assume the width of support equal to 400 mm and a side cover x'= 40 mm.

$$\therefore \quad L_0 = \frac{l_s}{2} - x' + 13\Phi$$

$$= \frac{400}{2} - 40 + (13 \times 30) = 550 \text{ mm.}$$

$$L_d = \frac{\Phi . \sigma_{st}}{4\tau_{bd}} = \frac{\Phi \times 140}{4 \times 0.6} = 58.3 \times \Phi$$

$$= 58.3 \times 30 = 1749.$$

Now $1.3\frac{M_1}{V} + L_0 = 1.3 \times \frac{135.3 \times 10^6}{136176} + 550$

$\therefore \quad = 1292 + 550 = 1792$ mm.

Thus $\left(\frac{M_1}{V} + L_0\right) > L_d$

Hence the anchorage requirements are satisfied.

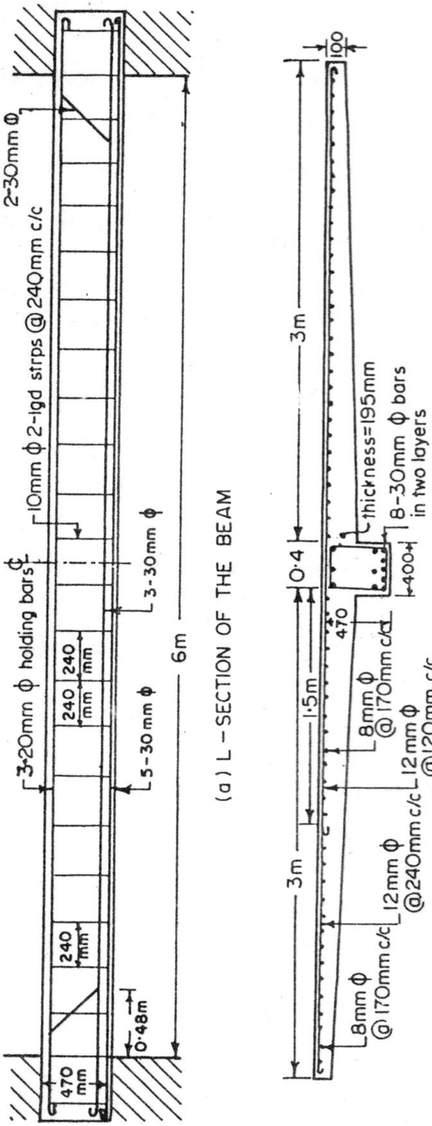

(a) L-SECTION OF THE BEAM

(b) SECTION AT THE MID-SPAN

FIG. 7.23

The details of reinforcement etc. are shown in Fig. 7.23.

7.11 DESIGN OF T-BEAM ROOF

Design Example 7.12. *A hall measures 10 m × 6 m from inside, and has walls 400 mm thick. Design a suitable R.C. T-beam roof to carry a super-imposed load of 2000 N/m². Use M 15 concrete. Take unit weight of concrete as 24 kN/m³.*

Solution. Let us keep end spans of slab slightly less than the middle span so that approximately equal maximum B.M. in induced in all the three spans. Assuming the T-beams to be 300 mm wide, the following spans of the slab are selected :

(*i*) End spans = 3 m (clear)

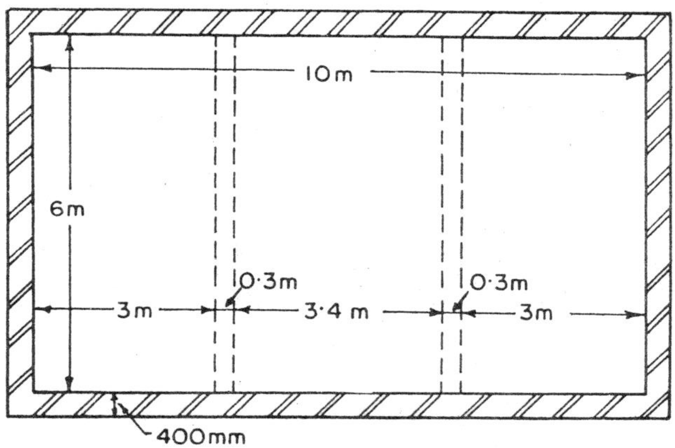

FIG. 7.24

(*ii*) Intermediate span = 3.4 m (clear).

The clear span of each T-beam will be 6 m, while the R.C slab monolithic with the beam will be continuous over the three spans.

For M 15 concrete, $\sigma_{cbc} = c = 5$ N/mm².
and $m \approx 19$. Taking $\sigma_{st} = t = 140$ N/mm²,
we have $k = 0.404$; $j = 0.865$ and $R = 0.874$.

(*A*) Design of continuous slab

Let us assume total thickness of slab = 1000 mm and effective thickness = 80 mm.

1. ***Effective spans***

The width of the supports is more than $\frac{1}{12}$ the clear span.

Effective end span= clear spans + $\frac{1}{2}$ (effective thickness of slab)

$= 3 + 0.04 = 3.04$ m.

Effective middle span = clear distance between supports
= 3.4 m.

2. ***Maximum B.M.***

Dead load/m^2 $= 0.1 \times 1 \times 24000 = 2400 = w_d$

Superimposed load = 2000 N/m^2$=w_s$

For end span : Near the middle of end span

$$M \text{ due to dead load } = \frac{w_d L^2}{12} = \frac{2400\,(3.04)^2}{12} \times 1000$$

$$\approx 1.85 \times 10^6 \text{ N-mm.}$$

$$M \text{ due to superimposed load } = \frac{w_s L^2}{10} = \frac{2400\,(3.04)^2}{10} \times 1000$$

$$= 1.85 \times 10^6 \text{ N-mm}$$

Total $M = 3.70 \times 10^6$ N-mm (+)

For intermediate span (at the middle)

$$M_d = \frac{w_d L^2}{24} = \frac{2400\,(3.4)^2}{24} \times 1000 = 1.16 \times 10^6 \text{ N-mm}$$

$$M_s = \frac{w_s L^2}{12} = \frac{2000\,(3.4)^2}{12} \times 1000 = 1.93 \times 10^6 \text{ N-mm}$$

Total $M = 3.09 \times 10^6$ N-mm (+)

At the support next to end support

$$M_d = \frac{w_d L^2}{10} = \frac{2400\,(3.04)^2}{10} \times 1000 = 2.22 \times 10^6 \text{ N-mm}$$

$$M_s = \frac{w_s L^2}{9} = \frac{2000\,(3.04)^2}{9} \times 1000 = 2.05 \times 10^6 \text{ N-mm}$$

Total $M = 4.27 \times 10^6$ N-mm

Hence $M_{max} = 4.27 \times 10^6$ N-mm.

3. ***Design of section***

$$d = \sqrt{\frac{M_{max}}{Rb}} = \sqrt{\frac{4.27 \times 10^6}{0.874 \times 1000}} \approx 70 \text{ mm,}$$

Let us check this depth from stiffness (deflection) point of view. Span/effective depth ratio = 26.

For a balanced section, % reinforcement

$$= \frac{kc}{2t} \times 100 = \frac{0.404 \times 5}{2 \times 140} \times 100 = 0.7\%$$

for M 15 concrete. Hence from Fig. 7.1, modification factor ≈ 1.6.

$$\therefore \qquad \frac{\text{span}}{d} = 26 \times 1.6 = 41.6$$

$$\therefore \qquad d = \frac{\text{span}}{41.6} = \frac{3040}{41.6} = 73 \text{ mm for end spans}$$

and $$d = \frac{3400}{41.6} = 82 \text{ mm for intermediate span.}$$

Let us keep total D = 100 mm. Using 8 mm Φ bars, and a clear cover of 15 mm, $d = 100 - 4 - 15 = 81$ mm.

Since the difference between the various bending moments is not large, reinforcement found for max. B.M. will be provided for other bending moments also.

$$A_{st} = \frac{M}{\sigma_{st} \cdot jd} = \frac{4.27 \times 10^6}{140 \times 0.865 \times 81} = 435 \text{ cm}^2$$

For 8 mm Φ bars, $A_\Phi = \frac{\pi}{4}(8)^2 = 50.3 \text{ mm}^2$.

$$\therefore \quad \text{Spacing} \qquad s = \frac{1000\,A_\Phi}{A_{st}} = \frac{1000 \times 50.3}{435} = 116 \text{ mm}$$

Hence provide 8 mm Φ bars @ 110 mm c/c at the middle of each span. Bend up alternative bars from the bottom at a distance of $\frac{L}{5} = \frac{3040}{5} \approx 600$ mm from the centre of intermediate support and at $L/7 \approx 450$ mm from the end support. Thus, reinforcement available at each intermediate support will be 8 mm Φ bars @ 110 mm c/c.

Distribution reinforcement

$$A_{sd} = \frac{0.15 \times 1000 \times 100}{100} = 150 \text{ mm}^2$$

Using 6 mm Φ bars, having $A_\Phi = \frac{\pi}{4}(6)^2 = 28.3 \text{ mm}^2$

$$s = \frac{1000 \times 28.3}{150} = 188 \text{ mm}$$

Hence provide these at 180 mm c/c.

4. ***Check for shear***

$$V = 0.4\, w_d \,.\, L + 0.45\, w_s \,.\, L \text{ at the outer support}$$

$$= (0.4 \times 2400 + 0.45 \times 2000)\, 3.04 = 5654 \text{ N}$$

$$\therefore \quad \tau_v = \frac{V}{bd} = \frac{5654}{1000 \times 81} = 0.07 \text{ N/mm}^2.$$

This is much less than the permissible value of 0.22 N/mm² even for minimum reinforcement of 0.25% (see Table 3.1). Hence safe.

5. ***Check for development length at supports.***

The code envisages that the diameter of the reinforcing bars should be such that the following equation is satisfied at the end supports:

$$1.3 \frac{M_1}{V} + L_0 \geq L_d$$

where $M_1 = \sigma_{st} \,.\, A_{st} \,.\, jd$

$$A_{st} \text{ at ends} = \frac{1000 \times 50.3}{220} = 228.7 \text{ mm}^2$$

$$\therefore \quad M_1 = 228.7 \times 140 \times 0.865 \times 81 = 2.243 \times 10^6 \text{ N-mm}$$

$$V = 5654 \text{ N}$$

Let us provide a side cover of 30 mm and assume a support width of 400 mm.

$$\therefore \quad L_0 = \frac{l_s}{2} - x' + 13\,\Phi = \frac{400}{2} - 30 + 13 \times 8$$

$$= 274 \text{ mm.}$$

$$L_d \approx 58.3\Phi = 58.3 \times 8 = 466 \text{ mm.}$$

$$\text{Now} \quad 1.3 \frac{M_1}{V} + L_0 = 1.3 \times \frac{2.243 \times 10^6}{5654} + 274$$

$$= 516 + 274 = 790 \text{ mm.}$$

$$\therefore \quad 1.3 \frac{M_1}{V} + L_0 > L_d$$

Hence the code requirements are satisfied.

(*B*) Design of T-beams

Let width of web $= b_w = 300$ mm.

Effective span of beam (l_0) = c/c of bearing = 6.4 m.

$$b_f = \frac{l_0}{6} + b_w + 6D_f = \frac{6400}{6} + 300 + 6 \times 100 = 1966 \text{ mm}$$

1. ***Loading per metre run***

Load from slab $=(w_d + w_s)\left[\frac{3}{2} + 0.3 + \frac{3.4}{2}\right]$

$= (2400 + 2000)(3.5) \approx 15400$ N/m.

Weight of web (assuming depth of rib = 300 mm)

$= 0.3 \times 0.3 \times 24000 \approx 2160$ N/m.

$\therefore$ Total $w = 15400 + 2160 = 17560$ N/m.

2. ***Design of section***

$$M = \frac{wL^2}{8} = \frac{17560\,(6.4)^2}{8} \times 1000 = 89.91 \times 10^6 \text{ N-mm}$$

For balanced section, we have from Eq. 5.17.

$$M = 0.45\, b_f\, \sigma_{cbc} \cdot D_f \left(\frac{2kd - D_f}{k}\right)$$

$$\therefore \quad 89.91 \times 10^6 = 0.45 \times 1966 \times 5 \times 100 \left(\frac{2 \times 0.404\, d - 100}{0.404}\right)$$

From which $d \approx 225$ mm

3. ***Area of steel***

$$a \approx d - \frac{D_f}{2} = 225 - \frac{100}{2} = 175 \text{ mm.}$$

Alternatively, $a \approx 0.9\,d = 0.9 \times 225 = 202$ mm

Adopt $a = 190$ mm for preliminary calculations.

$$A_{st} = \frac{M}{\sigma_{st} \times a} = \frac{89.91 \times 10^6}{140 \times 190} = 3380 \text{ mm}^2$$

Using 20 mm Φ bars, $A_\Phi = \frac{\pi}{4}(20)^2 \simeq 314 \text{ mm}^2$

$$\therefore \quad \text{No. of bars} = \frac{3380}{314} = 10.76$$

Provide 11-20 mm Φ bars.

Accommodate these in two layers, having clear distance between them = 20 mm. Using 25 mm clear cover.

$$D = 225 + 10 + 20 + 25 = 280 \text{ mm}$$

Actual $A_{st} = 11 \times 314 = 3454$ mm^2.

Bend 2 bars up at $L/7 \approx 1$ m from support, to take up negative B.M.

4. ***Determination of actual N.A. and stresses***

Assuming the N.A. to fall below the flange, we have,

$$b_f \cdot D_f \left(n - \frac{D_f}{2}\right) = m\, A_{st}\,(d - n)$$

$$\therefore \quad 1966 \times 100(n - 50) = 19 \times 3454\,(225 - n)$$

From which , n = 93.8 mm

Thus, the N.A. falls in the flange

$$\therefore \quad \frac{b_f \cdot n^2}{2} = mA_{st}(d - n)$$

or
$$\frac{1966}{2} n^2 = 19 \times 3454\,(225 - n)$$

which gives n = 93.6 mm

$$\therefore \quad a = d - \frac{n}{3}$$

$$= 225 - \frac{93.6}{3} = 193.8$$

$$\therefore \text{ Average stress in steel } = \frac{M}{A_{st} \times a} = \frac{89.91 \times 10^6}{3454 \times 193.8}$$

$$= 134.3 \text{ N/mm}^2$$

Corresponding stress in concrete is given by,

$$c = \frac{t}{m} \cdot \frac{n}{d - n} = \frac{134.3}{19} \times \frac{93.6}{225 - 93.6}$$

$$= 5.03 \text{ N/mm}^2$$

Stress in extreme layer of steel

$$= \frac{134.3}{225}(225 + 20) = 146 \text{ N/mm}^2$$

5. *Check for shear*

$$V = \frac{wl}{2} = \frac{17560 \times 6}{3} = 35120 \text{ N}$$

$$\tau_v = \frac{V}{b_w \cdot d} = \frac{35120}{300 \times 225} = 0.52 \text{ N/mm}^2$$

Assuming that 3 out of 11 bars will be bent up, $\frac{100 A_s}{bd}$ at the support $= \frac{100}{300 \times 225}\left(\frac{8}{11} \times 3454\right) = 3.72\%$. Hence from Table 3.1 permissible τ_c=0.44 N/mm^2. Thus, shear reinforcement is required. Shear resistance of concrete is,

$$V_c = \tau_c\, bd = 0.44 \times 300 \times 225 = 29700 \text{ N.}$$

Let us bend 3 bars, out of 11 bars of 20 mm Φ at a distance $= \sqrt{2}\, a = \sqrt{2} \times 194.1 \approx 275$ mm from the support.

$A_{sv} = 3\frac{\pi}{4}(20)^2 = 942$ mm. Hence shear resistance of bent up bar is $V_s = \sigma_{sv} A_{sv} \sin\alpha = 140 \times 942 \times 0.707 = 93280$ N. This itself is much more than the S.F. Hence only nominal stirrups will be required, the spacing of which is given by

$$s_v = \frac{2.5\,A_{sv}\,.\,f_y}{b}$$

Using 10 mm Φ 2-lgd stirrups,

$$A_{sv} = 2\frac{\pi}{4}(10)^2 = 157 \text{ mm}^2$$

$$\therefore \quad s_v = \frac{2.5 \times 157 \times 250}{300} = 327 \text{ mm.}$$

However, provide these @ 320 mm c/c throughout the length of the beam. Provide 2-20 mm Φ holding bars at the top.

6. ***Check for development length at support***

At the support, the following equation should be satisfied :

$$1.3\frac{M_1}{V} + L_0 \geq L_d$$

At the end only six bars are available, out of total 11 bars.

$$\therefore \quad A_{st} = 6 \times 314 = 1884 \text{ mm}^2$$

$$M_1 = 1884 \times 140 \times 0.864 \times (280 - 25 - 10)$$

$$= 54.83 \times 10^6 \text{ N-mm.}$$

$$V = 35120 \text{ N}$$

$$l_s = \text{length of support available} = 400 \text{ mm.}$$

Let the side cover $x' = 30$ mm.

$$\therefore \quad L_0 = \frac{l_s}{2} - x' + 13\,\Phi = \frac{400}{2} - 30 + 13 \times 20$$

$$= 430 \text{ mm.}$$

$$L_d \approx 58.3\,\Phi = 58.3 \times 20 = 1166 \text{ mm.}$$

Now $1.3\dfrac{M_1}{V} + L_0 = 1.3 \times \dfrac{55.83 \times 10^6}{35120} + 430$

$$= 2067 + 430 = 2497 \text{ mm.}$$

$$\therefore \quad 1.3\frac{M_1}{V} + L_0 > L_d$$

Hence the code requirements are satisfied. The details of reinforcement etc. are shown in Fig. 7.25.

7.12. DESIGN OF INVERTED T-BEAM ROOF

We have seen that if the size of a room is large, we may use monolithic beam-slab construction, forming the T-beam roof. However, in some cases, the provision of the beam (*i.e.* rib) below the slab may be undesirable from architectural or other point of view. In

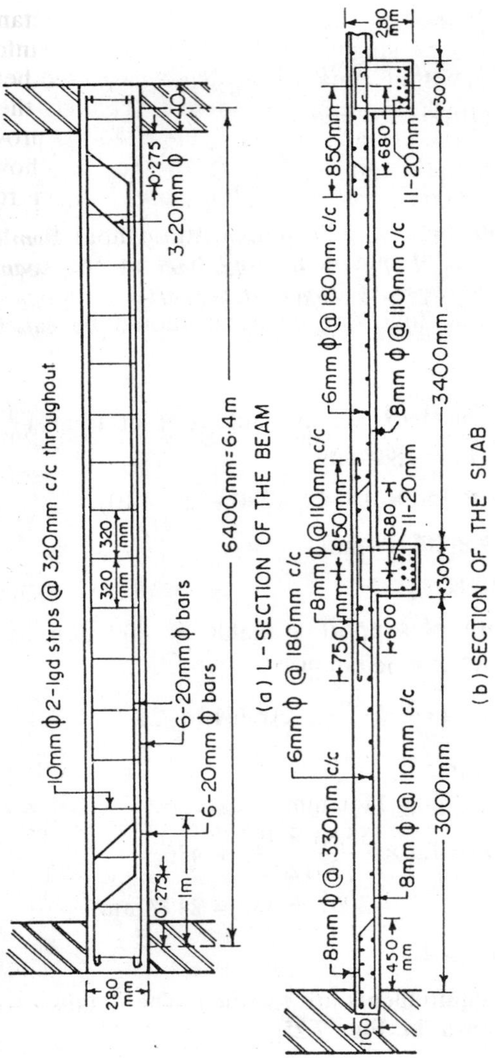
10mm ϕ 2-lgd strps @ 320mm c/c throughout
320 mm
320 mm
6-20mm ϕ bars
6-20mm ϕ bars
3-20mm ϕ
0·275
40
0·275
1m
280 mm
6400mm=6·4m
(a) L-SECTION OF THE BEAM
8mm ϕ @ 330mm c/c
6mm ϕ @ 180mm c/c
8mm ϕ @110mm c/c
6mm ϕ @ 180mm c/c
850mm
750 mm
850mm
8mm ϕ @ 110mm c/c
8mm ϕ @ 110mm c/c
450 mm
100
600
680
680
11-20mm
11-20mm
300
300
280 mm
3000mm
3400mm
(b) SECTION OF THE SLAB

FIG. 7.25

that case, the beam is provided *above* the slab, forming what is known as the inverted T-beam system. In such a case, the ceiling composed of the roof slab, is plane and the beam projects above the top of the slab. The slab is thus provided at the 'tension side' of the beam and is therefore, not helpful in resisting the tensile stresses with the beam. The beam, therefore, acts as a simple rectangular beam, which may either be singly reinforced or doubly reinforced. The depth of this beam will be equal to the depth measured between top of the beam and the bottom of the slab. Though the slab is cast monolithic with the beam it is tied to the beam by providing suitable vertical ties or stirrups. The design of the slab, however, is done in the same manner as for an ordinary T-beam roof.

Design Example 7.13. *A hall measures 10 m × 6 m from inside and has walls 400 mm thick. Design a suitable inverted T -beam roof to carry a super- imposed load of 2000 N/m². Assume suitable values of stresses. Use M 15 concrete. Take unit weight of concrete as 24 kN/m³.*

Solution. This example is the same as example 7.12 except that the system is inverted. However, since the slab is continuous over the two beams, its design will be the same as done in example 7.12. We will use the same set of stresses.

Design of rectangular beam

(i) Loading per metre run

Total thickness of slab = 100 mm

Let the width of the beam = 300 mm

$$\therefore \quad \text{Load from slab} = (2400 + 2000)\left(\frac{3}{2} + 0.3 + \frac{3.4}{2}\right)$$

$$= 15400 \text{ N}$$

Weight of rib (assuming d_w = 570 mm) = 0.3 × 0.57 × 24000

= 4105

$$\therefore \text{ Total load/m run} = 15400 + 4105 = 19505 \text{ N}$$

(ii) Design of section

Effective span L = c/c of bearing = 6.4 m.

$$\therefore \quad M = \frac{wL^2}{8} = \frac{19505\,(6.4)^2}{8} \approx 100000 \text{ N-m}$$

$$= 1 \times 10^8 \text{ N-mm}$$

$$\therefore \quad d = \sqrt{\frac{M}{Rb}} = \sqrt{\frac{1 \times 10^8}{0.874 \times 300}} = 618 \text{ mm}$$

Keep total depth = 660 mm. Using 20 mm Φ bars and a clear cover of 25 mm, available $d = 660 - 25 - 10 = 625$ mm.

(iii) Reinforcement

$$A_{st} = \frac{M}{\sigma_{st} \cdot jd} = \frac{1 \times 10^8}{140 \times 0.865 \times 625} = 1321 \text{ mm}^2$$

Use 20 mm Φ bars having A_Φ=314 mm²

$$\therefore \quad \text{No. of bars required} = \frac{1312}{314} = 4.2$$

Hence provide 5-20 mm Φ bars.

Actual A_{st} *provided* = 5 × 314 = 1570 mm².

Bend 2 bars up at a distance of $\sqrt{2}\, a = 1.414 \times 0.865 \times 625 = 764 \approx 800$ mm from the edge of support.

(iv) Check for development length at the supports

At the supports, $1.3 \dfrac{M_1}{V} + L_0 \geq L_d$

Available A_{st} at supports = 3 × 314 = 942 mm².

$$\therefore \quad M_1 = 942 \times 140 \times 0.865 \times 625 = 71.3 \times 10^6 \text{ N-mm}$$

$$V = \frac{wl}{2} = \frac{19505 \times 6}{2} = 58515 \text{ N.}$$

$$L_s = 400 \text{ mm} \; ; \; x' = 30 \text{ mm (say).}$$

$$\therefore \quad L_0 = \frac{L_s}{2} - x' + 13\Phi = \frac{400}{2} - 30 + 13 \times 20 = 430 \text{ mm.}$$

$$1.3 \frac{M_1}{V} + L_0 = 1.3 \times \frac{71.3 \times 10^6}{58515} + 430 = 1582 + 430 = 2012 \text{ mm}$$

$L_d = 58.3\,\Phi = 58.3 \times 20 = 1176$ mm.

Hence the code requirements are satisfied.

(v) Check for shear

$$\tau_v = \frac{V}{bd} = \frac{58515}{300 \times 625} = 0.312 \text{ N/mm}^2$$

$$\frac{100 A_{st}}{bd} = \frac{100 \times 942}{300 \times 625} = 0.5\%$$

$\therefore \quad \tau_c = 0.29$ N/mm² from Table 3.1, for M 15 concrete.

$$\therefore \quad V_c = 0.29 \times 300 \times 625 = 54375 \text{ N}$$

$$V_s = V - V_c = 58515 - 54375 = 4140 \text{ N.}$$

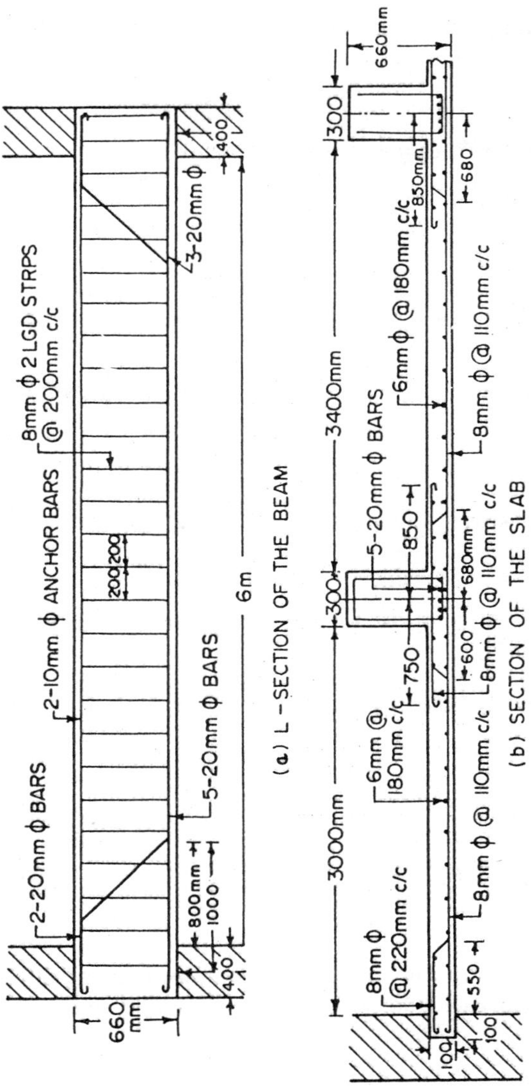
2-20mm ϕ BARS
2-10mm ϕ ANCHOR BARS
8mm ϕ 2 LGD STRPS @ 200mm c/c
3-20mm ϕ
5-20mm ϕ BARS
200 200
800mm
1000
400
400
660 mm
6m
(a) L-SECTION OF THE BEAM
3000mm
3400mm
300
300
660mm
8mm ϕ @ 220mm c/c
6mm @ 180mm c/c
750
5-20mm ϕ BARS
850
6mm ϕ @ 180mm c/c
850mm
550
600
680mm
680
100
100
8mm ϕ @ 110mm c/c
8mm ϕ @ 110mm c/c
8mm ϕ @ 110mm c/c
(b) SECTION OF THE SLAB

FIG. 7.26

The shear resistance of the bent up bars will be very much larger than this.

Hence only nominal stirrups are required, the spacing of which is given by

$$s_v = \frac{2.5\, A_{sv} f_y}{b} = \frac{2.5 \times 250}{300} A_{sv} = 2.083\, A_{sv}$$

Using 8 mm Φ two-legged stirrups, having

$$A_{sv} = 2\frac{\pi}{4}(8)^2 = 100.5 \text{ mm}^2$$

$$\therefore \quad s_v = 209 \text{ mm}.$$

Hence provide the stirrups @ 200 mm c/c throughout. These stirrups will also act as ties to transfer the load of the slabs to the beam.

Total load transferred by the slab to the beam per metre run

$$= 15400 \text{ N}.$$

$$\therefore \quad \text{Sectional area of ties required} = \frac{15400}{140} = 110 \text{ mm}^2$$

Actual area of stirrups provided per metre run

$$= \frac{1000}{200} \times 100.5 = 502.5 \text{ mm}^2.$$

Hence O.K. Use 2-10 mm Φ anchor bars at the top. The details are shown in Fig. 7.26.

7.13. DESIGN OF OVERHANGING T-BEAM ROOF

Design Example 7.14. *A hall measures 10 m × 6 m from inside and has walls 400 mm thick. The beams of the T-beam roof project out by 2 m on one side of the room. The slab is monolithic with the beam, and extends to the full length of the beam. Design a suitable roof system to carry a super-imposed load of 2 kN/m². Assume suitable values of the permissible stresses. Take unit weight of concrete as 24 kN/m³.*

Solution. The problem is similar to example 7.12, except that the beams overhang to one side. Let us arrange the beams as shown in Fig. 7.27.

Since the spans of the slab are the same as in example 7.12, its design will be the same as done in example 6.12. We will use the same set of stresses.

Hence thickness of the slab = 100 mm.

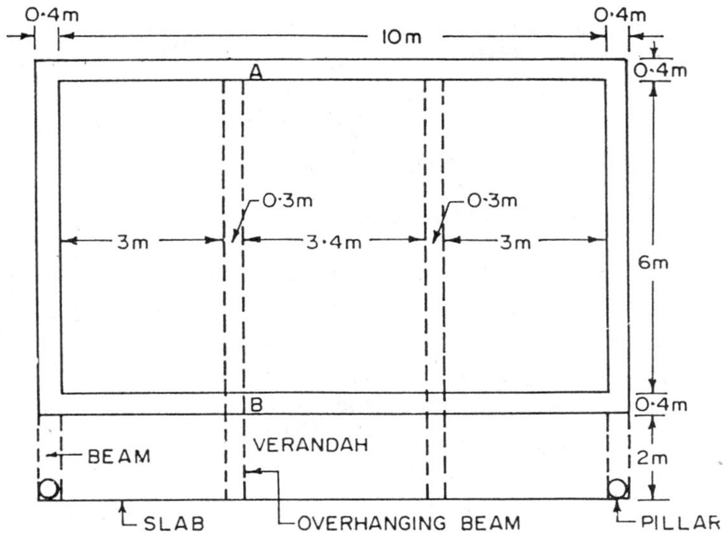

FIG. 7.27

Design of beam

The walls are 400 mm thick. Hence the following diagram showing the spans etc. may be drawn.

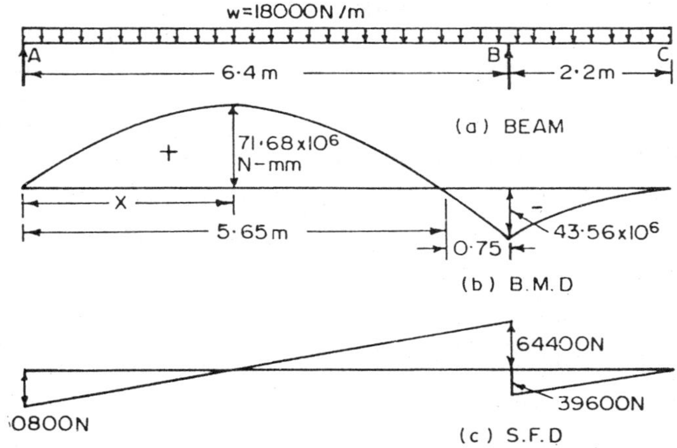

FIG. 7.28

Loading per metre run

Super-imposed load from slab = 15400 N

Let us assume the depth of rib = 300 mm and width of rib = 300 mm.

$\therefore$ Weight of rib = $0.3 \times 0.3 \times 24000 = 2160$ N

$\therefore$ Total load/m run $= 15400 + 2160 \approx 18000$ N (say)

Design of T-beam AB

1. *Bending moment.*

The bending moment diagram is shown in Fig. 7.28(*b*).

Maximum negative bending moment at support $B = \dfrac{18000(2.2)^2}{2}$

$$= 4.356 \times 10^4 \text{ N-m} = 43.56 \times 10^6 \text{ N-mm}.$$

The maximum positive bending moment will occur somewhere between A and B.

$$\text{Now, reaction } R_B = \left[\frac{18000(8.6)^2}{2}\right]\frac{1}{6.4} = 104000 \text{ N}$$

$$\text{Reaction } R_A = \frac{18000 \times 8.6\,(6.4-4.3)}{6.4} = 50800 \text{ N}$$

Measuring x from A, we have

$$M_x = 50800x - 18000\frac{x^2}{2}$$

$$\text{For maxima,} \quad \frac{dM_x}{dx} = 50800 - 18000x = 0$$

$$\therefore \quad x = \frac{50800}{18000} = 2.82 \text{ m}.$$

$$\therefore \quad M_{max} = \left[(50800 \times 2.82) - \frac{18000}{2}(2.82)^2\right] \times 10^3 \text{ N-mm}$$

$$= 71.68 \times 10^6 \text{ N-mm (sagging)}$$

The distance at which B.M. is zero is given by

$$50800x - \frac{18000x^2}{2} = 0.$$

$$\text{From which } x = \frac{50800}{9000} = 5.65 \text{ m}.$$

The B.M.D. and S.F.D. are shown in Fig. 7.28(*b*) and (*c*) respectively.

2. *Design of section*

The beam *AB*, for most of its length, has sagging bending moment, thus having tension at lower side. Hence it will work as a T- beam.

However, beam BC has hogging B.M. throughout its length, having tension at the upper fibres. Thus the slab will not be helpful, and the beam will act as a simple rectangular beam.

For the depth of T-beam, we have from Eq. 6.17,

$$M = 0.45\, b_f \,.\, \sigma_{cbc} \,.\, D_f \left[\frac{2kd - D_f}{k}\right] \qquad ...(i)$$

The width b_f of the flange will be the maximum of the following :

(*i*) $b_f = \frac{l_0}{6} + b_w + 6\, D_f$

where l_0 = distance between points of zero moments=5.65 m

$$\therefore \qquad b_f = \frac{5650}{6} + 300 + 6 \times 100 = 1842 \text{ mm}$$

(*ii*) $b_f = b_w + \frac{1}{2}$ [sum of clear distances to the adjacent beams on either side

$$= 300 + \tfrac{1}{2}[3000 + 3400] = 3500 \text{ mm}$$

Hence take b_f = 1842 mm.

Substituting the values in (*i*) we get

$$71.68 \times 10^6 = 0.45 \times 1842 \times 5 \times 100 \left[\frac{2 \times 0.404\, d - 100}{0.404}\right]$$

or $$0.808\, d = \frac{71.68 \times 10^6 \times 0.404}{0.45 \times 1842 \times 5 \times 100} + 100$$

From which d=210 mm.

Provide total depth = 250 mm. Using 25 mm Φ bars and a clear cover of 25 mm, available $d = 20 - 25 - \frac{25}{2} = 212.5$ mm.

3. *Area of steel*

Approximate L.A. $a = d - \frac{D_f}{2} = 212.5 - \frac{100}{2}$

$= 162.5$ mm

Alternatively, $a = 0.9d = 0.9 \times 212.5 = 191.2$ mm

Hence adopt a =180 mm for preliminary calculations.

$$A_{st} = \frac{M}{\sigma_{st} \times a} = \frac{71.68 \times 10^6}{140 \times 180} = 2844 \text{ mm}^2$$

Using 25 mm Φ bars having A_Φ = 490.9 mm²

$$\therefore \qquad \text{No. of bars} = \frac{2844}{490.9} = 5.8$$

Hence provide 6 No. of 25 bars

Actual A_{st} provided = 2945 mm².

At the simply supported end, bend two bars up one at a distance of $1.414\,a = 260$ mm and other at 320 mm from the edge of the support.

Similarly, bend two bars up near support B, one at a distance of 260 mm and the other at 520 mm from the face of the support. These bent up bars may be taken right up to end C, *to resist part of tension developed in beam BC.*

4. *Check for shear and development length at support A*

At A, $\tau_v = \dfrac{V}{bd} = \dfrac{50800}{300 \times 212.5} = 0.8 \text{ N/mm}^2$

$$\frac{100\,A_s}{bd} = \frac{100(4 \times 490.9)}{300 \times 212.5} = 3.08\ \%$$

Hence from Table 3.1, $\tau_c = 0.44$ N/mm^2

Hence shear reinforcement is necessary.

$$V_c = 0.44 \times 300 \times 212.5 = 28050$$

$$\therefore \quad V_s = V - V_c = 50800 - 28050 = 22750 \text{ N}$$

V_{s1} of one bent up bar $= 0.707 \times A_{sv} \times \sigma_{sv}$

$$= 0.707 \times 490.9 \times 140 = 48590 \text{ N}$$

Hence only nominal stirrups are required.

$$s_v \le \frac{2.5\,A_{sv} f_y}{b}$$

Using 8 mm dia. 2-lgd. stirrups, $A_{sv} = 2\dfrac{\pi}{4}(8)^2 = 100.5$

$$s_v \le \frac{2.5 \times 100.5 \times 250}{300} \le 209.$$

Hence provide these @ 200 mm c/c.

Also, $\dfrac{M_1}{V} + L_0 \ge L_d$

$$M_1 = (4 \times 490.9) \times 140 \times 180 = 49.48 \times 10^6 \text{ N-mm}$$

$$V = 50800 \text{ N}$$

$$L_s = 400 \text{ mm} \; ; \; x' = 30 \text{ mm}$$

$$L_0 = \frac{L_s}{2} - x' + 13\,\Phi = \frac{400}{2} - 30 + (13 \times 25) = 495 \text{ mm}$$

$$\therefore \quad \frac{M_1}{V} + L_0 = \frac{49.48 \times 10^6}{50800} + 495 = 974 + 495 = 1469 \text{ mm}$$

$$L_d \approx 58.3\,\Phi = 58.3 \times 25 = 1458 \text{ mm.}$$

Hence the Code requirements are satisfied.

5. *Check for actual stresses in beam AB*

Assuming the actual N.A. to fall below the flange, we have

$$b_f . D_f \left(n - \frac{D_f}{2}\right) = m A_{st} (d - n)$$

or $$1842 \times 100 \left(n - \frac{100}{2}\right) = 19 \times 2945 (212.5 - n)$$

From which $n = 87.9$ mm.

Hence the N.A. falls in the flange.

$$\therefore \quad b_f . \frac{n^2}{2} = m A_{st} (d - n)$$

or $$1842 \frac{n^2}{2} = 19 \times 2945 (212.5 - n)$$

From which $n \approx 87$ mm

$$\therefore \quad a = d - \frac{n}{3} \approx 212.5 - \frac{87}{3} \approx 183 \text{ mm}$$

$$\therefore \text{ Stress in steel } = \frac{M}{A_{st} . a} = \frac{71.68 \times 10^6}{2945 \times 183} = 133 \text{ N/mm}^2 \text{ (safe)}$$

Corresponding stress in concrete is given by

$$c = \frac{t}{m} . \frac{n}{d - n} = \frac{133}{19} \times \frac{87}{212.5 - 87} = 4.85 \text{ N/mm}^2 \text{ (safe)}$$

6. Shear reinforcement in beam AB at end B

S.F. $V = 64000$ N.

$$\tau_v = \frac{64000}{300 \times 212.5} = 1 \text{ N/mm}^2.$$

$$\frac{100 A_s}{bd} = \frac{100 (4 \times 490.0)}{300 \times 212.5} = 3.08\%$$

From Table 3.1, $\tau_c = 0.44$ N/mm^2.

Hence shear reinforcement is necessary.

$$V_c = 0.44 \times 300 \times 212.5 = 28050 \text{ N}$$

$$V_s = V - V_c = 64000 - 28050 = 35950 \text{ N}$$

V_{s1} of one bent up bar $= 0.707 \times A_{sv} . \sigma_{sv}$

$= 0.707 \times 490.9 \times 140 = 48590$ N

Hence only nominal stirrups are required.

Provide 8 mm dia. 2 lgd stirrups @ 200 mm c/c, (as provided at end A, step 4).

Design of beam BC

1. *Tensile reinforcement*

Beam BC will be designed as an ordinary rectangular beam with its tension side upwards. However, keep the depth and width of the beam the same as the T-beam.

Thus, width of beam = 300 mm
Total thickness of beam = 250 mm
Effective thickness = 212.5 mm (as before)

Moment of resistance of the beam $= M_1 = Rbd^2$

or $M_1 = 0.874 \times 300\,(212.5)^2 = 11.84 \times 10^6$ N-mm

Actual B.M. $= 43.56 \times 10^6$ N-mm

Hence the beam will have to be designed as a doubly reinforced section.

$$A_{st1} = \frac{M_1}{\sigma_{st}\, jd} = \frac{11.84 \times 10^6}{140 \times 0.865 \times 212.5} = 460 \text{ mm}^2$$

$M_2 = M - M_1 = (43.56 - 11.84)\,10^6 = 31.72 \times 10^6$ N–mm

This remaining B.M. has to be resisted by a couple provided by the tensile and compressive reinforcement. Let the centre of compressive reinforcement be placed at 40 mm below the top fibre, *i.e.*, $d_c = 40$ mm.

$$A_{st2} = \frac{M_2}{\sigma_{st}(d - d_c)} = \frac{31.72 \times 10^6}{140(212.5 - 40)} = 1313 \text{ mm}^2$$

$\therefore$ Total $A_{st} = A_{st1} + A_{st2} = 360 + 1313 = 1773 \text{ mm}^2$

Out of these, 2 bent-up bars are already available from span AB, having area $= 2 \times 490.9 = 981.8 \text{ mm}^2$.

$\therefore$ Remaining area $= 1773 - 981.8 = 791.5 \text{ mm}^2$.

Using 16 mm Φ bars, having $A_\Phi = 201 \text{ mm}^2$,

No. of bars = 4

Thus the main reinforcement consists of (*i*) 2-25 mm Φ bars available from AB, and (*ii*) 4-16 mm Φ bars. All these are provided at the top face of the beam BC, with a clear cover of 25 mm. Out of 4 bars of 16 mm Φ, continue 2 bars throughout span BA, to act as anchor bars, and curtail the remaining 2 bars at $L_d = 58.3 \times 16 \approx 950$ mm from the support B.

2. *Compressive reinforcement*

$$n = 0.404 \times 212.5 = 85.8 \text{ mm}$$

$$A_{sc} = \frac{m(d - n)}{(m_c - 1)(n - d_c)} A_{st2}$$

$$= \frac{19\,(212.5 - 85.8)}{(1.5 \times 19 - 1)(85.5 - 40)} \times 1313$$

$$= 2509 \text{ mm}^2$$

This is more than A_{st} which is 1773 mm² only. Hence redesign the beam, using steel beam theory.

Redesign. Let $A_{st} = A_{sc}$

Lever arm $= d - d_c = 212.5 - 40 = 172.5$ mm

$$A_{st} = A_{sc} = \frac{M}{\sigma_{st} \times a} = \frac{43.56 \times 10^6}{140 \times 172.5} = 1804 \text{ mm}^2$$

Now two bars of 25 mm Φ, having area = 981.8 mm² are already available from *AB*. Remaining area to provided = 1804−981.8 = 822.2 mm². Hence provide additional 1-25 Φ bar and 2-16 Φ bars, giving area = 490.9+2 × 201 = 892.9 mm². Curtail the 25 mm Φ bars at a distance of $L_d = 58.3 \times 25 \approx 1500$ mm from support *B*, inside span *BA*. However, continue 2-16 mm Φ bars throughout in span *BA*, to act as anchor bars.

Compression steel A_{sc} = 1804 mm². Now 4 bars of 25 mm Φ, giving area = 4 × 409.9 = 1963.6 mm², are already available from span *AB*. Hence these can be continued in span *BC* and no additional reinforcement is necessary.

At $L_d = 58.3\,\Phi = 58.3 \times 25 \approx 1500$ mm (=1.5 m) from end *B*, in beam *BC*,

$$M = \frac{18000\,(2.2 - 1.5)^2}{2} \times 1000 = 4.41 \times 10^6 \text{ N-mm}$$

This is much less than moment of resistance of singly reinforced beam. Hence discontinue the 3 - 25 mm Φ top bars and 2-25 mm Φ bottom bars at this location, and continue the remaining bars. At this point of curtailment,

$$\text{Required } A_{st} = \frac{4.41 \times 10^6}{140 \times 0.865 \times 212.5} = 171.4 \text{ mm}^2$$

Area available (*i.e.*, 2-16 mm Φ bars)= 2 × 201 = 402 mm²

Hence O.K.

3. *Shear reinforcement in BC at B*

$$V = 39600 \text{ N}$$

$$\tau_v = \frac{39600}{300 \times 212.5} = 0.621 \text{ mm}^2$$

$$\frac{100 A_s}{bd} = \frac{100(981.8 \times 490.9 + 402)}{300 \times 212.5} = 2.94\%$$

From Table 3.1, $\tau_c = 0.44$ mm²,

Hence shear reinforcement is necessary.

$V_c = 0.44 \times 300 \times 212.5 = 28050$ N

$V_s = V - V_c = 39600 - 28050 = 11550$ N.

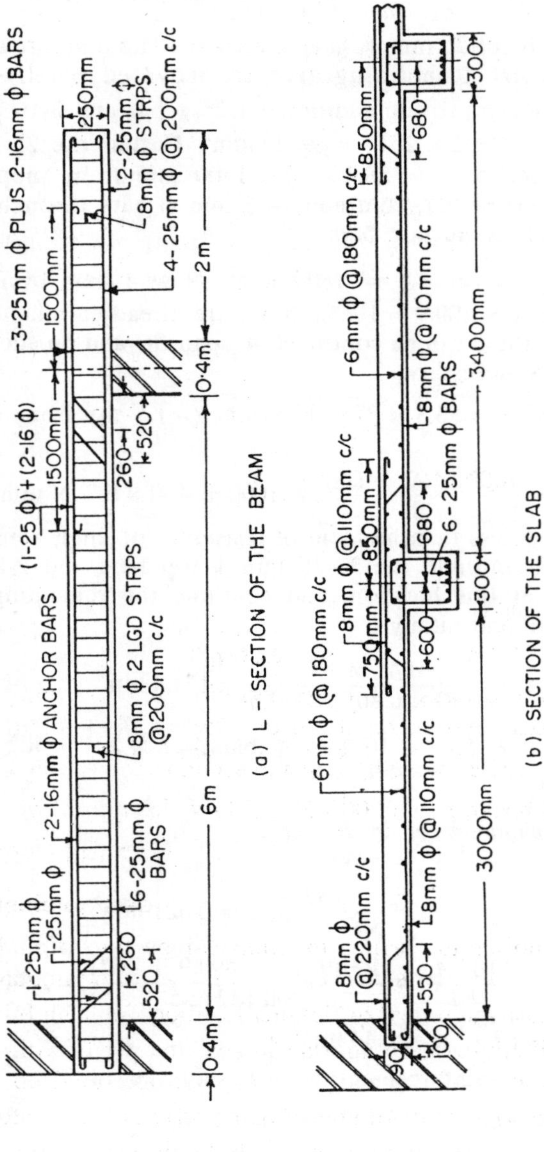

(a) L - SECTION OF THE BEAM

(b) SECTION OF THE SLAB

FIG. 7.29

Using 8 mm Φ 2-lgd stirrups, $A_{sv} = 2 \times \frac{\pi}{4}(8)^2 = 100.5$

$$s_v = \frac{\sigma_{sv} . A_{sv} . d}{V_s} = \frac{140 \times 100.5 \times 212.5}{11550}$$

$$= 259 \text{ mm.}$$

This should not be more than the spacing of nominal stirrups, given by

$$s_v \leq \frac{2.5 A_{sv} f_y}{b} \leq \frac{2.5 \times 100.5 \times 250}{300}$$

$$\leq 209 \text{ mm.}$$

Hence provide 8 mm Φ 2-lgd stirrups @ 200 mm c/c throughout the length.

The details of reinforcement etc. are shown in Fig. 7.29.

7.14 DESIGN OF CANTILEVER CANOPY

Design Example 7.15. *Design a R.C. cantilever canopy having 2.5 metre effective overhang and 6 metres wide as shown in Fig. 7.30. The canopy slab is supported on two beams cantilevered out from walls and spaced at 3.5 m centre to centre, and overhangs each beam by 1.25 m on each side. The beams and slab are cast monolithic, to give a flat soffit. The super-imposed load on the slab may be taken*

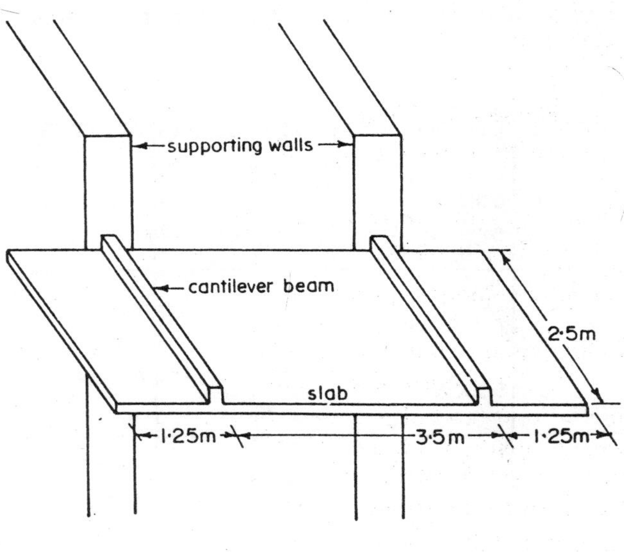

FIG. 7.30.

as 1500 N/m². If the height of the wall over the anchoring length of the beam is 5 m, calculate the anchoring length of the beam required for a factor of safety of 1.5. The width of the supporting walls is 600 mm. Use M 15 concrete and take $\sigma_{st} = 140$ N/mm².

Solution. (*a*) Design of Slab

1. ***The design constants***

For $c = \sigma_{cbc} = 5$ N/mm² ; $t = \sigma_{st} = 140$ N/mm² and $m \approx 19$, we have

$$k = 0.404 \; ; \; j = 0.865 \; ; \; R = 0.874$$

2. ***Bending moment***

Loads : Let the total thickness of slab be 100 mm.

Dead load per sq. metre $0.1 \times 1 \times 1 \times 25000 = 2500$ N

Superimposed load $= 1500$ N

Total $= 4000$ N

The B.M. and S.F. diagrams are shown in Fig. 7.31.

$$\text{Max. hogging B.M.} = \frac{4000\,(1.25)^2}{2} \times 1000 = 3.125 \times 10^6 \text{ N-mm}$$

$$\text{Max. sagging B.M.} = \left[12000 \times 1.75 - \frac{4000(3)^2}{2}\right] \times 1000$$

$$= 3 \times 10^6 \text{ N-mm}$$

Let the point of contraflexure occur at x from support B. Its value is given by

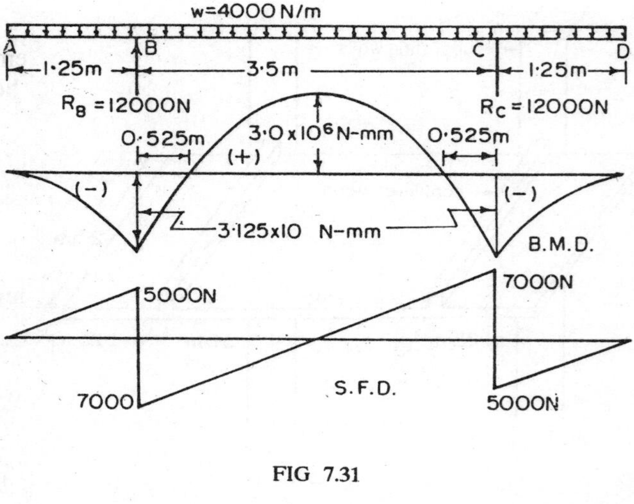

FIG 7.31

$$12000x - \frac{4000(1.25 + x)^2}{2} = 0$$

or $$x^2 - 3.5x + 1.5625 = 0$$

From which $x = 0.525$ m

The slab will be designed for the greater B.M. *i.e.* 3.125×10^6 N-mm.

3. ***Design of section***

$$d = \sqrt{\frac{M}{Rb}} = \sqrt{\frac{3.125 \times 10^6}{0.874 \times 1000}} = 60 \text{ mm}$$

From stiffness point of view, span/eff. depth = 7 for cantilever.

For a balanced section, % reinforcement

$$= \frac{k\,\sigma_{cbc}}{2\,\sigma_{st}} \times 100 = \frac{0.404 \times 5}{2 \times 140} \times 100 = 0.7\%, \text{ for M 15 concrete.}$$

Hence from Fig. 7.1, modification factor ≈ 1.6

$$\therefore \quad \frac{\text{span}}{d} = 7 \times 1.6 = 11.2$$

$$\therefore \quad d = \frac{\text{span}}{11.2} = \frac{1.25 \times 1000}{11.2} = 112 \text{ mm.}$$

However, since the slab is continuous, keep $D = 100$ mm. Using 8 mm Φ bars and a cover of 15 mm, available $d = 100 - 4 - 15 = 81$ mm.

4. ***Reinforcement***

For *AB*, there is hogging B.M. and hence reinforcement will be provided at the top. For *BC*, B.M. is sagging and hence the reinforcement will be provided at the bottom.

$$\text{For } AB,\ A_{st} = \frac{M}{\sigma_{st}\,jd} = \frac{3.125 \times 10^6}{0.874 \times 140 \times 81} = 315 \text{ mm}^2$$

$$\text{For } BC,\ A_{st} = \frac{M}{\sigma_{st}\,jd} = \frac{3.0 \times 10^6}{0.874 \times 140 \times 81} = 303 \text{ mm}^2$$

Minimum reinforcement $= 1.5\,D = 1.5 \times 100 = 150 \text{ mm}^2$.

Using 8 mm Φ bars, $A_\Phi = \frac{\pi}{4}(8)^2 = 50.27 \text{ mm}^2$

$$\therefore \quad \text{Spacing} \quad s = \frac{1000\,A_\Phi}{A_{st}} = \frac{1000 \times 50.27}{315} = 159.6 \text{ mm}$$

However, provide 8 mm Φ bars @ 150 mm c/c in all the three spans, for *AB* and *CD*, provide reinforcement at the top and take the bars inside the span for a distance equal to 525 mm or $L_d = 58.3\ \Phi$ ($= 58.3 \times 8 = 466$ mm), whichever is more. Hence take these bars inside by 600 mm . For *BC*, provide reinforcement at the bottom and take it upto the supports *B* and *C*. *This provision will meet the requirements of development length.*

5. *Distribution reinforcement*

$A_{sd} = 1.5\,D = 1.5 \times 100 = 150\ \text{mm}^2$

Using 6 mm Φ bars, having $A_{\Phi} = \frac{\pi}{4}(6)^2 = 28.4\ \text{mm}^2$

$$\therefore \quad s = \frac{1000 \times 28.4}{153} = 190\ \text{mm.}$$

6. *Check for shear*

S.F. = 7000 N.

$$\therefore \quad \tau_v = \frac{V}{bd} = \frac{7000}{1000 \times 81} = 0.0864\ \text{N/mm}^2.$$

This is much less than the permissible value of $0.22 \times 1.3 = 0.286\ \text{N/mm}^2$ even for min. reinforcement of 0.25% (Table 3.1 and 3.2). Hence safe.

(*b*) Design of beam

1. *Bending moment*

The beam and slab are cast monolithic. Since the compression is in the bottom fibres, flange action will be effective and the beam will act as a T-beam.

Span of cantilever = 2.5 m.

Super-imposed load from slab, per m run

$$= \frac{6}{2} \times 1 \times 4000 = 12000\ \text{N}$$

Self weight of web (assuming 300 mm width and 300 mm depth)

$$= 0.3 \times 0.3 \times 1 \times 25000 = 2250\ \text{N}$$

$\therefore$ Total load per m run $= 12000 + 2250 = 14250$ N

$$\therefore \ M = \frac{wL^2}{2} = \frac{14250(2.5)^2}{2} \times 1000 = 44.53 \times 10^6\ \text{N-mm}$$

2. *Design of section*

The width of flange available will be the minimum of the following :

(*i*) $b_f = \frac{l_0}{6} + b_w + 6\,D_f$

Taking $l_0 = 2.5$ m $= 2500$ mm,

$$b_f = \frac{2500}{6} + 300 + 6 \times 100 \approx 1316 \text{ mm}$$

(*ii*) b_f = cantilever projection of slab + $\frac{1}{2}$ spacing of beam

$$= 1250 + 1750 = 3000 \text{ mm}$$

Hence adopt $b_f = 1316$ mm.

For effective depth of T-beam, we have from Eq. 5.17,

$$M = 0.45\, b_f \,.\, \sigma_{cbc} \,.\, D_f \left[\frac{2kd - D_f}{k}\right]$$

or $44.53 \times 10^6 = 0.45 \times 1316 \times 5 \times 100 \left[\frac{2 \times 0.404\, d - 100}{0.404}\right]$

or $\qquad 0.808\, d = \frac{44.53 \times 10^6 \times 404}{0.45 \times 1316 \times 500} + 100$

From which $d \approx 200$ mm.

Provide total D = 250 mm. Using 20 mm Φ bars and a clear cover of 25 mm, available $d = 250 - \frac{20}{2} - 25 = 210$ mm

3. ***Area of steel***

Approximate L.A. $a = d - \frac{D_f}{2} = 210 - \frac{100}{2} = 160$ mm

Alternatively, $\qquad a = 0.9\, d = 0.9 \times 210 = 189$ mm

Adopt $a = 175$ mm for preliminary calculations

$$A_{st} = \frac{M}{\sigma_{st} \,.\, a} = \frac{44.53 \times 10^6}{140 \times 189} = 1683 \text{ mm}^2$$

Using 20 mm Φ bars, $A_\Phi \approx 314 \text{ mm}^2$

∴ $\quad$ No. of bars required $= \frac{1683}{314} = 5.3$

Hence provide 6 Nos. 20 mm Φ bars, providing $A_\Phi = 6 \times 314 = 1884 \text{ mm}^2$.

4. ***Determination of actual N.A. and stresses***

Assuming the N.A. to fall below the flange, we have

$$b_f \,.\, D_f \left(n - \frac{D_f}{2}\right) = m \,.\, A_{st}\, (d - n)$$

or $\quad 1316 \times 100 \left(n - \frac{100}{2}\right) = 19 \times 1884\, (210 - n)$

From which $n = 84$ mm.

Hence the N.A. falls in the flange.

$$\therefore \quad b_f \,.\, \frac{n^2}{2} = m\, A_{st}\, (d - n)$$

or $1316 \times \frac{n^2}{2} = 19 \times 1884\,(210 - n)$

from which $n = 83$ mm.

$\therefore \quad a = d - \frac{n}{3} = 210 - \frac{83}{3} = 182.3$ mm

$\therefore$ Stress in steel $= t = \frac{M}{A_{st}\,.\,a} = \frac{44.53 \times 10^6}{1884 \times 182.3}$

$= 129.6$ N/mm^2

Corresponding stress in steel is given by

$$c = \frac{t}{m}\cdot\frac{n}{d-n} = \frac{129.6}{19} \times \frac{83}{210 - 83} = 4.46 \text{ N/mm}^2.$$

5. ***Check for shear***

$V = wL = 14250 \times 2.5 = 35625$ N

$\tau_c = \frac{V}{bd} = \frac{35625}{300 \times 210} = 0.565$ N/mm^2

$\frac{100 A_s}{bd} = \frac{100 \times 1884}{300 \times 210} \approx 3\%$

Hence from Table 3.1, $\tau_c = 0.44$ N/mm^2

Hence shear reinforcement is necessary.

$V_c = 0.44 \times 300 \times 210 = 27720$ N

$\therefore \quad V_s = V - V_c = 35625 - 27720 = 7905$ N.

Using 8 mm Φ 2 lgd- stirrups, $A_{sv} = 2\frac{\pi}{4}(8)^2 = 100.5$

$$s_v = \frac{\sigma_{sv}\,.\,A_{sv}\,.\,d}{V_s} = \frac{140 \times 100.5 \times 210}{7905} = 374 \text{ mm.}$$

Spacing for nominal stirrups is given by

$$s_v \le \frac{2.5\,.\,A_{sv}\,.\,f_y}{V_s} \le \frac{2.5 \times 100.5 \times 250}{300} \le 209 \text{ mm}$$

Hence provide 8 mm Φ 2 lgd stirrups @ 200 mm c/c. Provide 2-10 mm Φ holding bars at the bottom.

6. ***Curtailment of main reinforcement***

Out of 6 main bars, let us curtail 3 bars at a distance of x from cantilever end.

$$M_x = \frac{wx^2}{2} = \frac{14250\,x^2}{2} \times 1000 = 7.125 \times 10^6 x^2 \text{ N-mm.}$$

This should be equal to half of the max. B.M.

$\therefore \quad 7.125 \times 10^6 x^2 = \frac{1}{2} \times 44.53 \times 10^6$

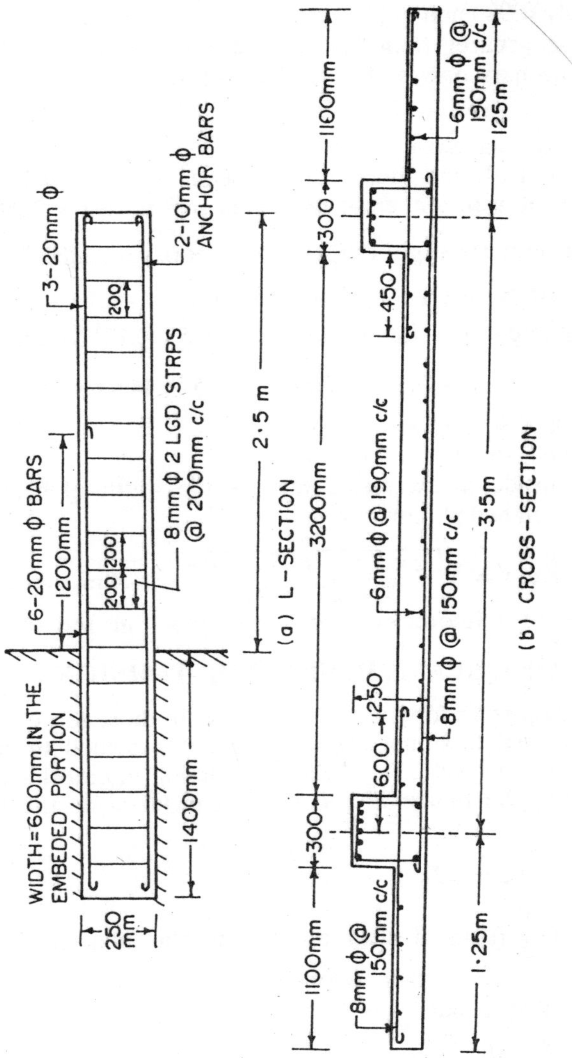
3-20mm ϕ
2-10mm ϕ ANCHOR BARS
200
8mm ϕ 2 LGD STRPS @ 200mm c/c
6-20mm ϕ BARS
1200mm
200 200
200
WIDTH=600mm IN THE EMBEDED PORTION
1400mm
250 mm
(a) L-SECTION
2·5 m
1100mm
300
3200mm
300
1100mm
6mm ϕ @ 190mm c/c
125m
450
6mm ϕ @ 190mm c/c
8mm ϕ @ 150mm c/c
250
600
8mm ϕ @ 150mm c/c
8mm ϕ @ 150mm c/c
3·5m
1·25m
(b) CROSS-SECTION

FIG. 7.32

$$\therefore\ x = \sqrt{\frac{44.53}{2 \times 7.125}} = 1.76 \text{ m} = 1760 \text{ mm}$$

Hence curtail two bars at a distance of (2500−1760)+12 Φ = 740+(12+20) = 980 mm ≈ 1000 mm from the fixed end. This distance should be greater than $L_d = 58.3\ \Phi = 58.3 \times 20 = 1160$ mm. Hence curtail the bars only at the distance of 1200 mm (say) from the fixed end.

7. ***Length of embedment of the beam***

Let the length of embedment of the beam in the wall be x m.

$\therefore$ Weight of wall on embeded beam $= 5 \times 0.3x \times 22500$ N

Its moment about the edge of support $= 5 \times 0.3 \times 22500\ \frac{x^2}{2}$ N–m

Factor of safety against overturning = 1.5

External moment causing rotation $= 44.53 \times 10^3$ N−m,

$$\therefore\quad 5 \times 0.3 \times 22500 \frac{x^2}{2} = 1.5 \times 44.53 \times 10^3$$

From which $x = 1.985$ m

Hence embed the beam by 2 m.

If this is considered excessive, increase the width of beam to 600 mm in the embeded portion.

This will give $x = \sqrt{\frac{2 \times 1.5 \times 44.53 \times 10^3}{5 \times 0.6 \times 22500}} = 1.4$ m.

The details of reinforcement etc. are shown in Fig. 7.32.

7.15 DESIGN OF L-BEAM : DESIGN FOR TORSION

The design principles for L-beam are similar to those for T-beam, except that additional torsional reinforcement is provided to resist torsional moment. The width b_f of the slab, acting monolithic with the beam, forming the flange of the L-beam is taken as the least of following :

(*a*) $b_f = \frac{l_0}{12} + b_w + 3\,D_f$

(*b*) $b_f = b_w + \frac{1}{2}$ (sum of clear distance to the adjacent beams on either side)

(*c*) For isolated beam,

$$b_f = \frac{0.5\,l_0}{\frac{l_0}{b} + 4} + b_w$$

(but in no case greater than the actual width of flange)

where l_0 is the distance between points of zero moments in the beam. For continuous beams and frames, l_0 may be assumed as 0.7 times of the effective span.

The location of N.A. and calculations of the moment of resistance of L-beam are done exactly in the same way as that for T-beam, and all the equations developed for T-beam are also applicable for L-beam. The design of L-beam is done in the following steps :

1. Assume some suitable value of width of rib.

2. Determine the effective width b_f of the flange, based on criteria given above.

3. Assume effective span of the beam and calculate bending moment and shear force.

4. Determine the effective depth of the beam on any one of the three considerations: (*i*) balanced section (*ii*) economy (*iii*) shear stress

For balanced section, the effective depth is given by Eq. 6.16.

$$M = b_f \,.\, \sigma_{cbc} \,.\, D_f \left[d - \frac{D_f(1+k)}{2k} + \frac{D_f^{\,2}}{3k\,d}\right]$$

Alternatively, taking lever arm to be equal to $0.9\,d$, we have, from Eq. 6.17 :

$$M = 0.45\, b_f \,.\, \sigma_{cbc} \,.\, D_f \left[\frac{2kd - D_f}{k}\right]$$

For economical depth, use Eq. 6.18

$$d = \sqrt{\frac{r \,.\, M}{\sigma_{st} \,.\, j \,.\, b_w}}$$

For shear considerations, use Eq. 6.21

$$d_{min} = \frac{V}{b_w\, \tau_v} = \frac{V}{b_w \,.\, \tau_{c\,max}}$$

where τ_v may by taken equal to $\tau_{c\,.\,max}$ given in Table 3.3.

The maximum value of d obtained from the above three criteria is taken as the effective depth of the beam.

5. Take the approximate lever arm $a = d - \dfrac{D_f}{2}$, or $a = 0.9\,d$, and determine the area of tensile reinforcement :

$$A_{st} = \frac{M}{t \times a}, \quad \text{where } t = \sigma_{st}$$

Arrange suitably steel reinforcement.

6. Determine the actual N.A. and check the stress in concrete and steel.

Assuming the N.A. to fall below the flange n may be determined from the expression :

$$b_f . D_f \left(n - \frac{D_f}{2}\right) = m\, A_{st}\, (d - n)$$

Also, $$\bar{y} = \frac{3n - 2D_f}{2n - D_f} \cdot \frac{D_f}{3}$$

$\therefore$ $$a = d - \bar{y}$$

If, however, N.A. falls in the flange,

$$b_f . \frac{n^2}{2} = m . A_{st}\, (d - n)$$

$\therefore$ $$a = d - \frac{n}{3}$$

Hence $$t = \frac{M}{A_{st} . a}$$

and $$c = \frac{t}{m} \cdot \frac{n}{d - n}$$

7. Calculate shear force V and torsional moment T. Compute the *equivalent shear* V_e from the expression :

$$V_e = V + 1.6 \frac{T}{b_w}$$

and $$\tau_{ve} = \frac{V_e}{b_w . d}$$

The value of τ_{ve} should not exceed $\tau_{c.max}$ given in Table 3.3. If it exceeds, the section should be redesigned by increasing the concrete area.

8. If equivalent shear stress τ_{ve} does not exceed τ_c (Table 3.1) provide nominal shear stirrups governed by the equation.

$$s_v \leq \frac{2.5\, A_{sv} f_y}{b}$$

9. If τ_{ve} exceeds τ_c , provide longitudinal reinforcement as well as transverse reinforcement as discussed in steps 10 and 11 respectively.

10. *Longitudinal reinforcement.* It is designed to resist an equivalent B.M. M_{e1} given by

$$M_{e1} = M + M_T$$

where M = B.M. at the section

and $$M_t = T\frac{\left(1+\frac{D}{b_w}\right)}{1.7}.$$

If M_T exceeds M, provide longitudinal reinforcement on compression face also, for a moment M_{e2} given by.

$$M_{e2} = M_T - M\ .$$

11. *Transverse reinforcement.* Find area of transverse reinforcement A_{sv} by

$$A_{sv} = \frac{T\,.\,s_v}{b_1 d_1 \sigma_{sv}} + \frac{V\,.\,s_v}{2.5\,d_1\,.\,\sigma_{sv}}$$

subject to a minimum of $\frac{(\tau_{ve} - \tau_c)\,b_w\,.\,s_v}{\sigma_{sv}}$

b_1 = centre to centre distance between corner bars in direction of width

d_1 = centre to distance between corner bars in direction of depth

The spacing s_v should not exceed x_1, $\frac{x_1+y_1}{4}$ and 300 mm, where s_1 and y_1 are respectively the short and long dimensions of the stirrup.

12. *Side face reinforcement.* Where the depth of web in a beam exceeds 450 mm, side face reinforcement shall be provided along the two faces. The total area of such reinforcement shall be not less than 0.1 percent of the web area and shall be distributed equally on two faces at spacing not exceeding 300 mm or web thickness whichever is less.

Design Example 7.16. *Fig 7.33 shows the plan and section of a canopy to be provided over an entrance opening of building. Design the beam and the slab. Use M 15 concrete. The superimposed load on the slab may be taken as 1200 N/m². Take unit weight of concrete as 24 kN/m³.*

Solution. Since the beam and slab are monolithic, the beam will act as L-beam and part of the slab will resist the compression of the beam. The breadth of the beam will be kept equal to the width of the column, *i.e.*, 400 mm. Using M 15 concrete, and taking $\sigma_{st} = t$ = 140 N/mm^2 , we have

$c = \sigma_{cbc} = 5$ N/mm^2 , m=19, k=0.404, j = 0.865 ; R = 0.874

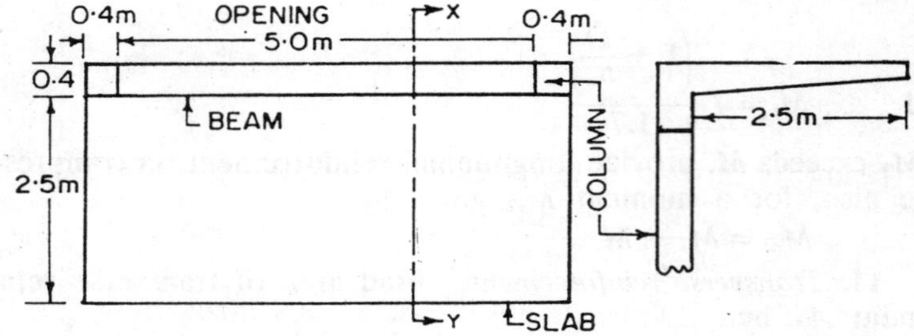

(a) PLAN (b) SECTION AT X-Y

FIG. 7.33

(*A*) Design of canopy slab

Let the average depth of slab = 120 mm.

1. ***Loading and B.M.***

Superimposed load per sq. m = 1200 N

Dead load due to self weight = 0.12×1×1×24000

= 2880 N

$\therefore$ Total $w = 4080\ \text{N/m}^2$.

$$M_{max} = \frac{w L^2}{2} = \frac{4080\,(2.5)^2}{2} \times 1000 = 12.75 \times 10^6\ \text{N–mm}.$$

2. ***Design of section***

$$d = \sqrt{\frac{M}{Rb}} = \sqrt{\frac{12.75 \times 10^6}{0.874 \times 1000}} \approx 120\ \text{mm}.$$

As per code requirement, $\frac{L}{d} = 7$. For M 15 concrete, taking $p \approx 0.72\ \%$ for balanced design, modification factor (Fig. 7.32) = 1.6.

$$\therefore \qquad \frac{L}{d} = 7 \times 1.6 = 11.2$$

$$\therefore \qquad d = \frac{L}{11.2} = \frac{2500}{11.2} = 223\ \text{mm}$$

Hence keep D = 250 mm at the support and reduce it to 00 mm at the edge. Average thickness = $\frac{250 + 100}{2}$ = 175 mm against an assumed value of 120 mm. Hence dead load due to revised dimension = 0.175 × 24000 = 4200 N.

$\therefore$ Total $w = 5400\ \text{N/m}^2$

$$M = \frac{5400\,(2.5)^2}{2} \times 1000 = 16.875 \times 10^6 \text{ N–mm}$$

3. ***Main reinforcement***

Using 10 mm Φ bars, available $d = 250 - 5 - 15 = 230$

$$A_{st} = \frac{M}{\sigma_{st}\, jd} = \frac{16.875 \times 10^6}{140 \times 0.865 \times 230} = 606 \text{ mm}^2$$

For 10 mm Φ bars, $A_\Phi = \frac{\pi}{4}(10)^2 = 78.5 \text{ mm}^2$

$$\therefore \quad \text{Spacing } s \quad \frac{1000\,A_\Phi}{A_{st}} = \frac{1000 \times 78.5}{606} = 130 \text{ mm.}$$

Hence provide 10 mm Φ bars @ 130 mm c/c.

Curtailment of reinforcement

Let us find the position where reinforcement can be reduced to half. Evidently, the B.M. at that point will be $= 8.44 \times 10^6$ N-mm. Measuring x from the free end, we have

$$8.44 \times 10^6 = \frac{5400\,(x)^2}{2} \times 1000$$

which gives $x = 1.77$ m from the free end, or 0.73 m from the fixed end. However, provide 10 mm Φ bars @ 130 mm c/c upto 1 m from the fixed end and thereafter, curtail half the bars so that 10 mm Φ bars are provided at 260 mm c/c.

4. ***Check for development length at the end***

The code envisages that the diameter of the reinforcing bar should be such that the following equation is satisfied at the end.

$$\frac{M_1}{V} + L_0 \geq L_d.$$

At the end, $A_{st} = 606 \text{ mm}^2$

$$\therefore\ M_1 = 606 \times 140 \times 0.865 \times 230 = 16.88 \times 10^6 \text{ N–mm}$$

$$V = 5400 \times 2.5 = 13500 \text{ N}$$

Support width $= 400 \text{ mm} = l_s.$

Let us provide side cover equal to 40 mm.

$$\therefore \quad L_0 = \frac{l_s}{2} - x' + 13\Phi = \frac{400}{2} - 40 + 13 \times 10 = 290 \text{ mm}$$

$L_d = 58.3\,\Phi = 58.3 \times 10 = 583$ mm.

Now $\frac{M_1}{V} + L_0 = \frac{16.88 \times 10^6}{13500} + 290 = 1250 + 290 = 1540$ mm

$\therefore \quad \frac{M_1}{V} + L_0 > L_d.$

Hence the anchorage requirements are satisfied.

5. *Distribution reinforcement*

$$A_{sd} = \frac{0.15}{100} \times \text{breadth} \times \text{average depth} = \frac{0.15}{100} \times 10000\,D$$

$$= 1.5D = 1.5 \times 175 = 262.5 \text{ mm}^2$$

Using 6 mm Φ bars, $A_\Phi = 28.2 \text{ mm}^2$

$$\therefore \quad s = 1000\frac{A_\Phi}{A_{sd}} = 1000 \times \frac{28.2}{262.5} = 107 \text{ mm.}$$

However, provide 6 mm Φ @ 100 mm c/c.

(*B*) Design of L-beam

Effective span $= 5 + 0.4 = 5.4$ m

Assume the total depth of beam = 600 mm

Breadth of the beam = 400 mm

1. *Loading and B.M.*

Dead load of beam/m $= 0.6 \times 0.4 \times 24000 = 5760$ N

Load from slab $= 2.5 \times 1 \times 5400 = 13500$ N

$\therefore$ Total $w = 19260$ N/m length.

Assuming partial fixidity at the ends,

$$M = \frac{wL^2}{10} = \frac{19260(5.4)^2}{10} \times 1000 = 56.16 \times 10^6 \text{ N-mm}$$

2. *Design of beam section*

For an isolated L-beam,

$$b = 400 + 2500 = 2900 \text{ mm}$$

$$b_f = \frac{0.5l_o}{\left(\frac{l_o}{b}\right) + 4} + b_w = \frac{0.5 \times 5400}{\frac{5400}{2900} + 4} + 400$$

$$= 860 \text{ mm.}$$

Thickness of slab at 0.86 m from the face of the beam

$$= 100 \times \frac{250 - 100}{2.5}(2.5 - 0.86) = 198.4 \text{ mm.}$$

$$\therefore \quad \text{Average} \quad D_f = \frac{250 + 198.4}{2} \approx 224 \text{ mm.}$$

For a balanced section, the effective depth is given by

$$M = 0.45\, b_f.\, \sigma_{cbc}.\, D_f\left(\frac{2kd - D_f}{k}\right)$$

$$\therefore \quad 56.16 \times 10^6 = 0.45 \times 860 \times 5 \times 224 \left(\frac{2 \times 0.404\, d - 224}{0.404}\right)$$

which gives $d \approx$ 342 mm.

Using 16 mm Φ bars and a cover of 25 mm,

$$D = 342 + 8 + 25 = 375 \text{ mm}$$

3. ***Depth from shear point of view***

The total thickness should be such that the beam is safe in shear due to both bending as well as torsion.

Torsional moment $T = \left\{5400 \times 2.5\left(\frac{2.5}{2} + \frac{0.4}{2}\right)\right\} \times \frac{5.4}{2}$ N–m

$$= 52850 \text{ N–m} = 52.85 \times 10^6 \text{ N– mm}$$

Also, $V = \frac{wl}{2} = 19260 \times \frac{5}{2} = 48150$ N

Hence equivalent shear is given by

$$V_e = V + 1.6\frac{T}{b_w} = 48150 + \frac{1.6 \times 52.85 \times 10^6}{400}$$

$$= 259550 \text{ N}$$

$$\therefore \quad \tau_{ve} = \frac{V_e}{b_w d}$$

From Table 3.3 for M 15 concrete, max. shear stress $(\tau_{c.max})$ = 1.6 N/mm².

Limiting τ_{ve} to $\tau_{c.max}$, we have

$$\tau_v = \tau_{c\,max} = 1.6 = \frac{V_e}{b_w d}$$

$$\therefore \quad d = \frac{V_e}{b_w \tau_{c\,max}} = \frac{259550}{400 \times 1.6} = 406 \text{ mm.}$$

Using 16 mm Φ bars and a cover of 25 mm,

$$D = 406 + 8 + 25 \approx 440 \text{ mm.}$$

However keep D = 480 mm so that available d = 480 – 8 – 25 = 447 mm, which is slightly more than 440 mm so that τ_{ve} is slightly less than $\tau_{c\,max}$.

4. ***Design of steel reinforcement***

Let $a = 0.9d = 0.9 \times 447 = 402$ mm (say)

$$A_{st} = \frac{M}{a\,\sigma_{st}} = \frac{56.16 \times 10^6}{402 \times 140} = 998 \text{ mm}^2.$$

Using 16 mm Φ bar, $A_\Phi = \frac{\pi}{4}(16)^2 = 201 \text{ mm}^2$

$$\therefore \text{ No. of bars } = \frac{998}{201} = 5.$$

Actual A_{st} provided $= 5 \times 201 = 1005 \text{ mm}^2$

Out of 5 bars, bend 2 bars upwards at a distance of $L/7$ = 80 cm from the supports.

5. *Determination of N.A. and actual stresses in T-beam*

Assuming the N.A. to fall in the flange,

$$b_f \frac{n^2}{2} = m A_{st} (d - n)$$

or $$\frac{860}{2} n^2 = 19 \times 1005(447 - n)$$

or $n^2 + 44.41\, n - 19850 = 0$

From which $n = 120$ mm

$$\therefore \quad a = d - \frac{n}{3} = 447 - \frac{120}{3} = 407 \text{ mm}$$

$$\therefore \quad t = \frac{M}{A_{st} \times a} = \frac{56.16 \times 10^6}{1005 \times 407} = 137.3 \text{ N/mm}^2 \text{ (safe)}$$

$$c = \frac{t}{m} \cdot \frac{n}{d - n} = \frac{137.3}{19} \times \frac{120}{447 - 120} = 2.65 \text{ N/mm}^2 \text{(Safe)}$$

6. *Design of shear reinforcement*

As found earlier, equivalent shear $V_e = 259550$

$$\therefore \quad \tau_v = \frac{V_e}{b_w . d} = \frac{259550}{400 \times 447} = 1.45 \text{ N/mm}^2$$

which is less than $\tau_{c.max} = 1.6 \text{ N/mm}^2$. However, shear reinforcement is required.

(a) Longitudinal reinforcement

Equivalent B.M. is given by

$$M_{e1} = M + M_T$$

where $M = 56.16 \times 10^6$ N–mm.

$$M_T = T \frac{(1 + D/b_w)}{1.7} = 52.85 \times 10^6 \left[\frac{1 + 480/400}{1.7}\right]$$

$$= 68.5 \times 10^6 \text{ N–mm}$$

$$\therefore \quad M_{e1} = (56.16 + 68.5)10^6 = 124.66 \times 10^6 \text{ N–mm}$$

$$\therefore \quad A_{st} = \frac{M_{e1}}{\sigma_{st} . jd} = \frac{124.66 \times 10^6}{140 \times 0.865 \times 447} = 2303 \text{ mm}^2$$

Using 16 mm Φ bars, $A_\Phi = \frac{\pi}{4}(16)^2 = 201$

$\therefore$ No. of bars $= \frac{2303}{201} \approx 12$. This number is large.

However, using 20 mm Φ bars, $A_\Phi = \frac{\pi}{4}(20)^2 = 314 \text{ mm}^2$

$\therefore$ No. of bars $= \frac{2303}{314} = 8.$

Hence provide 8 Nos. 20 mm Φ bars. This reinforcement is *inclusive* of main reinforcement found in step 4.

Since numerical value of M_T is more than M, the code specifies that longitudinal reinforcement shall be provided on the flexural compression face such that the beam can also withstand an equivalent moment given by

$$M_{e2} = M_T - M = (68.5 - 56.16)10^6 = 12.34 \times 10^6 \text{ N-mm}$$

$$A_{st} = \frac{M_{e2}}{\sigma_{st}(d - d_c)} = \frac{12.34 \times 10^6}{140(447 - 35)} = 214 \text{ mm}^2$$

This is quite small. However, provide 2 bars of 20 mm Φ at the top, which will also work as holding bars for shear stirrups.

(b) Transverse reinforcement

Transverse reinforcement will be provided in the form of vertical stirrups. Let the top 20 mm corner bars be provided at a clear cover of 25 mm.

b_1 = centre to centre distance between corner bars in the direction of width $= 400 - 2 \times 25 - 10 = 340$ mm

d_1 = centre to centre distance between corner bars in the direction of depth $= 400 - 2 \times 25 - 10 = 340$ mm.

Using 12 mm Φ stirrups, $A_{sv} = 2\frac{\pi}{4}(12)^2 = 226 \text{ mm}^2$

Now, $A_{sv} = \frac{T s_v}{b_1 d_1 \sigma_{sv}} + \frac{V s_v}{2.5 d_1 \sigma_{sv}}$

$$\therefore \quad 226 = \left[\frac{52.85 \times 10^6}{340 \times 340 \times 140} + \frac{48150}{2.5 \times 340 \times 140}\right] s_v$$

from which $s_v = 61.6$ mm.

This is too small. However, using two legged stirrups of 16 mm Φ, $A_{sv} = 2 \times 201 = 402 \text{ mm}^2$, and s_v will come out to be 110 mm. Hence provide 16 mm Φ two- legged stirrups @ 110 mm c/c at the supports. However, at the mid-section of the beam, where both V and T are zero, nominal shear stirrups may be provided. Provide 16 mm Φ 2 lgd stirrups @ 250 mm c/c at the middle 2.5 m length of the beam.

The details of the reinforcement etc. are shown in Fig. 7.34.

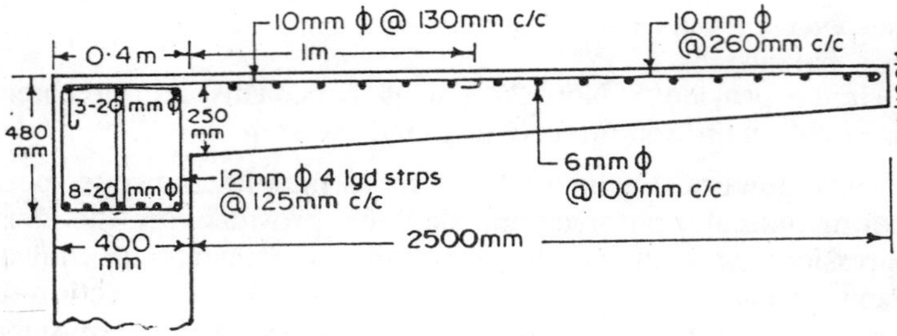

FIG. 7.34.

Design Example 7.17. *Solve example 7.16, if the canopy slab is flush with the bottom of the beam, so as to have a plane soffit.*

Solution. If the slab is attached to the lower end of the beam, it will be in the tension zone, and hence flange action will not be available. Thus, the beam will act as a simple rectangular beam, though the slab will be cast monolithic with it. *The design of the slab will remain unaltered.*

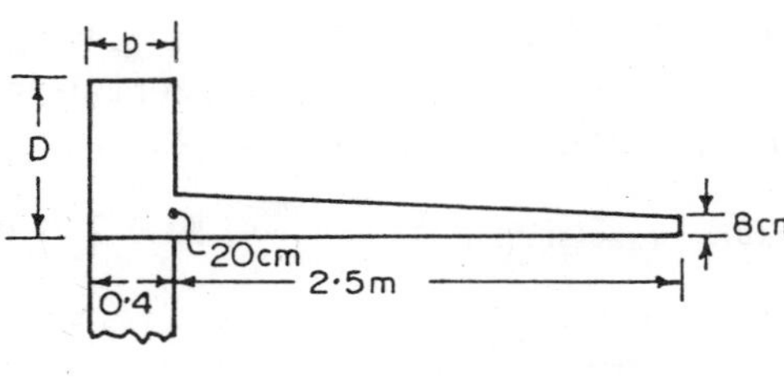

FIG. 7.35.

Design of the beam

1. *Loading and bending moment*

Width of beam = 400 mm

Let the overall depth of beam = 500 mm

Dead load of beam/m $= 0.4 \times 0.5 \times 1 \times 24000 = 4800$ N

Load from slab/m $= 2.5 \times 1 \times 5400 \qquad = 13500$

Total = 18300 N/m

Effective span = 5.4 m.

Assuming partial fixidity at the ends,

$$M = \frac{wL^2}{10} = \frac{18300(5.4)^2}{10} \times 1000 = 53.36 \times 10^6 \text{ N-mm}$$

2. *Design of section*

For balanced section,

$$d = \sqrt{\frac{M}{R\,b}} = \sqrt{\frac{53.36 \times 10^6}{0.874 \times 400}}$$

$$= 390.7 \text{ mm}$$

Let us keep $D = 420$ mm

3. *Check for shear*

Before designing reinforcement, let us check the section for shear because the beam is subjected to heavy torsional moment.

$$T = \left\{5400 \times 2.5\left(\frac{2.5}{2} + \frac{0.4}{2}\right)\right\} \times \frac{5.4}{2} \text{ N-m}$$

$$= 52850 \text{ N-m} = 52.85 \times 10^6 \text{ N-mm}$$

$$V = \frac{wl}{2} = \frac{18300 \times 5}{2} = 45750 \text{ N}$$

$$\therefore \quad V_e = V + 1.6\frac{T}{b_w} = 45750 + \frac{1.6 \times 52.85 \times 10^6}{400}$$

$$= 257150 \text{ N}$$

$$\tau_{ve} = \frac{V_e}{b_w . d}$$

Limiting τ_{ve} to τ_{cmax} (= 1.6 N/mm^2, Table 3.3),

$$d = \frac{V_e}{\tau_{c\,max} b_w} = \frac{257150}{1.6 \times 400}$$

$$= 402 \text{ mm}$$

Keep depth $D = 450$ mm

Using 16 mm Φ bars and providing 25 mm clear cover, available $d = 450 - 8 - 25 = 417$ mm, which is slightly more than 402 mm so that τ_{ve} is slightly less than $\tau_{c\,max}$.

4. *Design of main reinforcement*

$$A_{st} = \frac{M}{\sigma_{st} \times jd} = \frac{53.36 \times 10^6}{140 \times 0.865 \times 417}$$

Using 16 mm Φ bars, having $A_\Phi = \frac{\pi}{4}(16)^2 = 201$ mm^2

$$\therefore \quad \text{No. of bars} = \frac{1057}{201} = 5.3$$

Hence provide 6 bars of 16 mm Φ.

5. *Design for torsion and shear*

$$\tau_{ve} = \frac{V_e}{b_w d} = \frac{257150}{400 \times 417}$$

$= 1.54$ N/mm^2.

Though this is less than $\tau_{c\,max}$, it is more than τ_c and hence shear reinforcement is necessary.

(*a*) ***Longitudinal reinforcement***

$$M_{e1} = M + M_T$$

where $M = 53.36 \times 10^6$ N–mm.

$$M_T = T\frac{(1 + D/b_w)}{1.7} = 52.85 \times 10^6 \frac{1 + \left(\frac{450}{400}\right)}{1.7}$$

$$= 66.06 \times 10^6 \text{ N-mm.}$$

$$\therefore \quad M_{e1} = (53.36 + 66.06)10^6 = 119.42 \times 10^6 \text{ N–mm}$$

$$\therefore \quad A_{st} = \frac{119.4 \times 10^6}{140 \times 0.865 \times 417} = 2364 \text{ mm}^2.$$

If we use 16 mm Φ bars, No. of bars will be very large. Hence use 20 mm Φ bars, having

$$A_\Phi = \frac{\pi}{4}(20)^2 = 314 \text{ mm}^2.$$

$$\text{No. of bars} = \frac{2364}{314} \approx 8.$$

Hence provide 8 Nos. of 20 mm Φ bars. *This reinforcement is inclusive of main reinforcement found in step 4* (In fact, step 4 may be avoided altogether).

Out of these 8 bars, bend 3 bars up near supports, at a distance of $L/7 \approx 800$ mm from the support, to take negative B.M.

Since the numerical value of M_T is more than M, the code specifies that longitudinal reinforcement shall be provided at the flexural compression face such that the beam can also withstand an equivalent moment given by

$$M_{e2} = M_T - M = (66.06 - 53.36)10^6 = 12.7 \times 10^6 \text{ N-mm}$$

$$\therefore \quad A_{sc} = \frac{M_{e2}}{\sigma_{st}(d - d_c)} = \frac{12.7 \times 10^6}{140\,(417 - 35)} = 238 \text{ mm}^2.$$

This is quite small. However, provide 2 bars of 20 mm Φ at the top, which will also work as holding bars for shear stirrups.

(*b*) *Transverse reinforcement*

This will be provided in the form of vertical stirrups

Now b_1 = centre to centre distance between corner bars in the direction of width

$= 400 - 2 \times 25 - 10 = 340$ mm

d_1 = centre to centre distance between corner bars in the direction of depth
$= 450 - 2 \times 25 - 10 = 390$ mm.

Using 12 mm stirrups, $A_{sv} = 2\frac{\pi}{4}(12)^2 = 226$ mm^2

$$\text{Now } A_{sv} = \left[\frac{T}{b_1 d_1 \sigma_{sv}} + \frac{V}{2.5\, d_1 \sigma_{sv}}\right] s_v$$

$$\text{or} \qquad 226 = \left[\frac{52.85 \times 10^6}{340 \times 390 \times 140} + \frac{45750}{2.5 \times 390 \times 140}\right] s_v$$

from which $s_v = 71$ mm.

This is too small. Hence use 16 mm two legged stirrups, having, $A_{sv} = 2 \times 201 = 402$ mm^2. This will give $s_v = 126$ mm.

Hence use 16 mm Φ two legged stirrups @ 125 mm c/c near the supports. At the middle of beam, only nominal stirrups are required. However, increase the spacing of the stirrups to 250 mm c/c at the mid span of the beam.

The details of reinforcement etc. are shown in Fig. 7.36.

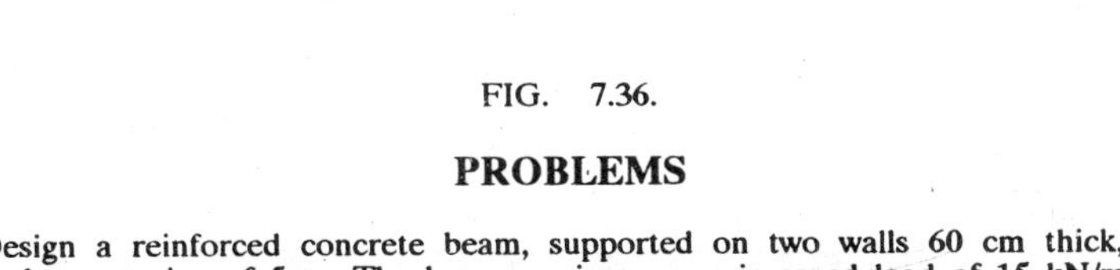

FIG. 7.36.

PROBLEMS

1. Design a reinforced concrete beam, supported on two walls 60 cm thick, over a clear opening of 5 m. The beam carries a superimposed load of 15 kN/m. Use M 15 mix and mild steel reinforcement.

2. Design a reinforced concrete lintel over the openings of two windows, each 2m wide separated by a wall of 50 cm. The height of wall over the lintel is 3 m, and thickness of wall is 40 cm. The lintel is to be provided with a sunshade projecting out by 80 cm and cast monolithically with the lintel. Use M 15 mix and mild steel reinforcement. The live load on the sunshade may be taken as 1500 N/m^2.

3. Design a simply supported reinforced concrete beam, carrying a superimposed load of 20 kN per metre run over a clear opening of 6 m. In addition to this the beam carries two concentrated loads of 30 kN each at a distance of 1.5 m from the face for each support. The width of supporting wall is 60 cm. Use M 15 concrete and Fe 415 steel.

4. Design a R.C. cantilever projecting out 3 m beyond the fixed end and carrying a superimposed load of 15 kN. The cantilever also carries a concentrated load of 5 kN at the free end. Use M 15 mix. Take $\sigma_{st} = 230$ N/mm^2.

5. Design a R.C.C. floor slab for a room having inside dimensions 4 m × 10 m and supported on all sides by a 40 cm thick brick wall. The superimposed load may be taken as 3 kN/m^2 . Use M 15 mix.

6. Fig. 7.37 shows the cross-section of a roof slab, the supports being 4 m part. Design the slab for a live load of 2000 N/m^2. The slab is to be finished with 30 mm thick granolithic flooring. Use M 20 concrete and HYSD bars.

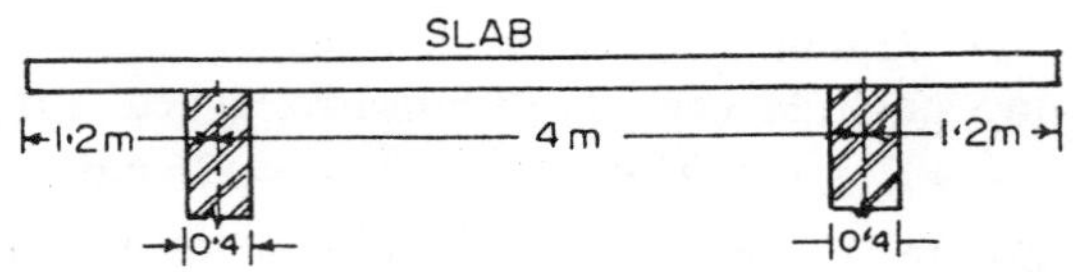

FIG. 7.37.

7. Design the roof for a hall, 6 m wide and 11 m long. The slab is supported on R.C.C. beams, 30 cm wide and arranged as shown in Fig. 7.38. The live load on the roof may be taken at 3500 N/m^2 and floor finishing load as 700 N/m^2. Use M 15 mix. and HYSD bars.

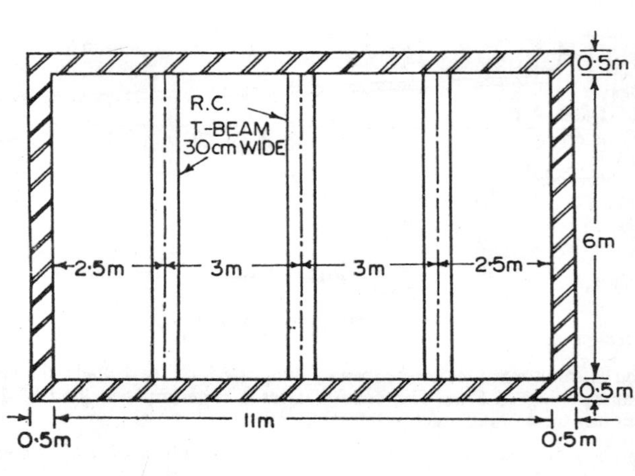

FIG. 7.38.

8. Design a doubly reinforced concrete beam, supporting a uniformly distributed superimposed load of 40 kN/m over a clear span of 6 m. The beam carries two concentrated loads of 20 kN each at a distance of 2 m from the face of the support. The overall depth and width of the beam are restricted to 90 cm and 40 cm respectively. Use M 15 mix. and Fe 250 steel.

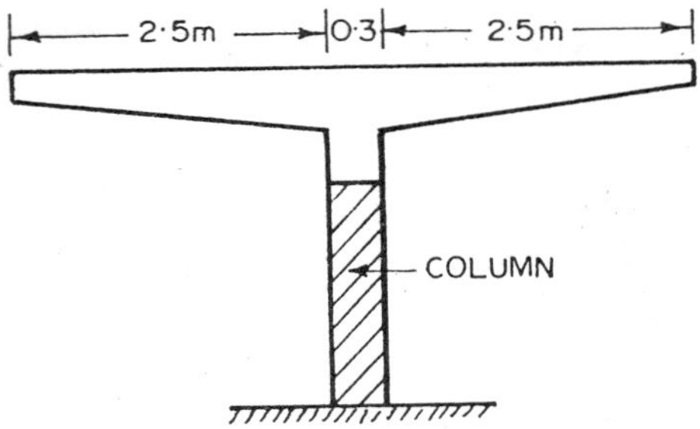

FIG. 7.39.

9. Design completely a R.C.C., roof of a railway platform, consisting of a R.C. slab which cantilevers 2.5 metres on each side of a central R.C. beam (Fig. 7.39). The slab is monolithic with the beam. The beam is supported on columns 5 metres apart, and 30 cm wide. The superimposed load may be taken as 2000 N/m^2. Use suitable values of permissible stresses.

10. The beam of a T-beam roof, over a room 6 m × 18 m projects out by 2 m on side of the room. The roof slab which is 12 cm thick extends to the full length of the beams and its monolithic with them. The beams are spaced 3 metres centre to centre and the web which projects below the slab is 30 cm wide. Design the T-beams completely, if the superimposed load on roof is 4000 N/m^2. The walls of the room are 40 cm thick. Use M 16 mix. and mild steel reinforcement.

11. Design a R.C.C., T-beam and a R.C.C. balcony for a lecture theatre having the following data:
    (*i*) Size of the lecture theatre....15 m × 9 m × 6 m high
    (*ii*) Size of the balcony.....1.5 m clear with 80 cm high and 1.5 cm thick R.C.C. railing.
    (*iii*) Height of balcony.....3.5 m from floor.
    (*iv*) Thickness of brick wall.....40 cm.
    (*v*) Live load for roof.....1500 N/m^2.
    (*vi*) Live load for balcony ....4000 N/m^2.
    Use M 20 mix. and Fe 415 reinforcement.

12. Design as inverted T-beam roof for data of problem 7.

13. A R.C. cantilevered canopy 3.5 metre effective overhang and 5.6 metre wide is supported on two beams cantilevered out from wall and spaced at 3.2 metres centre to centre so that the slab overhangs each beam by 1.2 metre from the centre of each beam. The ribs of the beams are integrated with the slab so as to have flat soffit. Design the canopy completely, for a superimposed load of 1500 N/m^2. Use M 15 mix. and Fe 415 steel.

14. Design the porch slab and the beam shown in Fig. 7.40 for a live load of 1500 N/m^2, inclusive of the floor finishes, but exclusive of the self-weight of the slab. Use M 15 mix. and Fe 415 steel.

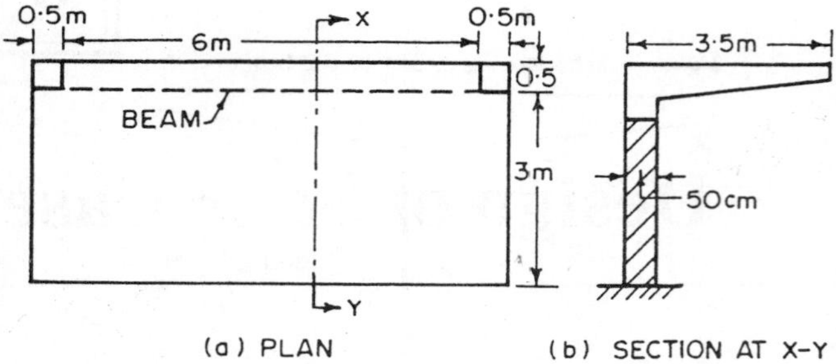

FIG. 7.40

15. A R.C.C. bent for a city bus stand is shown in Fig. 7.41. The total length of the bus stand is 11 m. The bents are spaced at 3 m centres so that the slab overhangs by 1 m on either side beyond the centres of bent ribs. Design the slab as well as the bents, for a live load of 2000 N/m^2 inclusive of the finishes on the slab. The ribs are to be provided above the roof. Use M 15 mix. and mild steel bars.

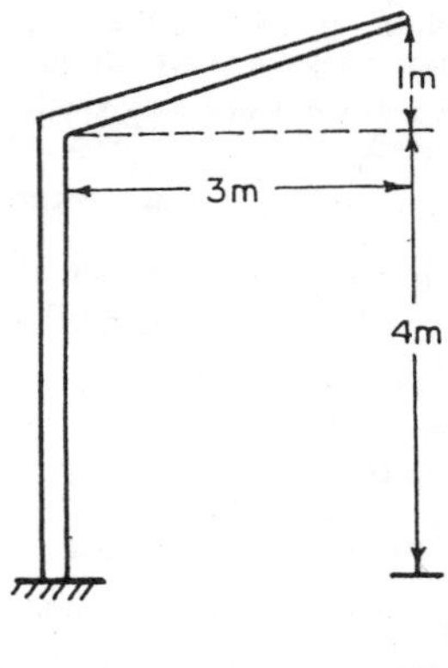

FIG. 7.41.

8

Design of Stair Cases

8.1 INTRODUCTION

Staircases are used in almost all buildings. A staircase consists of a number of steps arranged in a series, with landings at appropriate locations, for the purposes of giving access to different floors of a building. The width of a staircase may depend on the purpose for which it is provided, and may generally vary between 1 m for residential buildings to 2 m for public buildings. A *flight* is the length of the staircase situated between two landings. The number of steps in flight may vary between 3 to 12. The *rise* of a step and the *tread* should be so proportioned that it gives comfortable access. Generally, the sum of tread plus twice the rise is kept about 500 mm and the product of tread and rise is kept about 40000 to 42000. In residential buildings, the rise may vary between 150 mm to 180 mm and tread between 200 to 250 mm. In public buildings, rise is kept between 120 to 150 mm and tread between 200 to 300 mm.

Classification of stairs

Staircases may be broadly classified in the following heads (Fig. 8.1)

(1) Straight stair.

(2) Quarter turn stair.

(3) Half turn stair (open newel type or open well stair).

(4) Dog-legged stair.

(5) Opn newel stair with quarter space landing.

(6) Geometrical stairs such a circular stair, spiral stair etc .

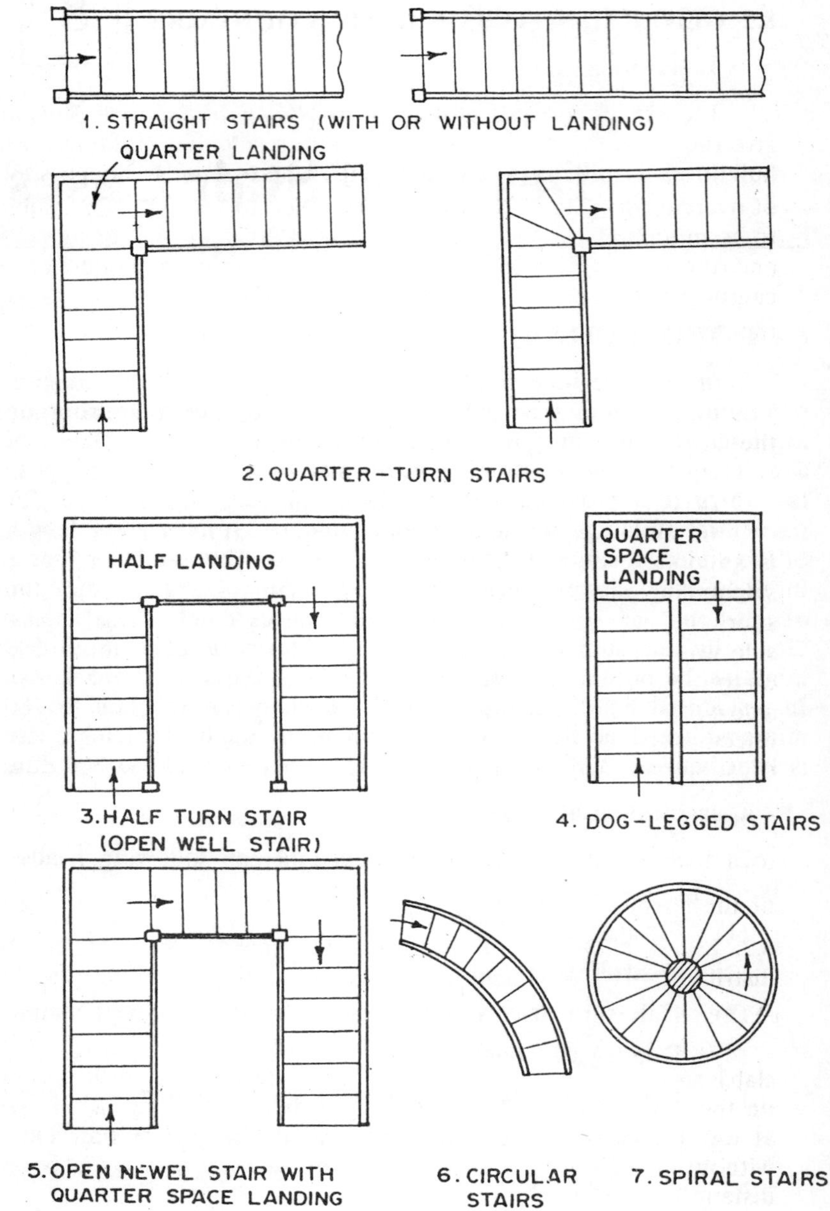

FIG. 8.1 DIFFERENT TYPES OF STAIRS.

8.2 GENERAL NOTES ON DESIGN OF STAIRS

1. Live load on stairs

I.S. 875-1964 (*Code of practice for Structural Safety of Buildings*) give the loads for staircases. For stairs in residential buildings, office buildings, hospital wards, hostels, etc., where there is no possibility of overcrowding, the live load may be taken to be 3000 N/m^2, subject to a minimum of 1300 N concentrated load at the unsupported end of each step for stairs constructed out of structurally independent cantilever step. For other public buildings liable to be overcrowded, the live load may be taken to be 5000 N/m^2.

2. Effective span of stairs

Stair slab may be divided into two categories, depending upon the direction in which the stair slab spans :

(*i*) Stair slab spanning horizontally

(*ii*) Stair slab spanning longitudinally.

(*i*) ***Stair slab spanning horizontally.*** In this category, the slab is supported on each side by side wall or stringer beam on one side and beam on the other side. Sometimes, as in the case of straight stair, the slab may also be supported on both the sides by the two side walls. The slab may also be supported horizontally by side wall on one side of each flight and the common newel on the other side between the backward and forward flights. In such a case the effective span L is the horizontal distance between centre-to-centre of supports. Each step is designed as spanning horizontally with a bending moment equal to $\frac{w L^2}{8}$. Each step is considered equivalent to a rectangular beam of width b (measured parallel to the slope of the stair) and an effective depth equal to $\frac{D}{2}$ as shown in Fig 8.2. Main reinforcement is provided in the direction of L, while distribution reinforcement is provided parallel to the flight direction. A waist of about 8 cm is provided.

(*ii*) ***Stair slab spanning longitudinally.*** In this category, the slab is supported at bottom and top of the flight and remain unsupported on the sides. Each flight of stairs is continuous, supported on beams at top and bottom or on landings. The effective span of such stairs, without stringer beams, should be taken as the following horizontal distances :

(*a*) where unsupported at top and bottom risers by beams spanning parallel with the risers, the distance centre-to-centre of beam ;

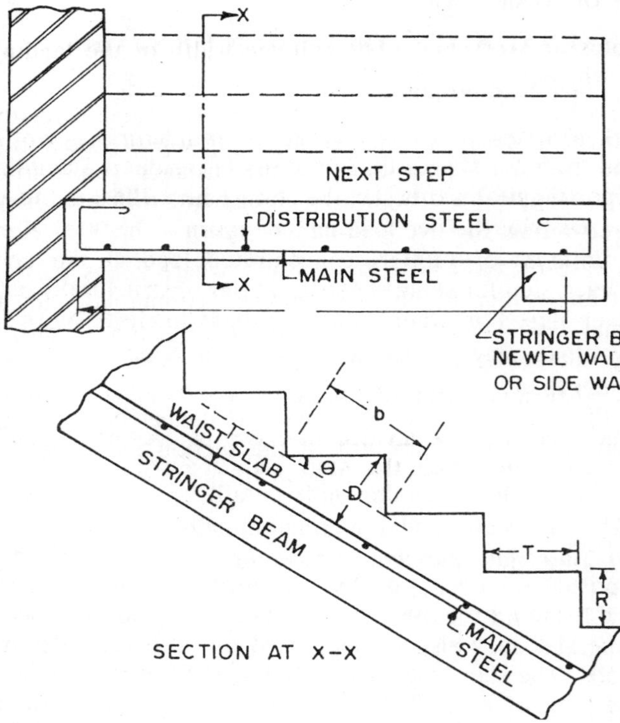

FIG. 8.2. STAIR SLAB SPANNING HORIZONTALLY.

(*b*) where spanning on the edge of a landing slab which spans parallel to the risers (Fig.8.3), a distance equal to the 'going' of the

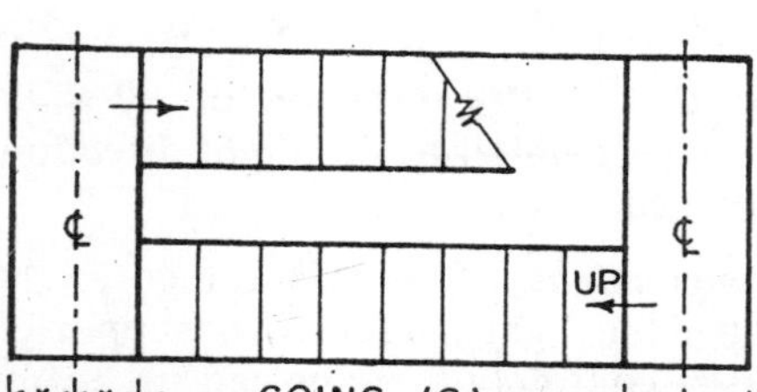

x (m)	y (m)	SPAN L (m)
<1	<1	G + x + y
<1	≥1	G + x + 1
≥1	<1	G + y + 1
≥1	≥1	G + 1 + 1

8.3 EFFECTIVE SPAN FOR STAIRS SUPPORTED AT EACH END BY LANDINGS SPANNING PARALLEL WITH THE RISERS.

stairs plus at each end either half the width of the landing or one metre, whichever is smaller ; and

(*c*) where the landing slab spans in the same direction as the stairs, they should be considered as acting together to form a single slab and the span determined at the distance centre-to-centre of the supporting beams or walls, the going being measured horizontally.

3. Distribution of loading on stairs

In case of stairs with open walls, where spans partly crossing at right angles occur, the load on areas common to any two such spans may be taken as one-half in each direction as shown in Fig. 8.4 (*a*). Where flights or landings are built into walls at a distance of not less than 110 mm and are designed to span in the direction of the flight, a 150 mm strip may be deducted from the loaded area and the effective breadth of the section increased by 75 mm for the purposes of design [Fig. 8.4(*b*)].

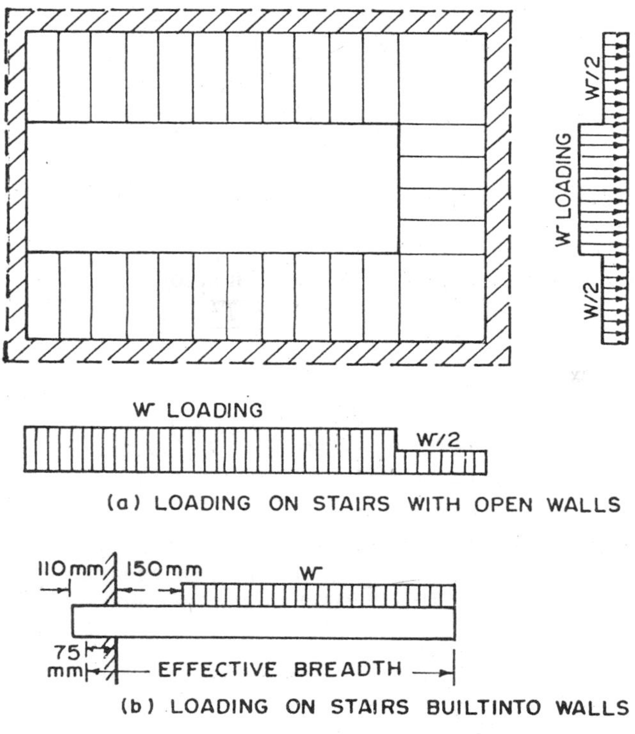

FIG. 8.4. DISTRIBUTION OF LOADING ON STAIRS.

4. Estimation of dead weight

The dead weight of stair consists of (*i*) dead weight of waist slab and (*ii*) dead weight of steps.

(i) Dead weight of waist slab

The dead weight w', per unit area, is first calculated at right angles to the slope. The corresponding load per unit horizontal area is then obtained by increasing w' by the ratio $\sqrt{R^2 + T^2}/T$ where $R =$ rise and $T =$ tread. Thus, if $t =$ thickness of waist in mm, then $w' = \frac{t \times 1 \times 1}{1000} \times 25000 = 25\,t$ N/m^2 of inclined area.

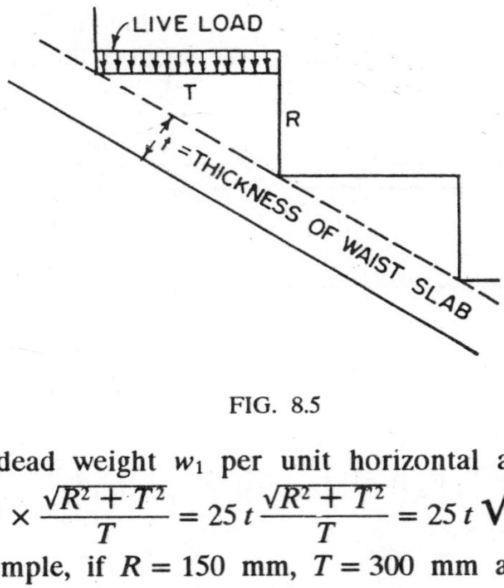

FIG. 8.5

Hence dead weight w_1 per unit horizontal area is given by

$$w_1 = w' \times \frac{\sqrt{R^2 + T^2}}{T} = 25\,t\,\frac{\sqrt{R^2 + T^2}}{T} = 25\,t\sqrt{1 + (R/T)^2}$$

For example, if $R = 150$ mm, $T = 300$ mm and $t = 80$ mm.

Then $w_1 = 25\,t\sqrt{1 + \left(\frac{150}{300}\right)^2} = 27.95\,t$

$= 2236$ N/m^2 of horizontal area.

(ii) Dead weight of steps

The dead weight of the steps is calculated by treating the step to be equivalent *horizontal* slab of thickness equal to half the rise $(R/2)$. Thus, if w_2 is the weight of step per unit horizontal area, we have

$$w_2 = \frac{R}{2 \times 1000} \times 1 \times 1 \times 25000 = 12.5\,R \text{ N/m}^2$$

where R is rise in mm.

Total $w = w_1 + w_2$ per unit horizontal area.

8.3 DESIGN OF STAIRS SPANNING HORIZONTALLY

In this type of stairs, the effective span is the horizontal distance between centre to centre of supports. Each step is designed as spanning horizontally. Each step is considered equivalent to a horizontal beam of width $b = \sqrt{R^2 + T^2}$ measured parallel to the slope of the stair, and an effective depth $D/2$, as shown in Fig. 8.2. Main reinforcement is provided in the direction of span. The distribution reinforcement, provided in the form of 6 mm diameter bars at 30 mm c/c is normally adequate.

Design Example 8.1. *A straight stair in a residential building in supported on wall on one side and stringer beam on the other side. The risers are 150 mm and treads are 250 mm, and the horizontal span of the stairs may be taken as 1.2 metres. Design the steps. Use M 15 concrete and HYSD bars.*

Solution. ***(i) Design constants***

For M 15 concrete, $\sigma_{cbc} = 5\ \text{N/mm}^2$ and $m = 19$. For HYSD bars, $\sigma_{st} = 230\ \text{N/mm}^2$, Hence we have $k = 0.292$, $j = 0\ 903$ and $R = 0.659$.

(ii) Step dimensions (Fig. 8.2)

Rise $R = 150$ mm ; Tread $T = 250$ mm

$$b = \sqrt{R^2 + T^2} = \sqrt{150^2 + 250^2} = 292 \text{ mm}$$

Let us keep waist thickness = 80 mm

$$D = 80 + \frac{RT}{b} = 80 + \frac{150 \times 250}{292} = 208 \text{ mm}$$

Hence the effective depth of equivalent beam $= \frac{D}{2} = 104$ mm

Width $= b = 292$ mm ; Span $= L = 1.25$ m.

(iii) Loading and B.M.

Each step spans horizontally.

Dead load of each step per metre

$$= \left(\frac{1}{2} \times \frac{150}{1000} \times \frac{250}{1000} \times 25000\right) \approx 469 \text{ N/m}$$

Dead load of waist $= \frac{80 \times 292}{10^6} \times 25000 = 584$ N/m

Loading of finishing = 70 N/m (say)

Total = 1123 N/m

Live load @ 3000 N/m² $= \frac{250}{1000} \times 3000 \times 1 = 750$ N/m

Total = 1873 N/m

$$\therefore\ M = \frac{wL^2}{8} = \frac{1873\,(1.25)^2}{8} = 362\ \text{N}-\text{m} = 36.2 \times 10^4\ \ \text{N - mm}$$

(iv) Design of section

$$d = \sqrt{\frac{M}{Rb}} = \sqrt{\frac{36.2 \times 10^4}{0.659 \times 292}} = 43.4\ \text{mm}$$

But available $d = 104$ mm

$$\therefore\quad A_{st} = \frac{M}{\sigma_{st}\, jd} = \frac{36.2 \times 10^4}{230 \times 0.903 \times 104} = 16.8\ \text{mm}^2$$

However, provide minimum steel of one bar of 6 mm diameter per step giving $A_{st} = 50\ \text{mm}^2$.

Provide distribution reinforcement in the form of 6 mm Φ bars @ 300 mm c/c. The reinforcement is arranged as shown in Fig. 8.2.

8.4 DESIGN OF DOG-LEGGED STAIR

Design Example 8.2 *Design a dog-legged stair for a building in which the vertical distance between floors is 3.6 m. The stair hall measures 2.5 m × 5 m. The live load may be taken as 2500 N/m². Use M 15 concrete, and HYSD bars.*

Solution.

(i) General arrangement

Fig. 8.6 shows the plan of stair hall. Let the rise be 150 mm and tread be 250 mm. Let us keep width of each flight =1.2 m.

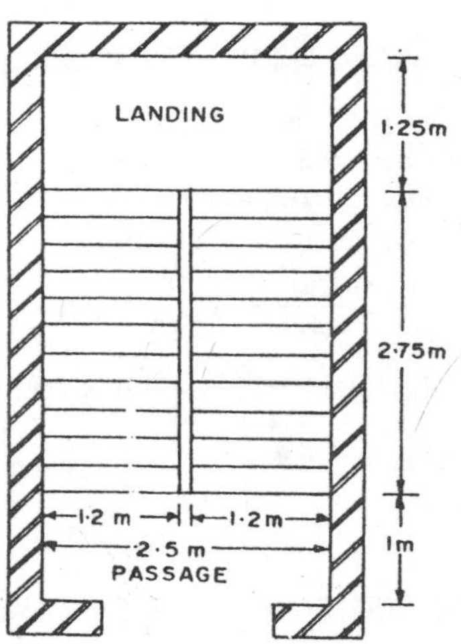

FIG. 8.6.

Height of each flight $= \frac{3.6}{2} = 1.8$ m

No. of risers required $= \frac{1.8}{0.15} = 12$ in each flight

No. of tread in each flight $= 12 - 1 = 11$.

Space occupied by treads $= 11 \times 25 = 275$ cm.

Keep width of landing equal to 1.25 m. Hence space left for passage $= 5.0 - 1.25 - 2.75 = 1$ m.

(ii) ***Design constants***

For M 15 concrete, and HYSD bars having $\sigma_{st} = 230\ \text{N/mm}^2$ we have $k = 0.292$; $j = 0.903$ and $R = 0.659$

(iii) ***Loading on each flight***

The landing slab is assumed to span in the same direction as the stairs, and is considered as acting together to form a single slab. Let the bearing of the landing slab in the wall be 160 mm.

Then effective span $= 2.75 + 1.25 + \frac{0.160}{2} = 4.08 \approx 4.1$ m

Let the thickness of waist slab be equal to 200 mm (assumed at the rate of 40 to 50 mm per metre span).

$\therefore$ Weight of slab w' on slope $= \frac{200}{1000} \times 1 \times 1 \times 25000$

$= 5000\ \text{N/m}^2$

Dead weight of horizontal area $w_1 = w' \frac{\sqrt{R^2 + T^2}}{T}$

$$= 5000 \times \frac{\sqrt{(150)^2 + (250)^2}}{250} = 5830\ \text{N/m}^2$$

Dead weight of steps is given by

$$w_2 = \frac{R}{2 \times 1000} \times 1 \times 1 \times 25000$$

$$= \frac{150}{2000} \times 25000 = 1875\ \text{N/m}$$

$\therefore$ Total dead weight per metre run

$= 5830 + 1875 = 7705$ N

Weight of finishing etc. $= 100$ N (assumed)

Live load $= 2500$ N

Total $w = 10305$ N/m

Note. The w on the landing portion will be $10305 - 1875 = 8505$ N, since weight of steps will not come on it. However, a uniform value of w has been adopted here.

(iii) Design of waist slab

$$M = \frac{wL^2}{8} = \frac{10305\,(4.1)^2}{8} = 21653 \text{ N - m}$$

$$= 21.65 \times 10^6 \text{ N - mm}$$

$$\therefore \quad d = \sqrt{\frac{M}{Rb}} = \sqrt{\frac{21.65 \times 10^6}{0.659 \times 1000}} = 181.3 \text{ mm}$$

Adopt 210 mm overall depth. Using 20 mm clear cover, and 12 mm Φ bars, efffective depth =210 – 20 – 6 = 184 mm.

(iv) Reinforcement

$$A_{st} = \frac{M}{\sigma_{st}\, jd} = \frac{21.65 \times 10^{6}}{230 \times 0.903 \times 184} = 566.5 \text{ mm}^2$$

Using 12 mm Φ bars $A_{\Phi} = \frac{\pi}{4}(12)^2 = 114.0 \text{ mm}^2$

No. of bars needed in 1.2 m width$= \frac{1.20 \times 566.5}{114} \approx 6$ (say)

$\therefore$ Spacing of bars $= \frac{1200}{6} = 200$ mm.

Hence use 12 mm Φ bars @ 200 mm c/c.

Distribution reinforcement @ 0.12% of area

$$A_{sd} = \frac{0.12 \times 210 \times 1000}{100} = 252 \text{ mm}^2$$

Using 8 mm Φ bars, $A_{\Phi} = \frac{\pi}{4}(8)^2 = 50.3 \text{ mm}^2$

$$\therefore \quad s = \frac{1000 \times 50.3}{252} \approx 200 \text{ mm}$$

Hence provide 8 mm Φ bars @ 200 mm c/c.

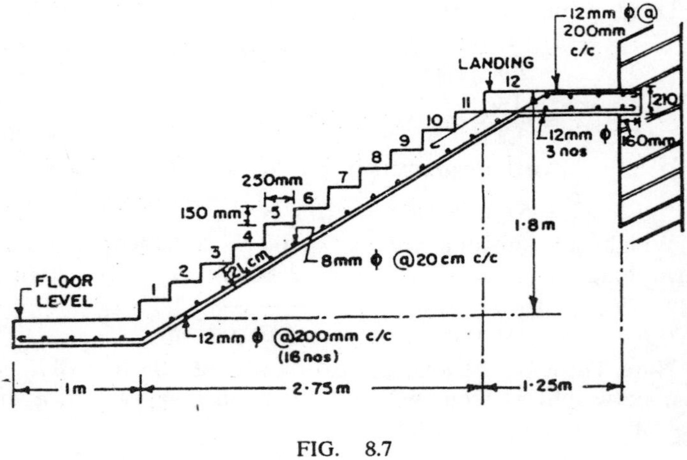

FIG. 8.7

The main reinforcement should be bent to follow the bottom profile of the stair. However, if this pattern is followed near the landing, an angle will be formed in the bar. When the bar is stressed, it will try to throw off the concrete cover. Hence near the landing, the bars are taken straight up and then bent in the compression zone of landing. In order to take tensile stresses in the loading portion, it is desirable to use separate set of bars as shown in Fig. 8.7. However, since the bending moment is very much reduced near the landing only half the number of bars may be provided, *i.e.* provide 6 nos. of 12 mm Φ bars. The length of each type of bar on either side of crossing should be at least $58\,\Phi = 58 \times 12 = 700$ mm.

The details of reinforcement etc. are shown in Fig. 8.7.

8.5 DESIGN OF STAIRS WITH QUARTER SPACE LANDING

Design Example 8.3. *Fig. 8.78 shows the general arrangement of a stair case for an office building. The tread is 300 mm and rise is 150 mm. The stairs is built in the side wall along the flights for a distance of 120 mm. Design the stair case for a live load of 3000 kN/ m*^{2} *taking the span in the direction of the flight. Use M 15 concrete and take* $\sigma_{st} = 140\ N/mm^2$.

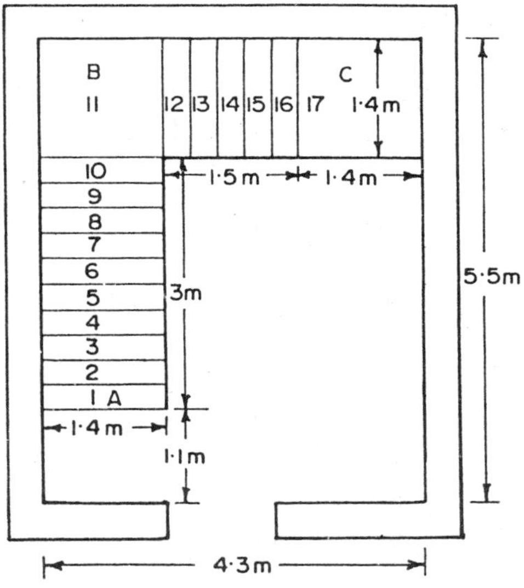

FIG. 8.8.

Solution.

(i) Design constants :

For the given set of stresses, we have

$$k = 0.404\,;\, j = 0.865 \text{ and } R = 0.874$$

(ii) Effective span

Assume 200 mm bearing of the landing in the walls.

Effective span of flight $AB = 3 + 1.4 + 0.10 = 4.5$ m.

Effective span of flight $BC = 0.1 + 1.4 + 1.5 + 1.4 + 0.1 = 4.5$ m

Thus, effective span of both the flights is equal. Hence any one flight (say flight BC) may be designed and the same reinforcement may be adopted for the other flight.

(iii) Loading on each flight

Let the thickness of the waist slab be 200 mm.

$$\therefore \text{ Weight of } w' \text{ on slope } = \frac{200}{1000} \times 1 \times 1.4 \times 25000 \text{ per metre run}$$

$\therefore$ Weight w_1 per horizontal metre run

$$= \frac{200}{1000} \times 1.4 \times 25000 \frac{\sqrt{(150)^2 + (300)^2}}{300}$$

$$= 7825 \text{ N/m run}$$

$$\text{Weight of each step } = 1.4 \frac{150}{2} \times \frac{300}{10^6} \times 25000 = 788 \text{ N}$$

$\therefore$ Weight w_2 of steps per horizontal metre run

$$= 788 \times \frac{1000}{300} = 2625 \text{ N}$$

Alternatively,

$$w_2 = \frac{R}{2 \times 1000} \times 1 \times 1.4 \times 25000 = \frac{150}{2} \times 1.4 \times 25$$

$$= 2625 \text{ N}$$

$\therefore$ Total dead weight per metre run	$= 7825 + 2625$
	$= 10450$ N
Weight of finishing etc.	$= 150$ N (say)
Total	$= 10600$ N/m

For the computation of live load, consider Fig. 8.9. Since the flight is built into the side wall by a distance 120 mm (> 110 mm), the loaded width $= 1.4 - 0.15 = 1.25$ m.

$\therefore$ Live load/m $= 3000 \times 1.25 = 3750$ N

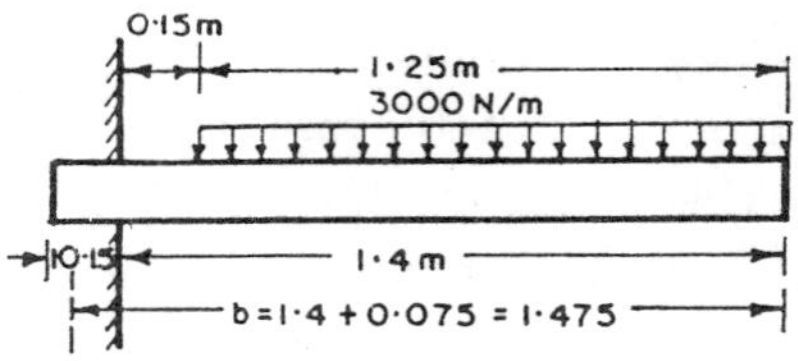

FIG. 8.9.

Effective breadth b = 1.4 + 0.075 = 1.475 m

= 1475 mm

Hence total w per horizontal metre run

= 10600 + 3750 = 14350 N/m

(iv) Design of flight BC

Landing B is common to both the flights. Hence w for landing $B = \frac{1}{2} \times 14350 = 7175$ N/m, while w for landing C will be taken as 14350 N/m. The loading, B.M.D. and S.F.D. are shown in Fig. 8.10.

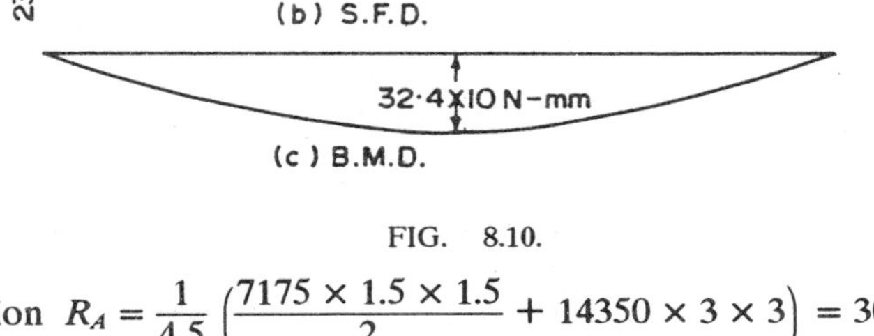

FIG. 8.10.

$$\text{Reaction } R_A = \frac{1}{4.5}\left(\frac{7175 \times 1.5 \times 1.5}{2} + 14350 \times 3 \times 3\right) = 30494 \text{ N}$$

$$R_B = 7175 \times 1.5 + 14350 \times 3 - 30494 = 23318 \text{ N}$$

$$\text{S.F. is zero at a distance } = \frac{30494}{14350} = 2.125 \text{ m from } C.$$

B.M. will be maximum where S.F. is zero.

$$M_{max} = \left[30494 \times 2.125 - \frac{14350\,(2.125)^2}{2}\right] 1000 = 32.4 \times 10^6 \text{ N-mm}$$

Breadth b of slab for design = 1475 mm

$$d = \sqrt{\frac{M}{Rb}} = \sqrt{\frac{32.4 \times 10^6}{0.874 \times 1475}} = 159 \text{ mm}$$

Keep total depth = 180 mm

Using 12 mm Φ bars and clear cover of 15 mm,

$$d = 180 - 15 - 6 = 159 \text{ mm.}$$

$$A_{st} = \frac{32.5 \times 10^6}{140 \times 0.864 \times 159} = 1685 \text{ mm}^2$$

Using 12 mm Φ bars, $A_\Phi = \frac{\pi}{4}(12)^2 = 113.1 \text{ mm}^2$

∴ No. of bars required in 1475 mm width $= \frac{1685}{113.1} = 14.9$

Hence provide 15 bars of 12 mm Φ

∴ Spacing $s = \frac{1475}{15} = 100$ mm

Distribution reinforcement, $A_{st} = 1.5 \times 180 = 270 \text{ mm}^2$

Using 8 mm Φ bars, $A_\Phi = \frac{\pi}{4}(8)^2 = 50.3 \text{ mm}^2$

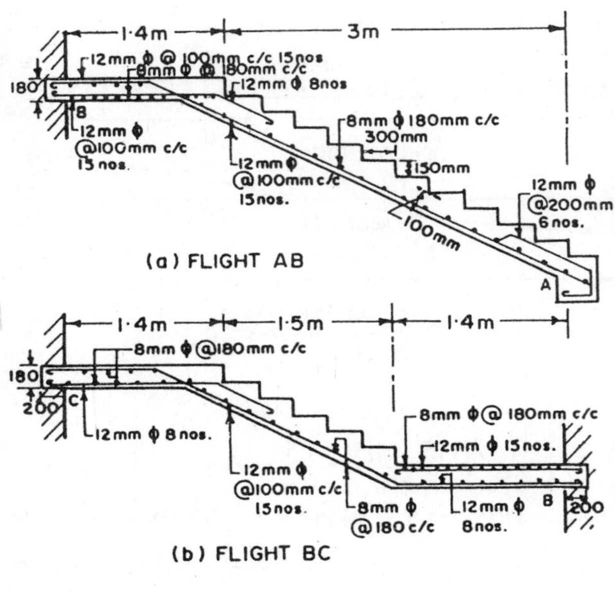

FIG. 8.11.

$\therefore$ Spacing $= \dfrac{1000 \times 50.3}{270} = 186$ mm

Hence provide 8 Φ mm bars @ 180 mm c/c.

The same reinforcement may be provided for both the flights. At the landing, provide reinforcement both at top as well as at bottom. The details of reinforcement are shown in Fig. 8.11.

8.6 DESIGN OF OPEN NEWEL STAIR WITH QUARTER SPACE LANDING

Design Example 8.4. *Fig 8.12 shows the general arrangement of a staircase of a building. The risers are 150 mm and the treads are 250 mm. Design the staircase for a live load of 3000 N/m². The width of stair is 1.5 m and the width of wall is 400 mm. Take* $c = \sigma_{cbc} = 5\ N/mm^2$, $t = \sigma_{st} = 140\ N/mm^2$ *and* $m = 18$. *Adopt unit weight of concrete as 2400 N/m³.*

Solution

(i) Design constants

For the given set of stresses, we have

$$k = 0.39\,; j = 0.87\,; \quad \text{and } R = 0.85$$

(ii) Effective span

Assume a bearing of 200 mm in the walls.

For flight AB and CD, $L = 0.1 + 2.25 + 1.5 + 0.1 = 3.95$ m

For flight BC, $L = 0.1 + 1.5 + 1.25 + 1.5 + 0.1 = 4.45$ m

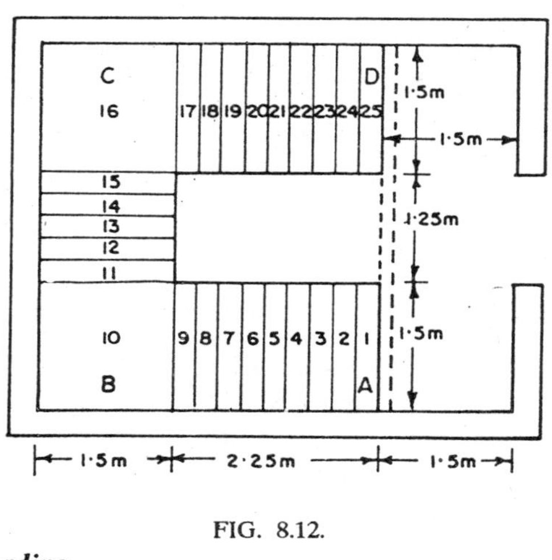

FIG. 8.12.

(iii) Loading

(a) For the flight portion:

Let the waist slab be 200 mm thick. Also, let the landing slab be 200 mm thick. Consider a strip of slab of 1 m wide.

Weight of slab w' on slope $= \dfrac{200 \times 1 \times 1}{1000} \times 24000 = 4800 \text{ N/m}^2$

Weight on horizontal area $= 4800 \dfrac{\sqrt{150^2 + 250^2}}{250} = 5600 \text{ N/m}^2$

Dead weight of steps is given by

$$w_2 = \frac{R}{2 \times 1000} \times 1 \times 1 \times 24000 = \frac{150}{2000} \times 24000 = 1800 \text{ N}$$

Weight of finishing $= 100 \text{ N/m}^2$ (assumed)

Live load $=$ 3000 N/m^2.

Hence total load $= 5600 + 1800 + 100 + 3000 = 10500 \text{ N/m}^2$

(b) For the landing portion :

Dead load $= \dfrac{200}{1000} \times 1 \times 1 \times 24000 = 4800 \text{ N/m}^2$

Weight of finishing $=$ 100 N/m^2

Live load $=$ 3000 N/m^2

Total $=$ $7900 = 8000 \text{ N/m}^2$

However, since each quarter space landing is common to both the flights, only half of the above loading, *i.e.* 4000 N/m will be taken.

(iv) Design of flight AB

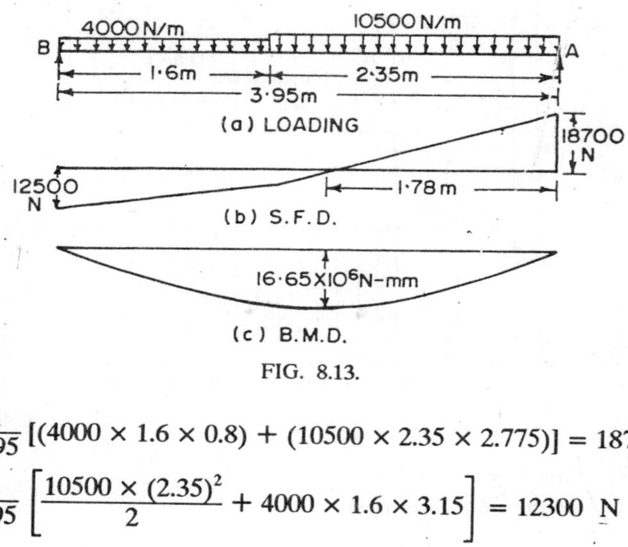

FIG. 8.13.

$$R_A = \frac{1}{3.95}\,[(4000 \times 1.6 \times 0.8) + (10500 \times 2.35 \times 2.775)] = 18700 \text{ N}$$

$$R_B = \frac{1}{3.95}\left[\frac{10500 \times (2.35)^2}{2} + 4000 \times 1.6 \times 3.15\right] = 12300 \text{ N}$$

S.F. is zero at $\frac{18700}{10500} = 1.78$ m from A.

$$M_{max} = \left[(18700 \times 1.78) - \frac{10500\,(1.78)^2}{2}\right] = 16.65 \times 10^6 \text{ N - mm}$$

$$\text{Depth } d = \sqrt{\frac{M}{Rb}} = \sqrt{\frac{16.65 \times 10^6}{0.85 \times 1000}} = 140 \text{ mm}$$

Provide overall depth of 170 mm so that available d=150 mm

$$A_{st} = \frac{M}{\sigma_{st}\, jd} = \frac{16.65 \times 10^6}{140 \times 0.87 \times 150} = 911.3 \text{ mm}^2$$

Using 12 mm Φ bars, $A_\Phi = 113.1$ mm^2

$$\therefore \quad \text{Spacing} = \frac{1000 \times 113.1}{911.3} = 124.1 \text{ mm}$$

$$\text{In width of 1.5 m, No. of bars} = \frac{1500}{124.1} \approx 12$$

$$\therefore \quad \text{Actual } s = \frac{1500}{12} = 125 \text{ mm.}$$

Distribution reinforcement $A_{sd} = 1.5D = 1.5 \times 170 = 255$ mm^2

Provide 8 mm Φ @ 180 mm c/c.

$$\text{Nominal shear stress } \tau_v = \frac{V}{bd} = \frac{18700}{1000 \times 150} = 0.124 \text{ N/mm}^2$$

This is less than the permissible shear stress.

(v) ***Design of flight BC***

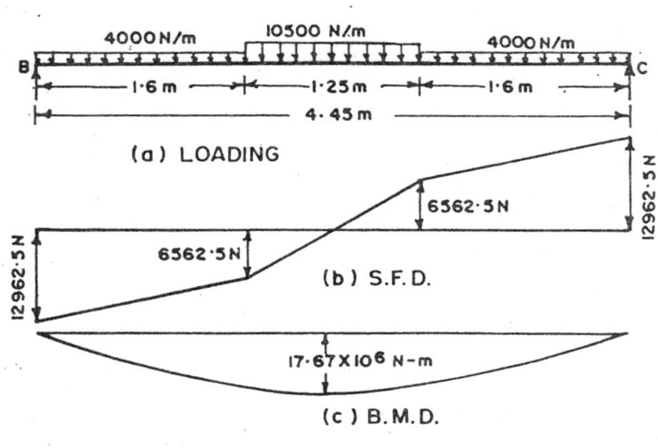

FIG. 8.14.

$$R_B = R_C = \frac{1}{2}[4000 \times 1.6 + 10500 \times 1.25 + 4000 \times 1.6] = 12962.5 \text{ N}$$

$$M_{max} = \left[(12962.5 \times 2.225) - (4000 \times 1.6 \times 1.425) - \frac{10500\,(0.625)^2}{2}\right] \times 1000 = 17.76 \times 10^6 \text{ N-mm}$$

$$d = \sqrt{\frac{M}{Rb}} = \sqrt{\frac{17.67 \times 10^6}{0.85 \times 1000}} = 144 \text{ mm}$$

Keep $D = 170$ mm so that $d = 150$ mm

$$A_{st} = \frac{M}{\sigma_{st}\, jd} = \frac{17.67 \times 10^6}{140 \times 0.87 \times 150} = 967 \text{ mm}^2$$

Using 12 Φ bars having $A_\Phi = 113.1$ mm^2.

$$\text{Spacing} = \frac{1000 \times 113.1}{967} = 117 \text{ mm.}$$

Keep $s = 100$ mm

In width of 1.5 m, No. of bars $= \dfrac{1500}{100} = 15$

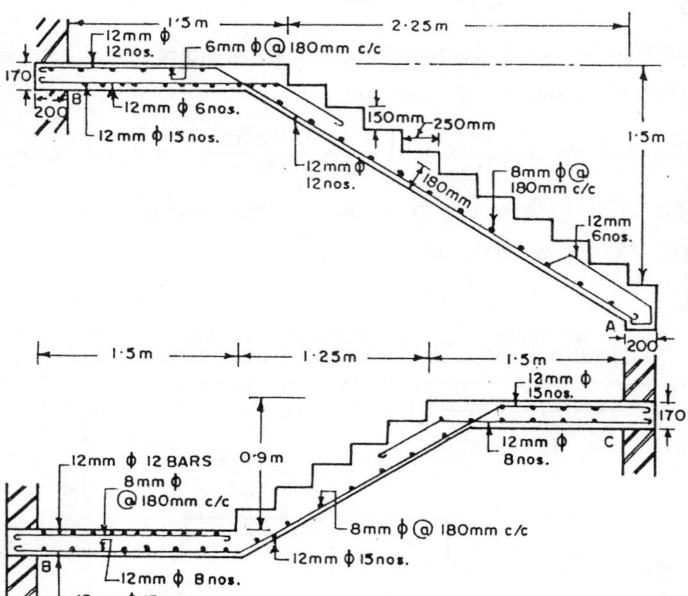

FIG. 8.15.

Distribution reinforcement

Provide 8 mm Φ bars @ 180 mm c/c.

The details of reinforcement for both the flights are shown in Fig. 8.15.

8.7 DESIGN OF STAIRCASE WITH CENTRAL STRINGER BEAM

Design Example 8.5. *Fig. 8.16 shows the general layout of a staircase with central stringer beam supported on columns at B, C and D. The rise and tread of the stairs are 100 mm and 250 mm respectively. The width of the steps is 1.40 m. Design the staircase for a live load of 4000 N/m². Take* $c = \sigma_{cbc} = 5 N/mm^2$, $t = \sigma_{st} =$, $140 N/mm^2$ *and* $m = 18$. *Adopt unit weight of concrete as 24000 N/m³.*

Solution. (*i*) ***Design constants***

For the given set of stress,

$k = 0.39$; $j = 0.87$ and $R = 0.85$

(ii) Loading on waist slab

Let the waist slab be 80 mm thick. The weight of waist slab on the slope should be multiplied by the factor $\frac{\sqrt{R^2 + T^2}}{T}$

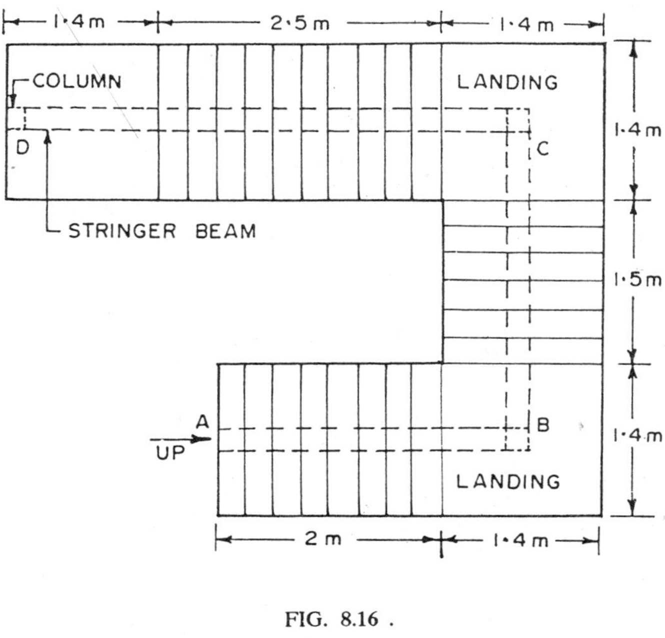

FIG. 8.16 .

$= \frac{\sqrt{(160)^2 + (250)^2}}{250} =$ 1.185 to get the equivalent weight on horizontal plane. Consider 1 m width of slab.

Load per metre horizontal run will be as follows:

(*a*) Self weight$= \left(\frac{80}{1000} \times 1 \times 1 \times 24000\right) 1.185 = 2280$ N

(*b*) Wt. of steps$= \frac{160}{2 \times 1000} \times 1 \times 1 \times 24000 \quad = 1920$ N

(*c*) Finishing etc. $= 100$ N (say)

(*d*) Live load $= 4000$ N

Total $= 8300$ N

The loading on the landing will be lesser ; however, for simplicity, we will take the same loading throughout.

(iii) Design of waist slab

The waist slab is supported on the central stringer beam. Hence the worst condition may be when we consider concentrated live load of 1300 N to act to one side only.

Dead weight $= 2280 + 1920 + 100 = 4300$ N

Let the width of stringer beam be 200 mm

Projection of slab beyond the rib of beam $= \frac{1.4 - 0.2}{2} = 0.6$ m

B.M. due to dead load $= \frac{4300\,(0.6)^2}{2} = 774$ N-m

B.M. due to U.D.. live load $= \frac{4300\,(0.6)^2}{2} = 720$ N-m

B.M. due to concentrated live load $= 1300 \times 0.6 = 780$ N-m

$\therefore$ Max. B.M. $M = 774 + 780 = 1554$ N-m 1.554×10^6 N-mm

$$d = \sqrt{\frac{M}{Rb}} = \sqrt{\frac{1.554 \times 10^6}{0.85 \times 1000}} = 42.8 \text{ mm}$$

However, keep a minimum total depth $=$ 80 mm

Effective depth $= 80 - 20 = 60$ mm

$$A_{st} = \frac{1.554 \times 10^6}{140 \times 0.86 \times 60} = 215 \text{ mm}^2$$

Using 10 mm Φ bars, having $A_\Phi =$ 78.5 mm^2.

$$s = \frac{1000 \times 78.5}{215} = 365 \text{ mm}$$

However, keep spacing $=$ 250 mm, *i.e.* one bar per step.

Distribution reinforcement $= 1.5 \times 80 = 120$ mm^2.

$\therefore$ Provide 6 mm Φ bars @ 180 mm c/c.

(ii) Design of stringer beam

The stringer beam will act as a T-beam. Flight *CD* is the longest, and hence we will design the stringer beam *CD*.

$$\text{Effective span} = \left(1.4 - \frac{0.2}{2}\right) + 2.5 + \frac{1.4}{4} = 4.5 \text{ m}$$

The loading on the stringer beam will be as follows, assuming the web to be 200 mm width and 200 mm deep.

(*a*) Weight of rib/m run $= \left(\dfrac{200 \times 200}{10^6} \times 24000\right) \times 1.185$

$= 1140$ N

(*b*) Load from waist slab $= 8300 \times 1 \times 1.4 = 11620$ N

Total $= 12760$ N/m

Assuming partial fixidity at the ends.

$$M = \frac{w L^2}{10} = \frac{12760\,(4.5)^2}{10} \times 1000 = 25.84 \times 10^6 \text{ N-mm}$$

Taking lever arm $= 0.9d$, balanced depth is given by Eq. 6.17

$$M = 0.45\, b_f\, \sigma_{cbc}\, D_f \left(\frac{2kd - D_f}{k}\right)$$

where $b_f =$ flange width of isolated T-beam given by

$$b_f = \frac{l_0}{\left(\dfrac{l_0}{b}\right) + 4} + b_w$$

where $l_0 \approx L = 4.5$ m ; $b =$ actual width of flange $= 1.4$ m

$b_w = 0.20$ m

$$b_f = \frac{4.5}{\left(\dfrac{4.5}{1.5}\right) + 4} + 0.20 = 0.824 \text{ m} = 824 \text{ mm}.$$

Hence from Eq. 4.17

$$25.84 \times 10^6 = 0.45 \times 824 \times 5 \times 80 \left[\frac{2 \times 0.39\, d - 80}{0.39}\right]$$

or $\quad 0.78\, d = \dfrac{25.84 \times 10^6 \times 0.39}{0.45 \times 825 \times 5 \times 80} + 80$

From which $d = 190$ mm

From shear considerations,

$$d = \frac{V_e}{b_w\, \tau_{c\,max}}$$

where $\tau_{c\,max} = 1.6$ N/mm^2 for M 15 concrete (Table 3.3)

$$V_e = V + 1.6\frac{T}{b_w}$$

$$V = \frac{wL}{2} = \frac{12760 \times 4.5}{2} = 28700 \text{ N}$$

T = torsional moment, which will be induced due to live load acting only to one side of step.

$$\therefore \quad T = \left[\frac{4000\,(0.6)^2}{2}\right] \times \frac{4.5}{2} \times 1000 \text{ N-mm} = 1.62 \times 10^6 \text{ N-mm}$$

or $$T = (1300 \times 0.6)\,\frac{4.5}{2} \times 1000 \text{ N-mm} = 1.755 \times 10^6 \text{ N-mm}$$

whichevere is more.

$$\therefore \quad T = 1.755 \times 10^6 \text{ N-mm.}$$

and $$V_e = 28700 + 1.6\,\frac{1.755 \times 10^6}{200} = 42740 \text{ N}$$

Hence $d = \dfrac{42740}{200 \times 1.6} = 134$ mm

However, keep total depth = 240 mm. Using 20 mm Φ bars and 25 mm clear cover, available $d = 240 - 10 - 25 = 205$ mm.

$$A_{st} = \frac{25.84 \times 10^6}{140 \times 0.9 \times 205} = 1000 \text{ mm}^2$$

For 20 mm Φ bars, $A_\Phi = 314$ mm^2

No. of bars $= \dfrac{1000}{314} = 3.2$

However, provide 4 bars of 20 mm Φ. Actual A_{st} provided = 314 × 4 = 1256 mm^2.

Note. The above reinforcement is for bending requirements only. There will be additional longitudinal reinforcement for torsion, as computed later.

Location of N.A.

Assuming the N.A. falls within the flange, we have

$$\frac{b_f n^2}{2} - m\,A_{st}\,(d - n) = 0$$

$$\therefore \quad \frac{824}{2} n^2 = 18 \times 1256\,(205 - n)$$

From which $n = 82$ mm.

Hence the resultant falls outside the flange.

Thus, $b_f D_f \left(n - \dfrac{D_f}{2}\right) = m\,A_{st}\,(d - n)$

or $$824 \times 80\left(n - \frac{80}{2}\right) = 18 \times 1256\,(205 - n)$$

From which $n = 82$ mm.

$$\bar{y} = \frac{3n - 2D_f}{2n - D_f} \times \frac{D_f}{3}$$

$$= \frac{3 \times 82 - 2 \times 80}{2 \times 82 - 80} \times \frac{80}{3} = 27.3$$

$\therefore$ L.A. $a = d - \bar{y} = 205 - 27.3 = 177.7$ mm

Stress in steel $= \dfrac{M}{A_{st}\, a} = \dfrac{25.84 \times 10^6}{1256 \times 177.7} = 115.8 \text{ N/m}^2$

(safe)

Corresponding stress in concrete is given by

$$c = \frac{t}{m}\frac{n}{d - n} = \frac{115.8}{18} \times \frac{82}{205 - 82} = 4.29 \text{ N/mm}^2$$

(safe)

(v) Design for torsion

A computed earlier, $T = 1.755 \times 10^6$ N–mm

$V = 28700$ N

and $V_e = 42740$ N.

$\therefore$ $$\tau_{ve} = \frac{V_e}{b_w d} = \frac{42740}{200 \times 205} = 1.042 \text{ N/mm}^2$$

$$\frac{100 A_{st}}{bd} = \frac{100 \times 1256}{200 \times 205} = 3\%$$

Hence from Table 3.2, $\tau_c = 0.44$ N/mm^2.

Since $\tau_{ve} > \tau_c$, shear reinforcement is necessary.

(a) Longitudinal reinforcement

$$M_{e1} = M + M_T$$

where $M = 25.84 \times 10^6$ N - mm

$$M_T = T\frac{(1 + D/b_w)}{1.7}$$

$$= 1.755 \times 10^6 \frac{(1 + 240/200)}{1.7} = 2.271 \times 10^6 \text{ N - mm}$$

$\therefore$ $$M_{e1} = 25.84 \times 10^6 + 2.271 \times 10^6 = 28.11 \times 10^6 \text{ N-mm}$$

$\therefore$ $$A_{st} = \frac{M_{e1}}{\sigma_{st}\, jd} = \frac{28.11 \times 10^6}{(140 \times 1177.7)} = 1130 \text{ mm}^2$$

Hence the provision of 4 bars of 20 mm Φ, giving A_{st}= 1130 mm^2 is O.K.

Near the column D, take the bars straight up. Provide 2 - 20 mm Φ bars at the lower face under the landing as shown.

(b) Transverse reinforcement

Transverse reinforcement will be provided in the form of vertical stirrups. Let us provide 25 mm clear cover all round.

b_1 = centre to centre distance between corner bars in the direction of width $= 200 - 2 \times 25 - 10 = 140$ mm

d_1 = centre to centre distance between corner bars in the direction of depth $= 240 - 2 \times 25 - 10 = 180$ mm

Using 10 mm Φ 2 - lgd stirrups, $A_{sv} = 2\frac{\pi}{4}(10)^2 = 157$ mm^2

$$\text{Now, } A_{sv} = \frac{T.s_v}{b_1 d_1 \sigma_{sv}} + \frac{V\, s_v}{2.5 d_1\, \sigma_{sv}}$$

or
$$157 = \left[\frac{1.755 \times 10^6}{140 \times 180 \times 140} + \frac{28700}{2.5 \times 180 \times 140}\right] s_v$$

From which $s_v = 165$ mm.

However, the spacing should not exceed the least of x_1, $\frac{x_1 + y_1}{4}$ and 300 mm, where

x_1 = short dimension of stirrup $= 140 + 120 + 10 = 170$

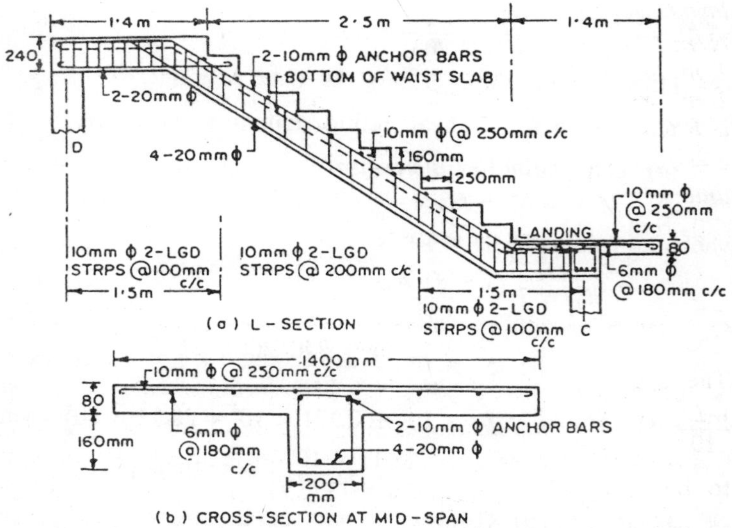

FIG. 8.17

$$y_1 = 180 + 20 + 10 = 210$$

$$\frac{x_1 + y_1}{4} = \frac{170 + 210}{4} = 95$$

Hence $s_v =$ 165 mm is not permissible.

Keep $s_v =$ 95 mm ≈ 100 mm c/c.

Increase the spacing to 200 mm c/c in the mid span where both transverse shear as well as torsional shear are minimum. Provide 2-10 mm Φ holding bars. Keep the same section for other flights. The details of reinforcement etc. are shown in Fig. 8.17.

8.8. DESIGN OF CINEMA BALCONY

A cinema balcony is a huge structure designed to carry the superimposed load of furniture and audience. The furniture is arranged on a number of successive steps so that view of the audience is not obstructed. These steps along with the waist slab are supported on *rackers* which on turn are supported on wall on one side and the *fulcrum girder* on the other side. The size of the huge fulcrum girder is sometimes restricted from the point of view of obstruction in the line of sight and other head-room requirements. The treads are normally kept between 900 to 1100 mm and the risers between 100 to 125 mm. The superimposed load may vary between 4000 to 5000 N/m^2.

Design Example 8.6. *Design a cinema balcony with the general layout shown in Fig. 8.18. The live load may be taken equal to 4500 N/m^2, inclusive of furniture etc. The horizontal tread of each step is 1 m and rise is 120 mm. The depth of fulcrum girder is limited to 1 m. At the end of the last step, there is a gangway 1 m wide and a R.C. balustrade 1 m high. Assume any other data not given. Take $c = \sigma_{cbc} = 5$ N/mm^2, $t = \sigma_{st} = 140$ N/mm^2 and $m = 18$ and adopt unit weight of concrete as 24000 N/mm^3.*

Solution.

For the given set of stresses, we have

$k = 0.3$; $j = 0.87$ and $R = 0.85$

(a) Design of waist slab or slab decking

The decking will be a continuous slab supported on the rackers. The slab may be designed for a maximum bending moment of $\frac{wL^2}{10}$. Since the deck slab is inclined, its weight on inclination has to be multiplied by a factor $\frac{\sqrt{R^2 + T^2}}{T} = \frac{\sqrt{(0.12)^2 + (1)^2}}{1} = 1.01$ to get the weight per unit horizontal area.

Let the thickness of slab be 100 mm

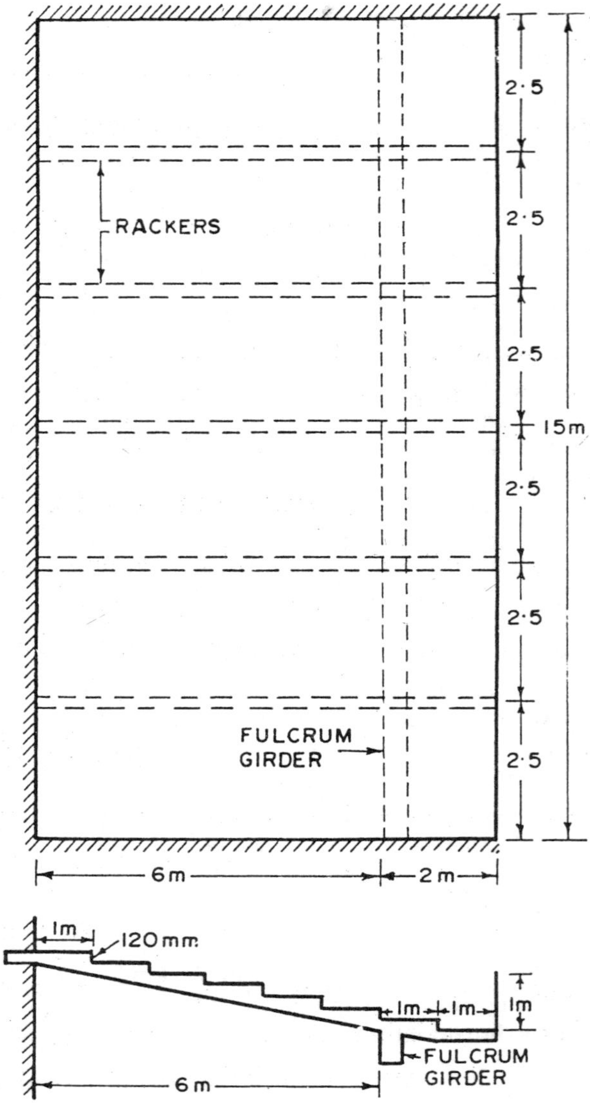

FIG. 8.18.

$\therefore$ Self weight per m^2 of slab$=(0.1\times1\times1\times 24000)\ 1.01 = 2420$ N

Wt of steps $= \frac{1}{2} \times 1 \times 0.12 \times 24000$ $= 1440$ N

Live load $= 4500$ N

Total $= 8360$ N

Effective span = 2.5 m.

$$\therefore \quad M = \frac{wL^2}{8} = \frac{8360\,(2.5)^2}{8} \times 1000 = 6.53 \times 10^6 \text{ N-mm.}$$

$$\therefore \quad d = \sqrt{\frac{6.53 \times 10^6}{0.85 \times 1000}} = 87.6 \text{ mm}$$

Provide total depth = 110 mm

Using 10 mm Φ bars, and 15 mm cover,

$$d = 110 - 5 - 15 = 90 \text{ mm.}$$

$$A_{st} = \frac{6.53 \times 10^6}{140 \times 0.87 \times 90} = 596 \text{ mm}^2$$

Using 10 mm Φ bars having $A_\Phi = 78.54$ mm^2,

$$s = \frac{1000 \times 78.54}{596} = 131.8 \text{ mm.}$$

However, provide 10 mm Φ bars @ 125 c/c. Bend half the bars from each side up near the raker beams, to resist the hogging bending moment of equal amount.

Check for shear

$$V = \frac{8360 \times 2.5}{2} = 10400 \text{ N}$$

$$\therefore \quad \tau_v = \frac{V}{bd} = \frac{10400}{1000 \times 90} = 0.116 \text{ N/mm}^2$$

At support, available $A_{st} = \frac{1000\,A_\Phi}{s} = \frac{1000 \times 78.54}{250}$

$$= 314.2 \text{ mm}^2$$

$\therefore \quad \frac{1000\,A_{st}}{bd} = \frac{100 \times 314.2}{1000 \times 90} = 0.35\%$. Hence from Table 3.2 ,

$\tau_v = 0.25$ N/mm^2. Hence $\tau_v < \tau_c$. (safe)

(*b*) Design of rackers

Let us design one of the interior rackers.

(*i*) ***Loading***

Load due to deck slab, steps and live load

$$= 8360 \times 2.5 = 20900 \text{ N.}$$

The racker beam will be a T-beam. Let the width of web be 300 mm and depth 300 mm.

$\therefore$ Self weight $= (0.3 \times 0.3 \times 24000)\,1.01 = 2160$ N.

$\therefore$ Total load $= 20900 + 2160 = 23100$ N

Assume the same load intensity in the gang way.

Weight of 100 mm thick balustrade, assumed as a concentrated load acting at the end of the racker= $1 \times 0.1 \times 2.5 \times 24000 = 6000$ N.

(ii) Effective span and bending moment

Let the bearing of the racker in the wall be 600 mm. Also, let width of the fulcrum girder be 600 mm. Then the effective span of the racker (T-beam) between the wall and the fulcrum girder will be $= 6 + 0.6 = 6.6$ m. Over-hang $= 1 + 1 - 0.3 = 1.7$ m.

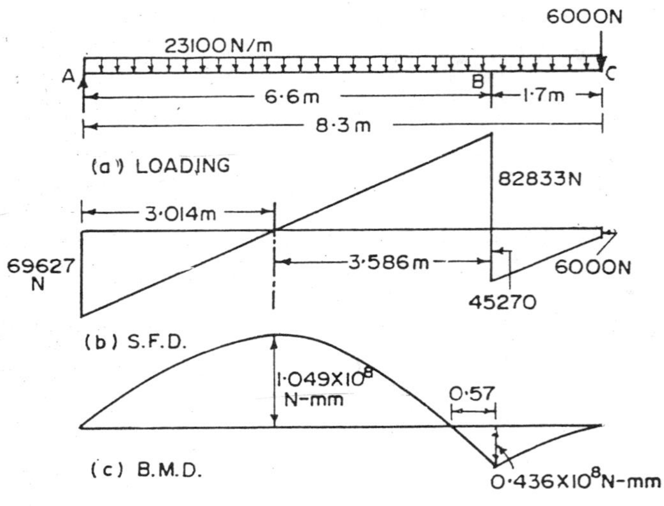

FIG. 8.19 RACKER BEAM

The racker with loading is shown Fig. 8.19(*a*). The S.F.D. and B.M.D. are shown in Fig. 8.19 (*b*) and (*c*) respectively.

$$R_B = \frac{1}{6.6}\left\{\frac{23100\,(8.3)^2}{2} + 6000 \times 8.3\right\} = 128103 \text{ N}$$

$$R_A = (8.3 \times 23100 + 6000) - 128103 = 69627 \text{ N.}$$

Zero S.F. is at a distance $= \dfrac{82833}{23100} = 3.586$ m from B or at 3.014 m from A. Hence maximum B.M. where S.F. is zero is given by

$$M = \left[69627 \times 3.02 - \frac{23100\,(3.02)^2}{2}\right] \times 1000$$

$= 1.049 \times 10^8$ N-mm

For this bending moment; the racker behaves as a T-beam. Also, for overhang *BC*,

$$M_B = \left[\frac{23100(1.7)^2}{2} + (6000 \times 1.7)\right] \times 1000 = 0.436 \times 10^8 \text{ N-mm.}$$

For this position, the racker will behave as an ordinary rectangular beam. Fig. 8.19 (*c*) shows the B.M.D. The point of contraflexure will occur at a distance 0.57 m from support *B*.

(iii) Design of section

(a) *For overhang portion BC.* In this portion the beam is rectangular in section having width $b =$ 300 mm .

$$\therefore \quad d = \sqrt{\frac{M}{Rb}} = \sqrt{\frac{0.436 \times 10^8}{0.85 \times 300}} = 413 \text{ mm} \qquad ...(1)$$

(*b*) *For portion AB.* For this portion, the beam will act as a T-beam. The width $B = b_f$ of the flange acting with the beam will be the minimum of the following:

(*i*) $b_f = b_w +$ clear distance between adjacent beams

= c/c distance between the beams

= 2.5 m

(*ii*) $b_f = \dfrac{l_0}{6} + b_w + 6D_f$

where $l_0 =$ distance between points of zero moments

$= 6.6 - 0.57 = 6.03$ m

$b_w = 300$ mm $= 0.3$ m.

$D_f = 110$ mm 0.11 m.

$$\therefore \quad b_f = \frac{6.03}{6} + 0.3 + 6 \times 0.11 = 1.97 \text{ m}$$

Hence $b_f = 1.97$ m = 1970 mm.

To find the effective depth for balanced section, we have

$$M = 0.45\, b_f\, c\, D_f\left[\frac{2kD - D_f}{k}\right]$$

$$\therefore 1.049 \times 10^8 = 0.45 \times 1970 \times 5 \times 110\left[\frac{2 \times 0.39d - 110}{0.39}\right]$$

or $$0.78\,d = \frac{1.049 \times 10^8 \times 0.39}{0.45 \times 1970 \times 5 \times 110} + 110$$

For which $d \approx$ 249 mm.

At the end and overhang portion, the beam will behave as rectangular section. Keep total $D =$ 350 mm. Using 20 mm Φ bars, a cover of 25 mm and arranging them in two layers with a clear vertical distance of 20 mm, we have effective depth = 350 – 25 – 20 – 10 = 295 mm.

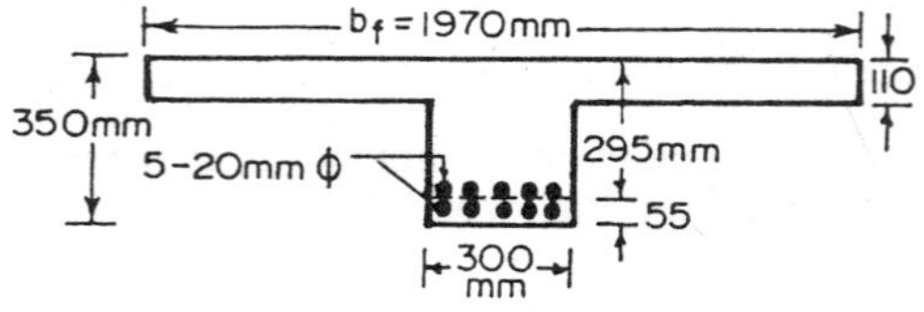

FIG. 8.20.

(iv) Design of reinforcement

(a) For the T-beam portion AB

Lever arm $a = 0.9d = 0.9 \times 295 = 265$ mm

Alternatively, $a = d - \frac{D_f}{2} = 295 - \frac{100}{2} = 240$ mm

However, use $a =$ 250 mm for preliminary calculations.

$$A_{st} = \frac{M}{\sigma_{st} \times a} = \frac{1.049 \times 10^8}{140 \times 250} = 2997 \text{ mm}^2$$

Using 20 mm Φ bars, $A_\Phi = \frac{\pi}{4}(20)^2 = 314$ mm^2

$$\text{No. of bars} = \frac{2997}{314} = 9.5$$

However, use 10 bars and arrange 5 bars in each layer as shown in Fig. 8.20. Bend two bars up at a distance equal to 1 m in given support A. Actual $A_{st} = 314 \times 10 = 3140$ mm^2.

(b) For the rectangular portion BC

Effective depth = 350 – 25 – 10 = 315 mm

$\therefore$ $M_r = Rbd^2 = 0.85 \times 300\,(315)^2 = 0.253 \times 10^8$ N-mm

This is less than the actual B.M. of 0.436×10^8 N-mm.

Let us design beam BC as a doubly reinforced beam, using steel beam theory. Let the compression reinforcement be placed at a cover $d_c = 35$ mm.

$$\therefore \; A_{st} = A_{sc} = \frac{M}{\sigma_{st}\,(d - d_c)} = \frac{0.436 \times 10^8}{140\,(315 - 35)} = 1112 \text{ mm}^2$$

Using 20 mm Φ bars, No. of bars $= \frac{1112}{314} \approx 4.$

Thus, we have to provide 4 bars of 20 mm Φ at the top and 4 bars at the bottom. These will be arranged as follows :

For the top. Bend 2 bars up from the portion *AB*, at the point of contraflexure, *i.e.* 0.57 m from the support *B*; the 2–20 mm Φ anchor bars of the T-beam portion will be continued in portion *BC*, to serve as the main reinforcement of the rectangular beam.

For the bottom. Out of the remaining 8 bars of the span *AB*, continue 3 bars straight into the portion *BC*. The remaining 4 bars may be curtailed at the support *B*.

(v) Check for stresses in T-beam

Assuming the N.A. to fall below the flange, we have

$$b_f D_f \left(n - \frac{D_f}{2}\right) = m A_{st} (d - n)$$

$$1970 \times 110 \left(n - \frac{110}{2}\right) = 18 \times 3140 (295 - n)$$

From which $n = 104.6$ mm

Hence the N.A. falls in the flange for which

$$b_f \frac{n^2}{2} = m A_{st} (d - n)$$

or $$1970 \frac{n^2}{2} = 18 \times 3140 (295 - n)$$

From which $n = 104.5$ mm.

$$\therefore \quad a = d - \frac{n}{3} = 295 - \frac{104.5}{3} = 260 \text{ mm}$$

Stress in steel, $t = \dfrac{M}{a\,\sigma_{st}} = \dfrac{1.049 \times 10^8}{260 \times 3140} = 128.5 \text{ N/mm}^2$

and $c = \dfrac{t}{m} \dfrac{n}{d - n} = \dfrac{128.5}{18} \times \dfrac{104.5}{295 - 104.5} = 3.92 \text{ N/mm}^2$

(safe).

(vi) Check for shear

(a) Portion AB

For the T-beam portion, $V_{max} = 82833$ N

$$\therefore \quad \tau_v = \frac{V}{b \times a} = \frac{82833}{300 \times 260} = 1.06 \text{ N/mm}^2$$

Hence shear reinforcement is necessary

$$\frac{100 A_s}{bd} = \frac{100 \times (8 \times 314)}{300 \times 295} = 2.84\%$$

Hence from Table 3.2, $\tau_c = 0.44 \text{ N/mm}^2$.

$$\therefore \quad V_c = 0.44 \times 300 \times 225 = 38940 \text{ N}.$$

Let us neglect the shear resistance of two bent-up bars.

$\therefore \quad V_s = V - V_c = 82833 - 38940 = 43893$ N

The spacing of the stirrups is given by

$$s = \frac{\sigma_{sv}\, d\, A_{sv}}{V_s} = \frac{140 \times 295}{43893} A_{sv}$$

$$= 0.941\, A_{sv} \text{ mm.}$$

Using 10 mm Φ 2-legged stirrups,

$$A_{sv} = 2\frac{\pi}{4}(10)^2 = 157 \text{ mm}^2$$

$$\therefore \quad s_v = 0.941 \times 157 = 148 \text{ mm}$$

Maximum spacing is given by

$$s_v = \frac{2.5\, A_{sv} f_y}{b} = \frac{2.5 \times 157 \times 250}{300}$$

$$= 327 \text{ mm}$$

Subject to maximum of $0.75d = 0.75 \times 295 = 221$ mm

Hence provide 10 mm Φ 2-lgd stirrups @ 140 mm c/c near the supports, and increase the spacing to 210 mm c/c in the mid poriton

(b) Portion BC

$$V = 45270 \text{ N}$$

$$\therefore \quad \tau_v = \frac{45270}{300 \times 315} = 0.48 \text{ N/mm}^2$$

$$\frac{100 A_s}{bd} = \frac{100\,(4 \times 314)}{300 \times 315} = 1.4\ \%$$

Hence from Table 3.2, $\tau_c = 0.41$ N/mm².

$\therefore \quad V_c = 0.41 \times 300 \times 315 = 38745$ N

$\therefore \quad V_s = V - V_c = 45270 - 38745 = 6525$ N

This is too small. Hence nominal stirrups will be sufficient. Hence provide 10 mm Φ 2-lgd stirrups @ 210 mm c/c throughout the length *BC*.

(c) Design of Fulcrum Girder

(i) Loading on the girder

The loading on the fulcrum girder will be in the form of concentrated loads transferred from the rackers. The magnitude of each concentracted load $= R_B$ (Fig. 6.19)= 128103 N.

Let the width of girder = 600 mm

Depth of girder= 1 m (given) = 1000 mm

$\therefore$ Self-weight per metre run= $0.6 \times 1 \times 1 \times 24000 = 14400$ N

The loading diagram, S.F.D. and B.M.D. for the fulcrum girder are shown in Fig. 8.21. Assuming bearing of 800 mm on each side, the effective span of the girder will be $= 15 + 0.8 = 15.8$ m.

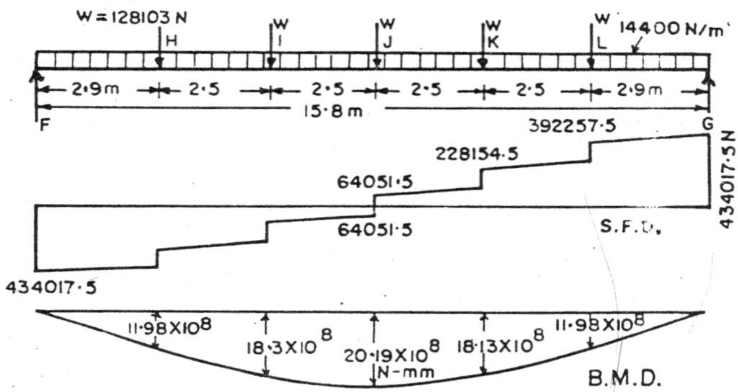

FIG. 8.21

(ii) Bending moment and S.F.

Due to symmetrical loading, $R_F = R_G = 434017.5$ N

$\therefore$ $F_{max} = 434017.5$ N

Maximum bending moment will occur at the centre, its magnitude bending equal to

$$M = (434017.5 \times 7.9) - \frac{14400(7.9)^2}{2} - 128103(5) - 128103(2.5)$$

$$= 2.019 \times 10^6 \text{ N-m}$$

$$= 20.19 \times 10^6 \text{ N-mm}$$

(iii) Design of section at midspan

Let us provide the main reinforcement in two layers. Using 40 mm diameter bars, a clear cover of 25 mm and vertical distance of 40 mm between the two layers, available effective depth $d = 1000 - 25 - 40 - 20 = 915$ mm.

$\therefore$ $M_r = Rbd^2 = 0.85 \times 600\,(915)^2 = 4.27 \times 10^8$ N-mm

This is much less than the external bending moment of 20.19×10^8 N-mm. Hence the beam will be designed as a doubly reinforced section.

Let the compression steel be also provided in two layers, so that its centre is at a distance $d_c = 85$ mm below the top fibres.

Hence from the steel beam theory,

$$A_{sc} = A_{st} = \frac{M}{\sigma_{st}(d - d_c)} = \frac{20.19 \times 10^8}{140\,(915 - 85)} = 17375 \text{ mm}^2$$

Using 40 mm Φ bars, $A_\Phi = 1256.6$ mm²

∴ No. of bars for each reinforcement $= \dfrac{17375}{1256.6} \approx 14$

Hence provide 14 bars in two tiers of 7 bars each, both near the top as well as bottom face of the beam.

Provide 12 mm Φ 4-legged nominal stirrups at the spacing given by

$$s_v = \frac{2.5\, A_{sv} f_y}{b}$$

where $A_{sv} = 4\,\frac{\pi}{4}\,(12)^2 = 452.4 \text{ cm}^2$

∴ $s_v = \dfrac{2.5 \times 425.4 \times 250}{600} = 471$ mm

Subject to a maximum of 450 mm

Hence provide the stirrups @ 450 mm c/c.

(*iv*) ***Design of section at the supports***

Minimum reinforcement required at the support $= \frac{1}{3}$ of main tensile reinforcement = 5 bars (say). Since there is practically no bending moment at the support, the main tensile reinforcement will be provided in one layer only, and out of the compression reinforcement, only 5 bars will be retained.

Then $d = 1000 - 25 - 20 = 955$ mm

Check for shear

$$V = 434017.5 \text{ N}$$

∴ $\tau_v = \dfrac{V}{bd} = \dfrac{434017.5}{600 \times 955} = 0.76 \text{ N/mm}^2$

$$\frac{100\,A_s}{bd} = \frac{100\,(5 \times 1256.6)}{600 \times 955} = 1.1\ \%.$$

Hence from table 3.1, $\tau_c = 0.38$ N/mm²

Hence shear reinforcement is necessary.

$V_c = 0.38 \times 600 \times 955 = 217740$ N.

∴ $V_s = V - V_c = 434017.5 - 217740 = 216277.5$ N

Using 12 mm Φ 4-legged stirrups, having $A_{\Phi} = 4\frac{\pi}{4}(12)^2$

$= 452.4 \text{ mm}^2.$

$$s = \frac{\sigma_{sv}\, d\, A_{sv}}{V_s} = \frac{140 \times 955 \times 425.4}{216277.5} = 280 \text{ mm}$$

Maximum spacing is given by

(*i*) $s_v = \frac{2.5 A_{sv} f_y}{b} = \frac{2.5 \times 452.4 \times 250}{600} = 471 \text{ mm}$

(*ii*) $0.75\, d = 0.75 \times 955 = 716 \text{ mm}$

or (*iii*) 450 mm

whichever is less.

Hence adopt $s =$ 280 mm c/c at the supports. Increase this to 450 mm c/c at the mid-span.

Check for development length

The code envisages that the diameter of the reinforcing bars should be such that the foll ›wing equation is satisfied at the end supports :

$$1.3\frac{M_1}{V} + L_0 \geq L_d$$

where $M_1 = \sigma_{st} A_{st}\, jd = 140\,(5 \times 1256.6) \times 87 \times 955$

$= 7.31 \times 10^8$ N-mm

$V = 434017.5$

Providing a side cover of 30 mm and assuming a support width 800 mm,

$$L_0 = \frac{l_s}{2} - x' + 13\,\Phi = \frac{800}{2} - 30 + 13 \times 40 = 890 \text{ mm}$$

$$\therefore \quad 1.3\frac{M_1}{V} + L_0 = 1.3 \times \frac{7.31 \times 10^8}{434017.5} + 890 = 2190 + 890 = 3080 \text{ mm}$$

$$L_d = 58.3\,\Phi = 58.3 \times 40 = 2332 \text{ mm.}$$

$$\therefore \quad 1.3\frac{M_1}{V} + L_0 > L_d.$$

Hence the code requirements are satisfied.

(*v*) ***Design of section at 2.9 m from the support***

$$M = \left[(434017.5 \times 2.9) - \frac{14400\,(2.9)^2}{2}\right] \times 1000 = 11.98 \times 10^8 \text{ N-mm}$$

$V = 434017.5 - (14400 \times 2.9) = 392257.5$ N

Hence from the steel beam theory,

$$A_{st} = A_{sc} = \frac{11.98 \times 10^8}{140\,(915 - 85)} = 10310 \text{ mm}^2$$

No. of 40 mm Φ bars required $= \dfrac{10310}{1256.6} = 8.2$

Hence provide 9 bars as follows : for tensile steel, provide 7 bars in the bottom layer and 2 bars in the second layer (*i.e.*, rest of the 5 bars of the second layer may be discontinued) ; for compressive steel also, provide 7 bars in the top layer and 2 bars in the second layer. The remaining 2 bars of second layer may however be curtailed at a distance 2 m from the support where the B.M. is approx. 7/8 of that at *H*. Similarly, at 1 m from the support, curtail 2 bars out of 7 bars in the first layer of each tension and compression reinforcement, where they are no longer required from B.M. of view.

$$\tau_v = \frac{V}{bd} = \frac{392257.5}{600 \times 915} = 0.71 \text{ N/mm}^2$$

$$\frac{100 A_s}{bd} = \frac{100\,(7 \times 1256.6)}{600 \times 915} = 1.6\%$$

Hence from Table 3.1, $\tau_c = 0.425$ N/mm^2

Hence shear reinforcement is necessary.

Use 12 mm Φ 4-lgd stirrups having $A_{sv} = 452.4$ mm^2

$$V_c = 0.425 \times 600 \times 915 = 233325 \text{ N}$$

$$\therefore \quad V_s = V - V_c = 392257.5 - 233325 = 158932.5 \text{ N}$$

$$\therefore \quad s = \frac{\sigma_{st}\, d\, A_{sv}}{V_s} = \frac{140 \times 915 \times 452.4}{158932.5} = 365 \text{ mm}$$

Hence provide the stirrups @ 360 mm c/c.

(iv) Design of section at 5.4 m from the support

B.M. at *K* is given by

$$M = \left[(434017.5 \times 5.4) - \frac{14400\,(5.4)^2}{2} - 128103\,(2.5) \right] \times 1000$$

$$= 18.13 \times 10^8 \text{ N-mm}$$

$$V = 434017.5 - (14400 \times 5.4) - 128103 = 228154.5 \text{ N}$$

Hence from steel beam theory :

$$A_{st} = A_{sv} = \frac{18.13 \times 10^8}{1490\,(915 - 85)} = 15602.4 \text{ mm}^2$$

$\therefore$ No. of 40 mm bars required $= \dfrac{15602.4}{1256.6} \approx 12$

Hence provide 7 bars in the first layer and 5 bars in the second layer, *i.e.*, curtail 2 bars of each second layer at this section.

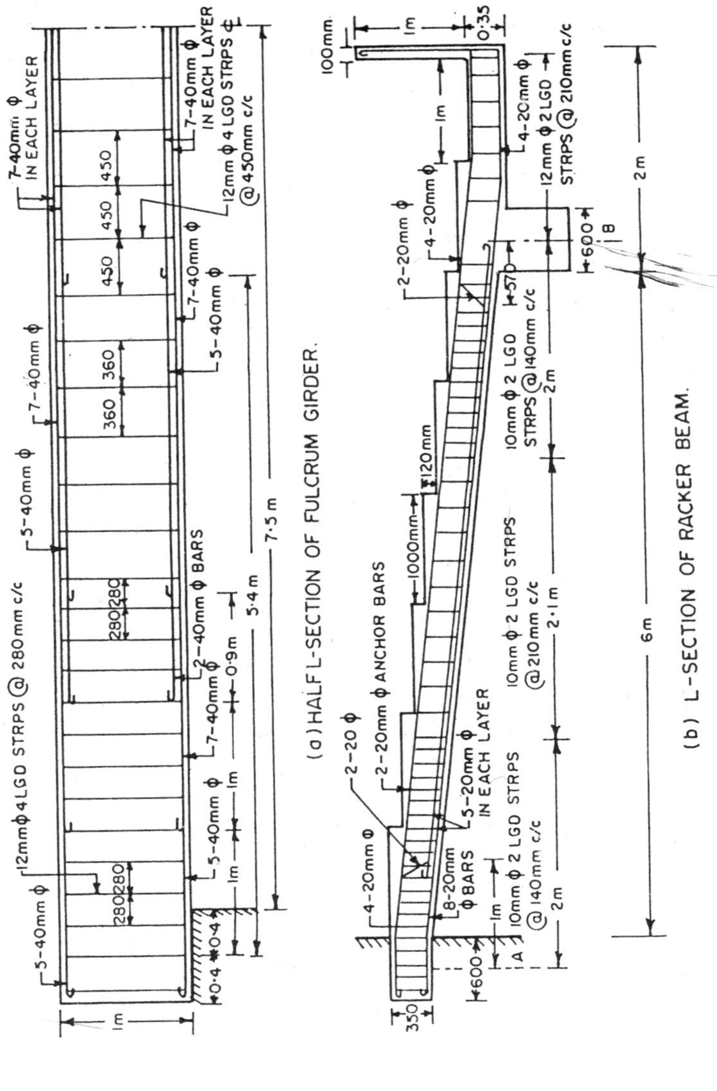

FIG. 8.22.

$$\tau_v = \frac{V}{bd} = \frac{228154.5}{600 \times 915} = 0.42 \text{ N/mm}^2.$$

Hence only nominal shear reinforcement is necessary. Provide 12 mm Φ 4-legged stirrups at a nominal spacing (maximum spacing) of 450 mm c/c

The details of reinforcement etc. are shown in Fig. 8.22.

PROBLEMS

1. Design the stairs for a public building, supported on wall on one side and stringer beam on the other side. The horizontal span of stairs if 1.4 m. The risers are 120 mm and tread are 300 mm. Use M 15 mix.
2. Design a suitable dog-legged stair in a public building, to be located in a staircase 6 metre long, 3.2 m wide and 3.7 m high, with a door of 1.1 m wide in each of the longitudinal walls. The doors face each other and are located with their centres at a distance of 0.9 metres from the respective corners of the staircase. Use M 15 mix.
3. A two storeyed building is to have a R.C. staircase from ground floor to first floor roof. The size of the staircase is 4.3 m × 4.3 m and there is one door opening in one wall and a window opening on the opposite wall. Design the staircase, giving the details of formation, R.C. slab arrangement of building, risers and treads with their top finishing with suitable sketch. The width of stair is 1.2 m and height of each storey is 3.4 m.
4. Fig 8.23 shows the arrangement of a staircase for an office building.

   The rise of steps is 150 mm. Design completely the stair alongwith the beams, landing etc. The live load may be taken as 4000 N/m². Use M 15 mix.

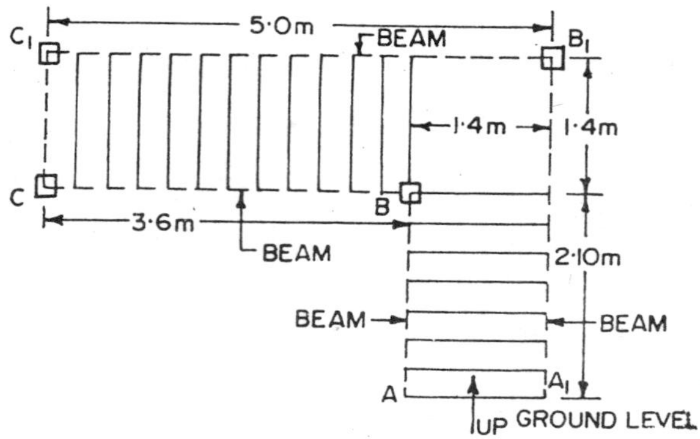

FIG. 8.23.

5. Fig. 8.24 shows the plan of stair hall for a public building. Design staircase with the central stringer beam. The tread is 300 mm and rise 150 mm. The live load may be taken as 4000 N/m^2. Use M 15 mix.

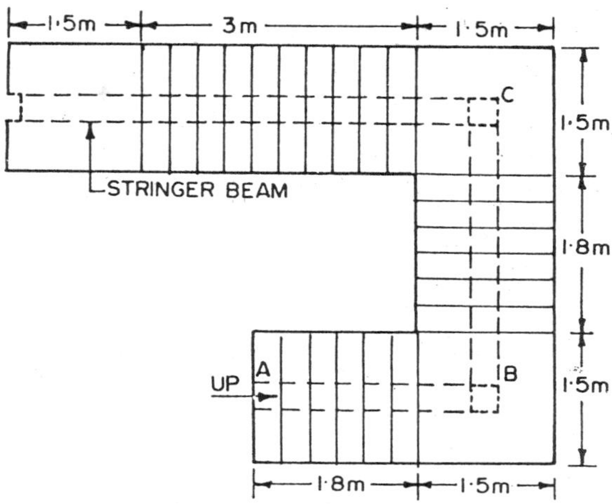

FIG. 8.24.

9

Reinforced Brick and Hollow Tile Roofs

9.1. REINFORCED BRICK WORK

Reinforced brickwork is a typical type of construction in which the compressive strength of bricks is utilised to bear the compressive stress and steel bars are used to bear the tensile stresses in a slab. In other words, the usual cement concrete is replaced by the bricks. However, since the size of a bricks is limited, continuity in the slab is obtained by filling the joints between the bricks by cement mortar. The reinforcing bars are embedded in the gap between the bricks, which is filled with cement mortar. Such type of construction is quite suitable and cheap for small span floor slabs carrying comparatively lighter loads. Fig. 9.1 shows a typical section of reinforced brick slab.

The depth of reinforced brick roof is governed by the thickness of the bricks available. Normally, bricks are 100 mm thick, inclusive of plaster to be used when the brick masonry is constructed. Hence thickness of slab may be kept as 100 mm or 200 mm. If 150 mm thickness is required from design point of view, 50 mm thick tiles are used on the 100 mm thick bricks to make a total thickness of 150 mm. The joint between the two layers of tile and bricks is filled with cement mortar. Before use, the bricks should be thoroughly soaked in water. The reinforcing bars put in the joint should not come in contact with bricks.

When two layers of bricks are used, vertical joints in the bricks should be broken so that slab does not shear along the joint. The bricks near the edge should rest half on the bearing wall so that vertical joint above the edge of the wall is avoided.

First class bricks should be used for such a work. Cement mortar used to fill the joints etc. should be of 1 : 3 ratio (*i.e.*, 1 part cement and 3 parts of clean coarse sand), with proper water cement ratio, to make the mortar workable. The width of joint between

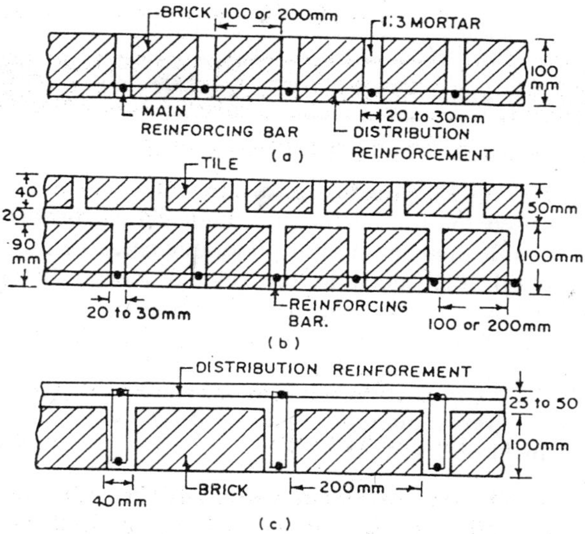

FIG. 9.1. REINFORCED BRICK WORK.

adjacent bricks is generally kept equal to 20 mm. The compressive strength of reinforced brickwork is sometimes increased by providing wider gap(say about 40 mm) between the bricks, and providing 25 to 50 mm thick layer of cement concrete on the top of the bricks, as shown in Fig. 9.1(*c*). The construction then becomes similar to *hollow tile roof*, the design principles of which are discussed in § 9.3.

9.2 DESIGN OF REINFORCED BRICKS SLAB

The principles of design of reinforced brick slab are similar to those of reinforced concrete slab. The following points should be kept in mind while designing such a slab.

1. ***Permissible stresses and design constants***

The permissible value of compressive stress in brick depends upon the quality of brick as well as that of mortar used to fill the gap. The usual range of permissible stress may be between 2

N/mm^2 to 3 N/mm^2. In absence of suitable data, a permissible value of $c = \sigma_{cbc} = 2.5\ N/mm^2$ may be taken. For mild steel usual value of $t = \sigma_{st} = 140\ N/mm^2$ may be utilised. The modular ratio m is usually taken as 40. Thus adopting $c = 2.5\ N/mm^2$, $t = 140\ N/mm^2$ and $m = 40$, we have the following design constants :

$$k = \frac{mc}{mc + t} = \frac{40 \times 2.5}{40 \times 2.5 + 140} \approx 0.42$$

$$j = 1 - k/3 = 1 - 0.42/3 = 0.86$$

$$R = \frac{1}{2}cjk = \frac{1}{2} \times 2.5 \times 0.86 \times 0.42 \approx 0.45$$

2. *Steel reinforcement*

The bars used for tensile reinforcement should be of as small diameter as possible, and should not preferably exceed 12 mm. These should be properly embedded in the cement mortar filled in the gap between the bricks. Half the bars are usually bent upwards at 1/7 of the span from the ends.

3. *Clear cover*

The clear cover should be equal to 15 mm or the diameter of the bar whichever is more.

4. *Distribution and temperature reinforcement*

The distribution reinforcement is provided @ 0.15% of the area of section. Sometimes, it is provided @ 20% of the area of the main reinforcement corresponding to maximum bending moment.

5. *Shear and bond*

Reinforced brick work is very weak against diagonal tension. The permissible shear stress may vary between 0.07 to 0.14 N/mm^2. An average value of 0.1 N/mm^2 may be adopted for design purposes. Normally, shear stress should not exceed this value. In exceptional cases, if the diagonal tension exceeds this value, suitable shear reinforcement should be provided. In such cases, however, the value of shear stress should not exceed 0.3 N/mm^2, otherwise the section should be redesigned, providing greater thickness. For development length, the same rules may be adopted as followed for reinforced concrete slab.

The design procedure is explained in designed example 9.1.

Design Example 9.1. *Design a reinforc d brick slab for a room 3 metre wide. The slab is paved with 20 mm thick mosaic flooring, and is to be designed to carry a live load 1500 N/m².*

Solution.

1. ***Permissible stresses and design constants***

Let $c = \sigma_{cbc} = 2.5$ N/mm^2 ; $t = \sigma_{st} = 140$ N/mm^2 and $m = 40$

Hence $k = 0.42$; $j = 0.86$ and $R = 0.45$.

2. ***Loading***

Let the thickness of slab be 150 mm. Assuming unit weight of brickwork (including) mortar etc. as 22000 N/m^3 and that of mosaic as 23000 N/m^3, the load on the slab per sq. metre will be as follows :

(*i*) Self weight $= 0.15 \times 1 \times 1 \times 22000 = 3300$ N

(*ii*) Weight of flooring $= 0.02 \times 1 \times 23000 = 460$ N

(*iii*) Live load @ 1500 N/m^2 $= 1500$ N

5260 N/ m^2

3. ***Effective span and bending moment***

Assuming effective depth = 130 mm, effective span= 3 + 0.13 = 0.13 m.

$$\therefore \quad M = \frac{wL^2}{8} = \frac{5260\,(3.13)^2}{8} \times 1000 = 6.44 \times 10^6 \text{ N-mm}$$

4. ***Effective depth***

$$d = \sqrt{\frac{M}{Rb}} = \sqrt{\frac{6.44 \times 10^6}{0.45 \times 1000}} = 120 \text{ mm.}$$

However, since total depth is governed by the thickness of brick let us keep the total thickness equal to 150 mm. Using 10 mm diameter bars and 15 mm clear cover, available effective depth will be equal to $150 - 5 - 15 = 130$ mm.

As per IS 1077, the working size of bricks is 200 mm × 100 mm × 100 mm and that of tiles 200 mm × 100 mm × 50 mm, the modular dimensions being 10 mm less in each direction. Hence we will use one tile over one brick to get a total thickness of 150 mm. The horizontal joint between the two layers will evidently be equal to 20 mm(since the modular thickness of brick is 90 mm and that of tile is 40 mm). Leave 30 mm gap between adjacent tiles. However both the joints should not be in same vertical line.

5. ***Main reinforcement***

$$A_{st} = \frac{M}{\sigma_{st}\, jd}$$

Since the effective depth provided is slightly more than that required for the balanced design, the section will be slightly unbalanced. However, we will use $j = 0.86$.

$$\therefore \quad A_{st} = \frac{6.44 \times 10^6}{140 \times 0.86 \times 130} = 411.4 \text{ mm}^2$$

Using 8 mm Φ bars, $A_\Phi = \frac{\pi}{4}(8)^2 = 50.3 \text{ mm}^2$

$$\therefore \quad s = \frac{1000 A_\Phi}{A_{st}} = = \frac{1000 \times 50.3}{411.4} = 122 \text{ mm}$$

Using 30 mm gap and noting that the modular width of brick is 90 mm, centre to centre spacing of bars will be 90 + 30 = 120 mm.

Bend half the bars up at a distance of $L/7 = \frac{3130}{7}$ = 450 mm from the edge of the support.

6. ***Distribution reinforcement***

$$A_{sd} = 0.15\,\% \text{ of area}$$

$$= \frac{0.15}{100} \times 1000 \times 150 = 225 \text{ mm}^2$$

Using 8 mm Φ bars

$$s = \frac{1000 A_\Phi}{A_{sd}} = \frac{1000 \times 50.3}{225} = 223 \text{ mm}$$

Using 3 mm gap and noting that modular length of brick is 190 mm, the centre to centre spacing of the bars will be 220 mm.

7. ***Check for shear***

$$V = \frac{wl}{2} = \frac{5260 \times 3}{2} = 7890 \text{ N}$$

$$\therefore \quad \tau_v = \frac{V}{bd} = \frac{7890}{1000 \times 130} = 0.06 \text{ N/mm}^2 \text{ (safe)}$$

8. ***Check for development length***

At the supports, $1.3\frac{M_1}{V} + L_o \geq L_d$

Here $M_1 = \sigma_{st} A_{st}\, jd = 140\left[\frac{1000 \times 50.3}{2 \times 120}\right] \times 0.86 \times 130$

$$= 3.28 \times 10^6 \text{ N-mm}$$

$$V = 7890 \text{ N.}$$

$$L_0 = \frac{l_s}{2} - x' + 13\Phi$$

Using $l_s = 200$ mm and $x' = 20$ mm.

$$L_0 = \frac{200}{2} - 20 + 13 \times 8 = 184 \text{ mm.}$$

$$\therefore \quad 1.3\frac{M_1}{V} + L_0 = 1.3 \times \frac{3.28 \times 10^6}{7890} + 184 = 540 + 184 = 724\text{mm}$$

$L_d \approx 58.3\,\Phi = 58.3 \times 8 = 466.4$ mm.

Hence code requirements are satisfied.

The details of reinforcement etc. are shown in Fig. 9.2.

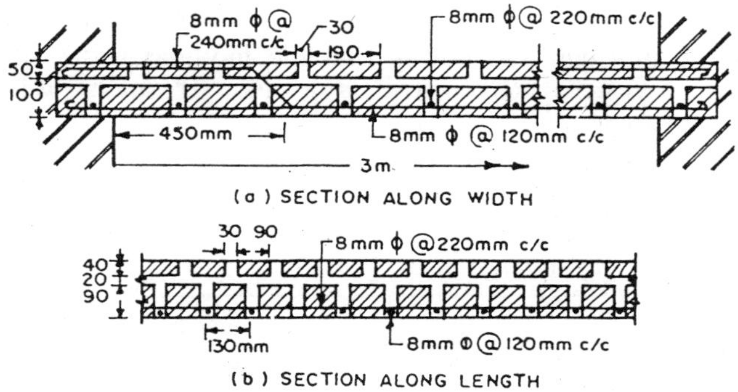

FIG. 9.2. REINFORCED BRICK ROOF.

9.3 HOLLOW TILE ROOF

We have seen earlier that the area of concrete below the N.A. in a rectangular beam, does not carry any appreciable moment of resistance and is neglected. From the point of view of economy, and also for reducing the self-weight, we reduce the width of the beam below the N.A., thus giving rise to T-beam. However, from architectural or hygienic considerations, sometimes plane ceiling is required. In that case, the space between the ribs of adjacent T-beams may be filled with hollow blocks or tiles of light weight concrete and ceramics of light fillers of wood, cane etc. Such a type of floor is known as *hollow tile floor (roof) or close-ribbed floor.* Some of the forms of hollow tile floors and close-ribbed floors are shown in Fig. 9.3.

Unlike T-beam floor construction, the ribs of hollow tile construction are closely spaced. The clear spacing between the ribs depends upon the size of hollow blocks available but it should normally not exceed 500 mm. The width of ribs may vary between 60 to 100 mm. The span of the ribs may be as much as 7 metres. However, when the span exceeds 5 metres, lateral ribs of the same width

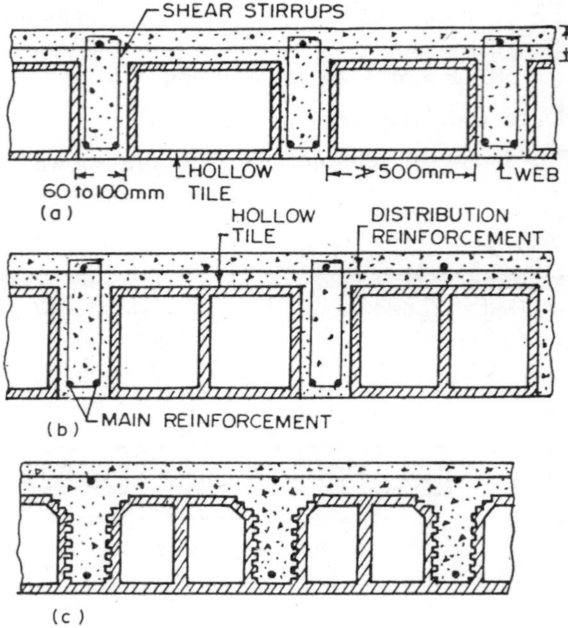

FIG. 9.3. HOLLOW TILE AND CLOSE-RIBBED FLOORS.

as the main longitudinal ones are provided at intervals between 1 to 3 metres. In that case, the main longitudinal ribs are designed as a continuous beam. Main reinforcement is provided at the bottom of the rib. To resist support moment(negative), an additional bar is placed at the top of rib section. A minimum cover of 25 mm is provided. Sometimes, ribs are reinforced with welded mats, usually one to a rib. The depth of rib is calculated on the basis of bending moment as well as on cost ratio of steel and concrete. Depth of ribs is usually taken as at least $\frac{L}{20}$ with free support and at least $\frac{L}{25}$ with fixed support.

Due to small span, the slab is normally not analysed. Slab thickness of 40 to 50 mm is generally provided. To check its cracking and to distribute the load properly, shrinkage and temperature reinforcement is provided in the slab, in both the directions. Sometimes, a welded fabric is arranged approximately along the middle of the thickness of the slab.

Hollow tiles are available in different widths and different depths. Sometimes, to suit the requirements of the depth of rib hollow tiles of required depth may be manufactured at the site. Various forms and types of hollow tiles are available, to suit the clear distance between the ribs.

9.4 DESIGN OF HOLLOW TILE ROOF

Design Example 9.2. *Design a hollow tile roof for a room 4 m wide. The live load may be taken as 2000 N/m². The roof supports lime concrete terracing 120 mm thick. Assume unit weight of concrete as 24000 N/m³.*

Solution.

1. ***Design constants***

Let us use the following values :

$t = \sigma_{st} = 140$ N/mm^2 ; $c = \sigma_{cbc} = 5$ N/mm^2 ; $m = 18$.

Unit weight of lime terracing = 19500 N/m^3

Width of tile =225 mm, height=200 mm and thickness=13 mm.

Weight of tile = 220 N per metre length of the tile.

2. ***Loading and bending moment***

Let us provide slab thickness of 50 mm and the width of rib equal to 80 mm. Hence the centre to centre spacing of the ribs will be equal to 225 + 80 = 305 mm.

Assuming the effective depth of rib equal to 230 mm, the effective span of the rib = 4 + 0.23 = 4.23 m. The total depth of floor will be equal to 200 + 50 = 250 mm.

The loading on the rib, per metre run will be as follows :

1. Wt. of slab $= \dfrac{305 \times 50}{10^6} \times 24000 = 365$ N
2. Wt. of rib $= \dfrac{80 \times 200}{10^6} \times 24000 = 384$ N
3. Terracing $= \dfrac{305 \times 120}{10000} \times 19500 = 712$ N
4. Live load $= \dfrac{305}{1000} \times 1 \times 2000 = 610$ N
5. Wt. of tiles @ 220 N/m =220 N

Total =2291 N ≈ 2300 N

Effective span =4.23 m.

$$\therefore \quad M = \frac{2300\,(4.23)^2}{8} \times 1000 = 5.14 \times 10^6 \text{ N-mm}$$

3. *Effective depth*

The beam is to be designed as a T-beam. Assuming the cost ratio of steel and concrete as 70, we have,

$$d = \sqrt{\frac{rM}{\sigma_{st}\, j\, b_w}} \quad \text{(Taking } j \approx 0.9\text{)}$$

$$d = \sqrt{\frac{70 \times 5.14 \times 10^6}{140 \times 0.9 \times 80}} = 190 \text{ mm.}$$

However, since the height of tiles = 200 mm, available total depth= 200 + 50 = 250 mm and hence available effective depth = 250 – 25 – 8 ≈ 217 mm, using 16 mm dia. bars. Thus for the T-beam section, $b_f = 305$ mm, $D_f = 50$ mm, $b_w = 80$ mm and $d = 217$mm.

4. *Steel reinforcement*

$$A_{st} = \frac{M}{\sigma_{st}\, jd} = \frac{5.14 \times 10^6}{140 \times 0.9 \times 217} = 188 \text{ mm}^2$$

Using 16 mm Φ bars, $A_\Phi = \frac{\pi}{4}(16)^2 = 201 \text{ mm}^2$

Hence provide one bar per rib.

5. *Check for stresses*

To locate the N.A. we have

$$b_f D_f \left[n - \frac{D_f}{2}\right] = m A_{st}(d - n)$$

or
$$305 \times 50 \left[n - \frac{50}{2}\right] = 18 \times 201\,(217 - n)$$

From which n = 618 mm.

The stress c_1 at the bottom of the slab is given by

$$c_1 = \frac{n - D_f}{n} c = \frac{618 - 50}{618} c = 0.19\,c$$

The distance $\bar{y}$ of the centroid of compressive force is given by

$$\bar{y} = \left(\frac{c + 2c_1}{c + c_1}\right)\frac{D_f}{3} = \left(\frac{10 + 38}{10 + 19}\right)\frac{50}{3} = 19.3 \text{ mm}$$

$$\therefore \text{ L.A.} = a = d - \bar{y} = 217 - 19.3 = 197.7 \text{ mm}$$

$$M_r = b_f \cdot D_f \cdot \frac{c + c_1}{2} \times a$$

or
$$5.14 \times 10^6 = 305 \times 50 \left(\frac{1 + 0.19}{2}\right) c \times 197.7$$

From which c = 2.85 N/mm² (safe)

The corresponding stress in steel is given by :

$$t = \frac{mc\,(d-n)}{2} = \frac{18 \times 2.85\,(217 - 61.8)}{61.8}$$

$$= 128 \text{ N/mm}^2 \text{ (safe)}$$

6. *Check for shear*

$$V = \frac{wl}{2} = \frac{2300 \times 4}{2} = 4600 \text{ N}$$

$$\tau_v = \frac{V}{bd} = \frac{4600}{80 \times 217} = 0.26 \text{ N/mm}^2. \text{ (safe)}$$

Hence provide nominal shear stirrups, given by

$$s_v = \frac{2.5\, A_{sv} \cdot f_y}{b}$$

Using 6 mm Φ bars, $A_{st} = 2 \times \frac{\pi}{4}(16)^2 = 56.5 \text{ mm}^2$

$$\therefore \quad s_v = \frac{2.5 \times 56.5 \times 250}{80} = 441 \text{ mm}$$

subject to a max. of 0.75 d = 0.75 × 217 = 163 mm

Hence provide the nominal stirrups @ 160 mm c/c.

7. *Temperature and distribution reinforcement*

Provide nominal reinforcement in the form of 6 mm Φ bars @ 305 mm c/c, with a clear cover of 20 mm from the top of the slab, in both the directions.

The details of the reinforcement etc are shown in Fig. 9.4.

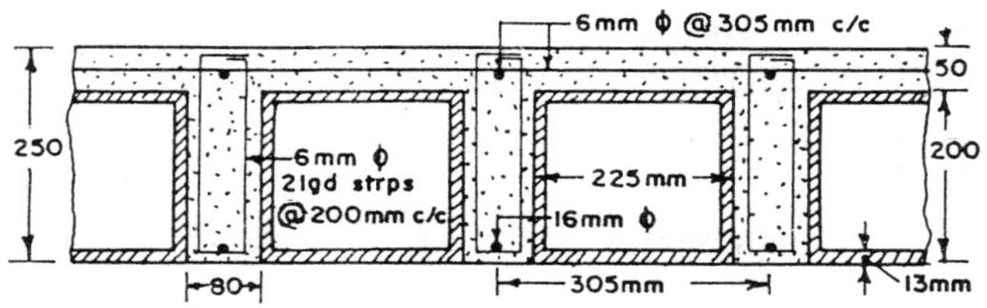

FIG. 9.4

10

Two-Way Slabs

10.1 INTRODUCTION

In Chapters 2 and 7, we have discussed the theory and design of slabs spanning in one direction, *i.e.* slabs supported on two parallel edges. In such a slab, bending takes place along the short span, and the B.M. carried along the longer span is usually a very small fraction of the B.M. carried along the short span. When the slab is supported on all the four edges, and when the ratio of long span to short span is small(say less than 2), bending takes place along both the spans. Such a slab is known as a two-way slab or a 'slab spanning in two directions. The maximum bending moment and deflection for such a slab is much smaller than that of a one-way slab and hence a thinner slab is required. However, reinforcement has to be provided in both the directions. When such a slab is loaded, the corners get lifted up. If the corners are held down, by fixidity at the wall support etc., the bending moment and deflection are further reduced, thus requiring still thinner slab. In that case, special torsional reinforcement at the corner has to be provided to check the cracking of corners.

We will divide the two-way slabs in the following three heads :

1. Slabs simply supported on the four edges, with corner not held down, and carrying uniformly distributed load.

2. Slabs simply supported on the four edges, with corners held down and carrying uniformly distributed load.

3. Slabs with edges fixed or continuous and carrying uniformly distributed load.

10.2 SLAB SIMPLY SUPPORTED ON THE FOUR EDGES, WITH CORNERS NOT HELD DOWN AND CARRYING U.D.L.

Such slabs are commonly used in single storey buildings, where the supporting walls do not cause any fixidity. When such a slab is loaded to failure, a crack pattern at the bottom of slab, shown in Fig. 10.1 is observed. As evident from the crack pattern, the bending takes place in the directions *c d* for the portion *EF* of the slab, while near the corners(say *A*) bending takes place along *ab*, perpendicular to the crack direction *AE*. However, the analysis of such a slab done on the basis of directions of cracks is tedious. Grashoff-Rankine gave the following simplified method based on 'working stress method'.

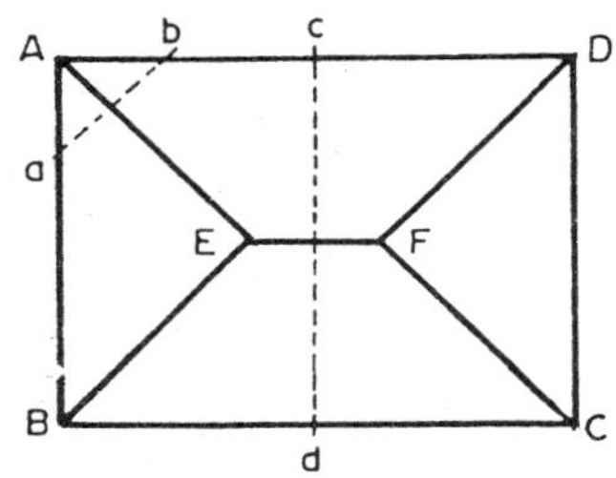

FIG. 10.1

GRASHOFF -RANKINE METHOD

Let $L =$ Long span of a slab

$B =$ Short span of slab

$w =$ Unformly distributed load per unit area.

In this method, the bending moments carried in the long and short directions are determined on the basis of maximum deflection

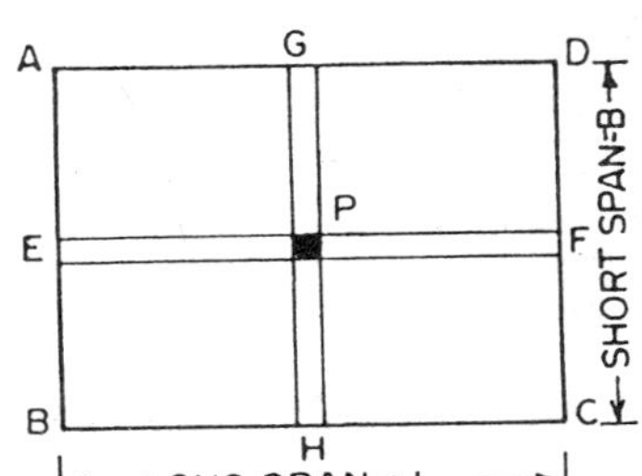

FIG. 10.2.

at the centre of the slab. Consider two middle strips *EF* and *GH* of unit width, parallel to long and short spans respectively.

Let w_L = Load carried by the slab in long direction L

w_B = Load carried by the slab in short direction B.

$$\therefore \quad w = w_L + w_B \qquad ...(1)$$

Consider the deflection of point P, at the intersection of both the strips.

For strip EF, $\delta_P = \dfrac{5}{384}\dfrac{w_L L^4}{EI}$

For strip GH, $\delta_P = \dfrac{5}{384}\dfrac{w_B B^4}{EI}$

Equating the two, we get

$$\frac{5}{384}\frac{w_L L^4}{EI} = \frac{5}{384}\frac{w_B B^4}{EI}$$

or

$$w_L = w_B\left(\frac{B}{L}\right)^4 \qquad ...(2)$$

Substituting in (1), we get

$$w = w_B\left(\frac{B}{L}\right)^4 + w_B$$

From which, $w_B = \dfrac{w}{1+\left(\dfrac{B}{L}\right)^4}$ ...[10.1(*a*)]

Hence $w_L = \dfrac{w}{1+\left(\dfrac{B}{L}\right)^4}\left(\dfrac{B}{L}\right)^4 = \dfrac{w}{1+\left(\dfrac{L}{B}\right)^4}$ ...[10.2(*b*)]

Since $\dfrac{L}{B} > 1$, we note that $w_B > w_L$.

Putting $\dfrac{L}{B} = r$, we have

$$w_L = \frac{w}{1+r^4} = r_L\, w \qquad ...(10.1)$$

and

$$w_B = \frac{w\, r^4}{1+r^4} = r_B\, w \qquad ...(10.2)$$

Where $r_L = \dfrac{1}{1+r^4}$ and $r_B = \dfrac{r^4}{1+r^4}$, and depend upon the value of L/B ratio. Table 10.1 gives the values of r_L and r_B for various values of r. From the Table it will be seen that when $L/B = r$ exceeds 2, the load carried in the longer direction is only 6% while that carried in the shorter direction is 94%. Hence for the slabs in which $\dfrac{L}{B} > 2$, one may design it as a slab supported on two edges, without any serious error.

It should be clearly noted that the above relations hold good only at the middle of the slab, where conditions of maximum B.M. and maximum deflection exist. However, at other points, this proportion of w_B and w_L does not exist, and the design is done by approximately assuming the same ratio as given above.

TABLE 10.1
VALUES OF r_B AND r_L

$r = L/B$	r_B	r_L	$r = L/B$	r_B	r_L
1.0	0.5	0.5	1.4	0.79	0.21
1.05	0.55	0.45	1.5	0.84	0.16
1.1	0.59	0.41	1.6	0.87	0.13
1.15	0.64	0.36	1.75	0.90	0.10
1.20	0.68	0.32	2.0	0.94	0.06
1.25	0.71	0.29	2.5	0.97	0.03
1.3	0.74	0.26	3.0	0.98	0.02

The maximum bending moments in the two directions are :

$$M_L = w_L \frac{L^2}{8} = \frac{w}{1+r^4}\frac{L^2}{8} = w\, r_L \frac{L^2}{8} \quad \text{...(10.3)}$$

$$M_B = w_B \frac{B^2}{8} = \frac{w\, r^4}{1+r^4}\frac{B^2}{8} = w\, r_B \frac{B^2}{8} \quad \text{...(10.4)}$$

Normally, the bending moment in the short span will be more than the bending moment on the long span, and hence the slab thickness will be determined on the basis of M_B. However, reinforcement in the short span is placed *below* the reinforcement in the long span. As per IS 456:1978 minimum total thickness of slab should satisfy the stiffness requirements. The area of steel in both the directions are calculated as follows :

$$(A_{st})_B = \frac{M_B}{\sigma_{st}\, jd}$$

$$(A_{st})_L = \frac{M_L}{\sigma_{st}\, j\,(d - \Phi_B)}$$

where d = effective depth slab determined corresponding to M_B

Φ_B = diameter of bars in short span.

σ_{st} = permissible tensile stress in steel.

Alternate bars can be bent up at 1/7th span from the centre of the supports, in each direction.

If both the spans are equal (for square slab), the effective depth should be taken up to the centre of the top layers of reinforcement.

Shear force

The relations for w_L and w_B developed above give reasonably good results from bending moment point of view only. The reactions at the supports obtained by the above values of w_L and w_B differ very much from those observed experimentally. This is because near the corners, the slab spans in a direction *ab* at right angles at the 45° lines *AE*, *EB* etc. It has been observed that almost the entire load of the triangle *AEB* is transferred to the short support *AB*, while the load in trapezium *AEFD* is transferred to the long support *AD*.

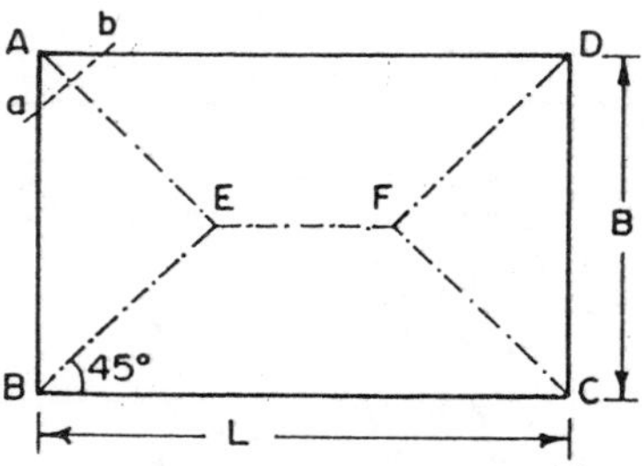

FIG. 10.3.

Thus total reaction on the edge *AB* will be equal to $\frac{w}{2} B \frac{B}{2} = w \frac{B^2}{4}$. The *average* reaction per unit width, along *AB*, will therefore be $\frac{w\,B}{4}$. However, the *maximum* reactions, per unit width along *AB* will occur near the centre of *AB*, and its value may be taken as $\frac{w\,B}{3}$ for all practical purposes.

Similarly, the S.F. along the long span *AD* or *BC*, per unit width may be taken as $w.B. \frac{w}{2+r}$, subject to a maximum of 0.5 $w.B.$ when r exceeds 2.

Design Example 10.1. *Design a R.C. slab for a room measuring 4 m × 5m from inside. The slab carries a live load of 2000 N/m², and is finished with 20 mm thick granolithic topping. Take* $c = \sigma_{cbc} = 5\ N/mm^2$, $t = \sigma_{st} = 140\ N/mm^2$ *and* $m=19$. *The slab is simply-supported at all the four edges, with corners free to lift.*

Solution.

1. ***Design constants***

For the given set of stresses, we have

$k = 0.404$; $j = 0.865$ and $R = 0.874$

2. ***Loading and bending moments***

Let the thickness of the slab be 120 mm for the purpose of calculating the self-weight. The load per square metre of the slab will be as under :

1. Self weight $= \frac{120}{1000} \times 1 \times 1 \times 25000 = 3000$ N
2. Wt. of flooring $= \frac{20}{1000} \times 1 \times 1 \times 2400 = 480$ N
3. Live load @ 2000 N/m² $= 2000$ N

Total $= 5480$ N

Taking effective depth ≈ 100 mm, we have

$$L = 5 + 0.1 = 5.1 \text{ m}$$

$$B = 4 + 0.1 = 4.1 \text{ m}$$

$$r = \frac{L}{B} = \frac{5.1}{4.1} = 1.245 \; ; r^4 = 2.4$$

Hence

$$w_B = w\frac{r^4}{1+r^4} = 5480\frac{2.4}{1+2.4} = 3868 \text{ N}$$

$$w_L = w\frac{1}{1+r^4} = 5480\frac{2.4}{1+2.4} = 1612 \text{ N}$$

$$\therefore \quad M_B = w_B\frac{B^2}{8} = 3868\frac{(4.1)^2}{1} \times 1000 = 8.13 \times 10^6 \text{ N-mm.}$$

$$M_L = w_L\frac{L^2}{8} = 1612\frac{(5.1)^2}{8} \times 1000 = 5.24 \times 10^6 \text{ N-mm.}$$

3. ***Design of section***

The effective depth of slab will be found on the basis of greater moment M_B.

$$\therefore \quad d = \sqrt{\frac{M_B}{Rb}} = \sqrt{\frac{8.13 \times 10^6}{0.874 \times 1000}} = 97 \text{ mm}$$

Adopt a total thickness = 120 mm

Using 10 mm Φ bars and a cover of 15 mm, we get $d = 120 - 15 - 5 = 100$ mm for short span. The effective depth for long span = 100 – 10 = 90 mm, since the reinforcement in long span is placed *above* the reinforcement in short span.

$$(A_{st})_B = \frac{M_B}{\sigma_{st} jd} = \frac{8.13 \times 10^6}{140 \times 0.865 \times 100} = 671.3 \text{ mm}^3$$

Using 10 mm Φ bars, $A_\Phi = \frac{\pi}{4}(10)^2 = 78.5 \text{ mm}^2$

$$\therefore \quad \text{Spacing } s_B = \frac{1000 A_\Phi}{A_{st}} = \frac{1000 \times 78.5}{671.3} = 117 \text{ m}$$

Hence provide 10 mm Φ bars @ 100 mm c/c along short span.

Bend-up alternate bars at $\frac{B}{7} = \frac{4.1}{7} \approx 60$ cm from the centre of each support.

For the long span, the area of steel is given by

$$(A_{st})_L = \frac{5.24 \times 10^6}{140 \times 0.865 \times 90} = 481 \text{ mm}^2$$

Hence spacing of 10 mm Φ bars is given by

$$s_L = \frac{1000 \times 78.5}{481} = 163 \text{ mm}$$

However, provide 10 mm Φ @ 150 mm c/c.

Bend up alternate bars at $\frac{L}{7} = \frac{5100}{7} \approx 700$ mm from the centre of each support.

4. ***Check for shear***

S.F. on the short edge per unit length is given by

$$V_B = \frac{1}{3} w . B = \frac{1}{3} \times 5480 \times 4.1 = 7489 \text{ N}$$

S.F. on the long edge, per unit length is

$$V_L = w B \frac{r}{2 + r} = 5480 \times 4.1 \frac{1.245}{2 + 1.245} = 8620 \text{ N}$$

Note carefully that the effective depth of slab at short edge is 90 mm while that at long edge is 100 mm.

$$\therefore \quad \tau_{VB} = \frac{V_B}{bd} = \frac{7489}{1000 \times 90} = 0.08 \text{ N/mm}^2$$

and

$$\tau_{VL} = \frac{8620}{1000 \times 100} = 0.086 \text{ N/mm}^2.$$

Permissible shear stress τ_c (Tables 3.1 and 3.2) is much more than these. Hence safe.

5. *Check for development length*

Let the slab have a bearing of 20 mm at the ends. Let us check for development length at the end of *short span.*

$$M_1 = \sigma_{st}\ A_{st}\ .jd = 140\left[\frac{100 \times 78.5}{200}\right] \times 0.865 \times 100$$

$$= 4.75 \times 10^6 \text{ N-mm}$$

$$V = V_L = 8620 \text{ N}$$

$$L_0 = \frac{l_s}{2} - x' + 13\,\Phi = \frac{200}{2} - 20 + 13 \times 10 = 210 \text{ mm}$$

$$\therefore\ 1.3\frac{M_1}{V} + L_o = 1.3 \times \frac{4.75 \times 10^6}{8620} + 210 = 716 + 210 = 926 \text{ mm}$$

$$L_d = 58.3\,\Phi = 58.3 \times 10 = 583 \text{ mm}$$

Thus code requirements are satisfied. Similar check can be applied at the end of long span.

The sections of the slab along the short and long spans are shown in Fig. 10.4.

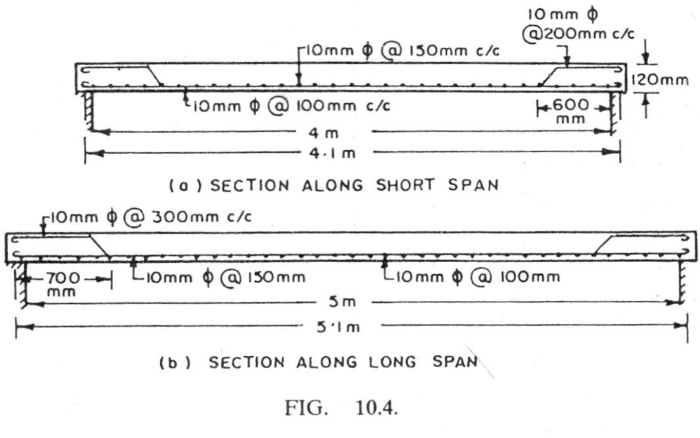

FIG. 10.4.

10.3 SLAB SIMPLY SUPPORTED ON THE FOUR EDGES WITH CORNERS HELD DOWN AND CARRYING U.D.L.

Such a case arises when the slab is built into brick walls or when they are cast monolithically into thin beam, provided the slab is not continuous over its edges. The exact analysis of such slab, based on theory of elasticity, is difficult and complicated. We shall discuss in brief the following three methods of analysis of such slabs :

1. Pigeaud's method.
2. Marcus's method.
3. I.S. Code method.(see § 10.6)

1. PIGEAUD'S METHOD

When corners are held down, the maximum bending moment and maximum deflection are further reduced. However, holding down of corners super-imposes a twisting moment in the slab, for which special reinforcement is needed at top and bottom faces of the slab at each corner. The behaviour of such a slab is similar to that of a uniformly loaded thin elastic plate, and has been analysed by Pigeaud. According to this method the bending moments M_B and M_L along the short and long spans, per unit width are given by the following expressions :

$$M_B = r_B' \frac{w B^2}{8} = \frac{w_B B^2}{8} \quad ...(10.5)$$

where $w_B = r_B' w$ ...[10.5(*a*)]

$$M_L = r_L' \frac{w L^2}{8} = \frac{w_L L^2}{8} \quad ...(10.6)$$

where $w_L = r_L' w$ ...[10.6(*a*)]

The coefficients r_B' and r_L' depend on $\frac{L}{B}$ ratio and are given in Table 10.2

TABLE 10.2.
VALUES OF r_B' AND r_L'

$r = \frac{L}{B}$	r_B''	r_L'	$r = \frac{L}{B}$	$'r_B'$	r_L
1	0.3	0.3	1.4	0.53	0.13
1.05	0.33	0.27	1.5	0.58	0.11
1.1	0.36	0.24	1.6	0.63	0.09
1.15	0.39	0.22	1.75	0.68	0.07
1.2	0.42	0.19	2.0	0.76	0.05
1.25	0.45	0.17	2.5	0.87	0.03
1.3	0.48	0.15	3.0	0.94	0.02

2. MARCUS'S METHOD

Dr. Marcus evolved a simplified but approximate method by which maximum bending moment in a slab with corners held down can be determined. Accordinig to this method, the bending moments M_B and M_L calculated by Rankine-Grashof method are multiplied by a reduction factor C, the value of which is always less than unity since the positive bending moments in the slab are reduced due to corner restraint. The value of C depends upon L/B ratio, and is given by

$$C = 1 - \frac{5}{6} \frac{r^2}{1 + r^4} \qquad \text{...(10.7)}$$

The value of C for various values of L/B ratio are tabulated in Table 10.3.

TABLE 10.3
VALUE OF FACTOR C

$r = \frac{L}{B}$	C	$r = \frac{L}{B}$	C
1	0.861	1.4	0.888
1.05	0.862	1.5	0.898
1.1	0.864	1.6	0.907
1.15	0.866	1.75	0.919
1.2	0.871	2.0	0.935
1.25	0.874	2.5	0.957
1.3	0.879	3.0	0.970

Thus, the midspan bending moment per unit width are given by

$$M_B = \left(1 - \frac{5}{6}\frac{r^2}{1 + r^4}\right) \frac{r^4}{1 + r^4} \frac{w B^2}{8} \qquad \text{...[10.8(a)]}$$

$$M_B = C \frac{r^4}{1 + r^4} \frac{w B^2}{8} = C\, r_B \frac{w B^2}{8} \qquad \text{...(10.8)}$$

and

$$M_L = \left(1 - \frac{5}{6}\frac{r^2}{1 + r^4}\right) \frac{1}{1 + r^4} \frac{w L^2}{8} \qquad \text{...(10.9(a)]}$$

or $$M_L = C \frac{1}{1+r^4} \cdot \frac{w L^2}{8} = C \cdot r_L \cdot \frac{w L^2}{8} \quad \text{...(10.9)}$$

The resulting bending moments calculated by Marcus's method and by the exact theory, with Poisson's ratio equal to zero, are almost identical. If, however Poisson's ratio is taken equal to 0.15, the resulting bending moments are given by the following expressions:

$$M_B = C r_B \left(1 + \frac{0.15}{r^2}\right) \frac{w B^2}{8} \quad \text{...[10.10(}a\text{)]}$$

and $$M_L = C r_L \left(0.15 + \frac{1}{r^2}\right) \frac{w L^2}{8} \quad \text{...[10.10(}b\text{)]}$$

Torsional reinforcement at corners

To resist torsional moments induced at the held-down corners, it is necessary to provide a mesh reinforcement at both the faces of the slab, at each corner. The quantity and direction of such mesh reinforcement required is not definitely known either by experimental evidence or by theoretical analysis. According to common practice both top and bottom reinforcement should consist of two layers bars parallel to the sides of the slab and extending in these directions for a distance of one-fifth of the shorter span. The area of the bars in each of the four layers, per unit width of the slab should be three quarters of the area required for the maximum positive moment in the slab.

To provide such mesh reinforcement effectively it is advisable to use the same diameters of bars, as used at the mid-span, and then bend these bars up through 180° in the corner strips.

Shear Force

The shear force, per unit width, along the long and short span is calculated by the same method explain in § 10.2.

Thus, $$F_B = \tfrac{1}{3} w B \quad \text{...[10.11(}a\text{)]}$$

$$F_L = w B \frac{r}{2+r} \quad \text{...[10.11(}b\text{)]}$$

(subject to maximum of $0.5\, w\, B$)

Design Example 10.2. *Solve example 10.1 assuming that the corners are held down. Use Marcus's method.*

Solution.

1. ***Design constants.*** $k = 0.404$, $j = 0.865$, $R = 0.874$

2. ***Loading and bending moments***

Assuming an overall depth of 120 mm the loading will be the same as in example 10.1.

Thus $w = 5480 \text{ N/mm}^2$

As before, $L = 5 + 0.1 = 5.1$ m

$B = 4 + 0.1 = 4.1$ m

$$r = \frac{L}{B} = \frac{5.1}{4.1} = 1.245 \; ; \; r^4 = 2.4 \; ; \; r^2 = 1.55$$

The bending moment, according to Marcus's method, are given by Eqs. 10.8 (a) and 10.9(a). Thus

$$M_B = \left(1 - \frac{5}{6}\frac{r^2}{1 + r^4}\right)\frac{r^4}{1 + r^4}\frac{wB^2}{8}$$

$$= \left(1 - \frac{5}{6} \times \frac{1.55}{1 + 2.4}\right) \cdot \frac{2.4}{1 + 2.4} \cdot \frac{5480\,(4.1)^2}{8} \times 1000$$

$$= 0.62 \times 0.706 \times 11515 \times 1000 = 5.04 \times 10^6 \text{ N-mm}$$

$$M_L = \left(1 - \frac{5}{6}\frac{r^2}{1 + r^4}\right) \cdot \frac{1}{1 + r^4} \cdot \frac{wL^2}{8}$$

$$= \left(1 - \frac{5}{6} \times \frac{1.55}{1 + 2.4}\right)\frac{1}{1 + 2.4}\frac{5480\,(5.1)^2}{8} \times 1000$$

$$= 0.62 \times 0.294 \times 17817 \times 1000 = 3.25 \times 10^6 \text{ N-mm}$$

3. ***Design of section***

$$d = \sqrt{\frac{M_B}{Rb}} = \sqrt{\frac{5.04 \times 10^6}{0.874 \times 1000}} = 76 \text{ mm}$$

However, from I.S. code of practice, minimum total depth required from the stiffness point of view is

$$D = \frac{B}{35} = \frac{4.1}{35} \times 1000 = 117 \text{ m}$$

Hence adopt a total depth = 120 mm.

Using 8 mm Φ bars and 15 mm cover, the effective depth in the short span = 120 − 15 − 4 = 101 mm and that in the long span will be = 101 − 8 = 93 mm.

4. ***Reinforcement***

Area of steel in the short span is given by

$$(A_{st})_B = \frac{M_B}{\sigma_{st}\, jd} = \frac{5.04 \times 10^6}{140 \times 0.865 \times 101} = 412 \text{ mm}^2$$

Using 8 mm Φ bars, $A_\Phi = \frac{\pi}{4}(8)^2 = 50.3$ mm^2

$$\therefore \quad s_B = \frac{1000 A_\Phi}{A_{st}} = \frac{100 \times 50.3}{412} = 122 \text{ mm}$$

However, provide these @ 120 mm c/c. Bend alternate bars up at $\frac{B}{7} = \frac{4100}{7} = 600$ mm from the centre of each support.

For the long span,

$$(A_{st})_L = \frac{M_L}{\sigma_{st} jd} = \frac{3.25 \times 10^6}{140 \times 0.865 \times 93} = 289 \text{ mm}^2$$

Hence spacing of 8 mm Φ bars is given by

$$s_L = \frac{1000 A_\Phi}{A_{st}} = \frac{1000 \times 50.3}{289} = 174 \text{ mm}$$

Provide these @ 170 mm c/c. Bend alternate bars up at $L/7 = \frac{5100}{7} = 700$ mm from centre of each support.

5. ***Check for shear***

As found in example 10.1,

Shear along long edge, $V_L =$ 8620 N/m
and shear along short edge $V_B =$ 7489 N/m

$$\tau_{VB} = \frac{V_B}{bd} = \frac{7489}{1000 \times 93} = 0.08 \text{ N/mm}^2$$

$$\tau_{VL} = \frac{V_L}{bd} = \frac{8620}{1000 \times 101} = 0.085 \text{ N/mm}^2.$$

Hence safe.

6. ***Check for development length at supports***

At supports, $1.3\frac{M_1}{V} + L_0 \geq L_d$..

For the ends of short span,

$$M_1 = \sigma_{st} A_{st} jd = 140 \left[\frac{1000 \times 50.3}{2 \times 120}\right] \times 0.865 \times 101$$

$$= 2.56 \times 10^6 \text{ N-mm}$$

$$V = V_L = 8620 \text{ N.}$$

$$L_0 = \frac{l_s}{2} - x' + 13\,\Phi = \frac{150}{2} - 20 + 13 \times 8 = 159 \text{ mm}$$

$$\therefore \quad 1.3\frac{M_1}{V} + L_0 = 1.3 \times \frac{2.56 \times 10^6}{8620} + 159 = 386 + 159 = 545 \text{ mm}$$

$$L_d = 58.3\,\Phi = 58.3 \times 8 = 466 \text{ mm.}$$

Hence the code requirements are *satisfied*

Similarly, at the ends of the long span,

$$M_1 = \sigma_{st} A_{st} jd = 140 \left[\frac{1000 \times 50.3}{2 \times 170} \right] \times 0.865 \times 93$$

$$= 1.67 \times 10^6 \text{ N-mm}$$

$$V = V_B = 7489 \text{ N}$$

$$L_0 = \frac{l_s}{2} - x' + 13\,\Phi = \frac{150}{2} - 20 + 13 \times 8 = 159 \text{ mm}$$

$$\therefore\ 1.3\frac{M_1}{V} + L_0 = 1.3 \times \frac{1.67 \times 10^6}{7489} + 159 = 290 + 159 = 449 \text{ mm}$$

$$L_d = 58.3\ \Phi = 58.3 \times 8 = 466 \text{ mm.}$$

Hence the code requirements *not satisfied* in this direction.

In order to satisfy code requirements there are three alternatives: (*i*) to decrease the diameter of bars (*ii*) to decrease the spacing of 8 mm Φ bars, to (*iii*) to carry all the bars at the bottom of the slab right upto the edge and then bend them up through 180° so that all the bars are available at the top face of the slab, near the edges. We will adopt the third alternative, in which case full reinforcement will be available at the bottom face of the slab, at each support.

Hence for the ends of long span,

$$M_1 = 140 \left[\frac{1000 \times 50.3}{170} \right] \times 0.865 \times 93 = 3.34 \times 10^6$$

and $1.3\dfrac{M_1}{V} + L_0 = 1.3 \times \dfrac{3.34 \times 10^6}{7489} + 159 = 605 + 159 = 764$ mm

which is more than $L_d = 466$ mm.

The bent up bars may extend to a distance of $\dfrac{L}{7} + 75$ $= \dfrac{5100}{7} + 75 = 800$ mm from the outer edge of the slab in the long span. For the end strip of 1000 mm, the bent up bars will be extended to a distance of 1000 mm from the outer edge of the slab so that these bars may serve as torsional reinforcement (see next step).

Use the same pattern in the short span also, *i.e.* continue 8 mm Φ bars @ 120 mm c/c upto the support and bend these up through 180° so that these bars are available at the top face of the slab near the edges, to serve as torsional reinforcement.

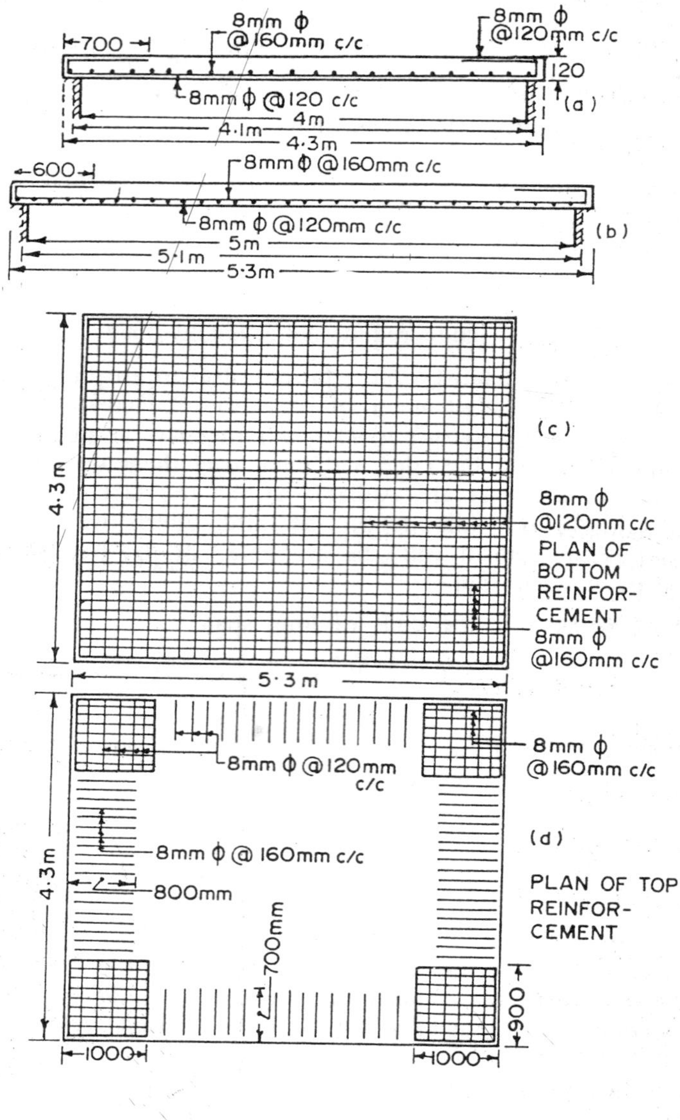

8mm ϕ @160mm c/c
8mm ϕ @120mm c/c
700
120
8mm ϕ @120 c/c
4m
4·1m
4·3m
(a)
600
8mm ϕ @160mm c/c
8mm ϕ @120mm c/c
5m
5·1m
5·3m
(b)
(c)
4·3m
8mm ϕ @120mm c/c
PLAN OF BOTTOM REINFORCEMENT
8mm ϕ @160mm c/c
5·3m
8mm ϕ @120mm c/c
8mm ϕ @160mm c/c
8mm ϕ @ 160mm c/c
800mm
700mm
(d)
PLAN OF TOP REINFORCEMENT
4·3m
900
1000
1000

FIG. 10.5.

7. ***Torsional reinforcement at corners***

Area of torsional reinforcement$=\frac{3}{4}\times 412 = 309$ mm^2

Size of mesh-reinforcement=1/5 × 4100 = 820 mm from the edge of support, or 820 + 150 = 1000 mm from the outer edge of the slab. It is always preferable to provide the same diameter of bar as that provided for the main reinforcement, and at spacing which is a convenient fraction or multiple of the spacing provided for the main reinforcement for the positive bending moments.

Using 8 mm Φ bars, spacing $=\frac{1000\times 50.3}{30.9} = 163$ mm

The main reinforcement has been provided @ 120 mm c/c in the short span, and @ 170 mm c/c in the long span. Thus at the bottom face of the corner mesh-reinforcement, no additional reinforcement is necessary, if the spacing in the long span direction is reduced to 160 mm. Similarly, at the top face of the slab, bent-up bars are available in the respective spacings of 120 mm and 160 mm in both the directions, and hence no additional reinforcement is required. However, these bent up bars are extended to a distance of 1000 mm both ways. The details of reinforcement etc. are shown in Fig. 10.5.

10.4 SLAB WITH EDGES FIXED OR CONTINUOUS AND CARRYING U.D.L.

The following three methods may be used for the analysis of slabs with fixed edges, or continuous slabs :

1. Pigeaud's coefficients method
2. Marcus's method
3. I.S. code method (see § 10.6).

1. PIGEAUD'S COEFFICIENTS METHOD

In this method, the load w_B and w_L causing bending moments are calculated by Eqs. 10.5(*a*) and 10.6(*a*), using coefficients r_B' and r_L' from Table 10.2 for the value of r based on the basis of reduced values of long and short span lengths depending on the end conditions of the slab. Table 10.4 gives the span reduction factors f_L and f_B by which r should be multiplied for use in Table 10.2. The maximum positive and negative bending moment in the slab are also given in Table 10.4.

TABLE 10.4. PIGEAUD'S VALUES FOR SLAB WITH FIXED OR FREE EDGES

Case	Condition of support	Reduction factors for r		Positive moments		Negative moments	
		f_L	f_B	M_L	M_B	M_L	M_B
1.	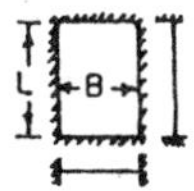	$\frac{2}{3}$	$\frac{2}{3}$	$\frac{1}{10} w_L L^2$	$\frac{1}{10} w_B B^2$	$\frac{1}{8} w_L L^2$	$\frac{1}{8} w_L L^2$
2.	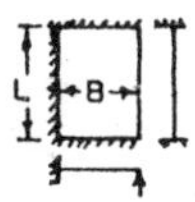	$\frac{2}{3}$	$\frac{4}{5}$	$\frac{1}{10} w_L L^2$	$\frac{1}{9} w_B B^2$	$\frac{1}{8} w_L L^2$	$\frac{1}{8} w_L L^2$
3.	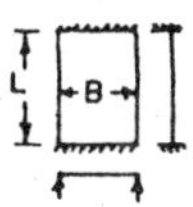	$\frac{2}{3}$	1	$\frac{1}{10} w_L L^2$	$\frac{1}{8} w_B B^2$	$\frac{1}{8} w_L L^2$	Zero, but design moment is taken as $1/8\, w_B B^2$
4.	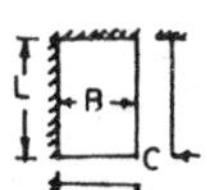 CORNER C HELD DOWN.	$\frac{4}{5}$	$\frac{4}{5}$	$\frac{1}{9} w_L L^2$	$\frac{1}{9} w_B B^2$	$\frac{1}{8} w_L L^2$	$\frac{1}{8} w_L L^2$
5.	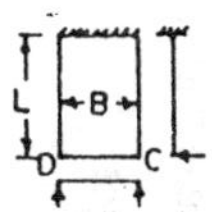 CORNER C & D HELD DOWN.	$\frac{4}{5}$	1	$\frac{1}{9} w_L L^2$	$\frac{1}{8} w_B B^2$	$\frac{1}{8} w_L L^2$	Zero, but design moment is taken as $1/8\, w_B B^2$

Table 10.4 describes 5 cases of end conditions: (1) all the four edges fixed(or continuous) (2) three edges fixed or continuous while one edge along long span is freely supported (3) two edges along short span are fixed while the other two edges along long span are free (4) two adjacent edges are fixed while the other two are free, with the common corner held down, and (5) one edge along short span is fixed while the three edges are free, with the opposite corners held down. For cases (1), (2) and (3), corners are always held down. For cases (4) and (5), if the free corners are not held down, the value of r may be calculated using reduction factors f_B and f_L, but the values of w_B and w_L are to determined using Grashoff-Rankine formulae (or Table 10.1). For case (4), the positive moments at the centre of span and negative moment at the fixed edges will be equal to $\frac{1}{10} w_B B^2$ for short span and $\frac{1}{10} w_L L^2$ for long span, when the free corner is not held down. Similarly, in case (5), if the free corners are not held down, the positive and negative moments each will be equal to $\frac{1}{10} w_L L^2$ for long span, while the positive moment for span will be equal to $\frac{1}{8} w_B B^2$.

2. MARCUS'S METHOD

In this method, the loads w_B and w_L along the two spans are calculated exactly in the same manner as done in Grashoff-Rankine method, except that proper expressions for central deflection are used. After having determined w_B and w_L, span moments M_B and M_L are calculated as per support conditions. However, due to torsion of corners, the span moments are further reduced by multiplying factors given by Marcus (Table 10.5).

For example, let us take the case of slab which has edges of short span fixed, while the long span is freely supported. Then the deflection at the middle of long span will be $\frac{5}{384} \frac{w_L L^4}{EI}$ while that at the middle of short span will be $\frac{1}{384} \frac{w_B B^4}{EI}$. Equating the two, we get

$$\frac{5}{384} w_L \frac{L^4}{EI} = \frac{1}{384} \frac{w_B \,.\, B^4}{EI}$$

or $$w_B = 5 w_L \left(\frac{L}{B}\right)^4 = 5 w_L r^4 \quad \text{...(1)}$$

TABLE 10.5 MARCUS'S VALUES FOR SLAB WITH FIXED OR FREE EDGES

Case	Condition of support	Load coefficient		Span moments (+)		Multiplying factors for span moment		Support moments	
		w_B/w	w_L/w	M_B	M_L	For M_B	For M_L	M_B'	M_L'
1.	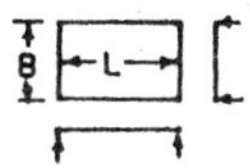	$\frac{r^4}{1+r^4}$	$\frac{1}{1+r^4}$	$\frac{1}{8}w_B.B^2$	$\frac{1}{8}w_L.L^2$	$1-\frac{5}{6}\cdot\frac{r^2}{1+r^4}$	$1-\frac{5}{6}\cdot\frac{r^2}{1+r^4}$	zero	zero
2.	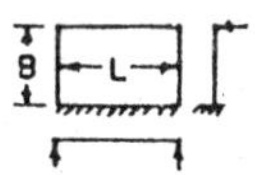	$\frac{5r^4}{2+5r^4}$	$\frac{2}{2+5r^4}$	$\frac{9}{128}w_B.B^2$	$\frac{1}{8}w_L.L^2$	$1-\frac{75}{32}\cdot\frac{r^2}{2+5r^4}$	$1-\frac{5}{3}\cdot\frac{r^2}{2+5r^4}$	$\frac{1}{8}w_B.B^2$ and zero	zero
3.	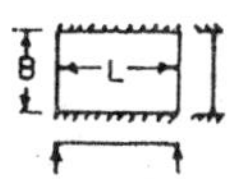	$\frac{5r^4}{1+5r^4}$	$\frac{1}{1+5r^4}$	$\frac{1}{24}w_B.B^2$	$\frac{1}{8}w_L.L^2$	$1-\frac{25}{18}\cdot\frac{r^2}{1+5r^4}$	$1-\frac{5}{6}\cdot\frac{r^2}{1+5r^4}$	$\frac{1}{12}w_B.B^2$ each	zero
4.	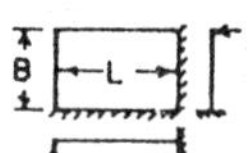	$\frac{r^4}{1+r^4}$	$\frac{1}{1+r^4}$	$\frac{9}{128}w_B.B^2$	$\frac{9}{128}w_L.L^2$	$1-\frac{15}{32}\cdot\frac{r^2}{1+r^4}$	$1-\frac{15}{32}\cdot\frac{r^2}{1+r^4}$	$\frac{1}{8}w_B.B^2$ and zero	$\frac{1}{8}w_L.L^2$ and zero
5	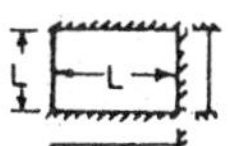	$\frac{2r^4}{1+2r^4}$	$\frac{1}{1+2r^4}$	$\frac{1}{24}w_B.B^2$	$\frac{9}{128}w_L.L^2$	$1-\frac{5}{9}\cdot\frac{r^2}{1+2r^4}$	$1-\frac{15}{32}\cdot\frac{r^2}{1+2r^4}$	$\frac{1}{12}w_B.B^2$ each	$\frac{1}{8}w_L.L^2$ and zero
6.	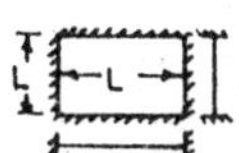	$\frac{r^4}{1+r^4}$	$\frac{1}{1+r^4}$	$\frac{1}{24}w_B.B^2$	$\frac{1}{24}w_L.L^2$	$1-\frac{5}{18}\cdot\frac{r^2}{1+r^4}$	$1-\frac{5}{18}\cdot\frac{r^2}{1+r^4}$	$\frac{1}{12}w_B.B^2$ each	$\frac{1}{12}w_L.L^2$ each

Also $\quad w_B + w_L = w$ ...(2)

Hence from (1) and (2),

$$w_L = \frac{1}{1 + 5r^4} w \qquad ...(9.12)$$

and
$$w_B = \frac{5r^4}{1 + 5r^4} w \qquad ...(9.13)$$

Similarly, values of w_L and w_B for other end conditions may be calculated. Table 10.5 gives the values of $\frac{w_L}{w}$ and $\frac{w_B}{w}$ for various end conditions, in terms of factor r, the span ratio. The values of span moments (positive moments) along with the multiplying factors for the span moments are also given. However, the support moments (fixed end moments) can be calculated. When a support is free, support moment will be $\frac{1}{12} w_B B^2$ or $\frac{1}{12} w_L L^2$ as the case may be. If one end of a span is fixed while the other end is free, the fixed end moment will be $\frac{1}{8} w_B B^2$ or $\frac{1}{8} w_L L^2$, as the case may be.

10.5 OTHER CASES OF SLABS

1. Non-rectangular slabs

Slabs with non-rectangular panels may be approximately analysed by treating the slab as an equivalent two way slab or a circular slab, depending upon the shape of the panel. If the shape of the panel is trapezoidal, the slab may be designed as an equivalent rectangular slab having spans as the centre to centre distance between the parallel sides, and the perpendicular distance between the parallel sides, as marked in Fig. 10.6(a).

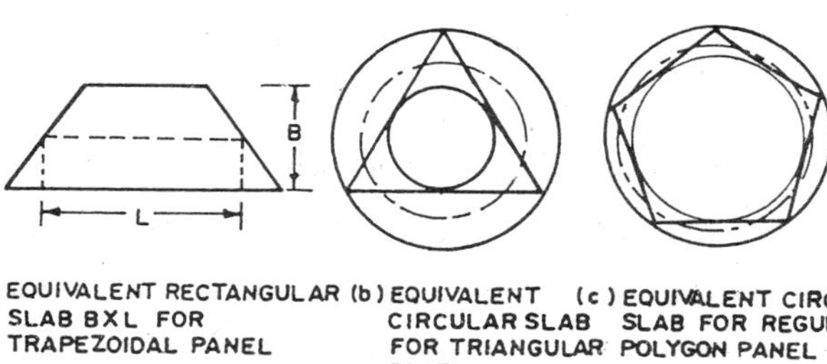

(a) EQUIVALENT RECTANGULAR SLAB BXL FOR TRAPEZOIDAL PANEL (b) EQUIVALENT CIRCULAR SLAB FOR TRIANGULAR PANEL (c) EQUIVALENT CIRCULAR SLAB FOR REGULAR POLYGON PANEL

FIG.10.6. EQUIVALENT SLABS FOR NON-RECTANGULAR PANELS.

If a panel is triangular or regular polygon in shape, it may be treated as an equivalent circular slab having a diameter equal to the mean of circumscribed and inscribed circles, as shown in Fig. 10.6 (*b*) and (*c*).

2. Loads other than U.D.L.

If the slab does not carry uniformly distributed load, but carries concentrated load, spread on a small specified areas or locations of slab, it may be analysed by using curves based on Pigeud's theory. A detailed treatment of various such cases have been dealt with separately in chapter on R.C.C. bridges. (Vol II).

Example 10.3 *Fig. 10.7 shows the plan of the roof of a building. The super- imposed load of the slab may be taken as 3000 N/m². Design all the slab panels. Also, design beam EFGH. Use Marcus's method for design. Take* $c = \sigma_{cbc} = 5\ N/mm^2$, $t = \sigma_{st} = 140\ N/mm^2$ *and* $m = 18$, *and unit weight of concrete* $= 24\ kN/m^3$.

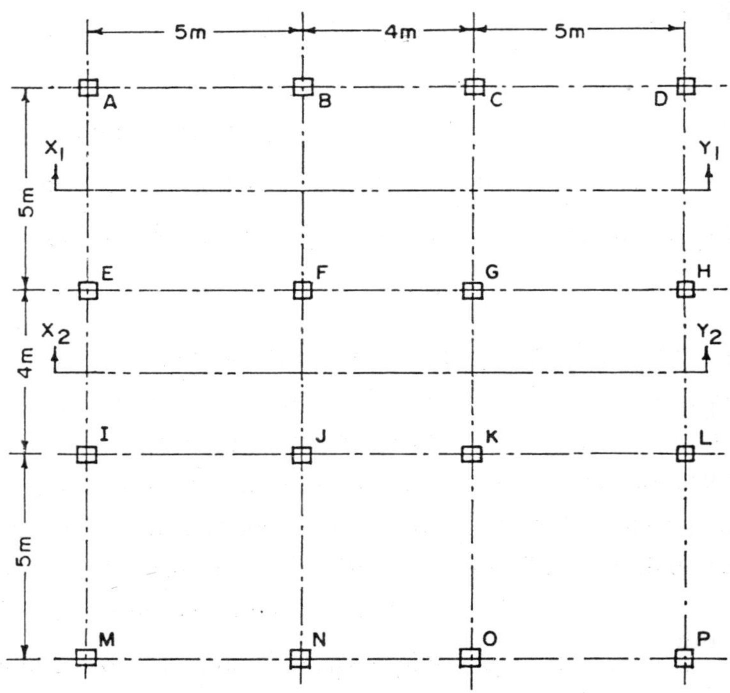

FIG. 10.7.

Solution. The four corner slabs *ABFE, CDHG, KLPO* and *IJNM* have the same edge conditions, *i.e.*, each slab is free at two adjacent edges, with the corner held down, and fixed at the other two adjacent edges. Hence design of only one slab out of these four need be done.

Similarly, the four slabs *EFJI, BFGC, GHLK* and *JKON* have the same edge conditions, *i.e.* free at one edge and fixed at the other. Hence we will design only one slab out of the four. Then remains the slab *FGKJ*, which is fixed on all the four edges.

For the given set of stresses, the design constant will be : $k = 0.39$, $j = 0.87$ and $R = 0.85$.

1. *Design of slab ABFE*

This is a square panel of effective size 5 m × 5 m.

Let $AB = L = 5$ m.

$AE = B = 5$ m.

Edges *BF* and *EF* are continuous, while *AB* and *AE* are freely supported, with corner *A* held down to the column at *A*(Fig. 10.8)

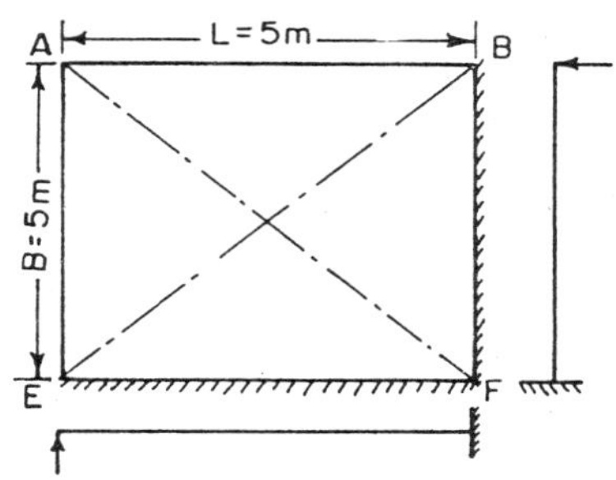

FIG. 10.8.

Let us assume total thickness of the slab = 140 mm.

∴ Self-weight per sq. metre = 140 × 24 = 3360 N/m^2

Super-imposed load = 3000 N/m^2

Total = 6360 N/m^2

$$r = \frac{L}{B} = \frac{5}{5} = 1 \,;\, r^4 = 1$$

From Table 10.5 (case 4), we have

$$w_B = \frac{r^4}{1+r^4} \times 6360 = \frac{1}{2} \times 6360 = 3180 \text{ N/m}$$

$$w_L = \frac{1}{1+r^4} \times 6360 = \frac{1}{2} \times 6360 = 3180 \text{ N/m}$$

Multiplying factor of span moment $= 1 - \frac{15}{32}\frac{r^2}{1+r^4}$

$= 1 - \frac{15}{32} \times \frac{1}{2} = 0.766$

$$M_B = M_L = \frac{9}{128} \; w_B B^2 \times \text{ Multiplying factor}$$

$$= \frac{9}{128} \times 3180(5^2) \times 0.766 \text{ N-m}$$

$$= 4.28 \times 10^6 \text{ N-mm}$$

Support moments are

$$M_{B'} = M_{L'} = \frac{w_B B^2}{8} = \frac{3180(5)^2}{8} \times 1000$$

$$= 9.94 \times 10^6 \text{ N-mm}$$

$$\therefore \qquad d = \sqrt{\frac{M_{B'}}{R\,b}}$$

$$= \sqrt{\frac{9.94 \times 10^6}{0.85 \times 1000}} = 108 \text{ mm.}$$

Let us use 8 mm Φ bars and provide 15 mm cover. Hence keeping total depth = 140 mm, available depth of the centre of the upper layer of reinforcement = 140 − 15 − 4 − 8 = 113 mm. Since the span moments in both the directions are the same, we will provide the same reinforcement, though the effective depth in one direction will be 113 mm and in other direction 121 mm.

∴ Reinforcement at centre of each span $= \frac{4.28 \times 10^6}{140 \times 0.87 \times 113}$

$= 311 \text{ mm}^2$

Reinforcement at fixed ends $= \frac{9.94 \times 10^6}{140 \times 0.87 \times 113} = 722 \text{ mm}^2$

∴ Spacing of 8 mm Φ bars at centre of span $= \frac{1000 \times 50.3}{311}$

$= 162 \text{ mm}$

Hence provide 8 mm Φ bars @ 160 mm c/c in the centre of each span. Bend alternate bars up at a distance of $\frac{L}{5} = 1$ m from the fixed end at a distance of $\frac{L}{7} = \frac{5000}{7} \approx 700$ mm from the free edge.

∴ Area of steel available from half bent up bars, at the fixed edges $= \frac{1000 \times 50.3}{320} = 157 \text{ mm}^2.$

∴ Remaining area to be provided at the fixed edges, to take up negative bending moments $= 722 - 157 = 565\ \text{mm}^2$

∴ Spacing of 8 mm Φ bars $= \dfrac{1000 \times 50.3}{565} = 89$ mm

However keep spacing @ 80 mm c/c. Thus, bent up bars are available @ 160 × 2 = 320 mm c/c, while these additional bars are to be provided @ 80 mm c/c. These bars are to be provided for a length $= \dfrac{L}{4} = \dfrac{5}{4} = 1.25$ m.

Torsional reinforcement has to be provided at corner A, since it is held down.

$$A_{st} = \tfrac{3}{4} \times 311 = 233\ \text{mm}^2$$

This is to be provided for a length $\dfrac{B}{5} = 1$ m, in both directions and at both the faces.

Using 8 mm Φ bars, $s = \dfrac{1000 \times 50.3}{223} = 226$ mm.

However, it is preferable to keep a spacing equal to 160 mm. It should be noted that the bent up bars @ 320 mm c/c are already available. The remaining bars need be provided @ 320 mm c/c, so that the overall spacing will be 160 mm c/c.

Torsional reinforcement will also be provided at corners E and B, the area being half of that provided at A. Hence provide 8 mm Φ bars @ 320 mm c/c at these corners, for a mesh size of 1 m × 1 m. No torsional reinforcement is needed at F, because of symmetrical fixidity conditions there.

Check for shear

Since it is a square panel, the distribution of S.F. on all the four panels will be triangular, as shown in Fig. 10.8.

Total reaction on fixed edge $= \dfrac{5}{8} w_B B L$

$$= \frac{5}{8} \times 3180(5)^2 \doteq 49688 \text{ N}$$

Total reaction on free edge $= \dfrac{3}{8} w_B B L$

$$= \frac{3}{8} \times 3180(5)^2 = 29812 \text{ N}$$

The distribution of each of these is triangular.

Now maximum F = Total reaction $\times \dfrac{2}{B}$

$\therefore$ Max. S.F. at fixed edge $= 49688 \times \frac{2}{2} = 19875$ N

Max. S.F. at free edge $= \frac{29812 \times 2}{2} = 11925$ N

$\therefore$ Nominal Shear stress at fixed edge $= \frac{19875}{1000 \times 113}$

≈ 0.176 N/mm^2

Shear stress at free edge $= \frac{11925}{1000 \times 113} \approx 0.101$ N/mm^2

Hence safe.

Check for development length at free edge.

The code envisages that the following equation should be satisfied at the free edge.

$$1.3 \frac{M_1}{V} + L_o \geq L_d$$

where $M_1 = \sigma_{st} A_{st} jd = 140\,[157] \times 0.87 \times 113$

$= 2.16 \times 10^6$ N-mm

$V = 11925$ N

$L_o = \frac{l_s}{2} - x' + 13\,\Phi$ (where $l_s = 150$ mm, say)

$= \frac{150}{2} - 20 + 13 \times 8 = 159$ mm

$$\therefore \quad 1.3 \frac{M_1}{V} + L_o = 1.3 \times \frac{2.16 \times 10^6}{11925} + 159 = 235 + 159 = 394 \text{ mm}$$

$L_d = 58.3\,\Phi = 58.3 \times 8 = 466$ mm.

Hence code requirements are not satisfied. In order to satisfy code requirements, we will not bend half the bars up, but instead we will carry them straight upto the edge and then bend them up through 180°. In that case $M_1 = 2 \times 2.16 \times 10^6$.

$$\therefore \quad 1.3 \frac{M_1}{V} + L_0 = 1.3 \times \frac{4.32 \times 10^6}{11925} + 159 = 470 + 159 = 629 \text{ mm} > L_d$$

2. *Design of slab EFJI*

The slab is freely supported on the short edge *EI* and is continuous on the remaining three edges. (Fig. 10.9).

$L = 5$ m and $B = 4$ m

$\therefore \quad r = \frac{L}{B} = \frac{5}{4} = 1.25$

$r^2 = 1.56\,; \; r^4 = 2.44.$

We take the same loading

i.e. $w = 6360 \text{ N/m}^2$

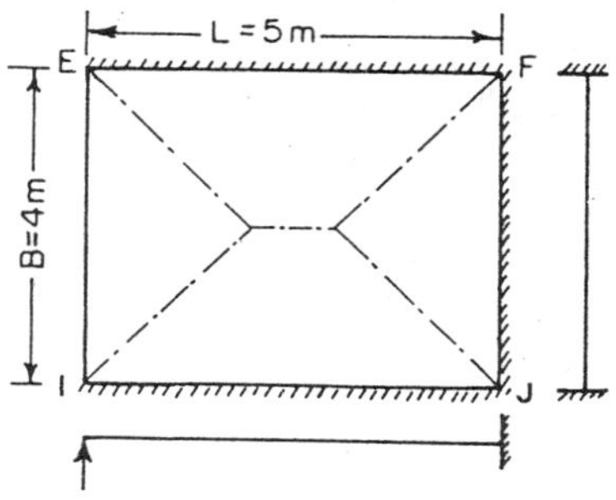

FIG. 10.9.

From the Table 10.5, Case 5,

$$w_B = \frac{2r^4}{1+2r^4} w = \frac{2 \times 2.44}{1 + 2.44 \times 2} \times 6360 = 5280 \text{ N/m}$$

$$w_L = \frac{1}{1+2r^4} w = \frac{1}{1 + 2.44 \times 2} \times 6360 = 1080 \text{ N/m}$$

Multiplying factor for $M_B = 1 - \frac{5}{9} \frac{r^2}{1+2r^4}$

$$= 1 - \frac{5}{9}\left(\frac{1.56}{1+4.88}\right) = 0.852$$

Multiplying factor for $M_L = 1 - \frac{15}{32} \frac{r^2}{1+2r^4}$

$$= 1 - \frac{15}{32}\left(\frac{1.56}{1+4.88}\right) = 0.876$$

$$\therefore \quad M_B = \frac{1}{24} w_B B^2 \times \text{ multiplying factor}$$

$$= \frac{1}{24} \times 5280\,(4)^2 \times 0.852 \times 1000 = 3 \times 10^6 \text{ N-mm}$$

$$M_L = \frac{9}{128} w_L L^2 \times \text{ multiplying factor}$$

$$= \frac{9}{128} \times 1080\,(5)^2 \times 0.876 \times 1000$$

$$= 1.66 \times 10^6 \text{ N-mm}$$

Fixed end moment at each end of short span is

$$M_B' = \frac{1}{12} w_B B^2$$

$$= \frac{1}{12} \times 5280\,(4)^2 \times 1000 = 7.05 \times 10^6 \text{ N-mm}$$

Fixed end moment at end *FJ* of long span is

$$M_L' = \frac{1}{8} w_L L^2 = \frac{1}{8} \times 1080\,(5)^2 \times 1000 = 3.38 \times 10^6 \text{ N-mm}$$

The fixed end moment at edge *EI* will be zero.

The thickness will be designed for the maximum of the above moments, *i.e.* 7.05×10^6 N-mm.

$$d = \sqrt{\frac{7.05 \times 10^6}{1000 \times 0.85}} = 91 \text{ mm.}$$

However, provide overall depth of 140 mm as provided in the panel *ABFE*. Using 8 mm Φ bars and 15 mm cover, the available minimum $d = 140 - 15 - 4 - 8 = 113$ mm in long span and $140 - 15 - 4 = 121$ mm is short span.

Area of steel for positive moment for short span

$$(A_{st})_B = \frac{3 \times 10^6}{140 \times 0.87 \times 121} = 204 \text{ mm}^2$$

Minimum reinforcement $= 1.5\,D = 1.5 \times 140 = 210 \text{ mm}^2$

Spacing of 8 mm Φ bars $= \dfrac{1000 \times 50.3}{210} = 239.5$ mm

However, provide 8 mm Φ bars @ 220 mm c/c. Bend half the bars up near the supports at a distance $= \dfrac{B}{5} = \dfrac{4}{5} = 0.8$ m from each fixed end.

Reinforcement for negative moment for short span

$$= (A_{st})_B' = \frac{7.05 \times 10^6}{140 \times 0.87 \times 121} = 478 \text{ mm}^2$$

Out of this, area available from half bend-up bars.

$$= \frac{1}{2} \times \frac{1000 \times 50.3}{220} = 114 \text{ mm}^2$$

∴ Remaining area $= 478 - 114 = 364 \text{ mm}^2$

Spacing of 8 mm Φ bars $= \dfrac{1000 \times 50.3}{364} = 138$ mm

However, provide these @ 110 mm c/c, for distance of $\dfrac{B}{4}$ = 1 m from each fixed edge.

For long span,

$$(A_{st})_L = \frac{1.66 \times 10^6}{140 \times 0.87 \times 113} = 120.6 \text{ mm}^2$$

Min. reinforcement $= 1.5\,D = 1.5 \times 140 = 210 \text{ mm}^2$, *i.e.* 8 mm Φ bars @ 220 mm c/c. Bend half the bars up near the support at a distance of $\frac{L}{7} = \frac{5}{7} \times 100 =$ 700 mm from free edge and at $\frac{L}{5}$ m $= 1$ from fixed edge.

$$(A_{st})_{L'} = \frac{3.38 \times 10^6}{140 \times 0.87 \times 113} = 264 \text{ mm}^2$$

Out of this, 114 mm^2 area is available from bent up bars. Hence provide remaining area $= 246 - 114 = 132$ mm^2. But we have to provide a minimum reinforcement of 210 mm^2. Hence provide 8 mm Φ bars @ 220 mm c/c additionally. Provide these for $L/4 =$ 1.25 m from each fixed support, by turning the positive reinforcement by 180° at the ends.

Torsional reinforcement

Both ends E and I are continuous. Hence provide area of reinforcement of half of that provided at A. However, provide 8 mm Φ bars @ 220 mm c/c, so that some of the bent up bars may be utilised for torsional reinforcement. This is to be provided as a mesh of dimension 1 m × 1 m in each direction.

Check for shear

Reaction R at each edge $= \frac{1}{2} w_B BL = \frac{1}{2} \times 5280 \times 4 \times 5$

$= 52800$ N

The distribution of these is trapezoidal. Let the maximum shear force per unit width be V

Then $$\frac{VB}{2} + V(L - B) = R$$

$$\therefore \quad V = \frac{2R}{2L - B} = \frac{2R}{L + (L - B)} = \frac{2 \times 52800}{5 + (5 - 4)} = 17600 \text{ N} \quad ...(i)$$

Reaction on the fixed short edge$= \frac{5}{8} w_L BL = \frac{5}{8} \times 1080 \times 4 \times 5$

$= 13500$ N

Assuming this to be distributed triangularly, $V = \frac{2 \times 13500}{4}$

$= 6750$ N ...(*ii*)

Reaction at the free short edge$=\frac{3}{8} w_L B \times L = \frac{3}{8} \times 1080 \times 5 \times 4$
$= 8100$ N

$\therefore \quad V = \frac{8100 \times 2}{4} = 4050$

$\therefore \quad$ Max. $V = 17600$ N

$\therefore \quad \tau_v = \frac{17600}{1000 \times 121} = 0.145 \text{ N/mm}^2$

From tables 3.1 and 3.2, it will be seen that the permissible shear stress τ_c is much greater than this. Hence safe.

Check for development length

Along the long edge (*i.e.* fixed supports of span B), $V = 17600$ N. Hence apply the criteria $\left(1.3 \frac{M_1}{V} + L_0 \geq L_d\right)$ for reinforcement of 8 mm Φ provided @ 110 mm c/c.

$$M_1 = \sigma_{st} A_{st} jd = 140 \left(\frac{1000 \times 50.3}{110}\right) \times 0.87 \times 121$$

$$= 6.74 \times 10^6 \text{ N-mm}$$

$$L_0 = \frac{l_s}{2} - l_0 + 13\,\Phi = \frac{150}{2} - 20 + 13 \times 8 = 159 \text{ mm}$$

$$\therefore \quad 1.3 \frac{M_1}{V} + L_0 = 1.3 \times \frac{6.74 \times 10^6}{17600} + 159 = 383 + 159 = 542 \text{ mm}$$

$L_d = 58.3\,\Phi = 58.3 \times 8 = 466$ mm. Hence safe.

Along the fixed short edge (*i.e.* fixed supports of long span L) $V = 6750$ N and $A_{st} = 8$ mm Φ @ 220 mm c/c.

$$\therefore \quad M_1 = 140 \left[\frac{1000 \times 50.3}{220}\right] \times 0.87 \times 113 = 3.15 \times 10^6 \text{ N-mm}$$

$$L_0 = \frac{l_s}{2} - l_0 + 13\,\Phi = 159 \text{ mm (as before)}$$

$$\therefore \quad 1.3 \frac{M_1}{V} + L_0 = 1.3 \times \frac{3.15 \times 10^6}{6750} + 159 = 607 + 159$$

$$= 766 \text{ mm}$$

$$L_d = 466 \text{ mm}$$

Hence code requirements are satisfied.

Along the free short edge (*i.e.* Free support of long span)
$V = 4050$ N and $A_{st} = 8$ mm @ 440 mm c/c.

$$\therefore \quad M_1 = 140 \left[\frac{1000 \times 50.3}{440}\right] \times 0.87 \times 113 = 1.58 \times 10^6 \text{ N-mm}$$

$$L_0 = 159 \text{ mm (as before)}$$

$$\therefore \qquad 1.3\frac{M_1}{V} + L_0 = 1.3 \times \frac{1.58 \times 10^6}{4050} + 159 = 666 \text{ mm}$$

$$L_d = 466 \text{ mm}$$

Hence safe.

3. Design of slab FGKJ

This slab is continuous on all the four edges (Fig. 10.10)

$$L = B = 4 \text{ m}$$

$$\therefore \qquad r = 1.$$

We take the same loading *i.e.*

$$w = 6360 \text{ N/m}^2$$

From Table 10.5, case 6,

$$w_B = w_L = \frac{1}{2} w = 3180 \text{ N/m}^2$$

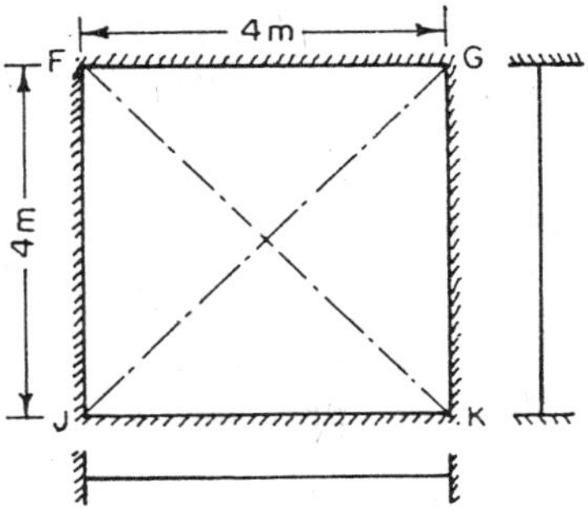

FIG. 10.10

Multiplying factor for each span moment

$$= 1 - \frac{5}{18}\frac{r^2}{1+r^4} = 1 - \frac{5}{18}\frac{1}{2} = 0.861$$

Span moments $M_B = M_L = \frac{1}{24} w_B B^2 \times$ multiplying factor

$$= \frac{1}{24} \times 3180\,(4)^2 \times 0.861 \times 1000 = 1.83 \times 10^6 \text{ N-mm}$$

Support moments $M_B' = M_L' = \frac{1}{12} w_B B^2$

$$= \frac{1}{12} \times 3180\,(4)^2 \times 1000 = 4.24 \times 10^6 \text{ N-mm}$$

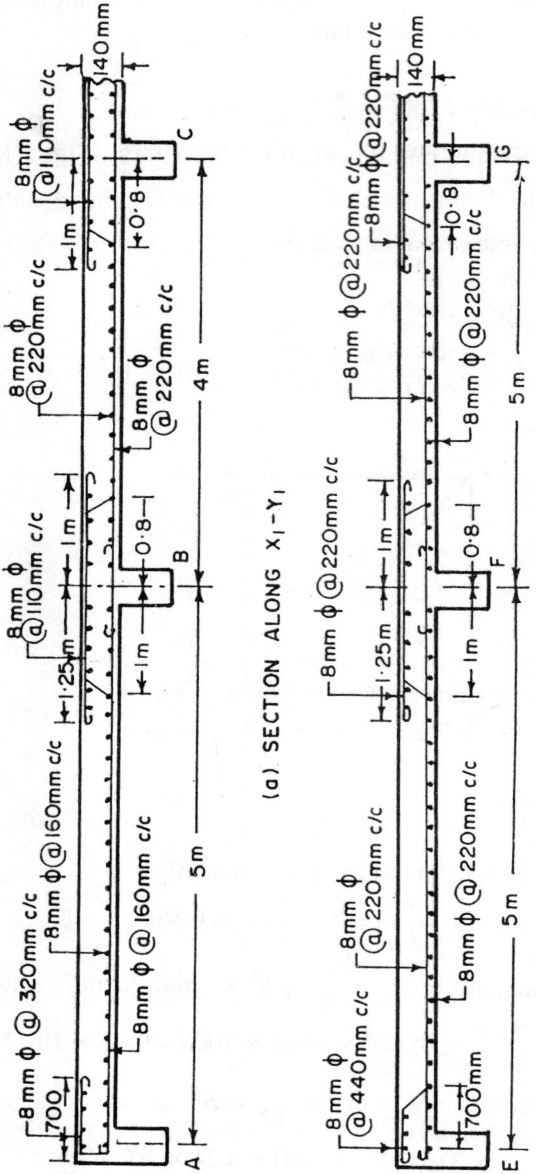

(a) SECTION ALONG $X_1 - Y_1$

(b) SECTION ALONG $X_2 - Y_2$

FIG. 10.11

Keep overall depth = 140 mm as before. Using 8 mm diameter bars and providing a cover of 15 mm, the minimum available $d = 110$ mm.

$$(A_{st})_B = (A_{st})_L = \frac{1.83 \times 10^6}{140 \times 0.87 \times 113} = 133 \text{ mm}^2$$

However, minimum reinforcement $= 1.5 \times 140 = 210 \text{ mm}^2$

Hence provide 8 mm Φ bars @ 220 mm c/c. Bend half the bars up near the support at a distance $= \frac{B}{5} = 0.8$ m from the edge.

For fixing moments,

$$(A_{st})_B' = (A_{st})_L' = \frac{4.24 \times 10^6}{140 \times 0.87 \times 113} = 308 \text{ mm}^2$$

Out of this, area available from bent up bars = 114 mm^2

$\therefore$ Remaining area $= 308 - 114 = 194 \text{ mm}^2$. However a minimum area of 210 mm^2 is to be provided . Hence provide 8 mm Φ bars @ 220 mm, for a distance $\frac{B}{4} =$ 1 m from each edge. No torsional reinforcement is required for this panel.

Check for shear

$$\text{Total reaction on any edge} = \frac{1}{2} w (B . L)$$

$$= \frac{1}{2} \times 6360 (4 \times 4) = 25440 \text{ N}$$

$$\therefore \quad V_{max} = \frac{25440 \times 2}{4} = 12720 \text{ N}$$

$$\therefore \quad \tau_v = \frac{12720}{1000 \times 113} = 0.113 \text{ N/mm}^2$$

This is less than permissible shear stress even at the minimum reinforcement (see Table 3.1 and 3.2).

Since it is an interior panel, it is not necessary to check for development length.

The sections of the panels along $X_1 - Y_1$ and $X_2 - Y_2$ are shown in Fig. 10.11.

10.6 INDIAN STANDARD CODE METHOD (IS : 456-1978)

Indian Standard Code (IS : 456-1978) lays down design rules for two cases of slabs spanning in two direction :

(*a*) restrained slabs (*b*) simply supported slabs.

TABLE 10.6 BENDING MOMENT COEFFICIENTS FOR RECTANGULAR PANELS SUPPORTED ON FOUR SIDES WITH PROVISION FOR TORSION AT CORERS (IS : 456 - 978)

Case No.	Type of panel and moment	Short span coefficient α_x for values of l_y/l_x								long span co-efficient α_y for all values of l_y/l_x
		1.0	1.1	1.2	1.3	1.4	1.5	1.75	2.0	
1.	**Interior Panels**									
	Negative moment at continuous edge	0.032	0.037	0.043	0.047	0.051	0.053	0.060	0.065	0.032
	Positive moment at mid-span	0.024	0.028	0.032	0.036	0.039	0.041	0.045	0.049	0.024
2.	**One short edge discontinuous**									
	Negative moment at continuous edge	0.037	0.043	0.048	0.051	0.055	0.057	0.064	0.068	0.037
	Positive moment at mid-span	0.028	0.032	0.036	0.039	0.041	0.044	0.048	0.052	0.028
3.	**One long edge discontinuous**									
	Negative moment at continuous edge	0.037	0.044	0.052	0.057	0.063	0.067	0.077	0.085	0.037
	Positive moment at mid-span	0.028	0.033	0.039	0.044	0.047	0.051	0.059	0.065	0.028
4.	**Two adjacent edges discontinuous**									
	Negative moment at continuous edge	0.047	0.053	0.060	0.065	0.071	0.075	0.084	0.091	0.047
	Positive moment at mid-span	0.035	0.040	0.049	0.049	0.053	0.056	0.064	0.069	0.035

Table 10.6 (*Continued*)

Case No.	Type of panel and moment	Short span coefficient α_x, for values of l_y/l_x								long span co-efficients α_y for all values of l_y/l_x
		1.0	1.1	1.2	1.3	1.4	1.5	1.75	2.0	
5.	**Two short edges discontinuous**									
	Negative moment at continuous edge	0.045	0.049	0.052	0.056	0.059	0.060	0.065	0.069	—
	Positive moment at mid-span	0.035	0.037	0.040	0.043	0.044	0.045	0.049	0.052	0.035
6.	**Two long edges continuous**									
	Negative moment at continuous edge	—	—	—	—	—	—	—	—	0.045
	Positive moment at mid-span	0.035	0.043	0.051	0.075	0.063	0.068	0.080	0.088	0.035
7.	**Three edges discontinuous (one long edge continuous)**									
	Negative moment at continuous edge	0.057	0.064	0.071	0.076	0.080	0.084	0.091	0.097	—
	Positive moment at mid-span	0.043	0.048	0.053	0.057	0.060	0.064	0.069	0.073	0.043
8.	**Three edges discontinuous (one short edge continuous)**									
	Negative moment at continuous edge	—	—	—	—	—	—	—	—	0.057
	Positive moment at mid-span	0.043	0.051	0.059	0.065	0.071	0.076	0.087	0.096	0.043
9.	**Four edges discontinuous**									
	Negative moment at continuous edge	0.056	0.064	0.072	0.079	0.085	0.089	0.100	0.107	0.056

(*a*) Restrained slabs

When the corners of a slab are prevented from lifting, the slab may be designed as specified in *a* – 1 to *a* – 11 below.

(a–1). The maximum bending moments per unit width in a slab are given by the following equation:

$$M_x = \alpha_x w \, l_x^2$$

$$M_y = \alpha_y \, . \, w \, . \, l_x^2$$

where α_x and α_y are coefficients given in Table 10.6.

w = total load per unit area

M_x, M_y = moments on strips of unit width spanning l_x and l_y respectively, and

l_x and l_y = lengths of shorter span and long span respectively.

(a–2). Slabs are considered as divided in each direction into middle strips and edge strips as shown Fig. 10.12, the middle strips being three quarters of the width and each edge strip one-eighth of the width.

(a–3). The maximum moments calcuiated as in *a*–1 apply on to the middle strips and no redistribution shall be made.

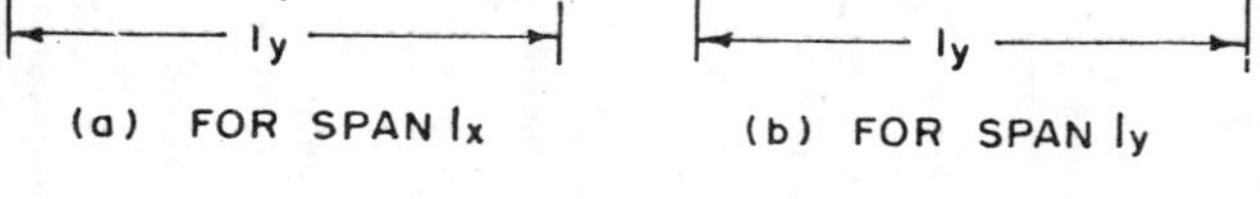

FIG. 10.12. DIVISION OF SLAB INTO MIDDLE AND EDGE STRIPS.

(a–4). Tension reinforcement provided at mid-span in the middle strip shall extend in the lower part of the slab to within 2.25 *l* of a continuous edge, or 0.15 *l* of a discontinuous edge.

(a–5). Over the continuous edge of the middle strip, the tension reinforcement shall extend in the upper part of the slab a distance of 0.15 *l* from the supports, and at least 50 percent shall extend a distance of 0.3 *l* .

(a–6). At a discontinuous edge, negative moments may arise. They depend on the degree of fixidity at the edge of the slab but, in general, tension reinforcement equal to 50 percent of that provided at mid span extending 0.1 l into the span will be sufficient.

(a–7). Reinforcement in edge strip, parallel to that edge shall comply with minimum given in chapter 7 (*i.e.* @ 0.15%) and the requirements for torsion given in $a-8$, $a-9$ and $a-10$.

(a–8). Torsion reinforcement shall be provided at any corner when the slab is simply supported on both edges meeting at that corner. It shall consist of top and bottom reinforcement, each with layers of bars placed parallel to the sides of the slab and extending from the edges to a minimum distance of one fifth of the shorter span. The area of reinforcement in each of these four layers shall be three quarters of the area required for the maximum mid-span moment in the slab.

(a–9). Torsion reinforcement equal to half that described in $a-8$ shall be provided at a corner contained by edges over only one which the slab is continuous.

(a–10). Torsion reinforcement need not be provided at any corner contained by edges over both of which slab is continuous..

(a–11). Where l_y/l_x is greater than 2, the slabs shall be designed as spanning one way.

(*b*) Simply supported slabs

(b–1). When simply supported slabs do not have adequate provision to resist torsion at corner and to prevent the corners from lifting, the maximum moments for unit width are given by the following equation :

$$M_x = \alpha_x w l_x^2$$

$$M_y = \alpha_y w l_x^2$$

Where, M_x, M_y, l_x, l_y are the same as those in $a-1$, and α_x and α_y are moment coefficients given in Table 10.8.

(b–2). At least 50 percent of the tension reinforcement provided at mid-span should extend to the supports. The remaining 50 percent should extend to within 0.1 l_x or 0.1 l_y of the support, as appropriate.

TABLE 10.8

BENDING MOMENT COEFFICIENTS FOR SLABS SPANNING IN TWO DIRECTIONS AT RIGHT ANGLES, SIMPLY SUPPORTED ON FOUR SIDES (IS : 456 –1978)

l_y/l_x	1.0	1.1	1.2	1.3	1.4	1.5	1.75	2.0	2.5	3.0
α_x	0.062	0.074	0.084	0.093	0.099	0.104	0.113	0.118	0.122	0.124
α_y	0.062	0.061	0.059	0.055	0.051	0.046	0.037	0.029	0.020	0.014

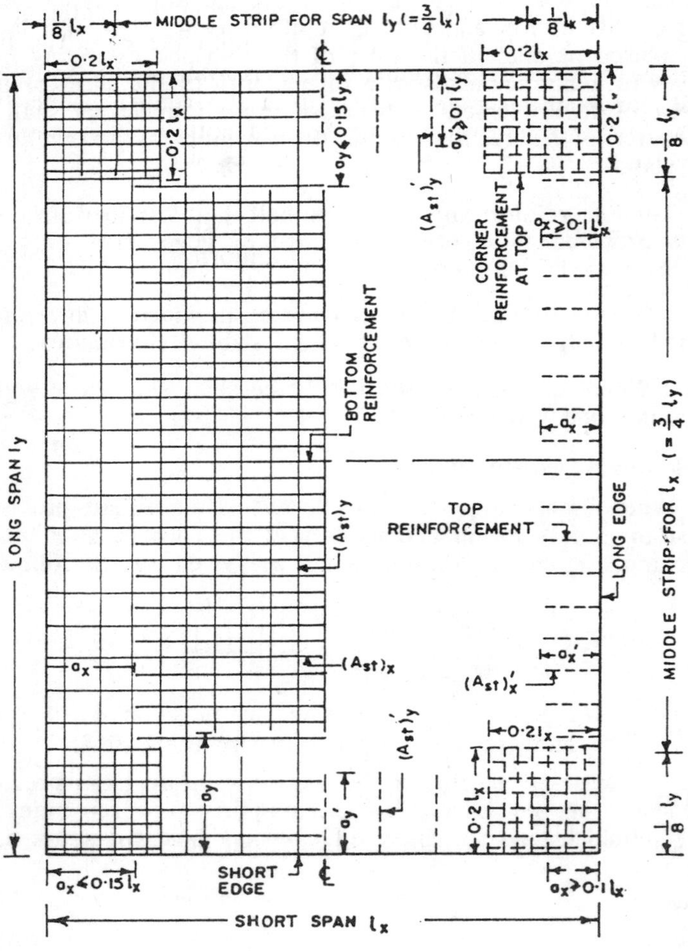

FIG. 10.13.

Details of Reinforcement

Fig. 10.13 shows the details of reinforcement. Half the positive moment reinforcement $(A_{st})_x$ is bent upward s, at a distance $a_x \leq 0.15 l_x$ from the centre of the support while the remaining half is continued upto the end. The bent up bars or the negative reinforcement(top reinforcement) $(A_{st})_x'$ should be available for distance $a_x' \geq 0.1\, l_x$ from the centre of the support.

Similarly, in the span l_y, half the positive moment reinforcement $(A_{st})_y$ is bent up, at a distance $a_y \leq 0.15\, l_y$ from the centre of the support. The remaining half is continued upto the end. The bent bars or the negative reinforcement $(A_{st})_y'$ is available at the top of the slab for a distance $a_y' \geq 0.1\, l_y$ from the centre of the slab.

The corner reinforcement, at top and bottom of each corner is provided in the form of square mesh of size $0.2\, l_x \times 0.2\, l_x$.

Load on supporting beams

The loads on the beams supporting solid slabs spanning in two directions at right angles and supporting uniformly distributed loads, may be assumed as marked in Fig. 10.14.

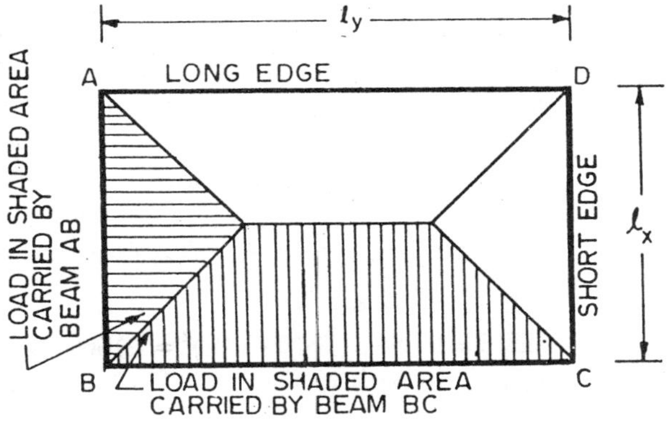

FIG. 10.14. LOADS CARRIED BY SUPPORTED BEAM.

Thus total reaction on short edge *AB*(or *CD*) will be equal to . $\frac{1}{2} w\, l_x \frac{l_x}{2} = \frac{w\, l_x^2}{4}$. The *average reaction* per unit width along short edge will therefore be $\frac{w\, l_x}{4}$, However the *maximum reaction*(hence max. S.F.) per unit width along *AB* will occur near the centre of *AB* and its value may be taken as $\frac{w\, l_x}{3}$ for all practical purposes.

Similarly, S.F. along the edge *AD* or *AC*, (*i.e.* ends of short span of slab) per unit width may be taken as $w\, l_x \dfrac{r}{2+r}$, subject to a maximum of 0.5 $w\, l_x$ when r exceeds 2.

Design Example 10.4. *Design a R.C. slab for a room measuring 5 m × 6 m size. The slab is simply supported on all the four edges, with corners held down, and carries a superimposed load of 3500 N/m², inclusive of floor finishes etc. Use M 15 mix, and IS code method. Take unit weight of concrete as 24 kN/m³.*

Solution

(1) *Design constants*

For M 15 concrete, taking $\sigma_{st} = 140$ N/mm², we have $k = 0.404$, $j = 0.865$ and $R = 0.874$.

(2) *Loading and bending moments*

Assume an overall depth of 180 mm

(*i*) Weight of slab per m² $= \dfrac{180}{1000} \times 1 \times 1 \times 24000 = 4320$

(*ii*) Super-imposed load @ 3500 N/m² $= 3500$

Total $w = 7820$ N/m²

Taking effective depth = 160 mm, we have effective $l_y = 6 + 0.16 = 6.16$ m and $l_x = 5 + 0.16 = 5.16$ m

$\therefore \quad r = l_y / l_x = 6.16/5.16 \approx 1.2$

This is case 9 of Table 10.6, from which $\alpha_x = 0.072$ and $\alpha_y = 0.056$

$$\therefore \quad M_x = \alpha_x w\, l_x^2 = 0.072 \times 7820\,(5.16)^2 = 14990 \text{ N-m}$$
$$= 14.99 \times 10^6 \text{ N-mm}$$
$$M_y = \alpha_x w\, l_x^2 = 0.056 \times 7820\,(5.16)^2 = 11660 \text{ N-m}$$
$$= 11.66 \times 10^6 \text{ N-mm}$$

(3) *Design of section for short span*

$$d = \sqrt{\frac{14.99 \times 10^6}{1000 \times 0.874}} = 131 \text{ mm}$$

From stiffness(deflection) point of view, span/effective depth ratio = 20. For a balanced design, percentage reinforcement $\dfrac{k\,\sigma_{cbc}}{2\,\sigma_{st}} \times 100 = \dfrac{0.404 \times 5}{2 \times 140} \times 100 = 0.72\,\%$ for M 15 concrete. Hence from Fig. 7.1, modification factor, using mild steel bars $\approx$ 1.6.

$$\therefore \quad \frac{\text{span}}{d} = 20 \times 1.6 = 32$$

$$\therefore \qquad d = \frac{\text{span}}{32} = \frac{5160}{32} = 160 \text{ mm.}$$

Hence provide total thickness= 180 mm. Using 10 mm Φ bars and a cover of 15 mm, available effective depth for short span = 180 – 15 – 5 = 160 mm for the short span, and 160 – 10 = 150 mm for the long span.

For short span, middle strip $= \frac{3}{4} l_y = \frac{3}{4} \times 6.16 \approx 4.62$ m

Width of edge strip $= \frac{1}{2}(6.16 - 4.62) = 0.77$ m

$$(A_{st})_x = \frac{14.99 \times 10^6}{140 \times 0.865 \times 160} = 774 \text{ mm}^2$$

Using 10 mm Φ bars, $A_\Phi = 78.5$ mm^2

$$\text{spacing } s = \frac{1000 \times 78.5}{774} = 101 \text{ mm}$$

However, use 10 mm Φ bars @ 100 mm c/c for the middle strip of width 4.62 m. Bend half the bars up at a distance $= 0.15 l_x = 0.15 \times 5160 \approx 770$ mm from the centre of the support, or at a distance of 770 + 80 = 850 mm from the edge of the slab(assuming a bearing of 160 mm on the wall). Available length of bars at the top = 770 – (160 – 20) = 630 mm from the centre of support, assuming bending of bar at 45°. The length is more than $0.1\, l_x$ (= 0.1 × 5160 = 516 mm) required by the code. Hence length of top bars from the edge of slab = 630 + 80 = 710 mm.

Edge strip is of length 0.77 m. The reinforcement in the edge strips is given by

$$A_{st}(@\, 0.15\,\%) = \frac{0.15}{100} \times 1000 \times 180 = 270 \text{ mm}^2.$$

$$\text{Using 10 mm } \Phi \text{ bars, } s = \frac{1000 \times 78.5}{270} = 290 \text{ cm c/c.}$$

However, it is more convenient to keep the spacing as the simple fraction of the edge strip or simple multiple of spacing for the middle strip. Hence keep $s =$ 200 mm.

4. ***Design of section for long span***

$$\therefore \qquad (A_{st})_y = \frac{11.66 \times 10^6}{140 \times 0.865 \times 150} = 642 \text{ mm}^2$$

$$\text{Using 10 mm } \Phi \text{ bars, } s = \frac{1000 \times 78.5}{642} = 122 \text{ mm.}$$

Hence provide 10 mm Φ bars @ 120 mm c/c in the middle strip of width. $\frac{3}{4} l_x = \frac{3}{4} \times 5.16 = 3.87$ m. For the edge strip of width 5.16/8 = 0.645 m, provide 10 mm Φ bars @ 240 mm c/c.

Bend half the bars of the middle strip, up at a distance of $0.15\, l_y = 0.15 \times 616 \approx 920$ mm from the centre of support or $920 + 80 = 1000$ mm from the edge of the slab. Available lengths of bars of top $= 920 - (150 - 20) = 790$ mm from the centre of the support or $790 + 80 = 870$ mm from the edge of the slab.

(5) *Check for shear and development length in short span*

$$\text{S.F. at long edges} = w\, l_x \frac{r}{2 + r}$$

$$= 7820 \times 5.16 \frac{1.2}{2 + 1.2} = 15132 \text{ N/m.}$$

$\therefore$ Nominal shear stress at long edges

$$= \frac{15132}{1000 \times 160} \approx 0.095 \text{ N/mm}^2 \text{ (safe)}$$

At the long edges, the diameter of the bars should be so restricted that the following requirement is satisfied :

$$1.3 \frac{M_1}{V} + L_0 \geq L_d$$

$$A_{st} \text{ at supports} = \frac{1000 \times 78.5}{200} = 392.5 \text{ mm}^2.$$

$\therefore$ $M_1 = 392.5 \times 140 \times 0.865 \times 610 = 7.605 \times 10^6$ N-mm.

$V = 15132$ N.

Assume that support width $l_s = 250$ mm and a side cover of 20 mm.

$$\therefore \quad L_0 = \frac{l_s}{2} - x' + 13\,\Phi = \frac{250}{2} - 20 + 13 \times 10$$

$$= 235 \text{ mm}$$

$$\therefore \quad 1.3 \frac{M_1}{V} + L_0 = 1.3 \times \frac{7.605 \times 10^6}{15132} + 235 = 888 \text{ mm}$$

$$L_d \approx 58.3\,\Phi = 58.3 \times 10 = 583 \text{ mm.}$$

$$\therefore \quad 1.3 \frac{M_1}{V} + L_0 \geq L_d$$

Note. The code requires that the positive reinforcement should extend into the support at least by $\frac{L_d}{3}$. Hence minimum support width $= \frac{L_d}{3} + x' = \frac{583}{3} + 20 \approx 215$ mm. In our case, the support width is 250 mm.

(6) *Check for shear and development length in long span*

S.F. at short edges $= \frac{1}{3} w\, l_x = \frac{1}{3} \times 7820 \times 5.16$

$= 13450$ N/m

$\therefore$ Nominal shear stress $= \dfrac{13450}{1000 \times 150} \approx 0.09$ N/m^2

(safe).

$$A_{st} = \frac{1000 \times 78.5}{240} = 327 \text{ mm}^2$$

$\therefore$ $M_1 = 327 \times 140 \times 0.865 \times 150 = 5.94 \times 10^6$ N-mm

$V = 13450$ N/m

$$L_0 = \frac{l_s}{2} - x' + 13\,\Phi = \frac{250}{2} - 20 + 13 \times 10$$

$= 235$ mm

$$1.3\frac{M_1}{V} + L_0 = 1.3 \times \frac{5.94 \times 10^6}{13450} + 235 = 809 \text{ mm}$$

$L_d \approx 58.3\,\Phi = 58.3 \times 10 = 583$ mm

$$\therefore \quad 1.3\frac{M_1}{V} + L_0 > L_d$$

Hence code requirements are satisfied.

(7) *Torsional reinforcement at corners*

Size of torsional mesh $= l_x/5 = 5.16/5 = 1.03$ m or $1.03 + 0.08 \approx 1.10$ m from the edge of the slab.

Area of torsional reinforcement $= \frac{3}{4}(A_{st})_x = \frac{3}{4} \times 774 \approx 580$ mm^2

Using 10 mm Φ bars, $s = \dfrac{1000 \times 78.5}{580} \approx 135$ mm.

However, it is preferable to use the same spacings as provided for main reinforcement. In this short span, the main reinforcement in the middle strip has been provided @ 100 mm c/c while for edge strip, it is provided @ 200 mm c/c. Hence provide 10 mm Φ bars @ 100 mm c/c in the short span direction. In the long span, main reinforcement is @ 120 mm c/c. Hence provide torsional reinforcement @ 120 mm c/c. This reinforcement is to be provided at both faces of the slab at each corner.

The details of the reinforcement are shown in Fig. 10.15.

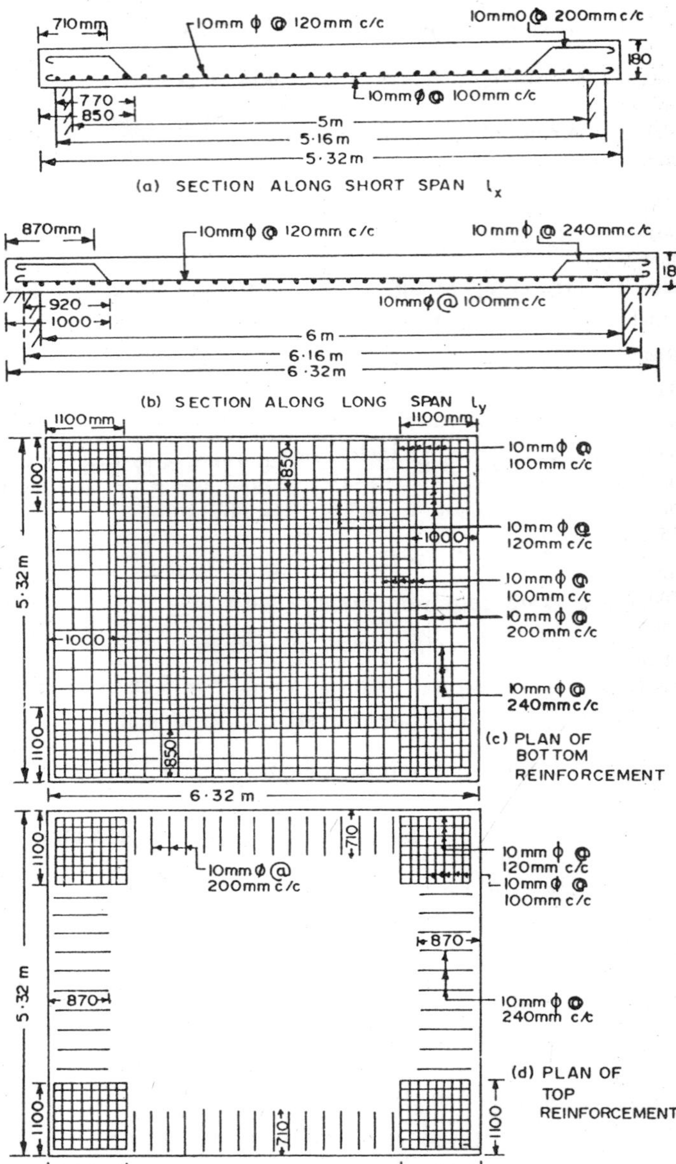
710mm
10mm ϕ @ 120mm c/c
10mm ϕ @ 200mm c/c
180
770
850
10mm ϕ @ 100mm c/c
5m
5·16 m
5·32 m
(a) SECTION ALONG SHORT SPAN l_x
870mm
10mm ϕ @ 120mm c/c
10 mm ϕ @ 240mm c/c
180
920
1000
10mm ϕ @ 100mm c/c
6 m
6·16 m
6·32 m
(b) SECTION ALONG LONG SPAN l_y
1100mm
1100mm
850
1100
1000
10mm ϕ @ 100mm c/c
10mm ϕ @ 120mm c/c
10mm ϕ @ 100mm c/c
10mm ϕ @ 200mm c/c
10mm ϕ @ 240mm c/c
5·32 m
1000
1100
850
(c) PLAN OF BOTTOM REINFORCEMENT
6·32 m
1100
710
10mm ϕ @ 200mm c/c
10 mm ϕ @ 120mm c/c
10mm ϕ @ 100mm c/c
870
5·32 m
870
10mm ϕ @ 240mm c/c
(d) PLAN OF TOP REINFORCEMENT
1100
710
1100
1100
1100

FIG. 10.15

PROBLEMS

1. Design a R.C. slab for a room 4 m × 4 m measuring from inside. The thickness of wall is 400 mm. The superimposed load, exclusive of the self weight of the slab, is 2 kN/m^2. The slab may be assumed to be simply supported at all the four edges, with corners free to lift. Use M 15 mix.
2. Redesign problem 1 assuming that the corners are held down. Use Marcus's method.
3. Design a simply supported R.C. slab over a room 4.5m × 6m from inside, assuming that the corners are not free to lift. The thickness of all the four walls is 400 mm. The live load on the floor is 2 kN/m^2. The floor carries a floor finish which weighs 8.5 kN/m^2. Use M 15 mix. Use Marcus's method.
4. Redesign Problem 3 using I.S. code method.
5. Redesign problem 1, assuming that the slab is fixed on all the four edges.
6. A R.C. floor is supported on R.C. columns 250 mm square at the corners of a rectangle 6 m ×7.5 m with R.C. beams connecting the columns along the perimeter of the rectangle. Design the floor as a two-way reinforced slab. Also design one of the 6 m span beams as an L-beam. Live load on floor = 8 kN/m^2. Use M 15 mix.
7. An underground room of a dimensions 10 m × 12 m between the existing walls 500 mm thick is to be provided with a R.C.C. roof using one column at the centre of the room and four beams, so that the slab is divided in four panels of 5m × 6m. The roof is to carry a live load of 5 kN/m^2. Design the slab and beams. Use M 15 mix.

11

Circular Slabs

11.1. INTRODUCTION

Circular slabs are used for the following purposes ; (*i*) roof of a room or hall circular in plan (*ii*) floor of circular water tanks or towers (*iii*) roof of pump houses constructed above tube wells (*iv*) roof of a traffic control post at the intersection of roads, etc.. The bending of such a slab is essentially different from a rectangular slab where bending takes place in distinctly two perpendicular directions along the two spans. When a circular slab simply supported at the edge is loaded with uniformly distributed load, it bends in the form of a saucer, due to which stresses are developed both in the radial as well as the in circumferential directions. The tensile radial and circumferential stresses develop towards the *convex* side of the saucer, and hence reinforcement need be provided at the convex face of the slab. Theoretically, reinforcement should be provided both in the radial as well as circumferential directions, but this arrangement would cause congestion and anchoring problem at the centre of the slab. Hence an alternative method of providing reinforcement is adopted : reinforcement is provided in the form of a mesh of bars having equal area of cross-section in both the directions, the area being equal to that required for the bigger of the radial and circumferential moments. However, if the stresses near the edge are not negligible, or if the edge is fixed, radial and circumferential reinforcement near the edge becomes essential.

The exact analysis of slab, based on theory of elasticity and assuming Poisson's ratio equal to zero, is beyond the scope of the present book ; only final equations are given here. Sometimes, empirical formulae are used for bending moments and shear force etc. We shall discuss in brief the following cases :

1. Slab freely supported at edges and carrying U.D.L.

2. Slab fixed at edges and carrying U.D.L.

3. Slab simply supported at the edges, with load W uniformly distributed along the circumference of a concentric circle.

4. Slab simply supported at edges, with U.D.L. inside a concentric circle.

5. Slab simply supported at the edges with a central hole and carrying U.D.L.

6. Slab simply supported at the edges, with a central hole and carrying W distributed along the circumference of a concentric circle.

11.2 SLAB FREELY SUPPORTED AT EDGES AND CARRYING U.D.L.

Let w = uniformly distributed load, kg/m^2.

a = radius of slab, m.

M_r = radial bending moment at any point at radius r from the centre of the slab.

$(M_r)_c$ &$(M_r)_e$ = radial moments at centre and edge respectively.

M_θ = circumferential bending moment at any point at radius r from the centre of the slab.

$(M_\theta)_c$ & $(M_\theta)_e$ = circumferential moments at centre and edge respectively.

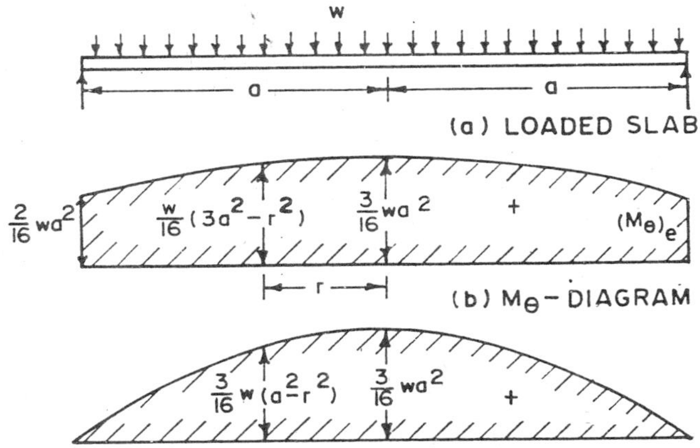

FIG. 11.1

The circumferential moment (M_θ) distribution diagram along any diameter is shown in Fig. 11.1(b). The moment varies parabolically with a maximum value of $\frac{3}{16}wa^2$ at the middle to a minimum value of $\frac{2}{16}wa^2$ at the edges. Similarly, the radial moment (M_r) distribution along any diameter is shown in Fig. 11.1 (*c*). The moment varies from $\frac{3}{16}wa^2$ at the middle to zero at the edges.

The values of various moments, per unit width are given below :

$$(M_\theta)_c = +\frac{3}{16}wa^2 \qquad ...(11.1)$$

$$(M_\theta)_e = +\frac{2}{16}wa^2 \qquad ...(11.2)$$

$$M_\theta = +\frac{w}{16}(3a^2 - r^2), \text{ at any radius } r \qquad ...(11.3)$$

$$(M_r)_c = +\frac{3}{16}wa^2 \qquad ...(11.4)$$

$$(M_r)_e = 0 \qquad ...(11.5)$$

$$M_r = \frac{3}{16}w(a^2 - r^2) \qquad ...(11.6)$$

The radial shear force F_r at any radius r is given by

$$F_r = \frac{1}{2}w \cdot r \text{ (per unit width)} \qquad ...(11.7)$$

The circumferential shear force is zero every where. C.E. Reynolds assumes that there is redistribution of moments in the slab due to plasticity of materials, during failure, and therefore recommends maximum values of M_θ and M_r as $\frac{1}{9}wa^2$ instead of $\frac{3}{16}wa^2$ given above.

11.3. SLABS FIXED AT EDGES AND CARRYING U.D.L.

Fig. 11.2 (*b*) shows the circumferential moment distribution diagram, where, M_θ varies parabolically from a maximum value at the centre to zero at the edge. The moment is *positive* throughout. Fig. 11.2(*c*) shows the radial moment distribution diagram. M_r varies from a maximum positive value at the centre to zero at a distance $\frac{a}{\sqrt{3}} = 0.577\,a$ from the centre, and then becomes negative at the edges. The various values of moments and shear per unit width are as under :

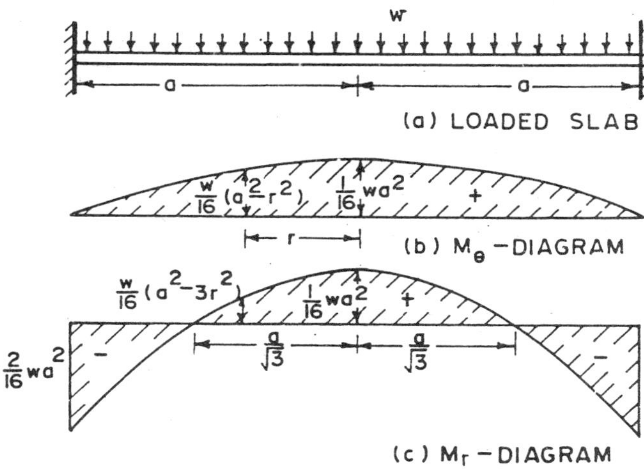

FIG. 11.2.

$$(M_\theta)_c = +\frac{1}{16}wa^2 \qquad ...(11.8)$$

$$M_\theta = +\frac{1}{16}w(a^2-r^2) \qquad ..(11.9)$$

$$(M_\theta)_e = 0 \qquad ...(11.10)$$

$$(M_r)_c = +\frac{1}{16}wa^2 \qquad ...(11.11)$$

$$M_r = +\frac{1}{16}w(a^2-3r^2) \qquad ...(11.12)$$

$$(M_r)_e = -\frac{2}{16}wa^2 \qquad ...(11.13)$$

$$F_r = \frac{1}{2}w\,.\,r \text{ (per unit width)}$$

Slabs partially fixed at the edges

This case is intermediate between the case of a freely supported slab and a fixed slab. Hence the moments may be taken as the average of the corresponding moment of the two cases. The values of various moments per unit width are as follows. For radial moment the point of contraflexure occurs at a radius $r = a\sqrt{2/3}$.

Thus, $$(M_r)_c = (M_\theta)_c = +\frac{2}{16}wa^2 \qquad ..(11.15)$$

$$(M_r)_e = -\frac{1}{16} wa^2 \qquad ...(11.16)$$

$$(M_\theta)_e = +\frac{1}{16} wa^2 \qquad ...(11.17)$$

11.4. SLAB SIMPLY SUPPORTED AT THE EDGES WITH LOAD W UNIFORMLY DISTRIBUTED ALONG THE CIRCUMFERENCE OF A CONCENTRIC CIRCLE

Both M_r as well as M_θ are constant from $r = 0$ to $r = b$. Where r is greater than b (*i.e.* outside the load circle), both M_θ and M_r decrease parabolically, to values $(M_\theta)_e$ and zero respectively at the edges . Variou values of moments etc. per unit width are as follows :

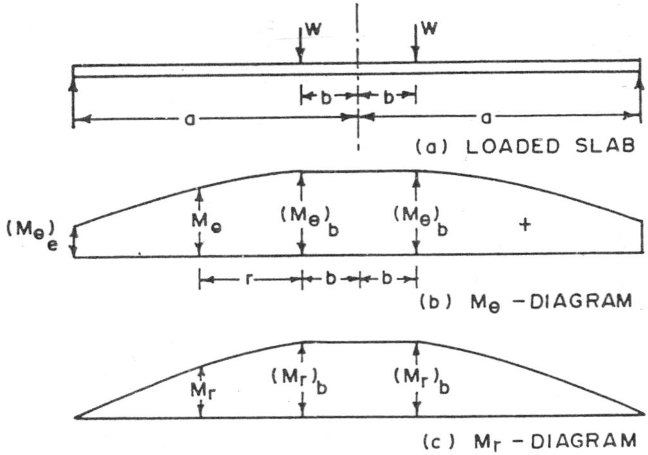

FIG. 11.3.

For $r < b$

$$M_r = (M_r)_b = M_\theta = (M_\theta)_b = \frac{W}{8\pi}\left[2\log_e\left(\frac{a}{b}\right) + 1 - \left(\frac{b}{a}\right)^2\right] \qquad ...(11.18)$$

$$F_r = 0 \qquad ...(11.19)$$

For $r > 0$

$$M_\theta = \frac{W}{8\pi}\left[2\log_e\left(\frac{a}{r}\right) - \left(\frac{b}{r}\right)^2 + 2 - \left(\frac{b}{a}\right)^2\right] \qquad ...(11.20)$$

$$M_r = \frac{W}{8\pi}\left[2\log_e\left(\frac{a}{r}\right) - \left(\frac{b}{a}\right)^2 + \left(\frac{b}{r}\right)^2\right] \quad ...(11.21)$$

$$F_r = \frac{W}{2\pi r} \quad ...(11.22)$$

11.5. SLAB SIMPLY SUPPORTED AT EDGES, WITH U.D.L. INSIDE A CONCENTRIC CIRCLE

Let b the radius of concentric circle carrying U.D.L. per unit area. The bending moments M_θ and M_r diagrams are shown in Fig. 11.4(b) and (c) respectively. The various values of moments per unit width are as follows :

For $r < b$

$$M_r = -\frac{3}{16}wr^2 + \frac{1}{4}wb^2\left[1 - \log_e\left(\frac{b}{a}\right) - \frac{b^2}{4a^2}\right] \quad ...(11.23)$$

$$M_\theta = -\frac{1}{16}wr^2 + \frac{1}{4}wb^2\left[1 - \log_e\left(\frac{b}{a}\right) - \frac{b^2}{4a^2}\right] \quad ...(11.24)$$

$$(M_r)_c = +\frac{1}{4}wb^2\left[1 - \log_e\left(\frac{b}{a}\right) - \frac{b^2}{4a^2}\right] \quad ...[11.23(a)]$$

$$(M_r)_b = (M_r)_c = \frac{3}{16}wb^2 \quad ...[11.23(b)]$$

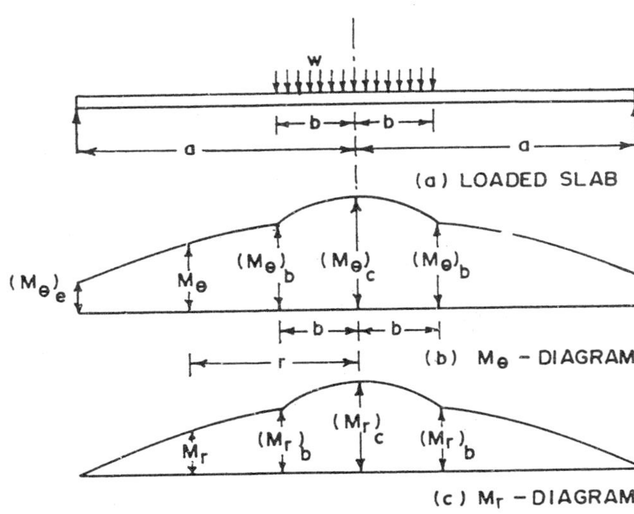

FIG. 11.4

$$(M_\theta)_c = +\frac{1}{4}wb^2\left[1 - \log_e\left(\frac{b}{a}\right) - \frac{b^2}{4a^2}\right] \qquad ...[11.24(a)]$$

$$(M_\theta)_b = (M_\theta)_c - \frac{1}{16}wb^2 \qquad ...[11.24(b)]$$

$$F_r = \frac{wr}{2} \qquad ...(11.25)$$

For $r > b$

$$M_r = -wb^2\left[\frac{1}{4}\log_e\left(\frac{r}{a}\right) + \frac{b^2}{16}\left(\frac{1}{a^2} - \frac{1}{r^2}\right)\right] \qquad ...(11.26)$$

$$M_\theta = -wb^2\left[\frac{1}{4}\log_e\left(\frac{r}{a}\right) - \frac{1}{4} + \frac{b^2}{16}\left(\frac{1}{a^2} + \frac{1}{r^2}\right)\right] \qquad ...(11.27)$$

$$\therefore \quad (M_r)_e = 0 \qquad ...(11.28)$$

$$(M_\theta)_e = -wb^2\left(-\frac{1}{4} + \frac{b^2}{8a^2}\right) \qquad ..(11.29)$$

$$F_r = \frac{wb^2}{2r} \qquad ...(11.30)$$

11.6 SLAB SIMPLY SUPPORTED AT EDGES, WITH A CENTRAL HOLE AND CARRYING U.D.L.

Let b be the radius of the hole. The B.M. diagrams are shown in Fig. 11.5. The various values of the moments per unit width are as under :

$$M_\theta = -\frac{1}{16}wr^2 + \frac{wb^2}{4}\left(\log_e\left(\frac{r}{a}\right) + \frac{3}{4}\left(-\frac{1}{3} + \frac{a^2}{b^2} + \frac{a^2}{r^2}\right)\right.$$
$$\left. - \frac{a^2 + r^2}{r^2}\,\frac{b^2}{a^2 - b^2}\log_e\frac{a}{b}\right) \qquad ...(11.31)$$

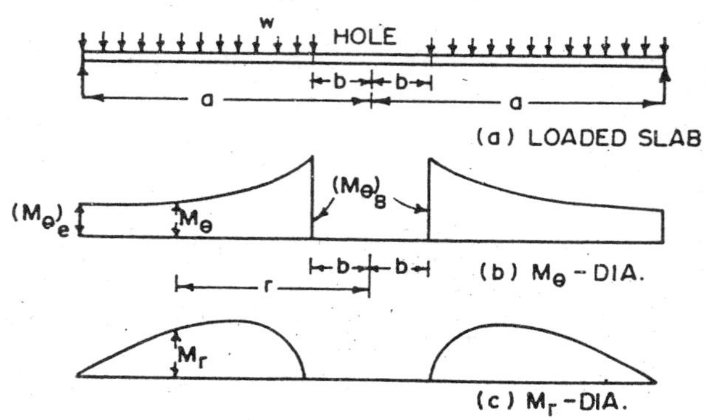

FIG. 11.5

$$M_r = -\frac{3}{16}wr^2 + \frac{wb^2}{4}\left[\log_e\frac{r}{a} + \frac{3}{4}\left(1 + \frac{a^2}{b^2} - \frac{a^2}{r^2}\right) + \frac{a^2 - r^2}{r^2}\,\frac{b^2}{a^2 - b^2}\log_e\frac{a}{b}\right] \quad ...(11.32)$$

$$F_r = \frac{wr}{2} - \frac{w\,b^2}{2r} \quad ...(11.33)$$

11.7. SLAB SIMPLY SUPPORTED AT THE EDGES WITH A CENTRAL HOLE AND CARRYING W DISTRIBUTED ALONG THE CIRCUMFERENCE OF A CONCENTRIC CIRCLE

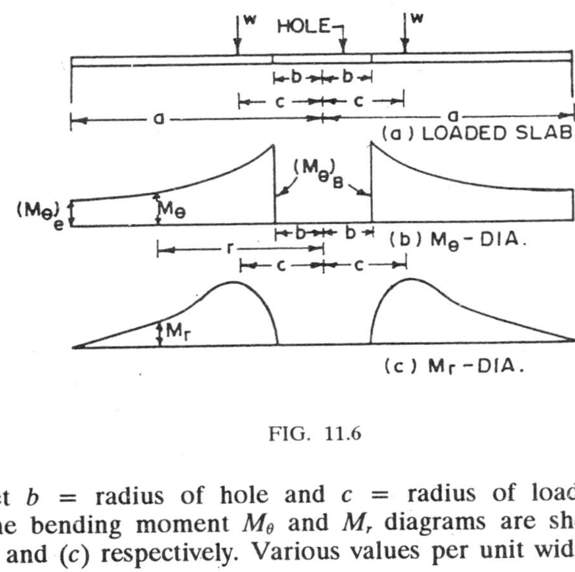

FIG. 11.6

Let b = radius of hole and c = radius of load circle.

The bending moment M_θ and M_r diagrams are shown in Fig. 11.6(b) and (c) respectively. Various values per unit width are given below :

For $r < c$

$$M_\theta = \frac{W}{4\pi}\cdot\frac{a^2}{a^2 - b^2}\left(1. + \frac{b^2}{r^2}\right)\left[\log_e\frac{a}{c} + \frac{1}{2} - \frac{c^2}{2a^2}\right] \quad ...(11.34)$$

$$M_r = \frac{W}{4\pi}\cdot\frac{a^2}{a^2 - b^2}\left(1 - \frac{b^2}{r^2}\right)\left[\log_e\frac{a}{c} + \frac{1}{2} - \frac{c^2}{2a^2}\right] \quad ..(11.35)$$

$$F_r = 0 \quad ...(11.36)$$

For $r > c$

$$M_\theta = \frac{W}{4\pi}\left[\log_e \frac{c}{r} + \frac{1}{2} + \frac{a^2}{a^2 - b^2} \cdot \frac{r^2 + b^2}{r^2}\left(\log_e \frac{a}{c} + \frac{1}{2} - \frac{c^2}{2a^2}\right) - \frac{c^2}{2r^2}\right] \quad ...(11.37)$$

$$M_r = \frac{W}{4\pi}\left[\log_e \frac{c}{r} - \frac{1}{2} + \frac{a^2}{a^2 - b^2} \cdot \frac{r^2 - b^2}{r^2}\left(\log_e \frac{a}{c} + \frac{1}{2} - \frac{c^2}{2a^2}\right) + \frac{c^2}{2r^2}\right] \quad ...(11.38)$$

$$F = \frac{W}{2\pi r} \quad ...(11.39)$$

Example 11.1. *A circular room has 5 m diameter from inside. Design a circular roof slab for room, to carry a superimposed load of 3800 N/m². Assume that the slab is simply supported at the edges. Use M 15 mix. and (a) Mild steel bars (b) HYSD bars.*

Solution. **(a) Mild Steel bars**

1 *Design constants :*

Taking $\sigma_{st} = 140$ N/mm², we have $k=0.404$; $j=0.865$ and $R=0.874$.

2. *Loading and B.M.*

Let the total depth of slab be 120 mm.

$\therefore$ Self weight $= \frac{120}{1000} \times 1 \times 1 \times 25000 = 3000$ N/m²

Super-imposed load $= 3800$ N/m²

Total $w = 6800$ N/mm²

$$(M_r)_c = (M_\theta)_c = +\frac{3}{16}wa^2 = \frac{3}{16} \times 6800\,(2.5)^2 = 7769 \text{ N--m}$$

$$= 7.769 \times 10^6 \text{ N}-\text{mm}$$

$$(M_\theta)_e = \frac{2}{16}wa^2 = \frac{2}{16} \times 6800(2.5)^2 = 5308 \text{ N}-\text{m}$$

$$= 5.308 \times 10^6 \text{ N}-\text{mm}$$

3. *Design of section*

The slab will be designed for the greater of the two moments.

$$\therefore \quad d = \sqrt{\frac{7.769 \times 10^6}{1000 \times 0.874}} = 94.3 \text{ mm}$$

Let us keep total depth = 125 mm. Using 10 mm Φ bars and 15 mm clear cover, available $d = 125 - 15 - 5 = 105$ mm for one layer and $105 - 10 = 95$ mm for the other layer.

4. *Steel reinforcement*

Radial and circumferential reinforcement required at the centre is

$$A_{st} = \frac{7.769 \times 10^6}{140 \times 0.865 \times 95} = 675 \text{ mm}^2$$

Using 10 mm Φ bars, $A_\Phi = 78.5 \text{ mm}^2$

$$\text{Spacing } s = \frac{1000 \times 78.5}{675} = 116 \text{ mm.}$$

Hence provide both the reinforcements in the form of a mesh consisting of 10 mm diameter bars spaced @ 110 mm c/c in each of the two layers at right angles of each other. Near the edges, the bars do not have proper anchorage since they are free. There will be slipping tendency and hence bars will not be capable of taking any tension. At the edges, radial tensile stress is zero, but circumferential tensile stresses exist because of $(M_\theta)_e = 5.308 \times 10^6$ N-mm. The tendency of slipping can be avoided by providing extra circumferential reinforcement, in the form of rings placed in a width equal to the development length of the mesh bars. Available d for the rings $= 95 - 10 = 85$ mm.

$$\therefore \qquad (A_{st})_\theta = \frac{5.308 \times 10^6}{140 \times 0.865 \times 85} = 516 \text{ mm}^2$$

$$\text{Spacing} \qquad s = \frac{1000 \times 78.5}{516} \approx 150 \text{ mm}$$

$(M_\theta)_e$ is $\frac{2}{3}$ of maximum moment at the centre. The circumferential steel is tobe provided for a length$= \frac{2}{3}(45\Phi) = \frac{2}{3} \times 45 \times 10 = 300$ mm (*i.e.* the length required to develop a tensile stress of $\frac{2}{3} \times 140 = 93.3 \text{ N/mm}^2$ by bond) Hence provide rings of 10 mm Φ bars @ 150 mm c/c, total rings being $\frac{300}{150} + 1 = 3$.

Since the slab is quite thin, no temperature or distribution reinforcement at the top faces of the slab is necessary. *However*, for a thicker slab (say greater than 200 mm) such reinforcement @ 0.15% of cross-sectional area of concrete, may be provided at the top face of the slab.

5. *Check of shear*

Maximum S.F. at the edges is given by

$$V_r = \tfrac{1}{2}wa = \tfrac{1}{2} \times 6800 \times 2.5 = 8500 \text{ N/m width}$$

$$\therefore \qquad \tau_r = \frac{V_r}{bd} = \frac{8500}{1000 \times 95} \approx 0.09 \text{ N/mm}^2 \text{(safe)}$$

The arrangement of the reinforcement will be as shown in Fig. 11.7.

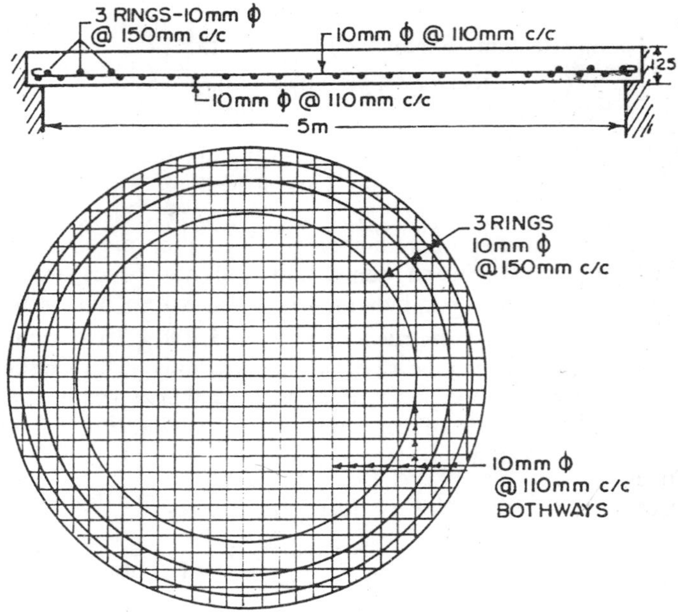

FIG. 11.7.

(b) HYSD bars

(i) Design constants

For HYSD bars, $\sigma_{st} = 230$ N/mm^2. With $\sigma_{cbc} = 5$ N/mm^2, we have k=0.2923, j=0.9026 and $R \approx 0.66$.

(ii) Loading and B.M.

Let the total depth be 130 mm.

$$\therefore \text{ Self weight} = \frac{130}{1000} \times 1 \times 25000 = 3250 \text{ N/m}^2$$

$$\text{Super-imposed load} = 3800 \text{ N/m}^2$$

$$\text{Total} \quad w = 7050 \text{ N/m}^2$$

$$(M_r)_c = (M_\theta)_c = \frac{3}{16} wa^2 = \frac{3}{16} \times 7050(2.5)^2 = 8262 \text{ N-m}$$

$$= 8.262 \times 10^6 \text{ N-mm}$$

$$(M_\theta)_e = \frac{2}{16} wa^2 = \frac{2}{16} \times 7050(2.5)^2 = 5508 \text{ N-m}$$

$$= 5.508 \times 10^6 \text{ N-mm}$$

(iii) ***Design of section***

Design moment $= 8.262 \times 10^6$ N-mm

$$\therefore \qquad d = \sqrt{\frac{8.262 \times 10^6}{1000 \times 0.66}} = 112 \text{ mm}$$

Let us keep total depth $= 145$ mm. Using 8 mm dia bars and 15 mm clear cover, $d = 145 - 15 - 4 = 126$ mm for one layer and $126 - 8 = 118$ mm for the other layer.

(iv) ***Steel reinforcement***

Radial and circumferential reinforcement required at the centre is

$$A_{st} = \frac{8.262 \times 10^6}{230 \times 0.9026 \times 118} = 337.3 \text{ mm}^2$$

Using 8 mm dia bars having $A_\Phi = 50.27$ mm^2

$$\text{Spacing} \qquad s = \frac{1000 \times 50.27}{337.3} = 149 \text{ mm.}$$

Hence provide a mesh 8 mm dia. bars spaced @ 140 mm c/c. To avoid slipping at the edges, provide extra circumferential reinforcement in the form of rings.

Available depth $= 118 - 8 = 110$ mm

$$(A_{st})_\theta = \frac{5.508 \times 10^6}{230 \times 0.9026 \times 110} = 241.2 \text{ mm}^2$$

$$\therefore \qquad \text{Spacing} \qquad s = \frac{1000 \times 50.27}{109.3} = 208 \text{ mm}$$

$(M_\theta)_e$ is $\frac{2}{3}$ of maximum moment at the centre. The circumferential steel is to be provided for a length of $\frac{2}{3}(68\,\Phi) = \frac{2}{3} \times 68 \times 8 = 363$ mm (*i.e.*, a length required to develop a tensile stress of $\frac{2}{3} \times 230 = 153.3$ N/mm^2 by bond). Hence provide rings of 8 mm dia bars @ 208 mm c/c, total rings being $= \frac{363}{208} + 1 \approx 3$,

i.e., spacing of rings $= \frac{363}{2} = 180$ mm.

(v) ***Check for shear***

Maximum S.F. at the edges is given by

$$V_r = \tfrac{1}{2} wa = \tfrac{1}{2} \times 7050 \times 2.5 = 8812 \text{ N/m width}$$

$$\tau_r = \frac{V_r}{bd} = \frac{8812}{1000 \times 118} = 0.075 \text{ N/mm}^2 \qquad \text{(safe)}$$

Design Example 11.2. *Solve example 11.1, assuming the slab to be partially restrained at the edges.*

Solution.

1. *Design constants :* $k = 0.404$; $j = 0.865$; $R = 0.874$

2. *Loading and B.M.*

Let the total depth of slab = 110 mm

$$\therefore \quad w = (110 \times 25) + 3800 = 6550 \text{ N/m}^2$$

Hence from Eq. (11.15),

$$(M_r)_c = (M_\theta)_c = +\frac{2}{16} wa^2 = +\frac{2}{16} \times 6550(2.5)^2 \times 1000$$

$$= +5.12 \times 10^6 \text{ N-mm}$$

$$(M_\theta)_e = +\frac{1}{16} wa^2 = +2.56 \times 10^6 \text{ N-mm}$$

$$(M_r)_e = -\frac{1}{16} wa^2 = -2.56 \times 10^6 \text{ N-mm}$$

The radial moment changes sign, the point of inflection being at a radius $a\sqrt{2/3} = 2.5\sqrt{2/3} = 2.04$ m = 2040 mm and or at a distance of 2500 − 2040 = 460 mm from the edge of the support.

3. *Design of section*

The thickness of the slab will be designed for a B.M. of 5.12×10^6.

$$d = \sqrt{\frac{5.12 \times 10^6}{1000 \times 0.874}} = 76.5 \text{ mm}$$

Let us keep total thickness = 110 mm. Using 10 mm Φ bars and a clear cover of 15 mm, available $d = 110 - 15 - 5 = 90$ mm for the bottom layer and $90 - 10 = 80$ mm for the top layer.

4. *Steel reinforcement at the centre*

The area of steel required per metre width at the centre of the slab is

$$A_{st} = \frac{5.12 \times 10^6}{140 \times 0.865 \times 80} = 528 \text{ mm}^2$$

Using 10 mm Φ bars, $A_\Phi = 78.5 \text{ mm}^2$

$$\therefore \qquad \text{Spacing} = \frac{1000 \times 78.5}{528} \approx 150 \text{ mm}$$

Hence provide a mesh of reinforcement consisting of 10 mm Φ bars placed ≈ 150 mm c/c in two mutually perpendicular directions.

5. *Circumferential reinforcement at the edges*

At the edges, $(M_\theta)_e = +2.56 \times 10^6$ N-mm and area of steel required$= \frac{1}{2} \times 528 = 264 \text{ mm}^2$. Since the mesh bars will not be in a position to bear it, additional ring reinforcement will have to be provided near the edges. Using 10 mm Φ bars for the rings,

$$s = \frac{1000 \times 78.5}{264} = 300 \text{ mm}$$

Since the circumferential moment at the edges is half of that at the centre, the required development length $= \frac{1}{2} \times 58.3\Phi = 291$ mm, *i.e.*, the length required to develop a tensile steel $= \frac{1}{2} \times 140 = 70 \text{ N/mm}^2$ by bond. Since $s = 300$ mm, only two rings spaced at 300 mm c/c will be required at the edges.

6. *Radial reinforcement at the edges*

Since there is negative radial moment of magnitude $= 2.56 \times 10^6$ N-mm, radial reinforcement has to be provided at the top face of the slab at the edges. The point of inflection is at a distance 460 mm from the edge of the support. Hence the radial distance, measured from the face of the support, upto which the radial reinforcement is to be provided will be the greater of the following :

(*i*) 460 mm $+ d = 460 + 90 = 550$ mm.

(*ii*) 460 mm $+ 12\Phi = 460 + 12 \times 10 = 580$ mm.

(*iii*) $L_d = 53.3\,\Phi = 583$ mm.

Hence provide these upto radial distance of 600 mm from the face of the support. The code specifies that at least one third of the total negative moment reinforcement at the support shall extend beyond the point of inflection for a distance not less than one six-teenth of clear span $\left(= \frac{1}{16} \times 5000 = 312.5 \text{ mm}\right)$. Hence provide alternate radial bars for a distance of $460 + 312.5 = 772.5 \approx 800$ mm from the face of support. Available $d = 110 - 15 - 5 = 90$ mm

$$A_{st} = \frac{2.56 \times 10^6}{140 \times 0.865 \times 90} = 234 \text{ mm}^2$$

$\therefore$ Spacing of 10 mm Φ bars$= \frac{1000 \times 78.5}{234} \approx 330$ mm

7. ***Check for shear***

$$F_r = \frac{1}{2} wa = \frac{1}{2} \times 6650 \times 2.5 = 8321.5 \text{ N}$$

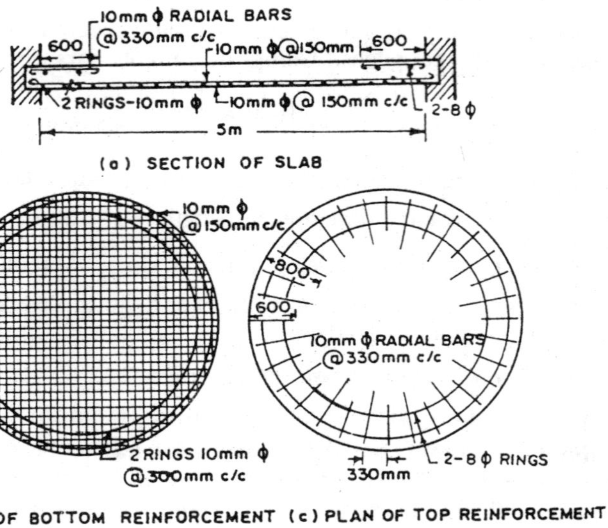

FIG. 11.8

$$\therefore \qquad \tau_v = \frac{8321.5}{1000 \times 90} = 0.092 \text{ N/mm}^2$$

This is much less than permissible shear stress. (Table 3.1 and 3.2) even at the minimum reinforcement.

The details of reinforcement etc. are shown in Fig. 11.8.

Example 11.3 *A traffic control post, 2 m in diameter is supported centrally by a reinforced concrete column, 30 cm in diameter. Design the circular slab for a super-imposed load of 1500* N/m^2*. Take* $c = \sigma_{cbc} = 5\ N/mm^2$, $t = \sigma_{st} = 140\,N/mm^2, m=18$ *and unit weight of concrete as* $2.4\,k\,N/m^3$.

Solution. Fig. 11.9 shows the slab supported centrally on the column and carrying U.D.L. w N/m^2. Let the column reaction be w_1 N/m^2 . Hence the slab may be considered to be simply supported at the edge and subjected to (*i*) a downward U.D.L. w and (*ii*) an upward U.D.L. w_1 over a concentric circle of radius $b = 150$ mm.

Hence we can use § 11.2 and § 11.5 for determining net moments on the slabs.

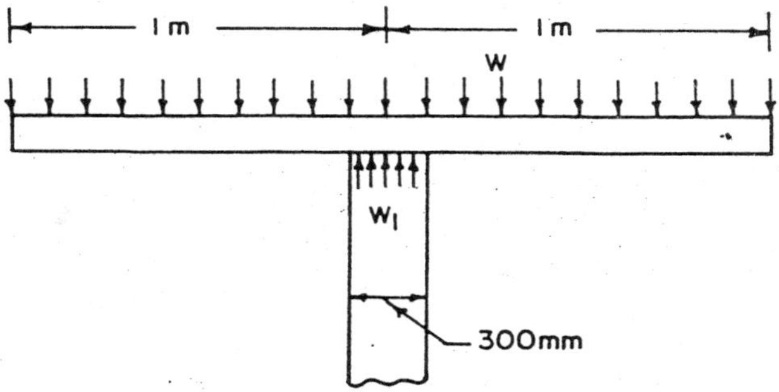

FIG. 11.9

1. ***Design constant***

For the given set of stresses, we have

$$k = 0.39\,;\, j = 0.87\,;\, R = 0.85$$

2. ***Loading***

Let the total thickness of slab = 100 mm

$$w = (100 \times 24) + 1500 = 3900 \text{ N/m}^2$$

$$w_1 = \frac{3900\,\pi\,(1)^2}{\pi\,(0.15)^2} = -\ 173000 \text{ N/m}^2$$

(minus sign has been used since w_1 acts in an upward direction).

3. ***Bending moments due to downward load w***

Refer §11.2

$$(M_r)_e = 0$$

$$(M_\theta)_e = +\frac{2}{16}wa^2 = \frac{2}{16} \times 3900(1)^2 \times 1000 = +4.88 \times 10^5 \text{ N-mm}$$

Since the slab is cast monolithic with the column, the critical section will be at the edge of the column, where $r = 0.15$ m. Hence from Eq. 11.3

$$M_\theta = +\frac{w}{16}(3a^2 - r^2) = +\frac{3900}{16}\{3 \times 1^2 - (0.15)^2\} \times 1000$$

$$= +\ 7.25 \times 10^5 \text{ N–mm}$$

Also, from Eq. 11.6

$$M_r = \frac{3}{16}w(a^2 - r^2) = +\frac{3 \times 3900}{16}\{1^2 - (0.15)^2\} \times 1000$$

$$= +\ 7.15 \times 10^5 \text{ N-mm}$$

4. Bending moments due to upward load w_1

Refer § 11.5

From Eq. 11.23, at $r=0.15\ m=b$

$$M_r = -\frac{3}{16}wr^2 + \frac{1}{4}wb^2\left[1 - \log_e\left(\frac{b}{a}\right) - \frac{b^2}{4a^2}\right]$$

Putting $w = w_1 = -173000$ and $r = b = 0.15$

$$M_r = +\frac{3}{16} \times 173000(0.15)^2 - \frac{1}{4} \times 173000(0.15)^2\left[1 - \log_e\left(\frac{0.15}{1}\right) - \left(\frac{0.15}{2}\right)^2\right]$$

$$= 730 - 2810 = -2080 \text{ N-m} = -20.8 \times 10^5 \text{ N-mm}$$

Similarly from Eq. 11.24,

$$M_\theta = -\frac{1}{16}wr^2 + \frac{1}{4}wb^2\left[1 - \log_e\left(\frac{b}{a}\right) - \frac{b^2}{4a^2}\right]$$

Putting $w = w_1 = -173000$ and $r = b = 0.15$ m,

$$M_\theta = \frac{1}{16} \times 173000(0.15)^2 - \frac{1}{4} \times 173000(0.15)^2\left[1 - \log_e\left(\frac{0.15}{1}\right) - \left(\frac{0.15}{2}\right)^2\right]$$

$$= 243 - 2810 = -2567 \text{ N-m} = -25.67 \times 10^5 \text{ N-mm}$$

The radial moment at the edge is zero.

$\therefore \quad (M_r)_e = 0.$

The circumferential moment at the edge is given by Eq. 11.29.

$$(M_\theta)_e = -wb^2\left(-\frac{1}{4} + \frac{b^2}{8a^2}\right)$$

$$= 173000(0.15)^2\left[-\frac{1}{4} + \frac{1}{8}\left(\frac{0.15}{1}\right)^2\right] = -962 \text{ N-m}$$

5. *Net moments and thickness of slab*

The net moments are tabulated below.

Case	*Radial moments (N-mm)*		*Circumferential moments (N-mm)*	
	r = 0.15 m	*r = 1 m*	*r = 0.15 m*	*r = 1m*
(i)	$+7.15 \times 10^5$	0	$+7.25 \times 10^5$	$+4.88 \times 10^5$
(ii)	-20.8×10^5	0	-25.67×10^5	-9.62×10^5
Net	-13.65×10^5	0	-18.42×10^5	-4.74×10^5

Thus the moments throughout are negative, as expected, since the slab will bend having convexity upwards. Hence both radial and circumferential reinforcement will be placed at the top face of the slab.

The thickness will be designed for a bending moment of 18.42×10^5 N–mm.

$$d = \sqrt{\frac{18.42 \times 10^5}{1000 \times 0.85}} = 46.5 \text{ mm.}$$

Keep overall thickness = 80 mm.

Using 8 mm Φ bars and 15 mm clear cover, available d = 80 – 4 – 15 = 61 mm for one layer and 61 – 8 = 53 mm for the second layer of reinforcement.

6. *Steel reinforcement*

Let us design the reinforcement for the greater of the two moments.

$$\therefore \quad A_{st} = \frac{18.42 \times 10^5}{140 \times 0.87 \times 53} = 285 \text{ mm}^2$$

Using 8 mm bars, $A_\Phi = 50.3 \text{ mm}^2$

$$\therefore \quad \text{Spacing } s = \frac{1000 \times 50.3}{285} = 176 \text{ mm}$$

Hence provide both the reinforcement in the form of a mesh consisting of 8 mm bars spaced at 175 mm c/c in two directions at right angles to each other. This mesh will not be in a position to resist circumferential tensile stress at the edge. The effective depth available for the rings reinforcement = 53 – 8 = 45 mm

$$\therefore \quad A_{st} = \frac{4.74 \times 10^5}{140 \times 0.87 \times 45} = 87 \text{ mm}^2, \text{ for 1000 mm width.}$$

Since the circumferential moment at the edge is about $\frac{1}{4}$ of the moment at the centre, the circumferential reinforcement is to be provided for a width $= \frac{1}{4} \times 58.3\Phi = \frac{1}{4} \times 58.3 \times 8 = 117$ mm.

Hence area of steel required for this width $= \dfrac{87 \times 117}{1000}$ $\approx 10 \text{ mm}^2$. Using 10 mm dia. bar, area $A_\Phi = 50.3 \text{ mm}^2$. *Hence one ring is sufficient.*

7. *Check for shear*

Shear force at the edge of the column support is given by Eq. 11.30 and 11.7 :

$$F = \frac{w_1 b^2}{2r} - \tfrac{1}{2} wr = \frac{w_1 b^2}{2b} - \tfrac{1}{2} wb$$

$$= \frac{w_1 b}{2} - \tfrac{1}{2} wb = \frac{b}{2}(w_1 - w)$$

$$= \frac{0.15}{2}(173000 - 3900) = 12680 \text{ N/m}^2$$

$$\therefore \quad \tau_v = \frac{12680}{1000 \times 53}$$

$$= 0.24 \text{ N/mm}^2.$$

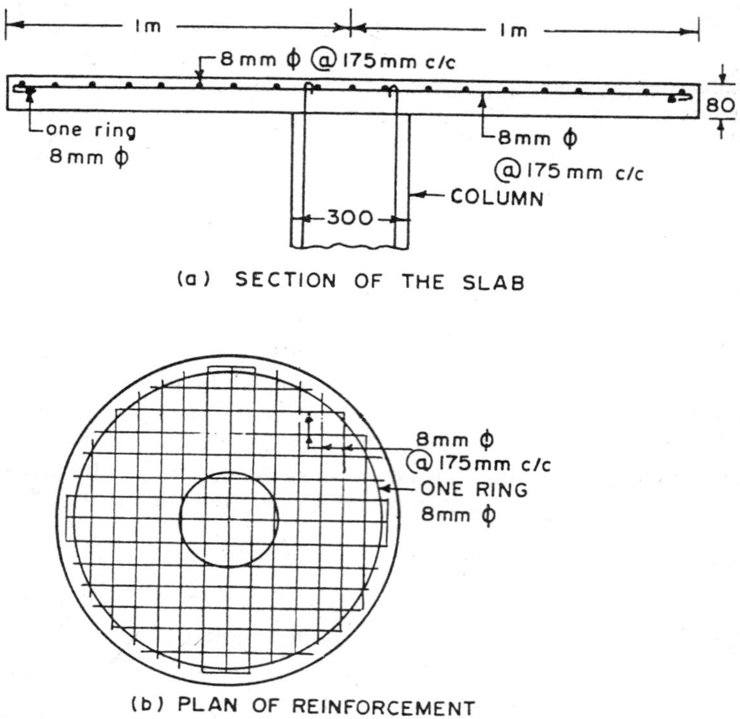

FIG. 11.10

The permissible shear stress, even at minimum reinforcement $= 1.3 \times 0.22 = 0.296$ N/mm^2 (Tables 3.1 and 3.2) . Hence safe. The details of reinforcement etc. shown in Fig. 11.10

Design Example 11.4. *The base slab of an overhead circular tank is 6 m in diameter and supported centrally over a ring beam of 2 m diameter. The slab has a central hole of 1.2 m diameter. At the periphery of the central hole, the inner walls of the tank transmit a total load of 30 kN. At the outer periphery of the slab, the outer walls of the tank transmit a total load of 150 kN. In addition to this, the slab supports a uniformly distributed water load of intensity 18350 N/mm², over the annular area. Design the base slab. Use M 15 concrete.*

Solution. Assuming the total thickness of the slab to be 450 mm its weight per sq. metre $= 450 \times 25 = 11250$ N. Hence total $w = 18350 + 11250 = 29600$ N/m². The loading on the base slab is shown in Fig. 11.11. The total load (reaction) on the support is $= 30000 + 150000 + \frac{\pi}{4}\{(6)^2 - (1.2)^2\} \times 29600 = 982000$ N.

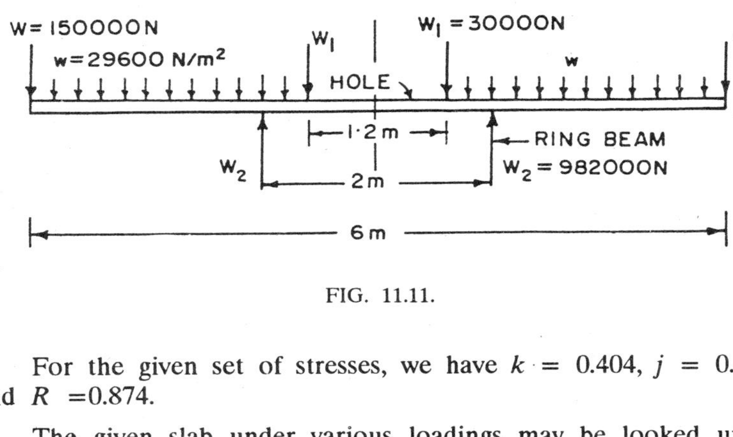

FIG. 11.11.

For the given set of stresses, we have $k = 0.404$, $j = 0.865$ and $R = 0.874$.

The given slab under various loadings may be looked upon a slab (with central hole), simply supported at the periphery and loaded as follows :

1. A downward load $W_1 = 30000$ N at the periphery of the hole.

2. A uniformly distributed load $w = 29600$ N/m² over the annual area.

3. An upward load $W_2 = 982000$ N at the outer periphery of a circle of radius 1 m.

4. A downward load $W = 150000$ N at the outer periphery of the slab. Since the imaginary supports are at the outer periphery,there

will be no B.M. or S.F. due to this load W. Hence this case has not been considered.

1. ***Bending moments due to W_1 (case 1 of loading)***

Refer §·11.7. The radial and circumferential moments are given by Eqs. 11.38 and 11.37 respectively. Substituting $W = W_1 = 30000$ N, $a = 3\,m, b = c = 0.6$ m, these equations reduce as follows:

$$M_\theta = \frac{W}{4\pi}\left[\log_e \frac{c}{r} + \frac{1}{2} + \frac{a^2}{a^2 - b^2} \cdot \frac{r^2 + b^2}{r^2} \times \left(\log_e \frac{a}{c} + \frac{1}{2} - \frac{c^2}{2a^2}\right) - \frac{c^2}{2r^2}\right]$$

$$= \frac{30000}{4\pi}\left[\log_e \frac{0.6}{r} + \frac{1}{2} + \frac{3^2}{3^2 - (0.6)^2} \cdot \frac{r^2 + (0.6)^2}{r^2} \times \left(\log_e \frac{3}{0.6} + \frac{1}{2} - \frac{(2.6)^2}{2(3)^2}\right) - \frac{1}{2}\left(\frac{0.6}{r}\right)^2\right]$$

$$= 2390\left[\log_e \left(\frac{0.6}{r}\right) + \frac{1}{2} + 2.18\left(\frac{r^2 + 0.36}{r^2}\right) - \frac{0.18}{r^2}\right] \quad ...(i)$$

$$M_r = \frac{W}{4\pi}\left[\log_e \frac{c}{r} - \frac{1}{2} + \frac{a^2}{a^2 - b^2} \cdot \frac{r^2 - b^2}{r^2} \times \left(\log_e \frac{a}{c} + \frac{1}{2} - \frac{c^2}{2a^2}\right) + \frac{c^2}{2r^2}\right]$$

$$= \frac{30000}{4\pi}\left[\log_e \frac{0.6}{r} - \frac{1}{2} + \frac{3^2}{3^2 - (0.6)^2} \cdot \frac{r^2 - (0.6)^2}{r^2} \times \left(\log_e \frac{3}{0.6} + \frac{1}{2} - \frac{(0.6)^2}{2(3)^2}\right) + \frac{1}{2}\left(\frac{0.6}{r}\right)^2\right]$$

$$= 2390\left[\log_e \left(\frac{0.6}{r}\right) - \frac{1}{2} + 2.18\left(\frac{r^2 - 0.36}{r^2}\right) + \frac{0.18}{r^2}\right] \quad ...(ii)$$

The values of M_θ and M_r at the salient locations are tabulated below :

r(m)	0.6	1	1.5	2	2.5	3
M_θ (N-m)	+10400	+6970	+4850	+3870	+3200	+2700
M_r (N-m)	0	+1720	+1170	+765	+381	0

2. ***Bending moments due to U.D.L. w(case 2 of loading)***

Refer § 11.6. The circumferential and radial moments are given by Eqs. 11.31 and 11.32 respectively.

$$M_\theta = -\frac{1}{16}wr^2 + \frac{wb^2}{4}\left[\log_e\left(\frac{r}{a}\right) + \frac{3}{4}\left(-\frac{1}{3} + \frac{a^2}{b^2} + \frac{a^2}{r^2}\right) - \frac{a^2+r^2}{r^2}\cdot\frac{b^2}{a^2-b^2}\log_e\frac{a}{b}\right]$$

$$= -\frac{29600}{16}(r^2) + \frac{29600}{4}(0.6)^2\left[\log_e\left(\frac{r}{3}\right) + \frac{3}{4}\left(-\frac{1}{3} + \left(\frac{3}{0.6}\right)^2 + \left(\frac{3}{r}\right)^2\right) - \frac{3^2+(r)^2}{r^2}\cdot\frac{(0.6)^2}{3^2-(0.6)^2}\log_e\left(\frac{3}{0.6}\right)\right]$$

$$= -1850\,r^2 + 2660\left[\log_e\left(\frac{r}{3}\right) + \frac{3}{4}\left(24.7 + \frac{9}{r^2}\right) - \frac{9+r^2}{r^2}\times 0.067\right] \quad ...(iii)$$

$$M_r = -\frac{3}{16}wr^2 + \frac{wb^2}{4}\left[\log_e\frac{r}{a} + \frac{3}{4}\left(1 + \frac{a^2}{b^2} - \frac{a^2}{r^2}\right) + \frac{a^2-r^2}{r^2}\,\frac{b^2}{a^2-b^2}\log_e\frac{a}{b}\right]$$

$$= -\frac{3}{16}\times 29600\,r^2 + \frac{29600(0.6)^2}{4}\left[\log_e\frac{r}{3} + \frac{3}{4}\left(1 + \left(\frac{3}{0.6}\right)^2 - \frac{9}{r^2}\right) + \frac{9-r^2}{3}\cdot\frac{(0.6)^2}{9-(0.6)^2}\log_e\left(\frac{3}{0.6}\right)\right]$$

$$= -5550\,r^2 + 2660\left[\log_e\frac{r}{3} + \frac{3}{4}\left(26 - \frac{9}{r^2}\right) + \frac{9-r^2}{r^2}\times 0.067\right] \quad ...(iv)$$

The values of M_θ and M_r for various for r are tabulated below:

r(m)	0.6	1	1.5	2	2.5	3
M_θ (N-m)	+89730	+60750	+50340	+44400	+39650	+34350
M_r (N-m)	0	+26850	+30020	+24200	+13950	0

3. ***Bending moments due to upward load $W_2 = 982000\,N$.***

Refer § 11.7. The moments are given by Eqs . 11.34, 11.35, 11.37 and 11.38, where $W = W_2 = -98200$ N, $b = 0.6$ m, $c = 1$ m and $a = 3$ m

For $r<c$

$$M_\theta = \frac{W}{4\pi} \times \frac{a^2}{a^2-b^2}\left(1+\frac{b^2}{r^2}\right)\left[\log_e\left(\frac{a}{c}\right)+\frac{1}{2}-\frac{c^2}{2a^2}\right]$$

$$= -\frac{98200}{4\pi} \times \frac{9}{9-0.36}\left(1+\frac{0.36}{r^2}\right)\left[\log_e\left(\frac{3}{1}\right)+\frac{1}{2}-\frac{1}{2\times 9}\right]$$

$$= -125000\left(1+\frac{0.36}{r^2}\right) \quad ...(v)$$

$$M_r = \frac{W}{4\pi} - \frac{a^2}{a^2-b^2}\left(1-\frac{b^2}{r^2}\right)\left[\log_e\frac{a}{c}+\frac{1}{2}-\frac{c^2}{2a^2}\right]$$

$$= -125000\left(1-\frac{0.36}{r^2}\right) ...(vi)$$

For $r>c$

$$M_\theta = \frac{W}{4\pi}\left[\log_e\frac{c}{r}+\frac{1}{2}+\frac{a^2}{a^2-b^2}\cdot\frac{r^2+b^2}{r^2}\left(\log_e\frac{a}{c}+\frac{1}{2}-\frac{c}{2a^2}\right)-\frac{c^2}{2r^2}\right]$$

$$= -\frac{982000}{4\pi}\left[\log_e\frac{1}{r}+\frac{1}{2}+\frac{9}{9-0.36}\cdot\frac{r^2+0.36}{r^2}\right.$$

$$\left.\times\left(\log_e 3+\frac{1}{2}-\frac{1}{2\times 9}\right)-\frac{1}{2r^2}\right]$$

$$= -78000\left[\log_e\frac{1}{r}+\frac{1}{2}+1.6\left(\frac{r^2+0.36}{r^2}\right)-\frac{1}{2r^2}\right] \quad ...(vii)$$

$$M_r = \frac{W}{4\pi}\left[\log_e\frac{c}{r}-\frac{1}{2}+\frac{a^2}{a^2-b^2}\cdot\frac{r^2-b^2}{r^2}\times\left(\log_e\frac{a}{c}+\frac{1}{2}-\frac{c^2}{2a^2}\right)+\frac{c}{2r^2}\right]$$

$$= -78000\left[\log_e\frac{1}{r}-\frac{1}{2}+1.6\left(\frac{r^2-0.36}{r^2}\right)+\frac{1}{2r^2}\right] \quad (viii)$$

The values of M_θ and M_r are calculated for various values of r, from Eqs *(v)*, *(vi)*, *(vii)* and *(viii)* and tabulated below :

r(m)	***0.6***	***1***	***1.5***	***2***	***2.5***	***3***
M_θ (N-m)	–250000	– 170000	–136000	– 110000	– 93600	– 78000
M_r (N-m)	0	– 80000	– 52200	–29200	– 132500	0

4. ***Net moments and S.F.***

The combined moments (N–m) are tabulated below :

Moment	*Loading case*	*r (metres)*					
		0.6	*1*	*1.5*	*2*	*2.5*	*3*
M_θ	1.	+10400	+6970	+4850	+3870	+3200	+2700
	2.	+89730	+65750	+50340	+44400	+39650	+34350
	3.	–250000	–170000	–136000	–110000	–93600	–78000
	Net	+149870	–102280	–80810	–61730	–50750	–40950
M_r	1.	0	+1720	+1170	+765	+381	0
	2.	0	+26850	+30020	+24200	+13950	0
	3.	0	– 80000	–52200	–29200	–13250	0
	Net	0	– 51430	–21010	–4235	+1081	0

5. ***Shear force***

Maximum shear will occur at the outer edge of the ring beam, its magnitude per metre width is given by Eqs. 11.39 and 11.33.

$$F = \frac{W_1}{2\pi r} + \left(\frac{wr}{2} - \frac{wb^2}{2r}\right) - \frac{W_2}{2\pi r}$$

$$= \frac{30000}{3\pi \times 1} + \frac{29600 \times 1}{2}\left\{1 - \left(\frac{0.6}{1}\right)^2\right\} - \frac{982000}{2\pi \times 1}$$

$= -$ 150770 N per meter width

6. ***Design of thickness***

From B.M. point of view, $d = \sqrt{\dfrac{149870 \times 1000}{1000 \times 0.874}} = 414$ mm

From S.F. point of view, permitting $\tau_v = \tau_c = 0.3$ N/mm^2 corresponding to about 0.6% reinforcement(Table 3.1), we get

$$d = \frac{V}{b \cdot \tau_v} = \frac{150770}{1000 \times 0.3} = 503 \text{ mm.}$$

Let us keep total thickness= 550 mm. Using 20 mm Φ bars and a cover of 20 mm, available d for circumferential reinforcement = 550 – 20 – 10 = 520 mm. Using 10 mm Φ radial bars, available d for radial reinforcement = 520 – 10 – 5 = 505 mm.

7. ***Design of reinforcement***

Circumferential reinforcement

For circumferential reinforcement, to be provided in the form of rings, near the inner edge of slab :

$$A_{(st)\theta} = \frac{149870 \times 1000}{140 \times 0.865 \times 520} = 2380 \text{ mm}^2 \text{ per meter width}$$

Using $20 \text{ mm } \Phi, A_\Phi = 314 \text{ mm}^2$,

$$\text{Spacing} \quad s = \frac{1000 \times 314}{2300} = 132 \text{ mm}$$

However, provide the rings @ 100 mm c/c near the inner edge. At the outer edge.

$$(A_{st})_\theta = \frac{40950 \times 1000}{140 \times 0.865 \times 520} = 650 \text{ mm}^2$$

$$\therefore \quad s = \frac{8000 \times 314}{650} = 483 \text{ mm}$$

But maximum permissible spacing is 450 mm as per IS code. However, the spacing of the rings can be increased gradually from 100 mm c/c at the inner edge to 400 mm c/c at the outer edge.

Radial reinforcement

$$(A_{st})_r = \frac{51430 \times 1000}{140 \times 0.865 \times 505} = 841 \text{ mm}^2$$

Using $10 \text{ mm } \Phi \; A_\Phi = 78.5 \text{ mm}^2$

$$\therefore \quad s = \frac{1000 \times 78.5}{841} = 93 \text{ mm}$$

Hence provide radial bars @ 90 mm c/c at the ring beam support. These radial bars may bent downwards near the inner edge.

8. ***Check for development length***

The radial reinforcement should be checked for development length at the ring beam support, where their spacing is @ 90 mm c/c.

$$\therefore \quad (A_{st})_r = 841 \times \frac{93}{90} = 869 \text{ mm}^2$$

$$\therefore \quad M_1 = \sigma_{st} \,.\, (A_{st})_r \,.\, jd = 140 \times 869 \times 0.865 \times 505$$

$$= 53.15 \times 10^6 \text{ N-mm}$$

$$V = 150770 \text{ N}$$

As per conditions of support (Fig. 11.12),

$$l_0 = \left(\frac{2000 - 1200}{2} - 20\right) + 16\,\Phi = 380 + 16 \times 10 = 540 \text{ mm}$$

$$\therefore \quad 1.3\frac{M_1}{V} + l_0 = 1.3 \times \frac{53.15 \times 10^2}{150770} + 540 = 1350 \text{ m}$$

$L_d = 58.3\ \Phi = 58.3 \times 10 = 583$ mm.

Hence code requirements are satisfied. However, bend the radial bars downwards, by 90°, at the perimeter of the central hole.

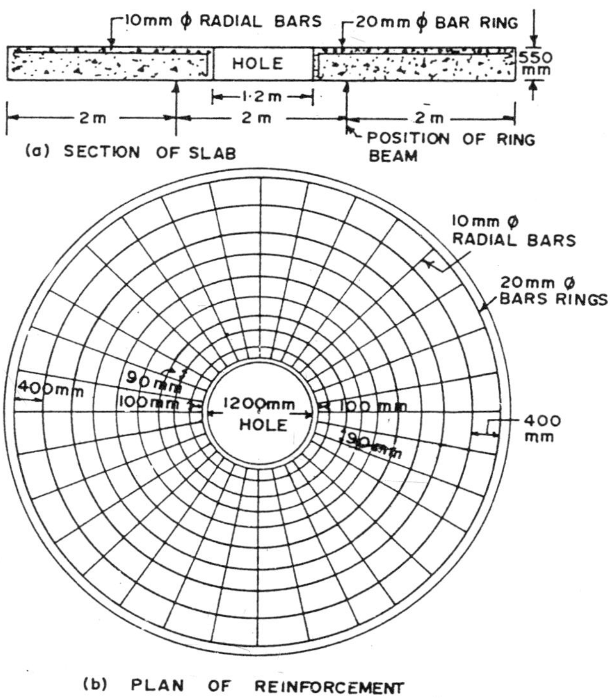

Fig. 11.12.

The details of reinforcement etc are shown in Fig. 11.12.

PROBLEMS

1. Design the roof slab for a circular room 6 metre in diameter from inside and carrying a super imposed load of 5 kN/m^2. Assume that the slab is simply supported at the edges. Use M 15 mix. The thickness of wall is 400 mm.
2. Redesign problem 1 assuming that the slab is restrained at the edge.
3. A circular room 8 metre in diameter from inside carrries a 500 mm dia. column at its centre. Design a circular slab, simply supported at the outer edge and supported on the column at its middle. The slab carries a total uniformly distributed load of 3 kN/m^2. including its own weight. Use M 15 mix.
4. Design the slab of a traffic control post, 1.5 m diameter and supported centrally on a column of 200 mm diameter. The total superimposed load, inclusive of its own weight may be taken as 1.8 kN/m^2. Use M 15 mix.

12

Flat Slabs

12.1 INTRODUCTION

A flat slab is a typical type of construction in which a reinforced

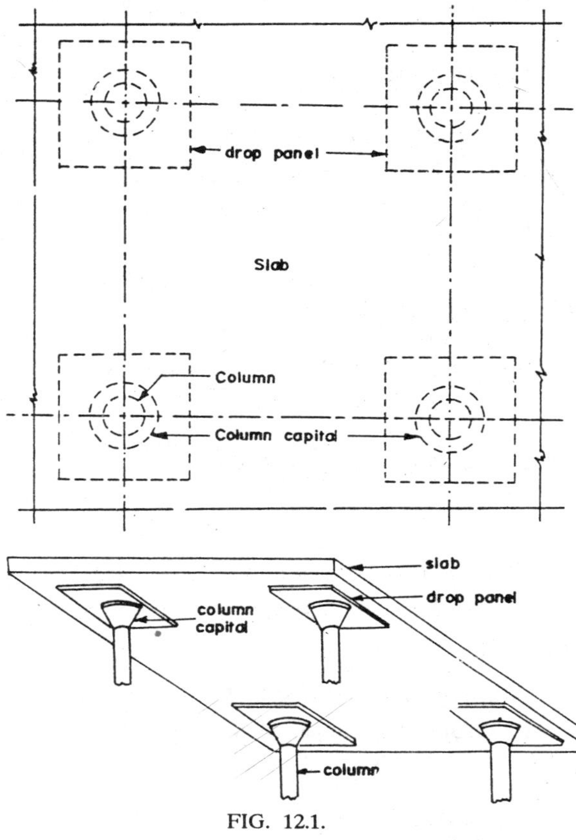

FIG. 12.1.

slab is built monolithically with the supporting columns and is reinforced in two or more directions, without any provision of beams. The flat slab thus transfers the load directly to the supporting columns suitably spaced below the slab. Because of exclusion of beam-system in this type of construction, a plain ceiling is obtained, thus giving attractive appearance from architectural point of view. The plain ceiling diffuses the light better and is considered less vulnerable in the case of fire than the usual beam slab construction. The flat slab is easier to construct and requires cheaper formwork. Concrete is more logically used in this type of construction, and hence in the case of large spans and heavy load, the total cost is considerably less.

12.2. COMPONENTS OF FLAT SLAB CONSTRUCTION

Some of the components of flat slab construction, shown in Fig. 12.1 and 12.2 are defined below.

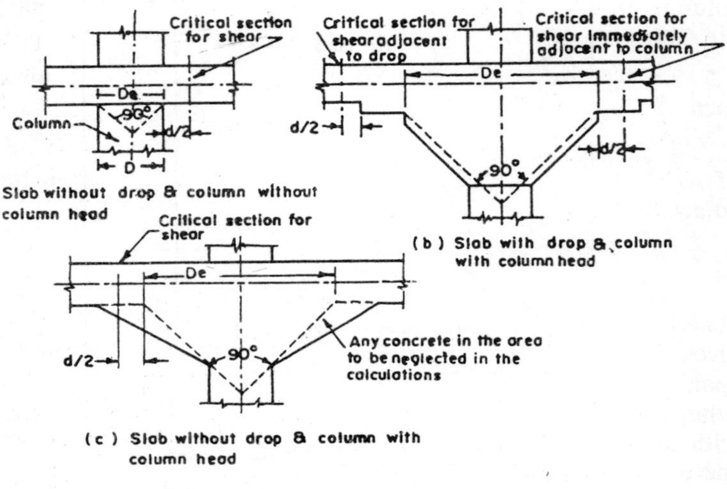

FIG. 12.2

1. ***Drop of flat slab.*** The slab in a flat slab construction may be either with drop or without drop. *Drop* is that part of the slab around the column, which is of greater thickness than the rest of the slab. The drop may have 25% to 50% more thickness, than the rest of the slab, as shown in Fig. 12.2 *(b)*. Flat slab without drop is shown in Fig. 12.2 (*a*) and (*c*).

2. ***Capital or column head.*** Sometimes, the diameter of a supporting column is increased below the slab. This part of column with increased diameter is called 'column head'. The column head increases rigidity of the slab and 'column connection' increases resistance of the slab to shear and decreases the effective span of the slab. Fig. 12.2*(c)* shows a flat slab without drop but with column head.

3. ***Panel.*** A panel of a flat slab construction is the area enclosed between the centre lines connecting adjacent columns in two directions and the outline of the column heads as shown in Fig. 12.3.

12.3 INDIAN CODE RECOMMENDATIONS (IS : 456-1978)

(*a*) General. Fig. 12.2 shows flat slab with or without drops. The following definitions shall apply :

1. ***Column strip.*** Column strip means a design strip having a width of 0.25 l_2, but not greater than 0.25 l_1 on each side of the column centre line, where l_1 is the span in the direction moments are being determined, measured centre to centre of supports.

2. ***Middle strip.*** Middle strip means a design strip bounded on each of its opposite sides by the column strip.

3. ***Panel.*** Panel means that part of the slab bounded on each of its four sides by the centre-line of a column or centre lines of adjacent spans.

(*b*) Proportioning

1. ***Thickness of Flat slab.*** The thickness of flat slab shall be generally controlled by considerations of span to effective depth ratios given in § 7.1. For slabs with drops conforming to clause *(b-2)*, span to effective depth ratios given in § 7.1 shall be applied directly; otherwise the span to effective depth ratios obtained in accordance with provisions of § 7.1 shall be multiplied by 0.9. For this purpose, the longer span shall be considered. The minimum thickness of slab shall be 125 mm.

2. ***Drops.*** The drops when provided shall be rectangular in plan, and have a length in each direction not less than one third of the panel length in the direction. For exterior panels, the width of drops at right angles to the non-continuous edge and measured from centre line of the columns shall be equal to one half the width of drop for interior panels.

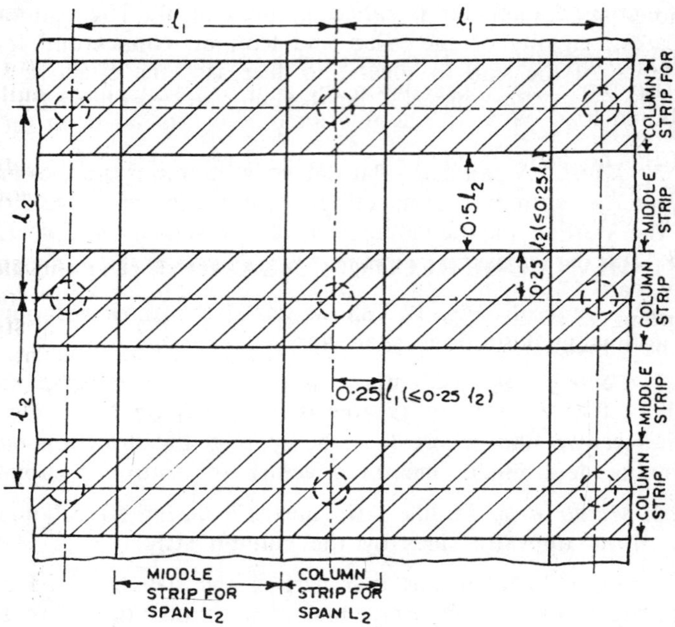

FIG. 12.3. DIVISION OF FLAT SLAB PANEL INTO STRIPS.

3. ***Column heads.*** When column heads are provided, that portion of a column head which lies within the largest circular cone or pyramid that has a vertex angle of 90° and can be included entirely within the outlines of the column and the column head, shall be considered for design purposes (see Fig. 12.2).

(c) Determination of bending moment

1. ***Methods of analysis and design.*** It shall be permissible to design the slab system by one of the following design methods.

(*i*) The direct design method.

(*ii*) The equivalent frame method.

2. ***Bending moments in panels with marginal beams or walls.*** Where the slab is supported by a marginal beam with depth greater than 1.5 times the thickness of the slab, or by a wall, then :

(*i*) The total load to be carried by the beam or wall shall comprise those loads directly on the wall or beam plus a uniformly distributed load equal to one-quarter of the total load on the slab.

(*ii*) The bending moments on the half column strip adjacent to the beam or wall shall be one-quarter of the bending moments for the first interior column strip.

3. ***Transfer of bending moments to columns.*** When unbalanced gravity load, wind, earthquake, or other lateral loads cause transfer of bending moment between slab and column, the flexural stresses shall be investigated using a fraction, α of the moment given by

$$\alpha = \frac{1}{1 + \frac{2}{3}\sqrt{\frac{a_1}{a_2}}}$$

where

a_1 = overall dimension of the critical section for shear in the direction in which moment acts, and

a_2 = overall dimension of the critical section for shear transverse to the direction in which moment acts.

A slab width between lines that are one and one-half slab or drop panel thickness, 1.5 D, on each side of the column or capital may be considered effective, D being the size of the column.

Concentration of reinforcement over column head by closer spacing or additional reinforcement may be used to resist the moment on this section.

12.4 DIRECT DESIGN METHOD

1. Limitations. Slab systems designed by the direct design method shall fulfil the following conditions :

(*i*) There shall be minimum of three continuous spans in each direction.

(*ii*) The panel shall be rectangular, and the ratio of the longer span to the shorter span within a panel shall not be greater than 2.0.

(*iii*) It shall be permissible to offset columns to a maximum of 10 percent of the span in the direction of the offset notwithstanding the provision in (*ii*).

(*iv*) The successive span lengths in each direction shall not differ by more than one third of the longer span. The end spans may be shorter but not longer than the interior spans, and

(*v*) The design live load shall not exceed three times the design dead load.

2. Total design moment for a span

(*i*) In the direct design method, the total design moment for a span shall be determined for a strip bounded laterally by the centre line of the panel on each side of the centre line of the supports.

(*ii*) The absolute sum of the positive and average negative bending moments in each direction shall be taken as :

$$M_0 = \frac{W l_n}{8}$$

M_0 = total moment

W = design load on area $l_2 . l_n$

l_n = clear span extending from face to face of columns, capitals, brackets or walls, but not less than $0.65\, l_1$

l_1 = length of span in the direction of M_0 ; and

l_2 = length of span transverse to l_1.

(*iii*) Circular supports shall be treated as square supports having the same area.

(*iv*) When the transverse span of the panels on either side of the centre line of support varies, l_2 shall be taken as the average of the transverse spans.

(*v*) When the span adjacent and parallel to an edge is being considered, the distance from the edge to the centre line of the panel shall be substituted for l_2 in (*ii*) above.

3. Negative and positive design moments

(*i*) The negative design moment shall be treated at the face of rectangular supports, circular supports being treated as square supports having the same area.

(*ii*) In an interior span, the total design moment M_0 shall be distributed in the following proportions :

Negative design moment : 0.65

Positive design moment : 0.35

(*iii*) In an end span, the total design moment M_0 shall be distributed in the following proportions :

Interior negative design moment :

$$0.75 - \frac{0.10}{1 + \frac{1}{\alpha_c}}$$

Positive design moment :

$$0.65 - \frac{0.28}{1 + \frac{1}{\alpha_c}}$$

Exterior negative design moment :

$$\frac{0.65}{1 + \frac{1}{\alpha_c}}$$

α_c is the ratio of flexural stiffness of the exterior columns to the flexural stiffness of the slab at a joint taken in the direction moments are being determined and is given by

$$\alpha_c = \frac{\Sigma K_c}{K_s}$$

where ΣK_c = Sum of the flexural stiffness of the columns meeting at the joint, and

K_s = flexural stiffness of the slab, expressed as moment per unit rotation.

(*iv*) It shall be permissible to modify these design moments by upto 10 percent, so long as the total design moment M_0 for the panel in the direction considered is not less than that required by 2 (*ii*) above.

(*v*) The negative moment section shall be designed to resist the larger of the two interior negative design moments determined for the spans forming into a common support unless an analysis is made to distribute the unbalanced moment in accordance with the stiffness of the adjoining parts.

4. Distribution of bending moments across the panel width :

Bending moments at critical cross-section shall be distributed to the column strips and middle strips as stated below :

(i) Column strip : Negative moment at an interior support. At an interior support, the column strip shall be designed to resist 75 percent of the total negative moment in the panel at that support.

(ii) Column strip : Negative moment at an exterior support.

(*a*) At an exterior support, the column strip shall be designed to resist the total negative moment in the panel at the support.

(*b*) Where the exterior support consists of a column or a wall extending for a distance equal to or greater than three quarters of the value of l_2, the length of span transverse to the direction moments are being determined, the exterior negative moments shall be considered to be uniformly distributed across the length l_2.

(iii) Column strip : Positive moment for each span: For each span, the column strip shall be designed to resist 60 percent of the total positive moment in the panel.

(iv) Moment in the middle strip : The middle strip shall be designed on the following basis :

(*a*) The portion of the design moment not resisted by the column strip shall be assigned to the adjacent middle strips.

(*b*) Each middle strip shall be proportioned to resist the sum of the moment assigned to its two half middle strips.

(*c*) The middle strip adjacent and parallel to an edge supported by a wall shall be proportioned to resist twice the moment assigned to half the middle strip corresponding to the first row of interior columns.

5. Moments in columns

(*i*) Columns built integrally with the slab system shall be designed to resist moments arising from loads on the slab systems.

(*ii*) At an interior support, the supporting members above and below the slab shall be designed to resist the moment M given by the following equation, in direct proportion to their stiffness unless general analysis is made.

$$M = 0.08 \frac{(w_d + 0.5\, w_l)\, l_2\, l_n^2 - w_d'\, l_2'\, l_n^{2\,\prime}}{1 + \dfrac{1}{\alpha_c}}$$

where w_d, w_l = design dead and live loads respectively, per unit area

l_2 = length of span transverse to the direction of M.

l_n = length of the clear span in the direction of M, measured face to face of supports.

$$\alpha_c = \frac{\Sigma K_c}{\Sigma K_s}$$

W_d' , l_2' and l_n' refer to the shorter span.

6. Effect of pattern of loading. In the direct design method, when the ratio of the live load to dead load exceed 0.5 :

(*i*) the sum of the flexural stiffness of the columns above and below the slab, ΣK_c shall be such that α_c is not less than the appropriate minimum value $\alpha_{c.min}$ specified in table 12.1, or

(*ii*) if the sum of flexural stiffnesses of the column ΣK_c does not satisfy (*i*) above, the positive design moments for the panel shall be multiplied by the coefficient β_s given by the following equation :

$$\beta_s = 1 + \left(\frac{2 - \dfrac{w_d}{w_l}}{4 + \dfrac{w_d}{w_l}}\right)\left(1 - \frac{\alpha_c}{\alpha_{c.min}}\right)$$

α_c is the ratio of flexural stiffness of the columns above and below the slab to the flexural stiffness of the slabs at the joint taken in the direction moments are being determined and the given by :

$$\alpha_c = \frac{\Sigma K_c}{\Sigma K_s}$$

where K_c and K_s are flexural stiffness of column and slab respectively. The method of computing flexural stiffnesses of various members has been illustrated in the design example.

TABLE 12..1

MINIMUM PERMISSIBLE VALUES OF α_c (IS : 456-1978)

Live load/Dead load	*Ratio l_2/l_1*	*Value of $\alpha_{c\,(min.)}$*
0.5	0.5 to 2.0	0
1.0	0.5	0.6
1.0	0.8	0.7
1.0	1.0	0.7
1.0	1.25	0.8
1.0	2.0	1.2

2.0	0.5	1.3
2.0	0.8	1.5
2.0	1.0	1.6
2.0	1.25	1.9
2.0	2.0	4.9
3.0	0.5	1.8
3.0	0.8	2.0
3.0	1.0	2.3
3.0	1.25	2.8
3.0	2.00	13.0

11.5 EQUIVALENT FRAME METHOD

1. Assumptions

The bending moments and shear forces may be determined by an analysis of the structure as a continuous frame and the following assumptions may be made :

(*i*) The structure shall be considered to be made up of equivalent frames on column lines taken longitudinally and transversely through the building. Each frame consists of a row of equivalent columns or supports, bounded laterally by the centre line of the panel on each side of the centre line of the columns or supports. Frames adjacent and parallel to an edge shall be bounded by the edge and the centre line of the adjacent panel.

(*ii*) Each such frame may be analysed in its entirelty, or for vertical loading–each floor thereof and the roof may be analysed separately with columns being assumed fixed at their remote ends. Where slabs are thus analysed separately, it may be assumed in determining the bending moment at a given support that the slab is fixed at any support two panels distant therefrom provided the slab is continuous beyond that point.

(*iii*) For the purpose of determining relative stiffness of members, the moment of inertia of any slab or column may be assumed to be that of the gross cross-section of the concrete alone.

(*iv*) Variations of moment of inertia along the axis of the slab on account of provision of drops shall be taken into account. In case of recessed or coffered slab which is made solid in the region of the columns, the stiffening effect may be ignored provided the solid part of the slab does not extend more than 0.15 l_e into the span measured from the centre line of the columns. The stiffening effect of flared column heads may be ignored.

2. Loading Pattern

(*i*) When the loading pattern is known, the structure shall be analysed for the load concerned.

(*ii*) When the live load is variable but does not exceed three quarters of the dead load, or the nature of the live load is that all panels will be loaded simultaneously, the maximum moments may be assumed to occur at all sections when full design live load is on entire slab system.

(*iii*) For other conditions of live load/dead load ratio and when all panels are not loaded simultaneously :

(*a*) maximum positive moment near mid-span of a panel may be assumed to occur when three-quarters of the full design live load is on the panel and on alternate panels ; and

(*b*) maximum negative moment in the slab at a support may be assumed to occur when three-quarters of the full design live load is on the adjacent panels only.

(*iv*) In no case shall design moments be taken to be less than those occuring with full design live load on all panels.

3. Negative Design Moment

(*i*) At interior supports, the critical section for negative moment in both the column strip and middle strip, shall be taken at the face of rectilinear supports but in no case at a distance greater than 0.175 l_1 from the centre of the column where l_1 is the length of the span in the direction moments are being determined, measured centre-to-centre of supports.

(*ii*) At exterior supports provided with brackets or capitals, the critical section for negative moment in the direction perpendicular to the edge shall be taken at a distance from the face of the supporting element not greater than one-half the projection of the bracket or capital beyond the face of the supporting element.

(*iii*) Circular or regular polygon shaped supports shall be treated as square supports having the same area.

4. Modification of Maximum Moment

Moments determined by means of the equivalent frame method, for slabs which fulfills the limitations of *direct design method* may be reduced in such proportion that the numerical sum of the positive and average negative moments is not less than the value of total design moment M_0 specified in para 2 of 12.4.

5. Distribution of Bending Moment Across the Panel width

(i) Column strip : Negative moment at an interior support. At an interior support, the column strip shall be designed to resist 75 percent of the total negative moment in the panel at that support.

(ii) Column strip : Negative moment at an exterior support

(*a*) As an exterior support, the column strip shall be designed to resist the total negative moment in the panel at the support.

(*b*) Where the exterior support consists of a column or a wall extending for a distance equal to or greater than three-quarters of the value of l_2, the length of span transvere to the direction moments are being determined the exterior negative moment shall be considered to be uniformaly distributed across the length l_2.

(iii) Column strip : Positive moment for each span: For each span, the column strip shall be designed to resist 60 percent of the total positive moment in the panel.

(iv) Moments in the middle strip: The middle strip shall be designed on the following bases :

(*a*) The portion of the design moment not resisted by the column strip shall be assigned to the adjacent middle strips.

(*b*) Each middle strip shall be proportioned to resist the sum of the moments assigned to its two half middle strips.

(*c*) The middle strip adjacent and parallel to an edge supported by a wall shall be proportioned to resist twice the moment assigned to half the middle strip corresponding to the first row of interior columns.

12.6 SHEAR IN FLAT SLAB

1. The critical section : The critical section shall be at a distance $d/2$ from the periphery of the column/capital/drop panel, perpendicular to the plane of the slab where d is the effective depth of the section (See Fig. 12.2). The shape in plan is geometrically similar to the support immediately below the slab. See Fig. 12.4*(a)* and 12.4 *(b)*.

Note : For column sections with re-entrant angles the critical section shall be taken as indicated in Fig. 12.4*(c)* and 12.4*(d)*.

(*a*) In the case of columns near the free edge of a slab, the critical section shall be taken as shown in Fig. 12.5.

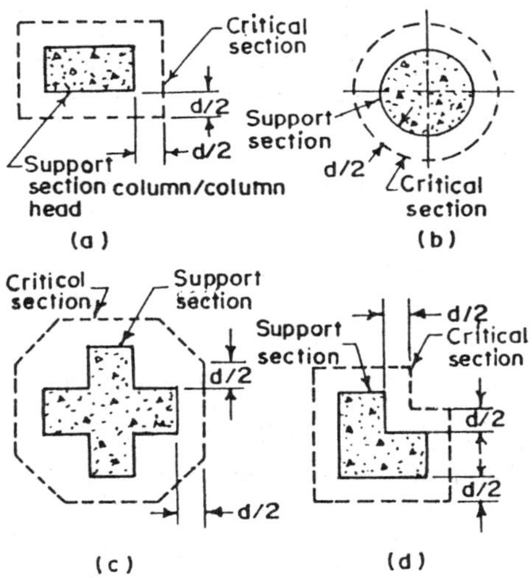

FIG. 12.4 CRITICAL SECTIONS IN PLAN FOR SHEAR.

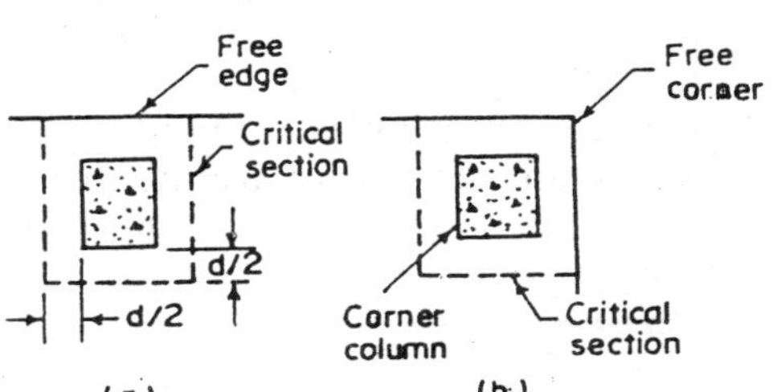

FIG. 12.5. EFFECT OF FREE EDGES ON CRITICAL SECTIONS FOR SHEAR

(*b*) When the openings in flat slabs are located at a distance less than ten times the thickness of the slab from a concentrated reaction or when the openings are located within the column strips, the critical sections specified in *(a)* shall be modified so that the part of the periphery of the critical section which is enclosed by radial projection of the openings to the centroid of the reaction area shall be considered ineffective (See Fig. 12.6), and openings shall not encroach upon column head.

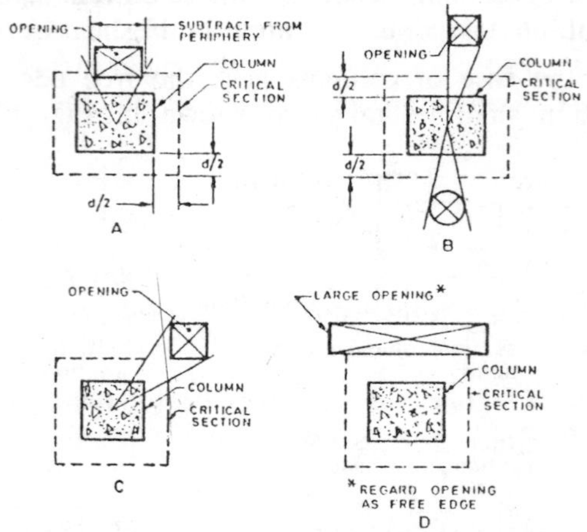

FIG. 12.6. EFFECTS OF OPENINGS ON CRITICAL SECTION FOR THE SHEAR.

2. Calculation of shear stress. The shear stress τ_v shall be the sum of the values calculated according to (*i*) and (*ii*) below:

(*i*) The nominal shear stress in flat slabs shall be taken as $V/b_0 d$ where V is the shear force due to design load, b_0 is the periphery of the critical section and d is the effective depth.

(*ii*) When unbalanced gravity load, wind, earthquake or other forces cause transfer of bending moment between slab and column, a fraction $(1-\alpha)$ of the moment shall be considered transferred by eccentricity of the shear about the centroid of the critical section. Shear stress shall be taken as varying linearly about the centroid of the critical section. The value of α shall be obtained from the equation given in clause C-3.

3. Permissible shear stress.

(*i*) When shear reinforcement is not provided, the calculated shear stress at the critical section shall not exceed $k_s \,.\, \tau_c$, where

$k_s = (0.5 + \beta_c)$ but not greater than 1, β_c being the ratio of short side to long side of the column/capital ; and

$\tau_c = 0.25\sqrt{f_{ck}}$ in limit state method of design and

$= 0.16\sqrt{f_{ck}}$ in working stress method of design.

(*ii*) When the shear stress at the critical section exceeds the value given in (*i*) above, but less than 1.5 τ_c, shear reinforcement shall be redesigned. Shear stresses shall be investigated at successive sections more distant from the support and shear reinforcement shall be provided up to a section where the shear stress does not exceed 0.5 τ_c. While designing the shear reinforcement, the shear stress carried by the concrete shall be assumed to be $0.5\tau_c$ and reinforcement shall carry the remaining shear.

12.7 SLAB REINFORCEMENT

1. Spacing. The spacing of bars in a flat slab, shall not exceed 2 times the slab thickness except where a slab is of cellular or ribbed construction.

2. Area of reinforcement. When drop panels are used the thickness of drop panel for determination of area of reinforcement shall be the lesser of the following :

(*i*) Thickness of drop, and

(*ii*) Thickness of slab plus one quarter the distance between edge of drop and edge of capital.

The reinforcement in flat slabs should have minimum lengths specified in Fig. 12.7.

12.8. OPENINGS IN FLAT SLAB

Openings of any size may be provided in the flat slab if it is shown by analysis that the requirement of strength and serviceability are met. However, for openings conforming to the following, no special analysis is required :

(1) Openings of any size may be placed within the middle half of the span in each direction provided the total amount of reinforcement required for the panel without the opening is maintained.

(2) In the area of common to two column strips, not more than one-eighth of the width of strip in either span shall be interrupted by the openings. The equivalent of reinforcement interrupted shall be added on all sides of the openings.

(3) In the area common to one column strip and one middle strip not more than one-quarter of the reinforcement in either strip shall be interrupted by the openings. The equivalent of reinforcement interrupted shall be added on all sides of the openings.

(4) Shear requirements of § 12.6 shall be satisfied.

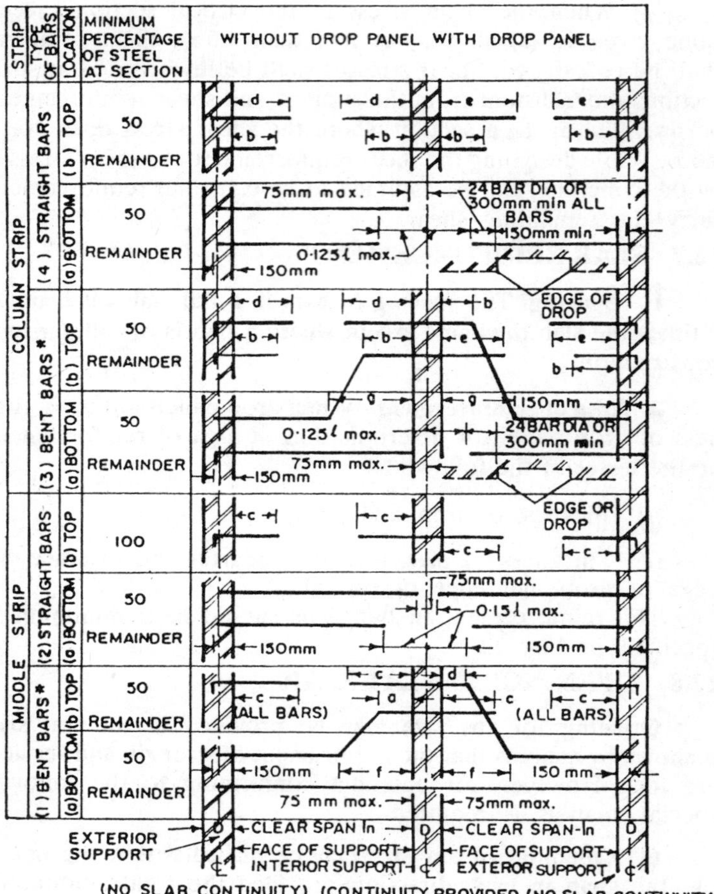

	BAR LENGTH FROM FACE OF SUPPORT						
	MINIMUM LENGTH					MAXIMUM LENGTH	
MARK	a	b	c	d	e	f	g
LENGTH	$0 \cdot 14\, l_n$	$0 \cdot 20\, l_n$	$0 \cdot 22\, l_n$	$0 \cdot 30\, l_n$	$0 \cdot 33\, l_n$	$0.20\, l_n$	$0 \cdot 24\, l_n$

FIG. 12.7. THE REINFORCEMENT DETAILS

Note. D is the dia. of column and the dimension of rectangular column in the direction under consideration.

Design Example : *Design the interior panel of a flat slab 5.6 m × 6.6 m in size, for a super-imposed load of 7.75 kN/m². Provide two-way reinforcement. Use M 15 concrete.*

Solution.

1. *Design constants*

For M15 concrete, we have $\sigma_{cbc} = 5\ \text{N/mm}^2$ and $m = 19$. Taking $\sigma_{st} = 140\ \text{N/mm}^2$ for mild steel bars, we have

$$k = 0.404\ ;\ j = 0.865 \text{ and } R = 0.874$$

2. *Panel dimensions*

Length of panel $L = 6.6$ m

Width of panel $B = 5.6$ m

Widths of strips will be as follows :

Along length $L : l_1 = L = 6.6$ and $l_2 = B = 5.6$ m

Width of column strip $= 0.25\, l_2 = 0.25 \times 5.6 = 1.4$ m, with an upper limit of $0.25\, l = 0.25 \times 6.6 = 1.65$ m. Hence keep width of column strip $= 1.4$ m on each side of column center line.

Width of middle strip $= 5.6 - 2.8$ m $= 2.8$ m

Along with B : $l_1 = B = 5.6$ and $l_2 = L = 6.9$ m

Width of column strip $= 0.25\, l_2 = 0.25 \times 6.6 = 1.65$, with an upper limit of $0.5\, l_1 = 0.25 \times 5.6 = 1.4$ m. Hence keep width of column strip $= 1.4$ m on each side of column centre line.

Width of middle strip $= 6.6 - 2.8 = 3.8$ m.

Let us provide drops also. The drops should be rectangular in plan, having a length in each direction not less than one third the panel length in that direction.

Thus, in the direction of length :

Along length L : Min length of drop $= \frac{1}{3} L = \frac{1}{3} \times 6.6 = 2.2$ m. However, keep it equal to the total width of column strip along $B = 2.8$ m

Along width B : Min. length of drop $= \frac{1}{3} B = \frac{1}{3} \times 5.6 = 1.87$ m. However, keep it equal to total width of column strip along $L = 2.8$ m.

Let the column have a column head of diameter one fifth of average span.

$$l = \frac{1}{2}(L + B) = \frac{1}{2}(6.6 + 5.6) = 6.1 \text{ m.}$$

$$\therefore \qquad D = \frac{1}{5} \approx 1.2 \text{ m.}$$

The panel details are shown in Fig. 12.8.

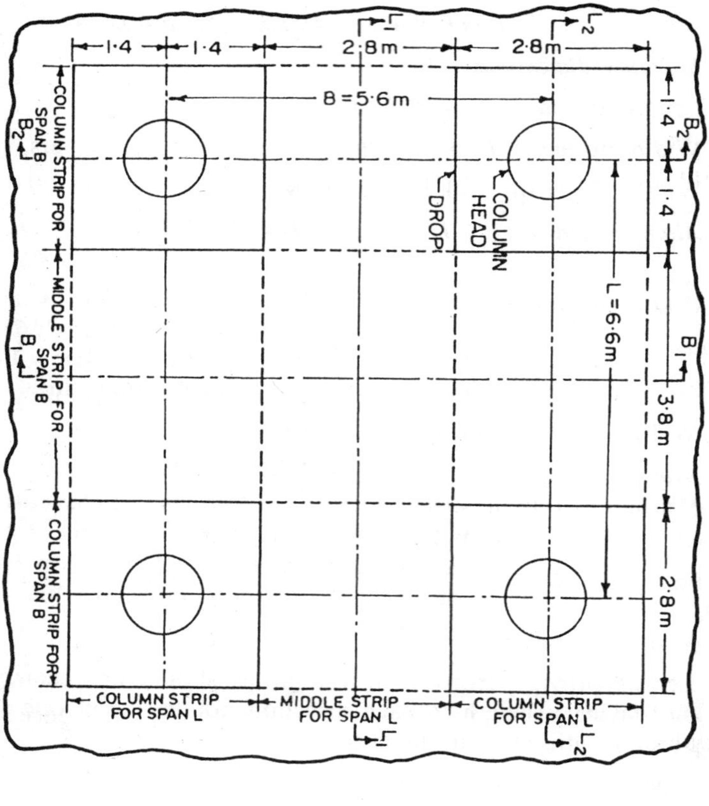

FIG. 12.8.

3. *Loading*

The thickness of flat slab is generally controlled by considerations of span to effective depth ratios given in § 7.16, clause 2.

$$\frac{\text{span}}{\text{effective depth}} \not> 26$$

Assuming balanced section, percentage reinforcement for M 15 concrete =0.72%. Hence from Fig. 7.1, modification factor for mild steel≈ 1.6.

$$\therefore \quad \frac{\text{span}}{d} = 26 \times 1.6$$

$$\therefore \quad d = \frac{\text{span}}{26 \times 1.6}$$

$$\therefore \quad d = \frac{6.6}{26 \times 1.6} = 0.159 \text{ m.}$$

$$\approx 160 \text{ mm.}$$

Assuming a cover of 15 mm, and using 12 mm Φ bars, total thickness = 160 + 6 + 12 + 15 = 193 mm. Let us provide a total thickness equal to 200 mm.

The thickness of drops is normally kept 25% to 50% more than the thickness of the slab.

Let us assume an *average* thickness of 250 mm for the calculation of dead load.

$\therefore$ Weight of slab/m^2 $= \frac{250}{1000} \times 1 \times 1 \times 25000 = 6250$ N

Super–imposed load/m^2 $= 7750$ N

Total w $= 14000$ N/m^2.

4. Moments along longer span

For the longer span $(L = 6.6 \text{ m})$,

$$l_1 = L = 6.6 \text{ m.}$$

$$l_2 = B = 5.6 \text{ m.}$$

The column head is circular, of diameter 1.2 m. Hence size of equivalent square supports

$$= \sqrt{\frac{\pi}{4}(1.2)^2} = 1.06 \text{ m}$$

$$\therefore \quad L_{nL} = l_1 - 1.06 = 6.6 - 1.06 = 5.54 \text{ m,}$$

(Subject to a minimum of $0.65\, l_1 = 0.65 \times 6.6 = 4.29$ m)

$$W_L = \text{ total design load on area } l_2 \,.\, l_{nL}$$

$$= (5.6 \times 5.54)\, 14000 \approx 434340 \text{ N}$$

$$\therefore \quad M_{OL} = \frac{W_L \,.\, l_{nL}}{8} = \frac{434340 \times 5.54}{8}$$

$$\approx 300800 \text{ N–m.}$$

Apportionment of moments

Total negative moment $= 0.65\ M_{OL}$

$= 0.65 \times 300800 = 195520$ N–m.

Total positive moment $= 0.35\ M_{OL}$

$= 0.35 \times 300800 = 105280$ N–m.

Effect of pattern of loading

In the direct design method, the ratio of live load to dead load exceeds 0.5, the factor $\alpha_c = \dfrac{\Sigma K_c}{\Sigma K_s}$ should be greater than $\alpha_{c.min}$ given in Table 12.1. If not, the positive moments for panel should be multiplied by coefficient β_0.

In order to compute α_c the flexural stiffness K_c for columns and K_s for the slab are needed. The flexural stiffness K for any member of *uniform section* is given by

$$K = \frac{4\,E_c\,I}{l}$$

where I is the moment of inertia of concrete section, neglecting reinforcement and l is the appropriate length of the member. For slab of uniform thickness h, span l_1 and transverse span l_2 :

$$K_s = \frac{4\,E_c\,l_2\,h^3}{12\,l_1}$$

Similarly, for uniform circular column of size $D \times D$ and storey height L,

$$K_c = \frac{4\,E_c\,D^4}{12\,l}$$

If, however, the member(*i.e.* slab or column) has variable moment of inertia, the flexural stiffness of the member is computed from the expression

$$K = \frac{k\,E_c\,I}{l}$$

where $k =$ stiffness factor for the section, which is more than 4.

The values of stiffness factors *(k)* for slab without drop, slab with drop and columns are given in Tables 12.2, 12.3 and 12.4 respectively, for some idealised cases.

TABLE 12.2
STIFFNESS COEFFICIENTS FOR SLAB WITH VARIABLE MOMENT OF INERTIA (WITHOUT DROP)

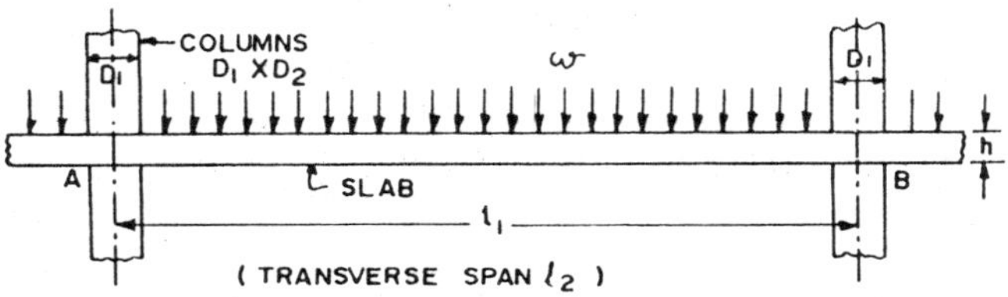

FIG. 12.9

Ratio D_1/l_1	*Stiffness factor k*
0.0	4.00
0.05	4.05
0.10	4.18
0.15	4.40
0.20	4.72
0.25	5.14

Note : The above table is valid for case $\frac{D_1}{l_1} = \frac{D_2}{l_2}$. For other cases the values are only approximate.

TABLE 12.3
STIFFNESS COEFFICIENTS FOR SLAB WITH VARIABLE MOMENT OFINERTIA (WITH DROP)

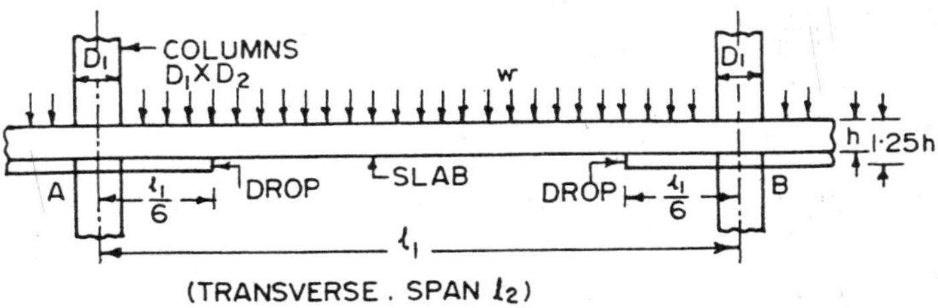

FIG. 12.10.

Ratio D_1/l_1	*Stiffness factor* k
0.0	4.78
0.05	4.84
0.10	4.98
0.15	5.22
0.20	5.55
0.25	5.98

Note. The above table is valid for case $\frac{D_1}{l_1} = \frac{D_2}{l_2}$. For other cases, the values are only approximate.

TABLE 12.4
STIFFNESS COEFFICIENTS FOR COLUMNS WITH VARIABLE MOMENT OF INERTIA

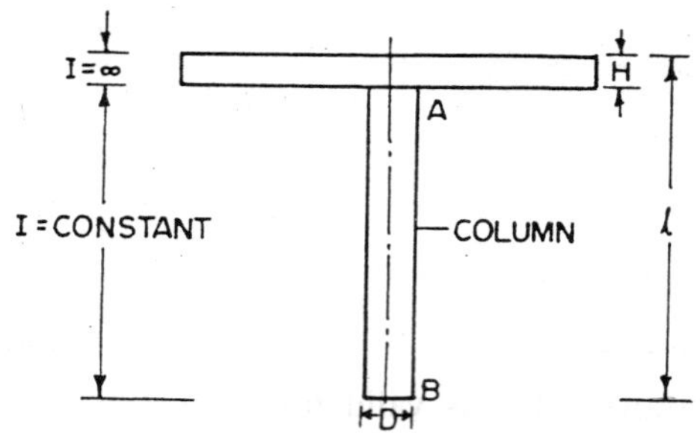

FIG. 12.11

Ratio H/l	k_{AB}	k_{BA}
0.0	4.00	4.00
0.05	4.91	4.21
0.10	6.09	4.44
0.15	7.64	4.71
0.20	9.69	5.00
0.25	12.44	5.33

For the present case, thickness of slab $h = 200$ mm $= 0.2$ mm, $l_1 = 6.6$ m and $l_2 = 5.6$ m. Let the thickness of drops be 300 mm $= 0.3$ m. Also, assume that column diameter=400 mm, height of column head =500 mm and diameter of column head =1200 mm.

Hence $D_1 = D_2 = D =$ Size of equivalent square column $= \sqrt{\frac{\pi}{4}(0.4)^2} = 0.35$ m ; $D_1/l_1 = D/l_1 = 0.35/6.6 = 0.053$.

Table 12.2 is applicable for slabs without drops. For most of the practical cases, D/l_1 will be between 0.05 to 0.1, and hence k will vary between 4.05 to 4.18. *In the absence of the table, therefore, the value of k may be taken as* 4.1

Table 12.3 is applicable for slab with drops of width $l_1/6$ on either side. For $D/l_1 = 0.05, k \approx 4.84$, against $k = 4.05$ for $D/l_1 = 0.05$ when the slab does not have drops. For width of drops more than $l_1/6$ on either side, the value of k will be more than 5. However, in absence of any table, one can adopt the value of $k = 5$ for all common cases for slabs with drops.

For the present case when drops are more than $l_1/6$ on either side, assume $k = 5$.

Now
$$k_s = \frac{k E_c I}{l} = \frac{k E_c l_2 h^3}{12 l_1}$$

$$= \frac{5 \times 5.6(0.2)^3}{12 \times 6.6} E_c = 2.83 \times 10^{-3} E_c$$

where E_c is the modulus of elasticity for concrete.

For computing the stiffness of columns meeting at the slab, Table 12.4 may be used, where H may be taken approximately as the thickness of the drop (or the slab) plus half the height of the padestal.

$$\therefore \quad H = 0.3 + 0.25 = 0.55 \text{ mm.}$$

The moment of inertia for this height H is taken to be infinite.

Let us assume that the height of storey is 4 m from top of one slab to top of other slab.

Hence l=4m. for H/l=0.55/4=0.1375, we have $k_{AB} \approx 7.25$ and $k_{BA} \approx 4.64$

For column $I = \frac{1}{12} D^4 = \frac{1}{12}(0.35)^4 = 1.25 \times 10^{-3}$

Hence for lower column,

$$K_c = \frac{k_{AB} E_c l}{l}$$

$$= \frac{7.25 \times 1.25 \times 10^{-3}}{4} E_c = 2.33 \times 10^{-3} E_c$$

and for upper column, $K_c = \dfrac{k_{BA} E_c I}{l}$

$$= \frac{4.64 \times 1.25 \times 10^{-3}}{4} E_c = 1.46 \times 10^{-3} E_c.$$

$$\therefore \qquad \alpha_c = \frac{\Sigma K_c}{\Sigma K_s} = \frac{(2.33 + 1.45)\, 10^{-3}}{2 \times 2.83 \times 10^{-3}} = 0.67$$

Now ratio $\dfrac{w_l}{w_d} = \dfrac{8000}{6000} = 1.33.$

Also, ratio $\dfrac{l_2}{l_1} = \dfrac{5.6}{6.6} = 0.85.$

For these ratios, the value of $\alpha_{c\,min.}$ is obtained from Table 12.1, which comes out to be 0.97.

Thus, $\alpha_c < \alpha_{c\,min}$.

Hence positive moment has to be multiplied by coefficient β, given by

$$\beta_0 = 1 + \frac{\left(2 - \frac{w_d}{w_l}\right)}{\left(4 + \frac{w_d}{w}\right)} \left(1 - \frac{\alpha_c}{\alpha_{c \cdot min \cdot}}\right)$$

or

$$\beta_0 = 1 + \left(\frac{2 - \frac{6}{8}}{4 + \frac{6}{8}}\right) \left(1 - \frac{0.67}{0.97}\right) = 1.081$$

Hence positive design moment

$$M_{PL} = 1.081 \times 105280$$

$$\approx 113800 \text{ N-m}$$

$\therefore$ Negative design moment

$$M_{NL} = 300800 - 113800 = 187000 \text{ N-m}.$$

These moments will be apportioned to the column strips and middle strip as under :

For column strips

– ve moment $M_{1L} = 0.75\, M_{NL}$

$$= 0.75 \times 187000$$

$$= 140250 \text{ N-m}$$

+ ve moment $M_{2L} = 0.60\, M_{PL}$
$= 0.6 \times 113800$
$= 68280$ N-m

For middle strip

– ve moment $M_{3L} = M_{NL} - M_{1L}$
$= 18700 - 140250 = 46750$ N-m

+ ve moment $M_{4L} = M_{PL} - M_{2L}$
$= 113800 - 68280 = 45520$ N-m

5. Moments along shorter span

For shorter span (B=5.6 m),

$$l_1 = B = 5.6\text{ m}\,;\, l_2 = L = 6.6\text{ m}$$

$$l_{nB} = 5.6 - 1.06 = 4.54\text{ m}$$

(Subject to a minimum of $0.65\, l_1 = 0.65 \times 5.6 = 3.65$ m)

W_B = total design load on area $l_2 \,.\, l_{nB}$

$= (6.6 \times 4.54) \times 14000 \approx 419500$ N

$$M_{0B} = \frac{W_B \,.\, l_{nB}}{8} = \frac{419500 \times 4.54}{8} \approx 238100\text{ N-m.}$$

Apportionment of moments

Total negative moment $= 0.65\ M_{0B}$
$= 0.65 \times 238100 = 154765$ N-m.

Total positive moment $= 0.35\ M_{0B}$
$= 0.35 \times 238100 = 83335$ N-m.

Effect of pattern of loading

$$\frac{D}{l_1} = \frac{0.35}{5.6} = 0.0625.$$

Hence from Table 12.3, $k \approx 4.9$. However take k =5 since width of drop is more than $l_1/6$ on each side of column centre.

$$\therefore \qquad K_s = \frac{k\, E_c \,.\, l_2\, h^3}{12\, l_1} = \frac{5 \times 6.6\, (0.2)^3\, E_c}{12 \times 5.6}$$

$$= 3.93 \times 10^{-3}\, E_c$$

$$\Sigma K_c = (2.33 + 1.45)10^{-3}\, E_c, \text{ as before.}$$

$$\therefore \qquad \alpha_c = \frac{\Sigma K_c}{\Sigma K_s} = \frac{(2.33 + 1.46)10^{-3}}{2 \times 3.93 \times 10^{-3}} = 0.48$$

Now ratio $\dfrac{w_l}{w_d} = \dfrac{8000}{6000} = 1.33$

Ratio $\frac{l_2}{l_1} = \frac{6.6}{5.6} = 1.18$

Hence from Table 12.3, value of $\alpha_{c\ min}$ comes out to be 1.09

$\therefore \quad \alpha_c < \alpha_{c\ min}.$

$$\therefore \quad \beta_0 = 1 + \frac{\left(2 - \frac{w_d}{w_l}\right)}{\left(4 + \frac{w_d}{w_l}\right)} \left(1 - \frac{\alpha_c}{\alpha_{c\ min}}\right)$$

$$= 1 + \frac{\left(2 - \frac{6}{8}\right)}{\left(4 + \frac{6}{8}\right)} \left(1 - \frac{0.48}{1.18}\right) = 1.156$$

$\therefore$ Positive design moment

$$M_{PB} = 1.156 \times 83335$$

$$= 96344 \text{ N-m}$$

Negative design moment

$$M_{NB} = 238100 - 96344$$

$$= 141756 \text{ N-m}$$

These moments will be apportioned to the column strips and middle strip as follows :

Column strips

$-$ ve moment $\quad M_{1B} = 0.75\, M_{NB} = 0.75 \times 141756$

$$= 106317 \text{ N-m}$$

$+$ ve moment $\quad M_{2B} = 0.6\, M_{PB} = 0.6 \times 96344$

$$= 57806 \text{ N-m}$$

Middle strips

$-$ ve moment $\quad M_{3B} = M_{NB} - M_{1B}$

$$= 141756 - 106317 = 35439 \text{ N-m}$$

$+$ ve moment $\quad M_{4B} = M_{PB} - M_{2B}$

$$= 96344 - 57806 = 38538 \text{ N-m.}$$

6. Thickness of slab and drops

The thickness of slab will be designed on the basis of maximum positive moment M_{2L} in the column strip of longer span.

Width of column strip of span L = 2.8 m.

$$\therefore \quad d = \sqrt{\frac{M_{2L}}{R.b}} = \sqrt{\frac{68280 \times 1000}{0.874 \times 2800}} = 167 \text{ mm}$$

Provide total thickness of 200 mm, as fixed from the deflection criterion. Using 12 mm of bars and a clear cover of 15 mm, available d for longer span $= 200 - 6 - 15 = 179$ mm and for shorter span (upper layer)$=179 - 12=167$ mm.

The thickness of drops d_d will be designed on the basis of maximum negative bending moment $M_{1L} = 140250$ N-m.

$$\therefore \qquad d_d = \sqrt{\frac{140250 \times 1000}{0.874 \times 2800}} = 239.4 \text{ mm.}$$

Keep total thickness of drop = 280 mm.

Using 12 mm Φ bars, available $d = 280 - 15 - 6 = 259$ mm for long span and $259 - 12 = 247$ mm for shorter span.

7. Check for shear

The critical section for the slab will be at a distance of $d/2$ from the face of the drop. Fig. 12.12(a) shows the plan for the periphery of critical section for slab.

Shear in slab is given by

$$V_s = (w_d + w_l)\,[l_1 l_2 - (a_1 + d)\,(a_2 + d)]$$

$$b_0 = \text{periphery of critical section in slab}$$

$$= 2(a_1 + a_2 + 2d)$$

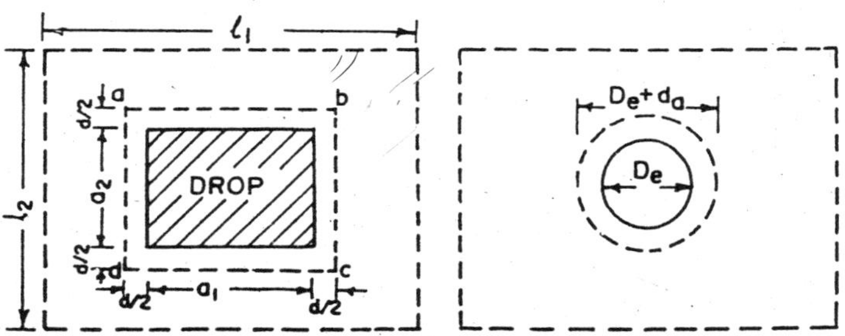

(a) SHEAR IN SLAB (b) SHEAR IN DROPS

FIG. 12.12 PERIPHERY OF CRITICAL SECTION.

For the present case,

$$a_1 = a_2 = 2.8 \text{ m} = 2800 \text{ mm}$$

$$d = \text{effective depth of slab} = 179 \text{ mm.}$$

$$\therefore \qquad V_s = [6000 + 8000]\,[6.6 \times 5.6 - (2.8 + 0.179)(2.8 + 179)]$$

$$\approx 393200 \text{ N}$$

$$b_0 = 2(2800 + 2800 + 2 \times 179) = 11916 \text{ mm}$$

$$\therefore \qquad \tau_v = \frac{V_s}{b_0 d} = \frac{393200}{11916 \times 179} = 0.184 \text{ N/mm}^2$$

Again, for shear in the drops, the critical section is at a distance of $d_d/2$ from the periphery of the column head, as shown in Fig. 12.12(*b*), where $D_c = 1.2$ m and $d_d = 0.259$ m $= 259$ mm.

Shear in drops is given by

$$V_d = (w_d + w_l)\left[l_1 l_2 - \frac{\pi}{4}(D_e + d_d)^2\right]$$

$$= (6000 + 8000)\left[6.6 \times 5.6 - \frac{\pi}{4}(1.2 + 0.259)^2\right]$$

$$= 494034 \text{ N}$$

and b_0 = periphery of critical section in slab

$$= \pi(D_e + d_d)$$

$$= \pi(1.2 + 0.259) = 4.583 \text{ m} = 4583 \text{ mm}$$

$$\therefore \quad \tau_V = \frac{V_d}{b_0 d_d} = \frac{494034}{4583 \times 259} = 0.416 \text{ N/mm}^2$$

Thus shear stress is more critical in the drops than in slabs.

Permissible shear stress $= k_s \,.\, \tau_c$

where $k_s = (0.5 + \beta_c)$ but $\not> 1$

β_c = ratio of short side to long side of the column capital

$= 1$, for the present case.

$\therefore \quad k_s = (0.5 + 1) = 1.5$, but $\not> 1$

Hence $k_s = 1$

$$\tau_c = 0.16\sqrt{f_{ck}}$$

For M-15 concrete, $f_{ck} = 15$ N/mm²

$$\therefore \quad \tau_c = 0.16\sqrt{1.5} = 0.62 \text{ N/mm}^2$$

$\therefore$ Permissible shear stress

$$= k_s \,.\, \tau_c = 1 \times 0.62 = 0.62 \text{ N/mm}^2$$

which is more than the actual shear stress in the drops and in the slab. Hence *safe*. Thus no shear reinforcement is required.

8. Reinforcement along span L

$$l_1 = L = 6.6 \text{ m}.$$

The spacing of the reinforcement is limited to twice the slab thickness, *i.e.*, maximum spacing $= 2 \times 179 = 358 \approx 350$ mm.

For determining area of reinforcement in drops, the thickness of drop panel will be the maximum of the following : (*i*) thickness of drop = 259 or (*ii*) thickness of slab $+ \frac{1}{4}$ the distance between edge of drop and edge of capital $= 179 + \frac{1}{4}(1400 - 600)$

$$= 179 + 200 = 379.$$

Hence thickness adopted for drops =259 mm (effective).

The reinforcements are to be provided for the following moments:

$$\left.\begin{aligned} M_{1L} &= -\ 140250\ \text{N–m} \\ M_{2L} &= +\ 68280\ \text{N–m} \end{aligned}\right\}\ \text{for column strip of width 2800 mm.}$$

$$\left.\begin{aligned} M_{3L} &= -\ 46750\ \text{N–m} \\ M_{4L} &= +\ 45520\ \text{N–m} \end{aligned}\right\}\ \text{for middle strip of width 2800 mm.}$$

(*i*) ***Middle strip***

For the *middle strip*, the effective thickness of slab = 179 mm

$$\therefore \qquad A_{st4L} = \frac{M_{4L}}{\sigma_{st}\, jd} = \frac{45520 \times 1000}{140 \times 0.865 \times 179}$$

$$= 2100\ \text{mm}^2\ (+\ \text{ve reinforcement})$$

Using 12 mm Φ bars, $A_\Phi = \frac{\pi}{4}(12)^2 = 113\ \text{mm}^2$.

$$\therefore\ \text{No. of bars required} = \frac{2100}{113} \approx 20$$

$$\therefore \quad \text{Average spacing} = \frac{2800}{20} = 140\ \text{mm c/c}.$$

$$\text{Again,} \qquad A_{st3L} = \frac{M_{3L}}{\sigma_{st}\, jd} = \frac{46750 \times 1000}{140 \times 0.865 \times 79}$$

$$= 2157\ \text{mm}^2\ (-\ \text{ve reinforcement})$$

$$\therefore\ \text{No. of 12 mm } \Phi \text{ bars} = \frac{2157}{113} \approx 20.$$

$$\therefore \quad \text{Average spacing} = \frac{2800}{20} = 140\ \text{mm c/c}.$$

Half of the positive reinforcement is continued upto 75 mm away from the centre of the support while the remaining half may be bent up at a distance of $f = 0.20\, l_{nL} = 0.2 \times 5.54 = 1.108$ m (max.) ≈ 1100 mm from the face of the column. The bent up bars should continue upto a distance $c = 0.22\, l_h = 0.22 \times 5.54 = 1.22$ m(min.) ≈ 1250 mm to the other face of the column. Since bent up bars will be available from both the adjacent spans, reinforcement for negative moment M_{3L} will be available as 12 mm Φ bars @ 140 mm c/c. It is to be ensured that such bending of bars provide a min. length $a = 0.14\, l_n = 0.14 \times 5.54 \approx 0.78$ m is available from the first face of the column.

The above reinforcement has been provided according to item (1) Fig. 12.7.

(ii) Column strips.

The column strips of span L have also width equal to 2800 mm. The effective thickness for +ve reinforcement is 179 mm (slab) while the effective thickness for –ve reinforcement is 159 mm (drops).

$$A_{st2L} = \frac{M_{2L}}{\sigma_{st}\ jd} = \frac{68280 \times 1000}{140 \times 0.865 \times 179}$$

$$= 3150\ \text{mm}^2 \quad (+\ \text{ve reinforcement})$$

and

$$A_{st1L} = \frac{M_{1L}}{\sigma_{st}\ jd_d} = \frac{140250 \times 1000}{140 \times 0.865 \times 259}$$

$$= 4470\ \text{mm}^2 \quad (-\text{ve reinforcement})$$

No. of 12 mm Φ bars for + ve reinforcement

$$= \frac{3150}{113} = 28$$

$$\therefore \text{ Average spacing } = \frac{2800}{28} = 100 \quad \text{mm c/c.}$$

Continue half the positive reinforcement to a distance of $0.125L = 0.125 \times 6.6 = 0.825$ m (max.) ≈ 800 mm from the column centre. Available length of bar from the edge of drop = 1400 − 800 = 600 mm which is more then 24 Φ (or 300 mm). The remaining positive reinforement bars will be bent up at a distance $= g = 0.24\,l_n = 0.24 \times 5.54 = 1.33$ m (max) ≈ 1300 mm from the face of the column. The bent up bars should continue upto a distance $d = 0.3\,l_n = 3.3 \times 5.54 \approx 1.662$ m (min.) ≈ 1700 mm beyond the outer face of the column (see item 3 of Fig. 12.7).

Thus, available negative reinforcement is 12 mm Φ @ 100 mm c/c, giving a total reinforcement of $\frac{2800}{100} \times 113 = 3164\ \text{mm}^2$ against a requirement of $A_{st1L} = 4470\ \text{mm}^2$. The balance $= A_{st1L} = 4470 - 3164 = 1306\ \text{mm}^2$ is to be provided in the form of *straight bars* at the top.

No. of straight bars required $= \frac{1306}{113} \approx 12$. However, provide 14 bars to have an even spacing$= \frac{2800}{14} = 200$ mm c/c. These bars should continue to a distance $e = 0.33\ l_n = 0.33 \times 5.54 = 1.828$ m (min.) $\approx$ 1900 mm on either side of column face, as required in item 4(*b*) of Fig. 12.7.

9. Reinforcement along span *B*

$l_1 = B = 5.6$ m.

(i) Middle strip

Width of the middle strip =3.8 m=3800 mm. The middle strip carries a positive moment of M_{4B}= 38538 N-m at its middle and negative moment of M_{3B} = 35439 N-m at its ends. Available d = 167 mm.

$$\therefore \quad A_{st4B} = \frac{M_{4B}}{\sigma_{st} jd} = \frac{38538 \times 1000}{140 \times 0.865 \times 167}$$

$$= 1906 \text{ mm}^2 \text{ (+ ve reinforcement)}$$

$\therefore$ No. of 12 mm Φ bars required $= \frac{1906}{113} \approx 17.$

$\therefore$ Spacing $= \frac{3800}{17} \approx 224$ mm c/c.

$$\text{Also, } A_{st3B} = \frac{M_{3L}}{\sigma_{st} jd} = \frac{35439 \times 1000}{140 \times 0.865 \times 167}$$

$$= 1752 \text{ mm}^2 \text{ (– ve requirement).}$$

No. of 12 mm Φ bars required $= \frac{1732}{113} \approx 16.$

However, provide 17 bars, as for +ve reinforcement, so that bent up bars serve the purpose.

l_n for span $B = l_{nB} = 4.54$ m.

Half the positive reinforcement is continued upto 75 mm away from the centre of the column, while the remaining half may be bent up at a distance of $f = 0.2\, l_n = 0.2 \times 4.54 = 0.908$ m (max.) $\approx$ 900 mm from the face of the column. The bent up bars should continue upto a distance $c = 0.22\, l_n = 0.22 \times 4.54 = 0.999$ m (min.) $\approx$ 1000 mm to the other face of the column.

(ii) Column strips

The width of column strips =2800 mm. The available effective depth =167 mm(slab) for positive reinforcement, and =247 mm(drops) for negative reinforcement. The strip carries as positive moment of M_{2B} = 57806 N-m at its middle and negative moment of M_{1B} = 106317 N-m at the ends.

$$\therefore \quad A_{st2B} = \frac{M_{2B}}{\sigma_{st} jd} = \frac{57806 \times 1000}{140 \times 0.865 \times 167}$$

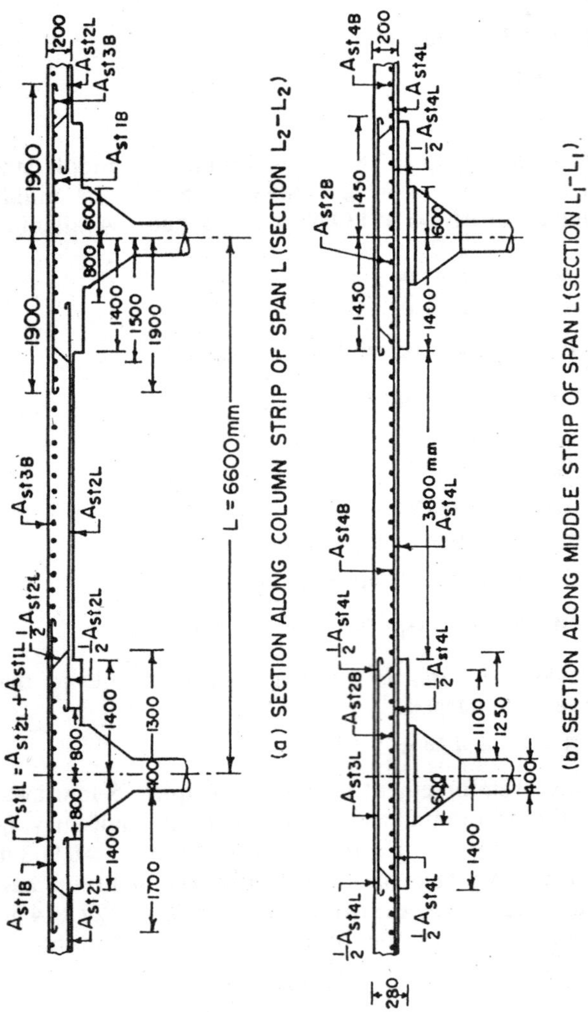

(a) SECTION ALONG COLUMN STRIP OF SPAN L (SECTION L_2-L_2)

(b) SECTION ALONG MIDDLE STRIP OF SPAN L (SECTION L_1-L_1)

FIG. 12.13

$$= 2858 \text{ mm}^2 \text{ (+ ve reinforcement)}$$

and

$$A_{st1B} = \frac{M_{1B}}{\sigma_{st}\, jd} = \frac{106317 \times 1000}{140 \times 0.865 \times 247}$$

$$= 3554 \text{ mm}^2 \text{ (− ve reinforcement)}$$

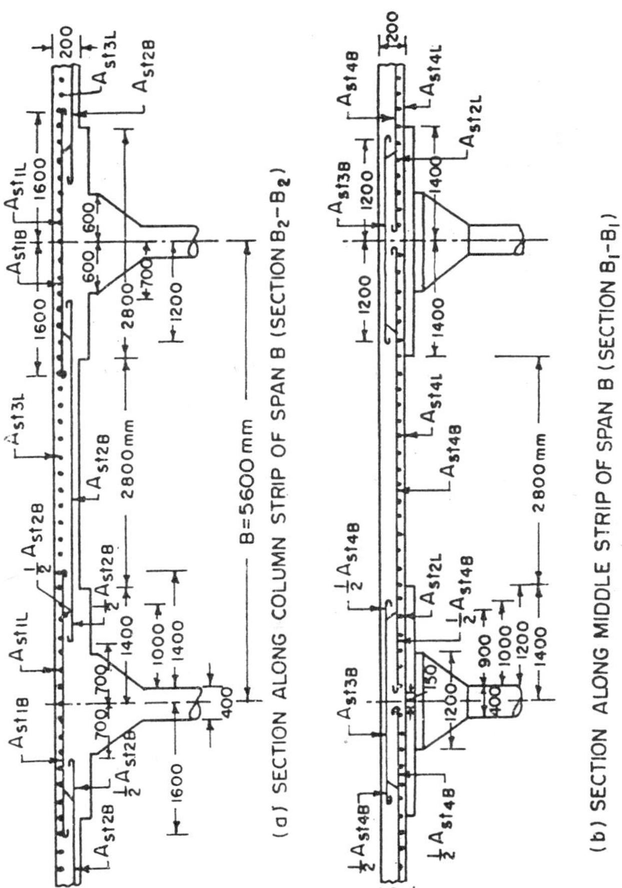

(a) SECTION ALONG COLUMN STRIP OF SPAN B (SECTION B_2-B_2)

(b) SECTION ALONG MIDDLE STRIP OF SPAN B (SECTION B_1-B_1)

FIG. 12.14

No. of 12 mm Φ bars required for + ve reinforcement

$$= \frac{2858}{113} \approx 26.$$

$\therefore$ Average spacing $= \dfrac{2800}{26} = 108$ mm c/c.

Continue half the positive reinforcement to a distance of $0.125\,B = 0.125 \times 5.6 = 0.7$ m (mix.) = 700 mm from the column centre. Available length of bar from the edge of drop =1400–700=700 mm which is more than 24 Φ (or 300 mm). The remaining +ve reinforcement bars will be bent up at a distance $g = 0.24\,l_n$ $= 0.24 \times 4.54 = 1.09$ m (mix.) $\approx$ 1000 mm from the face of the column. The bent up bars should continue upto a distance $d = 0.3\,l_n$ $= 0.3 \times 5.54 = 1.362$ m (min.) $\approx$ 1400 mm beyond the other face of the column(see item 3 of Fig. 12.7).

Thus, available negative reinforcement is 12 mm Φ @ 108 mm c/c, giving a total reinforcement of $\dfrac{2800}{108} \times 113$ $= 2930\text{ mm}^2$ against a requirement of $A_{st1B} = 3554$ mm^2 the balance $= 3554 - 2930 = 624\text{ mm}^2$ is to be provided in the form of *straight bars* at top.

No. of straight bars required $= \dfrac{624}{113} \approx 6$. Spacing of bars $= \dfrac{2800}{6} = 467$ mm c/c. However keep spacing= 4×108 = 432 cm c/c for facility in construction. These bars should continue to a distance $e = 0.33\,l_n = 0.33 \times 4.54 = 1.5$ m (min.) = 1500 mm on either side of column face, as required in item 4(*b*) of Fig. 12.7.

10. Details of reinforcement

The details of reinforcement etc. are shown in Figs. 12.13 and 12.14.

PROBLEMS

1. Design an interior panel of a flat slab, 6 m × 6 m, for a live load of 7 kN/m^2. Use M 15 concrete and provide two-way reinforcement.

2. Redesign problem 1 by providing four way reinforcement as per I.S. code of practice.

3. Redesign problem 1 if the slab has an opening of 1 m × 1 m at its middle.

4. Design an interior panel of a flat slab. 4 m × 4 m for a live load of 6 kN/m^2 Take the following values of stresses $\sigma_{cbc} = 5$ N/mm^2 ; $\sigma_{st} = 140$ N/mm^2 and m=19. Provide two-way reinforcement.

5. The roof of a hall measures 26 m × 34 m, and consists of flat slab construction dividing it into 16 panels. Design all the panels of the roof for a live load of 6.5 kN/m^2. Use M 15 concrete. Provide two way reinforcement.

PART II

COMPRESSION MEMBERS

Chapters :

13

Axially Loaded Columns

13.1 INTRODUCTION

A column is an important component of R.C. structures. A column, in general may be defined as a member carrying direct axial loads which causes compressive stresses of such magnitude that these stresses largely control its design. A column or strut is a compression member, the effective length of which exceeds three times the least lateral dimension. When a member carrying mainly axial load is vertical, it is termed as 'column' while if it is inclined or horizontal, it is termed as a 'strut'. Depending upon structural or architectural requirements, columns may be of various shapes, *i.e.* circular, rectangular, square, hexagonal, etc.

Concrete is strong in compression. However, longitudinal steel rods are always provided to assist in carrying the direct loads. A minimum area of longitudinal steel is provided in the column, whether it is required from load point of view or not.This is done to resist tensile stresses caused by some eccentricity of the vertical loads. There is also an upper limit of amount of reinforcement in R.C. columns, because higher percentage of steel may cause difficulties in placing and compacting the concrete. Longitudinal reinforcing bars are 'tied' laterally by 'ties' or stirrups' at suitable interval so that the bars do not buckle.

13.2 LONG AND SHORT COLUMNS

Columns are divided into two types : (*i*) short column and (*ii*) long column. When the ratio of *effective length* of the column to its least lateral dimension does not exceed 12, it is termed as a short column. A long column is the one whose ratio of effective length to its lateral dimension exceeds 12. In long columns, the permissible values of stresses in concrete and steel, used for short columns, should be multiplied by a coefficient C_r given by the following formula :

$$C_r = 1.25 - \frac{l_{ef}}{48b} \qquad ...(13.1)$$

C_r = reduction coefficient

l_{ef} = effective length of column (Table 13.1)

b = least lateral dimension of column.

When in a column, having helical reinforcement, the permissible load is based on the core area, the least lateral dimension should be taken as the diameter of the core.

For more exact calculations, the maximum permissible stresses in a reinforced concrete column or part there of having a ratio of effective column length to least lateral radius of gyration above 50 should not exceed those which result from the multiplication of the appropriate maximum permissible stresses by the coefficient C_r given by the following formula :

$$C_r = 1.25 - \frac{l_{ef}}{160\, i_{min}} \qquad ...(13.2)$$

where i_{min} = last radius of gyration.

Effective length of columns

The effective length (l_{ef}) of a column of length (l) may be determined with the help of Table 13.1. (Page 525)

13.3 TYPES OF COLUMNS

Columns can mainly be divided into the following three types :

1. Column with longitudinal steel and lateral ties
2. Column with longitudinal steel and spirals.
3. Composite column.

Fig. 13.1(a) shows a concrete column reinforced with longitudinal bars, but with no laterals or ties. When load is applied on such a column, the concrete bulges out laterally, as shown. The bars themselves act as long slender columns and therefore tend to buckle away from the column's axis. Due to this, tension is caused in the outside shell of the concrete which opens out. The failure usually takes place suddenly. In order to check this tendency, the longitudinal reinforcement is tied transversely, at suitable intervals, with the help of ties, as shown in Fig 13.1(b) and (c). These ties check the bars from buckling and also restrain the concrete from bulging action. When the number of bars are more than four, it is preferable to use two kinds of ties [(Fig.13.1(*b*)], placing them alternatively, so that one (set *a*) holds the corner rods while the other (set *b*) holds the intermediate rods.

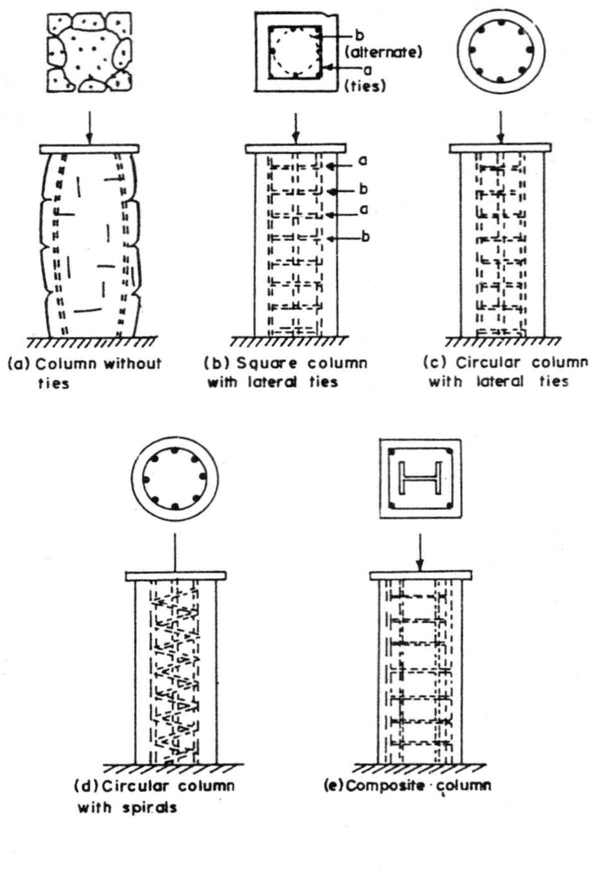

FIG. 13.1.

Each tie has to be spliced by lapping or by bending its ends around the main rod, which is quite troublesome. In order to overcome this difficulty, the longitudinal bars are tied continuously together with the held of 'spirals' shown in Fig. 13.1(*d*). The spiral so provided serve an additional purpose of laterally supporting the concrete inside and thus has confining effect on it.

Fig. 13.1(*e*) shows a typical composite column, reinforced with a centrally placed joist and with four or more longitudinal bars. Other steel sections may also be used. However, composite columns are used only for heavy loads.

13.4. LOAD CARRYING CAPACITY OF SHORT COLUMNS

(*a*) Short columns with lateral ties

According to the classical *elcstic theory* or the *compatible strain theory*, when a reinforced concrete column is loaded, both concrete and steel have equal strains during initial stage of loading, because they are well bonded. Hence if P is the load on the column, the loads carried by steel bars (P_s) and that carried by surrounding concrete(P_c) bears the relation

$$P = P_c + P_s \qquad ...(i)$$

From compatibility requirements,

$$\varepsilon_s = \varepsilon_c$$

$$\therefore \quad \frac{P_s}{E_s \cdot A_{sc}} = \frac{P_c}{E_c \cdot A_c}$$

or
$$\frac{\sigma_{sc}}{E_s} = \frac{\sigma_{cc}}{E_c}$$

or
$$\sigma_{sc} = \frac{E_s}{E_c} \cdot \sigma_{cc} = m \cdot \sigma_{cc} \qquad ...(ii)$$

where m is the modular ratio.

Hence from (i)

$$P = \sigma_{cc} \cdot A_c + \sigma_{sc} \cdot A_{sc}$$

or
$$P = \sigma_{cc} \cdot A_c + m\sigma_{cc} \cdot A_{sc} \qquad ...(13.3)$$

where A_{sc} = Area of steel reinforcement
A_c = Area of concrete
σ_{sc} = Compressive stress in steel
σ_{cc} = Compressive stress in concrete.

It has been found that the steel reinforcement is restrained against lateral expansion, due to concrete surrounding it, reducing the problem to be a plain strain problem rather than a plain stress problem. It is therefore essential to moderate the compatible stress in steel by restrained Poissons effect, using modified Poisson's ratio $m_c = 1.5\,m$. Thus stress in steel is given by

$$\sigma_{sc} = m_c \cdot \sigma_{cc} = 1.5\, m\, \sigma_{cc}$$

and equation (*iii*) reduces to the form

$$P = \sigma_{cc} \cdot A_c + 1.5\, m\, \sigma_{cc} \cdot A_{sc} \quad ...(13.4)$$

The above theoretical approach, based on strain compatibility, has been proved to be highly unrealistic due to creep and shrinkage behaviour of concrete. The above formula therefore gives very conservative results. It is therefore recommended to use *equilibrium approach* based on allowable stresses in the two materials. Such an equilibrium approach, however, violates the compatibility of strains. According to this approach, the load carrying capacity of a short column, axially loaded and provided with lateral ties, is given by

$$P = \sigma_{cc} \cdot A_c + \sigma_{sc} \cdot A_{sc} \quad ...(13.5)$$

where P = Permissible axial load

σ_{sc} = permissible compressive stress in column bars

σ_{cc} = permissible compressive stress in concrete

(*b*) **Short column with helical reinforcement**

For a column having longitudinal reinforcement tied with spirals (*i.e.* helical reinforcement), the load carrying capacity is taken as 1.05 times the strength of similar member with lateral ties, provided the requirements laid down in para 8*b* of § 13.5 are satisfied.

(*c*) **Composite columns**

The allowable load P on a composite column, consisting of a structural steel or cast-iron column thoroughly encased in concrete reinforced with both longitudinal and spiral reinforcement is given by the following expression :

$$P = \sigma_{cc} \cdot A_c + \sigma_{sc} \cdot A_{sc} + \sigma_{mc} \cdot A_m \quad ...(13.6)$$

where A_c = net area of concrete section

A_{sc} = cross-sectional area of longitudinal bar reinforcement

σ_{mc} = allowable unit stress in metal core, not to exceed 125 N/mm^2 for a steel core, or 70 N/mm^2 for a cast iron core

A_m = Cross-sectional area of steel or cast iron core.

13.5. INDIAN STANDARD RECOMMENDATIONS (IS : 456-1978)

(1) Longitudinal reinforcement

(*a*) The cross-sectional area of longitudinal reinforcement shall be not less than 0.8 percent nor more than 6 percent of the gross cross-sectional area of the column.

Note. The use of 6 percent reinforcement may involve practical difficulties in placing and compaction of concrete ; hence lower percentage is recommended. Where bars from the columns below have to be lapped with those in the column under consideration, the percentage of steel shall usually not exceed 4%.

(*b*) In any column that has a larger cross-sectional area than that required to support the load, the minimum percentage of steel shall be based upon the area of concrete required to resist the direct stress and not upon the actual area.

(*c*) The minimum number of longitudinal bars provided in a column shall be four in rectangular columns and six in circular columns.

(*d*) The bars shall not be less than 12 mm in diameter.

(*e*) A reinforced concrete column having helical reinforcement shall have at least six bars of longitudinal reinforcement.

(*f*) In a helically reinforced column, the longitudinal bars shall be in contact with the helical reinforcement and equidistant around its inner circumference.

(*g*) Spacing of longitudinal bars measured along the periphery of the column shall not exceed 300 mm.

(*h*) In case of padestals in which the longitudinal reinforcement is not taken into account in strength calculations, nominal longitudinal reinforcement not less than 0.15% of the cross- sectional area shall be provided.

Note : Padestal is a compression member, the effective length of which does not exceed three times the least lateral dimension.

(2) Transverse reinforcement

(a) General

A reinforced concrete compression member shall have transverse or helical reinforcement so disposed that every longitudinal bar nearest to the compression face has effective lateral support against buckling subject to provisions in 2(*b*). The effective lateral support is given by transverse reinforcement either in the form of circular rings capable of taking up circumferential tension or by polygonal links(lateral ties) with internal angles not exceeding 135°. The ends of the transverse reinforcement shall be properly anchored.

(*b*) ***Arrangement of transverse reinforcement***

(*i*) If the longitudinal bars are not spaced more than 75 mm on either side transverse reinforcement need only to go round corner and alternate bars for the purpose of providing effective lateral supports [Fig. 13.2(*a*)].

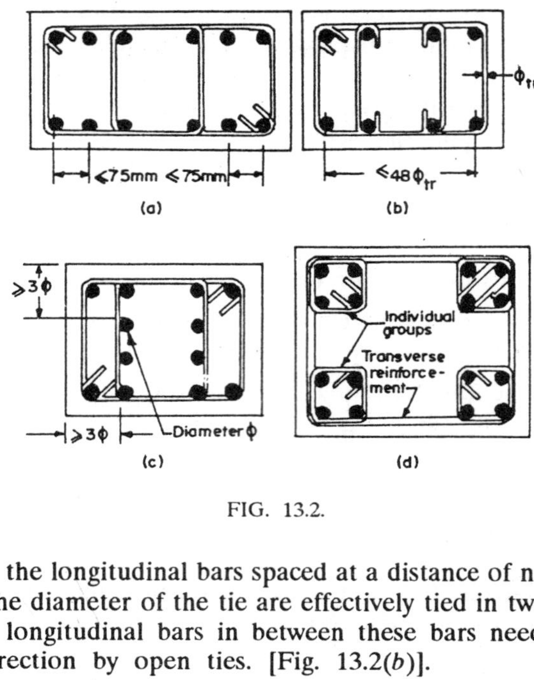

FIG. 13.2.

(*ii*) If the longitudinal bars spaced at a distance of not exceeding 48 times the diameter of the tie are effectively tied in two directions, additional longitudinal bars in between these bars need to be tied in one direction by open ties. [Fig. 13.2(*b*)].

(*iii*) Where the longitudinal reinforcing bars in a compression member are placed in more than one row, effective lateral support to the longitudinal bars in the inner rows may be assumed to have been provided if (*a*) transverse reinforcement is provided for outermost row in accordance with (2) above, and (*b*) no bar of the corner row is closer to the nearest compression face than three times the diameter of the largest bar in the inner row [Fig. 13.2(*c*)].

(*iv*) Where the longitudinal bars in a compression member are grouped(not in contact) and each group adequately tied with transverse reinforcement in accordance with clause (2), the transverse reinforcement for the compression member as a whole may be provided on the assumption that each group is a single longitudinal bar for purpose of determining the pitch and diameter of the transverse reinforcement in accordance with clause (2). The diameter of such transverse reinforcement need not, however, exceed 20 mm [see Fig. 13.2(*d*)].

(*c*) *Pitch and diameter of lateral ties*

(*i*) *Pitch.* The pitch of the transverse reinforcement shall be not more than the least of the following distances : (1) the least lateral dimension of the compression member ; (2) sixteen times the smallest diameter of the longitudinal reinforcement bar to be tied ; (3) Forty eight times the diameter of the transverse reinforcement.

(*ii*) *Diameter.* The diameter of the polygonal links or lateral ties shall not be less than one-fourth of the diameter of the largest longitudinal bar, and in no case less than 5 mm.

(*d*) *Helical reinforcement*

(*i*) *Pitch.* Helical reinforcement shall be of regular formation with the turns of the helix spaced evenly and its ends shall be anchored properly by providing one and a half extra turns of the spiral bar. Where an increased load on the column on the strength of the helical reinforcement is allowed for, the pitch of the helical turns shall be not more than 75 mm, nor more than one sixth of the case diameter of the column, nor less than 25 mm, nor less than three times the diameter of the steel bar forming the helix.

(*ii*) *Diameter.* The diameter of the helix reinforcement shall be in accordance with clause 2 (*c ii*) given above.

(3) Cover to reinforcement

For a longitudinal reinforcing bar in a column, the cover shall not be less than 40 mm, nor less than the diameter of such bar. In the case of columns of minimum dimension of 200 mm or under, whose reinforcing bars does not exceed 12 mm, a cover of 25 mm may be used.

(4) Effective length of columns

(*a*) In the absence of more exact analysis, the effective length of columns in framed structures may be obtained from the ratio of effective length to unsupported length l_{ef}/l given in Fig. 13.3 when relative displacement of the ends of the column is prevented

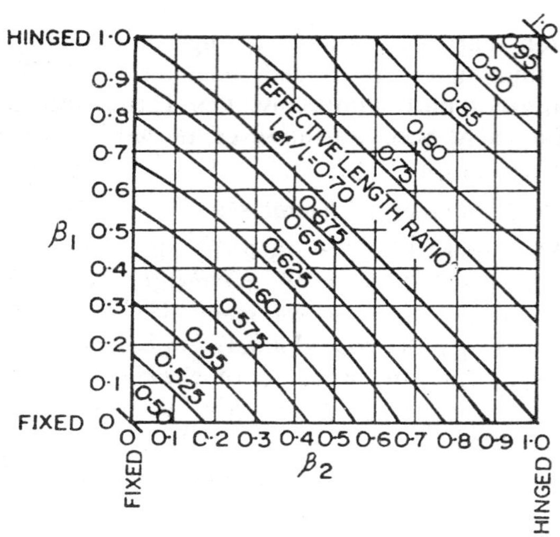

FIG. 13.3. EFFECTIVE LENGTH RATIOS FOR A COLUMN IN A FRAME WITHOUT RESTRAINT AGAINST SWAY.

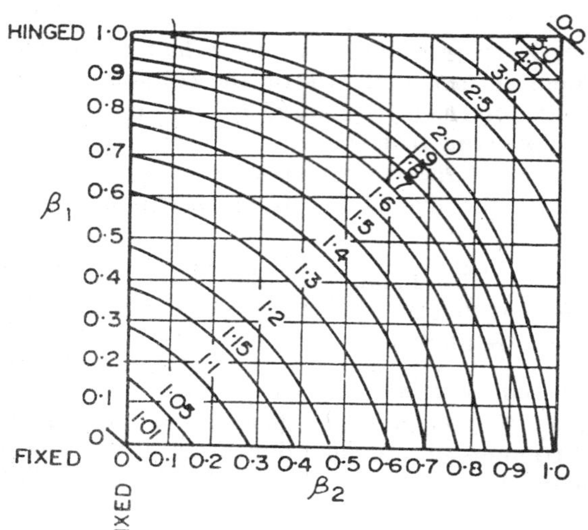

FIG. 13.4. EFFECTIVE LENGTH RATIOS FOR A COLUMN IN A FRAME WITH NO SWAY.

and in Fig. 13.4 when relative lateral displacement is not prevented. In the latter case, it is recommended that the effective length ratio l_{ef}/l may not be taken less than 1.2.

Note. In Figures 13.3 and 13.4, β_1 and β_2 are equal to $\dfrac{\Sigma k_c}{\Sigma K_c + \Sigma k_b}$ where the summation is done for the members framing into a joint at top and bottom respectively ; k_c and k_b being the flexural stiffness for column and beam respectively.

(*b*) For a normal usage, assuming idealised conditions, the effective length l_{ef} in a given plane may be assessed on the basis of Table 13.1. In the table, l is the unsupported length of the column.

TABLE 13.1. EFFECTIVE LENGTH OF COMPRESSION MEMBERS (IS : 456-1978)

	Degree of end restraint of compression member (*1*)	*Theoretical value of effective length* (*2*)	*Recommended value of effective length* (*3*)
1.	Effectively held in position and restrained against rotation at both ends.	$0.5l$	$0.65l$
2.	Effectively held in position at both ends, restrained regainst rotation at one end,	$0.7l$	$0.80l$
3.	Effectively held in position at both ends, but not restrained against rotation.	$1.00l$	$1.00l$
4.	Effectively held in position and restrained against rotation at one end, and at the other restrained against rotation but not held in position.	–	–
5.	Effectively held in position and restrained against rotation at one end, and at the other partially restrained against rotation but not held in position.	–	$1.50l$
6.	Effectively held in position at one end but not restrained against rotation, and at the other end restrained against rotation but not held in position.	$2.00l$	$2.00l$
7.	Effectively held in position and restrained against rotation at one end but not held in position nor restrained against rotation at the other end.	$2.00l$	$2.00l$

(5) Short and slender compression members

A compression member may be considered as short when both the slenderness ratios $\frac{l_{ex}}{D}$ and $\frac{l_{ey}}{b}$ are less than 12 :

where l_{ex} = effective length in respect of the major axis.

D = Depth in respect of the major axis.

l_{ey} = effective length respect of the minor axis, and

b = width of member.

It shall otherwise be considered as a slender compression member.

(6) Slenderness limits for columns

(*i*) The unsupported length between end restraints shall not exceed 60 times the least lateral dimension of a column.

(*ii*) If, in a given plane, one end of a column is unrestrained, its unsupported length l, shall not exceed $100b^2/D$.

where : b = width of that cross-section, and

D = depth of the cross-section measured in the plane under consideration

(7) Minimum eccentricity

All columns shall be designed for minimum eccentricity equal to the unsupported length of column/500 plus lateral dimension/30, subject to a minimum of 20 mm.

(8) Permissible loads in compression members

(*a*) *Padestal and short columns with lateral ties.* The axial load P permissible on a pedestal or short column reinforced with longitudinal bars and lateral ties shall not exceed that given by the following equation :

$$P = \sigma_{cc} A_c + \sigma_{sc} . A_{sc}$$

Note. The minimum eccentricity mentioned in clause 7 may be deemed to be incorporated in the above equation.

(*b*) *Short columns with helical reinforcement.*

The permissible load for columns with helical reinforcement shall be 1.05 times the permissible load for member with lateral ties, or rings, This similar clause is valid only if the ratio volume of helical reinforcement to the volume of core is not less than $0.36\ (A_g/A_c - 1) f_{ck}/f_y$

where A_g = gross area of section

A_e = area of core helically reinforced column measured to the outside diameter of the helix.

f_{ck} = characteristic compressive strength of the concrete

f_y = characteristic strength of helical reinforcement but not exceeding 415 N/mm².

(*c*) ***Long columns.*** The maximum permissible stress in a reinforced concrete column or part thereof having a ratio of effective column length to least later dimension above 12 shall not exceed that which results from the multiplication of the appropriate maximum permissible stresses(in concrete and steel) by the coefficient C_r given by the following formula :

$$C_r = 1.25 - \frac{l_{ef}}{48b}$$

where C_r = reduction coefficient

l_{ef} = effective length of column, and

b =least lateral dimension of column ; for column with helical reinforcement, b is the diameter of the core.

For more exact calculations, the maximum permissible stresses in a reinforced concrete column or part thereof having a ratio of effective column length to least lateral radius of gyration above 40 shall not exceed these which result from the multiplication of the appropriate maximum permissible stresses in concrete and steel, by the coefficient C_r given by the following formula :

$$C_r = 1.25 - \frac{l_{ef}}{160\, l_{min}}$$

where l_{min} is the least radius of gyration.

(9) Composite column

(*a*) ***Allowable load.*** The allowable axial load P on a composite column consisting of structural steel or cast-iron column thoroughly encased in concrete reinforced with both longitudinal and spiral reinforcement shall not exceed that given by the following formula :

$$P = \sigma_{cc} . A_c + \sigma_{sc} . A_{sc} + \sigma_{mc} . A_m$$

where σ_{cc} = permissible stress in concrete in direct compression

A_c = net area of concrete section, which is equal to the gross area of concrete section= $A_{sc} - A_m$

σ_{sc} = permissible compressive stress of column bars

A_{sc} = cross-sectional area of longitudinal bar reinforcement

σ_{mc} = allowable unit stress in metal core, not to exceed 125 N/mm^2 for a steel core , or 70 N/mm^2 for a cast iron core ;

A_m = the cross-sectional area of the steel or cast iron core.

(*b*) ***Metal core and reinforcement.*** The cross-sectional area of the metal core shall not exceed 20% of the gross area of the column. If a hollow metal core is used, it shall be filled with concrete. The amount of longitudinal and spiral reinforcement and the requirements as to spacing of bars, details of splices and thickness of protective shell outside the spiral, shall conform to the requirements of clause 1. A clearance of at least 75 mm shall be maintained between the spiral and the metal core at all points, except that when the core consists of a structural steel H-column, the minimum clearance may be reduced to 50 mm.

(*c*) ***Splices and connections of metal cores.*** Metal cores in composite columns shall be accurately milled at splices and positive provisions shall be made for alignment of one core above another. At the column base, provisions shall be made to transfer the load to the footing at safe unit stresses.

(*d*) ***Allowable load on metal core only.*** The metal core of composite columns shall be designed to carry safely any constructions or other loads to be placed upon them prior to their encasement in concrete.

Permissible stresses

The permissible compressive stress for various grades of concrete is given in Table 1.15. The values of σ_{cc} for M 15, M 20 and M 25 grades of concrete are 4 N/mm^2, 5 N/mm^2 and 6 N/mm^2 respectively.

The permissible compressive stress in longitudinal bars is given in Table 1.18. For mild steel bars σ_{sc} is 130 N/mm^2. For Fe 415 steel, the corresponding value is 190 N/mm^2.

13.6. REINFORCED CONCRETE WALLS

(*a*) **General.** Where reinforced concrete walls are intended to carry vertical loads, they should be designed generally in accordance with the recommendations given for columns. The provisions with regard to transverse reinforcement to restrain the vertical bars against buckling need not be applied to walls in which the vertical bars are not assumed to assist in resisting compression. The minimum thickness of wall should not be less than 100 mm.

(*b*) **Load carrying capacity.** The load carrying capacity of the walls shall be calculated as for columns ; however the strength of the wall may be increased by the values given below :

Ratio of storey height to length of wall	1.5 or more	1.0	3.5 or less
% increase in strength	0	10	20

Note. The length of the wall is the overall length or, where openings occur, the length between adjacent openings.

(*c*) **Slenderness effects.** When the effective height of wall exceeds 12 times the wall thickness, the slenderness effects shall be considered as in columns. Where the wall is stiffened by cross-walls, the slenderness ratios may be determined as in IS: 1905-1969 (Code of practice for structural safety of buildings : Masonry walls).

(*d*) **Reinforcement.** The reinforcement for walls shall be provided below :

(*i*) the minimum reinforcement for walls shall be provided below :

(1) 0.0012 for deformed bars not larger than 16 mm in diameter and with a characteristic strength of 415 N/mm^2 or greater.

(2) 0.0015 for other types of bars

(3) 0.0012 for welded wire fabric not longer than 15 mm in diameter

(*ii*) Vertical reinforcement shall be spaced not further apart than three times the wall thickness not 450 mm.

(*iii*) The minimum ratio of horizontal reinforcement to gross concrete area shall be :

(1) 0.0020 for deformed bars not larger than 16 mm in diameter and with a characteristic strength of 415 N/mm^2 or greater.

(2) 0.0025 for other types of bars

(3) 0.0020 for welded wire fabric not larger than 16 mm in diameter

(*iv*) Horizontal reinforcement shall be spaced not farther apart than three times the wall thickness nor 450 mm.

Note. The minimum reinforcement may not always be sufficient to provide adequate resistance to the effects of shrinkage and temperature.

13.7 DESIGN PROCEDURE

A R.C. column may be designed in the following steps :

1. Depending upon the grade of concrete to be used, determine the permissible stresses in concrete, longitudinal bars and temperature.

2. Find the super-imposed load the column is required to carry. To this, add dead weight(assumed) of the column to get the total load the column has to carry at its base.

3. Assume some suitable value of reinforcement A_{sc} say between 0.8 to 2% of gross area A of column. Determine the area A from the following expression :

$$P = \sigma_{cc} . A_c + \sigma_{sc} . A_{sc}$$

or
$$P = \sigma_{cc} (A - p . A) + \sigma_{sc} . pA \quad ...(13.7)$$

or
$$A = \frac{P}{\sigma_{cc}(1 - p) + p . \sigma_{sc}}$$

where A = gross area of column.

p = ratio of steel to total area $= \frac{A_{sc}}{A}$.

4. After having known the area A, determine the dimensions of the column. If it is a square of side b, then $b = \sqrt{A}$. If a circular column of diameter D is to be used, D (= b) will be equal to $\sqrt{4A/\pi}$.

5. For the given end conditions, determine the effective length l_{ef} of the column, and hence calculate $\frac{l_{ef}}{b}$ ratio to find whether it is short column or long column.

6. If $\frac{l_{ef}}{b} < 12$, it will be designed as a short column for which dimensions have already been found in steps 3 and 4. Determine the area of steel A_{sc} and distribute the bars suitably around the periphery of the column, keeping suitable cover.

7. If $\frac{l_{ef}}{b} > 12$, it will be designed as a long column for which the reduction factor C_r is determined from the expression.

$$C_r = 1.25 - \frac{l_{ef}}{48\, b}$$

Calculate the load P' for an equivalent short column :

$$P' = P \times C_r \quad ...(13.8)$$

Using the revised value of P' in Eq. 13.9, recalculate the area A and hence determine the size of the column. Also, calculate the area of steel $A_{sc} = pA$.

8. Find the diameter of bars used as ties, and determine its pitch as per rules discussed in the previous article.

Example 13.1. *(a) A reinforced concrete column 4 m long(effective) and 400 mm in diameter is reinforced with 8 bars of 20 mm diameter. Find safe load of columns can carry. Take* $\sigma_{cc} = 4\ N/mm^2$ *and* $\sigma_{sc} = 130\ N/mm^2$. *The column carries lateral ties.*

(b) If the effective length is increased to 8 m, what will be the safe load of column can carry ?

Solution. (*a*) $l_{ef} = 4000$ mm ; $b = D = 400$ mm

Hence $\frac{l_{ef}}{b} = \frac{4000}{400} = 10 < 12$. Thus the column is short.

$$P = \sigma_{cc} . A_c + \sigma_{sc} . A_{sc}.$$

where $A_{sc} = 8\frac{\pi}{4}(20)^2 = 2512\ mm^2 ; A = \frac{\pi}{4}(400)^2 = 125664\ mm^2$

$$A_c = 125664 - 2512 = 123152\ mm^2$$

$\therefore$ $P = (4 \times 123152) + (130 \times 2512) = 819170$ N

$\approx$ **819.2 kN**

(*b*) $l_{ef} = 8000$ mm ; $D = 400$ mm

$\frac{l_{ef}}{b} = \frac{8000}{400} = 20 > 12$. Hence the column is a long column for which the reduction factor is given by

$$C_r = 12.5 - \frac{l_{ef}}{48b} = 1.25 - \frac{8000}{48 \times 400} = 0.833$$

$P = 0.833 \times 819170 \approx 682600$ N $\approx$ **682 kN.**

Example 13.2. *Solve example 13.1 if the column carries spiral (helical)reinforcement of 10 mm Φ rod wound around the 20 mm Φ bars at a pitch of 60 mm. The 20 mm Φ longitudinal bars are placed with a clear cover of 40 mm. Use M 15 concrete. Take the characteristic strength of helical reinforcement as* $250\ N/mm^2$.

Solution.

Outside diameter of helix= $400 - (2 \times 40) + (2 \times 10) = 340$ mm

A_c = Area of core helically reinforced column measured to the outside diameter of the helix $= \frac{\pi}{4}(340)^2 = 90792\ mm^2$.

A_g = gross area of section $= \frac{\pi}{4}(400)^2 = 125664\ mm^2$

For M 15 concrete, $f_{ck} = 15\ N/mm^2$

Also $f_y = 250\ N/mm^2$

$\therefore$ Factor $0.36\ (A_g/A_c - 1) f_{ck}/f_y = 0.36\ (125664/90792 - 10)\ 15/250$
$= 0.0083$ ...(*i*)

Volume of helical reinforcement (V_h) per unit length of column

$$= \frac{\text{circumference of spiral} \times \text{its area of cross-section}}{\text{pitch of spiral}}$$

$$= \left(\frac{\pi d}{s}\right)\left(\frac{\pi}{4}\Phi_s^2\right)$$

where d = diameter upto centre of helix

$= 400 - (2 \times 40) + 10 = 330$ mm

Φ_s = dia. of spiral wire = 10 mm

s = pitch = 60 mm

$\therefore$ V_h per 1 mm length of column $= \left(\frac{\pi \times 330}{60}\right)\frac{\pi}{4}(10)^2 \times 2$

$= 1357\ mm^2$

Volume of core $V_c = A_c \times 1 = 90792 \times 1 = 90792\ mm^2$

$\therefore$ Ratio $\frac{V_h}{V_c} = \frac{1357}{90792} = 0.015$ ...(*ii*)

Since this ratio (V_h/V_c) is more than the factor calculated in (*i*), the load carried by column with helical reinforcement will be 1.05 times the permissible load for column with lateral ties or rings.

Hence for case (*a*) of short column (example 13.1),

$P = 1.05 \times 819.2 =$ **860 kN**

and for case (*b*) of long column (example 13.1),

$P = 1.05 \times 682.5 =$ **716 kN**

Example 13.3 *(a) Design a short square column to carry an axial load of 1200 kN. Use M 25 concrete mix. Take* $\sigma_{sc} = 130\ N/mm^2$.

Solution.

1. Design of section. Minimum steel = 0.8%

Let us use 1% steel *i.e.*, $p = 0.01$

σ_{cc} for M 25 mix = 6 N/mm^2

The load carrying capacity of a short column is given by

$$P = \sigma_{cc} . A_c + \sigma_{sc} . A_{sc} = \sigma_{cc}(A - pA) + \sigma_{sc}\, pA$$

From which $A = \dfrac{P}{\sigma_{cc}(1-p) + p\,\sigma_{sc}}$

or $$A = \frac{1200 \times 1000}{6(1-0.01) + 0.01 \times 130} = 165746 \text{ mm}^2$$

$\therefore$ size of square $= \sqrt{165746} = 407$ mm

However, provide square column 410 mm × 410 mm

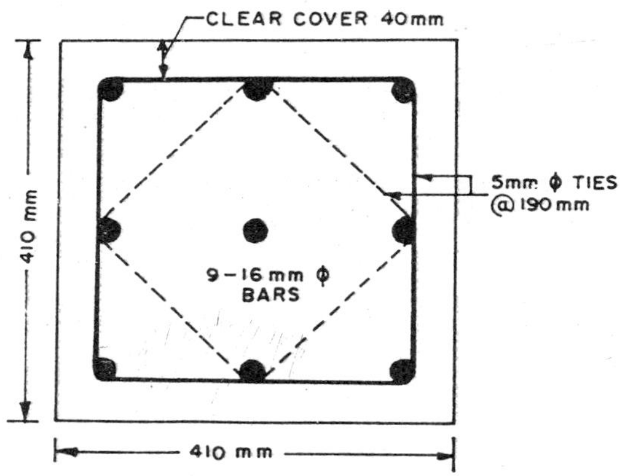

FIG. 13.5

2. *Longitudinal reinforcement*

$A_{sc} = pA = 0.01 \times 165746 \approx 1658 \text{ mm}^2$

Using 16 mm Φ bars, $A_\Phi = \frac{\pi}{4}(16)^2 = 201 \text{ mm}^2$

$\therefore$ No. of bars required $= \frac{1658}{201} \approx 9$

Arrange the bars as shown in Fig. 13.5 at a clear cover of 40 mm.

3. *Design of ties*

Diameter of ties should be 1/4 the diameter of longitudinal reinforcement subject to a minimum of 5 mm. Hence use 5 mm Φ bars of ties. The c/c spacing of the ties should not exceed the least of the following :

(*i*) least lateral dimension = 410 mm

(*ii*) 16 × diameter of main bar = 16 × 16 = 256 mm

(*iii*) 48 × dia. of ties = 48 × 5 = 240 mm.

Hence provide the ties @ 240 mm c/c. The ties will be square in shape in two size as shown in Fig. 13.6, using them alternately, so that longitudinal bars pass through the corners of the ties. Keep pitch of each set of ties at 240 mm.

Example 13.3(*b*) *Redesign column of example 13.3(a) using HYSD bars having σ_{sc} = 190 N/mm².*

Solution.

(i) Design of section

Use minimum steel = 0.8%

$\therefore \quad p = 0.008$; σ_{cc} for M 25 concrete $= 6 \text{ N/mm}^2$

The load carrying capacity of a short column is given by

$$p = \sigma_{cc} . A_c + \sigma_{sc} . A_{sc} = \sigma_{cc} (A - pA) + \sigma_{sc} . pA$$

From which $A = \dfrac{P}{\sigma_{cc} (1-p) + p\,\sigma_{sc}}$

$$= \frac{1200 \times 1000}{6\,(1 - 0.008) + 0.008 \times 190} = 160600 \text{ mm}^2$$

$\therefore$ Size of square $= \sqrt{160600} \approx 400$ mm.

(ii) Longitudinal reinforcement

$A_{sc} = pA = 0.008 \times 160600 = 1284.8 \text{ mm}^2$

Using 16 mm dia. bars, $A_\Phi = \dfrac{\pi}{4} (16)^2 = 201 \text{ mm}^2$

$\therefore \quad$ No. of bars $= \dfrac{1284.8}{201} = 6.39$

However, for the purpose of symmetrical arrangement of bars in a square column, provide 8 bars of 16 mm dia. and arrange them as shown in Fig. 13.5 (except for the central bar).

(iii) Design of ties

Same as in Example 13.3(*a*), except that plain bars are used for ties.

Example 13.4 *Design a short R.C. column to take an axial load of 5000 kN. The size of the column is not to be more than 700 mm. Use spiral reinforcement. Mix of concrete to be used is M 25. Take permissible stress in longitudinal steel as 130 N/mm². The characteristic strength of steel may be taken as 250 N/mm².*

Solution.

1 . ***Steel reinforcement***

Assuming that the conditions of helical reinforcement mentioned in para 8 (*b*) (§ 13.5) are fulfilled, the carrying capacity of the column may be taken to be equal to 1.05 times the load carried by a column with a lateral ties.

$\therefore \quad P = 1.05\,[\sigma_{cc}\,.\,A_c + \sigma_{sc}\,.\,A_{sc}\,]$

where $A = \frac{\pi}{4}(700)^2 = 385000 \text{ mm}^2 \,; A_c = A - A_{sc} = (385000 - A_{sc})$

For M 25 concrete, $\sigma_{cc} = 6 \text{ N/mm}^2$.

$\therefore \quad 5000{,}000 = 1.05\,[6(385000 - A_{sc}) + 130\,A_{sc}]$

From which $A_{sc} = 19773 \text{ mm}^2$.

This is less than 6% and more than 0.8% of area of cross-section of the column, and hence satisfactory.

Using 40 mm Φ bars, No. of bars required will be equal to 16. Keeping 40 mm clear cover, arrange the bars on a circle of diameter equal to $700 - 3 \times 40 = 580$ mm. Hence circumferential spacing of the bars $= \frac{\pi \times 580}{16} = 114$ mm which is less than 300 mm.

2. Spiral reinforcement

Let the helix bar be of 10 mm Φ.

Outside dia. of helix $= 700 - (2 \times 40) + (2 \times 10) = 640$ mm

A_c = area of core helically reinforced column measured to the outside diameter of helix $= \frac{\pi}{4}(640)^2 = 321700 \text{ mm}^2$

A_g = gross area of section $= \frac{\pi}{4}(700)^2 = 384845 \text{ mm}^2$

For M 25 concrete, $f_{ck} = 25 \text{ N/mm}^2$; Also, $f_y = 250 \text{ N/mm}^2$

$$\therefore \quad \text{Factor } 0.36\left(\frac{A_g}{A_c} - 1\right)\frac{f_{ck}}{f_y} = 0.36\left(\frac{384845}{321700} - 1\right)\frac{25}{250}$$

$$= 0.0071 \qquad \text{...}(i)$$

Dia. of core upto centre of helix $= 700 - (2 \times 40) + 10$
$= 630$ mm.

Let the pitch of the spiral be s mm.

$\therefore$ Volume of spiral V_h per 1 mm length of column

$$= \frac{\pi\, d}{s}\left(\frac{\pi}{4}\Phi_s^2\right)$$

$$= \frac{\pi \times 630}{s} \times \frac{\pi}{4}(10)^2 = \frac{155442}{s}$$

Volume of core per mm length, $V_c = A_c \times 1 = 321700 \times 1$

$$\therefore \text{ Ratio } \frac{V_h}{V_c} = \frac{155442}{s \times 321700} = \frac{0.4832}{s} \quad ...(ii)$$

Equating (*i*) and (*ii*), we get

$$\frac{0.4832}{s} = 0.0071$$

From which $s = 70$ mm

However the pitch should not be more than 75 mm, nor more than 1/6 core diameter (= 1/6 × 636 = 106). Also, it should not be less than 3 Φ (= 3 × 10 = 30 mm).

Hence keep the pitch equal to 70 mm. The details of reinforcement etc. are shown in Fig 13.6.

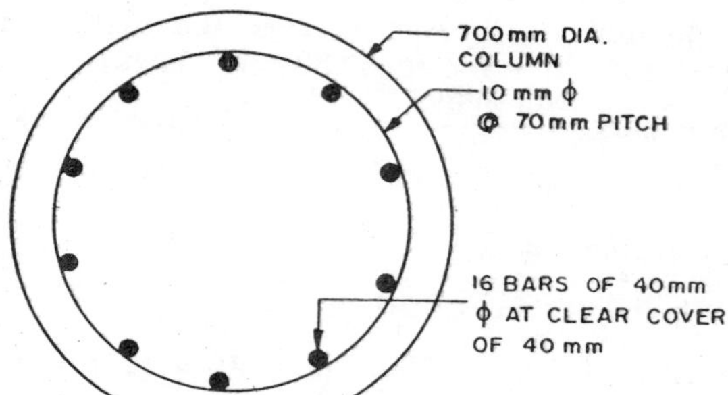

FIG. 13.6

Example 13.5. *Design a column 10 m long to carry an axial load of 600 kN. The column is restrained at the ends. Use M 25 concrete.*

Solution

1. ***Size of column***

From Table 13.1, $l_{ef} = 0.65l = 0.65 \times 10000 = 6500$ mm

Assume the column to be long one. Let the size of the column be $D \times D$ mm(square)

$$\therefore \text{ Reduction factor } C_r = 1.25 - \frac{6500}{48 \times D} = 1.25 - \frac{135.42}{D}$$

$$\text{Hence design load } = \frac{P}{C_r} = \left(\frac{600000}{1.25 - \dfrac{135.42}{D}}\right)$$

Providing only lateral ties, we have

$$\left(\frac{600000}{1.25 - \dfrac{135.42}{D}}\right) = \sigma_{cc} A_c + \sigma_{sc} A_{sc}$$

Let us provide reinforcement @ 1% of area.

$A = D^2 ; A_{sc} = 0.01\, D^2 ;$

$A_c = A - A_{sc} = D^2 - 0.01\ D^2 = 0.99\, D^2.$

For M 25 concrete , $\sigma_{sc} = 6\ \text{N/mm}^2$.

$$\therefore \quad \frac{600000}{1.25 - \dfrac{135.42}{D}} = 6 \times 0.99\, D^2 + 130 \times 0.01\, D^2 = 7.24\, D^2$$

or $D^2 - 108.2\, D - 66298 = 0$

From which $D = 317$ mm

Keep $D = 320$ mm.

2. *Longitudinal reinforcement*

$A_{sc} = 0.01\,(317 \times 317) = 1005\ \text{mm}^2$

Provide 8 bars. Area of each bar $= \dfrac{1005}{8} = 125.6$ mm

$\therefore$ Dia. of bar $= \sqrt{\dfrac{125.6 \times 4}{\pi}} = 12.6$ mm

Provide 14 mm bars. Keep a clear cover of 40 mm.

8. *Design of ties*

Use 5 mm Φ ties at a pitch not exceeding the minimum of the following :

(*i*) least lateral dimension = 320 mm.

(*ii*) $16\,\Phi = 16 \times 14 = 224$ mm.

(*iii*) $48 \times$ dia. of ties $= 48 \times 5 = 240$ mm.

However, provide the ties at 220 mm c/c.

Example 13.6. *Find the safe load for the composite column shown in Fig. 13.7. Take $\sigma_{cc} = 5\ N/mm^2$, $\sigma_{sc} = 130\, N/mm^2$ and $\sigma_{mc} = 125\, N/mm^2$. The joist has area of cros-section of 3500 mm².*

Solution

The safe load is given by

$$P = \sigma_{cc} A_c + \sigma_{sc} A_{sc} + \sigma_{mc} A_m$$

where, $\quad A_{sc} = 8\,\frac{\pi}{4}\,(20)^2 = 2512\ \text{mm}^2$

$\quad A_m = 3550\ \text{mm}^2 ; A = 500 \times 500 = 250000$ mm

$$\therefore \quad A_c = 250000 - 2512 - 3500 \approx 244000 \text{ mm}^2$$

$$P = 5 \times 244000 + 130 \times 2512 + 125 \times 3500$$

$$\approx 1984000 \text{ N} = \mathbf{1984\ kN}.$$

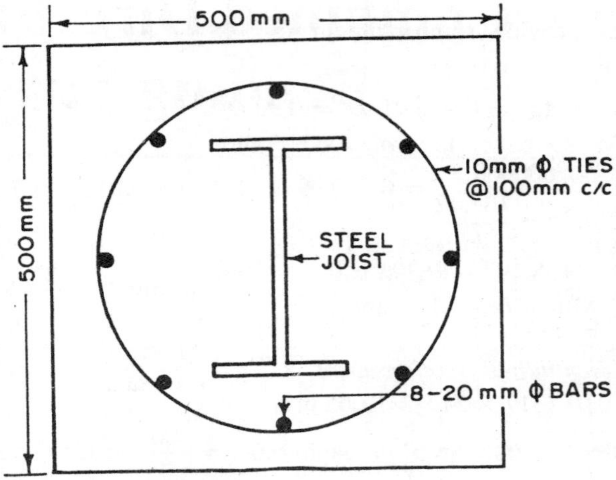

FIG. 13.7.

PROBLEMS

1. A reinforced concrete column, 3 m long (effective) and 240 mm × 240 mm in section is reinforced with 4 bars of 20 mm Φ . Find the safe load the column can carry. Take $\sigma_{cc} = 4 \text{ N/mm}^2$ and $\sigma_{sc} = 130 \text{ N/mm}^2$.
2. What will be the safe load for the column of problem 1 if the length of the column is 6 m.
3. A circular column of 500 mm diameter has an effective length of 5 m. The column is reinforced with 8 longitudinal bars of 20 mm Φ and spiral of 10 mm Φ rod wound around the 20 mm bars at a pitch of 60 mm. The clear cover for the main reinforcement is 40 mm. Determine the safe load for the column. Take $\sigma_{sc} = 130 \text{ N/mm}^2$; $\sigma_{cc} = 4 \text{ N/mm}^2$, $f_{ck} = 15 \text{ N/mm}^2$ and $f_y = 250 \text{ N/mm}^2$.
4. Design a short square column to carry an axial load of 1000 kN. Use M 25 mix. Take $\sigma_{sc} = 130 \text{ N/mm}^2$.
5. Design a circular column to carry an axial load of 500 kN. The diameter of column is limited to 500 mm. Use spiral reinforcement also. Take suitable permissible values.

14

Combined Direct and Bending Stresses

14.1. INTRODUCTION

In many cases, a reinforced concrete member is subjected to both direct load and bending moment, or to a direct load acting eccentrically. Exterior columns of a framed structure, arches, chimneys, silos, bunkers etc. are some of the common examples of such combination of loading. In some cases, the forces may be compression and bending, while in other cases, the forces may be tension and bending. The effect of the centrally applied load P and a moment M is the same as that of a load P acting at an eccentricity $e = M/P$. Hence the ratio of moment (M) to the direct load (P) is generally termed as *eccentricity*. The analysis and design of such members evidently depend upon the value of eccentricity. If the eccentricity is small so that section is subjected to the same kind of stress (*i.e.* either compressive or tensile), the analysis is slightly simplied. If the eccentricity is large, stress reversal may take place in the section, necessitating the determination of neutral axis. Based on the value of eccentricity, and also upon the nature of direct force (*i.e.* compressive or tensile), *six cases* may arise.

(*a*) Combined compression and bending

Case 1 *Eccentricity small.* when eccentricity is smaller than one–fourth the depth of the section : In such a case, no reversal of stresses takes place, or even if tension is induced, the resultant tension in concrete is not greater than 35 percent and 25 percent of the resultant compression for biaxial and uniaxial bending respectively, or exceeds three-fourths the 7-day modulus of rupture of concrete. The design of such a section is based on uncracked section.

Case 2. *Eccentricity large.* When the eccentricity is larger than 1.5 times the depth of the section.

Case 3. *Intermediate eccentricity.* When the eccentricity is between 0.25 and 1.5 times the depth of the section.

(*b*) **Combined tension and bending**

Case 4. *Eccentricity small.* The section is subjected to tensile stresses throughout.

Case 5. *Eccentricity large.*

Case 6. *Intermediate eccentricity.*

14.2. CASE 1. COMPRESSIVE LOAD AT ECCENTRICITY SMALLER THAN D/4.

Fig. 14.1 shows a R.C. section of size $b \times D$, subjected to a compressive load P at a distance e from the middle of the section. This eccentric load is equivalent to a direct load P acting through the C.G. of the section, and a bending moment equal to $P \times e'$ where e' is the eccentricity of the load measured from the centre of gravity of the section. It is assumed that eccentricity is small so that no stress reversal takes place, or even if tension is developed the whole area of concrete is effective.

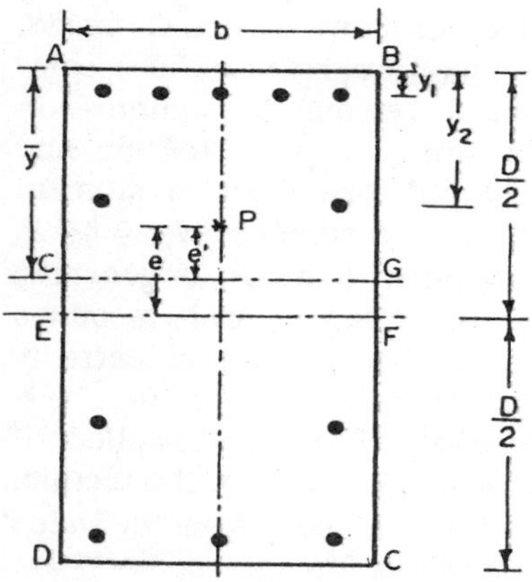

FIG. 14.1

Let A_{sc1}, A_{sc2}, A_{sc3} be the steel areas at various levels y_1, y_2, y_3, etc. from the top fibre (AB) as marked in Fig. 14.1

The equivalent area of section is given by

$$A_e = b\,D + \Sigma\,(m_e - 1)\,A_{sc} \qquad ...(14.1)$$

where, $m_c = 1.5\,m$

The distance $\bar{y}$ of the centre of gravity, measured from AB, is

$$\bar{y} = \frac{(m_c - 1)\,(A_{sc1}\,y_1 + A_{sc2}\,y_2 + ...) + b\,D^2/2}{(m_c - 1)\,(A_{sc1} + A_{sc2} + ...) + b\,D}$$

or

$$\bar{y} = \frac{b\,D^2/2 + \Sigma\,(m_c - 1)\,A_{sc}\,y}{b\,D + \Sigma\,(m_c - 1)\,A_{sc}}, \text{ where } m_c = 1.5\,m$$

The equivalent moment of inertia of the section is given by

$$I_e = (m_c - 1)\left(A_{sc1}\, a_1^2 + A_{sc2}\,.a_2^2 + \ldots\right) + \frac{b\,D^3}{12} + BD\left(\bar{y} - \frac{D}{2}\right)^2$$

$$I_e = \Sigma(m_c - 1)\,A_{sc}\,.\,a^2 + \frac{b\,D^3}{12} + BD\left(\bar{y} - \frac{D}{2}\right)^2 \qquad \ldots(14.3).$$

where $a_1, a_2, a_3, \ldots$ etc. are the distances of the steel areas from the centre of gravity of the section, and $m_c = 1.5\ m$

Now $\qquad M$ = load × eccentricity

$$\therefore \qquad M = P \times e' = P\left\{e - \left(\frac{D}{2} - \bar{y}\right)\right\}$$

The stress distribution in the column may be determined from the expression:

$$f = \frac{P}{A_e} \pm \frac{M\,.\,y}{I_e}$$

Hence the maximum compressive stress is given by

$$f_1 = \frac{P}{A_e} + \frac{P\left\{e - \left(\frac{D}{2} - \bar{y}\right)\right\}}{I_e}\,.\,\bar{y}$$

$$f_2 = \frac{P}{A_e} + \frac{P\left\{e - \left(\frac{D}{2} - \bar{y}\right)\right\}}{I_e}\,.\,(D - \bar{y})$$

Fig. 14.2 shows a symmetrically reinforced section, where $\bar{y} = D/2$.

$$A_e = b\,D + (m_c - 1)\,A_{sc} \qquad \ldots(14.4)$$

$$I_e = \frac{1}{12}bD^3 + (m_c - 1)\,A_{sc}\left(\frac{D}{2} - d_c\right)^2 \qquad \ldots(14.5)$$

where $m_c = 1.5\ m$

Hence maximum compressive stress at AB is given by

$$f_1 = \frac{P}{A_e} + \frac{P\,.e}{I_e}\,.\,\frac{D}{2}$$

Minimum stress at DC is

$$f_2 = \frac{P}{A_e} - \frac{P\,.e}{I_e}\,.\,\frac{D}{2}$$

If no tension is to develop in the section, f_2 must be equal to zero.

Hence $\quad \dfrac{P}{A_e} - \dfrac{P\,.e}{I_e}\,.\,\dfrac{D}{2} = 0$

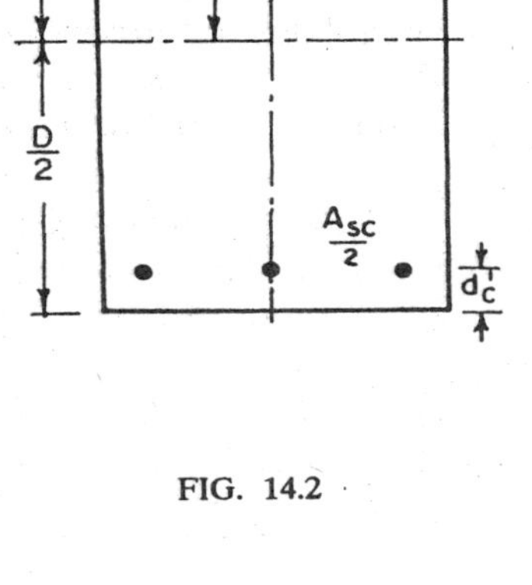

FIG. 14.2

which gives $e = \dfrac{2I_e}{A_e . D}$

A member subjected to axial load and bending(due to eccentricity of load, monolithic construction, lateral forces, etc.) should be considered safe provided the following condition is satisfied :

$$\frac{\sigma_{cc}'}{\sigma_{cc}} + \frac{\sigma_{cbc}'}{\sigma_{cbc}} \leq 1 \qquad ...(14.7)$$

where σ_{cc} = permissible direct compressive stress in concrete

σ_{cc}' = calculated direct compressive stress in concrete

σ_{cbc} = permissible bending compressive stress in concrete

σ_{cbc}' = calculated bending compressive stress in concrete.

$$\text{Stress in steel} = \sigma_s' = 1.5\, m \left[\frac{P}{A_e} + \frac{P . e}{I_e}\left(\frac{D}{2} - d_c\right)\right] \qquad ...(14.8)$$

14.3 BENDING ABOUT TWO AXES

Consider a symmetrically reinforced column subjected to an axial load P, a bending moment M_x about x-axis and a moment M_y about y-axis.

Let I_{ex} = equivalent moment of inertia of the section about x-x axis.

I_{ey} = equivalent moment of inertia of the section about y-y axis.

The stress f at any point (x, y) can be determined from the expression

$$f = \frac{P}{A_e} \pm \frac{M_x . y}{I_{ex}} \pm \frac{M_y . x}{I_{ey}} \qquad ...(14.9)$$

The maximum direct stress

$$\sigma_{cc}' = \frac{P}{A_e}$$

Maximum bending stress

$$\sigma_{cbc}' = \frac{M_x}{I_{ex}} . \frac{D}{2} + \frac{M_y}{I_{ey}} . \frac{B}{2}$$

The section will be safe if

$$\frac{\sigma_{cc}'}{\sigma_{cc}} + \frac{\sigma_{cbc}'}{\sigma_{cbc}} \leq 1$$

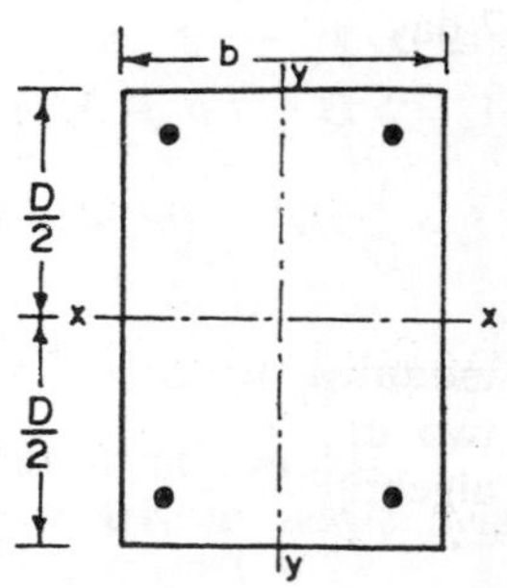

FIG. 14.3

13.4. DESIGN OF COLUMNS SUBJECTED TO COMBINED BENDING AND DIRECT STRESSES (IS : 456-1978)

The design of a column subjected to combined bending and direct stresses is that of trial and check. A suitable size of the column and steel reinforcement are first chosen. The preliminary section so selected is then checked for stresses in concrete and steel. The following notes refer to the design of such sections.

1. **Design based on uncracked section.** A member subjected to axial load and bending (due to eccentricity of load, monolithic construction, lateral forces etc.) shall be considered safe provided the following conditions are satisfied :

(*a*) $$\frac{\sigma_{cc}'}{\sigma_{cc}} + \frac{\sigma_{cbc}'}{\sigma_{cbc}} \leq 1$$

where σ_{cc}' = calculated direct compressive stress in concrete

σ_{cc} = permissible axial compressive stress in concrete

σ_{cbc}' = calculated bending compressive stress in concrete

σ_{cbc} = permissible bending compressive stress in concrete.

(*b*) The resultant tension in concrete is not greater than 35 percent and 25 percent of the resultant compression for biaxial and uniaxial bending respectively, or does not exceed three- fourths the 7-day modulus of rupture of concrete.

Note. (*i*) $\sigma_{cc}' = \dfrac{P}{A_c + 1.5\, m\, A_{sc}}$, for columns with ties.

(*ii*) $\sigma_{cbc}' = \dfrac{M}{Z}$, where M equals the moment and Z equals modulus of section. In the case of sections subject to moments in two directions, the stress shall be calculated separately and added algebraically.

2. **Design based on cracked section.** If the requirements specified in clause 1 above are not satisfied, the stresses in concrete and steel shall be calculated by theory of cracked section in which tensile resistance of concrete is ignored. If the calculated stresses are within the permissible stresses, the section may be assumed to be safe.

Note. The maximum stress in concrete and steel may be found from tables and charts based on the uncracked section theory or directly by determining the no-stress line which should satisfy the following requirements :

(*a*) The direct load should be equal to the algebraic sum of the forces on concrete and steel.

(*b*) The moment of the external loads about any reference line should be equal to the algebraic sum of the moment of the forces in concrete(ignoring the tensile force in concrete) and steel about the same line, and

(*c*) The moment of the external loads about any other reference lines should be equal to the algebraic sum of the moment of the forces in concrete(ignoring the tensile force in concrete) and steel about the same line.

3. **Members subjected to combined direct load and flexure.** Members subjected to combined direct load and flexure and designed by the methods based on elastic theory should be further checked for their strength under ultimate load conditions to ensure the desired margin of safety ; this check is specially necessary when the bending moment is due to horizontal loads.

Example 14.1. *A R.C. column 400 mm × 400 mm is reinforced with 4 bars of 25 mm diameter, placed at a cover of 50 mm to the centre of steel basrs. Determine the maximum and minimum stresses in concrete if the column is subjected to a load of 400 kN at an eccentricity of 50 mm about one of the axes. Also,check whether the section is safe or not. Use M 15 concrete, taking m = 19.*

Solution.

For M 15 concrete, $\sigma_{cc} = 4\ \text{N/mm}^2$ and $\sigma_{cbc} = 5\ \text{N/mm}^2$. Since the eccentricity is small, the analysis of § 14.2 can be used. However while calculating the equivalent area and moment of inertia, modular ratio m_c ($= 1.5\ m$) should be used with the steel area in compression.

$$A_{sc} = 4\frac{\pi}{4}(25)^2 = 1960\ \text{mm}^2$$

$$A_e = A_c + 1.5\,m\,A_{sc} = (A - A_{sc}) + 1.5\,m\ A_{sc} = A + (1.5\,m - 1)\,A_{sc}$$

$$= (400 \times 400) + (1.5 \times 19 - 1)\,1960 = 213900\ \text{mm}^2$$

$$I_e = \frac{1}{12}\,b\,D^3 + (m_c - 1)\,A_{sc}\left(\frac{D}{2} - d_c\right)^2$$

$$= \frac{1}{12} \times 400(400)^3 + (1.5 \times 19 - 1) \times 1960\,(200 - 50)^2$$

$$= 33.46 \times 10^8\ \text{mm}^4$$

$$\sigma_{cc}' = \frac{P}{A_e} = \frac{400000}{213900} = 1.87 \text{ N/mm}^2$$

$$\sigma_{cbc}' = \frac{P.e}{I_e} \times \frac{D}{2} = \frac{400000 \times 50}{33.46 \times 10^8} \times 200 \approx 1.20 \text{ N/mm}^2$$

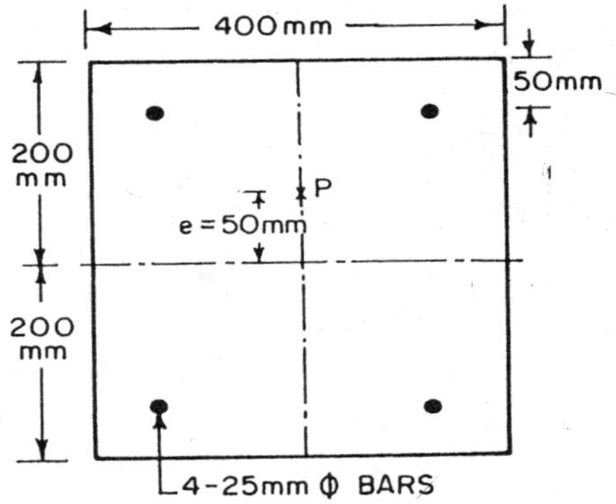

FIG. 14.4

$\therefore$ Max. stress $= \sigma_{cc}' + \sigma_{cbc}' = 1.87 + 1.20 = 3.07$ N/mm^2

Min. stress $= \sigma_{cc}' - \sigma_{cbc}' = 1.87 - 1.20 = 0.67$ N/mm^2

For the section to be safe, we have

$$\frac{\sigma_{cc}'}{\sigma_{cc}} + \frac{\sigma_{cbc}'}{\sigma_{cbc}} \leq 1$$

or
$$\frac{1.87}{4} + \frac{1.20}{5} \leq 1$$

or
$$0.71 \leq 1, \text{ Hence safe.}$$

Example 14.2. *Determine the maximum eccentricity of load in Example 14.1, if the column is to be safe as uncracked section.*

Solution. For safety, we have

$$\frac{\sigma_{cc}'}{\sigma_{cc}} + \frac{\sigma_{cbc}'}{\sigma_{cbc}} \leq 1$$

Hence $\dfrac{\sigma_{cbc}'}{\sigma_{cbc}} = 1 - \dfrac{\sigma_{cc}'}{\sigma_{cbc}} = 1 - \dfrac{1.87}{8} = 1 - 0.4675 = 0.5325$

$\therefore \quad \sigma_{cbc}' = 0.5325\, \sigma_{cbc} = 0.5325 \times 5 = 2.66$ N/mm^2

$\therefore \quad \sigma_{cbc}' = \dfrac{P.e}{I_e} \times \dfrac{D}{2} = 2.66$

$$e = 2.66 \times \frac{2\,I_e}{P.D} = \frac{2.66 \times 2 \times 33.46 \times 10^8}{400000 \times 400}$$

$$= \mathbf{111 \text{ mm.}}$$

Example 14.3 *If the column of Example 14.1 is subjected to an axial load of 400 kN along with a bending moment at 25 kN-m, determine the maximum and minimum stresses in concrete. Will the section be safe as an uncracked section ?*

Solution. $A_e = 213900$ mm^2 and $I_e = 33.46 \times 10^8$ mm^4, as found in Example 14.1.

$$\therefore \ \sigma_{cc}' = \frac{P}{A_e} = \frac{400000}{213900} = 1.87 \text{ N/mm}^2.$$

B.M. $M = 25$ kN-m $= 25 \times 10^6$ N-m

$$\therefore \ \sigma_{cbc}' = \frac{M}{I_e} \cdot \frac{D}{2} = \frac{25 \times 10^6}{33.46 \times 10^8} \times \frac{400}{2} = 1.49 \text{ N/mm}$$

Max. Stress $= \sigma_{cc}' + \sigma_{cbc}' = 1.87 + 1.49 =$ **3.36 N/mm²**.

Min. Stress $= \sigma_{cc}' - \sigma_{cbc}' = 1.87 - 1.49 =$ **0.38 N/mm²**.

For the section to be safe as an uncracked section,

$$\frac{\sigma_{cc}'}{\sigma_{cc}} + \frac{\sigma_{cbc}'}{\sigma_{cbc}} \le 1$$

or
$$\frac{1.87}{4} + \frac{1.49}{5} \le 1$$

or $0.766 \le 1$. Hence Safe.

Example 14.4. *A rectangular section 300 mm × 400 mm is reinforced with 8 bars of 20 mm Φ as shown in Fig. 14.5. Taking $\sigma_{cc} = 4$ N/mm², $\sigma_{cbc} = 5$ N/mm² and $m = 19$, determine (i) maximum eccentricity about x-axis at which load can be applied without developing tension in the section, (ii) the magnitude of the load.*

Solution

$$A_{sc} = 8 \times \frac{\pi}{4}(20)^2 = 2512 \text{ mm}^2$$

$$A_e = b\,D + (1.5\,m - 1)\,A_{sc} = (300 \times 400) + (1.5 \times 19 - 1) \times 2512$$
$$= 189080 \text{ mm}^2$$

$$I_e = \frac{1}{12} \times 300(400)^3 + (1.5 \times 19 - 1) \times 6 \times \frac{\pi}{4}(20)^2(150)^2$$
$$= 27.66 \times 10^8 \text{ mm}^4$$

For no tension to develop, we have

$$\frac{P}{A_e} - \frac{P \cdot e}{I_e} \cdot \frac{D}{2} = 0$$

or
$$e = \frac{2\,I_e}{A_e D} = \frac{2 \times 27.66 \times 10^8}{189080 \times 400} = 73.2 \text{ mm}$$

Again, if P is the maximum load, we have

$$\sigma_{cc}' = \frac{P}{A_e} = \frac{P}{189080}$$

$$\sigma_{cbc}' = \frac{P.e}{I_e} \cdot \frac{D}{2} = \frac{P}{I_e}\left[\frac{2I_e}{A_e.D}\right] \times \frac{D}{2} = \frac{P}{A_e} = \frac{P}{189080}$$

(**Note.** Both σ_{cc}' and σ_{cbc}' will be equal at the critical value of e)

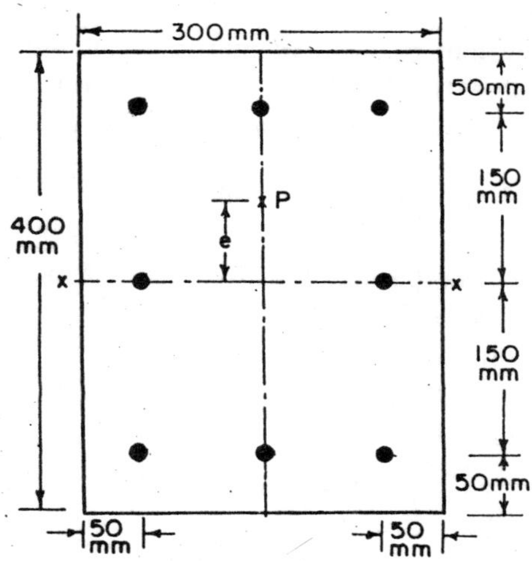

FIG. 14.5

$$\therefore \qquad \text{Now,} \quad \frac{\sigma_{cc}'}{\sigma_{cc}} + \frac{\sigma_{cbc}'}{\sigma_{cbc}} \leq 1$$

or
$$\frac{P}{189080}\left[\frac{1}{4} + \frac{1}{5}\right] = 1$$

or
$$P = \frac{189080 \times 4 \times 5}{9} = 420180 \text{ N}$$

Example 14.5 *Fig. 14.6 shows the cross-section of an arch rib. Determine the normal thrust and bending moment on the arch so that no tension is developed anywhere and the section is safe as uncracked section. Given : $\sigma_{cc} = 4\ N/mm^2$, $\sigma_{cbc} = 5\ N/mm^2$ and $m = 19$.*

Solution. Fig. (14.6)

$$A_{sc1} = \text{area of steel near face } AB = 6\frac{\pi}{4}(20)^2 = 1884 \text{ mm}^2.$$

$$A_{sc2} = \text{area of steel near face } CD = 6\frac{\pi}{4}(25)^2 = 2940 \text{ mm}^2.$$

Total $A_{sc} = 1884 + 2940 = 4824\ \text{mm}^2$

$\therefore \quad A_e = (400 \times 700) + (1.5 \times 19 - 1)\ 4824 = 412660\ \text{mm}^2$

Let $\bar{y}$ be the distance of centroidal axis from the face *AB*,

$$\bar{y} = \frac{(400 \times 700 \times 350) + (19 \times 1.5 - 1)1884 \times 50 + (19 \times 1.5 - 1) \times 2940 \times 650}{(400 \times 700) + (19 \times 1.5 - 1)1884 + (19 \times 1.5 - 1)\ 2940}$$

$= 371\ \text{mm}.$

The equivalent moment of inertia about the axis *EF* through the centre of gravity of the section is given by,

$$I_e = \frac{1}{12} 400(700)^3 + 400 \times 700\ (371 - 350)^2$$

$$+ (19 \times 1.5 - 1)\ [1884\ (367 - 50)^2 + 2940\ (700 - 371 - 50)^2]$$

$= 230.57 \times 10^3\ \text{mm}^4$

Let P be the permissible load and M be the moment.

Hence $e = M/P$

For no tension to develop, we have

$$\frac{P}{A_e} - \frac{M}{I_e}(D - \bar{y}) = 0$$

$$\therefore \quad \frac{P}{A_e} - \frac{P.e}{I_e}(D - \bar{y}) = 0$$

Hence $e = \dfrac{I_e}{A_e\,(D - \bar{y})} = \dfrac{230.57 \times 10^8}{412660\ (700 - 371)} = 170\ \text{mm}$

Again for an uncracked section, we have

$$\frac{\sigma_{cc}'}{\sigma_{cc}} + \frac{\sigma_{cbc}'}{\sigma_{cbc}} \leq 1$$

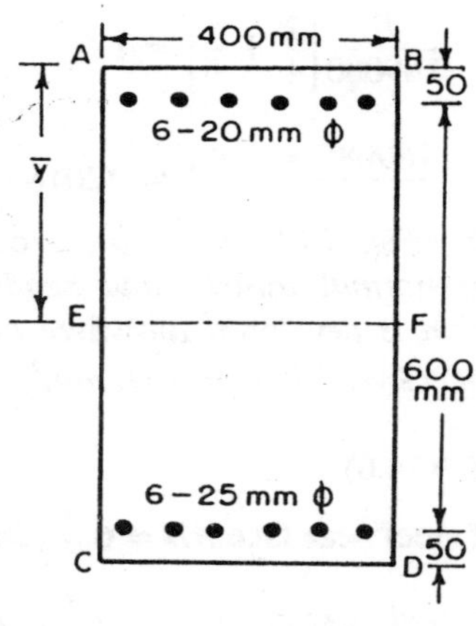

FIG. 14.6.

where $\sigma_{cc}' = \frac{P}{A_e} = \frac{P}{412660}$; $\sigma_{cc} = 4\text{ N/mm}^2$; $\sigma_{cbc} = 5\text{ N/mm}^2$

$$\sigma_{cbc}' = \frac{P.e}{I_e}.\bar{y} = P\left(\frac{170}{230.57 \times 10^8} \times 371\right) = \frac{P}{365578}$$

Hence $$\frac{\left(\frac{P}{412660}\right)}{4} + \frac{\left(\frac{P}{365578}\right)}{5} = 1.$$

or $$P\left[\frac{1}{412660 \times 4} + \frac{8}{365578 \times 5}\right] = 1$$

From which $P = 867400$ N = **867.4 kN**

Hence $M = P.e = 867.4 \times 170 = 147458$ kN-mm $\approx$ **147.5 kN-m**

Example 14.6. *Fig. 14.7 (a) Shows a reinforced concrete circular section, subjected to an axial load of 500 kN and bending moment of 25 kN-m. Taking the permissible stresses $\sigma_{cc} = 4\text{ N/mm}^2$ and $\sigma_{cbc} = 5\text{ N/mm}^2$, check whether the section is safe or not. Take m=19.*

Solution.

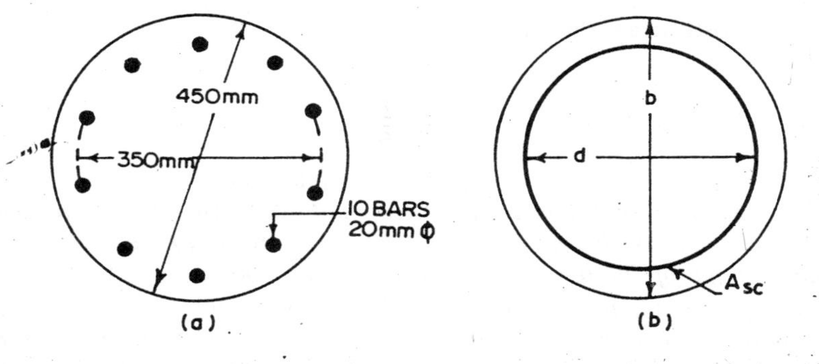

FIG. 14.7

$$A_{sc} = 10 \times \frac{\pi}{4}(20)^2 \approx 3140 \text{ mm}^2.$$

The steel bars can be replaced by angular ring of mean diameter 350 mm, having the same area.

$$A_e = \frac{\pi}{4}(450)^2 + (1.5 \times 19 - 1)\,3140 = 245390 \text{ mm}^2$$

Before finding the moment of inertia of the section, let us find the moment of inertia of the steel ring.

Polar moment of inertia (I_P) of ring is

$$I_p = \Sigma\, \delta A_{sc} \,.\, r^2 = \frac{A_{sc} \,.\, d^2}{4}$$

where d is the diameter of the circle joining the centres of the rods.

$$\therefore \qquad I_{xx} = I_{yy} = \frac{I_p}{2} = \frac{A_{sc} \,.\, d^2}{8}$$

Hence the moment of inertia of equivalent section is

$$\therefore \qquad I_e = \frac{\pi}{64}(450)^4 + (19 \times 1.5 - 1)\frac{3140(350)^2}{8}$$

$$= 33.35 \times 10^8 \text{ mm}^4$$

Now the section will be safe if

$$\frac{\sigma_{cc}'}{\sigma_{cc}} + \frac{\sigma_{cbc}'}{\sigma_{cbc}} \le 1$$

Here $\qquad \sigma_{cc}' = \frac{P}{A_e} = \frac{500 \times 1000}{245390} = 2.04 \text{ N/mm}^2$

$$\sigma_{cc} = 4 \text{ N/mm}^2$$

$$\sigma_{cbc}' = \frac{M}{I_e} \times 225 = \frac{25 \times 10^6}{33.35 \times 10^8} \times 225 = 1.69$$

$$\sigma_{cbc} = 5 \text{ N/mm}^2.$$

Substituting the values we get

$$\frac{\sigma_{cc}'}{\sigma_{cc}} + \frac{\sigma_{cbc}'}{\sigma_{cbc}} = \frac{2.04}{4} + \frac{1.69}{5}$$

$$= 0.51 + 0.338$$

$$= 0.848 < 1 \text{ Hence safe.}$$

Example 14.7 *The section of a reinforced concrete two hinged arch is 300 mm wide and 700 mm deep at the crown, and is reinforced with 6 bars of 25 mm Φ each at top and bottom at an effective cover of 50 mm, as shown in Fig. 14.8. The section is subjected to a resultant thrust of 800 kN inclined to the horizontal tangent to the centre of the arch at 3° and acting on the vertical axis at a distance of 60 mm from the centre line of the arch section. Determine the stress in concrete and steel at the top and bottom edge of the crown. If* $\sigma_{cc} = 4 N/mm^2$, $\sigma_{cbc} = 5 N/mm^2$ *and m = 19, find whether the section is safe or not.*

Solution

$$A_{sc} = 12\frac{\pi}{4}(25)^2 = 5890 \text{ mm}^2$$

$$A_e = (300 \times 700) + (1.5 \times 19 - 1)\, 5890 = 371980 \text{ mm}^2$$

$$I_e = \frac{1}{12} \times 300\,(700)^3 + (1.5 \times 19 - 1)\,5890\,(350 - 50)^2$$
$$= 2.135 \times 10^{10} \text{ mm}^4$$

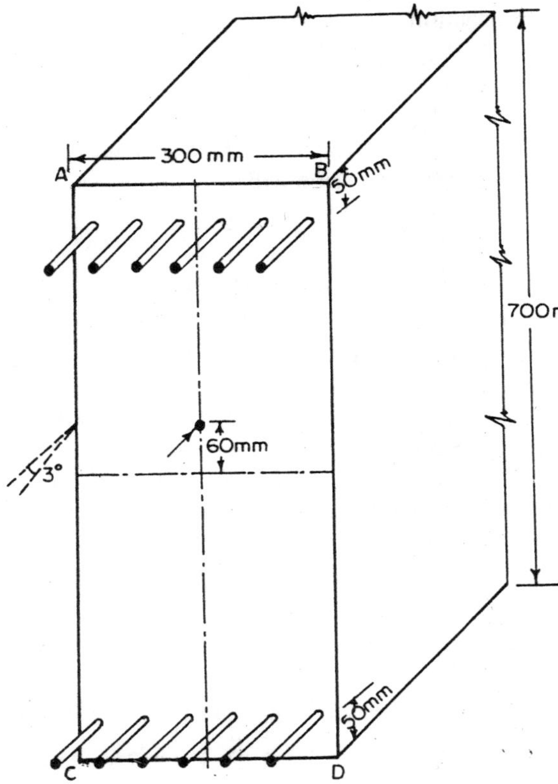

FIG. 14.8

Normal thrust $P = W \cos 3°$

$$= 800 \times 1000 \times 0.9986$$
$$= 7.989 \times 10^5 \text{ N}$$

$$\sigma_{cc}' = \frac{P}{A_e} = \frac{7.989 \times 10^5}{371980} = 2.149 \text{ N/mm}^2$$

$$\sigma_{cbc}' = \frac{P \cdot e}{I_e} \cdot \frac{D}{2} = \frac{7.989 \times 10^5 \times 60}{2.315 \times 10^{10}} \times 350$$
$$= 0.725 \text{ N/mm}^2$$

$\therefore$ Max. stress in concrete $= 2.149 + 0.725$

$= \mathbf{2.874\ N/mm^2}$ (compressive)

Min. stress in concrete $= 2.149 - 0.725$

$= \mathbf{1.424\ N/mm^2}$ (compressive)

For the section to be safe

$$\frac{\sigma_{cc}'}{\sigma_{cc}} + \frac{\sigma_{cbc}'}{\sigma_{cbc}} = \frac{2.149}{4} + \frac{0.725}{5}$$

$$= 0.537 + 0.145$$

$$= 0.682 < 1. \text{ Hence safe.}$$

Stress in steel near face AB

$$= 1.5 \times 19 \left\{2.149 + \frac{7.989 \times 10^5 \times 60}{2.315 \times 10^{10}} \times 300\right\}$$

$$= 28.5\,[2.149 + 0.621] = \mathbf{78.95\ N/mm^2} \text{ (comp.)}$$

Stress in steel near face CD

$$= 1.5 \times 19 \times 2.149 - 19 \times 0.621 = \mathbf{49.45\ N/mm^2} \text{ (comp.)}$$

Example 14.8 *A R.C. section 200 mm × 300 mm is reinforced with 4 bars of 20 mm Φ arranged as shown in 14.9. The section is subjected to (i) axial load of 200 kN (ii) bending moment of 3 kN-m about x-axis, and (iii) bending moment of 2 kN-m about y-axis. Taking $\sigma_{cc} = 4\ N/mm^2$ and $\sigma_{cbc} = 5\ N/mm^2$, find whether the section is safe or not. Take $m = 19$.*

Solution.

$$A_{sc} = 4 \times \frac{\pi}{4}(20)^2 = 1256 \text{ mm}$$

$$A_e = (200 \times 300) + (1.5 \times 19 - 1)\,1256 = 94540 \text{ mm}^2$$

$$I_{ex} = \frac{1}{12} 200(300)^3 + (1.5 \times 19 - 1) \times 1256\,(150 - 40)^2$$

$$= 8.679 \times 10^8 \text{ mm}^4$$

$$I_{ey} = \frac{1}{12} 300\,(200)^3 + (1.5 \times 19 - 1)\,1256\,(100 - 40)^2$$

$$= 3.243 \times 10^8 \text{ mm}^4$$

$$\sigma_{cc}' = \frac{P}{A_e} = \frac{200000}{94540} = 2.12 \text{ N/mm}^2$$

$$(\sigma_{cbc})_x' = \frac{M_x}{I_{ex}}\frac{D}{2} = \frac{3000 \times 1000}{8.679 \times 10^8} \times 150 = 0.52 \text{ N/mm}^2$$

$$(\sigma_{cbc})_y' = \frac{M_y}{I_{ey}}\frac{b}{2} = \frac{2000 \times 1000}{3.243 \times 10^8} \times 100 = 0.62 \text{ N/mm}^2$$

$\therefore$ Total bending stress $\sigma_{cbc}' = (\sigma_{cbc})'_x + (\sigma_{cbc})'_y = 0.52 + 0.62$
$= 1.14 \text{ N/mm}^2$

Now $$\frac{\sigma_{cc}'}{\sigma_{cc}} + \frac{\sigma_{cbc}'}{\sigma_{cbc}} \leq 1$$

or $$\frac{2.12}{4} + \frac{1.14}{5} < 1$$

or $$0.757 < 1 \text{ Hence safe.}$$

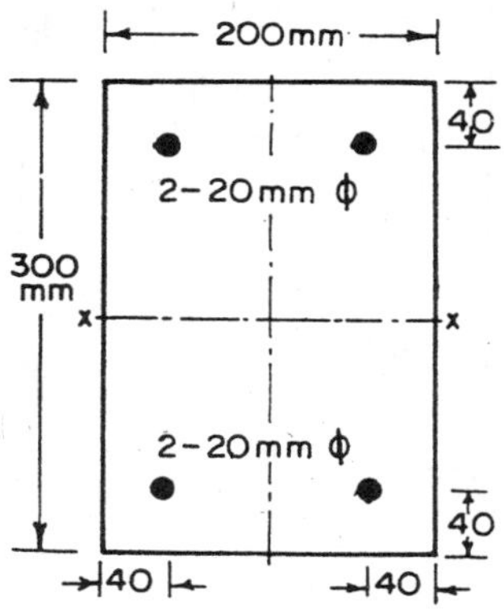

FIG. 14.9

Example 14.9. *Design a suitable column section to carry an axial load of 430 kN and bending moment of 8 kN-m. The allowable stresses may be taken as follows:* $\sigma_{cc} = 4\ N/mm^2$ $\sigma_{cbc} = 5\ N/mm^2$ *and* $m = 19$.

Solution. Let us assume a section of 320 mm × 320 mm, reinforced symmetrically with 4 bars of 25 mm Φ as shown in Fig. 14.10.

$$A_s = 4\frac{\pi}{4}(25)^2 = 196 \text{ mm}^2$$

$$A_c = (320 \times 320) + (1.5 \times 19 - 1)\, 1960 = 156300 \text{ mm}^2$$

$$I_e = \frac{1}{12} \times 320(320)^3 + (1.5 \times 19 - 1)\, 1960 \left(\frac{320}{2} - 50\right)^2$$

$$= 15.26 \times 10^8 \text{ mm}^4$$

$$\therefore \quad \sigma_{cc}' = \frac{P}{A_e} = \frac{430000}{156300} = 2.75 \text{ N/mm}^2$$

$$\sigma_{cbc}' = \frac{M}{I_e} \cdot \frac{D}{2} = \frac{8000 \times 1000}{15.26 \times 10^8} \times 160 = 0.84 \text{ N/mm}^2$$

Now $\frac{\sigma_{cc}'}{\sigma_{cc}} + \frac{\sigma_{cbc}'}{\sigma_{cbc}} = \frac{2.75}{4} + \frac{0.85}{5} = 0.86 < 1.$ Hence safe.

Provide 8 mm Φ ties @ 300 mm c/c.

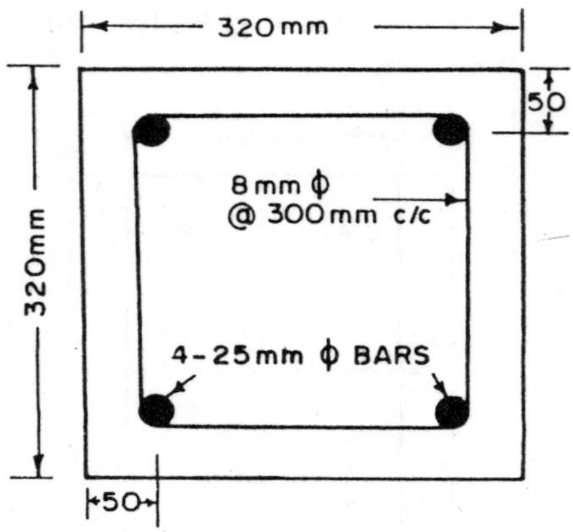

FIG. 14.10.

14.5. CASE 2. COMPRESSIVE LOAD AT LARGE ECCENTRICITY(e >1.5 D)

In this case, the compressive load acts at large eccentricity, resulting in large tensile stresses. Due to this, the section below neutral axis cracks and hence the concrete below it is neglected. The compressive stresses are borne by concrete above the neutral axis and the tensile stresses below the neutral axis are entirely borne by tensile steel. In this case, it is assumed that the eccentricity is large so that the bending stresses are predominant. The super-position of the direct stresses, found by dividing the load with the effective area of the section, does not therefore materially change the position of neutral axis located by the usual method adopted in the case of pure bending. This approximation is reasonable when eccentricity e is greater than 1.5 times the depth of the section.

Fig. 14.11 (*a*) shows a R.C. section subjected to a compressive force P acting at a large eccentricity e measured from the axis passing through the mid-depth.

Let A_{sc} = area of steel in compressive zone.

A_{st} = area of steel in tensile zone.

n = depth of neutral axis, based on bending stresses alone

c_a = compressive stress in concrete, due to bending

c_a' = compressive stress in concrete surrounding compressive steel

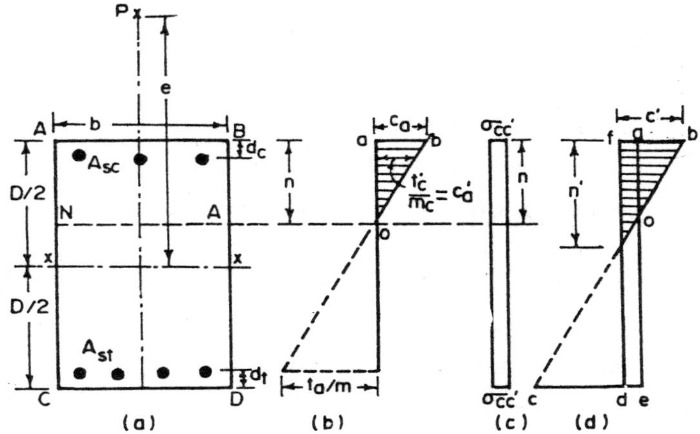

FIG. 14.11

σ_{cc}' = direct compressive stress in concrete

t_a = tensile stress in tensile steel

t_c' = compressive stress in compressive steel $= m_c \cdot c_a'$

$= 1.5\, m\, c_a'$

d_c = cover to compressive steel

d_t = cover to tensile steel

d = effective depth of the section

D = total depth of the section

e = eccentricity

e' = distance of P from N.A.

Fig. 14.11 (*b*) shows the bending stress diagram. The depth of neutral axis, neglecting the direct stresses, can be found by equating the moments of effective area about the neutral axis.

Thus, $\frac{b\,n^2}{2} + (m_c - 1)\,A_{sc}\,(n - d_c) = m.\,A_{st}\,(D - d_t - n)$...(14.9)

From this equation, the unknown n can be found. In order to determine the compressive stress c_a is concrete, due to bending, equate the moment of resistance of the section to the external bending moment. Thus :

$$\frac{b \cdot nc_a}{2}\left(D - d_t - \frac{n}{3}\right) + (m_c - 1) \cdot A_{sc} \cdot c_a' \,(D - d_c - d_t) = P \cdot e' \quad ..(14.10)$$

where e' is the distance of point of application of load P from the N.A.

In the above equation c_a' has the following value [Fig. 14.11(b)]

$$c_a' = \frac{c_a}{n}(n - d_c) \quad ...[14.11(a)$$

The stress in compressive steel is,

$$t_c' = 1.5\, mc_a' = 1.5\, m\,\frac{c_a}{n}\,(n - d_c) \quad ...[14.11(b)]$$

The tensile stress t_a in steel is given by

$$t_a = m \cdot \frac{c_a}{n}\,(D - d_t - n) \quad ...[14.11(c)]$$

The effective area of section is given by

$$A_c = b \cdot n + (m_c - 1) A_{sc} + m A_{st} \quad ...(14.12)$$

Hence the direct stress in concrete is given by

$$\sigma_{cc}' = \frac{P}{A_e} = \frac{P}{b \cdot n + (m_c - 1) A_{sc} + m A_{st}} \quad ...(14.13)$$

The final stresses can be found by superposition as under:

c' = stress in concrete $= c_a + \sigma_{cc}'$ ....[14.14(a)]

t_c = stress in compressive steel $= t_c' + m_c \cdot \sigma_{cc}'$

$= m_c \cdot c_a' + m_c \cdot \sigma_{cc}' = m_c\,(c_a' + \sigma_{cc}')$

$= 1.5\ m\,(c_a' + \sigma_{cc}')$ ...[14.14(b)]

t = stress in tensile steel $= t_a - m\,\sigma_{cc}'$ ...[14.14(c)]

Fig. 14.11(d) shows the super-position of both the stresses. It will be seen that the actual neutral axis, after superposition, shifts downwards increasing the value of n and hence of the effective area. However, the error involved is negligible if $e > \frac{3}{2}D$, and is on the safer side. A more exact method of locating the neutral axis is discussed in the next article, where e lies between $D/4$ and $\frac{3}{2}D$.

Example 14.10. *Fig. 14.12(a) shows a column subjected to a load of 25 kN acting at an eccentricity of 600 mm from the centre of the column. The column section is reinforced as shown in Fig. 14.12(b). Determine the stress in concrete and steel. Take m = 19.*

Solution.

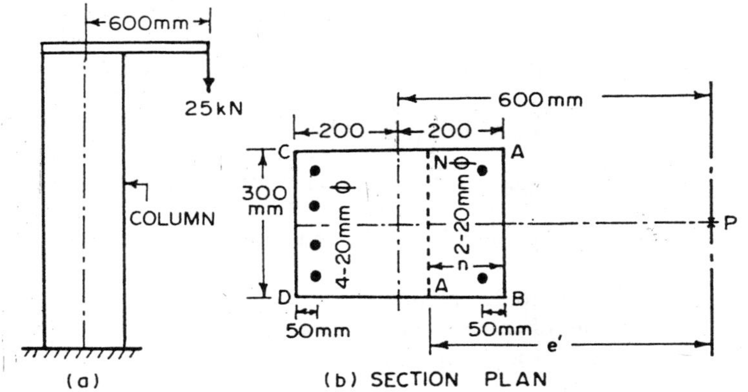

FIG. 14.12.

$$A_{st} = 4\frac{\pi}{4}(20)^2 \approx 1256 \ \text{mm}^2 \ ; \ d_t = 50 \text{ mm}$$

$$A_{sc} = 2\frac{\pi}{4}(20)^2 = 628 \ \text{mm}^2 ; \ d_c = 50 \text{ mm}$$

Let n be the depth of N.A. from compression face AB. Equating moment of the effective areas about N.A., we have

$$\frac{b\,n^2}{2} + (m_c - 1)\,A_{sc}\,(n - d_c) = m\,.A_{st}\,(D - d_t - n)$$

$$\therefore \quad \frac{300\,n^2}{2} + (1.5 \times 19 - 1)628(n - 50) = 19 \times 1256\,(400 - 50 - n)$$

or $\quad n^2 + 274.2 - 61440 = 0$

From which $n = 146.2$ mm

$\therefore \quad e' = 600 - (200 - n) = 600 - (200 - 146.2) = 546.2$ mm.

To find the compressive stress c_a in concrete, due to bending, equate the moment of resistance of the section to the external moment (Eq. 14.10) :

$$\frac{b.n.c_a}{2}\left(D - d_t - \frac{n}{3}\right) + (m_c - 1)\,A_{sc}.c_a'\,(D - d_c - d_t) = P.e'$$

where $\quad c_a' = \frac{c_a}{n}(n - d_c) = \frac{c_a}{146.2}(146.2 - 50) = 0.658\,c_a$

Substituting the values, we have

$$\frac{300 \times 146.2}{2}\,c_a\left[400 - 50 - \frac{146.2}{3}\right] + (1.5 \times 19 - 1)\,628 \times 0.648\,c_a \times (400 - 50 - 50) = 25000 \times 546.2$$

or $\quad 6606780\, c_a + 3409100\, c_a = 13655000$

or $\quad c_a = 1.36 \text{ N/mm}^2$

Stress in compression steel, $t_c' = 1.5\, m \frac{c_a}{n} (n - d_c)$

$$= 1.5 \times 19 \frac{1.36}{146.2} (146.2 - 50) = 25.5 \text{ N/mm}^2$$

Stress in tensile steel $= t_a = m \frac{c_a}{n} (D - d_t - n)$

$$= 19 \times \frac{1.36}{146.2} (400 - 50 - 146.2)$$

$$= 36.02 \text{ N/mm}^2.$$

Again, the effective area is given by

$$A_e = b.n + (m_c - 1) A_{sc} + m A_{st}$$

$$= (300 \times 146.2) + (1.5 \times 19 - 1)628 + 19 \times 1256$$

$$= 84994 \text{ mm}^2.$$

Hence the compressive stress due to direct load is

$$\sigma_{cc}' = \frac{25000}{84994} = 0.29 \text{ N/mm}^2.$$

The final stress will be as follows :

Compressive stress in concrete, $c' = c_a + \sigma_{cc}' = 1.36 + 0.29$

$$= \mathbf{1.65 \text{ N/mm}^2}$$

Compressive stress in steel $= t_c = t_c' + 1.5\, m\, \sigma_{cc}' = 25.5 + 1.5 \times 19 \times 0.29$

$$= \mathbf{33.8 \text{ N/mm}^2.}$$

Tensile stress in steel, $t = t_a - m\, \sigma_{cc}' = 36.02 - 19 \times 0.29$

$$= \mathbf{30.45 \text{ N/mm}^2.}$$

(See example 14.12 for comparison of results)

14.9. CASE 3. COMPRESSIVE LOAD AT MODERATE ECCENTRICITY $\left[\frac{D}{4} < e < \frac{3}{2} D\right]$

This is the most general case. The eccentricity is such that tensile stresses are set up in the concrete below the neutral axis, but at the same time both bending stresses as well as direct stress are of equal importance. The concrete below the neutral axis is neglected and the neutral axis is located taking into account the effects of both bending moments as well as direct load.

Let us use the same notations as in the previous article. Fig 14.13 shows the stress diagram. If c' is the compressive stress in concrete, the stresses in compressive and tensile steels are given by the following expressions :

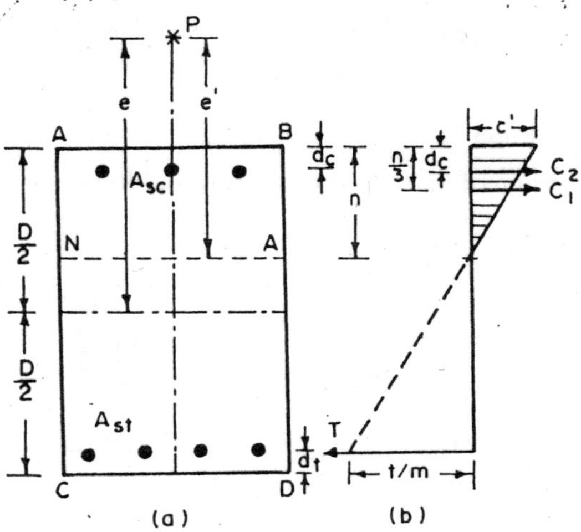

FIG. 14.13.

$$t_c = m_c \frac{c'}{n}(n - d_c) = 1.5\, m \frac{c'}{n}(n - d_c) \qquad ...(1)$$

$$t = m \frac{c'}{n}(D - d_t - n) \qquad ...(2)$$

Since the section is stable, the total force acting on the section must balance. That is, the algebraic sum of internal compressive and tensile forces must be equal to the external load P. Hence

$$\frac{b.n.c'}{2} + (m_c - 1) A_{sc.} \frac{c'}{n}(n - d_c) - m A_{st} \frac{c'}{n}(D - d_t - n) = P \qquad ...(14.15)$$

Similarly, the moments of all forces acting on the section must balance Taking the moments about the centre of tensile steel, we get

$$\frac{b.n.c'}{2}\left(D - d_t - \frac{n}{3}\right) + (m_c - 1) A_{sc.} \frac{c'}{n}(n - d_c)(D - d_t - d_c)$$
$$= P\left(e + \frac{D}{2} - d_t\right) \qquad ...(14.16)$$

From Eq. 14.15 and 14.16, the two unknowns c' and n can be determined. Dividing Eq. 14.16 by 14.15, we get

$$\frac{\frac{b.n}{2}\left(D - d_t - \frac{n}{3}\right) + (m_c - 1) A_{sc} \frac{1}{n}(n - d_c)(D - d_t - d_c)}{\frac{b.n}{2} + (m_c - 1) A_{sc}\left(\frac{n - d}{n}\right) - m A_{st}\left(\frac{D - d_t - n}{n}\right)}$$
$$= \left(e + \frac{D}{2} - d_t\right) \qquad ...(14.17)$$

This is a quadratic equation in n which can be solved either by trial and error, or graphically. Knowing n, c' can be determined either from Eq. 14.15 or from Eq. 14.16. The stresses in steel can then be found from (1) and (2) above.

Alternative Method

The value of n can alternatively be found as under. Let n

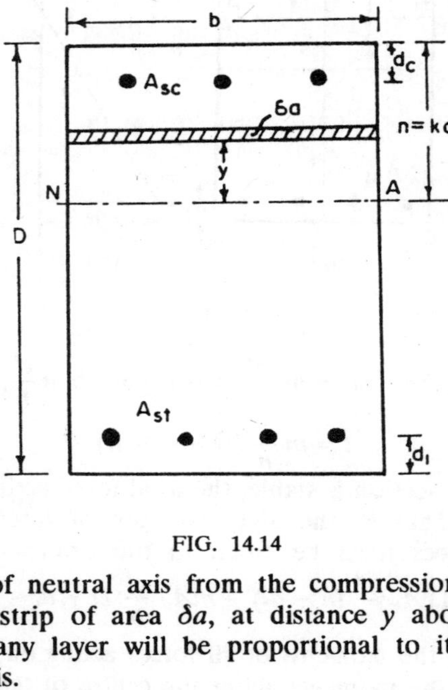

FIG. 14.14

be the depth of neutral axis from the compression face. Consider as elementary strip of area δa, at distance y above neutral axis. The stress at any layer will be proportional to its distance from the neutral axis.

Hence the stress at elementary strip $= r.y$ where r is a constant.

Elementary force $= r\,y\,\delta a$.

The total force on the whole area must be equal to the external force.

Hence $\Sigma r.y.\delta a = T$

or $r\,\Sigma y.\delta a = T$

Again moment of elementary force about the neutral axis will be equal to $ry^2.\delta a$. The sum of such elementary moments must be equal to the moment of the external force about the neutral axis.

Hence

$$\Sigma r.y^2.\delta a = T\left(e - \frac{D}{2} + n\right) = T.e'$$

or $$r\Sigma y^2 \,.\, \delta a = T\left(e - \frac{D}{2} + n\right) = T \,.\, e' \qquad ..(2)$$

Dividing (2) by (1), we get

$$\frac{\Sigma\, y^2\, \delta a}{\Sigma\, y.\, \delta a} = \left(e - \frac{D}{2} + n\right) = e' \qquad ...(14.18)$$

Equation 14.18 is very important conclusion. $\Sigma\, y^2 \,.\, \delta a$ is the moment of inertia of the effective section about the neutral axis, while $\Sigma\, y \,.\, \delta a$ is the moment of the effective area of the section about the neutral axis, the term $\left(e - \frac{D}{2} + n\right)$ or e' is the distance of the load from the neutral axis. *Hence the moment of inertia of the effective section about the neutral axis divided by the moment of area of the effective section about the N.A. is equal to the eccentricity of the load from the neutral axis.*

Again, the stress at any point is given by

$$f = r \,.\, y$$

or $$r = \frac{f}{y}$$

Substituting this is (2), we get

$$\frac{f}{y} \,.\, \Sigma\, y^2 \,.\, \delta\, a = T \,.\, e'$$

Putting $\Sigma\, y^2 \,.\, \delta a = I_{en}$ = moment of inertia of equivalent section about N.A.

and $$T \,.\, e' = M$$

We get $$\frac{f}{y} I_{en} = M$$

or $$f = \frac{M}{I_{en}}\, y \qquad ...(14.19)$$

Now $$I_{en} = \frac{bn^3}{3} + (m_c - 1)\, A_{sc}(n - d_c)^2 + mA_{st}(D - d_t - n)^2 \qquad ...(14.20)$$

$$\Sigma\, y.\, \delta a = \frac{bn^2}{2} + (m_c - 1)A_{sc}(n - d_c) - m\, A_{st}(D - d_t - n) \qquad ...(14.21)$$

Substituting these values in Eq. 14.18, we get

$$\frac{\frac{bn^3}{3} + (m_c - 1)A_{sc}(n - d_c)^2 + m \,.\, A_{st}(D - d_t - n)^2}{\frac{bn^2}{2} + (m_c - 1)A_{sc}(n - d_c) - m \,.\, A_{st}(D - d_t - n)} = e' \qquad ...(14.18)$$

which is a cubic equation in n and can be solved by trial and error.

Example 14.11. *The section of an arch rib, 400 mm × 800 mm is reinforced as shown in Fig 14.15. Determine maximum stress in concrete and steel when the section is subjected to a thrust of 400 kN and a bending moment of 200 kN-m. Take m = 19.*

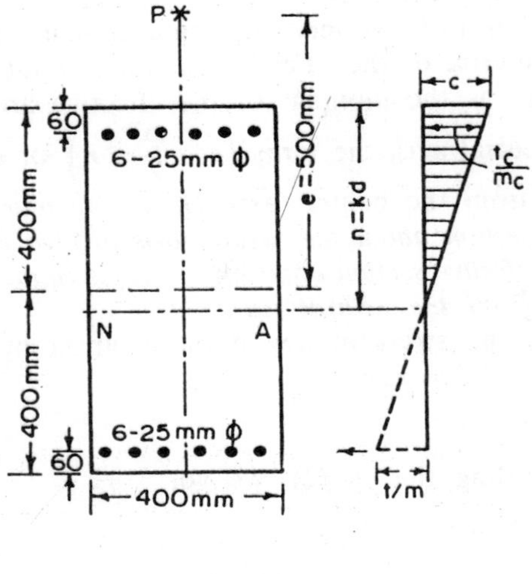

FIG. 14.15

Solution. The given axial thrust and bending moment is equivalent to an eccentric load acting at eccentricity $e = \dfrac{200 \times 10^6}{400 \times 10^3} = 500$ mm. This is more than $\dfrac{D}{4}$ and less than $\dfrac{3}{2}D$ and hence falls under case 3. The process of determining n is that of trial and error. As a first trial, k may be assumed to be between 0.4 to 0.7, for case 3. *Higher the value of eccentricity, lower will be the value of k.*

Let us assume $n = 450$ mm, so that $k = \dfrac{450}{740} \approx 0.61$

For this value of n, let us calculate the stress in steel and concrete, and then find the value of k, which should be near to the assumed value. If not, then an intermediate value of k is adopted and then calculations are repeated.

$$A_{st} = A_{sc} = 6 \times \frac{\pi}{4}(25)^2 = 2945 \text{ mm}^2.$$

Taking the moments of internal and external forces about the centre of tensile steel, we have, from Eq. 14.16.

$$\frac{bnc'}{2}\left(d-\frac{n}{3}\right) + (m_c-1)A_{sc}\,.\,\frac{c'}{n}(n-d_c)(d-d_c) = P\left(e+\frac{D}{2}-d_t\right) \quad ...(1)$$

or $$\frac{400 \times 450\,c'}{2}\left(740-\frac{450}{3}\right) + (1.5\times 19 - 1)\,2945\frac{c'}{450}(450-60) \times (740-60)$$

$$= 400 \times 10^3\left(500+\frac{800}{2}-60\right)$$

or $$53.1 \times 10^6 c' + 47.73 \times 10^6 c' = 336 \times 10^6$$

From which $c' = 3.33$ N/mm^2

Again, equating the sum of internal forces to the external forces, we have

$$\frac{bnc'}{2} + (m_c - 1)A_{sc}\,.\,\frac{c'}{n}(n-d_c) - A_{st}\,.\,t = P \quad ...(2)$$

or $$\frac{400\times450\times3.33}{2}+(1.5\times19-1)2945\times\frac{3.33}{450}(450-60)-2945\times t = 400 \times 10^3$$

or $$3 \times 10^5 + 2.338 \times 10^5 - 2945\,t = 4 \times 10^5$$

From which $t = 45.43$ N/mm^2

From these values of c and t, we have from stress diagram,

$$n = \frac{1}{1+\dfrac{t}{m\,.\,c'}} \times d = \frac{740}{1+\dfrac{45.43}{19\times 3.33}} = 430.8 \text{ mm}$$

This is less than the assumed value of 450 mm. Hence adopt an intermediate value of n = 436 mm(say). Substituting in Eq. (1) above, we get

$$\frac{400 \times 436\,c'}{2}\left(740-\frac{436}{3}\right) + (1.5\times 19 - 1)\,2945\,\frac{c'}{436}(436-60) \times (740-60)$$

$$= 400 \times 10^3\left(500+\frac{800}{2}-60\right)$$

which gives $c' = 3.38$ N/mm^2

Substituting in Eq. (2) above we get

$$\frac{400 \times 436 \times 3.38}{2} + (1.5\times 19-1)\,2945\,\frac{3.38}{436}(436-60) - 2946\,t = 400 \times 10^3$$

which gives $t = 44.55$ N/mm^2

Hence from stress diagram,

$$n = \frac{1}{1+\dfrac{t}{m \, . \, c'}} \, d = \frac{740}{1+\dfrac{44.55}{19 \times 3.38}} = 437 \text{ mm.}$$

This is practically the same as the assumed value, and no further trial is necessary.

Thus, $c' = \mathbf{3.38 \ N/mm^2}$ (compressive)

$t = \mathbf{44.55 \ N/mm^2}$ (tensile)

and $$t_c = 1.5 \, m \frac{c'}{n} (n - d_c) \quad = 1.5 \times 19 \frac{3.38}{436} (436 - 60)$$

$$= \mathbf{83.1 \ N/mm^2} \text{ (compressive)}$$

Alternative Solution

Let us solve this example by alternative method in which the neutral axis is first located. Let the N.A. be at a distance n mm below the compression face.

Moment of inertia of effective section about N.A. is given by Eq. 14.20.

$$I_{en} = \frac{bn^3}{3} + (m_c - 1) A_{sc} (n - d_c)^2 + mA_{st} (D - d_t - n)^2$$

$$= \frac{400 \times n^3}{3} + (1.5 \times 19 - 1) 2945 (n - 60)^2 + 19 \times 2945 (800 - 60 - n)^2$$

$$= 133.3 \, n^3 + 80988 \, (n - 60)^2 + 55955 \, (740 - n)^2 \qquad ...(1)$$

Moment of effective area about N.A is given by Eq. 13.21

$$\Sigma y \, . \, \delta a = \frac{bn^2}{2} + (m_c - 1) A_{sc} (n - d_c) - m \, A_{st} (D - d_t - n)$$

$$= \frac{400 \, n^2}{2} + (1.5 \times 19 - 1) \times 2945 (n - 60) - 19 \times 2945 (800 - 60 - n)$$

$$= 200 \, n^2 + 80988 \, (n - 60) - 55955 \, (740 - n) \qquad ...(2)$$

Distance of P from N.A. is given by

$$e' = \left[e - \frac{D}{2} + n \right] = 500 - \frac{800}{2} + n = 100 + n \qquad ...(3)$$

Hence from Eq. 14.18

$$\frac{I_{en}}{\Sigma y \, . \, \delta a} = e'$$

or $$\frac{133.3 \, n^3 + 80988 \, (n - 60)^2 + 55955 \, (740 - n)^2}{200 \, n^2 + 80988 \, (n - 60) - 55955 \, (740 - n)} = 100 + n$$

Solving this by trial and error, trying the value of n between 0.4 d (=296 mm) to 0.7 d(=518 mm), we get n=436.5 mm.

Hence $I_{en}=133.3(436.5)^2+80988(436.5-60)^2+55955(740-436.5)^2$
$= 2.772 \times 10^{10}$ mm^4

$$e' = e - \frac{D}{2} + n = 500 - \frac{800}{2} + 436.5 = 536.5$$

Moment of thrust about N.A. $= 400000 \times 536.5$

$$\therefore \quad \text{Stress in concrete, } c' = \frac{400000 \times 536.5}{2.772 \times 10^{10}} \times 436.5$$

$$= \mathbf{3.38\ N/mm^2}$$

$$\text{Stress in tensile steel} = \frac{400000 \times 536.5}{2.772 \times 10^{10}}(800 - 436.5 - 60) \times 19$$

$$= \mathbf{44.64\ N/mm^2}.$$

and stress in compressive steel,

$$t_c = 1.5 \times 19 \frac{400000 \times 536.5}{2.772 \times 10^{10}}(436.5 - 60)$$

$$= \mathbf{83.1\ N/mm^2}.$$

Example 14.12. *Solve Example 14.10 by the exact method.*

Solution. (Fig. 14.12).

Since the eccentricity was more than 1.5 D, example 14.10 was solved by method of superposition. Let us now solve it by first determining the neutral axis, by method of case 3.

Let the depth of neutral axis be n mm below the compression face AB. The moment of inertia of effective section about N.A. is given by Eq. 14.20 :

$$I_{en} = \frac{bn^3}{3} + (m_c - 1)A_{sc}(n - d_c)^2 + mA_{st}(D - d_t - n)^2$$

where $A_{sc} = 628$ mm^2 ; $A_{st} = 1256$ mm^2 ; $d_t = d_c = 50$ mm

$$\therefore\ I_{en} = \frac{300\,n^3}{3} + (1.5 \times 19 - 1)628(n-50)^2 + 19 \times 1256(400 - 50 - n)^2$$

$$= 100\,n^3 + 17270\,(n - 50)^2 + 24864\,(350 - n)^2 \qquad ...(1)$$

Moment of effective are about N.A. is given by Eq. 14.21.

$$\Sigma y\,.\,\delta a = \frac{bn^2}{2} + (m_c - 1)A_{sc}(n - d_c) - mA_{st}(D - d_t - n)$$

$$= \frac{300n^2}{2} + (1.5 \times 19 - 1)628(n - 50) - 19 \times 1256(400 - 50 - n)$$

$$= 150\,n^2 + 17270\,(n - 50) - 23864\,(350 - n) \qquad ...(2)$$

Distance of P from N.A. is

$$e' = \left(e - \frac{D}{2} + n\right) = 600 - 200 + n = 400 + n \qquad ...(3)$$

Hence from Eq. 14.18.

$$\frac{I_{en}}{\Sigma y \, . \, \delta a} = e'$$

or

$$\frac{100\, n^3 + 17270\,(n - 50)^2 + 23864\,(350 - n)^2}{150\, n^2 + 17270\,(n - 50) - 23864\,(350 - n)} = 400 + n$$

The above relationship can be solved by trial, assuming n between 0.4 d (= 140 mm) to 0.7 d (= 245 mm).

By trial and error, we get n = 176 mm.

Hence $I_{nn} = 100(176)^3 + 17270(176 - 50)^2 + 23864(350 - 176)^2$

$= 15.42 \times 10^8 \text{ mm}^4$

$$e' = e - \frac{D}{2} + n = 600 - 200 + 176 = 576 \text{ mm}$$

Moment of thrust about N.A.= $25000 \times 576 = 14.4 \times 10^6$ N/mm

$\therefore$ Stress in concrete $= c' = \dfrac{14.4 \times 10^6}{15.42 \times 10^8} \times 176$

$= \mathbf{1.64\ N/mm^2}$ (compressive)

(against a value of 1.65 N/mm^2 calculated by approximate method in Example 14.10)

Stress in tensile steel $= \dfrac{14.4 \times 10^6}{15.42 \times 10^8}(400 - 176 - 50) \times 19$

$= \mathbf{30.87\ N/mm^2}$ (tensile)

(against a value of 30.45 N/mm^2 in Ex. 14.10)

Stress in compression steel$= \dfrac{14.4 \times 10^6}{15.42 \times 10^8}(176 - 50) \times 1.5 \times 19$

$= \mathbf{33.53\ N/mm^2}$ (compressive)

(against a value of 33.8 N/mm^2 in Ex. 14.10).

The results shown that when , $e \geq \frac{3}{2}D$, the approximate method can be safely used without any appreciable error.

14.7 CIRCULAR SECTION SUBJECTED TO ECCENTRIC LOAD

The analysis of circular section subjected to load at small eccentricity has been done in example 14.6. Let us now take the case of moderate eccentricity, when tension is developed in the section.

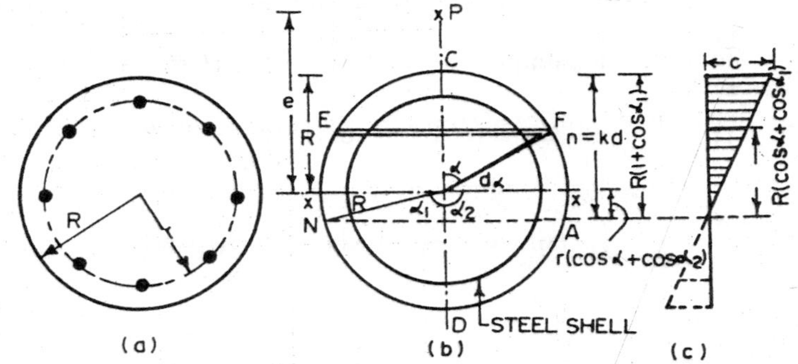

FIG. 14.16.

Fig 14.16(a) shows a column of circular section of radius R, having reinforcement area $=A_s$ arranged on a circle of radius r. The steel rods can be replaced by an equivalent steel shell of radius r, of equal area, as shown in Fig. 14.16(b). Let e be the eccentricity of the load with respect to the axis x-x passing through the centre of circle. Let the N.A. be at a distance $n = kd$ below the point C, so that angles, subtended at the centre, by its points of intersection with concrete extreme and steel shell are α_1 and α_2 respectively.

Consider a thin layer EF, subtending angle α at the centre, with the symmetrical axis CD, and parallel to N.A.

Length of strip $= 2R \sin\alpha$

Distance of strip from centre $= R\cos\alpha$

Width of strip $= \delta\,(R\cos\alpha) = R\sin\alpha \,.\, d\alpha$

Area of strip $= 2R^2\sin^2\alpha \,.\, d\alpha$.

$$n = R + R\cos\alpha_1 = R(1+\cos\alpha_1)$$

Distance of layer EF above N.A. $= R\cos\alpha + R\cos\alpha_1$

$$= R(\cos\alpha + \cos\alpha_1)$$

Let c' be the intensity of stress in concrete at point C.

$$\therefore \text{ Intensity of stress at } EF = \frac{c'\,R(\cos\alpha+\cos\alpha_1)}{R(1+\cos\alpha_1)}$$

$$= \frac{c'(\cos\alpha+\cos\alpha_1)}{1+\cos\alpha_1}$$

$\therefore$ Force on strip $= 2R^2 \sin^2\alpha \,.\, d\alpha \,.\, \dfrac{c'(\cos\alpha + \cos\alpha_1)}{1+\cos\alpha_1}$

Hence total compression C_1 in concrete is given by

$$C_1 = \int_0^{\pi-\alpha_1} \frac{2R^2 . c'}{1+\cos\alpha_1} . \sin^2\alpha\,(\cos\alpha + \cos\alpha_1)\, d\alpha$$

$$= \frac{2R^2 . c'}{1+\cos\alpha_1} \int_0^{\pi-\alpha_1} (\sin^2\alpha . \cos\alpha + \sin^2\alpha . \cos\alpha_1)\, dx$$

$$= \frac{2R^2 c'}{1+\cos\alpha_1} \left[\frac{\sin^3\alpha}{3} + \frac{\cos\alpha_1}{2}\left(\alpha - \frac{\sin 2\alpha}{2}\right)\right]_0^{\pi-\alpha_1}$$

$$= \frac{2R^2 c'}{1+\cos\alpha_1} \left[\frac{\sin^3(\pi-\alpha_1)}{3} + \frac{\cos\alpha_1}{2}(\pi-\alpha_1) - \frac{\cos\alpha_1}{4}\sin 2(\pi-\alpha_1)\right]$$

$$= \frac{2R^2 c'}{1+\cos\alpha_1} \left[\frac{\sin^3\alpha_1}{3} + \frac{\pi-\alpha_1}{2}\cos\alpha_1 - \frac{\cos\alpha_1}{4}\sin 2\alpha_1\right] \qquad ...(14.22)$$

Moment of compression C_1 in concrete, about the central axis x-x is :

$$M_1 = \int_0^{\pi-\alpha_1} 2R^2 \sin^2\alpha\, d\alpha\; \frac{c'(\cos\alpha + \cos\alpha_1)}{1+\cos\alpha_1} \times R\cos\alpha$$

$$= \frac{2R^3 c'}{1+\cos\alpha_1} \int_0^{\pi-\alpha_1} \sin^2\alpha\cos\alpha\,(\cos\alpha + \cos\alpha_1)\, d\alpha$$

$$= \frac{2R^3 c'}{1+\cos\alpha_1} \int_0^{\pi-\alpha_1} \sin^2\alpha\cos^2\alpha + \sin^2\alpha\cos\alpha\,\cos\alpha_1)\, d\alpha$$

$$= \frac{2R^3 c'}{1+\cos\alpha_1} \left[\frac{1}{8}\left(\alpha + \frac{\sin 4\alpha}{4}\right) + \cos\alpha_1 \frac{\sin^3\alpha}{3}\right]_0^{\pi-\alpha_1}$$

$$= \frac{2R^3 c'}{1+\cos\alpha_1} \left[\frac{1}{8}\left\{(\pi-\alpha_1) + \frac{\sin 4(\pi-\alpha_1)}{3}\right\} + \frac{\cos\alpha_1}{3}\sin^3(\pi-\alpha_1)\right]$$

$$= \frac{2R^3 c'}{1+\cos\alpha_1} \left[\frac{\pi-\alpha_1}{8} + \frac{\sin 4\alpha_1}{32} + \frac{\cos\alpha_1}{3}\sin^3\alpha_1\right] \qquad ...(14.23)$$

Similarly, consider steel area in compression. If A_{st} is the total steel area, the thickness z of steel is given by

$$2\pi r z = A_{st} \qquad ...(14.24)$$

Small area of compression steel at angle $\alpha = r\,\delta\alpha\, z$

Effective or equivalent area of compression steel element $= (m_c - 1)\, r\, \delta\alpha\, z$

Intensity of stress on elementary area $= c' \dfrac{r(\cos\alpha + \cos\alpha_2)}{R + r\cos\alpha_2}$

$\therefore$ Total compression C_2 in steel above neutral axis is given by

$$C_2 = 2\int_0^{\pi-\alpha_2} (m_c - 1)\, r\, d\alpha \,.\, z\, c' \frac{r(\cos\alpha + \cos\alpha_2)}{R + r\cos\alpha_2}$$

$$= \frac{2r^2 z(m_c - 1)c'}{r + r\cos\alpha_2}\int_0^{\pi-\alpha_2} (\cos\alpha + \cos\alpha_2)\, d\alpha$$

$$= \frac{2r^2 z(m_c - 1)c'}{R + r\cos\alpha_2}\Big[\sin\alpha + \alpha\cos\alpha_2\Big]_0^{\pi-\alpha_2}$$

$$= \frac{2r^2 z(m_c - 1)c'}{R + r\cos\alpha_2}\Big[\sin\alpha_2 + (\pi - \alpha_2)\cos\alpha_2\Big] \qquad ...(14.25)$$

Moment of total compression in steel about central axis x-x is given by

$$M_2 = 2\int_0^{\pi-\alpha_2} (m_c - 1)\, r\, d\alpha\, z\, c' \frac{r(\cos\alpha + \cos\alpha_2)}{R + r\cos\alpha_2}\, r\cos\alpha$$

$$= \frac{2r^3 z(m_c - 1)c'}{R + r\cos\alpha_2}\int_0^{\pi-\alpha_2} (\cos^2\alpha + \cos\alpha\cos\alpha_2)\, d\alpha$$

$$= \frac{2r^3 z(m_c - 1)c'}{R + r\cos\alpha_2}\left[\frac{\alpha}{2} + \frac{\sin 2\alpha}{4} + \cos\alpha_2\sin\alpha\right]_0^{\pi-\alpha_2}$$

$$= \frac{2r^3 z(m_c - 1)c'}{R + r\cos\alpha_2}\left[\frac{\pi - \alpha_2}{2} - \frac{\sin 2\alpha_2}{4} + \cos\alpha_2\sin\alpha_2\right]$$

$$= \frac{2r^3 z(m_c - 1)c'}{R + r\cos\alpha_2}\left[\frac{\pi - \alpha_2}{2} + \frac{\sin 2\alpha_2}{4}\right] \qquad ...(14.26)$$

where $m_c = 1.5\, m$

In the same manner total tension T is tension steel below the N.A. is given by

$$T = 2\int_0^{\alpha_2} r d\alpha\, z\, m\, c' \frac{r\cos\alpha - r\cos\alpha_2}{R + r\cos\alpha_2}$$

$$= \frac{2r^2 z\, m\, c'}{R + r\cos\alpha_2}\int_0^{\alpha_2} (\cos\alpha - \cos\alpha_2)\, d\alpha$$

$$= \frac{2r^2 z\, m\, c'}{R + r\cos\alpha_2}[\sin\alpha_2 - \alpha_2\cos\alpha_2] \quad ...(14.27)$$

Moment of total tension in steel, about the centre axis x-x is given by

$$M_3 = 2\int_0^{\alpha_2} r\, d\alpha\, .\, z\, m\, c' \frac{r(\cos\alpha - \cos\alpha_2)}{R + r\cos\alpha_2}\, .\, r\cos\alpha$$

$$= \frac{2r^3 .\, z. m\, c'}{R + r\cos\alpha_2}\int_0^{\alpha_2} (\cos^2\alpha - \cos\alpha\cos\alpha_2)\, d\alpha$$

$$= \frac{2r^3 .\, z. m_c'}{R + r\cos\alpha_2}\left(\frac{\alpha_2}{2} - \frac{\sin 2\alpha_2}{4}\right) \quad ...(14.28)$$

Since the section is in equilibrium, the algebraic sum of internal forces must be equal to the external force.

Hence $\qquad C_1 + C_2 - T = P \qquad ...(14.29)$

Similarly, total moment of internal force about centre line is equal to the moment of external force about the centre line

Hence $\qquad M_1 + M_2 + M_3 = P.e \qquad ...(14.30)$

Dividing Eq. 14.30 by Eq. 14.29, we get

$$\frac{M_1 + M_2 + M_3}{C_1 + C_2 - T} = e \quad ...(14.31)$$

Substituting the values of C_1, C_2, T, M_1, M_2 and M_3, we find that there are two unknowns α_1 and α_2 Another relationship between α_1 and α_2 is obtained by :

$$R\cos\alpha_1 = r\cos\alpha_2 \quad ...(14.32)$$

or $\qquad x_2 = \cos^{-1}\left(\frac{R}{r}\cos\alpha_1\right) \qquad ...[14.32(a)]$

Thus value of α_2 can be substituted in Eq. 14.31, which can then be solved by trial and error. However, it is better to assume some value of k (between 0.4 to 0.7) and then find the value of α_1 from the relation :

$$n = kd = R(1 + \cos\alpha_1) \quad ...(14.33)$$

Knowing α_1, α_2 is determined from Eq. 14.12(a). These values of α_1 and α_2 can then be substituted in Eq. 14.31 to see whether it is satisfied or not. If not satisfied, another intermediate value of n is chosen and procedure repeated.

Example 14.13. *A circular column of 500 mm diameter is reinforced with 8 bars of 20 mm diameter arranged on a circle of 400 mm diameter as shown in Fig. 14.17. Determine the maximum stresses in concrete and steel when the column is subjected to an axial load of 200 kN and a bending moment of 40 kN-m. Take m = 19.*

Solution . $A_{st} = 8\frac{\pi}{4}(20)^2 = 2513 \text{ mm}^2$

Let z be the thickness of equivalent steel shell placed at a radius of 200 mm.

$$2\pi r z = A_{st} = 2531$$

$$\therefore \quad z = \frac{2513}{2\pi r} = \frac{2513}{2\pi \times 200} = 2 \text{ mm}$$

$$e = \frac{40}{200} = 0.2 \text{ m}$$

$$= \mathbf{200 \ mm}.$$

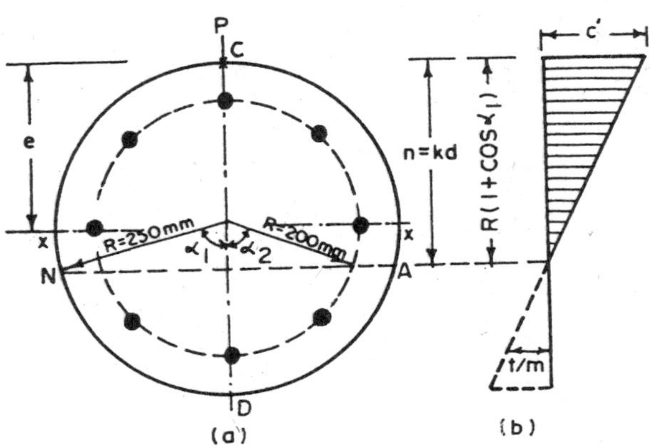

FIG 14.17.

Let $n = kd$ be the depth of N.A. The value of k may be assumed between 0.4 to 0.7, smaller value for larger eccentricity.

$$d = 500 - 50 = 450 \text{ mm}$$

Let us assume $k = 0.60$ so that $n = 450 \times 0.6 = 270$ mm

$R = 250$ mm ; $r = 200$ mm

Now $n = R\,(1 + \cos\alpha_1)$

$$\therefore\ \alpha_1 = \cos^{-1}\left(\frac{n}{R} - 1\right) = \cos^{-1}\left(\frac{270}{250} - 1\right) = \cos^{-1}(0.08)$$

$\therefore\ \alpha_1 = 85°\ 25' = 1.491$ radians

Again, from Eq. 14.32 (*a*),

$$\alpha_2 = \cos^{-1}\left(\frac{R}{r}\cos\alpha_1\right) = \cos^{-1}\left(\frac{250}{200} \times 0.08\right) = \cos^{-1}(0.1)$$

$\therefore\ \alpha_2 = 84°\ 16' = 1.47$ radians

The other functions useful in various equations are

$\sin\alpha_1 = 0.997$; $\cos\alpha_1 = 0.08$; $\sin 2\alpha_1 = 0.16$; $\sin 4\alpha_1 = -0.316$
$\sin\alpha_2 = 0.995$; $\cos\alpha_2 = 0.1$; $\sin 2\alpha_2 = 0.199$

Substituting these values in Eqs. 14.22, 14.23, 14.25, 14.26, 14.27 and 14.28, we get

$$C_1 = \frac{2\,(250)^2 c'}{1+0.08}\left[\frac{(0.997)^3}{3} + \frac{\pi - 1.491}{2} \times 0.08 - \frac{0.08}{4} \times 0.16\right]$$
$$= 46246\,c' \qquad \text{...(1)}$$

$$M_1 = \frac{2\,(250)^3 c'}{1 + 0.08}\left[\frac{\pi - 1.491}{8} - \frac{0.316}{32} + \frac{0.08}{3}(0.997)^3\right]$$
$$= 6448972\,c' \qquad \text{...(2)}$$

$$C_2 = \frac{2(200)^2 \times 2 \times 27.5\,c'}{250 + 200 \times 0.1}\left[0.995 + (\pi - 1.47)0.1\right] = 18939\,c' \qquad \text{...(3)}$$

$$M_2 = \frac{2\,(200)^3 \times 2 \times 27.5\,c'}{250 + 200 \times 0.1}\left[\frac{\pi - 1.47}{2} + \frac{0.199}{4}\right] = 288622\,c' \qquad \text{...(4)}$$

$$T = \frac{2\,(200)^2 \times 2 \times 19\,c'}{250 + 200 \times 0.1}[0.995 - 1.47 \times 0.1] = 9548\,c' \qquad \text{...(5)}$$

$$M_3 = \frac{2\,(200)^3 \times 2 \times 19\,c'}{250 + 200 \times 0.1}\left[\frac{1.47}{2} - \frac{0.199}{4}\right] = 1543082\,c' \qquad \text{...(6)}$$

Equating the sum of internal forces to the external forces, we have

$$C_1 + C_2 - T = P$$

$\therefore\ 46246\,c' + 18939\,c' - 9548\,c' = 200000$

From which $c' = 3.66$ N/mm^2 ...(*a*)

Again, equating the sum of internal moments to the external moment, we get

$$M_1 + M_2 + M_3 = M$$

or $$6448972\,c' + 288625\,c' + 1543082\,c' = 40 \times 10^2$$

From which $c' = 3.68\ \text{N/mm}^2$

Since the two values of c' differ very little, no second trial is necessary

Adopt $c' = \frac{1}{2}(3.66 + 3.68) = \mathbf{3.67\ N/mm^2}$

From stress diagram.

$$t = \frac{m\,c'}{n}(d - n) = \frac{19 \times 3.67}{270}(450 - 270)$$

$$= \mathbf{46.49\ N/mm^2}.$$

14.8. CASE 4 : TENSILE LOAD AT SMALL ECCENTRICITY

This case is similar to case 1, except that a tensile load acts in place of a compressive one. However, the load acts at such an eccentricity that only tensile stresses are set up in the section. This may happen when the point of application of pull happens to lie anywhere between the two areas. Because of tensile stresses, concrete becomes ineffective, and only the two steel areas bear the tensile load.

Fig. 14.18(*b*) shows a R.C. section reinforced with steel areas A_{st1} and A_{st2}. Let the directions of the C.G. of steel areas of the two steel areas be y_1 and y_2. The pull P acts at eccentricity e from both the steel areas. Let t_1 and t_2 be the tensile stresses set up in both steel areas, as shown in Fig. 14.18(*a*).

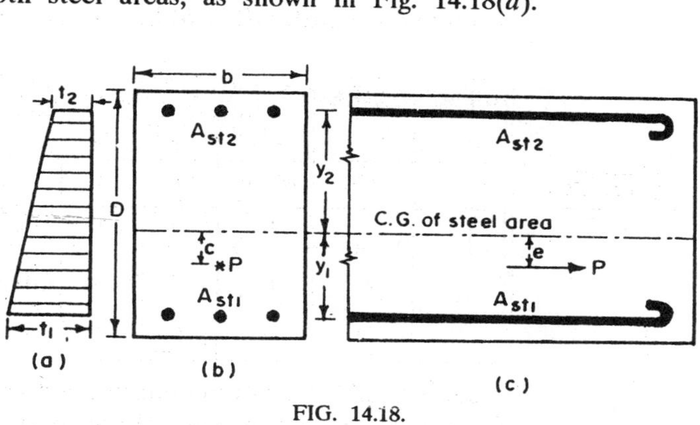

FIG. 14.18.

Taking moments about the centre of steel area A_{st2}, we get

$$t_1 A_{st1} (y_1 + y_2) = P (e + y_2)$$

$$t_1 = \frac{P (e + y_2)}{A_{st1} (y_1 + y_2)} \quad ...(14.34)$$

Similarly taking moments about the centre of steel area A_{st2} we get.

$$t_2 A_{st2} (y_1 + y_2) = P (y_1 - e)$$

$$t_2 = \frac{P (y_1 - e)}{A_{st2} (y_1 + y_2)} \quad ...(14.35)$$

Example 14.14. *The section of a bow-string girder suspender 200 m × 350 mm is reinforced with 3 bars of 16 mm Φ on each side as shown in Fig. 14.19. Determine the stress in steel reinforcement, when the suspender is subjected in a pull of 80 kN and a bending moment of 5 kN-m.*

Solution

$$A_{st1} = A_{st2} = 3\frac{\pi}{4}(16)^2 = 603 \text{ mm}^2$$

Since the section is symmetrically reinforced, the C.G. of steel areas will pass through the mid depth of the section.

Hence $y_1 = y_2 = 125$ mm

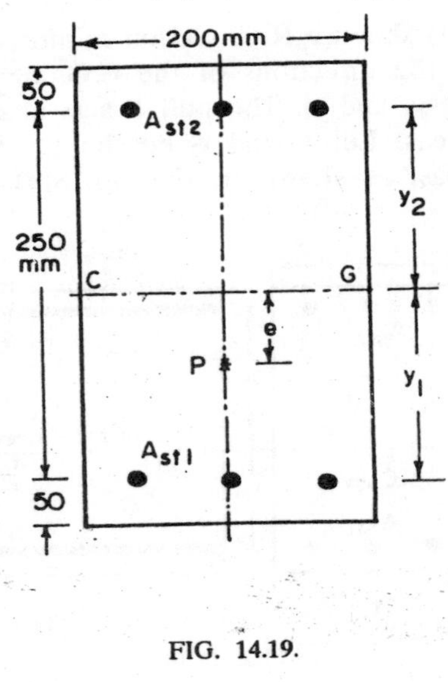

FIG. 14.19.

Also, eccentricity

$$e = \frac{5 \times 1000}{80} = 62.5 \text{ mm}$$

Taking moments about the centre of top steel area, we get

$$t_1 A_{st1} (250) = 80000 (125 + 62.5)$$

$$\therefore \quad t_2 = \frac{80000 \times 187.5}{603 \times 250} = \mathbf{99.5\ N/mm^2} \text{ (tension)}$$

Similarly, taking moments about the centre of bottom steel area, we get

$$t_2 A_{st2} (250) = 80000 (125 - 62.5)$$

$$\therefore \quad t_2 = \frac{80000 \times 62.5}{603 \times 250} = \mathbf{33.2\ N/mm^2} \quad \text{(tension)}$$

14.9. CASE 5. TENSILE LOAD AT LARGE ECCENTRICITY

This case is similar to case 2. The eccentricity is so large ($e > 3/2\,D$) that bending stresses become predominant. The neutral axis is located considering the section as doubly reinforced one, subjected to bending moment. The direct tensile stresses due to axial pull are then super-imposed to get final stresses. The super-imposition will shift the neutral axis towards the compression zone, as shown in Fig. 14.20, but the method gives satisfactory results when eccentricity is more than 1.5 times the depth of the section.

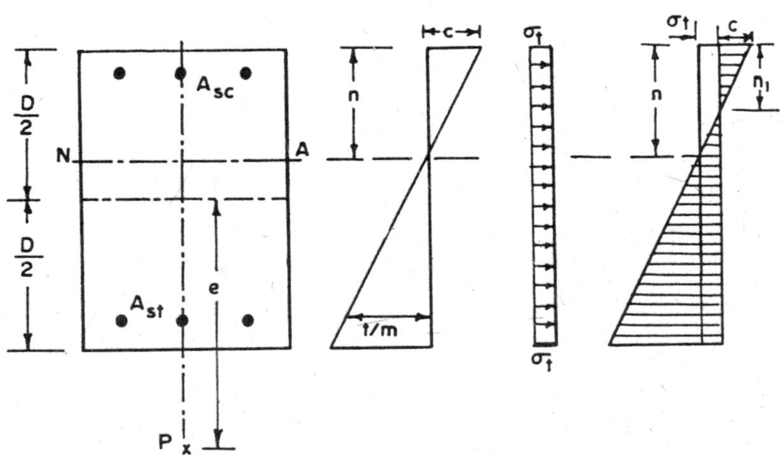

(a) (b) (c) (d)

FIG 14.20.

Fig. 14.20(b) shows the bending stress diagram, where c is the maximum compressive stress in concrete and t is the maximum tensile stress in steel due to bending. The concrete below the N.A. is neglected in the analysis. Fig 14.20(c) shows the uniform tensile stress in the section, due to direct pull. The superimposed stress diagram is shown in Fig. 16.20(d), where the N.A. shifts upwards $(n_1 < n)$.

Example 14.15. *The slab of a tank is reinforced with 8 bars of 16 mm Φ on each face, and is 200 mm thick, as shown in Fig. 14.21. The slab is subjected to a pull of 45 kN and a bending moment of 22.5 kN-m per metre width. Determine the maximum stresses in concrete and steel. Take m = 18.*

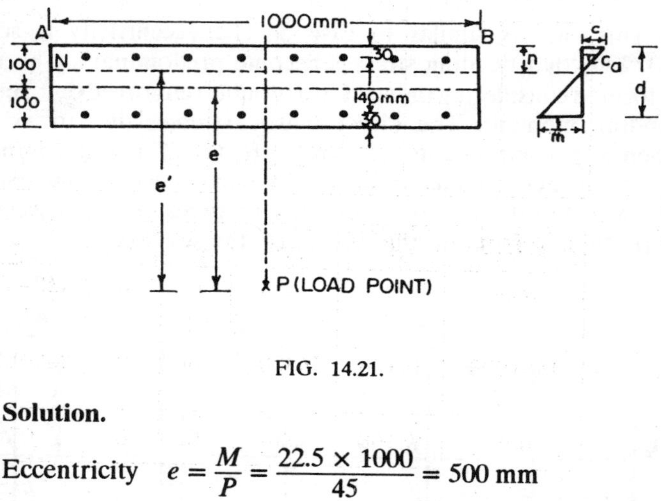

FIG. 14.21.

Solution.

$$\text{Eccentricity} \quad e = \frac{M}{P} = \frac{22.5 \times 1000}{45} = 500 \text{ mm}$$

Since this is more than 1.5D, case 5 will apply here, and bending stresses will be predominant.

Let the neutral axis be at a distance n below the compression face AB.

$$A_{sc} = A_{st} = 8 \frac{\pi}{4} (16)^2 = 1608 \text{ mm}^2$$

Taking the moments of effective areas about the N.A., we have

$$1000 \frac{n^2}{2} + (1.5 \times 19 - 1)\, 1608\, (n - 30) = 19 \times 1608\, (170 - n)$$

or
$$n^2 + 149.54\, n - 13041 = 0$$

From which $n = 61.7$ mm

Eccentricity of pull from N.A. $= e' = e + (100 - n)$

$= 500 + 100 - 61.7 = 538.8$ mm

Let the stress in concrete be c.

Let c_a = stress in concrete around compressive steel

$$= \frac{c}{n}(n - 30) = \frac{c}{61.7}(61.7 - 30) = 0.514\,c$$

∴ Moment of resistance of the section is

$$M_r = \frac{b\,n\,c}{2}\left(d - \frac{n}{3}\right) + (m_c - 1)\,A_{sc}\,(d - 30)c_a$$

$$= \frac{1000 \times 61.7 \times c}{2}\left(170 - \frac{61.7}{3}\right)$$
$$+ (1.5 \times 19 - 1)\,1608\,(170 - 30)\,0.514\,c$$

$$= 7.792 \times 10^6\,c.$$

Equating the moment of resistance to the moment of pull about N.A., we get

$$M_r = P\,e'$$

or $7.792 \times 10^6 c = 45000 \times 538.3$ mm

$$\therefore \quad c = \frac{45000 \times 538.3}{7.792 \times 10^6} = 3.11 \text{ N/mm}^2 \text{(comp.)}$$

Stress in compressive steel $= m_c \cdot c_a = (1.5 \times 19) \times 0.514 \times 3.11$

$= 45.56$ N/mm^2 (comp.)

Stress in tensile steel $= t = m\,\dfrac{c}{n}(d - n)$

$$= \frac{19 \times 3.11}{61.7}(170 - 61.7) = 103.72 \text{ N/mm}^2$$

Effective area of the section for direct tensile force is given by

$$A_e = 1000\,n + (m - 1)\,A_{sc} + m\,A_{st}$$
$$= 1000 \times 61.7 + (19 - 1)\,1608 + 19 \times 1608$$
$$= 121196 \text{ mm}^2$$

∴ σ_t = tensile stress in concrete due to direct pull

$$= \frac{P}{A_e} = \frac{45000}{121196} = 0.37 \text{ N/mm}^2 \text{ (tensile)}$$

Hence the final stresses will be as under :

Compressive stress in concrete $= 3.11 - 0.37 =$ **2.74 N/mm²**

Compressive stress in top steel $= 45.56 - 19 \times 0.37$

$=$ **38.51 N/mm²**

Tensile stress in bottom steel $= 103.72 + 19 \times 0.37$

$=$ **110.75 N/mm²**

14.10. CASE 6. TENSILE LOAD AT MODERATE ECCENTRICITY

This case is similar to case 3 in which both bending stresses as well as direct stresses are equally important. The point of application of pull is *outside* the section, but the eccentricity is less than 1.5 *D*. This is thus a general case, in which the neutral axis is located taking both B.M. and pull into consideration.

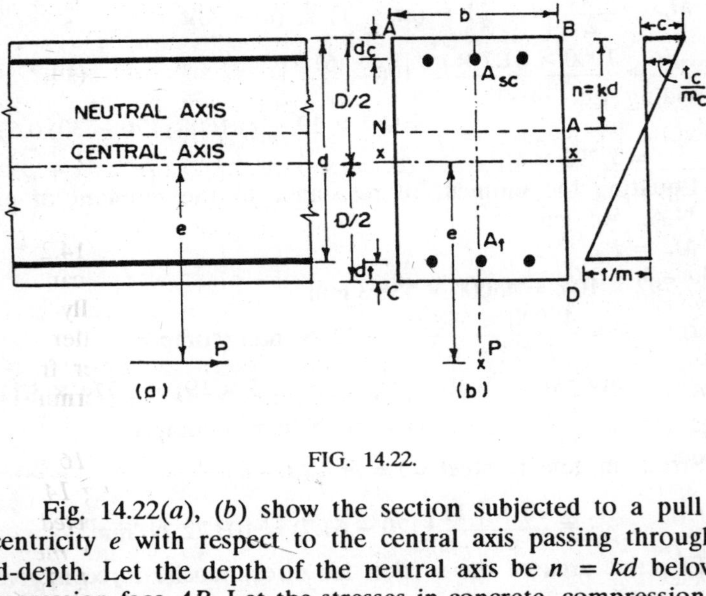

FIG. 14.22.

Fig. 14.22(*a*), (*b*) show the section subjected to a pull *P* at eccentricity *e* with respect to the central axis passing through the mid-depth. Let the depth of the neutral axis be $n = kd$ below the compression face *AB*. Let the stresses in concrete, compression steel and tensile steel be c', t_c, and t respectively. From the stress diagram[Fig. 14.22(*c*)], we have :

$$t_c = m_c \frac{c'}{n}(n - d_c) = 1.5\, m \frac{c'}{n}(n - d_c) \qquad ...(1)$$

$$t = m\frac{c'}{n}(D - d_t - n) = m\frac{c'}{n}(d - n) \qquad ...(2)$$

Since the section is stable, the total forces acting on the section must balance. Hence

$$t A_{st} - \frac{b\,n\,c'}{2} - (m_c - 1) A_{sc} \frac{c'}{n}(n - d_c) = P$$

$$\frac{mc'}{n}(d - n) A_{st} - \frac{bnc'}{2} - (m_c - 1) A_{sc} \frac{c'}{n}(n - d_c) = P \qquad ...(14.36)$$

where $m_c = 1.5\, m$

Similarly, the moments of all the forces acting on the section must balance. Taking moments about the centre of tensile steel, we get

$$\frac{bnc'}{2}\left(D-d_t-\frac{n}{3}\right)+(m_c-1)A_{sc}\frac{c'}{n}(n-d_c)(D-d_t-d_c)$$
$$=P\left(e-\frac{D}{2}+d_t\right)$$

or $$\frac{bnc'}{2}\left(d-\frac{n}{3}\right)+(m_c-1)A_{sc}\frac{c'}{n}(n-d_c)(d-d_c)=P\left(e-\frac{D}{2}+d_t\right)$$

...(14.27)

Dividing Eq. 14.37 by 14.36, we get

$$\frac{\frac{bnc'}{2}\left(d-\frac{n}{3}\right)+(m_c-1)A_{sc}\frac{c'}{n}(n-d_c)(d-d_c)}{\frac{mc'}{n}(d-n)A_{st}-\frac{bnc'}{2}-(m_c-1)A_{sc}\frac{c'}{n}(n-d_c)}=\left(e-\frac{D}{2}+d_t\right)$$

...(14.28)

This is a quadratic equation in terms of n, which can be solved by trial and error. The value of $k(=n/d)$ may generally vary between 0.2 and 0.4 ; smaller the value of eccentricity, smaller the value of k. Once n is known, c' may be determined either from Eq. 14.36 or 14.37. The values of t_o and t can then be determined from Eqs (1) and (2) above.

Example 14.16. *The walls of a tank is reinforced with 16 mm dia. bars spaced at 100 mm c/c on each face as shown in Fig 14.23. The total thickness of the wall is 200 mm. The wall is subjected to a pull of 160 kN and a bending moment of 19.2 kN-m per metre height. Determine the maximum stresses in concrete and steel. Take* $m = 19$.

Solution. Consider one metre height of the wall, in which there will be 10 bars on each face.

$$A_{sc}=A_{st}=10\times\frac{\pi}{4}(16)^2=2011\text{ mm}^2$$

$$d=170\text{ mm}\,;\,b=1000\text{ mm}\,;\,d_c=d_t=30\text{ mm}$$

$$e=\frac{M}{P}=\frac{19.2}{160}=0.12\text{ m}=120\text{ mm}$$

Equating the sum of internal forces to external forces, we have, from Eq. 14.36 :

$$m\frac{c'}{n}(d-n)A_{st}-\frac{bnc'}{2}-(m_c-1)A_{sc}\frac{c'}{n}(n-d_c)=P$$

or $$\frac{m}{n}(d-n)A_{st} - \frac{bn}{2} - (m_c - 1)A_{sc}\frac{n-d_c}{n} = \frac{P}{c'} \quad ...(1)$$

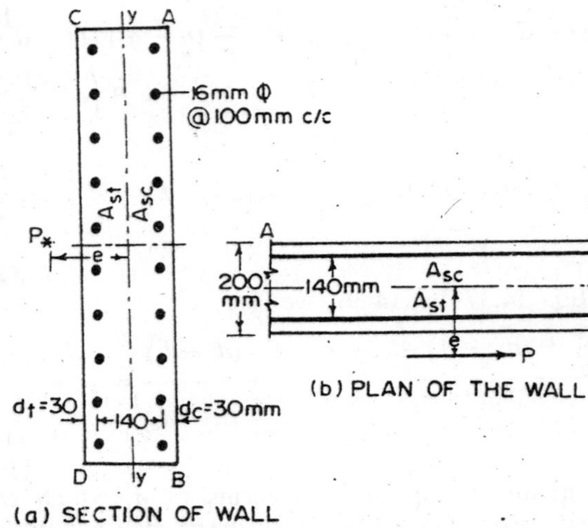

FIG. 14.23.

Similarly, equating the sum of internal moments about the centre of tensile steel to the moment of the pull, we have

$$\frac{bnc'}{2}\left(d-\frac{n}{3}\right) + (m_c-1)A_{sc}\frac{c'}{n}(n-d_c)(d-d_c) = P\left(e-\frac{D}{2}+d_t\right)$$

or $$\frac{bn}{2}\left(d-\frac{n}{3}\right) + (m_c-1)A_{sc}\frac{(n-d_c)(d-d_c)}{n} = \frac{P}{c'}\left(e-\frac{D}{2}+d_t\right) \quad ...(2)$$

Dividing the two and substituting the numerical values, we get

$$\frac{\frac{1000}{2}n\left(170-\frac{n}{3}\right) + (1.5\times19-1)\,2011\,\frac{(n-30)(170-30)}{n}}{\frac{19}{n}(170-n)\,2011 - \frac{1000n}{2} - (1.5\times19-1)\,2011\,\frac{n-30}{n}} = \left(120-\frac{200}{2}+30\right)$$

or $$\frac{500\,n\left(170-\frac{n}{3}\right) + 7742350\left(\frac{n-30}{n}\right)}{38209\left(\frac{170-n}{n}\right) - 500\,n - 55303\left(\frac{n-30}{n}\right)} = 50$$

Solving this by trial and error we get $n \approx 39.6$ mm. Substituting this in Eq. (1), we get.

$$\frac{19}{39.6}(170 - 39.6)\,2011 - \frac{1000 \times 39.6}{2} - (1.5 \times 19 - 1) \times \frac{2011}{39.6}(39.6 - 30) = \frac{160000}{c'}$$

or $$118086 = \frac{160000}{c'}$$

which gives $c' = 1.35$ N/mm^2(comp.)

The stress in tensile steel is given by

$$t = m\frac{c'}{n}(d - n) = \frac{19 \times 1.35}{39.6}(170 - 39.6)$$

$$= \mathbf{84.46\ N/mm^2}\ \text{(tensile)}$$

The stress in, compressive steel is given by

$$t_c = m_c\frac{c'}{n}(n - d_c) = 1.5 \times 19 \times \frac{1.35}{39.6}(39.6 - 30)$$

$$= \mathbf{9.33\ N/mm^2}\ \text{(comp.)}$$

14.11 SECTIONS OF IRREGULAR SHAPE

We have seen in cases 3 and 6 that the position of neutral axis is determined by trial and error, solving a cubic equation, even for a section of rectangular cross-section. When the shape of cross-section is not rectangular, such as sections of *L* or *T* shapes, rectangular section bending across a diagonal or any other irregular shape, the process of computing *n* becomes very tedius, though the basic principles of determination remain the same. In such cases, the position of neutral axis, and the maximum stresses in concrete and steel can be determined *step by step* in the following manner :

1. Assume any suitable value of *k* and draw the neutral axis.

2. Divide the compression zone into a number of thin strips, each parallel to the neutral axis.

3. Assume the maximum compressive stress in concrete as c', and determine the stresses at the middle of each strip, in terms of the maximum stress c', from the stress diagram. Multiply these stresses by the corresponding areas of the strip, each assumed to be rectangular, and find compressive forces C_1, C_2, C_3, etc., in the strips. Also, find the compressive force in the compressive steel if any.

4. Find the moment of these compressive forces about the centre of tensile steel. Find also the moment of the external eccentric load, about the centre of tensile steel.

5. Equate both the moments, to get the maximum compressive stress c'.

6. Equate the algebraic sum of the compressive and tensile forces to the external load. This will give the tensile stress t in tensile steel.

7. Knowing c' and t determine the values of k from the stress relationship. If this agrees with the assumed value of k within 5% then the stresses are correct. If not, asume some other suitable value of k and follow steps 1 to 6, till the two values of k do not differ by more than 5%.

14.12. DESIGN EXAMPLE

The design of eccentrically loaded sections, specially with moderate eccentricity $\left(\frac{D}{4} < e < \frac{3}{2}D\right)$, is done by trial and error. A suitable section is first assumed and then tested for stresses in concrete and steel. Alternatively, design may be done with the help of suitable design charts which give the value of coefficient k, eccentricity ratio $\frac{e}{d}$, steel percentage p, cover ratio a/d and compressive stress parameter. Such charts may be found in the book 'Reinforced Concrete Structures' by Peabody. See chapter 15.

Example 14.17. *Design a column section to carry a direct load of 300 kN and a bending moment of 90 kN-m. The width of column is limited to 300 mm. Take the following values of safe stresses* $\sigma_{cc} = 4\ N/mm^2$, $\sigma_{cbc} = c = 5\ N/mm^2$, $\sigma_{st} = t = 140\ N/mm^2$ *and* $m = 19$.

Solution.

$$\text{Eccentricity } e = \frac{90 \times 1000}{300} = 300 \text{ mm}$$

Let us keep total depth of column = 600 mm, so that eccentricity is equal to 0.5 D.

(a) ***Design based on permissible stresses.***

Let A_{sc} be the area of steel in compression and A_{st} be the area of steel in tension, each being provided at effective cover of 50 mm as shown in Fig. 14.24. Let n_c be the depth of critical neutral axis, when the permissible stresses in both the materials reach simultaneously [Fig. 14.24(*b*)].

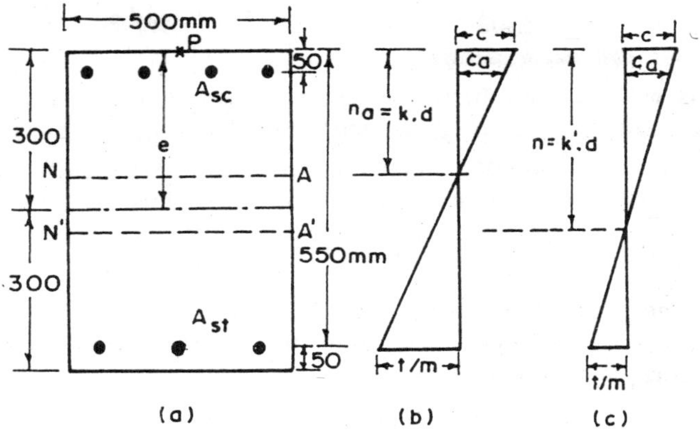

FIG. 14.24

Given : $t = \sigma_{st} = 140$; $c = \sigma_{cbc} = 5$ and $m = 19$

$$k = \frac{1}{1+\frac{t}{mc}} = \frac{1}{1+\frac{140}{19 \times 5}} = 0.404$$

$$d = 600 - 50 = 550 \text{ mm}$$

$$\therefore \quad n = 0.404 \times 550 = 222 \text{ mm}$$

Stress in concrete, surrounding compressive steel is given by

$$c_a = \frac{c}{n_c}(n_c - 50) = \frac{5}{222}(222 - 50) = 3.87 \text{ N/mm}^2$$

Equating the moment of compressive forces about the centre of tensile steel, to the moment of external forces, we have

$$(b \times n_c)\frac{c}{2}\left(d - \frac{n_c}{3}\right) + (m_c - 1)A_{sc}.c_a(d - d_c) = \text{P}\left(e + \frac{D}{2} - d_t\right)$$

or $(330 \times 221)\frac{5}{2}\left(550 - \frac{222}{3}\right) + (1.5 \times 19 - 1)A_{sc} \times 3.87(550 - 50)$

$$= 300000(300 + 300 - 50)$$

or $\quad 0.793 \times 10^8 + 53213\ A_{sc} = 1.65 \times 10^8$

From which $A_{sc} = 1610 \text{ mm}^2$

Similarly, equating the algebraic sum of internal forces to the external forces, we get

$$b\,.\,n_c\,.\,\frac{c}{2} + (m_c - 1)\ A_{sc}\,.\,c_a - m\,.\,A_{st}\,.\,t = P$$

or $\quad 300 \times 222 \times \frac{5}{2} + (1.5 \times 19 - 1)1610 \times 3.87 - 19 A_{st} \times 140 = 300000$

From which $A_{st} = 14.2 \text{ mm}^2$

$\therefore$ Total area of steel $= A_{sc} + A_{st} = 1610 + 14.2 = 1624.2 \text{ mm}^2$

(4) ***Alternative design***

It will be seen from the above design that very small steel area in tension is required. Hence if we lower down the value of t, keeping the value of $c = 5 \text{ N/mm}^2$, N.A. will be shifted down, thus increasing the compressive area. It will thus be seen that a section designed on the basis of reduced value of t is found to be more economical. When the value of eccentricity is about half the depth, the value of t is kept between 40 to 70 N/mm^2. Let us keep $t = 50 \text{ N/mm}^2$. The value of k' corresponding to this is given by [Fig 14.24(c)] :

$$k' = \frac{1}{1 + \dfrac{t'}{m \cdot c}} = \frac{1}{1 + \dfrac{50}{19 \times 5}} = 0.655$$

$$n = 0.655 \times 550 = 360 \text{ mm}$$

$$c_a = \frac{c}{n}(n - 50) = \frac{5}{360}(360 - 50) = 4.31 \text{ N/mm}^2$$

Equating the moment of compressive forces about the centre of tensile steel, to the moment of external force, we get

$$(300 \times 360)\frac{5}{2}\left(550 - \frac{360}{3}\right) + (1.5 \times 19 - 1)A_{sc} \times 4.31\,(550 - 50)$$
$$= 300000\,(300 + 300 - 50)$$

or $\quad 1.16 \times 10^8 + 59263\, A_{sc} = 1.65 \times 10^8$

From which $A_{sc} \approx$ **827 mm²**

Similarly, equating the algebraic sum of internal forces to the external forces, we get

$$(300 \times 360)\frac{5}{2} + (1.5 \times 19 - 1)827 \times 4.31 - 19 A_{st} \times 55 = 300000$$

From which $A_{st} \approx$ **72 mm²**

Hence total area of steel $= 827 + 72 = 999 \text{ mm}^2$ against 1624.2 mm² found above. Hence it is economical to design the section on the reduced value of t. However, with increase in eccentricity, higher value of t should be assumed. When eccentricity becomes more than 1.5 D, bending stress becomes predominent and hence $t = 140 \text{ N/mm}^2$ should be assumed.

PROBLEMS

1. A R.C. column 600 mm × 600 mm is reinforced with 8 bars of 20 mm Φ at a cover of 50 mm to the centre of steel bars as shown in Fig. 14.25. Determine the minimum and maximum stresses in concrete if the column is subjected to a load of 800 kN at an eccentricity of 50 mm about one of the axes. Also, if the permissible stresses in concrete are : $\sigma_{cc} = 4$ N/mm^2 and $c = \sigma_{cbc} = 5$ N/mm^2 check whether the section is safe or not. Take $m = 19$.

[Ans 2.72 N/mm^2; 1 N/mm^2; safe]

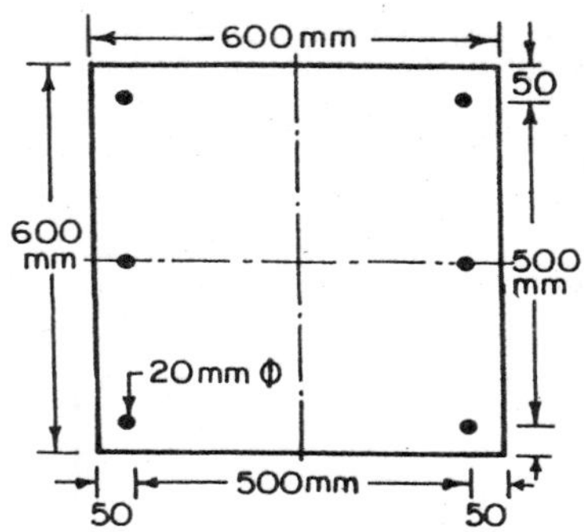

FIG. 14.25.

2. Determine the maximum eccentricity of load in Problem 1 if the column is to be safe as uncracked section.

[Ans. 156 mm]

3. If the column of problem 1 is subjected to an axial load of 800 kN along with a bending moment of 75 kN-m, determine the maximum and minimum stresses in concrete. Will the section be safe as uncracked section?

[Ans 3.47 N/mm^2(comp.) 0.27 N/mm^2(comp.)safe]

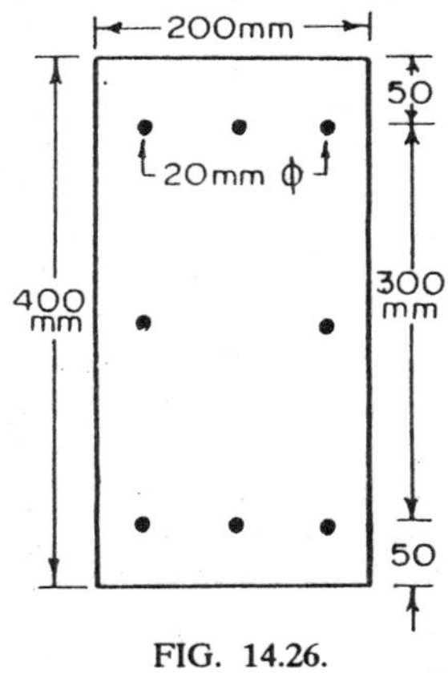

FIG. 14.26.

4. Given the column section of solved Example 14.1. If it carries an axial load of 400 kN, what is the additional safe load it can carry at 50 mm eccentricity from one of the axes ? Take $\sigma_{cc} = 4$ N/mm^2 and $c = \sigma_{cbc} = 5$ N/mm^2.
[Ans. 301 kN]

5. A rectangular section, 200 mm wide and 400 mm deep is reinforced with 8 bars of 20 mm Φ as shown in Fig. 14.26. Determine the maximum eccentricity about x-axis at which a load can be applied so that no tension is developed in the section. If $\sigma = 4$ N/mm^2 and $c = \sigma_{cbc} = 5$ N/mm^2 and $m = 19$. Determine the maximum value of the load. [**Ans.** 75 mm ; 331 kN]

6. Fig 14.27 shows the cross section of an arch rib. Determine the normal thrust and bending moment on the section so that no tension is developed any where and the section is safe as uncracked section. Take $\sigma_{cc} = 4$ N/mm^2 ; $\sigma_{cbc} = 5$ N/mm^2 and $m = 19$. [Ans. 555 kN, 79.4 kN-m]

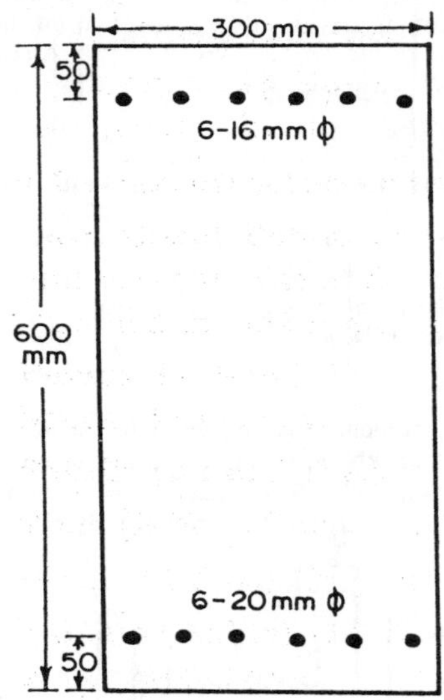

FIG. 14.27

7. A reinforced concrete column of circular section, 300 mm diameter is reinforced with 8 bars of 16 mm Φ, arranged on a circle of 200 mm diameter. It is subjected to an axial load of 150 kN and B.M. of 0.5 kN-m. Determine the maximum and minimum stresses in concrete. Taking the permissible stresses as $\sigma_{cc} = 4$ N/mm^2 and $c = \sigma_{cbc} = 5$ kN/mm^2. Check whether the section is safe or not. Take $m = 19$.
[Ans. 2.52 N/mm^2(comp.); 0.09 N/mm^2(comp).safe]

8. A two-hinged arch rib has a section 400 mm wide and 900 mm deep at the crown. The section is reinforced with 6 bars of 30 mm Φ at top and an equal

reinforcement at the bottom. The reinforcements are placed at an effective cover of 50 mm from the respective edges. If the resultant thrust on the section is 1200 kN at 3° with the tangent to arch centre line and acting on the vertical axis and 80 mm from the centre line of the section, determine the stresses in concrete and steel at the top and bottom of steel. Take $m = 18$, if $\sigma_{cc} = 4\ N/mm^2$ and $\sigma_{cbc} = 5\ N/mm^2$, determine whether the section is safe or not.

[Ans Max. stress in concrete $=2.78\ N/mm^2$; Min. Stress $=1.34\ N/mm^2$) (comp.) ; Max stress in steel $=73\ N/mm^2$(comp.) Min. stress $= 44.1\ N/mm^2$(comp.) ; safe].

9. Given the section of problem 1. If the section is subjected to (*i*) axial load of 800 kN (*ii*) bending moment of 40 kN-m about *x*-axis and (*iii*) B.M. of 30 kN-m about *y*-axis, check whether the section is safe or not. [Ans. safe]

10. Design a column section to carry an axial load of 300 kN and B.M. of 7.5 kN-m. Take $\sigma_{cc} = 4\ N/mm^2$, $c = \sigma_{cbc} = 5\ N/mm^2$ and $m = 18$.

11. A column section shown in Fig. 14.28 is subjected to a load of 50 kN acting at an eccentricity of 1 m from the centre line of the column section. Determine the stresses in concrete and steel. Take $m = 19$. Assume an effective cover of 50 mm to the centre of steel bars.

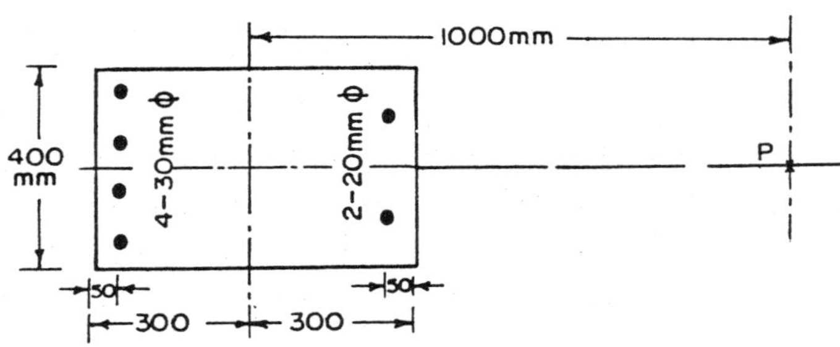

FIG. 14.28.

[Ans (*i*) Stress in concrete $=$ 2.06 N/mm^2(comp.)
(*ii*) Stress in comp steel $=$ 48.9 N/mm^2(comp.)
(*iii*) Stress in tensile steel $=$ 29.8 N/mm^2(tensile)

12. The section of an arch rib is 400 mm wide and 900 mm deep, and is reinforced with 6 bars of 30 mm Φ at the top and an equal reinforcement at from the bottom. The reinforcements are placed at an effective cover of 50 mm from the respective edges. Determine the maximum stresses in concrete and steel when the section is subjected to a thrust of 500 kN and a B.M. of 350 kN-m. Take $m = 19$. [Ans 0.64 N/mm^2(comp.), 33 N/mm^2(tensile)]

13. The suspender of a bow string girder, 200 mm × 300 mm is reinforced with 4 bars of 16 mm at each face placed at an effective cover of 40 mm. If it is subjected

to a pull of 100 kN and B.M. of 6 kN-m determine the stresses in steel reinforcements.
[Ans 96.1 N/mm^2(tensile) ; 28.3 N/mm^2(tensile)]

14. The slab of a tank is 240 mm thick and is reinforced with 8 bars of 20 mm Φ per meter length on each face as shown in Fig 14.29. The slab is subjected to a pull of 60 kN and a B.M. of 36 kN-m per meter width. Determine the stresses in concrete and steel. Take m = 19

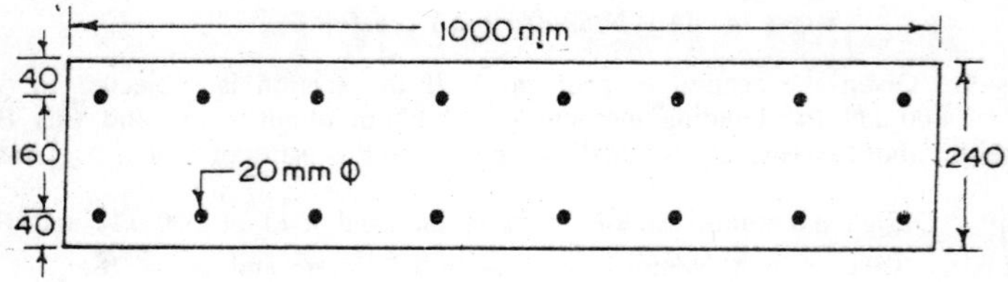

FIG. 14.29

[Ans. Comp. stress in concrete = 3.06 N/mm^2,
Comp. stress in steel = 41.41 N/mm^2.
Tensile stress in steel = 104.4 N/mm^2.]

PART III

FOUNDATIONS AND RETAINING STRUCTURES

Chapters :

15. CONTINUOUS AND ISOLATED FOOTINGS

16. COMBINED FOOTINGS

17. PILE FOUNDATIONS

18. RETAINING WALLS

15

Continuous and Isolated Footings

15.1. INTRODUCTION

Any structure is generally considered to have two main portions (*i*) the *superstructure* and (*ii*) the *substructure*. The substructure transmits the loads of superstructure to the supporting soil and is generally termed as the *foundation*. *Footing* is that portion of the foundation which ultimately delivers the load to the soil, and is thus in contact with it. The load of the superstructure is transmitted to the foundation or structure through either columns or walls. The object of providing foundation to a structure is to distribute the load to the soil in such a way that the maximum pressure on the soil does not exceed its permissible bearing value, and at the same time the settlement is within the permissible limits.

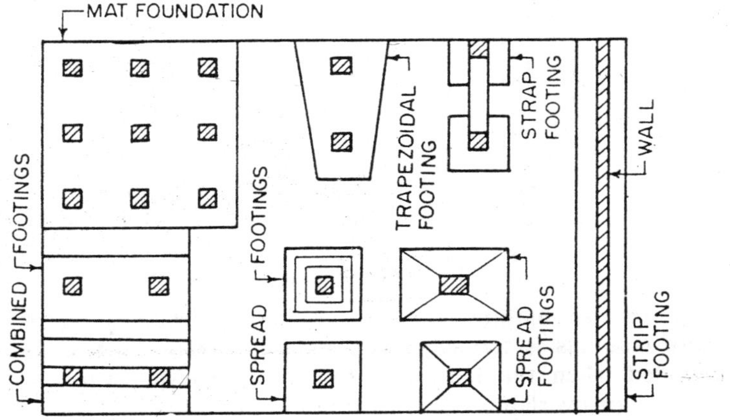

FIG. 15.1 VARIOUS TYPES OF SHALLOW FOOTINGS.

Foundations may be broadly classified under two heads : *shallow foundation and deep foundation.* According to Terzaghi, a foundation is shallow if its depth is equal to or less than its width. In the case of deep foundation, the depth is equal to or greater than the width. Apart from deep strip, rectangular or square foundations, other common forms of deep foundations are : *pier foundation, pile foundation and well foundation*. The shallow foundations are of the following types : Spread footing, strap footing, combined footing and mat or raft footing. Fig. 15.1 shows the common types of shallow foundations.

Spread footings. A spread footing or simply footing, is a type of shallow foundation used to transmit the load of an isolated column, or that of a wall, on the subsoil. In the case of wall, the footing is *continuous* while in the case of column, it is *isolated*. Fig. 15.2 shows some common types of reinforced concrete spread footing.

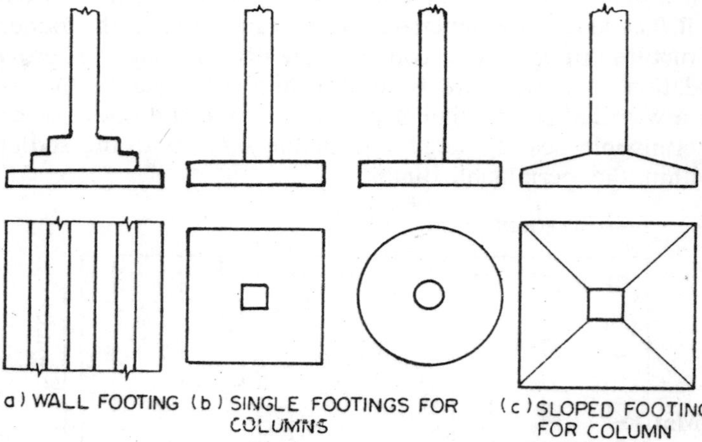

FIG. 15.2. TYPICAL SPREAD FOOTINGS.

Combined footings. A spread footing which supports two or more columns is termed as a *combined footing*. Such a footing, is provided when the individual footings are either very near to each other, or overlap. Combined footings may either be rectangular or trapezoidal (Fig. 15.2).

Strap or cantilever footing. A strap footing consists of spread footings of two columns connected by a strap beam. The strap beam does not remain in contact with soil, and thus does not transfer any pressure to the soil. Such a footing is generally used to combine the footing of outer column to the adjacent one so that the footing of the former does not extend in the adjoining property[Fig. 15.3(*c*)].

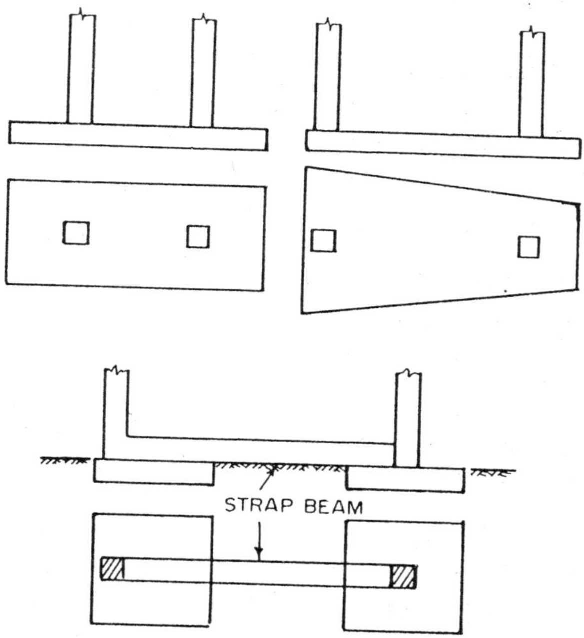

FIG. 15.3.

Mat or raft foundation. A mat or raft is a combined footing that covers the entire area beneath a structure and supports all the walls and columns. When the available soil pressure is low or the building loads are heavy, the use of spread footings would cover more than one-half of the area and it may prove more economical to use mat or raft foundation.

Pile foundation. Pile foundation is a deep foundation used where the top soil is relatively weak. Piles transfer the load to a lower stratum of greater bearing capacity, by way of end bearing, or to the intermediate soil through skin friction. This is most common type of deep foundation generally used for buildings where a group of piles transfer the load of the super-structure to the sub-soil.

15.2. PRESSURE DISTRIBUTION BENEATH FOOTINGS

Both from observations as well as the analytical studies from theory of elasticity, it is known that the pressure distribution beneath footings, symmetrically loaded is not uniform. The pressure intensities depend upon the rigidity of footing, soil type, and the condition of soil. Fig. 15.4(*a*) and (*b*) show the probable pressure distribution beneath a rigid footing on a loose cohesionless soil and cohesive soil. Fig. 15.4(*c*) shows the usually assumed uniform pressure distributions.

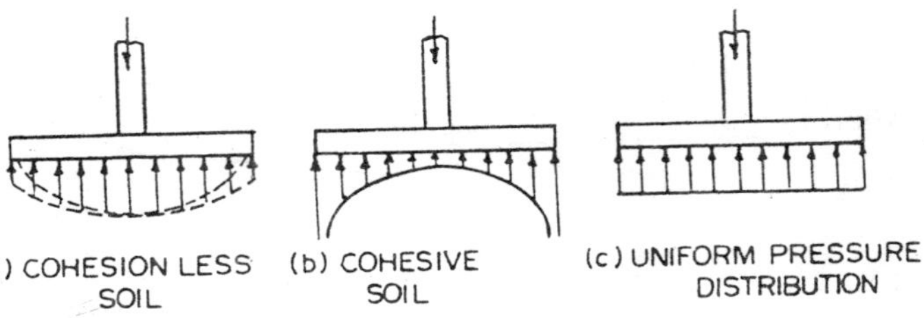

FIG. 15.4. PRESSURE DISTRIBUTION BENEATH FOOTINGS.

When a rigid footing rests on cohesionless soil, the soil grain at the outer edge have no lateral restraint, whereas in centre the soil is relatively confined resulting in a pressure distribution shown in Fig. 15.4(*a*). For the footings on cohesive soils, the edge stresses may be very large. However, the pressure distribution may be considered as *linear* as shown in Fig. 15.4(*c*), for the purpose of the design of reinforced concrete footings. Once the pressure distribution is known, the bending moments and shear force in the footing can be calculated, and the thickness of the structural member of the footing, alongwith the reinforcement etc., can be calculated using the usual principles of reinforced concrete.

15.3 BEARING CAPACITY OF SOIL AND SETTLEMENT OF FOOTINGS

In conventional design, the allowable bearing capacity should be taken as the smaller of the following two values :

(*i*) the safe bearing capacity based on ultimate capacity.

(*ii*) the allowable bearing pressure based on tolerable settlement.

The safe bearing capacity may be determined from a plate load test. Alternatively, the safe bearing capacity q_o may be computed from the following equations based on Terzaghi's analysis :

For strip footing

$$q_o = \frac{1}{F}[cN_c + \gamma D(N_q - 1)R_{w1} + 0.5\gamma BN_\gamma\, R_{w2}] + \gamma D \quad ...(15.1)$$

For square footing

$$q_0 = \frac{1}{F}[1.3\, cN_c + \gamma D\,(N_q - 1\,)R_{w1} + 0.4BN_\gamma R_{w2}\,] + \gamma D ...(15.2)$$

For circular footing

$$q_0 = \frac{1}{F}[1.3cN_c + \gamma D\,(N_q - 1) + 0.3\gamma BN_\gamma\;R_{w2}] + \gamma D \quad ...(15.3)$$

where D = depth of footing

B = width of footing(strip or square) or diameter of circular footing

F = factor of safety (2 to 3)

N_c, N_q, N_γ = Bearing capacity factors for general shear failure.

For local shear failure, N_c' , N_q' and N_γ' should be used.

R_{w1} and R_{w2} are the water reduction factors given by (Fig. 15.5).

$$R_{w1} = 0.5\left(1 + \frac{Z_{w1}}{D}\right) \quad ..[15.4(a)]$$

$$R_{w2} = 0.5\left(1 + \frac{Z_{w2}}{B}\right) \quad ..[15.4(b)]$$

The allowable bearing pressure q_ρ based on maximum settlement of individual footing to 25 mm, can be computed from the following empirical relation :

$$q_\rho = 35\,(N-3)\left(\frac{B + 0.3}{2B}\right)^2 . R_{w2}\, R_d \quad ...(15.5)$$

where q_ρ= allowable net increase in soil pressure over existing soil pressure for settlement of 25 mm, kN/m^2

N = standard penetration number with applicable corrections

B = width of footings, in metres

R_{w2} = water reduction factor

R_d = depth factor = $\left(1 + \frac{D}{F}\right) \le 2.0$

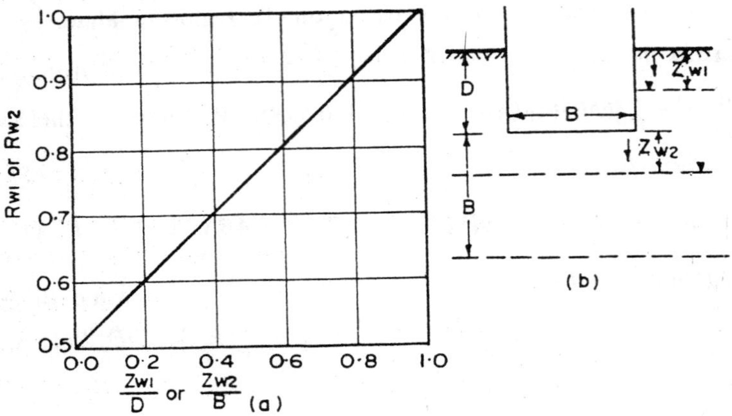

FIG. 15.5. WATER REDUCTION FACTORS.

If the safe bearing capacity is taken to be smaller of the two values discussed above the footings on granular soils will not suffer detrimental settlement. The total settlement of a footing in clay may be considered to consist of three components (Skempton and Bjerrum, 1957) :

$$S = S_i + S_c + S_s \qquad ...(15.6)$$

where S = total settlement

S_i = immediate elastic settlement

S_c = consolidation settlement

S_s = settlement due to secondary consolidation of clay.

The immediate settlement S_i is the elastic settlement, and can be computed from the following expression based on the theory of elasticity :

$$S_i = q\,B\left(\frac{1-\mu^2}{E_s}\right) I_w \qquad ...(15.7)$$

where q = intensity of contact pressure (kN/m^2)

B = least lateral dimension of footing, metres

E_s = modulus of elasticity for soil, (kN/m^2)

I_w = influence factor,

= 0.88 for rigid circular footing.

= 0.82 for rigid square footing.

= 1.06 for rigid rectangular footing with $L/B = 1.5$

= 1.70 for rigid rectangular footing width $L/B = 5$

It is difficult to determine μ and E_s for soils. However, the entire term $\frac{1-\mu^2}{E_s} I_w$ of Eq. 15.7 may be determined from plate load tests, by using two or three different size plates of the same shape.

A plot between S_i and $q.B$ give a straight line, the slope of which is equal to $\frac{1-\mu^2}{E_s} I_w$.

Janbu, Bjerrum and Kjaernisli (1966) have proposed the following equation for computing the immediate settlement :

$$S_i = \mu_0 . \mu_1 . q\, B \frac{1-\mu^2}{E_s} \qquad \text{...(15.8)}$$

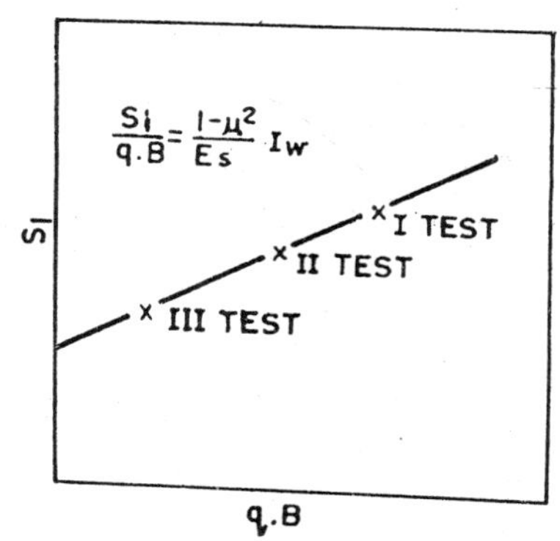

FIG. 15.6.

where the modulus of elasticity E_s is computed from the triaxial test data :

$$E_s = \frac{\sigma_1 - \sigma_3}{\Delta L/L} \qquad \text{...(15.9)}$$

and μ_0 and μ_1 are taken from Fig. 15.7.

The *consolidation settlement* S_c is computed from the following expression:

$$S_c = C \frac{C_c}{1+e_0} . H \log_{10} \frac{\sigma_0 + \Delta\sigma}{\sigma_0} \qquad \text{...(15.10)}$$

where $\sigma_0 =$ effective overburden pressure due to soil overburden

$\Delta\sigma =$ vertical stress on footing

$C_c =$ compression index

$e_0 =$ initial voids ratio

$H =$ thickness of compressible layer.

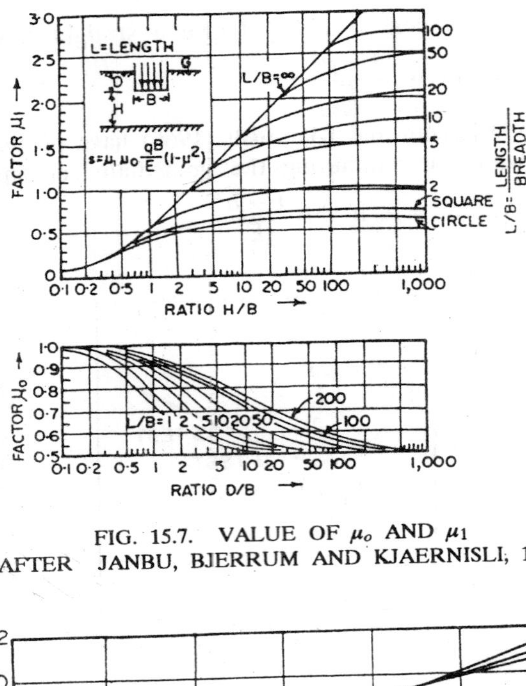

FIG. 15.7. VALUE OF μ_o AND μ_1
(AFTER JANBU, BJERRUM AND KJAERNISLI, 1966)

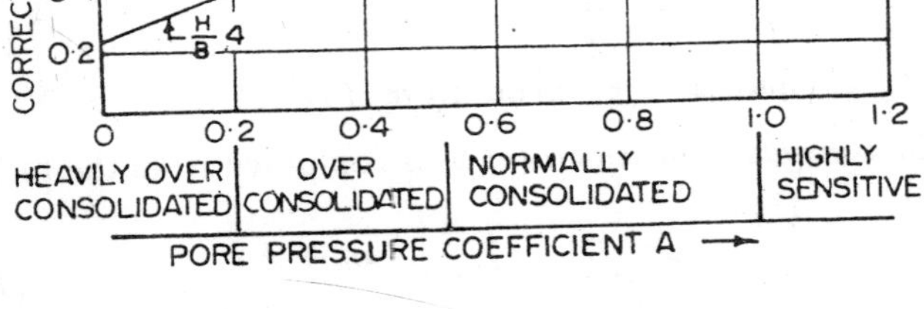

FIG. 15.8. VALUE OF FACTOR C.

C = a coefficient or correction factor (given by Fig. 15.8) depending upon the geometry of the footing and the history of loading on the clay(*i.e.*, on the pore pressure coefficient A).

15.4. INDIAN STANDARD CODE RECOMMENDATIONS FOR DESIGN OF FOOTINGS (IS: 456-1978)

(A) General

1. In sloped or stepped footings the effective cross-section in compression shall be limited by the area above the general plane, and the angle of slope or depth and location of steps shall be such that the design requirements are satisfied at every section. Sloped and stepped footings that are designed as a unit shall be constructed to assure action as a unit.

2. *Thickness at the edge of footing.* In reinforced and plain concrete footings, the thickness at edge shall be not less than 15 cm for footings on soils nor less than 30 cm above the tops of piles for footings on piles.

3. In the case of plain concrete pedestals, the angle α between plane passing through the bottom edge of the pedestal and the corresponding junction edge of the column with pedestal and the horizontal plane (Fig. 15.9) shall be governed by the expression :

$$\tan\alpha \nless 0.9\sqrt{\frac{100\,q_0}{f_{ck}}+1} \qquad ...(15.11)$$

where q_o = calculated maximum bearing pressure at the base of the pedestal in N/mm^2, and

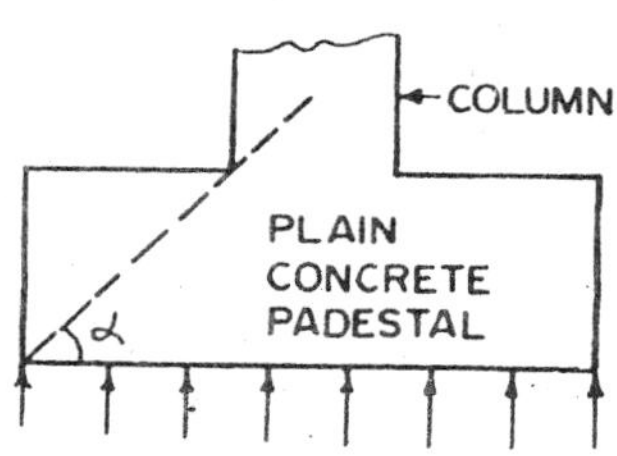

FIG. 15.9.

f_{ck} = characteristic strength of concrete at 28 days in N/mm^2.

(B) Moments and Forces

1. In the case of footings on piles, computation for moments and shears may be based on the assumption that the reaction from any pile is concentrated at the centre of the pile.

2. For the purposes of computing stress in footings which support a round or octagonal concrete column or pedestal, the face of the column or pedestal shall be taken as the side of a square inscribed within the perimeter of the round or octagonal column or pedestal.

3. *Bending Moment*

(*i*) The bending moment at any section shall be determined by passing through the section a vertical plane which extends completely across the footing and computing the moment of the forces acting over the entire area of the footing on the side of the side plane.

(*ii*) The greatest bending moment to be used in design of an isolated concrete footing which supports a column, pedestal or wall, shall be the moment computed in the manner prescribed in (*i*) at sections located as follows :

(*a*) At the face of the column, pedestal or wall, for footings supporting a concrete column, pedestal or wall,

(*b*) Half way between the centre line and the edge of the wall, for footings under masonry walls, and

(*c*) Half way between the face of the column or pedestal and the edge of the gussetted base, for footing under gusseted bases.

4. *Shear and bond*

(*i*) The shear strength of footings is governed by the more severe of the following two conditions :

(*a*) The footing acting essentially as a wide beam, with a potential diagonal crack extending in a plane across the entire width, the critical section for this condition shall be assumed as a vertical section located from the face of the column, pedestal or wall at a distance equal to the effective depth of the footing in case of footings on soils, and a distance equal to half the effective depth of footing on piles.

(*b*) Two-way action of the footing, with potential diagonal cracking along the surface of truncated cone or pyramid around the concentrated load : in this case, the footing shall be designed for shear in accordance with appropriate provisions discussed below (Fig 15.10)

ii) In computing the external shear on any section through a footing supported on piles, the entire reaction from any pile of diameter D_p whose centre is located $D_p/2$

or more outside the section shall be assumed as producing shear on the section ; the reaction from any pile whose centre is located $D_p/2$ or more inside the section shall be assumed as producing no shear on the section. For intermediate positions of the pile centre, the portion of the pile reaction to be assumed as producing shear on the section shall be based on straight line interpolation between full value at $D_p/2$ outside the section and zero value at $D_p/2$ inside the section.

(*iii*) The critical section for checking the development length in a footing shall be assumed at the same planes as described for bending moment in *B*(3) and also at all other vertical planes where abrupt changes of section occur. If the reinforcement is curtailed, the anchorage requirements should be checked.

Thus, according to the above provision, shear stress is to be checked for (*i*) *one way action* (*i.e.* , beam shear) for which the governing section *AB* is at a distance *d* from the face of column or pedestal [Fig. 15.10(*a*)] and (*ii*) *two way action,* (*i.e.*, punching shear), for which the governing section is along the perimeter *ABCD* situated at a distance *d*/2 from the face of the column or pedestal [Fig. 15.10(*b*)]

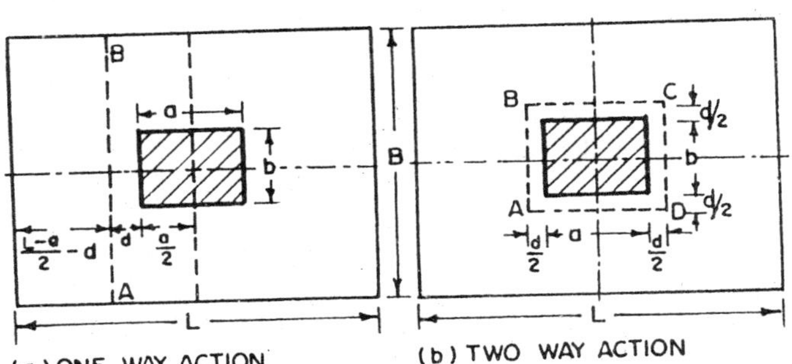

FIG. 1<.10. CRITICAL SECTION FOR SHEAR

For the *two way action,* the calculated shear stress τ_v should satisfy the following relation

$$\tau_v \leq k_s \cdot \tau_c \qquad ...[15.12(a)]$$

where $\tau_v = \dfrac{F}{2\,[\,(a+d)+(b+d)\,]\,d}$ ...[15.12(*b*)]

$k_s = (0.5 + \beta_c)$, but not greater than 1.0

$\beta_c = \dfrac{b}{a} = \dfrac{\text{short side of column}}{\text{long side of column}}$ ...[15.12(*c*)]

$\tau_c = 0.16\sqrt{f_{ck}}\ \text{N/mm}^2$ in working stress method ...[15.12(*d*)]

and F = net S.F. acting on the perimeter

For the *beam shear,* the nominal shear stress across *AB* should satisfy the relation

$$\tau_v \leq k\,\tau_c \qquad ...(15.13)$$

wheres τ_c = the permissible shear stress for the grade of the concrete, corresponding to the reinforcement $\left(100\dfrac{A_s}{bd}\right)$ as given in Table 3.1

k = factor for slabs, as given in Table 3.2.

(C) Tensile Reinforcement

The total tensile reinforcement at any section shall provide a moment of resistance at least equal to the bending moment on the section calculated in accordance with *B*(3). Total tensile reinforcement shall be distributed across the corresponding resisting section as given below.

(*i*) In one-way reinforced footing, the reinforcement shall be distributed uniformly across the full width of the footing ;

(*ii*) In two-way reinforced square footing, the reinforcement extending in each direction shall be distributed uniformly across the full width of the footing ;

(*iii*) In two-way reinforced rectangular footing, the reinforcement in long direction shall be distributed uniformly across the full width of the footing. For reinforcement in the short direction, a central band equal to the width of the footing shall be marked along the length of the footing and portion of the reinforcement determined in accordance with the equation given below shall be uniformly distributed across the central band :

$$\frac{\text{Reinforcement in central band width}}{\text{Total reinforcement in short direction}} = \frac{2}{\beta + 1}$$

where β is the ratio of the long side to the short-side of the footing.

The remainder of the reinforcement shall be uniformly distributed in the outer portions of the footing.

(D) Transfer of load at the base of column

The compressive stress in concrete at the base of a column or pedestal shall be considered as being transferred by bearing to the top of the supporting pedestal or footing. The bearing pressure (σ_{cbr}) on the loaded area shall not exceed the permissible bearing stress in direct compression multiplied by a value equal to $\sqrt{A_1/A_2}$ but not greater than 2.

where A_1 = supporting area for bearing, of footing, which in sloped or stepped footing may be taken as the area of the lower base of the largest frustrum of a pyramid or cone contained wholly within the footing and having for its upper base, the area actually loaded and having side slope of one vertical to two horizontal, and

A_2 = Loaded area at the column base.

For working stress method of design, the permissible bearing stress on full area of concrete shall be taken as 0.25 f_{ck} ; for limit state method of design the permissible bearing stress shall be 0.45 f_{ck}.

Thus $\sigma_{cbr} \leq 0.25 f_{ck} \sqrt{\dfrac{A_1}{A_2}}$ in working stress method ....[15.14(*a*)]

and $\sigma_{cbr} \leq f_{ck} \sqrt{\dfrac{A_1}{A_2}}$ in limit state method ....[15.14(*b*)]

The actual bearing pressure σ_{cbr} = column load divided by the area of column at the base.

Thus $$\sigma_{cbr} = \frac{W}{a+b}$$

where a and b are the sides of the column.

(1) Where the permissible bearing stress on the concrete in the supporting or supported member would be exceeded, reinforcement shall be provided for developing excess force, either by extending the longitudinal bars into the supporting members or by dowels (see 3 below).

(2) Where transfer of force is accomplished by reinforcement, the development length of the reinforcement shall be sufficient to transfer the compression or tension to the supporting member.

(3) Extended longitudinal reinforcement or dowels of at least 0.5 percent of the cross-sectional area of the supported column or pedestal and a minimum of four bars shall be provided. Where the dowels are used, their diameter shall not exceed the diameter of the column bars by more than 3 mm.

(4) Column bars of diameters larger than 36 mm, in compression only can be dowelled at the footings with bars of smaller size of the necessary area. The dowel shall extend into the column, a distance equal to the development length of the column bar and into the footing, a distance equal to the development length of the dowel.

15.5. DESIGN OF CONTINUOUS FOOTINGS

A continuous footing is provided under walls which may be either of masonry or of concrete. For design of reinforced concrete continuous footings, it is convenient to consider the forces on unit length.

1. CONTINUOUS FOOTING UNDER MASONRY WALL

Let b = width of wall(metres).

B = width of footing

q_o = safe bearing capacity of soil (kN/m^2)

W = load from wall, per metre run (kN/m)

W' = weight of footing per metre run.

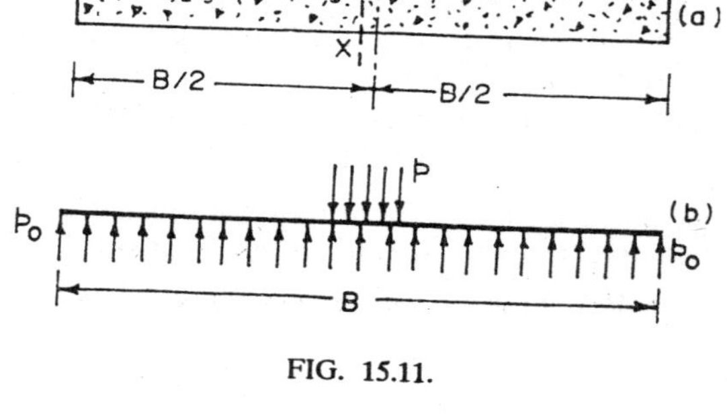

FIG. 15.11.

Then, the width of footing is given by

$$B = \frac{W + W'}{q_0} \qquad ...[15.15\ (a)]$$

Fig. 15.11(*b*) shows the pressure diagram. The *net* upward soil pressure p_0 on the footing will be

$$p_0 = \frac{W}{B \times 1} = \frac{W}{B} \qquad ...[15.15\ (b)]$$

The pressure intensity p at the contact of wall and footing is given by

$$p = \frac{W}{b \times 1} = \frac{W}{b} \qquad ...[15.15\ (c)]$$

The entire footing can be considered to consist of two cantilevers, each projecting out by a length $\frac{1}{2}(B - b)$, and having a fixed support length $\frac{b}{2}$. The maximum B.M. is considered to act at a section midway between the edge of the wall and centre of the wall *i.e.*, at a distance $\frac{b}{4}$ from the edge of the wall. Its magnitude per metre length is thus given by

$$M = p_0\, \frac{1}{2}\left[\frac{B-b}{2} + \frac{b}{4}\right]^2 - p\, \frac{1}{2}\left[\frac{b}{4}\right]^2$$

$$= \frac{W}{2B}\left[\frac{2B - b}{4}\right]^2 - \frac{W}{2b}\cdot\frac{b^2}{16}$$

$$= \frac{W}{2B}\left[\frac{4B^2 + b^2 - 4bB}{16}\right] - \frac{Wb}{32}$$

$$= \frac{W}{32}\left[4B - 4b + \frac{b^2}{B} - b\right] = \frac{W}{32}\left[4B - 5b + \frac{b^2}{B}\right]$$

$$= \frac{W}{32B}[4B^2 - 5Bb + b^2] = \frac{W}{32B}(4B - b)(B - b)$$

$$= \frac{W}{8B}(B - b)\left[B - \frac{b}{4}\right] \text{ kN-m}$$

$$= \frac{p_0}{8}(B - b)\left[B - \frac{b}{4}\right] \times 10^6 \text{ N-mm} \qquad ...(15.16)$$

$\therefore$ Effective depth $d = \sqrt{\dfrac{M}{1000\,R}}$

A minimum thickness of 150 mm should be provided at the edges.

Also $\quad A_{st} = \dfrac{M}{tjd},$ where $t = \sigma_{st}$

In addition to this, longitudinal reinforcement is provided @ 0.15% of the area of section of the footings.

Check for shear. The footing slab is checked for *one way action* (*i.e.*, beam shear) for which the critical section will be at a distance of effective depth from the face of the wall. The depth of footing at that section must be sufficient to keep nominal shear stress within safe limits.

2. CONTINUOUS FOOTING UNDER CONCRETE WALL

A concrete footing under a concrete wall is cast monolithic with it. Hence the maximum B.M. will occur at the face of the wall.

As found earlier

$$B = \frac{W + W'}{q_0};$$

$$B = \frac{W'}{b}\ ; \ p_0 = \frac{W}{B}$$

∴ Maximum B.M. per metre length is given by

$$M = p_0 \frac{1}{2}\left[\frac{B-b}{2}\right]^2 \text{ kN-m}$$

or $$M = p_0 \frac{(B-b)^2}{8} \times 10^6 \text{ N-mm} \qquad \text{...(15.17)}$$

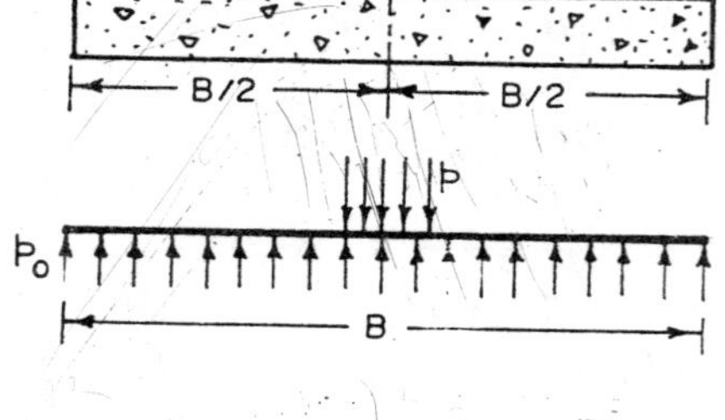

FIG. 15.12.

The effective depth and area of steel reinforcement are then given by

$$d = \sqrt{\frac{M}{1000\,R}} \quad \text{and} \quad A_{st} = \frac{M}{\sigma_{st}\,jd}$$

In the addition to this, longitudinal reinforcement is provided @ 0.15% of cross-sectional area of footing for mild steel bars and @ 0.12 % for HYSD bars.

The section is checked for beam shear at a distance of effective depth from the face of the wall.

15.6. ISOLATED FOOTING OF UNIFORM DEPTH

The total area at the base of such footing is determined from the point of view of the safe bearing capacity of soil. The depth of footing is determined both from the considerations of maximum bending moments as well as from punching shear.

Let W = weight transferred from the column (kN)

W' = weight of footing

q_o = safe bearing capacity of soil (kN/m^2)

A = area of footing at the base.

Then $$A = \frac{W + W'}{q_0}$$

If p_0 = net upward pressure at footing, then

$$p_0 = \frac{W}{A} = \frac{W q_0}{W + W'}$$

The weight W' of the footing may be assumed to the equal to 10% W, for preliminary calculations.

2. SQUARE FOOTING

Fig. 15.13 shows a square footing of uniform thickness(depth)D.

The width B of the footing will evidently be equal to

$$\sqrt{\frac{W + W'}{q_0}}$$

The thickness is calculated both from the considerations of bending moment as well as punching shear.

(a) Depth for bending moment

Since the column and footing are cast monolithic, maximum B.M. will take place about section X - X near the face of the column. The unit pressure p_o causing bending of cantilever of **length** $\frac{1}{2}(B-b)$ will be equal to $\frac{W}{B^2}$.

$$\therefore \quad M = p_0 \; B \left(\frac{B-b}{2}\right)\left(\frac{B-b}{4}\right) = p_0 \; \frac{B}{8}(B-b)^2 \text{ kN-m}$$

$$= \frac{p_0 B}{8}(B-b)^2 \times 10^6 \text{ N-mm} \quad ...(15.18)$$

The effective depth is given by

$$d = \sqrt{\frac{M}{BR}} = \sqrt{\frac{P_0 (B-b)^2 \, 10^6}{8R}}$$

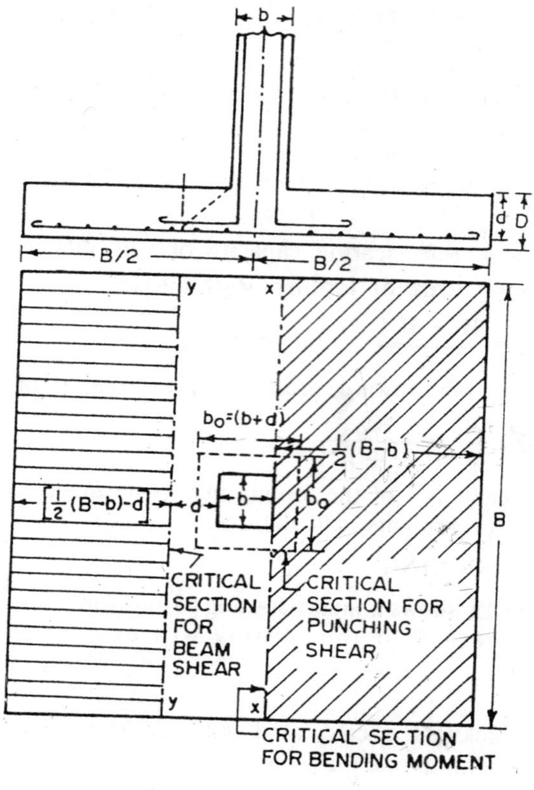

FIG. 15.13

(b) Depth for shear

The footing slab is checked both for *one way action* (*i.e.* beam shear) and *two way action* (*i.e.* punching shear).

For *one way action,* the critical section *Y-Y* is situated at a distance d from the face of the column. The nominal shear stress $\tau_v \left(= \frac{V}{B \cdot d}\right)$ is equated to the permissible shear stress $k\tau_c$ found from Table 3.1 for the grade of concrete corresponding to the reinforcement $\left(100 \frac{A_s}{bd}\right)$. Thus

$$\frac{V}{Bd} \leq k\,\tau_c \quad \text{...[15.19 (a)]}$$

From this the effective depth d is found.

For *two way action,* the calculated shear stress τ_v computed from Eq. 15.12 (*b*) should be less than $k_s\ \tau_c$.

Thus, $$\frac{F}{4\,[b + d]\,d} \leq k_s \cdot \tau_c$$

where $$k_s = (0.5 + \beta_c) \leq 1$$

$$= (0.5 + 1) \leq 1$$

$$= 1 \text{ (maximum value)}$$

$$\tau_c = 0.16\sqrt{f_{ck}}$$

and F = net S.F. acting on the periphery.

$$= p_0\,[B^2 - (b + d)^2]$$

Hence, $$\frac{F}{4(b + d)d} \leq 0.16\sqrt{f_{ck}} \quad \text{...[15.19 (b)]}$$

The greater of the depths found from Eqs 15.19(*a*) and 15.19(*b*) should be provided. A minimum thickness of 150 mm should be provided in case d comes out to be less than 150 mm.

(*c*) Steel reinforcement

Steel reinforcement is provided in both the directions, in the form of mesh, for the bending moment M.

Thus $A_{st} = \frac{M}{tjd}$, where $t = \sigma_{st}$

This area is to be uniformly distributed over a width B. If A_Φ is area of each bar chosen, the spacing of bars is given by

$$s = \frac{B \cdot A_\Phi}{A_{st}}$$

(*d*) Check for development length

The critical section for checking the development length in the footing is assumed at the same plane as used for maximum bending moment, *i.e.* at a section along the face of the column.

If the reinforcement is curtailed, the anchorage requirements should be checked.

2. RECTANGULAR FOOTING

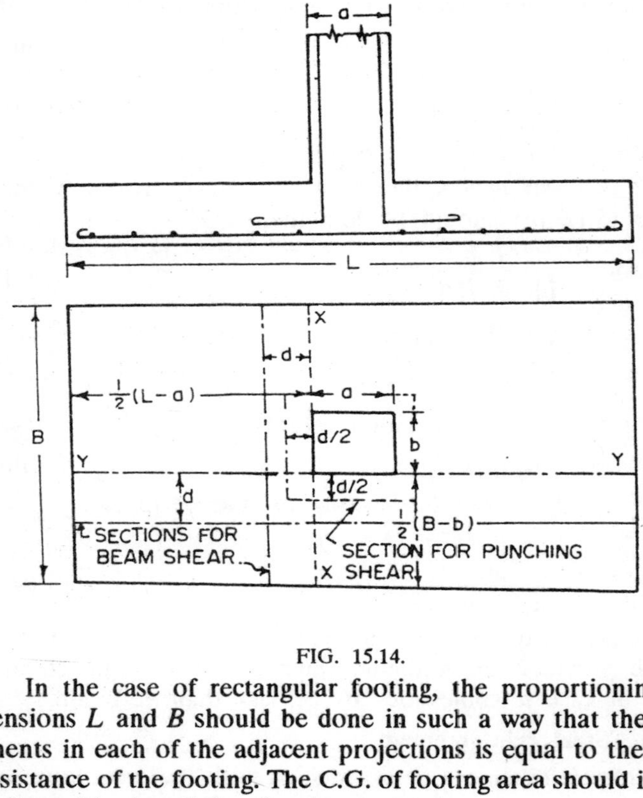

FIG. 15.14.

In the case of rectangular footing, the proportioning of the dimensions L and B should be done in such a way that the bending moments in each of the adjacent projections is equal to the moment of resistance of the footing. The C.G. of footing area should invariably coincide with the C.G. of column area.

Fig. 15.14 shows a column of size $a \times b$. Sometimes, a rectangular footing may be provided even for a square column, where there are some restrictions on the maximum value of the width of the footing. The thickness of the footing is determined both from the B.M. as well as shear point of view.

(a) Depth for bending moment

Bending moment M_1 about section XX is given by

$$M_1 = p_0\, B \left(\frac{L-a}{2}\right)\left(\frac{L-a}{4}\right) = \frac{p_0 B\,(L-a)^2}{8} \times 10^6 \text{ N-mm} \quad ...(15.20)$$

$$d = \sqrt{\frac{M_1}{RB}} = \sqrt{\frac{p_0 (L-a)^2 \times 10^6}{8R}} \qquad ...(15.21)$$

Bending moment M_2 about section Y-Y is given by

$$M_2 = p_0 \, L \left(\frac{B-b}{2}\right)\left(\frac{B-b}{4}\right) = \frac{p_0 L}{8}(B-b)^2 \times 10^6 \text{ N-mm} \qquad ...(15.22)$$

$$\therefore \quad d = \sqrt{\frac{M_2}{RL}} = \sqrt{\frac{p_0 (B-b)^2 \times 10^6}{8R}} \qquad ...(15.23)$$

(b) Depth for shear

From shear point of view, the depth is found for both *one way action (i.e.* beam section) as well as *two way action* (*i.e.* punching shear action).

For one way action, the sections are situated at a distance d from each face of the column. The nominal shear stress $\tau_v \left(= \frac{V}{BD} \text{ or } \frac{V}{LD}\right)$ should be less than permissible shear τ_c found from Table 3.1 for the grade of concrete corresponding to reinforcement ratio $\left(100 \frac{A_s}{bd}\right)$. To begin with, the reinforcement ratio is arbitrarily selected equal to the normal value.

Thus $$\frac{V}{Bd}\left(\text{or } \frac{V}{L} . d\right) \le \tau_c \qquad ...(15.24)$$

From which effective depth is found.

For *two way action,* the calculated shear stress(τ_v) should satisfy the following relation :

$$\tau_v \le k_s . \tau_c$$

where $$\tau_v = \frac{F}{2\,[(a+d)+(b+d)]\,d}$$

$$k_s = \left(0.5 + \frac{b}{a}\right) \le 1$$

$$\tau_c = 0.16\sqrt{f_{ck}} \quad \text{and}$$

F = net S.F. at the periphery = $p_0[(B \times L) - (a+d)(b+d)]$

Hence $$\frac{F}{2[(a+d)+(b+d)]d} \le 0.16\, k_s \sqrt{f_{ck}} \qquad ...(15.25)$$

The greater of the depths found from Eq. 15.22, 15.23, 15.24 and 15.25 should be provided.

(*e*) **Steel reinforcement**

Steel reinforcement is provided in both the directions for the bending moments M_1 and M_2.

The area of steel A_{st1}, of *long bars* parallel to directions L, is calculated for bending moment M_1 as under :

$$A_{st1} = \frac{M_1}{tjd} \quad \text{where } t = \sigma_{st} \qquad ...[15.26(a)]$$

This area should be uniformly distributed over a width B. If A_Φ is the area of each bar, the spacing is given by

$$s = \frac{B A_\Phi}{A_{st1}} \qquad ...[15.26(b)]$$

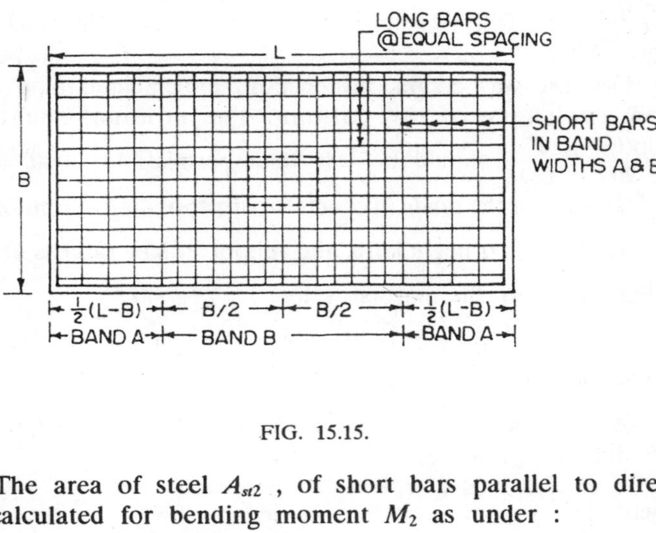

FIG. 15.15.

The area of steel A_{st2}, of short bars parallel to direction B, is calculated for bending moment M_2 as under :

$$A_{st2} = \frac{M_2}{t\,jd} \quad \text{where } t = \sigma_{st} \qquad ...[15.27(a)]$$

This area is provided in two distinct band widths : (*i*) the central band B of width B, and (*ii*) the end bands A, each of width $\frac{1}{2}(L-B)$.

The area $A_{st2(B)}$ in the central band B is given by

$$A_{st2(B)} = \frac{2 A_{st2}}{\beta + 1} \qquad ...[15.27(b)]$$

where $\beta = \frac{L}{B}$.

The area $A_{st2(A)}$ in each of the end bands A is given by

$$A_{st2(A)} = \frac{1}{2}(A_{st2} - A_{st2(B)}) \qquad ...[15.27(c)]$$

The reinforcement calculated above in the three bands should be distributed uniformly. (see Fig. 15.15)

(*b*) **Check for development length.** The section should be checked for development length at each face of the column, exactly in the same manner as for square column.

15.7. ISOLATED SLOPED FOOTING

In the previous article, we have seen that bending moment, beam shear and punching shear govern the thickness or depth of footing near the column face. This depth can be reduced toward the edges of the footing where the bending moment and shear decrease rapidly. If this decrease is achieved *linearly,* we get a *sloped footing* shown in Fig. 15.16 for a square footing. However, a minimum thickness of 150 mm should be kept at the edges. The areas at the base of footing and the net upward pressure p_o are found exactly in the same manner as for isolated footing of uniform depth.

1. SQUARE FOOTING

Fig. 15.16 shows a square footing of size $B \times B$, with variable depth.

Width $$B = \sqrt{\frac{W + W'}{q_0}}$$

and $$p_0 = \frac{W}{B^2} \qquad (\text{kN/m}^2)$$

The depth at the column face is found both from the consideration of bending moment as well as punching shear.

(*a*) **Depth for bending moment**

The maximum bending moment as the section XX through the column face is given by

$$M = p_0 B\left(\frac{B-b}{2}\right)\left(\frac{B-b}{4}\right) = p_0 \frac{B}{8}(B-b)^2 \times 10^6 \text{ N-mm} \quad ..(15.15)$$

This should be equated to the moment of resistance of the footing section at XX. Fig. 15.17(*a*) shows the section of the footing at XX. The top width is b, the bottom width is B and the effective depth is d.

The moment of resistance of the trapezoidal section consists of the moment of resistance of the central rectangular portion plus the moment of resistance of the two triangular portions.

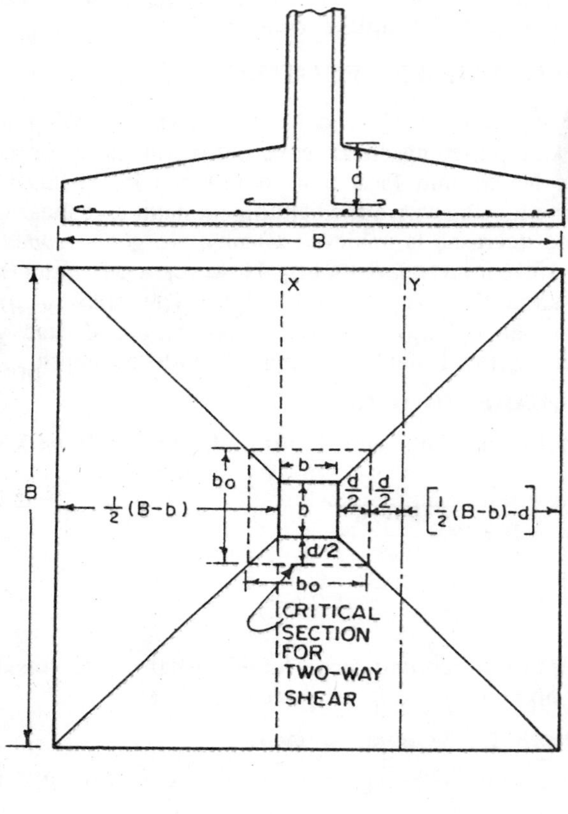

FIG. 15.16.

If α is the inclination of the footing top face with horizontal the moment of resistance M_{R1} of the central portion is given by

$$M_{R1} = Rbd^2 \cos^2 \alpha \qquad ...(15.28)$$

and compressive force $C_1 = \frac{1}{2} bkd \,.\, c \cos^2 \alpha$ ...(15.29)

for the central portion, where $c = \sigma_{cbc}$.

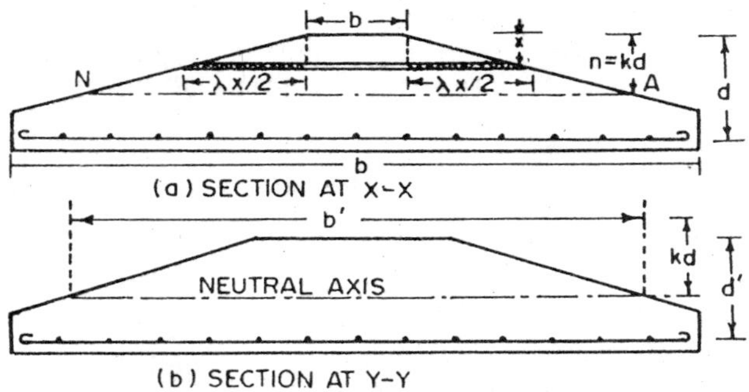

FIG. 15.17.

To find the moment of resistance M_{R2} of the two triangular portions consider an elementary strip at depth x below the apex, and let the total width of the strip be λx. If c is the compressive stress at the top fibre, the compressive stress at the level of the strip will be $\frac{kd-x}{kd}\,c$.

Hence the total compressive force in the triangular portion will be

$$C_2 = \int_0^{kd} \lambda x \,.\, dx \, \frac{kd-x}{kd}\, c.$$

$$= \lambda\, c \frac{k^2 d^2}{6} \qquad ...(15.22)$$

The moment of resistance of the triangular portion is :

$$M_{R2} = \int_0^{kd} \lambda x \,.\, dx \frac{kd-x}{kd} \,.\, c\,(d-x)$$

or

$$M_{R2} = \lambda \,.\, c \frac{k^2 d^3}{12} (2-k) \qquad ...(15.30)$$

Hence lever arm $a = \frac{M_{R2}}{C} = \frac{1}{2} d\,(2-k)$ ...(15.31)

$$\therefore \text{ Total } M_r = M_{r1} + M_{r2} = Rbd \cos^2\alpha + \lambda \,.\, c \frac{k^2 d^3}{12}(2-k) \qquad ...(15.32)$$

and combined level arm $= \dfrac{\text{Moment}}{\text{Force}}$

$$= \frac{Rbd^2 \cos^2 \alpha + \lambda \,.\, c\,\dfrac{k^2 d^3}{12}\,(2-k)}{\frac{1}{2}bkd\,c \,.\, \cos^2 \alpha + \lambda \,.\, c\,\dfrac{k^2 d^2}{6}} \qquad ...(15.33)$$

$= j'\,d$

where $j' =$ the lever arm factor

$$= \frac{Rbd \cos^2 \alpha + \lambda \,.\, c\,\dfrac{k^2 d^2}{12} \times (2-k)}{\frac{1}{2}\,bkc \cos^2 \alpha + \lambda \,.\, c\,\dfrac{k^2 d}{6}} \qquad ...[15.33(a)]$$

Equating Eqs. 15.15 and 15.32, the depth d can be calculated. In the above equations, the value of k and R are to be taken as for balanced section. The above method is a very tedious. As an approximation, the trapezoidal section can be replaced by an equivalent rectangular section of width $b + \frac{1}{8}(B - b)$ and of depth d to calculate the moment of resistance.

(*b*) **Check for shear**

For one way shear (beam action) the critical section is taken at a distance d from the face of column. The shear force V at the section is given by

$$V = [\tfrac{1}{2}(B - b) - d\,]\,B \times b_0$$

Hence nominal shear stress is given by

$$\tau_v = \frac{V}{b'\,d}$$

where b' = minimum width of section in the tensile zone
= width of the section at N.A. [Fig. 15.17(*b*)]

d' = effective depth of the section at the plane

Normally, the section is designed from the considerations of *two-way shear* (punching shear), and is later checked for beam shear.

For *two way action* (punching shear), the critical section located all round, is at a distance $d/2$ from the face of the column where $b_0 = (b + d)$

S.F. $F = p_0\,[B^2 - b_0^2] = p_0\,[B^2 - (b + d)^2]$

$$\tau_v = \frac{F}{4b_0 \,.\, d_o} = \frac{F}{4(b + d)d_0}$$

where d_o = effective depth of slab at the critical plane.

Hence $\tau_v \leq k_s \,.\, \tau_c \leq (0.5 + \beta_c) \,.\, \tau_c$

or $$\frac{F}{4(b + d)d_0} \leq (0.5 + \beta_c) \times 0.16\sqrt{f_{ck}}$$

where k_s or $(0.5 + \beta_c)$ has a maximum value of 1. ...(15.34)

(*c*) **Steel reinforcement**

Steel reinforcement is provided in both the directions, in the form of mesh, for the bending moment M.

Thus, $$A_{st} = \frac{M}{tj\,'d\,'} \quad \text{where } t = \sigma_{st}$$

where j' is the lever arm factor determined from Eq. [15.33(*a*)]

(*d*) **Check for development length**

The section should be checked for development length at the face of the column.

(2) **RECTANGUALAR FOOTING**

The procedure for design of rectangular footing is similar to that of a square sloped footing, except that each of the cantilever portion is analysed separately. The reinforcements A_{st1} and A_{st2} found in the two directions are provided as explained § 15.6.

15.8 ISOLATED STEPPED FOOTING

The construction of sloped footing is sometimes difficult and when the slope of the top face of footing is more, say more than 1 vertically to 3 horizontally, it may be difficult to finish the top without having concrete slump too much. In such cases, stepped footing shown in Fig. 15.18 is an alternative.

The design is first done at the section $X_1 - X_1$ passing through the face of the column. The depth d is determined both on the basis of bending moment and shear.

Thus $$M_1 = \frac{p_o B}{8}(B - b)^2 \times 10^6 \text{ N-mm} \qquad ...(15.15)$$

$$\therefore \quad d_1 = \sqrt{\frac{M_1}{RB_1}} \qquad ...(15.35)$$

where B_1 is the width of the footings at the top, at section $X_1 - X_1$.

The depth so found is checked both for one way shear (beam shear) as well as to way shear (punching shear) as explained in § 15.6.

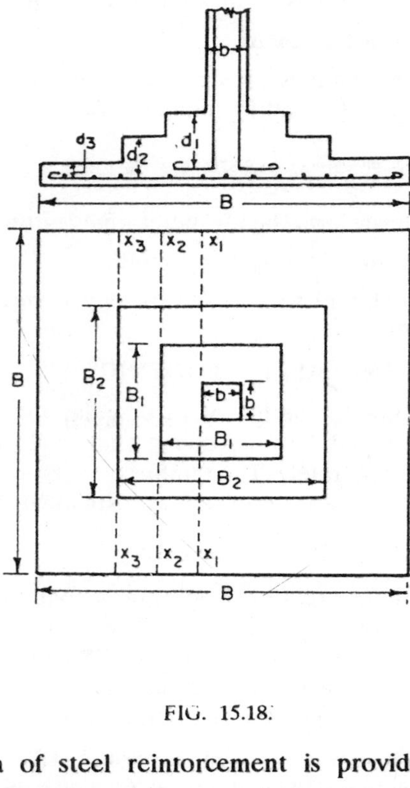

FIG. 15.18.

The area of steel reinforcement is provided as usual.

Let the size of footings at its top, at the next step be $B_2 \times B_2$ and its depth d_2. The critical section for design will be at section X_2 - X_2. The curtailed depth d_2 should be safe in compression, tension, shear and development length. The bending moment at X_2 - X_2 will be the moment of the total pressure acting on the rectangular area to one side of the section. The resisting width will be B_2.

Thus, from compression point of view,

$$d_2 = \sqrt{\frac{M_2}{RB_2}}$$, where M_2 is the bending moment at X_2 - X_2 .

From tension point of view, providing the *same* reinforcement A_{st} as at $X_1 - X_1$,

$$d_2 = \frac{M_2}{tj\,A_{st}} \quad \text{where } t = \sigma_{st}$$

The greater of the two values of d_2 is selected. This value of d_2 is checked both for beam shear(at a section distant d_2 from $X_2 - X_2$) and for punching shear(at perimeter section distant $d_2/2$ from the face of the step).

Similarly, depth d_3 at section $X_3 - X_3$ is determined.

15.9. ISOLATED FOOTING FOR CIRCULAR COLUMNS

Circular columns are usually preferred. The footing for a circular column may be either square or circular at the base.

1. SQUARE FOOTING

Fig. 15.19 show square stepped footing for a circular column for the purposes of design, the circular column of the same area as the circular column. The design of the square footing for the squivant square column is done exactly in the same manner as explained in § 15.7.

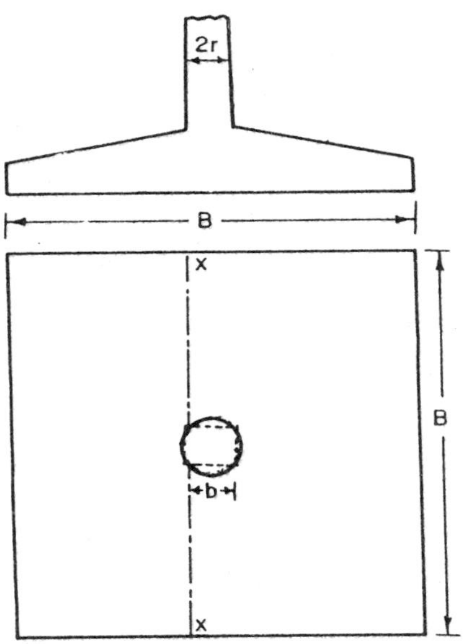

FIG. 15.19.

The critical section(XY) for bending moment will pass through the face of the equivalent square column.

2. CIRCULAR FOOTING

Fig. 15.20 shows a column of radius r with circular footing of radius R given by

$$R = \sqrt{\frac{W + W'}{\pi q_0}} \qquad \text{...(15.36)}$$

The net upward pressure intensity of footing is given by

$$p_0 = \frac{W}{\pi R^2} \qquad \text{...(15.37)}$$

1. *Depth for bending moment*

The maximum bending moment in the footing is calculated at the edge of the column for one quadrant such as $PQRS$ shown shaded.

The distance of $C.G.$ of the quadrant of a circle of radius R is at a distance of R from the centre. Hence if $\bar{x}$ is the distance of C.G. of shaded area from the centre of column, we have

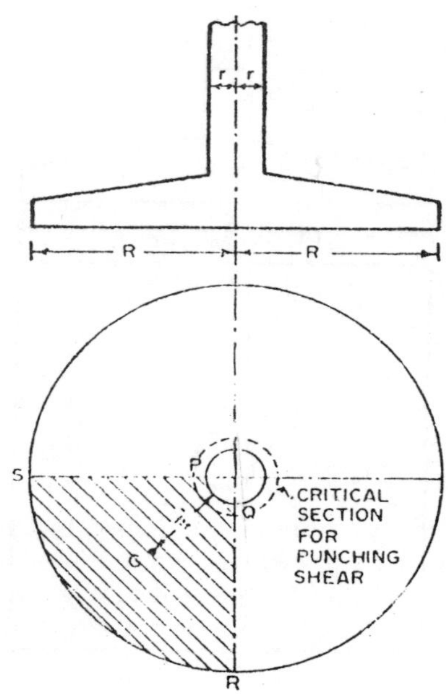

FIG. 15.20

$$\bar{x} = \frac{\frac{\pi}{4} R^2 \times 0.6\, R - \frac{\pi}{4} r^2 \times 0.6\, r}{\frac{\pi}{4} R^2 - \frac{\pi}{4} r^2} = \frac{0.6\,(R^3 - r^3)}{R^2 - r^2}$$

$$= \frac{0.6\,(R^2 + Rr + r^2)}{R + r} \qquad \text{...[15.38(a)]}$$

∴ Distanace of C.G. of shaded area from the face of the column is

$$\bar{x}' = \frac{0.6\,(R^2 + Rr + r^2)}{R + r} - r$$

$$= \frac{0.6\,(R^2 + Rr + r^2) - Rr - r^2}{R + r}$$

$$= \frac{0.6\, R^2 - 0.4 Rr - 0.4\, r^2}{R + r}$$

$$= \frac{0.2}{R + r}\,[3R^2 - 2Rr - 2r^2] \qquad \text{...(15.38)}$$

Upward pressure on shaded area is

$$= \frac{\pi}{4}\,(R^2 - r^2)\, p_0$$

∴ Bending moment at the face of the column is

$$M = \frac{\pi}{4}\,(R^2 - r^2)\, p_0 \times \frac{0.2}{R + r}\,(3R^2 - 2Rr - 2r^2)$$

$$= \frac{p_0\, \pi}{20}\,(R - r)\,(3R^2 - 2Rr - 2r^2) \qquad \text{...(15.39)}$$

Breadth b of quadrant $= PQ = \frac{\pi\, r}{2}$

Hence the effective depth is given by

$$d = \sqrt{\frac{M}{R \,.\, b}}$$

(where R in the above equation is equal to $\frac{1}{2}\sigma_{cbc} \,.\, jk$)

A minimum total thickness of 150 mm is provided at the outer edge of footing.

2. **Depth for two way shear (punching shear)**

Circular footing are checked for two way shear(punching shear) for which the critical circular plane exists at a radial distance of $d/2$ from the column face, as shown in Fig. 15.20. The radius r_o of this critical circle $= \left(r + \frac{d}{2}\right)$.

Net punching shear force $F = p_0\, \pi\,(R^2 - r_0^2)$

If d_0 is the effective depth of slab at that plane, the shear stress

$$\tau_v = \frac{F}{2\pi r_0 d_0} \qquad \text{...[15.40(a)]}$$

This should be less than $k_s \cdot \tau_c$.

$$\therefore \qquad = \frac{F}{2\pi r_0 d_0} \leq k_s \tau_c.$$

where $k_s = \left(0.5 + \frac{b}{a}\right) \leq 1 = 1$ (since $b = a = r$).

and $\tau_c = 0.16\sqrt{f_{ck}}$

Hence

$$\frac{F}{2\pi r_0 d_0} \leq 0.16\sqrt{f_{ck}} \qquad \text{...(15.40)}$$

The critical section for *one way shear* will occur at a radial distance d from the face of the column, *i.e.*, at AB (Fig. 15.21). The shear force V will be

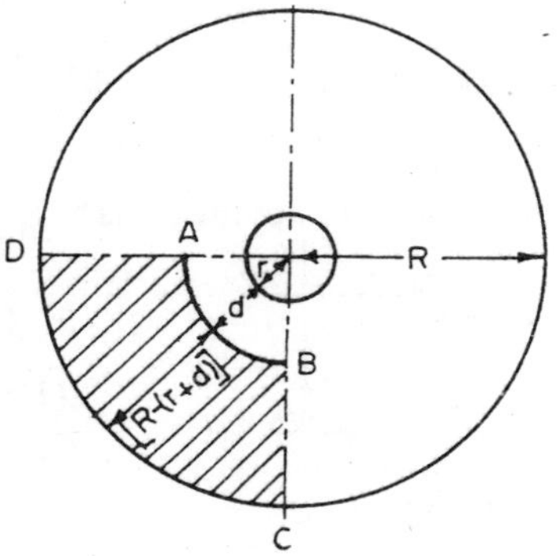

FIG. 15.21.

$$V = p_0 \frac{\pi}{4}[R^2 - (r + d)^2] \qquad \text{...[16.40(a)]}$$

The breadth b' of the footing at critical section

$$= \frac{\pi}{2}(r + d)$$

$$\therefore \quad \tau_v = \frac{V}{b'd'} = \frac{V}{\frac{\pi}{4}(r + d)\,d'}$$

where d' is the effective depth at the critical section.

3. ***Steel reinforcement***

Steel reinforcement is given by

$$A_{st} = \frac{M}{\sigma_{st}\, jd}.$$

Same reinforcement (A_{st}) is provided in two directions at right angles, *i.e.*, the bars are arranged in form of square mesh centrally located under the column, the width of the square being equal to the side of square that can be inscribed in the circular footing of radius R.

4. *Check for development length*

The section is checked for development length at the face of column.

Design Example 15.1 *A reinforced concrete wall 250 mm thick carries a load of 500 kN/m inclusive of its own weight. Design a reinforced concrete footing on soil having safe bearing capacity of 160 kN/m². Use M 15 concrete and take* $\sigma_{st} = 140\ N/mm^2$.

Solution.

1. *Design constants*

For M 15 concrete and for $\sigma_{st} = 140$ N/mm², we have

$$k = 0.404\ ;\ j = 0.865\ ;\ R = 0.874$$

2. *Size of footing*

$$W = 500 \text{ kN}\ ;\ q_0 = 160 \text{ kN/m}^2$$

Assume the self-weight of footing equal to 10% of the super imposed load.

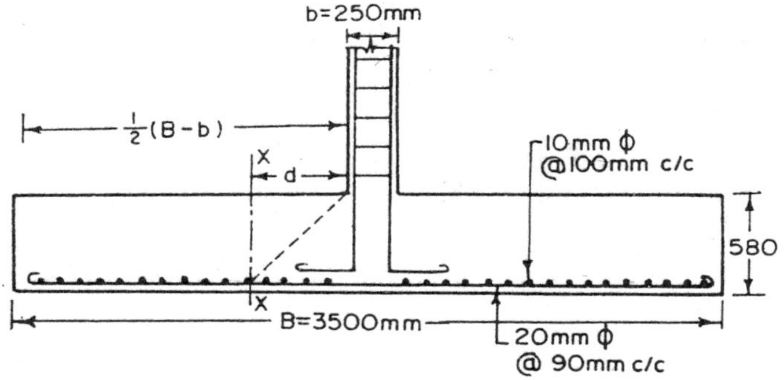

FIG. 15.22

$$\therefore \quad W' = 50 \text{ kN/m}$$

$$\therefore \quad B = \frac{W + W'}{q_0} = \frac{500 + 50}{160} = 3.43 \text{ m}$$

Adopt $B = 3.5$ m $= 3500$ mm .

$\therefore$ Net upward pressure $p_0 = \dfrac{500}{3.5} = 142.86$ kN/m²/m

3. *Design of section*

Maximum bending moment at the face of the wall is given by Eq. 16.14.

$$M = p_0 \frac{(B - b)^2}{8} \times 10^6 \text{ N-mm}$$

$$= \frac{142.86\,(3.5 - 0.25)^2}{8} \times 10^6 = 1.89 \times 10^8 \text{ N-mm}$$

$$d = \sqrt{\frac{M}{BR}} = \sqrt{\frac{1.89 \times 10^8}{1000 \times 0.874}} = 465 \text{ mm}$$

Using 60 mm effective cover to the centre of steel, total depth. $= 465 + 60 = 525$ mm. However, provide total depth = 530 mm so that available effective depth will be 470 mm.

$$A_{st} = \frac{M}{\sigma_{st}\, jd} = \frac{1.89 \times 10^8}{140 \times 0.865 \times 470} = 3321 \text{ mm}^2$$

Using 20 mm Φ bars having $A_\Phi = \frac{\pi}{4}(20)^2 = 314 \text{ mm}^2$

$$\therefore \quad \text{Spacing} \quad s = \frac{1000\,A_\Phi}{A_{st}} = \frac{1000 \times 314}{3321} = 94.5 \text{ mm}$$

However, provide 20 mm Φ bars @ 90 mm c/c.

$$\text{Longitudinal reinforcement} = \frac{0.15 \times 1000 \times 530}{100} = 795 \text{ mm}^2$$

Using 10 mm Φ bars, $A_\Phi = 78.5$

$$\therefore \quad \text{Spacing} = \frac{1000 \times 78.5}{795} \approx 100 \text{ mm}$$

Provide 10 mm Φ bars longitudinally @ 100 mm c/c

4. *Check for shear*

For *one way shear,* the critical section *XX* (Fig. 15.22) at distance *d* from the face of wall. Distance of section *XX* from the edge of footing $= \frac{1}{2}(B - b) - d = \frac{1}{2}(3.5 - 0.25) - 0.47 = 1.155$ m $= 1155$ mm.

$$\therefore \quad V = p_0 \times 1.155 = 142.86 \times 1.155 = 165 \text{ kN/m} = 165000 \text{ N/m}$$

$$\therefore \quad \tau_v = \frac{165000}{1000 \times 470} = 0.35 \text{ N/mm}^2$$

Permissible shear stress $= k\,\tau_c$
where $k = 1$ for total thickness of 530 mm (Table 3.2)

Take $\tau_c = 0.33$ N/mm² (Table 3.1) for reinforcement of 0.70%

Hence $\tau_v > k.\tau_c$.

Thus, the footing is not safe from shear point of view. In order to make the footing safe in shear, increase its depth. Taking maximum value of τ_v to equal τ_c $\left(= 0.32 \text{ N/m}^2 \text{ for } \frac{100 A_s}{bd} \text{ slightly less than } 0.7\%\right)$, we have

$$\tau_v = \frac{165000}{1000 \times d} = 0.32$$

which gives $d = \frac{16500}{1000 \times 0.32} = 516$ mm

However, provide total depth = 580 mm so that available $d = 580 - 60 = 520$ mm.

5. ***Check for development length***

$$L_d = \frac{\sigma_{st}}{4\,\tau_{bd}}\,\Phi = \frac{140}{4 \times 0.6}\,\Phi = 58.3\,\Phi$$

$$= 58.3 \times 10 = 583 \text{ mm}$$

Providing 60 mm side cover, length of bars available $\frac{1}{2}[B - b] - 60 = \frac{1}{2}[3500 - 250] - 60 = 1565$ mm, which is greater than L_d. Hence O.K.

The details of reinforcement etc. are shown in Fig 15.22.

Design Example 15.2. *A brick wall 300 mm thick carries a load of 180 kN/m length. Design a R.C.C. footing, if the safe bearing cpacity of soil is 120 kN/m². Assume the permissible stresses as given in example 15.1.*

Solution.

1. ***Design constants***

$$k = 0.404 \; ; \; j = 0.865 \text{ and } R = 0.874.$$

2. ***Size of footing***

$$W = 180 \text{ kN}$$

Assume $W' = 10\%$ of $W = 18$kN

$\therefore$ Width B of footing $= \frac{180 + 18}{120} = 1.65$ m

Adopt $B = 1.7$ m.

Net upward pressure $p_0 = \frac{180}{1.7} = 106$ kN/m²/m

3. ***Design of section***

Maximum bending moment occurs at section X - X distant $\frac{b}{4}$ from the centre of the wall, and its magnitude is given by Eq. 15.13.

$$M = \frac{p_0}{8}(B - b)\left(B - \frac{b}{4}\right) \times 10^6 \quad \text{N-mm}$$

$$= \frac{106}{8}(1.7 - 0.3)\left(1.7 - \frac{0.3}{4}\right) \times 10^6 = 30.14 \times 10^6 \text{ N–mm}.$$

$$\therefore \; d = \sqrt{\frac{30.14 \times 10^6}{1000 \times 0.865}} = 187 \text{ mm}$$

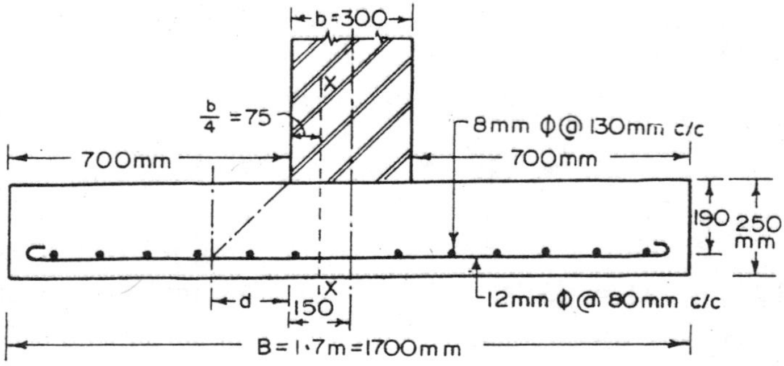

FIG. 15.23.

Provide a total depth of 250 mm and a cover to the centre of the steel equal to 60 mm, so that available $d = 250 - 60 = 190$ mm.

4. *Check for shear*

For balanced section, p = 0.72% for M 15 concrete. Hence from Table 3.1, $\tau_c \approx 0.33$ N/mm². Also, from Table 3.2, k = 1.10 for D = 250 mm. Hence permissible shear stress $= k\,\tau_c = 1.10 \times 0.33 = 0.36$ N/mm².

The critical section Y-Y lies at a distance of d (=190mm) from the face of the wall. Hence distance of Y-Y from the edge of footing $= \frac{1}{2}(B - b) - d = \frac{1}{2}(1.7 - 0.3) - 0.19 = 0.51$ m.

$$\therefore \quad V = 106000 \times 0.51 \approx 54000 \text{ N/m}$$

$$\therefore \; \tau_v = \frac{V}{bd} = \frac{54000}{1000 \times 190} = 0.284 \text{ N/mm}^2$$

This is less than the permissible shear stress. Hence safe.

5. *Design of reinforcement*

$$A_{st} = \frac{M}{\sigma_{st}\, jd} = \frac{30.14 \times 10^6}{140 \times 0.865 \times 190} = 1310 \text{ N/mm}^2$$

Using 12 mm Φ *bars*, $A_\Phi = \frac{\pi}{4}(12)^2 = 113 \text{ mm}^2$.

$$\therefore \quad \text{Spacing } s = \frac{1000\, A_\Phi}{A_{st}} = \frac{1000 \times 113}{1310} = 86 \text{ mm}.$$

However, provide 12 mm Φ bars @ 80 mm c/c.

Area of longitudinal reinforcement = 0.15% area of cross-section.

$$= \frac{0.15}{100} \times 1000\,D = 1.5\,D$$

$$= 1.5 \times 250 = 375 \text{ mm}^2$$

Use 8 mm Φ bars having $A_\Phi = \frac{\pi}{4}(8)^2 = 50 \text{ mm}^2$

$$\therefore \quad \text{Spacing } s = \frac{1000 \times 50}{375} = 133 \text{ mm.}$$

Provide 8 mm Φ bars @ 130 mm c/c.

6. ***Check for development length***

$$L_d = 58.3\,\Phi = 58.3 \times 12 \approx 700 \text{ mm.}$$

Providing 60 mm side cover, length of bars available

$$= \tfrac{1}{2}[B - b] - 60 = \tfrac{1}{2}[1700 - 300] - 60 = 640 \text{ mm.}$$

This is less than L_d. Hence sufficient development length is not available. There are three methods to satisfy the requirements : (*i*) decrease in the bar diameter (*ii*) increase in width of footing and (*iii*) provision of 90° bend on the bars at the edge of footing. The third alternative is the best. Hence bend the bar up by 90°, for a length of 6 Φ = 6 × 12 = 72 mm, so that effective length available =anchorage value of right angle bend = 12 Φ = 144mm. Hence total length available= 640 + 144 = 788 mm $> L_d$

Design Example 15.3. *Design an isolated footing of uniform thickness of a R.C. column bearing a vertical load of 600 kN and having a base of size 500 mm × 500 mm. The safe bearing capacity of soil may be taken as 120 kN/m². Use M 15 is concrete and taken* $\sigma_{st} = 140\ N/mm^2$.

Solution.

1. ***Design constants.***

For given set of stresses, we have k = 0.404 ; j =0.865 and R=0.874

2. ***Size of footing***

W = 600 kN

Let W' be equal to 10% W = 60 kN

$$\therefore \quad A = \frac{660}{120} = 5.5\text{m}^2$$

∴ Size B of square $\sqrt{5.5} = 2.345$ m.

Provide a square footing of 2.4 m × 2.4 m

Net upward pressure $p_0 = \frac{600}{2.4 \times 2.4} = 104.17 \text{ kN/m}^2$

3. ***Design of section***

The maximum bending moment acts at the face of column, and its magnitude is given by Eq. 15.18.

$$M = p_0 \frac{B}{8}(B - b)^2 \times 10^6 \text{ N–mm}$$

$$= 104.17 \times \frac{2.4}{8}(2.4 - 5)^2 \times 10^6 = 1.128 \times 10^8 \text{ N–mm}$$

$$\therefore \quad d = \sqrt{\frac{M}{RB}} = \sqrt{\frac{1.128 \times 10^8}{0.87 \times 2400}} = 232 \text{ mm}$$

4. ***Check for shear***

The depth found above should be checked for shear. For *one way shear,* the critical section *YY* is located at a distance $d(=232$ mm) from the face of the column (Fig 15.13), where shear force V is given by

$$V = p_0 B\left\{\frac{1}{2}(B - b) - d\right\}$$

$$= 104.17 \times 2.4\left\{\frac{1}{2}(2.4 - 0.5) - 0.232\right\} = 179.5 \text{ kN}$$

$$\therefore \quad \tau_c = \frac{B}{B.d} = \frac{179500}{2400 \times 232} = 0.322 \text{ N/mm}^2$$

Using 12 mm Φ bars 60 mm clear cover, total depth will be about $232 + 6 + 60 = 298$ mm ≈ 300 mm. Hence from Table 3.2, $k = 1$. Also, for a balanced section having $p = 0.72\%$ for M 15 concrete, $\tau_c \approx 0.33$ N/mm² from Table 3.1. Hence the permissible shear stress $= k\,\tau_c = 1 \times 0.33 = 0.33 \text{ N/mm}^2$. Hence the section is just safe.

Also, for *two way shear* (punching shear), the section lies at $d/2$ from the column face all round. The width p_0 of the section $= b + d = 500 + 232 = 732$ mm. Also, shear force around the section is

$$F = p_0[B^2 - b_0^2] = p_0[B^2 - (b + d)^2]$$

$$= 104.17[(2.4)^2 - (0.732)^2] = 544.2 \text{ kN}$$

$$\therefore \quad \tau_c = \frac{F}{4\,b_0\,d} = \frac{F}{4\,[b + d]\,d} = \frac{544200}{4 \times 732 \times 232}$$

$$= 0.801 \text{ N/mm}^2$$

Permissible shear stress $= k_s\,\tau_c$

where $k_s = (0.5 + \beta_c) = (0.5 + 1)$ with max. value of 1

Hence $k_s = 1$

Also, $\tau_c = 0.16\sqrt{f_{ck}} = 0.16\sqrt{15} = 0.62\ \text{N/mm}^2$

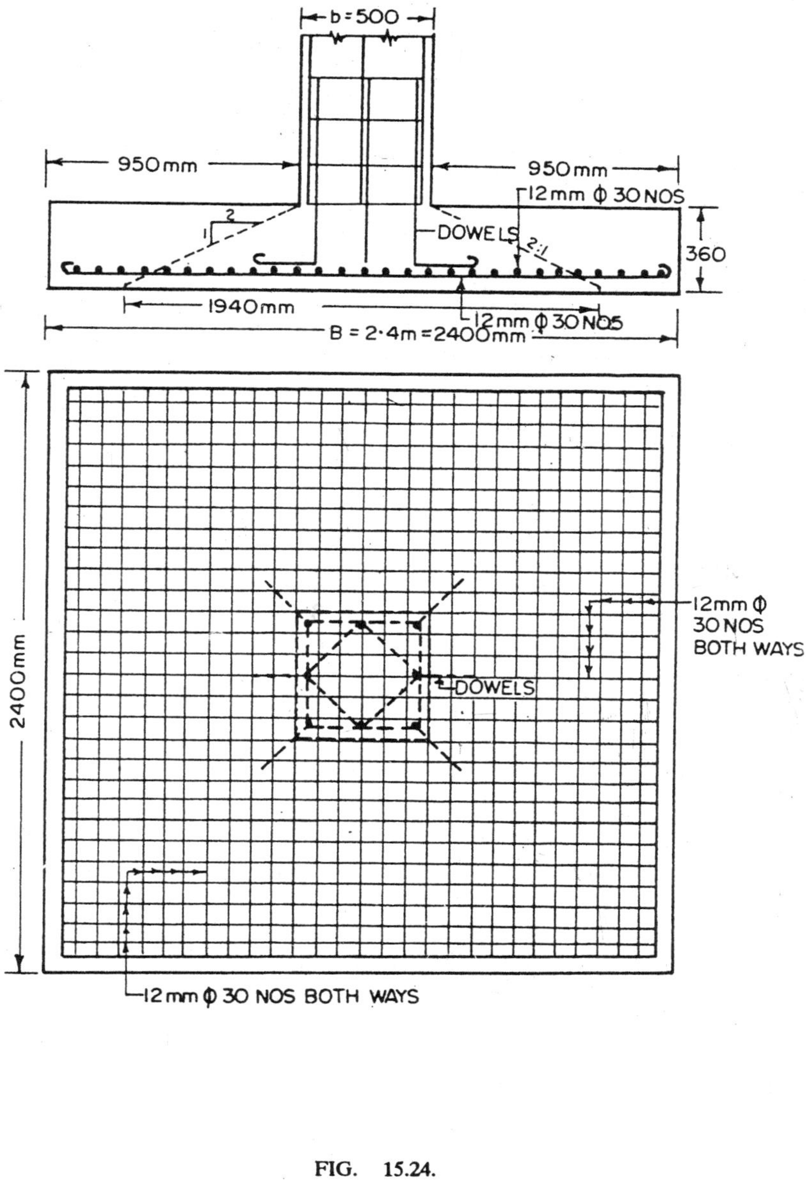

FIG. 15.24.

Hence permissible shear stress $= k_s \tau_c = 1 \times 0.62 = 0.62$ N/mm^2.

This is less that the punching shear stress. Hence the section is unsafe (*i.e.* depth is insufficient).

The required effective depth from punching shear point of view is found by equating τ_v to permissible shear stress of 0.62 N/mm^2.

$$\therefore \quad \tau_v = \frac{F}{4\,[b + d]\,d} \leq k_s \tau_c$$

or
$$\frac{104.17 \times 10^3 \,[(2400)^2 - (500 + d)^2] \times 10^{-6}}{4[500 + d]\,d} = 0.62$$

or
$$d^2 + 520.2\,d - 222092 = 0$$

which gives $d = 278$ mm.

Provide a total depth of 360 mm. Using 12 mm Φ bars and a clear cover of 60 mm, available $d = 360 - 6 - 60 = 294$ mm is one direction and $294 - 12 = 282$ mm in the other direction.

5. *Design of steel reinforcement*

The area of reinforcement in each direction is given by

$$A_{st} = \frac{M}{\sigma_{st}\, jd} = \frac{1.128 \times 10^8}{140 \times 0.865 \times 282} = 3303 \text{ mm}^2$$

Using 12 mm Φ bars, $A_\Phi = \frac{\pi}{4}(12)^2 = 113$ mm^2

$$\therefore \quad \text{No. of bars} = \frac{3303}{113} = 29.2$$

Provide 30 bars of 12 mm Φ uniformly distributed in each direction.

6. *Check for development length*

$$L_d = 58.3\,\Phi = 58.3 \times 12 \approx 700 \text{ mm}$$

Providing 60 mm side cover, length of bars available $= \frac{1}{2}[B - b] - 60 = \frac{1}{2}[2400 - 500] - 60 = 890$ mm $> L_d$. Hence safe.

7. *Transfer of load at column base*

$$A_2 = 500 \times 500 = 250000 \text{ mm}^2$$

$$A = [500 + 2\,(2 \times 360)]^2 \text{ for side spread rate of } 2 : 1$$

$$= (1940)^2 = 3763600 \text{ mm}^2$$

$$\therefore \quad \sqrt{\frac{A_1}{A_2}} = \sqrt{\frac{3763600}{250000}} = 3.88$$

Adopt max. value of $\sqrt{\frac{A_1}{A_2}} = 2.$

Hence permissible bearing stress $= 0.25 f_{ck} \sqrt{\dfrac{A_1}{A_2}}$

$$= 0.25 \times 15 \times 2 = 7.5 \text{ N/mm}^2.$$

Actual bearing pressure $= \dfrac{600000}{500 \times 500} = 2.4 \text{ N/mm}^2.$

Hence safe. Thus no separate dowal bars are required for the transfer of load. However, it is advisable to continue all the bars of the column, into the foundation.

The details of reinforcement etc. are shown in Fig 15.24.

Example 15.4. *Design a rectangular isolated footing of uniform thickness for R.C. column bearing a vertical load of 600 kN, and having a base size of 400 × 600 mm. The safe bearing capacity of soil may be taken as 120 kN/m². Use M 15 concrete. Take* $\sigma_{st} = 140 \text{ N/mm}^2$.

Solution

1. *Design constants*

For M 15 concrete, and for $\sigma_{st} = 140$, we have

$k = 0.404$; $j = 0.865$ and $R = 0.874$

2. *Size of footing*

$$W = 600 \text{ kN}$$

Let W' be equal to 10% W = 60 kN.

$$\therefore \quad A = \frac{660}{120} = 5.5 \text{ m}^2$$

Let ratio of B to $L = \dfrac{40}{60} = \dfrac{2}{3}$.

$$\therefore \quad \frac{2}{3} L \times L = 5.5$$

or $\quad L = 2.9$ m.

$$\therefore \quad B = \frac{2}{3} \times 2.9 \text{ m}$$

However, provide a footing of size 2 m × 3 m.

Net upward pressure $p_0 = \dfrac{600}{2 \times 3} = 100 \text{ kN/m}^2$

3. *Design of section*

Refer Fig. 15.14.

Bending moment M_1 about section X-X is given by

$$M_1 = \frac{p_0 B}{8} (L - a)^2 \text{ kN–m}$$

$$= \frac{100 \times 2}{8} (3 - 0.6)^2 \times 10^6 \text{ N-mm}$$

$$= 144 \times 10^6 \text{ N-mm}$$

$$\therefore \quad d = \sqrt{\frac{M_1}{RB}} = \sqrt{\frac{144 \times 10^6}{0.874 \times 2000}}$$

$$= 287 \text{ mm.}$$

Keep $d = 290$ mm and total depth = 350 mm. Provide uniform thickness for the entire footing. B.M. M_2 about section Y–Y is given by

$$M_2 = \frac{p_0 L}{8}(B - b)^2 \times 10^6 \text{ N-mm}$$

$$= \frac{100 \times 3}{8}(2 - 0.4)^2 \times 10^6 = 96 \times 10^6 \text{ N–mm}$$

Thus, $M_2 < M_1$.

The effective depth found above has to be checked for shear.

4. Check for shear

For the *beam action* total S.F. along section AB[Fig. 15.10(a)] is

$$V = p_0 B\left(\frac{L}{2} - \frac{a}{2} - d\right) = p_0\, B\left(\frac{L - a}{2} - d\right)$$

$$= 100 \times 2\left(\frac{3 - 0.6}{2} - 0.29\right) = 182 \text{ kN}$$

$$= 170000 \text{ N.}$$

$$\tau_v = \frac{V}{Bd} = \frac{182000}{2000 \times 290} = 0.314 \text{ N/mm}^2.$$

Fore M 15 concrete, $p = 0.72\%$ and hence $\tau_c = 0.33$ N/mm^2

Also for D > 300 mm , $k = 1$ from Table 3.2.

$$\therefore \quad \tau_c < \tau_c \text{ (safe).}$$

For the *two way action or punching shear action* along $ABCD$ [Fig. 16.10(b)],

$$\text{Perimeter } ABCD = 2\{(a + d) + (b + d)\}$$

$$= 2\{(0.6 + 0.29) + (0.4 + 0.29)\}$$

$$= 2\{0.89 + 0.69\} = 3.16 \text{ m}$$

$$= 3160 \text{ mm.}$$

Area $ABCD = 0.89 \times 0.69 = 0.6141$ m^2

$$\therefore \quad \text{Punching shear} = 100[(2 \times 3) - 0.6141]$$

$$= 538.59 \text{ kN}$$

$$\therefore \quad \tau_v = \frac{538.59 \times 1000}{3160 \times 290} = 0.59 \text{ N/mm}^2.$$

Allowable shear stress τ_c is given by

$$\tau_c = 0.16\sqrt{f_{ck}} = 0.16\sqrt{15} \approx 0.62 \text{ N/mm}^2.$$

$$k_s = (0.5 + \beta_c) = \left(0.5 + \frac{0.4}{0.6}\right) = 1.17$$

However, adopt max. $k_s = 1$.

$\therefore \quad k_s \tau_c = 1 \times 0.62 = 0.62 \text{ N/mm}^2.$

This is more than $\tau_v = 0.59 \text{ N/mm}^2$. Hence safe.

Thus the effective depth d = 290 mm is alright.

5. *Design for reinforcement*

Area A_{st1} of long bars calculated for moment M_1 is given by

$$A_{st1} = \frac{M_1}{\sigma_{st}\, jd} = \frac{144 \times 10^6}{140 \times 0.865 \times 290} = 4100 \text{ mm}^2$$

Using 12 mm Φ bars, $A_\Phi = 113 \text{ mm}^2$

$\therefore \quad$ No. of bars $= \dfrac{4100}{113} \approx 36.$

These are to be distributed uniformly in a width B = 2 m. Effecive depth for top layer of reinforcement =290−12=278 mm.

The area A_{st2} of short bars calculated for M_2 is given by

$$A_{st2} = \frac{M_2}{\sigma_{st}\, jd} = \frac{97 \times 10^6}{140 \times 0.865 \times 278} = 2880 \text{ mm}^2.$$

This area is to be provided in two distinct band widths. Area $A_{st2(B)}$ in central bend of width B = 2 m is given by

$$A_{st2(B)} = \frac{2A_{st2}}{\beta + 1} = \frac{2 \times 2880}{\frac{3}{2} + 1} = 2304 \text{ mm}^2.$$

$\therefore \quad$ No. of 12 mm Φ bars $= \dfrac{2304}{113} \approx 20$ to be provided in central bend width = 2 m.

Remaining area in each end band strip

$$= \tfrac{1}{2}(2880 - 2304) = 288 \text{ mm}^2.$$

No. of 12 mm Φ bars $= \dfrac{288}{113} \approx 3$, to be provided in each end band of width $\frac{1}{2}(L - B) = \frac{1}{2}(3 - 2) = 0.5$ m.

6. *Test for development length*

$$L_d = \frac{\sigma_{st}}{4\tau_{bd}}\Phi = \frac{140}{4 \times 0.6}\Phi = 58.3\,\Phi$$

$$= 58.3 \times 12 = 700 \text{ mm.}$$

Providing 60 mm side cover, length available

$$\frac{1}{2}[B - b] - 60 = \frac{1}{2}[2000 - 400] - 60 = 740 \text{ mm},$$

which is greater than L_d. Hence O.K.

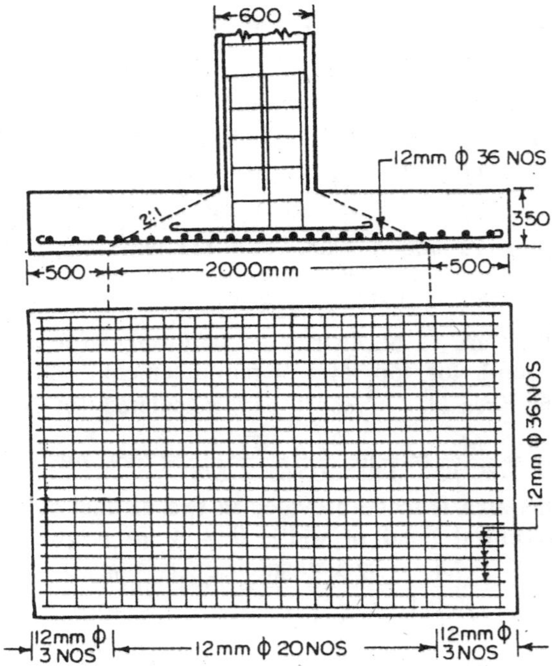

FIG. 15.25

7. Check for transfer of load at the base

$$A_2 = 600 \times 600 = 360000 \text{ mm}^2$$

At a rate of spread of 2:1,

$$A_1 = [600 + 2(2 \times 350)]^2 = (2000)^2 = 4 \times 10^6 \text{ mm}^2$$

$$\therefore \quad \sqrt{\frac{A_1}{A_2}} = \sqrt{\frac{4 \times 10^6}{360000}} = 3.33 > 2$$

Adopt max. value of $\sqrt{A_1/A_2}$ as 2.

$$\therefore \quad \text{permissible bearing stress} = 0.25 f_{ck} \sqrt{\frac{A_1}{A_2}}$$

$$= 0.25 \times 15 \times 2 = 7.5 \text{ N/mm}^2.$$

Actual bearing stress $= \frac{600000}{600 \times 600} = 1.67\ N/mm^2$

Hence satisfactory.

The details of reinforcement etc. are shown in Fig. 15.25.

Example 15.5. *Design an isolated footing for a column 500 mm × 500 mm, transmitting an axial load of 1200 kN. The column is reinforced with 8 bars of 20 mm diameter. The safe bearing capacity of soil is 120 tonnes/m². Use M 15 concrete. For mild steel bars, take* $\sigma_{st} = 140\ N/mm^2$.

Solution.

1. ***Design constants.***

For M 15 concrete, $\sigma_{cbc} = 5$ N/mm² and $m = 19$. Also, $\sigma_{st} = 140$ N/mm².

Hence $k = 0.404$; $j = 0.865$; $R = 0.874$.

2. ***Size of footing***

$$W = 1200 \text{ kN}$$

Assume $W' = 10\ \%\ W = 120$ kN

$\therefore$ Area of footing $A = \frac{1200 + 120}{12} = 11\ m^2$

Provide a footing of size 3.4 m × 3.4 m.

Actual upward pressure intensity $p_0 = \frac{1200}{3.4 \times 3.4}$

$$= 104\ kN/m^2.$$

3. ***Design of footing***

Maximum bending moment occurs at the face of the column and its magnitude is given by Fig. 15.15.

$$M = p_0 \frac{B}{8} (B - b)^2 \text{ kN-m}$$

$$= \frac{104 \times 3.4}{8} (3.4 - 0.5)^2 \times 10^6 \text{ N-mm}$$

$$= 372 \times 10^6 \text{ N-mm}.$$

The section of the footing at the columm face will be trapezoidal, as shown in Fig. 15.16. Let the effective depth at the column face be d and that at the edges be 0.2 d.

Slope $\lambda = \frac{B - b}{d - 0.2\ d} = \frac{3400 - 500}{0.8\ d}$

or $$\lambda = \frac{3600}{d} \qquad ...(i)$$

The moment of resistance of section is given by Eq. 15.32

$$M_r = Rbd^2 \cos^2 \alpha + \lambda \,.\, c \,\frac{k^2 d^3}{12}\,(2-k)$$

Adopting $\cos^2 \alpha = 0.85$ for practical purposes, we have

$$M_r = 0.875 \times 500\, d^2 \times 0.85 + \frac{3600 \times 5}{d}\,\frac{(0.404)^2}{12}\,(2-0.404)\, d^3$$

$$= 372\, d^2 + 391\, d^2 = 763\, d^2 \qquad ...(ii)$$

Total force of compression is

$$C = \tfrac{1}{2} kbd\, c \cos^2 \alpha + \lambda \,.\, c \,\frac{k^2\, d^2}{6}$$

$$= \tfrac{1}{2} \times 500 \times 0.404 \times 5 \times 0.85\, d + \frac{3600}{d} \times \frac{5}{6}\,(0.404)^2\, d^2$$

$$= 492\, d + 490\, d = 919\, d \qquad ...(iii)$$

L.A. for the section $= \dfrac{762\, d^2}{919\, d} = 0.83\, d$

$\therefore$ Lever arm factor $j = 0.83$

Alternatively taking the section to be an equivalent rectangle of width $= b + \frac{1}{8}(B - b) = 500 + \frac{1}{8}(3400 - 500) \approx 860$ mm.

$\therefore M_r = R\, bd^2 + 0.875 \times 860\, d^2 = 753\, d^2$ against $763\; d^2$ found in (*ii*). Since both the values are nearly equal, we can take the trapezoidal section to be an equivalent rectangle of width $b + \frac{1}{8}(B-b)$.

Adopting $M_r = 763\; d^2$, and equating it to the B.M., we get

$$763\, d^2 = 372 \times 10^6$$

or

$$d = \sqrt{\frac{372 \times 10^6}{763}} = 700 \text{ mm}.$$

Effective depth at end $= 0.2d = 140$ mm.

4. *Check for shear*

(*a*) *For beam shear*

The section is to be checked for beam shear is at a distance $d = 700$ mm from the column face, where shear force is given by

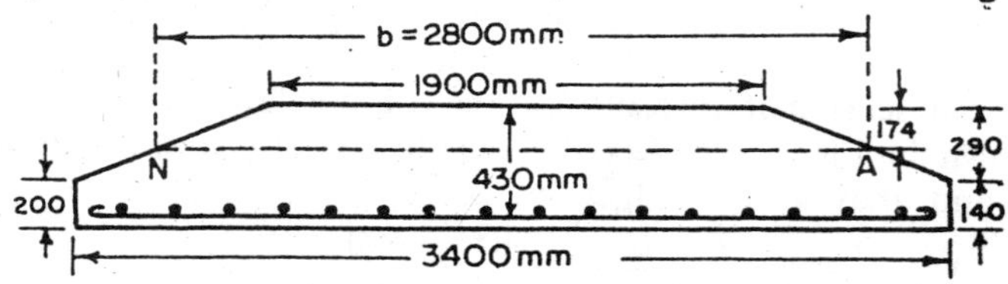

FIG. 15.26.

$$V = p_0 B \left\{ \tfrac{1}{2}(B - b) - d \right\}$$

$$= 104 \times 3.4 \left\{ \tfrac{1}{2}(3.4 - 0.5) - 0.7 \right\} = 266 \text{ kN.}$$

The section for footing is shown in Fig. 15.26.

Effective depth at that location $= 140 + \dfrac{700 - 140}{1450}(1450 - 700)$

$= 430$ mm.

Top width of section $= 500 + \dfrac{3400 - 500}{1450} \times 700 = 1900$ mm

Depth of N.A. $= 430 \times 0.404 = 174$ mm.

Width b' at N.A. $= 1900 + \dfrac{3400 - 1900}{290} \times 174 = 2800$ mm.

$$\therefore \quad \tau_v = \frac{V}{b'd} = \frac{266 \times 1000}{2800 \times 430} = 0.22 \text{ N/mm}^2;$$

For M 15 concrete, $p = 0.72\%$ and hence $\tau_c = 0.33$ N/mm^2

Thus, $\tau_v < \tau_c$. (Hence safe).

(b) Check for two way shear (punching shear) : [Fig. 15.10(*a*)]

Perimeter $ABCD = 2[(a + d) + (b + d)]$

$= 4(a + d) = 4(500 + 700) = 4800$ mm.

Area $ABCD = 1.2 \times 1.2 = 1.44$ m^2

Punching shear $= 104 [3.4 \times 3.3 - 1.44] = 1052.5$ kN.

$$\therefore \quad \tau_v = \frac{1052.5 \times 1000}{4800 \times 700} = 0.31 \text{ N/mm}^2.$$

Allowable shear stress τ_c is given by

$$\tau_c = 0.16\sqrt{f_{ck}} = 0.16\sqrt{15} \approx 0.62 \text{ N/mm}^2$$

$$k_s = (0.5 + \beta_c) = (0.5 + 1).$$

However adopt max. $k_s = 1$.

$\therefore \quad k_s \tau_c = 1 \times 0.62 = 0.62$ N/mm^2. Hence safe.

5. Steel reinforcement

$$A_{st} = \frac{M}{\sigma_{st}\, jd} \quad \text{where } j = 0.83$$

$$= \frac{372 \times 10^6}{140 \times 0.83 \times 700} = 4573 \text{ mm}^2$$

Using 16 mm Φ bars, $A_\Phi = 201$ mm^2.

$\therefore$ No. of bars $= \dfrac{4573}{211} \approx 23.$

Hence provide 16 mm Φ bars, 23 Nos., uniformly spaced in the width 3.4 m in each direction. Effective depth to the centre of bottom layer = 700 + 16 = 716 mm. Provide overall depth of 770 mm and reduce it to 200 mm at the edge.

6. *Check for development length*

$$L_d = 58.3 \times 16 = 932.8 \text{ mm.}$$

Providing 90 mm side cover, length, available $= \frac{1}{2}(B - b)$

$= \frac{1}{2}(3400 - 500) - 60 = 1390$ mm. Hence O.K.

The details of reinforcement etc. as shown in Fig. 15.27.

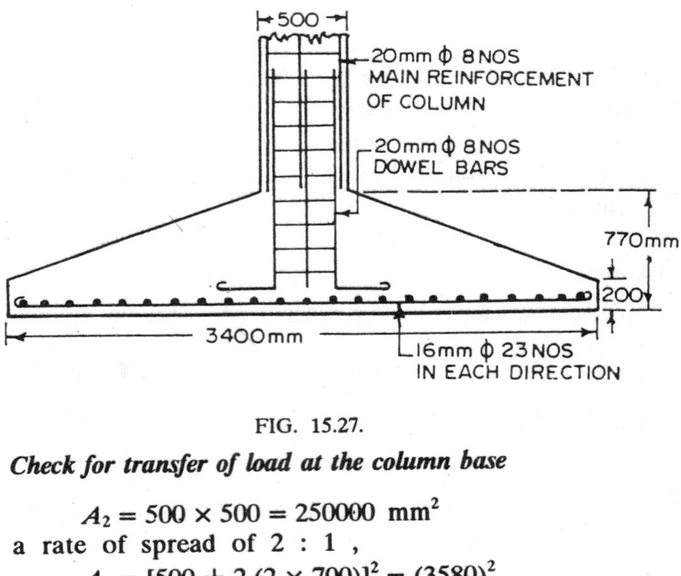

FIG. 15.27.

7. *Check for transfer of load at the column base*

$$A_2 = 500 \times 500 = 250000 \text{ mm}^2$$

At a rate of spread of 2 : 1 ,

$$A_1 = [500 + 2(2 \times 700)]^2 = (3580)^2$$

$$= 12816400 \text{ mm}^2$$

$$\therefore \quad \sqrt{\frac{A_1}{A_2}} = \sqrt{\frac{12816400}{250000}} = 7.16 > 2$$

Adopt a max. value of $\sqrt{\frac{A_1}{A_2}} = 2.$

$\therefore$ Permissible bearing stress $= 0.25 f_{ck} \sqrt{\frac{A_1}{A_2}}$

$$= 0.25 \times 15 \times 2 = 7.5 \text{ N/mm}^2$$

Actual bearing stress $= \frac{1200000}{500 \times 500} = 4.1 \text{ N/mm}^2$

Hence satisfactory.

Note. In case the permissible bearing stress is less than the actual bearing pressure, load transfer is done by the provision of *dowels bars.* For example, if the permissible bearing pressure were only 3.5 N/mm^2 (in place of 7.5 N/mm^2), a part of load transfer will be done by dowels. Assume that a *minimum* of 4 dowel bars are provided, each of 20 mm Φ . Hence capacity of concrete and dowels is equal to $3.5\,[500 \times 500 - 8 \times 314] + 4 \times 314 \times 130 = 866208 + 163280 = 1029488$ N against actual load of 1200 kN. Hence more dowel bars would be required. Load required to be transferred by dowels $= 1200000 - 866208 = 333792$ N. Providing 8 dowel bars, area of each bar $= \dfrac{333792}{8 \times 130} = 321\ mm^2$. Hence dia. of bar comes out to be 20.2 mm. The code *restricts* the dia. of dowel bars to be 3 mm more than the dia. of column bars. Hence 8 dowel bars to be required, each of 22 mm dia.

In the present case, the load transfer is there without exceeding the permissible bearing stress. However it is always a good practice to extend all the column bars into the foundation and anchored. If dowels bars are provided, development length requirements should be satisfied as required in § 15.4.

The details of reinforcement etc. are shown in Fig. 15.27.

Design Example 15.6 *Design an isolated stepped foundation of column of example 15.5.*

Solution.

1. ***Design constants***

$$k = 0.404 : j = 0.865 \text{ and } R = 0.874$$

2. ***Size of footing***

As determined in example 15.5,

$$B = 3.4 \text{ m} \ ; \ p_0 = 104 \text{ kN/m}^2$$

3. ***Design of footing near column face***

Refer Fig. 15.18.

Maximum bending moment at section $X_1 - X_1$ passing through the face of the column is given by

$$M_1 = \frac{p_0 B}{8} (B - b)^2 \text{ kN-m} = \frac{p_0 B}{8} (B - b)^2 \times 10^6 \text{ N-mm}$$

$$= \frac{104}{8} \times 3.4\,(3.4 - 0.5)^2 \times 10^6 = 3.72 \times 10^8 \text{ N-mm.}$$

Let us give first step at 450 mm from the column face and second step at 950 mm from the column face, as shown in Fig.15.28.

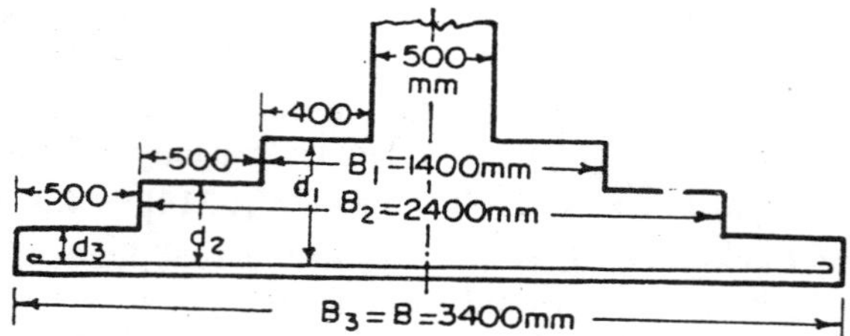

FIG. 15.28.

$$\therefore \quad B_1 = b + (2 \times 450) = 500 + 900 = 1400 \text{ mm}$$

and $$B_2 = b + (2 \times 950) = 500 + 1900 = 2400 \text{ mm}$$

The top width of footing at the column base $= B_1 = 1400$ mm. Hence the effective depth of footing near column face is given by

$$d_1 = \sqrt{\frac{M_1}{RB_2}} = \sqrt{\frac{3.72 \times 10^6}{0.874 \times 1400}} = 551 \text{ mm}$$

Let us check this depth for shear.

For one way shear, the critical plane is located at a distance of d = 551 mm from the column face which lies after the step where thickness d_2 is yet to be found.

For two way shear, the critical plane lies at a distance of $\frac{d}{2}$ $= \frac{551}{2} \approx 275$ mm from the column face where b_0 $= (500 + 2 \times 275) = 1050$ mm. Hence perimeter or the punching shear zone $= 4 \times 1050$. Also, punching shear force is given by

$$F = p_0 [B_3^2 - b_0^2] = 104 [(3.4)^2 - (1.05)^2] = 1088 \text{ kN}$$

$$\therefore \quad \tau_v = \frac{F}{4b_0 d} = \frac{1088 \times 1000}{4 \times 1050 \times 551} = 0.47 \text{ N/mm}^2.$$

Permissible shear stress $= k_s \tau_c$

where $k_s = (0.5 + 1)$ with max. value of 1.

and $\tau_c = 0.16\sqrt{15} = 0.62$ N/mm^2.

Hence actual shear stress is less than permissible shear stress and the depth is adequate.

Area of steel $= \dfrac{M_1}{\sigma_{st} jd} = \dfrac{3.72 \times 10^8}{140 \times 0.865 \times 551} = 5580 \text{ mm}^2$

Using 16 mm Φ bars, No. of bars $= \dfrac{5580}{201} \approx 28$

Hence provide 28 bars in each direction.

Actual $A_{st} = 28 \times 201 = 5628 \text{ mm}^2$

Let us check for development length of bars.

$L_d \approx 58.3\,\Phi = 58.3 \times 16 = 933$ mm.

Providing 50 mm side cover, available length of bar $= \frac{1}{2}(3400 - 500) - 50 = 1400$ mm. $> L_d$. Hence O.K.

5. ***Design of section after first step***

Let the effective depth be d_2 mm.

Width of footing at top $= B_2 = 2400$ mm

B.M. M_2 just at the face of the step is given by

$$M_2 = \frac{p_0 B}{8}(B - B_1)^2$$

$$= \frac{104 \times 3.4}{8}(3.4 - 1.4)^2 \text{ kN–m} = 1.77 \times 10^8 \text{ N–mm}$$

Hence from compression point of view,

$$d = \sqrt{\frac{M_2}{RB_2}} = \sqrt{\frac{1.77 \times 10^8}{0.874 \times 2400}} = 290 \text{ mm} \qquad ...(i)$$

From tension point of view, if the same reinforcement, as provided at the column face, is to be kept, we have

$$d_2 = \frac{M_2}{\sigma_{st}\, j\, A_{st}} = \frac{1.77 \times 10^8}{140 \times 0.865 \times 5628}$$

$$= 260 \text{ mm} \qquad ...(ii)$$

For *One way shear,* taking permissible shear stress $= k\,\tau_c = \tau_c$ (since $k = 1$ for $D > 300$ mm),

$$d_2 = \frac{F}{B_2\, \tau_c} \quad (\text{where } \tau_c = 0.33 \text{ N/mm}^2 \text{ for } p = 0.72\%)$$

where $F = p_0 B\left[\frac{1}{2}(B - B_1) - d_2\right]$; taking $d_2 \approx 290$ mm

$$= 104 \times 3.4\left[\tfrac{1}{2}(3.4 - 1.4) - 0.29\right] = 251 \text{ kN}$$

$$\therefore \quad d_2 = \frac{251 \times 1000}{1400 \times 0.33} = 317 \text{ mm} \qquad ...(iii)$$

For *punching shear,* taking $d_2 = 317$ mm (max. of the above three values),

$$b_0 = B_1 + \frac{d_2}{2} = 1400 + \frac{317}{2} \approx 1560 \text{ mm}$$

$$F = p_0 [B^2 - b_0^2] = 104 [(3.4)^2 - (1.56)^2] = 94.9 \text{ kN}$$

$$\tau_v = \frac{F}{4b_0 d_2} = \frac{949 \times 1000}{4 \times 1560 \times d_2}$$

Taking $\tau_v = k_c \tau_c = \tau_c = 0.16\sqrt{f_{ck}} = 0.16\sqrt{15} = 0.62 \text{ N/mm}^2$

We get $d_2 = \dfrac{949 \times 1000}{4 \times 1560 \times 0.62} = 245$ mm. ...(*iv*)

Hence maximum value of $d_2 = 317$ mm

6. *Design of section after second step*

Let the effective depth be d_3. This is provided upto the edge of the footing. The bending moment M_3 just at the face of the second steps is

$$M_3 = \frac{p_0 B}{8}(B - B_2)^2 = \frac{104}{8} \times 3.4\,(3.4 - 2.4)^2 \text{ kN–m}$$

$$= 44.2 \times 10^6 \text{ N–mm}$$

From compression point of view,

$$d_3 = \sqrt{\frac{M_3}{RB}} = \sqrt{\frac{44.2 \times 10^6}{0.874 \times 3400}} = 122 \text{ mm} \qquad ...(1)$$

From tension point of view, using the same steel,

$$d_3 = \frac{M_3}{\sigma_{st}\, j\, A_{st}} = \frac{44.2 \times 10^6}{140 \times 0.865 \times 5628} = 65 \text{ mm} \qquad ...(2)$$

From *one way shear* point of view,

$$d_3 = \frac{F}{B_3 \tau_c}$$

where $F = p_0 B \left[\frac{1}{2}(B - B_2) - d_3\right]$; taking $d_3 \approx 65$ mm

$$= 104 \times 3.4\left[\tfrac{1}{2}(3.4 - 2.4) - 0.065\right]$$

$$= 153.8 \text{ kN}$$

and $\tau_c = 0.33 \text{ N/mm}^2$ for $p = 0.72\%$ (Table 3.1)

$$\therefore \quad d_3 = \frac{153.8 \times 1000}{3400 \times 0.33} = 137 \text{ mm} \qquad ...(3)$$

For *punching shear,* taking $d_3 = 137$ mm(max. of the above three values), $b_0 = 2400 + \dfrac{137}{2} = 2469$ mm.

$$\therefore \quad F = p_0 [B^2 - b_0^2] = 104 [(13.4)^2 - (2.469)^2] = 568 \text{ kN}$$

$\therefore$ Now $\tau_v = \dfrac{F}{4b_0 d_3}$ or $d_3 = \dfrac{F}{4b_0 \tau_v}$;

Taking $\tau_v = k_s \tau_c = 1 \times 0.16\sqrt{15} = 0.62\ \text{N/mm}^2$,

$$d_3 = \frac{568 \times 1000}{4 \times 2469 \times 0.62} = 9.3\ \text{mm}$$

Hence maximum value of $d_3 = 137$ mm

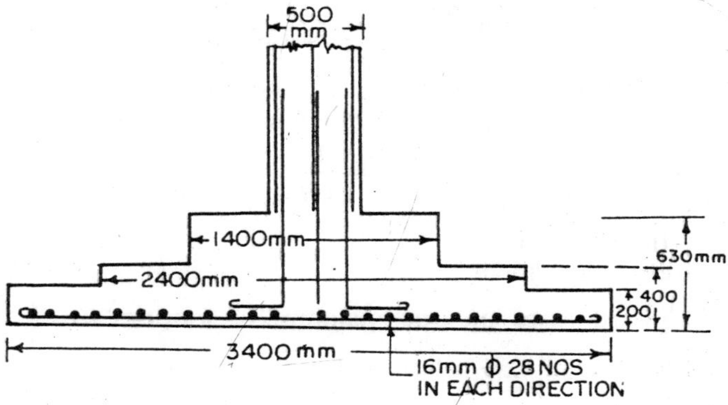

FIG. 15.29.

7. ***Total Thickness***

At the section before first step, $d_1 = 551$ mm. Hence effective depth in the other direction $= 551 + 16 = 567$ mm. Therefore $D_1 = 567 + 60 = 627$. Provide $D_1 = 630$ mm.

At the section after second step, $d_2 = 317$ mm. Hence effective depth in the other direction $= 317 + 16 = 333$ mm and $D_2 = 333 + 60 = 393$ mm. Provide $D_2 = 400$ mm.

For the outer edge $d_3 = 137$ mm for one direction and $137 + 16 = 153$ mm for the other direction. Hence $D_3 = 153 + 60 = 213$ mm. Provide $D_3 = 220$ mm.

Design Example 15.7. *Design a footing for circular column 560 mm diameter, transmitting an axial load of 1200 kN. The column is reinforced with 8 bars of 10 mm diameter. The safe bearing capacity of soil is 120 kN/m². Use M 15 concrete.*

Solution

(a) Circular footing

1. ***Design constants***

For M 15 concrete, we have $c = \sigma_{cbc} = 5\ \text{N/mm}^2$; $t = \sigma_{st} = 140\ \text{N/mm}^2$; $m = 19$; $k = 0.404$; $j = 0.865$; $R = 0.874$.

2. ***Size of footing***

$W = 1200$ kN

Assume $W' = 10\%\ W = 120$ kN

$\therefore$ Area of footing $A = \dfrac{1200 + 120}{120} = 11\ \text{m}^2.$

Radius $R = \sqrt{\dfrac{A}{\pi}} = \sqrt{\dfrac{11}{3.14}} = 1.87$ m.

Adopt radius $R = 1.9$ m or diameter $= 3.8$ m at the base.

Radius of column $r = 280$ mm $= 0.28$ m

Net upward pressure $p_0 = \dfrac{W}{\pi R^2} = \dfrac{1200}{\pi (1.9)^2} = 106\ \text{kN/m}^2$

3. ***Design of footing***

Refer Fig. 15.20.

Upward force on one quadrant of footing $= p_0 \dfrac{\pi}{4}(R^2 - r^2)$

Distance of C.G. of shaded area from the face of the column

$$= \frac{0.2}{R + r}[3R^2 - 2Rr - 2r^2]$$

Bending moment at the face of the column is given by

$$M = p_0 \frac{\pi}{4}(R^2 - r^2) \times \frac{0.2}{R + r}[3R^2 - 2Rr - 2r^2]$$

$$= p_0 \frac{\pi}{20}(R - r)(3R^2 - 2Rr - 2r^2)$$

$$= \frac{106\pi}{20}(1.9 - 0.28)[3(1.9)^2 - 2 \times 1.9 \times 0.28 - 2(0.28)^2]\ \text{kN–m}$$

$$= 2.592 \times 10^2\ \text{kN–m} = 2.592 \times 10^8\ \text{N–mm}$$

Breadth b of quadrant $= \frac{1}{2}\pi r = \frac{\pi}{2}(280) = 440$ mm.

$$\therefore \quad d = \sqrt{\frac{2.592 \times 10^8}{440 \times 0.874}} = 820\ \text{mm}.$$

4. Design for shear

The depth found above from bending compression should be adequate from shear point of view. Circular footings are checked for two way shear(punching shear) at a circular critical section distant $d/2(\approx 410$ mm) away from the face of the column.

$$\therefore \quad r_0 = r + \frac{d}{2} = 280 + 410 = 690\ \text{mm}$$

Punching shear $F = p_0 \pi (R^2 - r_0^2)$

$$= 106\pi[(1.9)^2 - (0.690)^2] = 1043.6\ \text{kN}$$

Hence from Eq. 15.40

$$\frac{F}{2\pi r_0 d_0} \le 0.16\sqrt{f_{ck}}$$

From which $$d_0 = \frac{F}{2\pi r_0 \times 0.16\sqrt{f_{ck}}}$$

$$= \frac{1043.6 \times 1000}{2\pi \times 690 \times 0.16\sqrt{15}} \approx 390 \text{ mm} \quad ...(3)$$

Let the effective depth d_1 at the outer edge be 140 mm. Then the available depth d_0 at the critical plane, corresponding to d =820 mm (found from bending compression) is given by (Fig. 15.30)

$$d_0 = d_1 + \frac{(d - d_1)}{R-r}\left(R - r - \frac{d}{2}\right)$$

$$= 140 + \frac{820-140}{1900-280}\left[1900 - 280 - \frac{820}{2}\right]$$

$$= 648 \text{ mm}$$

This is more than the one found above from punching shear of view.

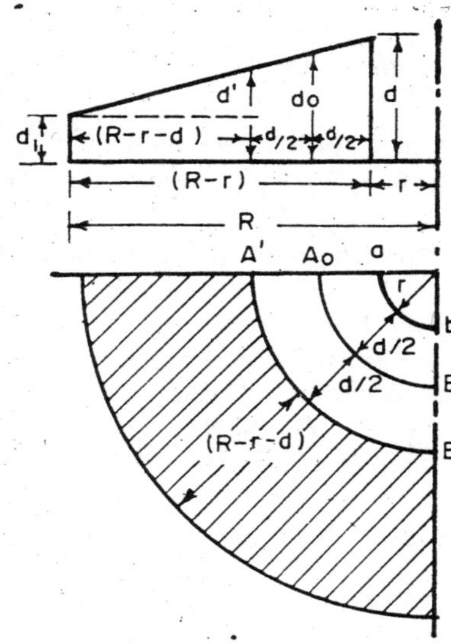

FIG. 15.30.

For *beam shear,* the critical plane $A'B'$ lies at a radial distance d from the column face(Fig. 15.30), where the effective depth d' is given by

$$d' = d_1 + \frac{(d - d_1)}{R - r}[R - r - d]$$

$$= 140 + \frac{820 - 140}{1900 - 280}[1900 - 280 - 820] = 476 \text{ mm}$$

Breadth b' at this section $= \frac{\pi}{2}(r + d)$

$$= \frac{\pi}{2}(280 + 820) = 1728 \text{ mm}$$

Shear force $V = p_0 \frac{\pi}{4}\left\{R^2 - (r + d)^2\right\}$

$$= 106 \times \frac{\pi}{4}[(1.9)^2 - (0.28 + 0.820)^2]$$

$$= 200 \text{ kN.}$$

$$\therefore \qquad \tau_v = \frac{V}{b'd'} = \frac{200 \times 1000}{1728 \times 476} = 0.243 \text{ N/mm}^2$$

Permissible shear stress at $p = 0.72\%$ is $\tau_c = 0.33$ N/mm^2 (Table 3.1).

Hence $d = 820$ mm found on the basis of bending compression is adequate.

5. *Design of reinforcement*

$$A_{st} = \frac{M}{\sigma_{st}\, jd} = \frac{2.592 \times 10^8}{140 \times 0.865 \times 820} = 2611 \text{ mm}^2$$

Using 16 mm Φ bars, $A_\Phi = 201$ mm^2

$\therefore$ No. of bars required in each direction $= \frac{2611}{201} = 13$

However provide 14 bars each, in two mutually perpendicular directions. These bars are to be accommodated in the form of square mesh, centrally located under the column. If B is the size of inscribed square, its magnitude $= \frac{D}{\sqrt{2}} = \frac{2R}{\sqrt{2}} = R\sqrt{2} = 1.9\sqrt{2} \approx 2.7\text{m}$.

Average spacing of bars $= \frac{2700}{13} = 208$ mm c/c

Provide two additional bars in each sector beyond this square. Provide 60 mm cover to centre of steel.

Total depth at edge $= 140 + 60 = 200$ mm.

Total depth at column face $= 820 + 60 = 880 \approx 900$ mm.

6. ***Check for Development Length***

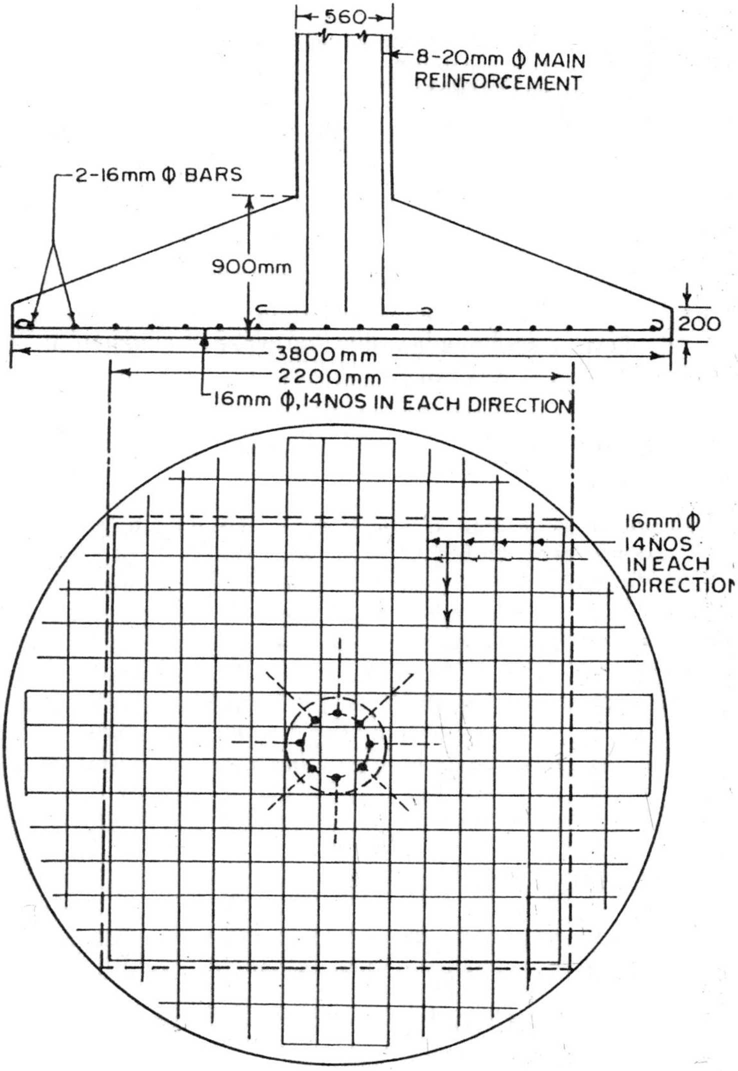

FIG. 15.31.

$$L_d \approx 58.3\,\Phi = 58.3 \times 16 = 933 \text{ mm}$$

Available length $= (R - r) -$ side cover

$$= 1900 - 280 - 60 = 1560 \text{ mm} \quad > L_d$$

Hence safe.

The details of reinforcement etc. are shown in Fig. 15.31.

(*b*) Square footing

For square footing, the width b of an equivalent square column is given by

$$b^2 = \pi r^2$$

$$b = r\sqrt{\pi} = 280\sqrt{\pi} = 496 \approx 500 \text{ mm}$$

Thus the problem reduces to the design of a square footing for a square column of size 500 mm ×500 mm carrying an axial load of 1200 kN. The footing design will be practically the same as worked out in Example 15.6.

15.10. ISOLATED FOOTING SUBJECTED TO ECCENTRIC LOAD

When a column is subjected to axial loading only, the C.G. of column area should coincide with the C.G. of footing area, so that the soil pressure distribution below the footing is *rectangular*. However, a column may be subjected to an axial load along with certain bending moment in one plane, or it may be subjected to an eccentric load. In both the cases, the soil pressure distribution will be trapezoidal if the C.G. of footing area coincides with the C.G. of column as shown in Fig. 15.31. If however, a rectangular pressure distribution is required, the C.G. of footing area may be placed eccentric with the C.G. of the column, by a distance $e = M/W$, or equal to eccentricity of axial load W, as shown in Fig. 15.32.

(*a*) Design of unsymmetrical footing

Fig. 15.32 shows such a case in which the footing is arranged eccentric with respect to column, *i.e.* the axis *Y-Y* of footing is placed at a distance $e\left(=\dfrac{M}{W}\right)$ from the axis of the column. The pressure distribuion will be rectangular.

As before, $B^2 = \dfrac{W + W'}{q_0}$ ...(*i*)

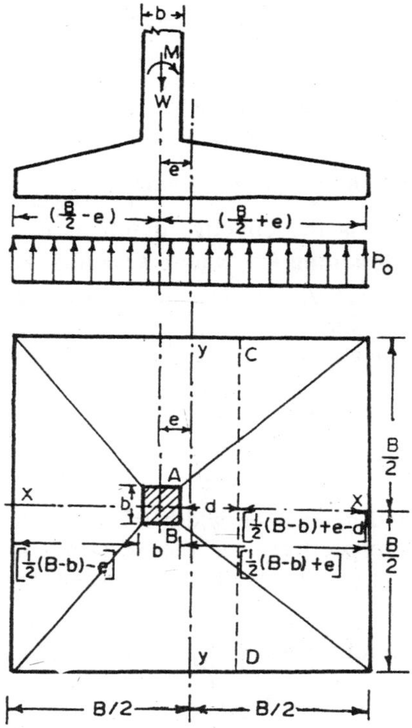

FIG. 15.32.

and $$p_0 = \frac{W}{B^2} \qquad ...(ii)$$

The bending moment will be maximum at face AB of the column, since the cantilever distance in front of this face is more, its value being equal to $\frac{1}{2}\{(B-b\}+e\}$.

$$\therefore \quad M = p_0 B\left\{\frac{1}{2}(B-b)+e\right\} \times \frac{1}{2}\left\{\frac{1}{2}(B-b)+e\right\}$$

$$= \frac{p_0 B}{2}\left\{\frac{1}{2}(B-b)+e\right\}^2 \qquad ...(iii)$$

The section of footing at plane AB will be trapezoidal and its moment of resistance can be determined by the method explained in section § 15.7. Alternatively, the trapezoidal section may be replaced by an equivalent rectangular section width $b + \frac{1}{8}(B-b)$.

Equating the moment of resistance to the external bending moment, the effective depth can be calculated.

The effective depth is also determined(or checked) from two way shear(punching shear), the critical section for which is at distance $d/2$ all round the column face. The width of the section = $b_0 = (b + d)$ and the shear stress will be $\frac{F}{4b_0 d_0}$ where F = punching shear force and d_0 is the effective depth at the section. This shear stress should be less than $k_s \tau_c$ where $k_s = (0.5 + \beta_c)$ but not greater than 1 and $\tau_c = 0.16 \sqrt{f_{ck}}$. The shear force $F = p_0 [B^2 - b_0^2]$.

For *one way shear,* the critical section is located at distance d from column face, where shear force V is given by

$$V = p_0 B \left[\tfrac{1}{2}(B - b) + e - d\right]$$

$$\therefore \qquad \tau = \frac{V}{b' d'}$$

where b' and d' are the width of the section at N.A. and the effective depth respectively, of the section CD (Fig. 15.32).

The shear stress τ_v should not exceed the permissible shear stress τ_c (Table 3.1).

The steel reinforcement , parallel to XX axis is given by

$$A_{st} = \frac{M}{\sigma_{st}\, jd}$$

The steel reinforcement is provided in the other direction, though the bending moment will be less in the other direction because of lesser cantilever projection.

(b) Design of unsymmetrical footing

Fig. 15.33 shows a symmetrically placed footing so that C.G. of footing coincides with that of column. The pressure distribution below the footing will evidently be trapezoidal.

Let B be the size of footing. The maximum stress is equated to the safe bearing capacity q_0 of the soil, to determine the magnitude of B :

$$\frac{W + W'}{B^2} + \frac{M}{B^3/6} = q_0$$

or $$6M + (W + W')B = q_0 B^3 \qquad \text{...(15.41)}$$

The above equation can be solved for B, by trial and error.

The net upward pressure distribution is given by

$$p_0 = \frac{W}{B^2} \pm \frac{M}{B^3/6} \qquad \text{...(15.42)}$$

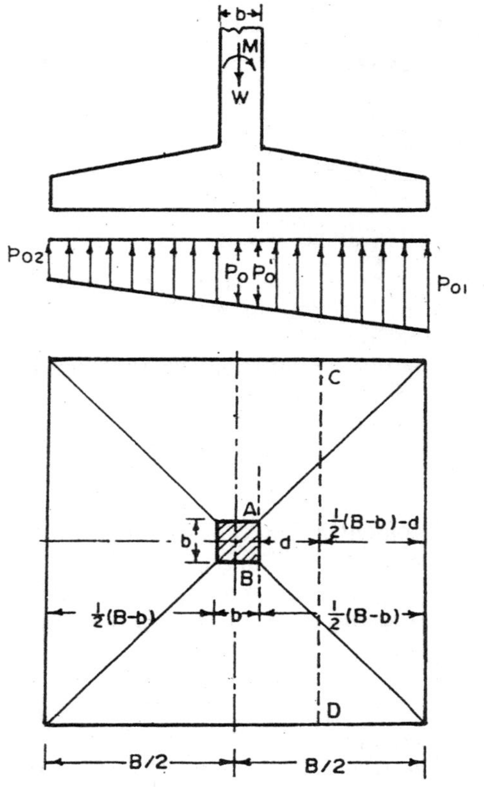

FIG. 15.33

so that pressure p_{01} and p_{02} at the two extremities are known. The lengths of cantilevers on all the four sides will be equal. However, the bending moment will be maximum at the face AB since the pressure intensity is maximum to the right of it. The total upward force F to the right AB is given by

$$F = B\left\{p_0'\left(\frac{B-b}{2}\right) + \frac{1}{2}(p_{01} - p_0')\left(\frac{B-b}{2}\right)\right\}$$

$$= \frac{B(B-b)}{2}\left\{\frac{p_0' + P_{01}}{2}\right\}$$

$$= \frac{B(B-b)}{4}(p_0' + P_{01}) \qquad \text{...(15.43)}$$

Hence B.M. M at AB is given by

$$M = B\left\{p_0'\left(\frac{B-b}{2}\right)\left(\frac{B-b}{4}\right)+\frac{1}{2}(p_{01}-p_0')\left(\frac{B-b}{2}\right)\frac{2}{3}\left(\frac{B-b}{2}\right)\right\}$$

$$= B(B-b)^2\left\{\frac{p_0'}{8}+\frac{p_{01}-p_0'}{12}\right\} = \frac{B(B-b)^2}{24}(2p_{01}+p_0') \quad ...(15.44)$$

where p_0' is the pressure intensity at AB, and is given by

$$p_0' = p_{02} + \frac{p_{01}-p_{02}}{B}\left\{\frac{1}{2}(B-b)+b\right\}$$

$$= p_{02} + \frac{p_{01}-p_{02}}{9B}(B+b) \quad ...(15.45)$$

The maximum bending moment can be equated to the moment of resistance of the section to get d. The effective depth should be sufficient so that punching shear stress is not exceeded. Similarly, effective depth should be sufficient so that nominal shear stress τ_v resulting from one way shear is not exceeded. Steel reinforcement at the plane of maximum B.M. is given by

$$A_{st} = \frac{M}{\sigma_{st}\, jd}.$$

The reinforcement is checked for the development length.

Design Example 15.8 *Design an isolated square footing for a column 500 mm × 500 mm, transmitting a load of 600 kN and a moment of 30 kN-m. The safe bearing capacity of soil is 120 kN/m². Use M 15 concrete.*

Solution.

1. *Design constants*

For M 15 mix, $k = 0.404$; $j = 0.865 : R = 0.874$

2. *Size of footing*

$$W = 600 \text{ kN}$$

Assume $W' = 10\%$ of $W = 60$ kN

Area of footing $= \dfrac{600+60}{120} = 5.5 \text{ m}^2$

$\therefore \quad B = \sqrt{5.5} = 235$ m,

Provide a square footing of size 2.4 m × 2.4 m.

Eccentricity $\quad e = \dfrac{M}{W} = \dfrac{30 \times 1000}{600} = 50$ mm

Hence place C.G. of footing of 50 mm away from the axis of column as shown in Fig. 15.34. Thus the load and moment is equivalent to a load W at the C.G. of the footing.

$$p_0 = \frac{W}{B^2} = \frac{600}{2.4 \times 2.4}$$

$$= 104 \text{ kN/m}^2$$

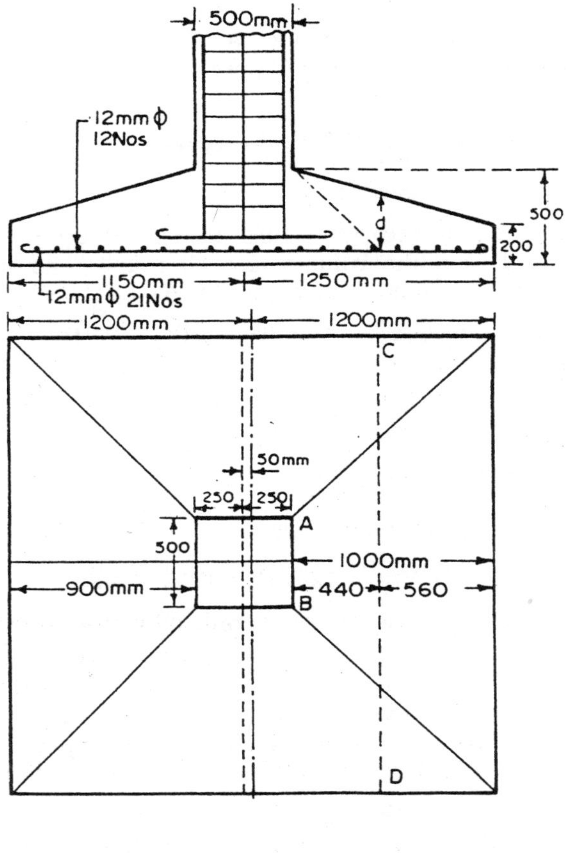

FIG. 15.34.

3. ***Design of footing***

Cantilever length to the right of AB

$$= \frac{1}{2}[2400 - 500] + 50 = 1000 \text{ mm} = 1 \text{ m}$$

Bending moment at the face AB of the column is given by

$$M = p_0 B \frac{(1)^2}{2} = 104 \times 2.4 \times \frac{1}{2} \text{ kN–m}$$

$$= 124.8 \text{ kN–m} = 1.248 \times 10^8 \text{ N–mm}$$

The section of footing at AB will be trapezoidal. However its moment of resistance can be approximately determined by considering it to be a rectangle of width $= b + \frac{1}{8}(B - b)$

$= 500 + \frac{1}{8}(2400 - 500) = 740$ mm.

$\therefore \quad M_r = Rbd^2 = 0.874 \times 740\, d^2$

Equating this to the B.M., we get

$$d = \sqrt{\frac{1.248 \times 10^8}{0.874 \times 740}} = 440 \text{ mm}$$

4. *Check for shear*

For two way shear(punching shear) the section is situated at distance $d/2$ from the column face all round. The side b_0 of the section is $b_0 = b + d = 500 + 440 = 940$ mm.

Shear force $F = p_0\,[B^2 - b_0^2] = 104\,[(2.4)^2 - (0.94)]^2$

$= 507$ kN

$$\therefore \quad \tau_v = \frac{F}{4\,b_0\,d_0}$$

From which $d_0 = \dfrac{F}{4\,b_0\,\tau_v}$

Now permissible shear stress $= k_s\,\tau_c$

$= 1 \times 0.16\sqrt{f_{ck}} = 1.6\sqrt{15} = 0.62$ N/mm²

Limiting τ_v to the above permissible shear stress,

$$d_0 = \frac{507 \times 1000}{4 \times 940 \times 0.62} = 218 \text{ mm.}$$

If an effective depth of 140 mm (total depth =200 mm) is provided at the outer edge, and if effective depth d=440 mm is kept at face AB (found on the basis of bending compression), the available effective depth at $\frac{d}{2}$ from column face will be $=140 + \frac{440 - 140}{1250 - 250} \times (1250 - 250 - 220) = 374$ mm, which is more than the required depth d_0 = 218 mm. Hence the section will be safe in punching shear.

Also, for one way shear, the critical section CD will be at a distance d = 440 mm from the column face. Cantilever length to the right of section CD = 1000 – 440 = 560 mm = 0.56 m. The shear force there is given by

$$V = p_0\,B \times 0.56$$

$$= 104 \times 2.4 \times 0.56 \approx 140 \text{ kN}$$

The section will be trapezoidal in shape as shown in Fig. 15.35. The width at the top is given by

$$b = 500 + \frac{2400 - 500}{100} \times 440 = 1336 \text{ mm}$$

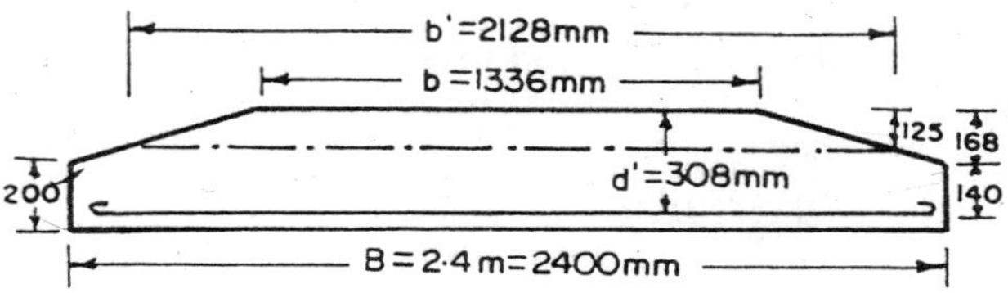

FIG. 15.35.

The effective depth $d' = 140 + \dfrac{440 - 140}{1000} \times 560 = 308$ mm

Depth of N.A. $= kd' = 0.404 \times 308 = 125$ mm

Hence the width of sectional N.A. is given by

$$b = 1336 + \frac{2400 - 1336}{168} \times 125 = 2128$$

Hence $\tau_v = \dfrac{V}{b'd'} = \dfrac{140 \times 1000}{2128 \times 308} = 0.214 \text{ N/mm}^2$.

This is less than the permissible shear stress 0.33 N/mm^2 at $p = 0.72\%$(Table 3.1) corresponding to a balanced section.

Hence $d = 440$ mm found on the basis of the bending compression is adequate.

5 ***Steel reinforcement***

$$A_{st} = \frac{M}{\sigma_{st} jd} = \frac{1.248 \times 10^8}{140 \times 0.864 \times 440} = 2345 \text{ N/mm}^2$$

Using 12 mm Φ bars $A_\Phi = 113 \text{ mm}^2$

$\therefore$ No. of bars $= \dfrac{2345}{113} \approx 21$

Arrange them uniformly spaced in width of 2400 mm. Use the same reinforcement in the perpendicular direction.

6. ***Check for the development length***

$L_d = 58.3\,\Phi = 58.3 \times 12 \approx 700$ mm

Using 60 mm side cover, available length $= 1250 - 50 - 60 = 1160 \text{ mm} > L_d$.

Hence O.K.

The details of reinforcement etc. are shown in Fig. 15.34.

2. ***Size of footing***

$$W = 600 \text{ kN}$$

$$W' = 60 \text{ kN} ; M = 30 \text{ kN-m}$$

Equating the maximum soil pressure to the safe bearing capacity of soil,

$$\frac{660}{B^2} + \frac{30}{B^3/6} = 120$$

or

$$660\,B + 180 = 120\,B^3$$

Solving this by trial and error, we get $B = 2.47$ m

However, provide the footing of size 2.5 m × 2.5 m

The maximum and minimum soil pressure are given by

$$p_{01} = \frac{W}{B^2} + \frac{6M}{B^3} = \frac{600}{(2.5)^2} + \frac{6 \times 30}{(2.5)^3} = 96 + 11.5$$

$$= 107.5 \text{ kN/m}^2.$$

$$p_{02} = \frac{W}{B^2} - \frac{6\,M}{B^3} = 96 - 11.5 = 84.5 \text{ kN/m}^2$$

Pressure intensity under the column axis is

$$p_0 = \frac{1}{2}[107.5 + 84.4] = 96 \text{ kN/m}^2$$

3. ***Design of section for bending compression***

Arrange the footing centrally under the column as shown in Fig. 15.36.

Intensity of net soil pressure below the column face is

$$p_0' = 96 + \frac{107.5 - 96}{1.25} \times 0.25 = 98.3 \text{ kN/m}^2$$

Maximum B.M. will occur at face AB. Cantilever length $\frac{1}{2}(B - b) = \frac{1}{2}(2.5 - 0.5) = 1$ m.

$$\therefore \text{ Total force under cantilever} = (2.5 \times 1)\left[\frac{98.3 + 107.5}{2}\right]$$

$$= 257 \text{ kN}$$

$$\text{Distance of its centroid from } AB = \left(\frac{98.3 + 2 \times 107.5}{98.3 + 107.5}\right)\frac{1}{3}$$

$$= 0.507 \text{ m}$$

$$\therefore \text{ B.M. } M = 257 \times 0.507 = 130 \text{ kN–m} = 1.3 \times 10^8 \text{ N–mm}$$

Alternatively, from Eq. 15.44,

$$M = \frac{B\,(B-b)^2}{24}\,(2\,p_{01} + p_0')$$

$$= \frac{2.5\,(2.5-0.5)^2}{24}\,[2 \times 107.5 + 98.3]$$

$$= 130 \text{ kN–m} = 1.3 \times 10^8 \text{ N–mm.}$$

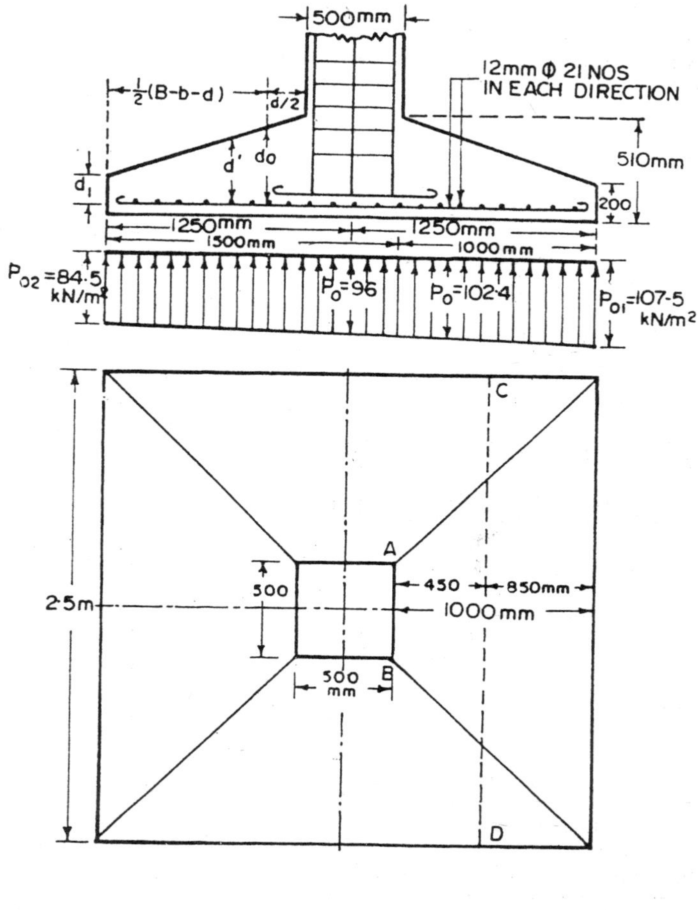

FIG. 15.36.

The section will be trapezoidal in shape. Width of equivalent rectangle $= b + \frac{1}{8}(B - b) = 500 + \frac{1}{8}(2500 - 500) = 750$ mm.

$$\therefore \qquad d = \sqrt{\frac{1.3 \times 10^8}{0.874 \times 750}} = 445 \text{ mm.}$$

Provide d = 450 mm and total depth of 510 mm. Reduce the total depth at edges to 200 mm so that effective depth(d_1) will be 140 mm there.

4. *Design for shear*

The depth found above should be safe for shear. For two way shear(punching shear), the critical plane lies at $\frac{d}{2} \approx 225$ mm from the column face for which width $b_0 = b + \frac{d}{2} = 500 + 225$ $= 725$ mm.

The punching shear stress

$$\tau_v = \frac{F}{4b_0 d_0}$$

where

$$F = \text{Punching shear} = W - p_0 b_0^2$$

$$= 600 - 96\,(0.725)^2 = 550 \text{ kN}$$

$$d_0 = d_1 + \frac{d - d_1}{(B - b)/2}\left(\frac{B - b - d}{2}\right) \qquad \text{...(see Fig. 15.36)}$$

or

$$d_0 = 140 + \frac{450 - 140}{2500 - 500}(2500 - 500 - 225)$$

$$= 415 \text{ mm}$$

$$\therefore \qquad \tau_v = \frac{550 \times 1000}{4 \times 725 \times 415} = 0.457 \text{ N/mm}^2.$$

Permissible shear stress $= k_s \times 0.16\sqrt{f_{ck}}$

$$= 1 \times 0.16\sqrt{15} = 0.62 \text{ N/mm}^2 > \tau_v$$

Hence the thickness found from point of view of bending comression is safe.

For one way shear, the critical plane CD lies at distance d (=450 mm) from column face. The cantilever length to the right of CD = 1.25 – 0.25 – 0.45 = 0.55 m.

Intensity of pressure p_0'' at CD is given by

$$p_0'' = 84.5 + \frac{107.5 - 84.5}{2.5}(2.5 - 0.55)$$

$$= 102.4 \text{ kN/m}^2$$

The section at *CD* will be trapezoidal as shown in Fig. 15.37. The width at top is given by

$$b = 500 + \frac{2500 - 500}{1000} \times 450 = 1400 \text{ mm.}$$

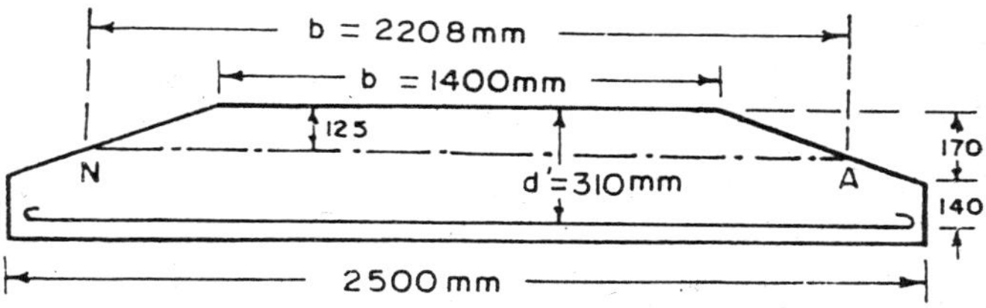

FIG. 15.37.

$$\text{Effective depth } d' = 140 + \frac{450 - 140}{1000} \times 550 = 310 \text{ mm}$$

$$\text{Depth of N.A.} = kd' = 0.404 \times 310 = 125 \text{ mm}$$

∴ width of the section at N.A. is given by

$$b' = 1400 + \frac{2500 - 1400}{170} \times 125 = 2208 \text{ mm}$$

∴ Shear force V' at *CD* is given by

$$V' = 2.5 \times 0.55 \times \tfrac{1}{2}[107.5 + 102.4]$$

$$= 144.3 \text{ kN}$$

$$\therefore \quad \tau = \frac{V'}{b'd'} = \frac{144.3 \times 1000}{2208 \times 310} = 0.21 \text{ N/mm}^2$$

Permissible shear stress at p =0.72%(balanced section)=0.33 N/mm^2 (Table 3.1).

Hence thickness provided from point of view of bending compression adequate.

5. *Design for bending tension*

$$A_{st} = \frac{1.3 \times 10^8}{140 \times 0.865 \times 450} = 2390 \text{ mm}^2$$

$$\therefore \text{ No. 12 mm } \Phi \text{ bars } = \frac{2390}{113} = 21$$

Provide the same reinforcement in the other direction.

6. *Check for development tension*

$$L_d = 58.3\,\Phi = 58.3 \times 12 = 700 \text{ mm}$$

Available length, providing a side a cover of 60 mm

$= 1000 - 60 = 940$ mm. Hence safe.

The details of reinforcement etc. are shown in Fig 15.36.

PROBLEMS

1. A steel stanchion carries 400 mm × 200 mm heavy beam section and carries a load of 1000 kN. Design a R.C.C. base for column. The safe bearing capacity of the soil may be taken as 100 kN/m^2 at a depth of 1 m below ground surface. Use M 15 concrete.

2. A R.C. column 400 mm × 400 mm in section, carries an axial load of 750 kN. Design sloping R.C. footing for the column for the following data , using M 15 concrete.

   (*a*) Permissible tensile stress in steel = 140 N/m^2
   (*b*) Modular ratio = 19
   (*c*) Safe bearing capacity of soil = 120 kN/mm^2

3. Redesign problem 2, if the column is circular in section having diameter = 40 cm. Provide square footings.

4. Redesign problem 3, by providing circular base for the footoing.

5. A rectangular R.C. column, 240 mm × 300 mm carries as axial load of 400 kN. Design a rectangular footing of uniform thickness, if the safe bearing capacity of the soil is 80 kN/m^2. Use M 15 concrete.

6. Design as isolated stepped footing for the column of problem 2.

7. Design a sloped footing for a rectangular column 400 mm × 500 mm carrying an axial load of 800 kN. The safe bearing capacity of soil is 150 kN/m^2. Use M 15 concrete. Take σ_{st} = 140 N/mm^2.

8. Redesign problem 2 if the column also carries a moment of 150 kN-m along with axial load of 750 kN. The footing should be located centraily under the column.

16

Combined Footings

16.1 INTRODUCTION

A combined footing supports the load of two or more adjacent columns. Such a footing is provided under the following circumstances: (*i*) when the columns are very near to each other so that their footings overlap (*ii*) when the bearing capacity of soil is less, requiring more area under individual footing (*iii*) when the end column is near a property line, so that its footing cannot be spread in that direction. A combined footing may be rectangular or trapezoidal in plan. The aim is to get uniform pressure distribution under the footing. For this, the C.G. of footing area should coincide with the C.G. of the combined loads of the two columns. There are four types of combined footings commonly used :

1. Combined rectangular footing
2. Combined trapezoidal footing
3. Strap footing

and 4. Raft footing

Rectangular footing is used when the two columns carry equal loads. If, however, one column near the property line carries heavier load, provision of trapezoidal footing becomes essential. A strap footing consists of spread footings of two columns connected by a strap beam. A raft or mat is a combined footing that covers the entire area beneath a structure and supports all the walls and columns.

16.2 COMBINED RECTANGULAR FOOTING

1. Proportioning of footing

Fig. 16.1 (*a*) shows two columns, with loads W_1 and W_2 spaced at l centre to centre. If W' is the weight of the footing, and q_0 is the safe bearing capacity, the footing area is given by

$$A = \frac{W_1 + W_2 + W'}{q_0} \qquad ...(16.1)$$

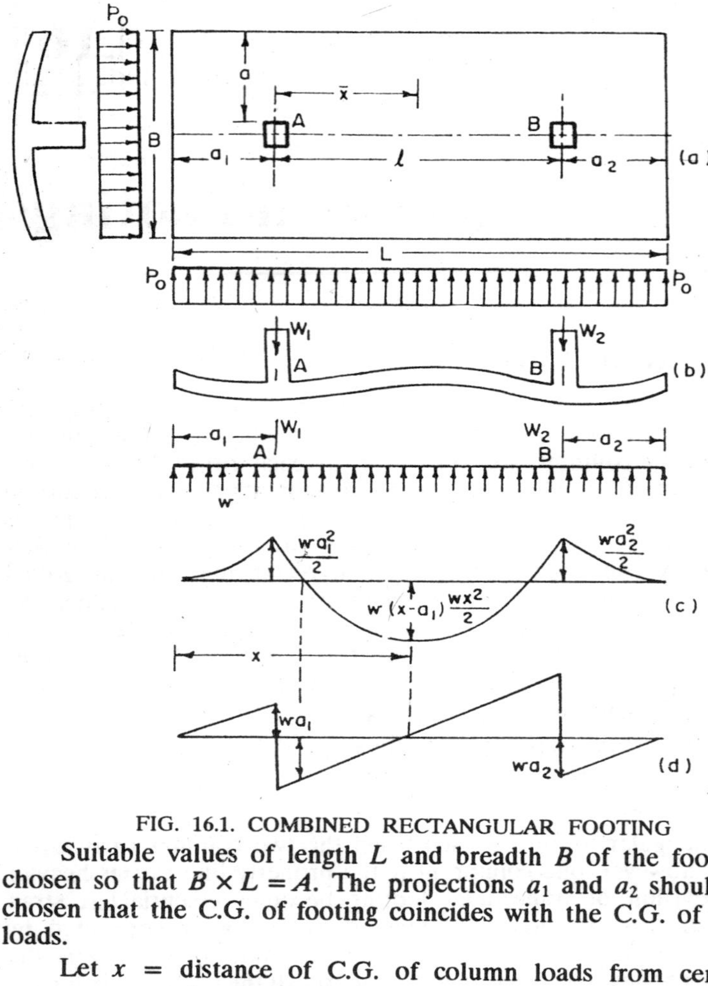

FIG. 16.1. COMBINED RECTANGULAR FOOTING

Suitable values of length L and breadth B of the footing are chosen so that $B \times L = A$. The projections a_1 and a_2 should be so chosen that the C.G. of footing coincides with the C.G. of the two loads.

Let x = distance of C.G. of column loads from centre of column A.

Then
$$a_1 + \bar{x} = \frac{L}{2} = a_2 + l - \bar{x} \qquad ...(16.2)$$

From which a_1 and a_2 can be determined.

The net upward pressure p_o is given by

$$p_0 = \frac{W_1 + W_2}{L.B} \qquad ...(16.3)$$

2. Bending pattern

The footing will bend both in longitudinal as well as transverse directions, as shown. In the longitudinal direction, the footing will

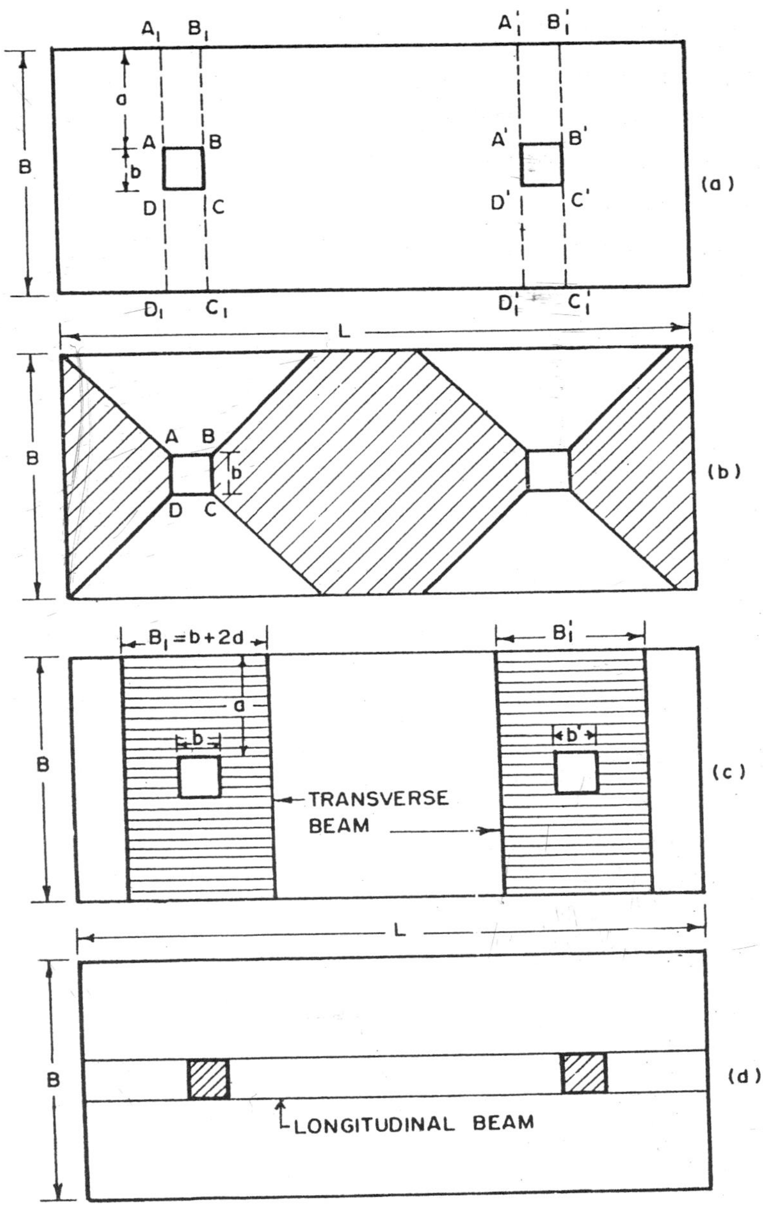

A1 B1
A'1 B'1
a
A B
b
D C
A' B'
D' C'
B
(a)
D1 C1
D'1 C'1
L
A B
b
D C
B
(b)
B1 = b+2d
B'1
a
b
b'
B
TRANSVERSE BEAM
(c)
L
B
(d)
LONGITUDINAL BEAM

FIG. 16.2

have sagging bending moment in the two cantilever portions, and hogging bending moment in some length of middle portion. In the transverse direction the footing will have sagging bending moment. Near the columns, therefore, the footing will have a tendency to bend in the form of saucer. The *transverse bending will decrease at distance away from the column, where the bending will be primarily longitudinal.* Hence sections near and around the column will be subjected to heavier bending stress.

(a) Design for longitudinal bending

The exact analysis of such a footing, having no transverse or longitudinal beam, is very difficult since the bending pattern is not precisely known at all points. The design is based primarily on empirical practices. One of the commonly adopted practice is to consider the footing as a longitudinal beam of width B. Thus, the footing will be a beam loaded with uniformly distributed load $w = p_0 B$ per unit length and supported on two columns with reactions W_1 and W_2 as shown in Fig 16.1 (*c*). The B.M. and S.F. diagrams are shown in Fig. 16.1(*d*) and (*e*) respectively. The maximum hogging bending moment will occur betweeen W_1 and W_2 where S.F. is zero. The maximum sagging moment at the column axis will be $\frac{w a_1^2}{2}$, while the maximum hogging bending moment will be $W_1 (x - a_1) - \frac{w a_1^2}{2}$, where x is the distance of the section, from the edge, where shear is zero, However, the footing will be designed for sagging moment at the *outer face* of each column, and for the maximum hogging moment. The reinforcement will be placed on the bottom face for sagging bending moment and on the top face for the hogging moment.

The above analysis is based on the assumption that full width B is available for longitudinal bending. However, in actual bending full width B will not be available for the whole length of the footing. Fig. 16.2(*a*) shows a critical strip $A_1 B_1 C_1 D_1$ for transverse bending.

The total shear force on the two planes $A_1 D_1$ and $B_1 C_1$ plus the upward soil reaction on the strip $A_1 B_1 C_1 D_1$ will be equal to the column load W_1. Thus, the strip $A_1 B_1 B A$ will bend up like a cantilever, with a total uniformly distributed upward force equal to half the column load. If p_0' is the net intensity of upward pressure on the strip, we have $p_0 = \frac{W_1}{\text{Area } A_1 B_1 C_1 D_1}$

If the strip $A_1 B_1 B A$ is designed as cantilever with an upward pressure p_0 only, it will be over-strained and will yield, and then the effective available *width* for longitudinal bending will be *AD* only and not the full width B. The available width however will increase from $AD=b$ at column face to B at distance B from the column face. Fig. 16.2 (*b*) shows the *probable* available widths shown shaded, for longitudinal bending all along the length.

There are three alternatives for the design of footing for longitudinal bending : (*i*) to design the footing for width equal to $AD=b$ at the column face, (*ii*) to provide a transverse beam of width B_1, of greater thickness then the rest of the slab, or (*iii*) to provide a longitudinal beam connecting the columns.

Alternative (i)

If the footing is designed as longitudinal beam with a width b only, at the column face, its thickness required for compression to resist bending moment at the face of the column will be excessive and uneconomical, specially when the cantilever length is more. For the hogging moment, however, full width L may be used. This method may be adopted only when the cantilever projections a_1 or a_2 are small, so that the sagging moments are much less than the hogging moment.

Alternative (ii) : Transverse beam

In order that strip $A\,B\,B_1A_1$ may not yield, another alternative is to provide a beam of width B_1 as shown in Fig. 15.39(*c*) to transfer the load W_1 of the column to the soil safely. In that case, the shaded portions in Fig. 16.2(*c*) will bend transversely, while the remaining portion will bend longitudinally as a beam utilizing full width B for compression resistance. The width B_1 of the transverse beam may vary from a value b to a value B. If the width of transverse beam $A_1B_1C_1D_1$ is kept equal to b only, its thickness required will be much more than the rest of the thickness of the footing. On the other hand, if the width B_1 is made equal to width B, each portion under the two transverse beams of columns A and B will act as individual footings. An intermediate value of B_1, between b and B, should therefore be chosen. Failure due to shear occurs approximately along 45° lines when footings are tested to destruction. This suggests that width of the transverse beam should be taken as $B_1 = b + 2d$, but not more than B. (Fig. 16.3).

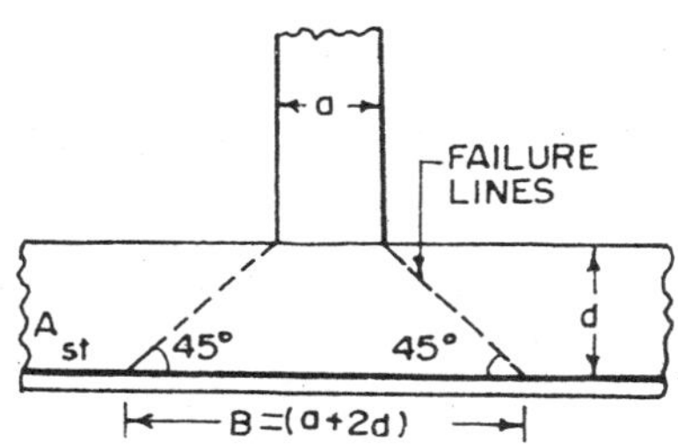

FIG. 16.3

After having determined B_1, the upward soil pressure p_0' for transverse bending will be $= \frac{W'}{B_1 \times B}$. The thickness of the transverse beam should be found for bending moment at column face equal to $(p_0 B_1)\, a^2/2$. This thickness will normally be greater than the thickness d for the rest of the footing. Even if both the thicknesses are equal, the reinforcement in the transverse direction should be found for the above B.M., and provided in width B_2. Similar reinforcement for width B_1 for the other column should be provided. For the rest of the portion of footing, minimum transverse reinforcement @ 0.15% of the area of cross-section of the footing should be provided. In the longitudinal direction, the footing will be designed for B.M. and S.F. shown Fig. 16.1(*d*) and (*c*).

Alternative (iii) Longitudinal beam

A third and more common alternative is to provide a longitudinal beam, all along the length, joining the two columns as shown in Fig. 16.2(*d*). In this case, the beam alone will be subjected to bending moments and shear forces shown in Fig. 16.1(*d*),(*e*) in the longitudinal direction. The slab of the footing on either side of the beam is designed as cantilever, for the transverse bending. The web of the longitudinal beam should project *below* the slab if the sagging moments in the beams are more, so that T-beam action is available in the cantilever portions of the beam. If, however, the hogging bending moment is more, the web of the beam should project at the top of the slab so that T-beam action is available for central portion.

3. Two way shear (punching shear)

The depth found on the basis of bending moment should always be tested for punching shear, so that punching shear stress does not exceed the permissible value. The critical section will be at distance *d*/2 from the column face.

4. One way shear

The footing should be tested for one way shear at a distance *d* from the two column faces, in the respective cantilever portions, as shown in Fig. 16.4. For the central portion, two locations, *F* and *G*, should be tried for diagonal tension. At *F*, distant *d* from column face, tension crack will occur only if sagging bending moment occurs at *F*. Another possibility of tension crack is at *G*, just to the right of point of inflexion, so that there is hogging bending moment giving rise to a crack on the top face. The corresponding S.F. at *G* can be determined. Out of the three points *E*, *F* and

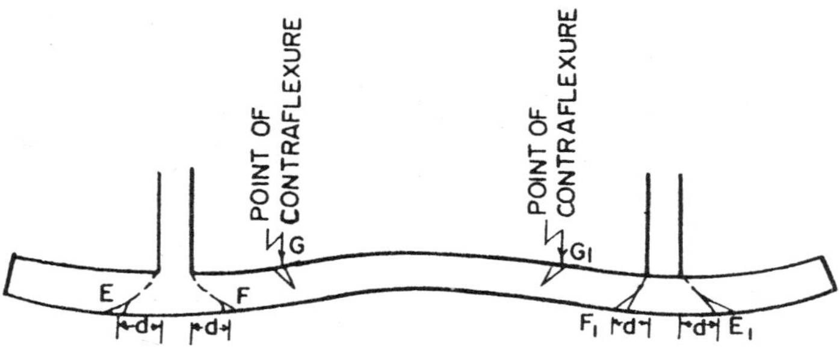

FIG. 16.4.

G, the diagonal tensional crack may occur at a point which is *nearer* to the column, face, or at which S.F. is more. If the shear stress comes out to be more than permissible one, shear stirrups should be provided for the required portion. However, nominal shear stirrups should always be provided throughout the length of the beam.

5. Development length

The reinforcement should be tested for development length.

16.3. COMBINED TRAPEZOIDAL FOOTING

Combined trapezoidal footing for two columns becomes essential when the heavily loaded column is near the property line, or when there is some restriction on the total length of the footing. The widths B_2 and B_1 are so proportioned that the C.G. of footing coincides with the C.G. of column loads.

Fig. 16.5 shows two columns with loads W_1 and W_2, spaced at distance l apart. Let L the length of the footing, and a_1 and a_2 be the cantilever projections. The widths B_1 and B_2 are unknown.

$$\therefore \quad \frac{(B_1 + B_2)\,L}{2} = \frac{W_1 + W_2 + W'}{q_0}$$

$$B_1 + B_2 = \frac{2\,(W_1 + W_2 + W')}{q_0\,L} \qquad ...(1)$$

$$\text{Distance } \bar{x} \text{ of C.G. of load} = \frac{W_2\,l}{W_1 + W_2}$$

$\therefore$ Distance of C.G. of load from short edge

$$= a_1 + \frac{W_2 l}{W_1 + W_2} \quad ...(i)$$

Distance C.G. of trapezium from the short edge

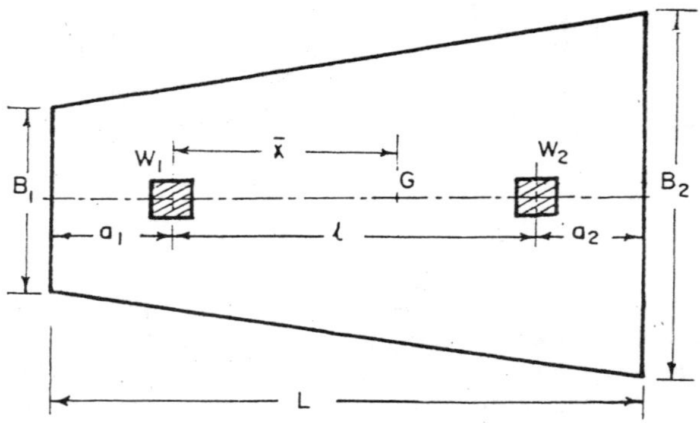

FIG 16.5.

$$= \frac{L}{3}\left(\frac{B_1 + 2B_2}{B_1 + B_2}\right) \quad ...(ii)$$

Equating the two, $a_1 + \dfrac{W_2 l}{W_1 + W_2} = \dfrac{L}{3}\left(\dfrac{B_1 + 2B_2}{B_1 + B_2}\right)$ ...(16.5)

From Eqs. (16.4) and (16.5), the unknowns B_1 and B_2 are determined. The net upward soil pressure intensity p_0 will be uniform throughout, and its magnitude is given by

$$p_0 = \frac{W_1 + W_2}{\frac{1}{2}(B_1 + B_2) L} \quad ...(16.6)$$

Knowing p_0, B.M. and S.F. diagrams may be drawn. The footing is designed exactly in the manner as for rectangular footing. Transverse reinforcement, for transverse bending, in provided near each column, in strip equal to width of column plus twice the effective depth of footing.

Design Example 16.1 *Design combined rectangular footing for two columns A and B, carrying loads of 500 and 700 kN respectively. Column A is 300 mm × 300 mm in size and column B is 400 mm×400 mm in size. The centre to centre spacing of the columns is 3.4 metres.*

The safe bearing capacity of soil may be taken as 150 N/m². Use M 15 mix. Take σ_{st} = 140 N/mm².

Solution.

1. *Design constants*

For M 15 mix. and $\sigma_{st} = 140$ N/mm², we have

$$k = 0.404\ ; j = 0.865 \text{ and } R = 0.874$$

2. *Size of footing*

$$W_1 = 500 \text{ kN and } W_2 = 700 \text{ kN}$$

Let the weight of footing= $W' = 10\%$ of $(W_1 + W_2) = 120$ kN

$$\therefore \quad A = \frac{500 + 700 + 120}{150} = 8.8 \text{ m}^2$$

Let the size of the footing be 1.8 m × 5 m

The projections a_1 and a_2 should be such that C.G. of footing coincides with the C.G. of column loads. The distance $\bar{x}$ of the C.G. of column loads from centre of column A is given by

$$\bar{x} = \frac{W_2 l}{W_1 + W_2} = \frac{700 \times 3.4}{500 + 700} \approx 2 \text{ m}$$

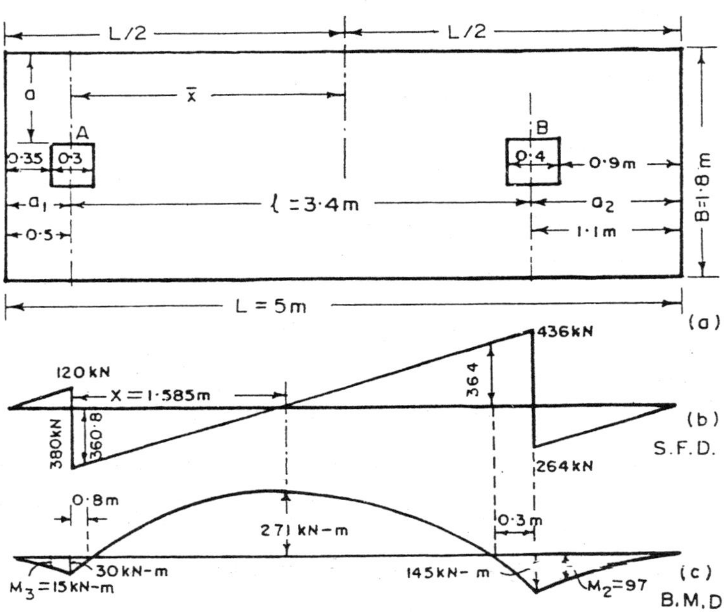

FIG. 16.6

$\therefore \qquad a_1 + \bar{x} = \frac{L}{2}$

or $\qquad a_1 = \frac{L}{2} - \bar{x} = \frac{5}{2} - 2 = 0.5 \text{ m}$

$$a_2 = L - (l + a_1) = 5 - (3.4 + 0.5) = 1.1 \text{ m}$$

The net upward pressure p_o is given by

$$p_0 = \frac{W_1 + W_2}{B \times L} = \frac{500 + 700}{1.8 \times 5} = \frac{400}{3} \text{ kN/m}^2$$

$\therefore$ pressure w per metre length

$= p_0 B = \frac{400}{3} \times 1.8 = 240 \text{ kN/m}^2$

3. *B.M. and S.F. diagrams*

The B.M. and S.F. diagrams in the longitudinal directions are shown in Fig. 16.6.

S.F. just to the left of $A = 240 \times 0.5 = 120$ kN

S.F. just to the right of $A = 500 - 120 = 380$ kN

S.F. just to the right of $B = 240 \times 1.1 = 264$ kN

S.F. just to the left of $B = 700 - 264 = 436$ kN

S.F. will be zero at distanace x from A, its magnitude being given by $x = \frac{380}{240} = 1.585$ m.

The maximum hogging B.M. M_1 will therefore occur at this section, its magnitude being given by

$$M_1 = (500 \times 1.585) - \frac{240}{2}(0.5 + 1.585)^2$$

$$\approx 271 \text{ kN-m} = 2.71 \times 10^8 \text{ N-mm}$$

Sagging B.M. at $B = \frac{240\,(1.1)^2}{2} = 145 \text{ kN-m} = 1.45 \times 10^8 \text{ N-mm}$

Sagging B.M. at $A = \frac{240}{2}(0.5)^2 = 30 \text{ kN-m} = 0.3 \times 10^8 \text{ N-mm}$

Sagging B.M. at the outer face of column B is

$M_2 = \frac{240}{2}(0.9)^2 = 97 \text{ kN-m} = 0.97 \times 10^8 \text{ N-mm}$

Sagging B.M. at the outer face of column A is

$M_3 = \frac{240}{2}(0.35)^2 = 15 \text{ kN-m} = 0.15 \times 10^8 \text{ N-mm}$

Let the point of contraflexure occur at x from the centre of column A.

Then $M_x = 500x - \frac{240}{2}(x + 0.5)^2 = 0$

which gives $x = 0.08$ or 3.08 m

Hence first point of contraflexure occurs at 0.08 m to the right centre of column A and the second one occurs at 3 m from A, or $3.4 - 3.1 = 0.3$ m to the left of centre of column B.

S.F. at the first point of contraflexure

$$= 380 - (240 \times 0.08) = 360.8 \text{ kN}$$

S.F. at the second point of contraflexure

$$= 436 - (240 \times 0.3) = 364 \text{ kN}$$

4. ***Effective depth from bending compression***

The effective depth d for hogging B.M. M_1 is given by

$$d = \sqrt{\frac{2.71 \times 10^8}{0.874 \times 1800}} = 416 \text{ mm}$$

5. ***Check for punching shear***

Let us check the above depth for punching shear, for which the critical plane lies at $d/2 = 208$ mm around column B. Width b_o of critical plane $= 400 + 208 = 608$ mm.

Punching shear force at column B $= 700 - \frac{400}{3} \times (0.608)^2$

$$\approx 651 \text{ kN.}$$

$\therefore$ Shear stress $= \frac{651 \times 1000}{4 \times 608 \times 416} = 0.643 \text{ N/mm}^2$.

Permissible shear stress $= k_s \tau_c = 1 \times 0.16 \sqrt{f_{ck}}$

$$= 1 \times 0.16 \sqrt{15} = 0.62 \text{ N/mm}^2$$

This is less than the actual shear stress. Hence the depth found from bending compression is inadequate.

The required depth $d = \frac{651 \times 1000}{4 \times 608 \times 0.62} = 432$ mm.

Using 60 mm cover to the centre of reinforcement, $D = 432 + 60 = 492$ mm.

However, provide $D = 500$ mm so that available $d = 500 - 60 = 440$ mm.

6. ***Design for bending tension in longitudinal direction***

(*i*) ***Reinforcement for hogging bending moment*** M_1

$$A_{st} = \frac{M_1}{\sigma_{st}\, jd} = \frac{2.71 \times 10^8}{140 \times 0.865 \times 440} = 5086 \text{ mm}^2$$

Using 16 mm Φ bars, having $A_\Phi = \frac{\pi}{4}(16)^2 = 201 \text{ mm}^2$

$\therefore$ No. of bars $= \frac{5086}{201} \approx 25.3 \approx 26.$

Thus provide 26 bars uniformly distributed over the width of 1.8 m over top face.

At the point of contraflexure near B, shear force $F = 364$ kN.

Let n =no. of 16 mm Φ bars required at the point of contraflexure, from the point of requirements of Code provisions inherent in the equation

$$\frac{M_1}{V} + L_0 \geq L_d$$

where M_1 = moment of resistance of the section $= A_{st}\, \sigma_{st}\, jd$

$= n \times 201 \times 140 \times 0.865 \times 440$

$= n \times 10.71 \times 10^6$ N-mm

V = S.F. at point of contraflexure $= 364 \times 10^3$ N

L_o = d for 12 Φ whichever is greater

= 440 or (12 × 16) which is greater

= 440 mm

$L_d \approx 58.3\, \Phi = 58.3 \times 16 = 933$ mm

$$\therefore \quad \frac{M_1}{V} + L_0 \geq L_d$$

or

$$\frac{n \times 10.71 \times 10^6}{364 \times 10^3} + 440 = 933$$

From which $n = 16.8 \approx 17$

However, all the 26 bars are taken up to the outer faces of both the columns. Out of these, 18 bars are stopped and 8 bars are taken straight upto either edge of footing. These bars will serve as anchorage for the stirrups.

(*ii*) ***Reinforcement for sagging B.M. M_2 is given by***

$$A_{st2} = \frac{M_2}{\sigma_{st}\, jd} = \frac{0.97 \times 10^8}{140 \times 0.865 \times 440} = 1820 \text{ mm}^2$$

Using 12 mm Φ bars, No. of bars $= \frac{1820}{113} = 16.1 \approx 17$

These no. of bars should be sufficient from development length point of view so as to satisfy the following criterion at the point of contraflexure near the inner face of column B.

$$\frac{M_1}{V} + L_0 \geq L_d$$

Let n = number of bars required to satisfy this criterion

$\therefore$ $M_1 = A_{st}\,\sigma_{st}\,jd = (n \times 113)\ 140 \times 0.865 \times 440$

$= n \times 6.02 \times 10^6$ N-mm

V = S.F. = 364 kN = 3.64×10^5 N

$L_0 = d$ or $12\ \Phi$ whichever is more = 440 mm.

$L_d = 58.3\ \Phi = 58.3 \times 12 = 700$ mm.

Substituting the values, we get

$$\frac{n \times 6.02 \times 10^6}{3.69 \times 10^5} + 440 = 700$$

which gives $n = 15.7 \approx 16$.

Hence 17 bars of 12 mm Φ will be sufficient.

These bars should be extended by l_o (=440 mm) beyond the point of contraflexure. At this point curtail 9 bars and continue 8 bars throughout the length to serve an anchor bars for stirrups.

(iii) Reinforcement for sagging B.M. M_3

$$A_{st3} = \frac{0.15 \times 10^8}{140 \times 0.865 \times 440} = 282 \text{ mm}^2$$

Minimum reinforcement @ 0.15% of cross-sectional area

$$= \frac{0.15 \times 500 \times 1800}{100} = 1350 \text{ mm}^2$$

$\therefore$ No. of 12 mm Φ bars $= \dfrac{1350}{113} = 12$

Let us find the reinforcement from development length point of view so that the following relation is satisfied at the point of contraflexure near the inner face of column A :

$$\frac{M_1}{V} + L_0 \geq L_d$$

Let n = number of bars required to satisfy the above criterion

$M_1 = A_{st}\,\sigma_{st}\,jd = (n \times 113) \times 140 \times 0.865 \times 440$

$= n \times 6.02 \times 10^6$ N-mm

V = shear force at point of contraflexure

$= 360.8$ kN $= 360.8 \times 10^3$ N

$L_0 = 12\ \Phi$ or d whichever is more = 440 mm

$L_d = 58.3\ \Phi = 58.3 \times 12 = 700$ mm.

Substituting the values,

$$\frac{n \times 6.02 \times 10^6}{360.8 \times 10^3} + 440 = 700$$

From which $n = 15.6 \approx 16$.

Continue these upto 440 mm beyond the point of contraflexure. At this point, curtail 9 bars and continue remaining 8 bars to serve as anchor bars for shear stirrups.

6. *Check for one way shear*

In the cantilever portion, test for one way shear (diagonal tension) is made at a distance $d = 0.44$ m from the column face or at $0.44 + 0.20 = 0.64$ m from the centre of column *B*. Shear $V = 264 - 240 \times 0.64 = 110.4$ kN.

$$\therefore \qquad \tau_v = \frac{110.4 \times 1000}{1800 \times 440} = 0.14 \text{ N/mm}^2$$

Permissible shear stress $= k\,\tau_c = \tau_c = 0.33$ N/mm² for $p = 0.72\%$ (balanced section), obtained from Table 3.1.

Hence safe.

Similarly, in the cantilever portion near column *A*, the footing will be safe in shear.

For diagonal tension between *A* and *B*, near column *B*, crack can occur at the bottom face of the footing(*i.e.* for sagging B.M.) at a distance of $d = 440$ mm, or at the top of footing (*i.e.*, for the hogging B.M.) at the point of contraflexure distant 300 mm from the centre of the column, whichever is nearer. Shear force at the point of contraflexure = 364 kN which is more.

$$\therefore \qquad \tau_v = \frac{364 \times 1000}{1800 \times 440} = 0.46 \text{ N/mm}^2$$

which is more than the permissible shear stress $\tau_c = 0.33$ N/mm². Hence shear reinforcement is necessary.

Using 12 mm Φ 8-legged stirrups $A_{sv} = 8 \times 113 = 904$ mm²

$$\text{Spacing} \quad s_v = \frac{\sigma_{sv} A_{sv}\, d}{V_s} = \frac{140 \times 904 \times 440}{364 \times 1000}$$

$$= 153 \text{ mm}$$

Hence provide there @ 150 mm c/c

The S.F. for which stress is 0.33 N/mm² is $= \dfrac{0.33}{0.46} \times 364$

$= 261$ kN

This occurs at a distance of $\frac{436 - 261}{240} = 0.73$ m from the centre of the column B or 0.53 m from the inner face of the column. Hence provide at least 5 stirrups @ 150 mm c/c, for a distance of 0.6 m from the face of the column B.

Similarly, near column A, tension crack can occur in the hogging portion at the inner face of the column since the point of contraflexure happens to fall under the column. S.F. at the inner face $= 380 - 240 \times 0.15 = 344$ kN

$$\therefore \qquad \tau_v = \frac{344 \times 1000}{1800 \times 440} = 0.434 \text{ N/mm}^2$$

against a permissible value of 0.33 N/mm². Hence shear reinforcement is necessary.

Hence provide 5 Nos. 12 mm Φ 8-legged stirrups @ 150 mm c/c, for a distance of 600 mm from the inner face of the column.

Spacing of nominal stirrups is given by

$$s_v = \frac{2.5\, A_{sv} f_v}{b} = \frac{2.5 \times 904 \times 250}{1800}$$

$$= 314 \text{ mm}$$

Hence for the rest of the length, provide 12 mm Φ 8-lgd nominal stirrups @ 300 mm c/c.

7. *Transverse reinforcement*

The footing will bend transversely near each column face. Projection a beyond the face of column $A = \frac{1}{2}(1800 - 300) = 750$ mm. Width B_1 of bending strip $= b + 2d = 300 + 2 \times 440 = 1180$ mm. However, width available to the left of outer face of column $A = 500 - 150 = 350$ mm only instead of 440 mm. Hence available $B_1 = 350 + 300 + 440 = 1090$ mm $= 1.09$ m.

Net upward pressure $p_0' = \frac{W_1}{B \times B_1} = \frac{500}{1.8 \times 1.09} = 254.8$ kN/m²

Maximum B.M. will occur the face of the column.

$$\therefore \qquad M = p_0' \frac{a^2}{2} = 254.8 \frac{(0.75)^2}{2} = 71.66 \text{ kN-m}$$

$$= 71.66 \times 10^6 \text{ N-mm}$$

$$\therefore \qquad d = \sqrt{\frac{71.66 \times 10^6}{0.874 \times 1000}} = 286 \text{ mm.}$$

Hence actual $d = 440$ mm provided earlier is sufficient. The transverse beam will thus be of the same thickness as the rest of the footing. Since transverse reinforcement will be provided *above* the longitudinal one, available $d = 440 - 12 =$ 428 mm.

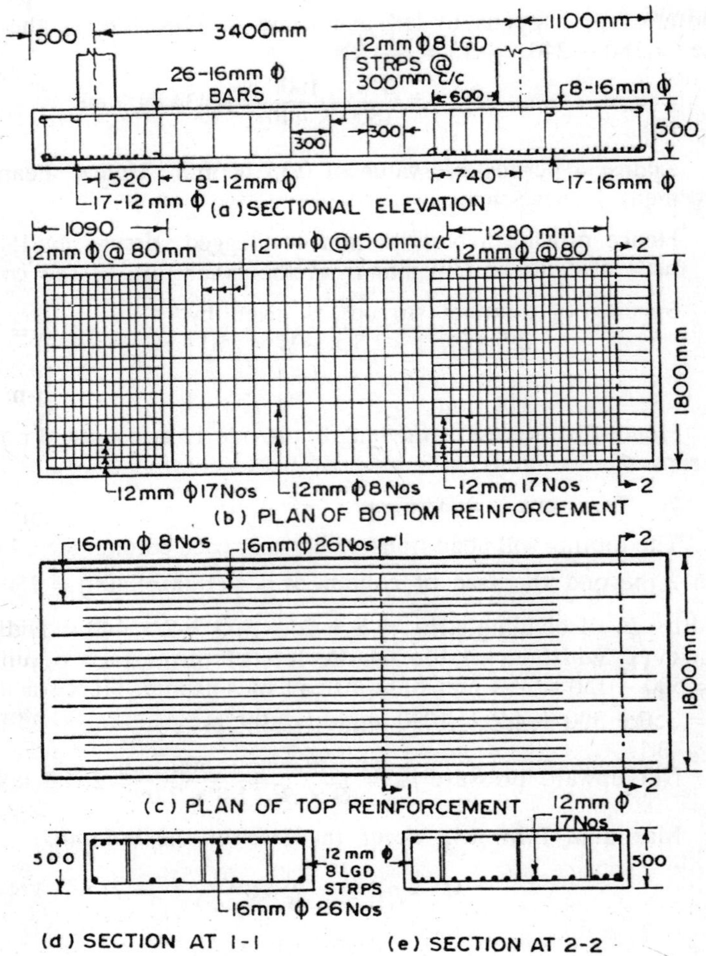

FIG. 16.7

The transverse reinforcement is given by

$$A_{st} = \frac{71.66 \times 10^6}{140 \times 0.865 \times 428} = 1383 \text{ mm}^2$$

Using 12 mm Φ bars, spacing is given by

$$s = \frac{1000\, A_\Phi}{A_{st}} = \frac{1000 \times 113}{1383} = 82 \text{ mm}$$

Provide these @ 80 mm c/c. The reinforcement provided above should have sufficient development length.

$L_d \approx 58.3\,\Phi = 58.3 \times 12 \approx 700$ mm.

Using 50 mm clear cover on the sides, length of bar available = 750 − 50 + anchorage value of hook. Hence safe.

Similarly, for column B, width B_1 for transverse beam = 400 + 2 × 440 = 1280 mm.

$$\therefore \quad p_0' = \frac{W}{B \times B_1} = \frac{700}{1.8 \times 1.28} = 304 \text{ kN/m}^2$$

Cantilever projection $a = \frac{1}{2}[1800 - 400] = 700 \text{ mm} = 0.7$ m

$$\therefore \quad M = p_0' \frac{a^2}{2} = 304 \frac{(0.7)^2}{2} = 74.5 \text{ kN-m}$$

$$= 75.5 \times 10^6 \text{ N-mm}$$

$$A_{st} = \frac{74.5 \times 10^6}{140 \times 0.865 \times 428} = 1437 \text{ mm}^2$$

Using 12 mm Φ bars, $s = \dfrac{1000 \times 113}{1437} \approx 80$ mm.

Hence provide these @ 80 mm c/c in the strip of width 1280 mm. The available length of the bars will just satisfy the requirements of the development length.

For the rest of the footing, provide transverse reinforcement @ 0.15% of the area of cross-section.

$$\therefore \quad A_{st} = \frac{0.15}{100}(1000 \times 113) = 750 \text{ mm}^2$$

$$\therefore \quad \text{spacing of 12 mm } \Phi \text{ bars} = \frac{1000 \times 113}{750} = 150 \text{ mm}$$

The details of reinforcement etc. are shown in Fig. 15.44.

Design Example 15.11. *Two reinforced concrete columns 400 mm × 400 mm in section carry a load of 1000 kN each, inclusive of* *the self-weight. Design a combined footing having central beam joining the columns. The centre to centre spacing of the columns is 4 metres. The safe bearing capacity of soil is 150 kN/m². Use M 15 mix.*

Solution.

1. ***Design constants***

For M 15 concrete, we have the following :

$\sigma_{cbc} = 5\ \text{N/mm}^2$; $\sigma_{st} = \sigma_{sv} = 140\ \text{N/mm}^2$; $k = 0.404$; $j = 0.865$; $R = 0.874$.

2. ***Proportioning of footing***

$$W_1 = W_2 = W = 1000 \text{ kN}$$

Let the weight of footing $W' = 0.1\,(1000 + 1000) = 200$ kN

$$\therefore \quad \text{Area of footing} = \frac{1000 + 1000 + 200}{150} = 14.65 \text{ m}^2$$

For the beam, upward load w per unit length

$$= \frac{1000 + 1000}{L} = \frac{2000}{L} \text{ kN/m}$$

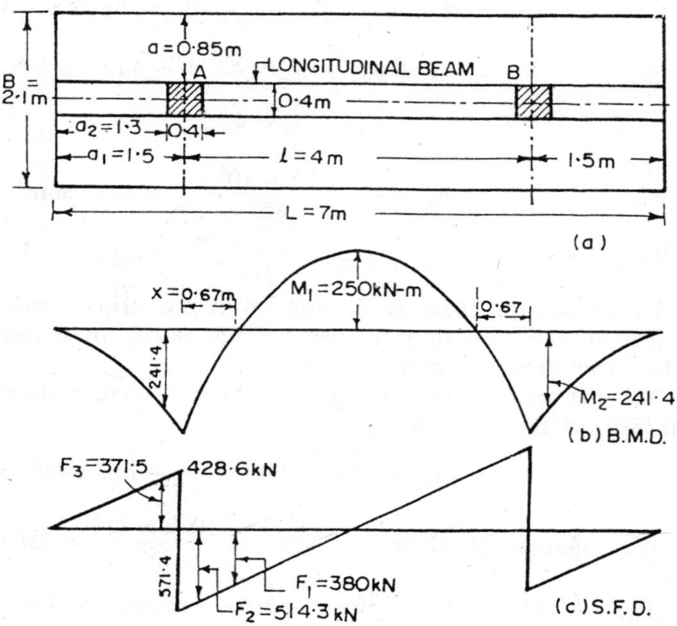

FIG. 16.8.

Since the footing is symmetrically loaded, the projection of the beam on side of either column will be equal. Projection a_1 measured from centre of column is

$$a_1 = \tfrac{1}{2}(L - l) = \tfrac{1}{2}(L - 4) = \frac{L}{2} - 2 \text{ metres.}$$

$\therefore$ Clear projection a_2 from outer face.

$$= \left(\frac{L}{2} - 2\right) - 0.2 = \left(\frac{L}{2} - 2.2\right) \text{ m.}$$

This projection a_2 should be such that maximum sagging bending moment at the *face* of column is equal to the maximum hogging bending moment at the middle of the beam.

$$\text{Maximum sagging B.M.} = \frac{w\,a_2^2}{2} = \frac{2000}{L}\frac{1}{2}\left(\frac{L}{2} - 2.2\right)^2$$

$$= \frac{1000}{L}\left(\frac{L}{2} - 2.2\right)^2 \qquad ...(i)$$

$$\text{Maximum hogging B.M.} = \frac{Wl}{2} - \frac{w}{2}\left(\frac{L}{2}\right)^2$$

$$= 1000 \times 2 - \frac{2000}{2L}\left(\frac{L}{2}\right)^2$$

$$= 2000 - 250\text{ L} \qquad ...(ii)$$

Equating the two, we get

$$\frac{1000}{L}\left(\frac{L}{2} - 2.2\right)^2 = 2000 - 250\text{ L}$$

or $$L^2 - 8.4L + 9.68 = 0$$

$\therefore$ $$L = 7.02 \text{ m}$$

Keep $L = 7$ m and $B = 2.1$ m, so that $A = 7 \times 2.1 = 14.7\ \text{m}^2$

$\therefore$ $$\text{Projection } a_1 = \frac{L}{2} - 2 = 3.5 - 2 = 1.5 \text{ m.}$$

$$a_2 = \frac{L}{2} - 2.2 = 3.5 - 2.2 = 1.3 \text{ m.}$$

Clear projection a of the slab from the face of the column is

$$a = \tfrac{1}{2}(B - b) = \tfrac{1}{2}(2.1 - 0.4) = 0.85 \text{ m.}$$

$$\text{Net upward soil reaction } p_0 = \frac{2000}{7 \times 2.1} = 136 \text{ kN/m}^2$$

$$\text{Upward load intensity on beam, } w = \frac{2000}{L} = \frac{2000}{7}$$

$$= 285.7 \text{ kN/m.}$$

3. ***Design of slab.***

Assuming width of beam equal to width of column, the cantilever projection of slab $= a = 0.85$ m.

$$\text{B.M. } M = p_0 \frac{a^2}{2} = \frac{136}{2}(0.85)^2$$

$$= 49.13 \text{ kN-m} = 49.13 \times 10^6 \text{ N-mm}$$

$$d = \sqrt{\frac{M}{Rb}} = \sqrt{\frac{49.13 \times 10^6}{0.874 \times 1000}} = 237 \text{ mm}$$

Provide $d = 240$ mm and total depth $= 300$ mm. Reduce the total depth to 200 mm at the edges so that an effective depth of 140 mm is available there.

Let us check the above depth of one way shear(diagonal tension), at the critical section situated at a distance $d = 240$ mm from the column face. The shear force F at the point is

$$V = p_0 (a - 0.24) = 136 (0.85 - 0.24)$$

$$= 83 \text{ kN} = 83000 \text{ N}$$

Effective depth of slab at their point

$$= 140 + \frac{240 - 140}{0.85}(0.85 - 0.24) = 212 \text{ mm}$$

$$\therefore \quad \tau_v = \frac{V}{bd} = \frac{83000}{1000 \times 212} = 0.39 \text{ N/mm}^2$$

Permissible shear stress $= k\,\tau_c$.

where $k = 1.05$ for $D = 214 + 60$ mm (Table 3.2)

and $\tau_c = 0.33$ N/mm^2 for $p = 0.72\%$ (balanced section)

$\therefore$ Permissible shear stress $= 1.05 \times 0.33 = 0.35$ N/mm^2.

Hence the section is *unsafe* for shear. In order to make it safe against diagonal tension, the revised depth

$$d_c = \frac{V}{b\,(k\,\tau_c)} = \frac{83000}{1000\,(1.05 \times 0.33)} = 240 \text{ mm.}$$

$\therefore$ Depth at the face of the column is given by the relation

$$240 = 140 + \frac{d - 140}{0.85}\left(0.85 - \frac{d}{1000}\right)$$

or

$$(d - 140)\left(0.85 - \frac{d}{1000}\right) = 85$$

which gives $d = 290$ mm

Hence keep $D = 290 + 60 = 350$ mm.

Area of steel reinforcement is given by

$$A_{st} = \frac{49.13 \times 10^6}{140 \times 0.865 \times 290} = 1399 \text{ mm}^2$$

Using 12 mm Φ bars, $A_\Phi = 113 \text{ mm}^2$

$$\therefore \quad \text{Spacing } s = \frac{1000 \times 113}{1399} = 80.1 \text{ mm}$$

Hence provide 12 mm Φ bars @ 80 mm c/c at the bottom of the slab.

$$L_d = 58.3\,\Phi = 58.3 \times 12 = 700 \text{ mm.}$$

Providing 50 mm side cover, available length of bar = 850 − 50 + anchorage value of hook, which is more than L_d. Hence safe.

Area of distribution reinforcement @ 0.15% of area of cross- section will be $= \frac{0.15}{100} \times 1000 \times 350 = 525 \text{ mm}^2$. Using 12 mm Φ bars ,

$$\text{Spacing} = \frac{1000 \times 113}{525} = 215 \text{ mm}$$

However, provide 12 mm Φ bars @ 200 mm c/c.

4. ***B.M. and S.F. diagrams for longitudinal beam.***

The B.M. and S.F. diagrams for the beam are shown in Fig. 16.8(*b*) and (*c*) respectively.

$$\text{Sagging B.M. } M_2 \text{ at column face} = \frac{w\,a_2^2}{2} = \frac{285.7}{2}(1.3)^2$$

$$= 241.4 \text{ kN-m} = 241.4 \times 10^6 \text{ N-mm}$$

$$\text{Hogging B.M. } M_1 = W\frac{l}{2} - \frac{w}{2}\left(\frac{L}{2}\right)^2$$

$$= 1000 \times \frac{4}{2} - \frac{285.7}{2}\left(\frac{7}{2}\right)^2$$

$$= 250 \text{ kN-m} = 250 \times 10^6 \text{ N-mm}$$

(Note. Both these moments are not exactly equal since the length L adopted as 7 m is slightly less than 7.22 m found on the basis of equal values of M_1 and M_2) .

The point of contraflexure occurs at x from the centre of column, its position is being given by

$$\frac{w\,(a_1 + x)^2}{2} - Wx = 0$$

or

$$\frac{285.7}{2}(1.5 + x)^2 = 1000x$$

From which $x = 0.67$ m

S.F. just to the left of $A = 285.7 \times 1.5 = 428.6$ kN

S.F. just to the right of $A = 1000 - 428.6 = 571.4$ kN

S.F. at the point of contraflexure is

$$F_1 = 571.4 - (285.7 \times 0.67) = 380 \text{ kN}$$

S.F. at the inner face of column is

$$F_2 = 571.4 - (285.7 \times 0.2) = 514.3 \text{ kN}$$

S.F. at the outer face of the column is

$$F_3 = 428.6 - (285.7 \times 0.2) = 371.5 \text{ kN.}$$

5. *Depth of Longitudinal beam*

Let us provide longitudinal beam projecting *below* the footing slab. In the cantilever portion, compression face is upwards and hence T-beam action will be available. In the central portion, however, T-beam action will not be available. Let us therefore, find the depth of beam for M_2 where T-beam action is not available. Keep width b of beam = 440 mm

$$d = \sqrt{\frac{M_2}{R\,b}} = \sqrt{\frac{241.4 \times 10^6}{0.874 \times 400}} = 830 \text{ mm.}$$

Let us check this depth for two way shear(punching shear). The critical plane occurs at a distance of half the effective depth from the column face. In this case, the effective depth d_b of the beam is larger than the effective depth d_s (= 290 mm) of the slab. The column will punch through both these depths, along its perimeter. Let us assume that the distance of critical plane $= \frac{d_b}{2}$ to one side and $\frac{d_s}{2}$ to the other side. Hence widths of the punching shear perimeter are

$$b_{01} = 500 + d_b$$

and

$$b_{02} = 400 + d_s = 400 + 290 = 690 \text{ mm}$$

$\therefore$ Punching shear area of concrete available along the perimeter for the critical plane.

$$= (2b_{01} \times d_s') + (2b_{02} \times d_b)$$

where d_s' = depth of slab at the critical plane.

$$= d_{se} + \frac{d_s - d_{se}}{a}\left[a - \frac{d_s}{2}\right]$$

$$= 140 + \frac{290 - 140}{0.85}\left[0.85 - \frac{0.29}{2}\right]$$

$$= 264 \text{ mm}$$

Punching shear F

$$= 1000 \times 10^3 - 136 \times 10^3 \, [b_{01} \times b_{02}]$$

$$= 1000 \times 10^3 - 136 \times 10^3 \, \frac{[(400 + d_b) \times 690]}{(1000)^2}$$

$$= 10^6 - 93.8\,(400 + d_b)$$

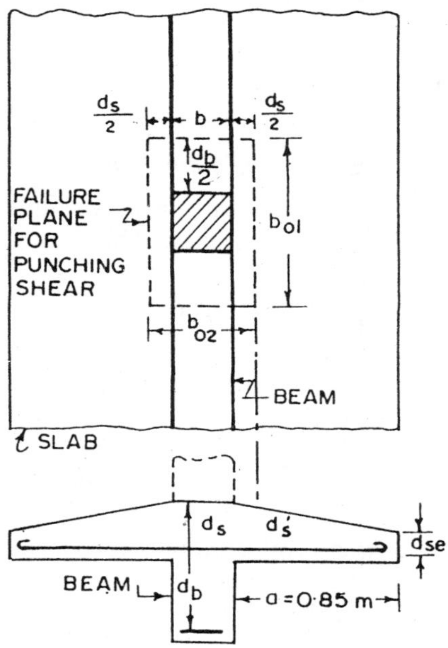

FIG. 16.9

$\therefore$ Punching shear stress

$$= \frac{F}{2\,[(b_{01} \times d_s') + (b_{02} \times d_b)]}$$

$$= \frac{10^6 - 93.8\,(400 + d_b)}{2\,[(400 + d_b)\,264 + 690\,d_b]}$$

$$= \frac{10^6 - 93.8\,(400 + d_b)}{2\,[105600 + 954\,d_b]} \qquad ...(1)$$

Permissible shear stress $= k_s \tau_c$

where $k_s = (0.5 + \beta_c) \not> 1 = 1$

$\tau_c = 0.16 \sqrt{f_{ck}} = 0.16 \sqrt{15} = 0.62 \text{ N/mm}^2$

Hence permissible shear stress $= 0.62 \text{ N/mm}^2$

Equating this to (1), we get

$$\frac{10^6 - 93.8 (400 + d_b)}{2 [105600 + 954 d_b]} = 0.62$$

From which $d_b = 650$ mm.

This is less than effective depth of 830 mm found from the bending compression.

Total depth $= 830 + 60 = 890$ mm.

Keep $D = 900$ mm so that $d = 900 - 60 = 840$ mm.

6. Reinforcement in longitudinal beam

For hogging B.M., the beam acts as a simple rectangular section with reinforcement at its top face. The area of steel is given by

$$A_{st1} = \frac{M_1}{\sigma_{st} jd} = \frac{250 \times 10^6}{140 \times 0.865 \times 840}$$

$$= 2458 \text{ mm}^2$$

$\therefore$ Use bars of 20 mm Φ, having $A_\Phi = 314 \text{ mm}^2$

$\therefore$ No. of bars $= \frac{2458}{1314} = 7.8$

Hence provide 8 bars of 20 mm Φ.

Let us check this reinforcement for development length at the point of contraflexure, as envisaged in the relation

$$\frac{M_1}{V} + L_0 \geq L_d$$

where $M_1 = \sigma_{st} A_{st} jd = 140 [8 \times 314] \times 0.865 \times 840$

$= 2.56 \times 10^8$ N-mm

V = shear force at point of contraflexure

$= 380 \text{ kN} = 380 \times 10^3$ N

$L_0 = 12\,\Phi$ or d whichever is greater

$= 840$ mm.

$L_d = 58.3 \times 20 = 1166$ mm

Now $\frac{M_1}{V} + L_0 = \frac{2.56 \times 10^8}{380 \times 10^3} + 840 = 1514$ mm.

which is more than L_d. Hence safe.

The above reinforcement (*i.e.* 8-20 Φ bars) may be curtailed at the face of the column. However, continue 3 bars in the cantilever portion, to serve as anchor bars for shear reinforcement.

The reinforcement in the cantilever portion is to be found for the bending moment M_2 = 241.1 kN-m. Here the beam will act as T- beam for which lever arm may be taken as 0.9 *d*.

$$\therefore \quad A_{st2} = \frac{M_2}{\sigma_{st}\, jd} = \frac{241.4 \times 10^6}{140 \times 0.9 \times 840} = 2280 \text{ mm}^2$$

$$\text{No. of 20 mm } \Phi \text{ bars} = \frac{2280}{314} = 7.3 \approx 8$$

This reinforcement should be safe from development length (L_d) point of view, to satisfy the following equation at the point of contraflexure near the inner face of the column :

$$\frac{M_2}{V} + L_0 \geq L_d.$$

S.F. at this point of contraflexure is 380 kN, and it has already been established in case of A_{st1} that the above criterion is satisfied. Since the point of contraflexure is the same, and since $A_{st2} = A_{st1},$ the bars of A_{st2} will also satisfy the Code requirements.

It is to be noted that A_{st2} bars should be extended beyond the point of contraflexure by a distance $= d$ or 12 Φ whichever is more (*i.e.* by 840 mm in this case) or to a distance of L_d(=1160 mm) from the centre of column (where bending moment M_2 occurs), whichever is more. Hence distance upto which A_{st2} bars are to be extended beyond centre of column $= 0.67 + 0.84 \approx 1.5$ m. At this point all the bars except 3 Nos, may be discontinued. The remaining 3 bars may be taken throughout the length to serve as anchor bars for shear reinforcement.

7. *Shear reinforcement*

Refer Fig. 16.4. Diagonal tension crack will occur at *F*, only if the zone of sagging B.M. extends to d = 900 mm beyond inner face of the column. However, point of contraflexure falls earlier. Hence there is a possibility of crack at *G*, at point of contraflexure where shear force V_1 = 380 kN.

$$\therefore \quad \tau_v = \frac{380 \times 1000}{840 \times 400} = 1.13 \text{ N/mm}^2.$$

This is more than the permissible value of $\tau_c = 0.33$ N/mm^2 for the balanced section. Hence shear reinforcement is required.

A_s at the point $= 8 \times 314 = 2512 \text{ mm}^2$

$$\frac{100 A_s}{bd} = \frac{100 \times 2512}{840 \times 400} = 0.75\%$$

$\therefore$ τ_c (for 0.75%, Table 3.1) $= 0.34 \text{ N/mm}^2$.

$\therefore$ $V_c = \tau_c bd = 0.34 \times 840 \times 400 = 114240 \text{ N}$

$\therefore$ $V_s = V_1 - V_c = 380000 - 114240 = 265760 \text{ N}$

Using 12 mm Φ 4-legged stirrups,

$$A_{sv} = 4 \times \frac{\pi}{4} (12)^2 = 452 \text{ mm}^2$$

$$\therefore \quad s_v = \frac{\sigma_{sv} . A_{sv} . d}{V_s} = \frac{140 \times 452 \times 840}{265 \times 760}$$

$$= 200 \text{ mm}$$

Distance upto which this is required (*i.e.*, at the point where S.F. $= V_c = 114240$ N) is given by

$$x_1 = \frac{V_2 - V_c}{285.7} = \frac{514.3 - 114.24}{285.7} = 1.4 \text{ m.}$$

Hence provide 12 mm Φ 4-legged stirrups, @ 200 mm c/c from the face of the column to a point 1.4 m, near each column.

In the cantilever portion, shear at the face of column =371.5 kN. Distance x_2 of point at which shear force is equal to V_c (=114.24 kN) is given by

$$x_2 = \frac{V_3 - V_c}{w} = \frac{371.5 - 114.24}{285.7} = 0.9 \text{ m}$$

Hence shear stirrups are to be provided for a distance of 0.9 m from the face of the column.

$$V_s = V_3 - V_c = 371.5 - 114.24 = 257.26 \text{ kN}$$

Using 12 mm Φ 4-legged stirrups, $A_{sv} = 452 \text{ mm}^2$

$$\therefore \quad s_v = \frac{452 \times 140 \times 840}{257.26 \times 1000} = 206 \text{ mm}$$

Hence provide these @ 200 mm c/c upto 0.9 m from the face of the column.

For the remaining of length of beam, provide 12 mm Φ 4-legged nominal stirrups at a spacing given by

$$s_v = \frac{2.5 A_{sv} . f_y}{b} = \frac{2.5 \times 452 \times 250}{400}$$

$$= 706 \text{ mm} \approx 700 \text{ mm}$$

The details of reinforcement etc. are shown in Fig. 16.10.

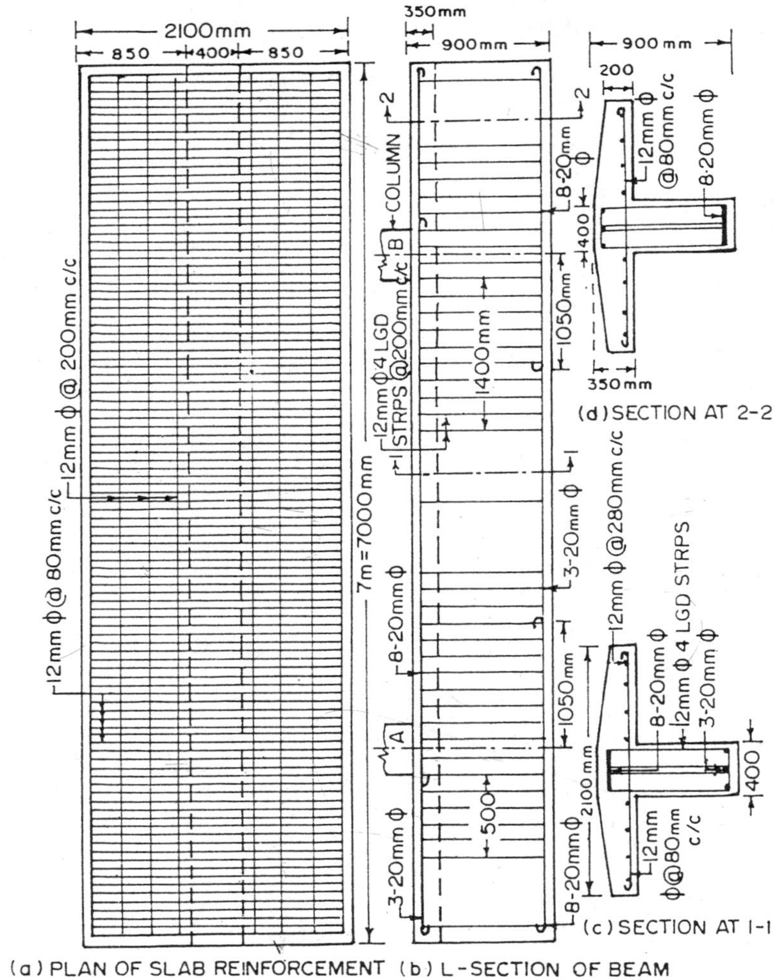

FIG. 16.10.

Design Example 16.3. *Design a combined trapezoidal footing for two columns A and B, spaced 5 metres centre to centre. Column A is 300 × 300 mm in size and transmits a load of 600 kN. Column B is 400 mm × 400 mm in size and carries a load of 900 kN.*

The maximum length of footing is restricted to 7 metres only. The safe bearing capacity of soil may be taken as 120 kN/m². Use M 15 mix for concrete.

Solution.

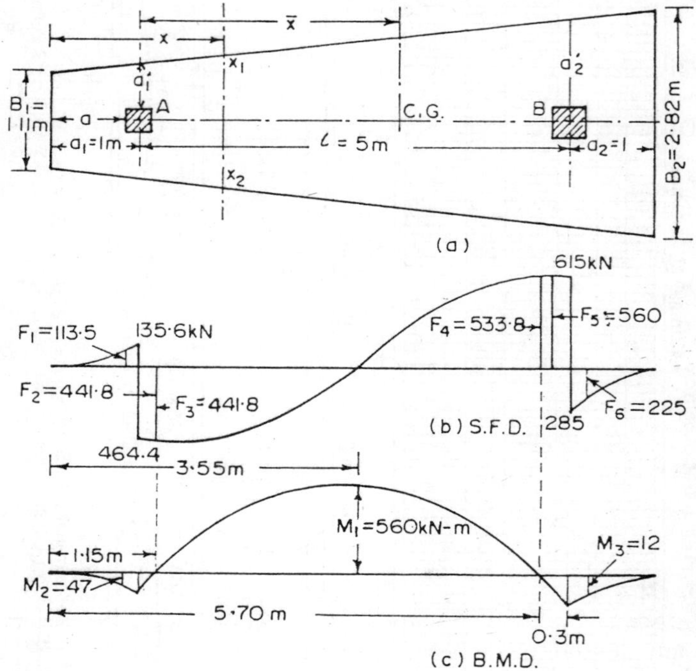

FIG. 16.11.

1. ***Design constants***

For M 15 mix, and mild steel bars

$\sigma_{cbc} = 4\ \text{N/mm}^2$; $\sigma_{st} = 140\ \text{N/mm}^2$; $k = 0.404$; $j = 0.865$; $R = 0.875$.

2. ***Size of footing***

Refer Fig. 16.11.

$L = 7$ m ; $l = 5$ m ; Keep $a_1 = a_2 = \frac{1}{2}(7 - 5) = 1$ m.

Let the width of footings be B_1 and B_2 as shown. The values of B_1 and B_2 should be such that C.G. of footing coincides with C.G. of the column loads.

Let weight of footing $W' = 0.1\,(W_1 + W_2)$
$= 0.1\,(600 + 900) \;\; = 150$ kN

$$\therefore \qquad \frac{(B_1 + B_2)L}{2} = \frac{W_1 + W_2 + W'}{q_0}$$

or
$$B_1 + B_2 = \frac{(600 + 900 + 150)}{120} \times \frac{2}{7} = 3.93 \text{ m} \quad ...(1)$$

Distance $\bar{x}$ of C.G. of load $= \dfrac{W_2 l}{W_1 + W_2} = \dfrac{900 \times 5}{600 + 900} = 3$ m.

Distance of C.G. of load from edge $B_1 = a_1 + \bar{x} = 1 + 3 = 4$ m. ...(*i*)

Distance of C.G. of footing from edge B_1

$$= \frac{L}{3}\left(\frac{B_1 + 2B_2}{B_1 + B_2}\right) = \frac{7}{3}\left(\frac{B_1 + 2B_2}{B_1 + B_2}\right) \quad ...(ii)$$

Equating (*i*) and (*ii*), we get

$$\frac{7}{3}\left(\frac{B_1 + 2B_2}{B_1 + B_2}\right) = 4 \quad ...(2)$$

From (1) and (2), we get

$B_1 = 1.11$ m and $B_2 = 2.82$ m.

Net upward soil pressure intensity is given by

$$p_0 = \frac{W_1 + W_2}{\frac{1}{2}(B_1 + B_2)\,L} = \frac{(600 + 900)}{\frac{1}{2}(1.11 + 2.82)\,7} \approx 110 \text{ kN/m}^2$$

3. *Bending moment and shear force*

Consider the bending of footing in the longitudinal direction. At any distance x from the face B_1, the width B_x of section X_1X_2 is given by

$$B_x = B_1 + \frac{B_2 - B_1}{L}x = 1.11 + \frac{2.82 - 1.11}{7}x$$

$$= 1.11 + 0.245\,x$$

$\therefore$ Area A_x to the left of section $X_1X_2 = \frac{1}{2}(B_1 + B_x)\,x$

$$= \frac{1}{2}\{1.11 + 1.11 + 0.245x\}\,x$$

$$= \frac{x}{2}\{2.22 + 0.245x\} \quad ...(i)$$

Distance of C.G. from section $X_1X_2 = \dfrac{x}{3}\left\{\dfrac{2B_1 + B_x}{B_1 + B_x}\right\}$

$$= \frac{x}{3}\left\{\frac{2.22 + 1.11 + 0.245x}{1.11 + 1.11 + 0.245x}\right\}$$

$$= \frac{x}{3}\left\{\frac{3.33 + 0.245x}{2.22 + 0.245x}\right\} \qquad ...(ii)$$

Hence total upward force to the left of X_1X_2

$$= \frac{p_o x}{2}(2.22 + 0.245x)$$

$$= \frac{110}{2}x\ (2.22 + 0.245\,x) \qquad ...(1)$$

$\therefore$ S.F. at outer face of column A, where $x = a = 0.85$ m, is

$$F_1 = \frac{110}{2} \times 0.85(2.22 + 0.245 \times 0.85) = 113.5 \text{ kN.}$$

$\therefore$ S.F. at x to the right of $A = 600 - \frac{110}{2}x\,(2.22 + 0.245x)$ ...(2)

$\therefore$ S.F. at inner face of column A, where $x = 1.15$ m, is

$$F_2 = 600 - \frac{110}{2} \times 1.15\,(2.22 + 0.245 \times 1.15) = 441.8 \text{ kN.}$$

To find the position of section wnere S.F. is zero, put (2) to zero.

$$\therefore \qquad 600 - \frac{110}{2}x\,[2.22 + 0.245\,x] = 0$$

From which $x = 3.55$ m

B.M. at section X_1X_2 is given by

$$M_2 = 600\,(x - 1) - (\text{Upward force} \times \text{Distance of its C.G.})$$

$$= 600\,(x - 1) - \left\{\frac{110}{2}x\,(2.22 + 0.245\,x)\right\} \times \frac{x}{3}\left\{\frac{3.33 + 0.245\,x}{2.22 + 0.245\,x}\right\}$$

$$= 600(x - 1) - \frac{110}{6}x^2\,(3.33 + 0.245\,x) \qquad ...(3)$$

Maximum hogging bending moment (M_1) will occur at $x = 3.55$ where S.F. is zero

$$\therefore \qquad M_1 = 6000(3.55 - 1) - \frac{110}{6}(3.55)^2\,(3.33 + 0.245 \times 3.55)$$

$$= 560 \text{ kN-m}$$

To find the value of x, where M_x is zero, equate (3) to zero.

$$\therefore \qquad 600\,(x - 1) - \frac{110}{6}x^2\ (3.33 + 0.245\,x) = 0$$

Solving this by trial and error, we get

$$x = 1.15 \text{ m and } 5.70 \text{ m}$$

Thus the position of both the points of contraflexure are known.

For getting the value of S.F. (F_3) at the first point of contraflexure, put $x = 1.15$ m in Eq. (2).

$$\therefore \quad F = 600 - \frac{110}{2}(1.15)\,(2.22 + 0.245 \times 1.15) = 441.8 \text{ kN}$$

Similarly S.F. (F_4) at $x = 5.70$ is

$$F_4 = 600 - \frac{110}{2}(5.7)(2.22 + 0.245 \times 5.7) = 533.8 \text{ kN}$$

(numerically)

S.F. at outer face of column B, at $x = 5.8$ m is

$$F_5 = 600 - \frac{110}{2}(5.8)(2.22 + 0.245 \times 5.8) = 560 \text{ kN(numerically)}$$

S.F. at inner face of column B, at $x = 6.2$ m is

$$F_6 = 600 - \frac{110}{2}(6.2)(2.22 + 0.245 \times 6.2) + 90 = 225 \text{ kN}$$

Sagging bending moment (M_2) at the outer face of column A is obtained by putting $x = 0.85$ m in the second term of Eq. (3) and neglecting the first term.

$$\therefore \quad M_2 = \frac{110}{6}(0.85)^2(3.33 + 0.245 \times 0.85) = 47 \text{ kN-m}$$

Sagging B.M. for the cantilever portion to the right of column B is given by

$$M_x = \frac{110}{6}x^2(3.33 + 0.245x) - 60(x - 1) - 90(x - 6)$$

Putting $x = 6.2$ m, we get B.M. (M_3) at the outer face of column B :

$$M_3 = \frac{110}{6}(6.2)^2(3.33 + 0.245 \times 6.2) - 60(5.2) - 90(0.2)$$

$$= 120 \text{ kN-m}$$

The B.M. and S.F. diagrams are shown in Fig. 16.11.

4. ***Effective depth of footing***

The effective depth is determined on the basis of hogging moment $M_1 = 560$ kN-m which is the maximum bending moment in the footing. It occurs at a section distant $x = 3.55$ m from the edge B_1.

The width B_x at that section is given by

$$B_x = B_1 + \frac{B_2 - B_1}{L}x$$

$$= 1.11 + 0.245x$$

$$= 1.11 + 0.245 \times 3.55 = 1.87 \text{ m} = 1870 \text{ mm}$$

$$\therefore \quad d = \sqrt{\frac{M_1}{RB_x}} = \sqrt{\frac{560 \times 10^6}{0.874 \times 1870}} = 586 \text{ mm}$$

This depth should be sufficient from punching shear point of view. Punching shear will be maximum near column B, where load $W_1 = 900$ kN.

Side $\qquad b_0 = b + d = (400 + d)$ mm

Punching shear $\qquad F = W_2 - p_0 b_0^2$

$$= 900 - \frac{110(400 + d)^2}{10^6} \text{ kN}$$

$$\therefore \qquad \tau_v = \frac{900 \times 10^3 - 0.11\,(400 + d)^2}{4\,(400 + d)\,d}$$

Permissible shear stress $= k_s \tau_c = 1 \times 0.16 \sqrt{f_{ck}}$

$$= 1 \times 0.16 \sqrt{15} = 0.62 \text{ N/mm}^2$$

$$\therefore \qquad \frac{900 \times 10^3 - 0.11\,(400 + d)^2}{4\,(400 + d)\,d} = 0.62$$

From which $d = 412$ mm which is less than the one found from bending compression.

Hence $d = 586$ mm . Using 60 mm cover, $D = 586 + 60 = 646$ mm. However, keep $D = 650$ mm so that available $d = 650 - 60 = 590$ mm.

5. *Reinforcement in the longitudinal direction*

(*i*) *Reinforcement for hogging bending moment* M_1 is given by

$$A_{st} = \frac{M_1}{\sigma_{st}\, jd} = \frac{560 \times 10^6}{140 \times 0.865 \times 590} = 7838 \text{ mm}^2$$

Using 20 mm Φ bars, $A_\Phi = \frac{\pi}{4}(20)^2 = 314 \text{ mm}^2$

No. of bars required $d = \frac{7838}{314} \approx 25$

Thus provide 25 bars at the top face of the footing.

These bars should be checked for development length at the point of contraflexure, to satisfy the criteria

$$\frac{M_1}{V} + L_0 \geq L_d.$$

Out of the two points of contraflexure in the hogging B.M. range, the one near column B, has maximum shear force

$$F_4 = 533.8 \text{ kN}.$$

$$M_1 = A_{st}\,\sigma_{st}\, jd = (25 \times 314) \times 140 \times 0.865 \times 590$$

$$= 5.61 \times 10^8 \text{ N-mm}$$

$$V = F_4 = 533.8 \text{ kN} = 5.33 \times 10^3 \text{ N}$$

$$L_0 = 12\,\Phi \text{ or } d \text{ whichever is more}$$
$$= 590 \text{ mm.}$$

$$L_d = 58.3\,\Phi = 58.3 \times 20 = 1166 \text{ mm.}$$

Substituting the values

$$\frac{M_1}{V} + L_0 = \frac{5.61 \times 10^8}{533.8 \times 10^3} + 590 = 1640 \text{ mm}$$

which is more than L_d, provided all the 25 bars are taken upto the point of contraflexure. These bars are to be taken beyond this point for a distance of 12 Φ (= 12 × 20 = 240 mm) or d (=500 mm), whichever is more, beyond this point. After that, curtail 17 bars and continue 8 bars straight upto either edge of footing so that these may serve as anchorage bars for shear stirrups.

(ii) Reinforcement for sagging bending moment M_3 near outer face of column B

$$A_{st3} = \frac{M_3}{\sigma_{st}\, jd} = \frac{120 \times 10^6}{140 \times 0.865 \times 590} = 1680 \text{ mm}^2$$

width of footing near the outer face of column B, where $x = 6.2$ m is

$$B_x = 1.11 + 0.245\,x = 1.11 + 0.245 \times 6.2 = 2.67 \text{ m}$$

$$\therefore \text{ Minimum reinforcement @ } 0.15\% = \frac{0.15}{100} \times 2670 \times 590$$

$$= 2363 \text{ mm}^2$$

Hence provide 2363 mm² area of reinforcement. Using 12 mm Φ bars, $A_\Phi = 133 \text{ mm}^2$.

$$\therefore \text{ No. of bars } = \frac{2363}{113} \approx 21$$

This reinforcement should be checked for development length at the point of contraflexure near the inner face column B, so as to satisfy the equation

$$\frac{M_1}{V} + L_0 \geq L_d$$

where $M_1 = \sigma_{st} A_{st}\, jd = 140\,(21 \times 113) \times 0.865 \times 590$

$$= 1.695 \times 10^8$$

$$V = F_4 = 533.8 \text{ kN} = 533.8 \times 10^3 \text{ N}$$

$$L_0 = 12\,\Phi \text{ or } d, \text{ whichever is greater } = 590 \text{ mm}$$

$$L_d = 58.3\,\Phi = 58.3 \times 12 = 700$$

$$\therefore \quad \frac{M_1}{V} + L_0 = \frac{1.695 \times 10^8}{533.8 \times 10^3} + 590 = 318 + 590 = 908 \text{ mm}$$

which is more than L_d; hence safe, provided all the 21 bars are available at the point of contraflerure. However, extend 21 bars upto a point distant 12 Φ (= 12 × 12 = 144 mm) or d (=590 mm) from the point of contraflexure or 590 + 150 ≈ 750 mm from the inner face of column B. After that discontinue 13 bars, but continue 8 bars throughout to serve as anchor bars for shear reinforcement.

Also, length of bars available, in the cantilever portion, beyond the centre of column B = 1m– side cover= 1000 – 60 = 940 mm > L_d. Hence safe.

(iii) Reinforcement for sagging B.M. M_2 near other face of column A.

$$A_{st2} = \frac{M_2}{\sigma_{st}\, jd} = \frac{47 \times 10^6}{140 \times 0.865 \times 590} = 658 \text{ mm}^2$$

Width of footing near the outer face of column A, where x =0.85 m is

$$B_x = 1.11 + 0.245\,x = 1.11 + 0.245 \times 0.85 \approx 1.32 \text{ m.}$$

$$\text{Minimum reinforcement @ } 0.15\% = \frac{0.15}{100} \times 1320 \times 590$$

$$= 1168 \text{ mm}^2$$

Using 12 mm Φ bars, having $A_\Phi = 113$ mm²

$$\therefore \text{ No. of bars } = \frac{1168}{113} \approx 11.$$

Let is check these from development length point of view, at the point of contraflexure near column A, so as to satisfy the criterion

$$\frac{M_1}{V} + L_0 \geq L_d$$

where $M_1 = \sigma_{st} A_{st}\, jd = 140\,(11 \times 113) \times 0.865 \times 590$

$= 88.8 \times 10^6$ N-mm

V = shear force = 441.8 kN at the point of contraflexure

L_0 = 12Φ or d whichever is more = 590 mm

$L_0 = 58.3\,\Phi = 58.3 \times 12 \approx 700$ mm.

$$\therefore \quad \frac{M_1}{V} + L_0 = \frac{88.8 \times 10^6}{441.8 \times 10^3} + 590 = 210 + 590 = 800 \text{ mm}$$

which is more than L_d. Hence safe.

However, extend the bars upto L_o = 590 mm from the point of contraflexure, *i.e.*, 590 + 1150 = 1740 mm from the edge of footing. After that, curtail 3 bars, but continue remaining 8 bars.

5. *Check for diagonal tension*

Test for diagonal tension should be made in the cantilever portion at a distance d from the column face. In the present case, d =590 mm =0.59 m. At 0.59 m from the outer face of column B, distance $x = 6.2 + 0.59 = 6.79$ m. At this section shear force is extremely small and available width B_x is large. Hence shear stress will be small. Similarly, for the cantilever portion to the left of column A, shear force, and hence shear stress will be small.

For diagonal tension between A and B, near column B, crack can occur at the bottom of the footing (*i.e.*, for hogging B.M.) at a distance d =590 mm, or at the top of footing (*i.e.* for hogging B.M.) at the point of contraflexure distant 300 mm from the centre of column B. Point of contraflexure being nearer, more S.F. will be at the point, S.F. at point of contraflexure= F_4 = 533.8 kN.

At the point of contraflexure $x = 5.70$ m.

$$\therefore \quad B_x = 1.11 + 0.245x = 1.11 + 0.245 \times 5.7 \approx 2.5 \text{ m}$$

$$\therefore \quad s_v = \frac{F_4}{bd} = \frac{533.8 \times 10^3}{2500 \times 590} = 0.362 \text{ N/mm}^2$$

$$\frac{100 A_s}{bd} = \frac{100\,(25 \times 314)}{2500 \times 590} = 0.53\%$$

Hence from Table 3.1, $\tau_c =$ 0.3 N/mm²

Thus, actual shear stress is more than the permissible one, and shear reinforcement is necessary.

$$V_c = \tau_c bd = 0.3 \times 2500 \times 590 = 442500 \text{ N} = 442.5 \text{ kN}$$

$$\therefore \quad V_s = V - V_c = 533.8 - 442.5 = 91.3 \text{ kN}$$

Using 10 mm Φ 8-lgd stirrups, $A_{sv} = 8 \times \frac{\pi}{4}(10)^2 = 628 \text{ mm}^2$

$$\therefore \quad s_v = \frac{628 \times 140 \times 590}{91.3 \times 1000} = 568 \text{ mm.}$$

Max. spacing of nominal stirrups is given by

$$s_v = \frac{2.5 A_{sv} f_y}{b} = \frac{2.5 \times 628 \times 250}{2500} = 157$$

This is too close. If however, 12 mm Φ bars are used for 8-lgd stirrups, $A_{sv} = 8\frac{\pi}{4}(12)^2 = 905 \text{ mm}^2$.

$$\therefore \quad s_v = \frac{2.5 \times 905 \times 250}{2500} = 226 \text{ mm.}$$

Hence provide these nominal stirrups @ 225 mm c/c throughout the length.

7. *Transverse Reinforcement*

The footing will bend transversely near column face. Projection a_2' beyond face of column $B = \frac{1}{2}(B_x - b_2)$ where B_x is the width of footing at the centre of the column B, where x=6 m.

$$\therefore \quad B_x = 1.11 + 0.245\,x = 1.11 + 0.245 \times 6 = 2.58 \text{ m}$$

$$\therefore \quad a_2' = \frac{1}{2}(2.58 - 0.4) = 1.09 \text{ m} = 1090 \text{ mm.}$$

Width of bending strip near column $B = d + b_2 + d$

$$= 590 + 400 + 590 = 1580 \text{ mm}$$

$\therefore$ Average area of bending strip $= 2.58 \times 1.58 = 4.076 \text{ m}^2$

$\therefore$ Net upward pressure $p_0' = \dfrac{900}{4.076} \approx 221 \text{ kN/m}^2$

(This is approximately double the value of average pressure p_0).

Maximum B.M. at the face of column is

$$M = p_0' \frac{(a_2')^2}{2} = \frac{221}{2}(1.09)^2 = 131.3 \text{ kN-m}$$

$$= 131.3 \times 10^6 \text{ N-mm}$$

$$\therefore \quad d \text{ required} = \sqrt{\frac{131.3 \times 10^6}{0.875 \times 1000}} = 388 \text{ mm.}$$

Actual d available =593 mm. Hence the transverse beam, of width = 1.58 m will be of the same thickness as the remaining footing. Since transverse reinforcement will be provided *above* the main reinforcement, available $d = 590 - 20 = 570$ mm (approx).

The transverse reinforcement is given by

$$A_{st} = \frac{131.3 \times 10^6}{140 \times 0.865 \times 570} = 1902 \text{ mm}^2$$

Using 16 mm Φ bars, $A_\Phi = 201 \text{ mm}^2$.

$$\therefore \quad \text{Spacing } s = \frac{1000 \times 201}{1902} = 106 \text{ mm.}$$

Hence provide 16 mm Φ bars @ 100 mm c/c.

Similarly, for column A, width B_x at the centre of column, where x =1 m is,

$B_x = 1.11 + 0.245\,x = 1.11 + 0.245 \times 1 = 1.36$ m.

Projection $a_1' = \frac{1}{2}(1.36 - 0.3) = 0.53$ m

Width of bending strip $= 2d + b_1 = 2 \times 590 + 300 = 1480$ mm

$\therefore$ Area bending strip $= 1.36 \times 1.48 = 2.013$ m^2

$\therefore$ Net upward pressure $p_0' = \dfrac{600}{2.013} = 298$ kN/m^2

$\therefore$ Max. B.M. at the face of the column is

$$M = p_0 \frac{(a_1')^2}{2} = 298 \frac{(0.53)^2}{2} = 41.8 \text{ kN-m}$$

$$= 41.8 \times 10^6 \text{ N-mm.}$$

The available depth d = 570 mm will be sufficient. Area of transverse reinforcement is

$$A_{st} = \frac{41.8 \times 10^6}{140 \times 0.865 \times 570} = 606 \text{ mm}^2$$

Minimum reinforcement $= \dfrac{0.15}{100}(1000 \times 570) = 855$ mm^2

Using 16 mm Φ bars, having $A_\Phi = 201$ mm^2

$\therefore$ Spacing $s = \dfrac{1000 \times 201}{855} = 235$ mm.

However provide 16 mm Φ bars @ 200 mm c/c in the strip of width 1.48 m.

For the remaining portion, provide 16 mm Φ bars @ 230 mm c/c, so that this reinforcement at any section is not less than 0.15% of area of cross-section.

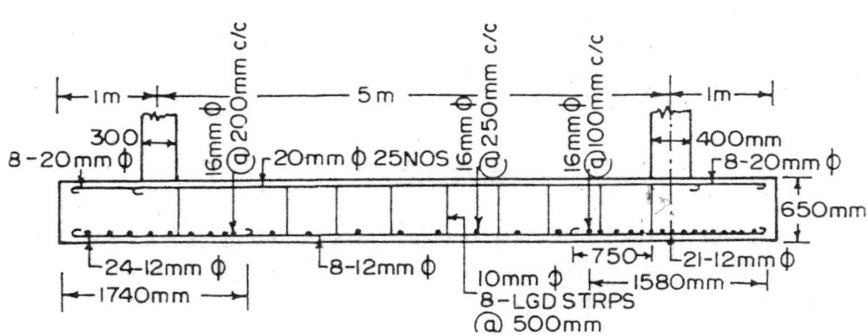

FIG. 16.12.

The details of reinforcement etc. are shown in Fig. 16.12.

Design Example 16.4. *Redesign example 16.3 by providing a longitudinal beam connecting the two columns.*

Solution. For the design of trapezoidal footing with a longitudinal beam, steps 1, 2 and 3 of example 16.3 will remain the same. The bending moment and shear force diagrams drawn in Fig. 16.4(*b*) and (*c*) will be applicable for the bending of the longitudinal beam which should be designed properly to resist these bending moments

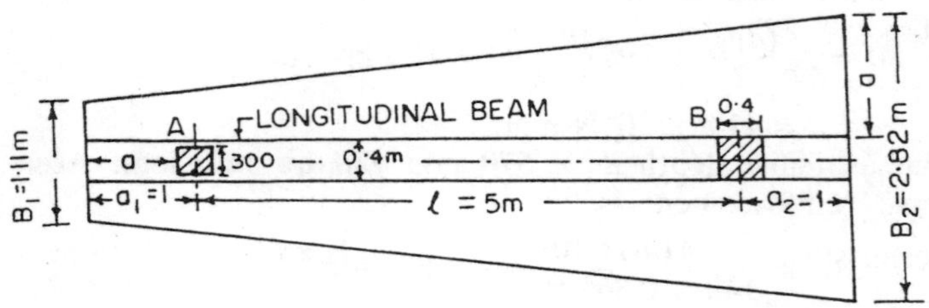

FIG. 16.13.

and shear forces at various sections. The slab of the footings will bend transversely.

The salient dimensions of the footing, determined in Example 16.3 are shown in Fig. 16.13. Let the beam project above the footing. Thus, for the hogging bending moment, T-beam action will be available. Let the width of beam = 400 mm.

1. ***Design of slab***

The slab will bend as cantilever on either side of the beam. The maximum cantilever projection of the slab will be at the end near width B_2, is magnitude being given by

$$a' = \frac{1}{2}(2.82 - 0.4)$$

$$= 1.21 \text{ m.}$$

Intensity of net upward pressure p_0 =110 kN/m², determined in the previous example. Maximum bending moment at the face of column B is given by

$$M = p_0 a'^2 = \frac{110\,(1.21)^2}{2} = 80.53 \text{ kN-m} = 80.53 \times 10^6 \text{ N-mm}$$

$\therefore$ Thickness of slab $= \sqrt{\dfrac{80.53 \times 10^6}{0.875 \times 1000}} = 304$ mm.

However, adopt $d =$ 310 mm, so that providing 60 mm cover, total $D = 310 + 60 = 370$ mm. Reduce it to 200 mm at the end, where available $d = 140$ mm.

The above depth should be safe against diagonal tension crack which may occur at $d = 310$ mm from the face of the column.

S.F. F' at that section $= 110\,(1.21 - 0.31) = 99$ kN

Effective depth d' at that section

$$= 140 + \frac{310 - 140}{1.21} \times (1.21 - 0.31) = 266 \text{ mm}$$

$$\therefore \quad \tau_v = \frac{F'}{bd'} = \frac{99 \times 1000}{1000 \times 266} = 0.372 \text{ N/mm}^2$$

Permissible τ_c (at $p = 0.72\%$) = 0.33 N/mm^2 (Table 3.1).

Hence unsafe.

To make it safe, increase the effective depth in the ratio $\dfrac{0.372}{0.33}$ approximately

$$\therefore \quad d \text{ at column face} = \frac{0.372}{0.33} \times 310 = 349 \text{ mm.}$$

Hence keep $d =$ 350 mm and $D =$ 450 mm at the face of the column.

The area of steel reinforcement is given by

$$A_{st} = \frac{M}{\sigma_{st}\, jd} = \frac{80.53 \times 10^6}{140 \times 0.865 \times 350} = 1900 \text{ mm}^2$$

Using 16 mm Φ bars, $A_\Phi = 201$ mm^2

$$\therefore \quad \text{Spacing } s = \frac{1000 \times 201}{1900} = 106 \text{ mm.}$$

Hence provide 16 mm Φ bars @ 100 mm c/c at the bottom of the section. Let us check this for development length.

$$L_d = 58.3\,\Phi = 58.3 \times 16 = 933 \text{ mm.}$$

Available length, using 60 mm side cover = 1210 – 60 = 1150 mm. Hence safe.

Similarly, at the end strip near column A, the cantilever projection of slab beyond the beam $= \frac{1}{2}(1.11 - 0.4) = 0.355$ m.

$$\therefore \quad \text{B.M. } M = 110\frac{(0.355)^2}{2} = 6.93 \text{ kN-m} = 6.93 \times 10^6 \text{ N-mm.}$$

This is extremely small, and reinforcement found on the basis of above B.M. will be extremely smail. However, for 16 mm Φ bars, $L_d = 58.3\,\Phi = 933$ mm while available length of bars(after providing 60 mm side cover $= 355 - 60 = 295$ mm only(plus anchorage value of hook, which has been ignored here). Hence maximum stress σ_s that can be developed $= \frac{295}{933} \times 140 = 44$ N/mm^2, so that bending failure is not there.

$$\therefore \quad A_{st} = \frac{M}{\sigma_s jd} = \frac{6.93 \times 10^6}{44 \times 0.865 \times 350} = 520 \text{ mm}^2$$

Also, minimum area of steel @ 0.15% of cross-sectional area is

$$A_{st} = \frac{0.15}{100} \times 1000 \times 410 = 615 \text{ mm}^2$$

Using 16 mm Φ bars, $s = \frac{1000 \times 201}{615} = 300$ mm

Hence the spacing of 16 mm Φ bars @ 100 mm c/c at the edge near column B can be gradually increased to 300 mm at the edge near column A.

Transverse reinforcement area @ 0.15% =615 mm^2

Using 10 mm Φ bars $A_\Phi =$ 78.5 mm^2

$$\therefore \quad \text{Spacing} = \frac{1000 \times 78.5}{615} = 125 \text{ mm c/c.}$$

2. *Depth of longitudinal beam*

Let us provide longitudinal beam of 400 mm width, projecting above the slab. In the hogging portion, therefore, T-beam action will be available. The hogging bending moment $M_1 = 500$ kN-m is much more than the sagging bending moment. Hence the depth of longitudinal beam will be designed for moment M_1, taking as a T-beam.

Average thickness of slab $= \frac{200 + 410}{2} = 305 \text{ mm} = D_f$

The maximum bending moment occurs at $x = 3.55$ m, where width B_x is given by

$$B_x = 1.11 + 0.245\,x = 1.11 + 0.245 \times 3.55$$

$= 1.87 \text{ m} = 1870 \text{ mm}.$

The width b_f of the flange will be corresponding to an isolated T-beam given by the expression

$$b_f = \frac{l_0}{\left(\frac{l_0}{b} + 4\right)} + b_w$$

where b = actual width of flange =1870 mm

l_0 = distance between points of zero moment in the beam

$= 5.70 - 1.15 = 4.55 \text{ m} = 4550$

b_w = width of web = 400 mm.

$$\therefore \quad b_f = \frac{4550}{\frac{4550}{1870} + 4} + 400 = 1106 \text{ mm}$$

Assume lever-arm =0.9 d, the moment of resistance is given by

$$M = 0.45\, b_f\, \sigma_{cbc}\, D_f \left[\frac{2kd - D_f}{k}\right]$$

$$560 \times 10^6 = 0.45 \times 1106 \times 5 \times 305 \left[\frac{2 \times 0.404\, d - 305}{0.404}\right]$$

or $$0.808\, d = \frac{560 \times 10^6 \times 0.404}{0.45 \times 1106 \times 5 \times 305} + 305$$

From width $d =$ 746 mm

Let us find the depth from punching shear point of view also. The column B will have to punch through depth d_b of the beam on two sides, and through depth d_s of the slab on the other two sides. The critical plane lies at a distance of $\frac{d}{2}$ from column face on two sides and at $\frac{d_s}{2}$ on the other two sides, where d_s is the depth of slab. The critical plane occurs at a distance of half the effective depth d_b of column on one side and $\frac{d_s}{2}$ to the other side. Hence widths of the punching shear zone are :

$$b_{01} = 400 + d_b$$

and $$b_{02} = 400 + d_s = 400 + 350 = 750 \text{ mm}$$

$\therefore$ Punching shear area of concrete available along the perimeter of the critical plane

$$= (2b_{01} \times d_s') + (2b_{02} \times d_b)$$

where d_s' = depth of slab at the critical plane (Fig. 65.46)

$$= d_{se} + \frac{d_s - d_{se}}{a}\left[a - \frac{d_s}{2}\right]$$

$$= 140 + \frac{350 - 140}{1000}\left[1000 - \frac{350}{2}\right] = 312 \text{ mm}$$

Punchig shear $F = 900 \times 10^3 - 110 \times 10^3 \left[\frac{(400 + d_b) \times 750}{(1000)^2}\right]$

$$= 9 \times 10^5 - 82.5\,[(400 + d_d)\,]$$

$\therefore$ Punching stress $= \dfrac{F}{2[(b_{01} \times d_s') + (b_{02} \times d_b)]}$

$$= \frac{9 \times 10^5 - 82.5\,(400 + d_b)}{2\,[(400 + d_b)\,312 + 750\,d_b]} \quad ...(i)$$

Permissible shear stress $= k_s\,\tau_c$

where $k_s = (0.5 + \beta_c) \not> 1 = 1$

$\tau_c = 0.16\sqrt{f_{ck}} = 0.16\sqrt{15} = 0.62 \text{ N/mm}^2$

$$\therefore \quad \frac{9 \times 10^5 - 82.5\,(400 + d_b)}{2[(400 + d_b)\,312 + 750\,d_b]} = 0.62$$

or

$$\frac{867000 - 82.5\,d_b}{124800 + 1062\,d_b} = 1.24$$

From which d_b = 510 mm.

This is less than the effective depth of 746 mm found from the bending compression.

3. *Steel reinforcement*

For hogging bending moment M_1, assuming lever arm = $0.9d$, the area of steel is given by

$$A_{st} = \frac{M_1}{\sigma_{st}\,(0.9d\,)} = \frac{560 \times 10^6}{140 \times 0.9 \times 746}$$

$$= 5958 \text{ mm}^2.$$

Using 25 mm Φ bars, $A_\Phi = 491 \text{ mm}^2$

$$\therefore \quad \text{No. of bars} = \frac{5958}{491} = 13.$$

Let us check this for development length of the point of contraflexure, so that the following equation is satisfied :

$$\frac{M}{V} + L_o \geq L_d$$

Assuming that all the 13 bars are available at the point of contraflexure near column B,

$$M = \sigma_{st.} A_{st.} jd = 140\,(13 \times 491) \times 0.9 \times 746$$
$$= 6 \times 10^8 \text{ N-mm.}$$
$$V = \text{shear force } F_4 \text{ at the point of contraflexure}$$
$$= 533.8 \text{ kN} = 533.8 \times 10^3 \text{ N}$$
$$L_o = 12\Phi \text{ or } d \text{ which ever is more}$$
$$= 746 \text{ mm}$$
$$L_d = 58.3\,\Phi = 58.3 \times 25 = 1458 \text{ mm}$$
$$\therefore \quad \frac{M}{V} + L_o = \frac{6 \times 10^8}{533.8 \times 10^3} + 746 = 1124 + 746 = 1870 \text{ mm}$$

which is more than L_d. Hence O.K.

Continue these 13 bars upto a point distant L_0 (=746 mm) beyond the point of contraflexure or to a point distant 746 − 300 − 200 = 246 mm from the outer face of the column B. Similarly, at the other side, continue these bars upto the outer face of column A. These bars are to be provided at the upper face of the beam, in two layers. Beyond the above two points, discontinue all bars except 3 bars which may be continued in the cantilever portions of the beam to serve as anchor bars. Using a cover of 60 mm, and keeping vertical distance between the two layers = 25 mm, total depth of beam = 746 + 12.5 + 12.5 + 60 = 831 mm. However, keep total depth of beam = 840 mm so that available d=840 − 12.5 − 12.5 − 60 = 755 mm.

The reinforcement in the cantilever portion near column B is to be found for B.M. M_3 = 120 kN-m. Here, the beam will act as simple beam.

$$A_{st3} = \frac{M_3}{\sigma_{st}\, jd} = \frac{120 \times 10^6}{140 \times 0.865\,(840 - 60)} = 1270 \text{ mm}^2$$

$\therefore$ No. of 12 mm Φ bars, having $A_\Phi = 113 \text{ mm}^2$ is

$$n = \frac{1270}{113} = 11.3 = 12.$$

Let us check this reinforcement for development length at the point of contraflexure near column B, so as to satisfy the equation

$$\frac{M}{V} + L_o \geq L_d$$

Assuming that all the 12 bars are available there,

$$M = \sigma_{st} A_{st}\, jd = 140(12 \times 113) \times 0.865 \times 780 = 1.28 \times 10^8 \text{ N-mm}$$

V = S.F. at point of contraflexure $= F_4 = 533.8$ kN

L_0 = 12 Φ or d whichever is more = 780 mm

$L_d = 58.3\,\Phi = 58.3 \times 12 = 700$ mm.

$$\frac{M}{V} + L_o = \frac{1.28 \times 10^8}{533.8 \times 10^3} + 780 = 240 + 780 = 1020 \text{ mm}$$

which is more than L_d.

However, continue all the 12 bars upto a distance of L_0(=780 mm) from the point of contraflexure, or upto a point distant 100 + 780 = 880 from the inner face of column B. Beyond this point, discontinue all bars, except 3 bars. These bars are to provided at the lower face of the beam in two layers.

Similarly, the steel reinforcement for the cantilever portion to the left of column A can be determined for $M_2 = 47$ kN-mm.

$$A_{st2} = \frac{M_2}{\sigma_{st}\, jd} = \frac{47 \times 10^8}{140 \times 0.865\,(840 - 60)} = 498 \text{ mm}^2$$

$$\therefore \text{ No. of 12 mm } \Phi \text{ bars} = \frac{498}{113} = 5.$$

Let us check this reinforcement for development length at the point of contraflexure near column A, so as to satisfy the equation

$$\frac{M}{V} + L_o \geq L_d$$

where $M = \sigma_{st} A_{st}\, jd = 140\,(5 \times 113) \times 0.865 \times 780$

$= 53.37 \times 10^6$ N-mm

V = Shear force $F_2 = 441.8$ kN.

L_o = 12 Φ or d whichever is greater = 780 mm

$L_d = 58.3\,\Phi = 58.2 \times 12 = 700$ mm

$$\therefore \frac{M}{V} + L_o = \frac{53.37 \times 10^6}{441.8 \times 10^3} + 700 = 120 + 700 = 820 \text{ m}$$

Hence the criterion is satisfied. However, continue the 5 bars of 12 mm Φ, upto a point distant L_0 (=780 mm) beyond point of contraflexure, or upto a point distant 1150+780=1930 mm from the free edge of the beam. At this point, 2 bars may be discontinued, while the remaining 3 bars may be continued throughout the length, to serve an anchor bars.

4. *Shear reinforcement*

Refer Fig. 16.4., Diagonal tension crack will occur at F only if the zero sagging B.M. extends to d = 780 mm beyond the inner face of column A (or B). However, point of contraflexure falls earlier.

Hence there is a possibility of crack at the point of contraflexure near column B where S.F. $F_4 = 533.8$ kN.

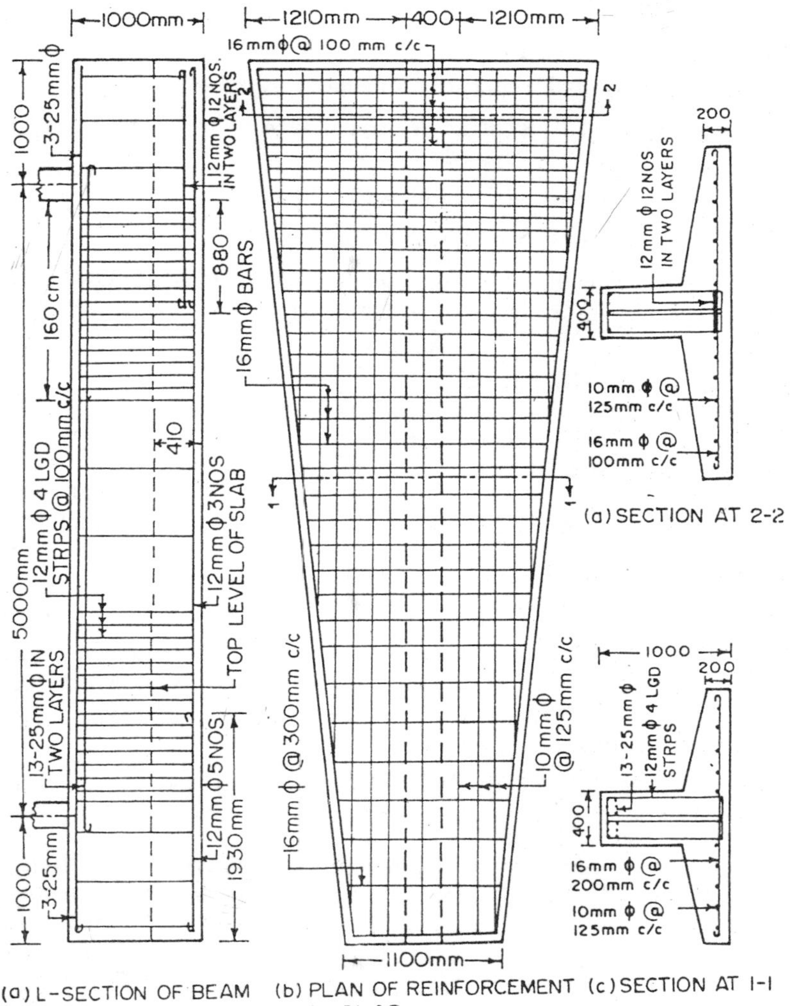

(a) L-SECTION OF BEAM (b) PLAN OF REINFORCEMENT IN SLAB (c) SECTION AT 1-1

FIG. 15.14

$$\therefore\ \tau_v = \frac{533.8 \times 10^3}{400 \times 780} = 1.71\ \text{N/mm}^2$$

$$100\frac{A_s}{bd} = 100 \times \frac{13 \times 491}{400 \times 780} = 2\%$$

In order to find the value of x, where the above S.F. occurs. Hence from Table 3.1, $\tau_c = 0.44$ N/mm^2.

Thus $\tau_v > \tau_c$ and shear reinforcement is necessary.

Hence crack will occur at the upper face of the beam if shear reinforcement is not provided.

$V = \tau_c . bd = 0.44 \times 400 \times 780 = 137280$ N $= 137.28$ kN

$V_s = 533.8 - 137.28 = 396.52$ kN

Using 12 mm Φ 4-lgd stirrups, $A_{sv} = 4 \times 113 = 452$ mm^2.

$$\therefore\quad s_v = \frac{\sigma_{sv}.A_{sv}.d}{V_s} = \frac{140 \times 452 \times 780}{396.52 \times 1000} = 124\ \text{mm}$$

$$= 120\ \text{mm c/c.}$$

Distance upto which this is required(*i.e.* at a point where S.F. $= V_c = 137.28$ kN) is obtained by putting $F_x = V_c = \pm 137.28$ in Eq. 2 of example 16.3.

$$\text{Hence}\quad F_x = \pm 137.28 = 600 - \frac{110}{2}x\,(2.22 + 0.245\,x\,)$$

Solving this, we get $x = 2.87$ m and 4.14 m.

These two points are correspondingly at distances of 2.87—1.15 =1.72 m from inner face of column A and 5.8—4.14=1.66 m from the inner face of column B. Between these two values of x, provide 12 mm Φ 4-lgd stirrups at nominal spacing of

$$s_v = \frac{2.5\,A_{sv}.f_y}{b} = \frac{2.5 \times 452 \times 250}{400} = 706\ \text{mm}$$

$$= 700\ \text{mm.}$$

From the inner face of column A to a distance of 1.72 m, the spacing of 12 mm Φ 4-lgd stirrups is given by

$$s_v = \frac{A_{sv}.\sigma_{sv}.d}{(F_3 - V_c)} = \frac{452 \times 140 \times 780}{(441.8 - 137.28)\,1000} = 162\ \text{mm}$$

However, for the sake of symmetry, provide these @ 120 mm c/c.

Thus, the shear reinforcement is summarised below :

1. From inner face of column A to 1.72 m distance, provide 12 mm Φ 4-lgd stirrups @ 120 mm c/c
2. From inner face of column B to a distance of 1.66 mm, provide 12 mm Φ 4-lgd stirrups @ 120 mm c/c.
3. For intermediate portion of beam, provide 12 mm Φ 4-lgd stirrups @ 700 mm c/c.
4. For the cantilever portions, where S.F. is less than V_c provide 12 mm Φ 4-lgd stirrups @ 700 m c/c, but at least 2 stirrups are necessary.

The details of reinforcement etc. are shown in Fig. 16.14

16.4 STRAP FOOTING

A strap footing consists of spread footings of two columns connected by a *strap beam*. This type of footing is useful when the external column(wall column) is very near to the property line so that its footing cannot be spread beyond the property line. If a combined trapezoidal footing is provided, the bending moment and shear force in the footing are very high since the footing is continuous under both the columns. In the case of strap footing, the individual spread footings have very large area under the columns, resulting in decrease in the maximum bending moment and shear force. The beam connecting the two spread footing does not transfer any load to the soil. The individual footing areas are so arranged that the C.G. of the combined loads of the two columns pass though the combined C.G. of the two footing areas. Once this criterion is achieved the pressure distribution below each individual footing will be uniform. The function of the strap beam is to transfer the load of heavily loaded outer column to the inner one. In doing so, the strap beam is subjected to bending moment and shear force, and it should be suitably designed to withstand these.

Fig.16.15 shows two columns A and B, transmitting axial loads W_1 and W_2 and are spaced l apart, centre to centre. Let W' be the total weight of both the individual footings. If A_1 and A_2 are the individual footing areas, and q_o is the safe bearing capacity of soil, we have

$$A_1 + A_2 = \frac{W_1 + W_2 + W'}{q_o}$$

or

$$B\,(L_1 + L_2) = \frac{W_1 + W_2 + W'}{q_o} \qquad ...(1)$$

where B is the common width of each footings and L_1 and L_2 are the individual lengths of the footings. Length L_2 is arranged centrally under column B.

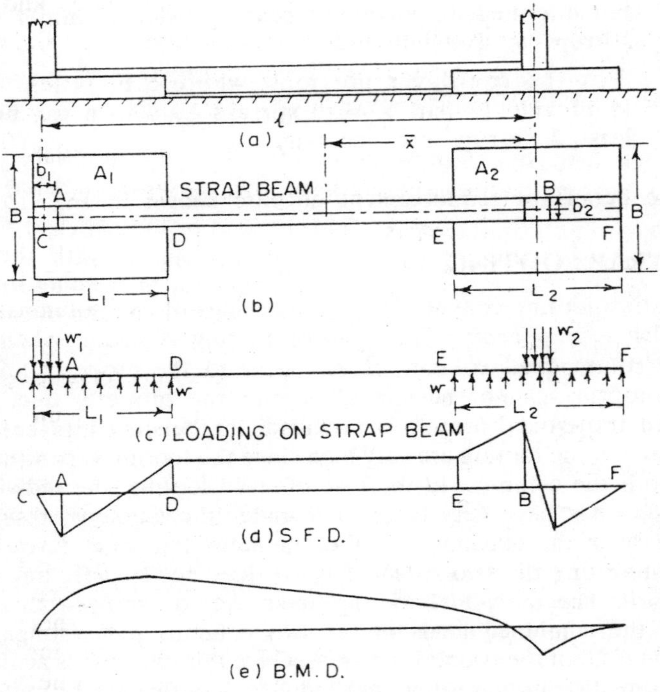

FIG. 16.15

The C.G. of the resultant load $W = W_1 + W_2$, falls at $\bar{x}$ from the centre of column B, so that

$$\bar{x} = \frac{W_1 l}{W_1 + W_2} \quad ...(i)$$

Let $(b_1 \times b_1)$ and $(b_2 \times b_2)$ be the size of the columns A and B respectively. Taking moments of the footing areas about centre of column B, we get

$$\bar{x} = \frac{(B \times L_1)\left(l + \frac{1}{2}b_1 - \frac{L_1}{2}\right)}{B(L_1 + L_2)} \quad ...(ii)$$

Equating (*i*) and (*ii*) , we get

$$\frac{L_1\left(l+\frac{1}{2}b_1-\frac{L_1}{2}\right)}{L_1+L_2}=\frac{W_1\,l}{W_1+W_2} \qquad ...(2)$$

From (1) and (2), the unknowns L_1 and L_2 can be known in terms of any suitable value of B.

Net upward soil pressure $p_o = \dfrac{W_1+W_2}{B(L_1+L_2)}$

Each individual footing is designed as cantilever slab, having sagging bending moment in each of the cantilever portion.

The strap beam transfers a part of load of footing A to footing B, in such a way that C.G. of two loads coincide with the C.G. of the footing areas. In doing so, it is subjected to bending moment and shear force all along its length, Fig. 16.15 (*c*) shows the loading diagram on the strap beam *CDEF*.

The upward uniform load w per unit length, under *CD* and *EF* will be equal to ($p_o \times B$). The downward w_1 per unit length in portion under column $A = \dfrac{W_1}{b_1}$ per unit length. Similarly downward load $w_2 = \dfrac{W_2}{b_2}$ per length. The characteristic bending moment diagram and S.F. diagram are sketched in Fig. 16.15 (*c*) and (*d*) respectively. For large portion of the strap beam, it will experience hogging bending moment, the maximum being at the section where S.F. is zero. For the cantilever portion beyond B, it will have sagging bending moment also, its maximum value being at the place where S.F. is zero.

Design Example 16.15 *Design a strap footing for two columns A and B, spaced 5 metres centre to centre. Column A, 300 mm × 300 mm in size carries a load of 600 kN and is on the property line. Column B, 400 mm×400 mm in size, carries a load of 900 kN. The bearing capacity of soil is 120 kN/m². Use M 15 mix. Take* $\sigma_{st} = 140\ N/mm^2$.

Solution.

1. *Design constants*

For M 15 mix we have the following

$c = \sigma_{cbc} = 5\ \text{N/mm}^2$; $t = \sigma_{st} = 140\ \text{N/mm}^2$. $m = 19$;

$k = 0.404$; $j = 0.865$; $R = 0.874$

2. ***Size of footings***

Fig.16.16 (*a*) show the general arrangement of the footing.

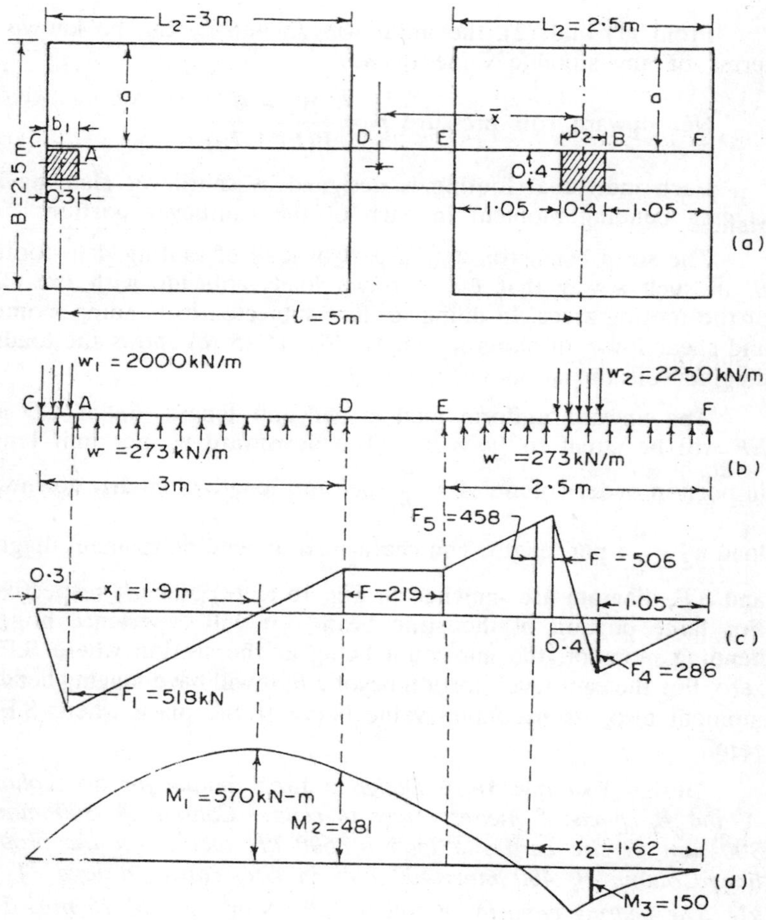

FIG. 16.16

Let the width of the two spread footings be B metres each. Length of footing under $A = L_1$ and length of footing under B be L_2 centrally arranged under B.

$$\therefore \quad W_1 = 600 \text{ kN} \; ; \; W_2 = 900 \text{ kN}.$$

Weight of footing, $W' = 0.1\,(600 + 900) = 150$ kN

$$\therefore \quad B(L_1 + L_2) = \frac{600 + 900 + 150}{120} = 13.75$$

or
$$L_1 + L_2 = \frac{13.75}{B} \qquad \text{...(1)}$$

Let $\bar{x}$ = distance of C.G. of loads from centre of column B,

$$\therefore \quad \bar{x} = \frac{W_1 l}{W_1 + W_2} = \frac{600 \times 5}{600 + 900} = 2 \text{ m}.$$

If $\bar{x}$ is also the distance of C.G. of areas, from the centre of column B, we have

$$\bar{x} = \frac{B \times L_1 \left(1 + \frac{b_1}{2} - \frac{L_1}{2}\right)}{B(L_1 + L_2)}$$

Substituting the values of $\bar{x}$, b_1 and l_1 we get

$$2 = \frac{L_1(5 + 0.15 - 0.5\,L_1)}{(L_1 + L_2)} \qquad \text{...(2)}$$

Substituting the value of $(L_1 + L_2)$ from (1), and choosing B=2.5 m, we get

$$\frac{L_1(5.15 - 0.5\,L_1)}{13.75/2.5} = 2$$

or
$$L_1^{\,2} - 10.3 L_1 + 22 = 0$$

From which L_1 = 3.025 m = 3 m (say).

$$\therefore \quad L_2 = \frac{13.75}{2.5} - 3.0 = 2.5 \text{ m}.$$

Net upward pressure $p_o = \dfrac{600 + 900}{2.5\,(3 + 2.5)} = 109$ kN/m^2

3. ***Design of footing slab***

Let the width of strap beam = 400 mm

∴ Cantilever projection, beyond the beam is

$$a = \frac{1}{2}(2.5 - 0.4) = 1.05 \text{ m}$$

Max. bending moment $M = p_o \dfrac{a^2}{2} = 109 \dfrac{(1.05)^2}{2}$

$$= 60 \text{ kN–m} = 60 \times 10^6 \text{ N–mm}.$$

$$\therefore \quad d = \sqrt{\frac{60 \times 10^6}{0.874 \times 1000}} = 262 \text{ mm}.$$

Keeping cover to centre of steel = 60 mm, D = 262+60=322 mm. However, keep D = 320 mm and d = 262 mm, so that effective cover available = 58 mm. Reduce to total depth to 200 mm at the edges, so that d = 200—60=140 mm there.

$$A_{st} = \frac{M}{\sigma_{st}\ jd} = \frac{60 \times 10^6}{140 \times 0.865 \times 262} = 1890 \text{ mm}^2$$

Using 16 mm Φ bars having A_Φ = 201 mm^2,

$$\text{Spacing} \quad s = \frac{1000 \times 201}{1890} = 106 \text{ mm.}$$

Hence provide 16 mm Φ bars @ 100 mm c/c.

Let us check this for development length.

$$L_d = 58.3\ \Phi = 58.3 \times 16 = 933 \text{ mm.}$$

Using 50 mm side cover, actual length available =(1050 – 50 + anchorage value of hook) > 933.

Hence O.K.

$$\text{Distribution reinforcement} = \frac{0.15}{100} \times 1000 \frac{(320 + 200)}{2} = 390 \text{ mm}^2.$$

$$\therefore \text{ Spacing of 10 mm } \Phi \text{ bars } = \frac{1000 \times 78.5}{390} \approx 200 \text{ mm c/c.}$$

4. *Bending moment and shear force diagrams for strap beam*

Upward load w per metre run, on the strap beam is given by

$$w = p_o B = 109 \times 2.5 = 273 \text{ kN/m.}$$

$$\text{Downward load of column } A = w_1 = \frac{600}{0.3} = 2000 \text{ kN/m.}$$

$$\text{Downward load of column } B = w_2 = \frac{900}{0.4} = 2250 \text{ kN/m.}$$

The loading diagram on the strap beam is shown in Fig. 16.16(b).

The shear force diagram for the strap beam is shown in Fig. 16.16(c).

S.F. at inner face of column $A = F_1 = (2000 \times 0.3) - (273 \times 0.3)$ = 518 KN.

S.F. at edge D = $F_2 = 273 \times 3 - 600 = 219$ kN.

S.F. at edge E = $F_2 = 219$ kN.

S.F. at inner edge of column $B = F_3 = 219 + 273 \left\{\frac{1}{2}(2.5 - 0.4)\right\}$ = 506 KN.

S.F. at outer edge of column $B = F_3 = 900 - 506 - (273 \times 0.4)$ = 286 KN.

or $= 273 \times \left\{\frac{1}{2}(2.5 - 0.4)\right\} = 286$ kN

In the range CD, shear force is zero at distance x_1 from the inner face of column A, its value being given by

$$x_1 = \frac{518}{273} = 1.9 \text{ m.}$$

Hogging bending moment will be maximum at this section, its values being given by

$$M_1 = (2000 \times 0.3)(1.9 + 0.15) - 273\frac{(0.3 + 1.9)^2}{2} = 570 \text{ kN–m}$$

Hogging B.M. at edge D is given by

$$M_2 = (2000 \times 0.3)(3 - 0.15) - \frac{273(3)^2}{2} = 481 \text{ kN–m}$$

Sagging B.M. at the outer face of column B is given by

$$M_3 = \frac{273}{2}\left[\frac{1}{2}(2.5 - 0.4)\right]^2 = 150 \text{ kN–m}$$

Let the point of contraflexure occur at x_2 from point F.

$$\therefore \; M_x = 0 = 273\left(\frac{x_2^2}{2}\right) - (2250 \times 0.4)(x_2 - 1.25)$$

which given $x_2 = 1.62$ m

Shear force at the point of contraflexure is given by

$$F_5 = (2250 \times 0.4) - (273 \times 1.62) = 458 \text{ kN}$$

Depth of strap beam

Let the width of strap beam be 400 mm. Since it projects above the footings, T-beam action will be available in range CD. Maximum bending moment is $M_1 = 570$ kN-m. However, since T-beam action is available here, the critical section will be at D, where the beam acts as rectangular beam. Bending moment at $D = M_2 = 481$ kN-m $= 481 \times 10^6$ N–mm.

$$\therefore \qquad d = \sqrt{\frac{481 \times 10^6}{0.874 \times 400}} = 1174 \text{ mm.}$$

Let us find moment of resistance of T-beam , with $d = 1174$ mm.

Average thickness of slab $= \dfrac{262 + 140}{2} = 200$ mm $= D_f$

For an isolated T-beam, width of flange is given by

$$b_f = \frac{l_o}{\left(\frac{l_o}{b} + 4\right)} + b_w$$

where $l_0 =$ distance between points of zero moments in beam$= (5 + 0.15 + 1.25) - 1.62 = 4.78$ m $= 4780$ mm.

$b =$ actual width of flange $= 2.5$ m $= 2500$ mm

$b_w =$ width of web $= 400$ mm.

$$\therefore \; b_f = \frac{4780}{\left(\frac{4780}{2500} + 4\right)} + 400 = 1208 \text{ mm.}$$

Taking lever arm $= 0.9\,d$, the moment of resistance of T-beam is given by

$$M_r = 0.45\, b_f.\, \sigma_{cbc}.\, D_f \left[\frac{2kd - D_f}{k}\right]$$

$$= 0.45 \times 1208 \times 5 \times 200 \left[\frac{2 \times 0.404 \times 1174 - 200}{0.404}\right]$$

$$= 1007 \times 10^6 \text{ N–mm} = 1007 \text{ kN–m}$$

which is much more than the hogging bending moment $M_1 = 570$ kN. Hence $d = 1174$ mm is safe.

6. *Reinforcement in strap beam*

For hogging bending moment M_1, lever arm $= 0.9\,d$ and the area of steel is given by

$$A_{st1} = \frac{M_1}{\sigma_{st}\, jd} = \frac{570 \times 10^6}{140 \times 0.9 \times 1174} = 3853 \text{ mm}^2$$

For hogging B.M. M_2, area of steel is given by

$$A_{st2} = \frac{481 \times 10^6}{140 \times 0.865 \times 1174} = 3383 \text{ mm}^2$$

Hence provide reinforcement equal to 3853 mm^2

Using 20 mm Φ bars, $A_\Phi = 314$ mm^2

$$\therefore \text{ No. of bars} = \frac{3853}{314} = 12.2 \approx 13.$$

Let us check the reinforcement for development of stresses, at point of contraflexure, so as to satisfy the criterion :

$$\frac{M}{V} + L_0 \geq L_d.$$

Assuming that the 13 bars, provided at the top face of the beam, are *available* at the point of contraflexure, we have

$$M = \sigma_{st} A_{st}\, jd = 140 \times (13 \times 314) \times 0.9 \times 1174$$

$$= 6.04 \times 10^8 \text{ N–mm}$$

V = S.F. at point of contraflexure $= F_5 = 458$ kN

L_0 = 12 Φ or d, whichever is more

= 1174 mm.

$L_d = 58.3\,\Phi = 58.5 \times 20 = 1166$ mm.

$$\therefore \quad \frac{M}{V} + L_0 = \frac{6.04 \times 10^6}{458 \times 1000} + 1174 = 1319 + 1174 = 2493 \text{ mm}$$

which is much more than L_d. This suggests that it is not necessary to take all the 13 bars upto the point of contraflexure. However, we will continue all the bars upto a distance of L_0 (= 1174 mm) from the point of contraflexure, or to a distance of 1620 − 1170 = 450 mm from the end of the beam. After that, discontinue all the bars, except 3 bars which will serve as anchor bars for shear stirrups.

Similarly at the face of support near B, the above reinforcement should satisfy the criterion

$$\frac{M}{V} + L_0 \geq L_d$$

where $M = 6.04 \times 10^8$ N-mm, as above

V = shear force = 518 kN.

$$L_0 = \frac{b}{2} - \text{cover} + \text{anchorage value of hook}$$

$$= 150 - 60 + (13 \times 20) = 350 \text{ mm.}$$

$$L_d = 58.3\,\Phi = 58.3 \times 20 = 1166 \text{ mm}$$

$$\therefore \quad \frac{M}{V} + L_0 = \frac{6.04 \times 10^8}{518 \times 10^8} + 350 = 1166 + 350 = 1516 \text{ mm}$$

which is more than L_d . Hence satisfactory.

Hence 13 bars of 20 mm Φ will be sufficient. These bars are arranged in two layers, keeping a clear distance of 20 mm between the two layers. Hence required total depth $D = d + 10 + \frac{\Phi}{2} + \text{cover}$

$$= 1174 + 10 + \frac{20}{2} + 60 = 1254 \text{ mm.}$$

However provide $D = 1260$ mm.

Maximum sagging B.M. occurs at the outer face of column B, where bending moment $M_3 = 150$ kN–m

$$\therefore \quad A_{st} = \frac{150 \times 10^6}{140 \times 0.865\,(1260 - 60)} = 1032 \text{ mm}^2$$

Using 16 mm Φ bars having $A_\Phi = 201$,

No. of bars $= \frac{1032}{201} = 5.14 = 6.$

Let us check whether this reinforcement is sufficient from the point of view of development of stress at the point of contraflexure so as to satisfy the criterion

$$\frac{M}{V} + L_0 \geq L_d$$

where $M = \sigma_{st} A_{st} jd = 140\,(6 \times 201) \times 0.865 \times 1200$

$= 1.75 \times 10^8$ N–mm

$V =$ S.F. at the point of contraflexure

$= 458$ kN $= 458 \times 10^3$ N

$L_0 = 12\,\Phi$ or d whichever is more $= 1200$ mm

$L_d = 58.3\,\Phi = 58.3 \times 16 = 933$ mm.

$$\therefore \quad \frac{M}{V} + L_0 = \frac{1.75 \times 10^8}{458 \times 10^3} + 1200 = 382 + 1200 = 1582 \text{ mm}$$

which is more than L_d, and hence safe.

Hence provide 6-16 mm Φ bars at the lower face, and continue these to a point distant L_0(= 1200 mm) from the point of contraflexure, or to a point distant 1200 + 1620 = 2800 mm from the end of the beam. After that, discontinue all the bars, except 3 bars which will serve as anchor bars.

7. ***Reinforcement for diagonal tension***

$$100\frac{A_s}{bd} = \frac{100 \times 13 \times 314}{400 \times 1174} = 0.87\%$$

Hence from Table 3.1, $\tau_c = 0.35$ N/mm^2

∴ Shear resistance of concrete, $V_c = 0.35 \times 400 \times 1174$

$= 164360$ N $= 164.36$ kN.

Hence shear reinforcement is necessary wherever shear force exceeds 164.36 kN

At the inner face of column A, shear force $F_1 = 518$ kN

∴ $V_s = F_1 - V_c = 518 - 164.36 = 353.64$ kN.

Using 10 mm Φ 4-lgd stirrups, $A_{sv} = 4 \times \frac{\pi}{4}(10)^2 = 314$ mm^2

$$\therefore \text{ Spacing } s_v = \frac{A_{sv}\,\sigma_{sv}\,d}{V_s} = \frac{314 \times 140 \times 1174}{353.64 \times 10^2} = 146 \text{ mm}$$

$= 140$ mm c/c.

Distance of point where shear force is 164.36 kN, from the inner face of column $= \dfrac{518 - 164.36}{273} \approx 1.3$ m.

Hence this distance from end of beam $= 1.3 + 0.3 = 1.6$ m.

Hence provide 10 mm Φ 4-lgd stirrups @ 140 mm c/c from end of the beam, upto a distance of 1.3 m.

At the inner face of column B_1 shear force $F_3 = 506$ kN. Hence spacing of 10 mm Φ 4-lgd stirrups is given by

$$s_v = \frac{314 \times 140 \times 1174}{(506 - 164.36)\, 10^3} = 150 \text{ mm}$$

However provide these stirrups @ 140 mm c/c from point E to the inner face of the column.

Similarly, at the outer face of column B, shear force $F_4 = 286$ kN.

$$\therefore \qquad s = \frac{314 \times 140 \times 1174}{(286 - 164.36)\, 10^3} = 424 \text{ mm.}$$

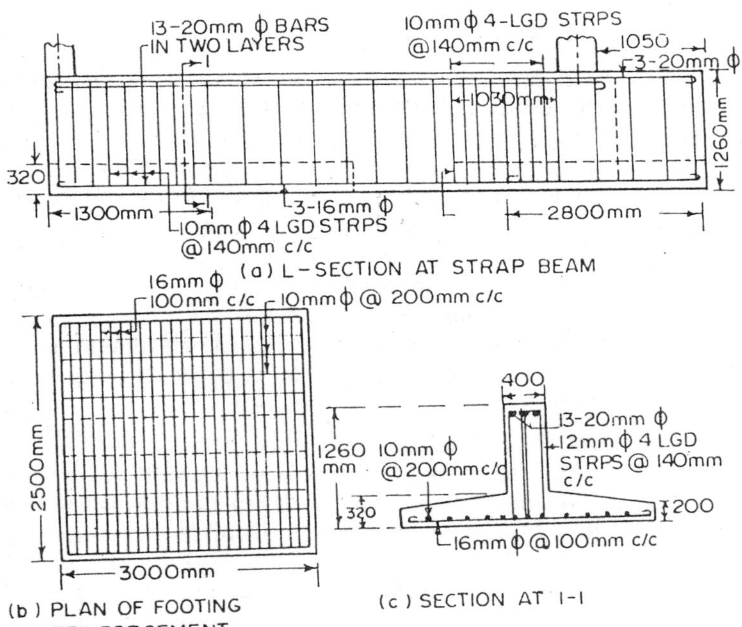

FIG 16.17.

Maximum spacing of stirrups is given by

$$s_v = \frac{2.5\,A_{sv}f_y}{b} = \frac{2.5 \times 314 \times 250}{400} = 490 \text{ mm.}$$

Hence keep s_v = 400 mm from the outer face of the column to the end of the beam.

At the point D, shear force $F_2 = 219$ kN.

$$\therefore \qquad s_v = \frac{314 \times 140 \times 1174}{(219 - 164.36)\,10^3} = 940 \text{ mm.}$$

subject to a maximum of 490 mm.

Hence for the remaining portion, provide stirrups of spacing of 490 mm c/c.

8. ***Check for punching shear***

The beam can be checked against punching shear at columns *A* and *B*, as explained in Example 16.2. The section will be found to be safe at both the locations.

16.5 RAFT FOOTING

A raft or mat is a combined footing that covers the entire area beneath a structure and supports all the walls and columns. When the allowable soil pressure is low, or the building loads are heavy, the use of spread footings would cover more than one-half of the area and it may prove more economical to use mat or raft foundation. They are also used where the soil mass contains compressible lenses or the soil is sufficiently erratic so that the differential settlement would be difficult to control. The mat or raft tends to bridge over the erratic deposits and eliminates the differential settlement. Raft foundation is also used to reduce settlement above highly compressible soil, by making the weight of structure and raft approximately equal to the weight of the soil excavated.

Ordinarily, rafts are designed as inverted beam slab system. If the C.G. of loads coincide with the centroid of the raft, the upward load is regarded as a uniform pressure equal to the downward load divided by the area of the raft. The weight of raft is not considered in the structural design because it is assumed to be carried directly by the sub-soil. When the column loads are equal, flat slab construction can be adopted. When raft foundation is also subjected to wind load moments, and when these are considered in design, the safe stresses can be increased by $33\frac{1}{3}$%. However, when the wind load reactions are less than $\frac{1}{3}$rd reactions due to other loads, wind is not considered in the design.

Example 16.6 illustrates the design procedure.

Design Example 16.6. *Fig. 16.18 shows the layout of the columns of a building. The outer columns are* 300 × 300 *mm in size and carry a load of 500 kN each. The inner columns are 400×400 mm in size and carry a load 800 kN each. In addition to this, each column carries a moment 160 kN-m due to wind load on the length of the building. Design a suitable raft foundation, if the bearing capacity of soil is 100 kN/m². Use M 15 mix. Take* $\sigma_{st} = 140$ *N/mm².*

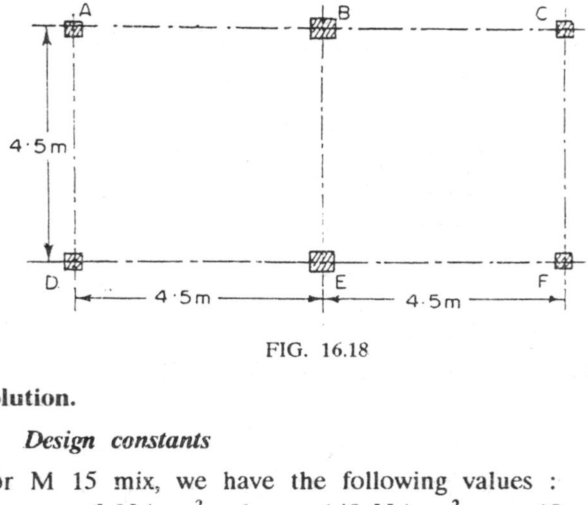

FIG. 16.18

Solution.

1. *Design constants*

For M 15 mix, we have the following values :

$\sigma_{cbc} = c = 5$ N/mm² and $\sigma_{st} = 140$ N/mm²; $m = 19$;

$k = 0.404$; $j = 0.865$; $R = 0.874$

2. *Size and general arrangement of mat*

Total weight $W = (4 \times 500) + (2 \times 800) = 3600$ kN

Self-weight of raft. $W' = 10\%\ W$ (say) $= 360$ kN

$\therefore$ Total load $= 3600 + 360 = 3960$ kN

Total moment $= 160 \times 6 = 960$ kN-m

$$\therefore \quad \text{Eccentricity } e = \frac{960}{3960} = 0.242 \text{ m}$$

Let the raft project 500 mm outside the column all round, so that length of raft $= (9 + 0.3 + 2 \times 0.5) = 10.3$ m, and width of raft $= (4.5 + 0.3 + 2 \times 0.5) = 5.8$ m.

Hence maximum and minimum intensities of pressures would be

$$= \frac{\Sigma W}{A}\left(1 \pm \frac{6\,e}{B}\right) = \frac{3960}{10.3 \times 5.8}\left(1 \pm \frac{6 \times 0.242}{5.8}\right)$$

$$\approx \mathbf{83\ N/m^2\ and\ 5\ N/m^2}$$

Since the maximum pressure is less than 100 kN/m^2 the size of the footing is satisfactory.

$$\text{Net upward soil reaction } p_0 = \frac{3600}{10.3 \times 5.8} \approx 60 \text{ kN/m}^2$$

Upward reaction due to wind moment is given by

$$p_w = \pm \frac{960}{\frac{1}{6} \times 10.3\,(5.8)^2} = \pm\ 16.6 \text{ kN/m}^2$$

i.e., +16.6 kN/m^2 on leeward side and −16.6 kN/m^2 on windward side.

Pressure intensity at inner face of longitudinal beams

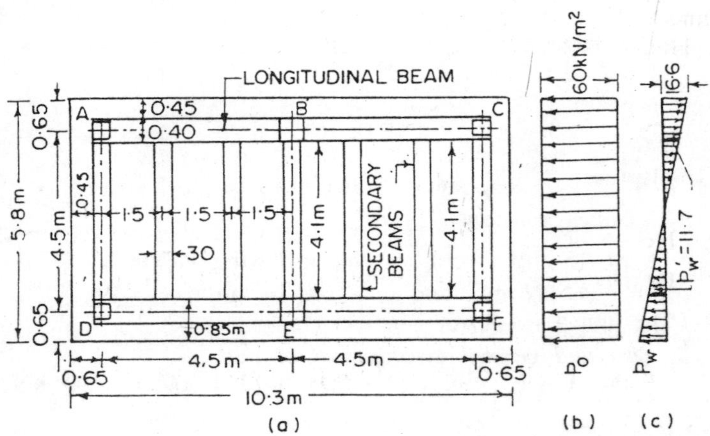

FIG. 16.19.

$$= \pm\ 16.6 \times \frac{4.1}{5.8} = \pm\ 11.7 \text{ kN/m}^2$$

Let us provide two longitudinal beams in direction *AC* and *DF*, joining three columns on each side. These two beams are joined by seven secondary beams. The general arrangement of beams etc. are shown in Fig. 16.19.

3. *Design of slab*

The soil pressure distribution due to load and wind moment are shown in Fig. 16.19 (*b*) and (*c*) respectively. Since the pressure p_w due to wind moments is less than $\frac{1}{3}$rd the pressure p_0 due to dead load, the effect of wind will not be considered in the design of slab.

The slab is to be designed for the greater of the following two moments :

(*a*) Bending moment in the cantilever portion of slab.

(*b*) Bending moment in the continuous slab.

Cantilever span $= 0.5 \text{ m} = a$ ...(*i*)

$$M_1 = p_0 \frac{a^2}{2} = \frac{60}{2}(0.5)^2 = 7.5 \text{ kN-m}$$

shear force $F_1 = p_0 a = 60 \times 0.5 = 30$ kN ...(*ii*)

The slab is continuous over the secondary beams. Centre to centre spacing of secondary beams=1.5 m. Taking width of secondary beams as 0.3 m, the clear span of slab, between two beams =1.5–0.3=1.2 m. Hence bending moment in interior spans is

$$M_2 = \frac{p_0 L^2}{12} = \frac{60(1.2)^2}{12} = 7.2 \text{ kN-m} \quad ...(iii)$$

$$\text{S.F.} \quad F_2 = p_0 \frac{L}{2} = \frac{60(1.2)}{12} = 36 \text{ kN} \quad ...(iv)$$

Hence design B.M. M =7.5 kN–m = 7.5×10^6 N-mm

Design shear force F = 36 kN.

The effective depth of slab is given by

$$d = \sqrt{\frac{7.5 \times 10^6}{0.874 \times 1000}} = 92.6 \text{ mm}$$

Keeping an effective cover of 50 mm and a total depth of 150 mm, available effective depth =150—50=100 mm.

Area of reinforcement is given by

$$A_{st} = \frac{7.5 \times 10^6}{140 \times 0.875 \times 100} = 612 \text{ mm}^2$$

Using 10 mm Φ bars, A_Φ = 78.5 mm^2

$$\therefore \text{ Spacing } s = \frac{1000 \times 78.5}{612} = 128 \text{ mm}$$

Hence provide 10 mm Φ bars @ 120 mm c/c.

Let us check the reinfoıcement for development length.

L_d required = 58.3 Φ = 58.3 × 10 = 583 mm.

Providing a side cover of 50 mm, available length of each bar = 500 − 50 + anchorage value of hook

= 450 + 13 Φ = 450 + 130 = 580 mm only.

This is quite near to the required L_d. Hence O.K.

For cantilever portion, and portion under the face of the secondary beams, this reinforcement is to be provided at the bottom face of the slab since bending moment is of sagging nature, while for intermediate portion between secondary beams, the bars are to be provided on the top face, as is done in a continuous slab.

$$\text{Area of transverse reinforcement} = \frac{0.15}{100}(1000 \times 150)$$

$$= 225 \text{ mm}^2$$

$$\text{Using 10 mm } \Phi \text{ bars, spacing} = \frac{1000 \times 78.5}{225} = 348 \text{ mm.}$$

For the portion of slab projecting from the face of the main slab, cantilever projection =0.45 m.

$$\therefore \quad M = \frac{60(0.45)^2}{2} = 6 \text{ kN–m}$$

Now, actual length of each bar in the cantilever portion, using 50 mm side cover =450 − 50 + 13 Φ = 400 + 130 = 530 mm.

Required L_d = 58.3 × 10 = 583 mm.

$$\text{Hence actual } \sigma_{st} \text{ carried by each bar} = \frac{530}{538} \times 140$$

$$= 127 \text{ N/mm}^2$$

Effective depth available =100 − 10 = 90 mm

$$\therefore \quad A_{st} = \frac{60 \times 10^6}{127 \times 0.875 \times 90} = 6000 \text{ mm}^2$$

$$\text{and spacing of 10 mm } \Phi \text{ bars} = \frac{1000 \times 78.5}{6000} \approx 130 \text{ mm}$$

For the sake of symmetry, provide these also at 120 mm c/c.

Also, the spacing of transverse reinforcement (found as 380 mm c/c) may be kept at 2 × 120 = 240 mm c/c, so that every alternate bar of cantilever reinforcement may be used as distribution bars.

Let us check the slab for shear.

$$\tau_v = \frac{V}{bd} = \frac{F_2}{bd} = \frac{36 \times 10^3}{1000 \times 100}$$

$$= 0.36 \text{ N/mm}^2$$

$$\frac{100 A_s}{bd} = 100\left(\frac{1000 \times 78.5}{120}\right) \div (1000 \times 100) = 0.65\%$$

$\therefore$ From Table 3.1, τ_c = 0.32 N/mm^2.

Also, from Table 3.2, k = 1.3 for D = 150 mm

$\therefore$ permissible shear stress $= k\ .\tau_c = 1.3 \times 0.32 = 0.42$ N/mm^2 which is more than τ_v. Hence safe.

4. *Design of intermediate secondary beam.*

There are in all seven secondary beams spaced at 1.5 m c/c out which the two outer secondary beams and the central secondary beams, join columns. The remaining four secondary beams are intermediate beams which join the longitudinal beams. These intermediate beams are subjected to reaction from the slab beneath. Since the reaction due to wind moment is less than $\frac{1}{3}$rd the dead load reactions, it will not be considered in design.

Span of the beam = 4.5 m.

However, the load on the beam will be the upward load acting on the slab on the clear span of 4.1 m. For the remaining poriton of the slab, load will be transferred directly to the longitudinal beams.

Load per metre run $w = p_0 \times 1.5 = 60 \times 1.5 = 90$ kN/m

The loading is shown in Fig. 16.20

End reactions transferred to each longitudinal beam

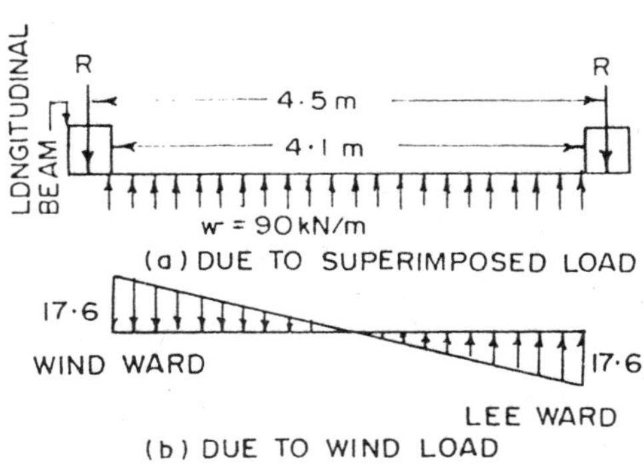

FIG. 16.20.

$$= R = \frac{1}{2}(4.1 \times 90) = 184.5 \text{ kN}$$

$$M = 184.5 \times \frac{4.5}{2} - \frac{90}{2}\left(\frac{4.1}{2}\right)^2 = 226 \text{ kN–m}$$

$$= 226 \times 10^6 \text{ N–mm}$$

$$F = R = 184.5 \text{ kN}$$

Since the beam experience hogging bending moment, it will act as T-beam. Let the width of beam $=b_w=300$ mm. The width of flange is given by

$$b_f = \frac{l_o}{6} + b_w + 6D_f$$

where $l_0 = 4.5 \text{ m} = 4500 \text{ mm}$

$b_w = 300 \text{ mm}$

$D_f = 150 \text{ mm}.$

$$\therefore \quad b_f = \frac{4500}{6} + 300 + 6 \times 150 = 1950 \text{ mm}$$

subject to maximum of centre to centre spacing(=1.5 m)

Hence $b_f = 1500$ mm. Taking lever arm $=0.9\,d$, effective depth is determined from the relation :

$$M = 0.45\, b_f .\, \sigma_{cbc} .\, D_f\left[\frac{2kd - D_f}{k}\right]$$

$$\therefore\ 226 \times 10^6 = 0.45 \times 1500 \times 5 \times 150 \left[\frac{2 \times 0.404\,d - 150}{0.404}\right]$$

$$\therefore\ 0.808\,d = \frac{226 \times 10^6 \times 0.404}{0.45 \times 1500 \times 5 \times 150} + 150 = 330.4$$

From which $d = 410$ mm

$\tau_{cmax} = 1.6 \text{ N/mm}^2$ for M 15 concrete (Table 3.3)

Hence maximum depth from considerations of shear is

$$d = \frac{F}{b.\,\tau_{c.\,ma}}$$

or
$$d = \frac{184.5 \times 10^3}{300 \times 1.6} = 384 \text{ mm.}$$

Hence adopt $d = 410$ mm.

$$A_{st} = \frac{226 \times 10^6}{140 \times 0.9 \times 410} = 4375 \text{ mm}^2$$

Using 25 mm Φ bars, No. of bars $= \dfrac{4375}{491} \approx 9.$

Provide these in two layers at the upper face of the beam. Keep a vertical distance of 25 mm between them and effective cover of 60 mm. Hence total depth = 410+12.5+12.5+60=495. Keep total depth of 500 mm.

Near ends, bend 3 bars downwards to take negative bending moment induced due to fixidity.

At the end, let us check for development length so as to satisfy the relation

$$\frac{M_1}{V} + L_0 \geq L_d$$

where $M_1 = \sigma_{st}.A_{st}.jd = 146(6 \times 491) \times 0.9 \times 410$

$$= 1.522 \times 10^8 \text{ N–mm}$$

$$V = 184.5 \text{ kN} = 184.5 \times 10^3 \text{ N}$$

$$L_0 = \frac{l_s}{2} - x' + 13\,\Phi = \frac{400}{2} - 40 + 13 \times 25 = 485 \text{ mm}$$

$$L_d = 58.3\,\Phi = 58.3 \times 25 = 1456 \text{ mm.}$$

$$\therefore \frac{M_1}{V} + L_0 = \frac{1.522 \times 10^8}{184.5 \times 10^3} + 485 = 1310 \text{ mm.}$$

This is less than required L_d.

Hence bend only two bars downwards, so that 7 bars are available at the top face.

Revised $M_1 = 1.522 \times 10^8 \times \frac{7}{6} = 1.78 \times 10^8$

$$\therefore \frac{M_1}{V} + L_0 = \frac{1.78 \times 10^8}{184.5 \times 10^3} + 485 \approx 1450 \text{ mm}$$

which is quite near to L_d

Also, $\tau_v = \frac{184.5 \times 10^3}{300 \times 410} = 1.5 \text{ N/mm}^2$

$$\frac{100 A_s}{bd} = \frac{100 \times 7 \times 491}{300 \times 410} = 2.8\%$$

Hence from Table 3.1, $\tau_c = 0.44 \text{ N/mm}^2$

Since $\tau_v > \tau_c$, shear reinforcements necessary.

$V_c = \tau_c.bd = 0.44 \times 300 \times 410 = 54120 \text{ N} = 54.12 \text{ kN}$

$\therefore V_s = V - V_c = 184.5 - 54.12 = 130.38 \text{ kN}$

Using 10 mm Φ 4-lgd stirrups, $A_{sv} = 4\frac{\pi}{4}(10)^2 = 314 \text{ mm}^2$

$$\therefore s_v = \frac{A_{sv}.\sigma_{sv}.d}{V_s} = \frac{314 \times 140 \times 410}{130.38 \times 1000} = 138 \text{ mm.}$$

The distance of the point where S.F. is equal to V_c is given by $x = \dfrac{184.50 - 54.12}{90} \approx 1.45$ m from the face of longitudinal beam.

Hence provide 10 mm Φ 4 lgd stirrups @ 130 mm c/c from the end to a point distant 1.45 m from the face of each longitudinal beam. For the remaining length, provide nominal stirrups at the maximum spacing given by

$$s_v = \frac{2.5.\,A_{sv}.f_y}{b} = \frac{2.5 \times 314 \times 250}{300} = 650 \text{ mm.}$$

Subject to a maximum of 450 mm or 0.75 d (=0.75 × 410 ≈ 300 mm). Hence provide 10 mm Φ 4-lgd nominal stirrups at a spacing of 300 mm c/c for the remaining length.

Let us calculate the reaction transferred by the secondary beam to main beam, due to wind load reaction p_x, since this will be useful for the design of main beam.

Load due to p_w, per metre run $= \pm\, 1.5\, p_w$

$= \pm\, 1.5 \times 11.7 = \pm\, 17.6$ kN/m

at the face of the main beam.

Couple due to this $= \left(\frac{1}{2} \times 17.6 \times 2.05\right)\left(\frac{2}{3} \times 4.1\right)$

$= 49.3$ kN–m

∴ End reactions $= \dfrac{49.3}{4.5} \approx 11$ kN.

∴ Net reaction at windward end =184.5—11=173.5 kN

Net reaction at leeward end = 114.5 + 11 = 195.5 kN.

5. *Design of central secondary beam*

The central secondary beam BE joins columns B and E and hence is subjected to end moments transmitted to it by the columns. Therefore, let us consider both p_o as well as p_w, along with the transferred end moments, to calculate the maximum bending moment and shear force. If the bending moment is greater than the intermediate secondary beam atleast by $33\frac{1}{3}\%$, then the design will change. Otherwise it will remain the same.

From Fig.16.19 (*c*), the pressure intensity p_w' due to wind, at the face of longitudinal beam = 11.7 kN/m².

∴ Net pressure intensity at $E' = 60 + 11.7 = 71.7$ kN/m².

Net pressure intensity at $B' = 60 - 11.7 = 48.3$ kN/m².

Width of slab transferring load = 1.5 m.

∴ Load intensity w_1 at end B' $=71.7 \times 1.5 = 107.6$ kN/m

Load intensity w_2 at end B' $=48.3 \times 1.5 = 72.5$ kN/m

The effective span of the beam =4.5 m while it will receive load only in a length of 4.1 m. The loading on the beam is shown in Fig. 16.21

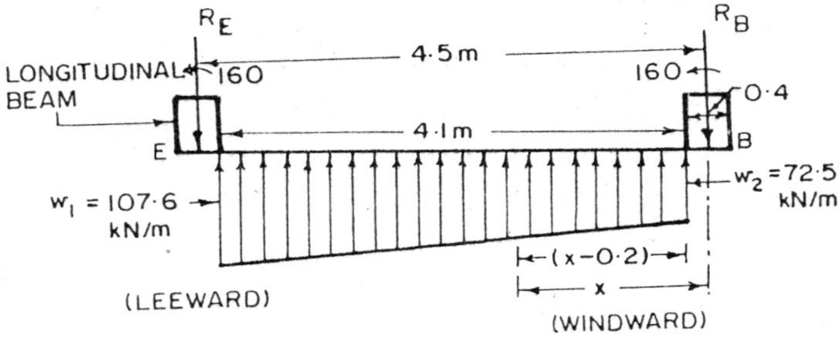

FIG. 16.21

To find reaction at E, take moments at B.

$$\therefore \; R_E = \frac{1}{4.5}\left[72.5 \times 4.1\left(\frac{4.1}{2} + 0.2\right) + \frac{107.6 - 72.5}{2} \times 4.1 \times \left\{\frac{2}{3}(4.1) + 0.2\right\} - 2 \times 160\right]$$

$$= 124.5 \text{ kN.}$$

Similarly, taking moments E,

$$R_B = \frac{1}{4.5}\left[72.5 \times 4.1\left(\frac{4.1}{2} + 0.2\right) + \frac{107.7 - 72.5}{2} \times 4.1 \times \left\{\frac{1}{3}(4.1) + 0.2\right\} + 2 \times 160\right]$$

$$= 244.8 \text{ kN}$$

Bending moment will be maximum where S.F. is zero. Let us write down equation of shear force at x from support B.

$$F_x = 244.8 - \left[72.5(x - 0.2) + \frac{107.6 - 72.5}{4.1}(x - 0.2)\,\frac{1}{2}\,(x - 0.2)\right]$$

Equating this to zero, we get $x = 3.1$ m

Hence maximum bending moment is given by

$$M = (244.8 \times 3.1) - 160 - \left[\frac{72.5(2.9)^2}{2} + \left(\frac{107.7 - 72.5}{4.1}\right) \times 3.1 \times \frac{1}{2} \times 3.1 \times \frac{3.1}{3}\right]$$

$= 251.5$ kN-m ...(*ii*)

It will be seen that maximum bending moment and shear force found above by taking the wind effect into account are *less* than $1\frac{1}{2}$ times the corresponding values for the intermediate secondary beam. *Hence the same depth and same main reinforcement will be used.* If however, the B.M. and S.F. found above were more than $1\frac{1}{2}$ times the B.M. and S.F. for the intermediate secondary beam, the central beam could be designed for the above increased B.M. and S.F., taking higher permissible values of stresses.

However, due to partial fixidity at the ends, the beam will be subjected to sagging B.M. $=\frac{WL}{24}$, in addition to wind moment of 160 kN-m.

$$\therefore \quad \text{Total sagging B.M.} = (124.5 + 244.8)\frac{4.5}{24} + 160$$

$$= 229.2 \text{ kN-m.}$$

Since the intermediate secondary beam has been designed for a B.M. of 226 kN-m, its depth $d = 410$ mm will be sufficient to take compression, but special reinforcement at the bottom face will have to be provided at the ends. However, provide the same reinforcement as the centre, *i.e.* 9 bars of 25 mm Φ provided in two layers at the bottom face for the portion under the columns for a distance at least equal to $L_d = 58.3 \times 2 \approx 1500$ mm.

6. *Design of end secondary beam*

The end secondary beam will get less load by way of reaction from the slab, because it will receive load from slab of width $= 0.65 + \frac{1}{2}(1.5) = 1.4$ m only against 1.5 m width for central secondary beam. However, provide the same section and reinforcement as for the central secondary beam. The ratio of load on the outer beam to the central one will be approximately equal to

$$\frac{1.4}{1.5} \approx 0.93$$

7. *Design of main beam*

The bending moments and shear forces in the main beam are to be found under the following three conditions :

(*a*) Ignoring wind effect.

(*b*) Taking wind effect : leeward beam.

(*c*) Taking wind effect ; windward beam.

(*a*) ***Ignoring wind effect***

When there is no wind acting, the reaction transferred to the main beam by all the intermediate as well as central beams will be equal, while the reaction transferred by the end beam will be 0.93 times that transferred by intermediate beam. In addition to this, the beam will get soil reaction from slab immediately below it and the slab cantilevering from its face.

Reaction transferred by each intermediate beam $= 184.5$ kN

Reaction transferred by end secondary beam $= 0.93 \times 184.5$

$= 172.5$ kN

Reaction transferred by slab, per metre run of beam $= 6 \times 0.85$

$= 60\,(0.4 + 0.45) = 51$ kN/m.

The loading diagram is shown in Fig. 16.22.

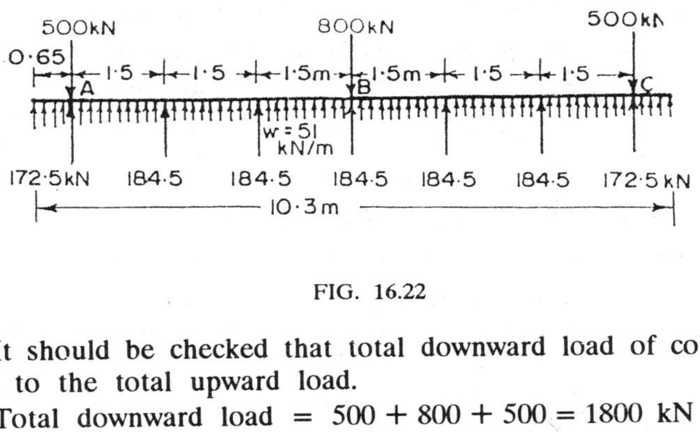

FIG. 16.22

It should be checked that total downward load of column is equal to the total upward load.

Total downward load $= 500 + 800 + 500 = 1800$ kN

Total upward load $= (184.5 \times 5) + (172.5 \times 2) + (10.3 \times 51)$

$= 1793$

Thus, both of these are approximately equal. The discrepancy is there because of rounding off the value of p_0 to 60 kN/m^2 in step 2.

Maximum sagging bending moment will occur at B, its value being given by

$$M_1 = 51\,\frac{(5.15)^2}{2} + (184.5 \times 1.5) + (184.5 \times 3.0) - (500 - 172.5)\,4.5$$

$= 32.8$ kN-m (sagging). ..(1)

Maximum hogging bending moment occurs somewhere between A and B, where shear force is zero. This is likely to occur between the two intermediate beams at distance x from A.

$\therefore \; F_x = 51\,(x + 0.65) + 172.5 + 184.5 - 500 = 0$

This gives x =2.15 m from A or 2.8 m from the edge. Hence the maximum hogging bending moment is given by

$$M_2 = (500 - 172.5)\,2.15 - 184.5\,(2.15 - 1.5) - \frac{51}{2}\,(2.8)^2$$

$$= 384.3 \text{ kN-m (hogging)} \qquad ...(2)$$

Shear force to the right of A is

$$F_1 = (500 - 172.5) - (51 \times 0.65) = 294.4 \text{ kN} \qquad ...(3)$$

Shear force to the left of B is

$$F_2 = \frac{1}{2}\,(800 - 184.5) = 307.8 \text{ kN}$$

(b) Wind effect : leeward beam

From step 4, reaction transferred by the intermediate secondary beam on the leeward end =195.5 kN.

From step 5, reaction transferred by the central secondary beam on the leeward end = 124.5 kN.

Reaction transferred by the end secondary beam, on the leeward

$$\text{end} = 0.93 \times 195.5 - \frac{1.60 + 160}{4.5}$$

$$= 182 - 71 = 111 \text{ kN}$$

Total downward load = 500 + 800 + 500 = 1800 kN

$\therefore$ Upward load from slab

$$= 1800 - (195.5 \times 4 + 124.5 + 111 \times 2)$$

$$= 671.50 \text{ kN}$$

$$\text{Uniform upward load } w = \frac{671.5}{10.3} = 65.2 \text{ kN/m}$$

The loading on the leeward beam DF is shown in Fig. 16.23.

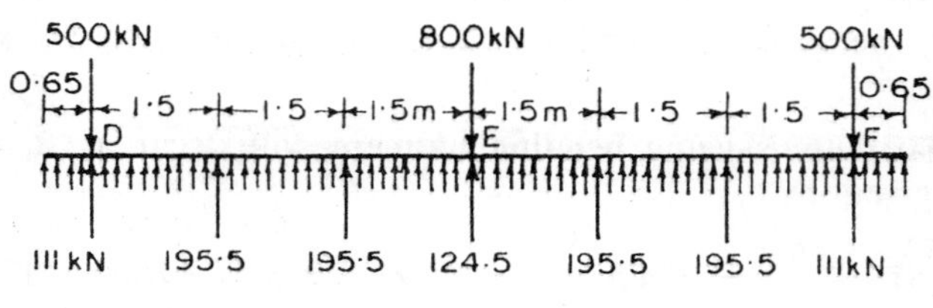

FIG. 16.23

Maximum sagging bending moment will occur at E, its value being

$$M_3 = \frac{65.2(5.15)^2}{2} + (195.2 \times 1.5) + (195.5 \times 3) - (500 - 111)4.5$$

$$= -6.1 \text{ kN-m } (i.e. \text{ hogging}) \qquad ...(5)$$

Maximum hogging bending moment occurs somewhere between D and E where shear force is zero. This is likely to occur between the two intermediate beams, at distance x from D.

$$F_x = 65.2(6.5 + x) + 111 + 195.5 - 500 = 0$$

This gives x = 2.32 m from D, or 2.97 m from the edge. Hence maximum hogging bending moment is given by

$$M_4 = (500 - 111) \times 2.32 - 195.5\,(2.67 - 1.5) - \frac{65.2}{3}(2.97)^2$$

$$= 327.5 \text{ kN-m (hogging)} \qquad ...(6)$$

S.F. to the right of D is

$$F_3 = (500 - 111) - (65.2 \times 0.65) = 346.6 \text{ kN} \qquad ...(7)$$

S.F. to the left of E is

$$F_4 = \frac{1}{2}(800 - 124.5) = 337.8 \text{ kN} \qquad ...(8)$$

(c) Wind effect : windward beam

From step 4, reaction transferred by the intermediate secondary beam on the windward side =173.5 kN

From step 5, reaction transferred by central secondary beam on the windward end =244.8 kN.

Reaction transferred by the end secondary beam on the windward end $= (0.93 \times 173.5) + \dfrac{160 + 160}{4.5} = 232.5$ kN

Total downward load =500 + 800 + 500 = 1800 kN

∴ Upward load from the slab = 1800 − (4 × 173.5 + 248.8 + 2 × 232.5) = 396.2 kN.

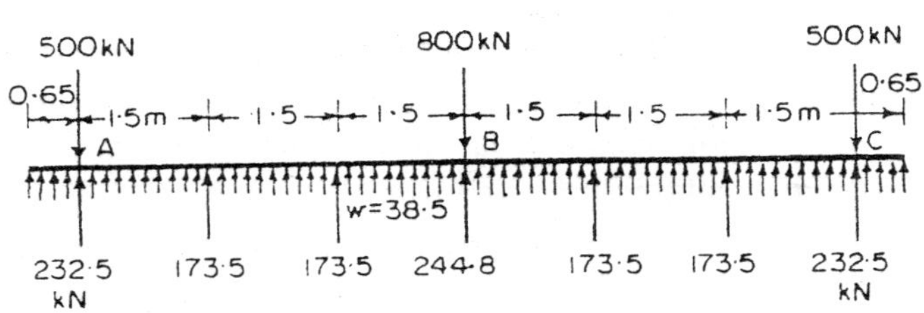

FIG. 16.24.

$\therefore$ Upward $w = \dfrac{396.2}{10.3} \approx 38.5$ kN/m.

The loading on the windward main beam AC is shown in Fig. 16.24.

Maximum sagging bending moment will occur at B :

$$M_5 = \frac{38.5(5.15)^2}{2} + (2 \times 173.5 \times 2.25) - (500 - 232.5) \times 4.5$$
$$= 87.6 \text{ kN-m}$$

Maximum hogging bending moment will occur where S.F. is zero, some where between A and B. Let it occur at x from A between the two intermediate beams.

$$F_x = 38.5\,(x + 0.65) + 173.5 + 232.5 - 500 = 0$$

where gives x = 1.79 m from A or 2.44 m from the edge.

Hence maximum hogging bending moment is given by

$$M_6 = (500 - 232.5) \times 1.79 - 173.5\,(1.79 - 1.5) - \frac{38.5}{2}(2.44)^2$$
$$= 314 \text{ kN-m (hogging)} \qquad \text{...(10)}$$

S.F. to the right of A is given by

$$F_5 = (500 - 232.5) - (0.65 \times 38.5) = 242.5 \text{ kN} \qquad \text{...(11)}$$

S.F. to the left of B is

$$F_6 = \frac{1}{2}(800 - 244.8) = 277.6 \text{ kN} \qquad \text{...(12)}$$

Design values

From the above cases, we get the following values for design purposes :

Without wind :	Maximum sagging moment	= 32.8 kN-m
	Maximum hogging moment	= 384.3 kN-m
	Maximum shear	= 307.8 kN
With wind :	Maximum sagging moment	= 87.6 kN-m
	Maximum hogging moment	= 327.5 kN-m
	Maximum shear	= 346.6 kN

Thus critical conditions are under 'no wind'. T-beam action will be available for maximum hogging bending moment = 384.3 kN-m. Let the width of rib =400 mm.

The flange width of T-beam is given by

$$b_f = \frac{l_0}{6} + b_w' + 6\,D_f$$

where l_0 = distance between points of zero moments in the beam

$= 10.3 - (2 \times 2.8) = 4.7 \text{ m} = 4700 \text{ mm}$

$b_w = 400 \text{ mm}$

$D_f = 150 \text{ mm}$

$$\therefore \quad b_f = \frac{4700}{6} + 400 + 6 \times 150 \approx 2083 \text{ mm}.$$

However, since the beam is situated nearly at the outer boundary, its half flange is restricted in its width to a value = 0.20 + 0.45 = 0.65 m = 650 mm. Hence maximum value of b_f is restricted to $2 \times 650 = 1300$ mm.

Hence adopt b_f = 1300 mm. Taking lever arm = 0.9 d, the effective depth is given by the relation :

$$M = 0.45\, b_f . \sigma_{cbc} . D_f \left(\frac{2kd - D_f}{k}\right)$$

or $$384.3 \times 10^6 = 0.45 \times 1300 \times 5 \times 150 \left(\frac{2 \times 0.404\, d - 150}{0.404}\right)$$

or $$0.808\, d = \frac{384.3 \times 10^6 \times 0.404}{0.45 \times 1300 \times 5 \times 150} + 150$$

From which d = 624 mm.

This depth will be sufficient to take the sagging bending moment when the beam will act as a simple rectangular beam.

$$A_{st} = \frac{384.3 \times 10^6}{140 \times 0.9 \times 624} = 4888 \text{ mm}^2$$

Using 25 mm Φ bars, No. of bars $= \dfrac{4888}{490} \approx 10.$

Hence provide these at the top face of the beam in two layers. This reinforcement should be safe to satisfy the development length criterion at the point of contraflexure, so as the satisfy the relation

$$\frac{M}{V} + L_0 \geq L_d .$$

Assuming that all the bars are continued even beyond the point of contraflexure.

$$M = \sigma_{st} . A_{st}\, jd = 140\,(10 \times 490) \times 0.9 \times 624$$

$$= 3.85 \times 10^8 \text{ N-mm}$$

V = Shear at the point of contraflexure

Assuming that the point of contraflexure falls just to the left of B, shear force $V = F_2$ = 307.8 kN

L_0 = 12 Φ or d whichever is more = 624 mm

$L_d = 58.3 \times 25 = 1458$ mm.

$$\therefore \ \frac{M}{V} + L_o = \frac{3.85 \times 10^8}{30.78 \times 10^3} + 624 = 1875 \text{ mm.}$$

Hence safe

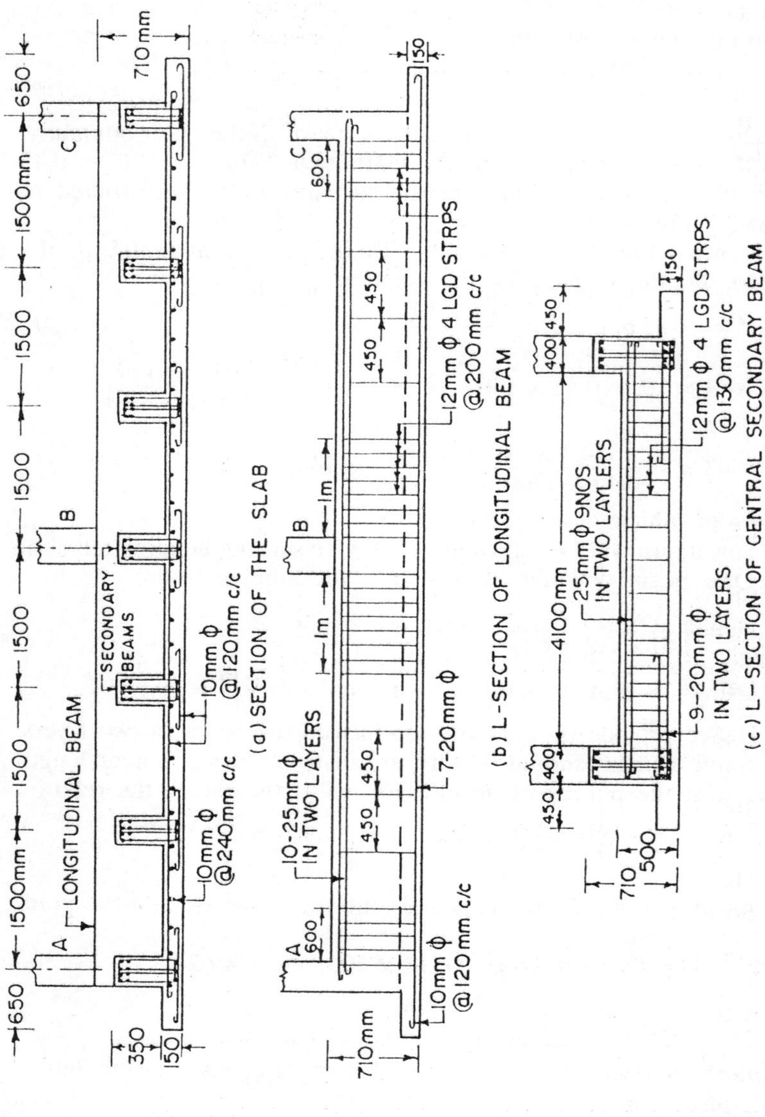

FIG. 16.25

Continue all the 10 bars throughout the length, in two layers. Total depth of beam = $624 + 24 + 60 \approx 710$ mm.

The beam should also have reinforcement at its bottom face to resist sagging bending moment at the centre and at the ends, for a bending moment of 87.6 kN-m under wind load conditions. The reinforcement required for bending moment will be less than 2 bars of 25 mm Φ. However, this reinforcement should satisfy the criterion of development length at the point of contraflexure near the centre of beam, where S.F. is maximum.

Let us use n number of 20 mm dia. bars each having $A_\Phi = 314 \text{ mm}^2$.

$$\therefore \quad M = \sigma_{st}.A_{st}\,jd = 140\,(314\,n) \times 0.9 \times 624$$

$$= n \times 24.69 \times 10^6 \text{ N-mm}$$

V = S.F. at the point of contraflexure

$\approx$307.8 kN (assuming this to be equal to S.F. at B)

$L_0 = 12\,\Phi$ or d, whichever is more = 624 mm.

$L_d = 58.3\,\Phi = 58.3 \times 20 = 1166$ mm.

$$\therefore \quad \frac{M}{V} + L_0 = \frac{n \times 24.69 \times 10^6}{307.8 \times 10^3} + 624$$

$$= (80.2\,n + 624)$$

Equating this to L_d,

$$80.2\,n + 624 = 1166$$

From which $n = 6.76 \approx 7$.

Hence provide 7 bars of 20 mm diameter at the bottom face, throughout the length of the beam.

Let us now check the beam for shear force of 307.8 kN.

$$\tau_v = \frac{307.8 \times 1000}{400 \times 624} = 1.23 \text{ N/mm}^2 \text{ (at } B\text{)}.$$

$$\frac{100\,A_s}{bd} = \frac{100(10 \times 490)}{400 \times 624} = 1.96\%$$

Hence $\tau_c = 0.44 \text{ N/mm}^2$ (Table 3.1)

Since $\tau_v > \tau_c$, shear reinforcement is necessary.

$$V_c = 0.44 \times 400 \times 624 = 109824 \text{ N} \approx 109.8 \text{ kN}$$

$$\therefore \quad V_s = 307.8 - 109.8 = 198 \text{ kN}.$$

Using 12 mm Φ 4-lgd stirrups, $A_{sv} = 452 \text{ mm}^2$.

$$\therefore \quad \text{Spacing} = \frac{\sigma_{sv}.A_{sv}.d}{V_s} = \frac{140 \times 452 \times 624}{198 \times 1000} \approx 200 \text{ mm}$$

Hence provide the stirups @ 200 mm c/c. for 1 m on either side of the faces of the column B.

Similarly, shear force at A = 294.4 kN. Hence provide 12 mm Φ 4-legged stirrups @ 200 mm c/c, for a distance of 0.6 m to the right of face of column A, and 0.6 m to left of column. C.

For the remaining length, provide nominal shear reinforcement at a maximum spacing of

$$s_v = \frac{2.5 A_{sv} f_y}{b} = \frac{2.5 \times 452 \times 250}{400}$$

$$\approx 700 \text{ mm.}$$

But not more than 450 mm nor more than 0.75 d (= 0.75 × 624 = 468 mm).

Hence provide 12 mm Φ 4-lgd nominal stirrups @ 450 mm c/c for the remaining length. The details of reinforcement etc. are shown in Fig. 16.25.

PROBLEMS

1. Design a combined footing for two columns, each 600 mm × 600 mm, 5 metres apart, and each carrying a load of 1800 kN. The available width is restricted to 2.4 metres. The safe bearing capacity of soil may be taken as 200 kN/m^2. Use M 15 conctrte. Take σ_{st} = 40 N/mm^2.

2. Redesign problem 1 by providing a longitudinal beam joining the column. There is no restriction on the width of the footing.

3. Design a combined footing for two R.C. columns A and B, separated by a distance of 4 m centre to centre:

   Column A is 500 mm square and carries a load of 1200 kN
   Column B is 600 mm square and carries a load of 1600 kN
   The safe bearing capacity of a soil is 200 kN/m^2. Use M 15 concrete. Take σ_{st} = 140 N/mm^2.

4. Redesign problem 3 if the property line is located 400 mm from centre of column A measured outwards along the centre line of the columns and runs at right angles to the column line. No encroachment beyond the property line is permissible.

5. Design a trapezoidal combined footing for the following requirements :
   Column A= 400 mm × 400 mm
   Column B= 600 mm × 600 mm
   Axial load on column A= 400 kN
   Axial load on column B= 800 kN
   Distance between centre of columns = 2.5 m
   Safe bearing capacity of soil = 200 kN/ m^2
   The footing is not to project more than 0.4 m beyond the outer face of the columns. Use M 15 concrete. Take σ_{st} = 140 N/mm^2.

6. Redesign problem 5 by providing a central longitudinal beam joining the columns.

7. Desing a strap footing for the following data:
   Column A, 400 mm× 400 mm carries a load of 800 kN
   Column B, 600 mm× 600 mm carries a load of 1000 kN
   Centre to centre distance between the two column is 8 metres.
   Column A is on the property line.
   Safe bearing capacity of soil= 150 kN/ mm^2
   Use M 15 mix. Take σ_{st} = 140 N/mm^2.

17

Pile Foundations

17.1 TYPES OF PILES.

The use of piles as foundation can be traced since olden times. The art of driving piles was well established in Roman times and the details of such foundations were recorded by Virtuvious in 59 A.D. Today, pile foundation is much more common than any other type of deep foundation. Based on foundation or the use, piles may be classified as : (1) end bearing pile (2) friction pile (3) compaction pile (4) tension pile or uplift pile (5) anchor pile (6) fender pile and dolphins (7) batter pile (8) sheet pile.

End *bearing piles* are used to transfer load through water or soft soil to a suitable bearing stratum. *Friction piles* are used to transfer loads to a depth of a friction load-carrying material by means of skin friction along the length of piles. *Compaction piles* are used to compact loose granular soils, thus increasing their bearing capacity. *Tension or uplift piles* anchor down the structures subjected to uplift due to hydrostatic pressure or due to overturning moment. *Anchor piles* provide anchorage against horizontal pull from sheet piling or other pulling forces. *Fender piles and dolphins* are used to protect water front structures against impact from ships or other floating objects. *Sheet piles* are commonly used as bulk heads or as impervious cut off to reduce seepage and uplift under hydraulic structures. *The batter piles* are used to resist large horizontal or inclined forces.

Piles can also be classified, based on materials composition, as follows :

1. *Concrete piles*
   (*a*) Precast.
   (*b*) Cast-in-situ :
      (*i*) driven piles ; cased or uncased.

(*ii*) bored piles ; pressure piles and under reamed pile.

2. *Timber piles*
3. *Steel piles*
   (*a*) H-pile.
   (*b*) Pipe pile.
   (*c*) Sheet pile.
4. *Composite piles*
   (*a*) Concrete and timber.
   (*b*) Concrete and steel.

The *precast concrete piles* are generally used for a maximum design load of about 800 kN except for large prestressed piles. They must be reinforced to withstand handling stresses. They require space for casting and storage, more time to set and cure before installation and heavy equipment for handling and driving. They also incur large cost in cutting off extra length or adding more length. The *cast-in-situ* piles are generally used for a maximum design load of 750 kN except for compacted, pedestal piles. They are installed by pre-excavation, thus eliminating vibration due to driving and the handling stresses. Cast-in-place piles may be classified into two classes : driven piles, (cased or uncased) and bored piles(pressure piles, pedestal piles and under-reamed piles). A variety of cast-in-situ piles are in use, each bearing the name of the manufacturer. The common types are as follows : (*i*) Raymond standard pile (*ii*) Raymond step-taper- pile (*iii*) Union metal pile or monotub, (*iv*) MacArthur compressed uncased pile (*v*) Mac Arthur cased pile. (*vi*) Franki standard pile. (*vii*) Western button bottom pile etc.

Under reamed pile is a special type of bored pile having an increased diameter or bulb at some point in its length, to anchor the foundation in expansive soil subjected to alternative expansion and contraction.

Concrete filled steel pipes and steel *H-piles* are used as long piles with high bearing capacity. They are rarely used unless they reach a stratum of exceptionally high supporting capacity, since their cost is very high. *Timber piles* have small bearing capacity and are not permanent unless treated. They are prone to damage by hard driving, and should not be driven through hard stratum or boulders. *Composite piles* are suitable where the upper part of the pile is to project above the water table. Such a pile consists of a lower portion of untreated timber and an upper portion of concrete. In the other type of composite pile, steel piles are attached to the lower end of cast in place concrete piles. This type is used in cases where the required length of pile is greater than that available for the cast-in-place type.

17.2 PILE DRIVING

Piles are commonly driven by means of a hammer supported by a crane or by a special device known as a *pile driver.* The hammer is guided between two parallel steel members known as *leads*. The leads are carried on a flange in such a way that they can be supported in a vertical position or an inclined position. Hammers are of the following types : (*i*) drop hammer (*ii*) single acting hammer (*iii*) double acting hammer (*iv*) diesel hammer and (*v*) vibratory hammer.

A *drop hammer* (ram or monkey) is raised by winch and allowed to fall by gravity on the top of a pile. A *single acting hammer* is raised by steam, compressed air or internal combustion, but is allowed to fall by gravity alone. The *double action hammer* employs steam or air for lifting the ram and for accelerating the downward stroke. It operates with succession of rapid blows. *Diesel hammer* is a small, light weight self contained and self-acting type, using gasoline for fuel. In the case of vibratory hammer, the driving unit vibrates at high frequency.

During pile driving, heads, helmets or caps are placed on the top of pile to receive the blows of hammer and to prevent damage to the head of the pile. The *cushion,* consisting of a pad of resilient material, hard wood or rope, is placed between the drive cap and the top of the pile to protect the pile head. Single acting hammers are advantageous when driving heavy piles in compact or hard soil while double acting hammers are generally used to drive piles of light or moderate weight in soils of average resistance against driving. Piles are ordinarily driven to a resistance measured by the number of blows required for the last 1 cm of penetration. Resistance of 3 to 5 blows per cm are commonly specified for concrete piles.

17.3 LOAD CARRYING CAPACITY OF PILES

The ultimate load carrying capacity of a pile is defined as the maximum load which can be carried by a pile, and at which the pile continues to sink without further increase of load. The *allowable load* is the load which the pile can carry safely, and is determined on the basis of : (*i*) ultimate bearing resistance divided by suitable factor of safety (*ii*) permissible settlement and (*iii*) overall stability of the pile foundation. The load carrying capacity of a pile can be determined by the following methods: (1) dynamic formulae (2) static formulae (3) pile load tests (4) penetration tests.

Dynamic formulae. When a pile hammer hits the pile, the total driving energy is equal to the weight of hammer times the height of drop of stroke. In addition to this, in the case of double acting hammers, some energy is also imparted by the steam pressure during the return stroke. This total downward energy is consumed by the work done in penetrating the pile and by certain losses. The various dynamic formulae are essentially based on this assumption. Following are some commonly used dynamic formulae :

(a) *Engineering News formula*

The Engineering News formula was proposed by A.M. Wellington(1888), in the following form :

$$Q_a = \frac{WH}{F(S+C)}$$

where Q = allowable load

W = weight of hammer

H = height of fall

F = factor of safety = 6

S = final set(penetration) per blow, usually taken as average penetration, cm per blow for the last 5 blows of drop hammer, or 20 blows of steam hammer.

C = empirical constant.

Denoting W in kg., H in cm and S in cm and C = 2.5 cm for drop hammer and C = 0.25 cm for single and double acting hammers the above formula reduces to the following forms :

(*i*) Drop hammers :

$$Q_a = \frac{WH}{6\,(S + 0.25)}$$

(*ii*) Single acting steam hammers :

$$Q_a = \frac{WH}{6\,(S + 0.25)}$$

(*iii*) Double acting steam hammers :

$$Q_a = \frac{(W + ap)\,H}{6\,(S + 0.25)}$$

where a = effective area of piston (cm^2) and p = mean effective steam pressure (kg/cm^2).

(*a*) Hiley's formula

Indian Standard IS : 2911 (Part II) 1964 gives the following formula based on original expression by Hiley :

$$Q_f = \frac{\eta_h\, W H\, \eta_b}{S + \frac{C}{2}}$$

where Q_f = ultimate load on pile

W = weight of hammer, in kg

H = height of drop of hammer, in cm

S = penetration or set, in cm, per blow

C = total elastic compression, $C_1 + C_2 + C_3$

C_1, C_2, C_3 = temporary elastic compression of dolly and packing, pile and soil respectively.

η_h = efficiency of hammer, variable from 65% for some double acting steam hammers to 100% for drop hammer released by triggers.

η = efficiency of hammer blow (*i.e.* ratio of the energy after impact to striking energy of ram)

$$\eta_b = \frac{W + e^2 P}{W + P} \quad \text{(for the case of } W > eP)$$

and

$$\eta_b = \frac{W + e^2 P}{W + P} - \left\{\frac{W - e P}{W + P}\right\}^2 \quad \text{(for the case when } W < e P)$$

P = weight of pile, anvil, helmet follower.

e = coefficient of restitution(variable from zero for a timber pile with poor condition of head or for excess packing in the driving cap to 0.5 for double acting hammer driven steel piles without helmet but with packing on top).

The allowable load is obtained by using a factor of safety of 2 or 2.5.

Static formulae

The static formulae are based on the assumption that the ultimate bearing capacity Q_f of a pile is the sum of the total ultimate skin friction R_f and total ultimate point or end bearing resistanc R_p :

$$Q_f = R_f + R_p$$

or

$$Q_f = A_s r_f + A_p r_p$$

where A_s = surface area of pile upon which the skin friction acts.

A_p = area of cross section of pile on which bearing resistance acts. For tapered piles, A_b may be taken as the cross-sectional area of the lower one-third of the embeded length.

r_f = average skin friction.

r_p = unit point or the resistance.

A factor of safety of 3 may be adopted for finding the allowable load.

Cohesive soil. For the pile in cohesive soil, the point bearing is generally neglected for individual pile action, since it is negligible as compared to friction resistance. The unit skin friction may be taken as equal to shear strength of the soil.

Hence $$r_f = \tau_f = c = \frac{q_u}{2}$$

$$r_p = c\,N_c = 9\,c$$

where q_u = unconfined compressive strength of soil and
c = unit cohesion.

Non-cohesive soil. For non-cohesive soil, the following values may be adopted :

$$r_f = K \tan \Phi(\gamma Z + q)$$

$$r_p = \frac{\gamma B}{2} N_{\gamma q} \text{ (for rectangular or square piles)}$$

or $$r_p = 0.3\,\gamma\,B\,N_{\gamma q} \text{ (for rectangular piles)}$$

where K = coefficient of lateral earth pressure*
γ = density of soil
q = surcharge on the ground
Z = depth of centre of gravity of the pile below ground surface
B = least lateral dimension of pile
Φ = angle of interal friction
$N_{\gamma q}$ = Meyerhof's non-dimensional factor.

17.4 GROUP ACTION IN PILE

When several closely spaced piles are grouped, together, it is reasonable to expect that the soil pressure developed in the soil as resistance, will overlap. The bearing capacity of a pile group may or may not be equal to the sum of the bearing capacity of individual piles constituting a group. Theory and test have shown that the bearing value Q_f of a group of *friction piles,* particularly in clay, is equal to bearing capacity of individual pile multiplied by the number of piles n in a group. However, no reduction due to grouping occurs in end bearing piles. For combined end bearing and friction piles, only the load carrying capacity of the friction portion is reduced. A method of estimating the bearing capacity of a group of friction piles is to multiply the quantity nQ_f by a reduction factor called the *efficiency of the pile group* η_g.

$$Q_g = nQ_f \cdot \eta_g \,.$$

*See Author's book 'Soil Mechanics and Foundations'.

The efficiency of pile group depends upon the following factors: characteristics of pile(*i.e.* length, diameter, material, etc.), spacing of piles, total number of piles (n) in a row, and number of rows (m). Out of a number of formulae for determining efficiency of a pile group, two are given below :

1. *Converse-Labarre formula*

$$\eta_g = 1 - \frac{\theta}{90}\left[\frac{(n-1)\,m + (m-1)\,n}{mn}\right]$$

where $\theta = \tan^{-1}\frac{d}{s}$ (degrees)

d = diameter of piles.

s = spacing of piles.

2. *Silver Keeney formula*

$$\eta_g = \left[1 - 0.479\left(\frac{s}{s^2 - 0.093}\right)\left(\frac{m+n-2}{m+n-1}\right)\right] + \frac{0.3}{m+n}$$

where s = average spacing, centre to centre in metres.

17.5. STRUCTURAL DESIGN OF R.C. PILE

A R.C. pile is designed for the following : (1) total load coming on it from the structure (2) handling stresses (3) driving stresses. While designing the pile as a column, it may be considered as fixed at one end hinged at the other. The length of the pile may be taken as 2/3 rd the length embedded in firm stratum. The cross-section of pile varies with its overall length.

The main longitudinal reinforcement in a pile should not be less than the following values :

(*i*) 1.25 percent of the cross-sectional area of the pile, for piles having length upto 30 times their least width.

(*ii*) 1.5 percent of the cross-sectional area of the pile, for piles having length between 30 to 40 times its least width.

(*iii*) 2 percent of the gross cross-sectional area of the pile, for piles having length greater than 40 times its least width.

Lateral reinforcement should be provided in form of links, of not less than 5 mm diameter wire. The volume of such lateral reinforcement should not be less than 0.2 percent of the gross volume of the pile, and their centre to centre spacing should not exceed half the least width of the pile. Near each end of pile, upto a distance of three times the least lateral dimension, lateral reinforcement should be provided @ 0.6% of the gross volume. If the pile has to penetrate hard stratum, the lateral reinforcement at the top of the pile for a distance of three times the width, should be in the form of helix. In addition to this, the longitudinal bars should be held apart by provision of *spacers* made of 12 to 16 mm wire and provided at the interval of not exceeding 1.5 m. In order to prevent the outward displacement of the bars, 6 mm Φ links should also be provided with the spacers (Fig. 17.5). The pile should be provided with a cast iron shoes at its lower end (Fig. 17.5).

Stresses during handling

Precast piles should be checked against handling stresses. When a pile is lifted by means of a derrick, it is subjected to bending moment due to its own weight. When the pile is of less than 12 metre length, it is suspended from one point at its middle. Piles longer than 12 metres are suspended at two or three points suitably spaced at its length so that handling moment is as small as possible.

(*i*) *Pile suspended at one point.* [Fig. 17.1(a)]

If w is the weight of the pile per unit length, and L is the length of the pile, it will be subjected to a minimum bending moment $= \frac{wL}{2} \times \frac{L}{4} = \frac{wL^2}{8}$ at the point of suspension.

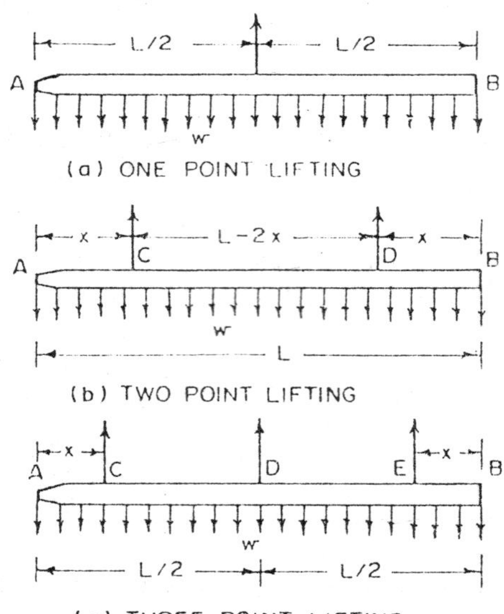

FIG. 17.1

(2) *Pile suspended at two points.* [Fig. 17.1(b)].

Let the distance of each point of suspension be x from the respective ends. The value of x should be such that maximum bending moment anywhere in the pile should be the least. This is possible when the hogging moment is equal to maximum sagging moment.

Reaction $R = \frac{wL}{2}$ at each point of suspension.

Maximum hogging moment at the end of each cantilever.

$$= \frac{w x^2}{2} \qquad ...(i)$$

Maximum sagging moment at the middle of the pile

$$= \frac{w L}{2}\left(\frac{L}{2} - x\right) - \frac{w L^2}{8} \qquad ...(ii)$$

Equating the two, we get

$$\frac{w x^2}{2} = \frac{w L}{2}\left(\frac{L}{2} - x\right) - \frac{w L^2}{8} = \frac{w L}{2}\left(\frac{L}{4} - x\right)$$

or $$x^2 + L x - \frac{L^2}{4} = 0$$

which gives $$x = 0.206\ L$$

Hence maximum B.M. $= \frac{w}{2}(0.206\ L)^2 = \frac{w L^2}{47}$

(3) ***Pile suspended at three points*** [Fig. 17.1(c)].

It can be shown that if the pile is suspended at three points, the end points should be located at $x = 0.15L$ from the corresponding ends so that B.M. is the least. The third point is located at the middle of the pile. The maximum bending moment in that case will be $\frac{w L^2}{90}$.

Erecting the pile from one end

Fig. 17.2 shows a pile being erected from one point C, the end of the pile being resting on the ground. Let the distance of the point be x from the end. In order that bending moment on the pile is the least, x should be equal to $0.293\,L$. The maximum bending moment will be equal to $\frac{w L^2}{23}$.

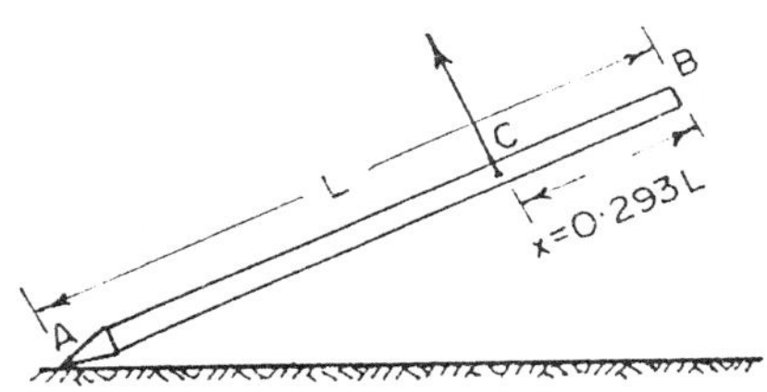

FIG. 17.2.

17.6 DESIGN OF PILE CAP

When a column or pier is supported on the pile only, the column should rest centrally on the pile. However, when the column is supported on more than one pile, the piles should be connected through a rigid pile cap, to distribute the load to the individual piles. The pile cap consists of a rigid, deep, reinforced concrete slab which acts monolithically with the group of piles. The piles should be arranged symmetrically about the axis of the column so that the load from column is distributed uniformly to all the columns. The pile cap slab is provided in uniform thickness. The pile cap should be extended beyond exterior piles by 10 to 15 cm. The pile should be embedded by at least 15 cm in the pile cap, and the reinforcement in the cap should be placed at least 10 cm above the pile head. Thus, effective depth of the pile cap will be equal to the total depth minus 25 cm. The pile cap, provided over the entire area of the piles, is considered to be divided into a framework of rectangular beams, along which main reinforcement is provided. The arrangement of these beams depends upon the number of piles, and the width of beam is taken equal to width of pile.

Pile cap for three piles

Fig. 17.3 shows the pile cap for three piles. The pile cap is considered to be composed of two beams *AB* and *CD* ; *A*, *B* and *C* being the three piles placed at distance *L* centre to centre. The column *W* is placed on the beam *DC*, at the centroid of the triangle *ABC*.

Let $W =$ total load of column.

$\therefore$ Load on each pile $= \dfrac{W}{3}$

Length of beam $CD = l = \dfrac{L\sqrt{3}}{2}$

Distance of W from $D = \frac{1}{3} l = \dfrac{L}{2\sqrt{3}}$

Beam *CD* is supported on *C* and *D*, and is loaded with *W*, such that reaction at *C* is *W*/3. Hence maximum bending moment is

$$M_1 = \frac{W}{3} \times \frac{2l}{3} = \frac{W}{3} \times \frac{2}{3}\left(\frac{L\sqrt{3}}{2}\right) = \frac{WL}{3\sqrt{3}}$$

Reaction transferred to $D = W - \dfrac{W}{3} = \dfrac{2W}{3}$

The beam AB is of length L, and is loaded at its middle point D with a load $\frac{2W}{3}$. Hence maximum bending moment is given by

$$M_2 = \frac{W}{3}\left(\frac{L}{2}\right) = \frac{WL}{6}.$$

The reinforcement for both beams can now be computed.

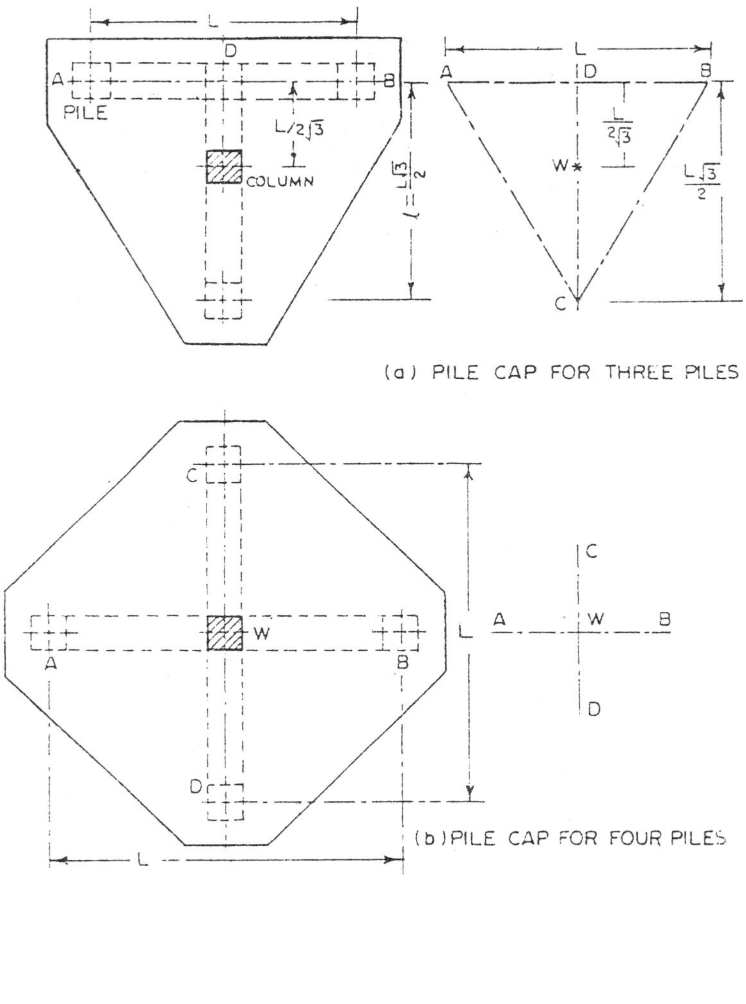

FIG. 17.3.

Pile cap for four columns [Fig. 17.3(*b*)]

Load on each pile $= \frac{W}{4}$

The pile cap is considered to be composed of two beams *AB* and *CD*.

$$\therefore \quad M_{max} = \frac{W}{4} \times \frac{L}{2} = \frac{WL}{8} \text{ for each beam.}$$

Secondary reinforcement

In order to prevent the outward splaying tendency of piles, secondary reinforcement should always be provided. The reinforcement is provided at the bottom of the pile cap, running round the longitudinal reinforcement projecting from the piles into the cap. It should be so bent that there is change of its direction at the head of every pile. The area of secondary reinforcement changing direction at every head of the pile should not be less than 20% of the tensile reinforcement.

Design Example 17.1. *Design a pile under a column transmitting an axial load of 600 kN. The pile is to be driven to a hard stratum available at a depth of 8 metres. Take* $\sigma_{cc} = 4\ N/mm^2$ *and* $\sigma_{sc} = 130\ N/mm^2$.

Solution.

1. ***Main reinforcement***

Let the length of the pile above ground, including pile cap, etc. = 0.6 m.

∴ Total length of pile = 8.6 m.

Let the size of the pile be 400 mm × 400 mm

$$\frac{l}{D} \text{ ratio } = \frac{8.6}{0.4} = 21.5$$

Since this is greater than 12, the pile behaves as long column.

Hence reduction coefficient $C_r = 1.25 - \frac{l_{ef}}{48\,D}$

$$= 1.25 - \frac{8.6}{48 \times 0.4} \approx 0.8$$

$$\therefore \text{ Design load for a short column } = \frac{600}{0.8} = 750 \text{ kN}$$

Load carrying capacity of column is given by

$$P = \sigma_{cc} A_c + \sigma_{sc} A_{sc}$$

where A_c = area of concrete= $(400 \times 400) - A_{sc} = 16 \times 10^4 - A_{sc}$

$\therefore \quad 750 \times 10^3 = 4\,(16 \times 10^4 - A_{sc}) + 130\,A_{sc}.$

From which $A_{sc} = 873\ \text{mm}^2$.

Since the length of pile is less than 30 times the width, minimum reinforcement @ 1.25% of gross cross-sectional area

$$= \frac{125}{100}(400 \times 400) = 2000\ \text{mm}^2.$$

However, provide 4 bars of 25 mm Φ giving total area of steel $= 4 \times 490 = 1960\ \text{mm}^2$ which is quite close to the minimum requirements. Provide an effective cover of 40 mm.

2. *Lateral reinforcement in the body of the pile*

Lateral reinforcement in the body of pile is provided @ 0.2% of gross volume.

$$\therefore \quad \text{Volume needed per mm length} = \frac{0.2}{100}(400 \times 400 \times 1) = 320\ \text{mm}^3.$$

Clear cover to main reinforcement $= 40 - 12.5 = 27.5$ mm

Using 8 mm Φ ties, length of each side of tie

$= 400 - 2 \times 27.5 + 8 = 353$ mm

Area $A_\Phi = \frac{\pi}{4}(8)^2 = 50.3\ \text{mm}^2$.

$\therefore$ Volume of each tie

$= 4 \times 353 \times 50.3$

$= 70975\ \text{mm}^3$

$\therefore \quad \text{Pitch} = \frac{70955}{320} = 222$ mm

Maximum pitch permissible

$= \frac{1}{2} \times 400 = 200$ mm.

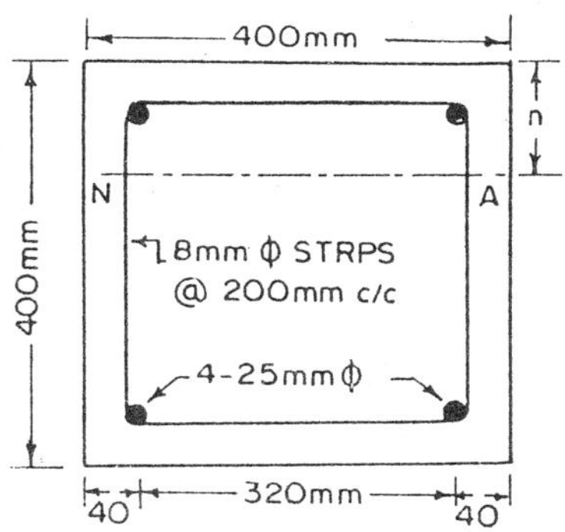

FIG. 17.4

Hence provide 8 mm Φ ties @ 200 mm c/c throughout the length of the pile.

3. *Lateral reinforcement near pile head*

Near pile head, special spiral reinforcement is to be provided for a length of $3 \times 400 = 1200$ mm. Volume of spiral, @ 0.6% of gross volume, per mm length is

$$= \frac{0.6}{100}(400 \times 400 \times 1) = 960\ \text{mm}^3$$

Using 8 mm Φ spiral, having $A_\Phi = 50.3\ \text{mm}^2$, pitch is given by

$$s = \frac{\text{circumference of spiral} \times A_\Phi}{960}$$

$$= \frac{\pi \times 353 \times 50.3}{960} = 58 \text{ mm}$$

Provide the spiral at 55 mm pitch. Provide 6 additional bars of 16 mm Φ vertically within the spiral. These spirals will be in addition to the normal ties.

4. *Lateral reinforcement near pile end*

Volume of ties per mm length @ 0.6% of gross volume = 960 mm^3.

$\therefore$ Volume of each tie = 70975 mm^3.

$$\therefore \quad \text{Pitch} = \frac{70975}{960} = 74 \text{ mm.}$$

Provide ties @ 70 mm c/c in a bottom length of 3 × 400 = 1200 mm.

5. *Spacer forks and links*

Provide two pairs of 12 mm Φ spacer fork with 6mm Φ links @ 1.5 m c/c along the length.

6. *Check for handling stresses*

Provide three holes in the pile as follows :

(*i*) One hole at $0.293L = 0.293 \times 8.6 \approx 2.5$ m from the pile for the purpose of hoisting it.

(*ii*) One hole each from either end, at a distance of $0.206L = 0.206 \times 8.6 \approx 1.75$ m for the purpose of stacking.

(*iii*) Weight of pile per meter run $= 0.4 \times 0.4 \times 1 \times 25000$
$= 4000$ N/m

$$\therefore M = \frac{4000\,(2.5)^2}{2} = 12500 \text{ N–m} = 125 \times 10^5 \text{ N–mm}$$

Effective depth of pile section $= 400 - 40 = 360$ mm. Let the neutral axis be situated at n below one face. Equating the moment of area about the N.A. we get (Fig. 17.4).

$$\frac{b\,n^2}{2} + (m_c - 1)\,A_{sc}\,(n - d_c) = m\,A_{sc}\,(d - n)$$

or
$$\frac{400}{2}n^2 + (1.5 \times 19 - 1)\,980\,(n - 40) = 19 \times 980\,(360 - n)$$

or
$$n^2 + 227.8\,n - 38900 = 0$$

which gives $n = 113$ mm.

Taking moment of forces about tensile steel, we get

$$b\,n\,\frac{c}{2}\left[d - \frac{n}{3}\right] + (1.5\,m - 1)\,A_{sc}\,c'\,(d - d_c) = M$$

where
$$c' = \frac{(n - d_c)\,c}{n} = \frac{113 - 40}{113} \times c = 0.65\,c$$

$$\therefore \quad 400 \times 113 \frac{c}{2}\left[360 - \frac{113}{3}\right] + 27.5 \times 980 \times 0.65\, c\,(360 - 40) = 125 \times 10^5$$

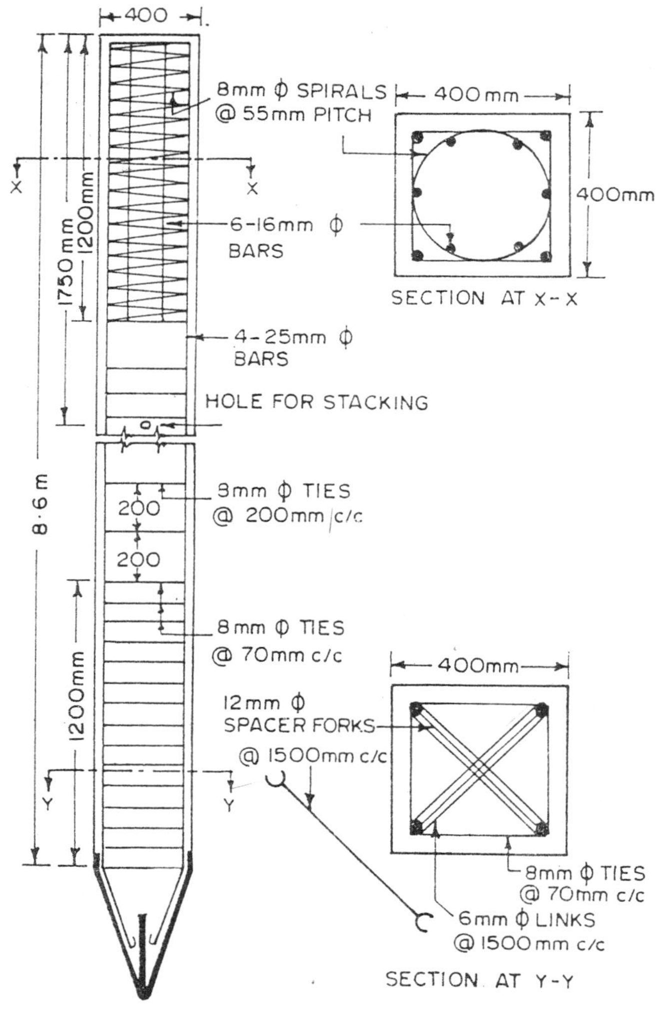

FIG. 17.5.

which gives $c = 0.97$ N/mm^2.

This is much less than 4 N/mm^2. Hence safe. The stress in steel will also be well within safe limit. The details of reinforcement are shown in Fig. 17.5.

Design Example 17.2 *A R.C. column, 400 mm × 400 mm carrying a load of 600 kN is supported on three piles 400 mm × 400 mm in section. The centre to centre distance between the piles is 1.5 m. Design a suitable pile cap. Use M 15 concrete.*

Solution. [Fig. 17.3 (*a*)]

1. ***Dimensions of pile cap***

Centre to centre spacing of piles $= L = 1.5$ m.

Keeping 200 mm clear projection of the cap beyond pile face. Overall length of the cap along the direction AB

$$= 1.5 + 0.4 + 0.4 = 2.3 \text{ m.}$$

Length of beam $CD = l = \dfrac{L\sqrt{3}}{2} = 1.5\dfrac{\sqrt{3}}{2} = 1.3$ m.

$\therefore$ Length of cap in the direction DC

$$= 1.3 + 0.4 + 0.4 = 2.1 \text{ m.}$$

2. ***Design of beam DC***

Load on each pile $= \dfrac{W}{3} = \dfrac{600}{3} = 200$ kN

Let the width of beam = width of column = 400 mm

$\therefore$ B.M. due to load $= \dfrac{WL}{3\sqrt{3}} = \dfrac{600 \times 1.5}{3\sqrt{3}}$

$$= 173 \text{ kN}-\text{m} = 173 \times 10^6 \text{ N}-\text{mm} \qquad ...(i)$$

In order to calculate the bending moment due to self weight of the beam plus weight of part of slab, let us assume total thickness of slab to be 800 mm. The self-weight of the beam is calculated on the assumption that weight of slab equal to two times the width of the beam acts with the beam.

$$\therefore \quad w = \left[\frac{3 \times 400 \times 800}{10^6}\right] \times 1 \times 25000 = 24000 \text{ N/m}$$

Length of beam $l = \dfrac{L\sqrt{3}}{2} = 1.3$ m.

$\therefore$ Totle load $= 1.3 \times 24000 = 31200$ N

$\therefore$ Reaction at $C = \dfrac{31200}{2} = 15600$ N

Distance of point of application of column load

$$= \frac{2}{3} l = \frac{2}{3} \times 1.3 = 0.87 \text{ m.}$$

$\therefore$ B.M. at the centre of column, due to self-weight is

$$= (15600 \times 0.87) - \frac{24000}{2} (0.87)^2$$

$$= 4489 \text{ N–m} = 4.489 \times 10^6 \text{ N–mm}$$

$\therefore$ Total B.M. $= 173 \times 10^6 + 4.489 \times 10^6 \approx 177.5 \times 10^6$ N–mm

$$\therefore \quad d = \sqrt{\frac{177.5 \times 10^6}{0.874 \times 400}} = 712 \text{ mm}$$

Keep $d = 720$ mm

$$\therefore \quad A_{st} = \frac{175.5 \times 10^6}{140 \times 0.865 \times 720} = 2036 \text{ mm}^2$$

$\therefore$ No. of 28 mm Φ bars $= \frac{2036}{616} = 3.3$

However, provide 4 bars of 28 mm Φ . Actual area of steel provided $= 4 \times 616 = 2464 \text{ mm}^2$.

3. *Design of beam AB*

Span $L = 1.5$ m

B.M. due to load from beam *CD*

$$= \frac{WL}{6} = \frac{600 \times 1.5}{6} = 150 \text{ kN–m} = 150 \times 10^6 \text{ N–mm}$$

B.M. due to self weight

$$= \frac{24000\,(1.5)^2}{8} = 6750 \text{ N–m} = 6.75 \times 10^6 \text{ N–mm}$$

$\therefore$ Total B.M. $= 150 \times 10^6 + 6.75 \times 10^6 = 156.75 \times 10^6$ N-mm

The reinforcement in direction *AB* will be placed below the reinforcement of *CD*. Hence available $d = 720 + 28 = 748$ mm.

$$\therefore A_{st} = \frac{156.75 \times 10^6}{140 \times 0.865 \times 748} = 1730 \text{ mm}^2.$$

However, provide the same reinforcement, *i.e.* 4 Nos. of 28 mm Φ bars. Keep total depth = 800 mm.

4. *Secondary reinforcement*

Area of secondary reinforcement running round each pile head $= 0.2 \times 2036 = 407 \text{ mm}^2$.

Using 12 mm Φ bars, $A_\Phi = 113$

$\therefore$ No. of bars $= \frac{407}{113} = 3.64 \approx 4.$

5. *Check for shear*

Shear is tested at a distance *d* from the beam. The dispersion lines(at 45°) transfer the load directly to the column. Hence there is no possibility of diagonal tension cracks.

The details of reinforcement etc. are shown in Fig. 17.6.

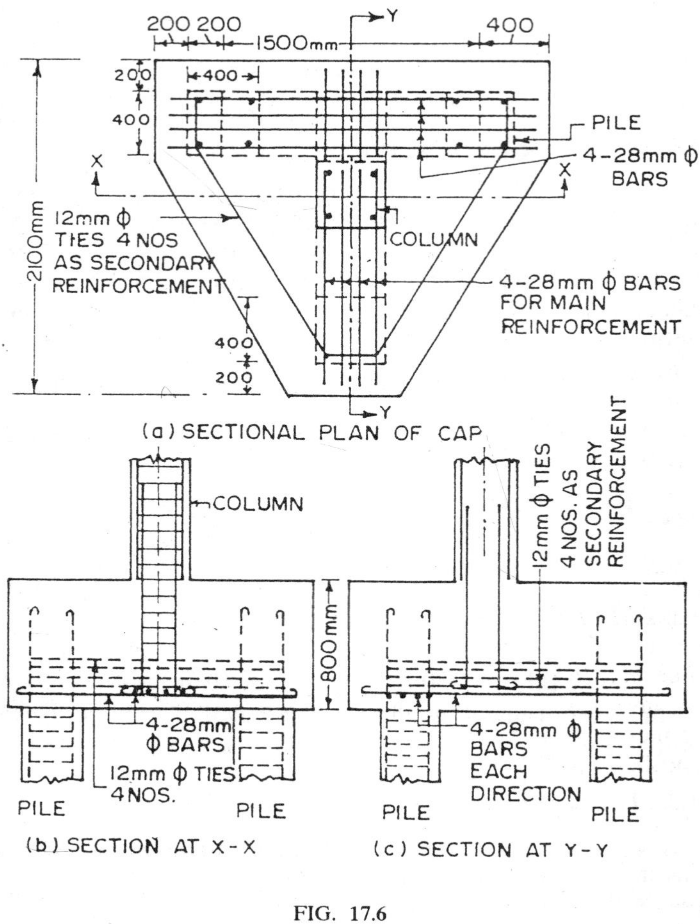

FIG. 17.6

PROBLEMS

1. A column 600 mm × 600 mm carries an axial load of 1200 kN and is supported on three pile. The piles are driven to hard strata available at the depth of 10 metres. Using M 15 concrete, design the pile and the pile cap. The column is placed at the centroid of the pile group. The centre to centre distance of the piles is restricted to 2 m.

18

Retaining Walls

18.1. INTRODUCTION

A *retaining wall* or retaining structure is used for maintaining the ground surfaces at different elevations on either side of it. Whenever embankments are involved in construction, retaining walls are usually necessary. In the construction, of buildings having basements, retaining walls are mandatory. Similarly in bridge work, the wing walls and abutments etc. are designed as retaining walls, to resist earth pressure along with superimposed loads. The material retained or supported by a retaining wall is called *backfill* which may have its top surface horizontal or inclined. The position of the backfill lying above the horizontal plane at the elevation of the top of a wall is called the *surcharge,* and its inclination to the horizontal is called the *surcharge angle* β.

In the design of retaining walls or other retaining structures, it is necessary to compute the lateral earth pressure exerted by the retaining mass of soil. The equation of finding out the lateral earth pressure against retaining wall is one of the oldest in the Civil Engineering field. The *plastic state of stress,* when the failure is imminent, was investigated by Rankine in 1860. A lot of theoretical and experimental work has been done in this field and many theories and hypothesis have been proposed.*

18.2. TYPES OF RETAINING WALLS

Retaining walls may be classified according to their mode of resisting the earth pressure, and according to their shape. Following are some of common types of retaining walls (Fig. 18.1).

---

*For detailed account of various earth pressure theories, the reader may refer to the Author's book 'Soil Machanics and Foundations'

(*i*) Gravity walls

(*ii*) Cantilever retaining walls

(*a*) T-shaped

(*b*) L-shaped

(*iii*) Counterfort retaining walls.

(*iv*) Buttressed walls.

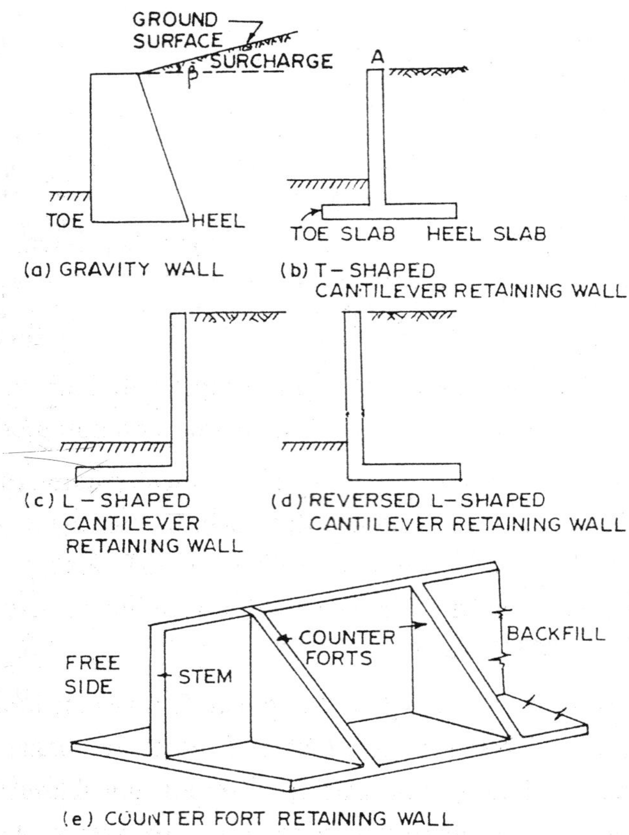

FIG. 18.1. VARIOUS TYPES OF RETAINING WALLS.

A gravity retaining wall shown in Fig. 18.1(*a*) is the one in which the earth pressure exerted by the backfill is resisted by dead weight of the wall, which is either made of masonry or of mass concrete. The stress developed in the wall is very low. These walls are so proportioned that no tension is developed anywhere, and the resultant of forces remain within the middle third of the base.

The cantilever retaining wall resists the horizontal earth pressure as well as other vertical pressures by way of bending of various components acting as cantilevers. A common form of cantilever retaining wall is the T-shaped wall shown in Fig. 18.1(*b*). The wall consists of stem *AB*, heel slab *BC* and toe slab *DB*. Each of these bend as cantilevers, about *B*. They are, therefore, reinforced on the tension face. Another form of cantilever retaining walls are the L-shaped walls shown in Fig. 18.1(*c*) and (*d*). They also resist the soil pressures by bending.

A counterfort retaining wall is shown in Fig. 18.1(*e*). The vertical stem and the heel slab are strengthened by providing counterforts at some suitable intervals. Because of provision of counterforts, the vertical stem as well as the heel slab acts as *continuous slab,* in contrast to the cantilevers of cantilever retaining wall. The toe slab however acts as cantilever bending upwards. This type of retaining wall is used when backfill of greater height is to be retained. A butteressed wall is a modification of the counterfort retaining wall in which the counterforts, called the buttresses, are provided to the other side of the backfill. However the buttresses reduce the clearance in front of the wall, and therefore these walls are commonly used.

18.3. ACTIVE EARTH PRESSURE : RANKINE'S THEORY

As originally proposed, Rankine's theory of lateral earth pressure is applied to uniform cohesionless soil only. Later, it was extended to include cohesive soils, by Resal and by Bell. The theory has been extended to stratified, partially immersed and submerged soils. Following are the assumptions of Rankine Theory : (1) The soil mass is semi-infinite, homogeneous, dry and cohesionless. (2) The ground surface is plane which may be horizontal or inclined. (3) The back of the wall is vertical and smooth. In other words, there are no shearing stresses between the wall and the soil, and the stress relationship for any element adjacent to the wall is the same as for any other element for away from the wall. (4) The wall yields about the base and satisfies the deformation conditions for plastic equilibrium.

However, the retaining walls are constructed of masonry or concrete, and hence the back of the wall is never smooth. Due to this, frictional forces develop. As a consequence of Rankine's assumption of non-existence of frictional forces at the wall face, the resultant pressure must be parallel to the surface of the backfill. The existence of friction makes the resultant pressure inclined to the normal to the wall at an angle that approaches the friction angle between the soil and the wall. We shall consider the following cases cohesionless backfill :

1. Dry or moist backfill with no surcharge.
2. Submerged backfill.
3. Backfill with uniform surcharge.
4. Backfill with sloping surfaces.

1. Dry or moist backfill with no surcharge

Fig. 18.2 (*a*) shows a retaining wall of height *h*, having dry or moist backfill, with no surcharge. According to Rankine's theory, the intensity of *active earth pressure*, p_a, trying to move the wall away from the fill, is given by

$$p_a = K_a \gamma H \quad ...(18.1)$$

where K_a = co-efficient of active earth pressure,

$$= \frac{1 - \sin \Phi}{1 + \sin \Phi} \quad ...(18.2)$$

Φ = angle of internal friction, for the backfill.

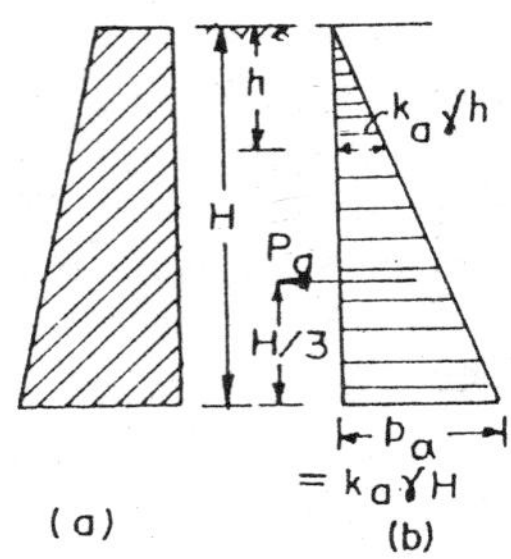

FIG. 18.2.

γ = unit weight of soil (backfill)

H = height of retaining wall.

At any height *h* below the horizontal ground surface, the intensity of active earth pressure is $K_a \gamma h$. Fig. 18.2(*b*) shows the distribution of active earth pressure p_a over the retaining wall. The total active earth pressure P_a or the resultant pressure per unit length of the wall is given by

$$P_a = \frac{1}{2} K_a \cdot \gamma \cdot H^2 \quad ...(18.3)$$

acting at *H*/3 above the base of retaining wall. If the soil is dry, γ is the dry weight of the soil, and if wet, γ is the moist weight, to be substituted in Eq. 18.1 or 18.3.

2. Submerged backfill

In this case, the sand fill behind the retaining wall is saturated with water. The lateral pressure is made up of two components.

(*i*) lateral pressure due to submerged weight γ' of soil, and

(*ii*) lateral pressure due to water.

Thus, at any depth *h* below the surface,

$$p_a = K_a\gamma'h + \gamma_w . h \quad ...(18.4)$$

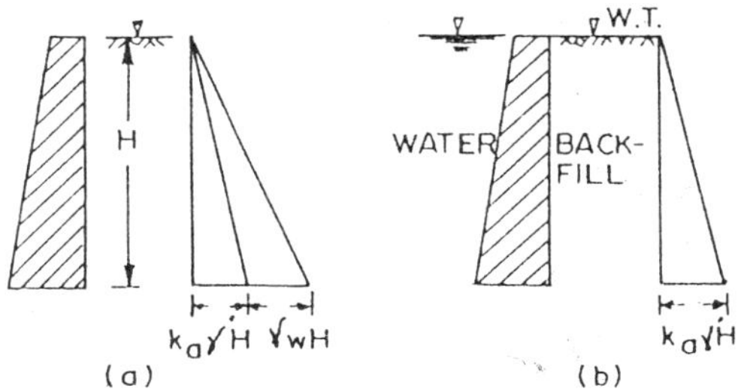

FIG. 18.3. SUBMERGED BACKFILL.

The pressure at the base of the retaining wall ($h=H$) is given by

$$p_a = K_a\gamma' H + \gamma_w H \quad ...(18.5)$$

If the free water stands to both the sides of the wall[Fig 18.3(*b*)], the water pressure need not be considered, and the net lateral pressure given by

$$p_a = K_a\gamma' H \quad ...(18.6)$$

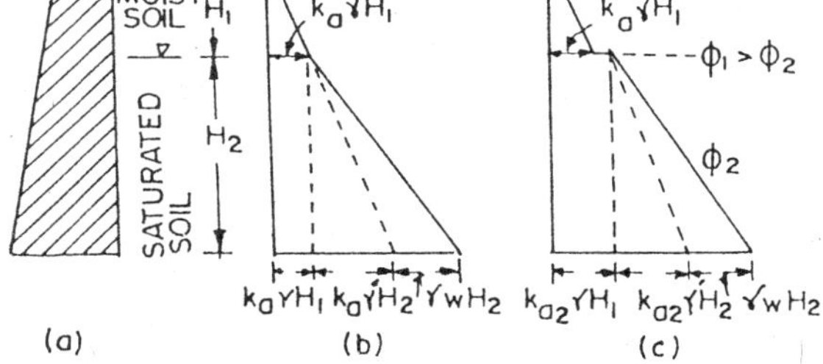

FIG. 18.4. PARTLY SUBMERGED BACKFILL.

If the backfill is partly submerged, *i.e.*, the backfill is moist to a depth H_1 below the ground level, and then it is submerged, the lateral pressure intensity at the base of the wall is given by

$$p_a = K_a \cdot \gamma H_1 + K_a \gamma ' H_2 + \gamma_w \cdot H_2 \qquad ...(18.7)$$

The above expression is bassed on the assumption that the value of Φ is the same for the moist as well as submerged soil. If it is different (say Φ_1 and Φ_2 respectively), the earth pressure coefficient K_{a1} and K_{a2} for both the portions will be different. As Φ decrease, K_a increases. The lateral pressure intensity [Fig. 18.4(*c*)] at the base of wall is

$$p_a = K_{a2} \gamma H_1 + K_{a2} \gamma' H_2 + \gamma_w H_2 \qquad ...(18.8)$$

3. Backfill with uniform surcharge

If the backfill is horizontal and carries surcharge of uniform intensity w per unit area, the vertical pressure increment, at any depth h, will increase by w. The increase in the lateral pressure due to this will be $K_a w$. Hence the lateral pressure at any depth h is given by

$$p_a = K_a \cdot \gamma \cdot h + K_a \cdot w \qquad ...(18.9)$$

At the base of the wall, the pressure intensity is :

$$p_a = K_a \gamma H + K_a w \qquad ...[18.9(a)]$$

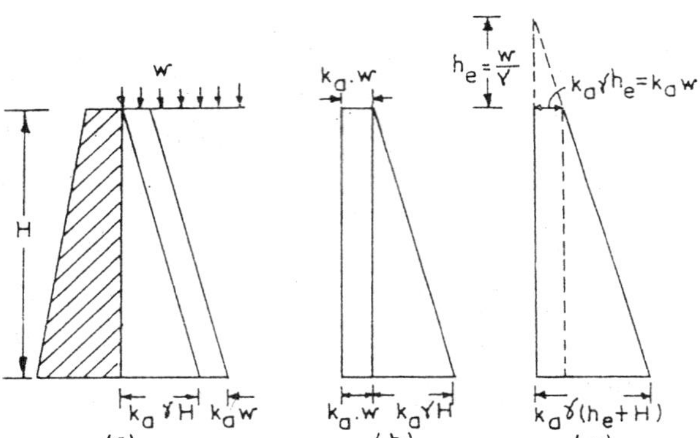

FIG. 18.5. BACK FILL WITH UNIFORM SURCHARGE

Fig. 18.5(*a*) and (*b*) show two alternative methods of plotting the lateral pressure diagram for this case. The lateral pressure increment $K_a \cdot w$ due to the surcharge is the same at every point of the back of the wall, and does not vary with height h. The height of fill h_e equivalent to the uniform surcharge intensity is given by the relation :

$$K_a \cdot \gamma \cdot h_e = K_a w$$

or
$$h_e = \frac{w}{\gamma} \qquad ...(18.10)$$

This means that the effect of the surcharge of intensity w is the same as that of a fill of height h_e above the ground surface[Fig. 18.5(*c*)].

4. Backfill with sloping surface

Let the sloping surface behind the wall be inclined at an angle β with the horizontal ; β is called the surcharge angle. In finding out the active earth pressure for this case by Rankine's theory, an additional assumption that the vertical and lateral stresses are *conjugate* is made. It can be shown that if the stress on a given plane at a given point is parallel to another plane, the stress on the latter plane at the same point must be parallel to the first plane.

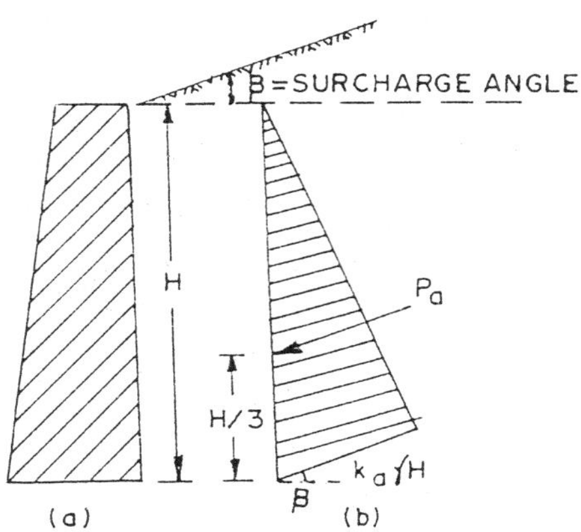

FIG. 18.6. LATERAL PRESSURE DISTRIBUTION FOR SLOPING SURCHARGE

Fig. 18.6(*a*) shows the retaining wall with a sloping backfill. The intensity of lateral earth pressure at the base wall is given by

$$p_a = \gamma H \cdot \cos\beta \frac{\cos\beta - \sqrt{\cos^2\beta - \cos^2\Phi}}{\cos\beta + \sqrt{\cos^2\beta - \cos^2\Phi}}$$

or
$$p_a = K_a \gamma H \qquad ...[18.11(a)]$$

where $$K_a = .\cos\beta\,\frac{\cos\beta - \sqrt{\cos^2\beta - \cos^2\Phi}}{\cos\beta + \sqrt{\cos^2\beta - \cos^2\Phi}} \quad ...[18.11(b)]$$

The pressure distribution is shown in Fig. 18.6(*b*). The pressure acts *parallel* to the sloping surface of the surcharge, *i.e.*, at β with the horizontal. The total active pressure P_a for the wall of height H is given by

$$P_a = \frac{1}{2} K_a \gamma H^2$$

The resultant acts at $H/3$ above the base, in direction parallel to the surcharge.

When $\beta = 0$ (*i.e.* horizontal ground surface), Eq. 18.11(*b*) reduces to

$$K_a = \frac{1 - \sin\Phi}{1 + \sin\Phi}$$ which is the same as Eq. 18.2.

18.4 PASSIVE EARTH PRESSURE

Passive earth pressure is exerted on a wall when it has a tendency to move towards the backfill. Such a condition may occur when

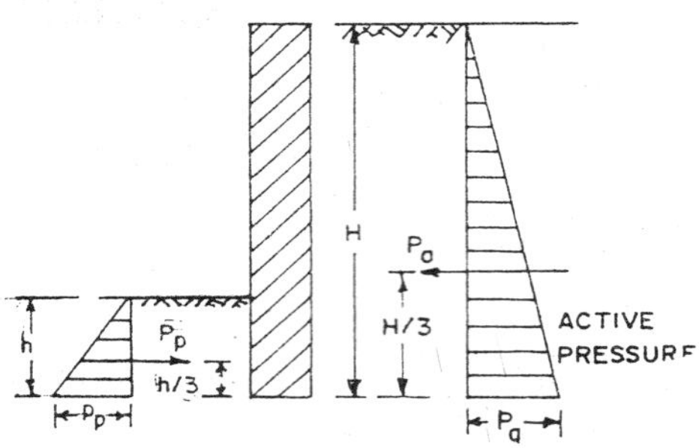

FIG. 18.7

the retaining wall supports an arch, and is subjected to arch thrust, moving it towards the fill. Another condition of passive earth pressure may be when the wall supports soil of different heights on both the sides as shown in Fig. 18.7. Due to active earth pressure from the right hand side, the wall moves to the left. The soil to the left is thus compressed and in turn exert passive earth pressure, resisting such moment.

If h is the height of fill, the intensity of passive pressure at height h is given by

$$p_p = K_p \, . \, \gamma \, h \quad ...(18.12)$$

where K_p = coefficient of passive earth pressure

$$= \frac{1+\sin\Phi}{1-\sin\Phi} = \frac{1}{K_a} \quad ...(18.13)$$

The passive pressure distribution will thus be a triangle. The total pressure is given by

$$P_p = K_p \cdot \frac{\gamma h^2}{2} \quad ...(18.14)$$

acting at $h/3$ above base.

It should be noted that K_p is very much greater than K_a. For example when $\Phi = 30°$, $K_a = \frac{1-\sin\Phi}{1+\sin\Phi} = \frac{1}{3}$ and $K_p = \frac{1+\sin\Phi}{1-\sin\Phi}$ $= 3$. Thus K_p, for this case, is 9 times K_a. In making calculations for active earth pressure behind retaining walls, the passive earth pressure due to other side of walls is generally neglected, since it is possible that this soil will erode away or that shearing will occur, thus eliminating the passive pressure.

If a uniform surcharge of intensity w per unit area acts over the surface of the backfill, the increase in the passive pressures will be equal to $K_p \cdot w$. The passive pressure intensity at depth h is given by

$$p_p = K_p\ (\gamma h + w) \quad ...(18.15)$$

If the backfill is having its top surface inclined at an angle β, the passive pressure is given by

$$p_p = \gamma h \cos\beta \frac{\cos\beta + \sqrt{\cos^2\beta - \cos^2\Phi}}{\cos\beta - \sqrt{\cos^2\beta - \cos^2\Phi}} \quad ...(18.16)$$

or

$$p_p = K_p \cdot \gamma h \quad ...[18.16(a)]$$

where

$$K_p = \cos\beta \frac{\cos\beta + \sqrt{\cos^2\beta - \cos^2\Phi}}{\cos\beta - \sqrt{\cos^2\beta - \cos^2\Phi}} \quad ...(18.16(b))$$

18.5. STABILITY OF CANTILEVER RETAINING WALL

Fig. 18.8 shows a cantilever retaining wall subjected to the following forces :

(1) Weight W_1 of the stem AB.

(2) Weight W_2 of the base slab DC.

(3) Weight W_3 of the column of soil supported on heel slab BC.

(4) Horizontal force P_a, equal to active earth pressure acting at $H/3$ above the base.

The following are the modes of failure of a retaining wall :

1. Overturning about the toe.

2. Sliding.

3. Failure of soil due to excessive pressure at toe or tension at the heel.

4 Bending failure of stem or base of slab or heel slat

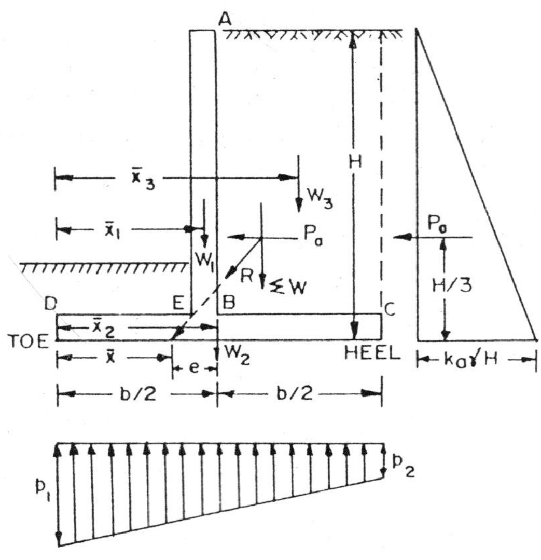

FIG. 18.8

1. Overturning

The most hazardous mode of failure of retaining walls is due to overturning because of unbalanced moments. Thus, in Fig 18.8, the overturning moment, due to active earth pressure, at toe is

$$M_o = P_a \frac{H}{3} = K_a \frac{\gamma H^2}{2} \cdot \frac{H}{3} = K_a \cdot \frac{\gamma H^3}{6} \qquad ...(18.17)$$

The resisting moment is due to the weights W_1, W_2 and W_3, neglecting the passive earth pressure and weight of soil above the toe slab.

Hence

$$M_R = W_1 \bar{x}_1 + W_2 \bar{x}_2 + W_3 \bar{x}_3 \qquad ...[18.17(a)]$$

Hence factor of safety due to overturning (F_1) is given by

$$F_1 = \frac{M_R}{M_o} \qquad ...(18.18)$$

A minimum factor of safety of 2 should be used.

2. Sliding

The horizontal force P_a tends to slide the wall away from the fill. The tendency to resists this is achieved by the friction at the base.

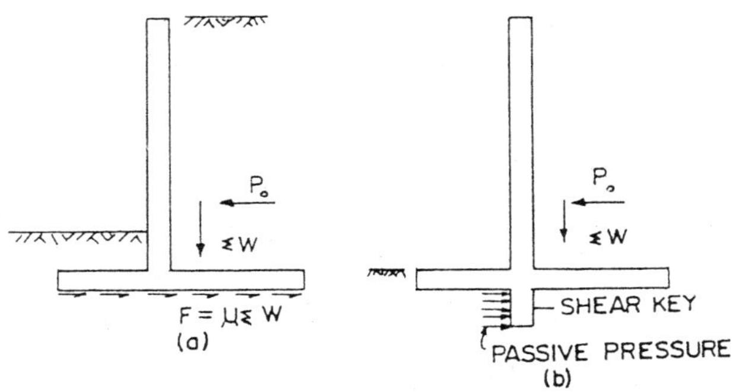

FIG. 18.9.

The force of resistance, F is given by

$$F = \mu \Sigma W \qquad ...(18.19)$$

where μ is the coefficient of friction(or friction factor) between soil and concrete, and ΣW is the sum of vertical forces. The factor of safety F_2 due to sliding is given by

$$F_2 = \frac{\mu \Sigma W}{H} \qquad ...(18.20)$$

If the wall is found to be unsafe against sliding, shear key below the base [Fig. 18.9(b)] should be provided. Such a key develops passive pressure which resists completely the sliding tendency of the wall. A factor of safety of 1.5 must be used against sliding.

In the absence of elaborate tests, the following values of μ may be adopted :

	μ
1. Coarse grained soil without suit	0.55
2. Coarse grained soil with silt	0.45
3. Silt	0.35

3. Soil pressure

Fig. 18.8. shows the various forces acting on the wall. If ΣW is the sum of all vertical forces, and P_a is the horizontal active earth pressure, the resultant R will strike the base slab at a distance e (say) from the middle point of the base.

Let $\Sigma M = W_1\bar{x}_1 + W_2\bar{x}_2 + W_3\bar{x}_3 - P_a \cdot \frac{H}{3}$

= net moment at the toe.

Then $\bar{x}$ = distance of point of application of resultant

$$= \frac{\Sigma M}{\Sigma W} \qquad ...(18.21)$$

Hence eccentricity $e = \frac{b}{2} - \bar{x}$ ...(18.22)

The pressure distribution below the base, is shown in Fig. 18.8. The intensity of soil pressure at the toe and heel is given by

$$p_1 = \frac{\Sigma W}{b}\left(1 + \frac{6e}{b}\right) \text{ at toe} \qquad ...(18.23)$$

and $$p_2 = \frac{\Sigma W}{b}\left(1 - \frac{6e}{b}\right) \text{ at heel} \qquad ...(18.24)$$

p_1 at toe should not exceed the safe bearing capacity of soil otherwise soil will fail. Similarly, p_2 at heel should be compressive. If p_2 comes to be tensile, the heel will be lifted above the soil, which is not permissible. In an extreme case, p_2 may be zero, where $e = b/6$. Hence in order that tension is not developed, the resultant should strike the base within the middle third.

4. *Bending failure*

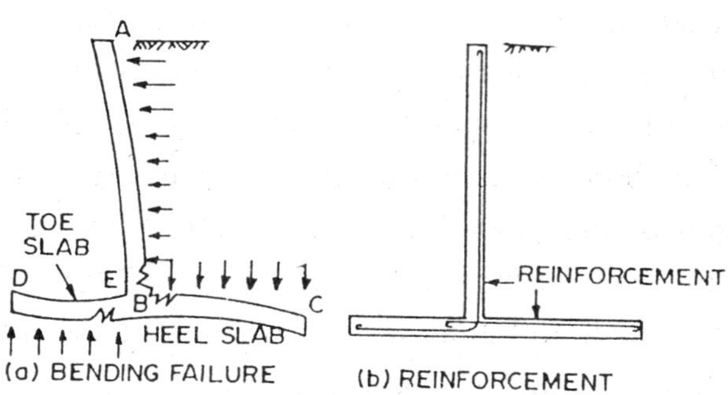

FIG. 18.10.

There are three distinct parts of T-shaped cantilever retaining wall ; the stem *AB*, the heel slab *BC* and toe slab *DE*. The stem *AB* will bend as cantilever, so that tensile face will be towards the backfill. The critical section will be at *B*, where cracks may occur at the inner face if it is not properly reinforced. The heel slab will have net pressure acting downwards, and will bend as a cantilever, having tensile face upwards. The critical section will be at *B*, where cracks may occur if it is not reinforced properly at the upper face as shown in Fig. 18.10(*b*). The net pressure on toe slab will act upwards , and hence it must be reinforced at the bottom face. The thickness of stem, heel slab and toe slab must be sufficient to withstand compressive stresses due to bending.

18.6. DESIGN PRINCIPLES OF CANTILEVER RETAINING WALL

The various dimensions of cantilever retaining wall should be so proportioned that the various failure criteria discussed above are taken care of. The design of a cantilever retaining wall consists of the following (*i*) fixation of base width *b* (*ii*) design of stem (*iii*) design of heel slab (*iv*) design of toe slab.

1. Fixation of base width (*b*)

The base width *b* of the retaining wall should be so chosen that the resultant of the forces remain within middle third, and

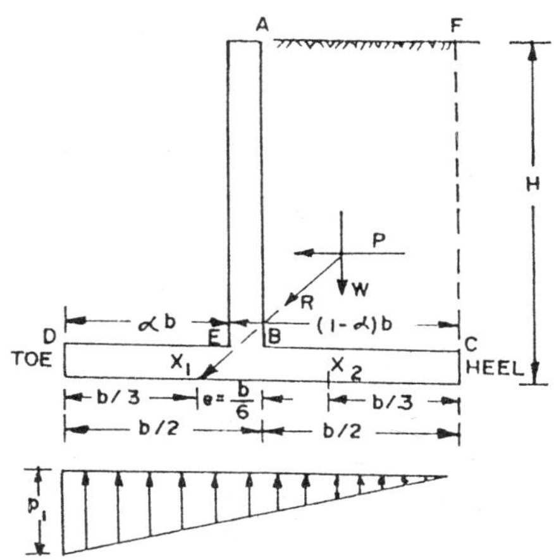

FIG. 18.11

the ratio of length of the toe slab to the base width should be such that the toe does not exceed the safe bearing capacity of soil. The method below gives an approximate method of finding the base width b and the width of toe slab $DE = \alpha b$.

Let us assume that the average unit weight of concrete and soil is equal to 1.1 γ where γ is the unit weight of soil. Neglecting the weight of toe slab DE, the total weight W of retaining wall plus the weight of soil in column $AFCB$ will be

$$W = (1-\alpha)b \times H(1.1\gamma) = 1.1\gamma bH(1-\alpha) \qquad ...(i)$$

This acts at a distance of $\frac{1}{2}(1-\alpha)b$ from E or C.

The horizontal earth pressure $p = K_a \dfrac{\gamma H^2}{2}$ ...(ii)

Let X_1 and X_2 be the outer and inner third points respectively. In order that no tension is developed at heel, the resultant should pass through the outer third point X_1. In that case, moment of resultant R will be zero, or the algebraic sum of moments of W and P about X_1 is zero. Hence, we have

$$W\left[\frac{2b}{3} - \frac{1}{2}(1-\alpha)b\right] - P \cdot \frac{H}{3} = 0$$

or
$$1.1\gamma bH(1-\alpha)\left\{\frac{2b}{3} - \frac{b}{2} + \frac{b\alpha}{2}\right\} - K_a \frac{\gamma H^3}{6} = 0$$

or
$$1.1\gamma b^2 H(1-\alpha)(1+3\alpha) = K_a \gamma H^3$$

which gives
$$b = 0.95\, H\sqrt{\frac{K_a}{(1-\alpha)(1+3\alpha)}} \qquad ...(18.25)$$

Thus b can be determined, if α, *i.e.* the ratio of heel slab DE to the total base width is known. The ratio α can be fixed on the basis of maximum pressure under the toe. When the resultant R passes through the outer third point X_1, eccentricity $e = \dfrac{b}{2} - \dfrac{b}{3}$ $= \dfrac{b}{6}$. Hence the maximum pressure p_1 under the toe is

$$p_1 = \frac{W}{b}\left(1 + \frac{6e}{b}\right) = \frac{W}{b}(1+1)$$

or
$$p_1 = \frac{2W}{b} = \frac{2}{b}\{1.1\gamma\, b\, H\,(1-\alpha)\} \qquad ...(iii)$$

The value of p_1 should not exceed the safe bending capacity q_0 of the soil. Hence,

$$q_0 = p_1 = 2.2\,\gamma H\,(1-\alpha)$$

$$\therefore \qquad (1-\alpha) = 1 - \frac{q_0}{2.2\gamma H}$$

or

$$\alpha = 1 - \frac{q_0}{2.2\,\gamma\,H} \qquad \text{...(18.26)}$$

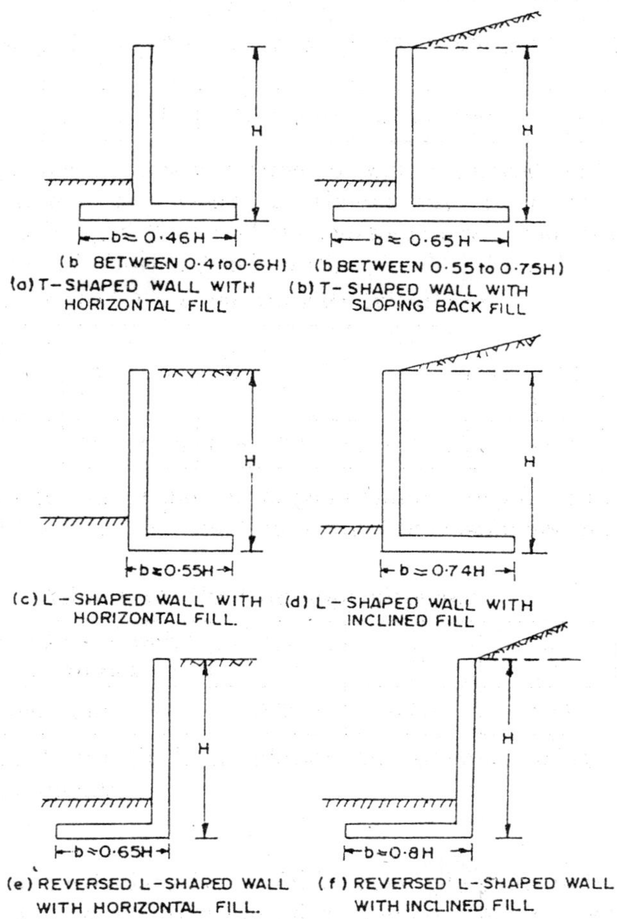

FIG. 18.12. FIXATION OF LENGTH OF BASE.

This value of α can be substituted in Eq. 18.25 to get the value of b. Eqs. 18.25 and 18.26 are valid, though as a rough guide, only when there is no surcharge. For the case of sloping backfill with surcharge angle β, the following approximate expression may be used :

$$\alpha = 1 - \frac{q_0}{2.7\gamma H} \quad \text{...(18.26(a)]}$$

and

$$b = H\sqrt{\frac{K_a \cos\beta}{(1-\alpha)(1+3\alpha)}} \quad \text{...[18.26(a)]}$$

Fig. 18.12 shows common values of b for various conditions of cantilever retaining walls.

Determination of base width from the considerations of sliding.

Eq. 18.25. gives the base width from the stress considerations. Let us find the base width from sliding consideration also. The factor of safety against sliding should at least be equal to 1.5. Thus,

$$\text{F.S.} = \frac{\mu W}{P} = 1.5$$

Substituting the values of $W = 1.1\gamma bH(1-\alpha)$

and $p = K_a \frac{\gamma H^2}{2}$, we get

$$\mu[1.1\gamma bH(1-\alpha)] = 1.5K_a \frac{\gamma H^2}{2}$$

$$b \approx \frac{0.7H}{1-\alpha} \cdot \frac{K_a}{\mu} \quad \text{...(18.27)}$$

Thus knowing α from Eq. 18.26, b can be determined from Eq. 18.27.

2. Design of stem

The vertical stem AB is designed as cantilever, for triangular loading. At any section h below the top point A, the force is equal to $K_a \frac{\gamma h^2}{3}$ and its bending moment about the section is $K_a \gamma \frac{h^3}{6}$. The thickness at B is maximum. The minimum thickness at A should vary from 20 to 30 cm depending upon the height of the wall. Reinforcement is provided towards the inner face of stem, *i.e.* towards

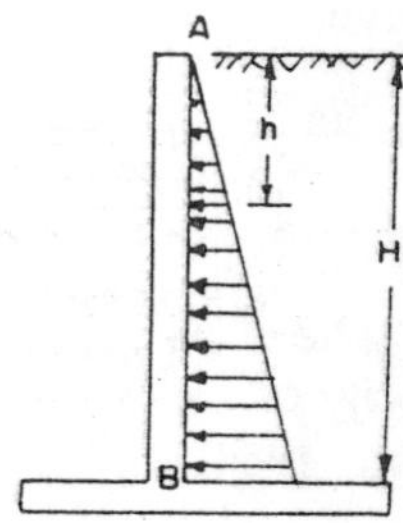

FIG. 18.13.

side of fill. The reinforcement towards the top of stem can be curtailed, since B.M. varies as h^3. Distribution reinforcement is provided @ 0.15% of the area of cross-section along the length of retaining wall at inner face. Similarly, at the outer face of the stem, temperature reinforcement is provided both in horizontal as well as in vertical direction, at the rate of 0.15% of the area of cross-section.

3. Design of heel slab

The heel is also to be designed as a cantilever. It has both downward pressure(due to weight of soil and self-weight) as well as upward pressure due to soil reaction. However, the *net* pressure is found to act *downward* and hence reinforcement is provided at the upper face *BC*.

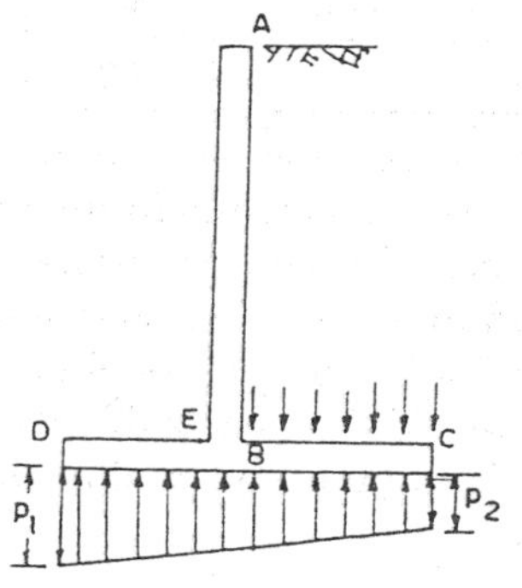

FIG. 18.14.

4. Design of toe slab

Neglecting the weight of the soil above it, the toe slab will bend upwards as a cantilever due to upward soil reaction. Hence reinforcement is placed at the bottom face. Normally, the thickness of both toe slab and heel slab is kept the same, determined on the basis of greater of the cantilever bending moments.

Depth of foundation

The height H_2 of the retaining wall, above ground level is fixed on the basis of height of the backfill to be retained. The depth of foundation y should be such that good quality of soil to bear the induced pressures is available. However, a minimum depth of foundation given below by Rankine's formula should be provided :

$$y_{min} = \frac{q_0}{\gamma}\left(\frac{1 - \sin \Phi}{1 + \sin \Phi}\right)^2 = \frac{q_0}{\gamma} K_a^{\ 2} \qquad \ldots(18.28)$$

where q_0 is the safe bearing capacity of soil, or equal to the maximum pressure likely to occur on soil.

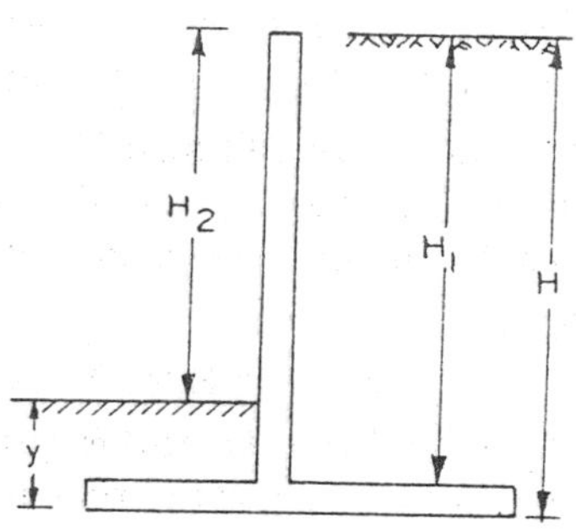

FIG. 18.15

18.7. DESIGN OF CANTILEVER RETAINING WALL WITH HORIZONTAL BACKFILL

Design Example 18.1. *Design a T-shaped cantilever retaining wall to retain earth embankment 3 m high above ground level. The unit weight of earth is* 18 kN/m^3 *and its angle of repose is 30°. The embankment is horizontal at its top. The safe bearing capacity of soil may be taken as 100 kN/m² and the coefficient of friction between soil and concrete as 0.5. Use M 15 mix. Take* $\sigma_{st} = 140\ N/mm^2$.

Solution

1. ***Design constants***

For M 15 concrete and mild steel reinforcement we have the following values :

$$c = \sigma_{cbc} = 5\ \text{N/mm}^2 : t = \sigma_{st} = 140\ \text{N/mm}^2 ; m = 19$$
$$k = 0.404 ; j = 0.865 \text{ and } R = 0.874$$

2. ***Depth of foundation***

$$\gamma = 18\ \text{kN/m}^3 = 18000\ \text{N/m}^3$$

The minimum depth of foundation is

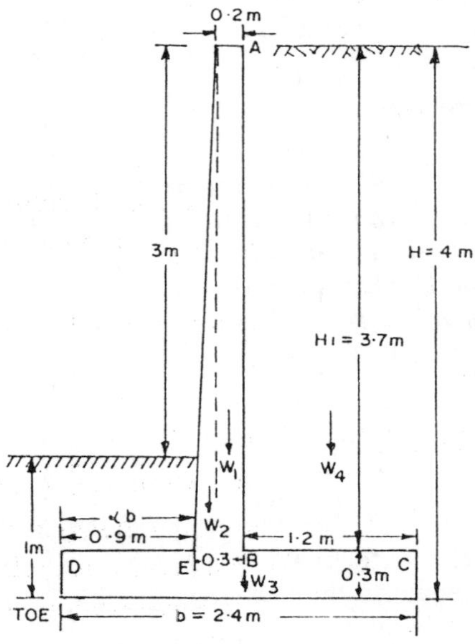

FIG. 18.16

$$y_{min} = \frac{q_o}{\gamma}\left(\frac{1-\sin\Phi}{1+\sin\Phi}\right)^2$$

$$= \frac{100}{18}\left(\frac{1-\sin 30°}{1+\sin 30°}\right)^2$$

$$= 0.62 \text{ m.}$$

However, keep depth =1 m to accommodate thickness of base wall below the ground surface.

Hence height of wall above its base $= H = 3 + 1 = 4$ m.

3. ***Dimensions of base***

The ratio of the length of toe slab (DE) to the base width b is given by Eq. 18.26 :

$$\alpha = 1 - \frac{q_o}{2.2\gamma H} = 1 - \frac{100}{2.2 \times 18 \times 4} = 0.365$$

Keep $\alpha = 0.37$ ...(*i*)

The width of base is given by Eq. 18.25.

$$b = 0.95\,H\sqrt{\frac{K_a}{(1-\alpha)(1+3\alpha)}}$$

$$k_a = \frac{1-\sin\Phi}{1+\sin\Phi} = \frac{1-\sin 30°}{1+\sin 30°} = \frac{1}{3}$$

$$b = 0.95 \times 4\sqrt{\frac{1}{3(1-0.37)(1+3\times 0.37)}} = 1.90\text{m.}$$

The base width from the considerations of sliding is given by Eq. 18.27.

$$b = \frac{0.7H\,K_a}{(1-\alpha)\mu} = \frac{0.7\times 4}{1-0.37} \times \frac{1}{3\times 0.5} = 2.96 \text{ m.}$$

This width is excessive. Normal practice is to provide b between 0.4 to 0.6 H. Taking maximum value of 0.6 H,

$$b = 0.6\,H = 0.6 \times 4 = 2.4 \text{ m.}$$

Hence provide $b = 2.4$ m. The wall will be unsafe against sliding. This will be made safe by providing a shear key at the base.

Width of toe slab $= 0.37 \times 2.4 = 0.89$ m.

Provide toe slab 0.9 m long.

Let the thickness of base be $\frac{1}{12}H \approx 0.3$ m for preliminary calculations.

4. ***Thickness of stem***

Height $AB = 4 - 0.3 = 3.7$ m.

Consider one meter length of retaining wall.

Maximum bending moment at B

$$= K_a \gamma \frac{H_1^3}{6} = \frac{1}{3}\,\frac{18}{6}(3.7)^3$$

$$= 50.65 \text{ kN-m } = 50.65 \times 10^6 \text{ N-mm}$$

Hence the effective depth is

$$d = \sqrt{\frac{50.65 \times 10^6}{1000 \times 0.874}} = 241 \text{ mm}$$

Keep d = 240 mm and total thickness =300 mm so that an effective cover of 60 mm is available. Reduce the total thickness to 200 mm at the top so that effective depth of 140 mm is available at the top.

5. *Stability of wall*

Fully dimensioned wall is shown in Fig. 18.16.

Let W_1 = weight of rectangular portion of stem

W_2 = weight of triangular portion of stem

W_3 = weight of base slab

W_4 = weight of soil on heel slab.

The calculations are arranged in Table 18.1

TABLE 18.1

S. No.	*Designation*	*Force (kN)*	*Lever arm (m)*	*Moment about toe (kN-m)*
1	W_1	$1 \times 0.2 \times 3.7 \times 25 = 18.5$	1.1	20.35
2	W_2	$\frac{1}{2} \times 0.1 \times 3.7 \times 25 = 4.63$	0.97	4.49
3	W_3	$1 \times 2.4 \times 0.3 \times 25 = 18.0$	1.2	21.60
4	W_4	$1 \times 1.2 \times 3.7 \times 18 = 79.97$	1.8	143.86
		$\Sigma W = 121.05$	M_R =	190.30

The total resisting moment $M_R = 190.30$ kN-m ...(1)

Earth pressure $P = K_a \gamma \dfrac{H^2}{2}$

$$= \frac{1}{3} \times \frac{18}{2}(4)^2 = 48 \text{ kN} \qquad ...(2)$$

Overturning

Overturning moment M_0 $= 48 \times \dfrac{4}{3} = 64$ kN-m ...(3)

$\therefore$ F.S. against overturning $= \dfrac{190.3}{64} = 2.97 > 2$. Hence safe.

Sliding

$$\text{F.S. against sliding} = \frac{\mu \Sigma W}{P} = \frac{0.5 \times 121.05}{48} = 1.26 < 1.5$$

Hence unsafe.

To make it safe against sliding, we will have to provide a shear key.

Pressure distribution

Net moment $\Sigma M = 190.3 - 64 = 126.3$ kN-m

$\therefore$ Distance $\bar{x}$ of the point of application of resultant, from toe is

$$\bar{x} = \frac{\Sigma M}{\Sigma W} = \frac{126.3}{122.05} = 1.04 \text{ m}$$

$$\text{Eccentricity} \qquad e = \frac{b}{2} - \bar{x} = 1.2 - 1.04 = 0.16 \text{ m}$$

This is less than $\frac{b}{6}\left(= \frac{2.4}{6} = 0.4 \text{ m}\right)$

$$\text{Pressure } p_1 \text{ at toe} = \frac{\Sigma W}{b}\left(1 + \frac{6e}{b}\right) = \frac{121.05}{2.4}\left(1 + \frac{6 \times 0.16}{2.4}\right)$$

$= 70.61$ kN/m$^2 < 100$. Hence safe.

$$\text{Pressure } p_2 \text{ at heel} = \frac{\Sigma W}{b}\left(1 - \frac{6e}{b}\right) = \frac{121.05}{2.4}\left(1 - \frac{6 \times 0.16}{2.4}\right)$$

$$= 30.26 \text{ kN/m}^2$$

Pressure p at the junction of stem with toe slab is

$$p = 70.61 - \frac{70.61 - 30.26}{2.4} \times 0.9 = 55.48 \text{ kN/m}^2$$

Pressure p' at the junction of stem with heel slab is

$$p' = 70.61 - \frac{70.61 - 30.26}{2.4} \times 1.2 = 50.44 \text{ kN/m}^2$$

6. *Design of toe slab*

The upward pressure distribution on the toe slab is shown in Fig. 18.17. The weight of the soil above the toe slab is neglected. Thus two forces are acting on it ;

(*i*) upward soil pressure.
(*ii*) downward weight of slab.

Downward weight of slab per unit area $= 0.3 \times 1 \times 1 \times 25$
$= 7.5$ kN/m^2

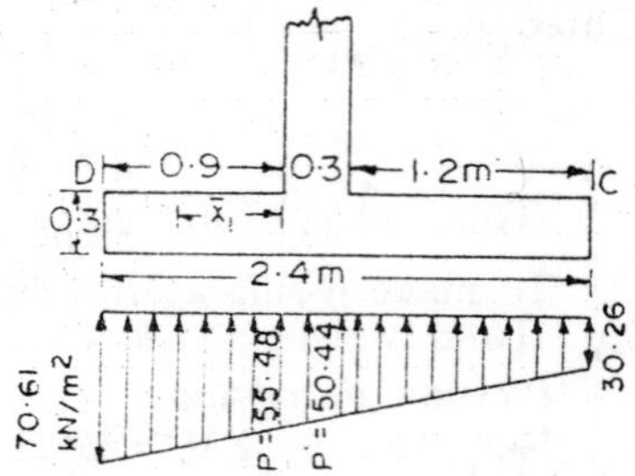

FIG. 18.17.

Hence net pressure intensities will be $= 70.61 - 7.5 = 63.11$ kN/m^2 under D and $55.48 - 7.5 = 47.98$ kN/m^2 under E.

Total force = S.F. at E = $\frac{1}{2}(63.11 + 47.98)0.9 = 50$ kN

$$\bar{x} \text{ from } E = \left(\frac{47.98 + 2 \times 63.11}{47.98 + 63.11}\right)\frac{0.9}{3} = 0.47 \text{ m.}$$

$\therefore$ B.M. at $E = 50 \times 0.47 = 23.52$ kN-m $= 23.52 \times 10^6$ N-mm

$$\therefore \quad d = \sqrt{\frac{23.52 \times 10^6}{1000 \times 0.874}} = 164 \text{ mm}$$

Let us keep total depth = 260 mm and effective depth of 200 mm so that 60 mm effective cover is available. Thickness can be reduced to 200 mm at the edge.

$$A_{st} = \frac{23.52 \times 10^6}{140 \times 0.865 \times 200} = 971 \text{ mm}^2$$

This reinforcement has to be provided at the bottom face. If alternate bars of stem reinforcement are bent and continued in the toe slab, area available $= \frac{1}{2} \times 2010 = 1005$ mm^2 (see step 8). This reinforcement will consist of 16 Φ bars @ 200 mm c/c. Let us check this reinforcement for development length.

$$L_d \approx 58.3\Phi = 58.3 \times 16 = 933 \text{ mm.}$$

Providing 50 mm clear side cover, actual length available $= 900 - 50 +$ anchorage value of hook $= 850 + 13 \times 16 = 850 + 208 = 1058 > L_d$. Hence safe.

$$\text{Distribution reinforcement} = \frac{0.15}{100} \times 1000\left[\frac{260 + 200}{2}\right]$$

$$= 345 \text{ mm}^2$$

Using 8 mm Φ bars, $A_\Phi = 50.3$ mm^2

$$\therefore \quad \text{Spacing} = \frac{1000 \times 50.3}{345} = 146 \text{ mm}$$

Hence provide these @ 140 mm c/c.

7. *Design of heel slab*

Three forces act on it : (*i*) downward weight of soil 3.7 m high
(*ii*) downward weight of heel slab
(*iii*) upward soil pressure

Total weight of soil $= 1.2 \times 3.7 \times 1 \times 18 = 80$ kN acting at 0.6 m from *B*.

Total weight of heel slab $= 1.2 \times 0.26 \times 1 \times 25 = 7.8$ acting at 0.6 m from *B*.

Total upward soil reaction $= \frac{1}{2}(50.44 + 30.26)1.2 = 48.42$ kN

$$\text{acting at } \left(\frac{50.44 + 2 \times 30.26}{50.44 + 30.26}\right)\frac{1.2}{3} = 0.85 \text{ m from } B$$

$\therefore$ Total force = S.F. at $B = 80 + 7.8 - 48.42 = 39.38$ kN

$\therefore$ B.M. at $B = (80 \times 0.6) + (7.8 \times 0.6) - (40.58 \times 0.85)$
$= 18.19$ kN – m $= 18.19 \times 10^6$ N–mm

$$\therefore \quad d = \sqrt{\frac{18.19 \times 10^6}{1000 \times 0.874}} = 144 \text{ mm.}$$

However keep the same total depth(=260 mm) as that of toe slab, so that available effective depth = 200 mm. The thickness is reduced to 200 mm at the edge.

$$A_{st} = \frac{18.19 \times 10^6}{140 \times 0.864 \times 200} = 752 \text{ mm}^2$$

Using 12 mm Φ bars, $A_\Phi = 113$ mm^2

$$\therefore \quad \text{Spacing} = \frac{1000 \times 113}{752} \approx 150 \text{ mm.}$$

Hence provide these @ 150 mm c/c. Take the reinforcement into the toe slab from a distance of $58.3\,\Phi = 58.3 \times 12 = 700$ mm to the left of *B*, and its ends should be hooked.

$$\text{Distribution steel} = \frac{0.15}{100} \times 1000\left[\frac{260 + 200}{2}\right] = 345 \text{ mm}^2$$

$$\therefore \quad \text{Spacing} = \frac{1000 \times 50.3}{345} = 146 \text{ mm.}$$

Hence provide three @ 140 mm c/c.

$$\text{Shear stress} \quad \tau_v = \frac{39.38 \times 1000}{1000 \times 200} = 0.20 \text{ N/mm}^2$$

This is much less than the permissible shear stress even at the minimum percentage of steel (Table 3.1).

8. *Reinforcement in the stem*

We had earlier assumed the thickness of heel slab as 0.3 m, while it has now been fixed as 0.26 m only.

Hence revised $H_1 = 4 - 0.26 = 3.74$ m

$$\therefore \quad M = K_a \gamma \frac{H_1^3}{6} = \frac{1}{3} \times \frac{18}{6}(3.74)^3$$

$$= 52.31 \text{ kN} - \text{m} = 52.31 \times 10^6 \text{ N} - \text{mm}$$

$$\therefore \quad d = \sqrt{\frac{52.31 \times 10^6}{1000 \times 0.874}} = 245 \text{ mm.}$$

Keep d = 250 mm so that D = 310 mm. Reduce the total thickness at the top.

$$A_{st} = \frac{52.31 \times 10^6}{140 \times 0.865 \times 250} = 1728 \text{ mm}^2$$

Using 16 mm Φ bars, $A_\Phi = 201 \text{ mm}^2$

$$\therefore \quad s = \frac{1000 \times 201}{1728} = 116 \text{ mm}$$

However provide 16 mm Φ bars @ 100 mm c/c.

Actual A_{st} provided $= 1000 \times \dfrac{201}{100} = 2010 \text{ mm}^2$.

Continue alternate bars in the toe slab to serve as tensile reinforcement there. Discontinue the remaining half bars after a distance of $58.3\Phi = 58.3 \times 16 \approx 930$ mm beyond B, in the toe slab.

Between A and B, some of the bars can be curtailed. Consider a section at depth h below the top of the stem. The effective depth d' at that section is

$$d' = 140 + \frac{250 - 140}{3.74} h = (140 + 29.4\,h) \text{ mm} \qquad \text{...(1)}$$

where h is in metres.

Now $A_{st} \propto \dfrac{H^3}{d}$

or $\quad H = (A_{st}\, d)^{1/3}$

$$\text{Hence} \quad \frac{h}{H_1} = \left(\frac{A_{st}'\, d'}{A_{st}\, d}\right)^{\frac{1}{3}} \qquad \text{...(2)}$$

where A_{st}' = reinforcement at depth h

d' = effective depth at depth h

A_{st} = reinforcement at depth H_1

d = effective depth at depth H_1

If $\quad A_{st}' = \frac{1}{2} A_{st}\,, \quad \dfrac{A_{st}'}{A_{st}} = \dfrac{1}{2}$

$$\therefore \quad \frac{h}{H_1} = \left(\frac{1}{2} \cdot \frac{d'}{d}\right)^{\frac{1}{3}}$$

Substituting $d = 250$ mm and $d' = (140 + 29.4\ h)$, we get

$$h = H_1\left[\frac{140 + 29.4\,h}{2 \times 250}\right]^{\frac{1}{3}} = 3.74\left[\frac{140 + 29.4\,h}{500}\right]^{\frac{1}{3}}$$

or $$h = 0.471[140 + 29.4\,h]^{\frac{1}{3}}$$

This can be solved by trial and error, noting that if the effective thickness of stem were constant, h would have been equal to $\frac{H_1}{(2)^{1/3}} \approx 0.79\,H_1 \approx 2.96$ m.

Sloving (3) by trial, we get $h = 2.86$ m.

Thus, half the bars can be curtailed at this point. However, the bars should be extended by a distance of $12\Phi\ (= 12 \times 16 = 192$ mm) or $d(= (250$ mm) which ever is more, beyond the point.

$\therefore \quad h = 2.86 - 0.25 = 2.61$ m. Hence curtail half the bars at a height 2.6 m below the top.

If we wish to curtail half of the remaining bars so that remaining reinforcement is one-fourth of that provided at B, we have

$$\frac{A_{st}'}{A_{st}} = \frac{1}{4}$$

Hence from (2), $\frac{h}{H_1} = \left(\frac{1}{4}\frac{d'}{d}\right)^{\frac{1}{3}}$

or $$h = H_1\left[\frac{140 + 29.4\,h}{4 \times 250}\right]^{\frac{1}{3}} = 3.74\left[\frac{140 + 29.4\,h}{1000}\right]^{\frac{1}{3}}$$

$$= 0.374\,[143 + 29.4\,h]^{\frac{1}{3}} \qquad ...(4)$$

This can be solved by trial and error, noting that if the effective thickness of stem were constant, h would have been equal to $\frac{H_1}{(4)^{\frac{1}{3}}} = 0.63\,H_1 \approx 2.36$ m.

Solving (4) by trial and error, we get $h = 2.2$ m. However the bars should be extended by 250 mm beyond this.

$\therefore\ h = 2.2 - 0.25 = 1.95$

Hence stop half the remaining bars by 1.95 m below the top of the stem. Continue rest of the bars to the top of the step.

Check for shear

Shear force $= p = K_a\gamma\frac{H^2}{2} = \frac{1}{3} \times \frac{18}{2}(3.74)^2$

$$= 41.96 \quad \text{kN}$$

$$\therefore \qquad \tau_v = \frac{41.96 \times 1000}{1000 \times 250} = 0.17 \text{ N/mm}^2 < \tau_c$$

Hence safe.

Distribution and temperature reinforcement

Average thickness of stem $= \frac{1}{2}(310 + 200) = 255$ mm

$$\therefore \quad \text{Distribution reinforcement} = \frac{0.15}{100} \times 1000 \times 255 = 383 \text{ mm.}$$

Using 8 mm Φ bars, $A_\Phi = 50.3$ mm²

$$\therefore \quad \text{Spacing} = \frac{1000 \times 50.3}{383} = 131 \text{ mm}$$

Hence provide 8 mm Φ bars @ 130 mm c/c at the inner face of the wall, along its length.

For temperature reinforcement, provide 8 mm Φ @ 260 mm c/c both, ways, in the outer face.

9. *Design of shear key*

The wall is unsafe in sliding, and hence shear key will have to be provided below the stem, as shown in Fig. 18.18.

Let the depth of key= a.

Intensity of passive pressure p_P developed in front of the key depends upon the soil pressure p in front of the key.

$\therefore \quad p_P = K_p p = 3 \times 55.48 = 166.4$ kN/m²

$\therefore \quad$ Total passive pressure $P_p = p_P\ a = 166.4\,a$ ..(1)

Sliding force at level D_1C_1

$$= \frac{1}{3} \times \frac{18}{2}(4 + a)^2$$

or

$$P_H = 3(4 + a)^2 \qquad ...(2)$$

Weight of soil between bottom of base and D_1C_1

$$= 2.4\,a \times 18 = 43.2\,a$$

$\therefore \quad \Sigma W = 121.05 + 43.2\,a$ (Refer Table 18.1)

Hence for equilibrium of the wall, permitting F.S. = 1.5 against sliding, we have

$$1.5 = \frac{\mu \Sigma W + P_p}{P_H}$$

or $$1.5 = \frac{0.5\,(121.05 + 43.2\,a) + 166.4\,a}{3\,(4 + a)^2}$$

or $$a^2 - 33.8\,a + 2.54 = 0$$

which gives $a \approx 0.09$ m = 90 mm

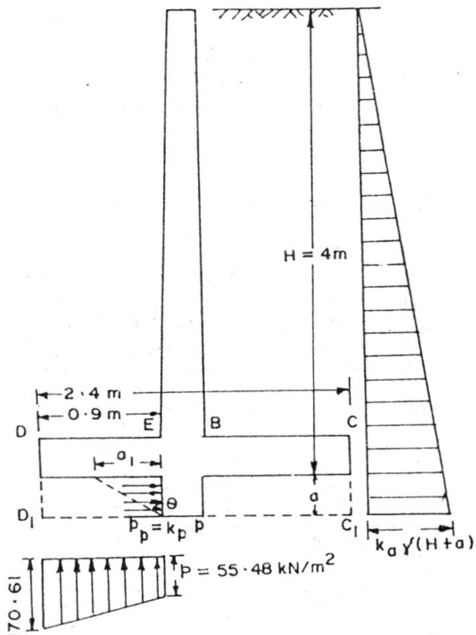

FIG. 18.18.

However, provide a minimum value of a = 0.3 m = 300 mm. Keep width of key = 300 mm. It should be noted that passive pressure taken into account above will be developed *only when* a length a_1 given below is available in front of the key :

$$a_1 = a \tan\theta = a \tan\left(45° + \frac{\Phi}{2}\right)$$

$$= a\sqrt{K_p}\,,\ \text{where}\ \left(45° + \frac{\Phi}{2}\right)$$

= shearing angle of passive resistance.

$\therefore \quad a_1 = 0.3\sqrt{3} = 0.52$ m.

Actual length of the slab available $= DE = 0.9$ m. Hence satisfactory.

Now size of key = 300 mm × 300 mm

$P_H = 3(4 + a)^2 = 3(4 + 0.3)^2 = 55.47$ kN

$P_P = 166.4\, a = 166.4 \times 0.3 = 49.92$ kN

$\therefore \quad \Sigma W = 121.05 + 43.2\, a = 121.05 + 43.2 \times 0.3 = 134.01$ kN

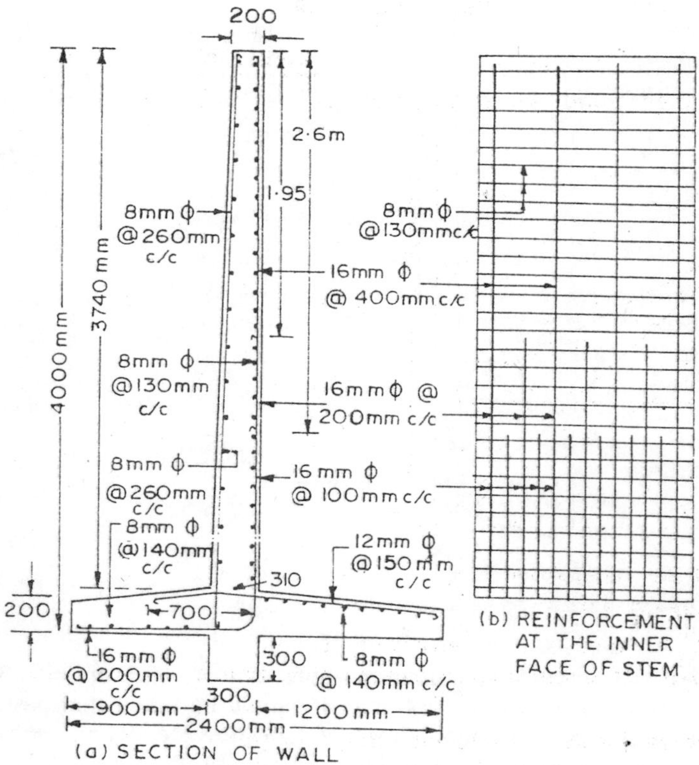

FIG. 18.19.

Actual force to be resisted by the key, at F.S.= 1.5 is

$$= 1.5\, P_H - \mu\, \Sigma W$$

$$= 1.5 \times 55.47 - 0.5 \times 134.01 = 16.2 \text{ kN}$$

$\therefore$ Shear stress $= \dfrac{16.2 \times 1000}{300 \times 1000} = 0.054\ \text{N/mm}^2$ (safe)

Bending stress $= \dfrac{16.2 \times 150 \times 1000}{\frac{1}{6} \times 1000\,(300)^2} = 0.16\ \text{N/mm}^2$ (safe)

The details of reinforcement etc. are shown in Fig. 18.19.

18.8. DESIGN OF CANTILEVER RETAINING WALL WITH HORIZONTAL BACKFILL AND TRAFFIC LOAD

Design Example 18.2. *Design a cantilever retaining wall for a road for the following requirements :*

1. *Height of wall from the bottom of base to top of stem=6m*
2. *Superimposed load due to road traffic = 18 kN/m*^{2}
3. *Unit weight of fill = 18 kN/m*^{3}
4. *Angle of internal friction for fill material = 30°*
5. *Allowable bearing pressure on ground = 160 kN/m*^{2}
6. *Coefficient of friction between concrete and ground = 0.4*

Also, provide a parapet wall 1 m high on the top of stem. Use M 15 concrete. Take $\sigma_{st} = 140\ N/mm^2$.

Solution.

1. *Design constants*

For M 15 concrete, and for mild steel reinforcement we have the following :

$c = \sigma_{cbc} = 5\ \text{N/mm}^2\,;\ t = \sigma_{st} = 140\ \text{N/mm}^2\,;\ m = 19$

$k = 0.404\,;\quad j = 0.865\quad \text{and}\quad R = 0.874$

2. *Dimensions of base*

Assume that a horizontal force $H = 2$ kN/m length of parapet wall will act because of person standing near the parapet.

Due to surcharge, equivalent height of fill is given by

$$h_e = \frac{w}{\gamma} \quad \text{(Eq. 18.10)}$$

$$\therefore \quad h_e = \frac{18}{18} = 1$$

Hence, in determining the values of b and α etc., we will use a height $H_2 = H + h_e = 6 + 1 = 7$ m. The ratio of the length of toe slab EF to the base width b may be determined by Eq. 18.28.

$$\alpha = 1 - \frac{q_0}{2.2\,\gamma\,H_2} = 1 - \frac{160}{2.2 \times 18 \times 7} = 0.42\,.$$

The width of base may be found from Eq. 18.25

$$b = 0.95\, H_2 \sqrt{\frac{K_a}{(1-\alpha)(1+3\alpha)}}$$

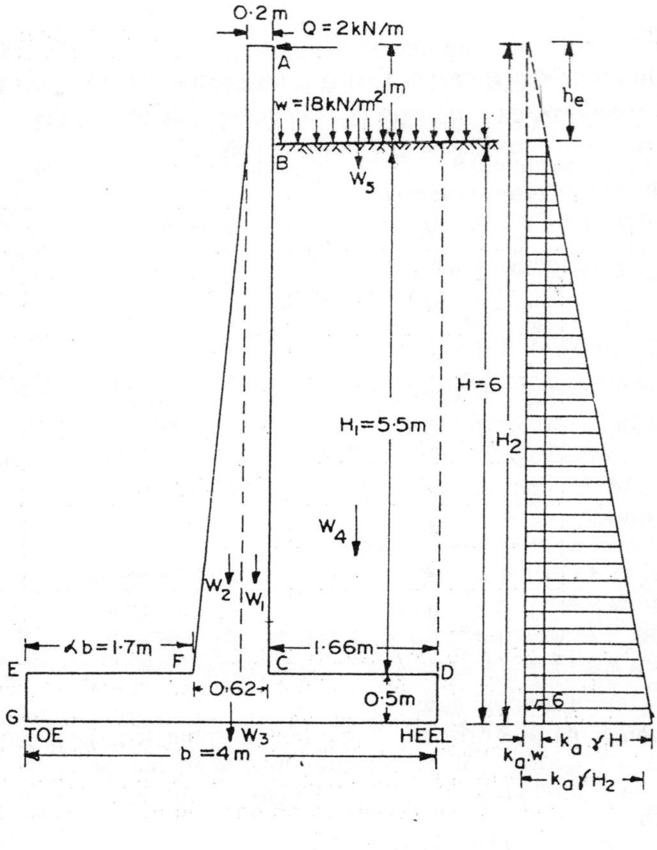

FIG. 18.20.

For $\Phi = 30, K_a = \dfrac{1-\sin 30°}{1+\sin 30°} = \dfrac{1}{3}$.

$$\therefore \quad b = 0.95 \times 7\sqrt{\frac{1}{3(1-0.42)(1+1.26)}} = 3.36 \text{ m}$$

However, keep minimum b between 0.5 to 0.6 H_2.

Hence adopt $b = 4$ m.

$\therefore$ Length of toe slab $= \alpha\, b = 0.42 \times 4 = 1.68$ m.

Keep it equal to 1.7 m.

Let the thickness of base = 50 cm for preliminary calculations.

3. *Thickness of stem*

Height $H_1 = 6 - 0.5 = 5.5$ m

Due to retained soil, the earth pressure diagram will be a triangle, having an ordinate equal to $K_a \gamma h$, at h below B. Due to surcharge w, there will be a uniform horizontal pressure $K_a w = \frac{1}{3} \times 18 = 6$ kN/m^2 throughout the height. The total bending moment at C will be due to moment of horizontal forced $Q =$ 2 kN acting at A, plus moment of rectangular pressure distribution, plus moment of triangular pressure distribution.

$$\therefore \quad M = Q\,(H_1 + 1) + K_a w \frac{H_1^2}{2} + K_a \gamma \frac{H_1^3}{6}$$

$$= 2\,(5.5 + 1) + \frac{1}{3} \times 18 \frac{(5.5)^2}{3} + \frac{1}{3} \times \frac{18}{6} (5.5)^2$$

$$= 13 + 90.75 + 166.35$$

$$= 270.1 \text{ kN-m} = 270.1 \times 10^6 \text{ N-mm}$$

$$\therefore \quad d = \sqrt{\frac{270.1 \times 10^6}{1000 \times 0.874}} = 556 \text{ mm}$$

Using an effective cover of 60 mm, total thickness of stem ≈ 620 mm. Reduce the total thickness to 200 mm at B, so that effective depth at $B = 200 - 60 = 140$ mm. Continue uniform thickness of 200 from B to A.

4. *Stability of wall*

Fully dimensioned wall is shown in Fig. 18.20.

Length of heel slab $CD = 4 - 1.7 - 0.62 = 1.68$ m.

Let W_1 = weight of rectangular portion of stem.

W_2 = weight of triangular portion of stem.

W_3 = weight of base slab.

W_4 = weight of soil above heel slab.

W_5 = total superimposed road-traffic load, over heel slab.

The calculations are arranged in Table 18.2.

TABLE 18.2

S. No.	Designation	Force (kN)	Level arm (m)	Moment about toe (kN-m)
1	W_1	$0.2 \times 6.5 \times 1 \times 25 = 32.5$	2.22	72.15
2	W_2	$\frac{1}{2} \times 0.42 \times 5.5 \times 1 \times 25 = 28.9$	1.98	57.17
3	W_3	$0.5 \times 4 \times 1 \times 25 = 50$	2.00	100.0
4	W_4	$1.68 \times 5.5 \times 1 \times 18 = 166.4$	3.16	525.82
5	W_5	$1.68 \times 1 \times 18 = 30.2$	3.16	95.56
		$\Sigma W \approx 308$	$M_R = 850.7$	

Total earth pressure at base $= K_a \dfrac{\gamma H^2}{2} = \dfrac{1}{3} \times \dfrac{18}{2} (6)^2$

$= 108$ kN

Overturning moment due to earth pressure $= 108 \times \dfrac{6}{3}$

$= 216$ kN-m

Overturning moment at toe, due to horizontal force Q

$= 2 \times 7 = 14$ kN-m

Overturning moment due to horizontal pressure caused by live load

$$= K_a w \frac{H^2}{2} = \frac{1}{3} \times 18 \times \frac{(6)^2}{2} = 108 \text{ kN-m}$$

$\therefore$ Total overturning moment $M_0 = 216 + 14 + 108$

$= 338$ kN-m

$\therefore$ F.S. against overturning $= \dfrac{850.7}{338} = 2.52 > 2.$ Hence safe.

Total horizontal pressure $= 2 + 108 + (6 \times 6) = 146$ kN $= \Sigma P$

$\therefore$ F.S. against sliding $= \dfrac{\mu \Sigma W}{\Sigma P} = \dfrac{0.4 \times 308}{146} = 0.84 < 1.5$

Hence unsafe. Special shear key will have to be designed to make the wall safe against sliding.

Pressure distribution

Algebraic sum of moments $= \Sigma M = M_R - M_0$

$= 850.7 - 338 = 512.7$ kN-m

Algebraic sum of vertical forces $= \Sigma W =$ 308 kN

Hence distance $\bar{x}$ from toe, of the point of application of resultant is given by

$$\bar{x} = \frac{\Sigma M}{\Sigma W} = \frac{512.7}{308} = 1.66 \text{ m}$$

$\therefore$ Eccentricity $e = \frac{b}{2} - \bar{x} = \frac{4}{2} - 1.66 = 0.34$ m

This is less than $\frac{b}{6} = \frac{4}{6} = 0.67$ m

Hence no tension is developed.

Pressure p_1 at toe $= \frac{\Sigma W}{b}\left(1 + \frac{6e}{6}\right)$

$$= \frac{308}{4}\left[1 - \frac{6 \times 0.34}{4}\right]$$

$= 116.3$ kN/m^2 < 160 Hence safe.

Pressure p_2 at heel $= \frac{\Sigma W}{b}\left(1 - \frac{6e}{b}\right)$

$$= \frac{308}{4}\left(1 - \frac{6 \times 0.34}{4}\right)$$

$= 37.7$ kN/m^2

Pressure p at the junction of stem with toe slab is

$$p = 116.3 - \frac{116.3 - 37.7}{4} \times 1.7 = 82.9 \text{ kN/m}^2$$

Pressure p' at the junction of stem with heel slab is

$$p' = 116.3 - \frac{116.3 - 37.7}{4} \times 2.32 = 70.7 \text{ kN/m}^2$$

5. *Design of toe slab*

The upward pressure distribution on the toe slab is shown in Fig. 18.21. The weight of the soil above the toe slab is neglected. Thus two forces are acting on it :

(*i*) upward soil pressure

(*ii*) downward weight of slab.

Downward weight of slab per unit area $= 0.5 \times 1 \times 1 \times 25 = 12$ kN/m^2 Hence net pressure intensities will be $116.3 - 12 = 104.3$ kN/m^2 under D and $82.9 - 12 = 70.9$ kN/m^2 under F.

$\therefore$ Total force = S.F. at $F = \frac{1}{2}(104.3 + 70.9) \times 1.7$

$= 148.92$ kN

$$\text{C.G. of force from } F = \bar{x}_1 = \left(\frac{70.9 + 2 \times 104.3}{70.9 + 104.3}\right)\frac{1.7}{3} = 0.904 \text{ m}$$

$\therefore$ B.M. at $F = 148.92 \times 0.904 = 134.63$ kN-m $= 134.63 \times 10^6$ N-mm

$$\therefore \qquad d = \sqrt{\frac{134.63 \times 10^6}{1000 \times 0.874}} = 392 \text{ mm.}$$

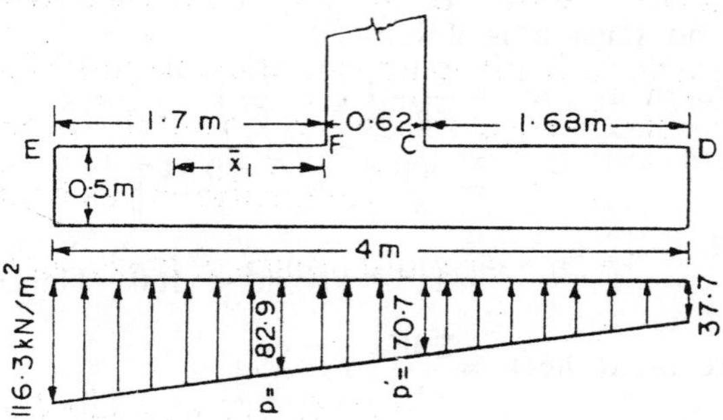

FIG. 18.21.

Keep $d =$ 400 mm and total depth equal to 460 mm so that 60 mm effective cover is available. Reduce the total depth to 200 mm at the edges.

$$A_{st} = \frac{134.63 \times 10^6}{140 \times 0.865 \times 400} = 2779 \text{ mm}^2$$

Bars available from stem reinforcement are 25 mm Φ @ 120 mm c/c giving A_{st} = 4092 mm^2 (step 7).

$$\tau_v = \frac{148.92 \times 1000}{1000 \times 400} = 0.37 \text{ N/mm}^2$$

$$100\frac{A_s}{bd} = \frac{100 \times 4092}{100 \times 400} = 1.02 \%$$

Hence from Table 3.2, $\tau_c = 0.375$ N/mm^2

Hence the slab is just safe in shear.

$$\text{Distribution steel} = \frac{0.15}{100}\left[\frac{460 + 200}{2}\right] \times 1000 = 495 \text{ mm}^2$$

Using 8 mm Φ bars, having $A_\Phi = 50.3$ mm^2

$$\text{spacing} = \frac{1000 \times 50.3}{495} = 102 \text{ mm}$$

Hence provide these @ 100 mm c/c.

6. *Design of heel slab*

Four forces act on it :

(*i*) downward weight of soil 5.5 m high

(*ii*) live load of traffic

(*iii*) weight of heel slab

(*iv*) upward soil pressure.

Weight of soil $= 1.68 \times 1 \times 5.5 \times 18 = 166.32$ kN acting at 0.84 m from C.

Live load $= 18 \times 1 \times 1.68 = 30.24$ kN acting at 0.84 m from C.

Weight of heel slab $= 0.5 \times 1.68 \times 25 = 21$ kN acting at 0.84 m from C

Upward soil reaction $= \frac{1}{2}(70.7 + 37.7) \times 1.68 = 91.06$ kN

acting at $\left(\dfrac{70.7 + 2 \times 37.7}{70.7 + 37.7}\right)\dfrac{1.68}{3} = 0.75$ m from C

$\therefore$ Total force = S.F. at $C = 166.32 + 30.24 + 21 - 91.06$
$= 126.5$ kN

B.M. at $C = (166.32 + 30.24 + 21)\, 0.84 - 91.06 \times 0.75$

$= 114.46$ kN-m $= 114.46 \times 10^6$ N-mm

Keep the same depth as that of toe slab, *i.e.* $d = 400$ mm and total depth = 460 mm. Reduce total depth to 200 mm at edges.

$$A_{st} = \frac{114.46 \times 10^6}{140 \times 0.865 \times 400} = 2363 \text{ mm}^2$$

Using 20 mm Φ bars, $A_\Phi =$ 31 mm^2

$$\therefore \quad \text{spacing} = \frac{1000 \times 314}{2363} = 132 \text{ mm.}$$

Hence, provide these @ 130 mm c/c. Take the reinforcement into the toe slab for a distance of $58.3\,\Phi = 58.3 \times 20 \approx 1170$ mm to the left of C, and its ends should be hooked. The length of heel slab = 1680 mm, hence sufficient development length will be available,

$$\tau_v = \frac{126.5 \times 10^3}{1000 \times 400} = 0.32 \text{ N/mm}^2$$

$$\text{Actual } A_{st} = \frac{1000 \times 2314}{130} = 2415 \text{ mm}^2$$

$$\therefore \quad 100\,\frac{A_s}{bd} = \frac{100 \times 2415}{1000 \times 400} = 0.6\ \%$$

Hence from Table 3.1, $\tau_c = 0.31$ N/mm², which is quite close to τ_v. Hence the slab can be considered to be safe against shear.

$$\text{Distribution steel} = \frac{0.15}{100}\left[\frac{460 + 200}{2}\right] \times 1000 = 495 \text{ mm}^2$$

$$\therefore \text{ Spacing of 8 mm } \Phi \text{ bars} = \frac{1000 \times 50.3}{495} \approx 100 \text{ mm.}$$

7. ***Reinforcement in the stem***

Revised value of $H_1 = 6 - 0.46 = 5.54$ m

Hence B.M. at C, calculated earlier will change slightly.

$$M = 2\,(5.54 + 1) + \frac{6\,(5.54)^2}{2} + \frac{1}{3} \times \frac{18}{6}\,(5.54)^3$$

$$= 13.08 + 92.08 + 170.03 \approx 275.2 \text{ kN-m}$$

$$= 275.2 \times 10^6 \text{ N-mm}$$

$$d = \sqrt{\frac{275.2 \times 10^6}{1000 \times 0.874}} \approx 560 \text{ mm.}$$

Keeping $D = 620$ and an effective cover of 60 mm available $d = 620 - 60 = 560$ mm. Hence O.K.

$$A_{st} = \frac{275.2 \times 10^6}{140 \times 0.865 \times 560} = 4058 \text{ mm}^2$$

Using 25 mm Φ bars, $A_\Phi = 491$ mm²

$$\therefore \quad s = \frac{1000 \times 491}{4058} = 121 \text{ mm}$$

Hence provide 25 mm Φ bars @ 120 mm c/c

$$\text{Actual } A_{st} \text{ provided} = \frac{1000 \times 491}{120} = 4092 \text{ mm}^2$$

Bend all the bars in the toe slab to serve as the reinforcement there. Due to this, sufficient bond length will be available to both the sides of point C (point of maximum bending moment)

$$\text{Total S.F. at } C = Q + K_a w H_1 + K_a \gamma \frac{H_1^2}{2}$$

$$= 2 + 6 \times 5.54 + \frac{1}{3} \times \frac{18}{2}\,(5.54)^2$$

$$= 127.3 \text{ kN}$$

$$\therefore \quad \tau_v = \frac{127.3 \times 1000}{1000 \times 560} = 0.23 \text{ N/mm}^2$$

$$\frac{100\,A_s}{bd} = \frac{100 \times 4092}{1000 \times 560} = 0.73\ \%$$

Hence $\tau_c = 0.33$ N/mm² $> \tau_v$ Hence safe.

Let us curtail reinforcement between C and B. If there were no other external force, except the earth pressure, and if the depth of stem were constant, half the bars could have been curtailed at a depth $\frac{H_1}{(2)^{1/3}} \approx 0.79 H_1$ below the point B. Because of presence of other forces, let us try at depth $= 0.65 H_1 = 0.65 \times 5.54 = 3.6$ m below B to see whether half the bars could be curtailed there or not.

Thus, depth of section below $B = h = 3.6$ m

$$\text{B.M.} = Q(h+1) + K_a w \frac{h^2}{2} + K_a \gamma \frac{h^3}{6}$$

$$= 2(3.6+1) + \frac{6(3.6)^2}{2} + \frac{1}{3}\frac{18}{6}(3.6)^3$$

$$= 9.2 + 38.9 + 46.7$$

$$= 94.8 \text{ kN-m} = 94.8 \times 10^6 \text{ N-mm}$$

$$d = 140 + \frac{560-140}{5.54} \times 3.6 = 413 \text{ mm}$$

$$\therefore \quad A_{st} = \frac{94.8 \times 10^6}{140 \times 0.865 \times 313} = 1895 \text{ mm}^2$$

This is slightly less than half of that provided at C. Hence half the bars can be curtailed at this depth. However, the bars should be extended by a distance of $12\Phi\,(= 12 \times 25 = 300$ mm) or $d\,(= 413$ mm) beyond this point, whichever is more. Hence $h = 3.6 - 0.413 = 3.18$ m. Hence curtail half the bars at a height of 3.1 m below B.

If we wish to curtail half of the remaining part, let us try it at a section at depth $h = 0.65 \times 3.6 \approx 2.2$ m below B.

$$\text{B.M.} = 2(2.2+1) + \frac{6(2.2)^2}{2} + \frac{1}{3} \times \frac{18}{6}(2.2)^3$$

$$= 6.4 + 14.5 + 10.6 = 31.5 \text{ kN-m} = 31.5 \times 10^6 \text{ N-mm}$$

$$d = 140 + \frac{500-140}{5.54} \times 2.2 = 307$$

$$A_{st} = \frac{31.5 \times 10^6}{140 \times 0.865 \times 307} = 874 \text{ mm}^2.$$

This is slightly less than one fourth of that provided at C. Hence we can curtail half of the remaining bars here. However, extend the bars by a distance of $d = 307$ mm beyond this, *i.e.* curtail half of the remaining bars at a distance of $2.2 - 0.3 = 1.9$ m below B. Rest of the bars will be continued upto the top of the parapet.

Distribution reinforcement

Average thickness of stem $= \dfrac{200 + 620}{2} = 410$ mm

$\therefore$ Area distribution reinforcement $= \dfrac{0.15}{100}(1000 \times 410)$

$= 615 \text{ mm}^2$

Using 10 mm Φ bars, $A_\Phi = 78.5 \text{ mm}^2$

$\therefore$ Spacing $= \dfrac{100 \times 78.5}{515} = 127$ mm

$\therefore$ Hence provide 10 mm bars @ 120 mm c/c at the inner face of the wall, along its length.

Temperature reinforcement

Total area $=$ 615 mm²

Provide 10 mm Φ bars @ 240 mm c/c both ways on the outer face.

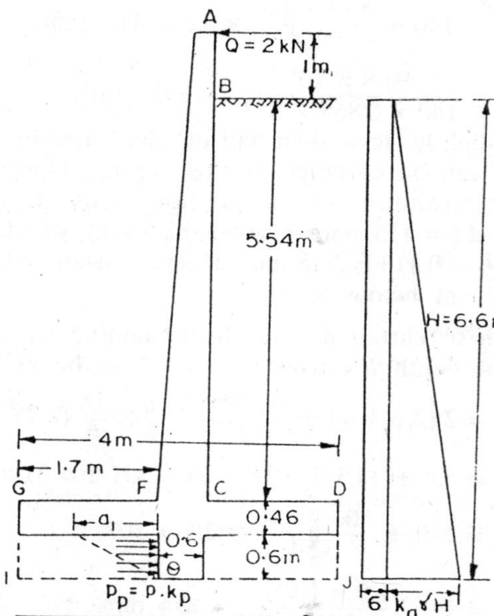

FIG. 18.22.

8. *Design of shear key*

The wall is unsafe against sliding.

Let us provide a shear key of depth a below the stem.

Let p_P be the intensity of passive pressure developed just in front of the shear key. This intensity p_P depends upon the soil pressure p just in front of shear key.

$$p_P = K_p \cdot p$$

where $$K_p = \frac{1 + \sin \Phi}{1 - \sin \Phi} = \frac{1}{K_a} = 3$$

and $$p = 82.9 \text{ kN/m}^2, \text{ from Fig. 18.21}$$

Hence $$p_P = 3 \times 82.9 = 248.7 \text{ kN/m}^2$$

This intensity may be considered to be constant along the depth of key, though there will be little increase in p_p, because of increase in p with depth. We will, however, consider the constant value of $p_P = 248.7$ kN/m².

$\therefore$ Total passive force $P_p = p_P\, a = 248.7\, a$ kN

Keeping $a = 600$ mm,

$$P_p = 248.7 \times 0.6 = 149.22 \text{ kN} \quad ...(1)$$

Total sliding force at the bottom level of key is

$$\Sigma P = 2 + (6 \times 6.6) + \frac{1}{3} \times \frac{18}{2} (6.6)^2$$

$$= 2 + 39.6 + 130.7 = 172.3 \text{ kN} \quad ...(2)$$

Weight of soil between the base and level JJ

$$= 0.6 \times 4 \times 1 \times 18 = 43.2 \text{ kN (approx.)}$$

$$\therefore \quad \Sigma W = 308 + 43.2 = 351.2 \text{ kN} \quad ...(3)$$

Hence F.S. against sliding is

$$= \frac{(\Sigma \mu W) + P_p}{\Sigma P} = \frac{0.4 \times 351.2 + 149.22}{172.3}$$

$$= 1.68 \text{ Hence safe.}$$

However, it should be noted that the above passive pressure will be developed *only when* a length a_1 given below is available in front of the key :

$$a_1 = a \tan \theta = a \tan \left(45 + \frac{\Phi}{2}\right) = a \sqrt{K_p}$$

$$= 0.6 \sqrt{3} = 1.04 \text{ m.}$$

where $\theta = 45 + \Phi/2$ = shearing angle of passive resistance.

Actual length of toe slab available $= GF = 1.7$ m

Hence satisfactory

Let us keep the width of key = 600 mm.

Actual force to be resisted by the key $= (1.5\,\Sigma P) - \mu\,\Sigma W$

$$= 1.5 \times 172.3 - 0.4 \times 351.2$$

$$\approx 118 \text{ kN}$$

Hence shear stress $= \dfrac{118 \times 1000}{1000 \times 600} \approx 0.2 \text{ N/mm}^2$. Safe

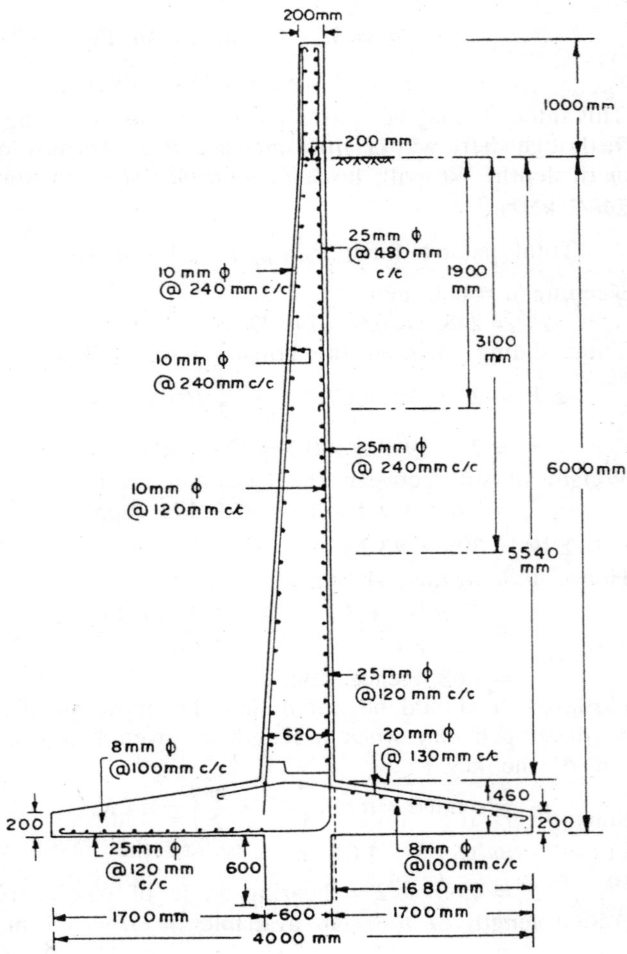

FIG. 18.23.

$$\therefore \text{ Bending stress} = \frac{118 \times 1000 \times 300}{1/6 \times 1000\,(600)^2} = 0.59 \text{ N/mm}^2$$

Since concrete can take this much of tensile stress no special reinforcement is necessary for the key. The key is to be cast monolithically with the base.

9. *Construction joint*

A construction joint, in the form of a key, is to be provided at the junction of stem with the base slab. The width of key is kept equal to $\frac{d}{4} = \frac{560}{4} = 140$ mm.

18.9. DESIGN OF CANTILEVER RETAINING WALL WITH SLOPING BACKFILL

Design Example 18.3. *Design a T-shaped cantilever retaining wall to retain earth embankment 3 m high above ground level. The embankment is surcharged at angle of 16° to the horizontal. The unit weight of earth is 18 N/m³ and its angle of repose is 30°. The safe bearing capacity may be taken as 100 kN/m² at a depth of 1m below the ground. The coefficient of friction between concrete and soil may be taken as 0.5. Use M 15 concrete. Take* $\sigma_{st} = 140$ *N/mm².*

Solution

1. *Design constants*

For M 15 concrete, and mild steel reinforcement we have the following :

$c = \sigma_{cbc} \doteq 5 \text{ N/mm}^2$; $t = \sigma_{st} = 140 \text{ N/mm}^2$; $m = 19$

$k = 0.404$; $j = 0.865$ and $R = 0.874$

2. *Dimensions of base*

$\sin\beta = 0.276$; $\cos\beta = 0.961$; $\tan\beta = 0.287$;

$\sin\Phi = \sin 30° = 0.5$; $\cos\Phi = 0.866$;

$$K_a = \cos\beta \, \frac{\cos\beta - \sqrt{\cos^2\beta - \cos^2\Phi}}{\cos\beta + \sqrt{\cos^2\beta - \cos^2\Phi}}$$

Substituting the values, we get $K_a = 0.38$

For a surcharged wall, the ratio of the length of the toe slab to the base width b may be fixed approximately by the expression :

$$\alpha = 1 - \frac{q_0}{2.7\,\gamma H} = 1 - \frac{100}{2.7 \times 18 \times 4} = 0.488.$$

Adopt $\alpha = 0.49$.

The base width can be determined from the expression

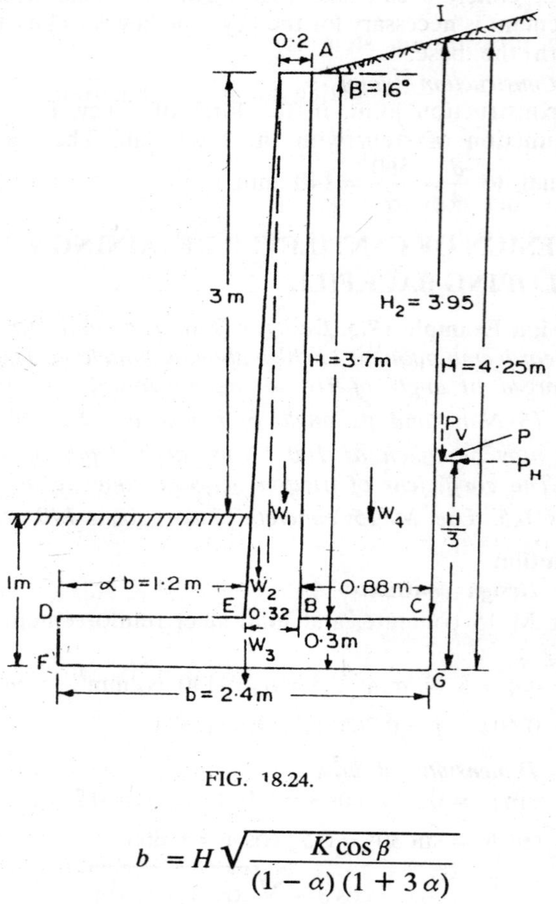

FIG. 18.24.

$$b = H\sqrt{\frac{K\cos\beta}{(1-\alpha)(1+3\alpha)}}$$

$$= 4\sqrt{\frac{0.38 \times 0.961}{(1-0.49)(1+3\times 0.49)}} = 2.15 \text{ m.}$$

From sliding point of view, b will be much more than this, as seen in example 18.1. However, a minimum base width $b = 0.6H = 2.4$ m should be adopted.

Hence keep $b = 2.4$ m

Length of toe slab $= 0.49 \times 2.4 \approx 1.2$ m.

Let the thickness of base slab be 300 mm for preliminary calculations.

3. *Thickness of stem*

Height $AB = H_1 = 4 - 0.3 = 3.7$ m.

$K_a = 0.38$

Earth pressure on stem is

$$P = K_a \frac{\gamma H_1^2}{2}, \text{ acting at } 16^\circ \text{ to horizontal}$$

$$= \frac{0.38}{2} \times 18\,(3.7)^2 = 46.82 \text{ kN}$$

Hence the horizontal earth pressure is

$$P_H = P\cos\beta = 46.82 \cos 16^\circ \approx 45 \text{ kN}$$

$$\text{B.M. at } B = P_H \frac{H_1}{3} = 45 \times \frac{3.7}{3} = 55.5 \text{ kN-m}$$

$$= 55.5 \times 10^6 \text{ N-mm}$$

$$\text{S.F. at } B = P_H = 45 \text{ kN}$$

$$d = \sqrt{\frac{55.5 \times 10^6}{0.874 \times 1000}} = 252 \text{ mm.}$$

Keep $d = 260$ mm and total thickness=320 mm so that an effective cover of 60 mm is available. Reduce total thickness to 200 mm at A, so that effective depth is available there = 140 mm.

$$\tau_v = \frac{45 \times 1000}{1000 \times 260} = 0.17 \text{ N/mm}^2 \text{ . Hence safe}$$

4. *Stability of wall*

Length of heel slab $= 2.4 - 1.2 - 0.32 = 0.88$ m

Height $H_2 = H_1 + 0.88 \tan 16^\circ = 3.7 + 0.25 = 3.95$ m

Height $H = 3.95 + 0.3 = 4.25$ m

The fully dimensioned wall is shown in Fig. 18.24.

Earth pressure $P = k_a \frac{\gamma H^2}{2}$, acting parallel to the surcharge

$$= \frac{0.38}{2} \times 18\,(4.25)^2 = 61.77 \text{ kN}$$

Its horizontal and vertical components are

$$P_H = P\cos\beta = 61.77 \times 0.961 = 59.36 \text{ kN}$$

$$P_V = P\sin\beta = 61.77 \times 0.276 = 17.05 \text{ kN}$$

Since P is acting on vertical face IG, at $H/3$ and hence P, will act along the vertical line IG(Fig. 18.24).

Let W_1 = weight of rectangular portion of stem

W_2 = weight of triangular portion of stem

W_3 = weight of base slab

W_4 = weight of soil on heel slab

TABLE 18.3

S. N.	Designation	Force (kN)	L.A. (m)	Moment about toe (kN-m)
1	W_1	$0.2 \times 3.7 \times 25 = 18.5$	1.42	26.27
2	W_2	$\frac{1}{2} \times 0.12 \times 3.7 \times 25 = 5.55$	1.28	7.10
3	W_3	$2.4 \times 0.3 \times 25 = 18$	1.20	21.60
4	W_4	$\frac{1}{2}(3.7 + 3.95)\, 0.88 \times 18 = 60.59$	1.965	119.06
5	W_5	$P_V = 17.05$	2.4	40.92
		$\Sigma W = 119.69$	$M_R = 214.95$	

$\therefore$ Total resisting moment $M_R = 214.95$ kN-m

Overturning moment $M_0 = P_H \dfrac{H}{3} = 59.36 \times \dfrac{4.25}{3} = 84.1$ kN-m

F.S. against overturning $= \dfrac{214.95}{84.1} = 2.56$. Hence safe.

F.S. against sliding $= \dfrac{\mu \Sigma W}{P_H} = \dfrac{0.5 \times 119.69}{59.36}$

$= 1 < 1.5$. Hence unsafe

To make it safe against sliding, will have to provide shear key.

Pressure distribution

Net moment $\Sigma M = M_R - M_0 = 214.95 - 84.1 = 130.85$ kN ...(1)

Distance $\bar{x}$ of the point of application of resultant from toe is

$$\bar{x} = \frac{\Sigma M}{\Sigma W} = \frac{130.85}{119.69} = 1.09 \text{ m.}$$

$\therefore$ Eccentricity $e = \dfrac{b}{2} - \bar{x} = 1.2 - 1.09 = 0.11$ m. This is much less than $b/6$.

Pressure p_1 at toe $= \dfrac{\Sigma W}{b}\left(1 + \dfrac{6e}{b}\right)$

$$= \frac{119.69}{2.4}\left(1 + \frac{6 \times 0.11}{2.4}\right)$$

$$= 63.59 \text{ kN/m}^2 < 100. \text{ Hence safe.}$$

Pressure P_2 at heel $= \frac{119.69}{2.4}\left(1 - \frac{6 \times 0.11}{2.4}\right) = 36.16 \text{ kN/m}^2$

Pressure p at the junction of stem with heel slab is

$$p = 63.59 - \frac{63.59 - 36.16}{2.4} \times 1.2 = 49.87 \text{ kN/m}^2$$

Presssure p' at the junction of stem with heel slab is

$$p' = 63.59 - \frac{63.59 - 36.16}{2.4} \times 1.52 = 46.22 \text{ kN/m}^2$$

5. *Design of toe slab*

The upward pressure distribution on the base is shown in Fig. 18.25. The weight of the soil above the toe slab is neglected. Thus two forces are acting on it :

(*i*) upward soil pressure

(*ii*) downward weight of the slab.

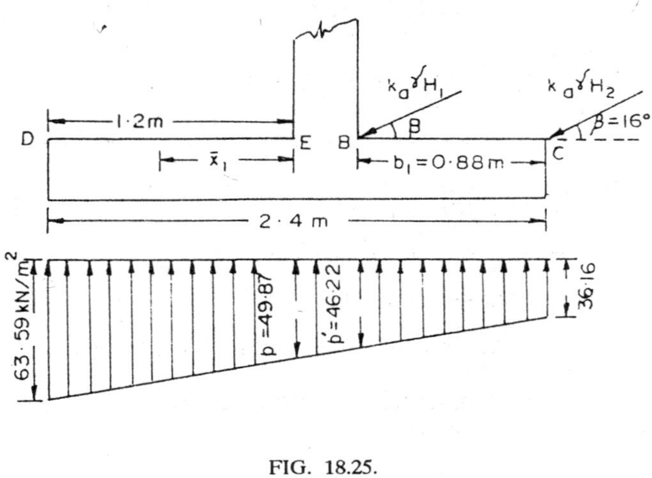

FIG. 18.25.

Downward weight of slab per unit area $= 0.3 \times 1 \times 1 \times 25$
$= 7.5 \text{ kN/m}^2$

Hence net pressure intensities will be $= 63.59 - 7.5$
$= 56.09 \text{ kN/m}^2$ under D and $49.87 - 7.5 = 42.37 \text{ kN/m}^2$ under E.

Total force $=$ S.F. at $E = \frac{1}{2}(56.09 + 42.47) \times 1.2$
$= 59.08$ kN.

C.G. of force from $E = \bar{x} = \left(\dfrac{42.37 + 2 \times 56.09}{42.37 + 56.09}\right)\dfrac{1.2}{3} = 0.63$ m

B.M. at $E = 59.08 \times 0.63 = 37.09$ kN-m

$= 37.09 \times 10^6$ N–mm

$\therefore \quad d = \sqrt{\dfrac{37.09 \times 10^6}{1000 \times 0.874}} = 206$ mm

Provide minimum total depth = 300 mm so that with an effective cover of 60 mm, $d = 240$ mm. Reduce the total depth to 200 mm at the edge.

$$\tau_v = \frac{59.08 \times 1000}{1000 \times 240} = 0.25 \text{ N/mm}^2$$

This is less than the permissible shear stress τ_c (Table 3.1)

$$A_{st} = \frac{37.09 \times 10^6}{140 \times 0.865 \times 240} = 1276 \text{ mm}^2$$

Hence 16 mm Φ bars bent down from the stem @ 110 mm c/c will be sufficient, providing $A_{st} =$ 1827 mm²(see step 7).

Distribution reinforcement $= \dfrac{0.15}{100} \times 1000 \left[\dfrac{300 + 200}{2}\right]$

$= 375$ mm²

Using 8 mm Φ bars, $A_\Phi = 50.3$ mm²

$\therefore \quad s = \dfrac{1000 \times 50.3}{375} = 134$ mm

Hence provide 8 mm Φ bars @, 134 mm c/c.

6. *Design of heel slab* (Fig. 18.25)

Let b_1 = 0.88 m = length of heel slab.
Four forces act on it :

(*i*) weight of soil over the heel slab
(*ii*) weight of heel slab
(*iii*) downward component of earth pressure
(*iv*) upward soil pressure

Weight of soil over heel slab $= \frac{1}{2}(3.7 + 3.95) \times 0.88 \times 18$

$= 60.59$ kN

Acting at $\left(\dfrac{3.7 + 2 \times 3.95}{3.7 + 3.95}\right)\dfrac{0.88}{3} = 0.445$ m from B.

Weight of heel slab $= 0.3 \times 0.88 \times 1 \times 25 = 6.6$ kN acting at 0.44 m from B.

Earth pressure intensity at $B = K_a \gamma H_1$ per unit inclined area, at β to horizontal.

$\therefore$ Earth pressure at B, on horizontal unit area

$$= K_a \gamma H_1 . \tan\beta$$

Vertical component of this, at B

$$= K_a \gamma H_1 \tan\beta . \sin\beta \quad ...(i)$$

Similarly, vertical component of earth pressure intensity at C

$$= K_a \gamma H_2 \tan\beta . \sin\beta \quad ...(ii)$$

Hence total force due to vertical component of earth pressure is

$$= K_a . \gamma \frac{(H_1 + H_2)}{2} \times b_1 \tan\beta . \sin\beta$$

$$= \frac{0.38 \times 18}{2}(3.7 + 3.95) \times 0.88 \times 0.287 \times 0.276$$

$$= 1.82 \text{ kN}$$

This acts at $\left(\dfrac{H_1 + 2H_2}{H_1 + H_2}\right)\dfrac{b_1}{3} = \dfrac{3.7 + 2 \times 3.95}{3.7 + 3.95} \times \dfrac{0.88}{3}$

$$= 0.445 \text{ m from } B.$$

Upward soil pressure $= \frac{1}{2}(46.22 + 36.16) \times 0.88 = 36.25$ kN

Acting at $\left(\dfrac{46.22 + 2 \times 36.16}{46.22 + 36.16}\right) \times \dfrac{0.88}{3} = 0.42$ m from B.

$\therefore$ Total force = S.F. at $B = 60.59 + 6.6 + 1.82 - 36.25$
$= 32.76$ kN.

$$\text{B.M. at } B = (60.59 \times 0.445) + (6.6 \times 0.44) + (1.82 \times 0.445) - (36.26 \times 0.42)$$

$$= 15.45 \text{ kN-m} = 15.45 \times 10^6 \text{ N-mm}$$

This is much less than the B.M. on the toe slab. However, we will keep the same depth, as that of toe slab, *i.e.* $d = 240$ mm and $D = 300$ mm, reducing it to 200 mm at the edges.

$$A_{st} = \frac{15.54 \times 10^6}{140 \times 0.865 \times 240} = 532 \text{ mm}^2$$

Using 12 mm Φ bars, having $A_\Phi = 113$ mm^2

$$\therefore \text{Spacing} = \frac{1000 \times 113}{532} = 212 \text{ mm.}$$

Provide them at 200 mm c/c. Take the reinforcement in the toe slab for a distance of $58.3\,\Phi = 58.3 \times 12 \approx 700$ mm to the left of B, and its ends should be hooked.

$$\tau_v = \frac{32.86 \times 1000}{1000 \times 250} = 0.13 \text{ N/mm}^2. \text{ Hence safe.}$$

$$\text{Distribution steel} = \frac{0.15}{100} \times 1000 \left(\frac{300 + 200}{2}\right) = 375 \text{ mm}^2$$

Hence provide 8 mm Φ bars @ 130 mm c/c as in the toe slab.

7. ***Reinforcement in the stem***

Revised $H_1 = 4 - 0.33 = 3.7$ m

$$\text{S.F. at } B = \frac{0.38}{2} \times 18\,(3.7)^2 \cos 16° = 45 \text{ kN} = P_H$$

$$\therefore \quad \text{B.M. at } B = 45 \times \frac{3.7}{3} = 55.5 \text{ kN-m} = 55.5 \times 10^6 \text{ N-mm}$$

$$d = 260 \text{ mm}$$

$$A_{st} = \frac{55.5 \times 10^6}{140 \times 0.865 \times 260} = 1763 \text{ mm}^2$$

Using 16 mm Φ bars, $A_\Phi = 201$ mm^2

$$\therefore \text{ Spacing} = \frac{1000 \times 201}{1763} = 114 \text{ mm}$$

Hence provide 15 mm Φ bars @ 110 mm c/c.

$$\text{Actual } A_{st} \text{ provided} = \frac{1000 \times 201}{110} = 1827 \text{ mm}^2$$

Bend these bars into the toe slab, to serve as reinforcement there. Sufficient development length is available.

Between A and B, some of the bars can be curtailed. Consider a section at a depth h below the top of the step. The effective depth d' at the section is

$$d' = 140 + \frac{260 - 140}{3.69} h = 140 + 32.4\, h \qquad \text{...(1)}$$

$$\text{Now } A_{st} \propto \frac{H^3}{d}$$

or $$H \propto (A_{st}\, d)^{\frac{1}{3}}$$

Hence $$\frac{h}{H_1} = \left(\frac{A_{st}'\, d'}{A_{st}\, d}\right)^{\frac{1}{3}} \qquad \text{...(2)}$$

where A_{st}' = reinforcement at depth h

A_{st} = reinforcement at depth H_1

d = effective thickness at depth H_1.

If $A_{st}' = \frac{1}{2} A_{st}$; $\frac{A_{st}'}{A_{st}} = \frac{1}{2}$

$$\therefore \quad \frac{h}{H_1} = \left(\frac{1}{2}\frac{d'}{d}\right)^{1/3}$$

Substituting $d = 260$ mm and $d' = 140 + 32.4h$, we have

$$h = H_1 \left(\frac{140 + 32.4h}{2 \times 260} \right)^{1/3}$$

or $$h = 3.7 \left(\frac{140 + 32.4h}{520} \right)^{1/3} = 0.46\,(140 + 32.4h)^{1/3} \quad ...(3)$$

This can be solved by trial and error, noting that if the effective thickness of stem were constant, h would have been equal to $\frac{H_1}{(2)^{1/3}} \approx 0.79\,H_1 \approx 2.92$ m. Solving (3) by trial, we get $h = 2.82$ m.

Thus half the bars can be curtailed at this point. However, bars should be extended by a distance of $12\Phi\,(= 12 \times 16)$ or $d\,(= 260)$ beyond this point, whichever is more.

$\therefore \quad h = 2.82 - 0.26 = 2.56$ m.

However, curtail half the bars at a height 2.5 m below the top.

If we wish to curtail half of the remaining bars, so that reinforcement is one-fourth of that provided at B, we have

$$\frac{A_{st}'}{A_{st}} = \frac{1}{4}$$

Hence from (2),

$$\frac{h}{H_1} = \left(\frac{1}{4} \frac{d'}{d} \right)^{1/3}$$

or $$h = H_1 \left(\frac{140 + 32.4h}{4 \times 260} \right)^{1/3} = 3.7 \left(\frac{140 + 32.4}{10210} \right)^{1/3}$$

or $$h = 0.365\,(140 + 32.4h)^{1/3}$$

This can be solved by trial, noting that if the effective thickness of the stem were constant, h would have been equal to $\frac{H}{(4)^{1/2}} \approx 0.63\,H_1 = 2.32$ m.

Hence, we get $h = 2.16$ m. However, bars should be extended by $d\,(= 260$ mm) beyond this

$\therefore \quad h = 2.16 - 0.26 = 1.90$ m

Hence stop half of the remaining bars at 1.90 m below the top of the stem. Continue rest of the bars to the top of the stem.

Distribution reinforcement

Average thickness $= \frac{1}{2}(320 + 200) = 260$ mm

$\therefore$ Distribution reinforcement $= \dfrac{0.15}{100}(1000 \times 260) = 390 \text{ mm}^2$

Using 8 mm Φ bars, $A_\Phi = 50.3 \text{ mm}^2$

$\therefore$ Spacing $= \dfrac{1000 \times 50.3}{390} = 128$ mm

Hence provide 8 mm Φ bars @ 120 mm c/c at the inner face of the stem.

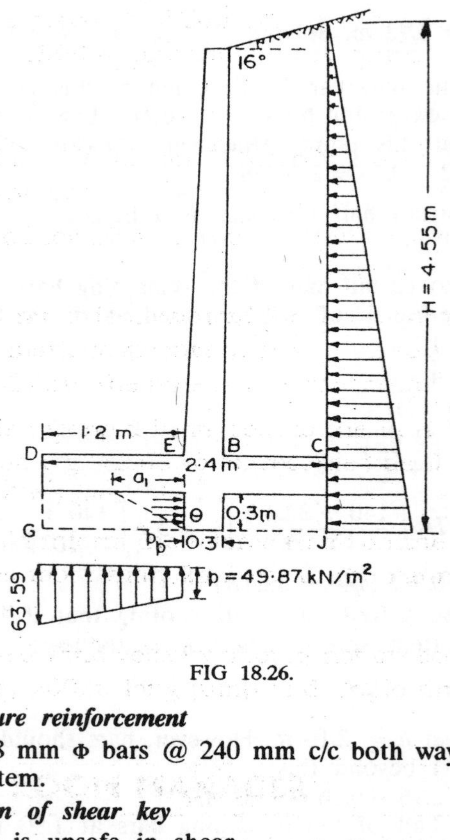

FIG 18.26.

Temperature reinforcement

Provide 8 mm Φ bars @ 240 mm c/c both ways on the outer face of the stem.

9. ***Design of shear key***

The wall is unsafe in shear.

Let us provide a shear key 300 mm × 300 mm.

Let p_p be the intensity of passive pressure developed just in front of the shear key. This intensity p_p depends upon the soil pressure p just in front of shear key.

$$p_p = K_p p \text{ where } K_p = \frac{1}{K_a} = \frac{1}{0.38} = 2.63$$

$\therefore \quad p_p = 2.63 \times 49.87 = 131.2 \text{ kN/m}^2$.

This intensity may be considered to be constant along the depth of key though there will be little increase in p_P because of increase in p with depth. We will, however, consider the constant value of $p_P = 131.2 \text{ kN/m}^2$.

Total passive force $p_P = 131.2 \times 0.3 = 39.36$ kN ...(1)

Let us calculate total sliding force at level *GJ*.

$$P_H = \frac{0.38 \times 18}{2} (4.55)^2 \cos 16° = 68.06 \text{ kN} \qquad ...(2)$$

Weight of soil between the base and level *GJ*.

$$= 0.3 \times 2.4 \times 18 = 12.96 \text{ kN (approx.)} \qquad ...(3)$$

$$\therefore \quad \Sigma W = 119.69 + 12.96 = 132.65 \text{ kN}$$

(There will be some more downward weight because of increasing vertical component of earth pressure).

$$\therefore \quad \text{F.S. against sliding} = \frac{\mu \Sigma W + P_p}{P_H}$$

$$= \frac{0.5 \times 132.65 + 39.36}{68.06} = 1.55 \quad >1.5 \quad \text{Safe.}$$

However, it should be noted that the above passive pressure will be developed *only* when a length a_1 given below is available in front of the key :

$$a_1 = a \tan\theta = a \tan\left(45 + \frac{\Phi}{2}\right) = a\sqrt{K_p}$$

$$= 0.3\sqrt{2.63} = 0.49 \text{ m.}$$

Actual length of toe slab available = 1.2 m. Hence satisfactory

Now, force to be resisted by key, at F.S. of 1.5 is

$$= 1.5\, P_H - \mu \Sigma W$$

$$= 1.5 \times 68.06 - 0.5 \times 132.65 = 35.8 \text{ kN}$$

$$\therefore \quad \text{Shear stress in key} = \frac{35.8 \times 10^3}{1000 \times 300} = 0.12 \text{ N/mm}^2 \text{ (Safe)}$$

$$\text{Bending stress} = \frac{35.8 \times 10^3 \times 150}{\frac{1}{6} \times 1000\,(300)^2} = 0.36 \text{ N/mm}^2$$

Since concrete can take this much of tensile stress, no special reinforcement is necessary for the key.

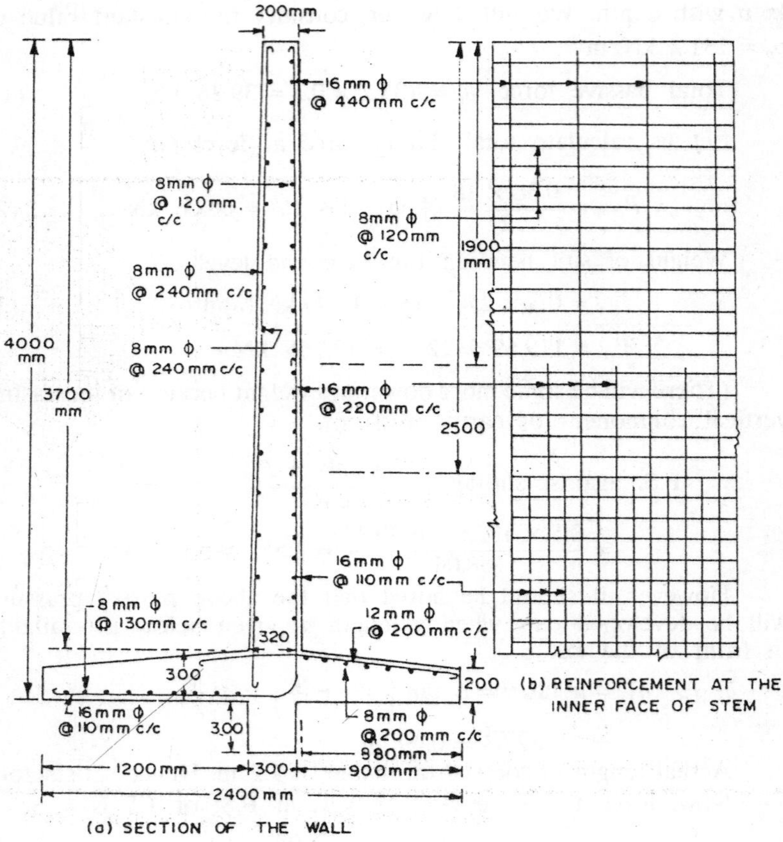

FIG. 18.27

The details of reinforcement etc. are shown in Fig. 18.27.

18.10. DESIGN OF COUNTERFORT RETAINING WALL

We have seen that in case of cantilever retaining wall, the bending moment in the vertical stem varies as h^3, where h is the height of the stem. When this height exceeds 6 m, a thick section is required for the stem, and the design becomes uneconomical In that case, counterfort retaining walls are used.

The counterforts support both the vertical stem as well as heel slab. Sometimes, counterforts are also provided over the toe slab, upto ground level, as shown in Fig. 18.28(*b*). Design principles for

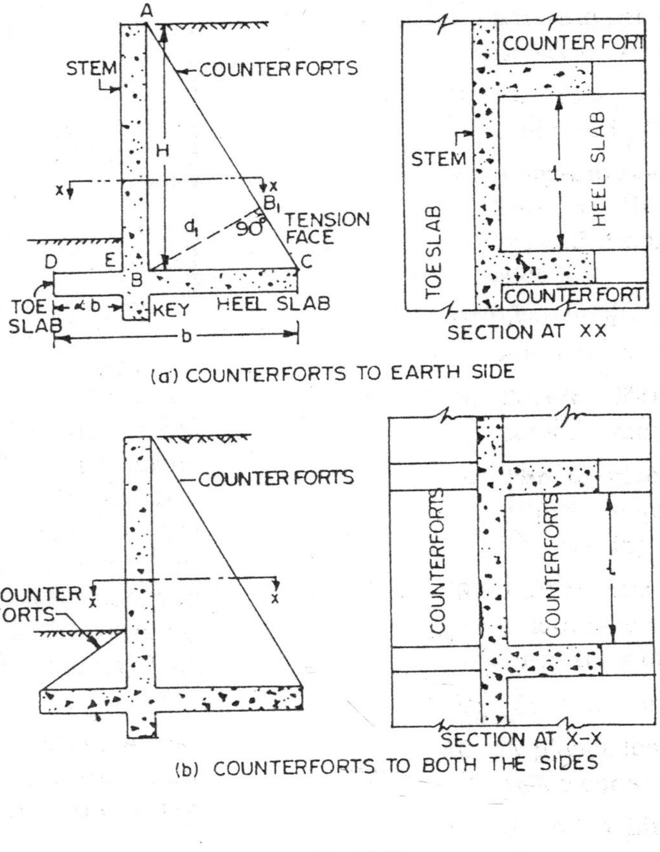

FIG. 18.28.

various component parts are discussed below in brief. The same criterion for fixing the length of base(b), and length of toe slab(αb) are used as cantilever retaining wall.

1. Design of stem

Unlike the stem of cantilever retaining wall, the stem of a counterfort retaining wall acts as a continuous slab supported on counterforts. Due to the varying earth pressure over the height of stem, the stem slab deflects outwards, and hence main reinforcement is provided along the length of the retaining wall, at the outer face of the stem between the counterforts, and at the inner face near the counterforts. The reaction of the stem is taken by the counterforts to which it is firmly anchored. The maximum bending moment occurs at B, where the uniformly distributed earth pressure load is calculated for unit height. If w is the load on the stem slab, at B, per unit length

$$w = p_a \times 1 \times 1 = K_a \gamma H.$$

If l is the clear distance between the counterforts, the negative bending moment in the stem at the counterforts may be taken as $w\,l^2/12$ and the positive bending moment in between the counterforts may be taken as $w\,l^2/16$, for the intermediate panels. The reinforcement may be varied along the height of the stem.

2. Design of heel slab

The action of heel slab is similar that of stem. The heel slab is subjected to the downward load due to weight of soil and self-weight, and upward load due to soil reaction. The net load is found to act downwards. The maximum net pressure is found to act on a strip of unit width near edge C, since the upward soil reaction is minimum there. If p is the net downward pressure per unit area, the maximum bending moment (negative) near counterforts will be $\frac{p\,l^2}{12}$ and positive bending moment will be $\frac{p\,l^2}{16}$. The total load (reaction) from the heel slab is transferred to the counterforts, and this load helps to provide a balancing moment against its overturning. The heel slab is firmly attached to the counterforts by means of vertical ties (two legged stirrups).

3. Design of toe slab

The toe slab bends as a cantilever due to upward soil reaction as in the case of cantilever retaining wall. Due to bending of the toe as cantilever, clockwise moments are induced at E, which are shared by both vertical stem and heel slab. The distribution of this moment between the stem and heel slab are not known. These transferred edge moments will cause bending of stem as well as heel slab, in a direction normal to their usual bending. These edge moments also cause shears in both the slabs which are not precisely known. Hence the design is done empirically, because of uncertain bending behaviour. However, if counterforts are also provided over the toe slab, upto the height of soil (*i.e.* upto ground level), their behaviour becomes certain. In such a case, the reaction is transferred directly to the counterforts, without affecting the bending behaviour of stem and heel slab.

4. Design of counterforts

The counterforts take reactions both from the stem as well as the heel slab. Since the active pressure on stem acts outwards, and net pressure heel slab acts downwards, the counterforts are subjected to tensile stresses along the outer face AC of the counterforts. The angle ABC between stem and heel slab has a tendency to increase from 90°, and this tendency is resisted by counterforts. Thus the counterfort may be considered to bend as a cantilever, fixed at BC. The counterfort acts as an inverted T-beam of varying rib depth. The maximum depth of this T beam is at the junction B. The depth is measured perpendicular to the sloping face AB, *i.e.*, depth $d_1 = BB_1$ at B. This depth thus goes on decreasing towards A_1 where the bending moment also decreases. The width b_1 of the counterfort is kept constant throughout its height. Main reinforcement is provided parallel to AC. The reinforcement may be varied along AC, maximum being at B_1. The faces AB and BC of the counterfort remain in compression. The compressive stresses on face AB are counter-balanced by the vertical upward reaction transferred by the slab. In addition to the main reinforcement, the counterforts are jointed firmly to the stem and heel slab by horizontal and vertical ties respectively, provided in the form of two legged stirrups. Normally, the spacing between the counterforts should not be kept less than 2 m. The spacing of the counterforts depends upon many factors, but usually ranges from $\frac{1}{3}$ to $\frac{1}{2}$ of the height of the wall. For an economical design the clear spacing l between the counterforts may be determined from the following approximate expression :

$$l = 3.5 \left(\frac{H}{\gamma}\right)^{\frac{1}{4}} \qquad ...(18.29)$$

where H is the height of wall mm and γ is the unit weight of soil in kN/m^3.

18.11. BACK ANCHORING OF RETAINING WALL

When the height of retaining wall is much more, it becomes uneconomical to even provide counterforts. In order to reduce the

section of stem etc. in the high retaining walls, the stem may be anchored at its back. The anchor practically takes all the earth pressure and B.M. and S.F. in the stem are thus greatly reduced.

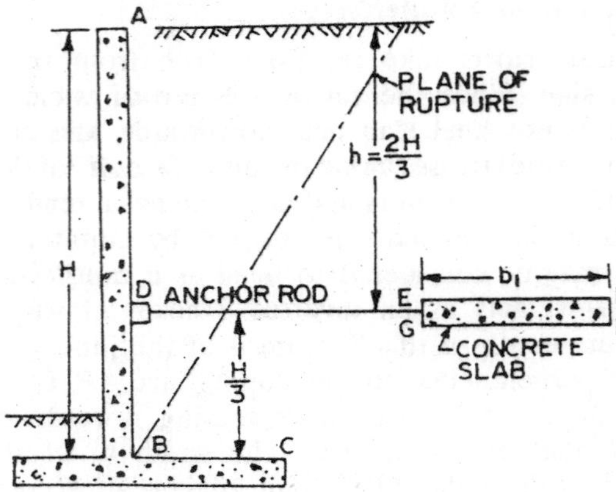

FIG. 18.29.

Fig. 18.29 shows such a wall anchored at its back, at a height $H/3$ where the resultant earth pressure acts. Due to the provision of the anchor at this location, bending moments in the stem at B and A would be zero, and the whole of the earth pressure will be resisted by the anchor rod which is provided at some suitable spacing l centre to centre along the length of the wall. The anchor bar is attached to a concrete slab which resists the pressure by way of development of frictional forces along both the surface EF and GH. The concrete slab is located well beyond the plane of rupture.

Let A = surface area of each face of slab, in a distance l.

μ = coefficient of friction between concrete and soil.

h = height to embedment of slab $= \frac{2}{3} H$.

l = centre to centre spacing of anchor bars.

A_s = area of cross-section of each anchor bars.

Horizontal sliding force $= \frac{1}{2} K_a \gamma H^2 \, l$ ...(*i*)

Resisting friction force $= 2 A\mu \, . \, \gamma h$ ...(*ii*)

Equating the two, $2A\mu\gamma h = \frac{1}{2} K_a \gamma H^2 l$

or $$A = \frac{K_a H^2 l}{4\mu h} = \frac{3}{8} \frac{K_a \, H \, l}{\mu} \qquad ...(18.30)$$

Hence width $b_1 = \frac{A}{l} = \frac{3K_a H}{\mu}$ ...(18.31)

$\therefore$ Area of rod $A_s = \frac{\frac{1}{2}K_a\gamma H^2 l}{\sigma_{st}} = \frac{K_a\gamma H^2 l}{2\sigma_{st}}$ ...(18.32)

If the height is more, two or more tie rods can used at different heights. The tie rod is fixed to the wall through horizontal or vertical beams cast monolithically with the stem. In case vertical beams are provided at interval l, the stem will act as a continuous slab as in the case of counterfort retaining wall. If horizontal beam is provided at height $H/3$, the portions of stem above and below it will act as cantilevers.

In some cases, such as in road or railway embankments, high retaining walls are provided to both the sides of the fill. In such a circumstances, concrete slab is not required to develop frictional force of resistance, but instead, both the walls are connected by common tie rods at suitable intervals, so that earth pressure on one is neutralised by the earth pressure on the other.

Design Example 18.4 *Design a counterfort retaining wall to retain 7 m high embankment above ground level. The foundation is to be taken 1 m deep where the safe bearing capacity of soil may be taken as 180 kN/m². The top of earth retained is horizontal, and soil weighs 18 kN/m³ with angle of internal friction Φ = 30°. Coefficient of friction between concrete and soil may be taken as 0.5.*

Solution.

1. ***Design constants***

For M 15 concrete and mild steel reinforcement we have the following :

$c = \sigma_{cbc} = 5 \text{ N/mm}^2;\ t = \sigma_{st} = 140 \text{ N/mm}^2\ ;\ m = 19$

$k = 0.404\ ;\ j = 0.865\ ;\ R = 0.874$

For the soil, $\Phi = 30°$

$$\therefore \quad K_a = \frac{1 - \sin 30}{1 + \sin 30} = \frac{1}{3}$$

$$K_p = \frac{1}{K_a} = 3.$$

2. *Dimensions of various parts*

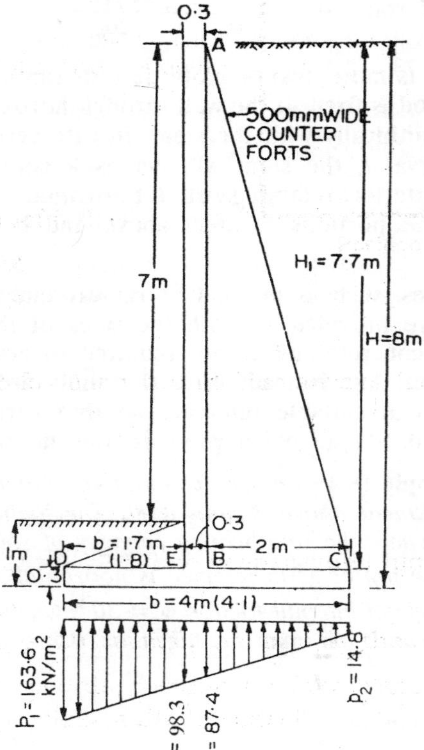

FIG. 18.30.

The ratio of length of toe slab(DE) to the base of width b may be found by the expression :

$$\alpha = 1 - \frac{q_o}{2.2\gamma H} = 1 - \frac{180}{2.2 \times 18 \times 8} = 0.43.$$

$$b = 0.95\, H\sqrt{\frac{K_a}{(1-\alpha)(1+3\alpha)}}.$$

$$= 0.95 \times 8\sqrt{\frac{1}{3(1-0.43)(1+3\times 0.43)}} = 3.64$$

Normal range of b is between 0.5 H to 0.6 H.

However, keep minimum $b = 0.5\ H = 4$ m.

$\therefore$ Length of toe slab $= 0.43 \times 4 = 1.72$ m.

Keep the length of toe slab $= 1.70$ m

Taking uniform thickness of stem $= 300$ mm, length of heel slab $BC = 4 - 1.70 - 0.3 = 2$ m.

Let the thickness of base slab $= 300$ mm.

Clear spacing of counterforts is

$$l = 3.5\left(\frac{H}{\gamma}\right)^{\frac{1}{4}} = 3.5\left(\frac{8}{18}\right)^{\frac{1}{4}} = 2.86 \text{ m.}$$

Keep them at 3 m apart. Let us also provide counterforts over toe slab, upto ground level, at 3 m clear distance.

3. *Stability of wall*

The preliminary dimensions of the wall are marked in Fig. 18.30.

Let W_1 = weight of stem, per metre length.

W_2 = weight of base slab.

W_3 = weight of soil on heel slab.

The calculations are arranged in Table 18.4.

TABLE 18.4

S. No.	*Designation*	*Force (kN)*	*L.A. (m)*	*Moment about toe (kN-m)*
1	W_1	$0.3 \times 7.7 \times 1 \times 25 = 57.8$	1.85	106.8
2	W_2	$0.3 \times 4 \times 1 \times 25 = 30$	2	60.0
3	W_3	$2 \times 7.7 \times 1 \times 18 = 277.2$	3	831.6
		$\Sigma W = 365$	$M_R = 998.4$	

$\therefore$ Resisting moment $M_R = 998.4$ kN-m ...(1)

Horizontal earth pressure $P_H = K_a \gamma \frac{H^2}{2} = \frac{1}{3} \times \frac{1}{2} \times 18\,(8)^2$

$= 192$ kN ...(2)

Overturning moment $M_o = P_H \cdot \frac{H}{3} = 192 \times \frac{8}{3}$

$= 512$ kN-m. ...(3)

$$\text{F.S. against overturning} = \frac{M_R}{M_0} = \frac{998.4}{512}$$

$$= 1.95. < 2. \quad \text{Hence unsafe.}$$

$$\text{F.S. against sliding} = \frac{\mu \Sigma W}{P_H}$$

$$= \frac{0.5 \times 365}{192} = 0.95$$

Hence the wall is unsafe against overturning and sliding. A shear key will have to be provided under the base. Also, width of the base will have to be increased.

Pressure distribution

Net moment $\Sigma M = M_R - M_0 = 998.4 - 512 = 486.4$ kN-m.

Distance $\bar{x}$ from toe, of the point of application of the resultant is given by

$$\bar{x} = \frac{\Sigma M}{\Sigma W} = \frac{486.4}{365} = 1.333 \text{ m}$$

$$\therefore \text{ Eccentricity} \quad e = \frac{b}{2} - \bar{x} = 2 - 1.333 = 0.667 \text{ m}$$

This is exactly equal to $\frac{b}{6} = \frac{4}{6} = 0.667$ m. However no tension will be developed. Pressure p_1 under the toe is given by

$$p_1 = \frac{\Sigma W}{b}\left(1 + \frac{6e}{b}\right) = \frac{365}{4}\left(1 + \frac{6 \times 0.667}{4}\right)$$

$$= 182.5 \text{ kN/m}^2 > 180 \text{ Hence unsafe.}$$

In order to make it safe, increase the length of toe slab DE to 1.8 m, so that total width $b = 4.1$ m.

The revised computations are arranged in table 18.4 (a)

TABLE 18.4 (*a*)

S. No.	*Designation*	*Force (kN)*	*L.A. (m)*	*Moment about toe (kN-m)*
1	W_1	$0.3 \times 7.7 \times 1 \times 25 = 57.8$	1.95	112.6
2	W_2	$0.3 \times 4.1 \times 1 \times 25 = 30.8$	2.05	63.0
3	W_3	$2 \times 7.7 \times 1 \times 18 = 277.2$	3.1	859.3
		$\Sigma W = 365.8$	$M_R =$	1034.9

Net moment $\Sigma M = M_R - M_0 = 1034.9 - 512 = 522.9$ kN-m

$$\bar{x} = \frac{\Sigma M}{\Sigma W} = \frac{522.9}{365.8} = 1.43 \text{ m}$$

$$\therefore \quad e = \frac{b}{2} - \bar{x} = 2.05 - 1.43 = 0.62 \text{ m} < \frac{b}{6}$$

$$p_1 = \frac{\Sigma W}{b}\left(1 + \frac{6e}{b}\right) = \frac{365.8}{4.1}\left(1 + \frac{6 \times 0.57}{4.1}\right)$$

$$= 163.6 \text{ kN/m}^2 < 180. \text{ Hence safe.}$$

Pressure p_2 under the heel is given by

$$p_2 = \frac{\Sigma W}{b}\left(1 + \frac{6e}{b}\right) = \frac{365.8}{4.1}\left(1 - \frac{6 \times 0.57}{4.1}\right) = 14.8 \text{ kN/m}^2$$

The pressure intensity p_1' under E is

$$p_1' = 163.6 - \frac{163.6 - 14.8}{4.1} \times 1.8 = 98.3 \text{ kN/m}^2$$

The pressure intensity p_2' under B is

$$p_2' = 163.6 - \frac{163.6 - 14.8}{4.1} \times 2.1 = 87.4 \text{ kN/m}^2$$

4. *Design of heel slab*

Clear spacing betwen counterforts =3 m. The pressure distribution below the heel slab is shown in Fig. 18.30. Consider a strip, 1 m wide, near the outer edge C. The upward pressure intensity is 14.8 kN/m², which is minimum at C.

Downward load due to weight of earth = 7.7 × 18 = 138.6 kN/m², and that due to self weight of heel slab = 0.3 × 25 = 7.5 kN/m². Hence net downward intensity of load p is given by

$$p = 138.6 + 7.5 - 14.8 = 131.3 \text{ kN/m}^2$$

∴ Maximum negative bending moment in heel slab, at counterforts is

$$M_1 = \frac{pl^2}{12} = \frac{131.3(3)^2}{12} = 98.5 \text{ kN-m} = 98.5 \times 10^6 \text{ N-mm}$$

$$\therefore \quad d = \sqrt{\frac{98.5 \times 10^6}{1000 \times 0.874}} = 336 \text{ mm}$$

$$\text{Shear force } V = \frac{131.3 \times 3}{2} = 197 \text{ kN.}$$

For a balanced section, having $p = 0.72\%$, $\tau_c = 0.33$ N/mm². Hence depth required from shear point of view is

$$d = \frac{V}{\tau_c b} = \frac{197 \times 1000}{0.33 \times 1000} = 597 \text{ mm}$$

This is excessive. However, keep $D = 500$ mm. Providing 60 mm effective cover, $d = 500 - 60 = 440$ mm.

$$\tau_v = \frac{197 \times 1000}{1000 \times 440} = 0.45 \text{ N/mm}^2 > \tau_c$$

Hence shear reinforcement will be necessary.
Area of steel at supports is given by

$$A_{st1} = \frac{98.5 \times 10^6}{140 \times 0.865 \times 440} = 1848 \text{ mm}^2$$

Providing 16 mm Φ bars, $A_\Phi = 201$ mm^2

$$\therefore \quad \text{Spacing} = \frac{1000 \times 201}{1848} = 108 \text{ mm}$$

Provide these @ 100 mm c/c.

$$\text{Actual } A_{st} \text{ provided} = \frac{1000 \times 201}{100} = 2010 \text{ mm}^2$$

Let us check this reinforcement for development length at the point of contraflexure, so as to satisfy the criterion

$$\frac{M}{V} + L_0 \geq L_d.$$

For a fixed beam or slab carrying U.D.L., the point of contraflexure is situated at a distance of 0.211 L. In our case, the slab is continuous, but we will assume the same position of point of contraflexure, *i.e.* at $0.211 \times 3 \approx 0.63$ m from the face of counterforts. Shear force at this point is given by

$$V = \frac{pL}{2}\left(\frac{l}{2} - x\right) \div \left(\frac{L}{2}\right) = p\left(\frac{l}{2} - x\right)$$

$$= 131.3\left(\frac{3}{2} - 0.63\right) = 114.2 \text{ kN.}$$

Assuming that all the bars will be available at the point of contraflexure,

$$M = \sigma_{st}.A_{st}jd = 140[2010] \times 0.865 \times 440$$

$$= 1.07 \times 10^8 \text{ N-mm}$$

$$L_0 = 12\Phi \text{ or } d, \text{ whichever is more } = 440 \text{ mm}$$

$$L_d = 58.3\Phi = 58.3 \times 16 = 932 \text{ mm.}$$

$$\therefore \quad \frac{M}{V} + L_0 = \frac{1.07 \times 10^8}{114.2 \times 10^3} + 440 = 1377 < L_d.$$

Hence safe.

Continue these bars by a distance l_0 ($= d = 440$ mm) beyond the point of contraflexure. After that, curtail half the bars, and continue the remaining half throughout the length. At the point of curtailment, length of each bar available = 630 + 440 = 1070 mm $>L_d$. These bars will be provided at the top face of the heel slab.

$$\text{Maximum positive B.M.} = \frac{pl^2}{16} = \frac{3}{4} M_1$$

$$\therefore \text{ Area of bottom steel } A_{st2} = \frac{3}{4} A_{st1} = \frac{3}{4} \times 1848 = 1386 \text{ mm}^2$$

$$\text{Spacing of 16 mm } \Phi \text{ bars} = \frac{1000 \times 201}{1386} \approx 140 \text{ mm}$$

$$\text{Actual } A_{st} = \frac{1000 \times 201}{140} = 1436 \text{ mm}^2$$

Let us check whether these bars satisfy the development length criterion at the point of contraflexure, inherent in the criterion :

$$\frac{M}{V} + L_o \geq L_d.$$

where V =shear at point of contraflexure =82.65 kN distant 0.63 m from the centre of support

Assuming that all the bars are available at the point of contraflexure,

$$M = 140 \times 1436 \times 0.865 \times 440 = 76.5 \times 10^6 \text{ N-mm}$$

$$L_0 = 440 \text{ mm and } L_d = 932 \text{ mm, as before.}$$

$$\therefore \quad \frac{M}{V} + L_0 = \frac{76.5 \times 10^6}{114.2 \times 10^3} + 440 = 670 + 440 = 1110 \text{ mm}$$

This is slightly more than L_d. Hence safe.

Thus continue all the bottom bars to a point distant L_0 (= 440 mm) from the point of contraflexure, *i.e.* upto a point distant 630 – 440 ≈ 190 mm from the centre of support. At this point, half the bars can be discontinued. Since this distance is quite small, it is better to continue these bars upto the centre of counterforts.

Reinforcement near B

The c/c spacing of reinforcement near *B* may be increased, because *p* decreases due to increase in upward soil reaction. Consider a strip 1 m wide near *B*. Upward soil reaction at *B* is = 87.4 kN/m², as found earlier.

$$\therefore \quad \text{Net downward load } p' = 138.6 + (0.5 \times 25) - 87.4$$
$$= 63.7 \text{ kN/m}^2$$

This is about $\frac{63.7}{131.3} = 0.48$ of load intensity at C.

Hence spacing of steel bars $= \frac{108}{0.48} \approx 200$ mm c/c. at the top face, near supports.

Spacing of steel bars at the bottom face, at midspan

$$= \frac{140}{0.48} = 280 \text{ mm c/.c}$$

Distribution steel

Area of steel $= \frac{0.15}{100}(1000 \times 500) = 750 \text{ mm}^2$.

Using 12 mm Φ bars, having $A_\Phi = 113 \text{ mm}^2$.

$\therefore$ Spacing $= \frac{1000 \times 113}{750} \approx 150$ mm c/c.

Shear reinforcement

Shear stress at $C = \tau_v = 0.45 \text{ N/mm}^2$.

$\frac{100A_s}{b\,d} = \frac{100 \times 2010}{1000 \times 440} = 0.46\%$

Hence from Table 3.1, $\tau_c = 0.28 \text{ N/mm}^2 < \tau_v$

Hence shear reinforcement is necessary.

$V_c = \tau_c bd = 0.28 \times 1000 \times 440 = 123.2 \times 10^8 \text{ N} = 123.2 \text{ kN}$.

Consider a section distant x_1 from the face of the counterfort, where shear force is 123.2 kN. The position is given by

$$\frac{123.2}{197} = \frac{1.5 - x_1}{1.5}$$

which gives $x_1 = 0.56$ m

Hence shear stirrups are required upto distance of 0.56 m on either side of each counterforts. The requirement is there for a strip of unit width passing through C. Let us consider a strip through B_1, distant y_1 from D, such that shear force at the counterforts is 123.2 kN

Net downward pressure at $D = 131.3 \text{ kN/m}^2$

Net downward pressure at $B = 63.7 \text{ kN/m}^2$

Let the net downward pressure at B_1 be w_1 kN/m^2

$\therefore$ S.F. at face of counterforts at $B_1 = w_1 \times \frac{3}{2} = 1.5\, w_1$. This should be equal to 123.2 kN.

$$\therefore \quad w = \frac{123.2}{1.5} = 82.1 \text{ kN/m}^2 \qquad ...(1)$$

However, at y_1 from C,

$$w_1 = 131.3 - \frac{131.3 - 63.7}{2} \times y_1$$

$$= 131.3 - 33.8 y_1 \qquad ...(2)$$

Equating the two, we get

$$131.3 - 33.8 y_1 = 82.1$$

From which $y_1 = 1.46$ m.

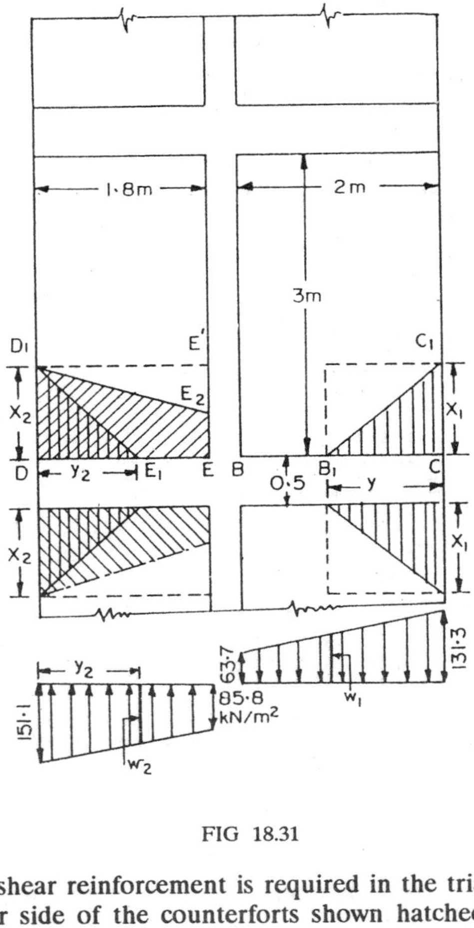

FIG 18.31

Hence shear reinforcement is required in the triangular portion on the other side of the counterforts shown hatched in Fig. 18.31.

However, we will provide shear stirrups in rectangular portions $x_1 \times y_1 = 0.56$ m $\times$ 1.46 m on either side of counterforts.

Let us provide 4-legged stirrups of 6 mm Φ wire.

$$\therefore \quad A_{sv} = 4\frac{\pi}{4}(6)^2 = 113 \text{ mm}^2$$

$$\therefore \quad s_v = \frac{A_{sv} \cdot \sigma_{sv} \cdot d}{V - V_c} = \frac{113 \times 140 \times 400}{(197 - 123.2)1000} = 94 \text{ mm}$$

Hence provide 7 stirrups on either side of each counterfort.

5. *Design of Toe slab*

Since the toe slab is also large, provide counterforts over it, upto ground level at 3 m clear distance face to face. The toe slab will thus bend like a continuous slab.

Assume total depth of toe slab = 500 mm.

$\therefore$ Downward weight per unit area, due to self weignt

$$= 0.5 \times 1 \times 1 \times 25 = 12.5 \text{ kN/m}^2.$$

Upward pressure intensity at $D = 163.6$ kN/m^2

$\therefore$ Net upward pressure intensity at $D = 163.6 - 12.5$

$$= 151.1 \text{ kN/m}^2.$$

Similarly, net upward pressure intensity at E

$$= 98.3 - 12.5 = 85.8 \text{ kN/m}^2$$

Consider strip of unit width at D.

Max negative B.M. is

$$M_1 = \frac{wl^2}{12} = \frac{151.1\,(3^2)}{12} = 113.3 \text{ kN-m}$$

$$= 113.3 \times 10^6 \text{ N-mm}$$

$$d = \sqrt{\frac{113.3 \times 10^6}{0.874 \times 1000}} = 360 \text{ mm}$$

$$\text{Shear force } V = \frac{151.3 \times 3}{2} = 226.8 \text{ kN.}$$

Taking a permissible stress $\tau_c = 0.33$ N/mm^2 for a balanced design (Table 3.1) the depth of slab required from shear point of view is given by

$$d = \frac{V}{b \cdot \tau_c} = \frac{226.8 \times 1000}{1000 \times 0.33} = 687 \text{ mm}$$

This is excessive. However, we will keep the same depth as that of heel, *i.e.*, d = 440 mm and D = 500 mm, and provide shear stirrups to take up excessive shearing stress.

Area of steel at supports, at bottom is

$$A_{st1} = \frac{113.3 \times 10^6}{140 \times 0.865 \times 440} = 2126 \text{ mm}^2$$

Using 16 mm Φ bars, $A_\Phi = 201 \text{ mm}^2$

$$s = \frac{1000 \times 201}{2126} = 94.5 \text{ mm}$$

However, provide these @ 90 mm c/c. Actual $A_{st1} = \frac{1000 \times 201}{90}$ $= 2233 \text{ mm}^2$.

Let us check these from point of view of development length, at the point of contraflexure, so as to satisfy the criterion.

$$\frac{M}{V} + L_0 \geq L_d .$$

The point of contraflexure occurs at a distance $x = 0.63$ m from the support. Hence S.F. at the point of contraflexure is

$$V = w\left(\frac{l}{2} - x\right) = 151.1\left(\frac{3}{2} - 0.63\right) = 131.5 \text{ kN}$$

$$M = 140 \times 2233 \times 0.865 \times 440 = 1.19 \times 10^8 \text{ N-mm}$$

$$L_0 = 12\Phi \text{ or } d, \text{ whichever is more } = 440 \text{ mm}$$

$$L_d = 58.3 \times 16 = 933 \text{ mm}$$

$$\therefore \quad \frac{M}{V} + L_0 = \frac{1.19 \times 10^8}{13.5 \times 10^3} + 440 = 905 + 440 = 1345 \text{ mm} > L_d$$

Hence satisfied. Continue these bars, at the bottom of toe slab, beyond the point of contraflexure by a distance of L_0(= 440 mm) *i.e.*, by a distance of 630 + 440 = 1070 mm from the face of the counterforts.

Again, positive B.M. $= \frac{3}{4} M_1 = \frac{3}{4} \times 113.3 \times 10^6$

$= 85 \times 10^6$ N-mm

Area of bottom steel $= \frac{3}{4} A_{st1} = \frac{3}{4} \times 2126 = 1595 \text{ mm}^2$

Spacing of 16 mm Φ bars $= \frac{1000 \times 201}{1595} = 126$ mm

Hence provide these @ 120 mm c/c. Actual area of steel $= \frac{1000 \times 201}{120} = 1675 \text{ mm}^2$. This steel should satisfy the following criterion at the point of contraflexure

$$\frac{M}{V} + L_o \geq L_d$$

where V = S.F. at the point of contraflexure = 131.5 kN

Assuming that all the bars, provided at the top face, are available at the point of contraflexure.

$$M = 140 \times 1675 \times 0.865 \times 440 = 89.25 \times 10^6 \text{ N-mm}$$

$$L_0 = 12\Phi \text{ or } d, \text{ whichever is more } = 440 \text{ mm.}$$

$$L_d = 58.3\Phi = 935 \text{ mm}$$

$$\therefore \quad \frac{M}{V} + L_0 = \frac{89.25 \times 10^6}{131.5 \times 10^3} + 440 = 678 + 440 = 1118 \text{ mm}$$

This is greater than L_d. Hence the criterion is satisfied. However, these bars are to be continued for a distance of $L_0 (= 440 \text{ mm})$ beyond the point of contraflexure, their distance from the face of counterforts being only $= 630 - 440 = 190$ mm. Hence it is better to continue these bars upto counterfort, at the top face of the toe slab.

Reinforcement at E

At a section distant 1m from E, upward soil pressure $= 163.6 - \frac{163.6 - 14.8}{4.1} \times 0.8 = 134.6 \text{ kN/m}^2$

$$\therefore \quad \text{Net upward pressure} = 134.6 - 12.5 = 122.1 \text{ kN/m}^2$$

This is $\frac{121.1}{151.1} \approx 0.81$ of w at D.

$$\therefore \quad \text{Spacing of bottom steel} = \frac{94.5}{0.81} = 117 \approx 110 \text{ mm c/c}$$

$$\text{Spacing of top steel} = \frac{126}{0.81} = 156 \approx 150 \text{ mm c/c}$$

Distribution steel

$$A_{sd} = \frac{0.15}{100}(1000 \times 500) = 750 \text{ mm}^2$$

Using 12 mm Φ bars, $A_\Phi \approx 113 \text{ mm}^2$

$$\therefore \quad \text{Spacing} = \frac{1000 \times 113}{750} \approx 150 \text{ mm}$$

Shear reinforcement

Shear force at D = 226.8 kN

$$\therefore \quad \tau_v = \frac{226.8 \times 1000}{1000 \times 440} = 0.52 \text{ N/mm}^2$$

$$\frac{100 A_s}{bd} = \frac{100 \times 2233}{1000 \times 440} \approx 0.5\%$$

Hence from Table 3.1, $\tau_c = 0.29 \text{ N/mm}^2$

Since $\tau_v > \tau_c$, shear reinforcement is necessary.

Now $V_c = \tau_c \,.\, bd = 0.29 \times 1000 \times 440 = 127600$ N

$= 127.6$ kN

Consider a section distant x_2 from the face of the counterfort (Fig. 18.31) where S.F. is 127.6 kN. Position x_2 is given by

$$\frac{127.6}{226.8} = \frac{1.5 - x_2}{1.5}$$

which gives $x_2 = 0.66$ m.

Hence shear stirrups are required upto a distance of 0.66 m on either side of counterforts. This requirement is there for a strip of unit width passing through D. Let us consider a strip through E_1, distant y_2 from D, such that shear force at the counterforts is 127.6 kN. To find the position of y_2, consider the net pressure distribution below the toe slab.

Self weight of toe slab = 12.5 kN/m^2. Hence net pressure intensity below D and E are respectively 163.6 – 12.5 = 151.1 kN/m^2 and 98.3 – 12.5 = 85.8 kN/m^2 respectively (Fig. 18.31). Let the net pressure intensity at E_1 be w_2 kN/m^2.

$\therefore$ S.F. at the counterforts at $E_1 = w_2 \times \frac{3}{2} = 1.5 w_2$ kN. This should be equal to 127.6 kN.

$$\therefore \quad w_2 = \frac{127.6}{1.5} = 85.1 \text{ kN/m}^2 \qquad \text{...(1)}$$

However, at y_2 from D,

$$w_2 = 151.1 - \frac{151.1 - 85.8}{1.8} y_2 = 151.1 - 36.3\, y_2 \qquad \text{...(2)}$$

Equating the two we get

$151.1 - 36.3\, y_2 = 85.1$

From which $y_2 = 1.82$ m.

This is more than $DE(= 1.8$ m).

Hence shear force at E is more than 127.6 kN.

Actual S.F. at $E = 1.5 \times 85.8 = 128.7$ kN.

Consider a section distant z from the face of counterforts (Point E), where S.F. is 127.6 kN. The position of z is given by

$$\frac{127.6}{128.7} = \frac{1.5 - z}{1.5}$$

which gives $z = 0.01$ m.

Hence shear stirrups are to be provided for a region DEE_1D_1, where $EE_1 = 0.01$ m only. However, we will provide stirrups for whole of rectangular area(shown dotted), for width $DD_1 = x_2 = 0.66$ and length DE = 1.8 m.

Let us provide 8 legged stirrups, of 6 mm Φ wire.

$$A_{sv} = 8\frac{\pi}{4}(6)^2 = 226 \text{ mm}^2$$

$$\therefore \quad \text{spacing} = \frac{A_{sv} \cdot \sigma_{sv}\, d}{V - V_c} = \frac{226 \times 140 \times 440}{(226.8 - 127.6)1000}$$

$$\approx 140 \text{ mm.}$$

Hence provide 6 stirrups on either side of each counterfort.

6. *Design of stem (vertical slab)*

The stem acts as a continuous slab. Consider 1 m strip at B. The intensity of earth pressure is given by

$p_h = K_a \gamma H_1 = \frac{1}{3} \times 18 \times 7.5 = 45$ kN/m^2

(where revised value of $H_1 = 8 - 0.5 = 7.5$ m)

∴ Negative B.M. in slab near counterforts is

$$M_1 = \frac{p_h \cdot l^2}{12} = \frac{45(3)^2}{12} = 33.75 \text{ kN-m}$$

$$= 33.75 \times 10^6 \text{ N-mm}$$

$$\therefore \quad d = \sqrt{\frac{33.75 \times 10^6}{1000 \times 0.874}} = 197 \text{ mm.}$$

Providing effective cover = 60 mm, total depth = 197 + 60 =257mm . However, provide total depth = 300 mm so that available $d = 300 - 60 = 240$ mm. This increased thickness will keep the shear stress within limits so that additional shear reinforcement is not required.

$$\text{Shear force } V = \frac{45 \times 3}{2} = 67.5 \text{ kN.}$$

$$\therefore \quad \tau_v = \frac{67.5 \times 1000}{1000 \times 240} = 0.28 \text{ N/mm}^2$$

This is less than $\tau_c = 0.29$ N/mm^2 at 0.5% reinforcement (see Table 3.11)

Area of steel near counterforts is

$$A_{st1} = \frac{33.75 \times 10^6}{140 \times 0.865 \times 240} = 1161.2 \text{ mm}^2$$

Using 12 mm Φ bars having $A_\Phi = 113$ mm^2.

$$s = \frac{1000 \times 113}{1161.2} \approx 100 \text{ mm}$$

Hence provide 12 mm Φ bars @ 100 mm c/c.

$$\text{Actual } A_{st} = \frac{1000 \times 113}{100} = 1130 \text{ mm}^2$$

and $$\frac{100 A_s}{bd} = \frac{100 \times 1130}{1000 \times 240} = 0.47\%.$$

From Table 3.1, $\tau_c = 0.28$ mm^2 corresponding to $p = 0.47\%$.

Let us check these bars for development length, near points of contraflexure, so as to satisfy the criterion

$$\frac{M}{V} + L_0 \geq L_{d.}$$

For a fixed beam or slab carrying U.D.L., the point of contraflexure is at a distance of 0.211 L = 0.63 m from the face of counterforts, S.F. at this point is given by

$$V = \frac{pL}{2}\left(\frac{l}{2} - x\right) \div \left(\frac{L}{2}\right) = p\left(\frac{l}{2} - x\right)$$

$$= 67.5\left(\frac{3}{2} - 0.63\right) = 58.53 \text{ kN}$$

Assuming that all the bars will be available at the point of contraflexure

$$M = \sigma_{st} \cdot A_{st} jd = 140 \times 1130 \times 0.865 \times 240$$

$$= 32.84 \times 10^6 \text{ N-mm}$$

$$L_d = 12\Phi \text{ or } d, \text{ whichever is more } = 240 \text{ mm}$$

$$L_d = 58.3 \times 12 \approx 700 \text{ mm}$$

$$\therefore \quad \frac{M}{V} + L_0 = \frac{32.84 \times 10^6}{58.53 \times 1000} + 240 = 561 + 240 = 800 \text{ mm}$$

Hence safe. It is thus essential to continue all the bars upto a point distant 240 mm beyond point of contraflexure, *i.e.* upto a point 630 + 240 = 870 mm ≈900 mm from the face of counterforts. These bars are to be provided at the inner face of the stem slab.

$$\text{Maximum positive B.M.} = \tfrac{3}{4}M_1 = \tfrac{3}{4} \times 33.75 \times 10^6$$

$$= 25.31 \times 10^6 \text{ N-mm}$$

Area of steel $= \frac{3}{4}A_{st1} = \frac{3}{4} \times 1130 = 847.5$ mm^2 and spacing of 12 mm Φ bars $= \dfrac{1000 \times 113}{847.5} = 133.0$ mm

However, provide these @ 130 mm c/c.

$$\text{Actual } A_{st} = \frac{1000 \times 113}{130} = 869 \text{ mm}^2.$$

Let us check these bars for development length at the point of contraflexure, so as to satisfy the criterion

$$\frac{M}{V} + L_0 \geq L_d$$

Assuming that all the reinforcement is extended upto point of contraflexure

$$M = 140 \times 869 \times 0.865 \times 240 = 25.26 \times 10^6 \text{ N-mm}$$

$$L_0 = 12\Phi \text{ or } d \text{ , whichever is more } = 240$$

$$L_d = 58.3 \times 12 \approx 700 \text{ mm}$$

$$V = 58.53 \text{ kN, as before,}$$

$$\therefore \quad \frac{M}{V} + L_0 = \frac{25.26 \times 10^6}{58.53 \times 10^3} + 240 = 432 + 240 = 672 \text{ mm}$$

This is less than L_d (= 700), and hence unsafe. However extend the bars upto the centre of counterforts so that L_0 is increased. These bars are to be provided at the outer face of the slab.

The spacing of the reinforcement at B, found above can be increased with height. The pressure p_h and hence the bending moment decreases linearly with height.

$$\therefore \quad A_{st} \propto h$$

Hence the spacing of the bars can be increased gradually to say 300 mm c/c near the top.

$$\text{Distribution steel} = \frac{0.15}{100}(1000 \times 300) = 450 \text{ mm}^2$$

Using 10 mm Φ bars, $A_\Phi = 78.5 \text{ mm}^2$

$$\therefore \quad \text{Spacing} = \frac{1000 \times 78.5}{450} = 174 \text{ mm.}$$

Hence provide these @ 170 mm c/c.

7. *Design of main counterfort*

Let us assume thickness of counterfort = 500 mm. The counterforts will thus be spaced @ 350 cm c/c. They will thus receive earth pressure from a width of 3.5 m and downward reaction from the heel slab for a width of 3.5 m.

At any section at depth h below the top A, the earth pressure acting on each counterfort will be

$$= \frac{1}{3} \times 18 \times h \times 3.5 = 21\,h \text{ kN/m} \qquad \text{...(1)}$$

Similarly, net downward pressure on heel at C is

$$= 138.60 + 0.5 \times 25 - 14.8 = 136.3 \text{ kN/m}^2$$

and that at B is $= 138.60 + 0.5 \times 25 - 87.4 = 63.7 \text{ kN/m}^2$

Hence reaction transferred to each counterfort will be

At C, $136.3 \times 3.5 = 477.1$ kN/m

At B, $63.7 \times 3.5 = 223$ kN/m.

The variations of horizontal and vertical forces on the counterfort are shown in Fig. 18.32.

The critical section for the counterfort will be at F, since below this, enormous depth will be available to resist bending.

Pressure intensity at $b = 7$ m is $= 21 \times 7 = 147$ kN/m

Shear force at $F = \frac{1}{2} \times 147 \times 7 = 514.5$ kN

$$\text{B.M.} \quad M = 514.5 \times \frac{7}{3} = 1200.5 \text{ kN-m}$$

$$= 1200.5 \times 10^6 \text{ N-mm.}$$

The counterfort acts as a T-beam. However, even as a rectangular beam, depth required is

$$d = \sqrt{\frac{1200.5 \times 10^6}{500 \times 0.874}} \approx 1660 \text{ mm}$$

$\therefore$ Total depth $= 1660 + 60 = 1720$ mm

Angle θ of face AC is given by

$$\tan \theta = \frac{2}{7.5} = 0.267$$

$\therefore \quad \theta = 14° 56'$

$\sin \theta = 0.2577$; $\cos \theta = 0.9662$

Depth $F_1C_1 = AF_1 \sin \theta = 7 \times 0.2577 = 1.8$ m $= 1800$ mm

$\therefore$ Depth $FG = 1800 + 300 = 2100$ mm

This is much more than required.

Assuming that steel reinforcement is provided in two layers and providing a clear cover of 50 mm and 20 mmΦ bars, the effective depth may be assumed approximately to be equal to 2020 mm.

$$\therefore \qquad A_{st} = \frac{1200 \times 10^6}{140 \times 0.9 \times 2020} = 4717 \text{ mm}^2$$

Using 20 mm Φ bars, $A_\Phi = 314$ mm^2

$$\text{No. of bars} = \frac{4717}{314} \approx 16$$

Provide these in two layers.

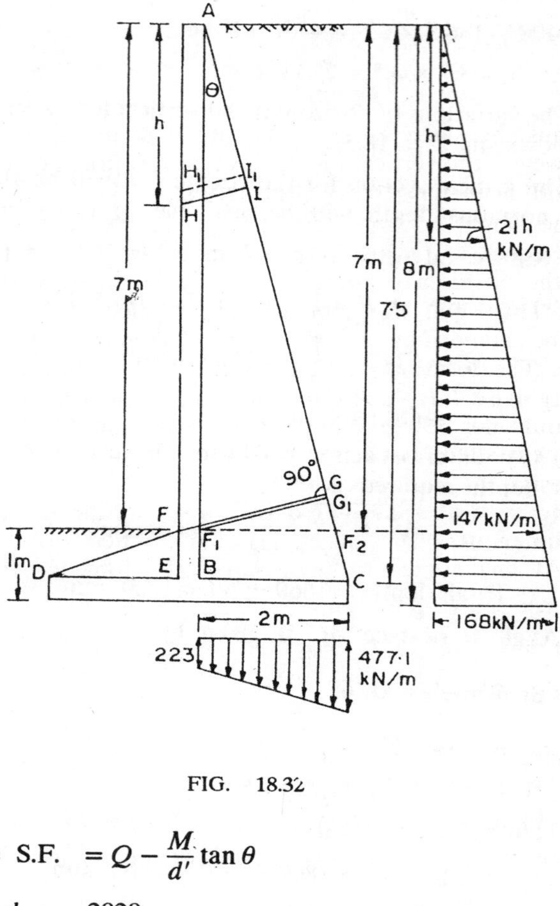

FIG. 18.32

$$\text{Effective S.F.} = Q - \frac{M}{d'} \tan \theta$$

vhere $d' = \frac{d}{\cos \theta} = \frac{2020}{0.9662} = 2090$ mm

$$\therefore \quad \text{Effective S.F.} = Q - \frac{M}{d'} \tan \theta$$

$$= 514.5 \times 10^3 - \frac{1200.5 \times 10^6}{2090} \times 0.267$$

$$= 361135 \text{ N}$$

$$\therefore \qquad \tau_v = \frac{361135}{500 \times 2090} = 0.346 \text{ N/mm}^2$$

$$100\frac{A_s}{bd} = \frac{100(16 \times 314)}{500 \times 2090} = 0.48\%$$

$\therefore \qquad \tau_c \approx 0.28\ \text{N/mm}^2.$

Thus the shear stress τ_v is more than permissible shear stress τ_c. However, the vertical and horizontal ties provided in the counterforts will bear the excess shear stress.

The height h where half of the reinforcement can be curtailed will be equal to $\sqrt{H} = \sqrt{8} \approx 2.8$ m below A, *i.e.* at point H. To locate the position of point of curtailment on AC, draw HI parallel to FG. Thus, half the bars can be curtailed at I. However, these should be extended by a distance $12\Phi = 240$ mm beyond I, *i.e.* extended upto I_1. The location of H, corresponding to I_1 can be located by drawing line I_1H_1 parallel to FG. It should be noted that I_1G should not be less than $58.3\,\Phi = 1166$ mm. Similarly, other bars can be curtailed, if desired.

Design of horizontal ties

The vertical stem slab has a tendency to separate out from the counterfort, and hence should be tied to it by horizontal ties. At any depth h below the top, force causing separation $\frac{1}{3} \times 18h \times 3 = 18\ h$ kN/m.

At $h = 7$ m, force $= 18 \times 7 = 126$ kN/m.

$\therefore$ Steel area required $= \dfrac{126 \times 1000}{140} = 900\ \text{mm}^2$ per metre height.

Using 10 mm Φ 2 legged ties, $A_\Phi = 2\frac{\pi}{4}(10)^2 = 157\ \text{mm}^2$

$\therefore$ Spacing $= \dfrac{1000 \times 157}{900} \approx 170$ mm c/c.

This spacing can be increased gradually to 300 mm c/c towards top.

Design of vertical ties

Similar to the stem slab, heel slab has also tendency to separate out from the counterfort, due to net downward force, unless tied properly by vertical ties. The downward force at C will be $477.1 \times \dfrac{3}{3.5} = 408.9$ kN/m at C and $223 \times \dfrac{3}{3.5} = 191.1$ kN/m at B (see Fig. 18.32). Near end C, the heel slab is tied to the counterforts with the help of main reinforcement of counterforts.

$\therefore$ Steel area at C $= \dfrac{408.9 \times 1000}{140} = 2921\ \text{mm}^2$

Using 16 mm Φ 2 legged ties, $A_\Phi = 2 \times \frac{\pi}{4}(16)^2 = 402\ \text{mm}^2$

$$\therefore \text{ Spacing of ties } = \frac{1000 \times 402}{2921} = 138 \text{ mm} \approx 130 \text{ mm c/c.}$$

$$\text{Steel area at } B = \frac{191 \times 1000}{140} = 1364 \text{ mm}^2$$

$$\therefore \text{ Spacing of ties } = \frac{1000 \times 402}{1364} \approx 290 \text{ mm c/c.}$$

Thus the spacing of vertical ties can be increased gradually from 130 mm c/c at C to 290 mm c/c at B.

8. *Design of front counterforts*

Refer Fig. 18.30. The upward pressure intensity varies from 163.6 kN/m^2 at D to 98.3 kN/m^2 at E. Downward weight of 500 mm thick toe slab = 0.5 × 2500 = 12.5 kN/m^2. Hence net w at D = 163.6–12.5 = 151.1 kN/m^2 and at E = 98.3–12.5 = 85.8 kN/m^2. The centre to centre spacing of counterforts, 500 mm wide is 3.5 m.. Hence upward force transmitted to counterfort at D= 151.1 × 3.5 = 528.9 kN/m and that at E = 85.8 × 3.5 = 300.3 kN/m.

The counterfort will bend up as cantilever about face FE. Hence DF will be in compression while $D_1 E_1$ will be in tension, and main reinforcement will be provided at bottom face $D_1 E_1$.

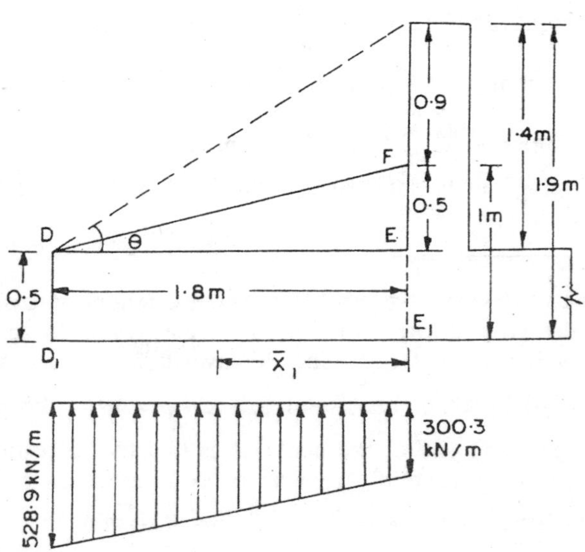

FIG. 18.33.

Total upward force $= \frac{1}{2}[528.9 + 300.3] \times 1.8 = 746.3$ kN acting

at $\bar{x} = \dfrac{300.3 + 2 \times 528.9}{300.3 + 528.9} \times \dfrac{1.8}{3} = 0.98$ m from E

$\therefore$ B.M. $= 746.3 \times 0.98 = 731.4$ kN−m

$= 731.4 \times 10^6$ N-mm

$$\therefore \qquad d = \sqrt{\frac{731 \times 10^6}{500 \times 0.856}} = 1307 \text{ mm}$$

Hence a total depth of 1307 + 60 = 1907 ≈ 1900 mm is required.

Thus, project the counterfort 900 mm above ground level, to point F_1 as shown.

$$A_{st} = \frac{731.4 \times 10^6}{140 \times 0.865 \times 1300} = 4646 \text{ mm}^2$$

Using 25 mm Φ bars, $A_\Phi = 491$ mm^2 .

$$\therefore \text{ No. of bars } = \frac{4646}{491} = 9.46 \approx 10.$$

These bars should be continued by a distance of 58.3 Φ = 58.3 × 25 ≈1460 mm beyond E_1 .

Net shear force $= F - \dfrac{M}{d}\tan\theta$.

From Fig 18.33, $\tan\theta = \dfrac{1.4}{1.8}$

$$\therefore \qquad V = 746.3 - \frac{731.4}{1.3} \times \frac{1.4}{1.8}$$

$$= 308.7 \text{ kN}$$

$$\tau_v = \frac{308.7 \times 1000}{500 \times 1300} = 0.475 \quad \text{N/mm}^2.$$

$$100\frac{A_s}{bd} = \frac{100(10 \times 491)}{500 \times 1300} = 0.75\%$$

Hence from Table, 3.1 $\tau_c = 0.34$ N/mm^2.
Since $\tau_v > \tau_c$, shear reinforcement is required.
Using 12 mm Φ 2-legged stirrups,

$$A_{sv} = 2 \times \frac{\pi}{4}(12)^2 = 226 \text{ mm}^2.$$

$$V_c = \tau_c bd = 0.34 \times 500 \times 1300 = 221000 \text{ N}$$

$$\therefore \qquad V_s = V - V_c = 308.7 \times 10^3 - 221 \times 10^3 = 87.7 \times 10^3 \text{ N}$$

$$\therefore \quad s_v = \frac{\sigma_{sv} \cdot A_{sv} \cdot d}{V_s} = \frac{140 \times 226 \times 1300}{87.7 \times 10^3} = 469 \text{ mm.}$$

However, provide these @ 400 mm c/c. Provide 2-16 mm Φ holding bars at the top.

9. *Fixing effects in stem, toe and heel slab*

At the junction of stem, toe and heel slabs fixing moments are induced, which are at right angles to their normal direction of bending. These moments are not determinate, but normal reinforcement given below may be provided :

(*i*) In stem @ 0.3% of cross section, to be provided at the inner face, in vertical direction, for a length =58.3 Φ.

$$\therefore A_{st} = \frac{0.3}{100} \times 1000 \times 300 = 900 \text{ mm}^2$$

Using 10 mm Φ bars, $A_\Phi = 78.5 \text{ mm}^2$

$$\therefore \text{ Spacing } = \frac{1000 \times 78.5}{900} \approx 80 \text{ mm c/c.}$$

Length of embedment in stem, above heel slab ≈580 mm.

(*ii*) In toe slab @ 0.15% to be provided at the lower face

$$A_{st} = \frac{0.15}{100}(1000 \times 500) = 750 \text{ mm}^2$$

$$\text{Spacing of 10 mm } \Phi \text{ bars } = \frac{1000 \times 78.5}{750} \approx 100 \text{ mm c/c.}$$

Length of embedment in toe slab = 580 mm.

(*iii*) In heel, @ 0.15% to be provided at the upper face.

∴ Provide 10 mm bars @ 100 c/c to be embeded by a distance of 580 mm.

Each of the above reinforcement should be anchored properly in the adjoining slab, as shown in Fig. 18.34

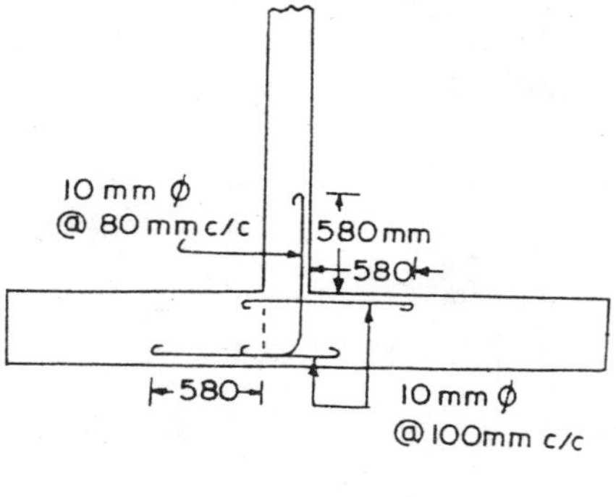

FIG. 18.34

10. ***Design of shear key***

The wall is unsafe in sliding, and hence shear key will have to be provided, as shown in Fig. 18.35. Let the depth of key=a.

Intensity of passive pressure p_P developed in front of the key depends upon the soil pressure p in front of key.

$$p_P = K_p p = 3 \times 98.3 = 294.9 \quad \text{kN/m}^2$$

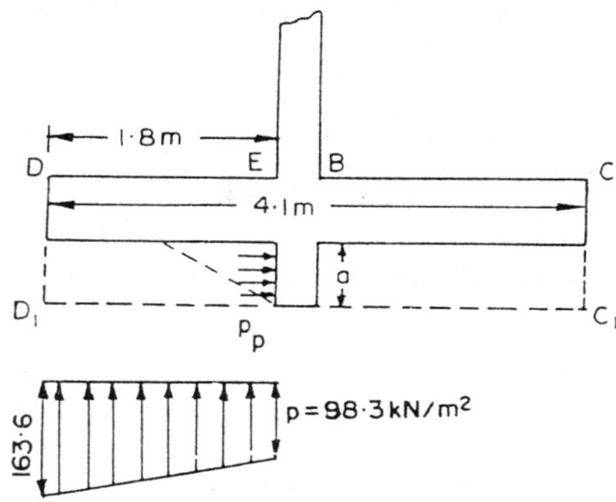

FIG. 18.35

$\therefore$ Total passive pressure $P_p = p_P \cdot a = 294.9\ a$ ...(1)

Sliding force at level $D_1C_1 = \frac{1}{3} \times \frac{18}{2}(8+a)^2$

or $$P_H = 3(8+a)^2 \quad ...(2)$$

Weight of soil between bottom of base and $D_1 C_1$

$$\approx 4.1 \times a \times 18 = 73.8\ a \text{ kg} \quad ...(3)$$

$\therefore$ $\Sigma W = 365.8 + 73.8\ a$ (Refer Table 18.4a)

Hence for equilibrium of wall, permitting F.S. =1.5 against sliding, we have

$$1.5 = \frac{\mu \Sigma W + P_P}{P_H}$$

or $$1.5 = \frac{0.5(365.8 + 73.8\,a) + 294\,a}{3(8+a)^2}$$

or $$a^2 - 57a + 23.3 = 0 \text{ which gives } a = 0.40 \text{ m}$$

Hence keep a = 400 mm and width of key = 400 mm.

Actual force to be resisted by key.

$$= 1.5 P_H - \mu \Sigma W$$

$$= 1.5\times3(8 + 0.4)^2 - 0.5(365.8 + 73.8\times0.4) = 119.8 \text{ kN}$$

$$\therefore \text{ Shear stress } = \frac{119.8 \times 1000}{400 \times 1000} \approx 0.3 \text{ N/mm}^2 \text{ (safe)}$$

$$\text{Bending stress } = \frac{119.8 \times 10^3 \times 200}{\frac{1}{6} \times 1000(400)^2} = 0.9 \text{ N/mm}^2 \text{ (safe)}$$

The details of reinforcement etc. are shown in Fig. 18.36.

PROBLEMS

1. Explain the methods of designing vertical stem, toe slab and heel slab of a T-shaped cantilever retaining wall. What will be the changes in the design if counterforts are provided at regular interval towards the side of backfil ?
2. Explain the method designing a shear key for a retaining wall.
3. Design a T-shaped cantilever retaining wall for the following data :

(*i*) Height of wall above ground, 4.5 m

(*ii*) Depth of foundation, 1.5 m

(*iii*) Unit weight of earth fill, 17 kN/m^2

(*iv*) Angle of internal friction, 20°

(*v*) Coefficient of friction between soil and concrete, 0.45

(*vi*) Safe bearing capacity of soil 130 kN/m^2.

Use M 15 mix.

4. Redesign problem 3 if the top surface of backfill is inclined at an angle of 15° with horizontal.
5. Redesign problem 3 if the embankment carries a load of 20 kN/m^2 over the backfill. Also provide a R.C. parapet 1 m high over the top of the stem of the retaining wall.
6. Design a suitable counterfort retaining wall for the data of problem 3.
7. Design a counterfort retaining wall for the following data :

(*i*) Height of wall above ground, 8 m.

(*ii*) Depth of foundation, 1.5 m

(*iii*) Safe bearing capacity, 200 kN/m^2

(*iv*) Unit weight of earthfill, 18 kN/m^2

(*v*) Surcharge angle 18°

(*vi*) Angle of internal friction for backfill, 30°

(*vii*) Face to face spacing of front counterforts, 2 m.

(*viii*) Face to face spacing of front counterforts, provided upon ground level, 2 m.

(*ix*) Coefficient of friction between soil and concrete, 0.55.

Use M - 15 mix.

PART IV

WATER TANKS AND MISCELLANEOUS STRUCTURES

Chapters :

19

Domes

19.1. INTRODUCTION

A dome may be defined as a thin shell generated by the revolution of a regular curve about one of its axes. The shape of the dome depends upon the type of the curve and the direction of the axis of revolution. When the segment of a circular curve revolves about its vertical diameter, a spherical dome is obtained. Similarly, conical dome is obtained by the revolution of a right angled triangle about its vertical axis, while an elliptical dome is obtained by the revolution of an elliptical curve about one of its axes. However, out of these, spherical domes are more commonly used. In the case of a spherical dome the vertical section through the axis of revolution in any direction is an arc of a circle.

Domes are used in variety of structures, such as (*i*) roof of circular areas (*ii*) circular tanks (*iii*) hangers, (*iv*) exhibition halls, auditoriums and planitoriums and (*v*) bottoms of tanks, bins and bunkers. Domes may be constructed of masonry, steel, timber and reinforced cement concrete. Stone and brick domes are one of the oldest architectural forms. However, reinforced concrete domes are more common now-a-days, since they can be constructed over large spans.

19.2. NATURE OF STRESSES IN SPHERICAL DOMES

A spherical dome may be imagined to consist of a number of horizontal rings placed one over the other. The diameters of the successive rings increase in the downward direction and the equilibrium is maintained independently of the rings above it. The circle of each ring is called 'latitude' while the circle drawn through two diametrically opposite points on a horizontal diameter and the crown is known as a *meridian circle.* All meridian circles converge at the crown(or top most point) of the spherical dome.

Fig. 19.1(*b*) shows the vertical section of a spherical dome. The successive horizontal rings subtend equal angle at the centre of the sphere. The joints between successive horizontal rings is *radial*. Every horizontal ring supports the load of the rings above it, and transmits it to the one below it. The reaction between the rings is tangential

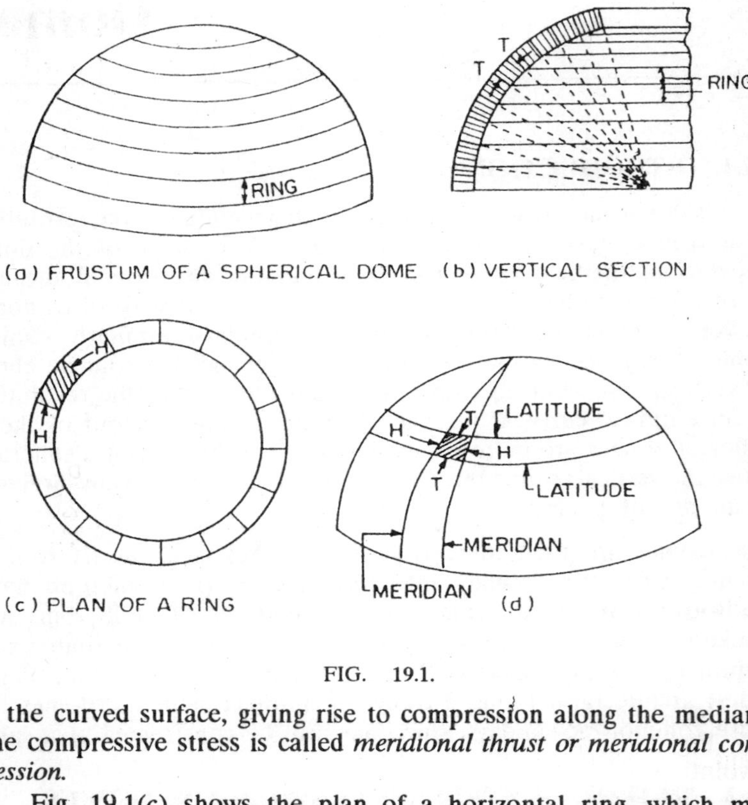

FIG. 19.1.

to the curved surface, giving rise to compression along the medians. The compressive stress is called *meridional thrust or meridional compression.*

Fig. 19.1(*c*) shows the plan of a horizontal ring, which may be imagined to consist of a number of *voussoirs.* The joints between adjacent voussoirs of the ring are radial. The tendency of separation of any voussoir will be prevented because of its wedge shape, and therefore, *hoop compression* will be caused in each ring.

To summarise, therefore, two types of stresses are induced in a dome [Fig. 19.1(*d*)].

(*i*) Meridional thrust (*T*) along the direction of meridian.

(*ii*) Hoop stress (*H*) along the latitudes.

19.3. ANALYSIS OF SPHERICAL DOMES

Let us now analyse stresses developed in a spherical dome of *uniform thickness.* Two cases of loading will be considered :

(*i*) uniformly distributed load,

(*ii*) concentrated load at the crown.

1. Uniformly distributed load

Let w = uniformly distributed load, inclusive of its own weight per unit area.

r = radius of the dome

t = thickness of dome shell

T = intensity of meridional thrust

H = intensity of hoop stress.

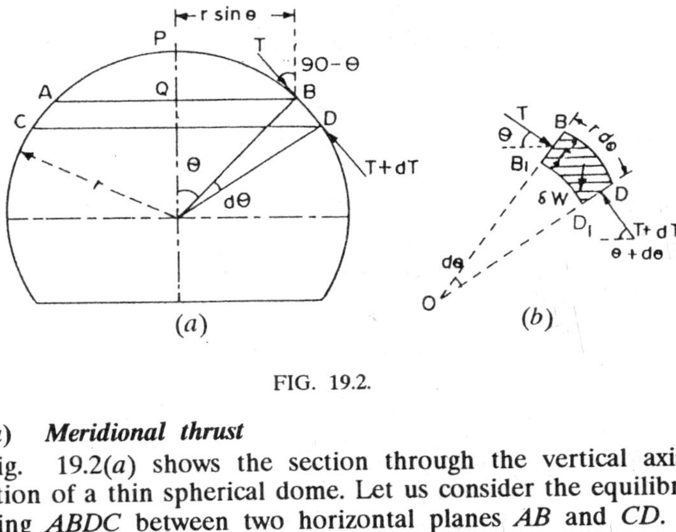

FIG. 19.2.

(*a*) *Meridional thrust*

Fig. 19.2(*a*) shows the section through the vertical axis of revolution of a thin spherical dome. Let us consider the equilibrium of a ring *ABDC* between two horizontal planes *AB* and *CD*. The extremity of the horizontal plane *AB* makes an angle θ with the vertical at the centre while the extremity of the horizontal plane *CD* makes angle $\theta + d\theta$. The ring thus subtends an angle $d\theta$ at the centre. The following are the forces acting on the unit length of the ring:

(*i*) The meridional thrust T per unit length of the circle of latitude *AB*, acting tangentially at *B* (or at right angles to the radial line *OB*).

(*ii*) The reaction or thrust $T + dT$ per unit length of circle on latitude *CD*, acting tangentially at *D*.

(*iii*) The weight δW of the ring itself, acting vertically down.

It should be noted that the reaction $T + dT$ will be greater than the thrust T due to the effect of the weight of the ring and due to the change in the inclination from θ of $(\theta + d\theta)$ of the radial lines. The meridional thrust T is caused due to the weight of the dome shell *APB* above the horizontal plane *AB*.

Surface area of dome shell $APB = 2\pi r \times PQ$

But $PQ = OP - OQ = r - r \cos\theta = r(1 - \cos\theta)$

$\therefore$ Weight of dome shell above $AB = 2\pi r \times PQ \times w$

$$= 2\pi r^2 w(1 - \cos\theta)$$

Since the sum of vertical components of thrust T acting along the circumference of the circle of latitude must be equal to the total weight of the dome shell *APB*, we have

$$T(2\pi \times QB)\sin\theta = 2\pi r^2 w(1 - \cos\theta)$$

or $$T \times 2\pi \times r\sin\theta \times \sin\theta = 2\pi r^2 w(1 - \cos\theta)$$

or $$T = \frac{wr(1-\cos\theta)}{\sin^2\theta} = \frac{wr}{1+\cos\theta} \quad ...(19.1)$$

This is thus the expression for the meridional thrust.

(b) *Hoop stresses*

We have seen that the meridional thrust T increases to $T + dT$ at the bottom of the ring. This difference in the meridional thrust T and $T + dT$ acting at θ and $\theta + d\theta$ respectively to the horizontal causes hoop stress.

Let H be the hoop force per unit length of surface measured on a great circle arc.

Breadth of ring $= r\,d\theta$,

$\therefore$ Hoop force $= H \times r\,d\theta$ ...(*iii*)

The horizontal components of T is $T\cos\theta$, and this horizontal component cause hoop tension, tending to increase the diameter of the ring, while the horizontal component of $T + dT$ will be $(T + dT)(\cos\theta + d\theta)$ and this horizontal component will cause hoop compression .

Now magnitude of hoop tension.

$$= T\cos\theta \times \text{Radius of ring } AB$$

$$= T\cos\theta\, r\sin\theta = Tr\sin\theta\cos\theta \quad ...(ii)$$

Magnitude of hoop compression

$$= (T + d\,T)\cos(\theta + d\,\theta) \times \text{radius of ring } CD$$

$$= (T + d\,T)\cos(\theta + d\,\theta)\, r\sin(\theta + d\,\theta) \qquad ..(iii)$$

The difference between *(ii)* and *(iii)* will cause the actual hoop stress. If *(iii)* is more than *(ii)*, hoop stress will be compressive, while if *(ii)* is more than *(iii)*, hoop stress will be tensile. Thus the hoop force, which is equal to the difference between *(ii)* and *(iii)*, is due to the change in the value of T when $d\theta$ increases by a small amount $d\theta$. Hence in the limiting case when $d\theta$ is extremely small,

$$H\,r\,d\,\theta = d\,[T\cos\theta\, r\sin\theta]$$

But $$T = \frac{w\,r(1-\cos\theta)}{\sin^2\theta} \text{ from Eq. 19.1.}$$

$\therefore$ $$H = \frac{1}{r}\cdot\frac{d}{d\theta}\left[r\sin\theta\cos\theta\,\frac{w\,r\,(1-\cos\theta)}{\sin^2\theta}\right]$$

$$= w\,r\,\frac{d}{d\,\theta}\left[\frac{1-\cos\theta}{\sin^2\theta}\sin\theta\cos\theta\right]$$

$$= w\,r\,\frac{d}{d\,\theta}\left[\frac{\cos\theta}{\sin\theta} - \frac{\cos^2\theta}{\sin\theta}\right]$$

or $$H = w\,r\left[\frac{-\sin^2\theta - \cos^2\theta}{\sin^2\theta} - \frac{-2\sin^2\theta\cos\theta - \cos^3\theta}{\sin^2\theta}\right]$$

$$= w\,r\left[\frac{-1 + 2\sin^2\theta\cos\theta + \cos^3\theta}{\sin^2\theta}\right]$$

$$= w\,r\left[\frac{-1 + 2\cos\theta - 2\cos^3\theta + \cos^3\theta}{\sin^2\theta}\right]$$

$$= w\,r\left[\frac{-1 + 2\cos\theta - \cos^3\theta}{\sin^2\theta}\right]$$

$$= w\,r\left[\frac{(1-\cos\theta)(\cos^2\theta + \cos\theta - 1)}{2-\cos^2\theta}\right]$$

$$= \frac{w\,r\,(\cos^2\theta + \cos\theta - 1)}{1+\cos\theta} \qquad ...(19.2)$$

The above expression gives the hoop stress in any horizontal ring the extremity of which subtends an angle θ with the vertical, at the centre. If the value of H obtained form Eq. 19.2 is positive, hoop force will be compressive ; otherwise it will be tensile.

At the crown, $\theta = 0$.

Hence from Eq. 17.2 $H = \frac{wr}{2}$

Intensity of *hoop stress at crown* $= \frac{H}{t} = \frac{wr}{2t}$ (compressive) [19.3(*a*)]

This is the maximum value of hoop stress. The hoop stress will go on decreasing as θ increases, till H becomes zero. After that H becomes tensile. To find the position of the plane where hoop stress becomes zero, we have

$$H = 0 = \frac{wr(\cos^2\theta + \cos\theta - 1)}{1 + \cos\theta}$$

or $$\cos^2\theta + \cos\theta - 1 = 0$$

or $$\cos\theta = 0.618$$

From which $\theta = 51° 49' 38''$

Hence round the circle of latitude at which the angle $\theta = 51° 49' 38''$, hoop stress is zero. For all portion of dome about this angle, hoop compression will be developed, while for the portion below this plane, hoop tension will be developed which will go on increasing further towards the base of the dome.

2. Concentrated load at the crown

Let us now find out the stresses developed in the spherical dome due to a concentrated load W situated at the crown P as shown in Fig. 19.3.

As before, let us consider a ring $ABCD$ bounded by horizontal planes AB and CD subtending angles θ and $\theta + d\theta$ with the vertical axis. Let T be the thrust per unit length of latitude AB acting tangentially and $T + dT$ be thrust per unit length of the circle of latitude CD.

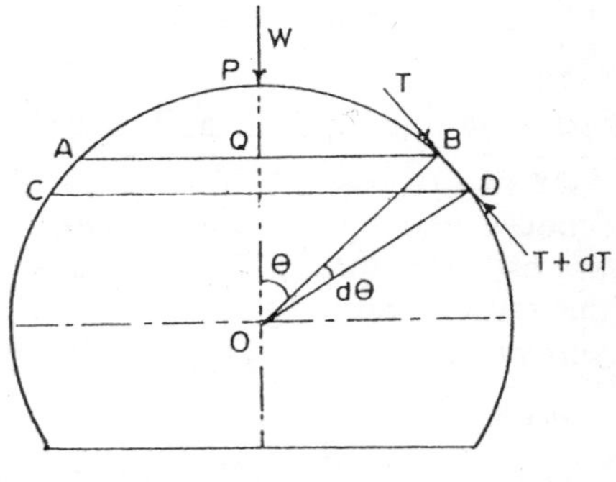

FIG. 19.3.

The sum of the vertical components of thrust T acting along the circumference of the circle of latitude must be equal to the load W ; hence

$$T \times 2\pi r \sin\theta \times \sin\theta = W$$

or $$T = \frac{W}{2\pi r \sin^2\theta} \qquad ...(16.4)$$

As discussed in the previous case, the hoop stress developed in any horizontal ring is due to the difference in the meridional thrust T and $T + dT$. Hence , as before,

$$H r.d\theta = d\,[T\cos\theta \times r\sin\theta]$$

Substituting the value of T from Fig. 19.4, we get

$$H = \frac{1}{r}\frac{d}{d\theta}\left[\frac{W}{2\pi r \sin^2\theta}\cos\theta\, r\sin\theta\right]$$

$$= \frac{W}{2\pi r}\frac{d}{d\theta}\left[\frac{\cos\theta}{\sin\theta}\right]$$

Differentiating the above, we get

$$H = \frac{W}{2\pi r}\left[\frac{-\sin^2\theta - \cos^2\theta}{\sin^2\theta}\right]$$

or $$H = -\frac{W}{2\pi r}\operatorname{cosec}^2\theta \qquad ...(19.5)$$

and hoop stress $= \frac{H}{t} = -\frac{W}{2\pi r t}\operatorname{cosec}^2\theta$ ...[19.5(a)]

where t is the thickness of the dome.

The minus signs shows that hoop stress will be tensile throughout, due to the concentrated load.

At the crown, $\theta = 0$. Hence H becomes infinite. Therefore, any concentrated load in the form of lantern or ornament etc., should always be distributed over sufficient area, to reduce the hoop stress at the crown. It is also desirable to thicken the dome at the top to spread the load over greater area.

Stresses due to combined uniformly distributed load and point load

The stresses due to point load applied at the crown should be added algebraically to the stresses developed due to uniformly distributed live load and self-weight of the dome, to get the final stresses.

Table 19.1 gives the coefficients for hoop stress and meridional stress due to uniformly distributed load and concentrated load at the crown, for different values of θ .

TABLE 19.1.
(+ for compression and – for tension)

θ (°)	*Uniformly distributed load w*		*Concentrated load W at the crown*	
	Coefficient of w.r. (for T)	*Coefficient of w. r. (for H)*	*Coefficient of $\frac{W}{r}$ (for T)*	*Coefficient of $\frac{W}{r}$ (for H)*
	$\frac{1-\cos\theta}{\sin^2\theta}$	$\frac{\cos^2\theta+\cos\theta-1}{1+\cos\theta}$	$\frac{\text{cosec}^2\theta}{2\pi}$	$-\frac{\text{cosec}^2\theta}{2\pi}$
0	+0.5	+0.5	∞	∞
5	+0.500	+0.496	+21.0	– 21.0
10	+0.505	+0.48	+5.3	– 5.3
20	+0.516	+0.425	+1.37	–1.37
30	+0.537	+0.33	+0.64	– 0.64
40	+0.566	+0.20	+0.38	– 0.38
50	+0.608	+0.034	+0.27	– 0.27
51° 49′ 38″	+0.618	+0.00	+0.26	– 0.26
60	+0.667	– 0.167	+0.21	– 0.21
70	+0.747	– 0.402	+0.18	– 0.18
80	+0.838	– 0.68	+0.16	– 0.16
90	+1.00	–1.00	+0.16	– 0.16
100	+1.21	– 1.38	+0.16	– 0.16
110	+1.52	– 1.86	+0.18	– 0.18
120	+2.00	– 2.50	+0.21	– 0.21
130	+2.79	– 3.45	+0.27	– 0.27
140	+4.27	– 5.05	+0.38	– 0.38
150	+7.48	– 8.33	+0.64	– 0.64
160	+16.60	– 17.5	+1.37	– 1.37
170	+66.60	– 66.60	+5.3	– 5.3
180	∞	∞	∞	∞

19.4. STRESSES DUE TO WIND LOAD

The analysis of stresses in domes due to wind load is very tedious since the dome has curvature in two directions. Since the thickness and reinforcement required for the vertical and uniformly distributed loads is much smaller than provided from practical considerations, exact calculation of stresses due to wind load and the effect of shrinkage and temperature variations are not necessary. Their effect is taken into account by adding and extra load of about 1000 to 1500 N per sq. m. of the surface of the dome.

19.5. DESIGN OF R.C. DOMES

The requirements of thickness of dome and reinforcement from the point of view of induced stresses are usually very small. However, a minimum thickness of 7.5 cm is provided to protect steel. Similarly, a minimum steel provided is 0.15% of the sectional area in each direction—meridionally as well as along the latitudes. This reinforcement will be in addition to the hoop tensile stresses. The steel reinforcement is provided in the middle of the thickness of the dome shell. Near the edges, some hogging bending moment may be developed, and hence meridional steel should be placed near the top surface.

Provision of ring beam.

If the dome is not hemishperical, the meridional thrust at the supporting circle of latitude(*i.e.*, at the base) will not be vertical. The inclined meridional thrust at the support will have horizontal component which will cause the supporting walls to burst outwards, causing its failure. In order to bear this horizontal component of meridional thrust, a ring beam is provided at the base of the dome. The reinforcement provided in the ring beam takes this hoop tension and transfer only vertical reaction to the supporting walls. The tensile stress on the equivalent area of concrete on the ring beam section should not exceed 12 N/mm^2.

Placement of main reinforcement in dome

As stated earlier, a minimum reinforcement of 0.15% of area is provided both in the direction of latitude as well as of the meridians. If the reinforcement along the meridians is continued upto crown, there will be congestion of steel there. Hence from practical considerations, the meridional reinforcement is stopped at any latitude circle near crown, and a separate mesh is provided as shown in Fig. 19.4. No separate reinforcement along latitude is provided in this area at the crown.

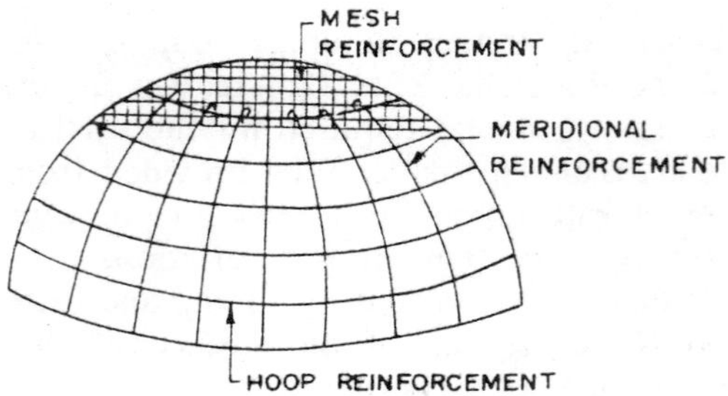

FIG. 19.4.

Provision of openings

Openings may be provided in the dome as required from other functional or architectural requirements. However, sufficient trimming reinforcement should be provided all round the opening, as shown in Fig. 19.5. The meridional and hoop reinforcement reaching the opening should be well anchored to the trimming reinforcement.

If there is an opening at the crown of the dome, and if there is any concentrated load of lantern etc. acting there, a ring beam

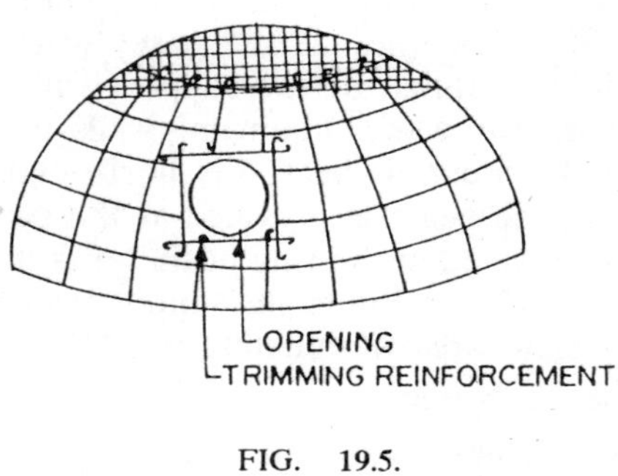

FIG. 19.5.

should be provided at the periphery of the opening, as shown in Fig. 19.7.

Design Example 19.1. *Design a spherical dome over a circular room, for the following data :*

(*i*) *Inside diameter of room* = 12 *m*

(*ii*) *Rise of dome* = 4 *m*

(*iii*) *Live load due to wind, ice, snow etc.* = 1.5 kN/m^2.

The dome has an opening of 1.6 m diameter at its crown. A lantern is provided at its top, which causes a dead load of 22 kN acting along the circumference of the opening.

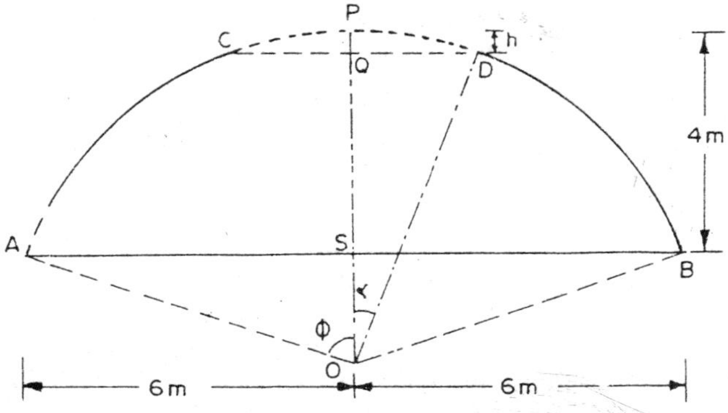

FIG. 19.6.

Solution.

1. ***Geometry of the dome***

Fig. 19.6. shows the dome. Let the radius of the dome be r. The diameter of base $= AB =$ dia. of room $=$ 12 m.

$$\text{Rise } PS = 4 \text{ m}$$

$$(2r - PS)\, PS = 6 \times 6$$

or

$$(2r - 4)\, 4 = 36$$

From which $r = AO = 6.5$ m

Again $CD =$ diameter of opening $=$ 1.6 m

$$PQ = \text{rise at opening} = h$$

$\therefore \quad h\,(2 \times 6.5 - h) = CQ^2 = (0.8)^2$

From which $\quad h \approx 0.05$ m

$\sin \alpha = \dfrac{0.8}{6.5} = 0.1231\,; \; \therefore \alpha = 7°\,4'\,; \cos \alpha = 0.9924$

$\sin \Phi = \dfrac{6}{6.5} = 0.9231\,; \; \therefore \; \Phi = 67°\,23'\,; \cos \Phi = 0.3840\,.$

2. ***Loading***

The various formulae derived earlier are valid when there is no opening at the crown. In our case, there is opening of 1.6 m diameter. However, for calculation purposes, we can assume that there is no opening, and the weight of the extra portion of the dome shell *CPD* can be accounted for by reducing the load of the lantern and taking into consideration only the effective weight of lantern. That is :

Effective wt. of lantern = Actual wt. of lantern – wt. of dome shell *CPD*.

Let the thickness of the dome be 100 mm.

The uniformly distributed loads per sq. m of surface area are

(*i*) Self-weight of dome shell $= 0.1 \times 25000 = 2500$ N.

(*ii*) Live load $= 1500$

$$\text{Total} = 4000 \text{ N/m}^2 = 4 \text{ kN/m}^2 = w$$

Weight of dome shell *CPD* $= w \times 2\pi r h$

$$= w \times 2\pi \times 6.5 \times 0.05 = 0.17 \text{ kN}$$

$\therefore$ Effective weight of lantern $= W = 22 - 8.17 = 13.83$ kN

3. *Calculation of stresses due to combined load*

The stress at any horizontal plane will be equal to the algebraic sum of stresses due to the two loading, and the dome will be designed for the maximum of these stresses.

$$\text{Total meridional stress} = \frac{w\,r(1-\cos\theta)}{t\sin^2\theta} + \frac{W}{2\pi r t \sin^2\theta}$$

$$= \frac{4 \times 6.5\,(1-\cos\theta)}{0.1\sin^2\theta} + \frac{13.83}{2\pi \times 6.5 \times 0.1 \sin^2\theta} \text{ kN/m}^2$$

$$= \left[260\,\frac{1-\cos\theta}{\sin^2\theta} + \frac{3.39}{\sin^2\theta}\right] \times 10^{-3} \text{ N/mm}^2$$

$$= 0.26\,\frac{1-\cos\theta}{\sin^2\theta} + \frac{0.00339}{\sin^2\theta} \text{ N/mm}^2 \qquad \text{...(I)}$$

$$\text{Hoop stress} = \frac{w\,r}{t}\left(\frac{\cos^2\theta + \cos\theta - 1}{1+\cos\theta}\right) - \frac{W}{2\pi r t}\,\frac{1}{\sin^2\theta}$$

$$= \frac{4 \times 6.5}{0.1}\left(\frac{\cos^2\theta + \cos\theta - 1}{1+\cos\theta}\right) - \frac{13.83}{2\pi \times 6.5 \times 0.1} \times \frac{1}{\sin^2\theta} \text{ kN/m}^2$$

$$= \left[260\left(\frac{\cos^2\theta + \cos\theta - 1}{1+\cos\theta}\right) - \frac{3.39}{\sin^2\theta}\right] \times 10^{-3} \text{ N/mm}^2$$

$$= 0.26 \frac{\cos^2\theta + \cos\theta - 1}{1 + \cos\theta} - \frac{0.00339}{\sin^2\theta} \text{ N/mm}^2 \quad ...(2)$$

The values of meridional stress and hoop stress for various values of θ are Tabulated in Table 19.2.

TABLE 19.2

θ	*Meridional stress (N/mm²)*			*Hoope stress (N/mm²)*		
	Due to w	*Due to W*	*Total*	*Due to w*	*Due to W*	*Total*
7° 4′	0.1305	0.223	0.354	0.127	− 0.225	− 0.098
10°	0.131	0.112	0.243	0.125	− 0.113	+ 0.012
20°	0.134	0.029	0.163	0.102	− 0.029	+ 0.073
30°	0.140	0.013	0.153	0.086	− 0.014	+ 0.072
40°	0.148	0.008	0.156	0.052	− 0.008	+ 0.044
50°	0.158	0.006	0.164	0.009	− 0.006	+ 0.003
60°	0.174	0.004	0.178	− 0.044	− 0.004	− 0.048
67° 23′	0.186	0.004	0.190	− 0.088	− 0.004	− 0.092

4. *Hoop stress in absence of live load*

Hoop stresses should also be found in absence of live load. This will increase the tensile stresses in the upper portion of the dome, specially near the periphery of the opening. However, meridional thrust will not increase by omitting the live load.

$w = 0.1 \times 25000 = 2500 \text{ N} = 2.5 \text{ kN}$

$\therefore$ Effective weight of lantern $= 22 - 2.5 \times 2\pi \times 6.5 \times 0.05$

$= 16.89 \text{ kN}$

Thus the hoop stresses due to w calculated above will be *decreased* in the ratio of $\frac{250}{400}$ (= 0.625) while the hoop stresses due to W will be *increased* by a ratio of $\frac{16.89}{13.84}$ (= 1.22). The results are tabulated in Table 19.3.

TABLE. 19.3

θ	*Hoop Stress* (N/mm^2)		
	Due to w	*Due to W*	*Total*
7° 4′	0.080	− 0.275	− 0.195
10°	0.078	− 0.138	− 0.060
20°	0.064	− 0.035	+ 0.029
30°	0.054	− 0.017	+ 0.037
40°	0.033	− 0.010	+ 0.023
50°	0.006	− 0.007	− 0.001
60°	− 0.027	− 0.005	− 0.032
67° 23′	− 0.055	− 0.005	− 0.060

Thus we see that the maximum hoop tension at the opening has been increased from 0.098 N/mm^2 to 0.195 N/mm^2.

5. ***Provision for reinforcement***

Max. compressive stress $= 0.354\ N/mm^2$ (safe)

Max. hoop tensile stress $= 0.195\ N/mm^2$.

$\therefore$ Max. hoop tension per metre length of meridian

$$= 0.195 \times 100 \times 1000 = 19500\ N$$

$\therefore$ Area of steel $= \dfrac{19500}{140} = 139\ mm^2$

Reinforcement for temp. etc. $= \dfrac{0.15}{100} \times 100 \times 1000 = 150\ mm^2/m$

$\therefore$ Total reinforcement $= 139 + 150 = 289\ mm^2$

Using 8 mm Φ bars, spacing $= \dfrac{1000 \times 50}{289}$

$$= 173\ mm$$

However, provide 8 mm Φ bars @ 160 mm c/c where hoop tension is developed. In the portion where no hoop tension is developed, minimum area of steel @ 0.15% will be 150 mm^2. Hence provide 8 mm Φ bars @ 300 mm c/c.

6. ***Design of lower ring beam***

Meridional thrust per metre length of the dome at its base $= 0.19 \times 100 \times 1000 = 19000$ N/m .

Horizontal component T per metre length

$$= 19000 \cos 67° 23' = 7307 \text{ N/m}$$

$\therefore$ Hoop tension, trying to rupture the beam

$$= 7307 \times \frac{12}{2} = 43840 \text{ N}$$

$\therefore$ Area of steel required $= \dfrac{43840}{140} = 313 \text{ mm}^2$

Using 10 mm Φ bars, No. of rings $= \dfrac{313}{78.5} = 4$

$\therefore$ Hence provide 4 rings of 10 mm Φ bars.

Equivalent area of composite section of beam of area of cross-section A.

$$= A + (m-1) A_{st} = A + 18 \times 78.5 \times 4$$

$$= A + 5652$$

Allowing tensile stress of 1.2 N/mm^2 in the composite section, we have

$$\frac{43840}{A + 5652} = 1.2$$

From which $A = 30880 \text{ mm}^2$

However, provide a ring beam of 200 mm × 200 mm. Provide 6 mm Φ stirrups @ 160 mm c/c to tie the rings in the ring beam.

7. ***Design of ring beam at the opening***

Hoop compression(horizontal) in the ring beam

$$= (0.345 \times 100 \times 1000) \cos \alpha \times \frac{1.6}{2}$$

$$= 35400 \times 0.9924 \times 0.8 = 28105 \text{ N}$$

The horizontal component will be provided for by a beam which will form the link between the lantern and the dome.

Provide a ring beam of size 160 mm × 160 mm either above the dome or below the dome, as required from architectural requirements.

Comp. stress in ring beam $= \dfrac{28105}{160 \times 160} = 1.1 \text{ N/mm}^2$ (safe)

Extend the 8 mm Φ rings in the ring beam also.

The details, of reinforcement are shown in Fig. 19.7.

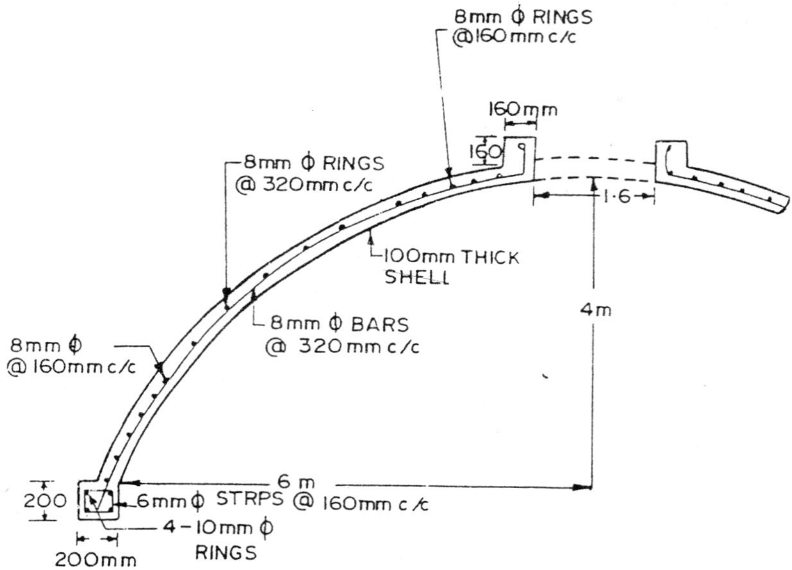

FIG. 19.7.

19.6. CONICAL DOMES

Let us now analyse a conical dome subtending an angle 2θ at the apex. Let w be the intensity of uniformly distributed load inclusive of its own weight, on unit area of the dome.

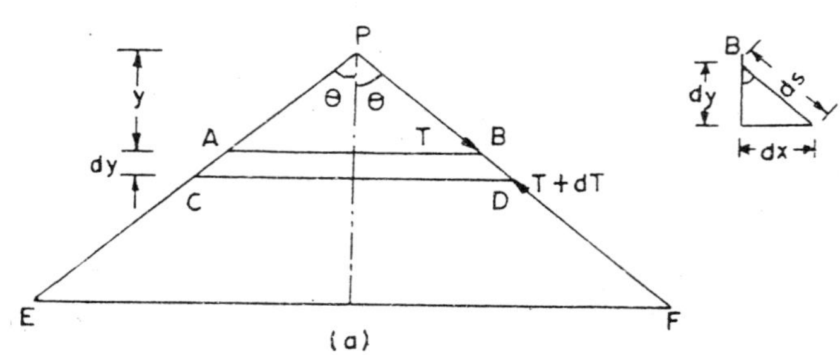

FIG. 19.8

Consider a horizontal ring *ABDC* as before. Let the meridional thrust per unit length of the ring be T at B and $T + dT$ at D, acting tangentially.

The vertical component of the total meridional thrust at B will evidently be equal to the load on the dome shell *APB*. Let the distance of horizontal planes *AB* and *CD* be y and $y + dy$ from apex P.

Diameter $AB = 2y \tan\theta$; length $AP = \dfrac{y}{\cos\theta}$

$\therefore$ Load on dome shell *APB*

$$= \frac{w}{2}(\pi\, 2y \tan\theta) \cdot \frac{y}{\cos\theta}$$

Vertical component of total $T = (\pi\, 2y\tan\theta)\, T\cos\theta$

$$\therefore \quad 2\pi y \tan\theta\, T\cos\theta = 2\pi y \frac{w}{2}\tan\theta \frac{y}{\cos\theta}$$

From which $$T = \frac{w}{2}\frac{y}{\cos^2\theta} \qquad ...(19.6)$$

$$\therefore \text{ Intensity of meridional stress } = \frac{w}{2t}\frac{y}{\cos^2\theta} \qquad ...[19.6(a)]$$

The horizontal component of T will cause hoop tension at B, while the horizontal component of $T + dT$ will cause hoop compression.

Magnitude of hoop tension $= T\sin\theta \times$ radius of ring at depth y

$$= T\sin\theta \times y\tan\theta = Ty\frac{\sin^2\theta}{\cos\theta}$$

$$\text{Magnitude of hoop compression } = (T + dT)(y + dy)\frac{\sin^2\theta}{\cos\theta}$$

The difference in these two horizontal components will give the value of hoop force. Let H be the hoop compression induced in the ring, per unit breadth. Let ds be the breadth of the ring of height dh [Fig. 19.8(b)]

Then $ds = \dfrac{dy}{\cos\theta}$

Hence we have

$$H\,ds = d\left(\frac{Ty\sin^2\theta}{\cos\theta}\right)$$

or $$\frac{H\,dy}{\cos\theta} = d\left(Ty\frac{\sin^2\theta}{\cos\theta}\right)$$

or $$H = \frac{d}{dy}(Ty \sin^2 \theta)$$

Substituting the value of T and differentiating, we get

$$H = \frac{d}{dy}\left[\frac{w}{2}\frac{y}{\cos^2}\cdot y \sin^2\theta\right] = \frac{w}{2}\frac{\sin^2\theta}{\cos^2\theta}\frac{d}{dy}(y)^2$$

Hence $H = wy \tan^2\theta$ ...(19.7)

Hence intensity of hoop stress $= \frac{wy}{t}\tan^2\theta$ ...[19.7(*a*)]

The hoop stress will be compressive throughout.

Design Example 19.2. *Design a conical roof for a hall having a diameter of 20 m. The rise of the dome has to be 4 m. Assume the live and other loads as 1500 N/m².*

Solution

(i) Geometry of the dome

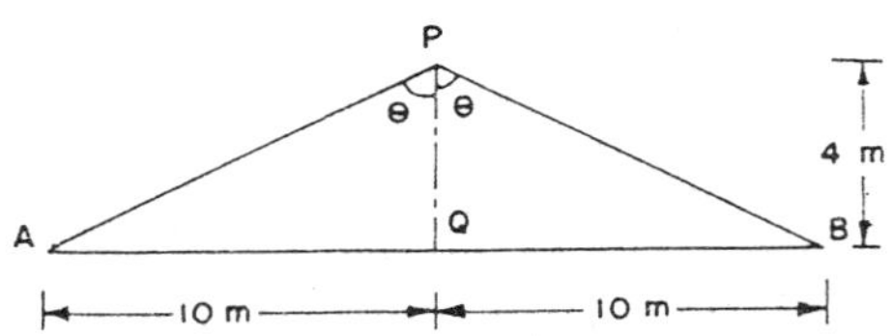

FIG. 19.9.

Diameter $AB = 20$ m.

Height $PQ = 4$ m.

$$\tan\theta = \frac{10}{4} = 2.5 \,;\, \theta = 68° \, 12'$$

$$\sin\theta = 0.9285 \,;\, \cos\theta = 0.3714$$

(ii) Loading

Let the thickness of shell be 10 cm.

Weight of dome shell /m² $= 0.1 \times 1 \times 25000 = 2500$ N

Live load $= 1500$ N

Total $w = 4000$ N/m²

(iii) Calculation of stress

Meridional thrust $T = \frac{wy}{2\cos^2\theta}$

Maximum T will occur at the base AB where $y = 4$ m.

$$\therefore \quad T_{max} = \frac{4000 \times 4}{2\,(0.3714)^2} = 58200 \text{ N (comp.)}$$

$$\therefore \quad \text{Meridional stresss} = \frac{58200}{100 \times 1000} = 0.582 \text{ N/mm}^2 \text{ (Safe)}$$

Hoop force $\quad H = w\,y \tan^2\theta.$

$$\therefore \quad H_{max} = 4000 \times 4\,(2.5)^2 = 100000 \text{ N (comp.)}$$

$$\therefore \quad \text{Hoop stress} = \frac{100000}{100 \times 1000} = 1 \text{ N/mm}^2 \text{ (Safe)}$$

(iv) Steel reinforcement

The stresses work out to be safe. Hence only nominal reinforcement has to be provided @ 0.15% of the area of concrete.

$$A_{st} = \frac{0.15}{100} \times 100 \times 10000 = 150 \text{ mm}^2$$

Using 8 mm Φ bars, $A_\Phi = 50$ mm^2

$$\text{Spacing} = \frac{1000\,A_\Phi}{A_{st}} = \frac{1000 \times 50}{150} = 333 \text{ mm}$$

However, provide 8 mm Φ bars @ 300 mm c/c both the ways.

The meridional bars may be discontinued near the apex, and a wire mesh may be provided there to avoid congestion of steel.

(v) Design of ring beam

Horizontal component of meridional thrust T will cause an outward force on the support, causing hoop tension. Hence a ring beam is necessary.

Hoop tension P in ring beam $= T \sin\theta$

$= 58200 \times 0.9285 = 54000$ N/m.

$$\therefore \quad \text{Total tensile force} = P \times \frac{D}{2} = 54000 \times \frac{20}{2} = 540000 \text{ N}$$

$$\therefore \quad \text{Area of steel to resist this, } A_{st} = \frac{540000}{140} = 3857 \text{ mm}^2$$

Using 25 mm Φ bars, $A_\Phi = 490$ mm^2

$$\therefore \quad \text{No. of bars} = \frac{3857}{490} \approx 8$$

Actual area of steel provided = 3920 mm^2.

Tie these by 8 mm Φ 2-lgd stirrups @ 300 mm c/c

Let A be the area of ring beam.

Equivalent area of composite section $= A + (m - 1)\,A_{st}$

$= A + 18 \times 3920 = A + 70560$

Assuming the allowable tensile stress in composite section to be 1.2 N/mm^2, we have

$$\frac{540000}{A + 70560} = 1.2$$

which gives $A = 379440$ mm^2

Provide ring beam of size 700 mm × 700 mm.

The details of the reinforcement is shown in Fig. 19.10.

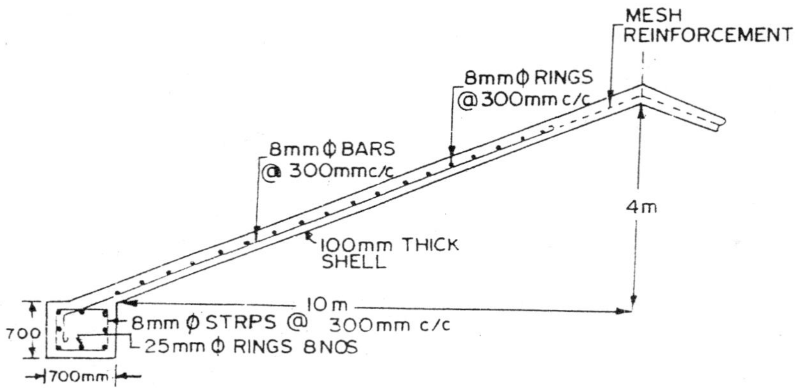

FIG. 19.10

PROBLEMS

1. A spherical dome, span 10 m and rise 2 m, has a shell which is 120 mm thick. It carries a lantern load of 5000 N at its apex. The wind load on the dome is estimated to the equivalent to 1200 N/m^2. Examine the stresses in dome and design suitable reinforcement. Also, design the ring beam.
2. Design a conical dome for the above, if there is no lantern.
3. Design a spherical dome over a circular room of 20 m diameter. The rise of the dome may be taken equal to 1/5 of the diameter. The dome carries a lantern load of 30 kN attached at the circumference of an opening of 2 m diameter at the apex. Take live load due to wind etc. as 1.5 kN/m^2 of the surface area of the dome. Use M 15 concrete.

20

Water Tanks-I

20.1 INTRODUCTION

A water tank is used to store water to tide over the daily requirements. In general, water tanks can be classified under three heads : (*i*) tanks resting on ground (*ii*) elevated tanks supported on staging, and (*iii*) underground tanks. From the shape point of view, water tanks may be of several types, such as (*i*) circular tanks (*ii*) rectangular tanks (*iii*) spherical tanks (*iv*) Intze tanks and (*v*) circular tanks with conical bottoms.

In the construction of concrete structures for the storage of water and other liquids, the imperviousness of concrete is most essential. The permeability of any uniform and thoroughly compacted concrete of given mix proportions is mainly dependent on the water-cement ratio. The increase in water-cement ratio results in increase in the permeability. The decrease in water-cement ratio will therefore be desirable to decrease the permeability, but very much reduced water-cement ratio may cause compaction difficulties and prove to be harmful also. For a given mix made with particular materials, there is a lower limit to the water-cement ratio which can be used economically on any job. It is essential to select a richness of mix compatible with available aggregates, whose particle shape and grading have an important bearing on workability, which must be suited to the means of compaction selected. Efficient compaction preferably by vibration is essential. It is desirable to specify cement content sufficiently high to ensure that thorough compaction is obtainable while maintaining a sufficiently low water-cement ratio. The quantity of cement should not be less than 330 kg/m^3 of concrete. It should also be less than 530 kg/m^3 of concrete to keep the shrinkage low. In thicker sections, where a reduction in cement content might be desirable to restrict the temperature rise due to cement hydration, a lower cement content is usually permissible. It is usual to use rich mix like M 30 grade in most of the water tanks.

Design of liquid retaining structure has to be based on the avoidance of cracking in the concrete having regard to its tensile strength. It has to be ensured in its design that concrete does not crack on its water face. Cracking may also result from the restraint to shrinkage, free expansion and contraction of concrete due to temperature and shrinkage and swelling due to moisture effects. Correct placing of reinforcement, use of small sized bars and use of deformed bars lead to a diffused distribution of cracks. The risk of cracking due to overall temperature and shrinkage effects may be minimized by limiting the changes in moisture content and temperature to which the structure as a whole is subjected. Cracks can be prevented by avoiding the use of thick timber shuttering which prevent the easy escape of heat of hydration from the concrete mass. The risk of cracking can also be minimized by reducing the restraints on the free expansion or contraction of the structure. For long walls or slabs founded at or below the ground level, restraints can be minimized by founding the structure on a flat layer of concrete with interposition of sliding layer of some material to break the bond and facilitate movement. However, it should be recognized that common and more serious causes of leakage in practice, other than cracking, are defects such as segregation and honey combing and in particular all joints are potential source of leakage.

20.2 GENERAL DESIGN REQUIREMENTS ACCORDING TO INDIAN STANDARD CODE OF PRACTICE (IS : 3370 - Part II, 1965)

1. Plain Concrete Structures

Plain concrete members of reinfored concrete liquid structures may be designed against structural failure by allowing tension in plain concrete as per the permissible limits for tension in bending specified in IS : 456 – 1978 (*i.e.*, permissible stress in tension in bending may be taken to be the same as permissible stress in shear, q measured as inclined tension). This will automatically take care of failure due to cracking. However, nominal reinforcement in accordance with the requirements of IS : 456 shall be provided for plain concrete structural members.

2. Permissible Stresses in concrete

(a) *For resistance to cracking*

Indian Standard code IS: 456-1978 does not specify the permissible stresses in concrete for its resistance to cracking. However, its earlier version (IS : 456-1964) included the permissible stresses in direct tension, bending tension and shear. These values are given in Table 20.1. The permissible tensile stresses due to bending apply to the face of the member in contact with the liquid. In members with thickness less than 225 mm and in contact with the liquid on one side, these permissible stresses in bending apply also to the face remote from the liquid.

TABLE 20.1.
PERMISSIBLE CONCRETE STRESSES IN CALCULATIONS RELATING TO RESISTANCE TO CRACKING

Grade of concerete	*Permissible stresses*		*Shear* $= \frac{Q}{b_j d}$
	Direct Tension (σ_{st})	*Tension due to bending* (σ_{cbt})	
	N/mm^2	N/mm^2	N/mm^2
M 15	1.1	1.5	1.5
M 20	1.2	1.7	1.7
M 25	1.3	1.8	1.9
M 30	1.5	2.0	2.2
M 35	1.6	2.2	2.5
M 40	1.7	2.4	2.7

(b) *For strength calculations*

In strength calculations the usual permissible stresses, in accordance with IS : 456-1978, Chapter 1, are used. Where the calculated shear stress in concrete above exceeds the permissible value, reinforcement acting in conjunction with diagonal compression in concrete shall be provided to take the whole of the shear.

3. Permissible Stresses in Steel Reinforcement

(a) *For resistance to cracking*

When steel and concrete are assumed to act together for checking the tensile stresses in concrete for avoidance of cracking the tensile stresses in steel will be limited by the requirement that the permissible tensile stress in concrete is not exceeded so that tensile stresses in steel shall be equal to the product of modular ratio of steel and concrete, and the corresponding allowable tensile stress in concrete.

TABLE 20.2.
PERMISSIBLE STRESSES IN STEEL REINFORCEMENT FOR STRENGTH CALCULATIONS

Types of stress in steel reinforcement	*Permissible stresses in N/mm²*	
	Plain round mild steel bars conforming to grade 1 of IS : 482 (Part 1) -1966	*High yield strength deformed bars (HYSD) conforming to IS :1789–1966 or IS : 1139 –1966.*
1. *Tensile stress in members under direct tension* (σ_s)	115	150
2. *Tensile stress in members due to bending* (σ_{st})		
(*a*) On liquid retaining face of members.	115	150
(*b*) On face away from liquid for members less than 225 mm.	115	150
(*c*) On face away from liquid for members 225 mm or more in thickness.	125	190
3. *Tensile stress in shear reinforcement* (σ_{sv})		
(*a*) For members less than 225 mm thickness	115	150
(*b*) For members 225 or more in thickness.	125	175
4. *Compressive stress in columns subjected to direct load* (σ_{sc})	125	175

(*b*) ***For strength calculations***

Though the Indian Standard Code IS : 456 had its *third revision* in 1978, the corresponding Codes IS : 3370 (Part I, II, III and IV) for concrete structures for the storage of liquids have not been revised since 1965. The main Code on concrete—IS : 456 is in SI units. However, the *fourth reprint* (May 1982) of IS : 3370 (Part II)-1965 incorporates the *amendment* regarding the permissible stresses in steel reinforcement. The revised values of permissible stresses are given in Table 20.2 converted into SI units, using the approximation 10 $kg/cm^2 \approx 1 N/mm^2$.

Note. Stress limitations for liquid retaining faces shall also apply to the following :

(*a*) Other faces within 225 mm of the liquid retaining face.

(*b*) Outside or external faces of structures away from the liquid but placed in water–logged soils upto the level of highest subsoil water.

4. Stresses due to drying shrinkage or temperature change

(*i*) Stresses due to drying shrinkage or temperature change may be ignored provided that :

(*a*) The permissible stresses specified in para (2) and (3), for concrete and steel respectively are not exceeded.

(*b*) Adequate precautions are taken to avoid cracking of concrete during the construction period and until the reservoir is put into use.

(*c*) The recommendations as regards the provision of joint and for suitable sliding layer(see § 19.3) are complied with, or the reservoir is to be used only for the storage of water or aqueous liquids at or near ambient temperature and the circumstances are such that the concrete will never dry out.

(*ii*) Shrinkage stresses may, however, be required to be calculated in special case, when a shrinkage coefficient of 300×10^{-6} may be assumed.

(*iii*) When the shrinkage stresses are allowed, the permissible stresses, tensile stresses in concrete (direct and bending) as given in Table 20.1 may be increased by $33\frac{1}{3}$ percent.

(*iv*) Where reservoirs are protected with an internal impermeable lining, consideration should be given to the possibility of concrete eventually drying out. Unless it is established on the basis of tests or experience that the lining has adequate crack bridging properties, allowance for the increased effect of drying shrinkage should be made in the design.

5. Steel Reinforcement

(a) Minimum reinforcement

(*i*) The minimum reinforcement in walls, floors and roofs in each of the two directions at right angles shall have an area of 0.3 percent of the concrete section in that direction for sections upto 100 mm thickness. For sections of thickness greater than 100 mm and less than 450 mm the minimum reinforcement in each of the two directions shall be linearly reduced from 0.3 percent for 100 mm thick section to 0.2 percent for 450 mm, minimum reinforcement in each of the two directions shall be kept at 0.2 percent. In concrete sections of thickness 225 mm or greater, two layers of reinforcing bars shall be placed one near each face of the section to make up the minimum reinforcement specified above.

(*ii*) In special circumstances, floor slabs resting directly on the ground may be constructed with percentage of reinforcement less than that specified above. In no case the percentage of reinforcement in any member be less than 0.15% of the concrete section.

(b) Minimum cover to reinforcement

(*i*) For liquid faces of parts of members either in contact with the liquid or enclosing the space above the liquid(such as inner faces of slab), the minimum cover to all reinforcement should be 25 mm or the diameter of the main bar, whichever is greater. In the presence of sea water and soils and water of corrosive character the cover should be increased by 12 mm but this additional cover shall not be taken into account for design calculations.

(*ii*) For faces away from the liquid and for parts of the structure neither in contact with the liquid on any face nor enclosing the face above the liquid, the cover should be the same as provided for other reinforced concrete sections.

20.3. JOINTS IN WATER TANKS

The various types of joints may be categorized under three heads :

(*a*) Movement joints

(*b*) Constructions joints

(*c*) Temporary open joints.

(*a*) **Movement joints**

These require the incorporation of special materials in order to maintain water-tightness while accommodating relative moment between the side of the joints. All movement joints are essentially flexible joints. Movement joints are of three types :

(*i*) Contraction joint

(*ii*) Expansion joint

(*iii*) Sliding joint.

(*i*) ***Contraction joint***

A contraction joint is a typical movement joint which accommodates the contraction of the concrete. The joint may be either a *complete contraction joint*([Fig. 20.1(*a*)], in which there is discontinuity of both concrete and steel, or it may be *partial contraction joint* in which there is discontinuity of concrete but the reinforcements run through the joint[Fig 20.1(*b*)]. In both cases, no initial gap is kept at the joint, but only discontinuity is given during construction. In the former type a water bar is inserted while in the later type, the mouth of the joint is filled with *joint sealing compound* and then *strip painted.* A *water bar* is a pre-formed strip of impermeable material (such as a metal, polyvinyl chloride or rubber). *Joint sealing compounds* are impermeable ductile materials which are required to provide a water-tight seal by adhesion to the concrete throughout the range of joint movement. The commonly used materials are based on asphalt, bitumen, or coal tar pitch with or without fillers such as limestone or slate dust, asbestos fibre, chopped hemp, rubber or other suitable material. These are usually applied after construction or just before the reservoir is put into service by pouring in the hot or cold state, by trowelling or gunning or as preformed strips ironed into position.

(*ii*) ***Expansion joint***

It is a movement joint with complete discontinuity in both reinforcement and concrete, and is intended to accommodate either expansion or contraction of the structure [Fig. 20.1(*c*)]. In general such a joint requires the provision of an initial gap between the adjoining parts of a structure which by closing or opening accommodates the expansion or contraction of the structure. The initial gap is filled with a joint filler. *Joint fillers* are usually compressible sheet or strip materials used as spacers. They are fixed to the face of the first placed concrete and against which the second placed concrete is cast. With an initial gap of 30 mm, the maximum expansion or contraction that the filler materials may allow may be of the order of 10 mm. Joint fillers, as at present available cannot by themselves function as water-tight expansion joints. But they can only be relied upon as spacers to provide the gap in an expansion joint when the gap is bridged by a water bar.

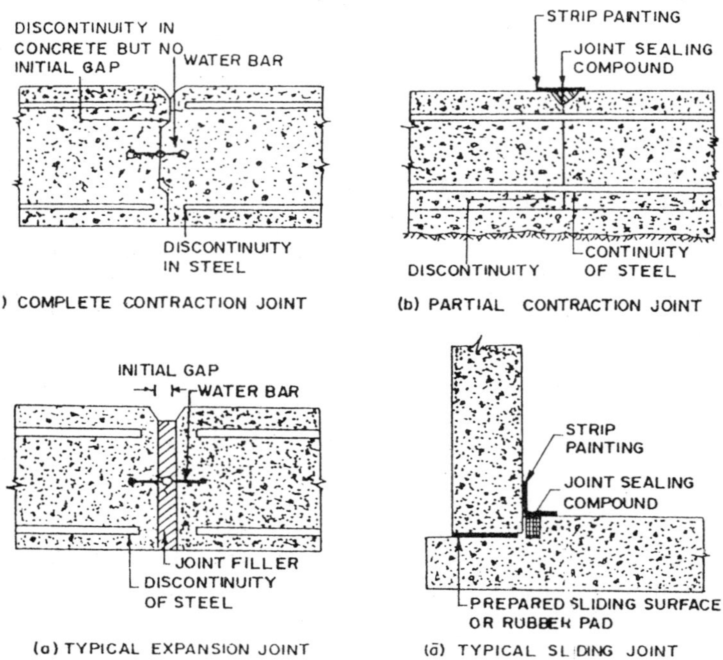

FIG. 20.1. TYPICAL MOVEMENT JOINTS

(iii) *Sliding joint*

Sliding joint is a movement joint with complete discontinuity in both reinforcement and concrete at which special provision is made to facilitate relative movement in place of the joint. A typical application of such a joint is between wall and floor in some cylindrical tank designs [Fig. 20.1(*d*)].

(*b*) Construction joints

A construction joint is a joint in the concrete introduced for convenience in construction at which special measures are takento achieve subsequent continuity without provision for further relative movement. It is, therefore, a rigid joint in contrast to a movement joint which is a flexible joint. Fig. 20.2 shows a typical construction joint between successive lifts in a reservoir wall. The position and arrangement of all construction joints should be predetermined by the engineer. Consideration should be given to limiting the number of such joints and to keeping them free from possibility of percolation in a manner similar to contraction joints.

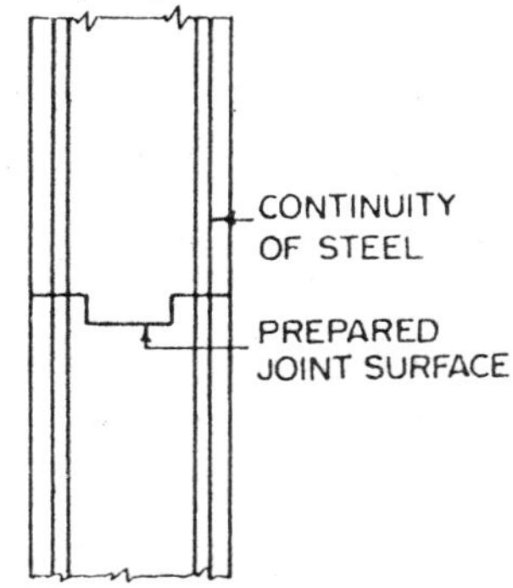

FIG 20.2 CONSTRUCTION JOINT.

(c) Temporary open joints

A temporary open joint is a gap temporarily left between the concrete of adjoining parts of a structure

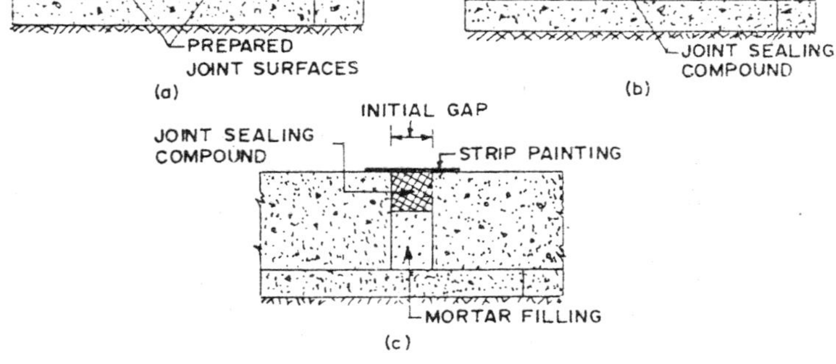

FIG. 20.3. TYPICAL TEMPORARY OPEN JOINTS.

which after a suitable interval and before the structure is put into use, is filled with mortar or concrete completely [Fig. 20.3(*a*)] or as provided below, with the inclusion of suitable jointing material [Fig. 20.3(*b*) and (*c*)]. In the former case the width of gap should be sufficient to allow the sides to be prepared before filling.

Where measures are taken for example, by the inclusion of suitable joining materials to maintain the water-tightness of the concrete subsequent to the filling of the joint, this type of joint may be regarded as being equivalent to a contraction joint (partial or complete) as defined above.

20.4. CIRCULAR TANK WITH FLEXIBLE JOINT BETWEEN FLOOR AND WALL

When water is filled in circular tank, the hydrostatic water pressure will try to increase its diameter at any section. However, this increase in the diameter all along the height of the tank will depend upon the nature of the joint at the junction *B* of the wall and bottom

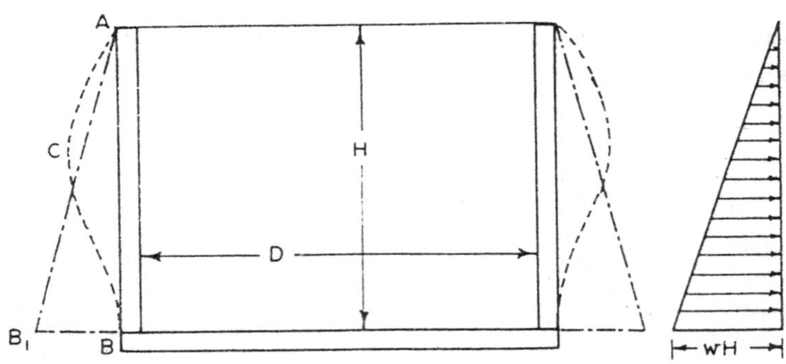

FIG. 20.4

slab. If the joint at *B* is flexible (*i.e.* sliding joint), it will be free to move outward to a position B_1. The hydrostatic pressure at *A* is zero, and hence there will be no change in the diameter at *A*. The hydrostatic pressure at *B* will be maximum, resulting in the maximum increase in diameter there, and hence maximum movement at *B* if the joint is flexible. Thus AB_1 will be the deflected position of wall *AB*. If however, the joint at *B* is fixed no movement is possible, and a fixing moment will be induced at *B*. In that case, *ACB* will be the deflected position.

To start with, we will consider the joint to be flexible so that the outward horizontal movement corresponding to the maximum horizontal pressure is possible. Due to this, hoop tension will be induced everywhere in the wall.

Let D = diameter of the tank

H = height of tank (ht. of water)

Maximum hoop tension at bottom, per unit height of wall is

$$= w H \frac{D}{2}.$$

Taking permissible stress in steel in direct tension as σ_s , area of steel per metre height at the base is given by :

$$A_{sh} = \frac{w H D}{2 \sigma_s} \qquad ...(20.1)$$

This area of steel may be provided at the centre of the wall if its thickness is small, or it may be provided on each face, keeping a minimum cover of 25 mm if the thickness is more than 225 mm. The above reinforcement is provided in the form of hoops suitably spaced, and the spacing may be increased towards the top. *However, the spacing should not be more than 3 times the thicknesss of the wall.*

Though the reinforcement has been provided to take the entire hoop tension, the concrete has not been prevented to take a part of this tension. The thickness of the wall should, therefore, be such that the tensile stress developed in the composite section is within safe limit. If σ_{ct} is the permissible tensile stress in the equivalent concrete section, and T is the thickness of the wall, we have

$$\sigma_{ct} = \frac{w H D/2}{1000 T + (m - 1) A_{sh}} \text{ (N/mm}^2\text{)} \qquad ...(20.2)$$

The value of σ_{ct} is taken as 1.2 N/mm^2 for M 20 concrete. From the above equation, the thickness T can be found. Alternatively, the thickness can be fixed from the following empirical formula and then stresses may be checked from Eq. 20.2.

$$T = 30 H + 50 \text{ mm} \qquad ...(20.3)$$

where H is the height of water (liquid) retained, in metres.

In addition to the main reinforcement, temperature and distribution reinforcement may be provided in the form of vertical bars @ 0.3% of the concrete section upto 100 mm thickness. For sections thicker than 100 mm and less than 450 mm, the minimum reinforcement may be reduced from 0.3% for 100 mm thick section to 0.2% for 450 mm thick section.

If the floor slab is *resting continuously on the ground,* a minimum thickness of 150 mm may be provided, with a nominal reinforcement of 0.3% in each direction. The slab should rest on a 75 mm thick layer of lean concrete (M 10 mix). The layer of lean concrete should first be cured, and then it should be covered with a layer of tar felt to enable the floor slab to act independent of the bottom layer of lean concrete.

Design Example 20.1. *Design a circular tank with flexible base for capacity of 400000 litres. The depth of water is to be 4 m, including a free board of 200 mm. Use M 20 concrete.*

Solution.

1. ***Dimensions of tank***

Effective depth of water from capacity point of view =3.8 m. If D is the inside diameter of the tank, we have,

$$\frac{\pi}{4} D^2 \times 3.8 = \frac{400000 \times 10^3}{10^6}$$

From which $D = \sqrt{\dfrac{400 \times 4}{\pi \times 3.8}} = 11.62$ m.

Provide a diameter of 11.7 m.

2. ***Design of section***

Take $w = 9800\ \text{N/m}^3$

Max. hoop tension $= wH\dfrac{D}{2} = 9800 \times 4 \times \dfrac{11.7}{2}$

$= 229320$ N per metre height at base.

Using $\sigma_s = 115\ \text{N/mm}^2$, area of hoop steel at base is

$$A_{sh} = \frac{229320}{115} = 1994\ \text{mm}^2/\text{m height.}$$

Using 20 mm Φ bars, $A_\Phi = 314\ \text{mm}^2$

Hence spacing of hoops $= \dfrac{1000 A_\Phi}{A_{sh}} = \dfrac{1000 \times 314}{1994} = 157$ mm

Hence provide hoops @ 150 mm c/c.

Actual $A_{sh} = \dfrac{1000 \times 314}{150} = 2093\ \text{mm}^2$

If T is thickness of wall, we have,

$$\sigma_{ct} = \frac{w H \frac{D}{2}}{1000\,T + (m-1)\,A_{sh}}$$

where σ_{ct} = permissible direct tensile stress in concrete

= 1.2 N/mm^2 for M 20 concrete

T = thickness of wall, in mm

m = modular ratio = 13 for M 20 concrete

$$\therefore \quad 1.2 = \frac{229320}{1000\,T + (13-1)\,2093}$$

$\therefore$ $1000\,T + 25116 = 191100.$

From which $T = 166$ mm.

Thickness from empirical formula $= 3H + 5 = 3 \times 4 + 5 = 17$ cm

$= 170$ mm.

Hence provide a thickness of 170 mm throughout the height. Spacing of hoops can be increased near the top. Providing a minimum reinforcement of 0.3% at the top,

$$A_{sh} = \frac{0.3}{100}(1000 \times 170) = 510 \text{ mm}^2$$

$$\text{Hence spacing of hoops} = \frac{1000 \times 314}{510} = 600 \text{ mm}$$

However, the spacing of hoops should not be more than 3 times of thickness of the wall. Hence keep a spacing of 500 mm c/c at the top. Since the thickness of wall is less than 225 mm, reinforcement will be provided at the centre of the thickness. If, however, the thickness were more than 225 mm, half the hoops would have been placed near the inner face and the other half at the outer face (*i.e.*, the spacing of the hoops would have been doubled). The spacing at the other heights can be similarly found. For example at the depth of 2 m below the top, the area of hoops is given by

$$A_{sh} = \frac{w\,h\,D/2}{115} = \frac{9800 \times 2 \times 11.7/2}{115} = 997 \text{ mm}^2$$

$$\therefore \quad \text{Spacing} = \frac{1000 \times 314}{997} = 315 \text{ mm}$$

However, keep spacing = 300 mm c/c.

3. *Vertical reinforcement*

Distribution and temperature reinforcement is provided in the vertical direction, and its area is

$$= 0.3 - 0.1 \left(\frac{170 - 100}{350} \right) = 0.28 \%$$

$$\therefore \quad A_{sd} = \frac{0.28}{100} \times 170 \times 1000 = 476 \text{ mm}^2$$

Provide 10 mm Φ bars having $A_\Phi = 78.5 \text{ mm}^2$.

$$\text{Spacing} = \frac{1000 \times 78.5}{476} = 165 \text{ mm.}$$

However, provide 10 mm Φ bars at 160 mm c/c in the vertical direction. This will also serve the purpose of tieing the hoop reinforcement.

4. ***Design of tank floor***

Since the tank floor is resting on ground throughout, provide a nominal thickness of 150 mm.

$$\text{Minimum } A_{st} = \frac{0.3}{100} \times 150 \times 1000 = 450 \text{ mm}^2 \text{ in each direction.}$$

Providing half the reinforcement near each face, $A_{st} = 225 \text{ mm}^2$

$$\text{Using 8 mm Φ bars, spacing} = \frac{1000 \times 50}{225} \approx 220 \text{ mm c/c.}$$

However, provide 8 mm Φ bars @ 200 mm c/c in both directions, at top and bottom of the floor slab. The floor slab will rest on 75 mm thick layer of lean concrete covered with a layer of tar felt.

5. ***Details of reinforcement***

The details of reinforcement etc. are shown in Fig. 20.5.

20.5. CIRCULAR TANK WITH RIGID JOINT BETWEEN FLOOR AND WALL

When the joint between the wall and floor is rigid, no horizontal displacement of the wall at the joint is possible. The deflected shape of the wall will be along *ABC* [Fig. 20.4 (*a*)]. The upper part of the wall will have hoop tension, while the lower part will bend like cantilever fixed at the joint *B*. For shallow tanks with large diameter, hoop stresses are very small and the walls act more like cantilever. For deep tanks of small diameter, the cantilever action due to fixidity at the base will be small and the hoop action is predominant. The exact analysis of the tank to determine the portion of the wall upto which hoop action will be predominant

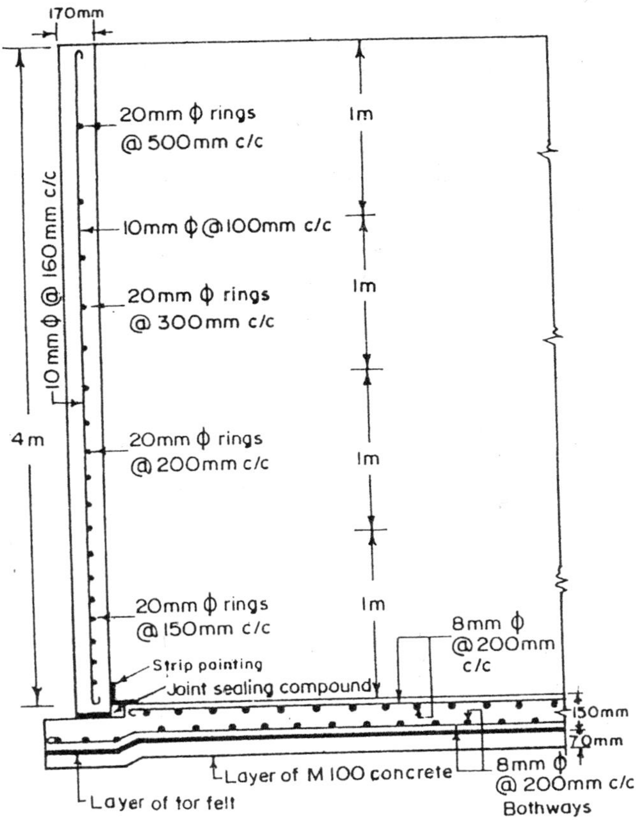

FIG. 20.5.

and to other portion in which cantilever action will be predominant, is difficult and is beyond the scope of the present volume. However, we will discuss the methods :

(1) Reissner's method

(2) Carpenter's simplified method

(3) Approximate method

(4) I.S. code method.

1. Reissner's method

Reissner gave tables from which the resultant moment M_f, the maximum tension and its location from the base are given in terms of a parameter K defined by

$$K = \frac{12H^4}{\left(\frac{D}{2}\right)^2 T^2} = \frac{48H^4}{D^2T^2} \qquad ...(20.4)$$

Thus, for a tank of given dimensions, and of assumed thickness T, the parameter K can be determined. Tables 20.3 and 20.4 give values of restraint moment M_f and the maximum ring tension, for both rectangular wall section as well as triangular wall section. To start with, T can be fixed from the following approximate formula :

$$T = (30H + 50) \text{ mm} \qquad ...(20.4)$$

Fig. 20.6(*a*) shows the deformation of the wall, Fig. 20.6(*b*) shows the load distribution along the height while Fig. 20.6(*c*) shows the bending moment diagram. In the diagram P_h denotes the load due to *hoop action* while P_c denotes the load due to cantilever action. Fig. 20.6(*c*) shows the approximate B.M. diagram. The maximum positive bending moment may be taken approximately equal to M_f / 4 to M_f / 3.

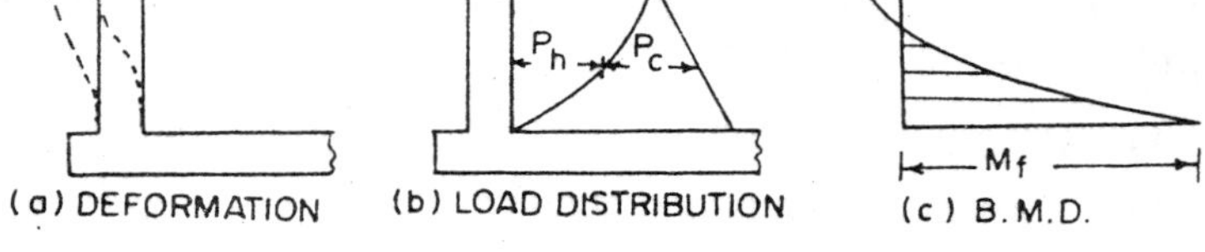

FIG. 20.6.

TABLE 20.3. REISSNER'S VALUE OF RESTRAINT MOMENT M_f
($p = wH$)

K	*Rectangular wall section*	*Triangular wall section*
0	$0.167\,pH^2$	$0.167\,pH^2$
10	$0.110\,pH^2$	$0.140\,pH^2$
100	$0.0582\,pH^2$	$0.0707\,pH^2$
1000	$0.024\,pH^2$	$0.026\,pH^2$
10000	$0.0085\,pH^2$	$0.009\,pH^2$
∞	0	0

TABLE 20.4 REISSNER'S VALUE FOR RING TENSION
($p = wH$)

K	*Rectanguar wall section*		*Triangular wall section*	
	Max. tension	*Height from base*	*Max. tension*	*Height from base*
0	0	—	—	—
10	$0.13\,p\left(\frac{D}{2}\right)$	1.0 H	$0.09\,p\left(\frac{D}{2}\right)$	0.65 H
100	$0.27\,p\left(\frac{D}{2}\right)$	1.0 H	$0.31\,p\left(\frac{D}{2}\right)$	0.58 H
1000	$0.47\,p\left(\frac{D}{2}\right)$	0.47 H	$0.52\,p\left(\frac{D}{2}\right)$	0.44 H
10000	$0.67\,p\left(\frac{D}{2}\right)$	0.31 H	$0.70\,p\left(\frac{D}{2}\right)$	0.30 H
∞	$1.0\,p\left(\frac{D}{2}\right)$	0	$1.0\,p\left(\frac{D}{2}\right)$	0

TABLE 20.5.
CARPENTER'S VALUES OF COEFFICIENTS *F* AND *K*

Factor		F				K			
	$\frac{H}{T}$ →	10	20	30	40	10	20	30	40
Values of H/D	0.2	0.046	0.028	0.022	0.015	—	0.50	0.45	0.40
	0.3	0.032	0.019	0.014	0.010	0.55	0.43	0.38	0.33
	0.4	0.024	0.014	0.010	0.007	0.50	0.39	0.35	0.30
	0.5	0.020	0.012	0.009	0.006	0.45	0.37	0.32	0.27
	1.0	0.012	0.006	0.005	0.003	0.37	0.30	0.24	0.21
	2.0	0.006	0.003	0.002	0.002	0.28	0.22	0.19	0.16
	4.0	0.004	0.002	0.002	0.001	0.27	0.20	0.17	0.14

2. Carpenter's simplified method

Carpenter simplified Dr. Reissner's method, and gave the values of maximum cantilever bending moment and the position and magnitude of maximum ring tension in terms of the following expressions :

(*i*) Position of max. hoop tension = KH above base . ...(20.6)

(*ii*) Max. hoop tension $= w\,(H - KH)\,\dfrac{D}{2} = w\,(H)\,\dfrac{D}{2}\,(1 - K)$ ...(20.7)

(*iii*) Maximum cantilever B.M. $= F\,w\,H^3$ ...(20.8)

The values of coefficient K and F depend upon H/D and H/T ratios, and may be taken from Table 20.5.

3. Approximate method

This method is followed when Reissner's table or Carpenter's tables are not available. In the approximate method, it is assumed that cantilever action will take place for a height $h = H/3$ or 1m (whichever is greater) above base for the value of $\dfrac{H^2}{DT}$ between 6 to 12 and for a height of $h = H/4$ or 1 m (whichever is more) for the value of $\dfrac{H^2}{DT}$ between 12 to 30. Above this height, hoop action will be predominant.

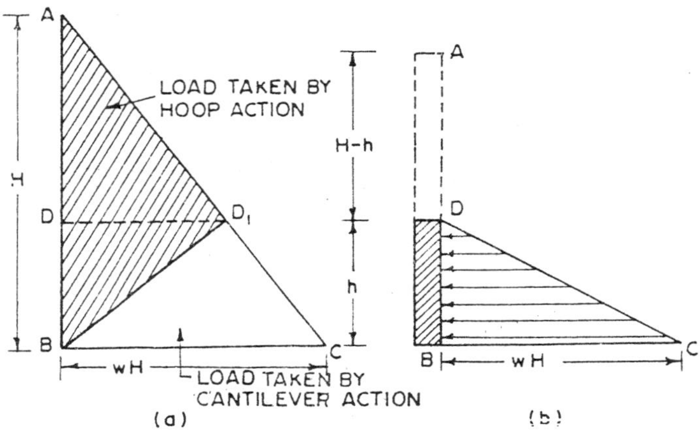

FIG. 20.7.

Thus, in Fig. 20.7, *ABC* is hydrostatic pressure diagram in which $BC = wH$. Cantilever action will take place upon the height $h = BD$ while hoop action will be predominant above this height. The maximum hoop tension is computed at level *D*, for height $(H - h)$, below *A*, where the water pressure will be $w(H - h)$. Fig. 20.7(*b*) shows the height *h* of the wall acting as cantilever fixed at base *B*. The load on the cantilever will be the area of triangular load D_1BC, having zero ordinate at *D* and maximum ordinate of *wH* at *B*. Cantilever reinforcement will be provided upto height *h*, at the *inner face*. Above this height, the spacing of the cantilever reinforcement can be increased.

The maximum hoop tension at *D*, per metre height

$$= w(H - h)\frac{D}{2} \quad ...(20.9)$$

Steel for this hoop tension is provided in form of rings provided at both the faces. The spacing of the bars for hoop tension will be kept equal upto height *h*(*i.e.,* for portion *BD*). Above this height, the spacing of the rings can be increased.

Example 20.2 *A circular tank has an internal diameter of 10 m and has maximum height of water as 4 m. The walls of the tank are restrained at the base. Determine the values of maximum hoop tension and its location, and the maximum cantilever bending moment by the following method :*

(*a*) *Reissner's method.*

(*b*) *Carpenter's method.*

Solution

(a) Reissner's method

Let thickness $T = 30H + 50 = (30 \times 4) + 50 = 170$ mm $= 0.17$m

$$\therefore \quad K = \frac{48 H^4}{D^2 T^2} = \frac{48 (4)^4}{(10 \times 0.17)^2} = 4250$$

From Table 20.3, we have

For $K = 1000, M_f = 0.024 \ pH^2$

For $K = 10000, M_f = 0.0085 \ pH^2$

Hence for $K = 4250, \quad M_f = \alpha pH^2$

where the value of α can be determined from the following logarithmic interpolation :

$$\alpha = 0.0085 + (0.024 - 0.0085) \frac{\log 10000 - \log 4250}{\log 10000 - \log 1000}$$

$$= 0.0085 + (0.024 - 0.0085) \times \frac{4 - 3.6284}{4 - 3} = 0.0143$$

$$\therefore \quad M_f = 0.0143 pH^2 = 0.0143 \ wH^3$$

$$= 0.0143 \times 9800 (4)^3 = \mathbf{8969 \ N\text{-}m}$$

For the position and value of maximum hoop tension, we have from Table 20.4 :

For $K = 1000$, Hoop tension $= 0.47 p \dfrac{D}{2}$

For $K = 10000$, Hoop tension $= 0.67 p \dfrac{D}{2}$

For $K \approx 4250$, Hoop tension $= \beta p \dfrac{D}{2}$, where the value of β may be determined from the following logarithmic interpolation :

$$\beta = 0.47 + (0.67 - 0.47) \frac{\log 4250 - \log 1000}{\log 10000 - \log 1000}$$

$$= 0.47 + (0.67 - 0.047) \frac{3.6284 - 3}{4 - 3} \approx 0.596$$

$$\therefore \quad \text{Max. hoop tension} = 0.596 p \left(\frac{D}{2}\right) = 0.596 w \frac{HD}{2}$$

$$= 0.596 \times 9800 \times 4 \times \frac{10}{2} = \mathbf{116816 \ N = 116.82 \ kN}$$

This occurs at γH above the base, where γ is given by

$$\gamma = 0.47 - (0.47 - 0.31) \times \frac{\log 4250 - \log 1000}{\log 10000 - \log 1000}$$

$$= 0.47 - 0.16\,(3.6284 - 3) \approx 0.37$$

$\therefore$ It occurs at $0.37 \times 4 = \mathbf{1.48\ m}$ above the base.

(*b*) *Carpenter's Method*

$$\frac{H}{T} = \frac{4}{0.17} = 23.5\,;\ \frac{H}{D} = \frac{4}{10} = 0.4$$

From Table 20.5. For $\frac{H}{T} = 20$ and $\frac{H}{D} = 0.4\,, F = 0.014$ and $K = 0.39$

For $\frac{H}{T} = 30$ and $\frac{H}{D} = 0.4\,, F = 0.010$ and $K = 0.35$

Hence, by linear interpolation, for $\frac{H}{T} = 23.5$ and $\frac{H}{D} = 0.4$, we have

$$F = 0.014 - \frac{(0.014 - 0.010)}{10} \times 3.5 = 0.0126$$

and

$$K = 0.39 - \frac{0.30 - 0.35}{10} \times 3.5 = 0.376$$

Hence (*i*) position of max. hoop tension $= KH = 0.376 \times 4$
$\approx$ **1.5 m** above base

(*ii*) Max. hoop tension $= wH\frac{D}{2}(1 - K)$

$$= 9800 \times 4 \times \frac{10}{2}(1 - 0.376)$$

$$= \mathbf{122300\ N} = \mathbf{122.3\ kN}$$

(iii) Max. cantilever B.M. $= F\,w\,H^3 = 0.0126 \times 9800 \times (4)^3$
$= \mathbf{7903\ N\text{-}m}$

Example 20.3. *Redesign the circular tank for the data of example 20.1, assuming that the point between the wall and the base is rigid. Use M 20 concrete. Follow the approximate method of analysis.*

Solution

1. *Dimensions of the tank*

As found earlier, $D = 11.7$ m and $H = 4$ m.

2. *Design constants*

For M 20 concrete, we have $\sigma_{cbc} = 7\ \text{N/mm}^2$ and $m = 13$. Taking $\sigma_{st} = 115\ \text{N/m}^2$, we have

$$k = \frac{m\,\sigma_{cbc}}{m\,\sigma_{cbc} + \sigma_{st}} = \frac{13 \times 7}{13 \times 7 + 115} = 0.442$$

$$j = 1 - \frac{k}{3} = 0.853$$

$$R = \tfrac{1}{2}\sigma_{cbc}\,j\,k = \tfrac{1}{2} \times 7 \times 0.853 \times 0.442 = 1.32.$$

3. *Determination of bending moment and hoop tension*

Let $T = (30\,H + 50)$ mm $= (30 \times 4) + 50 = 170$ mm $= 0.17$ m

$$\frac{H^2}{D\,T} = \frac{4 \times 4}{11.7 \times 0.17} = 8$$

The height h above base, upto which cantilever action will be there, is given by

$$h = \frac{H}{3} \text{ or } 1 \text{ m whichever is more}$$

$$\therefore \quad h = \frac{4}{3} = 1.33 \text{ m}$$

$\therefore$ Maximum ring tension at this level, per metre height

$$= w\,(H - h)\,\frac{D}{2}, \text{ where } w = 9800 \text{ N/m}^3$$

$$= 9800\,(4 - 1.33)\,\frac{11.7}{2} = 153070 \text{ N}$$

Water pressure at bottom $= w\,H = 9800 \times 4 = 39200$ N/m^2/m

$$\therefore \quad \text{Max. cantilever B.M.} = M_f = \frac{1}{2} \times 39200 \times 1.33 \times \frac{1.33}{3} \approx 11600 \text{ N-m/m}$$

4. *Design of section for cantilever action*

If d is the effective thickness of tank wall,

$$d = \sqrt{\frac{M_f}{1000\,R}} = \sqrt{\frac{11600 \times 1000}{1000 \times 1.32}} \approx 94 \text{ mm}$$

$\therefore$ Total thickness $= d +$ cover $= 94 + 35 = 130$ mm

However, provide a minimum thickness equal to the greater of the following :

(*i*) 150 mm

(*ii*) $30\,H + 50 = (30 \times 4) + 50 = 170$ mm

Hence provide $T = 170$ mm.

Providing 35 mm cover to the centre of reinforcement, available $d = 170 - 35 = 135$ mm.

Area of steel for cantilever bending is given by

$$A_{st} = \frac{M_f}{\sigma_{st}\, j\, d} = \frac{11600 \times 1000}{115 \times 0.853 \times 135} = 876 \text{ mm}^2$$

Using 12 mm Φ bars, $A_\Phi = 113 \text{mm}^2$.

$$\therefore \text{ spacing } = \frac{1000 \times 113}{876} = 129 \text{ mm}$$

However, provide 12 mm Φ bars @ 120 mm c/c up to a height of 1.33 m(or say 1.4 m) from the base. Above this height, curtail half the bars and continue the other half (*i.e.* @ 240 mm c/c) upto the top. Let us test this for development length.

$$L_d = \frac{\Phi\, \sigma_{st}}{4\, \tau_{bd}}$$

where $\Phi = 12$ mm ; $\sigma_{st} = 115 \text{ N/mm}^2$

and $\tau_{bd} = 0.8 \text{ N/mm}^2$ for M 20 concrete

$$\therefore \quad L_d = \frac{115}{4 \times 0.8}\Phi = 36\,\Phi = 36 \times 12$$

$$= 430 \text{ mm} = 0.48 \text{ m.}$$

Hence half the bars can be curtailed at 1.4 m height above the base.

These vertical bars are to be provided at inner face. Keep a clear cover of 25 mm.

5. *Design of section for hoop action*

Maximum hoop tension = 153070 N at 1.33 m above base

$$\therefore \quad \text{Area of rings, } A_{sh} = \frac{153070}{115} = 1330 \text{ mm}^2$$

Let us provide rings at both the faces. Hence area of rings on each face = 665 mm^2

Using 12 mm Φ rings, $A_\Phi = 113 \text{ mm}^2$

$$\therefore \quad \text{Spacing} = \frac{1000 \times 113}{665} \approx 170 \text{ mm.}$$

Hence provide 12 mm Φ rings @ 170 mm c/c on each face. This spacing is kept constant from the base to height $h = 1.33$ m(or say 1.4 m), and above this, the spacing may be increased.

$$\text{Actual } A_{sh} = 2 \times \frac{1000 \times 113}{170} = 1329.4 \text{ mm}^2$$

$$\therefore \quad \sigma_{ct} = \frac{153070}{1000 \times 170 + (13 - 1)\, 1329.4}$$

$$= 0.823 < 1.2 \text{ N/mm}^2 \text{. Hence safe.}$$

Hoop tension at 2 m below top $= 9800 \times 2 \times \frac{11.7}{2} = 114660$ N

$\therefore \quad A_{sh} = \frac{114660}{115} = 988 \text{ mm}^2.$

Area at each face $= 499 \text{ mm}^2.$

Spacing of rings $= \frac{1000 \times 113}{499} \approx 220$ mm

At the top, minimum $A_{sh} = \frac{0.3}{100} \times 170 \times 1000 = 510 \text{ mm}^2$, giving an area of 255 mm^2 on each face.

$\therefore \quad$ Spacing $= \frac{1000 \times 113}{225} = 443$ mm.

However, provide these @ 440 mm c/c at top.

6. *Distribution reinforcement*

Percentage area of distribution reinforcement is

$$= 0.3 - 0.1 \left(\frac{170 - 100}{350} \right) = 0.28 \%$$

$\therefore \quad A_{sd} = \frac{0.28}{100} \times 170 \times 1000 = 476 \text{ mm}^2.$

Area of steel on each face $= 238 \text{ mm}^2$. However, no additional reinforcement will be provided at the inner face since the vertical steel for cantilever action will serve this purpose.

Hence provide $A_{sd} = 238 \text{ mm}^2$ in vertical direction at the outer face. Using 8 mm Φ bars, $A_\Phi = 50.5 \text{ mm}^2$

$\therefore \quad$ Spacing $= \frac{1000 \times 50.5}{238} = 212$ mm

However, provide these @ 200 mm c/c.

7. *Provision of haunches*

It is customary to provide 150 mm × 150 mm haunches at the junction of the wall and the base. A haunch reinforcement of 8 mm Φ @ 220 mm c/c may be provided.

8. *Design of base slab*

Provide a 150 mm thick base slab with the reinforcement on both faces as indicated in Example 20.1.

The details of reinforcement etc. are shown in Fig. 20.8.

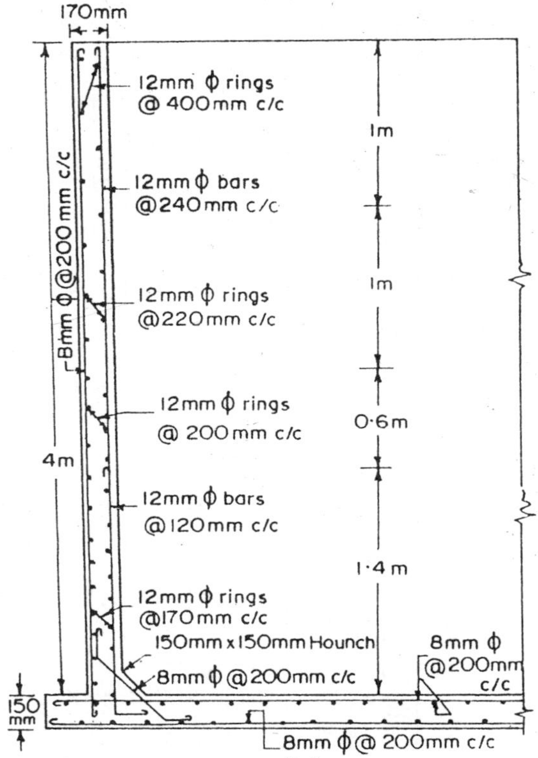

FIG. 20.8.

20.6. I.S. CODE METHOD FOR CIRCULAR TANKS

Indian Standard Code IS : 3370 (Part IV)-1967, gives design Tables for moment and hoop tension in circular tanks for various conditions of joint and various types of loading. However, we shall describe here the case of rigid joint between the wall and the base slab, subjected to triangular distributed water pressure. Table 20.6 gives the coefficient for tension at various height in the wall for various values of $\frac{H^2}{DT}$ ratio. The tension is given by the following expression :

TABLE 20.6 COEFFICIENTS FOR TENSION IN CYLINDRICAL WALL FIXED TO THE BASE

$\frac{H^2}{DT}$	Coefficient at point									
	0.0H	0.1H	0.2H	0.3H	0.4H	0.5H	0.6H	0.7H	0.8H	0.9H
0.4	+0.149	+0.134	+0.120	+0.101	+0.082	+0.066	+0.049	+0.029	+0.014	+0.004
0.8	+0.263	+0.239	+0.215	+0.109	+0.160	+0.130	+0.096	+0.063	+0.034	+0.010
1.2	+0.283	+0.271	+0.254	+0.234	+0.209	+0.180	+0.142	+0.099	+0.054	+0.016
1.6	+0.265	+0.268	+0.268	+0.265	+0.250	+0.226	+0.185	+0.134	+0.075	+0.023
2.0	+0.234	+0.251	+0.273	+0.285	+0.285	+0.274	+0.232	+0.172	+0.104	+0.031
3.0	+0.134	+0.203	+0.267	+0.322	+0.357	+0.362	+0.330	+0.262	+0.157	+0.052
4.0	+0.067	+0.164	+0.256	+0.339	+0.403	+0.429	+0.409	+0.334	+0.210	+0.073
5.0	+0.025	+0.137	+0.245	+0.346	+0.228	+0.477	+0.469	+0.398	+0.259	+0.092
6.0	+0.018	+0.119	+0.234	+0.344	+0.441	+0.504	+0.514	+0.447	+0.301	+0.112
8.0	−0.001	+0.104	+0.218	+0.355	+0.445	+0.534	+0.575	+0.530	+0.381	+0.151
10.0	−0.003	+0.098	+0.208	+0.323	+0.437	+0.542	+0.608	+0.589	+0.440	+0.179
12.0	−0.005	+0.097	+0.202	+0.312	+0.429	+0.543	+0.628	+0.633	+0.494	+0.211
14.0	−0.002	+0.098	+0.200	+0.306	+0.420	+0.539	+0.639	+0.666	+0.541	+0.241
16.0	−0.000	+0.099	+0.199	+0.304	+0.412	+0.531	+0.641	+0.687	+0.582	+0.265

Note : Positive sign indicates tension

TABLE 20.7 COEFFICIENT FOR MOMENT IN CYLINDRICAL WALL FIXED TO THE BASE

$\frac{H^2}{DT}$	Coefficient at point									
	0.1H	0.2H	0.3H	0.4H	0.5H	0.6H	0.7H	0.8H	0.9H	1.0H
0.4	+0.005	+0.0014	+0.0021	+0.0007	−0.0042	−0.0150	−0.0302	−0.0529	−0.0816	−0.1205
0.8	+0.0011	+0.0037	+0.0063	+0.0080	−0.0070	+0.0023	−0.0068	−0.0024	−0.0465	−0.0795
1.2	+0.0012	+0.0042	+0.0077	+0.0103	+0.0112	+0.0090	+0.0022	−0.0108	−0.0311	−0.0602
1.6	+0.0011	+0.0041	+0.0075	+0.0107	+0.0121	+0.0111	+0.0058	−0.0051	−0.0232	−0.0505
2.0	+0.0010	+0.0035	+0.0068	+0.0099	+0.0120	+0.0115	+0.0075	−0.0021	−0.0185	−0.0436
3.0	+0.0006	+0.0024	+0.0047	+0.0071	+0.0090	+0.0097	+0.0077	+0.0012	−0.0119	−0.0333
4.0	+0.0003	+0.0015	+0.0028	+0.0047	+0.0066	+0.0077	+0.0069	+0.0023	−0.0080	−0.0268
5.0	+0.0002	+0.0008	+0.0016	+0.0029	+0.0046	+0.0059	+0.0059	+0.0028	−0.0058	−0.0222
6.0	+0.0001	+0.0003	+0.0008	+0.0019	+0.0032	+0.0046	+0.0051	+0.0029	−0.0041	−0.0187
8.0	0.0000	+0.0001	+0.0002	+0.0008	+0.0016	+0.0028	+0.0038	+0.0029	−0.0022	−0.0146
10.0	0.0000	+0.0000	+0.0001	+0.0004	+0.0007	+0.0019	+0.0029	+0.0028	−0.0012	−0.0122
12.0	0.0000	−0.0001	+0.0001	+0.0002	+0.0003	+0.0013	+0.0023	+0.0026	−0.0005	−0.0104
14.0	0.0000	0.0000	0.0000	0.0000	−0.0001	+0.0008	+0.0019	+0.0022	−0.0001	−0.0090
16.0	0.0000	0.0000	−0.0001	−0.0002	−0.0001	+0.0004	+0.0013	+0.0019	+0.0001	−0.0079

Note : Positive sign indicates tension on outside

Tension $= \text{coefficient} \times w H \frac{D}{2}$ per metre ...(20.10)

Table 20.7 give the coefficients for bending moment at various heights in the wall for various values of $\frac{H^2}{DT}$ ratio. The moment is given by following expression :

Moment $= \text{coefficient} \times w H^2$ N–m/m ...(20.11)

The shear at the base of cylindrical wall for the case of triangular load can be determined from the following expression :

Shear $= \text{coefficient} \times w H^2$ kN ...(20.12)

where the coefficient can be taken from Table 20.8.

TABLE 20.8. COEFFICIENTS FOR SHEAR AT THE BASE

$\frac{H^2}{DT}$	*Coefficient*	$\frac{H^2}{DT}$	*Coefficient*
0.4	+0.436	5.0	+0.213
0.8	+0.374	6.0	+0.197
1.2	+0.339	8.0	+0.174
1.6	+0.317	10.0	+0.158
2.0	+0.299	12.0	+0.145
3.0	+0.262	14.0	+0.135
4.0	+0.236	16.0	+0.127

Example 20.4. *For the data of example 20.2, determine the maximum hoop tension and its position, moment at the base and shear at the base using I.S. code tables. Assume thickness of wall as 160 mm.*

Solution Thickness $T = 160$ mm $= 0.16$ m

Hence $\frac{H^2}{DT} = \frac{(4)^2}{10 \times 0.16} = 10$

From Table 20.6, for $\frac{H^2}{DT} = 10$, maximum tension may be assumed to occur at $0.6\,H = 2.4$ m from top, and coefficient may be taken as 0.608. Hence maximum tension $= 0.608 \times w H \frac{D}{2}$

$= 0.608 \times 9800 \times 4 \times \frac{10}{2} = 119168$ N. For exact location of maximum tension and its magnitude, the values of tension all along the height can be plotted by coefficients from Table 20.6. From the curve so obtained maximum value can be found.

From Table 20.7 for $\frac{H^2}{DT} = 10$, the moment coefficient for base (1.0H) is found to be -0.0122 (minus sign indicating tension at the inner face).

$\therefore$ Moment $= 0.0122 \times wH^3 = 0.0122 \times 9800\,(4)^3$

$= \mathbf{7652}$ **N-m/m.**

From Table 20.8, for $\frac{H^2}{DT} = 10$, the coefficient for shear is found to be 0.158.

Hence shear $= 0.158 \times wH^2 = 0.158 \times 9800\,(4)^2$

$= 24774$ N (acting inward)

20.7. RECTANGULAR TANKS

For tanks of smaller capacity, the cost of shuttering for circular tanks becomes high. Rectangular tanks are, therefore, used in such circumstances. However, rectangular tanks are normally not used for large capacities since they are uneconomical and also its exact analysis is difficult. For a given capacity, perimeter is least for circular tank.

The walls of a rectangular tank are subjected to bending moments both in the horizontal as well as in vertical direction. The analysis of moment in the wall is difficult since water pressure results in a triangular load on them. The magnitude of the moment will depend upon several factors such as length, breadth and height of tank, and the conditions of support of the wall at top and bottom edges. If the length of the wall is more in comparison to its height the moments will be mainly in the vertical direction *i.e.* the panel will bend as a cantilever. If, however, height is large in comparison to length, the moments will be in the horizontal direction, and the panel will bend as a thin slab supported on the edges. For intermediate condition bending will take place both in horizontal as well as vertical directions. In addition to the moments, the walls are also subjected to direct pull exerted by water pressure on some portion of side walls. The wall of the tank will thus be subjected to both bending moment as well as direct tension. The design of the walls is done on the premise that no cracks are developed in it. Though reinforcement is provided both for moments as well as direct tension, the maximum permissible value of tensile stresses for M 20 concrete may be taken as 1.2 N/mm^2 and 1.7 N/mm^2 respectively for direct tension and due to bending.

The exact analysis of rectangular tanks is beyond the scope of the present volume. An *approximate method* is discussed below

For *detailed analysis, the reader is referred to Author's Vol. II of this book.*

APPROXIMATE METHOD

For the design by approximate method, rectangular tanks may be divided into two categories : (*a*) tanks in which ratio of length to breadth is less than 2 ; (*b*) tanks in which ratio of length to breadth is more than 2.

(*a*) Ratio L/B less than 2

Let L be the length of the tank and B be the width of the tank from inside, and H be its height. If the ratio L/B is less than 2, the tank walls are designed as continuous frame subjected to a triangular load. Such a bending is assumed to take place from top to a height $h = \frac{H}{4}$ or 1 m above base (whichever is more). For the bottom height h, bending is assumed to act in vertical plane as cantilever, subjected to triangular load $D_1 \, B \, C$, having zero intensity at D_1 and wH at the base, as shown in Fig. 20.9(*a*).

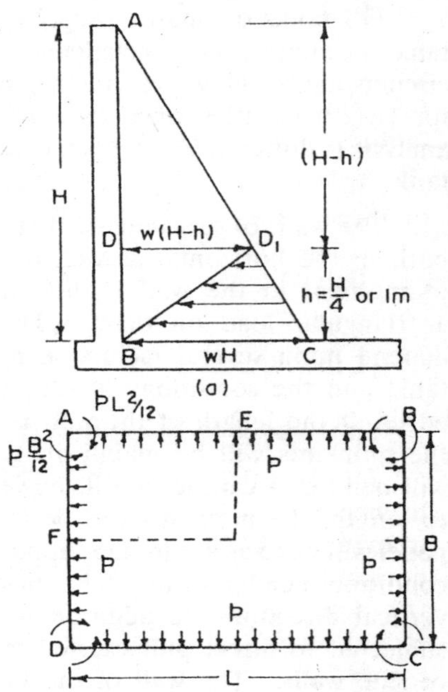

FIG. 20.9.

For horizontal bending, the maximum horizontal force per unit height, at D is taken as equal to $p = w\,(H - h)$ per metre run. The panels are first assumed to be fixed at the supports and fixed end moments are calculated. The final moments are then computed by moment distribution. Thus, in Fig. 20.9(*b*), if p is water pressure at level D [Fig. 20.9 (a)] the fixing moments for long slab AB will be $\frac{pL^2}{12}$ while fixing moments for short slab AD will be $\frac{pB^2}{12}$. Moment distribution can be carried out for the quarter frame EAF, with fixed corner at A. After having found the moment M_f at A, the moment at the centre of AB may be taken equal to $\left(\frac{pL^2}{8} - M_f\right)$. Similarly, moment at the centre of the short span may be taken equal to $\frac{pB^2}{8} - M_f$.

In the absence moment distribution, the bending moments may be computed by the following approximate expressions :

B.M. at centre of span $= \frac{pB^2}{16}$ (producing tension on outer face) ...(20.13)

B.M. at ends of span $= \frac{pB^2}{12}$ (producing tension on water face) ...(20.14)

The bottom portion of height h of each panel is designed as cantilever, in a manner similar to circular tank fixed at base.

In addition to the bending moments, the walls are subjected to direct tension of following values :

Direct tension on long walls $= P_L = w\,(H - h)\,\frac{B}{2}$ ...(20.15)

Direct tension on short walls $= P_B = w\,(H - h)\,\frac{L}{2}$ ...(20.16)

(b) Ratio L/B greater than 2

If the ratio L/B is greater than 2, the long walls are *assumed* to bend vertically as cantilever fixed at base and subjected to triangularly distributed load. The short wall is assumed to bend horizontally, supported on long walls, for the portion from top to the point D [Fig. 20.9(a)]. The load intensity for such a bending is taken as $p = w\,(H - h)$. The bottom portion of height $h = \frac{H}{4}$ or 1 m (whichever is more) the short wall is designed as cantilever subject to triangular load D_1BC of Fig. 20.9(a).

Thus, for long wall, maximum B.M. at the base, per unit length of the wall $\frac{w H^3}{6}$.

For short walls, the maximum bending moment at level D, at ends and centre may be taken equal to $\frac{w (H-h) B^2}{16}$

Similarly, the maximum cantilever B.M. for short wall is equal to $w H \frac{h}{2} \cdot \frac{h}{3} = \frac{w H h^2}{6}$

In addition to above bending moments, the long and short walls are subjected to direct pulls. Since the short walls are assumed to be supported on long walls, at its ends, the water pressure acting on the short walls, at the level D, is transferred to long walls thus causing pull in it. The direct tension, so developed. on long walls is given by

$$P_L = w (H-h) . \frac{B}{2} \qquad ...(20.15)$$

The long walls behave as cantilever, and hence they do not transfer any water load to short wall in the form of pull. However, it is observed that some portion of long walls near the corners transfer its water load to short walls. It is assumed that the end one metre width of long wall contributes to pull or direct tension in the short walls, its magnitude being given by

$$P_B = w (H-h) \times 1 .$$

Thus we see that long wall is subjected to cantilever bending moment as well as direct pull. The reinforcement due to direct pull is provided in the horizontal direction while the reinforcement due to cantilever action is provided in the vertical direction on the water face. However, short walls bend horizontally for major portion of its height and hence reinforcement provided for B.M. is in the horizontal direction. The reinforcement provided for direct tension is also in the horizontal direction. Hence the section is to be designed for combined bending and direct pull effect. The approximate method of design is given below.

Design of section subjected to combined effects of bending and pull

We have seen that walls of rectangular tanks of L/B ratio less than 2, and the short walls of rectangular tanks having L/B greater than 2, have reinforcements for B.M. and pull in horizontal direction.

Fig. 20.10 shows such a section of wall subjected to axial pull P and bending moment M. The nature of the B.M. is such that tension is induced to the right side of the axis. Let the distance of reinforcement from the central axis be x. Assume equal and opposite force P (pulls) acting along the reinforcement. The section may then be considered to be subjected to a net B.M. equal to $(M - Px)$ and a pull P in the reinforcement. Areas of steel are separately calculated for B.M. and pull by the following expression :

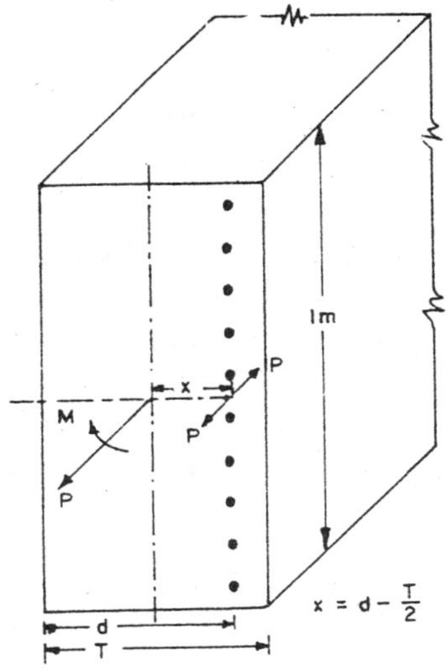

FIG. 20.10.

For B.M., $$A_{st1} = \frac{M - Px}{t_1 j d} = \frac{M - Px}{\sigma_{st} j d} \quad ...(20.17)$$

For pull, $$A_{st2} = \frac{P}{t_2} = \frac{P}{\sigma_{sh}} \quad ...(20.18)$$

where $t_1 (= \sigma_{st})$ is the safe stress in steel in bending while $t_2 = (\sigma_{sh})$ is the safe stress in steel in direct pull. The value of t_2 is taken as 115 N/mm^2 as usual. The value of tensile stress in bending (t_1) is taken equal to 125 N/mm^2 if it is placed on the water face and 125 N/mm^2 if it is used more than 225 mm away from water.

Total steel $A_{st} = A_{st1} + A_{st2}$

Example 20.5. *An open rectangular tank 4m×6m×3m deep rests on firm ground. Design the tank. Use M 20 mix.*

Solution.

1. ***Design constants***

For M 20 concrete, we have $\sigma_{cbc} = 7$ N/mm^2 and $m = 13$. Taking $\sigma_{st} = 115$ N/mm^2 ,we have,

$k = 0.442 \quad j = 0.853 \quad R = 1.32.$

2. ***Determination of B.M. for horizontal bending***

$$\frac{L}{B} = \frac{6}{4} = 1.5 < 2$$

Hence both long and short walls will bend horizontally for the upper portion, upto point D [Fig. 20.9(a)], where horizontal water pressure is $p = w\,(H - h)$.

Here $h = \frac{H}{4}$ or 1 m whichever is greater.

$\therefore \quad h = 1$ m.

Thus top 2 m height of walls will bend horizontally while the bottom 1m will bend as vertical cantilever. The bending moments for horizontal bending may be determined by moment distribution by considering tank as continuous frame or unit height at level of D.

Water pressure p at point D is given by

$$p = w\,(H - h) = 9800\,(3-1) = 19600 \text{ N-m}$$

The fixed end moments for long wall

$$= \frac{p\,L^2}{12} = \frac{p(6)^2}{12} = 3p \text{ N-m}$$

Fixed end moments for short wall

$$= \frac{p\,B^2}{12} = \frac{p\,(4)^2}{12} = \frac{4}{3}p \text{ N-m.}$$

Refer Fig. 20.9(b). Consider quarter frame FAE with joint A as rigid. Taking clockwise moment as positive and anticlockwise moment as negative, the fixed end moment M_{AF} for long wall will be $+3p$ while the fixed end moment M_{AF} for short wall will be $-\frac{4}{3}p$. Considering area A and moment of inertia I for both the walls to be the same, the stiffness of walls will be inversely proportional to these lengths. Thus we have following table :

Member	***Stiffness***	***Relative Stiffness***	***Sum***	***Distribution factor***
AE	$\frac{1}{3}$	$\frac{1}{3} \times 6 = 2$	5	$\frac{2}{5} = 0.4$
AF	$\frac{1}{2}$	$\frac{1}{2} \times 6 = 3$		$\frac{3}{5} = 0.6$

The moment distribution is carried out in the following Table :

Joint	*A*	
Member	*AE*	*AF*
Distribution Factors	0.4	0.6
Fixed end moments	$+3p$	$-\frac{4}{3}p$
Balancing moments	$-\frac{2}{3}p$	$-p$
Final moments (M_f)	$+\frac{7}{3}p$	$-\frac{7}{3}p$

Hence moment at supports, $M_f = \frac{7}{3}p = \frac{7}{3} \times 19600$

$$\approx 45740 \text{ N-m/m.}$$

This support moments will cause tension at the water force.

B.M. at the centre long span $= \frac{pL^2}{8} - M_f$

$$= \frac{19600\,(6)^2}{8} - 45740 = 42460 \text{ N-m/m.}$$

This B.M. will cause tension at the outer face. B.M. at the centre of short span $= \frac{pB^2}{8} - M_f$

$$= \frac{19600\,(4)^2}{8} - 45740 = -6540 \text{ N–m/m.}$$

This will cause tension at the water face.

$\therefore$ Max. design B.M. $= 45740$ N–m/m.

3. *Design of section*

Considering bending effect alone, effective depth

$$d = \sqrt{\frac{45740 \times 1000}{1.32 \times 1000}} = 185 \text{ mm}$$

Provide an overal depth $T = 220$ mm so that effective depth $= 220 - 3.5 = 185$ mm.

4. *Determination of pull*

Direct tension on long wall

$$= P_L = p \times \frac{B}{2} = 19600 \times \frac{4}{2} = 39200 \text{ N.}$$

Direct tension on short wall

$$= P_B = p\,\frac{L}{2} = 19600 \times \frac{6}{2} = 58800 \text{ N.}$$

5. ***Cantilever moment***

Cantilever moment at the base, per unit height

$$= w\,H\frac{h^2}{6} = 9800 \times 4\frac{(1)^2}{6} \approx 6535 \text{ N-m}$$

This will cause tension on the water face.

6. ***Reinforcement for long walls.***

The upper portion of long walls is subjected to both bending in horizontal direction as well as pull. The reinforcement for both will be in horizontal direction. Hence reinforcement has to be provided for a net moment $(M_f - P\,x)$, where M_f is the moment at ends(causing tension on the water face). Similarly reinforcement has to be provided for pull P_L in the horizontal direction.

Fig. 20.11 shows vertical section of unit height(1m) of long wall, at its end, at the level of 1m above the base, where reinforcement is provided at the water face.

$$x = d - \frac{T}{2} = 185 - \frac{220}{2} = 75 \text{ mm}$$

$$\therefore\ A_{st1} \text{ for B.M.} = \frac{M_f - P_L\,x}{\sigma_{st}\,j\,d}$$

$$= \frac{(45740 \times 1000) - 39200 \times 75}{115 \times 0.853 \times 185}$$

$$= 2359 \text{ mm}^2$$

$$A_{st} \text{ for pull} = \frac{P_L}{\sigma_s}$$

$$= \frac{39200}{115} = 341 \text{ mm}^2$$

$$\text{Total } A_{st} = A_{st1} + A_{st2}$$

$$= 2359 + 341 = 2700 \text{ mm}^2$$

per meter height.

Using 20 mm Φ bars, $A_\Phi = 314 \text{ mm}^2$

$$\therefore\ \text{spacing} = \frac{1000 \times 314}{2700} = 116 \text{ mm}$$

Hence provide 20 mm Φ bars @ 110 mm c/c.

The above reinforcement is to be provided at the inner face, near the corners, and at a height 1 m above the base.

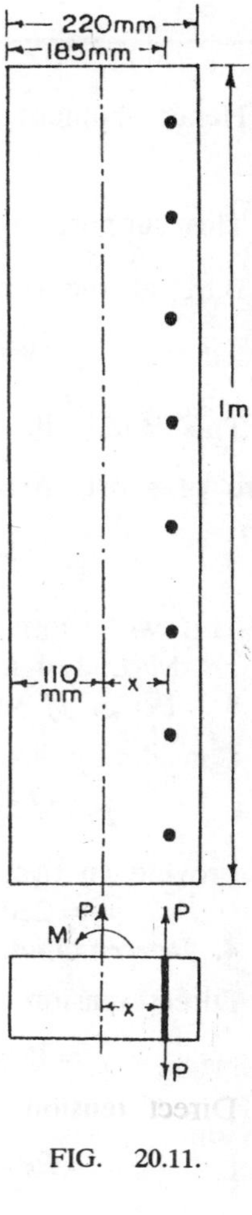

FIG. 20.11.

For other heights the above spacing may be varied, since bending moment will be reduced.

6. *Reinforcement at the middle of long walls*

Tension occurs at the outer face. However, since the distance of centre of steel from water face will be less than 225 mm, permissible stress in steel will be 115 N/mm^2 only. Hence the same design constants, found in step 1 will be used.

Design B.M. = 42460 N-m per metre height

Also P_L = 39200 N

$$A_{st1} = \frac{M - P_L x}{\sigma_{st} jd} = \frac{(42460 \times 1000) - 42460 \times 75}{115 \times 0.853 \times 185}$$

$$= 2164 \text{ mm}^2$$

$$A_{st2} = \frac{P_L}{\sigma_s} = \frac{39200}{115} = 341 \text{ mm}^2$$

Total $A_{st} = 2164 + 341 = 2505$ mm^2 per meter height.

This is very near to the reinforcement provided at the ends. Hence provide 20 mm Φ bars @ 110 mm c/c. Bend half the bars provided at ends, outwards, at distance of $\frac{L}{4} = 1.5$ m from ends. Provide additional 20 mm Φ bars @ 220 mm c/c. This reinforcement is to be provided at the outer face. The additional 20 mm Φ bars provided @ 220 mm c/c are to continued upto the end.

7. *Reinforcement for short walls*

B.M. at ends $= M_f = 45740$ N-m

Direct pull P_B = 58800 N

$$A_{st1} \text{ (for moment)} = \frac{M_f - P_B x}{\sigma_{st} jd} = \frac{45740 \times 1000 - 58800 \times 75}{115 \times 0.853 \times 185}$$

$$= 2278 \text{ mm}^2.$$

$$A_{st2} \text{ (for pull)} = \frac{58800}{115} = 511 \text{ mm}^2$$

$$\text{Total } A_{st} = A_{st1} + A_{st2} = 2278 + 511 = 2789 \text{ mm}^2$$

$$\text{Spacing of 20 mm } \Phi \text{ bars} = \frac{1000 \times 314}{2789} = 112 \text{ mm}$$

Hence provide 20 mm Φ bars @ 110 mm c/c at the inner face near the ends of short span.

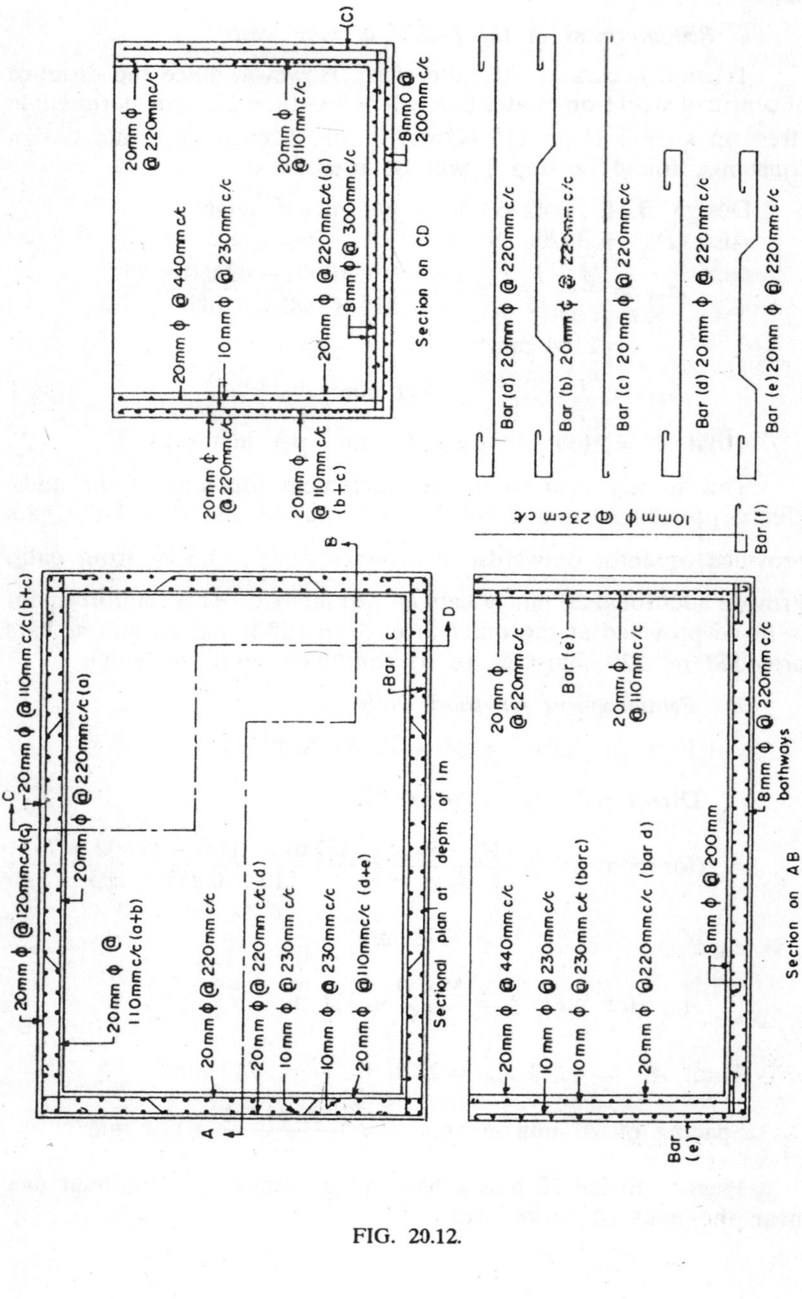
20mm φ @ 120mm c/c (a)
20mm φ @ 110mm c/c (b+c)
20mm φ @ 220mm c/c (a)
20mm φ @ 110mm c/c (a+b)
20mm φ @ 220mm c/c
20mm φ @ 220mm c/c (d)
10mm φ @ 230mm c/c
10mm φ @ 230mm c/c
20mm φ @ 110mm c/c (d+e)
Bar c
A
B
C
D
Sectional plan at depth of 1m
20mm φ @ 440mm c/c
10mm φ @ 230mm c/c
20mm φ @ 220mm c/c (a)
8mm φ @ 300mm c/c
20mm φ @ 220mm c/c
20mm φ @ 110mm c/c (b+c)
20mm φ @ 220mm c/c
20mm φ @ 110mm c/c
8mm φ @ 200mm c/c
(C)
Section on CD
Bar (a) 20mm φ @ 220mm c/c
Bar (b) 20mm φ @ 220mm c/c
Bar (c) 20mm φ @ 220mm c/c
Bar (d) 20mm φ @ 220mm c/c
Bar (e) 20mm φ @ 220mm c/c
10mm φ @ 23cm c/c
Bar (f)
20mm φ @ 440mm c/c
10 mm φ @ 230mm c/c
10mm φ @ 230mm c/c (bar c)
20mm φ @ 220mm c/c (bar d)
8mm φ @ 200mm
20mm φ @ 220mm c/c
Bar (e)
20mm φ @ 110mm c/c
8mm φ @ 220mm c/c bothways
Section on AB
Bar (e)

FIG. 20.12.

The B.M. at the centre of short walls causes tension at the water face (unlike that in the centre of long walls where tension is produced at the outer face). Since this B.M. is small, only nominal reinforcement is required. Similarly, we have to provide nominal reinforcement at the outer face. Hence bend half the bars outward at distance of $\frac{B}{4} = 1$ m from each end, and continue remaining half throughout. Thus, at the centre of span, the reinforcement on each face will consist of 20 mm Φ bars @ 220 mm c/c.

8. ***Reinforcement for cantilever moment and distribution reinforcement***

Max. cantilever moment $= 6535$ N-m

$$A_{st} = \frac{6535 \times 1000}{115 \times 0.853 \times 115} = 360 \text{ mm}^2.$$

But minimum reinforcement in vertical direction a 0.3% area

$$= \frac{0.3}{100} \times (220 \times 1000) = 660 \text{ mm}^2.$$

Since half of this area of steel can resist cantilever moment, we will provide 330 mm^2 steel area vertically on the inner face and the remaining area *i.e.* 330 mm^2 vertically at the outer face to serve as distribution reinforcement.

∴ Area of steel on each face $= 330 \text{ mm}^2$.

Using 10 mm Φ bars, spacing $= \frac{1000 \times 78.5}{330} = 238$ mm.

9. ***Design of base slab***

Since the tank rests on ground, provide a 100 mm thick base slab. Provide nominal reinforcement of 8 mm Φ bars @ 220 mm c/c in both directions, at top and bottom of base slab, provided in example 20.6.

The details of reinforcement etc. are shown in Fig. 20.12.

20.8. UNDERGROUND AND OVERHEAD TANKS

When circular and rectangular tanks are situated underground, the walls of the tanks are to be designed for earth pressure, as well as water pressure acting separately, and also acting simultaneously. Similarly, the floors of the tanks are to be designed for hydrostatic water pressure(if water table is higher) acting upwards.

If the tanks are supported on staging or on elevated walls, wind pressure etc. is to be taken into account. The detailed treatment of such tank, as well as for special types of tanks such as Intze tank etc. is given in Author's Vol. II of this book.

PROBLEMS

1. Design a circular water tank of capacaity 200000 litres. The depth of tank is limited to 3 m from inside. Keep the joint between the wall and base slab as flexible. The base slab rests on the ground.
2. Redesign the tank of problem 1 if the joint between wall and base is to be made rigid. Use approximate method.
3. A circular tank has 12 m diameter and 3 m water height. Determine (*i*) maximum hoop tension and its location and (*ii*) maximum bending moment. Use the following methods : (*a*) Reissner's method. (*b*) Carpenter's method (*c*) I.S. Code method.
4. Design a square water tank 5 m × 5 m × 3 m high. The tank is open at top and the walls are rigidly fixed to the base which rests on firm ground.
5. Design a rectangular tank 5 m × 8 m × 3 m deep. The tank is open at top and the walls are rigidly fixed to the base which rests on firm ground.
6. An open rectangular tank 3 m wide and 6 m long has a depth of 3 m from inside. The tank rests on firm ground. Design the tank. Use M 20 mix.

21

Formwork

21.1. INTRODUCTION

The *form work* or *shuttering* is a temporary ancillary construction used as a mould for the structure, in which concrete is placed and in which it hardens and matures. The construction of formwork involves considerable expenditure of time and material. The cost of formwork may be upto 15 to 25% of the cost of structure in building work and even higher in bridges. In order to reduce this expenditure, it is necessary to design economical types of formwork and to mechanize its construction. When the concrete has reached a certain required strength, the formwork is no longer needed and is removed. The operation of removing the formwork is commonly known as *stripping*. When stripping takes place, the components of formwork are removed and then reused for the forms of another part of the structure. Such forms, whose components can be reused several times are known as *panel forms. In contrast to this are stationary forms* which are made for individual non standard member and structures, which have no repeatable elements, and also for structural members the formwork of which cannot be stripped.

Forms are classified as wooden, plywood, steel, combined wood steel, reinforced concrete and plain concrete. Timber is the most common material used for formwork. The disadvantage of wooden formwork is the possibility of warping, swelling and shrinkage of the timber. However, these defects can be overcome by applying to the shuttering, water-impermeable coatings. This coating also prevents the shuttering from adhering to concrete and hence makes stripping easier. Steel shuttering is used for major work where every things is mechanised. Steel formwork has many advantages, such as follows (*i*) it can be put to high number of uses (*ii*) It provides ease of stripping (*iii*) it ensures an even and smooth concrete surface (*iv*) it posseses greater rigidity (*v*) it is not liable to shrinkage or distortion. However, steel formwork is comparatively more costly.

A good formwork should satisfy the following requirements :

(*i*) The material of the formwork should be cheap and it should be suitable for re-use several times.

(*ii*) It should be practically waterproof so that it does not absorb water from concrete. Also, its shrinkage and swelling should be minimum.

(*iii*) It should be strong enough to withstand all loads coming, on it, such as dead load of concrete and live load during its pouring, compaction and curing.

(*iv*) It should be stiff enough so that deflection is minimum.

(*v*) It should be as light as possible.

(*vi*) The surface of the formwork should be smooth. and it should afford easy stripping.

(*vii*) All joints of the formwork should be stiff so that lateral deformation under loads is minimum. Also, these joints should be leakproof.

(*viii*) The formwork should rest on non-yielding supports.

21.2. INDIAN STANDARD ON FORMWORK(IS : 456-1978)

1. ***General.*** The formwork shall conform to the shape, lines and dimensions as shown on the plans and be so constructed as to remain sufficiently rigid during the placing and compacting of the concrete, and shall be sufficiently tight to prevent loss of liquid from the concrete.

2. ***Cleaning and Treatment of forms.*** All rubbish, particularly chippings, shavings and sawdust, shall be removed from the interior of the forms before the concrete is placed and the formwork in contact with the concrete shall be cleaned and throughly wetted or treated with an approved composition. Care shall be taken that such approved composition is kept out of contact with the reinforcement.

4. ***Stripping time.*** In no circumstances shall forms be struck until the concrete reaches a strength of atleast twice the stress to which the concrete may be subjected at the time of striking.

The strength referred to shall be that of concrete using the same cement and aggregate, with the same proportions ; and cured under condition of temperature and moisture similar to those existing on the work. Where possible, the formwork should be left longer, as it would assist the curing.

In normal circumstances (generally where temperatures are above 20°C), and where ordinary cement is used, forms may be struck after expiry of the following periods :

(*a*) Walls, columns and vertical sides of beams	24 to 48 hours as may be decided by the engineer-in-charge.
(*b*) Slabs (props left under)	3 days.
(*c*) Beam soffits(props left under)	7 days.
(d) Removal of props to slab :	
(*i*) Spanning upto 4.5 m	7 days.
(*ii*) Spanning over 4.5 m	14 days.
(*e*) Removal of props to beams and arches :	
(*i*) Spanning upto 6 m	14 days.
(*ii*) Spanning over 6 m	21 days.

For rapid hardening cement 3/7 of the above period will be sufficient in all cases except vertical sides of slabs, between and columns which should be retained for 24 hours.

Note. The number of props, their sizes and disposition, shall be such as to be able to safely carry the full load of the slab, beam or arch as the case may be.

4. ***Procedure when Removing the Formwork.*** All formwork shall be removed without such shock or vibration as would damage the reinforced concrete. Before the soffit and struts are removed, the concrete surface shall be exposed, where necessary in order to ascertain that the concrete has sufficiently hardened. Proper precautions shall be taken to allow for the decrease in the rate of hardening that occurs with all cements in the cold water.

5. ***Camber.*** It is generally desirable to give forms an upward camber to ensure that the beams do not have a sag when they

have taken up their deflection, but this should not be done unless allowed for in design calculations of the beams.

6. *Tolerances.* Formwork shall be so constructed that the internal dimensions are within the permissible tolerance specified by the designer.

TABLE 21.1. PERMISSIBLE STRESSES IN TIMBER

Property	*Types of timber*			
	Fir	*Deodar*	*Kail*	*Chir*
Density (N/m²)	4500	5450	5150	5750
Modulus of elasticty E (N/mm²)	9400	9500	6800	9800
Permissible stresses in bending and tension (N/mm²)				
(*i*) Inside	7.8	10.2	6.6	8.4
(*ii*) Outside	6.6	8.8	5.6	7.0
(*iii*) Wet	5.6	7.0	5.0	6.0
Permissible stresses in shear (N/mm²)				
(*i*) Horizontal	0.6	0.7	0.6	0.4
(*ii*) Along grain	0.8	1.0	0.8	0.9
Permissible compresive stresses [Parallel] (N/mm²)				
(*i*) Inside	6	7.8	5.2	6.4
(*ii*) Outside	5.2	7.0	4.6	5.6
(*iii*) Wet	4.2	5.6	3.8	4.5
Permissible compressive stresses [perpendicular] (N/mm²)				
(*i*) Inside	1.6	2.6	1.7	2.2
(*ii*) Outside	1.2	2.1	1.3	1.7
(*iii*) Wet	1.0	1.7	1.0	1.4

21.3. LOADS ON FORMWORK

The formwork has to bear mainly the following loads apart from its own weight : (*i*) live load due to labour etc. (*ii*) dead weight of wet concrete (*iii*) hydrostatic pressures of the fluid concrete acting against the vertical or inclined faces of form, and (*iv*) impact due to pouring concrete. The temporary live loads of workmen and equipment, including the impact, may be taken equal to 3700 N/mm^2, for the design of planks and joists in bending and shear. The hydrostatic pressure due to fluidity of concrete in the initial stages of pouring depends upon several factors such as quantity of water in concrete, size of aggregates, rate of pouring and temperature. The hydrostatic pressure is maximum during pouring, and then decreases as concrete sets. Therefore, the *main* factor influencing this pressure is the depth of concrete poured before the concrete sets. The setting time may be taken between $\frac{3}{4}$ to 1 hour. Hence while computing the pressure, only the height of concrete poured in $\frac{3}{4}$ to 1 hour need only be taken into account. For heights of concrete upto 1.5 m, the equivalent fluid weight of concrete may be taken as 23000 N/m^3 . For higher heights, the equivalent fluid weight is reduced. When the height of concrete in one pour is 6 m, the equivalent fluid weight may be taken as only 12000 N/m^3. For intermediate heights between 1.5 to 6 m poured within the setting time of $\frac{3}{4}$ to 1 hour, linear interpolation of unit weight between 23000 to 12000 N/m^3 may be done.

Table 21.1 gives the safe values of stresses etc. for some common types of soft wood used for formwork. The maximum permissible deflection of sheathing and joists etc. should not exceed 2.5 mm.

21.4. SHUTTERING FOR COLUMNS

Shuttering for a column is probably the simplest. It consists of the following main components : (*i*) sheathing all round the column periphery (*ii*) side yokes and end yokes (*iii*) wedges and (*iv*) bolts with washers. Fig. 21.1 shows the formwork for a square column. The side yokes and end yokes consists of two number each, and are suitably spaced along the height of the column. The two-side yokes are comparatively of heavier section, and are connected together by two long bolts of 16 mm dia. Four wedges, one at each corner, are inserted between the bolts and the end yokes. The sheathing is nailed to the yokes.

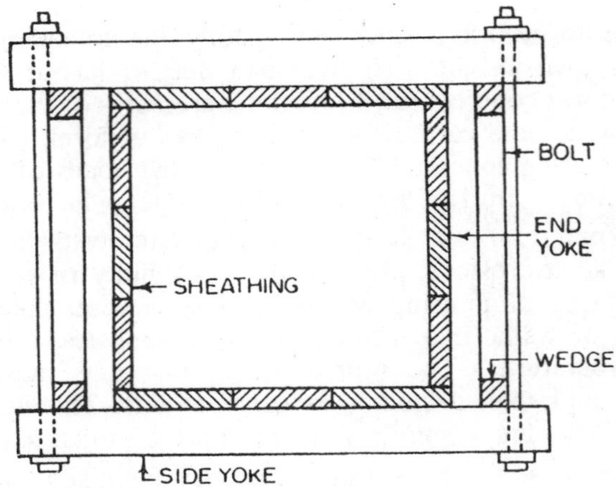

FIG. 21.1. FORMWORK FOR SQUARE OR RECTANGULAR COLUMN.

Spacing of yokes. The main design element for the formwork of a column is the spacing of yokes along the height of the columns. The sheathing may be considered to be a continuous slab over the yokes with its span(s) equal to the spacing of the yokes. Let b the width of the column. The fluid pressure of concrete over the width will be transferred to the yokes.

Let w = equivalent fluid weight of concrete(N/m^3)

h = depth of concrete poured within the setting time of $\frac{3}{4}$ to 1 hours (m).

W = total load on each yoke, uniformly distributed over the length b (N).

l = span of the side yoke, centre to centre between the bolts(mm).

s = centre to centre spacing of yokes(mm).

$$\therefore \quad W = \frac{wh\,.\,s\,.\,b}{1000 \times 1000} = wh\,.\,sb \times 10^{-6} \text{ N} \qquad ...(i)$$

The bending moment in sheathing $= \frac{W \times s}{10}$ (N–mm²)

If f is the permissible stress in sheathing (N/mm²) and d is the thickness of sheathing (in mm) we have

$$f \cdot \frac{b\,d^2}{6} = \frac{W.s}{10} \quad ...(ii)$$

Substituting the value of W from (i), we get

$$f\,\frac{b\,d^2}{6} = whbs^2 \times 10^{-7}$$

or $$s = 1291\sqrt{\frac{f}{wh}} \text{ mm} \quad ....(21.1)$$

The spacing should also be checked for shear and deflection in the sheathing. The above expression for s is based on permissible stress in sheathing. Let us also find the value of s based on permissible stress in the yoke. The span of the yoke is l mm centre to centre between the bolts, while it receives a total uniformly distributed load of W over length b.

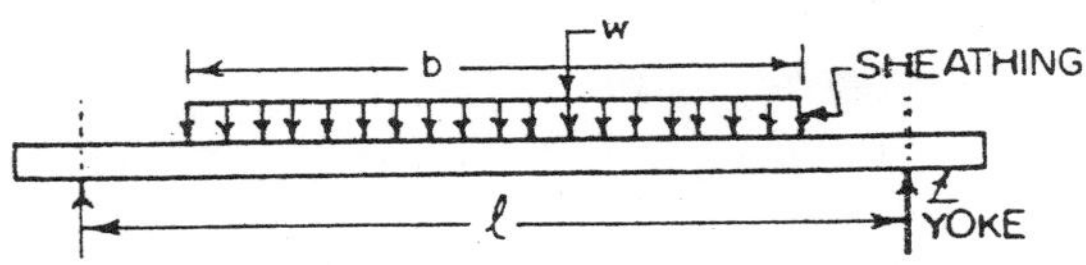

FIG. 21.2.

Hence B.M. $= \frac{W}{4}\left(l - \frac{b}{2}\right) = \frac{W}{2}(2l - b)$ N-mm

If Z is the section modulus of yoke, and f' is the permissible stress in bending (N/mm^2) for the timber of yoke, we have

$$\frac{W}{8}(2l - b) = f'Z \quad ...(iii)$$

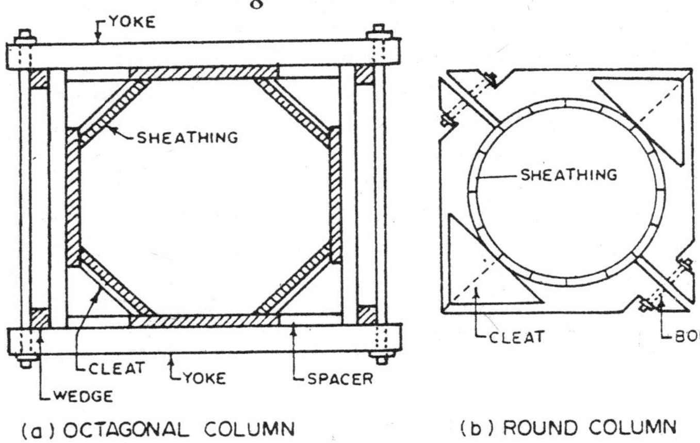

FIG. 21.3.

Substituting the value of W from (i), we get

$$\frac{whsb \times 10^{-6}}{8}(2l - b) = f' Z$$

$$s = \frac{8 \times 10^6 f' Z}{whb\,(2l - b)} \qquad \text{... (21.2)}$$

The lesser of spacing given by Eqs. 21.1 and 21.2 may be adopted

Fig. 21.3 shows shutterings for octagonal and round columns. Example 21.1 illustrates the method of designing the shuttering of a square column.

Example 21.1. *Design the formwork for a column* 350 mm × 350 *mm, having a height of 3 m. It is proposed to pour the entire concrete in one stage.*

Solution.

1. ***Equivalent liquid weight***

When the height of pour is 1.5 m, the equivalent fluid weight of concrete is taken as 23000 N/m^3 , while for height of pour of 6 m, equivalent fluid weight may be taken as 12000 N/m^3 . Hence by linear interpolation when the height of pour is 3 m the unit fluid weight may be taken as

$$= 23000 - \left(\frac{23000 - 12000}{6 - 1.5}\right)(3 - 1.5) = 23000 - 3667$$

$= 19333$ N/mm^3.

2. ***Bending stress in sheathing***

Let the spacing of yokes be s mm c/c.

Also, let the distance between the bolt centres = 600 mm.

From bending stress point of view, the spacing of yokes is given by Eq. 21.1. Taking the sheathing of chir wood, the permissible bending stress may be taken equal to 8.4 N/mm^2. Hence using 25 mm (=d) thick sheathing, we have

$$s = 1291\, d \sqrt{\frac{f}{wh}} = 1291 \times 25\sqrt{\frac{8.4}{19333 \times 3}} = 388 \text{ mm}$$

Keep spacing as 390 mm.

3. ***Check for shear and deflection in sheathing***

Let us check this spacing for shear and deflection

Permissible shear stress = 0.9 N/mm^2

$$\text{S.F.} = \tfrac{1}{2}(wh.b) \times \frac{390}{1000} = \tfrac{1}{2}(19333 \times 3 \times 0.35)\frac{390}{1000} = 3958 \text{ N}$$

$\therefore$ Shear stress $= \frac{3}{2}\frac{F}{bd} = \frac{3}{2} \times \frac{3958}{350 \times 25} = 0.68 \text{ N/mm}^2 < 0.9$

Hence safe.

Again $\delta = \frac{3}{384} \cdot \frac{(whb)\, s^4}{EI}$

where $EI = (9800)\frac{350(25)^3}{12} = 4.47 \times 10^9$ N-mm^2 units

whb = load per mm run of sheathing.

$= (19333 \times 3 \times 0.4)\frac{1}{1000} = 23.2$ N/mm

Substituting the values, we get

$$\delta = \frac{3}{384} \times \frac{23.2 \times (390)^4}{4.47 \times 10^9} \approx 1 \text{ mm.}$$

permissible deflection = 2.5 mm. Hence safe.

4. ***Design of yokes***

Let l = distance between bolt centres = 600 mm

Refer Fig. 21.2.

Load on yoke = W = $wh\, bs$ (where b and s are in metres).

$= 19333 \times 3 \times 0.35 \times 0.39 = 7917$ N

$\therefore$ Max. B.M. $= \frac{W}{8}(2l - b) = \frac{7917}{8}(2 \times 0.6 - 0.35)1000$

$= 841000$ N-mm

Taking permissible bending stress in yoke as 8.4 N/mm^2

We have $8.4z = 841000$

or $z = \frac{841000}{8.4} = 1 \times 10^5 \text{ mm}^3$

Using 80 mm × 100 mm section, $z = \frac{1}{6}\, 80\, (100)^2 = 13400 \text{ mm}^3$

Let us check this section for deflection.

$$\delta = \frac{5}{384} \cdot \frac{Wl^3}{EI} = \frac{5}{384} \frac{7917(600)^2}{9800 \times \frac{1}{12} 80(100)^3}$$

$= 0.34$ mm $<$ 2.5 mm. Hence safe.

5. ***Design of bolts***

Bolts are subjected to both direct pull as well as bending since the load is being transferred through wedges.

$W = 7917$ N

Load in each bolt $= \dfrac{7917}{2} = 3959$ N.

Assuming that the distance of load transmitted through wedges is 40 mm.

B.M. in bolt $= 3959 \times 40 = 158360$ N-mm.

Using 25 mm dia. bolts, $A = \dfrac{\pi}{4}(25)^2 = 491$ mm^2

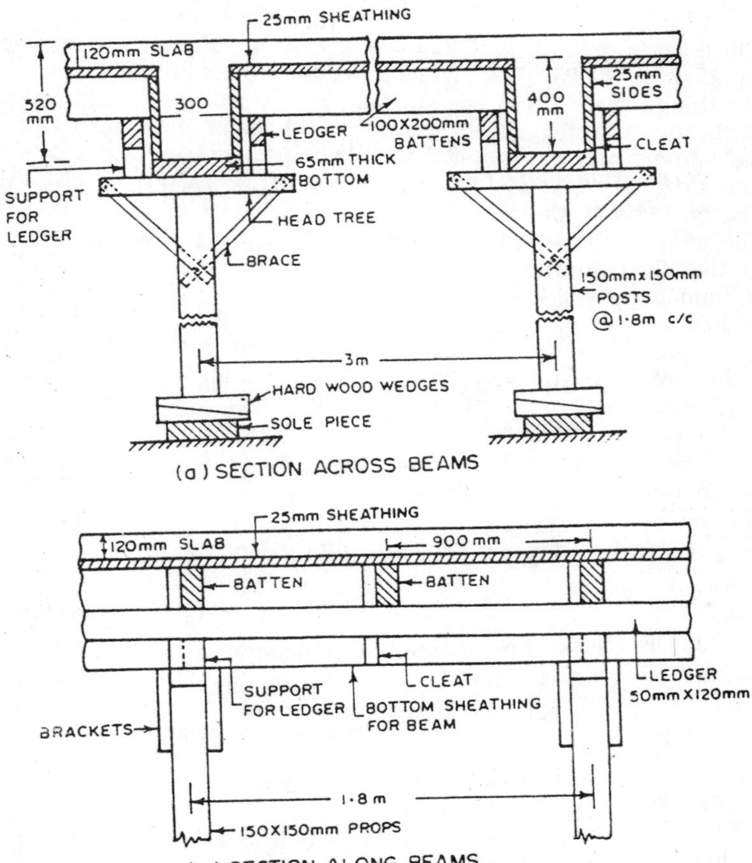

FIG. 21.4.

$$\therefore \text{ Tensile stress in bolt } = \frac{3959}{491} = 8.07 \text{ N/mm}^2$$

$$\text{Bending stress } = \frac{158360}{\frac{\pi}{64}(25)^4}\left(\frac{25}{2}\right) = 103 \text{ N/mm}^2$$

$$\therefore \text{ Total stress } = 103 + 8.07 = 111.07 \text{ N/mm}^2.$$

21.5. SHUTTERING FOR BEAM AND SLAB FLOOR

Fig. 21.4. shows the formwork for beam and slab floor. The slab is continuous over a number of beams. The slab is supported on 25 mm thick sheathing laid parallel to the main beams. The sheathing is supported on wooden battens which are laid between the beams, at some suitable spacing. In order to reduce the deflection, the battens may be propped at middle of the span through joists. The side forms of the beam consists of 30 mm thick sheathing. The bottom sheathing of the beam form may be 50 to 70 mm thick. The ends of the battens are supported on the ledger which is fixed to the cleats throughout the length. Cleats 100 mm × 20 mm to 30 mm are fixed to the side forms at the same spacing as that of battens, so that battens may be fixed to them. The beam form is supported on a head tree. The shore or post is connected to head tree through cleats. At the bottom of shore, two wedges of hard wood are provided over a sole piece.

Spacing of battens

Let s be the centre to centre spacing, in metres, of the battens which support the sheathing of the deck slab. The sheathing is continuous over battens.

Let δ = permissible deflection in metres

w = load on decking, N/mm^2

E = modulus of elasticity in N/m^2

Taking the deflection to be equal to $\frac{3}{384}\frac{ws^4}{EI}$. we have

$$\delta = \frac{3}{384}\frac{ws^4}{EI}$$

From which $s = \left(\frac{128\,E\,I\,\delta}{w}\right)^{1/4}$ metres ..(21.3)

The above spacing is based on deflection criterion. However, the sheathing should be thick enough to take up the bending stresses caused due to loading.

Example 21.2. *Design the formwork for the beam and slab floor, for the following data :*

1. *Thickness of floor : 120 mm*
2. *Centre to centre spacing of beams = 3 m*
3. *Width of beam = 300 mm and depth 400 mm below slab.*
4. *Height of ceiling of the roof =4 m above the floor.*

Take live load on sheathing equal to 4000 N/m² and dead weight of wet concrete as 26500 N/m³.

Solution

Sheathing for the slab

Let the sheathing be 25 mm thick, laid parallel to the beam. Let the spacing of battens be s metre centre to centre. Load per square metre of the decking will be as under :

Live load, including impact $= 4000 \text{ N/m}^2$

Dead load of wet concrete $= 26500 \times \dfrac{120}{1000} = 3180 \text{ N/m}^2$

Weight of sheathing $= 200 \text{ N/m}^2$ (say)

Total $w = 7380 \approx 7400 \text{ N/m}^2$

Let us use Deodar wood for the formwork.

Hence $EI = (9500 \times 10^6)\dfrac{1}{12} \times \left(\dfrac{25}{1000}\right)^3 \text{ N-m}^2$

$= 12370 \text{ N-m}^2$.

Taking permissible deflection = 3 mm,

$$\delta = \frac{3}{1000} = 0.003 \text{ m}$$

Substituting in Eq. 21.3,

Spacing of bettens $s = \left(\dfrac{128 \times 12370 \times 0.003}{7400}\right)^{1/4} \approx 0.9$ m.

Let us also find the spacing from bending stress point of view.

B.M. in sheathing $= \dfrac{ws^2}{10}$ N-m $= \dfrac{7400\, s^2}{10} = 740\, s^2$ N-m

Section modulus $= \dfrac{1}{6}\, bd^2 = \dfrac{1}{6} \times \left(\dfrac{25}{1000}\right)^2 \text{ m}^2$

Taking allowable stress f in bending $= 10.2 \text{ N/mm}^2$

$$= 10.2 \times 10^6 \text{ N/m}^2$$

We have $10.2 \times 10^6 \times \frac{1}{6}\left(\frac{25}{1000}\right)^2 = 740\, s^2$

From which $s = \sqrt{\dfrac{10.2 \times 10^6 \times 625}{6 \times 10^6 \times 740}} = 1.2$ m

Keep the spacing s equal to lesser of the two values. Hence provide battens at 900 mm centre to centre.

2. *Design of battens*

Centre to centre spacing of battens = 900 mm = 0.9 m.
Span of battens $= l = 3 - 0.3 = 2.7$ m.
Load on each batten $= 7400 \times 0.9 = 6660$ N/m.

$$\text{B.M.} = \frac{6660(2.7)^2}{8} \times 1000 = 6.07 \times 10^6 \text{ N-mm}$$

$$6.07 \times 10^6 = f.z = 10.2z$$

or $$z = \frac{6.07 \times 10^6}{10.2} = 5.95 \times 10^5 \text{ mm}^3$$

$$\therefore \quad bd^2 = 6 \times 5.95 \times 10^5 = 35.7 \times 10^5$$

Taking 100 mm wide battens, $d = \sqrt{\dfrac{35.7 \times 10^6}{100}} = 189$ mm

However, use 100 mm × 200 mm section.

$$\text{Max. S.F.} = \frac{5}{8}wl = \frac{5}{8} \times 6660 \times 2.7 \text{ N}$$

$$\therefore \text{ Shear stress} = 1.5\frac{F}{bd}$$

$$= \frac{1.5}{100 \times 200} \times \frac{5}{8} \times 6660 \times 2.7$$

$$= 0.84 \text{ N/mm}^2 < 1 \text{ N/mm}^2.$$

3. *Design of side forms of the beam*

The side forms of beam will be subjected to hydrostatic pressure of fluid concrete. Let the equivalent fluid weight of concrete be 23000 N/m³.

Total height of beam $= 120 + 400 = 520$ mm

Height of side form sheathing above bottom of beam

$$= 520 - (120 + 25) = 375 \text{ mm}$$

Pressure at top level of side form

$$= 23000\left(\frac{120 + 25}{1000}\right) = 23000 \times \frac{145}{1000}$$

Pressure at bottom level of side form $= 23000 \times \frac{520}{1000}$

Hence average pressure intensity on side form

$$= 23000\left(\frac{145+520}{2}\right) \times \frac{1}{1000} = 7650 \text{ N/mm}^2$$

The side form sheathing will span between battens.

Hence span of sheathing of side form may be taken as 0.9 m.

Height of side form = 375 mm = 0.375 m

Load w = 7650 × 0.375 × 1000 N/mm

B.M. $= (7650 \times 0.375) \times \frac{(0.9)^2}{10} \times 1000 = 232000$ N-mm

Let the thickness of side = 25 mm.

$\therefore \quad z = \frac{1}{6} \times 375\,(25)^2 = 39100 \text{ mm}^3$

$\therefore$ Bending stress $= \frac{232000}{39100} = 5.95 \text{ N/mm}^2 < 10.2$

Hence safe.

4. *Design of bottom of beams*

For the bottom of beams, the dead load of the full height of beam, including the thickness of slab, is to be taken into account. Similarly, the live load coming on the width of the beam alone is to be taken. Load per metre run of the bottom will be as follows :

Dead weight of concrete	= 26500 × 0.3 × 0.52	= 4140 N
Live load	= 4000 × 0.3	= 1200 N
Weight of boarding/m		= 400 N(say)
	Total w	= 5740 N/m.

The bottom of the beam is supported on props. Let the spacing of the props be double the spacing of the battens,

i.e. 2 × 900 = 1800 mm = 1.8 m c/c.

B.M. in bottom sheathing $= \frac{5740(1.8)^2}{8} \times 1000$

$= 2320000$ N-mm

Width of bottom = 300 + 2 × 25 = 350 mm

Let the thickness be t cm.

$\therefore \quad z = \frac{1}{6} \times 350\, t^2$

Allowable $f = 10.2 \text{ N/mm}^2$

$\therefore \quad \left(\frac{1}{6} \times 350\, t^2\right) \times 10.2 = 2320000$

From which $t = \sqrt{\dfrac{2320000 \times 6}{350 \times 10.2}} = 62.5$ mm

Hence provide 65 mm thick sheathing at the bottom.

5. ***Design of props***

Load from beam $= 5740 \times 1.8 = 10350$ N

Load from slab $= 6660 \times 2.7 \times 1.8 = 32350$ N

Total load $= 42700$ N

Taking allowable compressive stress perpendicular to grains as 2.1 N/mm^2.

$$\text{Area of section} = \frac{42700}{2.1} = 20300 \text{ mm}^2.$$

Use 150 mm × 150 mm section.

Let us test the prop against bucking,

$$\text{Safe stress} = f\left(1 - \frac{h}{50\,d}\right)$$

where f = safe compressive stress as short strut, parallel to grains = 7 N/mm^2

h = height of prop $\approx 4 - 0.4 = 3.6$ m

d = least dimension = 0.15 m

$$\therefore \text{Safe stress} = 7\left(1 - \frac{3.6}{50 \times 0.15}\right) = 3.64 \text{ N/mm}^2$$

$\therefore$ Allowable load $= 3.64 \times 150 \times 150 = 82000$ N > 42700

Hence safe.

The details of formwork are shown in Fig. 21.4.

22

Tests for Cement and Concrete

22.1. INTRODUCTION

The quantity of cement, and that of concrete manufactured from it, is judged from the tests conducted in the laboratory, as the work in the field progresses. The tests described in this chapter are conducted invariably in all big projects. There are a large number of tests for Portland cement, but we shall discuss the following important tests : (*i*) fineness test (*ii*) consistency test (*iii*) soundness test (*iv*) test for setting time (*v*) compressive strength test and (*vi*) tensile strength test.

Following are the common tests conducted on cement concrete:

(*i*) Workability tests

(*ii*) Test for compressive strength

(*iii*) Test for flexural strength

(*iv*) Test for modulus of elasticity

(*v*) Test for analysis of fresh concrete

(*vi*) Test for cement content of hardened Portland cement concrete.

(A) TESTS FOR CEMENT

20.2. TEST FOR FINENESS OF CEMENT

It has been found that the rate of hydration and hydrolysis and the subsequent setting of cement depends upon its fineness of particles. The rate of gain of strength is rapid for finer cement, though the final strength is not affected by fineness. However, the shrinkage and cracking of cement is greater for fine cement. The fineness of cement is measured in terms of its *specific surface*. There are several methods of measuring specific surface of cement, such as (*i*) particle size distribution method (*ii*) air permeability methods (*iii*) Wagner's turbidimeter method etc. Particle size distribution of cement can be determined by hydrometer test using kerosene as sedimentation medium and oleic acid as dispersing agent. The fineness can also be determined by sieve analysis but this is an unsatisfactory test and is not used much. Indian Standard specifies the last two methods. However, we shall describe here Blaine's air permeability method as specified in IS : 4031–1968.

Blaine's air permeability method

The Blaine's air permeability apparatus consists essentially of a means of drawing a definite quantity of air through a prepared bed of cement of definite porosity. The number and size of the pores in a prepared bed of cement of definite porosity is a function of the size of the particles and determines the rate of air flow through the bed.

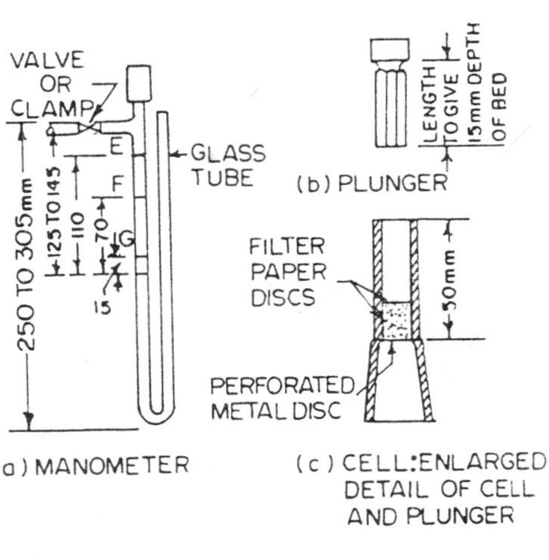

FIG. 22.1. BLAINE'S AIR PERMEABILITY APPARATUS.

Fig. 22.1. shows the parts of Blaine's air permeability apparatus as specified in IS : 4031–1968. The apparatus essentially consists of three parts : (*i*) the permeability cell (*ii*) the plunger and (*iii*) the manometer tube assembly. The permeability cell consists of a rigid cylinder of glass or non-corroding metal of 12.7 + 1 mm in diameter. The cell is fitted to the manometer tube assembly. The bottom of the cell forms an airtight connection with the top of the manometer. A perforated disc of 0.9 ± 0.1 mm thickness, perforated with 30 to 40 holes of 1 mm dia. equally distributed over its area, fits the inside of the cell snugly. The plunger is used to compact the bed of cement over the perforated disc. The height of the plunger is such that when the head of the plunger touches the top of the cell, the difference in height between the plunger bottom and the perforated disc is 15 mm. The top of one arm of the manometer forms an air tight connection with the permeability cell. A side outlet is provided at 250 to 305 mm above the bottom of the manometer used in the evacuation of the manometer arm connected to the permeability cell. A positive airtight valve or clamp is provided on the side out let. The manometer is filled to the mid-point with a non- volatile, non-hygroscopic liquid of low viscosity and density, such as dibutylphthalate or a light grade of mineral oil.

The test is conducted on two samples : (*i*) test on *standard sample* and (*ii*) test on sample of cement under investigation. For standard sample, *National Bureau of Standard's standard sample No. 114* is used.

Note : In the absence of NBS standard sample No. 114, samples of Indian cement having specific surface equivalent to NBS standard sample No. 114 may be used. Such samples may be obtained from the ACC Central Research Station, Bombay.

The test on the standard sample is commonly known as 'Calibration of Apparatus'. The complete test is done in the following steps :

Step 1. Determination of bulk volume of compacted bed of Cement powder.

A thin bed of cement, of height 15 mm is compacted in the cylinder. This bed of cement is actually the test specimen. In order to know exactly the bulk volume of cement compacted in the cell, mercury is used. The perforated metal disc is first placed in position in the cell and two filter paper discs are placed over it. The filter paper should be medium retentive, corresponding to No. 40 Whatman. The filter papers placed over the disc inside the cell are pressed with the help of a rod slightly smaller than the cell diameter until the filter discs are flat on the perforated disc. Mercury is then filled in the cell, to its top. Use a glass plate to level the mercury with the top of the cell. The mercury is removed from the cell and weighed (W_a gm). One of the filter discs is removed from the cell. Use a trial quantity of 2.80 g of cement and place over the filter paper in the cell. Tap lightly the side of the cell in order to level the bed of cement. Put the other filter paper disc on the top of the cement and compress it with the plunger until the plunger collar is in contact with the top of the cell. The plunger is then removed. Fill the space remaining in the top of the cell with mercury and level off the top as before. Remove the mercury from the cell and weigh it (W_b gm).

The bulk volume occupied by the cement is calculated by the following expression :

$$V = \frac{W_a - W_b}{D} \text{ cm}^3 \qquad ...(22.1)$$

where D = Density of mercury at temperature of test in g/cm^3, and its value may be taken from Table 22.1.

Step 2. ***Test on standard sample***

The air permeability is first conducted on the *standard sample.* The standard sample is generally available in the vial. The contents of the vial are enclosed in a 125 g jar and shaken vigorously. The weight (W) of standard sample used for the calibration test should be that required to produce a bed of cement having a porosity of 0.500 ± 0.005, and is calculated from the following expression :

$$W = \rho V(1 - e) \qquad ...(22.2)$$

Where ρ = specific gravity of test sample (for Portland cement, a value of 3.15 is used).

V = bulk volume of bed of cement, determined from Eq. 22.1

e = desired porosity of cement (0.500 ± 0.005).

Put a filter paper disc above the perforated disc and place the quantity of standard cement determined above. Place a filter paper disc over the cement bed and compress it with the plunger till the plunger collar is in contact with the top of the cell. Remove the plunger.

Attach the cell to the manometer tube, making certain that an airtight connection is obtained and taking care not to jar or disturb the prepared bed of cement. The air in one arm of the manometer U-tube is slowly evacuated until the liquid reaches the top mark (E) and the valve is then closed tightly. The stopwatch is started as the bottom of the meniscus of the manometer liquid reaches the second mark (F) and is stoped as the bottom of meniscus of liquid reaches the third mark (G). Measure the time interval between F and G and note the temperature of test.

Step 3. Test on the test sample

Repeat completely step 2 on the cement to be tested. The weight of sample to be compacted is calculated from Eq. 22.2, It is customary to use the same porosity(*i.e.*, 0.500 ± 0.005). However, while determining the fineness of high-early-strength cements, a porosity of 0.530 ± 0.005 should be adopted. Determine the time taken by the liquid to drop from point F to G as described in step 2. Measure the temperature during the test.

Step 4. Calculations

The specific surface of the cement is calculated from the following expression :

$$S = S_s \sqrt{\frac{T}{T_s}} \quad ...(22.3)$$

where S = specific surface in sq. cm per gram of the test sample.

S_s = specific surface of the standard sample.

T and T_s are the time intervals, in seconds, of manometer drop for the test sample and standard sample respectively.

The above expression in valid *only if* the temperature during both the tests is constant. If the temperature varies, the following expression is used :

$$S = S_s \sqrt{\frac{T}{T_s}} \sqrt{\frac{n_s}{n}} \quad ...(22.4)$$

where n_s = viscosity of air in poise at the temperature of test of the standard sample (Table 22.1).

n = viscosity of air in poise at temperature of test of the test sample (Table 22.1)

Both Eqs. 22.3 and 22.4 are valid only if the test sample is compacted at the same porosity as the standard sample. If the porosity is different, use the following expression :

$$S = S_s \sqrt{\frac{T}{T_s}} \sqrt{\frac{n_s}{n}} \sqrt{\frac{e^3}{e_s^3}} \cdot \frac{1-e_s}{1-e} \quad ...(22.5)$$

where e_s = porosity of bed of standard sample

e = porosity of bed of test sample.

TABLE 22.1. DENSITY OF MERCURY, VISCOSITY OF AIR (n) AND $\sqrt{n}$

Room Temperature °C	*Density of mercury g/cm³*	*Viscosity of air (n) poise*	$\sqrt{n}$
16	13.56	0.0001788	0.01337
18	13.55	0.0001798	0.01341
20	13.55	0.0001808	0.01344
22	13.54	0.0001818	0.01348
24	13.54	0.0001828	0.01352
26	13.53	0.0001837	0.01355
28	13.53	0.0001847	0.01359
30	13.52	0.0001857	0.01362
32	13.52	0.0001867	0.01366
34	13.51	0.0001876	0.01369

20.3. TEST FOR CONSISTENCY OF CEMENT PASTE

Many tests on cement, such as the determination of setting time, soundness, compressive strength and tensile strength etc. require the preliminary determination of the amount of mixing water required to produce a cement paste of *standard consistency*. The test for determination of water content to form the cement paste of standard consistency is conducted on Vicat's apparatus. The standard consistency of a cement paste is defined as that consistency which will permit the Vicat plunger (Fig. 22.2) to penetrate to a point 5 to 7 mm from the bottom of the Vicat mould.

Vicat's apparatus

Fig. 22.1 shows a Vicat's apparatus along with its components. It consists principally of a stand *D* in which a rod *B* weighing 300 g slides. At the end of the rod. *B*, either a needle (*C*) of 1 mm sq., 5 cm long is fitted (for determination of setting time) or a plunger (*G*) of 10 mm Φ can be fitted. A movable indicator is attached to rod *B*. The indicator travels over scale graduated from 0 to 40 mm to measure the travel of the rod. The cement paste is filled in the Vicat mould *E* which rests on a non-porous plate.

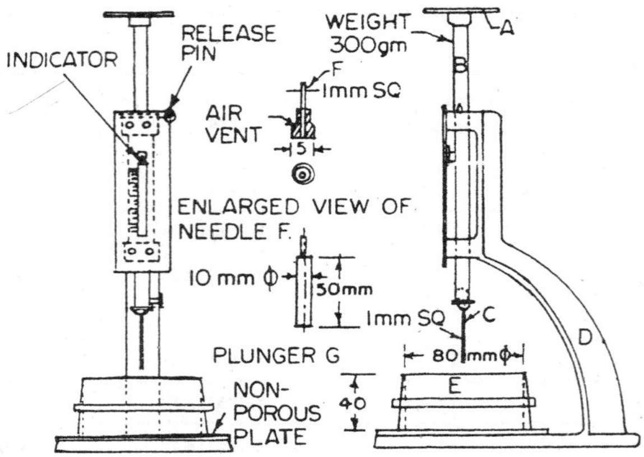

FIG. 20.2. VICAT'S APPARATUS.

Test procedure

The test consists of making a number of cement pastes with different amounts of mixing water, Each paste is placed in turn in the mould under the Vicat plunger which is lowered into contact with surface of the paste and quickly released. To prepare the paste, take weighed quantity(say about 300 g) of cement and place it in a crucible. Mix a weighed quantity of water to it and mix it with spatula. The time of mixing or gauging should not be less than 3 minutes, nor more than 5 min, and gauging time should be counted from the time of adding water to the dry cement until commencing to fill the mould. The Vicat mould is filled with the paste, which is levelled off at its top. The mould is placed under the rod bearing the plunger (*G*). The plunger is lowered gently to touch the surface of the test block and then quickly released, allowing it to sink in the paste. The reading against the indicator is noted. The test is repeated on a number of pastes prepared with different amounts of water. The paste of standard consistency is that which allows the Vicat plunger to settle to a point within 5 to 7 mm from the bottom of the Vicat mould. The amount of water required to prepare the paste of standard consistency is expressed as a percentage by weight of the dry cement. The test is performed at a temperature or 27° ± 2° C.

22.4 TEST FOR DETERMINATION OF SETTING TIMES

The setting time of cement may be divided into the *initial setting time* and the *final setting time.* It is difficult to say what exactly each time means as they are arbitrary points in relationship connecting the strength of cement with the time from adding water. The initial set is a stage in the process of hardening of cement after which any cracks that may occur or appear will not re-unite. The two setting times are defined *arbitrarily* in the Vicat's test, described below.

Preparation of Test Block. Take about 300 g of cement and prepare a cement paste by gauging it with 0.85 times the water required to give a paste of standard consistency. The paste is gauged in the manner described in the consistency test. Stop watch is started at the instant when water is added to the cement. The Vicat mould is filled with cement paste gauged above, the mould resting on a non-porous plate. The mould is filled completely and the surface of the paste is made level with the top of the mould. The cement block thus prepared in the mould is known as the *test block.*

Determination of Initial Setting Time. The mould along with the plate is placed under the rod bearing the needle(C). The needle is lowered slowly to bring it in contact with the surface of the test block. The needle is then released quickly, allowing it to penetrate into the test block. In the beginning, the needle will completely pierce the test block. The above procedure is repeated until the needle, when brought in contact with the test block and released, fails to pierce the block for 5 ± 0.5 mm measured from the bottom of the mould. *The period elasping between the time when water is added to the cement and the time at which the needle fails to pierce the test block by* 5 ± 0.5 *mm is taken as initial setting time.*

Determination of Final Setting Time. To determine the Final Setting Time, needle C is replaced by needle F having an annular attachment. This needle also has a diameter equal to 1 mm, but the needle projects out of annular attachment by 0.5 mm. The cement is considered as finally set when upon applying the needle gently to the surface of the test block, the needle makes an impression thereon, while the attachment fails to do so. *The period elasping between the time when water is added to the cement and the time at which the needle makes an impression on the surface of test block while the attachment fails to do so is taken as the final setting time.*

22.5. TEST FOR SOUNDNESS OF CEMENT

By *soundness* is meant the ability or otherwise of a cement to maintain a constant volume. The cement having some quantity of free lime and magnesia undergoes large changes of volume as the time elapses, tending to cause cracks. Such a cement is called an 'unsound cement'. The soundness of cement is determined either by 'Le Chatelier Method' or by means of an 'Autoclave Test'. We shall discuss here only the first method.

Le Chatelier Method

The test is carried out by means of Le Chatelier's apparatus which consists of small split cylinder of spring brass, or other suitable metal of 0.5 mm thickness, forming a mould 30 mm internal diameter and 30 mm high. On either side of the split are attached two indicators with pointed ends *AA*, the distance from these ends to the centre of the cylinder being 165 mm.

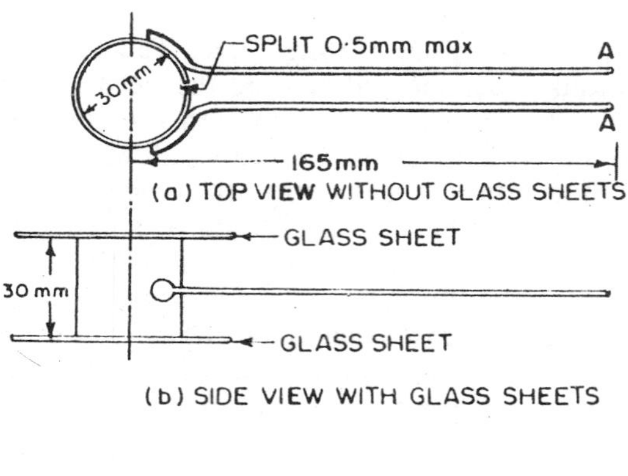

FIG. 22.3. APPARATUS FOR LE CHATELIER TEST.

Take about 100 g of cement. Mix it with 0.78 times the water required to give a paste of standard consistency. Gauge the paste thoroughly. The mould is placed on a glass plate and filled with the cement paste. The mould is covered with another piece of glass sheet and a small weight is placed over the glass sheet. The whole assembly is immediately placed in water bath having a temperature of 27° ± 0° C, and kept there for 24 hours. The assembly is taken out after 24 hours and the distance separating the indicator points is measured. The mould is then immersed in a water bath. The water of the bath is brought to boiling point, with the mould submerged in 25 to 30 minutes and kept boiling for 3 hours. The mould is taken out from the mould and allowed to cool. The distance between the points is then measured. The difference between these *two measurements* represents the expansion of cement. Un-aerated ordinary rapid hardening and low heat Portland cement should not have an expansion of more than 10 mm.

22.6 TEST FOR DETERMINATION OF COMPRESSIVE STRENGTH OF CEMENT

The compressive strength of concrete depends upon the strength of cement used for the manufacture of the concrete. Any test for the strength of cement should reflect the ability of the cement to make concrete of a given strength. The compressive strength test of cement is conducted on mortar cube, formed by mixing one part of cement with 3 parts of standard sand, compacted by means of a vibration machine. The complete procedure is outlined below :

1. ***Mix proportion and mixing***

Standard re-graded Ennore sand conforming to IS : 650–1966 is used for preparing the cement mortar. The following proportions are used :

Cement	200 g
Standard Sand	600 g
Water	$\left(\frac{P}{4} + 3.0\right)$ per cent of combined weight of cement and sand, where P is the percentage of water required to produce a paste of standard consistency.

However, if regraded Ennore sand is not available, single graded Ennore sand may be used. Such a sand should pass through 850 micron IS sieve and not more than 10 percent by weight should pass through 600 micron IS sieve. In that case mix proportions may be taken as under :

Cement	185 g	
Standard sand	555 g	
Water	$\left(\frac{P}{4}+3.5\right)$	percent of combined weight.

Cement and sand are first mixed dry on a non porous plate, with a trowel for one minute and then with water until the mixture is of uniform colour. The time of mixing should in any event not be less than 3 minutes and should the time taken to obtain a uniform colour exceed 4 min., the mixture should be rejected and the operation repeated with a fresh quantity of cement, sand and water.

2. *Moulding specimens*

Compressive strength test is done on mortar cubes having area of face equal to 50 cm^2. The moulds for the cube specimens of 50 cm^2 face area should be of metal which is not attacked by cement mortar. Normally, split moulds are used so that the moulded specimens could be removed easily. The parts of the mould when assembled together should be positively held rigidly together. In assembling the mould ready for use, the joints between the two halves of the mould are covered with a thin film of petroleum jelly. A similar coating of petroleum jelly between the contact surfaces of the bottom of the mould and its base plate is applied. The interior faces of the mould are treated with a thin coating of mould oil.

The assembled mould is mounted on the table of the vibrating machine and a hopper is attached to the top of it. Immediately after mixing the mortar, place the mortar in the cube mould and prod it with the standard packing rod made of non-absorptive, non-abrasive, non-brittle material, such as a rubber compound. The mortar should be proded 20 times in about 8 seconds to ensure the elimination of entrained air and honey combing. The remaining quantity of mortar is placed in the hopper of the cube, and it is proded again. The vibrating machine is then started and the mould is vibrated for 2 minutes. The vibrating machine should have a speed of 12000 ± 400 vibrations per minute. At the end of vibration, the mould assembly is removed from the machine, and its top is levelled with the blade of a trowel.

3. *Curing specimens*

Several such cubes are compacted in separate moulds. The mixture is made separately for each mould. The moulds are kept at a temperature of 27° ± 2° C and in an atmosphere of at least 90% relative humidity for 24 hours after the completion of vibration.

After that, the cube specimens are removed from the moulds and immediately submerged in clean fresh water and kept there until taken out just prior to breaking.

4. ***Compression testing***

Three such cubes are tested at the end of specified period, and the compressive strength should be the average of the strength of the three cubes for each period of curing. Normally 3 cubes are tested at the end of 30 days curing and 3 cubes at the end of 7 days curing. The testing can be done on any standard compression testing machine. The cubes should be tested on their sides without any packing between the cube and the steel plattens of the testing machine. One of the plattens should be carried on a base and should be self adjusting, and the load should be steadily and uniformly applied, starting from zero at a rate of 350 $kg/cm^2/min$. The compressive strength is calculated from the crushing load and the average area over which the load is applied. The minimum desirable values of compressive strengths at various curing periods, for different cements, are given in Chapter 1.

22.7. TEST FOR TENSILE STRENGTH OF CEMENT

Just like compressive strength, the tensile strength of cement is determined by testing specimens of cement sand mortar in direct tension.

1. ***Mix proportion and Mixing***

Standard sand is used for preparing the cement mortar. The following proportions are used :

Cement	250 g
Standard sand	750 g
Water	$\frac{P}{5} + 2.5$ percent of total weight of sand and cement, where P is the percent of water required for standard consistency.

The mortar is prepared exactly in the same way as for the compression test.

2. *Moulding specimens and curing*

Specimens are prepared in standard briquette moulds placed on non-porous plate. The moulds are of split type so that specimen can be easily taken out later. These moulds are open on both the faces. Fig. 22.4(a) shows the dimensions of standard briquette. The thickness of the briquette is 25.4 mm (1 inch) and it has a minimum sectional area of 645 sq. mm (1 sq. inch) at the central section. Before filling the mould, the two halves of the mould are lightly greased, and the inner surface of the mould is coated with oil. Mortar is then filled in the mould. To ensure complete filling of the mould the mortar should be heaped on the mould and the excess mortar is beaten down by means of standard spatula shown in Fig. 22.4(b), until the mortar is in level with the top of the mould and water appears on the surface. This process of heaping the motar and beating it down on the other-open face of the briquette mould is repeated. Finally, the faces are smoothened with the help of a trowel. Twelve specimens are prepared, six of which are tested after 3 days' curing, and the other six after 7 days. The briquette moulds are kept for 24 hours at $27° \pm 2° C$ and 90% humidity. The specimens are then taken out of mould and submerged in water for curing.

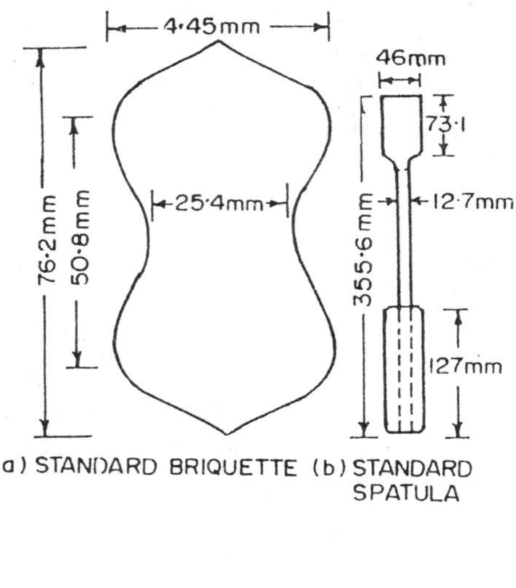

FIG. 22.4.

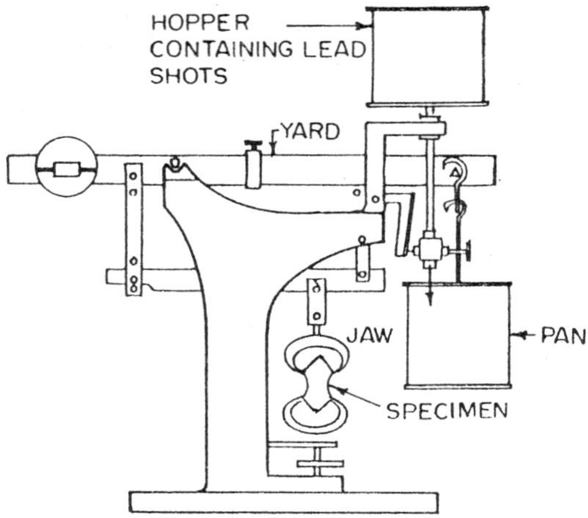

FIG. 22.5

3. *Test for tensile strength*

The briquettes so prepared are tested on a special *tension testing machine* shown in Fig. 22.5. The machine utilises the principle of compound lever for applying tensile force. The briquette is held in position in specially shaped jaws. A pan is attached to the free end of the lever. The loading is done by allowing lead shots to fall from a hopper to the pan. This causes tensile force in the briquette which ultimately breaks. When fracture occurs, the supply of lead shots is automatically cutoff. The lead shots collected in the pan are weighed by moving the counterpoise until balance is obtained. The reading on the yard gives the tensile load during fracture. The load divided by area of neck gives the tensile strength of the briquette.

(B) TESTS FOR CEMENT CONCRETE

22.8. TEST FOR WORKABILITY

Following are some of the tests carried for determining the workability of cement concrete : (*i*) slump test (*ii*) compacting factor test (*iii*) flow table test (*iv*) Vee-Bee consistometer method. We will discuss here the first two tests.

1. SLUMP TEST

Though it is a more common test, it gives only a rough idea of the workability of concrete. It can be adopted either in the laboratory or in the field during the progress of the work, where the maximum size of aggregate does not exceed 38 mm. The mould used for the test specimen is in the form of frustum of a cone having the following internal dimensions : bottom dia. = 20 cm ; top diameter = 10 cm and height = 30 cm. It is provided with suitable foot pieces and also handle to facilitate lifting it from the moulded concrete test specimen in vertical direction as required by the test. A mould provided with suitable guide attachment may be used, such as shown in Fig. 22.6. In addition to the mould a tamping rod of steel or other suitable material 16 mm in diameter, 0.6 m long and rounded at one end is used for tamping the concrete in the mould. Steel measure for measuring the slump is usually incorporated in the stand.

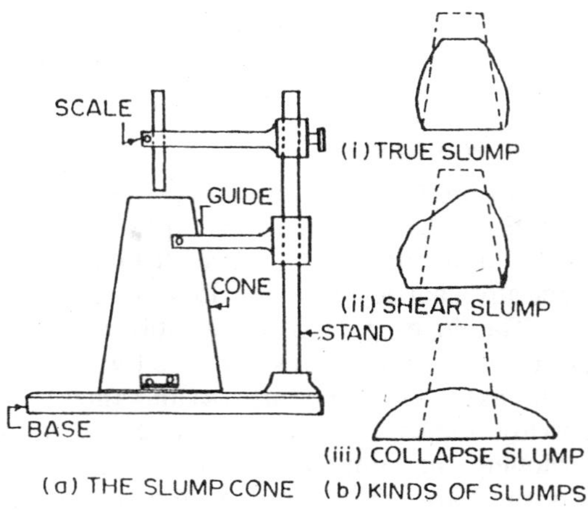

FIG. 22.6. THE SLUMP CONE.

To prepare the test specimen, the cleaned mould is placed on a smooth horizontal, rigid, non-absorbant surface(base). The mould is filled with freshly mixed concrete in *four* layers, each approximately one quarter of the height of the mould. Each layer is tamped with twenty five strokes of the rounded end of the tamping rod, the strokes being distributed in a uniform manner over the area. The strokes for the second and subsequent layers should penetrate into the underlying layer. The bottom layer should be tamped through out its depth. After the top layer has been rodded, the concrete is struck off level with a trowel or the tamping rod. The mould is then removed from the concrete immediately by raising it slowly and carefully in the vertical direction. This allows the concrete to subside and the slump is measured immediately by determining the difference between the highest point of the mould and that of the highest point of the slumped specimen. The slump measured is recorded in terms of millimeters of subsidence of the specimens during the test.

Fig. 22.6(*b*) shows three forms of slumps that may occur. The first is a *true slump*, the second is known as a *shear slump* and the third a *collapse slump* which is obtained with lean harsh or very wet mixes. Any slump specimen which collapses or shears off laterally gives incorrect results and if this occurs, the test should be repeated with another sample. If, in the repeat test also, the specimen should shear, the slump shall be measured and the fact that the specimen sheared, shall be recorded. Generally, if shear and collapse slumps are obtained, the concrete will be unsatisfactory for placing.

2. THE COMPACTING FACTOR TEST

The compacting factor test was introduced by the Road Research Laboratory, London. The compacting factor test gives more consistent and sensitive results than the slump test, and it is specially useful to dry mixes which give no slump. The test is designed to apply a given amount of work to a given amount of concrete and to reduce to a minimum the work lost in overcoming the friction between the concrete and the containing surfaces of the equipment. According to definition, workability of concrete is that property of concrete which determines the amount of work necessary to overcome the internal friction in order to obtain full compaction. Since a large proportion of the work is expanded usefully in overcoming the internal friction of concrete, the test gives a good measure of workability of concrete.

Fig. 22.7 shows two views of compacting factor apparatus. It consists of two conical hoppers *A* and *B* mounted above a cylindrical mould *C*, all made of cast brass or bronze or of stout sheet brass or steel. The lower ends of both the hoppers are closed with tightly fitting hinged trap-doors having quick release catches. The hoppers and the cylinder are mounted on a rigid frame. The hoppers and

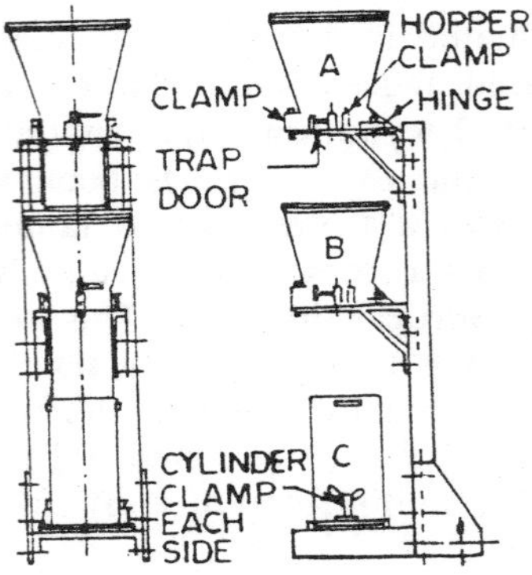

FIG. 22.7 COMPACTING FACTOR APPARATUS.

cylinder can be easily detached from the frame. In addition to the main apparatus, two ordinary bricklayers trowels, one hand scoop, a rod of steel and a balance to weigh upto 30 kg, to the nearest 10 g, are required.

Test procedure. The sample freshly mixed concrete is placed gently in the upper hopper, using the hand scoop. The hopper is filled level with its brim and the trap door of the first, hopper is opened so that concrete falls into the lower hopper. If concrete has a tendency to stick in hopper, the concrete may be helped through by pushing the rod gently into the concrete from the top. Immediately after the concrete has come to rest, the trap-door of the lower hopper is opened, so that the concrete falls into the lower hopper. The excess of concrete remaining above the level of the top of the cylinder is cut off by holding a trowel in each hand, with the plane of the blades horizontal and moving them simultaneously one from each side across the top of the cylinder, at the same time keeping them pressed on the top edge of the cylinder. After cleaning the cylinder from outside, the weight of the concrete in the cylinder is determined to nearest 10 g. This weight (w) is known as *the weight of partially compacted concrete*. The cylinder is refilled with concrete from the same sample in layers approximately 5 cm deep, the layers being heavily rammed or preferably vibrated so as to obtain full compaction. The top surface of the fully compacted concrete is carefully struck off level with the top of cylinder, and its weight W is determined. This weight is known as *the weight of fully compacted concrete*.

The compacting factor is defined as the ratio (w/W) of the weight of partially compacted concrete, to the weight of fully compacted concrete, and is stated to the nearest second decimal place.

22.9. TEST FOR COMPRESSIVE STRENGTH

1. *Shape and size of specimen*

The most important strength test for concrete is the compression test. It may be conducted either on specimens prepared in the laboratory or on specimens prepared in the field during the progress of the work. The test conducted on laboratory prepared specimen is called the *preliminary test,* while the test conducted on specimen made of concrete taken at the mixer is called the *works test.* The compression test can be conducted either on cubical specimens or on cylindrical specimens. If the maximum size of aggregate does not exceeds 30 mm, the size of the cube should be 15 × 15 × 15 cm. If the largest nominal size of the aggregate does not exceed 20 mm, 10 cm cubes may be used as an alternative. Cylindrical test specimens should have a length equal to twice the diameter. They should be 15 cm diameter and 30 cm long. The cube moulds, made of metal, should be such that moulded specimen can be removed easily. The moulds are of split type, and are provided with a metal base plate having a plane surface. The cylindrical mould should be capable of being opened longitudinally to facilitate removal of the specimen. Each cylindrical mould is provided with a metal base plate, and with a capping plate of glass or other suitable material. In addition to the moulds, tamping bar of steel, 16 mm in dia., 60 cm long and bullet pointed at the lower end is used for hand compaction of specimen.

2. *Preparation of specimen*

Before placing the concrete in the mould, its interior surface and the base plate should be lightly oiled. The test specimens should be made as soon as practicable after mixing, and in such a way as to produce full compaction of the concrete with neither segregation nor excessive liatance. The concrete is filled into the mould in layers of approximately 5 cm deep. Each layer is compacted either by vibration or by hand. Where the concrete at the actual work site is to be compacted by mechanical vibrator, the test specimen may be compacted with a mechanical vibrator. The mode and quantum of vibration of the laboratory specimen should be as nearly the ame as those adopted in actual concreting operation. When compacting by hand, the standard tamping bar should be used and the number of strokes should be evenly distributed. For cubical specimens, the number of strokes should not be less than 35 strokes per layer for 10 cm cubes. For cylindrical specimens, the number of strokes should not be less than 30 per layer. The strokes should penetrate into the underlying layer and the bottom layer should be rodded throughout its depth.

In the case of cylindrical samples, it becomes very necessary to have a uniform and smooth surface. The ends of all cylindrical test specimens should be capped. The cylinders may be capped with a thin layer of stiff, neat Portland cement paste after the concrete has ceased setting in the moulds generally for two to four hours or more after moulding. The cement for capping should be mixed to a stiff paste for about two to four hours before it is to be used in order to avoid the tendency of the cap to shrink. The stiff cement paste is placed at the top of the mould and the glass capping plate is worked on the cement paste. No gap is allowed to remain between the plate and the top surface of the mould. Adhesion of paste to the capping plate is avoided by coating the plate with a thin coat of oil or grease.

3. ***Curing of test specimens***

As soon as moulding is complete, the mould is stored in a place free from vibration, in moist air of at least 90% relative humidity and at a temperature of $27° \pm 2° C$ for 24 hours. After this period, the specimens are marked, removed from the moulds and immediately submerged in clean, fresh water, and kept there prior to testing.

4. ***Test for compressive strength***

The specimens are tested for compressive strength on any reliable *compression test machine* provided with two steel bearing platens with hardened faces. The test should be made at recognized ages of test specimen, the most usual being 7 and 28 days. At least three specimens, preferably from different batches, should be made for testing at each selected age. Specimens stored in water should be tested immediately on removal from the water and while they are in the wet condition. In the case of cubes, the specimen should be placed in the machine in such a manner that the load is applied to the opposite sides of the cubes as cast, that is, not to the top and bottom.The load should be applied without shock and increased continuously at a rate of approx. 140 $kg/cm^2/min$. until the resistance of the specimen to the increasing load breaks down and no greater load can be sustained. The measured compressive strength of the specimen is calculated by dividing the maximum load applied to the specimen during the test by the cross-sectional area, calculated from the mean dimensions of the section. Average of three values should be taken as the representative of the batch, provided the individual variation is not more than ± 15 percent of the average. Otherwise repeat test should be made.

22.10. TEST FOR FLEXURAL STRENGTH

For flexural strength, specimens of rectangualr beáms with square cross-sections are used. Following are the recommended sizes of the concrete beams, depending upon the maximum size of the aggregate :

Max. size of aggregate	*Cross-section*	*Length*
Less than 20 mm	10 cm × 10 cm	50 cm
Between 22 to 38 mm	15 cm × 15 cm	70 cm
Exceeding 38 mm	4 times max. size of aggregate	4 times the cross-sectional dimension

The specimen is prepared in steel or cast iron moulds constructed with the longer dimension horizontal. Each mould is provided with a metal base plate and two losse top plates of 4.0 × 0.6 cm cross-section and 5 cm longer than the width of the mould. The mould should be constructed in such a manner as to facilitate the removal of the moulded specimen without damage. For compacting concrete in the mould a *tamping bar* of steel, weighing 2 kg, 40 cm long and having a ramming face 25 mm square as used. The procedure of compaction is the same as adopted for preparing cubical specimens for compression test. The procedure for curing is also the same.

Fig. 22.8 shows arrangement for loading of flexure test specimens. The specimen is placed in the machine in such a manner

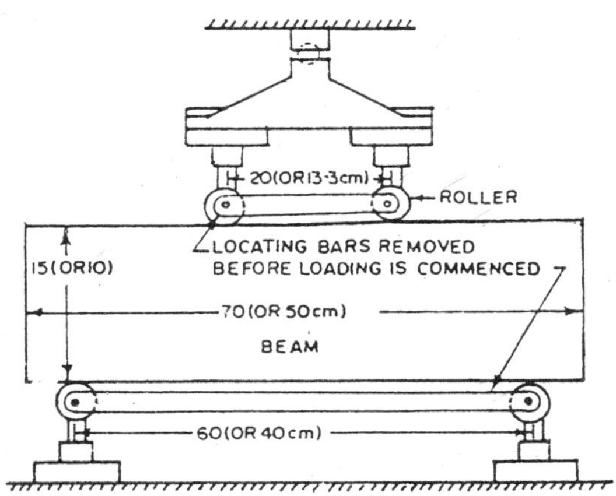

FIG. 22.8. ARRANGEMENT FOR LOADING OF FLEXURE TEST SPECIMEN

that load is applied to the uppermost surface as cast in the mould, along two lines spaced 22.0 cm for 70 cm long beam or 13.3 cm for the 50 long beam. The axis of the specimen is carefully aligned with the axis of the loading device. No packing should be used between the bearing surfaces of the specimen and the rollers. The load should be applied without shock and increasing continuously at a rate such that the extreme fibre stress increases at approximately $7\ kg/cm^2/min$, that is, at rate of loading of 400 kg/min. for 15.0 cm specimens and at a rate of 180 kg/min. for the 10.0 cm specimen. The load should be increased until the specimen fails, and the maximum load applied to the specimen during the test should be recorded. The appearance of the fractured faces of concrete and any unusual-features in the type of failure should be noted.

Let a = distance between the line of fracture and the nearer support, measured on the centre line of the tensile side of the specimen(cm).

f_b = modulus of rupture (kg/cm^2)

P = maximum load applied to the specimen (kg)

l = length of the span on which the specimen is supported (cm)

b = measured width of specimen (cm)

d = measured depth of specimen at the point of failure.

If a is *greater* than 22.0 cm for 15.0 cm specimen, or greater than 13.3 cm for 10.0 cm specimen, the modulus of rupture is calculated from the following expression :

$$f_b = \frac{P \times l}{bd^2} \qquad ...(22.6)$$

If a is *less* than 20 cm but greater than 17.0 cm for 15 cm specimen, or less than 13.3 cm but greater than 11.0 cm for a 10.0 cm specimen, the modulus of rupture is calculated from the following :

$$f_b = \frac{3\,P\,a}{bd^2} \qquad ...(22.7)$$

However, if a is less than 17.0 cm for a 15.0 cm specimen or less than 11.0 cm for a 10.0 cm specimen, the results of the test should be discarded.

PART V

LIMIT STATE DESIGN METHOD

Chapters

23

Design Concepts

23.1. METHODS OF DESIGN

The following design methods are used for the design of reinforced concrete structures :

(*a*) Working stress method.

(*b*) Ultimate load method, and

(*c*) Limit state method

(*a*) Working stress method of design

In this method, the structures are analysed by the classical *elastic theory.* The stresses in the members are considered for normal working load condition, and no attention is given do the conditions that arise at the time of structural collapse. The working loads are fixed by *limiting the stresses* in concrete and steel to a fraction of the stresses at which the material fails when tested as cubes and cylinders for concrete and bars in steel. The ratio which the yield stress of the steel or the cube strength of the concrete bears to the corresponding *permissible* or *working stress* is usually called *factor of safety*, which is actually a *stress factor of safety.* There is no guaranttee that a structure designed by this method, keeping the maximum stress in steel (or concrete) at working load to one half the yield stress will be able to carry the ultimate load of twice the working load.

Till recently, this method was in use in the English speaking country. In India, it is still widely used because of its simplicity in approach.

(*b*) Ultimate load method of design

An alternative method of design that was developed was the *ultimate load method* or the *load factor method,* in which the ultimate load is some known multiple of the maximum working load which the structure is likely to carry. The ratio of the collapse load to the working load is known as *load factor.* The load factor give exact margin of safety against collapse. At failure, the material ceases to behave elastically ; the strain increases many times beyond those considered for elastic theory and the stress distribution across the section adjust itself in such a way as to enable the member to develop its maximum capacity for resisting the conditions of applied loading.

There are three stages in ultimate load design :

1. Assessment of suitable value of load factor so that the value of ultimate load may be determined corresponding to a given working load ;

2. Analysis of structure under the ultimate load(or *factored load*) determined in stage 1 so that distribution of moments and loads may be determined at failure ; and

3. The design of members to fail at moments or forces so determined.

Since the method utilizes a large reserve of strength in plastic region (inelastic region) and of ultimate strength of member, the resulting section is very slender or thin. This gives rise to excessive deformations and cracking. Also, the method does not take into consideration the effects of creep and shrinkage. The method thus does not take into consideration serviceability requirements of avoidance of excessive deflection and cracking.

(*c*) Limit state method design

We have seen that while the *working stress method* gives satisfactory performance of the structure at working loads, it is unrealistic at ultimate state of collapse. Similarly, while the *ultimate load method* provides realistic assessment of safety, it does not guaranttee the satisfactory serviceability requirements at service loads. An ideal method is the one which takes into account not only the ultimate strength of the structure but also the serviceability and durability requirements. The newly emerging *'limit state method'* of design is oriented towards the simultaneous satisfaction of all these requirements. This new method makes a judicious combination of the working stress philosophy as well as the ultimate load philosophy, thus avoiding the demerits of both.

The acceptable limit of safety and serviceability requirements, before failure occurs is called a ***limit state.***

Three types of limit states are considered in the design

1. Strength limit state.
2. Serviceability limit state, and
3. Durability limit state.

In India, IS code has completely replaced the ultimate load method by the limit state method.

23.2. SAFETY AND SERVICEABILITY REQUIREMENTS (IS : 456-1978)

In the *Limit State Method*, the structure is designed to withstand *safely* all the loads liable to act on it throughout its life and also to satisfy the serviceability requirements, such as limitations to deflection and cracking.

The acceptable limit of safety and serviceability requirements before failure occurs is called a limit state.

A structure to be designed on the basis of *limit state* should satisfy the following limit states :

(a) Limit state of collapse (Safety requirements)

The limit state of collapse of the structure or part of the structure could be assessed from rupture of one or more critical sections and from buckling due to elastic or plastic instability or overturning. The resistance to bending, shear, torsion and axial loads at every section shall not be less than the appropriate value at that section produced by the probable most unfavourable combination of loads on the structure using the appropriate partial factors.

The following *limit states of collapse* are considered in design :

(*i*) Limit state of collapse in flexure

(*ii*) Limit state of collapse in compression

(*iii*) Limit state of collapse in shear

(*iv*) Limit state of collapse in torsion

(b) Limit state of serviceability

The limit states of serviceability consists of

(*i*) Excessive deflection

(*ii*) Premature or excessive cracking.

(*iii*) Corrosion

(*iv*) Excessive vibrations.

Out of these, the important limit states of serviceability are excessive deflection and cracking.

The deflection of a structure or part thereof shall not adversly affect the appearance or efficiency of the structure or finishes or partitions. Cracking of concrete should not adversely affect the appearance or durability of the structure.

The limit state of excessive deflection and crackwidth is applicable at *service loads* and is estimated on the basis of elastic analysis (working stress method). The limit state of collapse(or failure), however, depends upon *ultimate strength*.

Thus the *limit state method* is a combination of ultimate load and elastic methods for reinforced concrete structures. In fact *the working stress method* is a method of design of limit state of serviceability in which safety factors are applied to the characteristic strength of materials.

23.3. CHARACTERISTIC AND DESIGN VALUES AND PARTIAL SAFETY FACTORS

All relevant limit states should be considered in design to ensure adequate degree of *safety* and *serviceability*. The structure should be designed on the basis of the most critical limit state and then checked for other limits states. For this to be achieved, the design should be based on characteristic values of applied loads, which take into account the variations in the loads to be supported.

The design values are derived from the characteristic value through the use of partial safety factors.

(*a*) MATERIAL STRENGTH

(*i*) *Characteristic strength of materials*

The term characteristic strength means that value of strength of the material below which not more than 5 percent of the results are expected to fall. The characteristic strength of concrete is denoted by f_{ck} (N/mm^2) and its values for different grades of concrete are given in the Appendix. These values, however, may be modified regarding increase in strength with age. Until the relevant Indian Standard Specifications for reinforcing steel are modified to include the concept of characteristic strength, the characteristic value shall be assumed as minimum yield / 0.2 percent proof stress specified in the relevant Indian Standard Specifications.

(*ii*) *Design Values*

The design strength of the material, F_d is given by

$$F_d = \frac{f}{\gamma_m} \qquad ...(23.1)$$

where f = characteristic strength of the material (f_{ck} for concrete and f_y for steel).

γ_m = partial safety factor appropriate to the material and the limit state being considered.

(*iii*) ***Partial safety factor*** γ_m

When assessing the strength of a structure or structural member for limit state of collapse, the value of partial safety factor γ_m should be taken as 1.5 for concrete and 1.15 for steel.

Thus $\gamma_{mc} = 1.5$

and $\gamma_{mc} = 1.15$.

It should be noted that γ_m values are already incorporated in the equations and tables given in IS : 456-1978 for the limit state design.

A higher value of partial safety factor (γ_{mc}) for concrete has been adopted because there are greater chances of variation of strength of concrete due to improper compaction, inadequate curing, improper batching and mixing and variations in the properties of ingradients. The chances of variations in the strength of reinforcement are known to be small, and hence a lower value ($\gamma_m = 1.15$) has been adopted.

Thus, in the limit state method, the design stress for steel reinforcement $= \frac{f_y}{\gamma_{ms}} = \frac{f_y}{1.15} = 0.87 f_y$.

According to the IS code (IS : 456-1978) for design purposes, the compressive strength of concrete in the structure shall be assumed to be 0.67 times the characteristic strength. The partial safety factor $\gamma_{mc} = 1.5$ shall be applied in addition to this.

(*b*) IMPOSED LOADS

(*i*) *Characteristic loads*

The term *characteristic load* means that value of load which has a 95 percent probability of not being exceeded during the life of the structure. Since sufficient data are not available to express loads in statistical terms, the loads given in various relevant Codes shall be assumed as the characteristic loads.

(*ii*) ***Design loads***

The design load F_d is given by

$$F_d = F(\gamma_f) \qquad ...(23.2)$$

where F = characteristic load and

γ_f = partial safety factor appropriate to the nature of loading and the limit state being considered.

(iii) Partial safety factor (γ_f)

The partial safety factors (γ_f) for loads shall be as given in Table 23.1.

TABLE 23.1. PARTIAL SAFETY FACTOR γ_f FOR LOADS

Load Combinations	*Limit state of collapse*			*Limit state of serviceability*		
	DL	*LL*	*WL*	*DL*	*LL*	*WL*
(1)	*(2)*	*(3)*	*(4)*	*(5)*	*(6)*	*(7)*
DL+LL	1.5	1.5	—	1.0	1.0	—
DL+WL	1.5 or 0.9 *	—	1.5	1.0	—	1.0
DL+LL+WL	1.2	1.2	1.2	1.0	0.8	0.8

Note 1 : While considering earthquake effects, substitute *EL* for *WL*.

Note 2 : For the limit state of serviceability, the values of γ_f given in the above table are applicable for short term effects. While assessing the long term effects due to creep, the dead load and that part of live load likely to be permanent may only be considered.

(*This value is to be considered when stability against overturning or stress reversal is critical).

Comparison with working stress method

For the working stress method, we have

$$\gamma_f = 1.0 \text{ for all loads}$$

and

$$\gamma_{ms} = \frac{f_y}{\sigma_{st}}$$

$$= \frac{250}{140} = 1.79 \text{ for mild steel bars}$$

$$= \frac{415}{230} = 1.80 \text{ for HYSD bars of grade Fe 415.}$$

Thus average $\gamma_{ms} = 1.795$.

Also, $\gamma_{mc} = \frac{f_{cu}}{\sigma_{cbc}} \approx 3.0$ for all grades of concrete.

Hence $\gamma_f \times \gamma_{ms} = 1.0 \times 1.795 = 1.795$ for reinforcement and $\gamma_f \times \gamma_{mc} = 1.0 \times 3.0 = 3.0$ for concrete.

For the limit state design, taking maximum compressive stress in concrete $= 0.67 f_{cu}$,

and $\gamma_f = 1.5$; $\gamma_{ms} = 1.15$ and $\gamma_{mc} = 1.5$, we have

$\gamma_f \times \gamma_{ms} = 1.5 \times 1.15 = 1.73$ for reinforcement which is slightly less than 1.795 for working stress design.

and $\gamma_f \times \gamma_{mc} = 1.5 \times \frac{1.5}{0.67} = 3.36$ for concrete, which is more than 3.0 for working stress design.

PROBLEMS

1. Compare various methods of design of reinforced concrete structure.
2. What do you understand by 'limit state' ? Discuss various limit states.
3. Write notes on (*i*) characteristic strength, (*ii*) partial safety factors.

24

Singly Reinforced Sections

24.1. LIMIT STATE OF COLLAPSE IN FLEXURE

As stated in the previous chapter, the following limit state of collapse are considered in design :

(*i*) Limit state of collapse in flexure.

(*ii*) Limit state of collapse in compression.

(*iii*) Limit state of collapse in shear.

(*iv*) Limit State of collapse in torsion.

Design for the limit state of collapse in flexure is based on the following assumptions. These assumptions are valid only for shallow beams where the span to depth ratio is greater than 2.5.

Assumptions in limit state of collapse in flexure

(*a*) plane sections normal to the axis remain plane after bending.

(*b*) Maximum strain in concrete at the outermost compression fibre is taken as 0.0035 in bending.

(*c*) The relationship between the compressive stress distribution in concrete and the strain in concrete may be assumed to be rectangle, trapezoid, parabola or any other shape which results in the prediction of strength in substantial agreement with the results of tests. An acceptable stress-strain curve is given in Fig. 24.1. For design purposes, the *compressive strength of concrete in structure* shall be assumed to be 0.67 times the characteristic strength. The partial safety factor γ_{mc} = 1.5 shall be applied in addition to this.

(*d*) The tensile strength of the concrete is ignored.

(*e*) The stresses in the reinforcement are derived from representative stress-strain curve for the type of steel used. For design purposes, the partial safety factor $\gamma_{ms} = 1.15$ shall be applied.

(*f*) The maximum strain in the tension reinforcement in the section at failure shall not be less than

$$\frac{f_y}{1.15\,E_s} + 0.002$$

where f_y = characteristic strength of steel, and

E_s = modulus of Elasticity of steel.

Comments on the above assumptions

Assumption (*a*) implies linear variation of strain across the depth of the section (See Fig. 24.5). This has been verified by numerous tests on reinforced concrete beams. In assumption (*b*), the maximum strain in concrete is the strain at which the section reaches its maximum moment capacity (See Fig. 24.1). The value of such a strain is liable to large variations depending upon rate of deformation, loading time, grade of concrete, shape of cross- section, percentage of reinforcement etc. The code adopts a value of 0.0035 which will give conservative results in most cases of pure bending. Assumption (*c*) giving acceptable stress-strain curve in concrete (Fig.24.1) gives the shape of compressive stress distribution in concrete [See Fig. 24.5(*iii*)]. Such a stress distribution curve across the depth, popularly known as the *stress block* is a combination of a parabola and a rectangle. Assumption (*d*) is similar to the one made for the working stress method, in which the tensile strength of concrete is neglected. Assumption (*e*) is with reference to stress strain curve for steel reinforcement. For cold twisted bars which do not have definite yield point, the characteristic strength f_y is taken corresponding to 0.2% proof strain. Assumption (*f*) is intended to ensure ductile failure, that is, the tensile reinforcement at the critical section has to undergo a certain degree of inelastic deformation before the concrete fails in compression. In the Code, limit on maximum strain, the first term, $f_y/1.15\,E_s$, corresponds to the yield stress f_y for the case of mild steel and other types which have well defined yield point. However, the cold twisted steel (HYSD bars) do not have a well defined yield point and the yield stress f_y is taken as the conventional value of 0.2 percent proof stress. Hence a strain 0.002 is added so that there is sufficient yielding of steel before failure at a constant stress. It is to be noted that assumption (*b*) and (*f*) govern the maximum depth of neutral axis in flexural members.

24.2. STRESS STRAIN RELATIONSHIP FOR CONCRETE

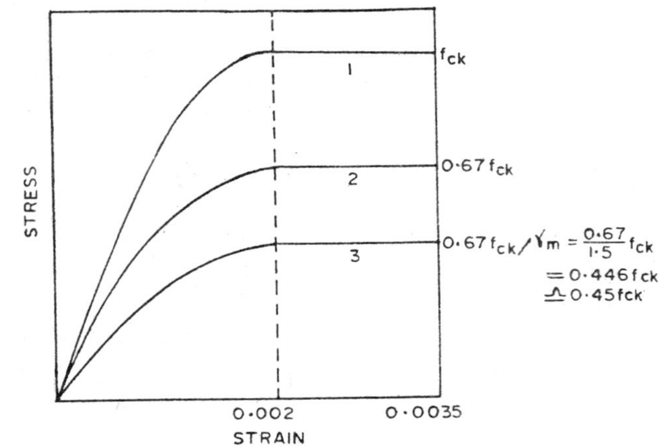

FIG. 24.1. STRESS STRAIN CURVE FOR CONCRETE.

The Code permits the use of any appropriate curve for relationship between the compressive stress and strain distribution in concrete, subject to the condition that it results in the prediction of strength in substantial agreement with test results. An acceptable stress-strain curve (Fig. 24.1) given in Fig.20 of the Code forms the basis of the design in this book. The compressive strength of concrete in the structure is asumed to be 0.67 f_{ck} (curve 2 of Fig. 24.1). The 0.67 factor is introduced to account for difference in strength indicated by a cube test and the strength of concrete in structure. The initial portion of the curve is parabolic. At a strain of 0.002(*i.e.* 0.2% strain) the stress remains constant with increasing load, until a strain of 0.35% is reached when the concrete is said to have failed. The design curve (curve 3 of Fig. 24.1) is obtained by using a partial safety factor $\gamma_m = 1.5$. Thus the maximum compressive stress in concrete for design purpose is equal to

$$\frac{0.67}{1.5} f_{ck} = 0.447 f_{ck} \approx 0.45 f_{ck}.$$

24.3. STRESS-STRAIN RELATIONSHIP FOR STEEL

The stress-strain curve for mild steel is shown in Fig. 24.2 while the stress strain curve for Fe 415 and Fe 500 steel are shown in Fig. 24.3 and Fig. 24.4 respectively.

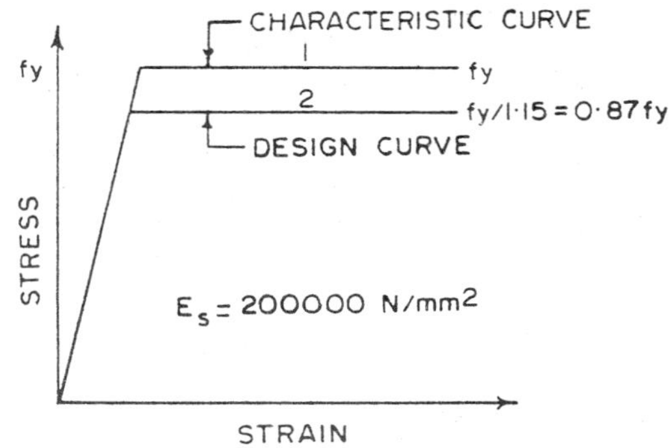

FIG. 24.2. STRESS STRAIN CURVE FOR MILD STEEL

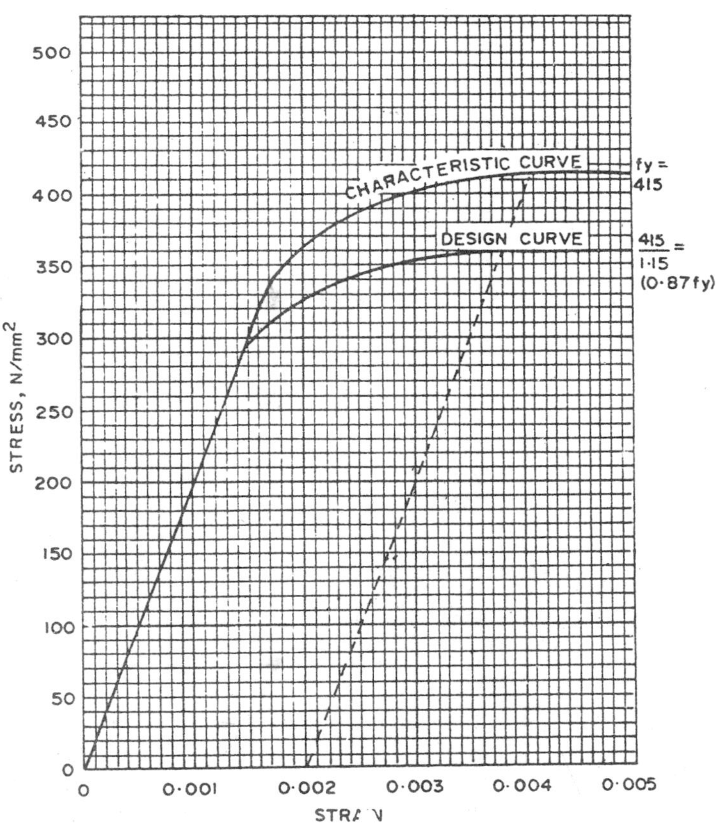

FIG. 24.3. STRESS STRAIN CURVE FOR Fe 415 STEEL.

The modulus of elasticity of all types of reinforcing steel, E_s is taken as 2×10^5 N/mm^2. Also, *the stress strain relationship for steel in tension and compression is assumed to be the same.*

For mild steel, the stress is proportional to strain upto yield point and thereafter the strain increases at constant stress. The design yield stress of steel is equal to f_y / γ_m. With a value of 1.15 for γ_m, the design yield stress f_{yd} becomes 0.87 f_y. For cold worked bars (Fe 415 and Fe 500 HYSD bars), there is no definite yield point ; hence yield stress is taken as 0.2 percent proof stress. The stress-strain curves (Fig. 24.3 and Fig. 24.4) for these two types of steels are linear upto a stress of 0.8 f_y and the strains are elastic. Thereafter, the stress-strain curves are defined as given below :

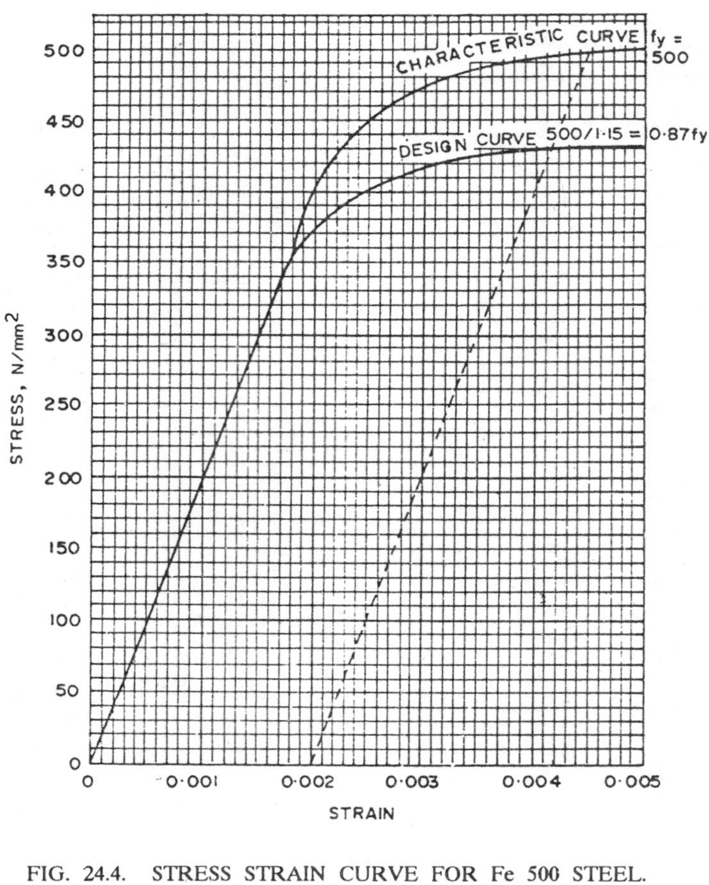

FIG. 24.4. STRESS STRAIN CURVE FOR Fe 500 STEEL.

Stress	*Inelastic strain*
0.80 f_y	Nil
0.85 f_y	0.0001
0.90 f_y	0.0003
0.95 f_y	0.0007
0.975 f_y	0.001
1.0 f_y	0.002

The stress-strain curve for design purposes is obtained by substituting f_{yd} for f_y in the above, where. $f_{yd}/\gamma_m = f_y/1.15 = 0.87 f_y$. The salient points on the design stress strain curve for two types of cold-worked steel are given in Table 24.4.

TABLE 24.4. SALIENT POINTS ON THE DESIGN STRESS-STRAIN CURVE FOR COLD WORKED BARS
(f_{yd} = Design stress ; f_y = Characteristic stress)

Stress level	*Fe 415 steel f_y = 415 N/mm²*		Fe 500 steel f_y = *500 (N/mm²)*	
	Strain	*Stress (N/mm²)*	*Strain*	*Stress (N/mm²)*
0.8 f_{yd}	0.00144	288.7	0.00174	347.8
0.85 f_{yd}	0.00163	306.7	0.00195	369.6
0.90 f_{yd}	0.00192	324.8	0.00226	391.3
0.95 f_{yd}	0.00241	342.8	0.00277	413.0
0.975 f_{yd}	0.00276	351.8	0.00312	423.9
1.0 f_{yd}	0.00380	360.9	0.00417	434.8

Note 1. $f_{yd} = \dfrac{f_y}{\gamma_{ms}} = \dfrac{f_y}{1.15} = 0.87 f_y$

Note. 2. Linear interpolation may be made for intermediate value.

24.4. STRESS BLOCK PARAMETERS

The various theories advanced for estimating the ultimate flexural resistance of reinforced concrete sections can be broadly classified into two categories. In the first category of theories (such as those of Suenson, Menaeh, Kempton, Dyson and Whitney) no assumptions are made regarding the distribution of strain across the section, and neutral axis is located empirically, for analysing compression failures. These theories are based on equilibrium only. The second category of theories (such as those by Stussi or Indian Standard) *include the strain compatibility.* A limiting strain in concrete is used as a criterion for the crushing of concrete in flexural compression.

The stress-strain behaviour of concrete under compression is generally obtained from cylinder or cube of concrete subjected to longitudinal compressive loading. Whereas the stress and strains are uniform for a cube(or for a cylinder), they vary across the depth of a flexural member. The ratio between the maximum flexural compression in a beam and the crushing strength of a cube is likely to be different from unity. The I.S. code (456-1978) recommends the compressive strength of concrete in the structure equal to 0.67 times the characteristic strength. The corresponding stress-strain diagram is shown in Fig. 24.5(*a*), which is the same as curve 2 of Fig. 24.5

Fig. 24.5(*bii*) shows the strain diagram, while Fig. 24.5(*biii*) shows the variation of stresses across the depth. The stress diagram *ABCDEA* [Fig. 24.5(*biii*)] has parabolic shape from *A* to *B* (corresponding to the curve *AB* of Fig. 24.5(*a*), and then linear from *B* to *C* at a constant stress of 0.67 f_{ck} corresponding to line *BC* of Fig. 24.5(*a*). The total compressive force C_u and its location below the top fibre can be expressed in terms of stress block factors k_1, k_2 and k_3 as follows :

k_1 =shape factor, defined as the ratio between the area of stress block *ABCD* and area of rectangle *AFCD*

$$= \frac{\text{Area } ABCD}{\text{Area } AFCD} \quad \text{[Fig 24.5 } (a)]$$

or $$k_1 = \frac{\text{Area } ABCD}{\text{Area } (x \times DC)} \quad \text{[Fig 24.5 } (biii)]$$

Now ultimate strain in concrete $\varepsilon_{cu} = AD = 0.0035$ [Fig. 24.5(*a*)] while the strain $\varepsilon_c = AE$. after which concrete yields at constant stress of 0.67 f_{ck} , is found to be equal to 0.002 for all grades concrete.

$$\text{Ratio } \frac{AE}{AD} = \frac{\varepsilon_c}{\varepsilon_{cu}} = \frac{0.002}{0.0035} = \frac{4}{7}\ .$$

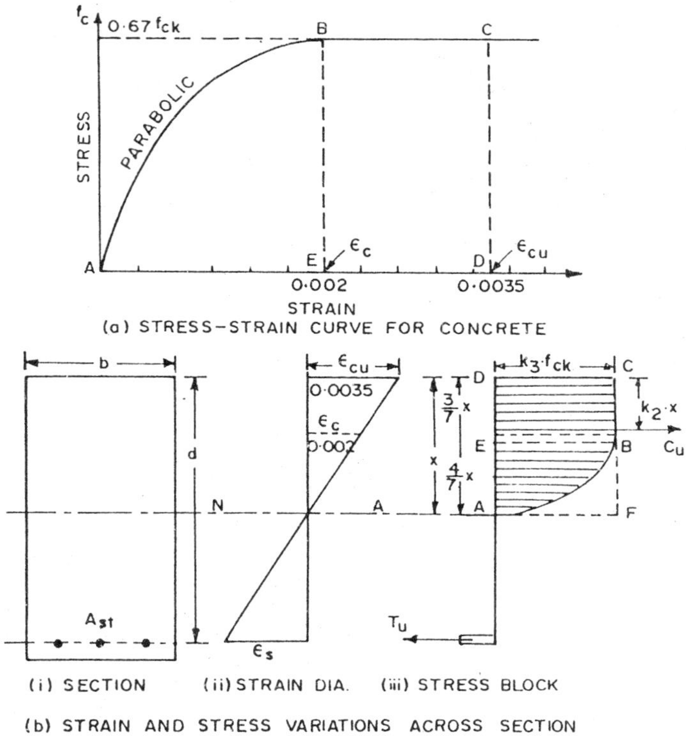

FIG. 24.5.

Also, $\quad \dfrac{ED}{AD} = \dfrac{0.0035 - 0.002}{0.0035} = \dfrac{0.0015}{0.0035} = \dfrac{3}{7}$

Hence $\quad AE = \dfrac{4}{7} AD.$

and $\quad ED = \dfrac{3}{7} AD.$

Now Area $ABCD$ = Area ABE + Area $BCDE$

$$= \left(\frac{2}{3} \times AE \times BE\right) + (ED \times CD)$$

$$= \left(\frac{2}{3} \times \frac{4}{7} AD \times CD\right) + \left(\frac{3}{7} AD \times CD\right)$$

$$= \frac{8}{21}(AD \times CD) + \frac{9}{21}(AD \times CD)$$

$$= \frac{17}{21} \times \text{Area } AFCD$$

$$\therefore\ k_1 = \frac{17}{21} = 0.8095 \approx 0.81 \qquad \text{...}(i)$$

The resultant compressive force is located at a depth $k_2 x$ below the top fibre, where x is the depth of the netural axis.

Now $$k_2.AD = \frac{(\text{Area } ABE) \times \bar{x}_1 + (\text{Area } BCDE) \times \bar{x}_2}{\text{Area } ABCD}$$

$$k_2.AD = \frac{\frac{8}{21}\left(ED + \frac{3}{8}AE\right) + \frac{9}{21}\left(\frac{1}{2}DE\right)}{\frac{17}{21}}$$

$$= \frac{\frac{8}{21}\left(\frac{3}{7} + \frac{3}{8} \times \frac{4}{7}\right)AD + \frac{9}{21}\left(\frac{1}{2} \times \frac{3}{7}\right)AD}{\frac{17}{21}}$$

or $$k_2 = \frac{\frac{36}{7} + \frac{27}{14}}{17} = \frac{99}{238} = 0.416 \approx 0.42.$$

Hence resultant compressive force is located at a depth of $k_2 x\ (=0.416x)$ below the top fibre.

Let the maximum ordinate (DC) of the stress block be $k_3 f_{ck}$ This is the maximum compressive stress in flexure on the structure, in relation to characteristic strength f_{ck} IS : 456-1978 recommends $k_3 = 0.67$.

The compressive force C_u in concrete as given by

$$C_u = b \times \text{area } ABCD \quad [\text{Fig. } 21.5\ (biii)]$$

$$= k_1 \times x \times DC \times b$$

or $$C_u = k_1 k_3 f_{ck}.bx \qquad \text{...(24.3)}$$

The moment of resistance is given by

$$M_u = C_u(d - k_2 x)$$

or $$M_u = k_1 k_3 f_{ck}.bd^2\left(\frac{x}{d}\right)\left(1 - k_2\frac{x}{d}\right) \qquad \text{...(24.4)}$$

24.5. DESIGN STRESS BLOCK PARAMETERS (IS : 456-1978)

The general expressions for C_u and M_u can be easily modified by incorporating partial safety factors γ_{mc} and γ_{ms} for concrete and steel. The stress strain curve for concrete is corresponding to curve 3 of Fig. 24.1 while the stress strain curve for steel are corresponding to curves 2 of Fig. 24.2, 24.3 and 24.4. Fig. 24.6(*b*) shows the strain diagram while Fig. 24.6(*c*) shows the stress block *for design* as recommended by IS : 456–1978. The maximum stress $DC = \frac{k_3 f_{ck}}{\gamma_{mc}}$ $= \frac{0.67 f_{ck}}{1.5} = 0.446 f_{ck.} \approx 0.45 f_{ck}$. *Factor* $k_2 = 0.416 \approx 0.42$ while depth of neutral axis is x_u. The strain ε_{cu} in concrete = 0.0035.

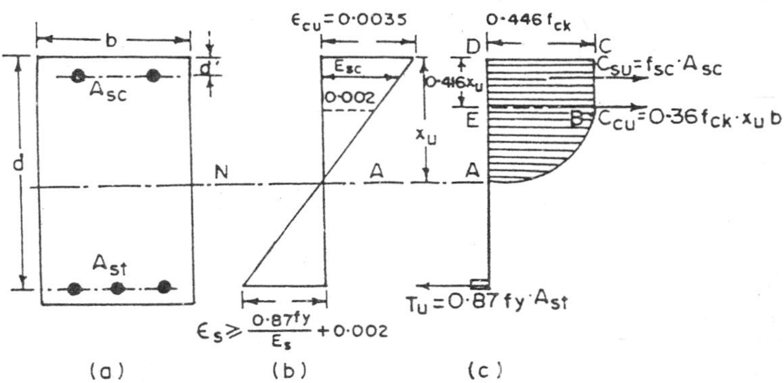

FIG. 24.6. STRAIN AND STRESS DISTRIBUTION FOR DESIGN. (IS : 456-1978)

The maximum stress in steel is limited to $\frac{f_y}{\gamma_{ms}} = \frac{f_y}{1.15} = 0.87 f_y$

Area of design stress block C_u is given by

$$C_u = \text{area } ABCD$$
$$= k_1 \times x_u \times DC$$
$$= k_1 \times x_u \times \frac{k_3 f_{ck}}{\gamma_{mc}}$$

Substituting $k_1 = 0.8095$. $k_3 = 0.67$ and $\gamma_{mc} = 1.5$,

we get $$c_u = 0.35 f_{ck.} x_u \quad \text{...[24.5(a)]}$$

Total compressive force for a rectangular beam of width b is given by

$$C_u = 0.36 f_{ck.} x_{u.} b \quad \text{...(24.5)}$$

The maximum strain ε_{cu} in concrete is taken equal to 0.0035. The Code requires that at failure the strain in reinforcement (ε_{su}) should not be less than, $\left(\frac{0.87 f_y}{E_s} + 0.002\right)$, thus ensuring a stress of $\frac{f_y}{1.15}$ ($= 0.87 f_y$). This limits the depth x_u of N.A. to its maximum of limiting value.

The depth x_u of the N.A. is given by

$$x_u = \frac{\varepsilon_{cu}}{\varepsilon_{cu} + \varepsilon_{su}} d \qquad ...(24.6)$$

Substituting min. value of $\varepsilon_{su} = 0.002 + \frac{0.87 f_y}{E_s}$, we get

$$\frac{x_{u\,(max.)}}{d} = \frac{0.0035}{0.0055 + \frac{0.87 f_y}{E_s}} \qquad ...(24.7)$$

or

$$\frac{x_{u\,(max.)}}{d} = \frac{0.0035\, E_s}{0.0055\, E_s + 0.87 f_y} \qquad ...[24.7(a)]$$

Substituting $E_s = 2 \times 10^5$ N/mm^2, we get

$$\frac{x_{u\,(max)}}{d} = \frac{700}{1100 + 0.87 f_y} \qquad ...[24.7(b)]$$

The limiting values of the depth of the neutral axis, for different grades of steel, based on the above expression are as follows

f_y	$\frac{x_{u(max)}}{d}$
250	0.53
415	0.48
500	0.46

The above limiting values are corresponding to assumptions (para f) of the IS code.

24.6. SINGLY REINFORCED RECTANGULAR BEAMS

The total compressive force C_u in a rectangular beam, singly reinforced, is given by Eq. 24.5., while the total tensile force T_u in reinforcement will be equal to $0.87 f_y . A_{st}$.

For equilibrium of forces [Fig. 24.6(c)].

$$C_u = T_u$$

$$\therefore \quad 0.36 f_{ck}\, x_u\, b = 0.87 f_y . A_{st}$$

or

$$\frac{x_u}{d} = \frac{0.87 f_y A_{st}}{0.36 f_{ck}\, bd} = 2.417\, p \frac{f_y}{f_{ck}} \qquad ...(24.8)$$

where $$p = \frac{A_{st}}{bd} = \text{reinforcement ratio.}$$

and the limiting or max. value of $\frac{x_u}{d}$ is given by Eq. 24.7.

Three cases may arise.

Case 1 : (x_u/d) less than limiting value

The case corresponds to under-reinforced section in working stress method, in which the reinforcement ratio $p \left(= \frac{A_{st}}{bd}\right)$ is less than p_{lim} corresponding to the limiting value $\frac{(x_{u,\,max})}{d}$, where

$$p_{lim} = \frac{1}{2.417} \cdot \frac{f_{ck}}{f_y} \frac{x_{u,\,max}}{d}$$

or $$p_{lim} = 0.414 \frac{f_{ck}}{f_y} \cdot \frac{x_{u,\,max}}{d}$$

The values of p_{lim} for various values of f_{ck} and f_y are given in Table 24.2.

TABLE 24.2. VALUE OF p_{lim} (%)

f_{ck} (N/mm^2)	*Values of 100 p_{lim} for*		
	f_y = 250 N/mm^2	f_y = 415 N/mm^2	f_y = 500 N/mm^2
15	1.316	0.718	0.571
20	1.755	0.958	0.762
25	2.194	1.197	0.952
30	2.633	1.436	1.112

The moment of resistance for such a case is conveniently expressed as

$$M_u = T_u (d - 0.416\, x_u)$$

where $$T_u = 0.87 f_y A_{st} = 0.87 f_y .\ pbd$$

$$\therefore \quad M_u = 0.87 f_y A_{st} (d - 0.416\, x_u)$$

$$M_u = 0.87 f_y A_{st} \cdot d \left(1 - 0.416 \frac{x_u}{d}\right) \quad ...[24.10(a)]$$

Substituting the value of $\frac{x_u}{d}$ from Eq. 24.8, we get

$$M_u = 0.87 f_y A_{st} . d \left(1 - 0.416 \frac{0.87}{0.36} \frac{f_y}{f_{ck}} . \frac{A_{st}}{bd}\right)$$

or
$$M_u = 0.87 f_y A_{st} . d \left(1 - \frac{f_y}{f_{ck}} . \frac{A_{st}}{bd}\right) \qquad ...(24.10)$$

Case 2 : (x_u/d) equal to limiting value

This case corresponds to *balanced design* in which the steel reinforcement reaches its yield stress at the same instant when ultimate strain is reached in concrete. The stress block is fully developed as shown in Fig. 24.6(c). The moment of resistance M_u reaches its limiting value $M_{u, limit,}$ and given by

$$M_{u, lim} = 0.36 f_{ck} . (x_u)_{max} b. (d - 0.416 x_{u\,(max)})$$

or
$$M_{u,lim} = 0.36 f_{ck} . \frac{x_{u\,(max)}}{d} \left(1 - 0.416 \frac{x_{u,\,max}}{d}\right) bd^2 \qquad ...(24.11)$$

Case 3 : (x_u/d) greater than the limiting value

The case corresponds to the beam in which the percentage of steel is sufficient to ensure that steel yield does not take place. Failure occurs when the strain in extreme fibres in concrete reaches its ultimate value. Thus the failure takes place due to crushing of concrete while strain ε_{su} in steel still remains below the yield strain ε_{su}.

It should be noted that compression failure is *sudden,* and therefore not desirable. The Code recommends that if (x_u/d) is found greater than the limiting value, *the beam should be redesigned.*

24.7. PROCEDURE FOR FINDING MOMENT OF RESISTANCE

The moment of resistance of rectangular sections without compression reinforcement should be obtained as follows (IS : 456-1978) :

(a) Determine the depth of neutral axis from the following equation :

$$\frac{x_u}{d} = \frac{0.87 f_y . A_{st}}{0.36 f_{ck}\, bd} = 2.417 \frac{f_y}{f_{ck}} . \frac{A_{st}}{bd} \qquad ...(24.8)$$

(b) If the value of x_u/d is less than the limiting(*i.e.* max.) value given by Eq. 24.7 calculate the moment of resistance by the following equation.

$$M_u = 0.87 f_y . A_{st} . d \left(1 - \frac{A_{st}}{bd} . \frac{f_y}{f_{ck}}\right) \qquad ...(24.10)$$

(*c*) If the value of x_u/d is equal to the limiting value, the moment of resistance of the section is given by the following expression :

$$M_{ulim} = 0.36 \frac{x_{u(max)}}{d} \left(1 - 0.416 \frac{x_{u(max)}}{a}\right) bd^2 f_{ck} \qquad ...(24.11)$$

(*d*) If x_u/d is greater than the limiting value, the section is *over-reinforced.* The moment of resistance of such a section is limited to $M_{u(lim)}$ given by Eq. 24.11. It should be noted that in the case of an over-reinforced section, not only that there is wastage of excess steel, the concrete reaches its ultimate capacity before steel yields resulting in a sudden failure. *The code therefore recommends that such a section should be redesigned.*

Variation of moment of resistance with p

Eq. 24.11 gives the variation moment of resistance M_u with reinforcement ratio $p = \left(\frac{A_{st}}{bd}\right)$. The equation can be rearranged in the form

$$\frac{M_u}{bd^2} = 0.87 f_y \frac{A_{st}}{bd} \left(1 - \frac{f_y}{f_{ck}} \frac{A_{st}}{bd}\right)$$

or

$$\frac{M_u}{bd^2} = 0.87 f_y p \left(1 - \frac{f_y}{f_{ck}} p\right) \qquad [24.10(a)]$$

Thus M_u increases with increase in the value of *p*. When *p* reaches limiting value $p_{lim.}$ given by Eq. 24.9(or by table 24.3), M_u also reaches the limiting or maximum value given by Eq. 24.11 which can be rearranged in the form :

$$\left(\frac{M_u}{bd^2}\right)_{lim} = 0.36 f_{ck} \frac{x_{u\,max}}{d} \left(0.146 \frac{x_{u\,max}}{d}\right) = R_u \qquad ...[24.11(a)]$$

For mild steel bars having $f_y = 250 \text{ N/mm}^2$, we have $\frac{x_{u\,max}}{d}$ $= 0.53$ and $p = 1.316 \times 10^{-2}$ when $f_{ck} = 15 \text{ N/mm}^2$. Hence

$$M_{u\,lim} = 0.36 \times 15 \times 0.53\,(1 - 0.416 \times 0.53)$$
$$= 2.231\, bd^2 \text{ N–mm.}$$

For other values of *p* , M_u can be found from Eq. 24.10(*a*). Fig. 24.7 shows the variation of $\frac{M_u}{bd^2}$ with *p*, for mild steel bars (having $f_y = 250 \text{ N/mm}^2$ and 15 concrete having$f_{ck} = 15 \text{ N/mm}^2$.

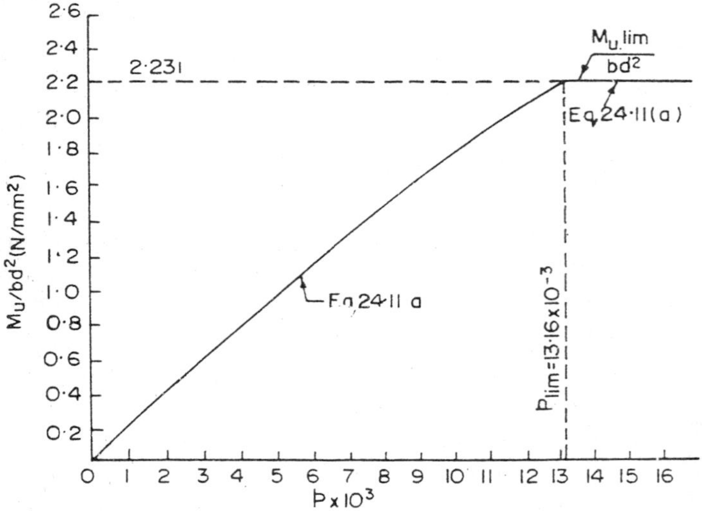

FIG. 24.7. VARIATION OF $\frac{M_u}{bd^2}$ WITH p FOR SINGLY REINFORCED SECTION.

Moment of resistance of balanced section

For a balanced design, the steel reinforcement reaches its yield stress at the same instant when ultimate strain is reached in concrete. The moment of resistance for such a beam is given by Eq. 24.11.

$$M_{u,\,lim} = 0.36 f_{ck} \frac{x_{u,\,max}}{d} \left(1 - 0.416 \frac{x_{u,\,max}}{d}\right) bd^2 \qquad ...(24.11)$$

In the above expression, the value of $\frac{x_{u,\,max}}{d}$ for mild steel, Fe 415 steel and Fe 500 steel are respectively 0.53, 0.48 and 0.46. Substituting these values, we get the following simplied expressions for moment of resistance, for various types of steel :

(*a*) ***For mild steel bars*** **(Fe 250)**

$$M_{u,\,lim} = 0.1487 f_{ck}\, bd^2 = R_u\, bd^2 \qquad ...[24.11(a)]$$

where $R_u = 0.1487 f_{ck}$.

(*b*) ***For Fe 415 Steel***

$$M_{u,\,lim} = 0.1383 f_{ck}\, bd^2 = R_u\, bd^2 \qquad ...[24.11(b)]$$

where $R_u = 0.1383 f_{ck}$

(c) ***For Fe 500 steel***

$$M_{u,\,lim} = 0.1339 f_{ck}\, bd^2 = R_u\, bd^2 \qquad ...[24.11(c)]$$

where $R_u = 0.1339 f_{ck}$.

Table 24.3 gives the values of R_u for various grades of concrete.

TABLE 24.3 VALUES OF R_u FOR BALANCED DESIGN

Concrete Grade	*Reinforcement*		
	Fe 250 ($R_u = 0.1487\ f_{ck}$)	*Fe 415* ($R_u = 0.1383\ f_{ck}$)	*Fe 500* ($R_u = 0.339\ f_{ck}$)
M 15	2.231	2.074	2.009
M 20	2.975	2.766	2.678
M 25	3.718	3.457	3.348
M 30	4.462	4.149	4.017
M 35	5.206	4.840	4.686
M 40	5.949	5.532	5.356

The readers are advised to compare these values of R_u with 1.5 R values of working stress method. It will be seen that R_u is greater than 1.5 R, resulting in smaller concrete section. Table 24.2 gives the limiting percentage ($p_{\lim}$) of steel. *It will thus be seen that the balanced design gives smallest concrete section and maximum area of reinforcement.* Since the cost of steel is quite high, an *under-reinforcement section,* which gives larger concrete area and smaller value of p, is always desirable in the limit state design. This can be achieved by (*i*) providing b or d or both greater than those required by balanced design, or (*ii*) using lesser values of $\frac{x_u}{d}$ than the limiting value or (*iii*) assuming lesser value of R_u than the limiting value given in Table 24.3.

24.8. DESIGN OF RECTANGULAR BEAM SECTION

The design of a section consists of determination of (*i*) cross-sectional dimensions b and d, and (*ii*) area of steel, so as to develop a given moment of resistance. Though the objective of a designer would be to design a *balanced section* so that the ultimate stresses in both the materials are developed simultaneously, such a design may not be the most economical. It has been stated in the previous para that a balanced design gives smallest concrete section and maximum area of reinforcement. Since the cost of steel is very high in comparison to that of concrete, a balanced design may not be economical. Also, for practical considerations, sometimes, it may be necessary to fix some uniform cross-sectional dimensions. In such a case, the design may result in a singly reinforced balanced section, under-reinforced section or doubly reinforced section. However, if the section has to be singly reinforced, an under-reinforced section is always more desirable in the limit state design.

1. Design to determine cross-sectional dimensions and reinforcement.

This is the most usual case of design in which the ultimate moment of resistance (M_u) to be developed by the section is given, and it is required to determine b, d and $A_{st.}$ The design is done in the following steps.

Step 1. Determine the limiting depth of N.A. by Eq. 24.7

$$x_{u\,(max)} = \frac{0.0035}{0.0055 + 0.87 f_y/E_s} d \qquad ...(24.7)$$

or

$$\frac{x_{u\,(max)}}{d} = \frac{700}{1100 + 0.87 f_y} \qquad ...[24.7(b)]$$

Step 2. Choose some suitable *ratio* of d and b. The value of d/b in the range of 1.5 to 3 is usually taken.

Step 3. Find d from the relation (Eq. 24.11).

$$M_u = 0.36 f_{ck} \frac{x_{u\,(max)}}{d} \left(1 - 0.416 \frac{x_{u\,max}}{d}\right) bd^2 \qquad ...(24.11)$$

or

$$M_u = R_u\, bd^2 \qquad ...[24.11(b)]$$

Step 4. Knowing b and d, determine area of reinforcement from the relation

$$M_u = 0.87 f_y A_{st} (d - 0.416 x_{u\,(max)}) \qquad ...(24.18\ (a)]$$

Alternatively, find $p_{lim.}$ from Eq. 24.9 :

$$p_{lim} = 0.414 \frac{f_{ck}}{f_y} \frac{x_{u\,(max)}}{d} \qquad ...(24.9)$$

and

$$A_{st} = p_{lim} \times bd$$

The values of $p_{lim.}$ are given in Table 24.2.

2. Design to determine area of reinforcement if *b* and *d* are known.

In this case M_u, b and d are given. To start with, we have to ascertain whether it will result in a singly reinforced section or a doubly reinforced section. The design is done in the following steps :

Step 1. Determine the limiting moment of resistance from Eq. 24.11

$$M_{u\,lim} = 0.36 f_{ck} \frac{x_{u\,(max)}}{d} \left(1 - 0.416 \frac{x_{u\,(max)}}{d}\right) bd^2$$

or $$M_{u\,lim} = R_u\, bd^2$$

Step 2. (*a*) If $M_u = M_{u\,lim}$, it will result in a balanced section, and area of steel (or p_{lim}) is found either from Eq. 24.9 or from Table 24.2.

(*b*) If $M_u > M_{u\,lim}$, , it will result in a doubly reinforced section (see chapter 25).

(*c*) If $M_u < M_{u\,lim}$, it will result in an *under-reinforced* section, and the design is done in the following steps :

Step 3. If $M_u < M_{u\,lim}$, determine actual x_u from relationship :

$$M_u = 0.36 f_{ck}\, b\, x_u\, (d - 0.416\, x_u)$$

This will result in a quadratic equation in terms of x_u which can be easily solved.

Step 4. Determine area of steel from the relation

$$M_u = 0.87 f_y A_{st}\, (d - 0.416\, x_u)$$

or $$A_{st} = \frac{M_u}{0.87 f_y\, (d - 0.416\, x_u)} \qquad ...(24.12)$$

Alternatively, find p from 24.8 :

$$p = 0.414 \frac{f_{ck}}{f_y} \frac{x_u}{d} \qquad ...[24.8(a)]$$

$\therefore$ $$A_{st} = p\, bd$$

Alternatively, steps 3 and 4 can be avoided and A_{st} can be found from Eq. 24.10 applicable for under-reinforced section.

$$M_u = 0.87 f_y A_{st}\, d \left(1 - \frac{A_{st}}{bd} \frac{f_y}{f_{ck}}\right) \qquad ...(24.10)$$

This gives a quadratic equation in terms of A_{st}, the solution of which works out as under :

$$A_{st} = \frac{0.5 f_{ck}}{f_y} \left[1 - \sqrt{1 - \frac{4.6\, M_u}{f_{ck}\, bd^2}}\right] bd \qquad ...(24.13)$$

Example 24.1. M.R. of singly reinforced section.

Find the M.R. of a singly reinforced concrete beam section of 200 mm width and 400 mm effective depth, reinforced with 4 bars of 20 mm Φ (mild steel). Use M 15 concrete. Use I.S. code method. Redesign the beam if necessary.

Solution

$$A_{st} = 4\frac{\pi}{4}(20)^2 = 1256.6 \text{ mm}^2$$

$$p = \frac{A_{st}}{bd} = \frac{1256.6}{200 \times 400} = 0.0157$$

For mild steel bars, $f_y = 250$ N/mm².

For M 15 concrete, $f_{ck} = 15$ N/mm².

Hence the depth of N.A. from Eq. 24.8 :

$$\frac{x_u}{d} = 2.417\,p\frac{f_y}{f_{ck}}$$

$$= 2.417 \times 0.0157 \times \frac{250}{15} = 0.6328 \qquad ...(i)$$

The limiting value of $\frac{x_u}{d}$ is given by Eq. 24.7(*b*) :

$$\frac{x_{u\,max}}{d} = \frac{700}{1100 + 0.87 \times 250} = 0.53$$

Thus, the actual N.A. depth is more than the limiting one. Such a beam is over-reinforced, and hence is *undesirable,* since the failure is sudden and without warning. The Code recommends that such a beam should be redesigned. However, the moment of resistance of such a beam will be taken equal to the limiting moment of resistance. The limiting moment of resistance of such a beam can be found by use of Eq. 24.11.

$$M_{u\,lim} = 0.36 f_{ck}\frac{x_{u\,max}}{d}\left(1 - 0.416\frac{x_{u\,max}}{d}\right)bd^2$$

$$= 0.36 \times 15\,(0.53)\,(1 - 0.416 \times 0.53)\,200\,(400)^2$$

$$= 71.4 \times 10^6 \text{ N–mm} = \mathbf{71.4 \text{ kN-m}}$$

The corresponding reinforcement for the balanced failure condition is given by Eq. 24.8.

$$\frac{x_{u\,max}}{d} = 2.417\,p\frac{f_y}{f_{ck}}$$

or
$$p = \left(\frac{x_{u\,max}}{d}\right)\frac{f_{ck}}{f_y} \times \frac{1}{2.417}$$
$$= \frac{0.53 \times 15}{250 \times 2.417} = 0.01316$$

$$\therefore \quad A_{st} = p\,bd = 0.01316\,(200)\,400 = \mathbf{1052.5\ mm^2}.$$

(Against 1256.6 mm^2 at present).

Example 24.2. Moment of resistance of singly reinforced beam.

For the beam of Example 24.1, find the M.R. if the reinforcement consists of 3-20 mm Φ bars.

Solution

$$A_{st} = 3\frac{\pi}{4}(20)^2 = 942.48\ \text{mm}^2$$

$$p = \frac{942.48}{200 \times 400} = 0.01178$$

$$\frac{x_u}{d} = 2.417\,p\,\frac{f_y}{f_{ck}} = 2.417 \times 0.01178\,\frac{250}{15}$$
$$= 0.4749$$

$$\frac{x_{u\,max}}{d} = 0.53\text{, as before.}$$

Hence $\frac{x_u}{d}$ is less than the limiting value and M.R. is given by Eq. 24.10.

$$M_u = 0.87 f_y A_{st}\,d\left(1 - \frac{f_y}{f_{ck}}\frac{A_{st}}{bd}\right)$$
$$= 0.87 \times 250 \times 942.48 \times 400\left(1 - \frac{250}{15} \times 0.01178\right)$$
$$= 65.9 \times 10^6\ \text{N–mm} = \mathbf{65.9\ kN\text{-}m.}$$

Example 24.3. Determination of actual stresses.

For the beam of Example 24.1, determine the actual stress in steel when the section is subjected to the limiting moment of resistance.

Solution. Given $A_{st} = 1256.6\ \text{mm}^2$ and $M_u = M_{u\,lim}$. As found earlier,

$$\frac{x_{u\,max}}{d} = 0.53 \quad \text{and} \quad \frac{x_u}{d} = 0.6328.$$

It is very clear that if $\left[\frac{x_u}{d}\right]$ is taken = 0.6328, the stress in concrete will exceed 0.45 f_{ck} . However, at $M_u = M_{u\,(\text{lim})}$, x_u will be equal to $x_{u\,(max)}$ and stress in concrete will be equal to 0.45 f_{ck}. In such a circumstance, the stress f_s in steel will be less than f_y.

From the equilibrium of internal forces, we have

$$C = T$$

or $$0.36 f_{ck}\, b\, x_u = A_{st} f_s$$

or $$f_s = \frac{0.36 f_{ck}\, b\, x_u}{A_{st}}, \text{ where } x_u = x_{u\,max}$$

or $$f_s = \frac{0.36 f_{ck}\, bd\, (x_{u\,max}/d)}{A_{st}}$$

$$= \frac{0.36 \times 15 \times 200 \times 400\ (0.53)}{1256.6}$$

$$= \mathbf{182.2\ N/mm^2} \text{ (against a value of } f_y = 250\ \text{N/mm}^2)$$

The corresponding strain in steel is

$$\varepsilon_s = \frac{0.0035\ (d - x_{u\,max})}{x_{u\,max}}$$

$$= \frac{0.0035\ (1 - x_{u\,max}/d)}{x_{u\,max}/d}$$

$$= \frac{0.0035\ (1 - 0.53)}{0.53} = 0.0031.$$

Check. Lever arm $jd = d - 0.416\, x_{u\,max} = d\left(1 - 0.416 \frac{x_{u\,max}}{d}\right)$

$$= 400\ (1 - 0.416 \times 0.53) \approx 312\ \text{mm}$$

$\therefore$ $$M_u = Cjd = (0.36 f_{ck}\, b\, x_{u\,max})\, jd$$

$$= 0.36 \times 15 \times 200\ (0.53 \times 400) \times 312$$

$$= 71.4 \times 10^6\ \text{N-mm}.$$

Also, $$M_u = T jd = A_{st} f_s\, jd$$

$$= 1256.6 \times 182.2 \times 312 = 71.4 \times 10^6\ \text{N-mm}$$

Each one of these is equal to $M_{u\,lim}$.

Example 24.4. Design of singly reinforced beam.

Design a balanced singly reinforced concrete beam section for an applied moment of 48 kN-m. The width of the beam is limited to 150 mm. Use M 15 concrete and mild steel bars.

Also, determine the reinforcement.

Solution.

$$M_D = 48 \text{ kN-m} = 48 \times 10^6 \text{ N-mm}.$$

Since $\gamma_f = 1.5$ for both dead and live loads,

$$M_{uD} = 48 \times 10^6 \times 1.5 = 72 \times 10^6 \text{ N-mm}.$$

For balanced design, we have, from Eq. 24.7 (*b*).

$$\frac{x_{u\,max}}{d} = \frac{700}{1100 + 0.87 f_y} = \frac{700}{1100 + 0.87 \times 250} = 0.53$$

The moment of resistance for such a beam is given by Eq. 24.11.

$$M_{u.limit} = 0.36 f_{ck} \frac{x_{u.max}}{d}\left(1 - 0.416 \frac{x_{u.max}}{d}\right) bd^2$$

Equating $M_{u.limit}$ and M_{uD}, we get

$$72 \times 10^6 = 0.36 \times 15\,(0.53)\,(1 - 0.416 \times 0.53)\,150\,d^2$$

From which $\boldsymbol{d = 465}$ **mm.**

The reinforcement is given by Eq. 24.8.

$$\frac{x_{u.max}}{d} = 2.417\, p_{lim.} \frac{f_y}{f_{ck}}$$

$$p_{lim.} = \frac{0.53 \times 15}{250 \times 2.417} = 0.01316$$

(This is same as given in Table 24.2 for $f_{ck} = 15$ and $f_y = 250$)

$$\therefore \quad A_{st} = 0.01316 \times 150 \times 465 = \mathbf{917.9\ mm^2}$$

Example 24.5. Design of singly reinforced beam.

Find the reinforcement for the beam of Example 24.4 if the depth of the beam is kept equal to the one obtained by working stress method.

Solution.

For M 15 concrete, with mild steel bars

$$k = \frac{m\,\sigma_{cbc}}{m\,\sigma_{cbc} + \sigma_{st}} = \frac{19 \times 5}{19 \times 5 + 140} = 0.404$$

$$j = 1 - \frac{0.404}{3} = 0.865$$

$$R = \frac{1}{2} \times 50 \times 0.865 \times 0.404 = 0.874$$

$$\therefore \qquad d = \sqrt{\frac{M_D}{Rb}} = \sqrt{\frac{48 \times 10^6}{0.865 \times 150}} \approx 600 \text{ mm}$$

The limiting ultimate moment for this beam is given by Eq. 24.11.

$$M_{u.lim.} = 0.36 f_{ck} \frac{x_{u.max}}{d} \left(1 - 0.416 \frac{x_{u.max}}{d}\right) bd^2$$

Taking $\frac{x_{u.max}}{d} = 0.53$, we have

$$M_{u.lim.} = 0.36 \times 15 \times 0.53 \times (1 - 0.416 \times 0.53)\, bd^2$$

$$= 2.231\, bd^2 \text{ (see Fig. 24.7 and Table 24.3 also)}$$

$$= 2.232 \times 150\, (600)^2 = 120.5 \times 10^6 \text{ N-mm}$$

$$M_{uD} = M_D \times \gamma_f = 48 \times 10^6 \times 1.5 = 72 \times 10^6 \text{ N-mm}$$

The reinforcement is given by Eq. 24.10.

$$M_u = 0.87 f_y A_{st}\, d \left(1 - \frac{f_y}{f_{ck}} \frac{A_{st}}{bd}\right)$$

or $$72 \times 10^6 = 0.87 \times 250 A_{st} \times 600 \left(1 - \frac{250}{15} \times \frac{A_{st}}{150 \times 600}\right)$$

or $$A_{st} \times (1 - 185 \times 10^{-6} A_{st}) = 551.7$$

or $$A_{st}^2 - 5402 A_{st} + 2980550 = 0$$

From which $A_{st} \approx$ **624 mm²** .

Alternatively, from Eq. 24.13.

$$A_{st} = \frac{0.5 \times f_{ck}}{f_y} \left[1 - \sqrt{1 - \frac{4.6 M_u}{f_{ck}\, bd^2}}\right] bd$$

or $$A_{st} = \frac{0.5 \times 15}{250} \left[1 - \sqrt{1 - \frac{4.6 \times 72 \times 10^6}{15 \times 150\, (600)^2}}\right] 150 \times 600$$

or $$A_{st} = 624 \text{ mm}^2$$

(Against 917.7 mm² provided in the previous example).

Let us compare this with working stress method in which the reinforcement is given by

$$A_{st} = \frac{M_D}{\sigma_{st}\, jd} = \frac{48 \times 10^6}{140 \times 0.865 \times 600} = 660.6 \text{ mm}^2$$

Thus the reinforcement by working stress method is only about 6% more.

Example 24.6. Design of singly reinforced beam.

A beam, simply supported over an effective span of 7 m carries a live load of 15 kN/m. Design the beam, using M 15 concrete and HYSD bars of grade Fe 415. Keep width equal to half the effective depth. Assume unit weight of concrete as $2.4\ kN/m^3$

Solution.

For the computation of dead load, let us assume that, the beam has size 250 mm × 550 mm.

$\therefore$ Self weight $w_d = 0.25 \times 0.55 \times 1 \times 24000 = 3300$ N/m.

Live load $w_l = 15000$ N/m

Total load $w_D = 3300 + 15000 = 18300$ N/m

Using $\gamma_f = 1.5$.

Ultimate design load $w_{uD} = 18300 \times 1.5$

$= 27450$ N/m

$$\therefore \quad M_{uD} = \frac{27450\,(7)^2}{8} = 168130 \text{ N-m.}$$

$$= 168.13 \times 10^6 \text{ N-mm.}$$

For M 15 concrete, $f_{ck} = 15\ N/mm^2$ and for *HYSD* bars of Fe 415 grade, $f_y = 415\ N/mm^2$. Hence the limiting depth of N.A. is [Eq. 24.7(*b*)].

$$\frac{x_{u.max}}{d} = \frac{700}{1100 + 0.87 f_y} = \frac{700}{1100 + 0.87 \times 415} \approx 0.48.$$

Let $d =$ effective depth, so that $b = 05\,d$

Also, from Eq. 24.11.

$M_{u.lim} = 0.36 \times 15 \times 0.48(1 - 0.416 \times 0.48)(0.5\,d)(d)^2 = 1.0372\,d^3$

Equating this to the design moment, we get

$168.13 \times 10^6 = 1.0372\,d^3$

From which $d = 647$ mm

and $b \approx \frac{1}{2} d = 273$ mm

The reinforcement is given by Eq. 24.9.

$$p_{lim} = 0.414 \frac{f_{ck}}{f_y} \frac{x_{u.max}}{d}$$

$$= 0.414 \times \frac{15}{415} \times 0.48 = 0.00718$$

$$A_{st} = 0.00718 \times 273 \times 546 = \mathbf{1070\ mm^2}.$$

Example 24.7. Design of singly reinforced beam.

A reinforced concrete beam has width equal to 300 mm and total depth equal to 800 mm, with a cover of 40 mm to the centre of reinforcement. Design the beam if it is subjected to a total bending moment of 140 kN-m. Use M 15 concrete and HYSD bars of grade Fe 415. Compare the design with that obtained by working stress method.

Solution.

$b = 300$ mm ; $d = 800 - 40 = 760$ mm

Using $\gamma_f = 1.5$; $M_{uD} = 140 \times 1.5 = 210$ kN-m

For M 15 grade concrete and *HYSD* bars,

$$\frac{x_{u.max}}{d} = \frac{700}{1100 + 0.87 f_y} = \frac{700}{1100 + 0.87 \times 415} \approx 0.48$$

Taking limiting moment is given by

$$M_{u.lim} = 0.36 \times 15 \times 0.48\,(1 - 0.416 \times 0.48) \times bd^2$$

$$= 2.074\, bd^2 = 2.074 \times 300 \times (760)^2$$

$$\approx 359 \times 10^6 \quad \text{N-mm} > M_{uD}.$$

Hence compression reinforcement is not needed.

$$M_{uD} = 0.87 f_y A_{st}\, d \left(1 - \frac{f_y}{f_{ck}} \frac{A_{st}}{bd}\right)$$

or $\quad 210 \times 10^6 = 0.87 \times 415 \times A_{st}\,(760) \left(1 - \frac{415}{15} \times \frac{A_{st}}{300 \times 760}\right)$

or $\quad A_{st}\,(1 - 121.3 \times 10^{-6} A_{st}) = 765.3$

or $\quad A_{st}^2 - 8244 A_{st} + 6.31 \times 10^6 = 0$

From which $A_{st} = \mathbf{853.8\ mm^2}$.

Alternatively, From Eq. 24.13, we have,

$$A_{st} = \frac{0.5 f_{ck}}{f_y} \left[1 - \sqrt{1 - \frac{4.6 M_u}{f_{ck}\, bd^2}}\right] bd$$

$$= \frac{0.5 \times 15}{415} \left[1 - \sqrt{1 - \frac{4.6 \times 210 \times 10^6}{15 \times 300\,(760)^2}}\right] 300 \times 760$$

$$= 854 \text{ mm}^2$$

Let us design the beam by working stress method. The moment of resistance of the singly reinforced beam is

$$M = Rbd^2$$

Now, $\quad k = \dfrac{m\,\sigma_{cbc}}{m\,\sigma_{cbc} + \sigma_{st}} = \dfrac{19 \times 5}{19 \times 5 + 230} = 0.2923$

$$j = 1 - \frac{0.2923}{3} = 0.9026$$

$$\therefore \quad R = \tfrac{1}{2} \times 5 \times 0.9026 \times 0.2923 = 0.66$$

$$\therefore \quad M = 0.66 \times 300\,(760)^2 = 114.3 \times 10^6 \text{ N-mm.}$$

Since $M < M_D$, compression reinforcement is required.

$$A_{st1} = \frac{M_1}{\sigma_{st}\, jd} = \frac{114.3 \times 10^6}{230 \times 0.9026 \times 760} = 724.5 \text{ mm}^2$$

$$M_2 = M_D - M_1 = (140 - 114.3)\, 10^6 = 25.7 \times 10^6 \text{ N-mm.}$$

$\therefore$ Assuming a cover of 40 mm for compressive reinforcement also,

$$A_{st2} = \frac{25.7 \times 10^6}{230\,(760 - 40)} = 155.2 \text{ mm}^2 \ .$$

Compressive steel to balance this is given by

$$A_{sc} = \frac{m\,(d - n)}{(m_c - 1)\,(n - d_c)} A_{st2}$$

where $m_c = 1.5 \times 19 = 28.5$; $n = kd = 0.2923 \times 760 = 222$ mm

$$\therefore \quad A_{sc} = \frac{19\,(760 - 222)}{(28.5 - 1)\,(222 - 40)} \times 155.2$$

$$\approx 317 \text{ mm}^2 \ .$$

$\therefore$ Total $A_{st} = A_{st1} + A_{st2} = 724.5 + 155.2 = 879.7 \text{ mm}^2$

$A_{sc} = 317 \text{ mm}^2$.

$\therefore$ Total steel $= 879.7 + 317 = 1196.7 \text{ mm}^2$

(Against 853.8 mm^2 required by limit state analysis)

PROBLEMS

1. A beam section 300 mm wide and 500 mm deep is reinforced with a tension reinforcement of 3000 mm^2 at an effective cover of 300 mm. Determine the ultimate moment of resistance of beam section. Use M 15 concrete and Fe 250 grade steel.
2. Solve problem 1 if the grade of concrete is M 20 and grade of steel is Fe 415.
3. Solve problem 1 for M 15–Fe 415 combination.
4. Design a balanced singly reinforced concrete beam with a span of 5 m to carry a dead load of 25 kN/m and a working live load of 20 kN/m. Use M 20 concrete and Fe 415 grade steel.
5. Redesign the beam of problem 2 if the cross-section of the beam is limited to 300 mm width and 600 mm effective depth. Use M 20 mix and Fe 415 grade steel.

25

Doubly Reinforced Sections

25.1. NECESSITY

We have seen in the previous chapter that a singly reinforced section has a limiting value of moment of resistance, corresponding to limiting value of steel reinforcement. However, if the applied moment M_{ua} is larger than $M_{u\,lim}$, two alternatives will be available: (*i*) to increase the depth of the section or (*ii*) to provide compression reinforcement. In many cases, the maximum value of depth of the section may be limited or restricted from architectural or other considerations. In that case, the only alternative that will be left will be to provide reinforcement in the compression zone, giving rise to a *doubly reinforced section.*

25.2. STRESS BLOCK AND N.A.

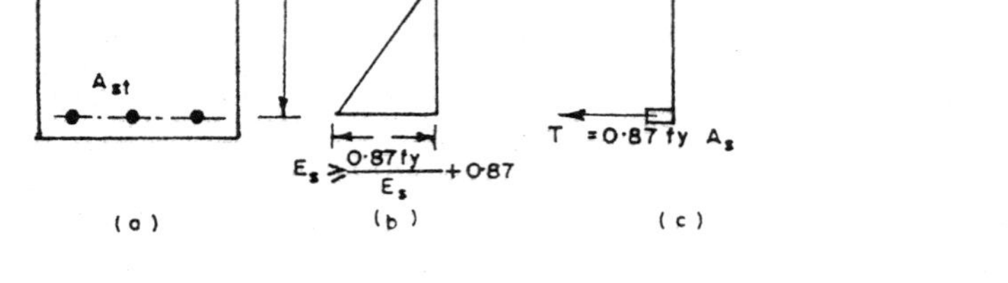

FIG. 25.1. DOUBLY REINFORCED SECTION.

Fig. 25.1(*a*) shows a doubly reinforced section having compression reinforcement at a depth of d' below the outermost compression fibre. Fig. 25.1(*b*) shows the strain diagram while Fig. 25.1(*c*) the stress block.

Let A_{st} = Total reinforcement at tension face

A_{sc} = Reinforcement in compression side.

x_u = Depth of N.A.

C_{cu} = Compressive force in concrete $= 0.36 f_{ck} x_u b$

C_{su} = Compressive force in compression steel $= f_{sc} A_{sc}$

f_{sc} = *Design stress* in compression reinforcement read off from the stress strain curve corresponding to the strain ε_{sc} in compression reinforcement.

ε_{sc} = Strain in compression reinforcement

$$= \frac{0.0035\,(x_u - d')}{x_u} \quad \text{...(25.1)}$$

Total compressive force is given by

$$C_u = C_{cu} + C_{su}$$

or
$$C_u = 0.36 f_{ck}\, b\, x_u + f_{sc} A_{sc} \quad \text{...(25.2)}$$

(neglecting the loss of concrete area occupied by compressive steel).

Total tensile force is given by

$$T_u = 0.87 f_y A_{st} \quad \text{...(25.3)}$$

In order to locate the N.A., equate the total compressive force to the total tensile force :

$$\therefore \quad 0.36 f_{ck}\, b\, x_u + f_{sc} A_{sc} = 0.87 f_y A_{st} \quad \text{...(25.4)}$$

From the above relation, x_u can be found. However, for the solution of the above equation, an iterative procedure will have to be adopted, since f_{sc} depends upon ε_{sc}, which in turn depends upon x_u.

If the loss of compressive area, occupied by the compressive steel is taken into account, Eqs. 25.2 and 25.4 are modified as under :

$$C_u = 0.36 f_{ck} x_u b + f_{sc} A_{sc} - 0.446 f_{ck} A_{sc}$$

or
$$C_u = 0.36 f_{ck} x_u b + (f_{sc} - 0.446 f_{ck}) A_{sc} \quad \text{...[25.2(a)]}$$

and
$$0.36 f_{ck} x_u b + (f_{sc} - 0.446 f_{ck}) A_{sc} = 0.87 f_y A_{st} \quad \text{...[25.2(b)]}$$

Note: Normally, the term $0.446 f_{ck} A_{sc}$ is very small, and can be neglected without causing any appreciable error.

Iterative procedure for computation of x_u.

Step 1. Compute the depth of N.A. of a balanced section given by strain compatibility [Fig. 25.1(*b*)] :

$$x_{u.max} = \frac{0.0035}{0.0055 + 0.87 f_y/E_s} \quad ...(25.3)$$

Taking $E_s = 2 \times 10^5$ N/mm^2 ,

$$x_{u.max} = \frac{700}{1100 + 0.87 f_y} \quad ...[25.3(a)]$$

It is to be noted that this expression is the same as Eq. 24.7(b) used for a singly reinforced section.

Step 2. For the given doubly reinforced section, *assume* x_u equal to $x_{u.max}$.

Step 3. Compute the value of ε_{sc} from Eq. 25.1.

Step 4 Compute value of f_{sc} from the stress strain curve of steel, corresponding to this value of ε_{sc} .

Step 5. Substitute the value of f_{sc} in Eq. 25.4 or 25.4(a) and compute the modified value of ε_{sc} .

Step 6. Repeat steps 3 to 5 till convergence for the value of x_u is achieved.

Note. A slightly different form of iterative procedure has been illustrated in Example 25.1.

24.3. DETERMINATION OF MOMENT OF RESISTANCE

For a given section,the depth of neutral axis, x_u can be determined as illustrated in the previous article. If this value of x_u is compared to the limiting depth ($x_{u.max}$) of the N.A. axis, three cases may arise:

Case 1. x_u/d less than limiting value

Case 2. x_u/d equal to the limiting value.

Case 3. x_u/d greater than limiting value.

Case 1. x_u/d less than limiting value

The case corresponds to the under reinforced-section in working stress method. The moment of resistance is given by

$$M_u = 0.36 f_{ck} x_u b (d - 0.416 x_u) + A_{sc} f_{sc} (d - d') \quad ...(25.4)$$

or

$$M_u = 0.36 f_{ck} \frac{x_u}{d} \left(1 - 0.416 \frac{x_u}{d}\right) bd^2 + A_{sc} f_{ck} (d - d')$$

The centroid $\bar{x}$ of the compressive forces, from the top fibre is given by

$$\bar{x} = \frac{0.36 f_{ck} x_u b \times 0.416 x_u + f_{ck} A_{sc} d'}{0.36 f_{ck} x_u b + f_{sc} A_{sc}} \quad ...(25.5)$$

$\therefore$ Lever arm $a = d - \bar{x}$

or $$a = d - \frac{0.36 f_{ck} x_u b \times 0.416 x_u + f_{sc} A_{sc} d'}{0.36 f_{ck} x_u b + f_{sc} A_{sc}} \quad ...(25.6)$$

Hence, with respect to tensile force,

$$M_u = 0.87 f_y A_{st} \times a \quad ...(25.7)$$

The moment of resistance found by Eqs 25.4 and 25.7 should be equal.

The above equations have been obtained by *neglecting* the loss of concrete occupied by compressive steel. The equations are modified as under if this loss is taken into account :

$$M_u = 0.36 f_{ck} x_u b\,(d - 0.416 x_u) + (A_{sc} f_{sc} - 0.446 f_{ck} A_{sc})\,(d - d') \quad ..(25.8)$$

where x_u is determined on the basis of Eq. 25.4(a)

$$\bar{x} = \frac{0.36 f_{ck} x_u b\,(0.416 x_u) + (f_{sc} - 0.446 f_{ck}) A_{sc} d'}{0.36 f_{ck} x_u b + (f_{sc} - 0.446 f_{ck}) A_{sc}} \quad ...(25.9)$$

$$a = d - \bar{x}$$

and $$M_u = 0.87 f_y A_{st} \times a$$

Case. 2. x_u/d equal to limiting value

This case corresponds to a *balanced design* in which the steel reinforcement reaches its yield stress at the same instant when ultimate strain is reached in concrete. The moment of resistance M_u reaches its limiting value $M_{u.lim.}$ and is given by

$$M_{u.lim.} = 0.36 f_{ck} x_{u.lim.} b\,(d - 0.416 x_u) + A_{sc} f_{sc}\,(d - d') \quad ...(25.10)$$

where f_{sc} = stress in compressive steel, found from stress strain curve, corresponding to ε_{sc} given by Eq. 25.1 in which $x_u = x_{u.lim.}$

If, however, loss of concrete area occupied by compressive steel in taken into account.

$$M_{u.lim} = 0.36 f_{ck} x_{u.max} b\,(d - 0.416 x_{u,max}) + (f_{sc} - 0.416 f_{ck}) A_{sc}\,(d - d') \quad ...[25.11(a)]$$

Case 3. x_u/d greater than the limiting value

This corresponds to an over-reinforced section. Failure occurs when strain in extreme fibres in concrete reaches its limiting value, while strain in tensile steel remain below the yield strain ε_{su}. The moment of resistance of such a section is limited to $M_{u.lim.}$ given by Eq. 25.10. It should be noted that compression failure is *sudden*, and therefore not desirable. The Code recommends that if x_u/d is found to be greater than the limiting value, the *section should be redesigned.*

25.4. DESIGN OF A DOUBLY REINFORCED SECTION

We have seen earlier that a doubly reinforced section is required when the applied moment M_{ua} is larger than $M_{u.lim.}$, given by Eq. 25.11 for a singly reinforced rectangular section. The difference between the applied moment M_{ua} and the limiting moment $M_{u.lim.}$ is carried by additional tensile reinforcement A_{st2} and compression reinforcement A_{sc}. Thus, a doubly reinforced section is equivalent to a singly reinforced balanced section and a section with additional tension and compression reinforcement, as illustrated in Fig. 25.2.

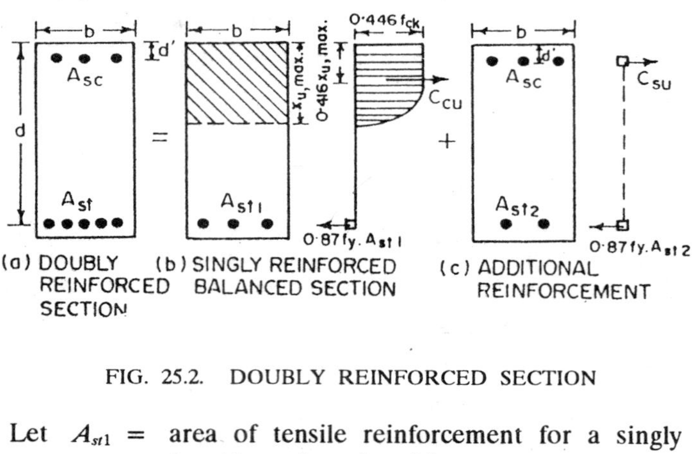

(a) DOUBLY REINFORCED SECTION (b) SINGLY REINFORCED BALANCED SECTION (c) ADDITIONAL REINFORCEMENT

FIG. 25.2. DOUBLY REINFORCED SECTION

Let A_{st1} = area of tensile reinforcement for a singly reinforced section, for $M_{u,\,lim.}$

$= p_{lim} \, . \, bd.$

A_{st2} = additional tensile reinforcement , given by

$$A_{st2} = \frac{M_{ua} - M_{u.lim.}}{0.87 f_y (d - d')} = \frac{A_{sc} f_{sc}}{0.87 f_y} \qquad ...(25.10)$$

The compression reinforcement A_{sc} is given by

$$A_{sc} = \frac{M_{ua} - M_{u.lim.}}{f_{sc} (d - d')} \qquad ...(25.11)$$

If however, loss of concrete area, occupied by compressive steel is taken into account, we have

$$A_{sc} = \frac{M_{ua} - M_{u.lim.}}{(f_{sc} - 0.446 f_{ck})(d - d')} \qquad ...[25.10(a)]$$

(where $d' < \frac{3}{7} x_u$)

where f_{sc} = *design stress* in compression reinforcement read off from the stress-strain curve corresponding to the strain ε_{sc} in the compression reinforcement.

ε_{sc} = Strain in compression reinforcement.

$$= \frac{0.0035 (x_{u.max} - d')}{x_{u.max}} \qquad ...(25.11)$$

and $x_{u.max}$ = limiting value of x_u given by Eq. 24.7(*b*).

Example 25.1. M.R. of doubly reinforced beam

Find the M.R. of a beam 200 mm wide and 400 mm effective depth, reinforced with 4 bars(mild steel) of 20 mm Φ for tension and 2 bars of 20 mm dia. for compression. The over to compressive reinforcement is 50 mm. Use M 15 concrete.

Solution

Given $f_{ck} = 15 \text{ N/mm}^2$; $f_y = 250 \text{ N/mm}^2$

$b = 200$ mm ; $d = 400$ mm ; $d' = 50$ mm ;

$$A_{st} = 4 \frac{\pi}{4} (20)^2 = 1256.6 \text{ mm}^2$$

$$A_{sc} = 2 \frac{\pi}{4} (20)^2 = 628.3 \text{ mm}^2$$

Let us assume that $d' < \frac{3}{7} x_u$.

$$T_u = 0.87 f_y A_{st} = 0.87 \times 250 \times 1256.6 = 273310 \text{ N} \qquad ...(i)$$

$$C_u = 0.36 f_{ck} x_u b + f_{sc} A_{sc} - 0.446 f_{ck} A_{sc}$$

$$= 0.36 \times 15 \times 200 x_u + f_{sc} \times 628.3 - 0.446 \times 15 \times 628.3$$

$$= 1080 x_u + 628.3 f_{sc} - 4203 \text{ N} \qquad ...(ii)$$

where f_{sc} = stress in the compressive reinforcement to be read off from the stress strain curve for strain

$$\varepsilon_{sc} = \frac{0.035 (x_u - d')}{x_u}$$

The depth x_u is best found by trial solution. We know that when $f_y = 250\ \text{N/mm}^2$, the value of $\frac{x_{u.max}}{d} = 0.53$, for singly reinforced sections. However, for most of the cases the value of x_u/d is less than this, and for doubly reinforced sections, x_u/d is still much less since the N.A. will be shifted up due to the presence of compressive steel.

As a first trial, let us assume $x_u = 0.4\,d = 0.4 \times 400 = 160$ mm

$$\therefore \qquad \varepsilon_{sc} = \frac{0.0035\,(160 - 50)}{160} = 0.0024$$

The stress-strain curve for mild steel bars is linear, with the yield point reaching at 0.2 percentstrain. Hence $f_{sc} = 0.87\ f_y = 0.87 \times 250 = 217.5\ \text{N/mm}^2$. Substituting in (*ii*), we get

$$C_u = (1080 \times 160) + (628.3 \times 217.5) - 4203$$

$$= 172800 + 136655 - 4203 = 309455 - 4203$$

$$= 305252\ \text{N}$$

(**Note :** It will be seen that $0.446 f_{ck} A_{sc} = 4203$ N only in comparison to total C_u , and hence can be neglected).

We find that $C_u > T_u$.

In the second trial, therefore, let us reduce x_u from the 160 mm to 135 mm.

(**Note:** f_{sc} is constant $= 0.87\ f_y$ when strain ε_{sc} is greater than 0.002).

$$\varepsilon_{sc} = \frac{0.0035\,(135 - 50)}{135} = 0.0022$$

$$\therefore \qquad f_{sc} = 0.87 f_y = 0.87 \times 250 = 217.5$$

$$\therefore \qquad C_u = 1080 \times 135 + 628.3 \times 217.5 - 4203$$

$$= 145800 + 136655 - 4203 = 278252\ \text{N}$$

This is quite near to $T_u = 273310$ N.

However, adopt $x_u = 130$ mm.

$$\therefore \qquad C_u = 1080 \times 130 + 136655 - 4203$$

$$= 140400 + 132452$$

$$= (C_{cu}) + (C_{su})$$

$$\therefore \qquad M_u = C_{cu}\,(d - 0.42\,x_u) + C_{su}\,(d - d\,')$$

$$= 140400\,(400 - 0.416 \times 130) + 132452\,(400 - 50)$$

$$= 48.58 \times 10^6 + 46.53 \times 10^6$$

$= 94.94 \times 10^6$ N-mm

$=$ **94.94 kN-m** $= M_{uD}$

$\therefore$ Design moment $M_D = 94.94/1.5 = 63.20$ kN-m

Note 1. The moment of resistance of the same beam without compression reinforcement was found to be 71.4 kN-m only, in Example 24.1.

Note 2. The moment of resistance of the doubly reinforced beam, by the working stress method will be as under :

For M 15 concrete, with mild steel bars,

$k = 0.404$; $j = 0.865$; $R = 0.874$.

For the N.A. location,

$$\frac{bn^2}{2} + (m_c - 1) A_{sc} (n - d') = m A_{st} (d - n)$$

or $$\frac{200}{2} n^2 + (1.5 \times 19 - 1) 628.3 (n - 50) = 19 \times 1256.6 \times (400 - n)$$

or $$100 n^2 + 17278 n - 863912 = 9550160 - 23875 n$$

or $$n^2 + 411.54 n - 104141 = 0$$

From which $n = 177$ mm

Critical depth $n_c = 0.404 \times 400 = 161.6$ mm.

The actual N.A. is below critical ; hence the stress in concrete will reach its maximum value first.

$\therefore \quad c = \sigma_{cbc} = 5$ N/mm^2.

$$c' = c \frac{b - d'}{n} = \frac{5}{177} (177 - 50) = 3.59 \text{ N/mm}^2$$

$$M_r = bn \frac{c}{2} \left(d - \frac{n}{3} \right) + (m_c - 1) A_{sc} . c' (d - d')$$

$$= 200 \times 177 \times \frac{5}{2} \left(400 - \frac{177}{3} \right) + (1.5 \times 19 - 1) 628.3 \times 3.59 \times (400 - 50)$$

$$= 30.179 \times 10^6 + 21.717 \times 10^6$$

$$= 51.896 \times 10^6 \text{ N-mm}$$

$$\approx 51.9 \text{ kN-m}$$

This is less than $M_D = 63.29$ kN-m found from the limit analysis. The ratio of the two moments is 1.23.

Example 25.2. M.R. of doubly reinforced beam.

Determine the moment of resistance of beam having the following data :

b = 350 mm ; d = 900 mm ; d ' = 50 mm ;

Tension reinforcement : 5-20 mm HYSD bars(Fe 415) ; compression reinforcement 2-20 mm HYSD bars(Fe 415). Grade of concrete : M 15.

Solution. $A_{st} = 5\frac{\pi}{4}(20)^2 = 1570.8 \text{ mm}^2$

$$A_{sc} = 2\frac{\pi}{4}(20)^2 = 628.3 \text{ mm}^2 .$$

$$T_u = 0.87 f_y A_{st} = 0.87 \times 415 \times 1570.8 = 567120 \text{ N} \quad ...(i)$$

$$C_u = 0.36 f_{ck} x_u b + f_{sc} A_{sc} - 0.446 f_{ck} A_{sc}$$

$$= 0.36 \times 15 \times 350 x_u + 628.3 f_{sc} - 0.446 \times 15 \times 628.3$$

$$= 1890 x_u + 628.3 f_{sc} - 4203 \quad ...(ii)$$

Let us assume $x_u = 230$ mm.

$$\frac{3}{7} x_u = \frac{3}{7} \times 230 = 98.6 \quad > d'.$$

$$\varepsilon_{cu} = \frac{0.0035 (x_u - d')}{x_u} = \frac{0.0035 (230 - 50)}{230}$$

$$= 0.00274.$$

Hence from stress-strain curve(Fig. 24.3 or Table 24.1), we get

$$f_{sc} = 315 \text{ N/mm}^2$$

$$\therefore \quad C_u = 1890 \times 230 + (628.3 \times 331) - 4203$$

$$= 434700 + 220533 - 4203 = 651030 \text{ N}$$

This is much more than $T_u = 567120$ N

Hence take $x_u = 190$ mm.

$$\varepsilon_{cu} = \frac{0.0035 (190 - 50)}{190} = 0.002580$$

$$\therefore \quad f_{sc} \approx 347 \text{ N/mm}^2 \text{ (Table 24.1)}$$

$$\therefore \quad C_u = 1890 \times 190 + (628.3 \times 347) - 4203$$

$$= 359100 + 218020 - 4203 = \underset{(C_{cu})}{359100} + \underset{(C_{su})}{213817}$$

$$= 572917 \approx T_u.$$

$$\therefore \quad M_u = C_{cu}(d - 0.416 x_u) + C_{su}(d - d')$$

$$= 359100\,(900 - 0.416 \times 190) + 213817\,(900 - 50)$$

$$= 294.8 \times 10^6 + 181.7 \times 10^6$$

$$= 476 \times 10^6 \text{ N-mm} = \mathbf{476.5 \text{ kN-m.}}$$

Example 25.3. Design of doubly reinforced beam.

A concrete beam has 350 mm breadth and 700 mm effective depth. Design the beam if it is subjected to a superimposed bending moment of 300 kN-m. Use HYSD bars of grade Fe 415 and concrete of grade M 15.

Solution.

Using $\gamma_f = 1.5$,

$$M_{uD} = 300 \times 1.5 = 450 \text{ kN-m.}$$

For $f_y = 415 \text{ N/mm}^2$, we have, from Eq. 24.7 (*b*),

$$\frac{x_{u.max}}{d} = \frac{700}{1100 + 0.87 \times 415} = 0.48$$

$$\therefore \quad x_{u.max} = 0.48 \times 700 = 336 \text{ mm}$$

The limiting reinforcement, from Eq. 24.9 :

$$p_{lim.} = 0.414\,\frac{15}{415} \times 0.48 = 0.00718$$

$$= 0.718\ \%$$

From Eq. 24.11,

$$M_{u.lim.} = 0.36 f_{ck}\,\frac{x_{u.max}}{d}\left(1 - 0.416\,\frac{x_{u.max}}{d}\right) bd^2$$

$$= 0.36 \times 15\,(0.48)\,(1 - 0.416 \times 0.48)\,350\,(700)^2$$

$$= 355 \times 10^6 \text{ N-mm}$$

$$A_{st1} = \text{reinforcement for } M_{u.lim.}$$

$$= p_{lim.}\,bd = 0.00718\,(350)\,(700)$$

$$= 1759 \text{ mm}^2.$$

$$A_{sc} = \frac{M_{uD} - M_u \text{ lim}}{0.87\,f_y\,(d - d')} = \frac{(453 - 355)10^6}{0.87 \times 415\,(700 - 50)}$$

$$= 405 \text{ mm}^2$$

$$\therefore \quad \text{Total } A_{st} = 1759 + 405 = \mathbf{2164 \text{ mm}^2}.$$

$$A_{sc} = \frac{M_{uD} - M_{u.lim.}}{f_{sc}\,(d - d')} = \frac{(450 - 355)\,10^6}{f_{sc}\,(700 - 50)}$$

$$= \frac{0.416 \times 10^6}{f_{sc}} \quad \text{(assuming } d' = 50 \text{ mm)}$$

$$\varepsilon_{sc} = \frac{0.0035\,(x_{u.max} - d')}{x_{u.max}} = \frac{0.0035\,(366 - 50)}{336}$$

$= 0.00298$

Hence from Table 24.1, for HYSD bars,

$$f_{sc} \approx 353.7 \text{ N/mm}^2 \text{ for } \varepsilon_{sc} = 0.00298$$

$$\therefore \quad A_{sc} = \frac{0.416 \times 10^6}{353.7} = \mathbf{413\ mm^2.}$$

If loss of concrete area, occupied by compressive steel is taken into account

$$A_{sc} = \frac{M_{uD} - M_{u.lim.}}{(f_{sc} - 0.446 f_{ck})(d - d')}$$

$$= \frac{(450 - 355)\,10^6}{(f_{sc} - 0.446 \times 15)(700 - 50)}$$

$$= \frac{0.416 \times 10^6}{f_{sc} - 6.69}$$

$$\varepsilon_{sc} = 0.00298 \text{ and } f_{sc} = 353.7, \text{ as before.}$$

$$\therefore \quad A_{sc} = \frac{0.416 \times 10^6}{353.7 - 6.69} \approx \mathbf{421\ mm^2.}$$

PROBLEMS

1. A rectangular beam 300 mm wide and 600 mm effective depth is reinforced with a tensile reinforcement of 9000 mm^2 and compressive reinforcement of 3000 mm^2. The compressive reinforcement has an effective cover of 50 mm. Determine its ultimate moment of resistance. Use M 15 grade concrete and Fe 250 grade steel.
2. Solve problem 1 for M 15–Fe415 combination.
3. Solve problem 1 for M 20–Fe415 combination.
4. A concrete beam has 300 mm breadth and 500 mm effective depth. Design the beam if it is subjected to a super-imposed bending moment of 200 kN-m. Use M 15 grade concrete and Fe 415 steel.

26

T and L-Beams

26.1. INTRODUCTION

In actual practice T-Sections are more common than the rectangular sections since the part of reinforced concrete slab, monolithic with the beam, participates with the structural behaviour of the beam. The slab, which forms the upper part of the T-beam, is stressed laterally in compression, but this does not reduce its capacity to carry longitudinal compression as a part of the beam.

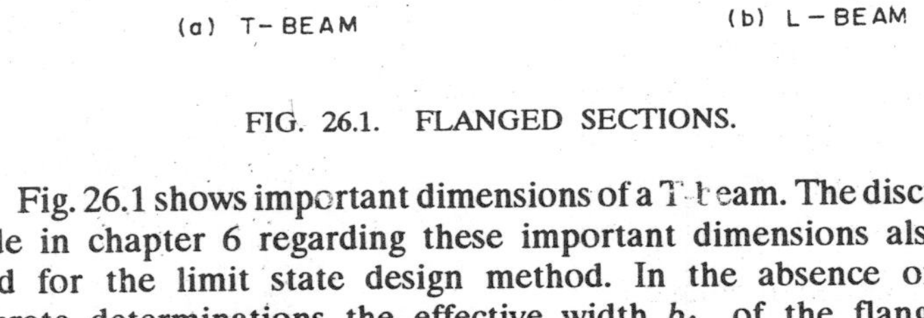

FIG. 26.1. FLANGED SECTIONS.

Fig. 26.1 shows important dimensions of a T-beam. The discussions made in chapter 6 regarding these important dimensions also hold good for the limit state design method. In the absence of more accurate determinations the effective width b_f of the flange of a T-beam may be taken as the least of the following.

(a) $b_f = \frac{l_0}{6} + b_w + 6\,D_f$

(b) $b_f = b_w + \frac{1}{2}$ (sum of clear distances to the adjacent beams on either side)

(c) For isolated beams, the flange shall be obtained as below but in no case greater than the actual width

$$b_f = \frac{l_0}{\frac{l_0}{b} + 4} + b_w$$

where l_0 is the distance between points of zero moments in the beam. For continuous beams and frames, l_0 may be assumed to be 0.7 times the effective span.

Similarly, for a L-beam.

$$b_f = \frac{l_0}{12} + b_w + 3\,D_f$$

For an isolated L-beam

$$b_f = \frac{0.5\,l_0}{\frac{l_0}{b} + 4} + b_w$$

26.2. STRESS BLOCK AND N.A.

Consider a T-beam having flange width b_f, web width b_w and flange thickness D_f, as shown in Fig. 26.2(*a*). The beam is singly reinforced. The stress block, shown in Fig. 26.2(*b*) is similar to the one for the rectangular section, with maximum compressive stress equal to $0.446\,f_{ck}$. The strain diagram, shown in Fig. 26.2(*b*) is triangular, as for the rectangular section, with a maximum strain in concrete equal to 0.0035.

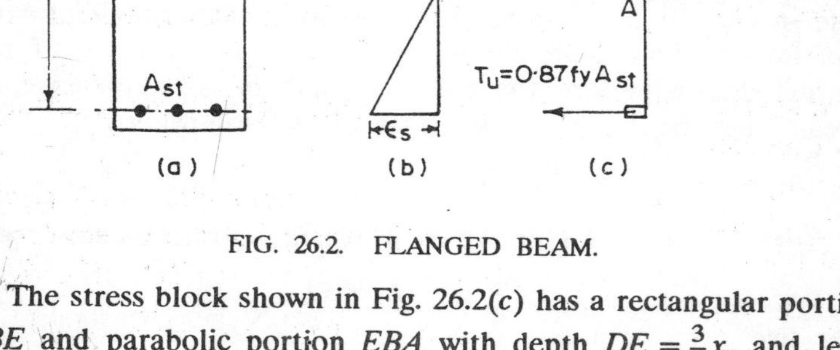

FIG. 26.2. FLANGED BEAM.

The stress block shown in Fig. 26.2(*c*) has a rectangular portion *DCBE* and parabolic portion *EBA* with depth $DE = \frac{3}{7}x_u$ and level

EB is corresponding to a strain of 0.002 in concrete. The remaining portion $EA = \frac{4}{7} x_u$.

Assuming that the N.A. falls in the flange, its position is given by equating the total compressive force to the total tensile force :

$$\therefore \quad 0.36 f_{ck}\, b_f\, x_u = 0.87 f_y A_{st}$$

or
$$\frac{x_u}{d} = \frac{0.87 f_y A_{st}}{0.36 f_{ck}\, b_f\, d} = 2.417\, p \frac{f_y}{f_{ck}} \qquad ...(26.1)$$

If this value of x_u comes out to be less than D_f , it will be treated as a rectangular section of width b_f . The section will than be under-reinforced, balanced or over- reinforced, depending upon whether $\frac{x_u}{d}$ is less than, equal to or more than $\frac{x_{u.max}}{d}$.

The depth of critical N.A. is given Eq. 24.7.

$$\frac{x_{u.max}}{d} = \frac{0.0035}{0.0055 + 0.87 f_y/E_s} = \frac{700}{1100 + 0.87 f_y}$$

If x_u found by Eq. 26.1 comes out to be more than D_f , the N.A. will fall outside the flange, and compression of the flange will have to be taken into account. The method of locating the depth of N.A. for such a case has been discussed in § 26.4.

26.3. MOMENT OF RESISTANCE WHEN $x_u < D_f$

If $x_u < D_f$, the section will behave as a rectangular section with width equal to b_f . However, if x_u is greater than D_f , the analysis will depend upon whether x_u is greater than $\frac{7}{3} D_f$ or lesser than $\frac{7}{3} D_f$, *i.e.*, whether the N.A. falls outside the rectangular portion of the stress block [Fig. 26.2(*c*)] or it falls inside the rectangular portion *DCBE*. Thus, in all, three cases may arise in the analysis of a T-beam.

Case 1. $x_u < D_f$.

Case 2. $D_f \le \frac{3}{7} x_u \left(\text{or } \frac{D_f}{x_u} \le 0.43 \right)$

Case. 3. $D_f > \frac{3}{7} x_u \left(\text{or } \frac{D_f}{x_u} > 0.43 \right)$

We will be considering the first case in this article and the other two cases in the next article.

Case 1. $x_u < D_f$

In this case, the neutral axis is entirely within the flange. Hence the moment of resistance can be found from the equations applicable for rectangular sections substituting b_f for b.

Here again three cases may arise depending upon whether x_u is less than, equal to or greater than the limiting value $x_{u,max}$.

When $\frac{x_u}{d} < x_{u.max}$, we have the case of under-reinforced section and the moment of resistance is given by Eq. 24.10 of rectangular section.

$$M_u = 0.87 f_y A_{st} d \left(1 - \frac{f_y}{f_{ck}} \frac{A_{st}}{b_f d}\right) \qquad ...(26.2)$$

When $\frac{x_u}{d}$ is equal to $\frac{x_{u.max}}{d}$, we have the *balanced* rectangular section, and the moment of resistance M_u, which reaches its $M_{u.lim.}$ is given by Eq. 24.11.

$$M_{u.lim.} = 0.36 f_{ck} \frac{x_{u.max}}{d} \left(1 - 0.416 \frac{x_{u.max}}{d}\right) b_f d^2 \qquad (26.3)$$

When $\frac{x_u}{d}$ exceeds $\frac{x_{u.max}}{d}$, we have the over-reinforced section.

The failure takes place due to crushing of concrete while strain ε_s in steel still remains below the yield strain $\varepsilon_{su.}$ *Therefore, the moment of resistance of such section is limited to $M_{u.lim.}$ given by Eq. 26.13.* However, the Code recommends that such a section should be *redesigned* because compression failure is *sudden* and therefore not desirable.

When $x_u = D_f$, the moment of resistance is given by

$$M_{u1} = 0.87 f_y A_{st} d \left(1 - 0.416 \frac{D_f}{d}\right)$$

or $$M_{u1} = 0.87 f_y A_{st}(d - 0.416\, D_f) \qquad ...(26.4)$$

or Alternatively, $$M_{u1} = 0.36 f_{ck}\, b_f D_f\, (d - 0.416\, D_f) \qquad ...[26.4(a)]$$

26.4. MOMENT OF RESISTANCE WHEN N.A. FALLS IN THE WEB

When the value of x_u, found from Eq. 26.1 exceeds D_f, the N.A. falls in the web, and *a new value of x_u will have to be found by equating the total compressive force to the total tensile force.*

Fig. 26.3(a) shows the flanged section in which the N.A. falls outside the flange. The total compressive force is equal to C_u which can be thought to be equal to the sum of compressive force

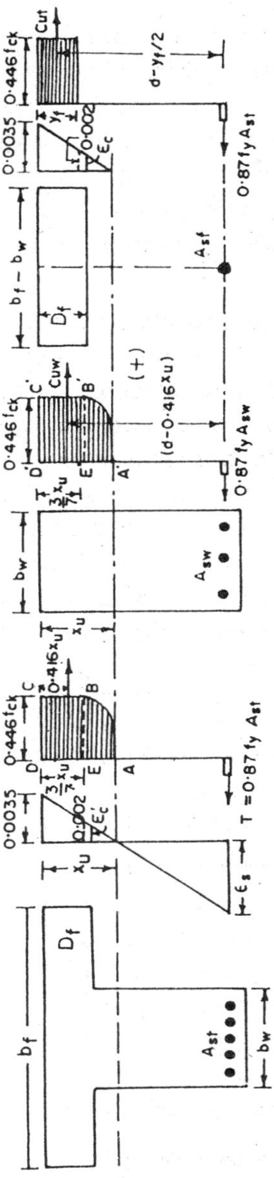

(a) FLANGED SECTION (b) WEB CONTRIBUTION (c) FLANGE CONTRIBUTION

FIG. 26.3. SINGLY REINFORCED FLANGED SECTION

C_{uw} developed by web contribution [Fig. 26.3(*b*)] and the compressive force c_{uf} developed by flange contribution [Fig. 26.3(*c*)]

The compressive force C_{uw} from the web contribution is given by

$$C_{uw} = 0.36 f_{ck} x_u b_w \qquad ...(26.5)$$

If out of the total tensile reinforcement A_{st}, a reinforcement A_{sw} is the component required to balance the compressive force C_{uw} and the remaining reinforcement A_{sf} is required to balance the compressive force C_{uf} in the flange, we have

$$A_{st} = A_{sw} + A_{sf}$$

and the tensile force T_{uw} to balance the compressive force C_{uw} will be equal to $0.87 f_y A_{sw}$. Hence

$$C_{uw} = 0.36 f_{ck} x_u b_w = 0.87 f_y A_{sw} \qquad ...[26.5(a)]$$

The flange contribution C_{uf} is shown in Fig. 26.3(*c*). The compressive force C_{uf} can be computed by approximating the actual stress block (rectangular-cum-parabolic) by a rectangular stress block of depth y_f, having the same peak stress value of $0.446 f_{ck}$.

The depth of equivalent rectangular stress block is given by

$$y_f = 0.15 x_u + 0.65 D_f \qquad ...(26.6)$$

(but not greater than D_f)

The compressive force contribution of the flange is given by

$$C_{uf} = 0.446 f_{ck} (b_f - b_w) y_f \qquad ...(26.7)$$

The tensile force T_{uw} to balance the compressive force C_{uf} will be equal to $0.87 f_y A_{sf}$.

The total compressive force C_u is then given by

$$C_u = 0.36 f_{ck} . x_u b_w + 0.446 f_{ck} (b_f - b_w) y_f \qquad ...(26.8)$$

The total tensile $T_u = 0.87 f_y A_{st}$ ...(26.9)

Hence equating the two, we get,

$$0.36 f_{ck} x_u b_w + 0.446 f_{ck} (b_f - b_w) y_f = 0.87 f_y A_{st} \qquad ...(26.10)$$

It should be clearly noted that the value of y_f in Eq .(26.10) in terms of x_u is given by Eq. 26.6. Hence the value of x_u found from Eq. 26.10 is valid only if the value of y_f, calculated from Eq. 26.6 by using this value of x_u, is equal to or less than D_f. If however, the value of y_f, comes out to be greater than D_f , the value of x_u should be *recalculated* from Eq. 26.10, taking the value of $y_f = D_f$. The modified expression is that case will be

$$0.36 f_{ck} x_u b_w + 0.446 f_{ck} (b_f - b_w) D_f = 0.87 f_y A_{st} \qquad ...[26.10(a)]$$

Thus, the value of x_u is known. This value of x_u should now be compared with the limiting value $x_{u.max}$. If $x_u < x_{u.max}$, we have *under reinforced section.* If $x_u = x_{u.max}$, we have the *balanced section* while if $x_u > x_{u.,max}$, we have *over-reinforced section.* We will take each one of these cases separately.

(*a*) UNDER-REINFORCED SECTION ($x_u < x_{u.max}$)

Here again, two cases may arise, depending upon the comparison of x_u with D_f.

Case 2. $D_f \le \frac{3}{7} x_u \left(\text{or } \frac{D_f}{x_u} \le 0.43 \right)$

Case 3. $D_f > \frac{3}{7} x_u \left(\text{or } \frac{D_f}{x_u} > 0.43 \right)$

(**Note.** Case 1 was corresponding to the situtation when $x_u < D_f$ as discussed in § 26.3).

Case 2. $D_f \le \frac{3}{7} x_u \left(\text{or } \frac{D_f}{x_u} \le 0.43 \right)$

If D_f is equai to or less than $\frac{3}{7} x_u$, the depth DE of the stress block will be greater than D_f. Hence the flange will lie completely within the rectangular portion of the stress block, and the strain

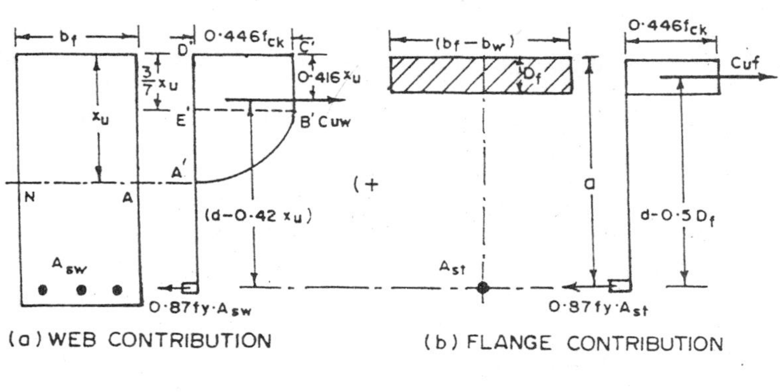

FIG. 26.:4. M_u FOR $\frac{D_f}{x_u} \le 0.43$.

$\varepsilon_c{}'$ in concrete, at the junction of web and flange will be less than 0.002.

The compressive forces C_{uw} and C_{uf} are given by

$$C_{uw} = 0.36 f_{ck} x_u b_w = 0.87 f_y A_{sw} \qquad ...[26.5(a)]$$

and

$$C_{uf} = 0.446 f_{ck} (b_f - b_w) D_f = 0.87 f_y A_{st} \qquad [26.7(a)]$$

The moment of resistance of web is given by

$$M_{uw} = C_{uw} (d - 0.416 x_u) = 0.36 f_{ck} \frac{x_u}{d} b_w \left(1 - 0.416 \frac{x_u}{d}\right) d^2 \qquad ...[26.11(a)]$$

The moment of resistance of the flange is given by

$$M_{uf} = C_{uf} (d - 0.5 D_f) = 0.446 f_{ck} (b_f - b_w) D_f (d - 0.5 D_f) \qquad ...[26.11(b)]$$

Hence

$$M_u = 0.36 \frac{x_u}{d} \left(1 - 0.416 \frac{x_u}{d}\right) f_{ck} b_w d^2 + 0.446 f_{ck} (b_f - b_w) D_f (d - 0.5 D_f) \qquad ...(26.11)$$

When $D_f = \frac{3}{7} x_u$ $\left(i.e., \text{ when } x_u = \frac{7}{3} D_f\right)$ the moment of resistance is given by

$$M_{u2} = 0.36 f_{ck} b_w \left(\frac{7}{3} D_f\right) \left(d - 0.416 \times \frac{7}{3} D_f\right) + 0.446 f_{ck} (b_f - b_w) \times D_f (d - 0.5 D_f) \qquad ...[26.11(c)]$$

Case 3. $D_f > \frac{3}{7} x_u$ $\left(\text{or } \frac{D_f}{x_u} > 0.43\right)$

Since D_f is greater than $\frac{3}{7} x_u$ in this case, the depth DE of the stress block will be smaller than the depth of the flange. Due to this, the distribution of compressive stress in the flange will not be rectangular. However, the distribution of stress in web will be the same as an Case 2. The total moment of resistance will be the sum of the contribution of the web and the flange. The actual distribution is shown in Fig. 26.5(*b*). This actual stress block can be replaced by a rectangular stress block of depth y_f, having the same peak stress value of $0.45 f_{ck}$. The depth of equivalent rectangular stress block is given by

$$y_f = 0.15 x_u + 0.65 D_f \qquad ...(26.6)$$

(but not greater than D_f).

The idealisation of actual stress block for the flange into a simple rectangular stress block of height y_f is based on the consideration that the compressive force and its centre of gravity of the idealised rectangular block and actual stress block compare within the acceptable limits of accuracy.

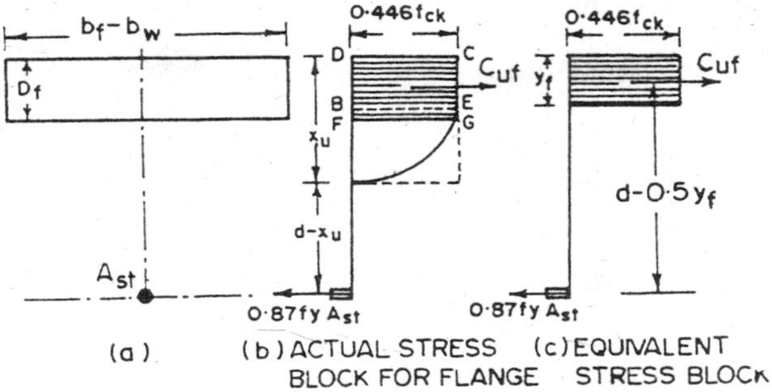

FIG. 26.5. FLANGE CONTRIBUTION WHEN $D_f > \frac{3}{7} x_u$.

The compressive forces C_{uw} and C_{uf} are given by

$$C_{uw} = 0.36 f_{ck} x_u b_w = 0.87 f_y A_{sw} \text{ (as before)} \quad ...(26.5(a)]$$

and
$$C_{uf} = 0.446 f_{ck} (b_f - b_w) y_f = 0.87 f_y A_{st} \quad ...(26.7)$$

The moment of resistance of the web is given by

$$M_{uf} = C_{uw} (d - 0.416\ x_u) = 0.36 f_{ck} \frac{x_u}{d} b_w \left(1 - 0.416\ \frac{x_u}{d}\right) d^2 \quad ...[26.12(a)]$$

The moment of resistance of the flange is given by

$$M_{uf} = C_{uf} (d - 0.5\, y_f) = 0.446 f_{ck} (b_f - b_w) y_f (d - 0.5\, y_f) \quad ...[26.12(b)]$$

$$\therefore \quad M_u = 0.36 \frac{x_u}{d}\left(1 - 0.416\ \frac{x_u}{d}\right) f_{ck} b_w d^2 + 0.446 f_{ck} (b_f - b_w) y_f (d - 0.5\, y_f) \quad ...(26.12)$$

(*b*) BALANCED SECTION ($x_u = x_{u.max}$).

When $x_u = x_{u.max}$ the moment of resistance is given by Eq. 26.12, by replacing x_u by $x_{u.max}$ Thus,

$$M_{u.lim} = 0.36 \frac{x_{u.max}}{d}\left(1 - 0.416\ \frac{x_{u.max}}{d}\right) f_{ck} b_w d^2 + 0.446 f_{ck} (b_f - b_w) y_f (d - 0.5\, y_f) \quad ...(26.13)$$

where $y_f = 0.15\, x_{u.max} + 0.65\, D_f$ [26.6(*b*)]

(subject to a maximum of D_f).

(*c*) OVER-REINFORCED SECTION ($x_u > x_{u.max}$)

When x_u is greater than $x_{u.max}$ the section will be *over-reinforced.* The failure takes place due to crushing of concrete while strain ε_s in steel still remains below the yield strain ε_{su}. The moment of resistance of such a section is limited to $M_{u.lim}$ given by Eq. 26.13.

The over-reinforced-section results in wastage of excess steel provided, and the concrete reaches its ultimate capacity before steel yields, resulting in a sudden failure. Such a section should therefore be redesigned.

26.4. IS CODE PROCEDURE FOR FINDING MOMENT OF RESISTANCE(IS : 456-1978)

IS Code procedure for finding moment of resistance of the flanged section depends upon the following *three cases.*

Case 1. When $x_u < D_f$

Case 2. when $(x_{u.max} = x_u) > D_f$

Case 3. When $x_{u.max} > x_u > D_f$

Case 1. When $x_u < D_f$

If $x_u < D_f$ the N.A. falls entirely in the flange. Hence the moment of resistance is computed, treating the section as a rectangular section, taking $b = b_f$.

(*a*) The value of x_u is found by Eq. 26.1 :

$$\frac{x_u}{d} = \frac{0.87 f_y A_{st}}{0.36 f_{ck} b_f d} \qquad ...{\prime}$$

(*b*) If the value of $\frac{x_u}{d}$ is less than $\frac{x_{u.max}}{d}$, we have

$$M_u = 0.87 f_y A_{st} d\left(1 - \frac{A_{st} f_y}{b_f d f_{ck}}\right) \qquad ...(26.14)$$

(c) If the value of $\frac{x_u}{d}$ is equal to $\frac{x_{u.max}}{d}$, the moment of esistance of the section is given by :

$$M_{u.lim.} = 0.36 \frac{x_{u.max}}{d}\left(1 - 0.416 \frac{x_{u.max}}{d}\right) bd^2 f_{ck} \qquad ...(26.15)$$

(*d*) If $\frac{x_u}{d}$ is greater than $\frac{x_{u.max}}{d}$, the section should be redesigned.

All these equations and recommendations are as per para *E*–1.1 of the I.S. code.

Case 2. $(x_{u.max} = x_u) > D_f$

This is the case when x_u is greater than D_f but at the same time, it is equal to $x_{u.max}$. In such a case, the moment of resistance of the section, commonly known as the *limiting moment of resistance* will depend upon whether the value of D_f/d is less than 0.2 or is greater than 0.2.

Case 2. (*a*) When $\frac{D_f}{d} \leq 0.2$

The moment of resistance is given by

$$M_{u.lim.} = 0.36 \frac{x_{u.max}}{d} \left(1 - 0.416 \frac{x_{u.max}}{d}\right) f_{ck}\, b_w\, d^2 + 0.446 f_{ck} (b_f - b_w) D_f \left(d - \frac{D_f}{2}\right) \quad ...(26.16)$$

Thus, in this case, the I.S. code has taken the depth y_f of the equivalent rectangular stress block equal to its maximum value of D_f . This equation is the same as given in para *E*-2.2 of the Code. Note that, this equation is same as Eq. 26.13. where y_f has been taken equal to D_f .

Case 2. (*b*) When $\frac{D_f}{d} > 0.2$

The moment of resistance is given by

$$M_{u.lim.} = 0.36 \frac{x_{u.max}}{d} \left(1 - 0.416 \frac{x_{u.max}}{d}\right) f_c\, b_w\, d^2 + 0.446 f_{ck} (b_f - b_w) y_f \left(d - \frac{y_f}{2}\right) \quad ...(26.17)$$

where $y_f = (0.15\, x_{u.max} + 0.65\, D_f)$

(But not greater than D_f).

The above equation is the same as given in para E-2.2.1 of the Code. Also, this equation is same as Eq. 26.13.

Case 3. $x_{u.max} > x_u > D_f$

This is a case of under-reinforced section in which the N.A. falls outside the flange. Here again, the moment of resistance will depend upon whether the value of $\frac{D_f}{x_u}$ is equal to or less than 0.43(or 3/7) or the value of $\frac{D_f}{x_u}$ is greater than 0.43.

Case 3. (*a*) $\frac{D_f}{x_u} < 0.43$

When D_f is equal to or less than $\frac{3}{7} x_u$ (or $0.43 x_u$), the depth *DE* of the stress block will be greater than D_f . Thus the flange will fall entirely within the rectangular portion of the stress block. Hence the moment of resistance is given by

$$M_u = 0.36\frac{x_u}{d}\left(1 - 0.416\frac{x_u}{d}\right) f_{ck}\, b_w\, d^2 + 0.446 f_{ck}\,(b_f - b_w)\, y_f \left(d - \frac{y_f}{2}\right) \quad ...(26.18)$$

This is same as Eq. 26.11. Also, this equation is the same as given in para E - 2.2 of the code except that $x_{u.max}$ has been replaced by

Case 3. (*b*) $\dfrac{D_f}{x_u} > 0.43$

When D_f is equal to or greater than $\frac{3}{7}x_u$ (or 0.43 x_u) the depth *DE* of the stress block will be smaller than the depth of the flange. Due to this, the distribution of compressive stress in the flange will not be rectangular. The moment of resistance is given by

$$M_u = 0.36\frac{x_u}{d}\left(1 - 0.416\frac{x_u}{d}\right) f_{ck}\, b_w\, d^2 + 0.446 f_{ck}\,(b_f - b_w)\, y_f \left(d - \frac{y_f}{2}\right) \quad ...(26.19)$$

where $y_f = (0.15\,x_u + 0.65\,D_f)$

(But not greater than D_f)

The above equation is the same as Eq. 26.12. Also, it is the same as given in para *E*-2.2.1 of the Code, except that $x_{u.max}$ has been replaced by x_u.

Note 1 : When the N.A. falls outside the flange, as concluded from the operation of Eq. 26.1, a new value of x_u should be found, by taking the compressive force of the web into account, as discussed in § 26.4. Though the Code has not specifically mentioned this, it is implied that this new value of x_u should be used while comparing it with $x_{u.max}$ or D_f . The value of x_u for such a case can be found either by Eq. 26.10 or by Eq. 26.10(*a*), as the case may be

Note 2. In para *E*-2.2.1, while computing the limiting value of M_u , the value of x_u has been erroneously printed in the placed of $x_{u.max}$. Hence the value of y_f should be taken as equal to, $(0.15\,x_{u.max} + 0.65\,D_f)$, subject to a maximum of D_f.

Note 3. The Code has not specifically dealt with the case when x_u is greater than D_f and is also greater than $x_{u.max}$. Hence, when $x_u > x_{u\,max} > D_f$, the section will be *over-reinforced.* The failure will take place due to the crushing of concrete while strain ε_s in steel still remains below the yield strain ε_{su}. Therefore, the moment of resistance of such a section is limited to $M_{u.lim}$, and is given by Eq. 26.16 $\left(\text{if } \frac{D_f}{d} \le 0.2\right)$ or by Eq. 26.17 if $\frac{D_f}{d} > 0.2$, as the case may be. Such a section should be redesigned because compression failure is *sudden* and therefore not desirable.

26.5. DESIGN OF T-BEAM

The design of T-beam of given dimensions (*i.e.* b_w, b_f, d known) and given design moment, is done in the following steps :

Step 1. Assuming $x_u = D_f$, compute the moment of resistance M_{u1} from Eq. 26.4(a)

$$M_{u1} = 0.36 f_{ck}\, b_f\, D_f\, (d - 0.416\, D_f) \qquad ...[26.4(a)]$$

Step 2. If $M_{u1} > M_{uD}$, the N.A. will fall inside the flange (*i.e.* $x_u \le D_f$). In that case the reinforcement is given by Eq. 26.2 :

$$M_u = M_{uD} = 0.87 f_y A_{st}\, d \left(1 - \frac{f_y}{f_{ck}} \frac{A_{st}}{b_f\, d}\right) \qquad ...(26.2)$$

Step 3. If $M_{u1} < M_{uD}$, the N.A. falls outside the flange. Assuming $x_u = \frac{7}{3} D_f$, compute M_{u2} from Eq. 26.11(c) :

$$M_{u2} = 0.36 f_{ck}\, b_w \left(\frac{7}{3} D_f\right) \left(d - 0.416 \times \frac{7}{3} D_f\right)$$

$$+ 0.446 f_{ck}\, (b_f - b_w) \times D_f\, (d - 0.5\, D_f) \qquad ...[26.11(c)]$$

Step 4. If $M_{u2} < M_{uD}$, $x_u > \frac{7}{3} D_f$.

Compute A_{sw} from Eq. 26.5(a) as under :

$$C_{uw} = 0.36 f_{ck}\, x_u\, b_w = 0.87 f_y A_{sw}$$

or $$A_{sw} = \frac{0.36 f_{ck}\, x_u\, b_w}{0.87 f_y} \qquad ...(26.20)$$

Similarly, compute A_{sf} from Eq. 26.7(a)

$$C_{uf} = 0.446 f_{ck}\, (b_f - b_w)\, D_f = 0.87 f_y A_{sf}$$

or $$A_{sf} = \frac{0.446 f_{ck}\, (b_f - b_w)\, D_f}{0.87 f_y} \qquad ...(26.21)$$

Total $$A_{st} = A_{sw} + A_{sf}$$

When $M_{uD} \gg M_{u2}$, it is always advisible to compute $M_{u,lim}$, since it is likely that $x_u > x_{u.max}$. If $M_{uD} > M_{u.lim.}$, the section will be doubly reinforced.

Step 5. If $M_{u2} > M_{uD}$, $x_u < \frac{7}{3} D_f$. The design procedure in this case will be that of trial and error. The following steps are recommended :

(*i*) Assume $x_u < \frac{7}{3} D_f$

(*ii*) Compute $y_f = 0.15\, x_u + 0.65\, D_f$
(Subject to a max. of D_f)

(*iii*) Compute M_u from Eq. 26.12 :

$$M_u = 0.36 \frac{x_u}{d}\left(1 - 0.416 \frac{x_u}{d}\right) f_{ck}\, b_w\, d^2 + 0.446 f_{ck}\, (b_f - b_w)\, y_f\, (d - 0.5\, y_f) \quad ...(26.12)$$

(*iv*) Compare M_u with M_{uD}.

If $M_u = M_{uD}$, assumed value of x_u is correct.

If $M_{uD} > M_u$, increase x_u in next trial.

If $M_{uD} < M_u$, decrease x_u in next trial.

Repeat till $M_u = M_{uD}$.

(*v*) knowing x_u, compute

$$C_{uw} = 0.36 f_{ck}\, x_u\, b_w \quad ...[26.5(a)]$$

$$C_{uf} = 0.446 f_{ck}\, (b_f - b_w)\, y_f \quad ...(26.7)$$

$$\therefore \quad C_u = C_{uw} + C_{uf}.$$

(*vi*) Compute $A_{st} = \dfrac{C_u}{0.87 f_y}$ ...(26.22)

Alternatively, x_u can be found by substituting the value of y_f in terms of x_u in Eq. 26.12. and then solving the quadratic equation in terms of x_u. This value of x_u will be acceptable only if the value of y_f obtained from this is less than (or equal to) D_f.

Example 26.1. M.R. of T-beam

Find the moment of resistance of T-beam having the following data :

b_f = 740 mm

d = 400 mm

b_w = 240 mm

A_{st} = 5-20 mm Φ mild steel bars

D_f = 100 mm.

Use M 15 concrete

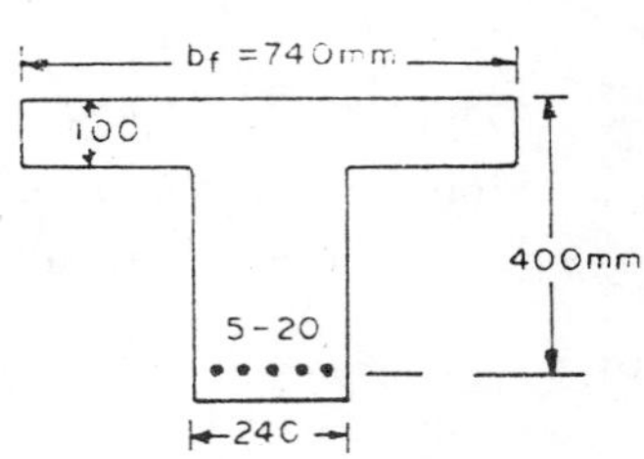

FIG. 26.6.

Solution

$$A_{st} = 5\frac{\pi}{4}(20)^2 = 1570 \text{ mm}^2$$

Assuming the N.A. to fall in the flange, we get, from Eq. 26.1

$$\frac{x_u}{d} = \frac{0.87 f_y A_{st}}{0.36 f_{ck} b_f d} = \frac{0.87 \times 250 \times 1570}{0.36 \times 15 \times 740 \times 400}$$

$$= 0.2136.$$

or $\quad x_u = 0.2136 \times 400 = 85.5 \text{ mm}$

$\therefore \quad x_u < D_f.$

Hence the N.A. lies in the flange.

Also, $\quad x_{u.max} = 0.53\, d$ for Fe 250 steel

$$= 0.53 \times 400$$

$$= 212 \text{ mm} \; > \; x_u$$

Hence, the moment of resistance is given by Eq. 26.2,

$$M_u = 0.87\, f_y \,.\, A_{st} \,.\, d \left(1 - \frac{f_y}{f_{ck}} \cdot \frac{A_{st}}{b_f \,.\, d}\right)$$

$$= 0.87 \times 250 \times 1570 \times 400 \left(1 - \frac{250}{15} \times \frac{1570}{740 \times 400}\right)$$

$$= 124.5 \times 10^6 \text{ N-mm}$$

$$= \mathbf{124.5 \text{ kN-m.}}$$

Example 26.2. M.R. of T-beam

Determine the M.R. of T-beam of example 26.1 if D_f=80 mm.

Solution

Assuming the N.A. to fall in the flange, we get x_u=85.5 mm, as obtained in example 26.1.

Since $x_u > D_f$ (=80 mm), the N.A. falls outside the flange. Let us obtain a new value of x_u by equating the total compressive force to the total tensile force.

Total tensile force $T_u = 0.87 f_y A_{st} = 0.87 \times 250 \times 1570$

$$= 341475 \text{ N}$$

Depth of equivalent stress block

$$= y_f = 0.15\, x_u + 0.65\, D_f$$

$$= 0.15\, x_u + 0.65 \times 80$$

$$= 0.15\, x_u + 52$$

Hence $C_u = 0.36 f_{ck} x_u b_w + 0.446 f_{ck} (b_f - b_w) y_f$

$$= (0.36 \times 15 x_u \times 240) + 0.446 \times 15\,(740 - 240)\,(0.15 x_u + 52)$$

$$= 1296 x_u + 501.8 x_u + 173940 = 1797.8 x_u + 173940$$

Equating $C_u = T_u$, we get

$1797.8\,x_u + 173940 = 341475$

From which $x_u = 93.2$ mm

This value will be acceptable only if $y_f \leq D_f$

Now $y_f = 0.15\,x_u + 0.65\,D_f = 0.15 \times 93.2 + 52 \approx 66 \text{ mm} < D_f$

Hence $x_u = 93.2$ mm is acceptable.

$$x_{u.max} = 0.48\,d \text{ (for Fe 415 steel)}$$
$$= 0.48 \times 400 = 192 \text{ mm} > x_u$$

Now $\frac{3}{7}x_u = \frac{3}{7} \times 93.2 = 39.9.$

$\therefore \quad D_f > \frac{3}{7}x_u =$ or $\frac{D_f}{x_u} > 0.43$ which corresponds to case 3 (*b*) of § 23.4 of IS Code procedure.

Hence, the moment of resistance is given by Eq. 26.19.

$$M_u = 0.36\frac{x_u}{d}\left(1 - 0.416\frac{x_u}{d}\right) f_{ck}\, b_w\, d^2 + 0.446\, f_{ck}\,(b_f - b_w)\, y_f \left(d - \frac{y_f}{2}\right)$$

where $y_f = 0.15\,x_u + 0.65\,D_f = 66$ mm (computed above)

$$\therefore \quad M_u = \frac{0.36 \times 93.2}{400}\left(1 - 0.416\,\frac{93.2}{400}\right) \times 15 \times 240\,(400)^2 + 0.446 \times 15\,(740 - 240) \times 66\,(400 - 0.5 \times 60)$$

$$= 43.63 \times 10^6 + 81.02 \times 10^6 = 124.65 \times 10^6 \text{ N-mm}$$

$$= \mathbf{142.65 \text{ kN-m.}}$$

Example 26.3. M.R. of T-beam.

Determine the M.R. of T-beam shown in Fig. 26.7. Use M 15 concrete and Fe 415 steel.

Solution

$A_{st} = 5 \times \frac{\pi}{4}(25)^2 = 2454.4 \text{ mm}^2$

Assuming the N.A. to fall in the flange, we have,

$$\frac{x_u}{d} = \frac{0.87\, f_y\, A_{st}}{0.36\, f_{ck}\, b_f\, d}$$

$$= \frac{0.87 \times 415 \times 2454.4}{0.36 \times 15 \times 700 \times 600} = 0.3907$$

$$x_u = 0.3907 \times 600$$
$$= 234.4 \text{ mm} > D_f.$$

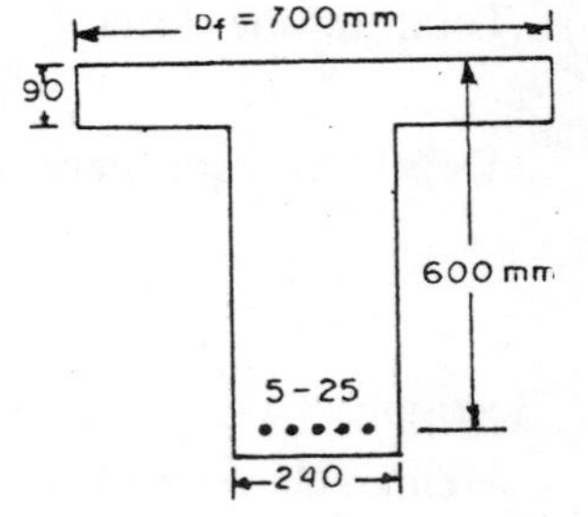

FIG. 26.7

Hence N.A. falls outside flange. To obtain a new value of x_u, equate total compressive force to the total tensile force.

$$T_u = 0.87 f_y A_{st} = 0.87 \times 415 \times 2454.4 = 886161 \text{ N}$$

$$y_f = 0.15 x_u + 0.65 D_f = 0.15 x_u + 0.65 \times 90 = 0.15 x_u + 58.5$$

with a maximum equal to D_f. Since approximate value of x_u found above is 234.4, $y_f \approx 0.15 \times 234.4 + 58.5 = 93.7 > D_f$.

Hence take $y_f = D_f = 90$ mm

$$C_u = 0.36 f_{ck} x_u b_w + 0.446 f_{ck} (b_f - b_w) D_f$$

$$= 0.36 \times 15 \times x_u (240) + 0.446 \times 15 (700 - 240) 90$$

$$= 1296 x_u + 276966.$$

Equating C_u and T_u, we get

$$1296 x_u + 276966 = 886161$$

$$\therefore \quad x_u = 470 \text{ mm.}$$

$$x_{u.max} = 0.48 d \text{ (for Fe 415 steel)}$$

$$= 0.48 \times 600 = 288 \text{ mm.}$$

Hence $x_u > x_{u.max}$

Thus the beam is *over-reinforced.* Though the Code recommends that such a beam should be redesigned, its moment of resistance is limited to $M_{u.lim.}$ corresponding to the balanced section.

$\therefore$ Take $x_u = x_{u.max} = 288$ mm.

Again, $\dfrac{D_f}{d} = \dfrac{90}{500} = 0.15 < 0.2$. Hence moment of resistance is given by Eq. 26.16 (corresponding to para E-2.2 of code) :

$$M_u = M_{u.lim.} = 0.36 \frac{x_{u.max}}{d} \left(1 - 0.416 \frac{x_{u.max}}{d}\right) f_{ck} b_w d^2 + 0.446 f_{ck} (b_f - b_w) D_f \left(d - \frac{D_f}{2}\right)$$

$$= 0.36 \times \frac{288}{600} \left(1 - 0.416 \frac{288}{600}\right) 15 \times 240 (600)^2 + 0.446 \times 15 (700 - 240) \times 90 \left(600 - \frac{90}{2}\right)$$

$$= 179.2 \times 10^6 + 153.7 \times 10^6 = 332.9 \times 10^6 \text{ N-mm}$$

$$= \mathbf{332.9 \text{ kN-m.}}$$

Example 26.4. Design of T-beam.

A T-beam has $b_f = 740$ mm, $b_w = 240$ mm, $d = 400$ and $D_f = 85$ mm. Find the reinforcement for an applied B.M. of 80 kN-m. Use M 15 concrete and Fe 250 steel.

Solution

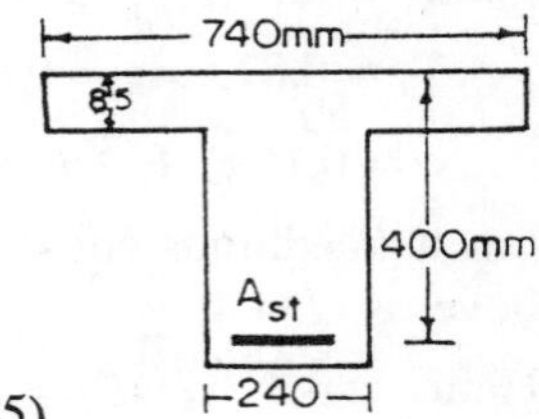

FIG 26.8

$M_D = 80$ kN-m ; $\gamma_f = 1.5$

$\therefore \quad M_{uD} = 80 \times 1.5 = 120$ kN-m

$= 120 \times 10^6$ N-mm.

Assuming $x_u = D_f$, we get

From Eq. 26.4(*a*)

$M_{u1} = 0.36 f_{ck} b_f D_f (d - 0.416 D_f)$

$= 0.36 \times 15 \times 740 \times 85 (400 - 0.416 \times 85)$

$= 123.85 \times 10^6$ N-mm $> M_{uD}$

$\therefore \quad x_u < D_f$

The reinforcement is found from Eq. 26.2.

$$M_{uD} = 0.87 f_y A_{st} d \left(1 - \frac{f_y}{f_{ck}} \frac{A_{st}}{b_f d}\right)$$

or $$120 \times 10^6 = 0.87 \times 250 \times 400 A_{st} \left(1 - \frac{250}{15} \times \frac{A_{st}}{740 \times 400}\right)$$

or $$A_{st}(1 - 56.3 \times 10^{-6} A_{st}) = 1379$$

or $$A_{st}^2 - 17762 A_{st} + 24.5 \times 10^6 = 0$$

From which A_{st} = **1507 mm²** .

Example 25.5. Design of T-beam.

Redesign the T-beam of example 26.4 if the applied B.M. is 124 kN-m.

Solution

$M_{uD} = 1.5 \times 124 = 186$ kN-m.

This is more than $M_{u1} = 123.7$ kN-m computed in Example 26.4.

Hence $x_u > D_f$.

Let us calculate compression force for $x_u = \frac{7}{3} D_f = \frac{7}{3} \times 85$

$= 198.3$ mm.

$\therefore \quad C_{uw} = 0.36 \times 15 \times 198.3 \times 240 = 256997$ N ...(26.5)

$C_{uf} = 0.446 \times 15 (740 - 240) 85 = 284325$ N ...(26.7(*a*)]

$M_{uw} = 256997 (400 - 0.416 \times 190.3) = 82.45 \times 10^6$ N-mm

$M_{uf} = 284325 (400 - 0.5 \times 85) = 101.65 \times 10^6$ N-mm

$\therefore \quad M_{u2} = 82.45 \times 10^6 + 101.65 \times 10^6 = 184.1 \times 10^6$ N-mm.

Since $M_{u2} < M_{uD}$, $\quad x_u > \frac{7}{3} D_f$.

Now M_{uf} does not depend on x_u, and thus remains constant.

$\therefore \quad M_{uw} = M_{uD} - M_{uf} = (186 - 101.65)10^6$ N-mm

Now, $M_{uw} = 0.87 f_y A_{st} d \left(1 - \frac{f_y}{f_{ck}} \frac{A_{sw}}{b_w d}\right)$ ...(26.23)

$$(186 - 101.65)\,10^6 = 0.87 \times 250\, A_{sw} \times 400 \left(1 - \frac{250}{15} \times \frac{A_{sw}}{240 \times 400}\right)$$

or $\quad 969.5 = A_{sw} (1 - 173.6 \times 10^{-6} A_{sw})$

or $\quad A_{sw}^2 - 5760\, A_{sw} + 5.585 \times 10^6 = 0$

From which $A_{sw} = 1234\ \text{mm}^2$

Also, $A_{sf} = \frac{C_{uf}}{0.87 f_y} = \frac{284325}{0.87 \times 250} = 1307\ \text{mm}^2$

$\therefore \quad$ Total $A_{st} = 1234 + 1307 = 2541\ \text{mm}^2$

It is now necessary to check that actual x_u is less than the critical (or limiting) value.

$$x_u = \frac{0.87 f_y A_{sw}}{0.36 f_{ck} b_w} = \frac{0.87 \times 250 \times 1234}{0.36 \times 15 \times 240} = 207\ \text{mm}$$

$$\therefore \quad \frac{x_u}{d} = \frac{207}{400} = 0.518.$$

For mild steel bars, $\frac{x_{u.max}}{d} = 0.53$.

Hence O.K.

Example 26.6. Design of T-beam.

Redesign T-beam of example 26.4 if the applied B.M. is 110 kN-m.

Solution

$$M_{uD} = 1.5 \times 110 = 165\ \text{kN-m}$$

In example 26.4, $M_{u1} = 123.85$ kN-m.

In example 26.5 $M_{u2} = 184.1$ kN-m.

$\therefore \quad M_{u1} < M_{uD} < M_{u2}$

Hence for this case $x_u < \frac{7}{3} D_f < 198.3$ mm.

The design procedure will be essentially that of trial and error.

Trial 1. Let us assume $x_u = 140$ mm.

Using I.S. Code method,

$$y_f = 0.15\,x_u + 0.65\,D_f$$
$$= 0.15 \times 140 + 0.65 \times 85 = 76.25 \text{ mm} < D_f$$

Hence O.K.

From Eq. 26.12,

$$M_u = 0.36\frac{x_u}{d}\left(1 - 0.416\frac{x_u}{d}\right) f_{ck}\, b_w\, d^2$$
$$+ 0.446 f_{ck}\,(b_f - b_w)\, y_f \times (d - 0.5\,y_f)$$
$$= 0.36 \times \frac{140}{400}\left(1 - 0.416\frac{140}{400}\right) \times 15 \times 240\,(400)^2$$
$$+ 0.446 \times 15\,(740 - 240)\,76.25\,(400 - 0.5 \times 76.25)$$
$$= 62.009 \times 10^6 + 92.298 \times 10^6 = 154.307 \times 10^6 \text{ N-mm}$$

$$M_{uD} = 165 \times 10^6 \text{ N-mm.}$$

Since $M_{uD} > M_u$, increase x_u.

Trial 2. Let $x_u = 160$ mm.

$$y_f = 0.15 \times 160 + 0.65 \times 85 = 79.25 \text{ mm} < D_f$$
$$M_u = 0.36 \times \frac{160}{400}\left(1 - 0.416 \times \frac{160}{400}\right) 15 \times 240\,(400)^2$$
$$+ 0.446 \times 15\,(740 - 240)\,79.25\,(400 - 0.5 \times 79.25)$$
$$= 69.142 \times 10^6 + 95.532 \times 10^6 = 164.674 \times 10^6.$$

$\therefore \quad M_u \approx M_{uD}.$

Hence adopt $x_u = 160$ mm.

Now $C_{uw} = 0.36 f_{ck}\, x_u\, b_w$ ...[26.5(*a*)]

$$= 0.36 \times 15 \times 160 \times 240 = 207360 \text{ N}$$
$$C_{uf} = 0.446 f_{ck}\,(b_f - b_w)\, y_f$$
$$= 0.446 \times 15\,(740 - 240)\,79.25 = 265091 \text{ N}$$

$\therefore$ Total compression $C_u = 207360 + 265091$
$= 472451$ N

$$\therefore \quad A_{st} = \frac{C_u}{0.87 f_y} = \frac{472451}{0.87 \times 250} = \mathbf{2172\ mm^2}.$$

Note. x_u can be alternatively obtained by combining the value of y_f in terms of x_u in Eq. 26.19 and then solving the quadratic equation.

Thus, $y_f = 0.15\,x_u + 0.65 \times 85 = 0.15\,x_u + 55.25$

Hence from Eq. 26.19, putting $M_u = M_{uD}$, we get

$$165 \times 10^6 = 0.36\frac{x_u}{400}\left(1 - \frac{0.416\,x_u}{400}\right) 15 \times 240\,(400)^2$$
$$+ 0.446 \times 15\,(740 - 240)\,[0.15\,x_u + 55.25]$$
$$\times [400 - 0.5\,(0.15\,x_u + 55.25)]$$

Simplifying and rearranging, we get

$$x_u^2 - 1199 x_u + 166797 = 0$$

From which $x_u = 160.63$ mm.

This value of x_u will be acceptable only if the value of y_f obtained from this is less than D_f.

$$y_f = 0.15 x_u + 0.65 D_f$$

$$= 0.15 \times 166.6 + 0.65 \times 85 = 79.35 \text{ mm} < D_f .$$

Hence the exact value of $x_u = 160.6$ mm.

$$\therefore \quad C_{uw} = 0.36 f_{ck} x_u b_w = 0.36 \times 15 \times 160.6 \times 240 = 208138 \text{ N}$$

$$C_{uf} = 0.446 f_{ck} (b_f - b_w) y_f = 0.446 \times 15 (740 - 240) 79.35$$

$$= 265426 \text{ N}$$

$$C_u = C_{uw} + C_{uf} = 208138 + 265426 = 473564$$

$$\therefore A_{st} = \frac{C_u}{0.87 f_y} = \frac{473564}{0.87 \times 250} = \mathbf{2177.3\ mm^2}.$$

Example 26.7. Design of T-beam.

Design the reinforcement for a T-beam for the following data :
Effective span : 8 m ; Ends simply supported.
Spacing of beams : 3.3 m centre to centre.
Thickness of slab : 130 mm
Width of web : 300 mm
Total Depth : 450 mm
Live load on the floor : 6 kN/m²
Floor finish load : 0.75 kN/m² .

The beam also supports a partition wall which transmits a load of 12 kN/m run.

Use M 15 concrete and Fe 415 steel

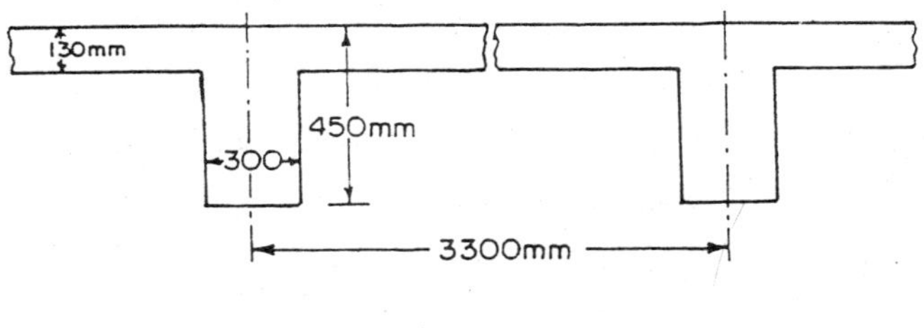

FIG. 26.9.

Solution

Weight of slab $= 25 \times 0.13 = 3.25$ kN/m^2

Weight of finishing $= 0.75$ kN/m^2

Live load $= 6.0$ kN/m^2

Total 10.0 kN/m^2.

$\therefore$ Load on beam, transferred from slab $= 3.3 \times 10 = 33$ kN/m

Load transferred from wall $= 12$ kN/m

Weight of rib $= 0.3\,(0.45 - 0.13) \times 25 = 2.4$ kN/m

Total working load $w = 47.4$ kN/m.

Load factor $= 1.5$.

$\therefore$ Ultimate load $w_u = 1.5 \times 47.4 = 71.1$ kN/m

$\therefore$ Ultimate design moment $M_{uD} = \dfrac{w_u L^2}{8}$

$$= \frac{71.1\,(8)^2}{8} = 568.8 \text{ kN-m.}$$

$$= 568.8 \times 10^6 \text{ N-mm.}$$

The flange width b_f given by,

$$b_f = \frac{L_0}{6} + b_w + 6\,D_f$$

$$= \frac{8}{6} + 0.3 + 6\,(0.13) = 2.413 \text{ m} = 2413 \text{ mm}$$

$$d = D - \text{cover} = 450 - 80 \text{ (say)} = 330 \text{ mm}$$

Step 1. Assuming $x_u = D_f$, we get from Eq.26.4(a)

$$M_{u1} = 0.36 f_{ck}\, b_f\, D_f\,(d - 0.416\, D_f)$$

$$= 0.36 \times 15 \times 2413 \times 130\,(330 - 0.416 \times 130)$$

$$= 467.4 \times 10^{-6} \text{ N-m.}$$

Step 2. Since $M_{uD} > M_{u1}$. N.A. falls outside the flange

Assuming $x_u = 7/3\, D_f \approx 303.3$ mm, we get from Eqs. 26.5 (a) and 26.5 (a).

$$C_{uw} = 0.36 f_{ck}\, x_u\, b_w = 0.36 \times 15 \times 303.3 \times 300$$

$$= 491346 \text{ N.}$$

$$C_{uf} = 0.446 f_{ck}\,(b_f - b_w)\, D_f = 0.446 \times 15\,(2413 - 300)\,130$$

$$= 1837676 \text{ N}$$

Hence $M_{uw} = C_{uw}\,(d - 0.416\, x_u) = 491346\,(300 - 0.416 \times 303.3)$

$$= 100.15 \times 10^6 \text{ N-mm}$$

and $M_{uf} = C_{uf}\,(d - 0.5\, D_f) = 1837676\,(330 - 0.5 \times 130)$

$$= 486.98 \times 10^6 \text{ N-mm}$$

$\therefore\ M_{u2} = M_{uw} + M_{uf} = (100.15 + 486.98)\ 10^6 = 587.13 \times 10^6$ N-mm

Since $M_{uD} < M_{u2},\ x_u < \frac{7}{3} D_f$

Hence the design procedure is done by trial and error.

Step 3. *Let us follow I.S. Code method :*

(*i*) Assume $x_u < \frac{7}{3} D_f < 303.3$ mm

Let us assume $x_u = 250$ mm.

Then $y_f = 0.15\,x_u + 0.65\,D_f$

$= 0.15 \times 250 + 0.65 \times 130 = 37.5 + 84.5 = 122$ mm

and $$M_u = 0.36 \frac{x_u}{d}\left(1 - 0.416 \frac{x_u}{d}\right) f_{ck}\, b_w\, d^2 + 0.446 f_{ck}\,(b_f - b_w) \times y_f\,(d - 0.5\,y_f) \quad \text{(Eq.26.12)}$$

$$= 0.36 \times \frac{250}{330}\left(1 - 0.416\ \frac{250}{330}\right) \times 15 \times 300\ (300)^2 + 0.446 \times 15\ (2413 - 300)\ 122\ (330 - 0.5 \times 122)$$

$$= 91.53 \times 10^6 + 463.91 \times 10^6 = 555.44 \times 10^6 \text{ N-mm}$$

This is quite close to $M_{uD} = 568.8 \times 10^6$ N-mm

However, in the second trial, let $x_u = 270$ mm.

$$y_f = (0.15 \times 270) + 0.65 \times 130 = 40.54 + 84.5 = 125$$

$$M_u = 0.36 \times \frac{270}{330}\left(1 - 0.416 \times \frac{270}{330}\right) \times 15 \times 300\ (330)^2 + 0.446 \times 15\ (2413 - 300)\ 125\ (330 - 0.5 \times 125)$$

$$= 95.21 \times 10^6 + 472.2 \times 10^2 = 567.91 \times 10^6 \approx \widetilde{M}_{uD}$$

Hence $x_u = 270$ mm

$$\therefore\quad C_{uw} = 0.36 f_{ck}\, x_u\, b_w = 0.36 \times 15 \times 270 \times 300 = 437400 \text{ N}$$

$$C_{uf} = 0.446 f_{ck}\,(b_f - b_w)\, y_f = 0.446 \times 15\ (2413 - 300)\ 125 = 1766996 \text{ N}$$

$$\therefore\quad C_u = C_{uw} + C_{uf} = 437400 + 1766996 = 2204396 \text{ N}$$

$$\therefore\quad A_{st} = \frac{C_u}{0.87 f_y} = \frac{2204396}{0.87 \times 415} = \mathbf{6106\ mm^2}$$

26.6. DOUBLY REINFORCED T-BEAMS

We have seen in the previous articles that a singly reinforced-section has limiting value of moment of resistance. However, if the applied moment is more than $M_{u.lim.}$ two alternatives will be available : (*i*) to increase the depth of the section, or (*ii*) to provide compressive reinforcement. In many cases the maximum value of depth of section may be limited or restricted from architectural or other considerations. In that case, the only alternative that will be left will be to provide reinforcement in the compression zone, giving rise to *doubly reinforced-section.*

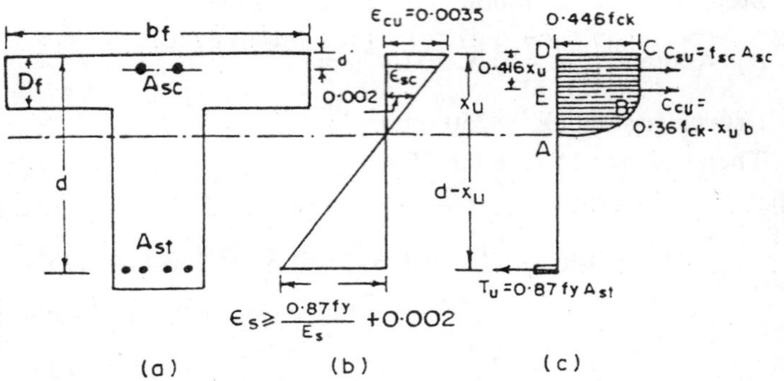

FIG. 26.10. DOUBLY REINFORCED T-BEAM.

Fig. 26.10(a) shows a doubly reinforced T-beam with compression reinforcement A_{sc} provided at a cover of d'. The strain in the compressive reinforcement is ε_{sc}, as shown in Fig. 26.10(b). The compressive force contribution of A_{sc} will be $C_{su} = f_{sc} A_{sc}$ [Fig. 26.10(c)] where f_{sc} is the compression stress in steel at a strain of ε_{sc}, and is found from the stress strain curve.

$$\text{Now,} \quad \varepsilon_{sc} = \frac{0.0035\,(x_u - d')}{x_u} \qquad ...(26.24)$$

DETERMINATION OF x_u

For a given section, x_u can be determined by equating the total tensile force to the total compressive force. Assuming that the N.A. lies in the flange.

$$0.36 f_{ck}\, b_f x_u + f_{sc} A_{sc} = 0.87 f_y A_{st} \qquad ...(26.25)$$

If the loss of concrete area occupied by compressive steel is taken into account, we have

$$0.36 f_{ck}\, b_f x_u + (f_{sc} - 0.446 f_{ck}) A_{sc} = 0.987 f_y A_{st} \qquad ...[26.25(a)]$$

(This is subject to the condition that $d' < \frac{3}{7} x_u$ so that stress in concrete at the level of compressive steel is equal to $0.446 f_{ck}$.)

Stress f_{sc} is found from stress-strain curve, corresponding to the value of ε_{sc} given by Eq. 26.24. Thus, f_{sc} indirectly depends upon x_u. Hence the following iterative procedure is adopted :

Step 1. Find $x_{u.max}$ for a balanced section from strain compatibility :

$$x_{u.max} = \frac{0.0035}{0.0055 + 0.87 f_y/E_s} = \frac{700}{1100 + 0.87 f_y}$$

Step 2. Assume $x_u = x_{u.max}$ or D_f (whichever is less) and compute ε_{sc} from Eq. 26.24.

Step 3. Find f_{sc} from stress-strain curve, corresponding to the above value of ε_{sc}.

(Alternatively, in the first trial, take f_{sc} slightly less than 0.87 f_y).

Step 4. Substitute this value of f_{sc} in Eq. 26.25 (or Eq. 26.25(*a*) and compute the modified value of x_u.

Step 5. Repeat step 2 to 4, till convergence of the value of x_u is achieved.

If $x_u < D_f$, the N.A. lies in the flange and hence Eq. 26.25 [or 26.25(*a*)] is valid. If however, x_u comes out to be greater than D_f, the N.A. will be outside the flange, *and a new value of x_u will have to be found* by equating the total compressive force to the total tensile force. The determination of x_u will be done exactly in the same manner as done for the singly reinforced T-beam discussed in § 26.4. The total compressive force C_u will consists of (*i*) web contribution, C_{uw} and (*ii*) flange contribution C_{uf}, where

$$C_{uw} = 0.36 f_{ck} x_u b_w + (f_{sc} - 0.446 f_{ck}) A_{sc} = 0.87 f_y A_{sw} \quad ...(26.26)$$

The compressive force C_{uf} can be computed by approximating the actual stress block into an equivalent rectangular block of height y_f given

$$y_f = 0.15 x_u + 0.65 D_f \text{ (subject to a max. of } D_f \text{)}$$

$$\therefore \quad C_{uf} = 0.446 f_{ck} (b_f - b_w) y_f = 0.87 f_y A_{sf} \quad ...(26.27)$$

$$\therefore \quad 0.36 f_{ck} x_u b_w + (f_{sc} - 0.446 f_{ck}) A_{sc} + 0.446 f_{ck} (b_f - b_w) y_f$$
$$= 0.87 f_y A_{st} \quad ...(26.28)$$

Since f_{sc} depends upon x_u , an iterative procedure will have to be adopted. Also, y_f depends upon x_u , and hence the final value of x_u found from Eq. 26.28 is valid only if the value of

y_f calculated by using this value of x_u, is less than or equal to D_f. If this value of y_f comes out to be greater than D_f, the value of x_u should be *recalculated* from Eq. 26.28 by taking $y_f = D_f$.

Thus, the value of x_u is known. This value of x_u should be compared with $x_{u.max}$ to find whether the section is under reinforced, balanced or over-reinforced.

MOMENT OF RESISTANCE WHEN $x_u < D_f$

If x_u comes out to be less than D_f, from Eq. 26.25 or 26.25(*a*), the moment of resistance can be found by treating the beam to be of a doubly reinforced rectangular section of width b_f.

Again, if $x_u < x_{u.max}$, the section is under-reinforced, and the moment of resistance is given by

$$M_u = 0.36 f_{ck} x_u b_f (d - 0.416 x_u) + (f_{sc} - 0.446 f_{ck}) A_{sc} (d - d')$$

When $x_u = D_f$, we get ...(26.29)

$$M_{u1} = 0.36 f_{ck} b_f D_f (d - 0.416 x_u) + (f_{sc} - 0.446 f_{ck}) A_{sc} (d - d') \quad ...[26.29(a)]$$

If $x_u = x_{u.max}$, we have the balanced section, for which,

$$M_{u.lim.} = 0.36 f_{ck} x_{u.max} b_f (d - 0.416 x_{u.max}) + (f_{sc} - 0.446 f_{sc}) A_{sc}(d - d') \quad ...(26.30)$$

If $x_u > x_{u.max}$, the section is *over-reinforced*, and its moment resistance M_u will be limited to $M_{u.lim.}$ given by Eq. 26.30. However, the Code recommends that such section should be redesigned.

MOMENT OF RESISTANCE WHEN $x_u > D_f$

The value of x_u obtained from Eq. 26.28 should be compared with $x_{u.max}$ to find whether the section is under-reinforced, balanced or over-reinforced.

The procedure for finding M_u will be the same as that discussed for singly reinforced section except that an additional moment of resistance contribution by compressive steel $= (f_{sc} - 0.446 f_{ck}) A_{sc}(d - d')$ will have to be taken into account.

(*a*) *Thus, for under-reinforced section* ($x_u < x_{u.max}$),

If $\frac{D_f}{x_u} \leq 0.43$, ,we have Case 2, for which

$$M_{uw} = 0.36 f_{ck} \frac{x_u}{d} b_w \left(1 - 0.416 \frac{x_u}{d}\right) d^2 + (f_{sc} - 0.446 f_{ck}) A_{sc}(d - d') \quad ...[26.31(a)]$$

$$M_{uf} = 0.446 f_{ck} (b_f - b_w) D_f (d - 0.5 D_f) \quad ...[26.31(b)]$$

and $M_u = 0.36 f_{ck} \frac{x_u}{d} b_w \left(1 - 0.416 \frac{x_u}{d}\right) d^2 + (f_{sc} - 0.446 f_{ck}) A_{sc} (d - d')$

$$+ 0.446 f_{ck} (b_f - b_w) D_f (d - 0.5 D_f) \quad ...(26.31)$$

When $D_f = \frac{3}{7} x_u \left(\text{or } x_u = \frac{7}{3} D_f\right)$, we have

$$M_{u2} = 0.36 f_{ck} b_w \left(\frac{7}{3} D_f\right) \left(d - 0.416 \times \frac{7}{3} D_f\right) + (f_{sc} - 0.446 f_{ck}) A_{sc} (d - d')$$

$$+ 0.446 f_{ck} (b_f - b_w) D_f (d - 0.5 D_f) \quad ...[26.31(c)]$$

(*b*) Similarly, when $x_u < x_{u.max}$ and $\frac{D_f}{x_u} > 0.43$, we have Case 3, for which.

$$M_{uw} = 0.36 f_{ck} \frac{x_u}{d} b_w \left(1 - 0.416 \frac{x_u}{d}\right) d^2 + (f_{sc} - 0.446 f_{ck}) A_{sc} (d - d') \quad ...[26.32(a)]$$

and $M_{uf} = 0.446 f_{ck} (b_f - b_w) y_f (d - 0.5 y_f)$ ...[26.32(*b*)]

and $M_u = 0.36 \frac{x_u}{d} \left(1 - 0.416 \frac{x_u}{d}\right) f_{ck} b_w d^2$

$$+ (f_{sc} - 0.446 f_{ck}) A_{sc} (d - d')$$

$$+ 0.446 f_{ck} (b_f - b_w) y_f (d - 0.5 y_f) \quad ...(26.32)$$

MOMENT OF RESISTANCE WHEN $(x_u = x_{u.max}) > D_f$

The limiting moment of resistance can be computed by modifying the equations (Eq. 26.16 and 26.17) suggested by IS Code, by including the moment of resistance term of the compression steel. Thus, we have the following modified equations.

(*a*) ***When*** $\frac{D_f}{d} \leq 0.2$.

$$M_{u.lim.} = 0.36 \frac{x_{u.max}}{d} \left(1 - 0.416 \frac{x_{u.max}}{d}\right) f_{ck} b_w d^2$$

$$+ (f_{sc} - 0.446 f_{ck}) A_{sc} (d - d') + 0.446 f_{ck} (b_f - b_w) D_f \left(d - \frac{D_f}{2}\right) \quad ...(26.23)$$

where f_{sc} is found from stress-strain curve corresponding to ε_{sc} given by Eq. 26.24 in which $x_u = x_{u.max}$.

(b) When $\dfrac{D_f}{d} > 0.2.$

$$M_{u.lim} = 0.36 \frac{x_{u.max}}{d}\left(1 - 0.416 \frac{x_{u.max}}{d}\right) f_{ck}\, b_w\, d^2$$

$$+ (f_{sc} - 0.446 f_{ck}) A_{sc} (d - d') + 0.446 f_{ck} (b_f - b_w) y_f \left(d - \frac{d_f}{2}\right) \quad ...(26.34)$$

where $y_f = (0.15\, x_{u.max} + D_f) \le D_f$

MOMENT OF RESISTANCE WHEN $(x_u > x_{u.max}) > D_f$

If $x_u > x_{u.max}$ the section is over-reinforced, and its moment of resistance M_u will be limited to $M_{u.lim.}$ given by Eq. 26.33 (when $D_f/d \le 0.2$) or by Eq. 26.34 (when $D_f/d > 0.2$). However, the Code recommends that such a section should be redesigned.

DESIGN OF DOUBLY REINFORCED T-BEAM

Refer § 26.5. A doubly reinforced section will be required when $M_{uD} > M_{u.lim.}$ For a T-beam of given cross-sectional dimensions, (*i.e.* given b_w, b_f, D_f and d), the design is done in the following steps.

Step 1. Assuming $x_u = D_f$, and assuming singly reinforced section, compute M_{u1} from Eq. 26.4(*a*).

Step 2. If $M_{u1} < M_{uD,}$ the N.A. will fall inside the flange. Taking $x_u = \dfrac{7}{3} D_f$ and taking singly reinforced section, compute M_{u2} from Eq. 26.11(*c*)

Step 3. If $M_{u2} < M_{uD}$, $x_u > \dfrac{7}{3} D_f$.

Compute $M_{u,lim}$ by taking $x_u = x_{u,max}$

If $M_{u,lim} < M_{uD}$, a doubly reinforced section will be required.

(*a*) Compute A_{st1} for $M_{u.lim.}$ as under :

$$A_{sw} = \frac{0.36 f_{ck}\, x_{u.max}\, b_w}{0.87 f_y}$$

and $$A_{sf} = \frac{0.446 f_{ck} (b_f - b_w) D_f}{0.87 f_y}$$

$$\therefore \quad A_{st1} = [0.36 f_{ck}\, x_{u.max}\, b_w + 0.446 f_{ck} (b_f - b_w) D_f] \div (0.87 f_y) \quad ...(26.35)$$

(*b*) Compute the additional tensile area A_{st2} from the relation

$$A_{st2} = \frac{M_{uD} - M_{u.lim.}}{0.87 f_y (d - d')} \quad ...(26.36)$$

(*c*) Compare the compressive area A_{sc} corresponding to A_{st2}, by the relation.

$$A_{sc} = \frac{M_u - M_{u.lim.}}{(f_{sc} - 0.446 f_{ck})(d - d')} \qquad ...(26.37)$$

Example 26.8. M.R. of doubly reinforced T-beam

Find the moment of resistance of T-beam of Example 26.3 if the beam also carries a compression reinforcement in the form of 3 bars of 25 mm dia. provided at a distance of 40 mm from the top fibre.

Solution

$$A_{st} = 2454.4 \text{ mm}^2 \text{ (as before)}$$

$$A_{sc} = 3 \times \frac{\pi}{4}(25)^2 = 1472.6 \text{ mm}^2.$$

Assuming the N.A. to fall in the flange, we get, from Eq. 26.25(*a*).

$$0.36 f_{ck}\, b_f\, x_u + (f_{sc} - 0.446 f_{ck}) A_{sc} = 0.87 f_y A_{st} \qquad ...[26.25(a)]$$

Take $f_{sc} < 0.87 f_y < 0.87 \times 415 < 361$, in the first trial.

Hence take $f_{sc} = 350$ N/mm² (say)

$$\therefore\ 0.36 \times 15 \times 700 x_u + (350 - 0.446 \times 15) \times 1472.6 = 0.87 \times 415 \times 2454.4$$

From which $x_u = 100.7$ mm $> D_f$

(Note that d' is less than $3/7\, x_u$ so that stress in concrete at A_{sc} level is $0.446 f_{ck}$)

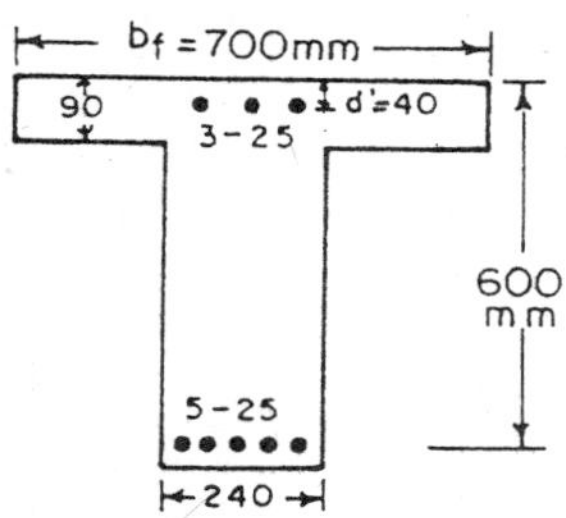

FIG. 26.11

Hence the N.A. lies outside the flange. To obtain a new value of x_u, equate total compressive force to the total tensile force.

$$T_u = 0.87 f_y A_{st} = 0.87 \times 415 \times 2454.4 = 886161 \text{ N}$$

$$y_f = 0.15 x_u + 0.65 D_f = 0.15 x_u + 0.65 \times 90 = 0.15 x_u + 58.5$$

with a maximum equal to D_f. Since the approximate value of x_u found above is 100.7 mm, $y_f = 0.15 \times 100.7 + 58.5 = 73.6$ mm $< D_f$.

$$\therefore \quad C_u = 0.36 f_{ck} x_u b_w + (f_{sc} - 0.446 f_{ck}) A_{sc} + 0.446 f_{ck}(b_f - b_w) y_f$$

or
$$C_u = 0.36 \times 15 x_u \times 240 + (f_{sc} - 0.446 \times 15)\, 1472.6 + 0.446 \times 15 (700 - 240) \times (0.15 x_u + 58.5)$$

or
$$C_u = 1296 x_u + 1472.6 f_{sc} - 9852 + 461.6 x_u + 180028$$

or
$$C_u = 1757.6 x_u + 170176 + 1472.6 f_{sc}$$

Equating C_u and T_u, we get

$$1757.6x_u + 170176 + 1472.6 f_{sc} = 886161$$

or $$x_u = 407.4 - 0.8378 f_{sc} \quad ...(i)$$

In the above equation, f_{sc} can be obtained from stress-strain curve, corresponding to ε_{sc} which itself depends upon x_u, as is evident from Eq. 26.24.

$$\varepsilon_{sc} = \frac{0.0035\,(x_u - d')}{x_u} \quad ...(ii)$$

In the first approximation, taking $x_u =$ 100.7 mm,

$$\varepsilon_{sc} = \frac{0.0035\,(100.7 - 40)}{100.7} = 0.0021$$

From stress-strain curve for Fe 415 (Fig. 24.3) or from Table 24.1, we get $f_{sc} \approx 328$ N/mm^2.

Hence from (*i*), $x_u = 407.4 - 0.8378 \times 328 = 132.6$ mm

For this new value of x_u, $y_f = 0.15 \times 132.6 + 58.5 = 78.4$ mm $< D_f$.

Hence O.K.

In the next trial, recalculate ε_{sc} by taking

$x_u = 132.6$ mm, from (*ii*)

$$\therefore \quad \varepsilon_{sc} = \frac{0.0035\,(132.6 - 40)}{132.6} = 0.0024$$

From Table 24.1 (or Fig. 24.3) , we get

$f_{sc} = 342$ N/mm^2 for $\varepsilon_{sc} = 0.0024$

Hence from (*i*) $x_u = 407.4 - 0.8378 \times 342 = 102.8$ mm.

In the next trial,

$$\varepsilon_{sc} = \frac{0.0035\,(120.8 - 40)}{120.8} = 0.00234$$

From Table 24.1 (or Fig. 24.3), $f_{sc} = 340$ N/mm^2. for $\varepsilon_{sc} = 0.00234$

Hence from (*i*), $x_u = 407.4 - 0.8378 \times 340 = 122.5$ mm

In the next trial, the solution will converage to a value of $x_u = 122$ mm, for which $\varepsilon_{sc} = 0.00235$ and $f_{sc} \approx 340.5$ N/mm^2

$$y_f = 0.15\,x_u + 58.5 = 0.15 \times 122 + 58.5 = 76.8 \text{ mm}$$

$$x_{u,max} = 0.48\,d \text{ (for Fe 415 steel)} = 0.48 \times 600 = 288 \text{ mm}$$

$\therefore$ $x_u < x_{u.max}$ and the section is under-reinforced.

Again, $\frac{3}{7}x_u = \frac{3}{7} \times 122 = 52.3$ mm $< D_f$

Hence $D_f > \frac{3}{7}x_u$

or $\frac{D_f}{x_u} > 0.43.$

Hence M_u is given by Eq. 26.32

$$M_u = 0.36\,\frac{x_u}{d}\left(1 - 0.416\,\frac{x_u}{d}\right)f_{ck}\,b_w\,d^2 + (f_{sc} - 0.446\,f_{ck})A_{sc}(d - d')$$
$$+ 0.446\,f_{ck}\,(b_f - b_w)\,y_f\,(d - 0.5\,y_f) \qquad ...(26.32)$$

or
$$M_u = 0.36\,\frac{122}{600}\left(1 - 0.416\,\frac{122}{600}\right) 15 \times 240\,(600)^2$$
$$+ (340.5 - 0.446 \times 15) \times 1472.6 \times (600 - 40)$$
$$+ 0.446 \times 15\,(700 - 240) \times 76.8\,(600 - 0.5 \times 76.8)$$
$$= 86.84 \times 10^6 + 275 \times 10^6 + 132.73 \times 10^6$$
$$= 494.57 \times 10^6 \text{ N-mm} = \mathbf{494.57 \text{ kN-m.}}$$

Example 26.9. M.R. of doubly reinforced T-beam

Find the moment of resistance of the T-beam of Example 26.3 if the beam also carries a compression reinforcement in the form of 2 bars of 20 mm dia. with d' = 40 mm and tensile reinforcement consists of 6 bars of 25 mm dia.

Solution.

$$A_{st} = 6 \times \frac{\pi}{4}\,(25)^2 = 2945.2 \text{ mm}^2$$

$$A_{sc} = 2\,\frac{\pi}{4}\,(20)^2 = 628.3 \text{ mm}^2$$

Assuming the N.A. to fall in flange, we have, from Eq. 26.25(*a*).

$$0.36\,f_{ck}\,b_f\,x_u + (f_{sc} - 0.446\,f_{ck})\,A_{sc} = 0.87\,f_y\,A_{st}$$

Taking $f_{sc} < 0.87\,f_y < 0.87 \times 415 < 361$ or equal to 350 N/mm^2 in the first trial, we get

FIG. 26.12

$$0.36 \times 15 \times 700\,x_u + (350 - 0.446 \times 15)\,628.3 = 0.87 \times 415 \times 2945.2$$

From which $x_u = 224.2 \text{ mm} > D_f$

(Note that d' is less than $\frac{3}{7}x_u$ so that stress in concrete at the level of compressive stress is equal to f_{ck}).

Hence the N.A. lies outside the flange. To obtain a new value of x_u, equate T_u and C_u.

$$T_u = 0.87\,f_y\,A_{st} = 0.87 \times 415 \times 2945.2 = 1063365 \text{ N}$$

$$y_f = 0.15\,x_u + 0.65\,D_f = 0.15\,x_u + 0.65 \times 90$$
$$= 0.15\,x_u + 58.5$$

with a maximum equal to D_f. Since the approximate value of x_u found above is 177.3 mm,

$$y_f = 224.2 \times 0.15 + 58.5 = 92.1 > D_f$$

Hence take $y_f = D_f$

$$\therefore \quad C_u = 0.36\,f_{ck}\,x_u\,b_w + (f_{sc} - 0.446\,f_{ck})\,A_{sc} + 0.446\,f_{ck}(b_f - b_w)D_f$$

or
$$C_u = 0.36 \times 15\,x_u \times 240 + (f_{sc} - 0.446 \times 15)\,628.3 + 0.446 \times 15\,(700 - 240)\,90$$

or
$$C_u = 1296\,x_u + 628.3\,f_{sc} - 4203 + 276966$$

or
$$C_u = 1296\,x_u + 272763 + 628.3\,f_{sc}$$

Equating C_u and T_u,

$$1296\,x_u + 272763 + 628.3\,f_{sc} = 1063365$$

or
$$x_u = 610 - 0.4848\,f_{sc} \quad \text{...}(i)$$

In the above equation, f_{sc} can be obtained from stress-strain curve, corresponding to ε_{sc} which itself depends upon x_u, as is evident from Eq. 26.24.

$$\varepsilon_{sc} = \frac{0.0035\,(x_u - d')}{x_u} \quad \text{...}(ii)$$

In the first approximation, taking $x_u \approx 242.2$ mm

$$\varepsilon_{sc} = \frac{0.0035\,(242.2 - 40)}{242.2} = 0.00292$$

From stress-strain curve for Fe 415(Fig. 24.3) or from Table 24.1, we get $f_{sc} \approx 353\ \text{N/mm}^2$.

Hence from (*i*), $x_u = 610 - 0.4848 \times 353 = 439$ mm

Substituting this value of x_u in (*ii*), we get

$$\varepsilon_{sc} = \frac{0.0035\,(439 - 40)}{439} = 0.0032$$

$\therefore$ $f_{sc} = 356\ \text{N/mm}^2$ from Table 24.1 (or Fig. 24.3)

$\therefore$ Next value of $x_u = 610 - 0.4848 \times 365 \approx 43$ mm

Since the value has converged, take $x_u = 437$ mm for which

$$f_{sc} = 356\ \text{N/mm}^2$$

Now, $x_{u.max} = 0.48\,d = 0.48 \times 600 = 288$ mm

Hence $x_u > x_{u.max}$.

Thus, the beam is over-reinforced. Though the Code recommends that such a beam should be *redesigned,* its moment of resistance is limited to $M_{u.lim.}$ corresponding to the balanced section.

$\therefore$ Take $x_u = x_{u.max} = 288$ mm.

For this value of x_u,

$$\varepsilon_{sc} = \frac{0.0035\,(288-40)}{288} = 0.003$$

$\therefore$ $f_{sc} \approx 354$ N/mm^2 .

Again $\frac{D_f}{d} = \frac{90}{600} = 0.15 < 0.2$. Hence moment of resistance is given by Eq. 26.33 (corresponding to para E-2.2 of code).

$$M_u = M_{u.lim.} = 0.36\frac{x_{u.max}}{d}\left(1-0.416\frac{x_{u.max}}{d}\right) f_{ck}\, b_w\, d^2$$

$$+ (f_{sc} - 0.446 f_{ck}) A_{sc}\,(d-d') + 0.446 f_{ck}\,(b_f - b_w)\, D_f \left(d - \frac{D_f}{2}\right)$$

$$= 0.36 \times \frac{288}{600}\left(1 - 0.416 \times \frac{288}{600}\right) 15 \times 240\,(600)^2$$

$$+ (354 - 0.446 \times 15) \times 628.3\,(600-40)$$

$$+ 0.446 \times 15\,(700-240) \times 90 \times (600 - 0.5 \times 90)$$

$$= 179.2 \times 10^6 - 122.2 \times 10^6 + 153.7 \times 10^6$$

$$= 455.1 \times 10^6 \text{ N-mm}$$

$= \mathbf{455.1\ kN\text{-}m.}$

PROBLEMS

1. Find the moment of resistance of a T-beam having the following data
   Width of flange : 800 mm
   Thickness of slab : 120 mm
   Width of rib : 200 mm
   Effective depth : 400 mm
   Tensile steel area : 3500 mm^2
   Use M 15 concrete and Fe 250 steel.
2. Solve problem 1 with M 15–Fe 415 combination.
3. Find the moment of resistance of a T-beam section having $b_w = 300$ mm, $b_f = 1600$ mm, $D_f = 100$ mm and $d = 510$ mm. The reinforcement consists of 4 bars of 25 mm dia. Use M 15 concrete and Fe 415 Steel.
4. Solve problem 3 if the reinforcement consists of 6 bars of 25 mm dia.
5. Solve problem 3 if the reinforcement consists of 8 bars of 25 mm dia.
6. Solve problem 3 if the beam also carries compressive reinforcement in the form of 2 bars of 16 mm dia.
7. Solve problem 4, if the beam also carries compressive reinforcement in the form of 4 bars of 16 mm dia.
8. For the T-beam of the dimensions of problem 1, find the reinforcement for an applied B.M. of 100 kN-m.

27

Shear, Bond and Torsion

27.1. LIMIT STATE OF COLLAPSE : SHEAR

1. *Nominal Shear Stress*

The nominal shear stress τ_v in beams of uniform depth is determined from the following expression :

$$\tau_v = \frac{V_u}{bd} \qquad \text{...(27.1)}$$

where V_u = shear force due to design load.

b = width of beam ($= b_w$ for flanged sections)

d = effective depth

In the case of beams of varying depth, the equation is modified as under :

$$\tau_v = \frac{V_u \pm \dfrac{M_u}{d} \tan \beta}{bd} \qquad \text{...(27.2)}$$

where M_u = bending moment at the section

and β = angle between the top and bottom edges of the beam.

The negative sign in the formula applies when the B.M. M_u increases numerically in the same direction as the effective depth d increases, and the positive sign when the moment decreases numerically in this direction.

2. *Design Shear Strength of Concrete*

The design shear strength of concrete in beams without shear reinforcement is given in Table 27.1.

TABLE 27.1. DESIGN SHEAR STRENGTH OF CONCRETE (N/mm^2)

$\frac{100A_s}{bd}$	*Grade of concrete*					
	M 15	*M 20*	*M 25*	*M 30*	*M 35*	*M 40*
0.25	0.35	0.36	0.36	0.37	0.37	0.38
0.50	0.46	0.48	0.49	0.50	0.50	0.51
0.75	0.54	0.56	0.57	0.59	0.59	0.60
1.00	0.60	0.62	0.64	0.66	0.67	0.68
1.25	0.64	0.67	0.70	0.71	0.73	0.74
1.50	0.68	0.72	0.74	0.76	0.78	0.79
1.75	0.71	0.75	0.78	0.80	0.82	0.84
2.00	0.71	0.79	0.82	0.84	0.86	0.88
2.25	0.71	0.81	0.85	0.88	0.90	0.92
2.50	0.71	0.82	0.88	0.91	0.93	0.95
2.75	0.71	0.82	0.90	0.94	0.96	0.98
3.00	0.71	0.82	0.92	0.96	0.99	1.01

Under no circumstances, even with shear reinforcement, shall the nominal shear in beams τ_v, exceed $\tau_{c.max}$ given in Table 27.2.

TABLE 27.2 MAXIMUM SHEAR STRESS $\tau_{c,max}$ (N/mm^2)

Concrete grade	M 15	M 20	M 25	M 30	M 35	M 40
$\tau_{c.max}$ *(N/mm^2)*	2.5	2.8	3.1	3.5	3.7	4.0

For solid slabs, the design shear strength of concrete shall be $k\,\tau_c$, where k has the following value : (Table 27.3.)

TABLE 27.3 VALUES OF *k*

Overall Depth of slab (mm)	300 or more	275	250	225	200	175	150 or less
k	1.00	1.05	1.10	1.15	1.20	1.25	1.30

Note. The above provision does not apply to flat slabs.

3. *Minimum Shear Reinforcement*

When τ_v is less than τ_c given in Table 27.1 minimum shear reinforcement in the form of stirrups shall be provided such that

$$\frac{A_{sv}}{b\,s_v} \geq \frac{0.4}{f_y} \qquad ...(27.3)$$

where A_{sv} = total cross sectional area of stirrup legs effective in shear,

s_v = stirrup spacing along the length of member.

b breadth of the beam or breadth of web of flanged beam, and

f_y = characteristic strength of the stirrup reinforcement in N/mm^2 which shall not be taken greater than 415 N/mm^2

Hence spacing based on minimum shear reinforcement is given by

$$s_v \leq \frac{2.5\,A_{sv} f_y}{b} \qquad ...[27.3(a)]$$

However, in members with minor structural importance, such as lintels or where the minimum shear stress calculated in less than half the permissible value, this provision need not be complied with.

4. *Design of Shear Reinforcement*

When τ_v exceeds τ_c given in Table 27.1 shear reinforcement is provided in any of the following forms :

(*a*) Vertical stirrups

(*b*) Bent up bars with stirrups, and

(*c*) Inclined stirrups.

Where bent up bars are provided, their contribution towards shear resistance shall not be more than half that of the total shear reinforcement.

Shear reinforcement shall be provided to carry a shear equal to $V_u - \tau_c bd$. The strength of shear reinforcement V_{us} shall be calculated as follows :

(*a*) For vertical stirrups :

$$V_{us} = \frac{0.87\,f_y A_{sv}\,d}{s_v} \qquad ...(27.4)$$

Alternatively, the spacing of vertical stirrups is given by

$$s_v = \frac{0.87 f_y A_{sv}}{(\tau_v - \tau_c)\, b} \qquad \text{...[27.4(a)]}$$

(*b*) For inclined stirrups or a series of bars bent up at different cross-sections.

$$V_{us} = \frac{0.87 f_y A_{sv} d}{s_v}(\sin\alpha + \cos\alpha) \qquad \text{...(27.5)}$$

(c) For single bar or single group of parallel bars, all bent-up at the same cross-section.

$$V_{us} = 0.87 f_y A_{sv} \sin\alpha \qquad \text{...(27.6)}$$

where α = angle between the inclined stirrup or bent-up bar and the axis of the member, not less than 45°.

Note 1. Where more than one type of shear reinforcement is used to reinforce the same portion of the beam, the total shear resistance shall be computed as the sum of the resistances for various types respectively.

Note 2. The area of stirrups shall not be less than the minimum specified in by Eq. 27.3,

Note 3. The shear resistance of minimum reinforcement envisaged in Eq. 27.3, is found by substituting the value $\frac{f_y A_{sv}}{s_v}$ $= 0.4\, b$ (Eq. 27.3) into Eq. 27.4

Thus, from Eq. 27.4.

$$V_{us.min.} = 0.87\left(\frac{f_y A_{sv}}{s_v}\right) d$$

or

$$V_{us.min.} = 0.87\,(0.4\, b)\, d = 0.348\, bd \qquad \text{...[27.3(a)]}$$

or

$$V_{us} \approx 0.35\, bd.$$

Thus, the shear carried by concrete and that carried by minimum stirrups is given by

$$V_{uc.min} = \tau_c\, bd + 0.348\, bd \qquad \text{...[27.3(b)]}$$

Note 4. The maximum spacing of shear reinforcement measured along the axis of the member shall not exceed 0.74 *d* for vertical stirrups and *d* for inclined stirrups at 45°, where *d* is the effective depth of the section. In no case shall the spacing exceed 450 mm.

Example 27.1. Design of shear stirrups

A reinforced concrete beam 250 mm wide and 400 mm effective depth is subjected to ultimate design shear force of 140 kN at the supports. The tensile reinforcement at the supports is 0.5 percent.

Design the shear stirrups at the supports. Also, design the minimum shear reinforcement at the mid span. Assume concrete of grade M 15 and mild steel bars of Fe 250 grade.

Solution

$$d = 400 \text{ mm.}$$

$$\tau_v = \frac{V_u}{bd} = \frac{140 \times 1000}{250 \times 400} = 1.4 \text{ N/mm}^2.$$

From Table 27.1, $\tau_c = 0.46 \text{ N/mm}^2$ for $M\,15$ concrete.

From Table 27.2, $\tau_{c.max} = 2.5 \text{ N/mm}^2$ for $M\,15$ concrete.

Thus τ_v is less than $\tau_{c.max}$, but greater than τ_c.

Hence shear reinforcement is necessary.

$$V_{uc} = \tau_c\, bd = 0.46 \times 250 \times 400 = 46000 \text{ N.}$$

$$\therefore \quad V_{us} = V_u - V_{uc} = (140 - 46)\,1000 = 94000 \text{ N.}$$

The shear resistance of nominal stirrups is given by

$$V_{usn} = 0.348\, bd = 0.348 \times 250 \times 400 = 34800 < V_{us}.$$

Hence nominal stirrups are not sufficient.

Again , from Eq. 27.4,

$$V_{us} = \frac{0.87 f_y A_{sv}\, d}{s_v}$$

or

$$s_v = \frac{0.87 f_y A_{sv}}{V_{us}}\, d$$

Using 10 mm dia. bars, $A_{sv} = 2 \frac{\pi}{4} (10)^2 = 157.1 \text{ mm}^2$.

$$\therefore \quad s_v = \frac{0.87 \times 250 \times 157.1}{94000} \times 400 = 145.4 \text{ mm}$$

Alternatively, from Eq. 27.4(*a*),

$$s_v = \frac{0.87 f_y A_{sv}}{(\tau_v - \tau_c)\, b} = \frac{0.87 \times 250 \times 157.1}{(1.4 - 0.46)\, 250}$$
$$= 145.4 \text{ mm}$$

Maximum spacing $= 0.75\, d = 300$ mm.

Hence provide 10 mm dia. two legged stirrups @ 145 mm c/c at the supports.

At the mid-span, the spacing of minimum shear reinforcement is given by

$$s_v = \frac{f_y A_{sv}}{0.4\, b}$$

Using 10 mm dia. 2-legged stirrups,

$$s_v = \frac{250 \times 157.1}{0.4 \times 250} = 393 \text{ mm.}$$

However, max. spacing is limited to $0.75\,d = 300$ mm.

Hence provide 10 mm dia. two-legged stirrups @ 300 mm c/c at the mid-span.

Example 27.2. Design of shear reinforcement

A simply supported beam, 300 mm wide and 600 mm effective depth carries a uniformly distributed load of 70 kN/m including its own weight, over an effective span of 6 m. The reinforcement consists of 5 bars of 25 mm diameter. Out of these, two bars can be safely bent up at 1 m distance from the support. Design the shear reinforcement for the beam.

Given : Grade of concrete : M 15 ; Grade of steel : Fe 415.
Assume width of support = 400 mm.

Solution

$$w_u = 1.5\,w = 1.5 \times 70 = 105 \text{ kN.}$$

Max. shear $$V_{u.max} = \frac{w_u L}{2} = \frac{105 \times 6}{2} = 315 \text{ kN.}$$

According to I.S. code, the critical section for shear occurs at a distance of $d(=600\text{mm})$ from the face of support, *i.e.* at a distance of $\frac{400}{2} + 600 = 800$ mm from the centre of support.

$$\therefore \qquad V_u = 315 - 105 \times 0.8 = 231 \text{ kN}$$

$$\tau_v = \frac{231 \times 1000}{300 \times 600} = 1.28 \text{ N/mm}^2.$$

Area of reinforcement $= (5-2)\frac{\pi}{4}(25)^2 = 1472.6 \text{ mm}^2$

$$\therefore \qquad \frac{100 A_s}{bd} = \frac{100 \times 1472.6}{300 \times 600} = 0.818\ \%.$$

Hence from Table 27.1, $\tau_c = 0.54 + \frac{0.60 - 0.54}{0.25}(0.818 - 0.75)$

$$= 0.556 \text{ N/mm}^2.$$

Also, from Table 27.2, $\tau_{c.max} = 2.5 \text{ N/mm}^2 > \tau_v$.

Since $\tau_v > \tau_c$, shear reinforcement is necessary.

Shear resistance of concrete $= V_{uc} = 0.556 \times 300 \times 600$

$= 100080 \text{ N} = 100.08 \text{ kN}.$

$\therefore$ Shear resistance to be provided by shear reinforcement

$= V_{us} = V_u - V_{uc} = 231 - 100.08 = 130.92 \text{ kN}.$

Area of cross-section of two bent-up bars $= A_{sv} = 2\frac{\pi}{4}(25)^2$

$= 490.0 \text{ mm}^2.$

Shear resistance of two bent-up bars

$= 0.87 f_y A_{sv} \sin \alpha$

$= 0.87 \times 415 (490.9) \times 0.707$

$\approx 125300 \text{ N} = 125.3 \text{ kN}.$

However, the maximum shear resistance to be assigned to the bent- up bars should not exceed half the resistance to be provided by shear reinforcement, *i.e.* it should not exceed $\frac{1}{2} \times 130.92 = 65.46$ kN.

Hence the remaining resistance to be provided by the vertical stirrups $= 130.92 - 65.46 = 65.46$ kN

$\therefore \quad V_{us} = 65.46$ kN for vertical stirrups

From Eq. 27.4, the spacing of vertical stirrups is given by

$$s_v = \frac{0.87 f_y A_{sv} d}{V_{us}}$$

Using 8 mm dia. 2-lgd stirrups, $A_{sv} = 2\frac{\pi}{4}(8)^2 = 100.5 \text{ mm}^2$

$$\therefore \quad s_v = \frac{0.87 \times 415 \times 100.5 \times 600}{65460}$$

$= 332.5 \text{ mm} < 450 \text{ mm},$

Maximum spacing is given by Eq. 27.35.

$$s_v = \frac{A_{sv} f_y}{0.4 b} = \frac{100.5 \times 415}{0.4 \times 300}$$

$= 348 \text{ mm} \approx 340 \text{ mm}.$

Hence provide 8 mm dia 2-lgd stirrups @ 330 mm c/c from the face of the support to the point at which the bars bent.

These two bars are bent at a distance of 1.414 × Lever arm $\approx 1.414 \times 0.9 \times 600 \approx 750\}$ mm from the face of support (*i.e.* at a distance of $750 + 200 = 950$ mm from the centre of the support). Beyond this bent up bars are not available.

At 0.95 m away from support,

$V_u = 315 - 105 \times 0.95 = 215.75 \text{ kN}$

$$A_s = 5\frac{\pi}{4}(8)^2 = 2454 \text{ mm}^2$$

$$\therefore \quad \frac{100 A_s}{bd} = \frac{100 \times 2454}{300 \times 600} = 1.36$$

$$\tau_c = 0.64 + \frac{0.68 - 0.64}{0.25}(1.36 - 1.25) \approx 0.658 \text{ N/mm}^2$$

$$\therefore \quad V_{uc} = 0.658 \times 300 \times 600 = 118440 = 118.44 \text{ kN}$$

$$\therefore \quad V_{us} = V_u - V_{uc} = 215.75 - 118.44 = 97.31 \text{ kN.}$$

The spacing of 8 mm dia. 2-*lgd* stirrups is given by

$$s_v = \frac{0.87 \times 415 \times 100.5 \times 600}{97310} = 223.7 \text{ mm.}$$

Hence provide these @ 220 mm c/c.

To locate the section where maximum spacing of 340 mm is provided, proceed as follows :

Let the distance of the section be x m from the mid- span.

$$V_u = \frac{315}{3}x = 105x \text{ kN} = 105000x \text{ N.}$$

Shear resistance of concrete $V_{uc} = 118440$ N (as before)

$$\therefore \quad V_{us} = 105000x - 118440$$

Now

$$s_v = \frac{0.87 f_y A_{sv} d}{V_{us}}$$

or

$$340 = \frac{0.87 \times 415 \times 100.5 \times 600}{105000x - 118440}$$

This gives $x \approx 1.73$ from mid-span (or 1.27 m from support).

27.2. DEVELOPMENT LENGTH

The development length in tension is given by

$$L_d = \frac{0.87 f_y}{4\,\tau_{bd}}\,\Phi \qquad \text{...(27.7)}$$

This is the same expression as used for working stress method except that σ_s has been replaced by 0.87 f_y. Values of the design bond stress (τ_{bd}) and development length ratio $\left(\frac{L_d}{\Phi}\right)$ for various grades of concrete and the two types of reinforcement, for bars in tension, are given in Table 27.4. For bars in compression, the values of bond stress for bars in tension shall be increased by 25 percent.

TABLE 27.4. DESIGN BOND STRESS AND DEVELOPMENT LENGTH IN TENSION

Grade fo Concrete	*HYSD bars Fe 415 (f_y = 415 N/mm²)*		*Mild steel bars (f_y = 250 N/mm²)*	
	τ_{bd} (N/mm²)	L_d / Φ	τ_{bd} (N/mm²)	L_d / Φ
M 15	1.60	57	1.00	55
M 20	1.92	47	1.20	46
M 25	2.24	41	1.40	39
M 30	2.40	38	1.50	37
M 35	2.72	34	1.70	32
M 40	3.04	30	1.90	29

Note. The values have been rounded off to the higher side.

Development length Requirements at the Supports

The Code stipulates that at the simple supports(and at the point of inflexion), the positive moment tension reinforcement shall be *limited* to a diameter such that :

$$L_d \leq \frac{M_1}{V} + L_0 \qquad ...[27.8(a)]$$

where L_d = development length, computed from Eq. 27.32.

M_1 = moment of resistance of the section at the support, assuming all the reinforcement at the section to be stressed to 0.87 f_y..

V = shear force at the section, due to ultimate load.

L_0 = sum of anchorages beyond the centre of support and the equivalent anchorage value of any hook or mechanical anchorage at the simple support (at the point of inflection, L_0 is limited to d or 12 Φ whichever is greater).

It should be noted that the method of computing the value of L_0 is the same as discussed for the working stress method of design.

The Code further recommends that the value of M_1/V in the above expression may be increased by 30 percent when the ends of the reinforcement are confined by a compressive reaction. This condition of *'confinement'* of reinforcing bar may not be available at all the types of simple supports.

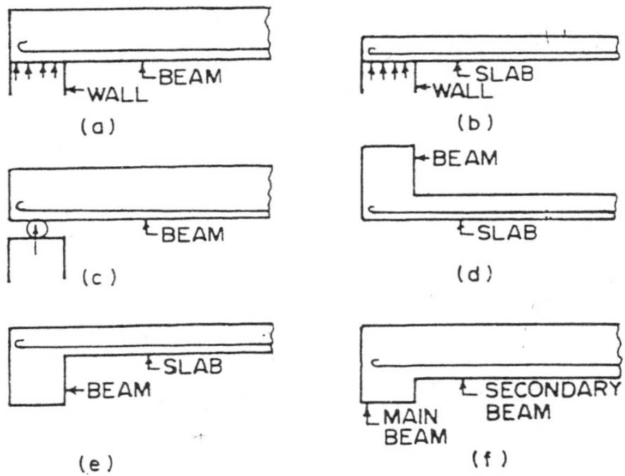

FIG. 27.1.

Four types of *simple supports* are shown in Fig. 27.1. In Fig. 27.1(*a*), a beam is simply supported on a wall which offers a compressive reaction which confines the ends of the reinforcement. Hence a factor 1.3 will be applicable with the term M_1/V. Similarly in Fig. 27.1(*b*), a slab is simply supported on a wall and hence factor 1.3 will be applicable. However, in Fig. 27.1(*c*) and 27.1(*d*) though a simple support is available, the reaction does not confine the ends of the reinforcement and hence the factor 1.3 will not be applicable with M_1/V term.

Similarly, for the case of a slab connected to a beam Fig. 27.1(*e*) or for the case of a secondary beam connected to main beam [Fig. 27.1(*f*)] tensile reaction is induced and hence factor 1.3 will not be available.

Thus, at simple supports, where the compressive reaction confines the ends of reinforcing bars, we have

$$L_d \leq 1.3\frac{M_1}{V} + L_0 \qquad ...[27.8\ (b)]$$

In Eq. 27.8(*a*) or 27.8(*b*), M_1 is the moment of resistance of the bars available at the support.

$\therefore \quad M_1 = 0.87 f_y A_{st1} z_u,$ where z_u is the lever arm, which may be assumed to be constant

Also, at the mid-span,

$$M = 0.87 f_y A_{st} z_u$$

$$\therefore \quad \frac{M_1}{M} = \frac{A_{st1}}{A_{st}} = \frac{N_1}{N}$$

or

$$M_1 = M \frac{N_1}{N} \qquad ...[27.8(c)]$$

where N_1 = No of bars at support
N = No. of bars at mid span
M = moment of resistance available

It should be clearly noted that Eq. 27.8(*c*) for the determination of M_1 is approximate, since it assumes a constant value of lever arm z_u. Its actual value is $(d - 0.416 x_u)$ where x_u is depth of N.A. of the section at support.

Hence $M_1 = 0.87 f_y A_{st1} (d - 0.416 x_u) \qquad ...(27.8(d)]$

where $x_u = \dfrac{0.87 f_y A_{st1}}{0.36 f_{ck} b} \qquad ...(27.8(e)]$

27.3. LIMIT STATE OF COLLAPSE : TORSION

Torsional reinforcement is not calculated separately from that required for bending and shear. Instead, the total longitudinal reinforcement is determined for a fictitious bending moment which is a function of actual bending moment and torsion. Similarly, shear reinforcement (web reinforcement) is determined for a fictitious shear which is a function of actual shear and torsion.

1. *Shear and Torsion*

Equivalent shear V_e is calculated from the expression

$$V_e = V_u + 1.6 \frac{T_u}{b} \qquad ...(27.9)$$

where V_e = equivalent shear
V_u = shear
T_u = torsional moment
b = breadth of the beam.

Equivalent nominal shear stress to be given by

$$\tau_{ve} = \frac{V_e}{bd} \qquad ...(27.10)$$

The values of τ_{ve} shall not exceed the values of $\tau_{c.max}$ given in Table 27.2.

2. ***Reinforcement in Members Subjected to Torsion***

If the equivalent nominal shear stress τ_{ve} does not exceed τ_c given in Table 27.1, minimum shear reinforcement shall be provided as per Eq. 27.3.

If τ_{ve} exceeds τ_c, reinforcement is required. Reinforcement consists of (*a*) longitudinal reinforcement and (*b*) transverse reinforcement :

(*a*) Longitudinal Reinforcement

The longitudinal reinforcement shall be designed to resist an equivalent bending moment, M_{e1} given by

$$M_{e1} = M_u + M_t \quad \text{...(27.11)}$$

where M_u = bending moment at the cross-section, and

$$M_t = T_u \left(\frac{1 + D/b}{1.7}\right) \quad \text{...(27.12)}$$

T_u = torsional moment

D = overall depth of beam.

If the numerical value of M_t given by Eq. 27.12 exceeds the numerical value of M_u, longitudinal reinforcement shall be provided on the flexural compression face, such that beam can also withstand an equivalent B.M. M_{e2} given by

$$M_{e2} = M_t - M_u \quad \text{...(27.13)}$$

In which M_{e2} is taken as acting in the opposite sense to the moment M_2.

(*b*) Transverse Reinforcement

Two-legged closed hoops enclosing the corner longitudinal bars shall have an area of cross-section A_{sv} given by

$$A_{sv} = \frac{T_u\, s_v}{b_1 d_1 (0.87 f_y)} + \frac{V_u\, s_v}{2.5\, d_1 (0.87 f_y)} \quad \text{...(27.14)}$$

However, the total reinforcement shall not be less than

$$\frac{(\tau_{vs} - \tau_c)\, b\, s_v}{0.87 f_y}.$$

where s_v = spacing of stirrups.

b_1 = centre to centre distance between corner bars in the direction of width.

d_1 = centre to centre distance between corner bars in the direction of depth.

Example 27.2. Design for Torsion.

Determine the reinforcement required for a rectangular beam section with the following data :

Width of section : 300 mm
Depth of section : 500 mm
Factored B.M : 80 kN-m
Factored torsional moment : 40 kN-m
Factored shear force : 70 kN
Use M 15 grade concrete and Fe 415 grade steel.

Solution.

1. ***Equivalent Shear***

Total depth D = 500 mm

Assume 20 mm dia. bars and 25 mm clear cover.

$$\therefore \quad d = 500 - 25 - \frac{20}{2} = 465 \text{ mm.}$$

$$V_e = V_u + 1.6\frac{T_u}{b}$$

$$= 70 + 1.6\frac{40}{0.3} = 283.333 \text{ kN.}$$

$$\therefore \quad \tau_{ve} = \frac{283.333 \times 1000}{300 \times 465} = 2.03 \text{ N/mm}^2$$

From Table 27.2, $\tau_{c.max} = 2.5$ N/mm^2

Hence the section does not require revision.

2. ***Longitudinal Reinforcement***

Equivalent B.M. is given by

$$M_{e1} = M_u + M_t$$

where $$M_t = T_u\left(\frac{1 + D/b}{1.7}\right) = 40\left[\frac{1 + 0.5/0.3}{1.7}\right] = 62.745 \text{ kN-m}$$

$$\therefore \quad M_{e1} = 80 + 62.745 = 142.745 \text{ kN-m}$$

The moment of resistance of a balanced section is given by

$$M_{u.lim.} = 0.36 f_{ck}\frac{x_{u.max}}{d}\left(1 - 0.42\frac{x_{u.max}}{d}\right) bd^2 \qquad ...(27.11)$$

where $\frac{x_{u.max}}{d} = 0.48$ for Fe 415 steel

$$\therefore \quad M_{u.lim.} = 0.36 f_{ck}(0.48)(1 - 0.42 \times 0.48) bd^2$$

or $\quad M_{u.lim.} = 0.138 f_{ck} bd^2$ [see Eq. 21.11 (*b*)]

$$= 0.138 \times 15\,(300)\,(500)^2 = 155.25 \times 10^6 \text{ N-mm.}$$

Since $M_{e1} < M_{u.lim.}$ the section is under-reinforced.
Hence from Eq. 24.10.

$$M_u = M_{e1} = 0.87 f_y A_{st}\, d \left(1 - \frac{f_y}{f_{ck}} \frac{A_{st}}{bd}\right)$$

or $\quad 142.745 \times 10^6 = 0.87 \times 415 A_{st} \times 500 \left(1 - \frac{415}{15} \times \frac{A_{st}}{300 \times 500}\right)$

or $\quad A_{st}\,(1 - 184.4 \times 10^{-6} A_{st}) = 790.72$

or $\quad A_{st}^2 - 5423 A_{st} + 4.288 \times 10^6 = 0$

From which $A_{st} = 961 \text{ mm}^2$

Hence provide 3 bars of 20 mm dia., and 1 bars of 12 mm dia. having total $A_{st} = 3 \times 314 + 113 = 1055 \text{ mm}^2$.

Again, as M_t is less than M_u , we need not consider M_{u2}. Therefore provide only bars of 12 mm dia. on the compression face to act as holding bars for transverse reinforcement.

3. ***Transverse Reinforcement***

$$\frac{100 A_s}{bd} = \frac{100 \times 1055}{300 \times 500} \approx 0.7\ \%$$

Hence from Table 27.1, $\tau_c = 0.46 + \dfrac{0.53 - 0.46}{0.25} \times 0.2$

$$= 0.524 \text{ N/mm}^2$$

Since $\tau_{ve} > \tau_c$, transverse shear reinforcement is essential. The area of two legs of the stirrups should satisfy the following (Eq. 27.14) :

$$A_{sv} = \frac{T_u\, s_v}{b_1 d_1 (0.87 f_y)} + \frac{V_u\, s_v}{2.5\, d_1 (0.87 f_y)}$$

Assuming dia. of stirrups as 10 mm,

$$d_1 = 500 - (25 + 10) - (25 + 6) = 434 \text{ mm}$$

$$b_1 = 300 - 2\,(25 + 10) = 230 \text{ mm.}$$

$$\therefore \quad \frac{A_{sv}(0.87 f_y)}{s_v} = \frac{40 \times 10^6}{230 \times 434} + \frac{70 \times 1000}{2.5 \times 434}$$

$$= 465.2 \text{ N/mm}^2.$$

Also, area of all the legs of the stirrups should satisfy the following condition

$$A_{sv} = \frac{(\tau_{ve} - \tau_c)\, b\, s_v}{0.87 f_y}$$

or $$\frac{A_{sv}\,(0.87 f_y)}{s_v} = (\tau_{ve} - \tau_c)\, b = (2.03 - 0.524)\, 300$$

$$= 451.8 \text{ N/mm.} \qquad ...(ii)$$

Choosing the greater one of the two, we have

$$\frac{A_{sv}}{s_v} = \frac{465.2}{0.87 f_y} = \frac{465.2}{0.87 \times 415} = 1.288.$$

Hence spacing of 10 mm dia. 2-legged stirrups is

$$s_v = \frac{A_{sv}}{1.288} = \frac{2 \times 75.5}{1.288} = 122 \text{ mm.}$$

Hence provide the stirrups @ 120 mm c/c.

As per Code requirements, this spacing should not exceed x_1, $(x_1 + y_1)/4$ and 300 mm.

where x_1 = short dimension of stirrup $= 300 - 2\,(25 - 5) = 260$

y_1 = long dimension of stirrup $= 500 - 2\,(25 - 5) = 460$

$$\frac{x_1 + y_1}{4} = \frac{260 + 460}{4} = 180 \text{ mm.}$$

Hence 10 mm dia. two legged stirrups @ 120 mm c/c satisfy the above Code requirements.

PROBLEMS

1. A rectangular beam, 240 mm wide and 450 mm effective depth is reinforced with 3 bars of 20 mm dia. at supports. Design the shear reinforcement if it carries a shear of 60 kN at service state. Use M 15 concrete and Fe 415 steel. Also, determine the nominal shear stirrups at mid-span.
2. Redesign beam of problem 1 if the S.F. is 200 kN at service state.
3. Check whether the beam of problem 1 is safe against the requirements of bond at the simple support. Assume that the width of support is 400 mm.

28

Design of Beams and Slabs

28.1 DESIGN OF BEAMS

(*a*) Basic Rules for Design (IS : 456-1978)

The basic design rules in the limit state design method are the same as given in chapter 7 for working stress design method.

(*b*) Design procedure

The design of a singly reinforced beam is done in the following steps :

Step 1. Determination of limiting depth of N.A.

Determine $\frac{x_{u,max}}{d}$ from Eq. 24.7. Alternatively, take the values of $\frac{x_{u,max}}{d}$ from following table.

f_y	$\frac{x_{u,max}}{d}$
250	0.53
415	0.48
500	0.46

Step 2. Computation of design B.M.

Assume suitable values of overall depth(D) and b of the beam, determine d and then the effective span.

This depth d should satisfy the requirements of *limit state of serviceability.*

Calculate self weight, and add to it the given super-imposed load, to get total U.D.L. w per metre run. Apply partial safety factors γ_f for loads, and compute factored load w_u per unit length of the beam. Compute max. B.M. M_u in the beam.

Step 3. Computation of effective depth d

Find d from the relation

$$M_u = 0.36 f_{ck} \cdot \frac{x_{u,max}}{d}\left(1 - 0.416\frac{x_{u,max}}{d}\right) bd^2$$

or $$M_u = R_u bd^2$$

(where b has already been chosen earlier)

Step 4. Steel Reinforcement

Knowing b and d, compute D by adding effective cover, and round off the value of D to slightly higher side. Find d by subtracting cover from D. Determine the area of reinforcement from the relation.

$$M_u = 0.87 f_y \cdot A_{st}(d - 0.416\, x_{u,max}) \qquad ...[28.18(a)]$$

or Alternatively, find p_{lim} from Eq. 24.9

$$p_{lim} = 0.414\frac{f_{ck}}{f_y} \cdot \frac{x_{u,max}}{d} \qquad ...(28.9)$$

and $$A_{st} = p_{lim} \times bd.$$

The reinforcement so computed should be equal to or more than minimum reinforcement given by

$$\frac{A_s}{bd} = \frac{0.85}{f_y}$$

Curtailment of reinforcement

Curtail the reinforcement at other locations where it is no longer required from the B.M. point of view, using Eq. 7.1

$$x = \frac{L}{2} \cdot \sqrt{\frac{N_x}{N}}$$

where x is the distance from the mid-span where N_x bars can be curtailed out of total bars N. However, the bars have to develop full tension at the centre of span, and hence x should be equal to or greater than the development length L_d of the bar which is to be curtailed. The Code also specifies that the reinforcement shall extend the point at which it is no longer required to resist flexure for a distance equal to effective depth d or 12 Φ whichever is greater. Hence the bar should be curtailed at a distance $x + d$ or (12Φ) from the centre, and this distance should be greater than L_d.

Note : If the depth d is fixed from other considerations and its value is more than the one found by bending compression(step

3), we have the *under-reinforced section,* for which the reinforcement is given by Eq. 24.13 :

$$A_{st} = \frac{0.5 f_{ck}}{f_y}\left[1 - \sqrt{1 - \frac{4.6 M_u}{f_{ck}\, bd^2}}\right] bd \qquad ...(24.13)$$

Step 5. Shear reinforcement

Compute max. shear force V_u at the critical section. The position of critical section is shown in Fig. 7.4. Compute nominal shear stress

$$\tau_v = \frac{V_u}{bd} \qquad ...(27.1)$$

Also, compute τ_c from Table 27.1.

(*i*) If $\tau \le \frac{1}{2}\tau_c$

or $\quad V_u = \frac{1}{2} V_{uc}$ where ($V_{uc} = \tau_c\, bd$),

No shear reinforcement is really necessary.

(*ii*) If $\frac{1}{2}V_{uc} \le V_u \le V_{uc}$

Provide nominal shear reinforcement given by Eq. 27.3 :

$$\frac{A_{sv}}{b \,.\, s_v} \ge \frac{0.4}{f_y}$$

Choosing bar dia., A_{sv} is known, and hence s_v can be computed from the above expression. However, max. spacing is restricted to 0.75 d or 450 mm, whichever is less.

(*iii*) If $V_{uc} \le V_u \le V_{uc\,,\,max} =$ (where $V_{uc\,,\,max} = \tau_{c\,max} \,.\, bd$)

Design the transverse reinforcement for the net S.F.

$$V_{us} = V_u - V_{uc}$$

For vertical stirrups,

$$V_{us} = \frac{0.87 f_y \,.\, A_{sv} \,.\, d}{s_v} \qquad ...(27.4)$$

(*iv*) If $V_u \ge V_{cu\,,\,max}$, redesign by increasing the depth such that $V_{cu\,,\,max}$ becomes equal to or more than the external shear V_u.

Step 6. Check for development length at end

For simple supports check if

$$L_d \le 1.3\,\frac{M_1}{V} + L_0$$

The factor 1.3 in the above equation should be chosen for only those simple supports which offer compressive reaction that confines the ends of reinforcing bars.

Example 28.1. Design of singly reinforced beam

A reinforced concrete beam is supported on two walls 750 mm thick, spaced at a clear distance of 6 m. The beam carries a super-imposed load of 9.8 kN/m. Design the beam using M 20 concrete mix and HYSD bars of Fe 415 grade.

Solution

Step 1. *Limiting depth of N.A.*

For Fe 415 steel, $f_y = 415$ N/mm^2.

For M 20 concrete, $f_{ck} = 20$ N/mm^2

$\frac{x_{u\,max}}{d}$ depends only a grade of steel and is independent on grade of concrete. For Fe 415 steel, $\frac{x_{u\,max}}{d} = 0.48$.

Step 2. *Computation of design B.M.*

The minimum depth of beam is based on limit state of serviceability requirements.

For simply supported beam, $\frac{L}{d} = 20$

For a balanced section, $p_{lim} = 0.96\%$ (Table 24.2) . Hence modification factor ≈ 1.

$$\therefore \qquad d = \frac{L}{20} \approx \frac{6000}{20} = 300 \text{ mm.}$$

This is the minimum value of d. Actual value of d, based on bending may be more than this. In working stress design method, d is kept equal to $\frac{l}{10}$ for preliminary computations. However, the limit state design method results in smaller depth. Hence take

$$d = \frac{l}{15} = \frac{6000}{15} = 400 \text{ mm.}$$

$$D = 400 + 30 = 430 \text{ mm (say).}$$

Keep $b = 300$ mm.

$\therefore$ Self weight of beam $= 0.43 \times 0.3 \times 1 \times 25000 = 3225$ N/m

External load $= 9800$ N/m.

$\therefore$ Total $w = 9800 + 3225 = 13025$ N/m

$$w_u = 1.5\,w = 1.5 \times 13025 = 19538 \text{ N/m}$$

Effective span $L = l + d = 6 + 0.4 = 6.4$ m, which is smaller than centre to centre distance of 6.75 m between the supports.

$$M_u = \frac{w_u . L^2}{8} = \frac{19538(6.4)^2}{8}$$

$$= 100035 \text{ N-m} \approx 100.04 \times 10^6 \text{ N-mm}$$

Step 3. Computation of effective depth d

$$M_u = 0.36 f_{ck} \frac{x_{u,max}}{d}\left(1 - 0.416 \frac{x_{u,max}}{d}\right) bd^2$$

or $$M_u = R_u . bd^2$$

where $$R_u = 0.36 \times 20 \times 0.48(1 - 0.416 \times 0.48) = 2.766$$

(Compare this with value given in Table 24.3)

$$\therefore \quad d = \sqrt{\frac{M_u}{R_u . b}} = \sqrt{\frac{100.04 \times 10^6}{2.766 \times 300}} = 347 \text{ mm.}$$

$$D = 347 + 25 + 10(\text{say}) = 382 \text{ mm.}$$

Keep D = 400 mm. Since this value is quite near to the one assumed earlier, no revised computations for self load are recommended.

Using 20 mm bars, $d = 400 - 25 - 10 = 365$ mm. Since available d(=365 mm) is more than the one required for balanced design, we will have an under-reinforced section.

Step 4. Steel reinforcement

The reinforcement, for an under-reinforced section, is given by Eq. 24.13 :

$$A_{st} = \frac{0.5 f_{ck}}{f_y}\left[1 - \sqrt{\frac{4.6 M_u}{f_{ck}\, bd^2}}\right] bd$$

or $$A_{st} = \frac{0.5 \times 20}{415}\left[1 - \sqrt{\frac{4.6 \times 100.04 \times 10^6}{20 \times 300 \times (365)^2}}\right] 300 \times 365$$

$$= 919.83 \text{ mm}^2$$

$$\text{No. of 20 mm dia. bars} = \frac{919.83}{314.15} = 2.93$$

Provide 3-20 mm dia. bars at mid span.

Actual $A_{st} = 3 \times 314.15 = 942.5 \text{ mm}^2$

Hence actual moment of resistance available at mid span is

$$M_{ur} = 0.87 f_y . A_{st} . d\left(1 - \frac{f_y}{f_{ck}} . \frac{A_{st}}{bd}\right) \quad \text{... (24.10)}$$

$$= 0.87 \times 415 \times 942.5 \times 365\left(1 - \frac{415}{20} \times \frac{942.5}{300 \times 365}\right)$$

$$= 102 \times 10^6 \text{ N/mm}$$

$$M_{u.lim} = 2.766\ bd^2 = 2.766 \times 300(365)^2 = 110 \times 10^6 \text{ N/mm}$$

Since $M_{ur} < M_{u,lim}$, the design is OK.

Since the Code recommends to provide at least $\frac{1}{3}A_{st}$ at the upper face of the beam, at supports, no curtailment of bars is proposed. Continue all the three bars, and then bend one bar upwards, at a distance x_1 from the support.

$$\text{B.M. at } x_1 = \frac{w_u \cdot L}{2}x_1 - \frac{w_u x_1^2}{2}$$

This should be 2/3 of the maximum B.M.

$$\therefore \quad \frac{w_u \cdot L}{2}x_1 - \frac{w_u \cdot x_1^2}{2} = \frac{2}{3}\frac{w_u L^2}{8}$$

which gives $x_1 = 0.211L = 0.211 \times 6.4 = 1.35$ m

However, the bars are to be taken further by a distance d or 12 Φ. Hence distance of point from the centre of support $= 135 - 0.365 = 0.985$ m, or at $0.985 - 0.75/2 = 0.61$ m. Hence bend one bar up, at a distance of 0.6 m from the face of the support, and continue the remaining two bars into the support.

Step 5. Shear reinforcement

The critical section for shear is at a distance of $d(=0.365 \text{ m})$ from the face of the support.

$$\therefore \quad V_{uD} = \frac{w_u\ L}{2} - w_u\left(\frac{0.75}{2} + 0.365\right)$$

$$\text{or} \quad V_{uD} = \frac{19538 \times 6.4}{2} - 19538[0.375 + 0.365]$$

$$= 48063 \text{ N}$$

$$\therefore \quad \tau_v = \frac{V_{uD}}{bd} = \frac{48063}{300 \times 365} \approx 0.439 \text{ N/mm}^2$$

$$\frac{100A_{st}}{bd} \text{ (at support)} = \frac{100}{300 \times 365}(314 \times 2) = 0.57\%.$$

Hence from Table 27.1, $\tau \approx 0.5$ N/mm^2.

Since $\tau_v < \tau_c$, no shear reinforcement is necessary. However, as per Code recommendations, provide nominal shear reinforcement, given by the expression

$$\frac{A_{sv}}{b \, . \, s_v} > \frac{0.4}{f_y}$$

or $$s_v = \frac{2.5 \, A_{sv} \, . \, f_y}{b}$$

Choosing 8 Φ bars, $A_{sv} = 2 \times \frac{\pi}{4}(8)^2 = 100.5 \text{ mm}^2$

$$\therefore \quad s_v = \frac{2.5 \times 100.5 \times 415}{300} = 348 \text{ mm}$$

Max. $s_v = 0.75\, d = 0.75 \times 365 = 274$ mm.

Hence provide 8 Φ *2-lgd stirrups* @ 260 mm c/c throughout the length of the beam. Provide 2-10 Φ holding bars at the top.

Step 6. Check for development length at the ends.

Since the beam is supported on wall, the compressive reaction will confine the reinforcement. Hence

$$L_d \leq 1.3 \frac{M_1}{V} + L_0$$

Only two bars are available at the supports.

$$x_u = \frac{0.87 f_y \, . \, A_{st1}}{0.36 f_{ck} \, . \, b} = \frac{0.87 \times 415 \; (2 \times 314.15)}{0.36 \times 20 \times 300}$$

$$= 105 \text{ mm}$$

$$\therefore \quad M_1 = 0.87 f_y A_{st1}(d - 0.42 x_u)$$

$$= 0.87 \times 415(2 \times 314.15)(365 - 105)$$

$$\approx 59 \times 10^6 \text{ N-mm}$$

$$V = \text{S.F.} = \frac{w_u \, . \, L}{2} = = \frac{19538 \times 6.4}{2} = 62522 \text{ N}$$

$$L_d = \frac{0.87 f_y}{4 \tau_{bd}} \times \Phi = \frac{0.87 \times 415}{4 \times 1.92} \times \Phi = 47\Phi$$

$$= 47 \times 20 = 940 \text{ mm}$$

L_0 = sum of anchorage beyond centre line of support and anchorage value of hook.

If no hook is provided,

$$L_0 = \frac{L_s}{2} - x' = \frac{750}{2} - 40 = 335 \text{ mm},$$

where x' = side cover assumed to be 40 mm. The Code further recommends that each bar should be extended by a distance of $L_d/3 = 940/3 = 313$ mm beyond the face of the support. Available distance, beyond the face of support $= L_0 + \frac{L_s}{2}$

$= 335 + \frac{750}{2} = 710$ mm, which is more than $\frac{L_d}{3}$. Hence no hook is required.

Hence $L_0 = 335$ mm.

$$\therefore\ 1.3\ \frac{M_1}{V} + L_0 = \left(1.3 \times \frac{59 \times 10^6}{62522}\right) + 335 = 1562 \text{ mm} > L_d$$

Hence O.K.

Step 7. Details of Reinforcement etc.

The details of reinforcement etc. are shown in Fig. 28.1.

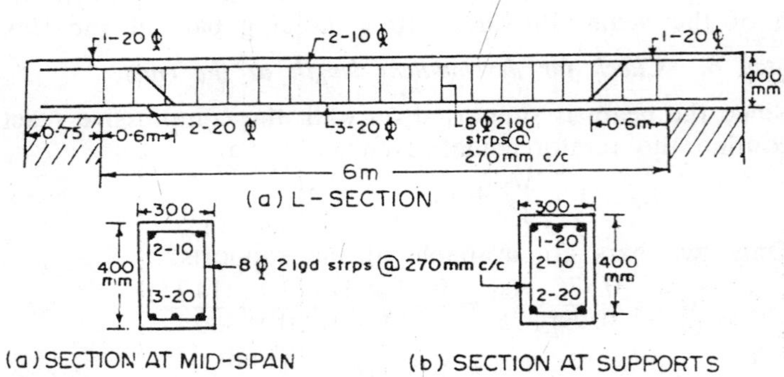

FIG. 28.1

28.2. DESIGN OF CANTILEVER

When a cantilever beam or slab is subjected to downward loads, it bends downwards with its convexity upwards. Due to this, the upper fibres of the section are subjected to tensile stress and the lower fibres are subjected to compressive stresses. Hence the main reinforcement is provided at the *upper face,* and not at the lower face. The B.M. varies from a zero value at the free section to maximum value at the fixed end(support). Hence the depth of the section and the steel reinforcement is varied as per requirements of bending moment. In order that the whole cantilever does not overturn, it should be properly fixed in the wall. The reinforcement should be embedded in the support for a distance of L_d from the edge of the support.

When curtailment of bar is done, it should be kept in mind that the embedment of the bar is not less than L_d from the point it is needed the most, nor less than 12 dia. (or d) from the point it can be spared from bending moment point of view.

Example 28.2. Design of a cantilever beam.

A cantilever beam projects 2.5 m beyond the fixed end, and carries a super-imposed load of 12 kN per metre run. Design the cantilever, using M 15 concrete and Fe 415 steel.

Solution.

Step 1. Design constants and limiting depths of N.A.

For Fe 415 steel, $f_y = 415\ \text{N/mm}^2$

For M 15 concrete, $f_{ck} = 15\ \text{N/mm}^2$.

$$\frac{x_{u,max}}{d}\ (\text{for Fe 415 steel}) = 0.48.$$

$$R_u = 0.36 f_{ck} \cdot \frac{x_{u,max}}{d}\left(1 - 0.416\ \frac{x_{u,max}}{d}\right)$$

(Also , see Table 24.3)

Step 2. *Computation of design B.M. and S.F.*

The minimum depth of beam is based on limit state of serviceability requirement.

For a cantilever, $\frac{L}{d} = 7$

$$\therefore \quad d = \frac{L}{7} = \frac{2500}{7} = 375\ \text{mm}.$$

Keep D = 400 mm at fixed end and 200 mm at free end.
Keep width b = 250 mm.

Self weight of beam $= \frac{1}{2}(0.4 + 0.2) \times 0.25 \times 2.5 \times 25000 = 4688$ N

acting at $\dfrac{0.4 + 2 \times 0.2}{0.4 + 0.2} \times \dfrac{2.5}{3} = 1.11$ m from fixed end.

$$\therefore \quad M = (4688 \times 1.1) + \frac{12000(2.5)^2}{2} = 42657\ \text{N-m}$$

$$\therefore \quad M_u = 1.5M = 1.5 \times 42657 = 63985\ \text{N-m}$$

$$\approx 63.99 \times 10^6\ \text{N-mm}$$

S.F. V at the edge of support $= 4688 + (12000 \times 2.5)$
$= 34688$ N

$$\therefore \quad V_u = 1.5\,V = 52032\ \text{N}$$

Step 3. Computation of effective depth

For a balanced section,

$$M_u = R_u bd^2 = 2.074\ bd^2$$

$$\therefore \quad d = \sqrt{\frac{M_u}{2.074b}} = \sqrt{\frac{63.99 \times 10^6}{2.074 \times 250}} = 351 \text{ mm.}$$

However, keep D = 400 mm, so that using 20 mm dia. bars, $d = 400 - 25 - 10 = 365$ mm. This is slightly more than the one required for the balanced section. Hence, we will have an under-reinforced section. Keep D = 200 mm at free end.

Step 4. Steel Reinforcement

The reinforcement for an under-reinforced section is given by Eq. 24.13 :

$$A_{st} = \frac{0.5 f_{ck}}{f_y}\left[1 - \sqrt{1 - \frac{4.6 M_u}{f_{ck}\, bd^2}}\right] bd.$$

$$= \frac{0.5 \times 15}{415}\left[1 - \sqrt{1 - \frac{4.6 \times 63.99 \times 10^6}{15 \times 250 (365)^2}}\right] \times 250 \times 365$$

$$= 592 \text{ mm}^2$$

$$\therefore \text{ No. of 16 mm dia. bars} = \frac{592}{201.06} = 2.94$$

Hence provide 3 bars of 16 mm dia.

Actual $A_{st} = 3 \times 201.06 = 603.2 \text{ mm}^2$

Hence actual moment of resistance available is

$$M_{ur} = 0.87 f_y . A_{st}\, d \left(1 - \frac{f_y}{f_{ck}} \cdot \frac{A_{st}}{bd}\right)$$

$$= 0.87 \times 415 \times 603.2 \times 365 \left(1 - \frac{415}{15} \times \frac{603.2}{250 \times 365}\right)$$

$$= 64.95 \times 10^6 \text{ N-mm} > M_u.$$

Also, $M_{u,lim,} = 2.075\, bd^2 = 2.074 \times 250\, (365)^2$

$$= 69.08 \times 10^6 \text{ N–mm.}$$

Since $M_{ur} < M_{u,lim}$, the design is OK.

Curtailment of reinforcement

Since the B.M. decreases to zero at the end, let us curtail one bar at a distance x mm.

Effective depth at free end $\approx 200 - 25 - 10 = 165$ mm

Effective depth d_x from x free end

$$= 165 + \left(\frac{365 - 165}{2500}\right) x = (165 + 0.08x) \text{ mm.}$$

Assuming B.M.D. to be parabolic, the B.M. at this section may be approximately taken to be equal to $\left(\frac{x}{2500}\right)^2 \times 63.99 \times 10^6$

$$= 10.24\, x^2 \text{ N-mm.}$$

Area of remaining 2 bars $= 2 \times 201.06 \approx 402.1 \text{ mm}^2$

$$M_{ur} = 0.87 f_y \,.\, A_{st}\left[d_x - \frac{f_y}{f_{ck}} \,.\, \frac{A_{st}}{b}\right]$$

$$= 0.87 \times 415 \times 402.1\left[(165 + 0.08x) - \frac{415}{15} \times \frac{402.1}{250}\right]$$

$$= 145178[120.5 + 0.08\, x] = 17.49 \times 10^6 + 11614\, x$$

$$\therefore \quad 10.24 x^2 = 17.49 \times 10^6 + 11614\, x$$

or $$x^2 - 1134x - 1.708 \times 10^6 \approx 0$$

From which $x \approx 1992$ mm

$$d_x = 165 + 0.08 \times 1992 \approx 324 \text{ mm.}$$

Hence one bar can be curtailed at a distance $=2500–1992 + 324 \approx 830$ mm away from the fixed end. However, minimum distance at which any bar can be curtailed from the fixed end is L_d so that all the bars can develop their full tensile strength at the fixed end.

Now $$L_d = \frac{0.87 f_y}{4\tau_{bd}} \,.\, \Phi = \frac{0.87 \times 415}{4 \times 1.6}\Phi = 56.4\ \Phi$$

$$= 56.4 \times 16 = 902 \text{ mm} > 830 \text{ mm.}$$

Hence one bar can be curtailed only at a distance of 902 mm from the fixed end. However , curtail one bar at 910 mm from the fixed end. Continue the remaining two bars upto the end. At the bottom, provide 2 — 10 Φ bar as holding bars.

Step 5. Shear reinforcement

$$V_u = 52032 \text{ N}$$

Nominal shear stress $$\tau_v = \frac{V_u - \frac{M_u}{d}\tan\beta}{bd}$$

where $$\tan\beta = \frac{400 - 200}{2500} = 0.06$$

$$\therefore \quad \tau_v = \frac{52032 - \frac{63.99 \times 10^6}{365} \times 0.06}{250 \times 365}$$

$$= 0.45 \text{ N/mm}^2.$$

$$p = \frac{100A_s}{bd} = \frac{100 \times 603.2}{250 \times 365} = 0.66\%$$

$$\therefore \qquad \tau_c \approx 0.5 \text{ N/mm}^2 > \tau_v.$$

Hence provide only nominal shear reinforcement, given by Eq. 27.3, rearranged below.

$$s_v = \frac{2.5\, A_{sv} \cdot f_y}{b}$$

Using 8 mm Φ 2 lgd. stirrups, $A_{sv} = 2 \times 50.3 = 100.6$

$$\therefore \qquad s_v = \frac{2.5 \times 100.6 \times 415}{250} = 417$$

However, max. $s_v = 0.75\, d = 0.75 \times 365 = 274$

Hence provide 8 Φ 2 lgd stirrups @ 270 mm c/c, at the supports. Reduce this spacing at the ends, to a value $= 0.75 \times 165 \approx 120$ mm

Step 6. Embedment in the support.

In order to develop full tensile strength at the face of the support, each of the three bars must be embedded into the support by a length equal to $L_d = 902$ mm. This could be best achieved by providing two bends of 90°, as shown in Fig. 28.2, where the anchorage value of each bend $= 12\Phi = 12 \times 16 = 192$ mm.

Step 7. Details of Reinforcement

Shown in Fig. 28.2.

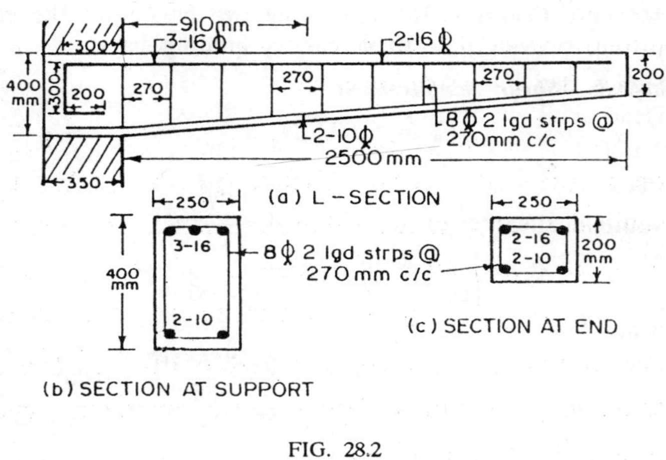

FIG. 28.2

28.3. DESIGN OF DOUBLY REINFORCED BEAM

Design Example 28.3

Design a reinforced concrete beam supported on two walls 500 mm thick, spaced at a clear distance of 6 m. The beam carries a super-imposed load of 30 kN/m. The size of the beam is restricted to 300 mm × 500 mm. Use M 15 concrete and Fe 415 steel.

Solution

Step 1. Computation of design constants and limiting depth of N.A.

For Fe 415 steel, $f_y = 415 \text{ N/mm}^2$ and $\frac{x_{u,max}}{d} = 0.48$

For M 15 concrete, $f_{ck} = 15 \text{ N/mm}^2$.

$$R_u = 0.36 f_{ck} \cdot \frac{x_{u,max}}{d}\left(1 - 0.416 \frac{x_{u,max}}{d}\right)$$

$$= 0.36 \times 15 \times 0.48(1 - 0.416 \times 0.48) = 2.074$$

$$\therefore \quad M_{u,lim} = R_u bd^2 \quad [\text{where } d \approx 500 - 25 - 10 = 465 \text{ mm}]$$

$$= 2.074 \times 300(465)^2 = 134.54 \times 10^6 \text{ N-mm}$$

$$x_{u,max} = 0.48\, d = 0.48 \times 465 = 223.2 \text{ mm.}$$

Step 2. Computation of design B.M.

$$L \approx 6000 + 465 = 6465 \text{ mm} < 6500 \text{ mm}$$

$$\therefore \quad \text{Available } \frac{L}{d} = \frac{6465}{465} = 13.9$$

Max. permissible value of $\frac{L}{d}$, even without modification factor = 20.

Hence deflection requirements are satisfied.

Dead load of beam $= 0.3 \times 0.5 \times 25000 = 3750$ N/m

Total $w = 3750 + 30{,}000 = 33750$ N/m.

Factored load $w_u = 1.5 \times 33750 = 50625 \text{ N/m}$.

$$M_{uD} \text{ (at midspan)} = \frac{w_u L^2}{8} = \frac{50625(6.465)^2}{8} = 264492 \text{ N-m}$$

$$\approx 264.5 \times 10^6 \text{ N-mm.}$$

Since $M_{uD} > M_{u,lim,}$ a doubly reinforced section will be required.

Step 3. Determinations of tensile reinforcement.

$$p_{lim} = 0.414 \frac{f_{ck}}{f_y} \cdot \frac{x_{u,max}}{d} \qquad \text{...(24.9)}$$

$$= 0.414 \times \frac{15}{415} \times 0.48 = 0.0071827$$

$$\therefore \quad A_{st1} = p_{lim} \times bd = 0.0071827 \times 300 \times 465 = 1002 \text{ mm}^2$$

Let cover to compressive reinforcement(d') = 25 + 10 = 35 mm

$$\therefore \quad A_{st2} = \frac{M_{uD} - M_{u,lim.}}{0.87 f_y (d - d')} = \frac{(264.5 - 134.54)10^6}{0.87 \times 415(465 - 35)}$$

$$= 837 \text{ mm}^2$$

$$\therefore \text{ Total } A_{st} = A_{st1} + A_{st2} = 1002 + 837 = 1839 \text{ mm}^2$$

(Max. reinforcement $= 0.04\, bD = 0.04 \times 300 \times 500$
$= 6000 \text{ mm}^2 > A_{st}$)

Hence O.K.

$$N - 20\,\Phi \text{ bars} = \frac{1839}{314.2} = 5.85 \approx 6$$

$$N - 25\,\Phi \text{ bars} = \frac{1839}{490.9} = 3.75 \approx 4.$$

Hence provide 6 bars of 20 mm dia.

Step 4. Determination of compressive reinforcement

$$A_{sc} = \frac{M_{uD} - M_{u,lim.}}{(f_{sc} - 0.446 f_{ck})(d - d')} \quad \text{...[25.10 } (a)]$$

or

$$A_{sc} = \frac{(264.5 - 134.5)10^6}{(f_{sc} - 0.446 \times 15)(465 - 35)} = \frac{302233}{(f_{sc} - 6.69)}$$

$$\varepsilon_{sc} = \frac{0.0035\,(x_{u,max} - d')}{x_{u,\max}} = \frac{0.0035(223.2 - 35)}{223.2} = 0.00295$$

Hence from Table 24.3, for Fe 415 steel

$$f_{sc} \approx 353.5 \text{ N/mm}^2 \text{ for } \varepsilon_{sc} = 0.00295$$

$$\therefore \quad A_{sc} = \frac{302233}{353.5 - 6.69} = 871.6 \text{ mm}^2$$

Provide 3-20 Φ bars having total $A_{sc} = 3 \times 314.2$
$= 942.2 \text{ mm}^2$.

Step 5. Curtailment of tensile and compressive reinforcement

Development length (L_d) of 20 dia. bars in tension.

$$= \frac{0.87 f_y}{4\,\tau_{bd}} \Phi = \frac{0.87 \times 415}{4 \times 1.60} \Phi = 56.4\,\Phi = 1128 \text{ mm}$$

Development length (L_d) of 20 mm dia. bars in compression.

$$= \frac{0.87 f_y}{4(1.25\,\tau_{bd})}\Phi = \frac{0.87 \times 415}{5 \times 1.60}\Phi = 45.1\,\Phi \approx 900 \text{ mm}$$

Hence the tension steel cannot be curtailed at less than 1.41 m from the centre of span, while the compression cannot be curtailed at less than 0.9 m from the centre.

Let us curtail 1 compression bar and two tension bars at a theoretical point of cut off distant 1.5 m from the centre of span.

∴ Actual point of cut-off from the centre of the span will be larger of the two values :

(*i*) $1.5 + d = 1.5 + 0.465 = 1.965$ m.

(*ii*) $1.5 + 12\,\Phi = 1.5 + 12 \times 0.02 = 1.74$ m

Keep the *actual* point of cut-off at 2 m from centre of span. Distance of theoretical point of cut off from centre of span = 2–0.465 = 1.535 m

∴ Distance of theoretical point of cut-off from centre of support

$$= \frac{6.465}{2} = 1.535 \approx 1.7 \text{ m}$$

It can be shown that the moment of resistance of the section at this point will be larger than the actual B.M. at this location.

Step 6. Shear reinforcement

The critical section for shear is at a distance of $d(=465$ mm) from the face of the support, or at a distance of $465 + \frac{465}{2} = 697.5$ mm ≈ 0.7 m from the centre of support.

$$V_u = \left(50625 \times \frac{6.465}{2}\right) - (50625 \times 0.7) = 128208 \text{ N}$$

$$\tau_v = \frac{V_u}{bd} = \frac{128208}{300 \times 465} \approx 0.92 \text{ N/mm}^2.$$

$$\frac{100 A_s}{bd} = \frac{100(4 \times 314.2)}{300 \times 465} = 0.9\%$$

Hence from Table 27.1. $\tau_c = 0.58$ N/mm^2.

Hence shear reinforcement is necessary.

Let us provide 8 mm Φ 2 lgd stirrups, having $A_{sv}=100.5\text{ mm}^2$

$$\therefore\ s_v = \frac{0.87 f_y . A_{sv}}{(\tau_v - \tau_c)b} = \frac{0.87 \times 415 \times 100.5}{(0.92 - 0.58) \times 300} = 365 \text{ mm}$$

Spacing based on minimum shear reinforcement

$$s_v = \frac{A_{sv} . f_y}{0.4\ b} = \frac{100 \times 415}{0.4 \times 300} = 348 \text{ mm}$$

Spacing based on effective depth of beam

$$s_v = 0.75\, d = 0.75 \times 465 = 348 \text{ mm}$$

Hence provide 8 mm Φ 2 lgd stirrups @ 340 mm c/c. Since this spacing is almost based on minimum shear reinforcement, provide this spacing throughout the length.

Step 7. Check for development length of support

At the support, the tensile reinforcement should satisfy the following relation

$$1.3\ \frac{M_1}{V} + L_0 \geq L_d$$

There are four tensile bars available at supports, having

$$A_{st} = 4 \times 314.2 = 1257 \text{ mm}^2$$

$$x_u = \frac{0.87 f_y A_{st1}}{0.36\ f_{ck} . b} = \frac{0.87 \times 415 \times 1257}{0.36 \times 15 \times 300} = 280 \text{ mm}$$

$$\therefore\ M_1 = 0.87 f_y A_{st}(d - x_u) = 0.87 \times 415 \times 1257\,(465 - 280)$$

$$= 83.96 \times 10^6 \text{ N-mm}$$

$$V = V_u = 50625 \times \frac{6.465}{2} = 163645 \text{ N}$$

$$L_d = 56.4\,\Phi = 1128 \text{ mm} \quad (\text{ as computed earlier})$$

Minimum embedment required from the face of the support

$$= \frac{L_d}{3} = \frac{1128}{3} = 376 \text{ mm.}$$

Width of support available (b_0) = 450 mm

Providing end cover x' of 40 mm, and providing 90° bend to the tensile bars,

$$L_0 = (500 - 40 - 232.5 + 12\,\Phi)$$
$$= 227.5 + 12 \times 20 = 467.5 \text{ mm}$$

$$\therefore \qquad 1.3\,\frac{M_1}{V} + L_0 = 1.3 \times \frac{83.96 \times 10^6}{163645} + 467.5$$

$$= 1134.5 \text{ mm} > L_d. \text{ Hence O.K.}$$

The compression reinforcement at the top has to be embedded into the support by a length $= b_0 -$ end cover $= 500 - 40 = 460$ mm, which is more than $L_d/3 (= 376$ mm).

Step 8. Details of reinforcement etc. As shown in Fig. 28.3.

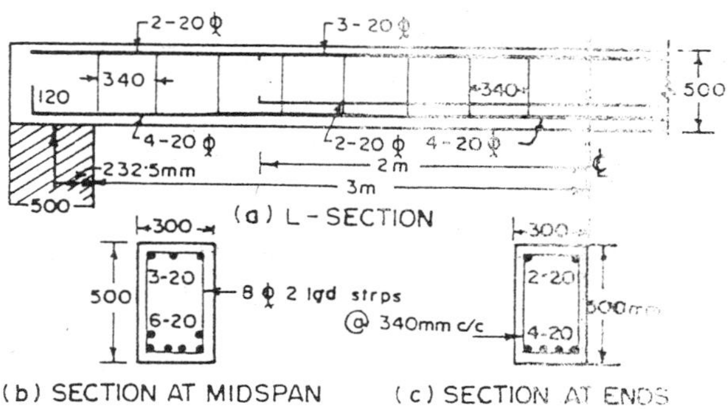

FIG. 28.3. DOUBLY REINFORCED BEAM.

28.4. DESIGN OF ONE WAY SLAB

The analysis of slab spanning in one direction is done by assuming it to be a beam of 1m width. The reinforcement etc. are calculated for 1 m width and the bars are distributed accordingly.

In addition to the main reinforcement (called *longitudinal reinforcement*), *transverse reinforcement* (also called distribution reinforcement) is provided in a direction at right angles to the span of the slab. The transverse reinforcement is provided to serve the following purposes :

(*i*) It distributes the effects of point load on the slab more evenly and uniformly.

(*ii*) It distributes the shrinkage and temperature cracks more evenly.

(*iii*) It keeps main reinforcement in position.

The reinforcement is either direction in slab shall not be 0.15% of the total cross-sectional area. However, this value can be reduced to 0.12% when high strength deformed bars or welded wire fabric are used.

The maximum diameter of the reinforcing bars shall not exceed one eighth of total thickness of slab.

As per IS 456-1978, atleast one third of the maximum positive reinforcement should extend along the same face of the slab into the support, to a length equal to $L_d/3$. Some positive reinforcement should also be bent up, near support, to take up negative B.M. which may develop due to partial fixidity. Unfortunately, IS : 456-1978 does not give any clear cut guidance about the bar bending and cut-off details, as given in CP 110- 1972. Due to lack of clear cut recommendations by IS code, different practices about bar bending is followed. One *practice* is to be bend alternate bars up at a distance of 0.15 L (or $L/7$) from the centre of support, so that bars are available at the top face, for atleast a length equal to 0.1 L (or $L/10$) from the centre of the support, The other *popular practice* is to bend alternate bars up at a distance of 0.15 l (or $l/7$) from the *face of the support*, where B.M. reduces to less than half its maximum value ; this will ensure, in majority of the cases, that length of bars available at the upper face is more than 0.1 L from the centre of support.

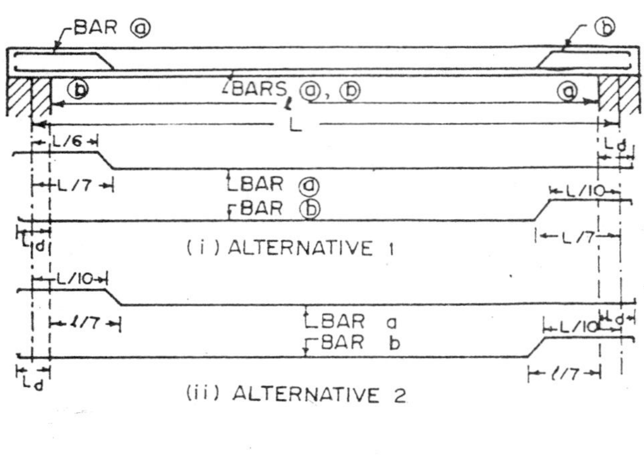

FIG. 28.4. BAR BENDING SCHEME IN ONE WAY SLAB.

For solid slabs, the design shear strength of concrete is taken as $k\tau_c$, where k may be taken from Table 27.3.

The other requirements and basic rules are the same as discussed in § 7.4 for design of one way slab by working stress method.

Example 28.4. Design of one way slab

Design a R.C. slab for a room having inside dimensions 3m × 7m. The thickness of supporting wall is 300 mm. The slab carries 75 mm thick lime concrete at its top, the unit weight of which may be taken as 20 kN/m^3. The live load on the slab may be taken as 2 kN/m^2. Assume the slab to be simply supported at the ends. Use M 15 concrete and Fe 250 steel.

Solution

Step 1. Design constants and limiting depth of N.A.

For Fe 250 steel, $f_y = 250$ N/mm^2
For M 15 concrete, $f_{ck} =$ N/mm^2

$$\frac{x_{u,max}}{d} \text{ (for Fe 250 steel) } = 0.53$$

$$R_u = 0.36 f_{ck} \frac{x_{u,max}}{d}\left(1 - 0.416 \frac{x_{u,max}}{d}\right)$$

$$= 0.36 \times 15 \times 0.53(1 - 0.416 \times 0.53)$$

$$= 2.23$$

Step 2. Computation of design B.M. and S.F.

The minimum depth of slab is based on serviceability requirement. For a simply supported slab

$$\frac{L}{d} = 20$$

In limit state design, p =1.32% for M 15–Fe 250 combination. However, assuming p = 0.8% for an *under-reinforced* section, we get modification factor ≈ 1.58 from Fig. 7.1.

Taking $L \approx 3.2$ m, we have

$$d = \frac{3200}{20 \times 1.58} = 101 \text{ mm}$$

Keep D = 120 mm. Using 15 mm clear cover and 8 mm dia. bars,

$$d = 120 - 15 - 4 = 101 \text{ mm}$$

Effective span $L = (l + d) = 3 + 0.101 = 3.1$ m

(or equal to c/c distance between supports, whichever is less)

Self weight of slab	$= 0.12 \times 25000$	$= 3000\ \text{N/m}^2$
Live load	$= 2\text{kN/m}^2$	$= 2000\ \text{N/m}^2$
Weight of lime terracing	$= 20000 \times 0.075$	$= 1500\ \text{N/m}^2$
Total		$w = 6500\ \text{N/m}^2$

$$\therefore \quad w_u = 1.5w = 1.5 \times 6500 = 9750\ \text{N/m}^2$$

$$M_u = \frac{w_u\, l^2}{8} = \frac{9750(3.1)^2}{8} = 11.712\ \text{N-m}$$

$$= 11.712 \times 10^6\ \text{N-mm}$$

$$V_u = \frac{w_u \,.\, L}{2} = 9750\ \frac{(3.1)}{2} = 15113\ \text{N}$$

Step 3. Computation of effective depth d

For a balanced section,

$$M_u = R_u \,.\, bd^2, \text{ where } R_u = 2.23 \text{ (Step 1)}$$

$$\therefore \quad d = \sqrt{\frac{M_u}{R_u \,.\, b}} = \sqrt{\frac{11.712 \times 10^6}{2.23 \times 1000}} = 72.5\ \text{mm}$$

Actual d available = 101 mm (based on the limit state of serviceability requirements).

Hence the section will be *under-reinforced.*

Step 4. Steel Reinforcement.

For an under-reinforced section,

$$A_{st} = \frac{0.5 f_{ck}}{f_y}\left[1 - \sqrt{1 - \frac{4.6\, M_u}{f_{ck}\, bd^2}}\right] bd \qquad \text{...(24.13)}$$

$$= \frac{0.5 \times 15}{250}\left[1 - \sqrt{\frac{4.6 \times 11.712 \times 10^6}{15 \times 1000\,(101)^2}}\right] 1000 \times 101$$

$$= 591\ \text{mm}^2$$

$$(A_{st})_{min} = \frac{0.15}{100} \times 1000 \times 120 = 180\ \text{mm}^2 = A_{sd}$$

Hence spacing of 8 Φ bars as main reinforcement is

$$s = \frac{1000\,A_\Phi}{A_{st}} = \frac{1000 \times 50.3}{591} = 85 \text{ mm c/c}$$

Spacing of 8 Φ bars as distribution reinforcement is

$$s = \frac{1000\,A_\Phi}{A_{sd}} = \frac{1000 \times 50.3}{180} = 279 \text{ mm} \approx 275 \text{ mm}$$

(which is less than max. spacing $5d = 5 \times 101$ mm = 505 or 450 mm whichever is smaller).

Bend alternate bars of main reinforcement up, at adistance of $\frac{l}{7} = \frac{3000}{7} = 430$ mm from the face of the support. Continue the remaining half bars into the support.

Step 5. Check for shear

The critical section for shear is at a distance of $d(= 101\text{mm})$ from the face of the support or at a distance of $d + \frac{d}{2} = 101 + \frac{101}{2} \approx 152$ mm from the centre of support.

$$V_{uD} = V_u - w_u(0.152) = 15113 - 9750(0.151)$$

$$= 13631 \text{ N.}$$

$$\tau_v = \frac{V_{uD}}{bd} = \frac{13631}{1000 \times 101} = 0.135 \text{ N/mm}^2$$

$$\frac{100\ A_s}{bd} \text{ (at supports)} = \frac{100(591/2)}{1000 \times 101} = 0.293\%$$

Hence $\tau_c \approx 0.37$ N/mm^2 from Table 27.1

Also $k = 1.30$ (from Table 27.3)

Since $\tau_v \ll k\tau_c$, the section is safe in shear.

Step 6. Check for development length and anchorage at ends

$$L_d = \frac{0.87 f_y}{4\tau_{bd}}\,\Phi = \frac{0.87 \times 250}{4 \times 1.0}\,\Phi = 54.4\Phi$$

$$= 54.4 \times 8 = 435 \text{ mm.}$$

The Code requires that the reinforcement should extend by a length equal to $\frac{L_d}{3} = \frac{435}{3} \approx 145$ mm beyond the face of support. This suggests that the width of support should not be less than $145 + 25 = 170$ mm. In the present case, the support width is 300 mm and hence the anchorage condition is satisfied.

Also, at the simple support like this, where compressive reaction confines the ends of bars,

$$1.3 \frac{M_1}{V} + L_0 >> L_d$$

Here $V = V_u = 15113$ N, at the centre of support.

M_1 = moment of resistance of the remaining bars. which is computed as under:

$$x_u = \frac{0.87 f_y A_{st1}}{0.36 f_{ck} \cdot b} = \frac{0.87 \times 250 \times (591/2)}{0.36 \times 15 \times 1000}$$

$$= 11.9 \text{ mm}$$

$$\therefore \quad M_1 = 0.87 f_y A_{st1}(d - 0.42 x_u)$$

$$= 0.87 \times 250 \times \frac{591}{2}(101 - 11.9)$$

$$= 5.78 \times 10^6 \text{ N-mm}$$

Providing U-hook and a clear cover = 25 mm at the side end,

$$L_0 = \frac{l_s}{2} - x' + 13\Phi = \frac{300}{2} - 25 + 13 \times 8 = 229 \text{ mm}$$

$$\therefore \ 1.3 \frac{M_1}{V} + L_0 = 1.3 \times \frac{5.78 \times 10^6}{15113} + 229$$

$$= 726 \text{ mm} > L_d$$

Hence the Code requirements are satisfied.

Step 7. Check for deflection.

At the mid-span,

$$\frac{100 A_{st}}{bd} = \frac{100 \times 591}{1000 \times 101} = 0.585\%$$

Hence from Fig. 7.1, modification factor ≈ 1.8

$$\therefore \quad \frac{L}{d} = 20 \times 1.8$$

or $$d = \frac{L}{20 \times 1.8} = \frac{3100}{20 \times 1.8} \approx 86 \text{ mm}$$

Actual d available = 101 mm.

Hence safe.

Note : 1. In limit state design, effective depth required from the point of view of bending will be very much less than the one required from deflection point of view. This will result in an *under-reinforced* section. Hence while taking the modification factor from Fig. 7.1, the value of p should be assumed equal to 40 to 50%

of p_{lim} given in Table 24.2. for mild steel bars, and equal to about 30% of p_{lim} for *HYSD* bars.

2. The above design may be *revised* by taking

D = 105 mm (or 110 mm), giving a value of

d = 105 – 15 – 4 = 86 mm (or = 110 – 15 – 4 = 91 mm).

Step. 8. Details of reinforcement. Shown in Fig. 28.5.

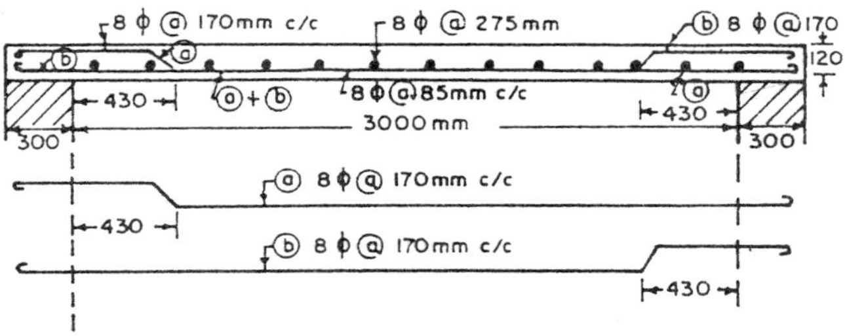

FIG. 28.5.

28.5. DESIGN OF ONE WAY CONTINUOUS SLAB

When a slab is continuous over several spans, negative(*i.e.* hogging) bending moment is induced over intermediate supports and hence reinforcement has to be provided at the top of the slab portion over the intermediate supports. These intermediate supports may either be monolithic with the slab, or may be of some different material such as rolled steel joist etc. The following *basic rules* for design, as per IS : 456-1978 are followed.

1. **Effective span.** Same as discussed in § 7.7.

2. **Limiting stiffeness** The ratio of span to effective depth should not exceed 26. Modification factor for tension reinforcement will be applicable extra as per Fig. 7.1.

3. **B.M. and S.F. coefficients.** Same rules, as are applicable for beams. See Table 7.2 and 7.3, and Fig. 7.16. Where a member is built into a masonry wall which develops only partial restraint, the slab should be designed to resist a negative moment at the face of the support of $\frac{WL}{24}$ where W is the total design load and L is the effective span. For such a condition, shear

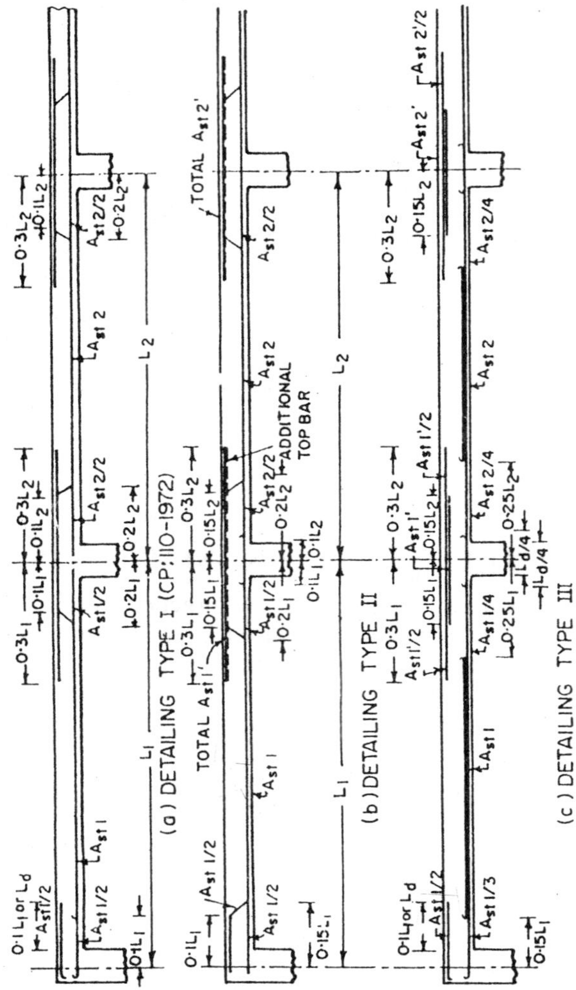

FIG. 28.6. ALTERNATIVE FORMS OF REINFORCEMENT DETAILING.

coefficient given in Table 7.3 at the end support may be increased by 0.05.

4. Reinforcement. The rules for main reinforcement and transverse/distribution reinforcement are the same as discussed for simply supported one way slab. However, the Code (IS : 456-1978) makes the following recommendations for continuation/curtailment of reinforcement

(a) Positive moment reinforcement At least one third the positive moment reinforcement in simple members and one fourth the positive moment reinforcement in continuous members shall extend along the same face of the member into the member, to a length equal to $\frac{L_d}{3}$.

(b) Negative moment reinforcement. At least one third of the total reinforcement provided for negative moment at the support shall extend beyond the point of inflexion for a distance not less than the effective depth of the member or 12 Φ or one sixteenth of the clear span whichever is greater.

Except for the above two recommendation IS : 456-1978 is silent in regard to simplified rules for curtailment of reinforcement in one-way continuous slabs. However, CP 110-1972 has given such simplified rules, as given in Fig. 28.6(*a*). In absence of any such recommendations from IS Code, various practices are followed, and one such practice is shown in Fig. 7.17.

Two more forms of reinforcement detailing are shown in Fig. 28.6(*b*) and (*c*) respectively. It may be noted that detailing type II, shown in Fig. 28.6(*b*) is similar to the one shown in Fig. 7.17, except that the bent-up point is at a distance of 0.25 L from centre of support (in place of $L/5$) and the negative reinforcement extends to a distance of 0.3 L to the either side of the support (in place of $L/4$). Detailing type II has been followed in Example 28.5. In detailing type III, no bar is bent up, and the positive reinforcement and negative reinforcement are provided separately at appropriate faces. The curtailment of positive reinforcement in type III detailing is as per IS Code recommendations.

Example 28.5. Design of one way continuous slab

Design a continuous R.C. slab for a hall 6.5 m wide and 13.5 m long. The slab is supported on R.C.C. beams, each 240 mm wide which are monolithic. The ends of the slab are supported on walls, 300 mm wide. Design the slab for a live load of 2 kN/m^2. Assume *the weight of roof finishing equal to 1.5 kN/m². Use M 15 concrete and Fe 415 steel.*

Solution.

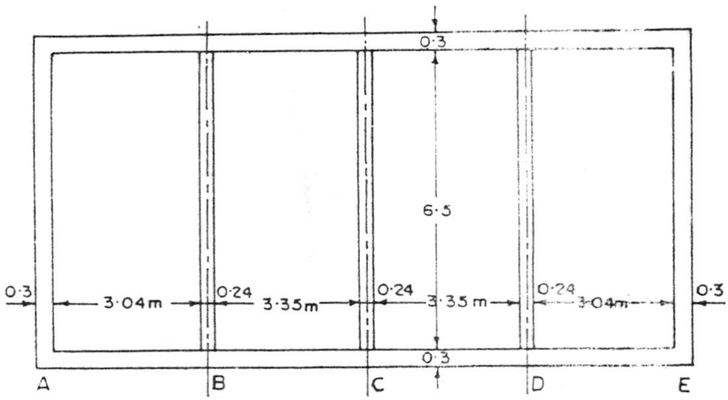

FIG. 28.7. ARRANGEMENT OF SPANS.

Step 1. Arrangement of spans

Let us arrange the spans as shown in Fig. 28.7. Assuming d = 100 mm. The width of intermediate supports is less than 1/12 of the clear span. Hence the effective span will be as under :

(*i*) *For end spans*

$$L = \text{centre to centre of supports} = 3.04 + \frac{0.3+0.24}{2} = 3.31 \text{ m}$$

or L = clear span + d = 3.04 + 0.10 =3.14 whichever is less

Hence L = 3.14 m

(*ii*) *For intermediate spans :*

L = centre to centre of supports = 3.35 + 0.24 = 3.59 m

or L = clear span + d = 3.35 + 0.10 = 3.45 whichever is less

Hence L = 3.45 m

Thus, in this case, the length of effective span is not equal. However, the difference is very much less than 0.15% of the longer spans ; hence they may be treated as being equal for the purposes of using the B.M. coefficients given in the I.S. Code.

Step 2. Fixation of d and D

From the limit state of serviceability, the effective depth is given by

$$\frac{L}{d} = 26 \text{ for continuous spans}$$

and

$$\frac{L}{d} = 20 \text{ for simply supported spans.}$$

hence for end span.

$$\frac{L}{d} \approx \frac{1}{2}(26 + 20) = 23$$

Assume a modification factor(corrresponding to $p \approx 0.3\%$) is equal to 1.4 for Fe 415 steel (Fig. 7.1).

Hence, on the basis of end span,

$$d = \frac{L}{23 \times 1.4} = \frac{3450}{23 \times 1.4} \approx 98 \text{ mm}$$

On the basis of intermediate spans,

$$d = \frac{L}{26 \times 1.4} = \frac{3450}{26 \times 1.4} = 95 \text{ mm}$$

Keep $D = 120$ mm, so that $d = 120 - 15 - 4 = 101$ mm.

Step 3. Design constants and limiting depth of N.A.

For Fe 415 steel, $f_y = 415 \text{ N/mm}^2$

For M 15 concrete, $f_{ck} = 15 \text{ N/mm}^2$.

$$= \frac{x_{u,max}}{d} \text{ (For Fe 415 steel)} = 0.48$$

$$\therefore R_u = 0.36 f_{ck} \frac{x_{u,max}}{d}\left(1 - 0.416\frac{x_{u,max}}{d}\right)$$

$$\therefore \quad = 0.36 \times 15 \times 0.48(1 - 0.416 \times 0.48) = 2.074$$

Step 4. Computation of design B.M. and effective depth.

Self weight of slab $= 0.12 \times 25000 = 3000 \text{ N/m}^2$

Weight of roof finishing $= 1500 \text{ N/m}^2$

$= 1500 \text{ N/m}^2$

Total $w_d = 4500 \text{ N/m}^2$

Live load $w_s = 2\text{kN/m}^2 = 2000\ \text{N/m}^2$

For end spans

$$M \text{ (near the centre)} = \frac{w_d L^2}{12} + \frac{w_s L^2}{10}$$

$$= \frac{4500(3.14)^2}{12} + \frac{2000(3.14)^2}{10} = 5669 \text{ N-m}$$

$$= 5.669 \times 10^6 \text{ N-mm}$$

$$M \text{ (at support next to end support)} = -\left[\frac{w_d L^2}{10} + \frac{w_s L^2}{9}\right]$$

$$= -\left[\frac{4500(3.14)^2}{10} + \frac{2000(3.14)^2}{9}\right]$$

$$= 6628 \text{ N-m} = 6.628 \times 10^6 \text{ N-mm}$$

For Intermediate spans

$$M \text{ (at the middle)} = +\frac{w_d L^2}{24} + \frac{w_s L^2}{12}$$

$$= \frac{4500(3.45)^2}{24} + \frac{2000(3.45)^2}{12} = 4215 \text{ N-m}$$

$$= 4.215 \times 10^6 \text{ N-mm}$$

$$M \text{ (at interior supports)} = -\left[\frac{w_d L^2}{12} + \frac{w_s L^2}{9}\right]$$

$$= -\left[\frac{4500\ (3.45)^2}{12} + \frac{2000\ (3.45)^2}{9}\right]$$

$$= 7108 \text{ N-m} = 7.108 \times 10^6 \text{ N-mm}$$

Out of the four values of moments, the effective depth will be determined for max. value.

$\therefore$ Design moment $= 7.108 \times 10^6$ N-mm

Hence $M_{uD} = 1.5 \times 7.108 \times 10^6 = 10.663 \times 10^6$ N-mm

$$\therefore \qquad d = \sqrt{\frac{M_{uD}}{R_u.b}} = \sqrt{\frac{10.663 \times 10^6}{2.074 \times 1000}} = 71.7 \text{ mm.}$$

Available d, determined from limit state of serviceability, is equal to 121 mm. Hence the section will be *under-reinforced.*

Step 5. Determination of reinforcement

It is proposed to provide type II detailing shown in Fig. 28.6(a), wherein there is a provision for different reinforcements at various critical locations.

(*a*) *At Middle of end span*

$M_{uD1} = 1.5 \times 5.669 \times 10^6 = 8.504 \times 10^6$ N-mm

$$\therefore \quad A_{st1} = \frac{0.5 f_{ck}}{f_y}\left[1 - \sqrt{1 - \frac{4.6 M_u}{f_{ck}\, bd^2}}\right] bd \qquad \ldots(24.13)$$

$$= \frac{0.5 \times 15}{415}\left[1 - \sqrt{1 - \frac{4.6 \times 8.504 \times 10^6}{15 \times 1000 (101)^2}}\right] 1000 \times 101$$

$$= 250.5 \text{ mm}^2$$

$$s\text{-}8\ \Phi \text{ bars} = \frac{1000 \times 50.3}{250.5} = 200.8 \text{ mm.}$$

Max. spacing $= 3d = 3 \times 120 = 360$ mm

Hence provide 8 Φ bars @ 200 mm c/c.

$$\text{Actual } A_{st1} \text{ provided} = \frac{1000 \times 50.3}{200} = 251.5 \text{ mm}^2$$

$$p = \frac{100 \times 251.5}{1000 \times 101} = 0.25\%$$

$$\left(A_{st,min} = \frac{0.12}{100} \times 1000 \times 140 = 168 \text{ mm}^2\right)$$

Bend half bars up at a distance of $0.15\,L = 0.15 \times 3140 = 471$ mm from the centre of support, or from $471 - d/2 = 471 - 101/2 \approx 420$ mm from the face of the support. Continue remaining half bars into the support for a distance $L_d/3$. At the other end, bend half bars up at a distance of $0.25\,L = 0.25 \times 3140 = 785$ from the centre of the support, and continue the remaining half into the support, for a distance or $L/10 \approx 320$ mm from the face of the support.

(*b*) *At middle of intermediate span*

$M_{uD2} = 1.5 \times 4.215 \times 10^6 = 6.323 \times 10^6$ N-mm

$$\therefore \quad A_{st2} = \frac{0.5 \times 15}{415}\left[1 - \sqrt{1 - \frac{4.6 \times 6.323 \times 10^6}{15 \times 1000\,(101)^2}}\right] 1000 \times 101$$

$$= 182.3 \text{ mm}^2$$

($A_{st,min} = 168$ mm^2).

However keep $A_{st2} = A_{st1}$ because of practical considerations, and also because of the reason that the section is already very much under-reinforced.

(c) *At support next to end support*

$M_{uD1}' = 1.5 \times 6.628 \times 10^6 = 9.942 \times 10^6$ mm

$$\therefore\ A_{st1}' = \frac{0.5 \times 15}{415}\left[1 - \sqrt{1 - \frac{4.6 \times 9.942 \times 10^6}{15 \times 1000(101)^2}}\right] 1000 \times 101$$

$= 297$ mm^2.

Area of steel available in the form of bent up bars from the adjacent spans $= \frac{1}{2}(A_{st1} + A_{st2}) = 251.5$ mm^2.

$\therefore$ Additional steel required $= 297 - 251.5 = 45.5$ mm^2

$$s\text{-}8\ \Phi \text{ bars } = \frac{1000 \times 50.3}{45.5} = 1105 \text{ mm.}$$

However, this has to be in multiple of earlier spacing of 200 mm.

Hence provide extra 8 mm Φ bars @ 800 mm c/c for length equal to $0.3L = 0.3 \times 3140 \approx 950$ mm to one side and $0.3 \times 3450 = 1050$ mm 1050 mm to the other side of the support.

(d) *At interior supports*

$M_{uD2}' = 1.5 \times 7.108 \times 10^6 = 10.662 \times 10^6$ N-mm.

$$\therefore\ A_{st2}' = \frac{0.5 \times 15}{415}\left[1 - \sqrt{1 - \frac{4.6 \times 10.662 \times 10^6}{15 \times 1000(101)^2}}\right] \times 1000 \times 101$$

$= 321$ mm^2.

Steel available in the form of bent-up bars from adjacent spans

$= 251.5$ mm^2.

$\therefore$ Additional steel required $= 321 - 251.5 = 69.5$ mm^2

$$s\text{-}8\ \Phi = \frac{1000 \times 50.3}{69.5} = 724 \text{ mm.}$$

However, this has to be in multiple of earlier spacing of 200 mm c/c. Hence provide 8 Φ extra bars @ 600 mm c/c, for a length $= 0.3L = 0.3 \times 3450 \approx 1050$ mm to the either side of the support.

Distribution reinforcement

$$A_{sd} = \frac{0.12}{100} \times 1000 \times 140 = 168 \text{ mm}^2.$$

$$\therefore\ s\text{-}8\ \Phi = \frac{1000 \times 50.3}{168} \approx 300 \text{ mm}$$

Hence provide 8 Φ bars @ 300 mm c/c.

Step 5. Check for anchorage and development length at end support.

The end support is a free support :

$$L_d = 56.4\Phi = 56.4 \times 8 \approx 450 \text{ mm}$$

$$L_d/3 = 450/3 \approx 150 \text{ mm.}$$

Width of support available= 300 mm. Length by which half the bars can be taken inside the support $= 300 - 25 = 275$ mm >150 mm. Hence safe.

Again, at the end support, the following relation should satisfied

$$1.3\frac{M_1}{V} + L_o \geq L_d$$

where $V = V_u = 1.5[0.4\ w_d L + 0.45 w_s . L]$

$$= 1.5[0.4 \times 4500 \times 3.14 + 0.45 \times 2000 \times 3.14]$$

$$= 12717 \text{ N.}$$

Area of steel available at supports $= A_{st1}/2 = 251.5/2$
$\approx 125.8 \text{ mm}^2$

$$x_u = \frac{0.87 f_y . A_{st}}{0.36 f_{ck} . b} = \frac{0.87 \times 415 \times 125.8}{0.36 \times 15 \times 1000} = 8.4 \text{ mm}$$

$$\therefore \quad M_1 = 0.87\ f_y A_{st}(d - 0.42 x_u)$$

$$= 0.87 \times 415(125.8)(101 - 8.4)$$

$$= 4.206 \times 10^6 \text{ N-mm}$$

$$\therefore \quad L_0 = \frac{b_s}{2} - x' = \frac{300}{2} - 25 = 125 \text{ mm.}$$

$$1.3\frac{M_1}{V} + L_o = 1.3 \times \frac{4.206 \times 10^6}{12717} + 125$$

$$= 555 \text{ mm.}$$

$$L_d = 450 \text{ mm.}$$

Hence safe.

Step 6. Check for shear.

Normally, slabs are found to be safe in shear. However, let us check for shear at the support next to the end support, where S.F. is maximum.

$$V_u = 1.5\ [0.6\, w_d L + 0.6\, w_s L]$$

$$= 1.5 \times 0.6 \times 3.14[4500 + 2000] = 18369 \text{ N}$$

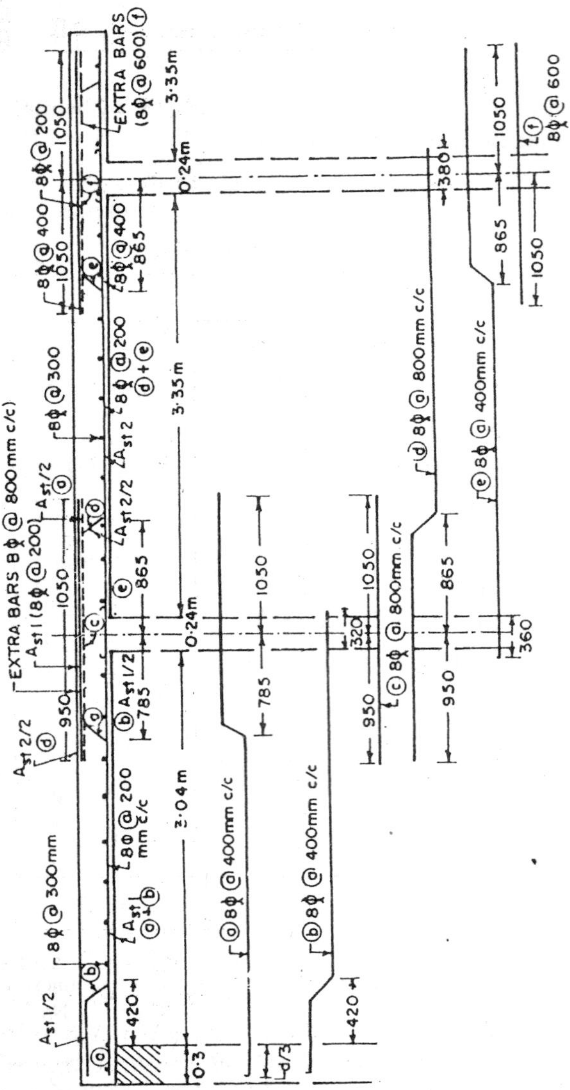
EXTRA BARS (8ϕ @ 600) (f)
EXTRA BARS 8ϕ @ 800mm c/c
EXTRA BARS A_{st}1 (8ϕ @ 200)
8ϕ @ 300mm
8ϕ @ 200 mm c/c
8ϕ @ 300
8ϕ @ 200
8ϕ @ 400
8ϕ @ 600
(a) 8ϕ @ 400mm c/c
(b) 8ϕ @ 400mm c/c
(c) 8ϕ @ 800mm c/c
(d) 8ϕ @ 800mm c/c
(e) 8ϕ @ 400mm c/c
3·04 m
3·35 m
3·35m
0·24m
0·3
d/3
420
785
950
1050
865
320
360
380

FIG. 28.8

Though the critical section will be at a distance of d from the support, where S.F. will be slightly less, we will adopt the same value of V_u as that the centre of the support.

$$\tau_v = \frac{18369}{1000 \times 101} = 0.182 \text{ N/mm}^2$$

$$\tau_{c,min.} = 0.35 \text{ N/mm}^2$$

$$\therefore \quad \tau_v < 1.3\ \tau_c$$

Hence safe.

Step 7. Check for deflection

For the end span, $p = 0.25\%$

$\therefore$ Modification factor ≈ 1.5

$$\therefore \quad d = \frac{L}{\frac{1}{2}(20 + 26) \times 1.5} = \frac{3140}{23 \times 1.5} = 91 \text{ mm}$$

Available $d = 101$ mm

Hence safe.

Step 8 Details of reinforcement etc.

Shown in Fig. 28.8.

28.6. DESIGN OF T-BEAM ROOF

A T-beam roof consists of two components :

(*i*) Continuous slab cast monolithic with beams.

(*ii*) T-beams.

The design of continuous slab has been given in the previous article. The design of T-beam has been thoroughly discussed and illustrated in Chapter 26.

PROBLEMS

1. Design a reinforced concrete beam, supported on two walls, 450 mm thick, over a clear span of 8 m. The beam carries a super-imposed load of 15 kN/m. Use M 15 mix. and Fe 415 reinforcement.

2. Design a reinforced concrete lintel over the openings of two windows, each 2 m wide, separated by a wall of 50 cm. The height of wall over the lintel is 3 m and the thickness of wall is 30 cm. The lintel is to be provided with a sunshade projecting out by 80 cm and cast monolithically with the lintel. Use M 15 mix and mild steel reinforcement. The live load on sunshade may be taken as 1 kN/m^2.

3. Design a R.C. cantilever beam projecting out 3 m beyond the fixed end and carrying a super-imposed load of 15 kN. The cantilever also carries a concentrated load of 5 kN at the free end. Use M 15 mix and Fe 415 steel.

4. Design a R.C.C. floor slab for a room having inside dimensions 4m × 10m and supported on all sides by a 40 cm thick brick wall. The super-imposed load may be taken as 3 kN/m^2. Use M 15 concrete and Fe 250 reinforcement.

29

Design of Columns

29.1. LIMIT STATE OF COLLAPSE : COMPRESSION

Assumptions

The limit state method of design of compression member is based on the following assumptions.

1. The plane section normal to the axis of column before deformation remain plane after deformation. This means that the strain at any point is proportional to its distance from the N.A.

2. The relationship between compressive stress distribution in concrete and strain in concrete may be assumed to be rectangle, trapezoid, parabola or any other shape which results in prediction of strength in substantial agreement with the results of tests. An acceptable stress-strain curve is given in Fig. 24.1. For design purposes, the *compressive strength of concrete in structure* is assumed to be 0.67 times the characteristic strength. The partial safety factor $\gamma_{mc} = 1.5$ is applied in addition to this. Thus the design strength of concrete is taken as $0.67 f_{ck}/1.5 = 0.446 f_{ck}$.

3. The tensile strength of concrete is ignored.

4. The stresses in the reinforcement are derived from representative stress-strain curve for the type of steel used. Typical curves are shown in Fig. 24.2, 24.3 and 24.4. For design purposes, partial safety factor $\gamma_{ms} = 1.15$ shall be applied.

5. The maximum compressive strain in concrete in axial compression is taken as 0.002.

6. The maximum strain in concrete, at the outer most compression fibre is taken as 0.0035 in bending when N.A. lies within the section.

7. The maximum compressive strain at the highly compressed extreme fibre in concrete subjected to axial compression and bending, and where there is no tension on the section shall be 0.0035 minus 0.75 times the strain at the least compressed extreme fibre.

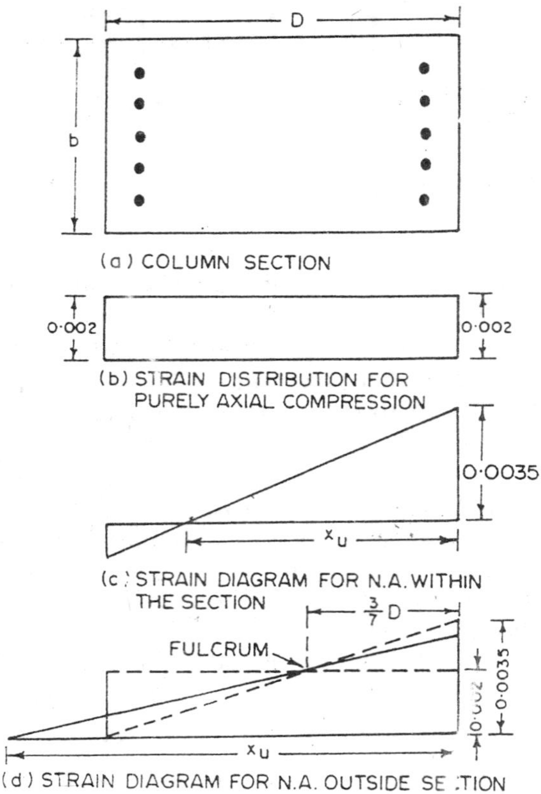

FIG. 29.1. STRAIN DISTRIBUTION ACROSS THE SECTION.

According to assumption 5, for purely axial compression, the strain is assumed to be uniform, equal to 0.002, across the section [Fig. 29.1(*b*)]. Also, for the case of axial compression along with bending, the maximum strain in concrete at the outer most fibre is 0.0035 [Fig. 29.1(*c*)] and this assumption is applicable when the N.A. lies within the section and in the limiting case when the N.A.

lies along the edge of the section. In the later case, the strain varies from 0.0035 at the highly compressed edge to zero at the opposite edge. The strain distribution for these two cases (of assumptions 5 and 6) intersect each other at a depth of $\frac{3D}{7}$ from the highly compressed edge. This point is assumed to act as fulcrum for the strain distribution line when the N.A. lies out side the section [Fig. 29.1(*c*)]. This leads to the assumption (No. 7) that the strain at highly compressed edge is 0.0035 minus 0.75 times the strain at least compressed edge.

29.2. SHORT AXIALLY LOADED MEMBERS IN COMPRESSION

When a short column is axially loaded, the strain distribution across the section will be rectangular. At failure, the strain in the concrete will be uniform at 0.002, The stress in concrete will be $0.67 f_{ck}/\gamma_{mc} = 0.67 f_{ck}/1.5 = 0.446 f_{ck}$.

The steel may develop full design stress in the case of mild steel reinforcement when the concrete attains a limiting strain of 0.002. However with cold twisted reinforcements ($f_y = 415$ N/mm^2 or $f_y = 500$ N/mm^2) the full design stress will not develop at a strain of 0.002.

The maximum stresses in the materials, at a strain of 0.002 will be as under :

Concrete : $0.446\ f_{ck}$

Mild steel : $0.87\ f_y$

Fe 415 Steel : $0.79\ f_y$

Fe 500 Steel : $0.75\ f_y$

Hence the capacity of member, subjected to only axial load, is given by

$$P = 0.446 f_{ck} A_c + \alpha f_y A_{sc} \quad \text{...[29.1}(a)\text{]}$$

Code has redesignated P as P_{uz} in section 38.6, and has given the following equation in the design aids (SP: 16-1980) :

$$P_{uz} = 0.446 f_{ck} \cdot A_c + 0.75 f_y \cdot A_{sc} \quad \text{...(29.1)}$$

Minimum eccentricity

According to IS : 456-1978, all compression members are to be designed for a minimum eccentricity of the load in two principal directions. Clause 24.4 of the Code specifies the following minimum eccentricity e_{min} for the design of columns :

$$e_{min} = \frac{l}{530} + \frac{D}{30}$$

subject to a minimum of 20 mm

where l = unsupported length of the column

D = lateral dimension of the column in the direction under consideration.

If the value of minimum eccentricity calculated as above is less than or equal to 0.05D, clause 38.3 of the Code permits the design of short axially loaded compression member by the following equation

$$P_u = 0.4 f_{ck} . A_c + 0.67 f_y A_{sc} \qquad ...(29.2)$$

where P_u = ultimate axial load

P_c = area of concrete

A_{sc} = area of reinforcement.

It should be noted that Eq. 29.2 is obtained by reducing the capacity of column, given be Eq. 29.1, by approximately 10% thereby allowing for the minimum eccentricity of 0.05D.

If, however, the minimum eccentricity is greater than 0.05D, the section is designed for combined axial load and bending, as discussed in § 29.3.

29.3. COMPRESSION MEMBERS WITH HELICAL REIN - FORCEMENT

The Code permits larger load in compression members with helical reinforcement because columns with helical reinforcement have greater ductility or toughness when they are loaded concentrically or with small eccentricity.

As per the Code, the strength of compression members with helical reinforcement satisfying the requirement given below shall be taken as 1.05 times the strength of similar members with lateral ties.

Requirement. The ratio of volume of helical reinforcement to the volume of core shall not be less than

$$0.36 \left(\frac{A_g}{A_c} - 1 \right) \frac{f_{ck}}{f_y}$$

where A_g = gross area of the section

A_c = Area of core of the helically reinforced column measured to the outside diameter of the helix.

f_y = characteristic strength of helical reinforcement but not exceeding 415 N/mm^2.

29.4. COLUMN WITH AXIAL LOAD AND UNI-AXIAL BENDING

When a column is subjected to direct compressive load as well as bending (or subjected to an eccentric compressive load), the direct design using analytical procedure is basically of a trial and error, and is extremely tedius. In such cases, use of interaction diagrams

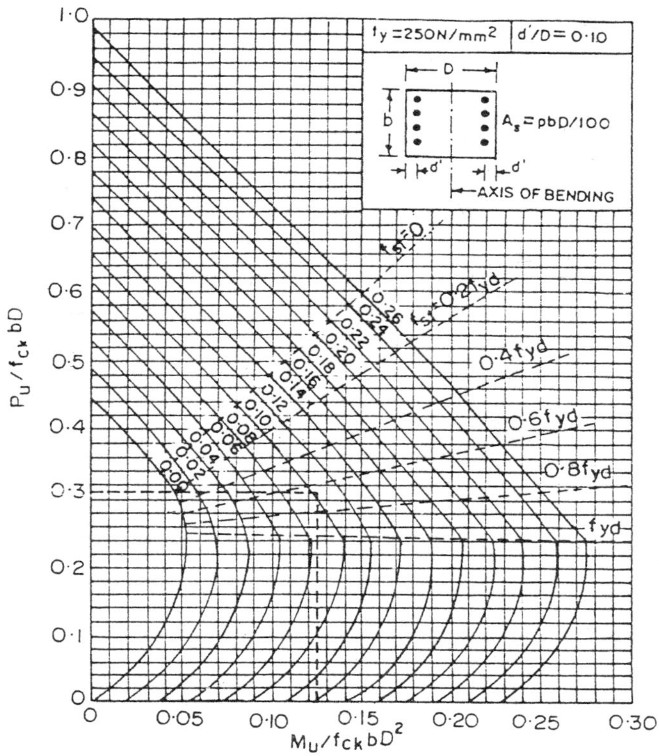

FIG. 29.2. INTERACTION CHARTS.

is very useful. I.S. Code recommends the use of these interaction diagrams, given in 'Design Aids for Reinforced Concrete to IS: 456-1978'. Two such interaction diagrams are reproduced here in Fig. 29.2 and 29.3. The interaction diagrams are presented using non-dimensional parameters $P_u/f_{ck} . bD$ and $M_u/f_{ck} . bD^2$. Each interaction diagram is for a given value for f_y and d'/D ratio, where d' is the distance of centre of steel from outer fibre. The use of these diagrams is illustrated in Example 29.3.

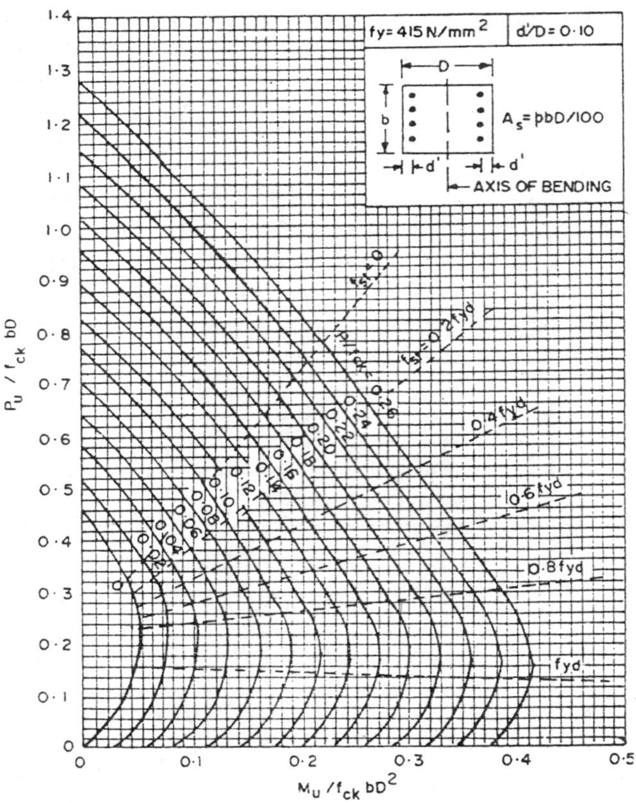

FIG. 29.3. INTERACTION CHARTS.

29.5. IS CODE RECOMMENDATIONS

1. Short and Slender Compression Members.

A compression member may be considered as *short* when both the slenderness ratios l_{ex}/D and l_{ey}/b are less than 12,

where l_{ex} = effective length in respect of the major axis

D = depth in respect of the major axis

l_{ey} = effective length in respect to minor axis

b = width of the member.

It shall otherwise be considered as a slender compression member. In case of slender columns, buckling effect gives rise to additional moments which are required to be taken into account.

2. Effective length of columns. The effective length (l_{ef}) of a column of length l may be determined with the help of Table 13.1.

3. Unsupported length of columns. The unsupported length of a column is defined as the clear distance between the end restraints. For a beam-slab floor construction, it is equal to floor height minus the depth of shallower beam forming into the column at its top.

4. Longitudinal and Transverse Reinforcement and cover to reinforcement. The same rules apply, as given in § 13.5.

Example 29.1. Load capacity of short column.

A reinforced concrete column, 300 mm square, is reinforced with 4 bars of 16 mm dia. Determine the ultimate load capacity of the column, using M 15 grade concrete and Fe 250 grade steel. What will be the allowable service load ?

Solution

$$A = 300 \times 300 = 90000 \text{ mm}^2$$

$$A_{sc} = 4 \times \frac{\pi}{4}(16)^2 = 804.2 \text{ mm}^2$$

$$\therefore \quad A_c = A - A_{sc} = 90000 - 804.2 = 89195.8 \text{ mm}^2.$$

Now from Eq. 29.2.

$$P_u = 0.4 f_{ck} \,.\, A_c + 0.67 f_y \,.\, A_s.$$

$$= 0.4 \times 15 \times 89195.8 + 0.67 \times 250 \times 804.2$$

$$= 535175 + 134703 = 669878 \text{ N}$$

$$= \mathbf{669.878 \text{ kN}}.$$

Allowable service load $= \frac{P_u}{1.5} = \frac{669.878}{1.5}$ = **446.585 kN.**

Example 29.2. Design of short column.

Design a short axially loaded column, 350 mm × 350 mm to support a service load of 820 kN. Use M 20 concrete and Fe 250 grade steel.

Solution

Service load = 800 kN
Load factor = 1.5

$\therefore$ Ultimate load $= 820 \times 15 = 1230$ kN

Now $P_u = 0.4 f_{ck} . A_c + 0.67 f_y . A_{sc}$ ...(29.2)

$\therefore$ $1230 \times 10^3 = 0.4 \times 20(350 \times 350 - A_{sc}) + 0.67 \times 250 A_{sc}$.

$167.5 A_{sc} - 8 A_{sc} = 250000$

From which $A_{sc} = 1567.4$ mm^2

$\therefore$ No. of 16 mm dia. bars $= \frac{1567.4}{201.4} = 7.8$

Hence provide 8 bars of 16 mm dia., having total area

$$= 201.06 \times 8 \approx 1608.5 \text{ mm}^2$$

Min. area of longitudinal reinforcement = 0.8 %

$$= \frac{0.8}{100} \times 350 \times 350 = 980 \text{ mm}^2.$$

Hence the reinforcement provided is O.K.

Using 6 mm dia. ties, spacing is the least of the following :

(*i*) Least lateral dimension = 350 mm,

(*ii*) $16 \times 16 = 256$ mm.

(*iii*) $48 \times 6 = 288$ mm.

Hence provide 6 mm dia. ties at 250 mm c/c.

Example 29.3. Design of column with eccentric load.

Design a reinforced concrete column, 400 mm square, to carry an ultimate load of 1000 kN at an eccentricity of 160 mm. Use M 20 grade concrete and Fe 250 grade steel.

Solution

Let cover to the centre of steel $= d' = 40$ mm

Here $b = D = 400$ mm

$d'/D = 40/400 = 0.1$

$P_u = 1000 \text{ kN} = 1000 \times 10^3$ N

$M_u = e . P_u = 1000 \times 10^3 \times 160$

The non-dimensional parametres are,

$$\frac{P_u}{f_{ck} \, . \, bD} = \frac{1000 \times 10^3}{20 \times 400 \times 400} = 0.3125$$

$$\frac{M_u}{f_{ck} \, . \, bD^2} = \frac{1000 \times 10^3 \times 160}{20 \times 400 \, (400)^2} = 0.125$$

Referring to the interaction diagram for $f_y = 250$ N/mm^2 and $d'/D = 0.1$ (*i.e.* Fig. 29.2), the corresponding value of p/f_{ck} is obtained as shown in dotted line.

$$\therefore \quad \frac{p}{f_{ck}} = 0.105$$

$$\therefore \quad p = 0.105 \times f_{ck} = 0.105 \times 20 = 2.1$$

$$p = \frac{100 A_s}{bD} = 2.1$$

$$\therefore \quad A_s = \frac{2.1 \, bD}{100} = \frac{2.1 \times 400 \times 400}{100} = 3360 \text{ mm}^2$$

Provide 2-25 mm dia. bars and 1-30 mm dia. bar on each face.

$$\text{Total } A_s = 4 \times \frac{\pi}{4}(25)^2 + 2 \times \frac{\pi}{4}(30)^2$$

$$= 1963.5 + 1413.7 = 3377.2 \text{ mm}^2.$$

Use 6 mm dia. lateral ties, at a spacing of the least of the following :

(*i*) Least lateral dimension = 400 mm,
(*ii*) 16 × 25 = 400 mm,
(*iii*) 48 × 6 = 288 mm.

Hence provide 6 mm dia. lateral ties at a spacing of 250 mm/cc.

PROBLEMS

1. Discuss various assumptions used in the limit state method of design of compression members.
2. A reinforced concrete column 200 mm × 300 mm is reinforced with 4 bars of 12 mm dia. one at each corner. Determine the service load which the column can carry, assuming it to be a short one. Use 15 grade concrete and Fe 250 steel bars.
3. Design a reinforced concrete column to carry a service load of 600 kN. The size of column is 250 mm × 300 mm. Use M 20 grade concrete and Fe 415 reinforcement.

30

Miscellaneous Designs

30.1. TWO WAY SIMPLY-SUPPORTED SLABS

When a slab is supported on all the four edges, and when the ratio of long span (l_y) to short span (l_x) is small, (say less than 2), bending takes place along both the spans. Such a slab is known as a '*two way slab*' or a 'slab spanning in two directions'. The maximum bending moment and deflection for such a slab is much smaller than that of a one-way slab and hence a thinner slab is required. However, reinforcement has to be provided in both the directions. When such a slab is loaded, the corners get lifted up. If the corners are held down, by fixidity at the wall support etc., the bending moment and deflection are further reduced, thus requiring still thinner slab. In that case, special torsional reinforcement at the corners has to be provided to check the cracking of corners.

We will divide the two-way slabs in the following three heads :

1. Slabs simply supported on the four edges, with corners not held down, and carrying uniformly distributed load.

2. Slabs simply supported on the four edges, with corners held down and carrying uniformly distributed load.

3. Slabs with edges fixed or continuous and carrying uniformly distributed load. (§30.2).

The analysis of a two-way slab may be done by the following methods.

1. Grashoff—Rankine method.
2. Pigeaud's method.
3. Marcus's method, and
4. I.S. Code method.

The two way slab, simply supported on the four edges with corners not held down can be analysed either by the first method or by the fourth method. The other cases of two-way slabs can be analysed by second, third or fourth method. For a detailed discussion, reference may be made to chapter 10. We will discuss here only the fourth method (*i.e.* I.S. Code Method).

Simply supported slab with corners free to lift (I.S. Code Method).

When simply supported slab does not have adequate provision to resist torsion at corners and to prevent the corners from lifting, the maximum moments per unit width are given by :

$$M_x = \alpha_x w l_x^{\ 2} \quad ...(30.1)$$

and

$$M_y = \alpha_y w l_x^{\ 2} \quad ...(30.2)$$

where α_x and α_y are moment coefficients given in Table 30.1.

TABE 30.1 VALUES OF α_x AND α_y

l_y / l_x	*1.0*	*1.1*	*1.2*	*1.3*	*1.4*	*1.5*	*1.75*	*2.0*	*2.5*	*3.0*
α_x	0.062	0.074	0.084	0.093	0.099	0.104	0.113	0.118	0.122	0.124
α_y	0.062	0.061	0.059	0.055	0.051	0.046	0.037	0.029	0.020	0.014

At least 50 percent of tension reinforcement provided at midspan should extend to the supports. The remaining 50 percent should extend to within 0.1 l_x or 0.1 l_y of the support, as appropriate.

When once the maximum B.M. is found, the rest of the design procedure is the same as given in § 28.4.

The shear force transferred to walls or supporting beams is computed as marked in Fig. 10.3. Thus, the average S.F., per unit width, along the short edge is equal to $\frac{w\,l_x}{4}$ with a maximum value equal to $\frac{w\,l_x}{3}$. The S.F., per unit width, along long edge is taken as $w\,l_x \frac{r}{2+r}$ wher e $r = \frac{l_y}{l_x}$ rati o, subject to a maximum of $0.5 w\,l_x$ when r exceeds 2.

Example 30.1. Design of simply supported two way slab.

Design a R.C. slab for a room measuring 4 m × 5 m from inside. The slab carries a live load of 2000 N/m², and is finished with 20 mm thick granolithic finishing, having unit wt. of 24 kN/m³. Use M 15 concrete and Fe 250 steel. The slab is simply supported at all the four edges, with corners free to lift. Take the width of supporting walls as 300 mm.

Solution

Step 1. Design constants and limiting depth of N.A.

For Fe 250 steel, $f_y = 250\ \text{N/mm}^2$

For M 15 concrete, $f_{ck} = 15\ \text{N/mm}^2$.

$$\frac{x_{u.max}}{d} \text{ (for Fe 250 steel) } = 0.53$$

$$R_u = 0.36 f_{ck} \frac{x_{u.max}}{d}\left(1 - 0.416 \frac{x_{u.max}}{d}\right)$$

$$= 0.36 \times 15 \times 0.53\,(1 - 0.416 \times 0.53) = 2.23.$$

Step 2. Minimum depth.

From the point of view of limit state of serviceability, we have for a simply supported slab

$$\frac{L}{d} = 20$$

where $L = l_x = 4 + 0.1 = 4.1\ \text{m (say)}$

Assuming an under-reinforced section, with $p = 0.6\%$, modification factor = 1.75 from Fig. 7.1.

$$\therefore \qquad d = \frac{l_x}{20 \times 1.75} = \frac{4100}{20 \times 1.75} = 117\ \text{mm}$$

This is too large for a two way slab. It may be noted that the same *d* will be required for a one way slab of 4 m span. Again, in para 23.1 of the Code, for a two way slab of span upto 3.5 m, *L*/*d* ratio for simply supported slab has been given as 35. For spans greater than 3.5 m of a two way slab, Code has not given separate recommendations for *L*/*d* ratio. The commentary on the Code recommends to use *L*/*d* ratio(for two way slab of span greater than 3.5 m) the same values as given in para 22.21 of the Code. This recommendation is quite strange because in two way slabs the deflections are small because of load sharing in two directions.

However, take $D = 135$ mm
Available $d = 135 - 15 - 4 = 116$ mm.

Step 3. Computation of design B.M.

$$l_x = 4 + 0.116 = 4.116 \text{ m}$$

$$l_y = 5 + 0.116 = 5.116 \text{ m}$$

Ratio $$r = \frac{l_y}{l_x} = \frac{5.116}{4.116} = 1.24$$

Self weight of slab $= 0.135 \times 25000 = 3375$ N/m^2
Wt. of flooring $= 0.02 \times 24000 = 480$ N/m^2
Live load @ 2000 N/m^2 $= 2000$ N/m^2

Total $w = 5855$ N/m^2

$$\therefore \quad w_u = 1.5\,w = 1.5 \times 5855 \approx 8783 \text{ N/m}^2$$

From Table 30.1

$$\alpha_x = 0.086$$

$$\alpha_y \approx 0.058$$

$$w_u l_x^2 = 8783\,(4.116)^2 = 148797$$

$$\therefore \quad M_x = M_{ux} = \alpha_x\, w_u\, l_x^2 = 0.086 \times 148797$$

$$= 12796 \text{ N-m} = 12.796 \times 10^6 \text{ N-mm}$$

$$M_y = M_{uy} = \alpha_y\, w_u\, l_x^2 = 0.058 \times 148797$$

$$= 8630 \text{ N-m} = 8.63 \times 10^6 \text{ N-mm}$$

Step 4. Computation of d from bending

$$\therefore \quad d = \sqrt{\frac{M_{ux}}{R_u\, b}} = \sqrt{\frac{12.796 \times 10^6}{2.23 \times 1000}} = 75.8 \text{ mm}$$

Using 8 mm dia. bars, available $d = 135 - 15 - 4 = 116$ mm
Hence we will have an under-reinforced section , as expected.
Depth in the other direction $= 116 - 8 = 108$ mm.

Step 5. Steel reinforcement

For an under-reinforced section ,

$$A_{st} = \frac{0.5\, f_{ck}}{f_y}\left[1 - \sqrt{1 - \frac{4.6\, M_u}{f_{ck}\, bd^2}}\right] bd \qquad ...(24.13)$$

$$\therefore \quad A_{stx} = \frac{0.5 \times 15}{250}\left[1 - \sqrt{1 - \frac{4.6 \times 12.796 \times 10^6}{15 \times 1000 (116)^2}}\right] \times 1000 \times 116$$

$$= 551 \text{ mm}^2$$

and $$A_{sty} = \frac{0.5 \times 15}{250}\left[1 - \sqrt{1 - \frac{4.6 \times 8.63 \times 10^6}{15 \times 10000(108)^2}}\right] \times 1000 \times 108$$

$$= 391 \text{ mm}^2$$

$$(A_t)_{min} = \frac{0.15}{100} \times 1000 \times 135 = 202.5 \text{ mm}^2.$$

The spacing of 8 mm dia. bars in the two directions are :

$$s_x = \frac{1000 A_\Phi}{A_{stx}} = \frac{1000 \times 50.3}{551} = 91.3 \text{ mm}$$

$$s_y = \frac{1000 \times 50.3}{391} = 128.6 \text{ mm}.$$

Hence provide 8 mm dia. bars @ 90 mm c/c in the short span direction. Discontinue half the bars at a distance of $0.1\, l_x = 0.1 \times 4116 = 411.6$ mm from the centre of support, or at a distance of $411.6 - 116/2 \approx 350$ mm from the face of the support, and continue remaining bars into the support.

Similarly, in the other direction, provide 8 mm dia. bars @ 125 mm c/c. Discontinue half the bars at a distance of $0.1\, l_y = 0.1 \times 5116 = 511.6$ mm from the centre of support, or at a distance of $511.6 - 116/2 \approx 450$ mm from the face of the support, and continue the remaining bars into the support.

Step 6. Check for shear

The critical section for shear is at a distance of d (= 116 mm, from the face of the support. However, because of complications in computing the exact shear force at this section in the case of two way slab, it is usual to provide this check at the centre of the support.

$$V_{uB} = \frac{1}{3} w_u\, l_x = \frac{1}{3} \times 8783 \times 4.116 = 12050 \text{ N}$$

Similarly,

$$V_{uL} = w_u\, l_x \frac{r}{2 + r} = 8783 \times 4.116 \times \frac{1.24}{2 + 1.24}$$

$$= 13836 \text{ N}$$

(Subject to a max. $= 0.5\, w_u\, l_x = 0.5 \times 8783 \times 4.116 = 18075$)

Thus, along the short edge, where $d = 108$ mm, we have

$$\tau_{VB} = \frac{V_{uB}}{bd} = \frac{12050}{1000 \times 108} = 0.11 \text{ N/mm}^2$$

and along the long edge, where $d = 116$ mm,

$$\tau_{VL} = \frac{V_{uL}}{bd} = \frac{13836}{1000 \times 116} \approx 0.12 \text{ N/mm}^2$$

Both of these are much less than the permissible of $k \cdot \tau_c$.

Step 7. Check for development length and anchorage at ends

$$L_d = \frac{0.87 f_y}{4 \tau_{bd}} \Phi = \frac{0.87 \times 250}{4 \times 1.0} \Phi = 54.4 \Phi$$

$$= 54.4 \times 8 = 435 \text{ mm.}$$

The Code requires that half the reinforcement should extend by a length equal to $L_d/3 = 435/3 = 145$ mm. beyond the face of the support. This suggests that width of support should not be less than $145 + 25 = 170$ mm. In the present case, the support width is 300 mm and hence the anchorage condition is satisfied.

Also, at the simple support like this, where compressive reaction confines the ends of the bars,

$$1.3 \frac{M_1}{V} + L_0 \geq L_d$$

Here $V = V_u = 13836$ N (along long edge)

$$A_{st1} = \frac{1000 \times 50.3}{180} = 279.4 \text{ mm}^2.$$

$$x_u = \frac{0.87 f_y A_{st1}}{0.36 f_{ck} b} = \frac{0.87 \times 250 \times 279.4}{0.36 \times 15 \times 1000}$$

$$= 11.3 \text{ mm}$$

$$\therefore \quad M_1 = 0.87 f_y A_{st1} (d - 0.42 x_u)$$

$$= 0.87 \times 250 \times 279.4 (116 - 11.3)$$

$$= 6.36 \times 10^6 \text{ N-mm}$$

Providing *U*-hook and a clear side cover of 25 mm,

$$L_0 = \frac{l_s}{2} - x' + 13 \Phi = \frac{300}{2} - 25 + 13 \times 8 = 229 \text{ mm}$$

$$\therefore \quad 1.3 \frac{M_1}{V} + L_0 = 1.3 \times \frac{6.36 \times 10^6}{13836} + 229$$

$$= 826 \text{ mm}$$

This is greater than $L_d = 435$ mm

Hence the Code requirements are satisfied.

Step 8. Check for deflection

At the mid-span,

$$\frac{100\,A_{st}}{bd} = \frac{100\,(2 \times 279.4)}{1000 \times 116} = 0.48\ \%$$

Hence from Fig. 7.1, modification factor $\approx$ 1.9

$$\therefore \qquad \frac{L}{d} = 20 \times 1.9$$

or
$$d = \frac{4116}{20 \times 1.9} = 108 \text{ mm.}$$

Available $d = 116$ mm. Hence safe.

Step 9. Details of reinforcement. Shown in Fig. 30.1.

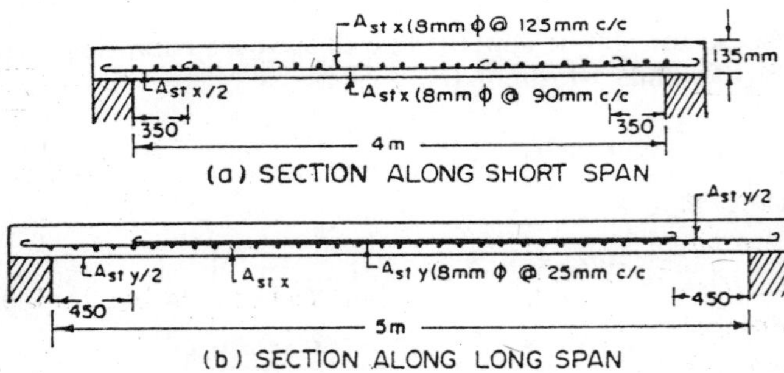

FIG. 30.1.

30.2. DESIGN OF STAIR CASES

Various types of stair cases, commonly used in buildings have been described in § 8.1. The general notes on the design of stairs are given in § 8.2 and the same notes apply to the limit state design method also.

Example 30.2. Design of dog-legged stair

Design a dog legged stair for a building in which the vertical distance between the floors is 3.6 m. The stair hall measures 2.5 m × 5m. The live load may be taken as 2500 N/m². Use M 15 concrete and Fe 250 reinforcement.

Solution

Step 1. General Arrangement

Fig. 8.6 shows the plan of stair hall. Let the rise be 150 mm and tread be 250 mm. Let us keep width of each flight = 1.2 m.

Height of each flight $= \frac{3.6}{2} = 1.8$ m.

No. of risers required $= \frac{1.8}{0.15} = 12$ in each flight

No. of tread in each flight $= 12 - 1 = 11$.

Space occupied by treads $= 11 \times 25 = 275$ cm.

Keep width of landing equal to 1.25 m. Hence space left for passage $= 5.0 - 1.25 - 2.75 = 1$ m.

Step 2. Design constants and limiting depth of N.A.

For Fe 250 steel, $f_y = 250$ N/mm^2

For M 15 concrete, $f_{ck} = 15$ N/mm^2

$$\therefore \quad \frac{x_{u.max}}{d} \text{ (for Fe 250 steel) } = 0.53$$

$$R_u = 0.36 f_{ck} \frac{x_{u.max}}{d}\left(1 - 0.416\frac{x_{u.max}}{d}\right)$$

$$= 0.36 \times 15 \times 0.53\,(1 - 0.416 \times 0.53) = 2.23$$

Step 3. Loading on each flight

The landing slab is assumed to span in the same direction as the stairs, and is considered as acting together to form a single slab. Let the bearing of the landing slab in the wall be 160 mm.

Then effective span $= 2.75 + 1.25 + \frac{0.160}{2} = 4.08 \approx 4.1$ m

From the point of view of limiting state for serviceability. $\frac{L}{d} = 20$ for a simply supported slab.

Assuming $p = 0.8\,\%$ for an *under-reinforced* section, we get a modification factor ≈ 1.58 from Fig. 7.1.

$$\therefore \quad d = \frac{L}{20 \times 1.58} = \frac{4100}{20 \times 1.58} \approx 130 \text{ mm}$$

Using 12 mm dia bars, $D = 130 + 15 + 6 = 151$ mm

However, keep $D = 150$ mm, so that $d = 150 - 15 - 6 = 129$ mm.

$\therefore$ Weight of slab w' on slope

$$= \frac{150}{1000} \times 1 \times 25000 = 3750 \text{ N/m}^2$$

$\therefore$ Dead weight on horizontal area

$$= w_1 = w' \frac{\sqrt{R^2 + T^2}}{T}$$

$$= 3750 \times \sqrt{\frac{(150)^2 + (250)^2}{250}} = 3750 \times 1.166$$

$$= 4372 \text{ N/m}^2$$

Dead weight of step is given by

$$w_2 = \frac{R}{2 \times 1000} \times 1 \times 1 \times 25000 = \frac{150}{2000} \times 25000$$

$$= 1875 \text{ N/m}^2$$

Weight of finishing etc. = 100 N (assumed)

Live load = 2500 N

∴ Total $w = 8847$ N/m

Hence $w_u = 1.5\,w = 1.5 \times 8847 = 13270$ N

Note : The value of w on the landing portion will be $8847 - 1875 = 6972$ N, since weight of steps will not come on it. However, an uniform value of w has been adopted here.

Step 4. Computation of design B.M. and S.F. and depth of waist slab

$$M_u = \frac{w_u L^2}{8} = \frac{13270\,(4.1)^2}{8} = 27885 \text{ N-m}$$

$$= 27.885 \times 10^6 \text{ N-mm}$$

$$V_u = \frac{w_u L}{2} = \frac{13270\,(4.1)}{2} = 27204 \text{ N}$$

$$\therefore \quad d = \sqrt{\frac{M_u}{R_u b}} = \sqrt{\frac{27.885 \times 10^6}{2.23 \times 1000}} = 111.8 \text{ mm.}$$

Available $d = 129$ mm.

Hence safe. However, the section will be slightly under-reinforced.

Step 5. Steel reinforcement

For an under-reinforced section ,

$$A_{st} = \frac{0.5 f_{ck}}{f_y}\left[1 - \sqrt{1 - \frac{4.6\,M_u}{f_{ck}\,bd^2}}\right] bd \qquad ...(24.13)$$

$$= \frac{0.5 \times 15}{250}\left[1 - \sqrt{1 - \frac{4.6 \times 27.885 \times 10^6}{15 \times 1000 (129)^2}}\right] \times 1000 (129)$$

$$= 1172 \text{ mm}^2$$

Using 12 mm Φ bars,

$$A_\Phi = \frac{\pi}{4}(12)^2 = 114 \text{ mm}^2$$

∴ No. of bars required in 1.2 m width

$$= \frac{1.2 \times 1172}{114} \approx 13 \text{ (say)}$$

∴ Spacing of bars $= \frac{1200}{13} = 92.3$ mm.

Actual A_{st} provided $= \frac{13 \times 114}{1.2} = 1235 \text{ mm}^2.$

Distribution reinforcement @ 0.15% of area :

$$A_{sd} = \frac{0.15 \times 150 \times 1000}{100} = 225 \text{ mm}^2$$

∴ s-8 Φ bars $= \frac{1000 \times 50.3}{225} = 224$ mm

Hence provide 8 mm Φ bars @ 220 mm c/c.

The main reinforcement should be bent to follow the bottom profile of the stair. However, if this pattern is followed near the landing, an angle will be formed in the bar. When the bar is stressed, it will try to throw off the concrete cover. Hence near the landing,

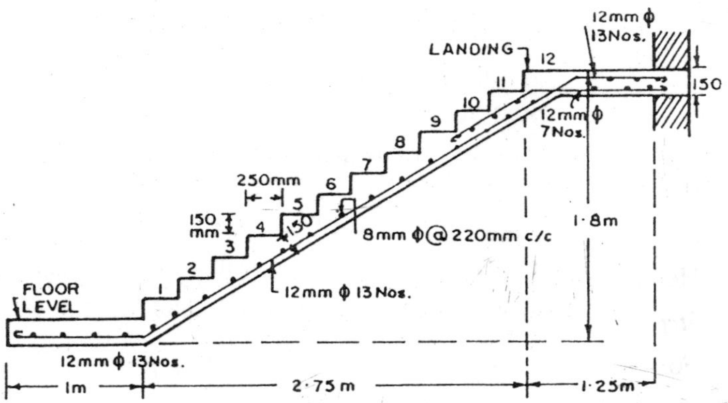

FIG. 30.2.

the bars are taken straight up and then bent in the compression zone of landing. In order to take tensile stresses in the landing portion, it is desirable to use separate set of bars as shown in Fig. 30.2. However, since the bending moment is very much reduced near the landing, only half the number of bars may be provided, *i.e.*, provide 7 nos. of 12 mm Φ bars. The length of each type of bar on either side of the crossing should be at least equal to

$$L_d = 54.4 \times 12 = 660 \text{ mm}$$

The details of reinforcement etc. are shown in Fig. 30.2.

Since all the bars of tensile reinforcement are taken into the support, the anchorage and development length requirements will be automatically satisfied.

30.3. ISOLATED FOOTING OF UNIFORM DEPTH

The Indian Standard Code recommendations for design of footings have been given in § 15.4. The recommendations hold good for design of footing by limit state method.

Example 30.3. Design of isolated square footing of uniform thickness for a R.C. column.

Design an isolated footing of uniform thickness of a R.C. column having a vertical load of 600 kN and having a base size 500 mm × 500 mm. The safe bearing capacity of soil may be taken as 120 kN/m². Use M 15 concrete and Fe 250 steel.

Solution.

Step 1. Size of footing

$$W = 600 \text{ kN}$$

Let W' be equal to 10% $W = 60$ kN.

$$\therefore \quad A = \frac{W + W'}{q} = \frac{660}{120} = 5.5 \text{ m}^2$$

$\therefore$ Size B of square $= \sqrt{5.5} = 2.345$ m.

Provide a square footing of size 2.4 m × 2.4 m.

Net upward pressure, under a factored load of 1.5 W is

$$p_{0u} = \frac{W_u}{B \times B} = \frac{1.5 \times 600}{2.4 \times 2.4} = 156.25 \text{ kN/m}^2.$$

Step 2. Design constants and limiting depth of N.A.

For Fe 250 steel, $f_y = 250 \text{ N/mm}^2$

For M 15 concrete, $f_{ck} = 15 \text{ N/mm}^2$

$\frac{x_{u,max}}{d}$ (for Fe 250 steel) $= 0.53$

$$\therefore \quad R_u = 0.36 f_{ck} \frac{x_{u.max}}{d}\left(1 - 0.416 \frac{x_{u.max}}{d}\right)$$
$$= 0.36 \times 15 \times 0.53\,(1 - 0.416 \times 0.53) = 2.23$$

Step 3. Effective depth from bending point of view

The maximum bending moment acts at the face of the column and it magnitude is given by Eq. 15.18.

$$M_u = p_{0u} \frac{B}{8}(B - b)^2 \times 10^6 \text{ N-mm}$$
$$= 156.25 \times \frac{2.4}{8}(2.4 - 0.5)^2 \times 10^6$$
$$= 169.22 \times 10^6 \text{ N-mm}$$
$$\therefore \quad d = \sqrt{\frac{M_u}{R_u B}} = \sqrt{\frac{169.22 \times 10^6}{2.23 \times 2400}} = 178 \text{ mm}$$

Before finding the reinforcement, let us check whether this thickenss is adequate from shear point of view or not.

Step 4. Check for shear

For *one way shear*, the critical section *Y-Y* is located at a distance d (= 178 mm) from the face of the column (Fig. 15.13) where shear force V_u is given by

$$V_u = p_{0u} B\left[\frac{1}{2}(B - b) - d\right]$$
$$= 156.25 \times 2.4\left[\frac{1}{2}(2.4 - 0.5) - 0.178\right]$$
$$= 289.5 \text{ kN}$$
$$\therefore \quad \tau_v = \frac{V_u}{B\,d} = \frac{289.5 \times 1000}{2400 \times 178} = 0.68 \text{ N/mm}^2$$

Using 12 mm Φ bars, and 60 mm clear cover, total depth will be about $178 + 6 + 60 = 244$ mm ≈ 250 mm(say). Hence $k = 1.10$ from Table 27.3. Also, for a balanced section, having $p = 1.316\,\%$, $\tau_c = 0.65$ N/mm². However, the section will be mostly *under reinforced* and hence p will be much less than this value. Keeping $p = 0.5\,\%$, $\tau_c = 0.46$ N/mm² (Table 27.1).

$$\therefore \quad k\,\tau_c = 0.46 \times 1.1 = 0.51 \text{ N/mm}^2$$

Since $\tau_v > k\,\tau_c$, the section is *unsafe.*

Also, for two way shear(punching shear), the section lies at $d/2$ from the face of the column round. The width b_0 of the section $= b + d = 500 + 178 = 678$ mm. The S.F. around the section is

$$F_u = p_{0u}\,[B^2 - b_0^2]$$
$$= 156.25\,[(2.4)^2 - (0.678)^2] = 828.2 \text{ kN}$$

$$\therefore \quad \tau_v = \frac{F_u}{4b_0 d} = \frac{828.2 \times 10^3}{4 \times 678 \times 178} = 1.716 \text{ N/mm}^2$$

Permissible shear stress $\tau_c' = k_s \tau_c$

where $k_s = (0.5 + \beta_c) = (0.5 + 1)$, with a max. value of 1

$\therefore \quad k_s = 1$

Also, for limit state method, $\tau_c = 0.25\sqrt{f_{ck}}$

$\therefore \quad \tau_c = 0.25\sqrt{15} = 0.968 \text{ N/mm}^2$

$\therefore \quad \tau_c' = \tau_c = 0.968 \text{ N/mm}^2$

Since $\tau_v > \tau_c'$, the section is not safe from two way shear also.

Let us, therefore, find a fresh value of d from *two way shear* point of view first, and check it for one way shear.

Keeping $\tau_v = \tau_c' = 0.968$, we have

$$\tau_v = \frac{F_u}{4\,b_0\,d}$$

where $b_0 = b + d = (500 + d)$

$$\therefore \; 0.968 = \frac{828.2 \times 10^3}{4\,(500 + d)\,d}$$

or $d^2 + 500\,d - 213895 = 0$

From which $d = 276$ mm

$\therefore \quad D = 276 + 6 + 60 = 342$ mm

However, keep D=350 mm, so that d=50−6−60=284 mm.

Let us check this depth for *one way shear.*

$$\tau_v = \frac{V_u}{Bd} = \frac{289.5 \times 1000}{2400 \times 284} = 0.425 \text{ N/mm}^2$$

Since the total thickness is more than 300 mm, $k = 1$. Hence from Table 27.1, even for $p = 0.5\,\%$, $\tau_c = 0.46 \text{ N/mm}^2$ which is more than τ_v. Thus $d = 284$ mm (and hence $D = 350$ mm) will be adequate.

Step 5. Steel Reinforcement

Effective depth in one direction = 284 mm.

Effective depth in other direction = 284 − 12 = 272 mm

Since the effective depth found from shear point of view is much more than the depth required for a balanced section, the section will be *under-reinforced.*

For an under-reinforced section , the area of steel is given by Eq. 24.13

$$A_{st} = \frac{0.5 f_{ck}}{f_y}\left[1 - \sqrt{1 - \frac{4.6\, M_u}{f_{ck}\, Bd^2}}\right] Bd$$

$$= \frac{0.5 \times 15}{250}\left[1 - \sqrt{1 - \frac{4.6 \times 169.22 \times 10^6}{15 \times 2400(272)^2}}\right] 2400 \times 272$$

$$= 3109 \text{ mm}^2$$

Using 12 mm Φ bars, $A_{\Phi} = \frac{\pi}{4}\,(12)^2 = 113 \text{ mm}^2$

∴ No. of bars $= \frac{3109}{113} = 27.5$

Hence provide 28 bars of 12 mm Φ uniformly distributed in each direction.

Step 6. Check for development length

$$L_d = 54.4\,\Phi = 54.4 \times 12 = 652.8 \text{ mm}$$

Providing 60 mm side cover, length of bars available

$$= \tfrac{1}{2}(B - b) - 60 = \tfrac{1}{2}(2400 - 500) - 60 = 890 \text{ mm} > L_d$$

Hence safe.

Step 7. Trasfer of load at column base

$$A^2 = 500 \times 500 = 250000 \text{ mm}^2.$$

$$A_1 = [500 + 2\,(2 \times 350)]^2 \text{ for side spread of } 2 : 1$$

$$= (1900)^2 = 3610000 \text{ mm}^2.$$

$$\therefore \quad \sqrt{\frac{A_1}{A_2}} = \sqrt{\frac{3610000}{250000}} = 3.8$$

Adopt max. value of $\sqrt{\frac{A_1}{A_2}} = 2.$

Hence permissible bearing stress $= 0.45 f_{ck} \sqrt{\frac{A_1}{A_2}}$

$$= 0.45 \times 15 \times 2 = 13.5 \text{ N/mm}^2$$

Actual bearing pressure $= \frac{1.5 \times 600000}{500 \times 500} = 3.6 \text{ N/mm}^2$

Hence safe. Thus, no separate dowel bars are required for the transfer of the load. However, it is advisable to continue all the bars of the column into the foundation.

Step 8. Details of Reinforcement. Shown in Fig. 30.3.

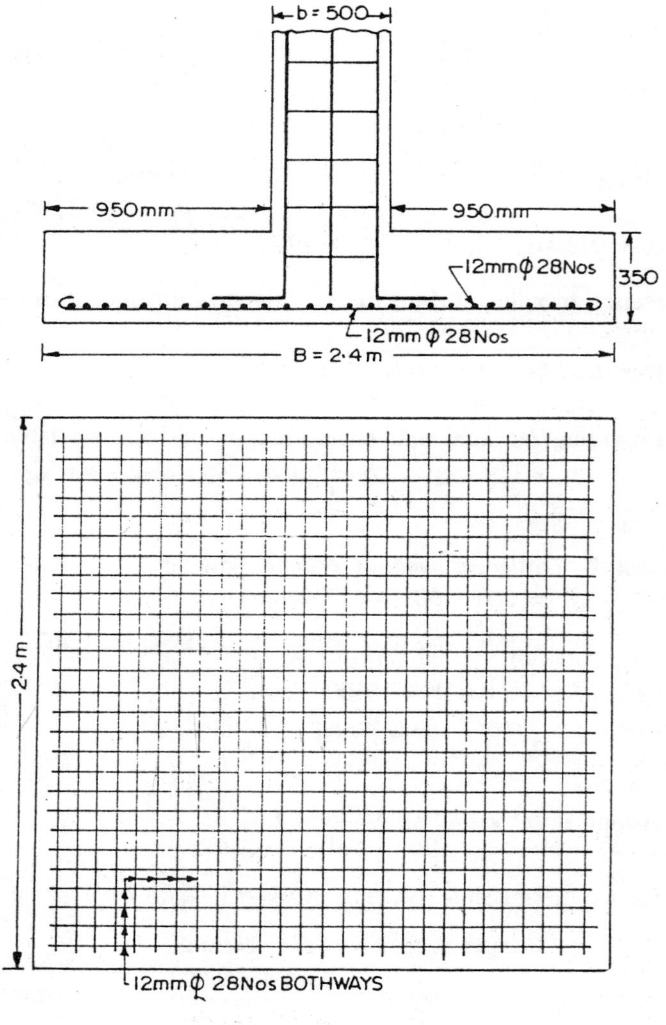

FIG. 30.3.

31

Yield Line Theory and Design of Slabs

31.1. INTRODUCTION

The *yield line theory* is one of the most important development in the analysis and design of slab systems. The yield line theory is the ultimate load theory for the design of R.C. slabs. In the case of slabs, the computation of ultimate load is quite complicated. In the yield line method, the computation of ultimate load is based on the pattern of *yield lines* that are developed in the slabs under conditions approaching collapse. The method was innovated by Ingerslav (1923) and was greatly extended and advanced by Johanssen. The most commonly used limit state or collapse method is based on Johanssen's yield line theory.

When a slab is loaded with increasing loads, the stresses in the reinforcement and the concrete increase more or less proportionately upto load level corresponding to the yield stress in the reinforcement. If the load is increased further, excessive deformations and rapid increase in strains will result. These deformations will be *elastoplastic* upto a load level called *limit load*. When limit load is reached, the slab will continue to deform without any additional load, leading to total collapse. At this stage, a pattern of cracks will form a set of lines known as *yield lines*, resulting into a mechanism leading to the total collapse of the slab. A *yield line* is defined as a line in the plane of slab across which reinforcing bars have yielded and about which excessive deformation (plastic rotations) under constant limit moment (ultimate moment) continues to occur leading to failure.

31.2. YIELD LINE PATTERNS

(a) One Way Slabs

The behaviour of one way slab regarding its ultimate moment capacity is the same as that of beam of unit width. Consider a one way slab of unit width, simply supported at the ends, and loaded with uniformly distributed load [Fig. 31.1 (*a*)]. The maximum positive B.M. will be at the mid-span of the slab, its value being equal to $+\frac{wL^2}{8}$. Similarly, Fig. 31.1(*b*) shows a one way slab of unit width fixed at the ends. The maximum negative B.M. will be $-\frac{wL^2}{12}$ at the ends and maximum mid-span positive B.M. will be $+\frac{wL^2}{24}$.

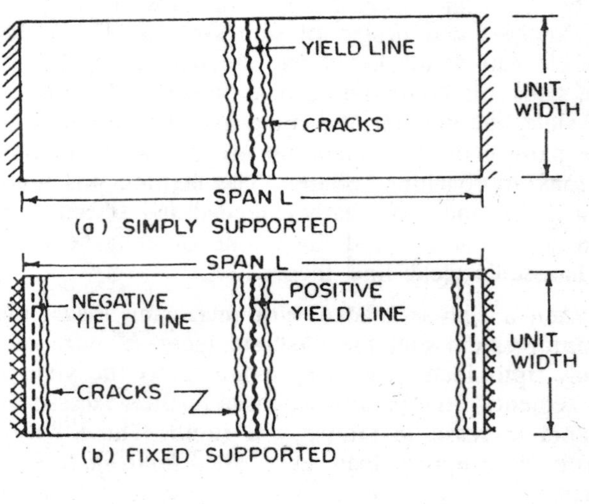

FIG. 31.1. YIELD LINES IN ONE WAY SLABS.

At service loads, the slab will be uncracked, but as the load is increased, cracks will be developed at the top of slab in negative moment regions, and at bottom in the positive moment regions. As the load is further increased, the negative and/or positive steel will start yielding. At the yield load w_y, the entire width of the slab will develop cracks, and *yield lines* will be formed. In the case of simply supported slab, such an yield line will be formed at mid-span, while for slab with fixed supports, *negative yield line* will be formed at the supports and *positive yield line* will be formed at the mid-span. At the yield lines, plastification takes place, creating *'plastic hinges'*. These plastic hinges withstand plastic moments. It is to be noted that the direction of yield line is perpendicular to the direction of reinforcement.

(*b*) Two Way Slabs

In one way slab, the direction of yield line is perpendicular to the direction of steel. But in two way slab (square or rectangular slabs) the direction of yield lines are not perpendicular to the two directions of reinforcement.

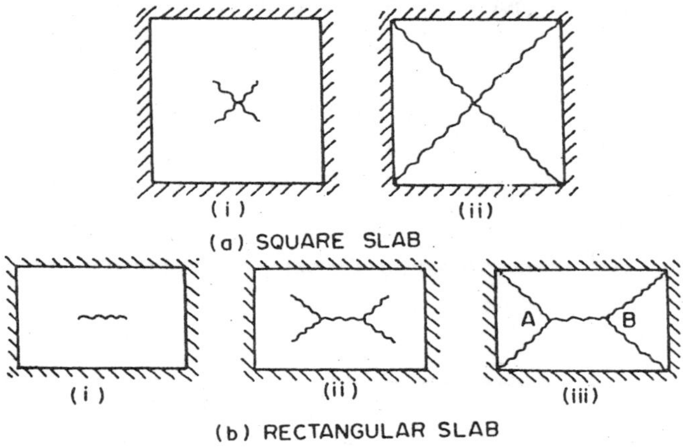

FIG. 31.2. INITIATION, GROWTH AND COMPLETION OF YIELD LINES IN SIMPLY SUPPORTED TWO WAY SLABS.

Fig. 31.2 (*a*) shows the initiation, growth and development of yield lines in square slabs. The cracks first develop along with small patches yield of diagonals at mid-span. These cracks spread towards the corners, till yield lines are fully developed. Similarly, in a rectangular slabs, [Fig. 31.2 (*b*)], the initial crack is developed in a direction perpendi- cular to the short-span, at the mid-region of the slabs. Further increase in the load results in continuous growth of cracks as shown in Fig. 31.2 (*b*) (*ii*) and (*iii*). It is seen that except for the length *AB,* the cracks are not transverse to the directions of reinforcements.

Characteristic Features of Yield Lines

The following characteristics are useful in selecting the yield line patterns for slab with various boundary conditions.

1. Yield lines are *straight lines* so that they may act as plastic hinges of a collapse mechanism.

2. Yield lines terminate at a slab boundary or at the intersection of other yield lines.

3. Yield lines act as axes of rotation for the movements of adjoining segments.

4. Each of the segments of the slab will tend to rotate in a rigid body motion. The axes of rotation generally lie along the lines of supports and pass over the columns.

5. If an edge is fixed or continuous, a yield line may form along the support.

6. For mechanism to develop (*i.e.* for compatibility of deformations) yield lines or yield lines produced, pass through the intersection of the axes of rotation of adjacent slab elements.

Sign Conventions for Yield Line Patterns and Supports

Fig. 31.3 Shows generally accepted sign conventions.

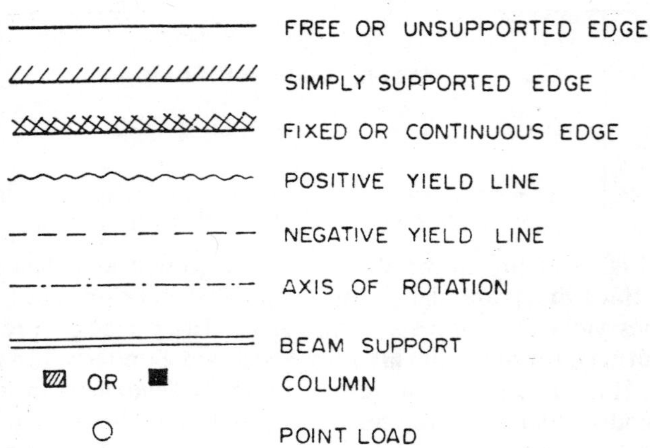

FIG. 31.3. SIGN CONVENTIONS.

Yield Line Patterns for Slabs

Fig. 31.4 shows the yield line patterns for slabs of various shapes, with various boundary conditions.

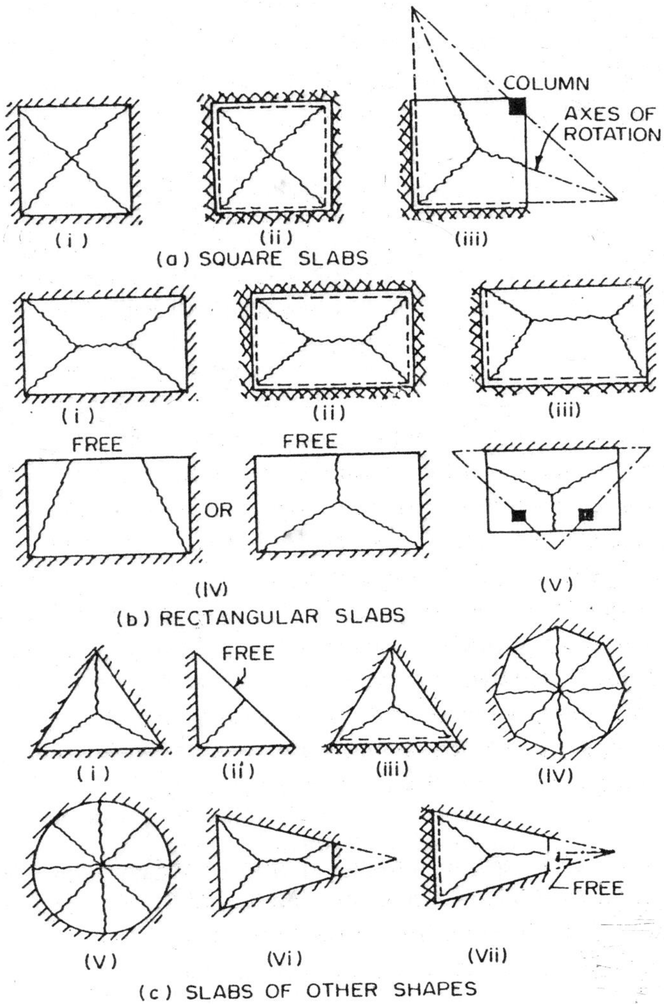

FIG. 31.4. TYPICAL YIELD LINE PATTERNS IN R.C. SLABS.

31.3. MOMENT CAPACITY ALONG AN YIELD LINE

In the case of one way slab, the yield line is perpendicular to the direction of reinforcement (Fig. 31.1). However, in the case of rectangular slabs having two-way reinforcement, the direction of the yield line may be inclined to the principal directions of reinforcement. Fig. 31.5 shows an inclined yield line *AB* in orthogonally reinforced slab. The normal to the yield line makes an angle α with the *x*-direction, *i.e.* the yield line makes an angle α with the *y*-direction

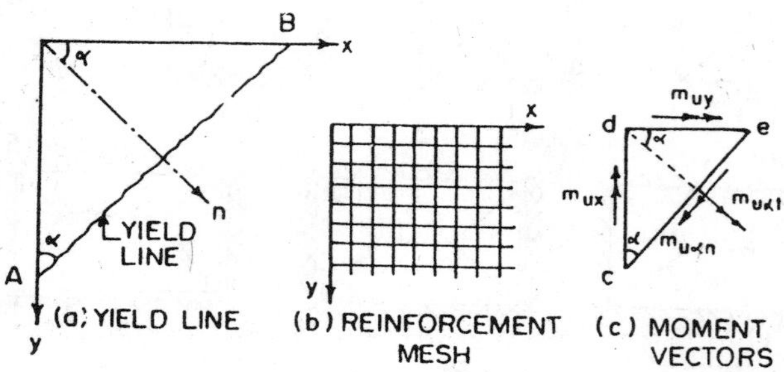

FIG. 31.5. INCLINED YIELD LINE IN ORTHOGONALLY REINFORCED SLAB.

Let m_{ux} = ultimate moment of resistance per unit length across a plane whose normal is in *x*-direction.

m_{uy} = ultimate moment of resistance per unit length, across a plane whose normal is in *y*-direction.

$m_{u\alpha n}$ = ultimate moment of resistance per unit length, across a plane whose normal makes an angle α with *x*-direction.

$m_{u\alpha t}$ = ultimate torsional moment of resistance per unit length of yield line.

Fig. 31.5 (*c*) shows the moment vectors. Taking moment about the side *ce* (parallel to the yield line),

$$m_{u\alpha n}\,(ce) = m_{ux}\,(cd)\cos\alpha + m_{uy}\,(de)\sin\alpha$$

or $$m_{u\alpha n} = m_{ux}\left(\frac{cd}{ce}\right)\cos\alpha + m_{uy}\left(\frac{de}{ce}\right)\sin\alpha$$

or $$m_{u\alpha n} = m_{ux}\cos^2\alpha + m_{uy}\sin^2\alpha \qquad ...(31.1)$$

If the slab is *isotropically reinforced* (*i.e.* equal reinforcement in the two directions), $m_{ux} = m_{uy} = m_u$

$$\therefore \quad m_{uan} = m_u(\cos^2\alpha + \sin^2\alpha) = m_u \qquad ...(31.2)$$

Thus the ultimate moment of resistance in an isotropically reinforced slab, in any direction, is the same.

If the reinforcement in the two directions is not the same, it is said to be *orthotropically reinforced*. In such a case, let $m_{ux} = m_u$ and $m_{uy} = \mu m_u$. Then from Eq 31.2, we get

$$m_{uan} = m_u \cos^2\alpha + \mu m_u \sin^2\alpha$$

or $$m_{uan} = m_u (\cos^2\alpha + \mu \sin^2\alpha) \qquad ...(31.3)$$

If $$\alpha = 45°, \; \cos^2\alpha = \sin^2\alpha = \frac{1}{\sqrt{2}}$$

$$\therefore \quad m_{uan} = \frac{m_u(1+\mu)}{2} \qquad ...(31.4)$$

Again, considering the moment vectors of Fig. 31.5 (*c*) and taking moments about an axis perpendicular to *ce*, to get torsional moment per unit length of yield line as :

$$m_{uat} = (m_{uz} - m_{uy}) \sin\alpha \cos\alpha \qquad ...(31.5)$$

For an isotropically reinforced slab, $m_{ux} = m_{uy}$, and hence torsional moment is zero.

For an orthotropically reinforced slab,

$$m_{uat} = m_u(1-\mu) \sin\alpha \cos\alpha \qquad ...(31.6)$$

31.4. ULTIMATE LOAD ON SLABS

There are two methods of determining the ultimate load capacity of reinforced concrete slabs.

1. Virtual work method.
2. Equilibrium method.

The *virtual work method* is based on the principle that the work done by the external forces in undergoing a small virtual displacement is equal to the internal work done (or energy dissipated) in rotation along the yield lines. The elastic deformations are however ignored in comparision to the plastic deformations that take place along the yield lines. In the equilibrium method, the collapse load is calculated from the equilibrium of the individual segments of a mechanism.

Both the virtual work method as well as the equilibrium method give the upper bound solution for the collapse load, *i.e.* the computed collapse load on the basis of an assumed yield line patterns is bound to be larger than the actual collapse load. Hence it is essential to examine several yield line patterns and a minimum value of collapse load has to be obtained. Test results have shown that the actual failure load of slab is greater than the one predicted by the upper bound solution of the yield line analysis, because of membrane action. Hence the results of yield line analysis can be used in design with reasonable degree of safety.

31.5. ANALYSIS BY VIRTUAL WORK METHOD

The principle of *virtual work* states that :

'If a deformable structure in equilibrium under the action of a system of external forces is subjected to a virtual deformation compatible with its condition of support, the work done by these forces on the displacements associated with the virtual deformation is equal to the work done by the internal stresses on the strains associated with this deformation'.

Thus, the work done during small motion of collapse mechanism is equal to work absorbed by the plastic hinges formed along the yield lines. The segments of the slab within the yield lines go through rigid body displacements with the collapse load acting on the structure.

Thus work done by external forces = work done by internal forces (or energy absorbed by hinges)

or $W_E = W_I$ ...(31.7)

Where W_E = work done by external forces

W_I = Energy absorbed by plastic hinges in undergoing corresponding rotations.

If w_u is the uniformly distributed external load, we have

$$W_E = \int\int w_u \, \delta_{x,y} \, dx \, dy = \Sigma W_u \Delta \quad ...(31.8)$$

where $\delta_{x,y}$ = virtual displacement at any point.

W_u = resultant load on each segment.

Δ = corresponding displacement at the centroid of load in each segment.

Again if m_{uan} = ultimate moment across and yield line.

l_0 = length of yield line.

and θ_n = relative rotation of two adjacent plates, perpendicular to the yield line.

Then, $M_I = \Sigma m_{uan} \theta_n l_0$ ...(31.9)

Equating M_E and M_I, we get

$$\Sigma W_u \Delta = \Sigma m_{uan} \theta_n l_0 \quad ...(31.10)$$

It is to be noted that the work done by the twisting moment (m_{uat}) has not been considered in Eq. 31.9 since its effect will cancel out when the entire slab is eventually considered.

We shall now illustrate the method by considering several cases.

1. Simply Supported Square Slab

A square slab is isotropically reinforced. Hence moment of resistance m_{uan}, is the same in all the directions. Fig. 31.6 shows such a slab of span L, having length of diagonal equal to $\sqrt{2}\,L$. If δ is the virtual displacement at the middle, the total rotation of the diagonal segments $= 2\,\theta_n = 2\dfrac{\delta}{\dfrac{\sqrt{2}\,L}{2}} = \dfrac{2\sqrt{2}\,\delta}{L}$

Length of each diagonal yield line $= \sqrt{2}\,L$

Hence for the two diagonal yield lines,

$$W_I = \Sigma\, m_{uan}\,\theta_n\,l_0 = 2\left(m_u \frac{2\sqrt{2}\,\delta}{L}\sqrt{2}\,L\right) = 8\,m_u\,\delta \qquad ...(1)$$

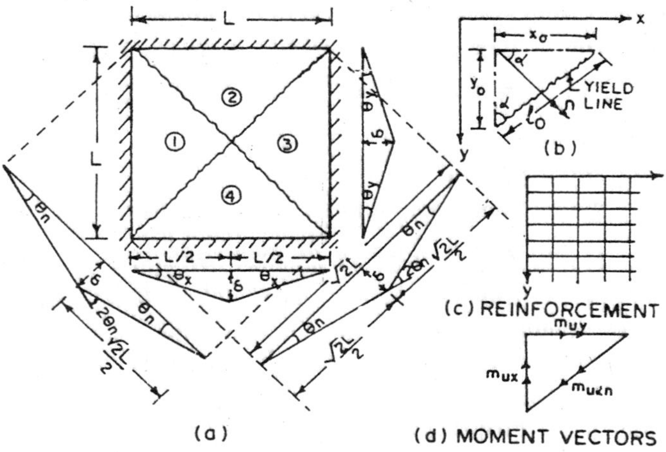

FIG. 31.6. SQUARE SLAB : FREELY SUPPORTED.

There are four segments, each segment carrying equal total load $W_u = w_u L \dfrac{L}{2} = w_u \dfrac{L^2}{4}$ acting through its centroid which undergoes a virtual displacement $\Delta = \dfrac{\delta}{3}$.

Hence $W_E = \Sigma W_u \Delta = 4\left(\frac{w_u L^2}{4}\frac{\delta}{3}\right) = w_u L^2 \frac{\delta}{3}$ ...(2)

Equating W_E and W_I, we have

$$w_u L^2 \frac{\delta}{3} = 8 m_u \delta$$

or $$w_u = \frac{24 m_u}{L^2} \quad ...(31.11)$$

or $$m_u = \frac{w_u L^2}{24} \quad ...[31.11(a)]$$

Alternative Solution

The problem can be solved alternatively by combination of components of moments and rotations in the co-ordinate directions x and y. Refer Fig. 31.6 (b).

From Eq. 31.1, we have

$$m_{uan} = m_{ux} \cos^2 \alpha + m_{uy} \sin^2 \alpha.$$

Hence $\Sigma m_{uan}\, \theta_n\, l_0 = \Sigma(m_{ux} \cos^2 \alpha + m_{uy} \sin^2 \alpha)\, \theta_n\, l_0$

or $\Sigma m_{uan}\, \theta_n\, l_0 = \Sigma m_{ux}\, y_0 \cos \alpha\, .\theta_n + \Sigma m_{uy}\, x_0 \cos \alpha\, .\theta_n$

(since $\cos \alpha = y_0/l_0$ and $\sin \alpha = x_0/l_0$)

Hence $\Sigma m_{uan}\, \theta_n\, l_0 = m_{ux}\, y_0\, \theta_x + \Sigma m_{uy}\, x_0\, \theta_y$ ...(31.12)

where θ_x = Component of θ_n in x *direction.*

θ_y = Component of θ , in y dirction.

x_0 = Projected length of yield line in x direction.

y_0 = Projected length of yield line in y direction.

It is to be noted that θ_x and θ_y are the sum of the rotations of segments adjacent to the yield line under consideration.

Substituting the value of $\Sigma m_{uan}\, \theta_n\, l_0$ in Eq 31.10, we get

$$\Sigma m_{ux}\, y_0\, \theta_x + \Sigma m_{uy}\, x_0\, \theta_y = W_u \Delta \quad ....(31.13)$$

Now, for the square slab of Fig. 31.6 (a), we have four segments. Segment 1 and 3 rotate about y-axis and there is no rotation about x-axis. Similarly, segments 2 and 4 rotate about x-axis and there is no rotation about y-axis.

Rotation $\theta_x = \theta_y = \frac{\delta}{L/2} = \frac{2\delta}{L}$

Also, $x_0 = y_0 = L$, and $m_{ux} = m_{uy} = m_u$

$\therefore$ For segment 1, $m_{ux}\, y_0\, \theta_x + m_{uy}\, x_0\, \theta_y = \left(m_u L \frac{2\delta}{L}\right) + 0$

For segment 2, $m_{ux} y_0 \theta_x + m_{uy} x_0 \theta_y = 0 + m_u L \dfrac{2\delta}{L}$

For segment 3, $m_{ux} y_0 \theta_x + m_{uy} x_0 \theta_y = m_u L \dfrac{2\delta}{L} + 0$

and for segment 4, $m_{ux} y_0 \theta_x + m_{uy} \theta_y x_0 = 0 + m_u L \dfrac{2\delta}{L}$

$$\therefore \ \Sigma m_{ux} \theta_x y_0 + \Sigma m_{uy} \theta_y x_o = 4 m_u L \frac{2\delta}{L} = 8 m_u \delta$$

which is the same as found earlier.

$$\Sigma W_u \Delta = 4 \left(\frac{w_u L^2}{4} \right) \frac{\delta}{3} = w_u L^2 \frac{\delta}{3}$$

$$\therefore \ 8 m_u \delta = w_u L^2 \frac{\delta}{3}$$

or $$m_u = \frac{w_u L^2}{24}$$

or $$w_u = \frac{24 m_u}{L^2}, \text{ as before.}$$

2. Square Slab Fixed on all Edges

Fig. 31.7 shows an isotropically reinforced square slab fixed at

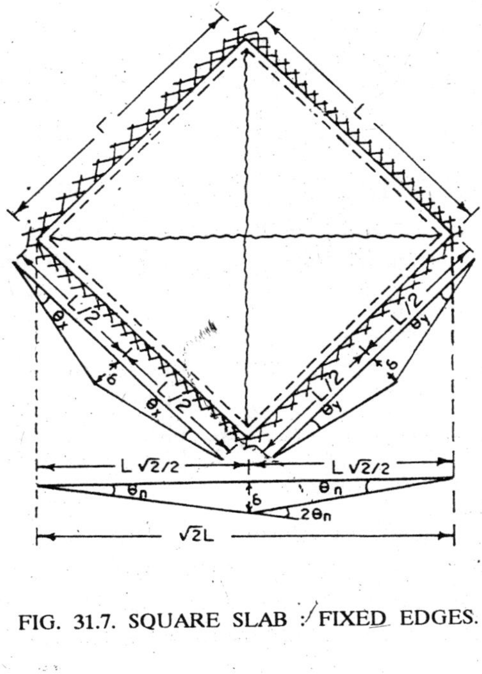

FIG. 31.7. SQUARE SLAB : FIXED EDGES.

the four supports. Due to this, negative yield lines will also be formed along the fixed edges. The rotation $\theta_x = \theta_y$ of these negative yield lines will be equal to $\frac{\delta}{L/2}$. The length of each negative yield line $= L$.

Hence internal work done by negative yield lines AB, BC, CD and DA is

$$\Sigma\, m_{u\alpha n}\, \theta_n\, l_0 \quad = 4\, m_u\, L \frac{\delta}{L/2} = 8\, m_u\, \delta \qquad ...(i)$$

Internal work done along positive yield lines along the diagonals AC and BD, as found in case 1, is

$$\Sigma\, m_{u\alpha n}\, \theta_n\, l_0 \quad = 8\, m_u\, \delta \qquad ...(ii)$$

Hence $\quad W_I = 8\, m_u\, \delta + 8\, m_u\, \delta = 16\, m_u\, \delta \qquad ...(a)$

It is to be noted that the energy absorbed by both negative yield line as well as positive yield line has positive sign.

Again, as found in the previous case,

$$W_E = \Sigma\, W_u\, \Delta = w_u\, L^2 \frac{\delta}{3} \qquad ...(b)$$

$$\text{Hence } 16\, m_u\, \delta \quad = w_u\, L^2 \frac{\delta}{3}$$

or $$m_u \quad = \frac{w_u\, L^2}{48} \qquad ...[31.14(a)]$$

or $$w_u \quad = \frac{48\, m_u}{L^2} \qquad ...(31.14)$$

3. Equilateral Triangular Slab Isotropically Reinforced

Fig. 31.8 shows isotropically reinforced triangular slab. The ultimate moment of resistance m_u will be same in all directions. When

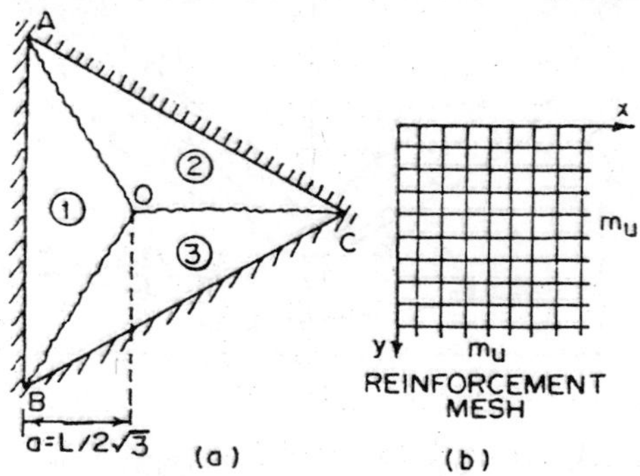

FIG. 31.8.

it is simply supported along all the three edges, the three yield lines will divide the slab in three symmetrically placed sectors. Hence we need consider only one sector. The distance of point O from edge AB will be $\dfrac{L}{2\sqrt{3}}$.

For sector 1, $W_I = m_u \theta l_0 = m_u \dfrac{\delta}{L/2\sqrt{3}.} L$

$$= 2\sqrt{3}\, m_u \delta \quad \text{...(1)}$$

Also, $W_E = \dfrac{1}{2} L \times \dfrac{L}{2\sqrt{3}} \times w_u \times \dfrac{\delta}{3} = w_u \dfrac{\delta L^2}{12\sqrt{3}}$ ...(2)

Equating the two, we get

$$\frac{w_u \delta L^2}{12\sqrt{3}} = 2\sqrt{3}\, m_u \delta$$

$$\therefore \quad w_u = \frac{72\, m_u}{L^2} \quad \text{...[31.15(a)]}$$

$$\text{or} \quad m_u = \frac{w_u L^2}{72} \quad \text{...(31.15)}$$

4. Generalised Triangular Slab

Let us now take the general case of a triangular slab, simply supported along two edges, third edge being free, and isotropically reinforced.

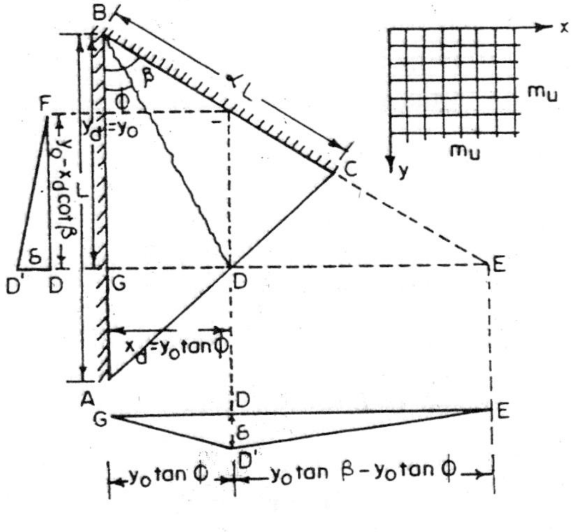

FIG. 31.9.

Let the side BA of length L be oriented along x-axis, and the other side BC be of length αL. Edge AC is free. Let $\angle ABC$ be β and the yield line BD make an angle Φ with BA (or y-axis). Let the point D, having coordinates (x_d, y_d) have a downward virtual displacement δ. Let us have proper identification of rotations. Also, $m_{ux} = m_{uy} = m_u$.

For segment AB (segment 1),

$$y_{01} = y_0$$

$$x_{01} = x_d = y_0 \tan \Phi$$

$$\theta_{x1} = \frac{\delta}{x_d} = \frac{\delta}{y_d \tan \Phi} = \frac{\delta}{y_0 \tan \Phi}$$

$$\theta_{y1} = 0$$

For segment BCD (segment 2)

$$\theta_{x2} = \frac{\delta}{DE} = \frac{\delta}{y_0 \tan \beta - y_0 \tan \Phi}$$

$$\theta_{y2} = \frac{\delta}{FD} = \frac{\delta}{y_0 - x_d \cot \beta} = \frac{\delta}{y_0 - y_0 \tan \Phi \cot \beta}$$

$$y_{02} = y_0$$

$$x_{02} = x_d = y_0 \tan \Phi$$

Now from Eq. 31.12

$$W_I = \Sigma m_{ux} y_0 \theta_x + \Sigma m_{uy} x_0 \theta_y$$

$$W_I = [m_u y_0 \theta_{x1} + m_u y_0 \theta_{x2}] + [m_u x_{01} \theta_{y1} + m_u x_{02} \theta_{y2}]$$

$$= m_u \left[y_0 \frac{\delta}{y_0 \tan \Phi} + y_0 \frac{\delta}{y_0 (\tan \beta - \tan \Phi)} \right]$$

$$+ m_u \left[y_0 \tan \Phi \times \text{zero} + y_0 \tan \Phi \frac{\delta}{y_0 (1 - \tan \Phi \cot \beta)} \right]$$

$$= m_u \left[\frac{\delta}{\tan \Phi} + \frac{\delta}{\tan \beta - \tan \Phi} + 0 + \frac{\delta \tan \Phi}{1 - \tan \Phi \cot \beta} \right]$$

$$= m_u \delta \left[\cot \Phi + \frac{1}{\tan \beta - \tan \Phi} + \frac{\tan \Phi . \tan \beta}{\tan \beta - \tan \Phi} \right]$$

or $$W_I = m_u \delta \left[\cot \Phi + \frac{1 + \tan \beta \tan \Phi}{\tan \beta - \tan \Phi} \right]$$

or $$W_I = m_u \delta [\cot \Phi + \cot (\beta - \Phi)] \quad ...(1)$$

Also, $W_E = \Sigma W_u \Delta = w_u \times$ area of segments $\times \frac{1}{3} \delta$

or $$W_E = w_u \frac{1}{2} L (\alpha L \sin \beta) \frac{\delta}{3} = w_u \delta \alpha L^2 \frac{\sin \beta}{6} \quad ...(2)$$

Now $$W_I = W_E$$

$$\therefore \quad m_u \delta [\cot \Phi + \cot (\beta - \Phi)] = w_u \delta \alpha L^2 \frac{\sin \beta}{6}$$

or
$$w_u = \frac{6\,m_u}{\alpha\,L^2} \cdot \frac{\cot\Phi + \cot(\beta-\Phi)}{\sin\beta}$$

$$= \frac{6\,m_u}{\alpha\,L^2}\,\frac{1}{\sin\Phi\sin(\beta-\Phi)} \qquad ...(31.16)$$

For w_u to be minimum, $\dfrac{\partial w_u}{\partial \Phi} = 0$

This gives $\cos\Phi\sin(\beta-\Phi) - \sin\Phi\cos(\beta-\Phi) = 0$

or $\tan\Phi = \tan(\beta-\Phi)$

$\therefore$ $\Phi = \beta/2$

Hence for minimum yield load, the yield line must bisect the angle β.

Substituting $\Phi = \beta/2$ in Eq. 31.16, we get

$$w_u = \frac{6\,m_u}{\alpha\,L^2}\,\frac{1}{\sin\beta/2\,\sin\beta/2}$$

$$= \frac{6\,m_u}{\alpha\,L^2\sin^2\beta/2} \qquad ...[31.17(a)]$$

or
$$m_u = \frac{w_u\,\alpha\,L^2}{6}\sin^2\beta/2 \qquad ...(31.17)$$

For an equilateral triangle, $\alpha = 1$ and $\beta = 60°$.

$$m_u = \frac{w_u\,L^2}{6}\sin^2 30° = \frac{w_u\,L^2}{24} \qquad ...(31.18)$$

5. Rectangular Slab, Simply Supported

Fig. 31.10 shows rectangular slab, orthotropically reinforced, and simply supported along all the four edges. Let the ultimate moment capacity in the short span direction be m_u and that in the long span direction be μm_u. The yield line pattern will be symmetrical, with point E at a distance βL from AD, where β is an unknown. Let us give a virtual displacement δ to the yield line EF.

Let L = Long span

αL = Short span

θ_x = Rotation in the x-direction $= \dfrac{\delta}{\beta L}$

θ_y = Rotation in the y-direction $= \dfrac{\delta}{\alpha L/2} = \dfrac{2\delta}{\alpha L}$

$m_{ux} = \mu\,m_u$

$m_{uy} = m_u$

For segment AED

$$m_{ux}\,y_0\,\theta_x + m_{uy}\,x_0\,\theta_y = (\mu\,m_u)(\alpha\,L)\left[\frac{\delta}{\beta L}\right] + 0$$

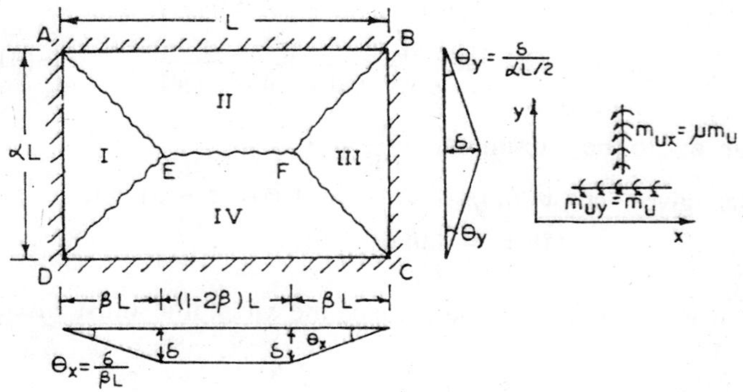

FIG. 31.10

For segment *ABFE,*

$$m_{ux}\,y_0\,\theta_x + m_{uy}\,x_0\,\theta_y \;= 0 + m_u\,L\frac{2\,\delta}{\alpha\,L}$$

For segment *BCF,*

$$m_{ux}\,y_0\,\theta_x + m_u\,y\,x_o\,\theta_y \;= (\mu m_u)\,(\alpha\,L)\left(\frac{\delta}{\beta\,L}\right) + 0$$

For segment *CDEF,*

$$m_u\,y_0\,\theta_x + m_{uy}\,x_0\,\theta_y \;= 0 + m_u\,L\frac{2\,\delta}{2\,L}$$

$$\therefore \quad \Sigma\,m_{ux}\,y_0\,\theta_x + \Sigma\,m_{uy}\,x_0\,\theta_y \;= \frac{2\,\mu\,m_y\,\alpha\,L\,\delta}{\beta\,L} + 2m_u\,L\,\frac{2\,\delta}{\alpha\,L}$$

$$= 2\,\delta\left[\frac{2\,m_u}{\alpha} + \frac{\alpha\,\mu\,m_u}{\beta}\right] \qquad \text{...(1)}$$

Work done by external load

For *AED,* $W_u\,\Delta \;= \frac{1}{2}\alpha\,L\,\beta\,L\,w_u\,\frac{\delta}{3} = \alpha\,\beta\,L^2\,\frac{\delta}{6}\,w_u$

For *ABEF,* $W_u\;\Delta \;= 2\left\{\frac{1}{2}\beta\,L\,\frac{\alpha\,L}{2}\,w_u\,\frac{\delta}{3}\right\} + (1 - 2\,\beta)\,L\,\frac{\alpha\,L}{2}\,w_u\,\frac{\delta}{2}$

$$= \alpha\,\beta\,L^2\,\frac{\delta}{6}\,w_u + (1 - 2\,\beta)\,\alpha\,L^2\,\frac{\delta}{4}\,w_u$$

For BCF, $W_u\,\Delta \;= \alpha\,\beta\,L^2\;\frac{\delta}{6}\;w_u.$

and for $CDEF$, $W_u \Delta = \alpha \beta L^2 \frac{\delta}{6} w_u + (1 - 2\beta) \alpha L^2 \frac{\delta}{4} w_u$

$$\therefore \quad \Sigma W_u \Delta = 2\left[\alpha\beta L^2 \frac{\delta}{6} w_u\right] + 2\left[\alpha\beta L^2 \frac{\delta}{6} w_u + (1 - 2\beta)\alpha L^2 \frac{\delta}{4} w_u\right]$$

$$= \delta w_u L^2 \left[\frac{\alpha\beta}{3} + \frac{\alpha\beta}{3} + \frac{\alpha}{2} - \alpha\beta\right]$$

$$= \delta w_u L^2 \left[\frac{\alpha}{2} - \frac{\alpha\beta}{3}\right] = \delta w_u L^2 \alpha \left(\frac{3 - 2\beta}{6}\right) \qquad ...(2)$$

Equating internal and external work, we get

$$= \alpha \delta w_u L^2 \left(\frac{3 - 2\beta}{6}\right) = 2\delta \left[\frac{2m_u}{\alpha} + \frac{2\alpha\mu m_u}{\beta}\right]$$

or $$\alpha w_u L^2 \left(\frac{3 - 2\beta}{6}\right) = 2m_u \left(\frac{2}{\alpha} + \frac{\alpha\mu}{\beta}\right) = 2m_u \left(\frac{2\beta + \alpha^2\mu}{\alpha\beta}\right)$$

$$\therefore \quad m_u = \frac{\alpha L^2 (3 - 2\beta)}{6} \times \frac{\alpha\beta}{2(2\beta + \alpha^2\mu)} \times w_u$$

or $$m_u = \frac{w_u \alpha^2 L^2}{12}\left[\frac{3\beta - 2\beta^2}{2\beta + \mu\alpha^2}\right] \qquad ...(31.19)$$

In the above equation, β is variable. For a maximum value of m_u (or minimum value of w_u), we should take the derivative with respect to β and equate it to zero.

In general, the work equation has the form

$$m_u = w_u \left[\frac{f_1(x_1, x_2)}{f_2(x_1, x_2)}\right]$$

For a maximum value of m_u, $\frac{\partial m_u}{\partial x_1} = 0$.

By differential process,

$$\frac{\partial m_u}{\partial x_1} = \frac{\partial}{\partial x_1}\left[\frac{f_1(x_1, x_2)}{f_2(x_1, x_2)}\right]$$

$$= \frac{f_2(x_1, x_2)\frac{\partial}{\partial x_1}[f_1(x_1, x_2)] - f_1(x_1, x_2)\frac{\partial}{\partial x_1}[f_2(x_1, x_2)]}{[f_2(x_1, x_2)]^2}$$

Equating this to zero, we get

$$f_2(x_1, x_2)\frac{\partial}{\partial x_1}[f_1(x_1, x_2)] - f_1(x_1, x_2)\frac{\partial}{\partial x_1} f_2[(x_1, x_2)] = 0$$

or $$\frac{f_1(x_1, x_2)}{f_2(x_1, x_2)} = \frac{\frac{\partial}{\partial x_1}[f_1(x_1, x_2)]}{\frac{\partial}{\partial x_1}[f_2(x_1, x_2)]} \qquad ...(31.20)$$

Eq. 31.20 is a very important governing equation for maximum value of m_u. Applying this to our present case, we get

$$\frac{3\beta - 2\beta^2}{2\beta + \mu\alpha^2} = \frac{3 - 4\beta}{2}$$

or
$$6\beta - 4\beta^2 = 6\beta + 3\mu\alpha^2 - 8\beta^2 - 4\mu\beta\alpha^2$$

or
$$4\beta^2 + 4\mu\alpha^2\beta - 3\mu\alpha^2 = 0$$

Solution of the above quadratic equation, we get,

$$\beta = \frac{1}{2}\left[\sqrt{\mu^2\alpha^4 + 3\mu\alpha^2} - \mu\alpha^2\right] \qquad ...(31.21)$$

Substituting this value of β in Eq 31.19, we get

$$m = \frac{1}{12} w_u\alpha^2 L^2 \frac{\left[\frac{3}{2}\left\{\sqrt{\mu^2\alpha^4 + 3\mu\alpha^2} - \mu\alpha^2\right\} - \frac{2}{4}\left(\mu^2\alpha^4 + 3\mu\alpha^2 + \mu^2\alpha^4 - 2\mu\alpha^2\sqrt{\mu^2\alpha^4 + 3\mu\alpha^2}\right)\right]}{\frac{2}{2}\left\{\sqrt{\mu^2\alpha^4 + 3\mu\alpha^2} - \mu\alpha^2\right\} + \mu\alpha^2}$$

Simplifying the above, we get

$$m_u = \frac{w_u\alpha^2 L^2}{24}\left[\sqrt{3 + \mu\alpha^2} - \alpha\sqrt{\mu}\right]^2 \qquad ...(31.22)$$

Special cases :

(*i*) *For an isotropically reinforced* square slab,

$$\alpha = 1 \text{ and } \mu = 1$$

$$\therefore \quad m_u = \frac{w_n\alpha^2 L^2}{24}\left[\sqrt{3 + 1} - 1\sqrt{1}\right]^2 = \frac{w_u L^2}{24} \qquad ...(31.23)$$

which is the same as Eq 31.11 (*a*)

(*ii*) *For an isotropically reinforced rectangular slab,*

$$\mu = 1$$

$$m_u = \frac{w_u\alpha^2 L^2}{24}\left[\sqrt{3 + \alpha^2} - \alpha\right]^2 \qquad ...(31.24)$$

6. Polygonal and Circular Slabs

Let us first consider polygonal slab, fixed along its edges, with isotropic positive moments m_u and isotropic negative moment m_u'. Let the length of each side be L, and let the perpendicular distance of each side from centre be r, which is also the radius of the inscribed circle. The positive yield lines will be along the radial lines as shown in Fig. 31.11 (*a*). The negative yield lines will be along the outer edges.

Let us give a virtual displacement δ at the apex O.

For any sector *ABO*,

$$M_I = (m_u + m_u')\, L\frac{\delta}{r} \qquad ...(1)$$

$$M_E = \frac{1}{2} L\, r \frac{\delta}{3} w_u \qquad ...(2)$$

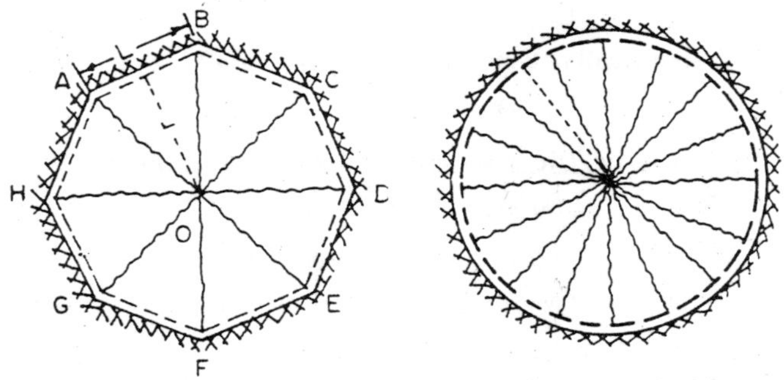

(a) POLYGONAL SLAB (b) CIRCULAR SLAB

FIG. 31.11

Equating the two, we get

$$w_u L\, r \frac{\delta}{6} = (m_u + m_u') L \frac{\delta}{r}$$

or

$$w_u = \frac{6\,(m_u + m_u')}{r^2} \qquad ...(31.25)$$

For a circular slab, there will be innumerable number of positive yield lines, such that number of sides n will tend to be infinity and L will tend to be zero. However, r will be the radius of the circle, and for any sector ABO ,

$$w_u = \frac{6\,(m_u + m_u')}{r^2}$$

Thus, Equation 31.25 is applicable to both polygonal as well as circular slabs.

For circular slabs simply supported along the edge, $m_u'=0$

$$\therefore \qquad w_u = \frac{6\, m_u}{r^2} \qquad ...[31.26]$$

or

$$m_u = \frac{w_u r^2}{6} \qquad ...[31.25(a)]$$

For hexagonal slab, $r = \sqrt{3}\, L/2$

Hence from Eq 31.25, $w_u = \dfrac{6\,(m_u + m_u')}{(\sqrt{3}L/2)^2} = \dfrac{8\,(m_u + m_u')}{L^2}$

If the hexagonal slab is simply supported.

$$w_u = \frac{8\, m_u}{L^2} \qquad ...(31.28)$$

$$\text{or} \qquad m_u = \frac{w_u L^2}{8} \qquad \text{...[31.27(a)]}$$

31.6. ANALYSIS BY EQUILIBRIUM METHOD

1. Isotropically Reinforced Square Slab

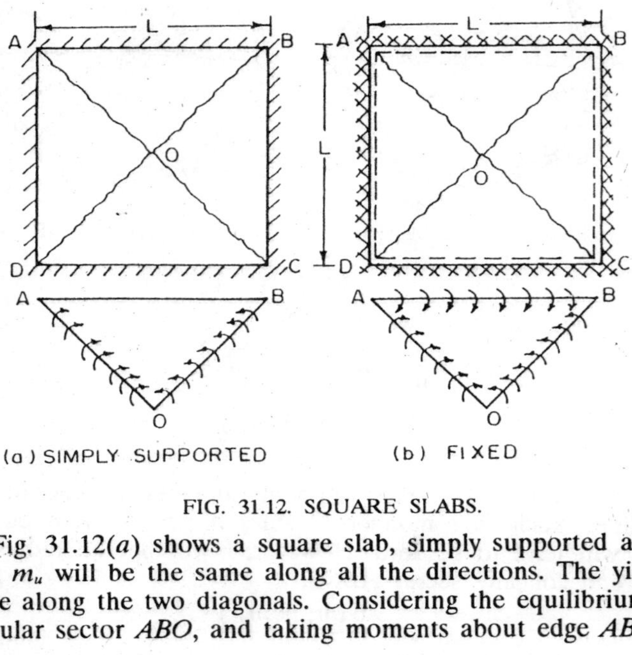

FIG. 31.12. SQUARE SLABS.

Fig. 31.12(a) shows a square slab, simply supported along the edges. m_u will be the same along all the directions. The yield lines will be along the two diagonals. Considering the equilibrium of the triangular sector ABO, and taking moments about edge AB, we get

$$m_u L = \frac{1}{2} L . \frac{L}{2} w_u \frac{L}{6}$$

$$\text{or} \qquad m_u = \frac{w_u L^2}{24} \qquad \text{...[31.11(a)]}$$

This is the same as Eq. 31.11 (a).

Again, for a fixed slab of Fig. 31.12 (b). positive yield lines will be along the diagonals, while the negative yield lines will be developed along the fixed edges. Considering the equilibrium of sector ABC, and taking moments about edge, AB, we get

$$m_u L + m_u L = \frac{1}{2} L \frac{L}{2} w_u \frac{L}{6}$$

$$\text{or} \qquad m_u = \frac{w_u L^2}{48} \qquad \text{...[31.14(a)]}$$

This is the same as Eq. 31.14(a).

2. Orthotropically Reinforced Rectangular Slab

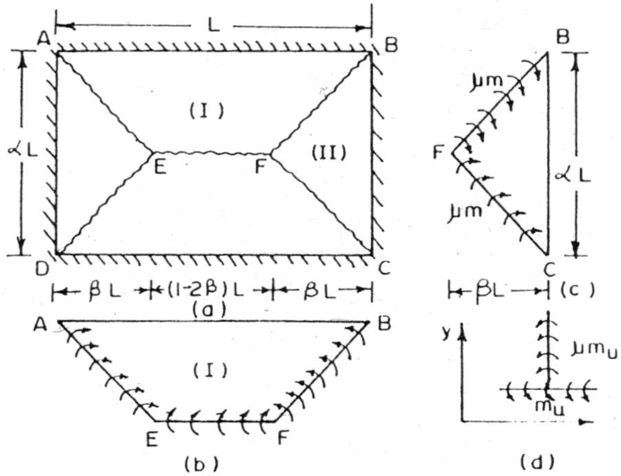

FIG. 31.13. RECTANGULAR SLAB.

Fig. 31.13 (a) shows a rectangular slab, with length of long span equal to L and that of short edge as αL. The yield line pattern will be symmetrical with point E at a distance βL from AD, where β is unknown. Let the slab be orthatropically reinforced with $m_{uy} = m_u$ and $m_{ux} = \mu m_u$.

Consider equilibrium of sector $ABFE$ [Fig. 31.13 (b)]. Taking moments about edge AB, we get.

$$m_u L = w_u \left[\left\{ (1-2\beta) L \frac{\alpha L}{2} \frac{\alpha L}{4} \right\} + 2 \left\{ \frac{1}{2} \beta L \frac{\alpha L}{2} \frac{\alpha L}{6} \right\} \right]$$

or $$m_u = w_u \alpha^2 L^2 \left[\frac{(1-2\beta)}{8} + \frac{\beta}{12} \right]$$

or $$m_u = \frac{w_u \alpha^2 L^2}{24} (3 - 4\beta) \qquad ...(1)$$

Similarly, consider equilibrium of sector FBC. Taking moments about edge BC, we get

$$\mu m_u \alpha L = \frac{1}{2} \alpha L \beta L \frac{\beta L}{3} w_u = \frac{\alpha \beta^2 L^3}{6} w_u$$

or $$m_u = \frac{\beta^2 L^2}{6\mu} w_u \qquad ...(2)$$

Equating the two equilibrium equations, we get

$$\frac{w_u \alpha^2 L^2}{24}(3 - 4\beta) = \frac{\beta^2 L^2}{6\mu} w_u$$

or $$\mu \alpha^2 (3 - 4\beta) = 4\beta^2$$

or $$4\beta^2 + 4\mu\alpha^2\beta - 3\mu\alpha^2 = 0$$

$\therefore$ $$\beta = \frac{1}{2}\left[\sqrt{\mu^2\alpha^4 + 3\mu\alpha^2} - \mu\alpha^2\right]$$

This is the same as Eq. 31.21

Substituting this value of β in Eq. (2), we get

$$m_u = \frac{w_u L^2}{6\mu} \times \frac{1}{4}\left[\sqrt{\mu^2\alpha^4 + 3\mu\alpha^2} - \mu\alpha^2\right]^2$$

or $$m_u = \frac{w_u L^2}{24\mu} \times \mu\alpha^2\left[\sqrt{3 + \mu\alpha^2} - \alpha\sqrt{\mu}\right]^2$$

or $$m_u = \frac{w_u \alpha^2 L^2}{24}\left[\sqrt{3 + \mu\alpha^2} - \alpha\sqrt{\mu}\right]^2$$

This is the same as Eq. 31.22

3. Isotropically Reinforced Hexagonal Slab

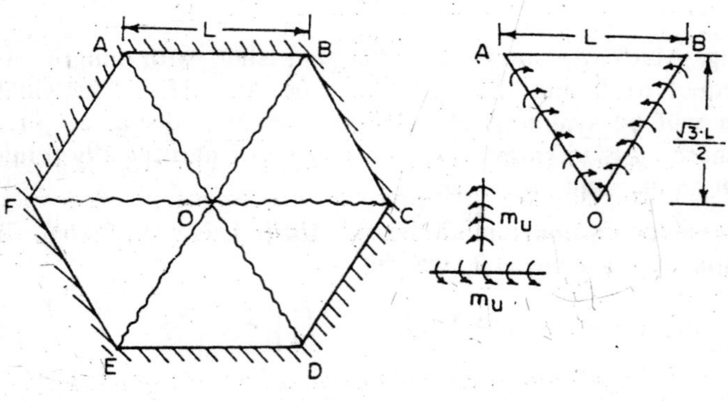

FIG. 31.14

Consider the equilibrium of sector *ABO*, and take moments about *AB*. The value of perpendicular $r = \dfrac{\sqrt{3}\,L}{2}$.

$\therefore$ $$m_u L = \frac{1}{2} L\, r\, \frac{r}{3} w_u$$

or $$m_u = \frac{w_u}{6} r^2 = \frac{w_u}{6}\left(\frac{\sqrt{3}\,L}{2}\right)^2$$

$\therefore$ $$m_u = \frac{w_u L^2}{8}$$

This is the same as Eq. 31.27 (*a*)

Example 31.1. *A square slab of side length 4 m is simply supported at the ends and carries a service live load of 3 kN/m². Design the slab. Use M 15 concrete and mild steel bars.*

Solution. Give : $L = 4$ m

$$f_{ck} = 15 \text{ N/mm}^2 ; f_y = 250 \text{ N/mm}^2.$$

Let us assume depth $= \dfrac{\text{span}}{35} = \dfrac{4000}{35} = 114$ mm

Let us assume an overall depth = 130 mm.

Using 15 mm clear cover and 10 mm dia. bars,

$$d = 130 - 15 - 10 = 105 \text{ mm.}$$

Self weight of slab $= 0.105 \times 25000 = 2625 \text{ N/m}^2$

Floor finishes 50 mm thick $= 0.05 \times 22000 = 1100 \text{ N/m}^2$ (say)

Live load $= 3000 \text{ N/m}^2$

Total service load $= 6725 \text{ N/m}^2$.

Load factor = 1.5

Ultimate design load $= 1.5 \times 6725 = 10090 \text{ N/m}^2$

∴ Ultimate capacity of the slab is given by Eq. 31.11 (*a*)

$$m_u = \frac{w_u L^2}{24}$$

$$= \frac{10090 \, (4)^2}{24} = 6726.7 \text{ N-m}$$

$$= 6.727 \times 10^6 \text{ N-mm}$$

The limiting or balancing moment capacity of the slab is given by

$$M_{u.lim} = 0.148 f_{ck} \, bd^2$$

$$= 0.148 \times 15 \, (1000) \, (105)^2$$

$$= 24.476 \times 10^6 \text{ N-mm}$$

Since $m_u < M_{u.lim}$, the slab is under-reinforced

The reinforcement in the slab is given by

$$M_u = m_u = 0.87 f_y A_{st} d \left[1 - \frac{A_{st} f_y}{b \, d \, f_{ck}} \right]$$

$$\therefore \; 6.727 \times 10^6 = 0.87 \times 250 A_{st} \times 105 \left[1 - \frac{A_{st} \times 250}{1000 \times 105 \times 15} \right]$$

$$A_{st}^2 - 6300 A_{st} + 1.717 \times 10^6 = 0$$

which gives $A_{st} = 285 \text{ mm}^2$

Minimum steel area $= \dfrac{0.15}{100} \times 1000 \times 130 = 195 \text{ mm}^2$

Using 8 mm dia. bars, spacing $= \dfrac{1000 \times 50}{285} = 175$ mm

Hence use 8 mm dia. bars @ 175 c/c both ways.

Example 31.2. *A reinforced concrete slab 5 m × 5 m is simply supported along the four edges and is reinforced with 10 mm dia. mild steel bars at 150 mm c/c both ways. The average effective depth of the slab is 100 mm and the overalll depth of the slab is 130 mm. The slab carries a flooring of 50 mm thick having unit weight of 2.2 kN/m². Determine the maximum permissible service load, if M 15 concrete is used.*

Solution. $A_{st} = \dfrac{1000 \times 78.54}{150} = 523.6 \text{ mm}^2/\text{m}$

Now, $M_u = 0.87 f_y A_{st} d \left[1 - \dfrac{A_{st} f_y}{b\, d\, f_{ck}}\right]$

$= 0.87 \times 250 \times 523.6 \times 100 \left[1 - \dfrac{523.6 \times 250}{1000 \times 100 \times 15}\right]$

$= 10.394 \times 10^6$ N-mm

$M_{u.lim} = 0.148 f_{ck}\, bd^2$

$= 0.148 \times 15\, (1000)\, (100)^2 = 22.2 \times 10^6$ N-mm

Since $M_u < M_{u.lim}$ the slab is under-reinforced.

Again, $m_u = M_u = 10.394 \times 10^6$ N-mm $= 10.394$ kN-m

But $m_u = \dfrac{w_u L^2}{24}$ [Eq. 31.11 (*a*)]

$\therefore \quad w_u = \dfrac{24}{L^2} m_u$

$= \dfrac{24}{(5)^2} \times 10.394 = 9.978 \text{ kN/m}^2.$

$\therefore$ Service load $= \dfrac{9.978}{1.5} \approx 6.650 \text{ kN/m}^2$

Dead load of slab $= 0.130 \times 25 = 3.25 \text{ kN/m}^2$

Dead load of finishing $= 0.05 \times 22 = 1.10 \text{ kN/m}^2$

Total dead load $= 3.25 + 1.10 = 4.35 \text{ kN/m}^2$

$\therefore$ Permissible service load $= 6.65 - 4.35 =$ **2.3 kN/m².**

Example 31.3. *A reinforced concrete slab, 4 m × 6 m is reinforced with 10 mm diameter bars at 150 mm spacing in the short direction and 200 mm spacing in the long direction. The slab is 100 mm thick with a average effective depth of 80 mm. If the yield lines are inclined at 45° to either direction of reinforcement, find the ultimate moments*

m_{uan} and m_{uat} for unit length along yield line. Use M 20 concrete and Fe 415 steel.

Solution. Refer Fig. 31.10 in which the yield lines *AE, DE, BF* and *CF* are inclined at 45° to the principal x and y directions. The reinforcement in y-direction is spaced at 150 mm c/c while that in x-direction is spaced at 200 mm c/c.

Now $M_{uy} = m_{uy} =$ moment across plane whose normal is in y-direction.

$$= 0.87 f_y A_{st}\, d \left(1 - \frac{A_{st} f_y}{bd\, f_{ck}}\right)$$

where $A_{st} = \dfrac{1000}{150} \times 78.6 = 524 \text{ mm}^2$

$$\therefore \quad m_{uy} = 0.87 \times 415 \times 524 \times 80 \left(1 - \frac{524 \times 415}{1000 \times 80 \times 20}\right)$$

$$= 13.078 \times 10^6 \text{ N-mm} = 13.078 \text{ kN-m}$$

Also, $M_{ux} = m_{ux} =$ moment across plane whose normal is in x-direction.

$$= 0.87 f_y A_{st}\, d \left(1 - \frac{A_{st} f_y}{bd\, f_{ck}}\right)$$

where $A_{st} = \dfrac{1000}{200} \times 78.6 = 393 \text{ mm}^2$

$$\therefore \quad m_{ux} = 0.87 \times 415 \times 393 \times 80 \left(1 - \frac{393 \times 415}{1000 \times 80 \times 20}\right)$$

$$= 10.194 \times 10^6 \text{ N-mm} = 10.194 \text{ kN-m.}$$

Now, from Figs, 31.1, and 31.6

$$m_{uan} = m_{ux} \cos^2 \alpha + m_{uy} \sin^2 \alpha$$

$$m_{uat} = (m_{ux} - m_{uy}) \sin \alpha \cos \alpha\,.$$

Here $\alpha = 45\,°$; $\cos \alpha = \sin \alpha = 0.707$

$$m_{uan} = 10.894\,(0.707)^2 + 13.078\,(0.707)^2 = \mathbf{11.636 \text{ kN-m}}$$

$$m_{uat} = (10.194 - 13.078)\, 0.707 \times 0.707 = \mathbf{-1.442 \text{ kN-m}}$$

(Negative sign shows that the direction of torsional moment is opposite to the one marked in Fig. 31.5 (*c*).

Example 31.4 *A rectangular slab 3.5 × 4.5 m is isotropically reinforced with 8 mm dia. bars spaced at 150 mm bothways. The average effective depth may be taken as 80 mm and the total depth of the slab is 100 mm. If Fe 415 steel and concrete of grade M 20 are used, determine the safe service live load. The dead load of floor finishing may be assumed as 1.5 kN/m^2.*

Solution. Given $L = 4.5$ m

$$\alpha L = 3.5 \text{ m}$$

$$\alpha = \frac{3.5}{4.5} = 0.7778$$

Since the slab is isotropically reinforced,

$$m_{ux} = m_{uy} = m_u$$

Hence $\mu = 1$

$$A_{st} = \frac{1000 \times 50.27}{150} = 335 \text{ mm}^2/\text{m}.$$

The ultimate moment is given by

$$M_u = m_u = 0.87 f_y A_{st} d \left[1 - \frac{A_{st} f_y}{bd f_{ck}}\right]$$

$$= 0.87 \times 415 \times 335 \times 80 \left[1 - \frac{335 \times 415}{1000 \times 80 \times 20}\right]$$

$$= 8.835 \times 10^6 \text{ N-mm} = 8.835 \text{ kN-m}$$

The yield moment of the slab is given by Eq. 31.24

$$m_u = \frac{w_u \alpha^2 L^2}{24}\left[\sqrt{3+\alpha^2} - \alpha\right]^2$$

or

$$w_u = \frac{24 m_u}{\alpha^2 L^2 \left[\sqrt{3+\alpha^2} - \alpha\right]^2}$$

$$= \frac{24 \times 8.835}{(0.7778)^2 \times (4.5)^2 [\sqrt{3 + (0.7778)^2} - 0.7778]^2}$$

$$= 13.76 \text{ kN/m}^2.$$

$\therefore$ Total serviced load $= \frac{13.76}{1.5} \approx 9.17$ kN/m^2

Dead load of slab $= 0.1 \times 25 = 2.5$ kN/m^2

Dead load of floor finishing $= 1.5$ kN/m^2

$\therefore$ Service live load $= 9.17 - (2.5 + 1.5) =$ **5.17 kN/m²**.

Example 31.5. *For the slab of example 31.3. determine the safe permissible service live load. The dead load of flooring may be assumed as 1 kN/m².*

Solution. Refer Fig. 31.10

Given :

$$L = 6 \text{ m}.$$

$$\alpha L = 4 \text{ m}.$$

$$\therefore \quad \alpha = \frac{4}{L} = \frac{4}{6} = 0.667$$

$$D = 100 \text{ mm} \; ; \; d = 80 \text{ mm}$$

As found in example 31.3.

$$m_{uy} = 13.078 = m_u.$$

$$m_{ux} = 10.194 = \mu m_u.$$

$$\therefore \quad \mu = \frac{10.194}{13.078} = 0.779.$$

Now from Fig. 31.22

$$m_u = \frac{w_u \alpha^2 L^2}{24}\left[\sqrt{3+\mu\alpha^2} - \alpha\sqrt{\mu}\right]^2$$

$$w_u = \frac{24\, m_u}{[\alpha^2 L^2 \sqrt{3+\alpha^2} - \alpha\sqrt{\mu}]^2}$$

$$= \frac{24 \times 13.078}{(0.667)^2 (6)^2 [\sqrt{3 + 0.779\,(0.667)^2} - 0.667\sqrt{0.779}]^2}$$

$$= \frac{24 \times 13.078}{0.445 \times 36 \times 1.539} = 12.729 \text{ kN/m}^2$$

$\therefore$ Service load $= \dfrac{12.729}{1.5} = 8.486$ kN/m^2

Dead load of slab $= 0.1 \times 25 = 2.5$ kN/m^2

Dead load of flooring $= 1.0$ kN/m^2

$\therefore$ Service live load $= 8.486 - 3.5 =$ **4.986 kN/m^2.**

Example 31.6 *A rectangular slab 3.5 × 5 m in size, simply supported at the edges. The slab is expected to carry a service live load of 3 kN/m^2 and a floor finishing load of 1 kN/m^2. Use M 15 concrete and Fe 415 steel. Design the slab if (a) it is isotropically reinforced (b) If it is orthotropically reinforced with $\mu = 0.75$.*

Solution. Given : $L = 5$ m.

$$\alpha L = 3.5\,; \ \therefore\ \alpha = 3.5/5 = 0.7$$

$$\left(\frac{\text{span}}{\text{depth}}\right) = 35$$

$\therefore$ Depth of slab $= \dfrac{350}{35} = 10$ cm.

Use 10 mm dia. bars both ways. Using 15 mm clear cover, assume average effective depth $= 100 - 15 - 10 = 75$ mm

Dead weight of slab $= 0.1 \times 25 = 2.5$ kN/m^2

Dead weight of flooring $= 1.0$ kN/m^2

Live load $= 3.0$ kN/m^2

$\therefore$ Total service load $= 6.5$ kN/m^2

$\therefore$ Ultimate design load $= w_u = 1.5 \times 6.5 = 9.75$ kN/m^2

(*a*) ***Isotropically reinforced slab*** : $\mu = 1$

Now, ultimate moment on slab is given by Eq 31.22 :

$$m_u = \frac{w_u \alpha^2 L^2}{24}\left[\sqrt{3+\mu\alpha^2} - \alpha\sqrt{\mu}\right]^2$$

$$= \frac{9.75\,(0.7)^2\,(5)^2}{24}\left[\sqrt{3+1(0.7)^2} - 0.7\sqrt{1}\right]^2$$

$$= 6.791 \text{ kN-m/m} = 6.791 \times 10^6 \text{ N-mm/m}$$

The reinforcement is given by

$$M_u = m_u = 0.87 f_y A_{st} d\left[1 - \frac{A_{st} f_y}{bd f_{ck}}\right]$$

or $$6.791 \times 10^6 = 0.87 \times 415 A_{st} \times 75\left[1 - \frac{A_{st} \times 415}{1000 \times 75 \times 15}\right]$$

which simplifies to

$$A_{st}^2 - 2711 A_{st} + 0.6798 \times 10^6 = 10$$

Hence $A_{st} = 280 \text{ mm}^2$

Using 8 mm dia. bars, spacing $= \dfrac{1000 \times 50.3}{280} = 179.5$ mm.

Hence provide 8 mm dia. bars @ 175 mm c/c both ways.

(*b*) ***Orthotropically reinforced slab***

$$\mu = 0.75$$

$$\therefore \quad m_u = \frac{w_u \alpha^2 L^2}{24}\left[\sqrt{3+\mu\alpha^2} - \alpha\sqrt{\mu}\right]^2$$

$$= \frac{9.75\,(0.7)^2\,(5)^2}{24}\left[\sqrt{3+0.75\,(0.7)^2} - 0.7\sqrt{0.75}\right]^2$$

$$= 8.541 \text{ kN-m/m} = 8.541 \times 10^6 \text{ N-mm/m}$$

But $m_u = 0.87 f_y A_{st} d\left[1 - \dfrac{A_{st} f_y}{bd f_{ck}}\right]$

or $$8.541 \times 10^6 = 0.87 \times 415 A_{st}\,(75)\left[1 - \frac{A_{st} \times 415}{1000 \times 75 \times 15}\right]$$

or $$A_{st}^2 - 2711 A_{st} + 0.855 \times 10^6 = 0$$

which gives $A_{st} = 364.4 \text{ mm}^2$.

$\therefore$ Spacing of 8 mm dia. bars in short direction $= \dfrac{1000 \times 50.3}{364.4}$

$= 138$ mm

and spacing of 8 mm dia. in long direction $= \dfrac{138}{0.75} = 184$ mm.

Hence provide 8 mm dia. bars @ 130 mm c/c in short direction and @ 180 mm c/c in long direction.

Example 31.7. *A right angled rectangular slab is simply supported along the two edges. The length of the supported edges are 3 m and 4 m respectively. The slab is isotropically reinforced with 8 mm dia. bars spaced at 100 mm centres both ways. The average effective depth of the slab is 95 mm and total depth is 120 mm. Using M 15 concrete and mild steel bars, determine the safe permissible service live load on the slab. Assume dead load of floor finishing at 1.5 kN/m*^{2}.

Solution

Given : $L = 4\text{ m}\,;\alpha L = 3\text{ m}$

$$\therefore \quad \alpha = \tfrac{3}{4} = 0.75$$

$$A_{st} = \frac{1000 \times 50.3}{100} = 503\text{ mm}^2$$

The ultimate moment is given by

$$m_u = M_u = 0.87 f_y A_{st}\, d \left[1 - \frac{A_{st} f_y}{bd\, f_{ck}}\right]$$

$$= 0.87 \times 250 \times 503 \times 95 \left[1 - \frac{503 \times 250}{1000 \times 95 \times 15}\right]$$

$$= 9.476 \times 10^6 \text{ N-mm} = 9.476 \text{ kN-m}$$

Now, from Eq. 31.17,

$$m_u = \frac{w_u\, \alpha L^2}{6} \sin^2 \beta/2$$

or

$$w_u = \frac{6\, m_u}{\alpha L^2 \sin^2 \beta/2}$$

Hence $\beta = 90°$

$$\therefore \quad w_u = \frac{6 \times 9.476}{0.75\,(4)^2\,(\sin 45°)^2} = 9.476 \text{ kN/m}^2$$

$$\therefore \quad \text{Total service load} = \frac{9.476}{1.5} = 6.317 \text{ kN/m}^2$$

Dead load of slab $= 0.12 \times 25 = 3 \text{ kN/m}^2$

Dead load of flooring $= 1.5 \text{ kN/m}^2$

Total dead load $= 4.5 \text{ kN/m}^2$

$\therefore$ Permissible service live load $= 6.317 - 4.5 = 1.82 \text{ kN/m}^2$.

Example 31.8. *A hexagonal slab is simply supported along the edges with a side length of 3.5 m. It is isotropically reinforced with 10 mm dia. bars at 125 mm centres both ways, at an average effective depth of 120 mm. The overall depth of the slab is 150 mm. Calculate (i) ultimate load capacity of slab, (ii) safe permissible service live load. Use M 20 concrete and Fe 415 steel. Take the load of floor finishing as 1.25 kN/m*^{2}.

Solution

$$L = 3.5 \text{ m.}$$

$$A_{st} = \frac{1000 \times 78.54}{125} = 628.3 \text{ mm}^2$$

$$m_u = M_u = 0.87 f_y A_{st} d \left[1 - \frac{A_{st} f_y}{bd f_{ck}}\right]$$

$$= 0.87 \times 415 \times 628.3 \times 120 \left[1 - \frac{628.3 \times 415}{1000 \times 120 \times 120}\right]$$

$$= 24.264 \times 10^6 \text{ N-mm} = \mathbf{24.264 \text{ kN-m.}}$$

Now, from Eq. 31.28

$$w_u = \frac{8 m_u}{L^2}$$

$$= \frac{8 \times 24.264}{(3.5)^2} = 15.845 \text{ kN/m}^2$$

Total service load $= \frac{15.845}{1.5} = 10.564 \text{ kN/m}^2$

Dead load of slab $= 0.15 \times 25 = 3.75 \text{ kN/m}^2$
Load of floor finishing $= 1.25 \text{ kN/m}^2$

Total dead load $= 5 \text{ kN/m}^2$
∴ Permissible service live load $= 10.564 - 5 =$ **5.564 kN/m²**.

Example 31.9 *Design a reinforced circular slab for the following data :*

(i) Diameter of slab : 5.5 m
(ii) Service live load : (4 kN/m²)
(iii) Floor finishing load : 1 kN/m²
(iv) Grade of concrete : M 20
(v) Grade of steel : Fe 415.

The slab is simply supported along the edge.

Solution

For a simply supported two way slab,

$$\frac{\text{Span}}{\text{Depth}} = 35$$

$$\text{Depth} = \frac{5500}{35} = 157 \text{ mm.}$$

Let us provide overall depth = 160 mm. Using 10 mm dia. bars and a clear cover of 15 mm, average effective depth = 160 − 15 − 10 = 135 mm.

Self load of slab = 0.16 × 25 = 4 kN/m²

Dead load of flooring = 1 kN/m²

Live load = 4 kN/m²

Total service load = 9 kN/m².

$$w_u = 9 \times 1.5 = 13.5 \text{ kN/m}^2$$

Now, from Eq. 31.26 (*a*),

$$m_u = \frac{w_u r^2}{6} = \frac{13.5\,(5.5/2)^2}{6}$$

$$= 17.016 \text{ kN-m/m} = 17.016 \times 10^6 \text{ N-mm/m.}$$

Limiting moment capacity :

$$M_{u.lim} = 0.148 f_{ck}\, bd^2$$

$$= 0.145 \times 20\,(1000)\,(135)^2$$

$$= 52.853 \times 10^6 \text{ N-mm.}$$

Since $m_u < M_{u.lim}$ the slab is under reinforced.

$$\therefore \quad m_u = M_u = 0.87 f_y A_{st}\, d \left(1 - \frac{A_{st} f_y}{bd\, f_{ck}}\right)$$

or $$17.016 \times 10^6 = 0.87 \times 415 A_{st} \times 135 \left(1 - \frac{A_{st} \times 415}{1000 \times 135 \times 20}\right)$$

or $$A_{st}^2 - 6506 A_{st} + 2.271 \times 10^6 = 0$$

which gives $A_{st} = 370.2 \text{ mm}^2$

Spacing of 10 mm dia. bars $= \dfrac{1000 \times 78.5}{370.2} = 212$ mm

However, use 10 mm dia. bars @ 200 mm centres both ways.

PROBLEMS

1. A reinforced concrete square slab, 3.5 m × 3.5 m is simply supported at the ends and is reinforced with 8 mm diameter bars spaced at 150 mm centres both ways. Determine the safe service live load if the average effective depth of slab is 100 mm and the total thickness of slab, inclusive of flooring, is 160 mm. Use M 15 concrete and mild steel bars.

2. Redesign the slab of problem 1 to carry a service load of 4 kN/m², inclusive of floor finishes, but exclusive of self load.

3. A rectangular slab 4 m × 5 m is simply supported at the ends. Design the slab to carry super-imposed service load of 5 kN/m^2, if the slab is to be isotropically reinforced. Use M 20 concrete and Fe 415 steel.

4. Redesign the slab of problem 3 if the slab is to be orthoropically reinforced, with $\mu = 0.8$.

5. Find the safe superimposed service load on the slab of problem 3, if it is reinforced with 8 mm diameter bars spaced at 125 mm centres both ways. The total thickness of the slab may be assumed as 125 mm.

6. A triangular slab has equal side lengths of 4.5 m is supported on two edges and is isotropically reinforced with 8 mm dia. bars of Fe 415 grade, spaced at 125 mm centres both ways. Determine (*i*) ultimate moment capacity, (*ii*) ultimate collapse load. The total thickness of slab is 125 mm, in M 20 Grade concrete.

7. A hexagonal slab, simply supported on all the sides has a side length of 3 m. Find the uniformly distributed load which would cause failure of the isotropically reinforced slab, if the ultimate moment of resistance of slab is 2.7 kN-m/m. Work from the first principles.

8. A rectangular slab 6 m × 4.5 m, simply supported at its edge is to be designed as an isotropically reinforced slab to support an uniformly distributed working load of 18 kN/m^2 which includes the self weight of the slab. Calculate the ultimate moment of resistance required for the slab section from first principles.

9. Design a circular slab of 5 m diameter, simply supported along the edge, to carry a service live load of 5 kN/m^2. Use M 20 concrete and Fe 415 steel.

Appendix-A

B.M. AND S.F. COEFFICIENTS

TABLE A-1 CONTINUOUS BEAMS

(a) B.M. Coefficients

B.M. = (Coefficient) × (Total load on one Span) × Span

Load	Dead load (All spans loaded)	Live load (Sequence of loaded spans to give max. B.M.)
Uniformly distributed	0·125 0·071, 0·071	0·125 0·096, 0·096
	0·100, 0·100 0·080, 0·025, 0·080	0·117, 0·117 0·101, 0·075, 0·101
	0·107, 0·072, 0·107 0·077, 0·056, 0·056, 0·077	(0·116) 0·121, (0·107) 0·107, (0·116) 0·121 0·099, 0·081, 0·081, 0·099
	0·105, 0·08, 0·08, 0·105 0·078, 0·033, 0·046, 0·053, 0·078	(0·116) 0·120, (0·107) 0·111, (0·107) 0·111, (0·116) 0·120 0·100, 0·080, 0·086, 0·080, 0·100
Concentrated at mid-span	0·188 0·156, 0·156	0·188 0·203, 0·203
	0·150, 0·150 0·175, 0·100, 0·175	0·175, 0·175 0·213, 0·175, 0·213
	0·161, 0·107, 0·161 0·164, 0·116, 0·116, 0·164	(0·174) 0·181, (0·160) 0·160, (0·174) 0·181 0·210, 0·183, 0·183, 0·210
	0·158, 0·119, 0·119, 0·158 0·171, 0·110, 0·130, 0·110, 0·171	(0·174) 0·174, (0·160) 0·167, (0·160) 0·167, (0·174) 0·174 0·211, 0·181, 0·141, 0·81, 0·211
Concentrated at third points	0·167 0·111, 0·111	0·167 0·139, 0·139
	0·133, 0·133 0·123, 0·054, 0·123	0·157, 0·157 0·145, 0·100, 0·145
	0·143, 0·075, 0·143 0·119, 0·056, 0·056, 0·119	(0·155) 0·160, (0·143) 0·144, (0·155) 0·160 0·143, 0·111, 0·111, 0·143
	0·141, 0·106, 0·106, 0·141 0·120, 0·050, 0·061, 0·050, 0·120	(0·155) 0·159, (0·142) 0·148, (0·142) 0·148, (0·155) 0·159 0·144, 0·108, 0·115, 0·108, 0·144

(For notes, see next page)

TABLE A-1 (contd.)

(*b*) S.F. coefficients

S.F. = Coefficient × (Total load on one span)

Load	Dead Load (All spans loaded)	Live Load (Sequence of loaded spans to give max. B.M.)
Uniformly distributed	0·38 0·62 0·62 0·38	0·44 0·62 0·62 0·44
	0·40 0·50 0·60 0·60 0·50 0·40	0·45 0·58 0·62 0·62 0·58 0·45
	0·39 0·54 0·46 0·61 0·61 0·46 0·54 0·39	0·45 0·60 0·57 0·62 0·62 0·57 0·60 0·45
	0·40 0·53 0·50 0·47 0·60 0·60 0·47 0·50 0·53 0·40	0·45 0·60 0·59 0·58 0·62 0·62 0·56 0·59 0·60 0·45

Note 1. B.M. coefficients above line apply to negative B.M. at supports.

B.M. coefficients below line apply to positive B.M. in span.

2. S.F. coefficients above line apply to S.F. at right hand side of support.

S.F. coefficients below line apply to S.F. at left hand side of support.

3. Coefficients apply when all spans are equal.

Loads on each loaded span are equal.

Moment of inertia are same throughout.

4. B.M. coefficients (live load) in brackets apply if two spans only are laoded.

TABLE A-2 . SIMPLY SUPPORTED BEAMS

(Reaction = r W , Max. B.M. = K_C . WL)

Loading	Total load W	Reaction coefficients r_A	Reaction coefficients r_B	Coefficient k_c for max B.M.
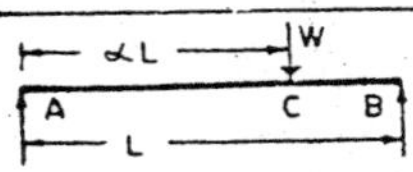	W	$1-\alpha$	α	$\alpha(1-\alpha)$
central load, $\alpha = \frac{1}{2}$	W	1/2	1/2	1/4
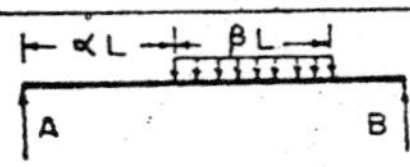	$\omega \beta L$	$1-(\alpha+\frac{\beta}{2})$	$\alpha+\frac{\beta}{2}$	$\frac{(\alpha+\frac{\beta}{2})(1-\alpha-\frac{\beta}{2})(2-\beta)}{2}$
span fully loaded ($\alpha=0$; $\beta=1$)	ωL	1/2	1/2	1/8
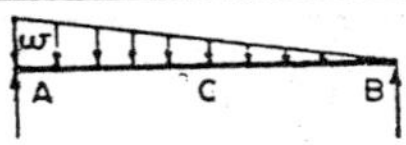	$\frac{\omega L}{2}$	2/3	1/3	1/7·81
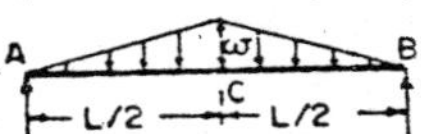	$\frac{\omega L}{2}$	1/2	1/2	1/6
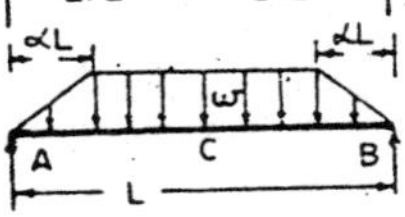	$\omega L(1-\alpha)$	1/2	1/2	$\frac{3-4\alpha^2}{24(1-\alpha)}$
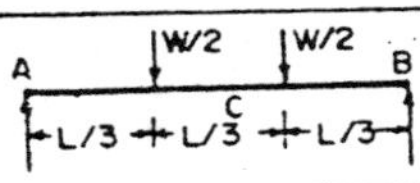	W	1/2	1/2	1/6

TABLE A-3. CANTILEVERS

(Max. B.M. = K_A . W L)

Loading	Total load W	B.M. coeff. K_A	Loading	Total load W	B.M. coeff. K_A
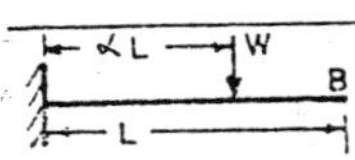	W	α	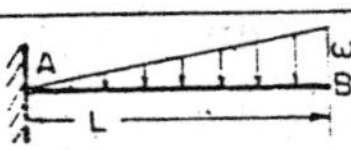	$\frac{\omega L}{2}$	$\frac{2}{3}$
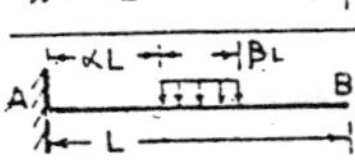	$\omega \beta L$	$(\alpha+\beta/2)$	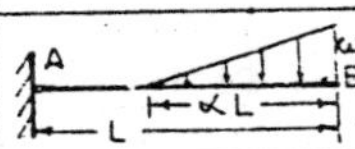	$\frac{\omega \alpha L}{2}$	$(1-\frac{\alpha}{3})$
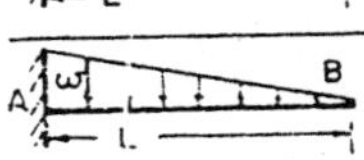	$\frac{\omega L}{2}$	$\frac{1}{3}$	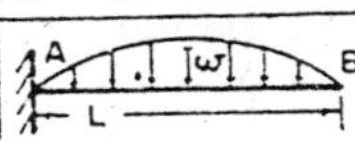	$\frac{2}{3}\omega L$	$\frac{1}{2}$
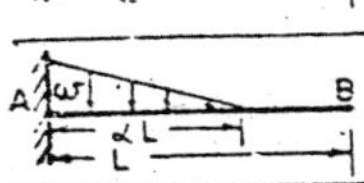	$\frac{\omega \alpha L}{2}$	$\frac{\alpha}{3}$	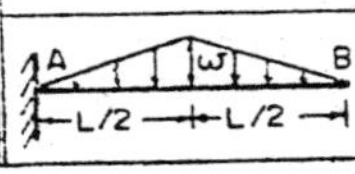	$\frac{\omega L}{2}$	$\frac{1}{2}$

TABLE A-4. FIXED BEAMS

Reaction at $B = r_B \cdot W$; Reaction at $A = r_A \cdot W$

$$M_A = -\,k_A \cdot WL$$

$$M_B = -\,k_B \cdot WL$$

$$M_C = +\,k_C \cdot WL$$

Position of plane of max. B.M. = $C\,L$ = distance CB

Loading	Total load W	Reaction coefficients r_B	B.M Coefficients k_A	k_B	k_C	C
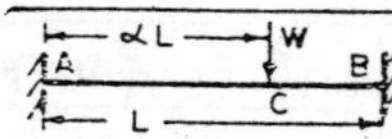	W	$\alpha^2(3-2\alpha)$	$\alpha(1-\alpha)^2$	$\alpha^2(1-\alpha)$	$2\alpha^2(1-\alpha)^2$	$1-\alpha$
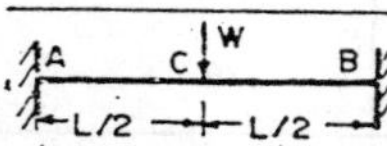	W	1/2	1/8	1/8	1/8	1/2
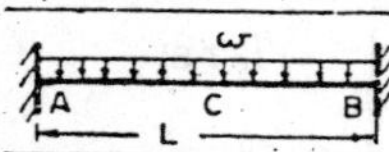	ωL	1/2	1/12	1/12	1/24	1/2
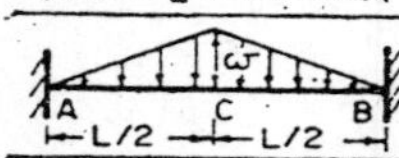	$\frac{\omega L}{2}$	1/2	5/48	5/48	1/16	1/2
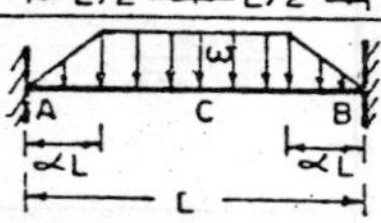	$(1-\alpha)\omega L$	1/2	$k_A = k_B = \frac{1-2\alpha^2+\alpha^3}{12(1-\alpha)}$		$\frac{1-2\alpha^3}{24(1-\alpha)}$	1/2
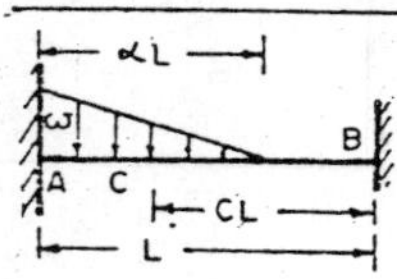	$\frac{1}{2}\,\omega\,\alpha L$	$\frac{\alpha^2}{10}(5-2\alpha)$	$\frac{\alpha}{3}-\frac{\alpha^2}{3}+\frac{\alpha^3}{10}$	$\frac{\alpha^2}{6}-\frac{\alpha^3}{10}$	$r_B\,C-k_B-\frac{1}{3\alpha^2}(\alpha+C-1)^3$	$1-\alpha+\alpha^2\sqrt{\frac{5-2\alpha}{10}}$
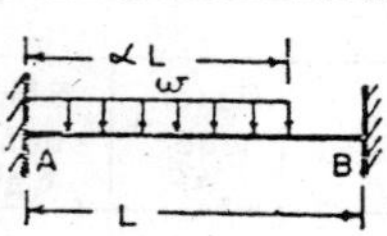	$\omega\alpha L$	$r_B = \frac{1}{2}(2\alpha^2-\alpha^3)$ $r_A = \frac{1}{2}(\alpha^3-2\alpha^2+2)$ $k_A = \frac{\alpha}{12}(3\alpha^2-8\alpha+6)$ $k_B = \frac{\alpha^2}{12}(4-3\alpha)$				
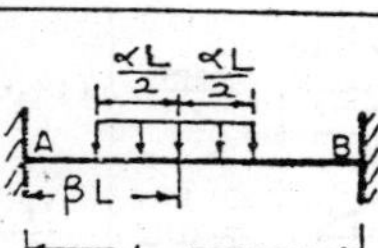	$\omega\alpha L$	$r_B = (-2\beta^3+3\beta^2-\frac{\alpha^2\beta}{2}+\frac{\alpha^2}{4})$ $r_A = (2\beta^2-3\beta^2+\frac{\alpha^2\beta}{2}-\frac{\alpha^4}{4}+1)$ $k_A = \beta(1-\beta)^2-\alpha^2(\frac{1}{6}-\frac{\beta}{4})$ $k_B = \beta^2(1-\beta)-\alpha^2(\frac{\beta}{4}-\frac{1}{12})$				

Appendix-B

PROPERTIES OF MATERIALS AND CONCRETE

TABLE B-1. GRADES OF CONCRETE

IS : 456-1964		*IS : 456-1978*	
Grade Designation	*Specified 28 days work cube compressive strength (kg/cm²)*	*Grade designation*	*Specified characteristic compressive strength f_{ck} at 28 days (N/mm²)*
M 100	100	M 10	10
M 150	150	M 15	15
M 200	200	M 20	20
M 250	250	M 25	25
M 300	300	M 30	30
M 350	350	M 35	35
M 400	400	M 40	40

TABLE B-2. OPTIONAL TESTS REQUIRMENTS OF CONCRETE

Grade of Concrete		*Compressive strength on 15 cm cubes, min. at 7 days*		*Modulus of rupture by beam test, min.*			
				at 72± 2h		*at 7 days*	
IS : 456-1964	*IS : 456-1978*	*kg/cm²*	*N/mm²*	*kg/cm²*	*N/mm²*	*kg/cm²*	*N/mm²*
M 100	M 10	70	7	12	1.2	17	1.7
M 150	M 15	100	10	15	1.5	21	2.1
M 200	M 20	135	13.5	17	1.7	24	2.4
M 250	M 25	170	17.0	19	1.9	27	2.7
M 300	M 30	200	20.0	21	2.1	30	3.0
M 350	M 35	235	23.5	23	2.3	32	3.2
M 400	M 40	270	27.0	25	2.5	34	3.4

TABLE B-3 PERMISSIBLE STRESSES IN CONCRETE (kg/cm^2)
(IS : 456—1964)

Grade of concrete	*permissible stress in compression*		*permissible stress in shear measured as inclined tension*	*Permissible bond stress*		*permissible bearing pressure on full area (plain concrete only)*
	Bending	*Direct*		*Average*	*Local*	
	(σ_{cbc})	(σ_{cc})	(q)	(σ_{ba})	(σ_{bl})	(σ_{br})
M 100	30	25	3.0	4.0	7.0	20
M 150	50	40	5.0	6.0	10.0	30
M 200	70	50	7.0	8.0	13.0	40
M 250	85	60	8.0	9.0	15.0	50
M 300	100	80	9.0	10.0	17.0	60
M 350	115	90	10.0	11.0	18.0	70
M 400	130	100	11.0	12.0	19.0	80

TABLE B-4 PERMISSIBLE STRESSES IN CONCRETE
(IS : 456-1978)

Grade of concrete	*Permissible stress in compression (N/mm^2)*		*Permissible stress in Bond (Average) for plain bars in tention (N/mm^2)* (τ_{bd})
	Bending (σ_{cc})	*Direct* (σ_{cc})	
M 10	3.0	2.5	—
M 15	5.0	4.0	0.6
M 20	7.0	5.0	0.8
M 25	8.5	6.0	0.9
M 30	10.0	8.0	1.0
M 35	11.5	9.0	1.1
M 40	13.0	10.0	1.2

TABLE B-5 DESIGN CONSTANTS FOR DIFFERENT VALUES OF σ_{cbc} AND m

σ_{st} = 1400 kg/cm² (140 N/mm²)

σ_{cbc}		m = 18.7					m = 13.3					m = 11.0				
		R					R					R				
kg/mm²	N/mm²	kg/cm²	N/mm²	p %	k	j	kg/cm²	N/mm²	p %	k	j	kg/cm²	N/mm²	p %	k	j
25	2.5	2.86	0.286	0.22	0.250	0.917	2.25	0.225	1.17	0.192	0.936	1.95	0.195	0.15	0.165	0.945
30	3.0	3.87	0.387	0.30	0.285	4.905	3.07	0.307	0.24	0.222	0.926	2.67	0.267	0.20	0.190	0.937
35	3.5	5.00	0.50	0.40	0.318	0.894	4.00	0.40	0.31	0.254	0.917	3.50	0.350	0.27	0.216	0.928
40	4.0	6.18	0.618	0.50	0.348	0.894	5.00	0.50	0.39	0.275	0.908	4.4	0.44	0.34	0.240	0.920
45	4.5	7.40	0.74	0.60	0.375	0.875	6.08	0.608	0.48	0.300	0.900	5.38	0.538	0.42	0.261	0.913
50	5.0	8.70	0.87	0.72	0.400	0.870	7.22	0.722	0.58	0.324	0.892	6.4	0.640	0.50	0.282	0.906
55	5.5	10.00	1.00	0.83	0.423	0.859	8.35	0.835	0.67	0.342	0.896	7.41	0.74	0.59	0.300	0.900
60	6.0	11.4	1.14	0.95	0.445	0.852	9.56	0.955	0.78	0.363	0.879	8.6	0.86	0.68	0.320	0.893
65	6.5	12.8	1.28	1.08	0.646	0.848	1.08	0.108	0.89	0.381	0.873	9.8	0.98	0.79	0.340	0.887
70	7.0	14.2	1.42	1.20	0.480	0.839	1.22	1.22	1.00	0.400	0.870	11.0	1.1	0.89	0.355	0.82
75	7.5	15.6	1.56	1.34	0.500	0.833	1.35	1.35	1.24	0.417	0.861	12.2	1.22	1.00	0.370	0.876
80	8.0	17.15	1.72	1.48	0.519	0.827	14.8	1.48	1.54	0.432	0.856	13.5	1.35	1.10	0.387	0.871
85	8.5	19.00	1.90	1.62	0.531	0.823	16.2	1.62	1.35	0.442	0.851	14.8	1.48	1.21	0.400	0.870

TABLE B-6 RECOMMENDED MIXES FOR VARIOUS TYPES OF CONSTRUCTION

Type of work	Recommended mix by volume	Max. size of Aggregate mm	Water in litres per bag of cement*	Best consistency
1. Long span arches ; heavily loaded columns	1 : 1 : 2	12 to 20	16 to 18	medium
2. Heavily stressed members ; small precast works such as posts and poles for fencing. telegraphs signals, garden furniture, and decorative and other items of work of very thin section ; watertight construction for high heads ; long piles.	1 : 1 : 2	12 to 20	20 to 23	medium
3. Columns and members subjected to medium loads : walls and floors of reservoirs and tanks ; cisterns, sewers, well kerbs, platforms and other watertight constructions for moderate heads : non surfaced roof slabs ; concrete deposited under water.	1 : 2 : 3 or $1 : 1\frac{2}{3} : 3\frac{1}{2}$	20	25	medium
4. General building work subjected to ordinary stresses such as beams, slabs, columns panel walls, basement and retaining walls, stairs lintals and sills; road, pavements : driveways and side walks ; floors ; steps ; bunker sand silos ; bridges, dams, piers and exposed of water and frost ; machine foundations subjected to vibrations ; footings piles.	1 : 2 : 4	12 to 40 as required	27 to 30	Stiff to medium
5. Mass concrete work in culverts, retaining walls, to action compound walls and ordinary machine bases : foundation walls which need not be water tight.	1 : 3 : 6	25 to 50	34	Stiff to medium
6. Mass concrete for heavy walls : foundations under column footings and under heavy duty floors.	1 : 4 : 8	40 to 50	40 to 48	medium

*The exact quantity will depend on the method of compaction *i.e.* whether the concrete is to be compacted by hand or vibrated, and also on the absorption of water by aggregate.

TABLE B-7 SPECIFIC GRAVITY, BULK DENSITY AND VOIDS OF DIFFERENT TYPES OF AGGREGATES

1. Specific Gravity			**3. Voids (percent)**		
Trap	...	2.9	River sands :		
granite	...	2.8	Fine	...	43
Gravel	...	2.66	Coarse	...	35
sand	...	2.65	Mixed and moist	...	38
			Mixed and dry	...	30
2. Bulk density (kg/l)			Broken stone graded :		
River sand :			25 mm max. size	...	46
Fine	...	1.44	50 mm max. size	...	45
Medium	...	1.52	63 mm max. size	...	41
Coarse	...	1.60	Stone screenings	...	48
Beach or river shingle	...	1.60			
Broken stone	...	1.60			
Stone screenings	...	1.44			
Broken Granite	...	1.68			

TABLE B–8
TOLERABLE CONCENTRATIONS OF SOME IMPURITIES IN MIXING WATER

	Impurity	*Tolerable concentration*
1.	Sodium and potassium carbonates and bicarbonate	1000 ppm (total). If this is exceeded, it is advisable to make tests both for setting time and 28 days strength.
2.	Chlorides	10,000 ppm.
3.	Sulphuric anhydride	3000 ppm.
4.	Calcium chloride	2% by wt. of cement in non prestressed concrete.
5.	Sodium iodate, Sodium phosphate, Sodium arsonate, Sodium borate	very low
6.	Sodium sulphate	Even 100 ppm warrants testing.
7.	Sodium hydroxide	0.5% by wt. of cement, provided quick set is not induced.
8.	Silt and suspended particles	2000 ppm. Mixing water with a high content of suspended solids should be allowed to stand in a setting basin before use.
9.	Total dissolved salts	1500 ppm.
10.	Organic material	3000 ppm. Water containing humic acid or such organic acids may adversely affect the hardening of concrete ; 780 ppm of humic acid are reported to have seriously impaired the strength of concrete. In the case of such waters, therefore, further testing in necessary.
11.	pH	4.5 to 8.5

Note : When a water of unknown performance contains impurities of possible adverse effect, it is certainly good practice to make for setting time strength in which known, acceptable water is used for compression. For the indications, test should be made using the particular cement involved. The procedure for the tests is set down in BS 3148 : 1959 'Tests for water for making concrete.' The water can be considered suitable for making concrete if :

(*i*) initial setting times of the cement do not differ by more than 30 minutes.

(*ii*) the average compressive strength of concrete test cubes using the suspected water is not less than 90% of he average strength of the control concrete test cubes using known good water or distilled water.

TABLE B-9 MASS OF SOME COMMON MATERIALS (IS : 1911—1967)

	Material	Mass (kg/m^3)
1.	*Bituminous substances*	
	Anthracite coal	... 1500
	peat (dry)	... 560 to 640
	Charcoal (light)	... 300
	Coke	... 1000
	Graphite	... 500
	Crude oil	... 880
	Pitch	... 1010
	Coal tar	... 1010
2.	*Excavated material*	
	Clay (dry, compact)	... 1440
	Clay (damp, compact)	... 1760
	Earth (dry)	... 1410 to 1840
	Earth (moist)	... 1600 to 2000
	Sand (dry)	... 1540 to 1600
	Sand	... 1660 to 2000
3.	*Liquids*	
	Alcohols	... 780
	Gasoline	... 670
	Ice	... 910
	Nitric Acid 91%	... 1510
	Sulphuric Acid 87%	... 1790
	Vegetable oil	... 930
	Water (Fresh)	... 1000
4.	*Building Materials*	
	Bricks	... 1600 to 1920
	Cement (ordinary)	... 1440
	Chalk	... 2240
	Glass	... 2400 to 2700
	lime stone	... 2400 to 2640
	Sand stone	... 2240 to 2400
	Steel	... 7850
	Timber	... 650 to 720
5.	*Structural items, ceilings, finishes etc.*	
	Asbestos cement sheets	... 12 to 15.6
	Brick masonry	... 1920
	Brick wall, 100 mm thick	... 192
	Bricks wall, 200 mm thick	... 384
	Bricks wall, 300 mm thick	... 576
	Cement plaster, 25 mm thick	... 52
	Concrete, plain	... 2300
	Concrete, reinforced	... 2400
	Galvanised iron sheet, 0.56 mm thick	... 5
	Galvanised iron sheet, 1.63 mm thick	... 13
	Mangalore tiles with bettens	... 63
	Dry rubble masonry	... 2080

Appendix-C

REINFORCEMENT

TABLE C-1 AREA , PERIMETER AND MASS OF ROUND BARS

Diameter of bar, mm	*Area (cm^2)*	*Perimeter (cm)*	*Mass (kg/m^3)*
5	0.20	1.57	0.15
6	0.28	1.88	0.22
8	0.50	2.51	0.39
10	0.79	3.14	0.62
12	1.13	3.77	0.89
16	2.01	3.03	1.58
20	3.14	6.28	2.47
22	3.80	6.91	2.98
25	4.91	7.85	3.85
28	6.16	8.80	4.80
32	8.04	10.05	6.31
36	16.18	11.31	7.99
40	12.57	12.57	9.86
45	15.90	14.14	12.49
50	19.64	15.71	15.41

TABLE C-2 REQUIREMENTS OF HOT ROLLED DEFORMED BARS (IS : 1139—1966)

Property	*Deformed mild steel bar*		*Deformed medium tensile steel bars*	
	kg/mm^2	*N/mm^2*	*kg/mm^2*	*N/mm^2*
1. Min. ultimate tensile strength. All sizes	42	420	55	550
2. Min. yield stress, Bar dia. upto and including 20 mm	26	260	36	360
Bar dia. over 20 mm upto and including 40 mm	24	240	34.5	345
Bar dia. over 40 mm	24	240	33	330
3. Min. Elongation (%)	23	23	20	20

*The gauge length = $5.65\sqrt{S_o}$ where S_o is the cross-sectional area of the test piece.

Appendix-D

LOADINGS

TABLE D-1 LIVE LOADS ON FLOORS (IS : 875—1964)

Loading class Number	*Types of floor*	*Min. live loads, kg/m² of floor areas*	*Alternate min. live load*
200	Floors in dwelling houses, tenements, hospital wards, bedrooms and private sitting rooms in hostels and dormitories.	200	Subject to a min. total load of 2-5 times the values in column 3 for and given slab panel and 6 times the values in column 3 for any given beam This total load shall be assumed uniformly distributed on the entire area of the slab panel or the entire length of the beam.
250	Office floor other than entrance halls, floors of light workrooms.	250*-400	
300	Floors of backing halls, office entrance halls and reading rooms.	300	
400	Shop floors used for the display and sale of merchandise; floors of workroom generally; floors of classrooms in schools; floor of place of assemby with fixed seating; restaurants, circulation space in machinery halls; power stations etc. where not occupied by plant or equipment.	400	
500	Floors of warehouse, work shops, factories and other buildings or parts of buildings of similar categories for light weight loads; office floors for storage and filling purposes; floors of places of assembly without fixed seating, public rooms in hostels, dance halls waiting halls etc.	500	
700	Floors of warehouses, workshops, factories and other buildings or part of buildings of similar category for medium weight, loads.	700	
1000	Floors of warehouses, workshops, factories and other buildings or parts of building of similar category for heavy weight loads, floors of book stores and libraries, roofs and payments projecting under the public footpath.	1000	

Loading class number	*Types of floor*	*Min live loads, kg/m^2 of floor area*	*Altenrative min. live load*
Garrage Light	Floors used for garages for vehicles not exceeding 2-5 metric tonnes gross weight slabs.	400	The worst combination of actual wheel loads, which ever is greater.
	beam	250	The worst combination of actual wheel loads, whichever is greater.
Garrage heavy	Floors used for garages for vehicles not exceeding 4 metric tonnes gross weight	750	Subject to min. of one and a half times max. wheel load but not less than 900 kg considered to be uniformly distributed over 76 cm square.
Stairs	Stairs, landings but not considered for class 200 loading but not liable to over crowding	300	Subject to a min. of 130 kg concentrated load at the unsupported end of each step for stairs constructed out of structurally independ ent cantilever steps
	Stairs, Landing and corridors for class 200 loading but liable to over crowding, and for all other classes	500	
Balcony	Bolconies not liable to over crowding :		
	For class 200 loading	500	—
	For all other classes	500	—
	Balconies liable to over crowding.	300	—

*The lower value of 250 kg/m^2 should be taken where separate storage facilities are provided and the higher value of 400 kg/m^2 should be taken where such provisions are lacking.

Note 1. In the above table, a reference to a 'floor' includes a reference to any part of that floor, and a reference to 'slabs' includes boarding and beams of ribs, spaced not further apart than one metre between centres, and reference to 'beams' means all other beams and ribs.

2. Under loading class No. 250, the reference to 'light workrooms' envisages room in which some light machines (for example, sewing mechines used by milliners or tailors) are operated without a central power driven unit, that is, the machines are independently operated, either by hand or by small motors. Under loading class No. 400, the reference to workrooms' generally envisages the installation of machines operated with a central power driven unit, with the individual units being belt driven.

3. 'Fixed seating' implies that the removal of seating and the use of space for other purposes is improbable. The maximum likely load in this case is, therefore, closely controlled.

4. The loading in workshops, warehouses and factories varies considerably and so three loadings under the terms 'light', medium' and 'heavy' are introduced in order to allow for more economical designs but the terms have no special meaning in themselves other than the live load for which the relevant floor is designed. It is, however, important, particularly in the case of heavy weight loads, to assess the actual loads to ensure that they are not in excess of 1000 kg/m^2 ; in case where they are in excess, the design shall be based on the actual loading.

5. The load classification for stairs, corridors, balconies and landings provide for the fact that these often serve several occupancies and are used for transporting the furniture and goods.

TABLE D-2 SAFE LOADS ON FOUNDATIONS (IS : 1904—1966)

Types of rock and soil	Max. Safe bearing capacity		Remarks
	t/m^2	kN/m^2	
I. Rock			
1. Rocks hard without lamination and defects for example, granite trap and diorite	330	3300	
2. Laminated rock (for example, Sand stone and limestone in sound condition	165	1650	
3. Residual deposits of shattered and broken bed rock and hard shale, cemented material.	90	900	
4. Soft rock.	45	450	See Note 2
II. Non–Cohesive soils			
5. Gravel, sand and gravel, compact and offering high resistance to penetration when excavated by tools	45	450	
6. Coarse sand, compact and dry	45	250	Dry means that ground water level is at a depth not less than the width of foundation below the base of the foundation
7. Medium sand, compact and dry	25	150	
8. Fine sand, silt (dry lumps pulverised by fingers)	15	150	
9. Loose gravel or sand gravel mixture, loose coarse to medium sand, dry.	25	100	
10. Fine sand, loose and dry.	10	100	See Note 2
III. Cohesive soils			
11. Soft shale, hard or stiff clay in deep bed, clay	45	452	This group is susceptible to long term consolidation settlement.
12. Medium clay, readily indented with a thumb nail	25	250	
13. Moist clay and sand clay mixture which can be indented with strong thumb pressure.	15	150	
14. Soft clay indented with moderate thumb pressure.	10	100	
15. Very soft clay which can be penetrated several cm with thumb	5	50	

16. Black cotton soil or other shrinkable or expansive clay in dry condition (50% saturation)	—	—	See Note 2. To be determined after investigation
IV. peat 17. Peat	—	—	See Note 3 and 4. To be determined after investigation
V. Made up ground 18. Fills or made up ground	—	—	See Notes 2 and 5. To be determined after investigation

Note 1. Increase or decrease the allowable bearing values as follows :

(*a*) The allowable bearing values may be increased by an amount equal to the weight of the material (virgin soil) removed from above the bearing level, that is, the base of the foundation.

(*b*) For non cohesive soils, the allowable bearing value shall be reduced by 50% if the water table is above or near the bearing surface of the soils. If the water table is below the bearing surface of the soils at a distance at least equal to the width of the foundation, no such reduction shall apply. For intermediate depth of water table, proportional reduction of the allowable bearing value may be made.

2. Compactness of looscness of non cohesive material may be determined by driving wooden picket of dimension 5× 5× 70 cm with a sharp point. The picket shall be pushed vertically into the coil by the full weight of a person weighting at least 70 kg. If the penetration of the picket exceeds 20 cm, the loose state shall be assumed to exist.

3. No generalised values for safe bearing pressures can be given for these types of soils. In such areas, adequate, site investigation (See IS : 1982-1962, Code of Practice for site investigation for foundations) shall be carried out and except advice shall be sought.

4. Peat may occur in very soft spongy condition may be quite firm and compact. While the ultimate bearings capacity may be high in the compact cases, very large consolidation settlements occur even under small pressure and the moments continue for decades.

5. The strength of made–up ground depends on the nature of the material, its depth the age, and the method use for consolidating it (if any).

TABLE D-3
LIVE LOAD ON ROOFS (IS : 875—1964)

Types of roofs	*Live load measured on plan*	*Min. Live load measured on plan*
1. Flat sloping or curved roof with slopes upto and including 10° (*a*) access provided	150 kg/m^2	375 kg uniformly disibuted over any span of one metre width of roof state and 900 kg. uni disti. over the span in case of all beams
(*b*) access not provided except for maintenance	75 kg/m^2	190 kg uniformly distributed over any span of one metre width of roof state and 450 kg uni. dist. over the span in the case of all beams.
2. Sloping roof with slope greater than 10°	(*i*) For roof memberanes sheets of purlins—75 kg/m^3 for every degree increase in slope over 10° (*ii*) For members suppor-ing the roof membrane and roof purlins, such as trusses, beams, girders etc— 2/3 of load in (*i*) (*iii*) Loads in (*i*) and (*ii*) do not include loads due to snow, rain, dust collection, etc. and the effects of such loads should be appropriately considered.	subject to a minimum of 40 kg/m^2
3. Curved roof with slope at spring greater than 10°	$75—345\sqrt{\gamma}$ kg/m^2 where $\gamma = \frac{h}{l}$ h = the height of the highest point of structure measureed from its spring and l = chord width of the roof singly curved, and shorter of the two sides if doubly curved	subject of to a minimum of 40 kg/m^2

Note. For special types of roofs with highly permeable and absorbent material, the contingency of roof material increasing weight due to absorption of moisture should be provided for.

$$[1 \text{ kg/m}^2 = 10 \text{ N/m}^2 = 0.01 \text{ kN/m}^2]$$

TABLE D-4 SI UNITS AND METRIC/IMPERICAL EQUIVALENTS
(Imposed Loads on Floors and Roofs)

Concentrated Load				Loading per unit length				Intensity of Load			
lb	*kg*	*Newtons N*	*kN*	*lb per foot*	*kg per metre*	*Newtons per metre*	*lb per metre*	*lb per sq. ft.*	*kg per sq.ft.*	*Newtons per sq. m.*	*kN per sq. m*
202	91.8	900	0.9	15.0	22.4	220	0.22	15.7	76.5	750	0.75
225	102	1000	1.0	24.6	36.7	360	0.36	20.9	102	1000	1.0
415	143	1400	1.4	50.7	75.5	740	0.74	31.3	253	1500	1.5
405	184	1800	1.8	103	153	1500	1.5	41.8	204	2000	2.0
603	275	2700	2.7	168	251	2460	2.46	52.6	255	2500	2.5
809	367	3000	3.6	206	306	3000	3.0	62.7	306	3000	3.0
1012	459	4500	4.5	280	418	4100	4.1	83.5	408	4000	4.0
2023	918	9000	9.0	308	459	4500	4.5	104	510	5000	5.0
				560	836	8200	8.9	136	663	6000	6.5
				672	1018	10000	10.0	157	765	7500	7.5
								209	1020	10000	10.0
								261	1275	12500	12.5
								313	1570	15000	15.0
								418	2040	20000	20.0

Conversion factors expressed to four significant figures but are calculated on more exact values

1 ft = 0.3048 m   1 m = 3.281 ft.

1 sq. ft. = 0.0929 sq. m   1 sq. m = 10.76 sq. ft.

TABLE D-4 (contd.)

SI : UNITS AND METRIC/IMPERICAL EQUIVALENTS

1 lb = 0 4536 kg = 4.448 N = 0.00448 kN 1 kg = 2.2046 =9.807 N = 0.009807 kN 1 N = 22243 lb = 0.1020 kg = 0.001020 kN 1kN = 224.8 lb = 102.0 kg = 1000N	1b/ft^2/ft 15.3 25.3 30.1	kg/m^2/m 245 408 510	N/m^2/m 2400 4000 5000	kN/m^2/m 2.4 4.0 5.0
lb per foot	kg per metre	Newtons per metre	Kg per metre	
1 0.6720 0.06542 68.52	1.488 1 0.1020 102.2	14.59 9.807 1 1000	0.01452 0.09807 0.001000 1	
lb per sq. ft.	kg per sq. m	Newtons per sq. m	kN per sq. m	
1 0.2048 0.02089 20.89	4.882 1 0.1020 102.0	47.88 9.807 1 1000	0.04781 0.009807 0.001000 1	
lb per sq.ft. per ft.	kg. per sq. m per m	N per sq. m per m	kN per sq. m per m	
1 0.06250 0.09375 63.75	1.600 1 0.1020 102.0	15.69 9.807 1 1000	0.01556 0.009827 0.001000 1	

Index